Principles of
ANIMAL
COMMUNICATION
Second Edition

Principles of
ANIMAL
COMMUNICATION
Second Edition

Jack W. Bradbury | Sandra L. Vehrencamp

*Cornell Lab of Ornithology and
Department of Neurobiology and Behavior
Cornell University*

 Sinauer Associates, Inc., Publishers
Sunderland, Massachusetts

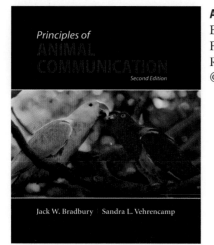

ABOUT THE COVER
Eclectus parrot (*Eclectus roratus*)–male (green) and female (red).
Females mate with multiple males who feed them, as shown here.
Rainforest Habitat, Cape York Peninsula, Australia.
© Gary Bell/OceanwideImages.com.

PRINCIPLES OF ANIMAL COMMUNICATION, Second Edition

For information, address:
Sinauer Associates Inc., 23 Plumtree Road, Sunderland, MA 01375 U.S.A.
FAX: 413-549-1118
email: publish@sinauer.com
internet: www.sinauer.com

Library of Congress Cataloging-in-Publication Data

Bradbury, J. W.
  Principles of animal communication / Jack W. Bradbury, Sandra L. Vehrencamp. -- 2nd ed.
    p. cm.
  Includes index.
  ISBN 978-0-87893-045-6 (hardcover)
  1. Animal communication. I. Vehrencamp, Sandra Lee, 1948- II. Title.
  QL776.B73 2011
  591.59--dc22

                    2011011920

Printed in China
6 5 4 3 2 1

To our own many mentors

Contents

Chapter 15
Communication Networks 611

Chapter 16
The Broader View: Microbes, Plants, and Humans 651

Preface to the Second Edition

More than a decade has passed since the publication of the First Edition of *Principles of Animal Communication*. That decade saw explosive growth in this field: the game theoretical underpinnings of signal evolution were formalized and classified, the study of visual signals caught up with and, in some respects, surpassed the prior dominance of acoustic and olfactory signals in animal communication research, and major advances occurred at the interfaces between animal communication and neurobiology, immunology, endocrinology, evolutionary genetics, economics, and conservation biology. Even the philosophers got in on the act. This growth created a massive increase in the primary and secondary literatures on animal communication that continues to this day. Studies involving animal communication currently account for 15–35% of the articles in the major journals devoted to animal behavior, and 20–30% of the talks and posters presented at national and international meetings on animal behavior and behavioral ecology.

Although we had not planned to write a second edition, the remarkable growth of research in this area, the continued enthusiasm for the topic, and lobbying by colleagues persuaded us to reconsider. A questionnaire circulated among faculty who had taught the material showed a desire for a new edition with fewer chapters (say 15–16 to match the number of weeks in a typical US semester), more even coverage of taxa and modalities, as thorough as possible an update on the relevant literature, and some sort of sequestering of the more mathematical aspects that tend to bog down undergraduates. Our publisher, Andy Sinauer, offered to produce the book in a larger format and in full color. So we set to work.

Five years later, you are holding the result. There are 16 chapters, but given the amount of updating needed and our attempts to cover all modalities and taxa more evenly, they are fairly long chapters. Faculty may want to "cherry pick" sections to assign to students (as many already do with the First Edition). To provide the requested updates, we read 500–1200 journal articles, chapters, and books *per chapter* in this Second Edition. Over 90% of the citations in this edition were published since the First Edition came out. While some recent textbooks—to our horror—now skip any textual citations, our new edition is fully referenced but uses citation numbers (instead of names and dates) to improve readability. The large chapter bibliographies and their corresponding citation numbers have been posted online where they can be easily downloaded. We have also written over 60 online Web Topics, which list links to enrichment resources, such as sample sound and video clips, and review the more difficult math that we have removed from the main undergraduate text but still want to make available for graduate students and faculty. We have tried to set the level of the physical text to that of college juniors or seniors who have had (but could probably survive in this course without) an introductory course in animal behavior using a text such as John Alcock's *Animal Behavior: An Evolutionary Approach*. While there will be some overlap in material and examples between the two courses, few students grasp all the concepts the first time, and the online Web Topics allows readers to go far beyond either text if they wish.

The opportunity to publish this edition in full color encouraged us to spend a significant amount of time searching for new and exciting photos of communicating animals. With a few notable and much appreciated exceptions, most of our colleagues no longer take photos of their research subjects and instead use online photos for their own lectures. We thus began perusing the website Flickr (http://www.flickr.com/) and discovered hundreds of superb photos including those of species and behaviors that have never before been illustrated in textbooks. Even more gratifying, most Flickr photographers were happy to include their photos in the book with no licensing fees, thus helping keep the cost of the book down. We owe the Flickr community a great deal of gratitude: they take superb photos and are very generous about sharing their work. In addition, nearly all of the professional wildlife photographers that we approached greatly reduced their licensing fees for this edition. Photo credits and contact information for our many photographic contributors are provided at the end of the book. The combination of these excellent photos and the lovely artwork and page design by the Sinauer staff has made this edition as visually appealing as we could have hoped.

As scientific fields reach middle age, one often sees polarization into opposing camps on theoretical issues. Current debates in behavioral ecology include the role of information in animal communication, the relative importance of inclusive fitness versus network topologies in shaping the evolution of cooperation, whether sexual selection is cooperative or competitive, and which type of evolutionary model best predicts observed patterns of animal behavior. Because we are writing a text about "principles," we have often had to review the evidence for alternative viewpoints and indicate which way we think the issue is leaning. Our judgments will surely gratify some colleagues and elicit dissent from others. Only time will determine how often we got it right. However, it is important to point out that taking stands does not reflect excessive hubris on our part, but stems from our conclusion after 40 years of undergraduate teaching that a clear structure is a better starting point for students than the typical journal style of playing Dr. Jones off against Dr. Smith. There is room for both forms of pedagogy, but we find the latter works better in a lab or discussion group setting than in an undergraduate textbook.

Many colleagues have helped us with this new edition. Any errors or omissions are of course our responsibility and not theirs. Among the many who have given us input and feedback, we particularly thank Thorsten Balsby, Andy Bass, Carlos Botero, Mark Briffa, Alexis Chaine, Ken Clifton, William Cooper, Tom Cronin, Selvino de Kort, Stephanie Doucet, Neville Fletcher, Carl Gerhardt, Judith Hall, Oren Hasson, Simo Hemila, Ron Hoy, Sönke Johnsen, Bob Johnston, André Kessler, Josh Kohn, Frederich Ladich, Jürgen Liebig, Charles Linn, Irby Lovette, Kevin McGraw, Colleen McLinn, Molly Morris, Sirpa Nummela, Tomasz Osiejuk, Julian Partridge, Richard Prum, David Puts, Rob Raguso, Kern Reeve, Tommy Reuter, Edwin Scholes, Mohan Srinivasarao, Kerry Shaw, Rod Suthers, John Swaddle, Marc Théry, Stuart West, Jerry Wilkinson, and Tristram Wyatt. Copy editor Carrie Compton and artist Elizabeth Morales made major contributions. The talented Sinauer Associates' staff has been fantastically supportive and they deserve major credit for the look and feel of the final book. In particular, we want to thank staff artist Joanne Delphia, book designer Janice Holabird, Joan Gemme for her work on the photographs, art director Chris Small, website director Jason Dirks, licensing guru David McIntyre, and our project editors Sydney Carroll, Kathy Emerson, Laura Green, and particularly Azelie Aquadro, who was our guiding light through the difficult stages between writing the chapter drafts and finalizing the last page proofs. Since 1972, Andy Sinauer has continued to believe in a text on this topic and our ability to deliver something useful. We hope we have lived up to his expectations.

Jack W. Bradbury
Sandra L. Vehrencamp

Chapter *1*

Signals and Communication

Overview

The evidence that animals communicate is all around us. But what accounts for the conspicuous diversity of signals that animals seem to be using? Is there any order or pattern to this diversity? Research on animal communication seeks to identify general principles from biology, the physical sciences, and economics that together can explain why one animal species relies on one type of signal and another animal species relies on a different type of signal. In this chapter, we provide an overview of the important principles and how they will be presented in more detail later in this book.

Why Study Animal Communication?

Do animals communicate?

Any observant person knows that animals communicate (**Figure 1.1**). When your dog hears a cat jump up onto your porch at night, it begins barking and soon, all the neighborhood dogs are also barking despite the fact that they cannot possibly have heard the cat's soft thump. It does not take musical training to notice that when one songbird in your backyard starts singing, its neighbors not only sing back, but may even match the theme sung by the first bird. If you are good at imitating bird songs by whistling, you can easily provoke a currently silent cardinal, mockingbird, or titmouse to start singing back at you, often with a similar theme. Given these widely experienced examples, most people presume that the roars of lions, the chirping of crickets, the deafening choruses of cicadas, and the songs of whales are also used by these animals for communication.

Sound production is just one clue that animals communicate. Anyone who has had to wait interminably while their dog meticulously sniffed the roots of a tree before finally adding its own urine cannot help but surmise that the dog is checking out odor signals left previously by other dogs and then leaving its own "message." Surely the bright red coloration of the male northern cardinal, which makes it extremely conspicuous to predators against a green background, has to serve some compensatory utility to the birds bearing it. The color might provide insurance that males and females can recognize members of their own species for mating; serve as an early morning advertisement to potential intruders that a territorial owner has survived another night; or create a plumaged "canvas" that helps females to assess the health of potential mates. Sound, odor,

(A)

(B)

FIGURE 1.1 **Commonly encountered examples of animals communicating** (A) Male Northern Mockingbird (*Mimus polyglottos*) singing to defend its breeding territory from other males and attract a female mate. (B) Domestic cat (*Felis catus*) performing a defensive visual display when threatened. (C) Domestic dog (*Canis lupus familiaris*) urinating on a tree to leave scent mark for other dogs in the neighborhood.

(C)

One of the expectations of modern science is that principles discovered in one discipline will be compatible with principles in other disciplines. Biologists invoke principles of acoustic physics to explain why body size ultimately constrains the pitch of animal communication sounds. Similarly, the evolutionary principle of kin selection, in which organisms favor cooperation with genetic kin, has proved to have significance in medicine [58]. Studies of animal communication must integrate and in some cases can help refine principles previously identified in other disciplines. As we shall see in this book, an understanding of the diversity observed in animal communication requires the melding of principles from physics, chemistry, genetics, physiology, evolutionary biology, taxonomy, behavioral ecology, community and population ecology, informatics, and economics. In turn, principles derived from studies of animal communication are providing new insights and tools for fields such as conservation biology and wildlife management, pest control, linguistics, developmental biology, immunology, epidemiology, neurobiology, and psychology (**Figure 1.3**).

Beyond the pursuit of scientific inquiry and tests of concordance across disciplines, even a minimal knowledge of animal communication principles can enrich anyone's daily life. It is hard to find a location on earth where one is not exposed to the signals of communicating animals. Knowing what they are doing, why they are doing it, and why they do it the way they do makes a walk in the woods or a snorkeling trip on a shallow reef a far richer experience. A birder so preoccupied with checking off newly sighted species that she fails to stop and attend to what the birds are doing is truly missing half the story. Even when one cannot see a forest bird, one can often hear it exchanging vocalizations with other birds. Why is the bird using such low-frequency sounds? Why does it use such a slow tempo? Is it defending a territory or attracting a mate? If one knows what questions to ask, it is amazing how much more one can get out of that walk or snorkeling dive. You do not need fancy equipment to eavesdrop on most of the communicating animals that you are likely to encounter. Human senses may not be as well tuned to every animal

and visual signals are only some of the various stimuli that we find animals using to communicate.

So, if most reasonable people have already concluded that animals communicate, what else is there left to say about the subject?

Diversity and principles

Diversity is a ubiquitous property of nature. The major task of science is making sense of this diversity by extracting and then verifying general principles that singly or in combination explain most of the variation in a particular aspect of nature. As demonstrated by the earlier examples and those in **Figure 1.2**, animals can show enormous diversity in whether, when, and how they communicate. Rabbits barely make a sound except when grabbed by a predator, whereas male hammer-headed bats devote two thirds of their anatomy and a quarter of their waking hours to producing honking calls that attract females. What basic principles can explain these differences? Or put another way, what principles might we discover by comparing the sound signals of rabbits and hammer-headed bats?

FIGURE 1.2 **Diversity in animal color signal patterns** A sample of the head and facial markings on males of different species of jumping spiders in the genus *Habronattus*. Although very closely related, these species show an amazing diversity of color patterns. A similar figure could be constructed for closely related species of crabs, butterflies, fish, lizards, or birds. (A) *H. pugillis*; (B) *H. tarsalis*; (C) *H. americanus*; (D) *H. sansoni*; (E) unnamed species 1; (F) unnamed species 2.

signal as those of the intended receivers, but we are amazingly well equipped to monitor a very broad range of stimuli, more than enough to make a knowledgeable person's walk in the park the highlight of the day.

Web Topic 1.1 *Animal communication and science education*
Because most students naturally like animals, and animal communication integrates so many disciplines, the topic can be used as an entry point for science education in middle and high school curricula. Here we provide some background and relevant links.

Cues, Signals, and Signal Evolution

Cues

All animals have **sense organs**. These provide current **information** about the physical, ecological, and social conditions surrounding the animal. This information is then used by the animal's brain and associated systems to adjust physiological states and refine decisions about subsequent actions. Most sense organs do not measure external conditions directly, but instead monitor **cues**. Cues are assessable properties that are at least partly correlated with a condition of interest. While one animal might rely on thermoreceptor neurons that directly measure ambient temperatures, others may attend to cues such as visible heat waves rising from the substrate or the dryness of an exposed tongue. Many conditions of interest, such as the health of a potential mate or the intentions of a nearby predator, are nearly impossible to measure directly. Instead, animals have evolved sense organs that are tuned

(A)   (B)

FIGURE 1.3 **Studying the interface between animal communication and cognition** (A) Dr. Irene Pepperberg and the late Alex, an African Grey Parrot (*Psittacus erithacus*). Alex was raised to respond to and reply with human speech in meaningful ways. In (B) Alex is shown being asked to identify the quantity of a specific set of objects defined by their shape and color. At his death, he could identify over 100 different items, including locations, foods, and objects made of various materials, colors, and shapes; assign items to labeled categories; understand concepts such as object permanence, same versus different, and bigger versus smaller; count to eight; recombine the elements in his labels to form other novel labels; and say "None" if asked to find something that was not there. This work demonstrated that parrots can use complex communication to perform cognitive tasks with a very high rate of accuracy [68]. How parrots might use these abilities in the wild remains to be studied.

to one or more cues correlated with those conditions. Few cues are perfectly correlated with conditions of interest, so an animal is often faced with a trade-off between relying on a cue that is easy to measure but imperfectly correlated with a condition of interest versus trying to measure the condition directly. In the majority of cases, animals opt for the most quickly evaluated cues despite their imperfect correlations with the properties of interest.

Humans monitor cues, just like other animals. When it is hot, and our sweat does not evaporate, we conclude that the humidity is probably high. We do not have direct humidity sensors and so rely on a related cue. We assume that a person with a wrinkled face and gray hair is probably old. However, there are diseases that can produce wrinkles and gray hair in younger people; the correlations between the cues (wrinkles and gray hair) and the property of interest (age) are imperfect. However, they are usually good enough for us to continue to monitor these traits as useful cues. We even make important behavioral decisions about how to interact with others based upon such imperfect correlations.

Animals behave similarly. Many species have sensory organs and associated brains that can track changes in ambient cues with great speed and accuracy. Simultaneous input from sensors monitoring different but related cues facilitates cross-checking to correct for imperfect correlations between any one cue and conditions. Mammalian predators combine olfactory, visual, and auditory cues to detect and locate prey. Interacting animals usually alter their behaviors more rapidly than the nonbiological environment changes. It thus pays for animals to monitor the cues that predict future actions of nearby conspecifics, predators, and prey more often and more accurately than they monitor atmospheric conditions. Most animals stop other activities, alter their posture, or otherwise prepare themselves before making a significant change in their actions. Observer animals can watch for these subtle antecedents and use them as cues to the subsequent behaviors of others. In fact, monitoring the behavioral cues generated by other nearby animals (including predators and prey) is the dominant task for the sensory organs and brains of most animal species.

Signals

Signals are stimuli produced by a **sender** and monitored by a **receiver**, to the average net benefit of both parties. Like cues, signals are correlated with conditions outside the receiver and thus provide potential information to it. Unlike cues, which are generated either inadvertently or for purposes other than communication, the function of most signals is to provide information to another animal. If this provision of information benefits both sender and receiver, mutations in either party that refine and improve the process will be favored over evolutionary time. We thus might expect that the correlations between signals and their referent conditions will usually be higher than the correlations between cues and conditions. As we shall see, this is often true, given sufficient evolutionary time and a commonality of interests between sender and receiver.

In practice, sender and receiver may not have identical interests, senders can err in their evaluation of the condition about which they are signaling, and noise and other factors may distort signals during propagation. This does not necessarily mean that communication is a waste of time. On the contrary, even a slight net benefit to one or both parties may favor the continued production and reception of imperfect signals. If the costs of further improvement by either party are higher than the benefits, evolution will not favor refinement, and the animals will continue to communicate imperfectly. This fact has generated some confusion in the study of animal communication about when senders are or are not deceiving receivers. True **deception** occurs when a sender produces a signal whose reception will benefit it at the expense of the receiver regardless of the condition with which the signal is supposed to be correlated. An observation that a sender produces a "wrong" signal, given the actual state of the referent conditions, could be an example of deceit, but it could also reflect economics that favor continued reliance on imperfect signals [7, 10, 34, 56, 75, 89]. Determining whether a misleading signal is a case of true deceit or imperfect signaling invariably requires more refined data, careful economic and sensory analysis, and often, a critical experiment based on the relevant principles. Research has shown that misleading signals are most often the outcome of economic constraints on signal perfection and only rarely due to deceitful intent on the part of the sender.

Many actions by animals have both signaling and non-signaling functions and thus are not easily assigned to discrete categories. While the song of a nightingale or the dance of a honeybee fit most scientists' definition of a signal, other behaviors do not fit the definition as neatly. Philosophers of science spend considerable time debating definitions for natural phenomena with the hope that everything can be clearly assigned to one discrete category or another. Evolution tends to favor economics over philosophy: if a single animal action can efficiently serve multiple functions, it is often favored by natural selection. A threat display at close range may function both to place the sender in a better tactical position for attacking its opponent and provide information to the opponent about the sender's estimation that it would win an escalated fight [84, 85]. Grooming of one primate by another provides hygienic benefits and information to both the recipient and any nearby observers about the groomer's perceived affiliations with the recipient [86]. Males of many birds provide food samples to courted females; this provides nourishment that may later contribute to egg production, but also provides information to the female about the courting male's future abilities as a provisioning parent [61]. These actions, which combine signaling and nonsignaling behaviors, are not easily assigned to tidy, discrete definitions. Not surprisingly, there is continued debate over suitable definitions of biological signals, information, and communication [4, 10, 23, 30–33, 35, 36, 38, 51–53, 55, 59, 60, 71, 76–79, 82, 83].

In this textbook, we shall invoke a broad and quantitative (as opposed to discrete) definition of signals. We accept that

actions such as threat displays, shared grooming, and courtship feeding can have both tactical and signaling functions, and that the impacts of these combined functions on both an actor and a recipient of the action can vary continuously. This approach has the advantage of expanding the range of phenomena that can be considered in trying to extract general principles. At the same time, it makes the detection and quantification of signal content in an action more challenging. A careful examination of the contexts in which an action is performed and its economic consequences, followed by informed experiments and manipulations, can be very helpful in evaluating signal content. For example, roosters often emit a specific call when they find food. This is not some uncontrollable expression of excitement [45, 71], but instead is given most often when hens are nearby [14, 15, 44]. Males will even pick up and present samples of food to nearby females [14, 44, 80, 81]. The economic benefit of this selective calling for the rooster is greater access to hens for mating [87]. Given this motivation, roosters will sometimes call falsely and proffer inedible objects to hens [24, 46]. Similar economic data have been used to confirm and quantify signal content in a variety of actions and a variety of taxa [2, 3, 9, 11, 12, 19, 40, 41, 47, 48, 57, 62, 63, 66, 67, 69, 70, 72]. We shall provide other examples in later chapters, where contexts, economics, and experiments combine to demonstrate and quantify the signal content in animal actions.

Web Topic 1.2 *Information and communication*
Some scientists feel that the role of information provision should be downplayed in definitions of animal communication. A few even recommend elimination of the term when applied to animal interactions. Here, we outline the case for information as a useful and even key concept in understanding the evolution and diversity of animal signals.

Signal evolution

Since most animals have already invested inordinately great amounts of time, energy, and anatomical specialization in monitoring cues generated by other animals, the evolution of signaling is relatively easy. Consider a female bird that routinely examines the plumage of potential mates for ectoparasite infestations. She might do so to avoid becoming infected during mating contact, or because she is looking for evidence of parasite-resistance genes in males that could be passed on to her offspring. Relevant cues that he is unhealthy might include excessive feather dust, missing vanes and elements in key feathers, lethargy, slow reaction times, or discolored skin. If a mutant male with low parasite infestations adopts a posture or activity that makes the female's assessment of his plumage easier or more accurate, he is more likely to be selected for mating. His many offspring will carry the genes that promote this display behavior as well as the genes of his mate, who responded to it. As a result, the trait could become increasingly common in successive generations. While males

(A)

(B)

(C)

(D)

FIGURE 1.4   Evolution of display from behavior with other functions   Most waterfowl drink by scooping up water in the bill and raising the beak high enough that the water runs into the throat. One puzzle is why waterfowl that have just spent extensive periods sitting on the water and filtering out food items appear to drink and to do so repeatedly. Why would they drink when it seems they have had plenty of opportunity for accumulating water during feeding? It turns out that many species have ritualized the drinking movements into a display that is used to mediate conflicts, courtship, and social integration [37]. Careful observation shows that the motions are similar but the contexts and functions are quite different. Here we see examples of ritualized drinking displays by a male (A) wood duck (*Aix sponsa*), (B) mallard (*Anas platyrhynchos*), (C) hooded merganser (*Lophodytes cucullatus*), and (D) bufflehead (*Bucephala albeola*).

with higher parasite loads penalize themselves when they perform the display, once enough males perform it, females should reject not only males that clearly have high infestations, but also males that refuse to perform the display. If the only way to obtain a mating is to display, the behavior becomes obligatory in all males if they wish to reproduce.

In this example, we assumed that the mutation generated a new posture or action. However, it is not necessary that the new behavior be entirely novel. As we noted earlier, most animals adopt postures or perform subtle actions that precede major changes in behavior. A careful watcher can use these cues to anticipate what the animal is likely to do

next. If it pays both parties to have the watcher anticipate the actor's next behavior, mutations that favor exaggeration of the actor's cue posture or action might be favored over evolutionary time. This is surely how many displays performed during aggressive encounters have evolved. In the case of the hypothetical interaction just discussed, the antecedent of the male's display might simply be normal preening activities. Birds spend considerable effort to keep their feathers cleaned, arranged, and oiled. Any of the normal hygiene activities of the male could be exaggerated slightly to make it easier for a nearby female to assess his ectoparasite load. Subsequent mutations might exaggerate this action further and shape the preening behavior such that the exaggerated form is used only when receptive females are nearby; when they are absent, the male might continue to use the original unexaggerated form of preening. Preening is just one of many examples of animal behaviors that have both nonsignaling and signaling versions (**Figure 1.4**). Such examples have proved very useful in understanding the process of signal evolution.

We shall discuss how signals evolve in Chapter 10. But at this stage, the general point is that the extensive monitoring of cues by animals sets the stage for the subsequent evolution of signals. Usually, the relevant cues are linked to the same condition that is subsequently the focus of the signals.

However, there are cases in which a mutant sender produces a stimulus that mimics some cue that is already of major interest to receivers but fails to provide any new information. Examples might be sender production of a color or sound that receivers typically associate with food or predators. A mutant sender could exploit this general sensitivity of the receiver to attract the latter's attention and then try to induce it to behave in ways that benefit the sender. This type of sensory exploitation usually leads to stable communication only if receipt of the signal provides some incidental benefit to the receiver, or if subsequent sender mutations cause the signal to become correlated with information that the receiver can use. For example, a male display that by chance exploits existing female sensory biases might benefit the male by catching the attention of more females, while ameliorating the task faced by females of finding and comparing potential mates. Whether an incipient signal exaggerates a cue already being monitored or instead exploits a sensory bias to get another animal's attention, prior cue monitoring is the key preadaptation for the evolution of communication.

Principles and Animal Communication

Luckily, explanations for the observed diversity in animal communication systems do not require that we invoke principles from all of the relevant disciplines at once. Instead, it is possible to divide up relevant principles into general topics, and then tackle the topics serially. Because signals evolve from cue monitoring, the **physiological mechanisms** with which senders develop signals and receivers process them are those that the animals are already using before signals evolve. These physiological precursors of communication have been shaped over prior evolutionary time by constraining principles of physics and chemistry. The **physical constraints** differ depending upon the animals' ambient medium (air, water, solid substrates); habitat (e.g., forest versus open plains); circadian rhythm (diurnal versus nocturnal); mobility; position in the food web; and body size. Different physiological preadaptations for monitoring cues in these different situations are a major source of diversity in animal communication systems and constitute the focus of Chapters 2–7 in this book.

A second source of diversity in animal communication systems arises from the taxonomic affiliations of each species. Without the effects of historical and phylogenetic constraints, we might expect all species, regardless of taxon, to utilize similar signals when the physical, chemical, and functional contexts are identical. In practice, this is not what is found. Most birds have wings and can fly (although a few groups have lost this ability), and most mammals have four legs but no wings (although one group has turned the front legs into wings, and a few others have turned theirs into flukes and flippers for swimming). As a result of these different heritages, flight displays are not an option for gorillas (at least over the time scales considered in this book), whereas they are a common signal among many bird taxa. It is thus important when considering the contributions of physical, chemical,

and physiological principles to animal signal diversity to also specify the taxonomic limitations on options. We have noted these taxonomic constraints throughout Chapters 2–7.

The third source of diversity in animal signal systems concerns the **economics of communication** (Figure 1.5). A signal emitted by a sender should be sufficiently correlated with conditions of interest to the receiver that it pays the receiver to attend to the perceived signal and incorporate the new information into its future decisions and physiological states. Similarly, a sender should only send a signal if its reception makes a receiver more likely to behave in ways that benefit the sender. In short, both sender and receiver should benefit, at least on average, by communicating.

The economics of communication can be both subtle and complicated. Both senders and receivers will pay costs for participating in communication: these costs include energetic, temporal, and anatomical investments; increased exposure to predators, disease, and parasites; and the risks of being deceived or manipulated by other parties. If the correlations between the condition being monitored and the signals perceived by the receiver are sufficiently poor that the average costs exceed the average benefits for either party, there will usually be no communication. The tightness of these correlations can vary with the accuracy of the sender's assessment of the conditions of interest, the modalities of the signals used (sound, light, odor, touch, electrical fields), the physics and chemistry of the environment in which the signals propagate between sender and receiver, and the accuracy with which a receiver can perceive and discriminate between alternative signals. All of these factors are constrained by the differing physiological heritages of each taxon.

Because it would be unrealistic for receivers to insist on perfect cue and signal correlations before making decisions, they usually choose what to do after receiving some intermediate amount of information. Refinement of the communication process above this level would only increase the costs on either or both parties with little if any benefit. The optimal level of signal accuracy can vary depending upon the context in which a given species is communicating. A population of birds living on islands, where predators generally tend to be less common, can likely afford to invest in a higher level of signal accuracy than can a population of the same species on the mainland, where the risks of a predator detecting the signal and attacking the sender or receiver are more severe. Basic principles of optimality economics and information theory have proved very useful in explaining the economic bases of animal signal diversity. These principles are the focus of Chapters 8 and 9.

The fourth source of signal diversity follows from the degree to which sender and receiver have commensurate interests in successful communication. At one extreme, fighting animals have minimal common interests, and one might think that it would not pay to engage in communication at all. In fact, a significant portion of most social species' signal repertoires is dedicated to conflict mediation. As we shall see, the distrust between sender and receiver in aggressive contexts places special constraints on the kinds of signals that are used in these contexts. Partners in cooperative groups such as mated pairs of birds raising offspring, pack hunters, or large flocks or schools sharing predator vigilance might be assumed to have identical interests. However, there is always a temptation to take more than one's fair share of the spoils or let a partner do more of the work. Thus even in apparently cooperative contexts, some conflict of interest is usually present. In fact, the only communication in which there are no potential conflicts of interest occurs when an animal "talks" to itself. An example of this is echolocation by bats and dolphins, in which the animal emits a sound, listens for the echo, and then uses differences between the two to infer the presence of nearby obstacles, predators, or prey.

When sender and receiver experience **conflicts of interest**, it would seem that the optimal level of signal accuracy might degenerate over evolutionary time until it did not pay to communicate. Despite this, animals continue to communicate across the entire spectrum of conflicts of interest. The answer to this puzzle is that receivers facing a conflict of interest often limit responses to those signals that have some **honesty guarantee**. How these guarantees are achieved has been clarified by invoking principles from **evolutionary game theory**. This is a discipline that merges classical game theory from human economics with basic principles from behavioral ecology and evolutionary biology. We introduce the basic ideas of evolutionary game theory and its links to signal evolution in Chapters 8–10, and then review the diversity of mechanisms discovered for ensuring signal honesty over the full range of potential conflicts of interest in Chapters 11–14.

The traditional approach to studies of animal communication focuses on a single sender and a single receiver. To the degree that natural communities of interacting animals can be described by summing up the interactions of each dyadic pair, such approaches to animal communication are sufficient. However, many animals broadcast signals expressly because it pays to contact many receivers at once. This can set off a wave of successive responses that both radiates away from the signalers and feeds back on them in complicated ways. The pooled interactions within the **network** of communicating animals can produce emergent properties that are not predictable from dyadic interactions alone (**Figure 1.6**). The nature of these emergent properties will vary with the physiological heritage, physical and ecological

Condition of interest (presence of eagle)

Cues   Prior information

Sender

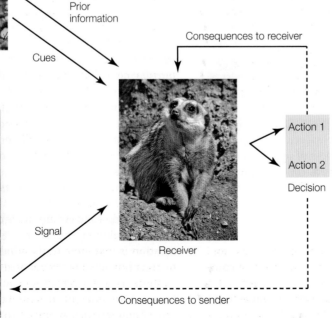

FIGURE 1.5   The process of communication   Here a meerkat (the receiver) needs information about a condition of interest (it doesn't know if an eagle is perched nearby or not) before it can make a decision about what to do next (flee or keep foraging). By itself, the receiver has to base its decision on prior information that it may have acquired (such as the frequency with which eagles occur in this site) and any cues that it can evaluate itself (e.g., whether other meerkats and small mammals have all fled, whether it can see an eagle, etc.). Suppose there is another meerkat nearby for whom the receiver's decision will also have consequences. For example, the receiver might be a close relative of the other meerkat. If this second meerkat, the sender, has access to cues that indicate the presence of an eagle, it might give an alarm call to the receiver; if the sender sees no evidence of an eagle, it could give a foraging call that encourages the receiver to keep feeding. In either case, the receiver can combine its prior information, its own cue assessment, and the nature of the signal given by the sender to make a better decision than it would by itself. If the benefits of the receiver making a good decision outweigh both parties' respective costs of giving the signal and attending to it, evolution will favor their participation in this communication.

Prior information

Cues

Signal

Consequences to receiver

Action 1

Action 2

Decision

Receiver

Consequences to sender

(A)

(B)

FIGURE 1.6 **Visual signal networks** (A) At low tide, each male fiddler crab (*Uca annulipes*) advertises himself and his burrow to females by waving his single grossly enlarged claw. When a female inspects the burrow of a particular male, other males cluster around the site and synchronize both their rate of waving and the phase of their wave display with those of the visited male. This synchrony provides no advantages in avoiding predation or attracting additional females, but is solely an emergent property of the local competition between males [5]. (B) Males of some species of fireflies (here *Photinus carolinus*) synchronize their flashes when attempting to attract female mates [8]. Like the fiddler crabs, the firefly males adjust their display behaviors according to those of their neighbors. In both species, the interactions between a single male and female cannot be considered independently of the presence and activities of neighboring males. The best approach is to study these systems as communication networks.

environment, and the appropriate economics for each taxon. Extraction of the principles governing the diversity of communication networks in animals is still in its infancy, but it is clearly a critical approach that is needed to complete the story. We review current principles of network communication in Chapter 15.

Finally, can the principles elucidated by studying the diversity of animal communication signals pass the test of taxonomic generality? For example, it is now widely recognized that communication occurs in bacteria and archaea, the diverse protist groups, and plants. Do these taxa follow similar rules? Humans are, of course, animals: to what degree are principles extracted by studying animal communication applicable to humans? In Chapter 16, we briefly review recent studies that have sought to extend the principles of animal communication to other taxa.

Principles of Evolutionary Biology

Evolution is increasingly seen as the core concept integrating all of the biological sciences. It is a *theory* in the scientific sense that although there is overwhelming evidence supporting it, scientists remain willing to refine or even refute aspects of the current version should new and persuasive evidence become available. The atomic basis of chemical reactivity and the well-known tenets advanced by Newton for object motion are theories that have been sufficiently tested and confirmed that they are now effectively referred to as *laws* or *principles*. Evolutionary theory has reached the same level of maturity: centuries of research, quantitative and experimental challenges, and tests of consistency across taxa and between disciplines have convinced nearly all scientists that the basic precepts of evolution are as likely to survive further tests as are the laws and principles of physics and chemistry [6, 18, 50, 73]. In this book, we thus refer to these precepts as *principles of evolutionary biology*.

The principles of evolutionary biology are relevant to every aspect of the study of animal communication. The physiological substrates that senders and receivers recruit for communication are considered by evolutionary biology to be **adaptations**: that is, they are likely to be those combinations of traits that in prior generations most effectively promoted their owners' survival and reproduction in their current contexts. Variants that were less effective resulted in early death or reduced reproduction of their owners. The process of differential contribution to future generations is called **sexual selection** when the relevant traits focus on competition for mates within a sex, and **natural selection** otherwise. Since most traits are to some degree **heritable** through **genetic transmission** to progeny, differential offspring production results in some trait combinations becoming more common over successive generations while other traits disappear. This is the process of evolution. Because new variants are continually appearing through **mutations** in the genes affecting traits, selection and evolution are continuing processes.

One evolutionary principle relevant to animal communication is that most traits suitable for signal production or reception are already adapted to specific functions and contexts. This why ears in aquatic animals like fish are likely to have very different structural designs from those in terrestrial animals such as birds. Evolutionary biology does not predict that every trait will be as adaptive as possible. Many traits have multiple effects and selection to improve a trait's effectiveness for the most crucial consequences may result in reduced suitability of the trait for other consequences. In addition, selection can only improve the effectiveness of a trait if mutation provides sufficient variants. Over evolutionary time, there may be adequate mutational variation to optimize the trait given the animal's basic anatomical constraints. But that is not necessarily the case for all traits. While we can expect most traits to be relatively adapted given a species' recent history, we should be alert to possible exceptions.

A second relevant evolutionary principle is that all organisms have arisen by descent. Ducks and geese appear to have evolved from a common ancestor, and humans and chimpanzees had a common ancestor. In increasing numbers of cases, extinct ancestors are now known from the fossil record. Because most traits have at least some heritable basis, it is possible to reconstruct a **phylogenetic tree** of organisms by examining fossil forms and looking for similarities and differences in the anatomy, physiology, and genetic structure of current taxa (**Figure 1.7**). Most phylogenetic trees show increased branching over evolutionary time, with no branches fusing and many branches arrested when a taxon goes extinct. However, there are notable exceptions. Many bacteria and archaea exhibit extensive gene transfer between species, and there are examples of taxa descended from quite different branches apparently combining into a new kind of organism. This is the most likely mechanism by which archaea evolved into eukaryotic organisms and by which some eukaryotic algae fused to create new algal taxa [39, 54]. There is little evidence that any animal taxa were created by organismal fusion, although one such claim remains controversial [29, 90]. On the other hand, specific genes can move between animal taxa, even between phyla, when transmitted by parasites with multiple host species [22].

This second principle of evolutionary biology means that closely related species are likely to have similar physiological substrates and physical constraints on the evolution of their signals. They are also likely to have similar ecologies, including diets and predators, and often relatively similar body sizes. All of these shared contexts favor similar economic constraints on their signal systems. The advantage is that once one understands the signaling economics of one member of a group, this often provides immediate hints as to the relevant economics of related taxa. Note that when initially related taxa move into different environments (e.g., some terrestrial mammals becoming aquatic), physiologies, ecologies, and signals are likely to diverge. Ecology and phylogeny then become two independent factors affecting signal diversity.

The third principle of evolutionary biology relevant to animal communication is a corollary of the first: behavior, like anatomy and physiology, is often an evolved and heritable trait. This means that communication behavior should largely be adaptive, and related species should perform similar communication tasks in similar ways. The courtship displays of dabbling ducks are strikingly similar, although some species use more of an ancestral repertoire than others [37]. The howls of dogs, wolves, and coyotes are quite similar, due to the similar preadaptations and functions of the howls [20, 21, 28, 65, 74].

The ubiquity of these principles should encourage us to ask of any animal signal system why the animal performs it the way it does. How much of the system is due to inheritance of signals evolved in immediate ancestors and shared with slight modification by related species? Why is a particular signal system adaptive, given the ecological contexts and phylogenetic constraints faced by the species? Answers will require us to compare the economics of the observed system with likely alternatives that may have been eliminated by selection in previous generations or may even be present in related species. At each of the stages of analysis outlined in this book—physiological and physical constraints, phylogenetic heritage, economics, and honesty guarantees—we can ask how particular aspects of animal communication are likely to have evolved and why the current form is adaptive compared to alternatives. As noted earlier, evolution provides a very powerful schema for examining any set of traits in organisms, and animal communication is no exception.

Classifying Communication Systems

Scientists classify natural phenomena in ways that reflect currently accepted principles. The taxonomy that defines species of living organisms and lumps them into genera, families, orders, and higher categories is based on the evolutionary principle of historical descent: species that share a common ancestor are combined into the same higher-level category. The utility of this classification is that the biology of a species about which little is known can often be inferred from knowledge of other species that, based on their anatomy, genes, and the fossil record, appear to be related by descent to the lesser-known form. Chemists classify atomic elements into the families of the periodic table based on the principle that the chemical reactivity of atoms is largely determined by the structure of their outer electronic shells. Classification is thus a way to summarize and predict similarities and differences among a diverse set of examples given underlying scientific principles.

It will be useful in this book to reduce the enormous diversity of animal communication systems down to a more manageable set of categories. If done properly, specific animal examples assigned to the same category would share many properties: knowing details about one example in that category may allow us to predict with reasonable confidence the properties of other yet-unstudied examples in that category. Our review of relevant principles above suggests four different schemes with which we might try to classify signal systems based on: (1) shared physiological and physical pre-adaptations for communication; (2) the informational economics of signals; (3) signal honesty guarantees; and (4) context. None of these schemes meets our goal perfectly by

FIGURE 1.7 **Bird phylogenetic tree** A recent study using the DNA of living bird taxa has reconstructed the most likely evolutionary relationships between major bird groups. (After [25].)

itself; but taken together, they do a pretty good job of sorting out the diversity seen in animal communication systems.

Classifying by preadaptation, modality, and medium

Animals monitor cues conveyed by sound, light, ambient chemicals, electric fields, hydrodynamics, and touch. Each of these modalities has also been recruited to convey signals by at least some animals. Although a remarkable number of species are capable of detecting magnetic cues [16, 17, 42, 43], so far no animals have been shown to use this modality for communication. The preadaptations required for the production and reception of signals vary markedly depending upon the modality. This is particularly true for receivers: for example, adaptations for monitoring sound cues are very unlikely to be recruited for detecting visual signals.

The medium (air, water, or solid substrates) in which animals monitor cues and exchange signals can have a dramatic effect on the adaptations required. Some modalities, such as electrical fields, simply do not propagate sufficiently between terrestrial animals to be useful for signaling. Vision is of little use for animals that spend their lives tunneling in the earth. Even where the same modality is feasible in multiple media, the constraints imposed by each medium on that modality can be markedly different. The many possible combinations of medium and modality thus provide a major source of diversity in animal communication systems. On the other hand, all animals trying to communicate with the same modality in the same medium face the same physical constraints. To the degree that each taxon's existing preadaptations allow, we should expect these shared constraints to produce at least some convergence in their signaling systems.

It thus seems that one useful way to classify animal signals would invoke the criteria of modality, medium, and taxonomic affiliation (in that order). As examples, we might expect fish and whales to show some convergence in the ways in which they produce and receive sounds, whereas both might be expected to show significant differences from terrestrial vertebrates such as toads or rats. Within the whales, the design of underwater hearing organs differs somewhat given the different evolutionary trajectories of the toothed whales (Odontocetes) and baleen whales (Mysticetes). As we shall see, a signal classification scheme based on modality and medium, with some taxonomic fine-tuning, can be very useful in interpreting the observed signal diversity of animals. This is the logic used to organize Chapters 2–7 in this book.

While many signals rely on only one modality, others are **multimodal**. Many vertebrates produce courtship displays that combine visual and acoustic components. Signals of insects, spiders, and aquatic invertebrates often combine visual and olfactory components; some even contain acoustic components as well. Multimodal signaling permits a wider variety of distinctive signals to be generated, but it also requires sufficient preadaptations in the relevant ancestors for manipulating and detecting more than one modality. Coordination of multiple modality components may also require additional nervous system integration not required for single modality signals

Classifying by informational focus

CODING RULES   The utility of communication to receivers hinges upon the ability of senders to correlate alternative signals with specific conditions of interest to receivers in a consistent way. As we have seen, this correlation need not be perfect, but it does have to be sufficiently reliable that attending to a signal improves a receiver's chances of making an appropriate decision or adjustment in its physiological state. To interpret signals, receivers must have prior access to the likely correlations that a sender will use to select signals. The correlations between a set of signals and conditions invoked by either party are called the **coding rules** for the exchange.

For communication to be effective, there must be some degree of concordance between the coding rules used by senders and those expected by receivers. There are several factors that can limit this concordance. One factor is the means by which each party acquires its coding rules. Species in which coding rules for both parties are largely heritable generally show a good match between sender and receiver coding. In contrast, species in which one or both parties largely acquire their coding rules by learning are prone to making mistakes. The most general procedure for acquisition combines some broad and heritable template for coding rules in both parties with subsequent refinement by learning and experience. For example, the canvas on which individual identity signals develop is largely heritable in most species, but the actual pattern in those signals depends on many different and unpredictable factors such as diet, injury, weather, local habitat conditions, and so on. Although the canvas is the same, individuals end up with at least slightly different signal variants. Receivers must then learn to associate specific signal variants with particular individuals to be able to discriminate identities.

A second factor limiting sender and receiver coding rule concordance is the degree to which signal propagation distorts the form of the signals. For close-range signals, this is usually not an issue. But for animals communicating over a distance, signal distortion can become a major problem. Clever receivers may thus use slightly different coding rules for long- and short-distance communication; receivers would continue to use the same rules. Alternatively, senders might only use signals that resisted propagation distortions, and then both parties could again share the same coding schemes. In Chapter 8, we outline a number of independent axes along which coding rules can vary depending upon functions, contexts, and taxa.

INFORMATION AND CLASSIFICATION   Sender and receiver coding rules must have a minimal level of concordance before signals can provide **information** to a receiver. The term information often has a different practical significance to animals than it does to literate humans. When we read a book or attend a lecture, we often learn of new facts that we might never have imagined before. The coding rules for animal signal systems rarely provide sufficient flexibility for this type of information acquisition. Instead, the possible *facts* have been discovered long ago during the species' evolution or more recently by the receiver's prior experience. The typical problem facing a receiver is which of several known alternatives is currently the case. Suppose the receiver is assessing whether or not there is a predator nearby. It recruits its own prior experience, any inherited biases favored over evolutionary time, and its monitoring of ambient cues to estimate the probability that a predator is present. If it then detects a conspecific's alarm call, which is usually correlated with predator presence, it will use this signal to **update** its estimates of predator risk before deciding what to do. As we shall discuss in more detail in Chapter 8, this change in probability estimates after receipt of a signal is what we mean by provision of information during animal communication. Properly scaled, it is also the best measure of how much information has been provided.

A consequent way to classify animal signals is by specifying the topic about which a sender is providing information to a receiver. A given topic might have several conditions of interest to receivers. To the degree that different species share interests in the same topics and conditions, we might expect relevant coding rules to show convergences. This provides a different perspective from which we might extract some general principles. Table 1.1 provides a hierarchical summary of topics and conditions that are most commonly encoded into animal signals. The broadest level divides signals according to three topic categories: (1) possible conditions of the sender; (2) possible conditions about the receiver that are uncertain or unknown to the receiver; and (3) conditions concerning a third party or entity.

For each topic, we list conditions that are of widespread interest across many animal taxa. We can further subdivide conditions into those that are relatively stable over time versus those that are more likely to vary over short time periods. We might expect signals reflecting fairly static conditions to require a different type of coding rule from those reflecting more rapidly varying conditions.

SIGNALS ABOUT THE SENDER   Since senders know themselves better than they know others, it is not surprising that the majority of the conditions about which they can provide information relates to themselves. Relevant signals have been described as "self-reporting" [30, 31, 49, 53]. Among the most common conditions of senders that are stable over long time periods are specifications of the sender's **identity**. These include its **species**, local **population**, current **social group** (if stable), **kin affiliation**, **pair bond** if mated, and own **individual identity**. Each of these aspects might be declared through one or more signals augmented by additional cues. Other relatively stable conditions that might be associated with self-reporting signals include the sender's **sex** (male, female, hermaphrodite, or asexual), **age**, **body size**, **toxicity** and **palatability** (to warn off predators), and **dominance status**. Some species wear visible "badges" that reflect their current **dominance rank** in their group.

A variable sender condition that a receiver often needs to know is the **location** of the sender. Senders can make the estimation of their location easy or difficult by careful

TABLE 1.1 *Informational classification of animal signals by topic and topic condition*

Signal topic	Condition stability	Specific examples

Sender conditions	Stable conditions	Identity:
		  Species
		  Population
		  Social group
		  Kin affiliation
		  Pair group
		  Individual

		Sex (male, female, hermaphrodite, asexual) Age Body size Dominance status Toxicity and palatability

	Variable conditions	Location of sender
		Sender's intended next action
		Physiological condition:
		  Hunger
		  Thirst
		  Reproductive receptivity
		  Health
		  Physical stamina
		  Coordination
		  Agility
		Learned skills (foraging, parental care, nesting)

Receiver conditions	Variable conditions	Detection of second party's presence Receiver selection: pointers Receiver selection: declaration of choice
Third party/ entity conditions	Variable conditions	Presence of third party/entity Location of third party/entity Quality or type of third party/entity Quantity of third party/entity Identity of third party/entity Urgency of response to third party/entity

adjustment of their signals' properties [64]. Another commonly signaled condition is the intended or threatened **next action** of the sender: this can be as obvious as exaggerated postures that anticipate that action or as subtle as an alerting note before a bird sings a complex song. Predators often seek to assess the **physiological condition** of alternative prey before selecting the most vulnerable to chase. Healthy senders may benefit by demonstrating their condition, and less healthy individuals will then have no choice but to do likewise or risk being hunted because they failed to signal. Receivers seeking mates may want to assess the **genetic quality** of potential candidates. While this is technically a stable condition, the most useful index of underlying genetic quality may be one or more current physiological conditions. In these and similar situations, senders often produce signals to indicate their current physiological states such as **hunger**, **thirst**, **reproductive receptivity**, **health**, **physical stamina**, **coordination**, and **agility**. A sender's level of **learned skills** with respect to foraging, parental care, or nest building can also be advertised using signals that mimic or demonstrate these abilities.

SIGNALS ABOUT THE RECEIVER   Many prey species produce signals that notify a stalking predator that it has been **detected**. The receiver (the predator) thus receives information about itself from the sender. Similarly, social animals may acknowledge the arrival of another with greeting signals that make it clear to the arriving animal that it has been noticed. The selective direction of broadcast signals to particular individuals is often difficult in crowded social situations: any nearby animal may mistakenly think a threat was directed at it. Some animals solve this risk by using **pointer** signals that specify the receiver to which subsequent signals will be directed. In a similar vein, a sender comparing several possible partners for some subsequent activity may use a signal to indicate to the favored individual that it has been selected (**choice**).

THIRD PARTY SIGNALS   Third party signals provide information about some individual or entity other than the sender or the receiver. The topic could be another conspecific, a member of another species, a predator, a nesting site, food, water, or a refuge. The simplest type of third party signal simply announces the **detection** of something of interest in the immediate environment. For example, many alarm signals announce the detection of predators, and food calls announce the discovery of a new food source. Information about the **location** of a third party or entity is often of great use to receivers. At one extreme, signals might only indicate whether the third party is nearby or far; at the other, they may actually indicate the direction and distance to be traveled to reach the third party or entity. Signals may also provide information about the **quantity** and **quality** of a food source, or the **type** of predator. In the case of a predator alarm, variant versions of the signal may indicate what the predator is doing and thus how **urgent** the situation may be.

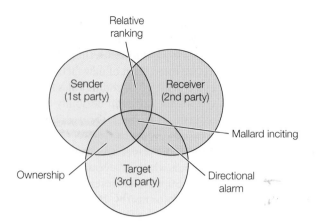

FIGURE 1.8   Examples in which information about more than one topic is provided by signals   In relative ranking situations, a sender signals its estimate of the relative values of itself and a receiver along some biologically important axis. This axis might involve dominance status, chances of winning a fight, or suitability for group leadership. The signal provides information both about the sender and about the receiver. When a sender uses a signal to declare ownership of a territory, food item, or a sleeping site, it is providing information about both itself and the target commodity that it claims to own. Consider a sentinel in a foraging group that spots a predator approaching a particular group member. If it can provide an alarm directed either acoustically or by its stare at the vulnerable group member, it will be providing information both about the target (the predator) and the receiver (vulnerable group member) at the same time. Finally, the inciting display of female mallard ducks indicates to her consort receiver that she accepts a sufficient level of bonding with him and expects him to attack the other (third party) male nearby. This signal provides information to her consort about herself, about her perceived relationship with the consort, and her evaluation of the other male as an intruder who should be chased away.

In the case of social interactions, an animal may use signals to **identify** for a receiver which other individual will be the target of the sender's subsequent displays or actions.

MIXED TOPIC SIGNALS   Some signals involve more than one topic and thus can be considered hybrids of the simpler examples listed in Table 1.1. Possible overlaps are summarized in **Figure 1.8**. For example, a sender may provide signals during a contest with a receiver that reflect the sender's estimate that it, rather than the receiver, will win the contest. It thus creates a **relative ranking** of the two parties. This type of signal provides a receiver with information about the sender's evaluation of itself and the sender's evaluation of the receiver.

Another type of hybrid signal occurs when a sender declares **ownership** of a territory, sleeping site, food item, or even another individual. The intended receivers are any possible intruders or competitors in the area, and the topics are both the sender's status as owner and the identity of the commodity that the sender claims to own. Alternatively, consider a sentinel watching for predators while its group's members forage. If it spots a predator approaching a particular group member, it might send a **directional alarm** signal toward a particular receiver to inform it that it is at risk from a predator. This type of signal provides information about both the receiver and a third party.

Finally, some animals produce signals that provide information on all three topic categories at the same time. One example is the inciting display of female mallard ducks. This display is performed as the female swims with her current male consort and approaches or is approached by a second male. The female's display is a repeated pointing of her bill at the intruder male, and her consort's response is often to attack the second male. This display indicates to the female's consort (the receiver) that she (the sender) feels sufficiently bonded to him that she expects him to ward off competitor males (the third parties).

MIXED CONDITION SIGNALS   A signal may provide information not only about multiple parties, but also about multiple conditions. Information about identity is often combined with information about other conditions. For example, the odor trail that a worker ant lays on the ground to guide its nestmates back to a food discovery is also likely to provide information about the species, population, nesting colony, and even individual identity of the worker. Body size may be indicated by a male frog's call at the same time that he uses the signal to declare ownership of a territory in the pond and his readiness to mate with receptive females. In some cases, multiple conditions are encoded using different modalities in a multimodal signal; in others, different aspects of the pattern in a single modality can be used to provide different kinds of information.

Classification by honesty guarantees

We noted earlier that senders and receivers may have conflicts of interest and thus disagree as to the optimal decision by the receiver. This provides an incentive for senders to emit signals that cause the receiver to make a decision that is optimal for the sender but not for the receiver. These are considered deceitful signals. Since receivers usually have the upper hand in making decisions, they can minimize this risk, at least over evolutionary time, by attending only to those signals that carry some guarantee of honesty. There are several types of guarantees possible and we shall discuss them in considerable detail in later chapters. For the purposes of signal classification, we can focus on five common cases [27, 30, 31, 35, 49, 53, 84, 85, 88]. **Index signals** are sufficiently constrained by physiology, body size, or access to information that they are unbluffable. Only those senders with appropriate properties can produce them. **Handicap signals** are available to all senders, but the costs of their production are higher (or the benefits lower) for deceitful senders. The encoding rules for **conventional signals** are arbitrary, and thus lack the direct link between signal and honest content seen in index and handicap situations. However, honesty can still be enforced if receivers test senders at intervals and punish those found being deceitful. **Proximity signals** are largely used during aggressive encounters and place a sender in a vulnerable position where the risks of injury keep deceit at a minimum. Finally, if senders and receivers have minimal conflicts of interest, honesty is favored by both parties and no guarantees are necessary. Note that honesty in one component of a multimodal signal might be guaranteed because it is an index, whereas another component might be kept honest because of handicap costs. As with the other classifications listed above, a given multimodal signal may fall into more than one category of honesty guarantee.

Classification by context

We can also classify signals by the social context in which they occur. Receiver decisions made in different social contexts result in different kinds of payoffs to each party. In some contexts, payoffs are associated with health, successful foraging, physical integrity and survival; in others, the key payoffs are related to successful reproduction. The signal repertoires of most species can be partitioned among four general categories according to social context. Usually, a given signal is used in only one of these contexts, making for a natural criterion for signal classification. These four categories form the basis for our treatments of signals in Chapters 11–14.

Most species have at least some signals that are used in **aggressive** contexts. Such signals function as threats, opponent assessments, appeasements, and indications of dominance status. The diversity of signals in this category is highly variable among species. A second contextual category involves **mating signals**. We often find separate signals in a species repertoire for mate attraction, courtship, solicitation, copulation/gamete release, and postmating announcements. **Social integration signals** are largely devoted to activity coordination. The group being coordinated can range from a mated pair performing parental care to a large group of individuals that arrives, forages, and flees as a unit. Finally, **environmental signals** provide information about the presence, and in some cases the location, of predators, food, water, refuges, or other resources.

The payoffs of exchanging signals in a given context may differ significantly depending upon who is involved in the exchange. Close **kin** have shared genetic interests: production of offspring by one individual helps promote genes shared by its relatives. The benefits and costs that determine the payoffs of a given exchange will vary depending upon whether sender and receiver are or are not relatives. Similarly, some animals may have long-term relationships involving **reciprocity**: aid by one partner to the other at one time is compensated by a reversal of the roles on some later occasion. Again, the economics of a given signal exchange may differ depending upon whether sender and receiver have already established a reciprocal relationship. Threaded throughout the four contexts outlined above are adjustments in the magnitude of payoffs depending on relationships between the communicating parties. Classifications of signals based on context usually require qualifications based on who is involved in the interaction.

Categorization of signals based on context is at best a fuzzy set exercise. Many male songbirds produce songs that are designed to repel male intruders and attract potential female mates at the same time. Social integration signals can become complicated by aggressive considerations when exchanged by animals with differing dominance status in the group. Environmental signals can play key roles in maintaining social integration of cohesive foraging groups. It is important to recognize that none of the signal classification schemas outlined in this chapter produce exclusive categories, and there are, in fact, no reasons why the underlying principles should require such exclusivity. While mixes may require a higher level of anatomical and physiological integration, these costs may be made up by the economy of using one multimodal signal for multiple functions and contexts. We review the various types of mixed-function signals in Chapter 8.

Cross-classifications

In practice, most researchers in animal communication invoke two or more of the above classifications concurrently. Thus some workers have combined informational focus as one classificatory axis and honesty guarantees as a second axis [30, 35]. Another obvious combination might invoke social context and honesty guarantees. We might expect both the degree of conflict of interest and the relevant payoffs to vary in different social contexts. The literature on animal communication often mixes classification schemes rather blithely. In the same issue of major scientific journals, one can find articles with titles referring to *mating solicitation* (a context-specific term with no topic or modality information), *signature whistles* (a mixture of modality and informational content, in this case, sender identity), and *predator alarm pheromone* (a mixture specifying the olfactory modality, a third party subject, and environmental context). Many scientists working on animal communication are not hampered by this inconsistent approach, because they are sufficiently familiar with the alternative modalities, focal topics, honesty guarantees, and contexts to be able to identify a given signal system with any of these names. However, it can be confusing to students and others new to the field, and we need to be honest at this point and confess that there is currently no consistency in how animal signals are named and classified. Be warned!

At the same time, the fact that no single classification system has gained dominance testifies to the complex nature of communication. By definition, communication involves at least two active participants: a sender and a receiver. Each is out to promote its own interests, which may or may not be congruent with those of the other. The overall payoffs of an exchange to each party may vary in complicated ways. Add in the interests of referenced third parties and eavesdroppers, and the classification of animal signals can become very complex, indeed. Animal communication is subject to coevolutionary processes: as one party in the system gains an advantage, selection pressures on others increase, leading to subsequent changes in their contributions. This makes for a very large number of potential evolutionary forces and potential trajectories, and one might worry that there cannot be order or general rules in such a situation. In fact there are, and as the field matures further, more and more predictable structure is becoming apparent. There will always be diversity in signal systems as long as there is a diversity of animals to use them. But luckily, general principles and useful classification schemes are emerging, and these are the main focus of this book.

The Signaling Sequence

While the opportunity to create different signal patterns might initially seem limitless, in practice, the modality, medium, topic, and need for honesty guarantees impose severe constraints on which patterns will work as signals and which will not. In each modality, there are only certain patterns that a given sender will be capable of generating. After emission of the signal, there will be only certain patterns that can propagate between sender and receiver without serious distortion. And once the signal arrives at the receiver, there will be only certain patterns in that modality that a given receiver is capable of recognizing. When the constraints imposed at each stage of the communication process are added up, the number of useful signal patterns can be quite limited. This winnows down the actual coevolutionary trajectories and outcomes significantly. It is thus very important to understand both the possibilities and the constraints for each signal modality.

In the following six chapters, we examine these possibilities and constraints for acoustic, visual, chemical, tactile, hydrodynamic, and electrical communication, respectively. Although we treat only one modality at a time, it should be remembered than many real animal signals are multimodal. We take up the issues of multimodality later in the book. For now, it will be sufficiently challenging to get a good grasp of which patterns are possible and which not in each modality separately.

Regardless of modality, the **signaling sequence** involves seven steps (**Figure 1.9**). The first three steps occur in the sender. These are the initial **generation** of a patterned stimulus, the subsequent **modification** and refinement of the stimulus pattern, and finally the **coupling** of the modified stimulus to some **medium** that links the sender to the receiver. Usually, senders must evolve special adaptations and tricks to accomplish each step of their part of the signaling sequence.

Once emitted by a sender, a signal must then **propagate** through the medium to the receiver. During this fourth and intermediate step, the pattern so carefully crafted into the signal by the sender can easily be distorted and, if propagated over a long enough distance, can disappear in the ambient noise. For sound, the medium linking sender and receiver can be air, water, or a solid such as a plant stem or the ground. For light, no medium is needed, since light can propagate in a vacuum; however, no animals live in a vacuum, so light signals propagating between senders and receivers must pass through

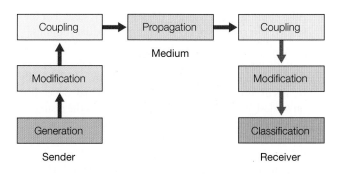

FIGURE 1.9 **The seven steps of the signaling process** Each modality will create its own challenges to communicating accurate information at any of these stages.

air, water, or even in certain cases, solids. Chemical signals propagate through air, water, or a few porous solids. Tactile, hydrodynamic, and electrical signals are all limited to short-range communication. Touch, of course, propagates directly from the surface of the sender to a surface on the receiver. Distortion is limited to interactions between the sender and receiver surfaces. Hydrodynamic signals tend to remain near to the location in which they were generated. However, they may persist long after the sender has moved away and thus may not be detected until the sender is at a considerable distance from the receiver. Air is a sufficiently poor conductor of electricity that electrical signals of animals are limited to aquatic environments. Distortion of electrical signals in water depends upon the nature of dissolved solutes and the proximity of objects with differing electrical properties.

The final three steps in the signaling sequence occur at the receiver and are the mirror images of those undertaken by the sender. First, the receiver must couple the incoming signal from the propagating medium into its sensory organs. Because the optimal propagation pattern may not be in the optimal form for information extraction, receivers often modify or transform the pattern of the signal once it enters their bodies. Finally, the receiver's sense organs and brain **identify** and **classify** the pattern that remains in the received and modified signal. At this point, the brain regions that oversee decision making step in and take over the remainder of the process.

SUMMARY

1. While most people acknowledge that animals communicate, they often lack the background needed to recognize common patterns and principles amidst the enormous diversity of animal signals.

2. The field of animal communication research, like other sciences, seeks to identify the basic principles that are required to explain and organize animal signal diversity. As is customary in modern science, these principles must be concordant with principles derived in other fields such as physics, chemistry, physiology, and economics, and they are proving general enough to provide insights for other disciplines such as psychology, medicine, conservation and wildlife management, and linguistics.

3. All animals monitor the ambient physical, ecological, and social conditions around them. Because many of the relevant conditions are not directly measurable, animals use their **sense organs** to monitor **cues** that are correlated with the conditions of interest. Because cues are only correlated with conditions, they typically provide imperfect information.

4. **Signals** are stimuli produced by one animal (the **sender**) and propagated to another animal (the **receiver**). Like cues, they are correlated with conditions of interest to the receiver, but unlike cues, which are generated inadvertently or as a by-product of some function other than communication, signals are generated expressly to provide additional **information** to receivers.

5. Because all animals have evolved the sense organs and sampling behaviors required to monitor ambient cues, the **evolution of signals** is often a relatively simple process. A mutant sender displays a slightly exaggerated anatomical feature or performs a slightly enhanced version of some action that another animal is already monitoring as a cue. If both parties benefit by this exaggerated or enhanced display, subsequent coevolution of the sender trait and receiver perception leads to further exaggeration and refinement and a new signal is created.

6. The relevant principles needed to explain animal signal diversity can be grouped into general categories: (A) those due to **physical constraints** imposed by the habitat in which each species lives and reflected in the **physiological mechanisms** that serve as **preadaptations** for signal generation and reception; (B) those due to different evolutionary histories for each taxon, that limit the kinds of preadaptations available for signal evolution; (C) those imposed by the **economics of communication**, which require that both senders and receivers benefit on average when exchanging signals; (D) those imposed by **conflicts of interest** between sender and receiver and the need for **honesty guarantees** in the design of signals; and (E) those imposed by the hordes of eavesdroppers that form the ambient **communications network** and may exploit the exchanges between any pair of animals.

7. As with all other fields of biology, evolutionary theory provides the backbone for studying all aspects of animal communication. Basic evolutionary principles such as the **genetic heritability** of traits, **sexual** and **natural selection** for some variants over others, **phylogenetic affinity**, and the perpetual introduction of new variation through

mutation are all critical to understanding the origins and maintenance of animal signal diversity.

8. Animal communication systems can be classified in several ways. One axis focuses on the **modality** (light, sound, chemicals, touch, hydrodynamics, or electricity) used to create signal stimuli, the **medium** (air, water, or solids) through which the signal propagates, and the preadaptations for that modality and medium that are present in each taxon.

9. A second axis for signal classification focuses on the alternative kinds of information that signals might provide. Since the sender knows more about its own status than any other animal, many signals provide information about sender conditions. Such information may include sender identity, sex, age, dominance status, and location. In some cases, the sender can provide information about the receiver that the latter cannot know on its own. Finally, the sender may provide receivers with information about other parties, such as predators or approaching conspecifics, or about objects, such as food.

10. A third method of classifying signals focuses on the mechanisms by which signals are kept honest when sender and receiver have a conflict of interest. **Index signals** are constrained in ways that make it impossible for a sender to be deceitful. **Handicap signals** impose higher costs or lower benefits on deceitful senders. Receivers can punish deceitful senders using **conventional signals**, and in aggressive contests, confident senders can confirm their self-assessment by making themselves vulnerable with **proximity signals**. Where there is minimal conflict of interest between sender and receiver, no honesty guarantees are needed.

11. A fourth axis classifies signals into one of four contextual categories that differ in the relevant payoffs that each party receives following communication exchanges. Usually, a species' signal repertoire can be divided into its **aggressive signals**, **mating signals**, **social integration signals**, and **environmental signals**. Different species may have more or fewer signals in each category, but most species have at least a few signals of each type.

12. Animals may use signals that mix categories both within and across the above classifications. **Multimodal signals** utilize two or more modalities at once, and each modality may invoke a different honesty guarantee. Birds often combine acoustic and visual displays, whereas insects combine visual and chemical components. Animals may also provide information about multiple conditions in the same signal: a frog's call can indicate the caller's body size and its species in the same signal. Finally, the same signal may be used in multiple contexts: the songs of many birds are designed both to repel male competitors and to attract potential female mates.

13. All communication requires the same seven steps. For the sender, this consists of (1) generating an initial stimulus, (2) modifying it to ensure proper pattern, and (3) coupling it to the propagation medium. These three sender steps are followed by (4) propagation in the medium, which usually results in some distortion of the released signal, depending on its design, the modality, the medium, and the distance between sender and receiver. The final three steps occur at the receiver and are the reverse of those undertaken by the sender: (5) coupling the propagated signal from the medium into the receiver's sense organs, (6) modifying it as necessary to improve detectability and resolution, and finally (7) identifying and classifying the perceived signal. Variations among species at each step contribute to the overall patterns of signal diversity seen in animals.

Further Reading

Good introductions to the evolution of animal behavior in general can be found in Alcock [1] and Dugatkin [13]. Signal classification and associated evolutionary processes are discussed by Hauser [33] who provides an in-depth review of the physiological preadaptations for signal evolution; Hailman [26], who classifies signals according to the complexity and redundancy of coding rules; and Searcy and Nowicki [77], and Maynard Smith and Harper [53], who review current thinking about conflicts of interest, honesty guarantees, and signal evolution.

COMPANION WEBSITE
sites.sinauer.com/animalcommunication2e

Go to the companion website for Chapter Outlines, Chapter Summaries, and References for all works cited in the textbook. In addition, the following resource is available for this chapter:

Web Topic 1.1 *Animal communication and science education*
Because most students naturally like animals, and animal communication integrates so many disciplines, the topic can be used as an entry point for science education in middle and high school curricula. Here we provide some background and relevant links.

Web Topic 1.2 *Information and communication*
Some scientists feel that the role of information provision should be downplayed in definitions of animal communication. A few even recommend elimination of the term when applied to animal interactions. Here, we outline the case for information as a useful and even key concept in understanding the evolution and diversity of animal signals.

Chapter 2

Sound and Sound Signal Production

Overview

Most humans are born into a world in which sound communication is a dominant activity. While we tend to take the mechanisms that allow vocal exchanges for granted, the production of useful sound signals presents significant hurdles, and most animals dispense with sound communication altogether. In this chapter, we outline the basic physical constraints within which we, and animals, must operate if we are to communicate with sound, and describe some of the techniques that different species use to mitigate and even take advantage of these constraints.

Properties of Sound

The media of sound communication

In contrast to the vacuum of outer space, every cubic meter of our planet is filled with matter. Matter consists of atoms that are largely bound together into molecules. When animals want to communicate with each other, their signals must pass through some type of intervening matter. We call the matter linking sender and receiver the **medium** of the communication process. Different media impose different constraints on the kinds of signals that senders can create and receivers can detect and process. For sound, the different states of matter constitute quite different media.

STATES OF MATTER We can initially categorize a medium according to whether it is a **fluid** or a **solid**. In turn, we can divide fluids into **gases** and **liquids**. To understand the differences, note first that all molecules on the Earth are moving at least slightly. A number of types of motion are possible, including internal vibrations within each a molecule, molecular rotations, or translational movements of a molecule from one location to another. The total energy in all such movements is the **kinetic energy** of the molecule. The **temperature** of a medium is a measure of the average kinetic energy of its molecules.

Molecules and atoms can be attracted to each other through a number of forces. Whereas kinetic movements tend to separate molecules, the attractive forces bring them closer together. When the kinetic energy of molecules is high relative to the attractive forces between them, the matter occurs as a gas (such as air). Air molecules occur at low densities, which means they can travel considerable distances before they collide with another molecule. When they do collide, they bounce and do not stick to each other. In a liquid (such as water), the attractive forces between molecules are high relative to kinetic energies; molecules thus remain much closer together and at higher densities than air. A water molecule can move, but not very far, before it collides with another, and water molecules tend to cohere (stick together) when they do collide. Air and water are both considered fluids because they can assume any shape. In contrast to fluids, solids tend to retain their shape when perturbed. The attractive forces in a solid are so much stronger than the kinetic energy of the molecules that molecules can make only tiny translational movements. Internal vibrations may also be limited by the proximity of nearby molecules, and molecular rotations are usually prevented by the strong linkage between neighbors.

PRESSURE   If we were to map the translational movements of individual air or water molecules, we would see that the directions moved in any small neighborhood were random. Suppose we place a small square surface inside such a neighborhood. When a molecule collides with that surface, it exerts a small force on it. Over a short period, many different molecules will collide with that surface, exerting a combined force on it. This total force divided by the area of the surface is known as the **ambient pressure** of the medium. Since molecular movements are random, similar numbers of collisions occur over time in any given direction. We would record the same pressure whether we aimed the surface upward, to the side, or even diagonally. Pressure is nondirectional.

In a neighborhood with a higher density of molecules than our first sample, there would be more molecules available to collide with the square, and the net force, and thus the measured pressure, would be higher. Similarly, if we examined another neighborhood with the same density of molecules as the first sample but at a higher temperature, the higher kinetic energy per molecule would result in a greater force per collision, and again, a higher recorded pressure. Ambient pressure is thus positively correlated with both molecular density and temperature.

The nature of sound

GENERATING SOUNDS   Suppose we insert a large flat circle of a hard material into a fluid medium such as air or water. After the molecules settle down to their usual random movements, we can use the disk to measure the ambient pressure in the medium. Once we have this measurement, we rapidly move the flat side of the disk forward by one centimeter (**Figure 2.1**). As the front side of the disk moves forward, it pushes all the air molecules in front of it forward as well. This creates a higher-than-average density of molecules (a **condensation**) on the front side of the disk, and a lower-than-average density of molecules (a **rarefaction**) on the back side of the disk. The condensation layer of molecules will show a pressure that is higher than the ambient pressure we measured earlier, and the rarefaction layer will show a lower-than-ambient pressure.

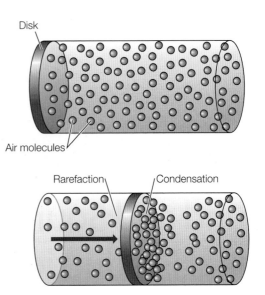

FIGURE 2.1   **Generation of a sound by movement of a disk in an air-filled tube** If the disk is immobile, air pressures will be identical throughout the tube. As we move the disk suddenly to the right, we collect air molecules and create a condensation ahead of it. At the same time, we create a rarefaction behind the disk. The disturbances created on each side of the disk are then transferred to successive layers of molecules further away from it even though individual molecules do not move very far. We will have created a propagating sound.

In addition to altering local pressures, the movement of the disk transfers additional kinetic energy and a relatively cohesive forward direction of motion to each of the molecules in the condensation. These molecules move forward and collide with the next layer of molecules. These collisions transfer both the forward direction and extra kinetic energy to the molecules in the second layer, while bouncing the first layer back toward the disk. This process continues to transfer the condensation to successively distant layers while individual molecules move no further than the 1-cm displacement caused by the disk movement. Behind the disk—the area of rarefaction—a similar process takes place. In this case, molecules just outside the rarefaction encounter more collisions from molecules farther from the disk than from those close to it. They thus get bounced into the rarefaction and fill it up. However, this leaves a rarefaction layer behind them, and this causes the next layer to move into that zone and fill it up. In this way, the rarefaction also radiates away from the disk.

The net effect of moving the disk forward is to generate two disturbances (a condensation and a rarefaction), both of which radiate away from the source over time. While the molecules closest to the disk are forced to move distances much larger than they would have traveled given no disk movement, successive layers of molecules acquire the forward kinetic energy not from the solid disk but from collisions with other molecules. Their subsequent direction of movement will be a combination of the random movement they would have undergone after any normal collision plus

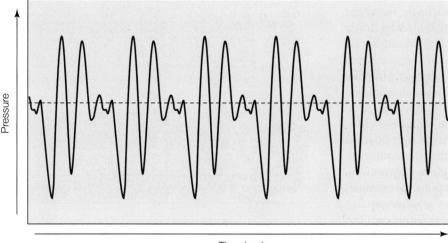

FIGURE 2.2 Waveform of English-speaking human child saying the word _are_ It is easy to see that even though there is considerable variation in successive peaks of this waveform, there is a clear pattern that repeats every fourth peak. This signal is thus periodic.

the forward motion of the disturbance. As a result, molecules farther from the disk pass on the forward disturbance, but their actual net displacement gets smaller and smaller with distance from the disk. Eventually at long distances from the disk, it is only the disturbances that change location, and not individual molecules. A radiating disturbance in the pressure of a medium is called a **sound**.

> ### Web Topic 2.1 _Measuring sound pressure_
> Microphones are used to measure the variations in pressure caused by a propagating sound. Specialized types of microphones exist for sound propagation in air, water, and solid substrates. These microphones work by converting pressure variation into electrical signals that can then be measured, stored, and characterized.

GENERATING WAVES IN A FLUID A single sudden disturbance in the pressure of a medium generates an **impulse** sound. What if, instead of creating a single impulse, we repeatedly move the disk back and forth along some fixed axis? Let us further ensure that the movement of the disk is **periodic**: that is, that it repeats the same motion over and over. This will generate successive condensations and rarefactions on both sides of the disk. Instead of a single radiating layer of condensations around the sound source, we now have a series of successive condensation (and intervening rarefaction) layers. The inner layers reflect the most recent disk movements and the outer layers reflect earlier movements.

If we use a fixed microphone to record the successive condensations and rarefactions as they radiate past us, we can generate a graph of pressure versus time. Depending on the way we move the disk back and forth, this graph might appear as in **Figure 2.2**. Because the pressure measurement in this graph rises and falls around the ambient pressure average, we say that the sound propagates as **waves**, and the graph of pressure versus time is thus called the sound's **waveform**. In our example, the shared direction of movement that

molecules pass on to each other as the disturbance spreads is the same direction in which the condensations and rarefactions are moving. When molecular movements and sound disturbances travel in the same direction, we say that the sound consists of **longitudinal waves**. This is the primary process for sound propagation inside large volumes of fluids such as air and water (**Figure 2.3A**).

> ### Web Topic 2.2 _Visualizing sound waves_
> The best way to understand the differences between different types of sound waves is to view an animation that shows how the molecules move as the sound propagates. We list a number of websites where you can watch visualizations of most of the basic acoustic processes described in this chapter.

GENERATING WAVES IN A STRING If we tie a rope to a tree, add just a bit of tension, and make a small flip in our end, a wave of displacement will move down the rope until it hits the tree. This is another example of a propagating wave. Note that each segment of the rope does not move toward or away from the tree, as would the molecules propagating a longitudinal wave, but instead oscillates along a line perpendicular to the rope axis. A similar type of oscillation is generated in the strings of a guitar or harp when they are plucked. This oscillation is called a **transverse wave** (**Figure 2.3B**). Light signals are propagated as transverse waves of electromagnetic energy. As we shall see, spiders and some insects communicate by sending transverse waves along strands of silk, and some fish species pluck tightened tendons like guitar strings to make sounds.

GENERATING WAVES ON A WATER SURFACE Unlike air, water is a relatively sticky medium: water molecules are strongly attracted to each other. Below the surface of a body of water, the forces experienced by each water molecule include the downward pull by gravity and the attractive forces from all directions exerted by nearby molecules. At the surface, gravity continues to pull down on the molecules; but now, attractive forces only pull the molecules downward and to the sides. Unless disturbed, the result is a flat surface with a strong surface tension. When we toss a rock into a pond, the ripples radiate out as circular waves from the impact point. These waves propagate with both longitudinal and transverse components (**Figure 2.3C**). In a ripple moving from left to right, a molecule near the surface moves in a clockwise trajectory, rising and moving in the same direction as the ripple when the crest passes, and then descending and moving opposite to the direction of ripple movement while in the trough (see Web Topic 2.2 for an animated example). When ripples are close together, surface tension (capillarity) dominates ripple behavior, and the resulting **capillary waves** radiate more slowly as the distance between them is increased [39, 119, 161, 230]. When the ripples are far apart (at least 1.7 cm or more), gravity affects the ripple behavior more than surface tension does. The resulting **gravity waves** radiate more quickly from the source as the distance between ripples is increased.

GENERATING WAVES IN A SOLID   In a solid, molecules cannot move very far before they collide with adjacent molecules. As a result, one often finds both longitudinal waves (also called compression waves) and transverse waves (also called shear waves) propagating inside a solid at the same time. In cylindrical solids such as plant stems, a third category, called **bending waves**, may also be present. These are similar to transverse waves but propagate in more complicated ways [254]. One may also find waves propagated on the surfaces of solids. These are similar to the ripples described above for water surfaces, except that surface molecules in solids move in an elliptical counterclockwise trajectory as the sound propagates from left to right (**Figure 2.3D**). In **Rayleigh waves**, the transverse component moves on a vertical line perpendicular to the surface, (see animation at Web Topic 2.2), and in **Love waves**, the transverse component is a side-to-side motion parallel to the surface. As one moves into the interior of a solid, the elliptical trajectories of molecules near the surface gradually lose their longitudinal component and become entirely transverse.

The characterization of sounds

Sound is a propagated disturbance in the density and pressure of a medium. This means that characterizing any given sound requires that we specify both how sound pressure might vary at a fixed location over time (temporal properties) and how sound pressure might vary over space at any given time (spatial properties). We take up each of these perspectives in turn.

BASIC TEMPORAL PROPERTIES OF SOUNDS   We now return to the use of a moving disk to generate sound waves in air or water, only this time we place a single microphone at some moderate distance from the disk to record pressure variations at that site over time. Let us begin with a sound source

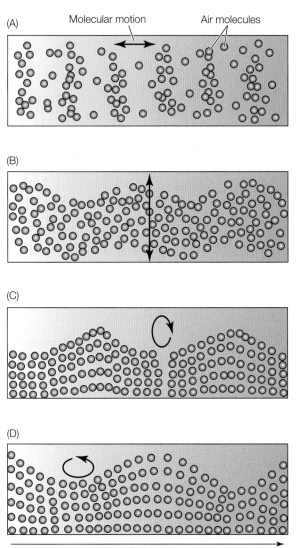

FIGURE 2.3   Types of propagating sound waves In all examples, sound propagates from left to right, and the dark arrows indicate the trajectory of motion of propagating molecules. (A) Longitudinal waves such as one finds in air or water far from any boundaries. Molecular motion is parallel with the direction of sound propagation. (B) Transverse waves as in a vibrating guitar string. Molecular motion here is perpendicular to the direction of sound propagation. (C) Surface waves on a body of water. Surface molecules propagate the ripples by moving in a clockwise (relative to direction of propagation) elliptical trajectory. (D) Rayleigh waves on the surface of a solid substrate. Molecules on the surface move along a counter-clockwise elliptical path. Love waves (not shown) are similar to Rayleigh waves except that they move the substrate side to side in a transverse motion, whereas Rayleigh waves move the substrate up and down. In solids, one may find any combination of these wave types as well as additional types such as bending waves.

that moves in a simple **sinusoidal pattern**. This is the movement pattern that we would record if we tracked the oscillating position of a child on a playground swing or a bouncing weight suspended from a spring. In each case, the object being tracked moves rapidly in one direction, gradually slows down, and then reverses direction and moves the other way, building up speed as it passes the midpoint, and then gradually slows down to reverse its movement again. If we move our disk in

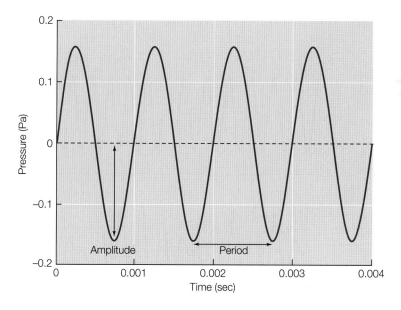

FIGURE 2.4   Time recording of varying pressure due to passage of sinusoidal sound waves past a microphone   Amplitude is a measure of the deviation of the passing sound pressures relative to the ambient level (here marked as 0). A typical songbird would produce a sound amplitude of about 0.16 Pa recorded 1 m away. The period is the amount of time it takes to record one complete cycle of the sound. Here, the period is about 0.001 sec. The frequency of the sound is the reciprocal of the period and in this example would be about 1000 Hz.

such a sinusoidal way, it will generate successive condensations and rarefactions in the medium that will pass by our microphone. A graph of the corresponding pressure recorded by the microphone over time would appear as in **Figure 2.4**.

There are several obvious measures we could make on this graph to characterize the sound. We see that the pattern of pressure variation repeats in a regular fashion. One obvious measure that we could extract from the graph is the interval we have to wait before the same part of the sound wave repeats. This interval, measured in seconds, is called the **period** of the wave. We could also measure how many complete cycles occur per second, which is called the **frequency** of the sound. Because wave frequencies are important measures throughout physics, the fundamental unit of frequency, one cycle per second, has been named the **Hertz** (abbreviated **Hz**), after the German physicist Heinrich Hertz. In practice, we need to measure only period or frequency, since one is the reciprocal of the other. For example, if we find the period of a sinusoidal wave to be 0.001 seconds (or 1 msec), the frequency can easily be computed as 1/0.001 = 1000 Hz (or 1 kHz). The only other time measurement we might want to make on this graph is *when*, relative to some reference time, the wave achieves a specific part of the cycle, such as a peak or valley. This is known as the **phase** of the wave. Phase is not very important when there is only a single frequency present in a sound. But as we shall see, it can become very important when multiple sine waves are present in the same sound or at the same location.

In addition to the time measurements, this graph allows us to measure the degree to which the passing sound waves differ from ambient pressure levels. One obvious measure is the difference between the peak (or valley) pressure and the ambient pressure (shown as the midline of the graph). This difference is one measure of the **amplitude** of the sound wave, measured as a standard unit of pressure, the **Pascal** (abbreviated **Pa**). While the difference between the peak and ambient pressure is a useful amplitude measure for single sine waves, it may not be very representative for more complicated sounds. The most common alternative is a weighted average of the differences between each part of the wave and the ambient level

called **RMS** amplitude. Details can be found at Web Topic 2.3. Finally, whether one measures peak or RMS amplitude, most publications on sound will not report these absolute values, but instead will report the ratio between the measured amplitude and a reference sound. The most common reference is the pressure difference between the measured sound and the softest sound an average human can hear (called SPL). Because humans can hear over a very wide range of amplitudes, these ratios can get very large. The practical solution is to use the logarithm of the ratio, properly scaled, to report amplitudes. This relative unit of amplitude is called the **decibel** (abbreviated **dB**). See Web Topic 2.3 for details on computing amplitudes in dB. However measured, the amplitude of a sound signal is usually referred to as its **sound pressure**.

Web Topic 2.3 *Quantifying and comparing sound amplitudes*

A variety of methods are available for measuring and comparing sound amplitudes. Here we define some of these methods, show how they are computed, and discuss when each might be most useful.

One might wonder why we don't use ambient pressure as the reference point and then refer to different sound levels as the percentage change from ambient pressures. There are several reasons why this is not done. The first is that ambient pressures change with altitude, the weather, local exposure to sunlight or wind, and many other factors. Having a fixed reference standardizes sound measurements. The second reason is that the change in pressure created by a passing sound wave is extremely small. Even a sound with an amplitude so high that it reaches the pain threshold in humans would only change local pressures by 0.02% as it passed. This is another reason why the instruments for measuring ambient pressure and sound pressures require quite different designs.

An additional temporal measure of sound propagation that will be important in later chapters is **particle velocity**. Velocity is defined as the distance an object moves per unit time. If we focus on an infinitesimal volume of fluid as a sound propagates, this "particle" of fluid will move back and forth at the same frequency as the sound. At low-fluid densities, this particle will be able to move some distance before it turns around and moves back in the other direction; at higher densities, it will encounter more resistance to its movements and the displacement will be less. Since the frequency is the same in both media, the time available for movement in a given direction has to be the same. Thus, the particle velocity in the lower-density medium will be higher than that in the higher-density medium. We shall see that this has important implications for how animals emit and capture ambient sounds.

TEMPORAL PATTERNS OF SOUNDS: INTERFERENCE AND BEATS   The prior example considered a sound source moving in a sinusoidal manner. What if the movement of the disk is best described as the sum of two sine waves? Alternatively, what if our microphone responds simultaneously to sine waves radiated from two different nearby disks? In either case, any medium molecule near the microphone is going to be receiving kinetic energy and directions of motion from two different sine waves at the same time. What would we record? In fact, what we record is simply the sum of the pressures predicted at each moment for the two waves. Suppose the two sine waves have exactly the same frequencies and amplitudes. If they arrive at the medium molecules near the microphone so that they are in exactly the same part of their cycle, we say that they are **in phase**. When both peak at the same time, the total pressure at the microphone is twice what it would be if only one sine wave were present. Now consider the opposite possibility that the two waves are completely **out of phase**: when one has its peak near the microphone, the other is in its valley. The condensation caused by one wave will be countered by the rarefaction caused by the other. The result will be no deviation from ambient sound pressures at all: the two waves simply cancel each other out. When two sine waves with the same frequencies and amplitude are in phase, we say that they experience **positive interference**; when they are out-of-phase, they experience **negative interference**.

What if the two sine waves do not have the same amplitude or frequency? They can still interfere with one another. As the two waves become more dissimilar in amplitude, the higher-amplitude wave will dominate the combination and the second wave will have a decreasing effect on the first (whether positive or negative). If they have the same amplitude, but differ slightly in frequency, then the behavior of the sum is a bit more complicated. At first, one may have its peaks when the other is having its valleys and the interference will be negative, reducing the amplitude of the sum. However, after several more cycles, the wave with a higher frequency will begin to catch up with the other wave, they may interfere more positively, and the amplitude of the sum will increase. The higher-frequency wave then moves ahead of the slower one again, and they return to the out-of-phase state.

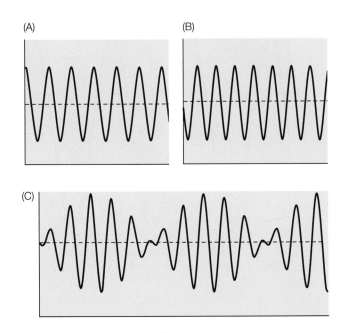

FIGURE 2.5   Generation of beats (A) Waveform of pure sinusoidal tone at 1 kHz. (B) Waveform of pure sinusoidal tone at 1.2 kHz. (C) Waveform of combination of first two tones producing beats. Combination waveform is periodic but nonsinusoidal; the pattern repeats at a rate equal to the frequency difference between the original two pure tones (1.2 kHz – 1.0 kHz = 200 Hz).

The summed result of two waves with slightly dissimilar frequencies is a sine wave whose frequency is the average of the two components, and whose amplitude rises and falls with a regular pattern (**Figure 2.5**). The frequency of this rise and fall in amplitude equals the difference between the frequencies of the two original sine waves, and the resulting variations in amplitude are called **beats**. An animated example can be found at Web Topic 2.2.

TEMPORAL PATTERNS OF SOUNDS: COMPLEX WAVEFORMS   In practice, it is very difficult for most physical systems (or animals) to produce a sinusoidal sound in air or water. The waveform recorded at a single microphone will not be sinusoidal but some other more complicated pattern. To complicate this further, many animal sounds are not really **periodic**: the same waveform pattern is not present throughout the entire sound, and so we see a series of different waveforms during the course of the sound signal. How can we possibly measure and compare such sounds?

Luckily, several mathematicians, particularly J. B. J. Fourier (1768–1830) and J. P. G. L. Dirichlet (1805–1859), showed that most periodic and continuous waveforms can be decomposed into an infinite sum of simple sine waves. Because the amplitudes of the high-frequency components tend to be very small, (see Web Topic 2.4 for reasons), we really need to consider only a limited range of low-frequency sine waves to adequately characterize most periodic signals. This characterization would specify the frequency, amplitude, and relative phase of each component of the sound signal, measures that we already know how to make. Because most animal ears perform this kind of **Fourier analysis** on sound signals before sending information to the brain, this approach also provides an excellent way to compare sounds similarly to the way an ear would do it. Even better, most animal ears ignore the relative phase information except when two components are very similar in frequency. This allows us to focus largely on the component differences in frequency and amplitude when we compare two animal sound signals.

How do we handle animal sounds that are not periodic throughout the signal? Fortunately, most animal signals can be divided into consecutive segments that are sufficiently periodic that we can perform a separate Fourier decomposition on each segment and then string the results together to see the overall pattern. The cost of this segmentation is that the ability to discriminate between adjacent frequencies declines as segments get shorter. But in practice, we can usually find a suitable segmentation of a complex sound that tells us most of what we need to know.

The most common way to characterize animal sound signals is with a **spectrogram** (also called a **sonogram**). This is a plot with time on the x-axis running from left (earlier) to right (later), and frequency on the y-axis (running from low frequencies near the origin and extending upward).

Each point in the plot corresponds to a specific time segment within the signal and a particular band of frequencies. If the Fourier decomposition indicates a high amplitude for frequencies in a given band at a given time, the plot shows one extreme of color or darkness; but if that band of frequencies is not present or is at low amplitude at that time, the plot shows an opposite color or darkness at that point. A variety of computer programs are now available (many for free) that generate spectrograms quickly and effectively. Since where the segmentation cuts are made is arbitrary, several sophisticated techniques are used to smooth out these plots so that junctions between adjacent segments are not visible. Some programs also allow the user to extract and view the Fourier composition of any segment that they wish to examine. Such a plot is called a **power spectrum** or **section**.

As an example of this approach, consider a single call of a male túngara frog (*Engystomops pustulosus*). The overall waveform of this call is shown in **Figure 2.6**. We can see that the call is fairly explosive, with a rapid onset at high amplitude and then a gradual diminution in amplitude during the call. Variation in amplitude during the course of a signal is called **amplitude modulation**. If we expand the waveform

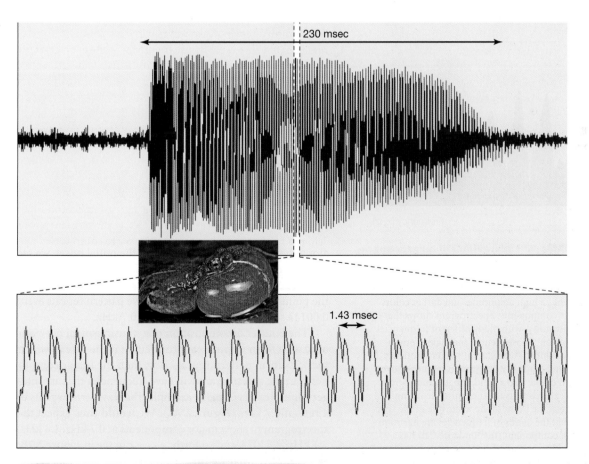

FIGURE 2.6 Waveform of single call by male túngara frog (*Engystomops pustulosus*) Both plots show varying sound pressure relative to ambient pressure (middle line) on the vertical axis and time on the horizontal axis (events on left occurred prior to events on right). The waveform in the dashed section of the top plot is expanded along the time axis in the lower plot to show the wave details. Inset shows a male túngara frog with his throat sac filled with air while calling.

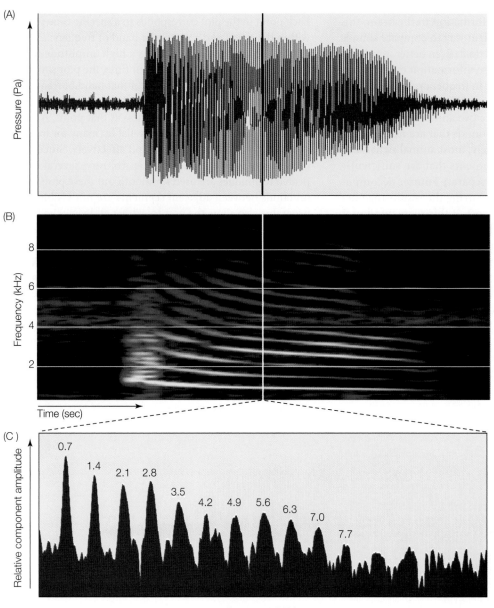

FIGURE 2.7 **Three different visualizations of túngara frog call** (A) Waveform (same as in Figure 2.6). (B) Spectrogram of same sound aligned and on same time scale as waveform in (A). Brighter (yellow) colors in the spectrogram indicate that a given frequency at that time has a high amplitude, and darker colors indicate low-amplitude components. Spectrogram shows that the beginning of call is quite noisy, with many different frequencies contributing to this portion. Subsequently, the major components are evenly spaced bands. These start at higher frequency values early in the call and gradually decrease in frequency over the course of the call. We note also a band of background noise between 4–5 kHz that extends through the entire recording. (C) A power spectrum slice at the point of the white line in the spectrogram. This shows that the bands of frequencies are harmonically related: the lowest component (the fundamental) has a frequency of about 0.7 kHz, and all higher frequency bands are integer multiples of this frequency. Note that the component amplitudes tend to decrease with frequency. This pattern of harmonically related bands is very common in the spectrograms of animal sound signals.

along the time axis at any point after the initial onset, we see that it is not a simple sine wave, but is periodic within short segments: the same basic shape repeats at regular intervals. At the point shown in Figure 2.6, the wave pattern repeats every 0.00143 seconds or at a rate of about 0.7 kHz.

The Fourier decomposition of a nonsinusoidal but periodic waveform is fairly easy to predict: if the rate at which the waveform pattern repeats is w, then the major components in the spectrogram will have frequencies that are integer multiples of w. In our example, the waveform repeats at a rate of 0.7 kHz (**Figure 2.7A**). We would thus expect the spectrogram to show major components at 0.7 kHz, 1.4 kHz, 2.1 kHz, 2.8 kHz, and so forth. As we can see in **Figure 2.7B**, this is exactly what we find. This spectrogram shows evenly spaced bands throughout the course of the call. The power spectrum for the segment indicated by the vertical white line in the spectrogram is shown in **Figure 2.7C**. Here we plot the relative amplitude of each frequency component on the vertical axis against the frequencies of possible components. We see that there is a series of peaks in this plot with each peak corresponding to the band at that frequency in the spectrogram. If we measure the frequency at the top of each peak, we find that these are all integer multiples of 0.7 kHz, the repeat rate for the waveform. Frequencies that are integer multiples of a reference frequency are called **harmonics** and the entire suite of them is called a **harmonic series**. The reference frequency is called the **fundamental**, the component at twice the reference is called the **second harmonic**, and so on. For most animal sounds, all harmonics will be present. In cases where the repeating wave pattern is very symmetrical, only the fundamental and the odd-numbered harmonics are present; however, this happens only rarely with animal sounds.

In the spectrogram in Figure 2.7B, we see that the fundamental and the harmonics all seem to decrease in frequency as the call proceeds. This is because the repeat rate of the basic pattern actually slows down during the call, and thus the frequency components must all decrease accordingly. We hear this as a slightly descending pitch during each frog's call. Variation in frequency composition during the course of a signal is called **frequency modulation**. Note that if we had performed the Fourier decomposition on the entire sound, we would not be able to detect these slow frequency modulations during the call. Segmenting the call into sections and performing Fourier analysis on each provides us with some ability to detect such changes during the course of a signal.

This is a very useful example because the majority of animal sounds in air or water are nonsinusoidal waves but are almost periodic within short segments. We should thus expect to see harmonics in the spectrograms of such animals' signals. Where we see only single pure frequencies, the animal has probably gone to a lot of work to filter out all but a few components in what was initially a harmonically rich sound. As we shall discuss later, animals have an easier time generating nearly sinusoidal sounds in solid media.

Web Topic 2.4 *Fourier analysis of animal sounds*
Here we provide an introduction to the logic behind Fourier decomposition of animal sounds, including links to several excellent software packages for creating spectrograms and introductions on how to use these packages. We also provide links to sites where one can use such methods to compare archived animal sounds.

SPATIAL PATTERNS OF SOUNDS: THE SPEED OF SOUND   As we have seen, a sound disturbance is passed from molecule to molecule at successive distances from the source. The speed with which a layer of molecules can hand off the pressure disturbance to the next layer, and thus the speed at which the sound propagates in that medium, depends on two properties of the medium (**Figure 2.8**). The first property is the density of the medium. This is a function of the mass of the

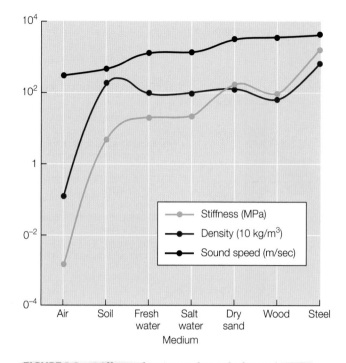

FIGURE 2.8   Stiffness, density, and speed of sound in some sample media Stiffness is measured as bulk modulus for fluids and Young's modulus for solids (Mpa), density in tens of kg/cubic meter, and sound speed in m/second. The speed of sound increases with stiffness, but decreases with medium density. Because both measures tend to increase together as one switches between media, the range of sound speeds is much less than might be expected given independent variation in stiffness and density.

molecules (larger masses result in higher densities), and the average distance between the molecules (more tightly packed molecules result in higher densities). All other factors being equal, the speed of sound is inversely related to medium density. This is because it takes longer for an activated layer to get the next layer moving if the second layer has many molecules in it or if those molecules have large masses. Water is about 800 times as dense as air; and a solid, such as steel, is about 8 times as dense as water. Based on density alone, we might expect sound speed to be highest in air, intermediate in water, and least in solids. In fact, the reverse is true. This is due to a second factor that must be considered.

The second factor affecting sound speed is the **stiffness** of the medium. For fluids such as air and water, stiffness is the relative resistance to compression: water resists being compressed much more than air does. For solids, stiffness refers both to incompressibility and to the solid's resistance to distortion in shape. All other properties held constant, the speed of sound in a medium is positively correlated with its stiffness. This is because a pressure disturbance passing through a stiff medium spends less time in compression and distortion than a disturbance moving through a non-stiff medium does. Water is about 10,000 times as stiff as air, and steel is about 100 times as stiff as water. Although the differences in density bias the speed of sound in these media in one

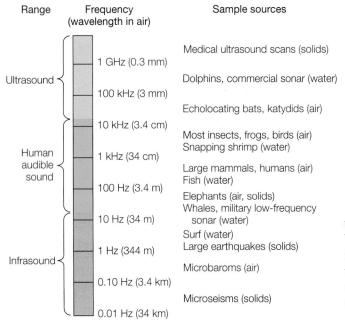

Range   Frequency   Sample sources
             (wavelength in air)

Ultrasound
                1 GHz (0.3 mm)   Medical ultrasound scans (solids)

                                   Dolphins, commercial sonar (water)
                100 kHz (3 mm)

                10 kHz (3.4 cm)   Echolocating bats, katydids (air)

Human   Most insects, frogs, birds (air)
audible   1 kHz (34 cm)   Snapping shrimp (water)
sound
                                   Large mammals, humans (air)
                100 Hz (3.4 m)   Fish (water)

                                   Elephants (air, solids)
                                   Whales, military low-frequency
                10 Hz (34 m)   sonar (water)

                                   Surf (water)
                1 Hz (344 m)   Large earthquakes (solids)
Infrasound
                                   Microbaroms (air)
                0.10 Hz (3.4 km)

                                   Microseisms (solids)
                0.01 Hz (34 km)

FIGURE 2.9   The acoustic spectrum The plot shows the sound frequencies used by different animals, including humans; the wavelengths of those frequencies in air; and whether the resulting sounds are propagated in air, water, or solid substrates. Note that wavelengths for a given frequency in water will be 4.4 times as long as those in air, and those in solids roughly 14–15 times as long as those in air. Microbaroms are low-frequency airborne sounds created by large-scale movements of the ocean that compress or rarefy the layer of air above. Microseisms are low-frequency vibrations propagated in the Earth from distant earthquakes, volcano action, fault movements, and explosions.

direction, the differences in stiffness more than compensate. As a result, the speed of sound is about 344 m/sec in air, 1500 m/sec in water, and 5100 m/sec in a bar of steel. Actual sound speeds will differ a bit from these values depending on local temperatures, ambient pressures, and the composition of the medium. Humidity can change the speed of sound in air, and salinity can alter the speed of sound in water. Although the speed of sound is independent of frequency within fluids like air or water, it can be very frequency-dependent inside solids and at boundaries between media. For example, the speed with which ripples spread on the surface of a pond will vary depending on the frequency being propagated. Sounds in solids can also show frequency-dependent propagation speeds.

SPATIAL PATTERNS OF SOUNDS: WAVELENGTH   The speed of sound plays a major role in determining the spatial pattern of a sound. Consider a sound wave with a frequency of 1 kHz in air. It will take 0.001 seconds (the period) to generate one complete cycle of this sound. By the time the sound source has finished generating this single cycle, the leading edge of the wave will have radiated 0.34 m (about 1 foot) away from the sound source. If we had created this same 1-kHz sound in water, the initial part of the wave would be 1.5 m away from the final part. The spatial length of a single cycle of a sound wave in a particular medium is known as its **wavelength**.

Because it takes longer to create a sound with a longer wavelength, we see that the wavelength and frequency of a sound are inversely related: high frequencies have small wavelengths, and low frequencies will have large wavelengths (**Figure 2.9**). Wavelength will also depend critically on the speed of sound of the medium. A given frequency of sound in water will have a wavelength about 4.4 times as long as that same frequency in air because of the difference in the speeds of sound in the two media. Sounds in solids will have

wavelengths 14–15 times as long as the same sounds in air. Why does wavelength matter? As we shall see later, it is difficult for most animals to generate an intense sound with a wavelength more than twice their body size. Given the effects of different speeds of sound, this constraint will be greater for aquatic animals than for terrestrial ones.

SPATIAL PATTERNS OF SOUNDS: DOPPLER EFFECTS   Consider a male bird that is flying toward its perched mate and emitting a call in flight (**Figure 2.10**). As successive waves radiate from its beak toward its mate, the male follows in the same direction. This means that the leading edge of a wave will travel a shorter distance by the time the trailing edge is emitted than would be the case if the caller were not moving. The wavelengths of all frequencies in the male's call will thus be shortened because of his motion toward his mate. Because successive peaks and valleys of the sound waves arrive at the female's ear with reduced delays, she perceives his call as being shifted to higher frequencies. If the male were flying away from the female, then the wavelengths of his emitted sounds would lengthen, and she would perceive his call as being shifted to lower frequencies. The same effects would be created if the male were to call from a perch and the female receiver were to fly toward or away from him during his calling. If both birds were to fly in the same direction, then the changes in wavelengths and frequencies would cancel out. Shifts occur only when the sender and receiver have different relative motions.

The change in propagated frequencies when either the sender or the receiver is moving relative to the other is called a **Doppler shift**. The magnitude of the shift in sound frequencies depends on the ratio of the speed of the moving individual to the speed of sound in the relevant medium. Birds and bats are among the fastest fliers in terrestrial environments. Their speeds rarely exceed 10 m/sec, although a stooping falcon might achieve a speed of nearly 50 m/sec. In air, a perched receiver attending to a call emitted by a sender flying directly toward or away from the receiver at 10 m/sec will experience a Doppler shift of about 3% in all song frequencies. The speed of movement of most animals sending and receiving sound signals in air—and thus the Doppler shifts they experience—will be less than this, and animals typically ignore the slight shifts in frequency. Foraging bats

FIGURE 2.10   Doppler shift when calling sender is moving relative to receiver Calling bird is flying from left to right. This causes successive waves in its emitted call to crowd together in front of it and spread out behind it. A receiver to the right of the flying bird will hear all components in the flyer's call at slightly higher frequencies. A receiver positioned behind the flying bird will hear the call at slightly reduced frequencies.

are an exception. They emit calls and listen for echoes from possible food targets. Slight shifts in the frequencies of the outgoing sound and returning echoes provide information to the bat on its speed relative to an edible target. We discuss this strategy further in Chapter 14. In water and solids, the speed of sound is so much higher than the velocities of moving animals that Doppler shifts are much less likely to be biologically relevant. See an animation of Doppler shifts at Web Topic 2.2.

SPATIAL PATTERNS OF SOUNDS: NEAR FIELD VERSUS FAR FIELD   We noted that near a sound source, medium molecules get moved over distances larger than they would normally travel if no sound were being propagated. However, as a disturbance radiates away through successive layers of molecules, the distances traveled by the molecules quickly become little more than they would have been without a disturbance present. This does not mean that molecular movements are the same whether a sound is being propagated or not. What differs during sound propagation is that adjacent molecules in a layer will all have a component of their movement in the same direction and in sufficient concert that they increase pressure in the next layer. It is not *how far* they move, but *where and when* they move that distinguishes the movements of molecules that are propagating sounds from their movements when no sound is present.

The zone around the sound source where the distances that propagating molecules travel are greater than usual is

called the **near field** of the sound, and the outer zone where molecular translocations return to normal is called the **far field** of the sound (**Figure 2.11**). In acoustics, the outer boundary of the near field is usually defined as a distance of 1/6 of a wavelength from the source or 1/3 the diameter of the source, whichever is larger. However, it should be remembered that the transition is actually gradual and not

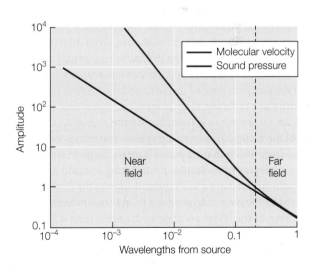

FIGURE 2.11   Plot of sound pressure and molecular velocity as a function of the distance (in wavelengths of the sound) from a point source Pressure values have been standardized relative to the characteristic acoustic impedance of the medium. Both pressure and velocity decrease with distance from the source, but velocity begins at a much higher value close to the source and decreases more rapidly with increasing distance. The two eventually converge at about one wavelength's distance. Although the transition is clearly gradual, the convention is to consider the region between the source and the distance marked with the dashed line the near field of the sound, and all greater distances the far field.

sharp, so this guideline is only approximate. In practice, a distance of 1/3 of a wavelength from the source or 2/3 the diameter of the source is required before near field effects become negligible. The ability of a receiver to detect a sound signal may depend significantly on whether it is located in the near or the far field of a sound source. A receiver in the near field could perceive the sound through some delicate sensor that is dragged by the tidal movements of the surrounding medium. In the far field, it is unlikely that tidal movements of the medium will be sufficient to cause stimulation of the sensor. In this situation, the receiver can at best try to monitor variations in pressure. Even this strategy can be challenging: by the time a sound propagates into the far field, the pressure variations due to the sound may only be about 0.02% or less of the ambient pressure of the medium.

The propagation of sound

One cannot hear a sound once one gets far enough away from the source. Sound signals get fainter with distance and eventually drop below the level of ambient noise. To understand this process, we now examine the various factors that erode a sound disturbance as it propagates and thus limit the range over which the sound can be detected by a receiver.

SPREADING LOSSES   Each time a sound source moves, it must expend energy to push the adjacent medium molecules ahead of it. The stiffer and denser the medium, the more energy the sound source has to expend to move the molecules a given distance. Because most animal sound sources move repeatedly to produce waves, the best measure of this process is the amount of energy expended per unit time. This is known as the **power** spent by the sound source. By forcing adjacent medium molecules to move in a given direction, the sound source transfers some of its kinetic energy to the molecules. This transfer is never very efficient: in animals, less than 1% of the energy expended by the sound source ends up in the pressure wave radiating away from the sound source [119].

A sound source first transfers kinetic energy to the first layer of molecules. This layer then passes this energy on to the second layer which in turn passes it on to the third layer. The successive layers of molecules propagating a sound are concentrically arrayed around the sound source: the first layer is enclosed by the second layer which is, in turn, enclosed by the third layer. If the layers are truly concentric, then it must be the case that an outer layer constitutes a larger surface area and consists of more molecules than an inner layer. It follows that as the sound radiates away from the sound source, the initial kinetic energy supplied to the first layer gets spread out over more and more molecules as it moves to successively more distant layers (**Figure 2.12**). The amount of energy each molecule acquires will thus decrease with distance from the sound source. As the kinetic energy per molecule decreases, the difference between the peak sound pressure and the ambient pressure must also decrease. This **spreading loss** causes the pressure of a sound signal propagating into three dimensions to decrease with the reciprocal of the distance

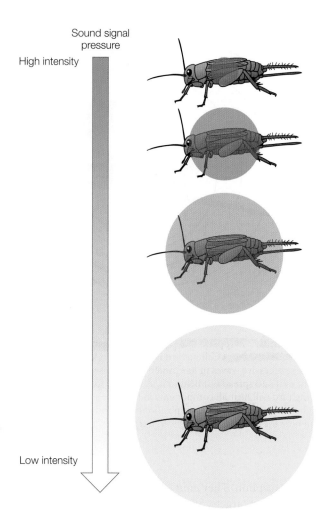

Sound signal pressure

High intensity

Low intensity

FIGURE 2.12   Spreading losses in a sound signal's pressure as it propagates away from its source   Reddish spheres show the location of the leading portion of a single call element at successive times after the cricket made the sound. Because the same initial energy is spread over larger and larger spherical surfaces as the sound radiates away from the source, the pressure at any point on a larger and later sphere is less than that on a smaller and earlier one.

between the sender and the receiver. Because spreading losses arise simply from geometrical considerations, they occur at the same rate for all frequencies.

Water striders and other aquatic insects signal by sending ripples outward on the surface of the water. In this case, the sound energy is largely spread in two dimensions instead of three. This reduces the rate of spreading loss; in fact, pressure decreases with the square root of the distance from the source for sound waves moving on a surface. Tree hoppers and spiders that live on thin-stemmed plants can communicate by introducing sound waves into the plant tissues. Receivers that are some distance away from the sender detect these waves as they pass under their feet. Because such stems and branches are essentially one-dimensional media, there is *no* spreading loss as these sounds propagate [32].

HEAT LOSSES   In addition to spreading losses, sound energy can be lost each time that molecules collide with each other. Such losses are largest when a molecule collides with another that has different physical properties. For example, air is largely composed of oxygen and nitrogen molecules. When one layer of these molecules is moved forward by a propagating sound wave and collides with the next layer, most (though not all) of the kinetic energy is passed on to the next layer, since nitrogen and oxygen gas molecules have similar properties. If, however, there are water vapor molecules in the next layer, they may convert the kinetic energy passed to them by nitrogen or oxygen molecules into internal vibrations and molecular rotations. This reduces some of the translational kinetic energy needed to sustain the sound wave and eventually converts it to heat (random movements of the molecules). The amount of heat loss in air is a complicated function of frequency, temperature, and relative humidity. At a temperature of 20° C, heat losses are maximal at relative humidities of 4–20% for frequencies of 1–10 kHz, respectively, and less for lower and higher humidities. As the frequency is increased, the relative humidity at which peak heat losses occur also increases. In seawater, dissolved salts such as magnesium sulfate and boric acid are very dissimilar to water molecules and thus have a similar effect on sound propagation. Here too, attenuation depends on frequency, temperature, dissolved salt concentrations, and pressure.

Since **heat losses** (also known as **excess attenuation**) occur with each successive collision between layers of molecules propagating the sound, their effect is cumulative. The drop in sound pressure during propagation due to heat losses alone will thus be roughly proportional to the distance between sender and receiver. Also, since adjacent layers collide more frequently when propagating a high-frequency sound wave than when propagating a low-frequency one, high frequencies lose more sound energy to heat losses than do low frequencies. In most media, heat losses increase with the square of the sound frequency. This means that the maximum distance over which two animals can communicate by sound will be shorter if they use high frequencies than if they use low frequencies. We discuss this principle in more detail in Chapter 3.

The type of medium significantly affects the rate of heat losses (**Figure 2.13**). For example, a 1-kHz sound propagating 500 m in fresh water will lose about 0.001% of its energy due to heat losses. The same sound propagating the same distance in seawater would lose 0.5% of its initial energy, and in air the heat losses would reduce the signal amplitude by 44%. Sounds are detectable at much greater distances in water than in air. Excess attenuation is very complicated in solids, in part because most solids have bounded surfaces that also affect sound propagation. We take up such boundary effects in the next sections.

ACOUSTIC IMPEDANCE   When a layer of medium molecules propagating a sound pushes against the adjacent layer, the latter will resist being compressed and distorted.

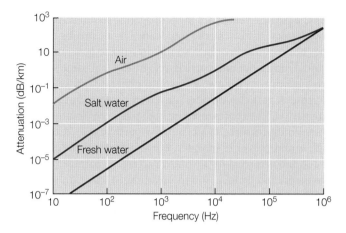

FIGURE 2.13   Heat losses of propagating sound as a function of frequency and medium   Heat losses need to be combined with spreading losses and any scattering effects to compute actual sound pressures at a distance from a source. As a rough rule, heat losses increase with the square of the frequency in all media. Deviations from this rule depend on humidity and temperature effects in air, and dissolved salts and temperature in water. Note that for a given frequency in the range typically used by animals for communication, heat losses in salt water are about 100 times as high as those in fresh water, and those in air are 100 times as high as those in salt water. (After [323].)

This resistance by the medium to having its current behavior altered is called its **acoustic impedance**. When a large volume of medium is far from sound sources and also from objects consisting of different media, its acoustic impedance is approximately equal to the product of its current density and speed of sound. This product is called the **characteristic acoustic impedance** of the medium. Since the speed of sound depends largely on the stiffness of the medium, characteristic acoustic impedance is essentially the product of medium density and stiffness. Both factors will make it harder for a given sound pressure to move a given volume of medium. In fact, the higher the characteristic acoustic impedance of a medium, the shorter the distance that one layer of medium can move the next when propagating a sound. As we noted earlier, water is about 800 times as dense as air, and also has a sound propagation speed that is 4.4 times as high. As a result, the characteristic acoustic impedance of water is about 3500 times as great as that of air. Most solids, such as wood, have characteristic acoustic impedances 5000 times as great as that of air. This has very important implications for how sound propagates in each medium, and what happens at boundaries between different states of matter.

We have seen that the energy imparted to a medium by a sound source becomes spread over larger and larger areas of medium as the sound propagates away from the source (spreading losses). It follows that the amount of energy per unit area of medium, known as the **intensity** of the sound, must decrease as the distance from the source increases. Sound intensity is equal to the product of the sound pressure and the particle velocity of the propagating sound wave. The low characteristic acoustic impedance in air means that it does not require very much sound pressure to generate substantial molecular and particle velocities. That same sound pressure in water, however, would generate only a small increase in particle velocities. As a result, a sound of a given intensity will be represented by high particle velocities and low sound pressures in air, but low particle velocities and high sound pressures in water. Put another way, the energy of a sound is carried for the most part by the particle velocities in air, and by local sound pressures in water. In solids, which usually have higher characteristic acoustic impedances than either air or water, the energy in sound waves is almost entirely conveyed by the sound pressure, and molecules are barely able to move at all.

In practice, the actual acoustic impedance for sound propagation is often different from that predicted by the medium's characteristic acoustic impedance value. For example, consider a tube with walls consisting of some solid material and a hollow filled with air. When sound waves propagate along the air-filled hollow of the tube, the air molecules in the center of the tube largely encounter only other air molecules, while those at the boundary of the hollow space will also collide with the incompressible and unmoving molecules that make up the tube wall. This will create a drag or friction that resists the passage of the sound waves close to the tube wall. The acoustic impedance will be closest to the characteristic value for air in the center of the hollow but higher at its margins. Overall, the average acoustic impedance inside a tube filled with air will be higher than that for an unbounded volume of air at the same temperature and humidity. Similarly, the acoustic impedance of air close to the surface of a body of water will be higher than the acoustic impedance of air high above this surface: air molecules close to the surface will not be able to move as freely as those with no water molecules nearby. Conversely, the acoustic impedance for water near a pond's surface will be lower than that deeper into the pond since water molecules on the surface will meet much less resistance to movement upwards than they would if surrounded on all sides by water molecules. While we can easily compute the characteristic acoustic impedance for a medium using its density and sound speed, we often have to modify this value up or down when considering that same medium near a boundary with another medium.

REFLECTION AND REFRACTION   What happens when sound traveling in one medium encounters a boundary with another medium? The answer depends critically on the relative acoustic impedances of the two media at the boundary. Because sound energy is conveyed largely by molecular velocities in air, but by high sound pressures in water, a sound wave traveling in one of these media will face a difficult time crossing a boundary into the other medium. Instead, nearly all of the incident sound energy will be **reflected** back into the initial medium at the surface of the boundary (**Figure 2.14**). The angle of travel of the reflected waves will be determined by

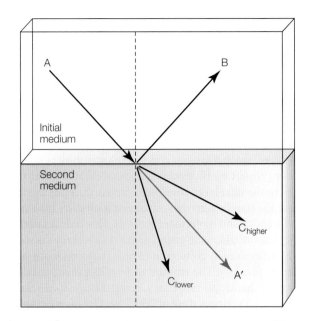

FIGURE 2.14   Reflection and refraction of sound at a boundary between two media In this example, sound initially traveling in the white medium along line A strikes the surface of the blue medium at an angle. If the acoustic impedances of the two media differ, some of the incident sound energy will likely be reflected from the surface and back into the white medium along line B. Note that the angles of the incident and reflected sound waves are the same relative to the boundary surface. If the acoustic impedances of the two media are not too different, some incident sound energy may also be transmitted through the boundary and into the blue medium. If the speed of sound (C) in the blue medium is higher than that in the white medium, the transmitted sound will be refracted (bent) away from the dashed line perpendicular to the surface as shown in line C_{higher}. If the blue medium has a slower sound speed, the transmitted sound's trajectory will be refracted toward the dashed perpendicular line as shown in line C_{lower}.

the angle of the incident waves. For example, if the incident sound waves were traveling from left to right at an angle of 30° relative to the surface of the boundary, then the reflected waves would also travel from left to right at an angle of 30° relative to the surface. Depending on the two media and the incident angle of the sound waves before reflection, the reflected sound waves may experience a phase shift: what were peaks in the incident sound wave can become valleys and vice versa in the reflected wave.

Reflections can have a major effect on sound propagation within a medium. For example, if two animals that are both in air are communicating by sound close to the ground, some of the sound waves arriving at the receiver will travel on a straight-line trajectory from the sender, but others will arrive at the receiver after being reflected off of the ground. If these are out of phase with the directly transmitted waves, the two can cancel and the receiver will hear nothing. Reflections within a small volume of medium such as inside a solid or small body of water can also produce very complicated interference patterns and thus modify the efficacy of sound communication significantly.

When the acoustic impedances of the two media are more similar than those of air and water, then some sound energy will traverse the boundary and continue propagating in the second medium. The direction of travel of the sound waves will likely be shifted: if the speed of sound is slower in the second medium, the travel direction of sound waves in the second medium will be bent closer to a line perpendicular to the surface than was the incident wave direction; if the speed of sound in the second medium is higher than the first, then the waves in the second medium will be bent upward so that their direction is more parallel to the surface than was the incident sound wave. This bending of the waves after crossing a boundary between media is called **refraction**. It is most likely to affect animal communication when sound is traveling in a single medium and encounters adjacent regions of medium with different temperatures, pressures, or compositions. Both water and air often exist in vertically stacked layers. These layers are usually insufficiently different in acoustic impedance to generate reflections, but are sufficiently different in sound speed to cause refraction. For useful animations demonstrating reflection and refraction, see Web Topic 2.2.

> **Web Topic 2.5** *Reflection and refraction*
> The fraction of sound energy reflected or refracted at a boundary is a complicated function of incident angle, relative acoustic impedances, and relative sound speeds. Here we present the equations for several different cases, and provide some real physical examples.

DIFFRACTION   Consider a solid wall separating two large air-filled spaces. Sound traveling in one of the air-filled spaces will be reflected at the wall and no sound will appear on the other side. Suppose we then open up a circular window in the wall. Now some of the incident sound can pass through the hole and propagate in the second space. One might expect the sound field in the second space to be cylindrical in shape, given the circular shape of the hole. This is true only when the hole is large relative to the wavelengths of the incident sound. As the hole is made smaller relative to the sound wavelengths, the sound waves emerging from the hole will form a cone of sound. The angle of the cone increases as the relative size of the hole decreases, and for a small enough hole, emerging sound waves will radiate into all parts of the second space. The bending of sound waves after passing through or around an obstruction is called **diffraction**. Diffraction is a consequence of the wave nature of sound and explains why we can hear someone speaking around the corner of a building or detect a sound with both ears when the source is located on only one side of our heads. See an animated example at Web Topic 2.2.

SCATTERING   A special case of reflection occurs when many objects consisting of one medium are distributed throughout a second one. Examples include trees and bushes sticking into the air space above the ground, and the air-filled swim bladders of fish scattered about in seawater (e.g., over a coral reef). If the acoustic impedances of the two media are sufficiently different, each object will constitute an individual reflector of incident sound waves. If the reflecting objects are situated between a sender and a receiver, this **scattering** of the sound during propagation can be another significant source of signal attenuation. The amount of scattering, and thus the amount of reduction in sound energy reaching the receiver, will depend on the abundance of such objects, their size relative to the wavelengths of the sound, and the location of the receiver relative to the source of sound and the objects. When wavelengths are much larger than the scattering objects, they simply bend around these obstacles, and very little sound energy is scattered. This is called **Rayleigh scattering** (Figure 2.15). As the wavelengths decrease in size, the fraction of incident sound energy that is scattered back toward the source gradually increases. Rayleigh scattering thus prevents higher-frequency components in a sound signal from reaching the receiver. Once wavelengths get down to about six times the size of the scattering objects, a more complicated behavior, **Mie scattering**, takes over. In this case, part of the incident sound energy arriving at a scattering object is directly reflected backward, whereas another part is diffracted around the object. Interference between the diffracted and reflected components can be positive or negative depending on the circumference of the object relative to the sound wavelengths and the relative location of the receiver. At even smaller wavelengths of sound, diffraction is negligible and all sound energy that strikes the object is reflected and lost to the receiver. See animation at Web Topic 2.2. We shall discuss the consequences of sound scattering in more detail in Chapter 3.

Sound Signal Generation

Sound signals begin with the production of some sort of vibration. There are dozens of ways that animals might generate a vibration, and nearly every mechanism has been used by at least one species. Given this, we might expect sound communication to be widespread among animal taxa. In fact, sound communication is largely restricted to arthropods and vertebrates, and even among these, modalities other than sound often predominate in their signaling repertoires. The major constraint that limits the use of sound for communication is not the generation of a vibration but the efficient coupling of that vibration into the propagating medium. In species that need to communicate over distances smaller than a body length, inefficient sound radiation can often be tolerated, and it is here that we see the widest variety of vibration mechanisms being used. However, animals that use sound to communicate over larger distances either use only those vibration sources that radiate efficiently or they feed the vibrations into secondary structures that are designed specifically for efficient sound radiation. In the following sections, we first examine the many ways in which animals

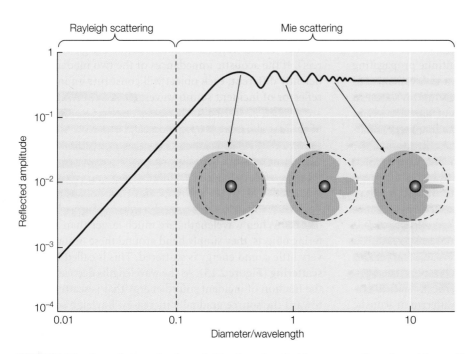

FIGURE 2.15 Sound scattering by a rigid sphere in a fluid medium Vertical axis is the ratio of reflected to incident sound pressure measured at a location about one-half wavelength away from the sphere along the line joining the sphere and the source. Horizontal axis indicates ratio of sphere diameter (here held fixed) to varying wavelengths of incident sound. At low ratios, sound waves sweep around the sphere and there is little reflection. Increasing the ratio (by decreasing the wavelengths) initially creates a smooth increase in reflected sound pressures, an effect called Rayleigh scattering. Above this ratio, reflection processes can become very complex, an effect called Mie scattering. For a sphere, reflection pressures rise and fall as the ratio is increased and eventually asymptote to a fixed value by a ratio of 6:1. While sound is reflected roughly equally in all directions for Rayleigh scattering, Mie scattering is increasingly asymmetric as the ratio is increased. Each blue plot shows the sum of incident and reflected sound pressure at the surface of the sphere for a given sphere/wavelength ratio; the sound source is located to the left of each plot, and the radius of the plot for any given angle around the sphere indicates the overall sound pressure in that direction. The dashed circle shows what that pressure would be if there were no sphere at that location. As the ratio increases, incident and reflected sound pressure add up positively in front of the sphere, whereas they interfere to form a "sound shadow" with lower-than-incident pressure on most of the rear side of the sphere.

generate vibrations and then look at the various devices that have allowed arthropods and vertebrates to improve the efficiency of sound radiation.

Producing vibrations

The production of a vibration usually entails an animal moving one part of itself relative to a nearby volume of medium or moving the nearby medium relative to some part of its body. There are four broad categories of movement that we encounter in animal sound production:

1. movement of a solid body part against another solid;
2. movement of a body part to create surface waves at a boundary between media;
3. movement of a body part to produce waves within a fluid medium; and
4. movement of a fluid medium against a body part.

We discuss each of these mechanisms in more detail below.

MOVING A BODY PART AGAINST ANOTHER SOLID   There are several ways in which an animal can generate a vibration by moving a body part against another solid. The simplest method is **percussion**, in which the animal strikes two solid objects together with a rapid motion (**Figure 2.16**). The two most commonly used targets for percussion are nearby substrates and another body part. In arthropods, substrate percussion is known in crabs (Decapoda), stoneflies (Plecoptera), grasshoppers (Orthoptera), katydids (Orthoptera), crickets (Orthoptera), termites (Isoptera), booklice (Psocoptera), true bugs (Hemiptera), lacewings (Neuroptera), alderflies (Megaloptera), snakeflies (Raphidioptera), scorpion flies (Mecoptera), beetles (Coleoptera), caddis flies (Trichoptera), caterpillars (Lepidoptera), ants and wasps (Hymenoptera), and spiders (Araneae) [1, 15, 23, 62, 66, 92, 189, 243, 303, 331, 390]. Body parts used by arthropods to strike the substrate include the head, antennae, mouthparts, two or more legs, wings, the abdomen, or the entire body. Vertebrate examples of substrate percussion include head slaps against the substrate by benthic fish such as the mottled sculpin [399], vocal sac and foot drumming against the substrate in frogs [51, 232, 233, 269], rapid bill drumming of many woodpecker species against anything that will reverberate [90, 94, 196, 351, 406], beating a drumstick against a hollow

FIGURE 2.16 Vibration generation by drumming on external substrates (A) Like most members of their taxon, Pileated woodpeckers (*Dryocopus pileatus*) drum their beaks against tree trunks. Typical drums consist of 10–30 strikes at rate of 15/sec. (B) Male palm cockatoo (*Probosciger aterrimus*) holding the branch he used to drum against an advertised nesting cavity in a tree. (C) Lacewings (*Chrysoperla agilis*) vibrate their body and strike the vegetation substrate with a curved abdomen. (D) American beavers (*Castor candadensis*) strike the water's surface with their tails as an alarm signal to conspecifics.

log by male palm cockatoos [407], foot stamping by a wide variety of mammals [311], thumping of the head against burrow walls by mole rats [272], and the loud drumming against tree buttresses by chimpanzees [6, 7].

Mouthparts and wings are the most common anatomical tools for body–body percussive actions (**Figure 2.17**). Among vertebrates, many birds pop or snap their bills together as a threat; storks, herons, and the roadrunner perform more elaborate bill clacking as an integral part of their displays [347, 401]. Arthropods such as caterpillars and grasshoppers also click their mandibles to make sounds [92]. Wings can be struck against each other or against some other part of the body to produce sounds. Avian examples include owls [402], nightjars [252], manakins [42-45, 306], and flappet larks [38, 293]. Arthropod examples of wing clapping include grasshoppers [92] and whistling moths; the latter have hardened "castanets" on each wing to increase the intensity of the percussive sound

(A)

(B)

(C)

(D)

FIGURE 2.17   Vibration generation by striking body parts together (A) Male ruddy duck (*Oxyura jamaicensis*) drumming bill against water trapped in feathers and air-filled sac on breast. (B) White stork (*Ciconia boyciana*) performing bill clacking display at nest. (C) Male red-capped manakin (*Pipra mentalis*) striking wing against leg to make sound during courtship display. (D) Eastern diamond-backed rattlesnake (*Crotalus adamanteus*) coiled and rattling dry buttons on tail tip as a threat. Rattle is not in focus in the photo because it was vibrating back and forth at about 60 times/sec.

[2]. Rattlesnakes whip the hard multiple segments at the ends of their tails back and forth at a very rapid rate, resulting in a percussive staccato [76, 325, 411]. The displaying male stiff-tailed duck drums its bill against an inflated air sac on its breast [202, 203], and gorillas beat air-filled sacs on their breasts as a threat.

Instead of percussive striking, some species rub two solids together to produce more continuous vibrations. This is called **stridulation.** In the simplest case, simple friction between the two surfaces generates a repetitive cycle of sticking and slipping (see analysis of relevant physics in [119]). Fish such as grunts (Haemulidae) grind their pharyngeal teeth together, producing noisy vibrations with little periodic structure [50]. However, most stridulating species have accentuated this process by providing bumps, hairs, spines,

ridges or ribs, on one or both of the rubbing surfaces. The most efficient system is a row of carefully spaced teeth on one surface (called a **file**) and a sharp edge or single tooth on the other (called a **plectrum**). The animal either moves the plectrum across the file or vice versa to generate a steady sequence of nonsinusoidal but relatively periodic vibrations. Thus spectrograms of stridulation signals can range from broadband smears across a wide range of frequencies (grunt tooth grinding) to stable harmonic series (orthopteran wing stridulation and catfish spine stridulation).

Given their hard exoskeletons and multi-jointed bodies, many arthropods (**Figure 2.18**) use stridulation to generate vibrations [99]. Almost every pair of juxtaposed or nearby body parts has been used for this purpose by some arthropod [92]. Insect examples include rubbing one segment of an antenna or leg against another segment of the same appendage, one wing against another wing, wings against legs, head against thorax, thorax against abdomen, legs against thorax, legs against abdomen, proboscis against thorax, etc. Crustaceans are a bit less inventive, but examples include rubbing

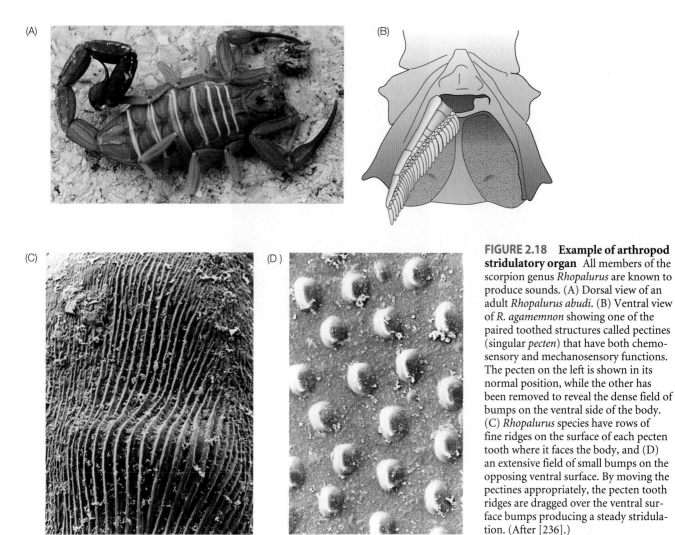

(A)

(B)

(C)

(D)

FIGURE 2.18 Example of arthropod stridulatory organ All members of the scorpion genus *Rhopalurus* are known to produce sounds. (A) Dorsal view of an adult *Rhopalurus abudi*. (B) Ventral view of *R. agamemnon* showing one of the paired toothed structures called pectines (singular *pecten*) that have both chemosensory and mechanosensory functions. The pecten on the left is shown in its normal position, while the other has been removed to reveal the dense field of bumps on the ventral side of the body. (C) *Rhopalurus* species have rows of fine ridges on the surface of each pecten tooth where it faces the body, and (D) an extensive field of small bumps on the opposing ventral surface. By moving the pectines appropriately, the pecten tooth ridges are dragged over the ventral surface bumps producing a steady stridulation. (After [236].)

antenna against cephalothorax, uropods against each other, one segment of a claw or leg against another segment of the same appendage, legs against carapace, claw against legs, and legs against legs. Spiders stridulate using pairs of mouthparts, legs against the body, or the cephalothorax against the abdomen. Scorpions use their claws and legs against each other or against the body, body segments against adjacent body segments, and some even use their sting as the plectrum against a file on a more basal segment of the tail. Other arachnids such as amblypygids, solpugids, and harvesters usually stridulate using mouthparts. Not surprisingly, millipedes and centipedes put their stridulation structures on adjacent legs. There is a very large literature on arthropod stridulation. Sample publications include studies on lobsters [290, 291], hermit, ghost, and fiddler crabs [1, 62, 195, 303, 331], orthopteran insects [82, 84, 103, 235, 264, 265, 298, 329], hemipteran insects [69, 242, 296, 322, 336, 418], beetles [77, 129, 193, 205, 237, 283, 284, 287], butterflies, moths, and caterpillars [75, 89, 117, 223, 409], caddis flies [197], ants and wasps [68, 158, 183, 189, 321, 382], spiders [23, 96, 98, 102, 201, 297, 308, 355, 379, 385], and scorpions [236].

Stridulation also occurs in vertebrates. Fish stridulate by grinding their teeth together [16, 50, 109, 222, 267], grating adjacent bones [73, 108, 378], or rubbing bony spines against adjacent bones, scales, or other spines [41, 105, 182, 335, 339]. Spine stridulation is used for sound production by hundreds of species of catfish [107, 217]. Some gekko lizards whip their tails so that adjacent scales rub and produce a rasping sound [130]. The elaborately barbed structure of bird feathers facilitates stridulatory rustling when plumes are rubbed or shaken together [147]. Male sage grouse rub the hard leading edge of the folded wings against rows of stiff breast feathers to produce swishing sounds during their display strut [188]. The small mammals called tenrecs can produce sound vibrations by rustling the spines on their back [55].

A third mechanism for vibration production involves the bending of a highly elastic plate of cuticle by body muscles until the plate buckles and snaps into a different configuration producing impulsive vibrations (**Figure 2.19**). The plate usually snaps back when the pressure on it is released, and this may produce an additional vibration. Such **buckling** is largely restricted to insects in the orders Hemiptera and

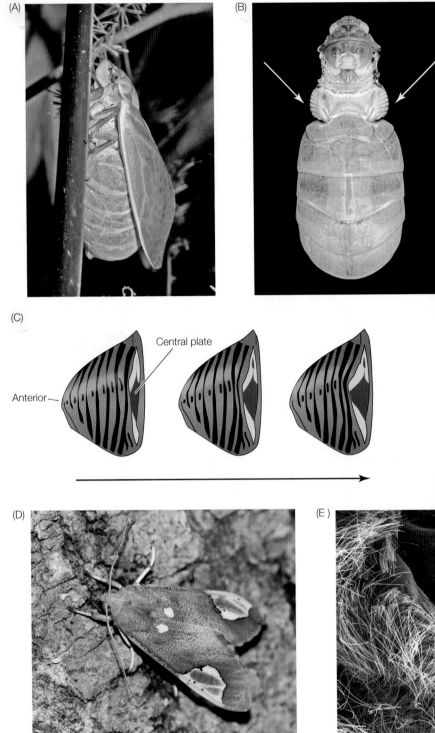

FIGURE 2.19 Insect tymbals (A) Bladder cicada (*Cystosoma saundersii*). (B) Dorsal view of male bladder cicada with wings removed to show ribbed tymbals on each side of body (arrows). (C) Successive stages in buckling of cicada tymbal (left to right). Each tymbal is viewed from the side with the anterior end on the left. Brown area is central plate. As plate is pulled towards anterior end by muscles, it, and then successive ribs, buckle and produce a sequence of sounds. (D) Arctiid moth (*Bertholdia trigona*) and (E) SEM photo of one of the tymbals located on its thorax. Note ridges along side of tymbal similar to ridges in cicada tymbal. Male moths use these tymbals to produce ultrasonic sounds that jam the sonar of attacking bats (see Chapter 14).

Lepidoptera. Auchenorrhynchan Hemiptera such as cicadas (Cicadidae), leafhoppers (Cicadellidae), froghoppers (Cercopidae), treehoppers (Membracidae), and planthoppers (Delphacidae) typically possess a convex cuticular plate called a **tymbal** covering an air space on each side of the abdomen [57, 66, 286]. Except for cicadas, tymbals are present in both sexes. Muscles rooted at the center of the body attach to each tymbal either directly or through a levered rod called an **apodeme**. As the muscles contract, a force is exerted on one portion of the convex tymbal plate until it suddenly pops into a concave state with a sudden generation of vibrations. Some cicadas have elaborated this scheme with a series of parallel ribs at successive distances from the apodeme insertion point. Instead of a single buckling of the entire tymbal, these ribbed tymbals buckle in steps, producing a longer series of initial vibrations [33, 35, 123, 178, 344, 354, 356, 416]. Cicadas typically buckle and relax the tymbals repeatedly at rates up to 100 times/sec. In some cicada species, the two tymbals are buckled repeatedly at this high rate but out-of-phase, producing even higher rates of buckling cycles [124, 231]. Tymbals also occur in Prosorrhynchan Hemiptera such as stink bugs (Pentotomidae), burrowing bugs (Cydnidae), and assassin bugs (Reduviidae) [390].

Lepidopteran tymbals are used to generate ultrasonic vibrations by a variety of nocturnal moths. Tiger moths (Arctiidae) have paired muscle-activated tymbals on the thorax; some owlet moths (Noctuidae), such as the Australian *Amyna natalis*, have tymbals on each forewing that are activated by twisting the wings; many waxmoths (Pyralidae) have tymbals at the base of each wing that are buckled by wing movements; and male *Symmoracma minoralis* (Pyralidae) have a single large tymbal on the ventral side of their abdomen [40, 75, 93, 160, 200, 268, 334, 346, 348, 349]. Many of these species have ribbed tymbals like those of the cicadas to produce longer bursts of vibrations. Neotropical butterflies in the genus *Hamadryas* (Nymphalidae) typically produce audible "cracking" sounds in flight. Although it was initially thought that this was due to percussion between the wings [263], recent work shows that a single wing can produce the sound and that some form of buckling vibration is the most likely source [408].

One group of fish, the croaking gouramis (Osphronemidae), has evolved a unique mechanism of sound production. Two enlarged tendons attached to each pectoral fin are stretched as the fin is moved in an anterior direction. The tendons are adjacent to bony processes of the fin rays that trap the tendons momentarily until further tension releases them with a snap. The effect is rather like the plucking of a guitar string. Each forward fin movement produces two short pulses of sound, and the two fins often move in opposite directions, producing long series of double pulses. Both sexes use these sounds when interacting with conspecifics [212, 213, 216, 221].

A final mechanism for generating vibrations using solids is **tremulation**. This involves rocking the entire body (or large parts of it) and transmitting the resulting vibrations into a solid substrate through the legs or some other appendage. It has been reported in insects in the orders Plecoptera, Orthoptera, Hemiptera, Megaloptera, Rhapidioptera, Neuroptera, Mecoptera, Trichoptera, Diptera, and Hymenoptera [14, 184, 390]. This is also the likely source of vibrations induced in plant stems by chameleons [21]. Such motions may also generate vibrations in the surrounding fluid medium (air or water), a mechanism that will be discussed in a later section.

Web Topic 2.6 *Sample animal sounds*
Visit this website to hear examples of the kinds of animal vibrations discussed in this and following sections of the chapter. You will also be able to see the waveform and spectrogram of each example.

MOVING A BODY PART TO CREATE SURFACE WAVES   As we have seen, the layer of medium closest to a boundary is likely to have somewhat different acoustic properties than a layer of the same medium farther from the boundary. Acoustic impedance is one property that is likely to vary in this manner. Various animal species have taken advantage of these differences to produce sound vibrations that then propagate on the boundary surface [230].

Some aquatic animals living on or near the surface of a body of water communicate by generating surface ripples that radiate away from the initial disturbance (**Figure 2.20**). Insect examples include Hemipteran water striders (Gerridae), which generate radiated ripples using vertical movements of the legs [403], and waterbugs (Belostomatidae), which pump the entire body up and down under water but

FIGURE 2.20 Calling signal of male water strider Repetitive motion of a male of this North American species (*Rhagadotarsus anomalus*) generates regular ripples that radiate out in concentric circles. The male is the tiny object at the center of the concentric ripples.

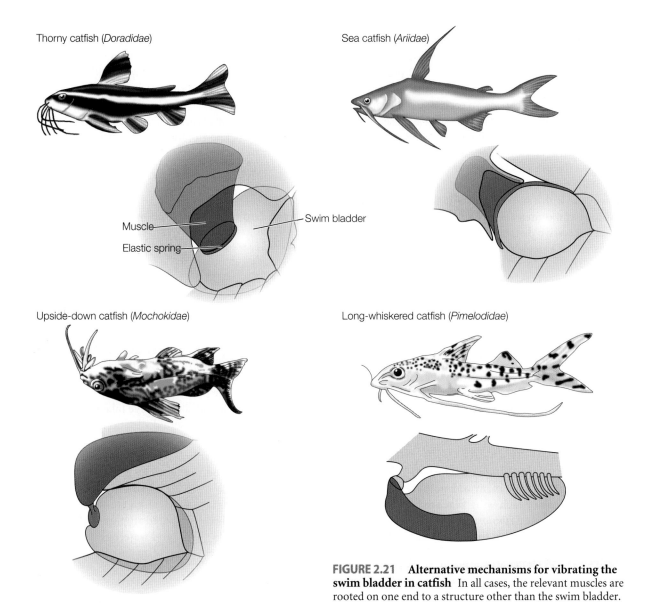

Thorny catfish (*Doradidae*)

Sea catfish (*Ariidae*)

Muscle

Swim bladder

Elastic spring

Upside-down catfish (*Mochokidae*)

Long-whiskered catfish (*Pimelodidae*)

FIGURE 2.21   Alternative mechanisms for vibrating the swim bladder in catfish   In all cases, the relevant muscles are rooted on one end to a structure other than the swim bladder. Examples from four catfish families are shown with details of representative swim bladder, relevant muscles, and elastic spring. (After [220].)

near the surface [214]. Male Siamese fighting fish signal to small fry feeding near the water's surface by pulsing water with their fins to generate surface ripples [215]. American alligators slap the water's surface with their heads [391], and a variety of whales and porpoises either perform tail slaps or leap out of the water and fall back to produce a massive splash [241, 395, 400]. It is not clear in either case whether the relevant vibrations are propagated at the surface or instead in the body of water as a whole.

Many of the percussive actions on substrates noted earlier are designed to create vibrations in the substrate instead of, or in addition to, vibrations in the nearby air or water. Where the wavelengths of the resulting vibrations are shorter than the thickness of the substrate, propagation will be limited largely to surface waves. Where wavelengths are larger than the thickness of the substrate (for example, when the substrate is a leaf or stem of a plant), the entire structure will propagate the vibrations. At intermediate wavelengths, both

surface and interior waves may be propagated. We discuss these processes in further detail in the next chapter.

MOVING A BODY PART INSIDE A FLUID MEDIUM   Many animals move a solid object to create vibrations and waves in their surrounding fluid medium. In some cases, these body parts are simply passing on vibrations generated elsewhere in the body. We take up this type of process in a later section. Here, we focus on movements of body parts that are the initial sources of sound vibrations in fluid media. There are at least four ways that animals initiate these sound vibrations.

The first method, **pulsation**, is to alternately contract and expand the surface of a closed but flexible object, so that medium is forced to move in concert with the surface. American lobsters [179] and mantis shrimp [292] expand

and compress their cephalothoracic carapace in this manner. Most bony (teleost) fishes maintain buoyancy using a gas-filled sac called a **swim bladder** in the center of their bodies. Because the gas is more compressible than any fluid-filled tissue, the swim bladder can be compressed, stretched, or struck to produce relatively large vibrations over its surface. These are then radiated into the surrounding medium as sounds [52, 81, 104, 175, 176, 218-220, 288, 289, 377, 378].

In many fish, the relevant **sonic muscles** are attached to the back of the skull, the backbone, or the sides of the body and are used to move, stretch, or compress the swim bladder relative to that base attachment point. There is an amazing diversity of designs (**Figure 2.21**). In the simplest case—naked catfish (Pimelodidae), some rockfish (Sebastidae), and tiger perch (Terapontidae)—the muscles attach directly to the swim bladder. In thorny catfish (Doradidae), ocean catfish (Ariidae), and upside-down catfish (Mochokidae), an intervening bony plate or lever, called an elastic spring, is attached both to the swim bladder and to the muscles. Tension on the muscles pulls the elastic spring and the bladder in one direction, and muscle relaxation allows the elastic spring to snap the bladder back to its original position. In squirrelfish (Holocentridae), the muscles move flattened ribs that are attached to the sides of the swim bladder and thus compress and expand the swim bladder walls indirectly. The muscles in some rockfish (Sebastidae) connect first to the pectoral girdle and then terminate on ribs or vertebrae, allowing for overall contractions and expansions of the cavity containing the swim bladder. Piranhas (Characidae) have a flattened tendon enveloping the lower half of the swim bladder that can be pulled upward by muscles attached to one of the ribs. The reverse system is used in drums and croakers (Sciaenidae), which have a flattened tendon over the top of the swim bladder attached to muscles in the ventral body walls that can pull downward on the swim bladder. Cusk eels (Ophidiidae) and pearl fish (Carapidae) have one pair of muscles that stretches the anterior end of the swim bladder perpendicular to the body axis while a second set pulls the anterior end forward. Because the posterior end of the swim bladder is attached to vertebrae, these muscles stretch the anterior end of the swim bladder in two directions at once. Some mormyrid electric fish make sounds with muscles attached to the posterior end of the swim bladder [81]. Finally, the relevant muscles in toadfish and midshipman fish (Batrachoididae), sea robins (Triglidae), flying gunards (Dactylopteridae), dories (Zeidae), wasp fishes (Apistidae), cod and haddock (Gadidae), and other mormyrid electric fish are unattached to any skeletal component (**Figure 2.22**). This allows for rapid and repetitive excitations of the swim bladder at rates up to 300 contractions/sec in toadfish and midshipman fishes. Interestingly, the sonic muscles in toadfish and midshipman fishes are among the fastest-oscillating muscles known in vertebrate animals [106, 325]. The vibrations generated by these species are usually periodic but nonsinusoidal in waveform. The resulting spectrograms typically show a harmonic series with

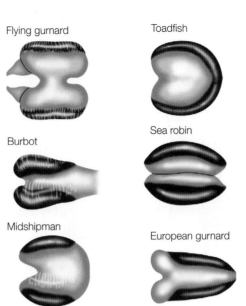

FIGURE 2.22 Examples of fish in which vibratory muscles are attached only to the swim bladder Swim bladders are drawn with anterior end to the left. Differences in geometry and shape reflect functional differences. For example, muscles (dark red) surround the sides and rear of the toadfish swim bladder, whereas they are found only on the sides of the midshipman's organ. This allows sufficient posterior flexibility for the midshipman to inflate its swim bladder before drumming. (After [220].)

a fundamental identical to the contraction rate of the sonic muscles [220].

A second mechanism, **fanning**, involves moving a flat solid object cyclically along a line perpendicular to its surface to produce a parallel movement of the nearby fluid medium. The wings of insects provide a classic example, as they may be vibrated (with or without flight) at rates up to several hundred Hz (**Figure 2.23**). In nearly all cases, little far-field sound is produced, but in the near field, air movements can be quite large and easily detected at distances of several centimeters by nearby receivers. Examples include the courtship dances of male *Drosophila* (Drosophilidae) [28, 32, 375, 376], large fruit flies (Tephritidae) [4, 48, 49, 240, 260] and sand flies (Psychodidae) [85, 86, 285]; the waggle dances of honeybees [256, 257]; and the wing-flicking displays of cave crickets (Phalangopsidae) [88, 177]. Male mosquitoes (Culicidae) are attracted to the near-field wing sounds of flying females, and females signal their interest by matching their wing-beat frequencies to that of the approaching male [146].

A third method, **fluid compression**, remains poorly understood but seems to involve such a rapid modification of local pressure in a fluid medium that it generates far-field sounds. In air, the rapid wing flicks of displaying male manakins produce loud snaps and pops that are easily heard in the far field [44, 45]. One possible mechanism is a sonic boom: this is produced whenever a solid object moves faster than the speed of sound in that medium, and is the cause of

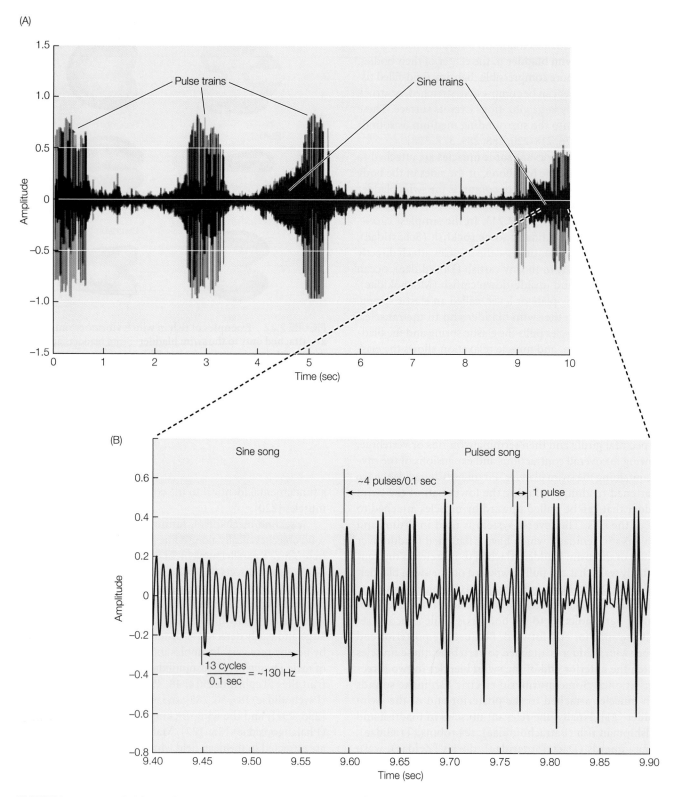

(A)

(B)

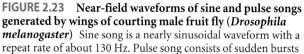

FIGURE 2.23 Near-field waveforms of sine and pulse songs generated by wings of courting male fruit fly (*Drosophila melanogaster*) Sine song is a nearly sinusoidal waveform with a repeat rate of about 130 Hz. Pulse song consists of sudden bursts of waves. The two kinds of songs are interspersed sequentially in normal courtship song. (A) Broad time scale showing 10 seconds of continuous courtship song. (B) Expanded portion of A showing 0.5-second portion. (After [204].)

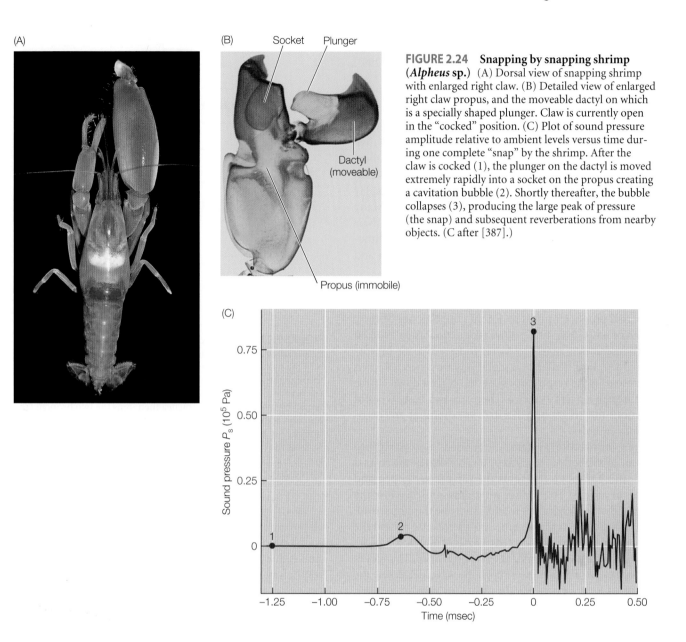

FIGURE 2.24 Snapping by snapping shrimp (*Alpheus* sp.) (A) Dorsal view of snapping shrimp with enlarged right claw. (B) Detailed view of enlarged right claw propus, and the moveable dactyl on which is a specially shaped plunger. Claw is currently open in the "cocked" position. (C) Plot of sound pressure amplitude relative to ambient levels versus time during one complete "snap" by the shrimp. After the claw is cocked (1), the plunger on the dactyl is moved extremely rapidly into a socket on the propus creating a cavitation bubble (2). Shortly thereafter, the bubble collapses (3), producing the large peak of pressure (the snap) and subsequent reverberations from nearby objects. (C after [387].)

the loud snap when one cracks a whip. Whether manakins can move the tips of their wings fast enough to produce a whip-crack remains unknown. Male ruffed grouse produce an accelerating series of low (40 Hz) thumping sounds by repeatedly bringing their partially extended wings toward each other very rapidly [8, 188]. High-speed films have eliminated the possibility of percussive effects in this behavior. The male sage grouse produces two loud "pops" during his strut display that are associated with a sudden expansion and oscillation of the two air sacs on his breast. While the pops might be generated in part by percussion between the two expanding sacs, it is also likely that each sac produces very rapid pressure changes near the sac surfaces [83]. Finally, very rapid movements in water can generate sufficiently extreme pressures around the moving object that nearby layers turn into a gas and form a bubble. The subsequent collapse of the bubble then generates a loud sound. This process, which can

occur only in liquids, is called **cavitation**, and is the source of bubbles behind boat propellers. Snapping shrimp (Alpheidae) have an enlarged claw in which one of the two terminal segments can be locked, stressed, and then released very rapidly (**Figure 2.24**). The loud sound is the result of cavitation and not percussion [101, 387]. The shrimp use snapping as a defense against predators, to stun prey, during intraspecific contests, and to rally neighbors to eject intruders from a group territory [180, 209, 338, 381]. Because snapping shrimp are common throughout most of the world's oceans, they constitute the major source of high-frequency noise in most marine environments [10, 37, 56, 304, 309, 312].

The fourth mechanism, **streaming**, occurs when the signaler moves sufficiently rapidly through a fluid medium that the flow over its appendages generates vibrations. Several groups of birds produce whistling, winnowing, or humming vibrations while in flight. Among shorebirds, the woodcock

(*Scolopax minor*) has special feathers in the tips of its wings that can be spread to produce vibrations while in flight [299], whereas several species of snipe (genera *Gallinago* and *Coenocorypha*) have similar adaptations in their tails [53, 259, 383, 386]. Wing feathers specialized to produce flight sounds also occur in honeyeaters (Meliphagidae) [78], and several nighthawks in the genus *Chordeiles* perform an aerial dive display that ends with a booming sound when wing feathers are angled into the airflow [166, 258]. Hummingbirds also have specialized tail or wing feathers that generate display sounds during particular modes of flight [59, 60, 192]. Many pigeons and doves produce a whistling sound with their wings when startled into sudden flight, and this sound functions as an alarm signal attended to by both conspecifics and heterospecifics [72, 186].

MOVING FLUID MEDIUM OVER BODY PARTS   Although some fish produce sounds by forcing bubbles out of their mouth or anus [393, 405], sound production using a forced flow of medium is largely restricted to terrestrial air-breathing animals. In the latter case, vibrations are generated when air present in the respiratory system is forced down a tube and through a narrow orifice. There are two basic types of vibrations that might be generated: vocalizations and aerodynamic sounds.

In a **vocalization system**, a tube used to move air into and out of the body is blocked by a valve. When the animal is breathing through this tube, the valve is kept open and out of the airflow. To produce vibrations, muscles close the valve, and other respiratory muscles increase the air pressure behind the valve. At some point, the air pressure behind the valve, (the upstream side) exceeds whatever muscle forces are holding the valve closed, and the valve pops open. This allows a sudden pulse of air to flow through the valve, decreasing the pressure difference between the upstream and downstream portions of the tube, and generating a suction on the valve lips called a **Bernoulli force**. The continued tension on the valve-closing muscles, the reduced pressure difference across the valve, the Bernoulli suction, and differential latencies in the movements of different masses in this system then cause the two lips of the valve to close and stop the airflow [87, 181, 380]. Upstream pressure starts to build up again, and the cycle repeats. Such a system will usually produce vibrations that are periodic but nonsinusoidal. A spectrogram of the resulting sound will show a harmonic series whose fundamental frequency is equal to the rate at which the valve pops open and closed. This rate will depend on the size, mass, and elastic properties of the valve lips, the muscle force used to close them, any tension applied to them by other muscles, and the pressure exerted by the upstream respiratory system prior to valve opening [119].

All terrestrial vertebrates have similar respiratory plumbing (**Figure 2.25**). Each of the two lungs is connected to a tube called a **bronchus**. The two bronchi join at some point to form a single tube called the **trachea**, which connects to the outside via the mouth and nasal cavities. Even though

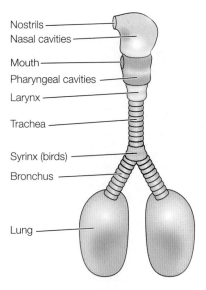

FIGURE 2.25   The basic elements of a terrestrial vertebrate respiratory system Air is inspired through the nostrils or mouth into the nasal or pharyngeal cavities, respectively. It then passes through the larynx (originally evolved to keep food out of the trachea) into the trachea, which consists of bone or cartilage rings joined by membranes. The trachea then joins the two bronchi (the junction elaborated into the syrinx in birds), which convey the air into the lungs. During sound emission, the reverse process occurs, with air moving through vibrating valves in the upper bronchi or syrinx junction (birds), or in the larynx (amphibians, reptiles, and mammals).

all terrestrial vertebrates share this basic design, their vocal sound production mechanisms are quite varied. This variation is due in part to different strategies for ventilating the respiratory system: those taxa with more sophisticated controls over ventilation are the ones most likely to produce vocal sounds, and the finer that control, the more complex the vocalizations.

Frogs and toads acquire air by expanding a throat sac while all other openings except the nostrils are closed. Once partially inflated, the sac is contracted, forcing air into the lungs. This is called **buccal pumping**. Air is expelled by the contraction of muscles in the body wall. Lizards and snakes, unlike frogs, have ribs and muscles that allow expansion of the thoracic cavity. Turtles' ribs are attached to their shells. Turtles, crocodiles, and alligators all have special muscles that move their viscera backward and forward to facilitate inhalation. Birds, like lizards and snakes, use muscles in their rib cages to expand and contract their thoracic cavities. Birds' lungs, unlike those of most other vertebrates, are not elastic and do not inflate or deflate with breathing. Instead, inhalation fills a set of air sacs that can subsequently pass the air through the lungs (**Figure 2.26**). This system stores large amounts of air needed for the high metabolic costs of flight. Mammals have augmented their thoracic musculature with a diaphragm that contracts to expand the thoracic cavity and draw air into the lungs. Whereas relaxed birds have relatively

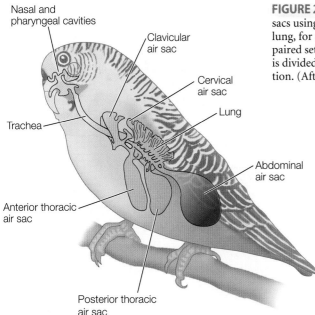

Nasal and pharyngeal cavities

Clavicular air sac

Cervical air sac

Lung

Trachea

Abdominal air sac

Anterior thoracic air sac

Posterior thoracic air sac

FIGURE 2.26 **Air sac system in birds** Diagram shows typical distribution of sacs using budgerigar (parrot) as example. There is a pair of sacs, one for each lung, for all but the clavicular air sac. Only one lung and one example of each paired set of sacs is shown in the figure. While there is a single clavicular air sac, it is divided into two lobes that encircle the syrinx and may be involved in vocalization. (After [100].)

As might be expected given their relatively simple ventilation systems, many modern reptiles do not vocalize. Exceptions include some turtles and tortoises, most gecko lizards, and most crocodilians [130, 266, 319, 326, 328]. In contrast, the more sophisticated ventilation system in mammals has made vocalization extremely common in this group. The roars of lions, rumbles of elephants, lowing of cattle, screams of the tree hyrax, howls of monkeys, and human speech are all examples of vocalization sounds. In a variation on this theme, some bats have a very thin ribbon of tissue just downstream from the free margin of each vocal cord. These membranes, which are thinner than

full air sacs and must work to *exhale*, relaxed mammals have empty lungs and must work to *inhale*. The consequent differences in neuromuscular control of breathing may be partly responsible for the greater abundance of complex vocalizations in birds than in mammals.

All terrestrial vertebrates except birds position the valve for production of vocal vibrations at the top of the trachea [270]. This valve, called the **glottis**, likely first served functions other than vocalization. Possibilities include preventing food from entering the trachea, generation of internal pressures for egg laying or defecation, or prevention of air loss during diving [208]. When used to produce sounds, the glottal lips are called **vocal cords** (**Figure 2.27**). The glottis, along with its associated cartilages and muscles, is part of the **larynx**. In most vocalizing mammals and reptiles, sound vibrations are produced when the laryngeal muscles move the vocal cords into a laryngeal cavity at the top of the tracheal passageway. Upstream air pressure is then increased by contracting the muscles used for exhalation. At some point, the vocal cords begin to open and close, letting small puffs of air flow through the larynx. Additional muscles are often present that can vary tension on the vocal cords. This tension will help to determine the rate at which the vocal cords oscillate open and closed: increased tension usually leads to more rapid rates of vibration, and a spectrogram of the sound will thus show a higher fundamental frequency and a greater spacing between harmonics. Varying the tension during a call is often used to modulate both the fundamental and the harmonic structure in the resulting sound.

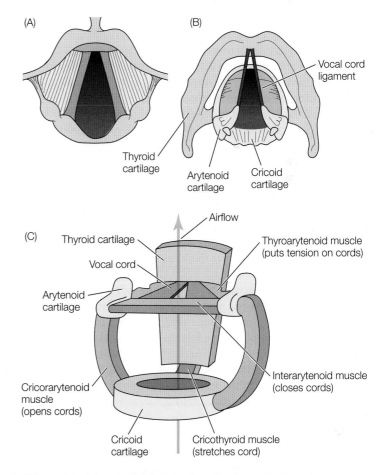

(A)

(B)

Vocal cord ligament

Thyroid cartilage

Arytenoid cartilage

Cricoid cartilage

(C)

Airflow

Thyroid cartilage

Vocal cord

Arytenoid cartilage

Thyroarytenoid muscle (puts tension on cords)

Interarytenoid muscle (closes cords)

Cricorarytenoid muscle (opens cords)

Cricoid cartilage

Cricothyroid muscle (stretches cord)

FIGURE 2.27 **Mammalian larynx** View from mouth of top of human larynx (A) and human larynx with soft tissues removed to show cartilages and ligaments (B). (C) Simplified diagram of essential working parts in mammalian larynx. Separate muscles open, close, and put tension on the vocal cords by moving them relative to attached cartilages. Cords are closed just prior to vocalization.

the vocal cord base, can vibrate independently and at much higher frequencies than the thicker part of the cords [164, 361]. Bats with these membranes typically echolocate by emitting short pulses of ultrasonic sounds in the dark and using the echoes to avoid obstacles and to locate food items (see Chapter 14). It has been proposed that the thicker bases of their vocal cords open and close to begin and end each pulse, and that airflow during the open phase causes the thinner membranes to vibrate independently at ultrasonic frequencies. Similar membranes are also known in some primates, cats, pigs, and llamas, but their function in these species is not yet clear [253].

Diving mammals present another special case. Because it is inefficient for these animals to release stored air when they are deep underwater, they need a way to trap and recycle air after it has passed through the vibratory valve during vocalization. Baleen whales (mysticete cetaceans) appear to generate vibrations in the larynx like other mammals, but pass the air into expandable nasal passages or a vocal sac attached to the larynx [9]. Toothed whales and porpoises (odontocete cetaceans), however, move air back and forth between nasal sacs situated between the larynx and the blowhole. A new valve, called **sonic lips**, has evolved in the passage between these sacs to produce the requisite vibrations [79, 80, 238, 239]. Seals, sea lions, and walruses also turn out to be highly vocal underwater (**Figure 2.28**). The loud and complex advertisement calls of male bearded seals (*Erignathus barbatus*), Weddell seals (*Leptonychotes weddellii*), ribbon seals (*Histriophoca fasciata*), and walruses (*Odobenus* spp.) are particularly striking [324, 345, 353]. Several of these species have special sacs attached to the trachea or the pharynx (the throat region between the larynx and the mouth), and the usual bony rings on one entire side of the trachea in the bearded seal have been replaced with a soft and expansible membrane [384]. These specializations contribute to underwater sound production, but the details are not yet understood. Tracheal sacs may provide air for long-duration vocalizations, and pharyngeal sacs, such as those found in frogs, may store air after vocalization for reuse.

Modern frogs and toads also produce sounds with a larynx at the top of their trachea, but like echolocating bats, most species use the glottis to turn airflow on and off. When on, the airflow passes over secondary membranes that are set into vibration and insert higher frequency components into the call (**Figure 2.29**). These membranes, unlike the ones in bats, are not attached to the margins of the glottis, but protrude into the airflow upstream (closer to the lungs) from the glottis [244–246, 337, 340]. In fire-bellied toads (*Bombina* spp.), the vibrating structures are thickened ridges on the sides of the larynx. These animals produce sounds on both inspiration and exhalation, and setting the ridges to vibrate with airflow in either direction requires symmetrical attachments. In contrast, common toads (*Bufo* spp.) have thin membranes, aimed downstream, that are set into vibration when the glottis pops open and airflow begins. In many toads, glottis oscillation is produced by mechanisms similar

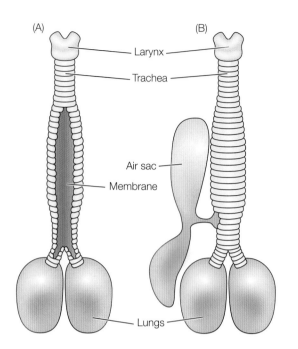

FIGURE 2.28   Modifications of trachea for vocalization by phocid seals (A) Incomplete tracheal and bronchial rings and associated membrane in bearded seal (*Erignathus barbatus*). (B) Tracheal air sac in male ribbon seal (*Histriophoca fasciata*). There is only one sac, located on the seal's right side. Females usually lack the sac and in males, it appears to get larger with body size or age. (After [384].)

to that in mammals: upstream pressure, airflow, elasticity, and muscle tension combine to produce repetitive glottal opening and closing at rates of several hundred Hz [244, 245]. When the glottis is open, the vocal membranes then oscillate at fundamental frequencies of 1–2 kHz. In some toads, the vocal membrane oscillations are nearly sinusoidal and display little energy at higher harmonics in spectrograms.

Amphibian genera with more complex laryngeal structures include common frogs (*Rana* spp.), cricket frogs (*Acris* spp.), and tree frogs (*Hyla* spp.), which have perpendicular ridges at the end of the vocal membranes (forming a T in cross section). The glottis is usually opened and closed by muscles in these species. When it opens, air pressure first pushes the flattened edges of the two vocal membranes together to close the airflow, and then forces them open again, setting them to vibrate [337, 340, 394]. The decrease in pressure and Bernoulli forces then pull them back together, and the cycle repeats as long as the glottis is held partially open. The result is a long call consisting of successive pulses of high-frequency vibrations [144]. The duration of the call (up to many seconds) is determined by how long the muscles hold the glottis open; the pulse rate (usually several hundred Hz) is set by the physical structure and properties of the vocal membranes and the airflow; and the vibration of the membranes when open (usually one to several kHz) is also determined by physical properties of the membranes and the

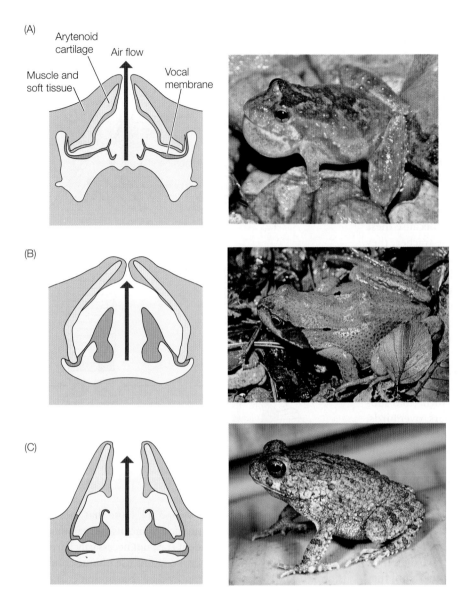

(A) Arytenoid cartilage
Air flow
Muscle and soft tissue
Vocal membrane

(B)

(C)

FIGURE 2.29 **Some common frogs and toads and their larynges** Glottal opening is at the top of each diagram, and arrow shows direction of airflow during vocalization. Blue area below the glottis indicates the tracheal and laryngeal air space; white area above the glottis is the pharyngeal cavity. (A) Cricket frog (*Acris crepitans*). Vocal membranes are T-shaped with a slight thickening at midpoint. (B) European common frog (*Rana temporaria*). T-shaped vocal membranes are considerably enlarged on their ends relative to cricket frog. (C) African toad (*Bufo regularis*). These vocal membranes are very enlarged before the filamentous tip and incorporate an ossicle that is heavy enough to depress the fundamental frequency considerably. Note that toads often have two pairs of vocal membranes, one upstream from the other. (After [145, 248, 340].)

airflow. Whether harmonics are present in spectrograms of the vibrations within a pulse depends on how sinusoidally the membranes can vibrate when not touching.

Many frogs and toads attach fibrous masses to the upstream side of the vocal membranes. These masses often lower the fundamental frequency of the membrane vibrations, and they can significantly affect the movement of the membranes and thus the presence or absence of harmonic structure in the sound [247, 248, 404]. In the túngara frog (*Engystomops pustulosus*), unusually large masses are present in cavities adjacent to the vocal membranes. Air flows not only between the edges of the vocal membranes, but also laterally through these cavities. The masses are also attached to the wall of the larynx and to the vocal cords in ways that allow them to vibrate at about half the fundamental frequency of the vocal cords. Male túngara frogs can vary laryngeal tensions and airflows either to prevent oscillation of the large masses, producing the obligatory "whine" component of the

call, or to allow the masses to vibrate interactively with the vocal membranes, producing the facultative "chuck" component after the whine [91, 162, 198, 327].

Perhaps because they have such sophisticated neuromuscular control over exhalation, birds exhibit the most diverse and complex vocalization patterns. Unlike frogs, reptiles, and terrestrial mammals, birds do not have a sound-producing valve at the top of the trachea, but instead have one or more vibrating valves near the junction of the trachea and the bronchi. The specialized cartilages and muscles associated with these valves are called the **syrinx**. Normally, the trachea and bronchi of vertebrates consist of bony or cartilaginous rings linked by connective tissue to form tubes. Many birds create a vibration-producing valve in the syrinx by expanding the soft tissues linking the bony rings into large **tympaniform membranes** [5, 26, 207]. In some cases, these membranes are created using incomplete C-shaped rings in which the missing bony tissue is replaced with soft tissues. In other cases, several adjacent rings are converted into flattened and flexible elements that are embedded in the softer tissues. As a third option, the distance between a pair of adjacent rings is increased and filled with a much larger patch of membrane. However created, pulling the syrinx toward the head or tail, rotating particular rings, or expanding the air pressure in the interclavicular air sac that surrounds the syrinx can cause the tympaniform membranes to buckle into the center of the tube where airflow will set them vibrating. Avian syringes have classically been divided into three categories based on where these membranes are located.

FIGURE 2.30   Example of bird tracheal syringes All figures are ventral views with mouth at top and lungs below. (A) Syrinx of blue-and-gold macaw (*Ara ararauna*). (B) Cross section of macaw syrinx showing lateral tympaniform membranes buckled into airflow (vertical arrows) by external muscles. (C) Syrinx of domestic chicken (*Gallus gallus*). Horizontal arrows indicate which membranes are buckled into tracheal cavity to create vibrating valve. (D) Syrinx of imperial pigeon (*Ducula latrans*). (After [26, 46, 207, 227].)

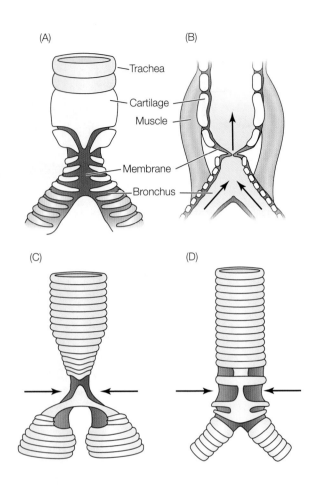

In birds with **tracheal syringes** the tympaniform membranes are located at the base of the trachea just before the junction with the two bronchi (**Figure 2.30**). Because these membranes form part of the outer wall of the trachea, they are called **lateral tympaniform membranes**. In many of these species, such as parrots and tyrannid flycatchers, but not pigeons and doves, the last 2–8 tracheal rings are fused into a single bony chamber called a **tympanum**. This appears to serve as a firm base against which the tension in adjacent membranes can be adjusted. In chickens and some relatives (Galliformes), doves and pigeons (Columbiformes), parrots (Psittaciformes), and some wading birds (Ciconiiformes), a tympaniform membrane exists on each side of the trachea just above the junction with the bronchi. In furnarioid suboscines (Passeriformes) such as woodcreepers, ovenbirds, gnateaters, antpipits, and antbirds, and in whistling ducks (genus *Dendrocygna*), a single tympaniform membrane is positioned centrally on the ventral side of the trachea. Physiological studies have demonstrated that these tracheal tympaniform membranes are indeed the source of vocal vibrations [46, 47, 95, 132–135, 138, 139, 153, 226, 227]. In pigeons and parrots, the tympaniform membranes are already partially buckled into the tracheal cavity during normal respiration. To create sounds, muscles move the syrinx toward the head, buckling the membranes further until they form a thin slit. Airflow through this slit generates the vocal vibrations, and adjacent muscles vary tension on the membranes to alter fundamental frequencies.

Birds with **bronchial syringes** have 1–2 tympaniform membranes on each bronchus below the junction of the trachea and bronchi (**Figure 2.31**). Examples include oilbirds, pauraques, and frogmouths (Caprimulgiformes), cuckoos, coucals, and guiras (Cuculiformes), penguins (Sphenisciformes), and some owls (Strigiformes). Oilbirds vocalize by relaxing the tension between the trachea and the syrinx, thus causing the lateral (external) tympaniform membranes in each bronchus to buckle into the bronchial cavity until each is very close to the corresponding medial (internal) tympaniform membrane. Airflow then generates vibrations in the

tympaniform membranes [363, 371]. Note that these birds have a vibration generator on each bronchus and can theoretically produce two independent (i.e., not necessarily harmonically related) sounds at once. Penguins use unique combinations of the two sounds for individual recognition [12].

FIGURE 2.31   Example of bird bronchial syringes (A) Syrinx of ground cuckoo (*Carpococcyx renauldi*). (B) Syrinx of oilbird (*Steatornis caripensis*). Note that there are tympaniform membranes on both sides of each bronchus in both species. (A after [26]; B after [207].)

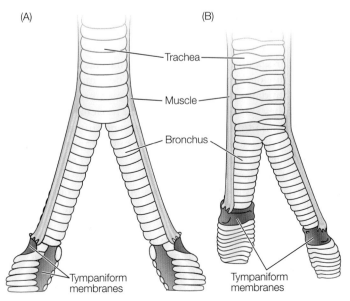

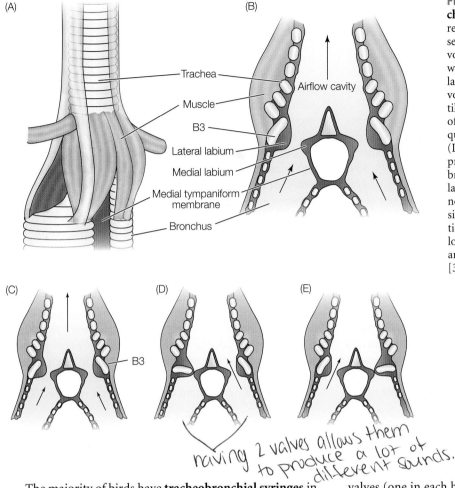

FIGURE 2.32 Example of bird tracheobronchial syrinx External view with relevant muscles attached (A) and cross section (B) of mockingbird syrinx when not vocalizing. Arrows show direction of airflow when exhaling and vocalizing. Medial and lateral labia are moved into airflow during vocalization by rotating third bronchial cartilage (B3). (C) Diagrammatic cross section of songbird syrinx when breathing, showing quiescent position of labia and B3 cartilages. (D) Configuration of songbird syrinx when producing high-frequency notes. Third bronchial cartilages (B3) are rotated forcing lateral labia into airflow. For high-frequency notes, left side is closed completely and right side pulses open and closed. (E) Configuration of songbird syrinx when producing low-frequency notes: right side is closed and left side pulses open and closed. (After [358].)

having 2 valves allows them to produce a lot of different sounds.

The majority of birds have **tracheobronchial syringes** in which the main expanses of elastic tissue are the **medial tympaniform membranes** located on the insides of each bronchus just below the junction of the trachea and the bronchus (**Figure 2.32**). Examples include ratites such as cassowaries (Struthioformes); most ducks and geese (Anseriformes); gulls and shorebirds (Charadriiformes); woodpeckers, barbets, and toucans (Piciformes); trogons, hornbills, and kingfishers (Coraciiformes); and perching birds (Passeriformes, with the exception of the furnarioid suboscines noted earlier). For many years, indirect evidence suggested that the medial tympaniform membranes were buckled into each bronchial cavity and set into vibration by concomitant exhalation [118, 139, 159, 352]. Subsequent work on songbirds (the Oscine Passeriformes) showed that vibrations are generated not by tympaniform membranes but by a pair of soft tissue pads, called the **lateral** (outside) and **medial** (inside) **labia**, located opposite each other in each bronchus [152, 155, 199, 226–228, 370]. A similar mechanism may operate with analogous structures in echolocating swiftlets [362]. In songbirds, each lateral labium is attached to a syringeal ring that can be rotated. Each medial labium is connected directly to the medial tympaniform membranes below it. Rotation of the ring behind a lateral labium and movement of the entire syrinx toward the head forces the two labia into the bronchial cavity. Such a syrinx thus has two independently vibrating

valves (one in each bronchus). As with the other two types of syringes, vibrations generated by each valve are typically periodic but nonsinusoidal.

Web Topic 2.7 *Animations of vocalizing birds*
Dr. Roderick Suthers and his team at Indiana University have pioneered our understanding about how the avian syrinx works. His lab has conveniently produced several animated clips demonstrating key steps in song production by northern cardinals and brown-headed cowbirds.

While all songbirds have labia and appear to use them to create vibrations, other taxa with a similar syrinx (e.g., gulls) appear to lack labia [132, 207]. All taxa with tracheobronchial syringes have medial tympaniform membranes, and it is possible that these contribute to sound production when labia are absent. Some species also show thinned membranes, (often called lateral tympaniform membranes but not to be confused with the lateral membranes in tracheal syringes), in the outer wall of the bronchi where songbirds would have a labium. Other species appear to have intermediate types of syringes in which tracheal lateral membranes extend down past the junction with the bronchi, and medial membranes that extend further toward the head than in songbirds [207]. Male sage grouse (*Centrocercus urophasianus*) have thin membranes on the external top sides of each bronchus and apparently can use these independently to produce two different and unrelated sounds at once [211]. An extreme case occurs in some geese and swans in which nearly the entire top section of each bronchus consists of membranous tissues. Sound production in geese is clearly an effective process, but the exact mechanisms remain to be studied.

Although both bronchial and tracheobronchial syringes allow for concurrent production of vibrations from two independent sources, the songbirds have taken most advantage of this opportunity [149–151, 154, 228, 274, 276, 278, 359, 366, 368–371, 373, 374]. With their sophisticated syringeal musculature, songbirds can move the labia into or out of the airflow at will. They can also use the labia to close off the airflow in a bronchus. This allows songbirds to produce rapid and complex songs by alternating between the two sides of the syrinx for successive syllables. Because the two sides of the syrinx are largely independent (except perhaps in very small songbirds; [277]), the successive syllables in a song can have quite different amplitude and frequency patterns. Many species have adopted a division of labor, producing low-frequency syllables on the left side of the syrinx and high-frequency syllables on the right. Male cardinals (*Cardinalis cardinalis*) combine these tricks by producing frequency-modulated syllables in which the first half is generated by one side of the

syrinx, and the second half is added nearly seamlessly by the other side (**Figure 2.33**). In addition to using the two sides of the syrinx to produce successive syllables or successive parts of syllables, species such as brown thrashers (*Toxostoma rufum*) may use the two sides to produce two concurrent notes that are not harmonically related. This, combined with a specialization of the right side of the syrinx for rapid modulations, allows this group to produce some very complicated sounds.

Many songbird males include trills in their songs. These are very rapid and long duration bursts of successive notes. At lower trill rates, the singers can take **mini-breaths** between syllables to maintain sufficient air to continue the trill over long periods; at higher rates, no inhalation is possible and the birds simply pulse the labial valve on one side of the syrinx until they run out of air [171, 172]. As a result, there is often an inverse relationship between trill duration and the trill rate, with larger birds having to switch to pulsatile trills at a lower trill rate than smaller birds [374]. A similar trade-off occurs between the frequency range over which a trill syllable can be swept and the trill rate [18]. Many songbirds thus appear to have pushed vibration production to a point where any further increase in one property (e.g., trill rate) degrades another concurrent property (e.g., trill duration or frequency range of each syllable). These

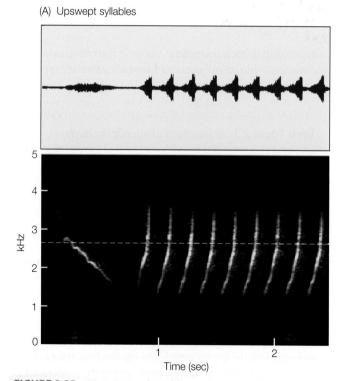

(A) Upswept syllables

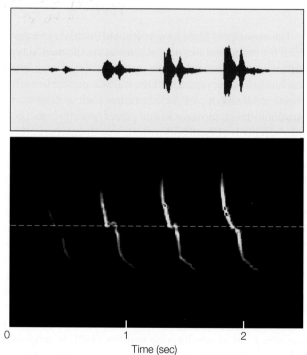

(B) Downswept syllables

FIGURE 2.33 Two examples of frequency-modulated syllables in cardinal song (*Cardinalis cardinalis*) In each example, the horizontal dashed white line in the spectrogram shows the break point (at about 2.7 kHz) at which one side of the syrinx takes over from the other to produce a nearly continuous single syllable. (A) Male song showing upswept syllables. Very faint break

can be seen in spectrogram at 2.7 kHz as the left side of the syrinx takes over from the right side, but no break is visible in the waveform. (B) Female song showing downswept syllables. Note the visible break in both spectrogram and waveform when the right side of syrinx takes over from the left side.

trade-offs will be of relevance later when we discuss the role of song in avian mate choice.

The final mechanism involving fluid flow over solid parts is called **aerodynamic vibration** [119]. In contrast to the vocalizations just discussed, the solid parts remain immobile as fluid passes over them and do not themselves vibrate. Instead, the flow of fluid through an aperture or against a hard obstacle will generate local turbulence in the medium. Turbulence takes the form of eddies or vortices that spiral away from the aperture or obstacle; their size determines the frequencies of any far-field sounds produced [119, 392]. There are three general types of aerodynamic sounds. Streaming of fluid through an aperture in a solid barrier will generate a wide range of vortex sizes; the resulting sounds will have many different frequencies present and be noisy and atonal like a hiss. This is an inefficient process, and the resulting sounds are likely to be useful at close range only. It is how humans generate many of their consonants. A second method involves aiming a steady stream of fluid against a small or sharp obstacle. This is more likely to produce vortices of similar size than is an aperture, and thus produces somewhat tonal sounds that can be heard at a distance. An example is the musical wail generated by wind against the outside corners of a building. Finally, positioning an aperture or obstacle to an airstream near a cavity can result in selective amplification (resonance; see next sections) of a limited number of vortex frequencies, resulting in a relatively pure-tone whistle. It is the most efficient of the three aerodynamic mechanisms and can generate sounds heard at significant distances. Humans whistle using the oral cavity in this manner.

A number of insects force air (or in grasshoppers, foam) out of their respiratory spiracles to produce hissing sounds [92, 271, 357]. In a similar manner, various reptiles force air through a partially closed and immobile glottis or through their mouths or nostrils to produce hisses [130, 410, 412–414]. Similar oral or nasal hisses are produced by some birds and mammals. More whistle-like mechanisms have been described in the death's-head hawkmoth (*Acherontia atropos*), which draws air into and out of its pharynx creating a squeaky sound in each direction [92, 330], and the king cobra (*Ophiophagus hannah*), which "growls" using tracheal air spaces to amplify vortices generated as air passes through its glottis [410]. While it was initially suggested that some bird vocalizations might be whistles [54, 136, 137, 275], this proposition has since gained no experimental support [155, 374].

Web Topic 2.8 *Linear versus nonlinear systems*
Although animal sound vibrators typically change frequency and amplitude linearly as the forces on them are changed, most will adopt less predictable nonlinear behavior at extreme force values. Some animals such as parrots, zebra finches, and canids use nonlinear vibration behavior to increase their vocal diversity.

Modification and coupling of sound signals

Clearly, there are several ways that arthropod and vertebrate animals can generate vibrations. It is not so easy, however, to convert these initial vibrations into useful sound signals. There are two problems. First, many of the initial vibrations are of such low amplitude that by the time they are coupled to the propagating medium, they are no longer detectable at the required distances. Second, many mechanisms for coupling vibrations from animals into a propagating medium can be hopelessly inefficient. The obvious solution is to amplify the vibrations before coupling, and then use those mechanisms that provide the most efficient coupling. As so often happens in physics and biology, the best amplifiers are not necessarily compatible with the best coupling mechanisms, and sound production is necessarily compromised. Below, we examine each of these options in principle, and conclude this chapter with some of the compromises that animals have adopted in practice.

MODIFICATION OF SOUND SIGNALS: RESONANCE AND FILTERING   The vast majority of vibrations generated by animals have periodic but nonsinusoidal waveforms, and thus will consist of many different frequencies. Noisy sounds (e.g., hisses) also contain many different frequency components. When a vibrational pattern is introduced into an object such as a wing, swim bladder, or pharyngeal cavity, some of the component frequencies will have wavelengths that fit within the dimensions of the object, and some will not. If the acoustic impedance of the object is similar to that of the adjacent medium, this fit does not matter, as all frequencies will simply propagate across the boundary and into the medium. If there *is* a significant impedance difference between the object and the medium, then much of the sound energy inside the object will be reflected back at its boundary and the object will store successive energy inputs cumulatively. The result is a series of superimposed waves for each frequency component in the vibration: some will have just entered the object from the vibrational source, some will have already traversed the object and bounced back the other way, and still others will have been reflected multiple times between opposite sides of the object. Given the right ratio between the wavelength of a frequency component and the dimensions of the object, both incoming and reflected waves will be in phase; successive waves will combine their sound pressures additively; and over time, the amplitude of the original vibrations will be amplified. This is known as **resonance** and those frequencies that fit and are amplified are called **normal modes** of the object. At the same time, those components that do not have the appropriate wavelength-to-object ratio, (e.g., do not fit), will produce reflected waves that are out-of-phase with incoming waves, and the resulting destructive interference will reduce the amplitudes of these components. This is known as **filtering**. See the examples of resonant systems at Web Topic 2.2.

Whether a given frequency component is amplified by resonance or reduced by filtering depends on the object's size, shape, speed of sound, and acoustic impedance relative

FIGURE 2.34 Frequency and time responses of resonators (A) Two resonators with a resonance at 2000 Hz but differing in their Q (quality) factor. The high-Q resonator is a better amplifier than the low-Q resonator at their shared resonant frequency. Q values can be estimated from such a graph by finding those frequencies on either side of the resonant frequency for which the pressure amplitude is 70.7% (–3 dB) of the peak value. (These will also be the frequencies at which the power is *half* of the peak value). The –3 dB bandwidth is the difference between these two frequencies and is marked with arrows on the graphs. Q is then the ratio of the resonant frequency to this bandwidth. (B) Damping properties of the same two resonators at their resonant frequency. Graphs show the amplitude of successively radiated cycles once the source of sound vibrations ends. The high-Q resonator is poorly damped, and continues to radiate for many cycles after stimulation ceases. The highly damped low-Q resonator, on the other hand, quickly loses all prior sound energy and becomes silent. Damping and Q are inversely related.

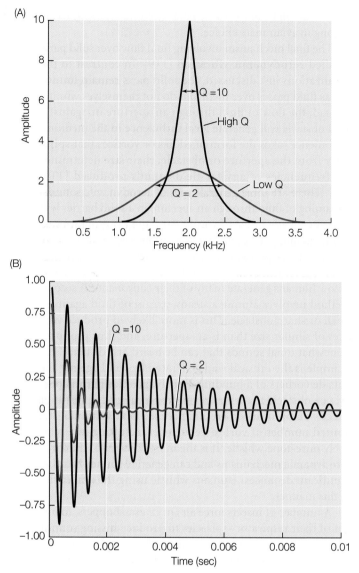

to the nearby medium. Small objects cannot accommodate large wavelengths; thus any resonant amplification will be limited to higher frequencies. However, if a given fundamental frequency of the introduced vibration fits within an object, most of the higher harmonics in a periodic nonsinusoidal vibration will also fit and can be amplified. The natural modes are then likely to be integer multiples of some fundamental value. If the fundamental frequency of the introduced vibrations is less than the lowest natural mode of the object, the latter will filter out the fundamental, amplify subsets of adjacent harmonics (called **formants**), and filter out intervening subsets. Nonsymmetrical objects may be able to accommodate several unrelated frequencies and thus can have normal modes that are not necessarily harmonics of the same fundamental frequency. The speed of sound in the object is important because this determines the wavelength of each frequency component in the introduced vibrations. Where the object is a solid, (e.g., a cricket wing), the relevant speed of sound is that of the relevant tissues. Many animals use gas-filled cavities either in their bodies or in a substrate for resonance. The relevant value in such **cavity resonators** is the speed of sound of gas in the cavity.

If the acoustic impedance of the object is sufficiently similar to that of the medium, introduced sound energy is quickly lost through the boundary. A sudden short pulse of vibrations will cause only a few cycles of successive wave overlap before the signal has completely moved into the medium. We say that such a system is **highly damped**. Because only a few recent waves will overlap in this object even with sustained stimulation, there is less opportunity to produce significant resonance or filtering, and thus the frequency selectivity of the object will be low (**Figure 2.34**). We say that the object supports a **wide bandwidth** (of frequencies). Objects with wide bandwidths and high damping are said to have a **low Q** (quality of resonance). In contrast, an object with an acoustic

impedance very different from the medium will be very slow to transfer sound energy to the medium. Many successive waves will build up inside it, allowing for very strong resonance and filtering. Such an object generally will have a very narrow bandwidth around each normal mode frequency. A prior signal may continue to reflect back and forth inside the object long after the input source has halted. This is called **ringing** and reflects very low damping. An object that loses sound energy slowly and has a narrow bandwidth and low damping is said to have a **high Q**.

While a high-Q resonator can significantly amplify a narrow bandwidth of frequencies if these are present in the introduced vibrations, it will smear and distort any rapid amplitude and frequency modulations of those vibrations. Similarly, a low-Q resonator can track and replicate rapid modulations accurately, but it will result in much lower levels of amplification. There is thus a trade-off between amplification and accurate temporal tracking when using secondary objects to amplify signals. In general, animals communicating over long distances, such as cicadas, are more likely to use high-Q resonators to maximize signal range. The cost is that they will lose any rapid temporal patterns in the signal. At shorter distances, animals need less amplification, and one expects—and finds—sound signals with greater temporal complexity.

COUPLING TO THE MEDIUM: LIMITS ON EFFICIENCY   At some point, an animal's vibrations must be radiated into the propagating medium (air, water, a solid substrate, or some boundary between two media). If the medium is a fluid (e.g., air or water) the surface of a vibrating radiator forces a layer of medium away from itself during part of each cycle, and causes medium to flow toward it during the rest of the cycle. If the medium is a solid, the pressure forces of the radiator generate corresponding pressure waves in the solid with little if any movement of medium molecules. The efficiency of this vibrational transfer is critically dependent on the size of the radiating organ, the way in which the organ moves, and the organ's acoustic impedance relative to that of the medium.

For fluid media, a larger radiator is better than a small one, because larger radiating surfaces will move more fluid per cycle. Another way to see this advantage is to consider two spheres differing in diameter by a factor of ten and both pulsating in air or water at the same rate. Suppose each sphere expands and contracts the same small distance per pulsation. The amount of medium moved *per unit area of sphere surface* will be very similar for the two spheres, and thus the initial energy provided per unit area will be similar. However, by the time the sound generated by the small sphere travels a distance from its center equal to the radius of the large sphere, its pressure will have decreased significantly due to spreading loss. It follows that at any further location equidistant from the two spheres, the larger sphere's sound pressures will always be greater. As a result, larger radiators produce more intense sounds for all frequencies.

Efficiency can also be affected by *how* the radiator moves, especially at low frequencies. A **monopole** radiator, like the spheres in the example above, simply expands and contracts equally in all directions. It thus changes its size to couple its vibrations to the medium. A **dipole** radiator, in contrast, is coupled to the medium by moving back and forth along an axis. The simple disk that we discussed at the beginning of this chapter is a good example of a dipole radiator. Finally, a **quadrupole** radiator maintains its location and size but changes its shape. A fish swim bladder that is compressed at one end and balloons out on the other is a good example of a quadrupole radiator. While there are additional radiators used by animals, nearly all of these behave like one or some combination of these three basic types. For example, the open mouth of a vocalizing bird or mammal radiates sound in a manner similar to a monopole placed close to a solid reflecting surface [119].

The coupling of vibrations to a medium becomes increasingly inefficient as the wavelengths of the sounds being radiated grow larger than the diameter of the radiator. This is true regardless of which type of motion is used. However, whenever the radiators are smaller than the wavelength, monopoles are much more efficient than dipoles, which in turn are more efficient than quadrupoles [119, 167]. In part, the increased inefficiency of dipoles and quadrupoles, when compared to monopoles, arises from acoustic **short-circuiting**. Because these radiators create rarefactions and condensations in different locations but at the same time, sufficiently slow oscillations (e.g., low frequencies) allow the condensation enough time to propagate to the rarefaction and cancel it out [255]. Once the radiator is at least one-third the size of the wavelength or larger, short-circuiting is minimal, efficiency becomes dependent only on the absolute size of the radiator (as noted earlier), and the type of motion is largely irrelevant.

> **Web Topic 2.9** *Radiation efficiency and sound radiator size*
> Monopoles, dipoles, and quadrupoles all show reduced efficiency when the wavelengths being produced are larger than the sound source. Here we explain in more detail why this happens.

These effects essentially limit sounds useful for long-distance communication to those with wavelengths no greater than 1–2 times the sender's body size. For animals smaller than 30 cm, this requires vibration frequencies of 0.5 kHz or more in air, and 2–3 kHz in water. The fastest muscles known in vertebrates are those used to generate repetitive vibrations in toadfish swim bladders; their maximal contraction rate is several hundred Hz [325]. Insect flight muscles just reach the minimal value of 1 kHz for 15 cm animals, but can barely generate the 5 kHz rate required for a 3 cm katydid [388]. Some cicadas achieve vibration rates twice that of their maximal muscle contractions by buckling their two tymbals out-of-phase with separate muscles [124]. Sea robins achieve doubling by alternating contractions of the muscles on each side of their swim bladder [74]. However, the more widespread solution to this problem is a **frequency multiplier**: a device that produces many vibratory cycles for every muscle stroke. Stridulation and vocalization are two vibration-generating mechanisms that act as frequency multipliers: each stroke of a katydid's wings causes many successive teeth in its file to pass over the corresponding plectrum, and each expelled breath by a vertebrate produces many successive vibrations of its glottis, labia, or tympaniform membranes. Without devices for frequency multiplication, most small animals would be unable to send sound signals or be limited to very short-range acoustic communication.

In addition to creating differences in low-frequency efficiency, different radiator motions also produce different distributions of sound pressure in the medium. The spatial pattern of pressure around a sound source is called its **sound field**. For example, a spherical monopole far from any boundaries produces the same pressure at all points equidistant from its center. A dipole, however, suffers negative interference between radiating condensations and rarefactions at points midway between its ends and perpendicular to its vibrational axis. Points closer to one or the other end of the vibrational axis show only minor interference and thus higher pressures. A graph connecting all locations at a given pressure around a dipole will show two large **lobes**, one radiating from each end of the vibrational axis. These lobes become narrower at higher frequencies. A quadrupole typically shows a sound field graph with four lobes (see Web Topic 2.2 for animated examples). Note that proximity between a radiator and a reflecting surface can change the sound-field shape. A monopole near to a reflecting surface will generate lobes similar to those of a dipole in open space; a dipole close to a surface will reduce acoustic short-circuiting and act more like a monopole. The significance of sound fields for communicating animals is that sounds from monopole radiators can be detected at equal distances in all directions, whereas the range of detection of a dipole can be highly dependent on the relative angle of sender and receiver.

The final factor affecting radiation efficiency is the relative acoustic impedance of the radiator and the propagating medium. Ideally, the impedance of the radiator would be identical to that of the medium to ensure complete transfer of the vibrations across the boundary. Aquatic organisms consist largely of water, and sounds are easily radiated into the medium. A hippopotamus lying at the water's surface is an interesting case: when it vocalizes to produce airborne signals, its submerged tissues also vibrate, coupling the same sounds to the surrounding water, where submerged conspecifics can detect them [19, 20]. More commonly, a terrestrial radiator has an acoustic impedance that is quite different from the medium. The acoustic impedance is much higher within a tracheal tube than in the outside air, causing sound vibrations to be reflected back into the trachea when they reach the mouth opening. The solution to this problem is to provide an **impedance matching device** at the boundary to ease the transition.

One impedance-matching device used by many terrestrial vertebrates is an inflated air sac. The thin membrane of the sac and the contained air create a combination that has an acoustic impedance intermediate between ambient air and solid living tissues. If sufficiently inflated, it can also produce a much larger radiating surface than any other part of the animal. Examples include the throat sacs of most frogs and toads, and the esophageal sacs of doves and pigeons.

Another common impedance-matching device is a **horn** like the bell on a trumpet: its narrow end matches the high acoustic impedance of the attached tube, and as the bell flares, it provides a gradual match to the acoustic impedance of open air. Horns can greatly improve the efficiency of radiation. The cost is that, like most acoustic tools, they can only radiate sound efficiently over a limited frequency range that depends on their shape and size. In addition, the resulting sound fields are usually very directional, with one or more lobes [119]. Despite these constraints, many birds and mammals use an opened mouth or beak as a horn, and some species even elaborate the opening to produce a flared bell like a trumpet (**Figure 2.35**).

Horns made of living tissue will not work underwater because sound will pass through the horn walls without any channeling. Toothed whales (Odontocete Cetacea) have

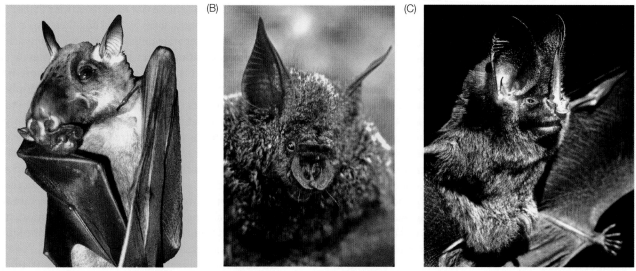

FIGURE 2.35   **Horns on bats for sound emission** (A) Flared bell around the mouth of a male hammer-headed bat (*Hypsignathus monstrosus*). These bats produce loud calls to attract females for mating. (B) African leaf-nosed bat (*Hipposideros cyclops*). This bat emits its echolocation calls through its nostrils, which are surrounded by a horseshoe-shaped horn. (C) Neotropical leaf-nosed bat (*Mimon crenulatum*). This bat also emits its echolocation calls through its nostrils, but in contrast with the prior species, hosts a vertically elongated horn that can be moved in concert with pulse emission.

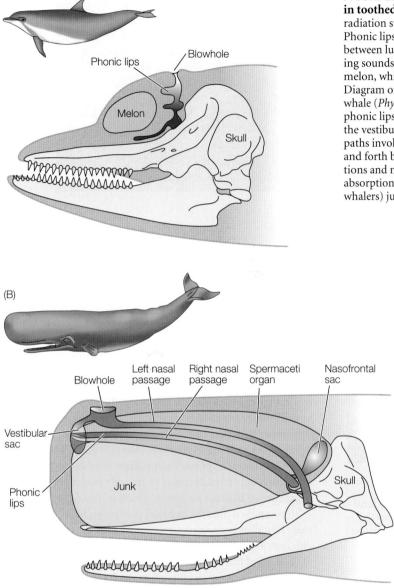

(A)

Phonic lips

Blowhole

Melon

Skull

(B)

Blowhole

Left nasal
passage

Right nasal
passage

Spermaceti
organ

Nasofrontal
sac

Vestibular
sac

Phonic
lips

Junk

Skull

FIGURE 2.36 **Resonance and impedance matching devices in toothed (Odontocete) cetaceans** (A) Diagram of acoustic radiation structures in the spinner dolphin (*Stenella longirostris*). Phonic lips are a presumed source of vibrations and are interposed between lungs and blowhole. The skull is a partial reflector aiming sounds forward, where they are focused and radiated by the melon, which consists of a layered mixture of oils and fats. (B) Diagram of acoustic radiation structures in adult male sperm whale (*Physeter macrocephalus*). Vibrations are thought to arise in phonic lips located at the junction of the right nasal passages and the vestibular air sac. They may then follow several concurrent paths involving direct radiation into the water, reflections back and forth between the vestibular and nasofrontal air sacs, reflections and modification inside the oil-filled spermaceti organ, and absorption and radiation from the fatty material (called junk by whalers) just below the spermaceti organ.

194, 254, 308, 320, 390] and elephants [165, 279–282] transfer vibrations through their appendages into solid substrates. Because the vibrating animal legs (and insect proboscises) are also solid, impedances between the animals and the medium are not sufficiently different to require elaborate modification and coupling devices. However, some filtering and resonance in the appendages themselves probably affect which frequencies are most apparent in the substrates.

Crickets and katydids produce stridulations by rubbing their forewings together. The file and plectrum are both small and thus inefficient sound radiators. However, they transfer the vibrations to adjacent areas of the forewings, (called the *harp* in crickets and the *mirror* in katydids), that act as resonant radiators [31, 34]. Cricket harps have low damping and radiate a highly amplified narrow band of frequencies. Katydid mirrors are more highly damped but produce a wider bandwidth of sound. In both groups, the sound-radiating wings act as dipoles, which makes them susceptible to acoustic short-circuiting at low frequencies. Because low frequencies propagate further, this poses a problem for long-distance signalers. Tree crickets (Oecanthinae) deal with this problem by calling near the edge of a leaf or putting their bodies in a hole or notch in a leaf (**Figure 2.37A,B**); by increasing the distance between concurrent condensations and rarefactions, they can radiate low-frequency sounds at reasonable amplitudes [125, 126]. The short-tailed cricket (*Anurogryllus muticus*) excavates a horn-shaped depression in the ground and positions itself in the cavity so that reflections of its stridulations from the cavity walls minimize short-circuiting and turn the cavity into a directional monopole [30, 125]. Mole crickets also create a horn in the ground surface, but then connect it through a narrow tunnel to a bulbous cavity below (**Figure 2.37C**). The combination of horn and cavity results in a complex

evolved an aquatic equivalent, however (**Figure 2.36**). The **melon** on the whale's forehead consists of multiple layers of special oils and fats that form a graded sequence of acoustic impedances. Sounds generated by phonic lips are reflected off of the bony skull and then focused and radiated forward through the melon [11, 61, 131, 190, 210, 260–262, 273, 419].

Balancing amplification and efficiency

Some animals do not require special modification or coupling devices for their vibrations. The beating wings of displaying fruit flies are dipoles, and the wavelengths of the sounds produced are several thousand times as large as the insects themselves. They are thus very inefficient as sources of far-field sounds. However, because female receivers are within a body length of a displaying male, the strong near-field flows are sufficient stimuli at that close distance. Certain arthropods [15, 23, 58, 62, 64, 65, 67, 69–71, 96, 97, 148, 185,

(A)

(B)

(C)

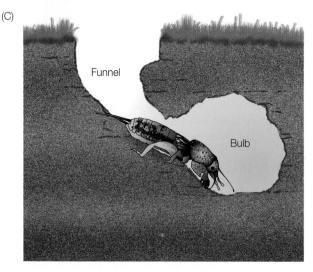

Funnel

Bulb

FIGURE 2.37 **Insect strategies to maximize radiated sound pressures** *Oecanthus fultoni* (A) and *Neoxabea bipunctata* (B), both tree crickets, reduce acoustic short-circuiting by placing their bodies in notches or holes that they have chewed in leaves. By the time a pressure condensation has propagated away from one side of the body and around both sides of the leaf, it is too attenuated to interfere significantly with a rarefaction. (C) Mole cricket (*Scapteriscus acletus*) stridulating at a critical location in its burrow that provides both a resonant amplification of its vibrations in the nearby bulb and enhanced radiation efficiency due to the funnel dug into the opening. (C after [29].)

resonance structure that is carefully tuned to the insect's fundamental stridulation frequency [27, 29, 31, 34].

Most cicadas have a large air-filled sac in the center of their abdomen that acts as a tuned cavity resonator for the vibrations produced by the two nearby tymbals [33–35, 305, 332, 333, 354, 389]. The amplified sound is radiated through two large eardrums located on either side of the body. In contrast, the gas-filled swim bladders of fish [106, 220] and inflated throat sacs of frogs and toads [145, 198, 307, 310] do not appear to function as cavity resonators. The walls of both types of structures do not differ sufficiently in acoustic impedance from the medium to retain successive waves

inside their respective cavities, and are thus heavily damped. However, the properties of the vocal sac tissues in frogs appear to be adjusted to act as filters that emphasize particular harmonics produced by the larynx [163]. In fish, the temporal pattern of acoustic signals is key to species identity, and low-Q swim bladders may thus be advantageous. Frog and toad throat sacs have acoustic impedances intermediate between solid tissue and free air, and thus should be good radiators. However, many frogs and toads also radiate sound from their entire bodies, and like cicadas, some species such as bullfrogs even radiate sound from their eardrums [307]. A number of frogs, like mole crickets, create burrows with cavity resonances tuned to their calls [13, 294, 295], and one Asian species, *Metaphrynella sundana*, adjusts the frequency and temporal characteristics of its calls to match the cavity resonance of the tree hollow in which it lives [224, 225].

The opened mouths or beaks of reptiles, birds, and mammals function as monopole radiators and are thus relatively efficient. However, most species exploit resonant properties of their vocal tracts to modify the patterns in the initial vibrations and to increase coupling further. Even when the beak or mouth is used as an impedance-matching horn, the acoustic impedance inside the vocal tract is sufficiently higher than outside that some sound energy is reflected at the opening back into the tract. The reflected sound waves then interfere with subsequent waves moving toward the mouth to produce resonance. Since the vocal tract is roughly tubular, its normal modes depend primarily on the length of the tract and whether the ends are open or closed. If the tube is open on both ends, the lowest normal mode will have a wavelength approximately equal to twice the length of the tube; if it is closed on one end, the first mode's wavelength will be four times the length of the tube. Thus closing one end decreases the frequency of the lowest normal mode by a factor of two. Since most birds and mammals produce vibrations with a fundamental frequency lower than the lowest normal mode of their vocal tract, the upper vocal tract acts as a resonator and filter to emphasize some vibrational frequencies over others. Attaching additional cavities to the vocal tract can both alter and accentuate the normal modes of the vocal

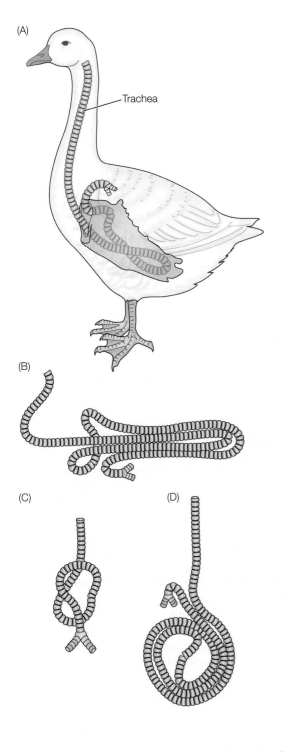

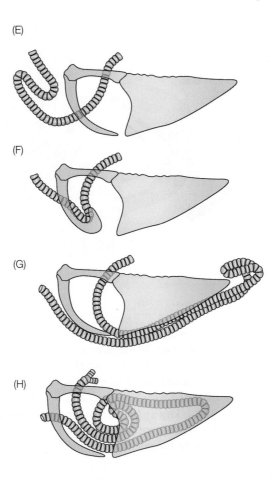

FIGURE 2.38 **Elongated tracheae in birds** (A) Trumpeter swan (*Cygnus buccinator*) showing coiling of elongated trachea inside hollowed breastbone. (B) Magpie goose (*Anseranas semipalmata*). Elaborate tracheal coils are located just under skin of breast. (C) Spoonbill (*Platalea leucorodia*). The tracheal coils lie above the breastbone (sternum). (D) Manucode bird-of-paradise (*Manucodia keraudrenii*). Highly convoluted trachea is also arrayed just under skin of breast. (E) Capercaillie grouse (*Tetrao urogallus*). Single loop lies just under skin of breast. (F) Crested guinea fowl (*Guttera edouardi*) with single loop stored in special hollow of clavicle bones. (G) Helmeted curassow (*Crax pauxi*). Elongated coil lies along breastbone just under skin. (H) Whooping crane (*Grus americana*). Sternum is hollowed out to accommodate extensive tracheal coils. (B after [113]; A, C–H after [251].)

tract. As a general rule, males are more likely than females to have such additional cavities or to have larger ones; however, there are exceptions, such as painted snipe (*Rostratula* spp.), in which females have the more elaborated vocal tracts [251].

The basic avian vocal tract consists of the beak, pharyngeal cavity, trachea, and syrinx when the vibrating valve(s) are closed. When the vibrating valve(s) are open, the upstream respiratory components, including the bronchi, lungs, and air sacs, are similar in acoustic impedance and must be included in the calculation of the length of the vibrating column of air. Thus a typical bird may have two sets of normal modes: one in which the vibrating valves are open, and another in which they are closed. As long as damping in the vocal tract

is low relative to the vibration rate of the valves, both sets of normal modes can contribute to the formants in the radiated sound [119].

Birds have multiples ways to adjust the normal modes of their vocal tracts (**Figure 2.38**). A permanent mechanism is to elongate the trachea beyond a direct connection between syrinx and mouth. This will lower the frequencies of all normal modes. Elongated tracheae are found in curassows and their allies, grouse, and guinea fowl (Galliformes); swans and magpie geese (Anseriformes); limpkins and cranes (Gruiformes); painted snipe (Charadriiformes); spoonbills and wood ibises (Ciconiiformes); and manucode birds-of paradise (Passeriformes) [26, 63, 110, 113, 128, 140, 251]. Oilbirds have

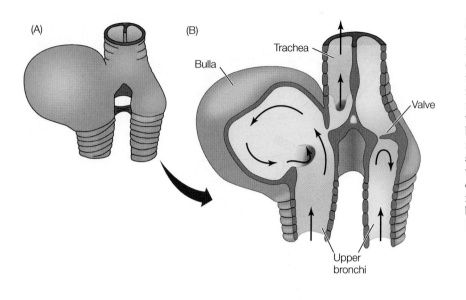

(A)

(B)

Trachea

Bulla

Valve

Upper bronchi

FIGURE 2.39 **Syringeal bullae in ducks** Males of many species of ducks have a cartilaginous bulla on the left side of their syrinx or just above the syrinx on their trachea. The shape and complexity differs among species, and the bulla is often a useful character for taxonomy. (A) Dorsal view of tracheal bulla, trachea, and upper bronchi of male mallard duck (*Anas platyrhynchos*). (B) Cut-away view of mallard syrinx showing presumed airflow during vocalization. A valve in the right bronchus can block airflow on that side while airflow on the left is routed through the bulla before rejoining the trachea (arrows). (After [207].)

asymmetrical bronchial syringes with the valves on the two bronchi situated at different distances from the junction with the trachea. This results in different normal modes for the two sides of the vocal tract [367]. An additional permanent mechanism is the attachment of a fixed resonant cavity at a location along the vocal tube. This is largely found in male ducks (Anseriformes), which have one or more bony **bullae** (**Figure 2.39**) attached just above or to the side of the syrinx [26, 202, 207].

A more customizable approach is to open or close the oral end of the vocal tract with the beak or tongue: closing lowers the frequencies of all the normal modes in the tract; opening increases these modes [187]. The greatest effect will be on higher modes and thus on the higher harmonics in the syringeal vibrations [120]. Many birds vary their beak and tongue positions in synchrony with the amplitude and frequency modulations of their syringeal vibrations [25, 156, 157, 191, 300–302, 398]. Cardinals (*Cardinalis cardinalis*) coordinate beak gape with a muscular expansion and contraction of the pharyngeal cavity and upper esophagus to remove all but the fundamental component in their syringeal vibrations before radiating the songs [122, 318]. Because many syllables of these songs are rapidly frequency modulated, the volume of the cavity must be altered just as quickly to track the changing fundamental frequency of the vibrations. See the animation at Web Topic 2.6.

Other birds use air-filled chambers to modify, amplify, and even radiate their vocalizations (**Figure 2.40**). Doves keep their beaks closed during vocalization; airflow toward the head picks up vibrations in the syrinx, and then passes through the glottis into an inflatable esophageal chamber. The wall of this chamber, enveloping the trachea and puffing up the breast of the bird, acts both as a filter/resonator and as the main radiator of the dove's vocalizations [24, 121]. Many species of grouse (Tetraonidae, Galliformes) inflate esophageal sacs and use them for modifying and radiating vocalization as well as for eversion of colored skin patches

on the sides of the head or body [188]. Other birds that use inflatable sacs for vocalization resonance and radiation include emus (Struthioniformes), bustards (Gruiformes), musk ducks (Anseriformes), the kakapo parrot (Psittaciformes), the American bittern (Ciconiiformes), and button quail (Charadriiformes) [83, 207, 250, 251].

In mammals, the acoustic impedance of the respiratory tract upstream from the glottis is sufficiently greater than that downstream that even when the valves are open, the upstream tubes have little effect on resonance and filtering [119]. This excludes the trachea as a possible resonant cavity, and limits most terrestrial mammals to the short space between the glottis and the mouth or nostrils for any resonance, filtering, or coupling devices. Despite these constraints, many mammalian species rely heavily on these processes during sound emission.

The open mouths of many mammals, as well as special nose-leaf structures around the nostrils in some echolocating bats, can easily function as impedance-matching horns [119, 168, 170]. Certain body positions may accentuate this effect and modify normal modes [174]. Pharyngeal and nasal cavities are often used as resonance- and impedance-matching devices [110–112, 114, 313-315, 317]. To make these cavities large enough to accommodate lower frequencies, some mammals depress their larynx away from the mouth, either permanently or when vocalizing, to increase the cavity dimensions. Examples include dogs, roaring cats (lions, tigers, jaguars, and leopards), pigs, goats, and tamarin monkeys [110, 114, 115, 173, 249, 397]. In humans, particularly adult males, the larynx is positioned away from the mouth to facilitate the resonant filtering of vowel sounds [234]. Pharyngeal and nasal cavities in many echolocating bats filter out all but certain harmonics in their echolocation calls [169, 170, 364, 365]. For those species emitting the sounds through the nostrils, nasal sacs may also function as impedance-matching devices [168, 169]. Males of the non-echolocating hammerheaded bat (*Hypsignathus monstrosus*) have exaggerated nasal

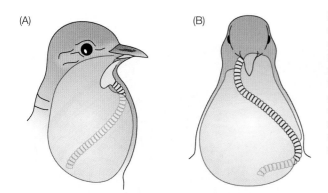

(A)

(B)

FIGURE 2.40 **Vocal air sacs in birds** (A) Male ring dove (*Streptopelia risoria*) with inflated esophagus (pink) enveloping trachea (blue) during emission of "coo" call. (B) Same bird viewed from the front. (C) Waveforms (top) and spectrograms (bottom) of ring dove "coo" recorded from outside the bird (left) and inside the bird at the syringeal end of the trachea (right). The inflated esophagus acts as a resonance device to filter out all but the fundamental frequency in the syringeal vibrations before emission. (D) Paired sacs on chest of strutting male sage grouse (*Centrocercus urophasianus*) during sound emission. (E) Male greater prairie chicken (*Tympanuchus cupido*, another species of grouse), showing one of paired vocal sacs on sides of head. (A–C after [316].)

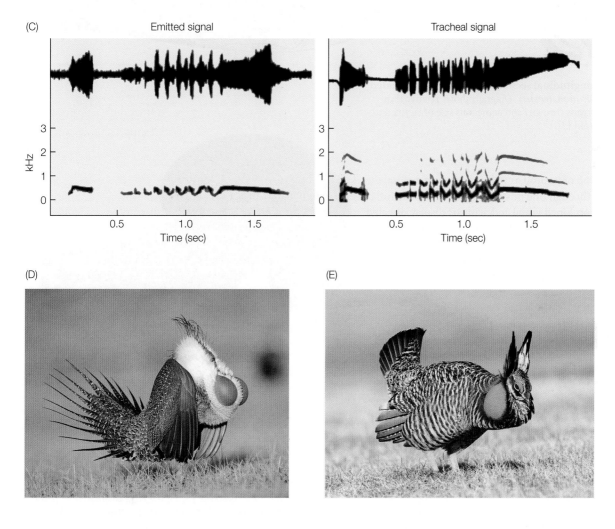

(C) Emitted signal   Tracheal signal

cavities that appear to function as resonators and filters of their sharp honking calls (**Figure 2.41**). These bats also have an enlarged and cartilaginous larynx that fills nearly two-thirds of their body cavity [342].

Other mammals have air-filled sacs located just downstream from the glottis or in adjacent tissues and cartilages [206, 270, 341]. Reindeer and muskoxen have a single inflatable sac in their throat used to amplify their calls [127, 270]. Other taxa such as bears and epomophorine bats have paired

pharyngeal pouches that function in a similar way [396, 417]. Resonant and impedance-matching air sacs are widespread in primates [110, 111, 141, 142, 206, 270, 350]. Males of forest species such as the African De Brazza's monkey (*Cercopithecus neglectus*) and the neotropical howler monkeys (*Alouatta* spp.) have particularly well-developed vocal sacs and produce calls that can be detected at great distances [141, 343]. Perhaps the most spectacular mammalian air sacs occur on the male hooded seal (*Cystophora cristata*), which can inflate a dark

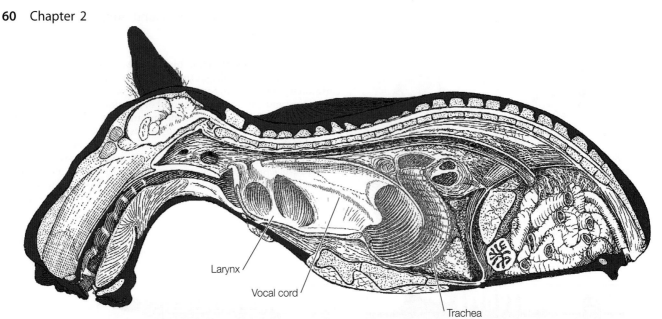

Larynx

Vocal cord

Trachea

FIGURE 2.41   Longitudinal section of male hammer-headed bat (*Hypsignathus monstrosus*) Diagram shows relative locations of internal organs (brown) and enormous size of larynx and trachea (blue). (From [3].)

nasal sac on the top of its head as well as a bright red septal sac that is extruded through one nostril (**Figure 2.42**). Sac inflation is part of a combined visual and acoustic display directed toward competing males during the breeding season. Interestingly, males perform the display in the same way whether in the air or underwater [17, 36, 229, 384]. How the sacs might modify sound radiation in each medium remains unknown.

FIGURE 2.42   Vocalizing male hooded seal (*Cystophora cristata*) Male with black nasal hood inflated and red nasal septum inflated so that it protrudes out of nostrils. Males can inflate either sac or both when performing combined visual and acoustic display to opponent males.

SUMMARY

1. Sound propagation requires a material **medium** containing atoms or molecules. The atoms or molecules may be in the form of a **fluid** (such as a gas or liquid), a **solid**, or an interface between two of these phases of matter. **Pressure** is a measure of the degree to which the motions of atoms and molecules affect the motions of other nearby atoms and molecules.

2. **Sound** is the propagation of a perturbation in local pressure away from an initial location. It thus consists of alternating regions of higher-than-average pressure (molecular **condensations**) and lower-than-average pressure (molecular **rarefactions**).

3. Inside fluids, the molecules that pass on a pressure disturbance tend to move back and forth along the same axis on which the sound is propagating. The pattern of successive

condensations and rarefactions moving away from the source of the disturbance is called a **longitudinal wave**. The molecules of a vibrating string propagate sounds along the string's length, but they do so by vibrating on a line perpendicular to the direction in which the sound is propagating. This is called a **transverse wave**. The molecules on the surface of a body of water that is propagating ripples move elliptically and thus have both longitudinal and transverse components in their motion. Solids can also propagate sounds, but the molecules can move only tiny distances, if at all. Solids can support longitudinal and transverse waves, as well as a variety of elliptical and more complex patterns such as **Rayleigh**, **Love**, and **bending waves**.

4. A **periodic wave** repeats the pattern of pressure variation over time. The **period** of the wave is the amount of time

(seconds) required to produce one complete cycle. The simplest type of periodic wave shows a **sinusoidal pattern** in pressure over time. The **frequency** of such a wave is equal to the number of times the pattern is repeated per second, and is measured in **Hertz** (abbreviated **Hz**). The part of the cycle that occurs at some reference time is called the **phase** of the wave. Finally, a wave's deviations of pressure from ambient levels provide a measure of its **amplitude**. Amplitudes are usually measured on a logarithmic scale relative to some reference value. The units are called **decibels** (abbreviated **dB**).

5. Two waves of the same sinusoidal frequency passing through the same location are **in phase** if they have maxima and minima at the same time and **out of phase** otherwise. Waves that are in phase will show positive **interference** by creating a composite wave of similar frequency but enhanced amplitude; waves that are out of phase interfere negatively and tend to cancel each other out. Two sinusoidal waves that are similar but not identical in frequency will drift in and out of phase, creating **beats**.

6. Most animal sounds are periodic but not sinusoidal. In most cases, they can be decomposed into the sum of a set of pure sine waves (called a **harmonic series**) with frequencies that are integer multiples (called **harmonics**) of the lowest frequency in the set (called the **fundamental**). The decomposition (called **Fourier analysis**) can also be performed on aperiodic signals, but the sine wave components then tend to be more numerous, and their frequencies are not integer multiples of a single fundamental. Where the periodicity varies within animal sounds, one can usually break the sound into roughly periodic segments and perform a Fourier analysis on each segment. A plot of the Fourier decomposition of successive segments in an animal sound is called a **spectrogram**.

7. Inside fluids, all frequency components of a complex sound propagate at the same speed. In air, the speed of sound is about 344 m/sec (with variation depending on temperature and humidity). The speed of sound in water is about 4.4 times as fast as that in air, and the speed of sound in solids is about 15 times as fast as that in air. On the surface of a body of water and in certain solids, different frequency components may propagate at different speeds.

8. The spatial distance between the beginning and end of one cycle of a propagating sinusoidal wave is called its **wavelength** and is measured in meters. Wavelengths are inversely related to the wave frequency and directly related to the speed of sound in the medium. Wavelengths for a given frequency are thus longer in water than in air. Wavelengths may also be decreased, (and the effective frequency increased), if a sender and receiver are moving toward each other; wavelengths will increase if they are moving apart. This is called a **Doppler shift**.

9. Close to a sound source, medium molecules flow back and forth in unison. This is called the **near field** around the source. At distances of about 1/3 of the wavelength of the sound (or 2/3 the diameter of the source), medium molecules pass on the pressure disturbance without taking part in a cohesive tidal flow back and forth. The propagating medium at this and further distances from the source constitutes the **far field** of the sound.

10. Sound amplitudes decrease with distance from the source. Inside fluids such as water or air, the pressure of a sound radiating away from the source decreases with the reciprocal of the distance from the source. This is called **spreading loss**. Spreading losses are less severe for ripples on the water's surface or inside the solid stems of plants. In addition to spreading losses, propagating high frequencies cause molecules to collide more often, and thus lose pressure to **heat losses** faster than propagating low frequencies. Spreading losses are the same in air and water, but heat losses are much higher in air.

11. **Acoustic impedance** is the resistance of a medium to a change in its molecular behavior. Away from interfaces between media, acoustic impedance depends on the density and speed of sound of that medium. Media with low acoustic impedances (e.g., air) propagate sound with weak pressures but significant molecular velocities; high-impedance media (e.g., water and solids) propagate sounds with high pressures and low molecular velocities. At an interface between two media with different acoustic impedances, most of the sound traveling in one medium will be **reflected** back into the same medium at the boundary. If the impedances are not too different, then some sound energy will pass into the second medium, but its direction of travel will likely be bent (**refraction**).

12. In addition to spreading and heat losses, a propagating sound can be attenuated by reflective **scattering** from objects that have acoustic impedances different from that of the medium, and by refraction that bends sound waves out of the path connecting sender and receiver.

13. Most animals produce sound signals in three steps: generation of vibrations, vibration modification, and coupling of the modified vibrations to the medium. Generation of vibrations can be achieved in many ways. Hard-bodied animals can use **percussion** (striking a body part against a substrate or another body part); **stridulation** (rubbing a file over a plectrum); **buckling** (of flat plates); or **tremulation** (vibrating the whole body on the surface of water or solid substrates). Animals in fluid media can also use **pulsation**, **fanning**, **fluid compression**, or **streaming** as vibrational sources. Finally, animals can force respiratory system air through openings to create **aerodynamic vibrations** in the form of hisses or single-frequency whistles, or through a valve like a **glottis** to create periodic but nonsinusoidal vibrations. Frogs, reptiles, and mammals have their vocalization valves in a **larynx** at the top of their trachea, whereas birds may have one valve at the bottom of the trachea (parrots and chickens), or a separate valve on each bronchus (songbirds, oilbirds, and woodpeckers). Wherever the bird valve is located, the associated tracheal-bronchial junction is called the **syrinx**. Birds with a valve on each bronchus can produce two different sounds at once, or more commonly, assign high frequencies to one valve and low frequencies to the other.

14. **Modification** usually involves linking the vibrational source to a flat surface or cavity with an acoustic impedance sufficiently different from the medium that successive vibrations overlap and can interfere. This facilitates positive interference and amplification (**resonance**), and negative interference (**filtering**) depending on the sound frequencies. **High-Q** modifiers provide strong amplification of a few select frequencies (**normal modes**), but have **low damping** that muddles rapid temporal patterning in the sound. **Low-Q** modifiers track temporal patterns accurately, but are much less selective for frequency and provide only minimal amplification.

15. Effective **coupling** of vibrations to the propagating medium depends in part on the size of the radiating surface, and in part on how it moves: **monopoles** expand and contract in all directions at once, **dipoles** oscillate along a single axis, and **quadrupoles** change shape by moving along two axes at the same time. Unlike monopoles, dipoles and quadrupoles produce directional **sound fields**. All three mechanisms are inefficient when the wavelengths of the radiated sounds are larger than the radiator, and the latter two are even more inefficient due to acoustic **short-circuiting**. These effects limit small animals communicating over moderate or greater distances to high frequencies. To create sufficiently high frequencies given limits on muscle contraction rates, small animals often resort to **frequency multipliers**. Examples include stridulation and vocalization mechanisms. Terrestrial animals can also improve radiation efficiency with acoustic horns or inflated sacs, which have acoustic impedances intermediate between that inside their bodies and that of the outside air.

16. Different species have evolved different compromises between elaborated sound structure and amplitude. Adaptations such as elongated tracheae in cranes and currasows, or cartilaginous vessels in ducks and howler monkeys, allow great amplitude at the cost of severely constraining what kinds of sounds can be produced and radiated. Species such as frogs, tree crickets, and mole crickets excavate or select calling sites that dramatically increase the resonance or radiation efficiency of the sounds they produce. Finally, species such as cardinals and humans actively vary the dimensions of their internal resonant cavities dynamically to amplify different frequency combinations during the production of a given sound.

Further Reading

Readers interested in more detailed treatments of biological acoustics should consult Fletcher [119] and Michelsen [255]. Additional explanations of monopole, dipole, and quadrupole radiator efficiencies can be found in Harris [167] and Kalmijn [204]. Most other reviews or general treatments of the topics in this chapter are taxon-specific. Suggested resources include: Bailey [14], Bennet-Clark [30-34], Gerhardt and Huber [145], Greenfield [161], and Virant-Doberlet and Čokl [390] for insects; Barth [22, 23] for spiders; Ladich [219] and Ladich and Fine [220] for fish; Gerhardt [143] and Gerhardt and Huber [145] for amphibians; Young [415], Kirchner [208] and Gans [130] for reptiles; Goller and Larsen [155, 227] and Suthers [360, 372, 374] for birds; Fitch and Hauser[111], Fitch et al. [116], and Frey et al. [127] for terrestrial mammals; and Tyack [384] for marine mammals.

COMPANION WEBSITE
sites.sinauer.com/animalcommunication2e

Go to the companion website for Chapter Outlines, Chapter Summaries, and References for all works cited in the textbook. In addition, the following resources are available for this chapter:

Web Topic 2.1 *Measuring sound pressure*
Microphones are used to measure the variations in pressure caused by a propagating sound. Specialized types of microphones exist for sound propagation in air, water, and solid substrates. These microphones work by converting pressure variation into electrical signals that can then be measured, stored, and characterized.

Web Topic 2.2 *Visualizing sound waves*
The best way to understand the differences between different types of sound waves is to view an animation that shows how the molecules move as the sound propagates. We list a number of websites where you can watch visualizations of most of the basic acoustic processes described in this chapter.

Web Topic 2.3 *Quantifying and comparing sound amplitudes*
A variety of methods are available for measuring and comparing sound amplitudes. Here we define some of these methods, show how they are computed, and discuss when each might be most useful.

Web Topic 2.4 *Fourier analysis of animal sounds*
Here we provide an introduction to the logic behind Fourier decomposition of animal sounds, including links to several excellent software packages for creating spectrograms and introductions on how to use these packages. We also provide links to sites where one can use such methods to compare archived animal sounds.

Web Topic 2.5 *Reflection and refraction*
The fraction of sound energy reflected or refracted at a boundary is a complicated function of incident angle, relative acoustic impedances, and relative sound speeds. Here we present the equations for several different cases, and provide some real physical examples.

Web Topic 2.6 *Sample animal sounds*
Visit this website to hear examples of the kinds of animal vibrations discussed in this and following sections of the chapter. You will also be able to see the waveform and spectrogram of each example.

Web Topic 2.7 *Animations of vocalizing birds*
Dr. Roderick Suthers and his team at Indiana University have pioneered our understanding about how the avian syrinx works. His lab has conveniently produced several animated clips demonstrating key steps in song production by northern cardinals and brown-headed cowbirds.

Web Topic 2.8 *Linear versus nonlinear systems*
Although animal sound vibrators typically change frequency and amplitude linearly as the forces on them are changed, most will adopt less predictable nonlinear behavior at extreme force values. Some animals such as parrots, zebra finches, and canids use nonlinear vibration behavior to increase their vocal diversity.

Web Topic 2.9 *Radiation efficiency and sound radiator size*
Monopoles, dipoles, and quadrupoles all show reduced efficiency when the wavelengths being produced are larger than the sound source. Here we explain in more detail why this happens.

Chapter 3

Sound Signal Propagation and Reception

Overview

In Chapter 2, we described the properties of sound and the various ways that senders create and modify vibrations before sending them on their way to a receiver. In this chapter, we pick up the communication sequence by following the emitted sound signals as they propagate through the medium. It turns out to be a complex journey in which the signal can be significantly altered. Once a sound signal arrives at a receiver, it must be captured and processed. We shall see that the physical processes underlying sound reception are generally the same as for sound signal production, but reversed in order. Finally, we look at how combining the physics of sound generation, propagation, and reception helps explain why different animals use different sound frequencies for communication.

Sound Propagation

As we have seen, animals can put a lot of effort into producing a sound with a particular set of properties: a well-equipped sender can fine-tune its signal's amplitude, frequency composition, temporal pattern, and directionality. However, once the sound enters the medium, the sender has no further control over it. Signals are invariably altered during propagation between sender and receiver; how they are changed and to what degree will depend on the medium, the distance traveled, and the original properties of the sound [117, 218, 219, 242, 243, 250, 279, 307, 309, 310, 364, 371].

Animals that can anticipate likely changes during propagation are able to modify their communication strategies accordingly. How they do so depends upon the context. For example, sound is often used for long-distance communication. The volume of medium within which a receiver can detect and recognize a sender's signal is called the **active space** of that signal. Sound production, propagation, and reception will each affect the size of a signal's active space. A sender can increase the active space of its signal by modifying its properties to minimize potential propagation effects. Receivers can increase active space by focusing on those signal properties that are least affected by propagation. The proposition that evolution favors those acoustic signals that suffer minimal propagation losses and noise overlap is called the **acoustic adaptation hypothesis** [132, 250, 313, 372]. However, while some species show some of the adaptations predicted by this hypothesis [6, 14, 62, 77, 81, 95, 120, 132, 151, 174, 202, 231, 259, 261, 327, 346, 356, 372], others show only minor adjustments, suggesting that in some cases, the potential benefits may be outweighed by other selective forces [31, 33, 44, 165, 197, 277, 318, 344]. Some senders (e.g., those living in the constant presence of predators) may not benefit by maximizing their active space, and may instead favor signals that are easily attenuated or degraded by propagation.

Most animal sound signals suffer some degradation during propagation. Many species exploit this fact by using the amount of degradation in a received signal to estimate the distance to calling senders. This information can then be a key determinant of subsequent receiver behavior. Estimation of sender distances from acoustic degradation is called **ranging (Figure 3.1)** and is widespread in birds [251, 254, 271, 373]. In this and in other ways, acoustic degradation can play much more complex roles in animal social interactions than simply limiting active spaces.

There are four types of degradation that a sound signal might experience during propagation: (1) overall attenuation; (2) distortion of its frequency composition; (3) distortion of its temporal pattern; and (4) masking by noise. While the basic principles are similar in all media, the details often differ depending on whether the sound is propagating in air, water, or a solid substrate. Below, we discuss each of these types of potential change, compare effects in the three types of media, and note strategies that senders and receivers might pursue to enhance or minimize that type of modification during propagation.

Overall attenuation

Two main processes that result in decreasing overall signal amplitudes with distance from the sender are spreading losses and refraction. Both are largely frequency-independent and over usual communication distances produce minimal changes in the temporal pattern of a signal. However, both processes play major roles in determining the active space of a sound signal.

SPREADING LOSSES   As we saw in Chapter 2, the pressure of a sound signal decreases with the reciprocal of the distance from the source due to spreading losses. In the far field, there will be a 6 dB decrease in signal amplitude for each doubling of distance. Spreading losses are the same for all frequencies of sound and for all media as long as no boundaries with other media are nearby. If there are nearby boundaries, then spreading losses may be altered. We noted in Chapter 2 that waves on the water's surface spread into two dimensions, not in three as in open air or deeper water, and thus sound pressures decrease much more slowly with distance.

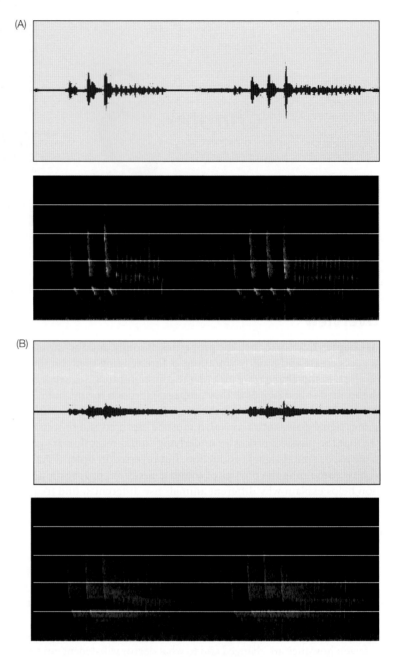

FIGURE 3.1   Degradation in song of male cardinal (*Cardinalis cardinalis*) due to long-distance propagation (A) Waveform (top) and spectrogram (bottom) of male song recorded close to bird. (B) Waveform and spectrogram of the same song recorded several hundred meters away in the forest. White lines in each spectrogram indicate 2 kHz intervals. Degradation includes overall reduction in amplitude, filtering out of higher frequencies, and addition of reverberations (echoes) that blur the temporal pattern. These effects can be used by other males to estimate their distance from the singer (ranging).

Similarly, sound propagating inside a narrow stem of a plant may show little spreading loss at all because the sound is essentially propagating along one dimension (**Figure 3.2**).

REFRACTION   Large bodies of air or water often contain gradients or vertically stacked layers in which the speed

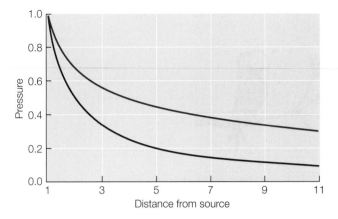

FIGURE 3.2 Spreading losses during sound propagation
Red line shows drop in sound pressure with increasing distance from the sound source in a free field with spherical (3-dimensional) spreading. Pressure decreases by 6 dB for every doubling of distance. Blue line shows slower drop for cylindrical spreading. This occurs for ripples on the surface of water, surface waves on the ground, and when sounds in air or water propagate between two or more boundaries. Pressure here decreases 3 dB for every doubling of distance.

of sound varies. Speeds of sound can differ between layers because of different temperatures (sound usually moves faster in warmer layers); different pressures (sound usually moves faster under higher pressure); different current velocities in the medium (sound moving with a current moves faster than sound moving against it); or some combination of these. Sound traveling in a layer with one sound speed will be refracted if it crosses a boundary into another layer with a different sound speed: if the second layer has a lower sound speed, the sound's trajectory will be bent into

the second layer (e.g., toward a line perpendicular to the layer surface). If the new layer has a higher sound speed, the trajectory of the propagating sound will be bent back toward the initial layer.

Refraction can exacerbate or reduce spreading losses by changing the volume into which a sound is radiating. Whether the active space of a sound decreases or increases depends on how the medium gradients or layers are arranged and upon the locations of the sender and receiver with respect to the layers. As an example, consider two birds communicating near the ground on a sunny day (**Figure 3.3**). The sunlight heats the ground and causes the air immediately above it to be warmer than the air higher above the ground. As a result, the speed of sound is highest close to the ground and decreases with increasing height above the ground. Sound waves traveling from the sender to the receiver will be refracted up and into the higher and cooler layers of air. At a sufficiently large distance, this can create a **sound shadow** close to the ground within which signal amplitudes are minimal. On hot and sunny days in open country, refraction can significantly reduce the active space of sound signals. This may explain why many birds sing only sporadically in such contexts.

The opposite situation can occur on a cold clear night or subsequent dawn (**Figure 3.4**). During the night, as the Earth radiates its accumulated heat back into the clear night sky, the ground becomes colder than the air. Air closest to the ground will then be cooled relative to the air somewhat higher above the ground. (This is called a **temperature inversion**.) Sound waves propagating between two animals close to the ground will now be refracted downward into the cooler layer and thus kept near the ground. Sound waves that otherwise would have radiated upward will be refracted and now arrive at the receiver. In this case, refraction due

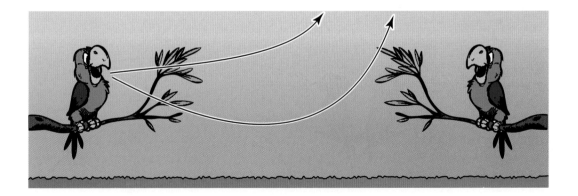

FIGURE 3.3 Refraction of sound waves for a receiver close to the ground on a hot day
The warm ground heats the immediately adjacent layer of air, and this generates a gradient of decreasing air temperatures, (and thus sound speeds), with increasing height above the ground. Sound traveling parallel to the ground is refracted up and away from potential receivers. At a sufficient distance from the sender, a sound shadow may be created within which receivers cannot detect the sound. The active space of a sender is reduced in this context.

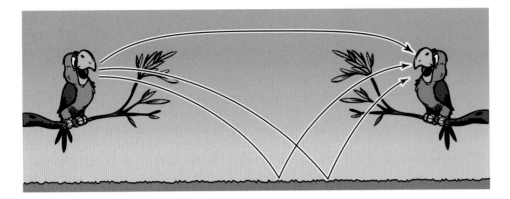

FIGURE 3.4 Refraction of sound waves for a receiver close to the ground on a clear night or at dawn The earth radiates its heat back into the sky on a clear night, generating a gradient of increasing temperatures, (and thus sound speeds), with increasing height above the ground. Sound that would have radiated above a receiver is refracted downward toward it. Sounds directed at the receiver are initially refracted downward away from the receiver, but are reflected by the ground and travel back up along curved refracted paths toward the receiver. The active space of a sender is increased in this context.

to a temperature gradient can increase the active space of sound signals. Many species of animals vocalize intensively at dawn, a phenomenon called the **dawn chorus** [142, 186]. One explanation for this intense period of sound emission at, or for nocturnal animals just prior to, dawn is the potential for a greater active space due to refraction.

Refraction can also be generated by currents in media, such as wind (**Figure 3.5**). Wind speeds are typically lowest near the ground and increase with height above it. If a receiver is upwind from a sender, the speed of a sound traveling between them is highest near the ground, (where reductions in sound speed due to wind are minimal), and decreases at successively greater heights, since sound propagation is opposed by the increasing wind velocities. This situation is analogous to that of a sunny day: sound will be refracted up and into higher layers of air where sound

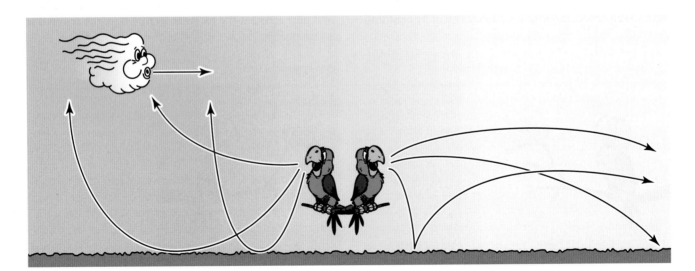

FIGURE 3.5 Refraction of sound waves near the ground due to wind In this figure, wind is blowing from left to right. When a receiver is upwind from the sender, wind slows down the propagation speed. Because wind speed is least near the ground and increases with increasing height above the ground, sound speeds will be highest near the ground and increasingly slowed down by wind at greater heights. Refraction of upwind sounds thus bends them up into the layers of slower sound speed, and away from distant receivers. It may even generate a sound shadow at sufficient distances. Sound speeds traveling downwind are lowest at the ground and highest higher above the ground. Refraction here bends radiating sounds back down toward the receiver and thus enhances signal range.

FIGURE 3.6 **Refraction of sound waves inside a forest** The canopy traps incident sunlight and heats up, whereas shade and evaporation leave the ground and understory cool. Air temperatures and thus sound speeds increase along a gradient between the ground and the underside of the canopy. Bird sounds emitted under the canopy are refracted down, as in propagation over cold ground. This extends the active space. Sounds emitted above the canopy are refracted up and into the atmosphere. Like sounds propagated near the ground on a hot day, their range is reduced.

speeds are low. Given a sufficient distance between sender and receiver, winds can generate an upwind sound shadow just as temperature layering does. If the receiver is downwind from the sender, sound speeds will be lowest near the ground (where wind has minimal effects) and increase at successively greater heights due to increasing wind speeds. In this case, sounds will be refracted back down toward the lower-speed layer near the ground, enhancing detection range and active space. The magnitude of wind effects will depend on wind speed relative to the speed of sound in air. Even on a blustery day, one might expect at most a change of 5% in sound speeds due to wind. As we know from experience, this can be sufficient to make it hard to hear a downwind speaker, but much easier to hear an upwind one. Note that mild winds may combine with temperature effects, but stronger winds create eddies that tend to mix air layers and thus eliminate contributions to refraction due to temperature differences.

Complex environments, such as mature forests, may consist of vertical stacks of layers with *alternating* sound speeds (**Figure 3.6**). For example, a forest canopy typically absorbs the sun's rays during the day and becomes hot on its top side, whereas the understory is kept cool not only by the shade of the canopy but also by the evaporation of

moisture in the forest litter. The air beneath a forest canopy thus tends to be coolest close to the ground and somewhat warmer as one approaches the canopy. The resulting refraction of sounds near the ground and in the understory favors long-range sound propagation. The warmed canopy vegetation heats the air layer immediately above it, so sound speeds are greater immediately above the canopy than in the higher air layers. Thus sounds generated in the canopy will refract up and away from a receiver in or on the canopy. Of course, the situation can change at dawn, when the canopy foliage may be cooler than the ambient air, or when the wind is blowing. Wind velocity is usually higher above the canopy than below it.

We find very similar principles at work in water (**Figure 3.7**). Near the surface, a critical factor influencing the speed of sound is the relative temperature of the air and the water. When the air temperature is higher than the average water temperature, (as in summer at intermediate and low latitudes), water closest to the surface will be warmed and have a higher sound speed than water just a bit deeper. The active space for sound-producing fish close to the surface will be reduced, because sound waves traveling between them will be refracted down and into the deeper and cooler water. If the air above the surface is colder than the water, (as in winter), then the reverse will be found: water near the surface will be colder and have a slower sound velocity than deeper water. Active spaces near the surface will be enhanced due to refraction. In principle, currents in streams, waves, or tidal flows could produce layered sound speeds in water just as wind does in air. However, the magnitude of these differences in water will be much smaller, because although the velocity of water flow is quite similar to that of air, the speed of sound is much greater in water than it is in air.

(A)

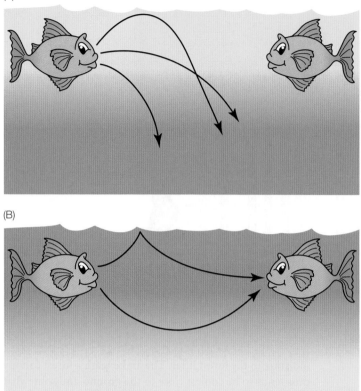

(B)

FIGURE 3.7 **Refraction in water near the surface** (A) Conditions in summer. The water nearest the surface is warmed by the atmosphere, while deeper waters remain cold. This creates a gradient of decreasing sound speed with depth. Sounds emitted near the surface are refracted down into the slower-speed regions of deeper water and lost to receivers near the surface. (B) Conditions in winter. Cold air cools the water surface to temperatures lower than deeper layers. Sound speeds now show a gradient of increasing speed with depth. Sounds emitted near the surface that begin to radiate downward are refracted back up toward the surface.

Like terrestrial forests, large bodies of water may consist of multiple layers with alternating sound speeds. As with most fluids, ocean sound speeds increase with pressure and with temperature. In the open ocean, temperature decreases with increasing depth until it reaches 3–4° C (in nonpolar regions, at a depth of about 1000 m); it then shows only minimal reductions in temperature for subsequent depths. Taken alone, this temperature gradient would result in decreasing sound speeds with increasing depth down to a minimal value and then remain constant at greater depths. However, pressure increases with depth continuously and by itself, would generate increasing sound speeds with increasing depth. The result of the interaction between the two is a layer at intermediate depths where the water is cold but pressure is still moderate. Sound speeds are low in this layer and higher either above or below this depth (**Figure 3.8**). Sounds propagated in this SOFAR (*Sound Fixing and Ranging*) channel are restrained by refraction from radiating into either upper or lower layers. The result is very long-range propagation for sounds emitted in the channel. The SOFAR layer typically begins at about 600–1200 m depths in the tropics and warmer temperate oceans, and varies from 100–1500 m in thickness. It moves closer to the surface in polar regions, and its thickness varies with temperature profiles, salinity, and ocean current patterns. In the North Atlantic in early spring, there may be two depths at which sound speeds are minimal: one at the usual SOFAR depth of 1000 m, and a second due to cold water at the ocean's surface [35]. While it has been suggested that baleen whales could exploit the deep SOFAR channel for very long-distance communication [276, 351, 375], current data suggest that most baleen whales actually vocalize at shallower depths (20–40 m) [340, 357].

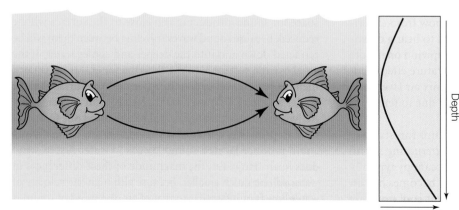

FIGURE 3.8 **Refraction in deep ocean SOFAR channel in summer** Graph on right shows typical deep ocean sound speed profile with increasing depth. Sound speed depends on water temperature (which is highest near surface and decreases to minimum value with depth) and water pressure (which is least at the surface, and increases with depth). Sound speed is high near the surface (because high temperature compensates for low pressure), and again high in very deep water (because high pressure compensates for low temperature). At intermediate depths, sound speed is low because neither temperature nor pressure is sufficiently high. Sound emitted in this channel can travel long distances due to refraction both above and below.

Depth

Sound speed

Frequency pattern distortion

Changes in the frequency composition of a sound result in changes in its waveform, and changes in a waveform imply changes in its frequency composition. Spectrograms and waveforms thus provide similar information about a sound, with the exception that phase relationships between frequency components contribute to waveforms but are ignored in spectrograms. As we shall see later, phase relationships are often ignored by animal ears unless components of a sound have very similar frequencies. In this section, we discuss changes in the spectrographic view of a sound signal as it propagates through different media.

TRANSFER FUNCTIONS   Let us begin with a sound that has just been emitted by a sender and characterize that sound with its spectrogram. For the moment, consider the block of medium between the sender and the receiver as a "black box." Sounds are introduced into one side of the black box and emerge after propagation from the other side (**Figure 3.9**). What do we need to know to be able to predict the spectrogram of the signal as it emerges from the far side of the black box and arrives at a receiver?

Luckily, each sound frequency propagates independently of the others in fluids such as air or water, a property called **linearity**; see Web Topic 2.8. If we know the initial amplitude of each frequency component in the emitted signal, and we can determine what happens to that amplitude during propagation in the black box, we can easily predict that

FIGURE 3.9   Frequency response of a linear black box This example shows the initial waveform, spectrogram, and frequency spectrum (at a point marked by the vertical white line in the waveform and spectrogram) for a parrot call (input). Horizontal lines in spectrograms indicate 5 kHz intervals, and vertical lines in frequency spectra indicate 2 kHz intervals. The call propagates through the air in the forest (the black box). During this passage, the amplitude of each frequency component in the original signal is decreased by the amount specified by the red line on the black box graph. This red line is the frequency response of this particular propagation path. Note that there is no resonance in this example and thus there are no increases in component frequency amplitudes (e.g., no positive values for the red line). The waveform, spectrogram, and frequency spectrum (for the same point in the signal) after propagation are shown as the output signal. The phase response of this system, which together with the frequency response forms the transfer function for this propagation path, is not shown here, but the same logic would apply.

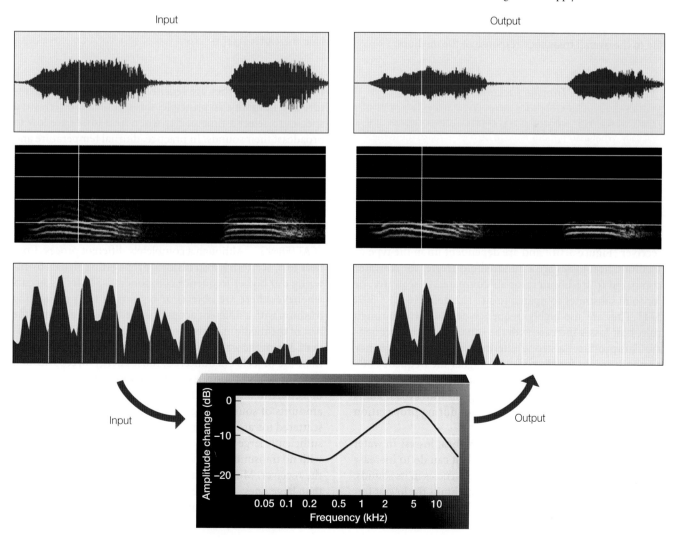

component's amplitude when it emerges. The spectrogram of the propagated signal will consist of each original frequency component in each time segment but with a new amplitude. A graph that displays the changes in the amplitude of a signal component caused by the box on the vertical axis and the frequency of that component on the horizontal axis is called the **frequency response** of the box. The vertical axis is usually scaled in dB and centered at a value of zero, which means no change in amplitude. If a given frequency in this graph has a positive vertical axis value, then its amplitude has been enhanced (probably by resonance, echoes, or addition of noise) during propagation; a negative value means that amplitude is being filtered out during propagation. The frequency-response graph is one of two graphs engineers use to characterize the way in which a black box converts an input signal into an output signal. The second is the **phase response** which shows, for each frequency, whether its phase is advanced or retarded as the signal passes through the black box. When combined into a three-dimensional graph, the frequency and phase responses define the **transfer function** of the box. As noted earlier, when dealing with animal sounds, we can usually ignore the phase-response graph and concentrate on the frequency-response graph. In the following sections, we shall examine several sources of frequency distortion during propagation and show how these contribute to the overall frequency response of the medium.

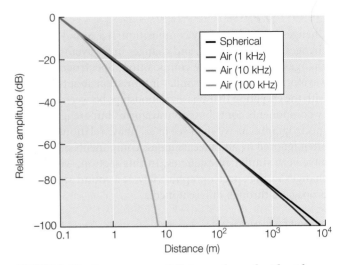

FIGURE 3.10 Frequency- and distance-dependent heat losses during sound propagation Black line shows decrease in dB of the amplitude of a sound due to spherical spreading alone during propagation over different distances. Red line shows expected decrease in amplitude due to spreading combined with heat losses for a 1 kHz sound in air. Blue line shows equivalent losses in air for a 10 kHz signal, and green line shows losses for a 100 kHz signal. In seawater, spreading losses are the same, but heat losses are lower. Amplitudes of frequencies in the range of 100 Hz to 10 kHz would decrease at rates similar to the red line; losses for ultrasonic sounds of 50–100 kHz in seawater would be more similar to the blue line. (After [112].)

> **Web Topic 3.1 Transfer functions**
> Here we discuss methods for measuring transfer functions of black boxes in general, and blocks of propagation medium specifically. What assumptions have to be met to make these measurements, and what happens if they are not met?

HEAT LOSSES   As we saw in Chapter 2, sound signals lose energy to heat losses during propagation. Heat losses increase monotonically with the distance between the sender and the receiver (**Figure 3.10**), and are dependent upon the type of medium; heat losses are a hundredfold greater for seawater compared to fresh water, for air compared to seawater, or for solid substrates compared to air [112, 185, 216, 217, 307, 371]. They are also highly frequency-dependent: as a rough rule, heat losses increase with the square of sound frequency. This produces a strong filtering effect in the frequency response graphs for all media: the higher the frequency of a component in a sender's signal, the lower its amplitude will be when it arrives at a receiver. For signals with broad ranges of component frequencies, heat losses during propagation can seriously distort the signal pattern.

What can a sender do to minimize heat losses? In water or within plant material, there is little it can do to increase active spaces except to favor lower frequencies in its sounds. As we have seen in Chapter 2, body size limits the lowest frequencies that can be radiated efficiently, so for small animals, this option is limited. In air, heat losses are more variable than in water, as they are sensitive to fluctuating temperature and humidity. In principle, this should allow terrestrial species to select a time of day when heat losses are low to broadcast their sounds. In practice, diurnal temperature and humidity often have opposite effects on heat losses [371], and the best times for minimizing heat losses often coincide with the maximal constraints on active space due to refraction. These complex interactions make prediction of optimal transmission times difficult.

SCATTERING   Although a completely different process from heat loss, scattering also distorts propagating signals by filtering out higher frequency components. Scattering occurs whenever there are objects or regions in the propagating medium with a different acoustic impedance from that of the rest of the medium. This process is highly frequency dependent. When the wavelengths of the incident sounds are larger than the objects encountered, most energy sweeps around the objects and continues on to the receiver. As the ratio of object size to incident sound wavelength increases, increasing amounts of sound energy striking the object will be backscattered toward the source and likely lost to the receiver. At sufficiently large object-to-wavelength ratios, sound shadows with no transmitted sound may exist between the objects and the receiver. Most media will contain more small reflectors than large ones: a summer forest consists of some tree trunks but many more leaves and branches. The result is that there are many more objects that scatter high frequencies than ones

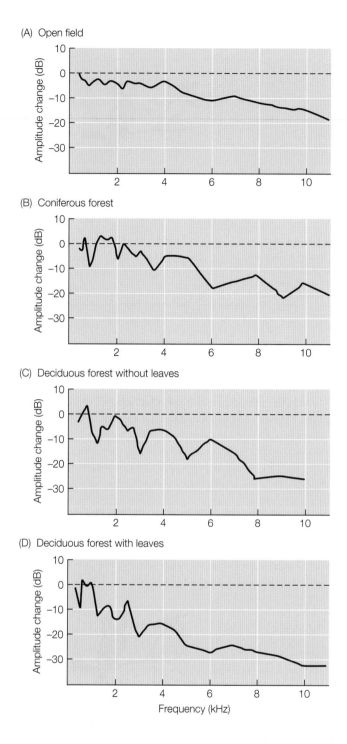

(A) Open field

(B) Coniferous forest

(C) Deciduous forest without leaves

(D) Deciduous forest with leaves

Frequency (kHz)

FIGURE 3.11   Scattering and heat losses in temperate terrestrial habitats Frequency response graphs for sound propagation in air across a distance of 100 m (A) over an open field, (B) in a coniferous forest, (C) in a deciduous forest without leaves, and (D) in a deciduous forest with leaves. All samples have spreading losses removed and are taken 10 m above the ground to exclude boundary effects. Note the significantly higher losses at high frequencies when a deciduous forest has leaves. This difference is largely due to scattering effects. (After [218].)

can also generate significant backscatter when aggregated in sufficient numbers. Surf turbulence near the shore can create vortices full of air bubbles that also serve as scattering sites. Because wavelengths for any frequency in water are 4.4 times as large as those of the same frequencies in air, whereas the size distributions of objects are similar in air and water, scattering in water only becomes important for component frequencies of 10 kHz or higher [146]. This exceeds the frequencies used by most marine animals, excepting toothed whales and porpoises, for sound communication. Over the same frequency ranges, scattering losses are thus much less of a problem for aquatic animals than they are for terrestrial ones. Inside a solid substrate such as a plant, the speed of sound is sufficiently high that potential obstacles would have to be larger than the entire plant to produce scatter. However, as we shall see in the next section, the plant boundaries themselves become important reflectors, and this leads to a similar filtering out of higher frequencies.

BOUNDARY EFFECTS   A boundary is a large expanse of interface between two media with significantly different acoustic impedances. There are three basic situations in which boundaries affect sound propagation: (1) when a sender and receiver exchange sound signals in one medium that is near a boundary with another medium; (2) when sender, receiver, and sounds exchanged between them are confined between two or more boundaries; and (3) when sender and receiver are located in one medium but communicate with signals propagating along or through a boundary with another medium. We take up each case in turn.

Sounds emitted by a sender close to a boundary with another medium may reach a receiver by multiple paths (**Figure 3.12**). Assuming no intervening obstacles, a **direct wave** will propagate along the straight line connecting the two animals. In addition, some of the sound that normally would not reach the receiver now encounters the boundary and is reflected toward it. This is a **reflected wave**. The arrival of two or more concurrent waves at the receiver sets the scene for interference between them [99, 100, 102, 371]. If the direct and reflected waves arrive at similar amplitudes but out of phase, they will cancel each other out, and the receiver will detect little sound. If they arrive at similar amplitudes but in phase, then the receiver will hear the sound at a greater amplitude than if there were no boundary nearby. Whether the two waves do or do not arrive with similar amplitudes depends on the strength of the reflection and on how far

that scatter low frequencies. The frequency response graph will again show a drop in amplitude that becomes more severe with increasing frequency.

In forested environments, trees and foliage are the main sources of scatter [13, 218-220]. Significant scattering with sound shadows can occur at frequencies as low as 1–2 kHz (**Figure 3.11**). In open terrestrial sites, rising pockets of heated air or wind vortices may generate similar frequency-dependent scatter [367, 368]. In the open ocean, the main sources of scatter are the air-filled swim bladders of fish and the air bubbles and fat globules carried by various zooplankton. Marine animals with acoustic impedances similar to but still not equal to that of water (e.g., shrimp and squid [159])

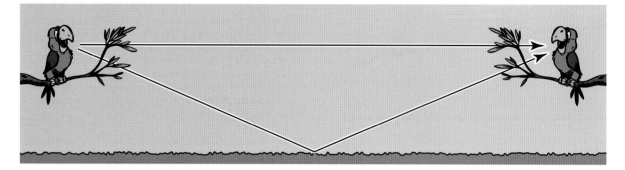

FIGURE 3.12 Interference between direct and reflected waves near a boundary
When sender or receiver are located in one medium, but one or both are close to a boundary with another medium of quite different acoustic impedance, sound waves that would otherwise have missed the receiver can now reach it due to reflections at the boundary. Such reflected waves can interfere (positively or negatively) with the direct wave that has traveled on a straight line between sender and receiver.

above the boundary the two parties are located (**Figure 3.13**). If one or both are high above the boundary, the reflected wave must travel much farther than the direct wave, suffer greater spreading, heat, and scattering losses, and thus arrive at a sufficiently lower amplitude that any interference will be negligible. When both parties are close to the boundary, the reflected wave will travel only slightly farther than the direct wave and interference can be significant.

Whether the direct and reflected waves are in or out of phase is more complicated. All other factors being equal, the

(A)

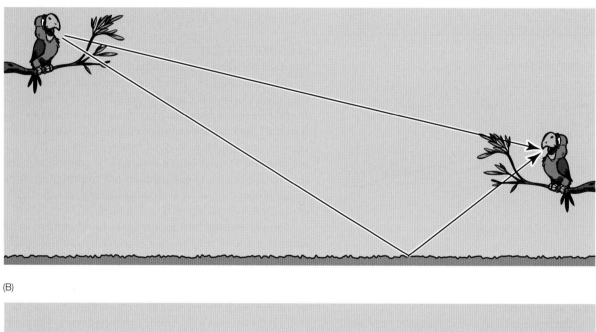

(B)

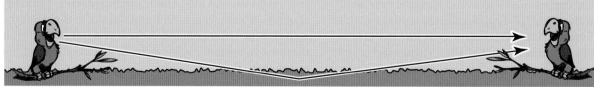

FIGURE 3.13 Effects of sender and receiver height on boundary interference If one or both parties are high above a nearby boundary (A), the path taken by the direct wave will be significantly shorter than that taken by the reflected wave. As a result, the amplitude of the reflected wave will be much lower than that of the direct wave, and interference will be minimal. When both animals are on the ground (B), the paths taken by the direct and reflected waves will be very similar and interference can be maximal.

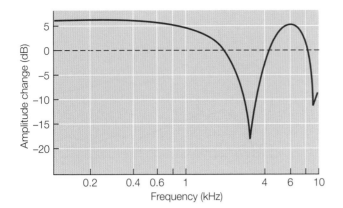

FIGURE 3.14 Boundary interference for airborne sounds over a hard surface For a very hard surface, such as asphalt, and any biologically likely incident angle, the reflected wave from the nearby hard surface will be large in amplitude and without any phase shifts at reflection. Reflected and direct waves thus interfere positively at low frequencies, but interfere negatively for that frequency where the reflected wave travels 1/2 of a wavelength further than the direct wave, (here, 3.2 kHz), and again when the reflected path length is 3/2 of a wavelength further than the direct wave (here, 9.6 kHz). (After [101].)

two waves will cancel each other out when the reflected wave has to travel an odd number of half-wavelengths (e.g., 1/2, 3/2, 5/2, etc., of a wavelength) farther than the direct wave. The two waves will interfere positively if the reflected wave travels 0, 1, or more full wavelengths farther than the directed wave (**Figure 3.14**). At very low frequencies, the difference in direct and reflected travel distances will be only a small fraction of a wavelength, and the two waves will be almost in phase. The frequency response graph for a signal propagating near a boundary would thus exhibit:

- high amplitudes for very low frequencies;
- a decline to a minimum at a frequency with a wavelength twice the difference in direct and reflected wave travel distances;
- an increase to a maximum at the frequency with a wavelength equal to the distance difference;
- a decline to a second minimum for a wavelength equal to 2/3 of the distance difference, etc.

When the boundary is between air and the ground, this oscillating pattern of negative and positive interference is usually lost at higher frequencies, because each blade of grass or pebble becomes a separate sound reflector once it is larger than the incident wavelengths of sound. This creates many out-of-phase reflected waves instead of one cohesive one, and thus no consistent pattern of interference with the direct wave.

All factors are rarely equal in natural contexts. If both the speed of sound and the acoustic impedance of the first medium are greater than those in the second medium, (as happens when sound is traveling in water and hits an air–water boundary), all component frequencies in the incident sound will be phase-shifted by one-half of a cycle at reflection. This generates a frequency-response graph that is the exact opposite of the one described above: low frequencies are now heavily cancelled out, a maximum arises for that frequency

with a wavelength twice the difference in direct and reflected wave distances, a minimum occurs for the frequency with a wavelength equal to this distance, and so on (**Figure 3.15**).

A similar phase shift can occur when the speed of sound is higher in the second medium, (as with sound in air striking the ground), if the angle at which the sound strikes the boundary is lower than a **critical angle**. The angle at which the sound strikes the boundary depends on the distance between the two parties (greater distances mean lower incident angles), the height of the sender above the boundary (lower heights mean lower incident angles), and the height of the receiver above the boundary (lower heights again mean lower incident angles). The value of the critical angle depends on the relative speeds of sound in the two media. If they are very different (as with sound in air striking an asphalt or cement surface), the critical angle is close to zero, and only animals very far apart or very close to the ground will experience a phase shift in reflected sound waves. If the sound speeds are not so different, as with sound in air striking soft ground, grassy turf, or new snow, the critical angle can be quite high and phase shifts at reflection will be the more common situation. As with sound in water striking the surface, phase shifts cause lower frequencies to be severely attenuated during propagation. Note that the angle of incidence can also affect the amplitude of the reflected wave, and this will also modulate the degree of interference. For more details on phase shifts, go to Web Topic 2.5.

Interference between direct and reflected waves can clearly impose major changes in the frequency composition of sound signals propagating near a boundary. For underwater animals near the water's surface, there is no recourse. For

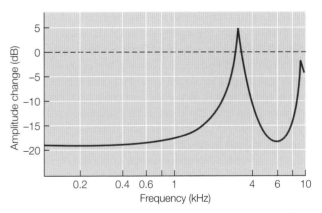

FIGURE 3.15 Boundary interference for airborne sounds at low incident angles The boundary surface is here assumed to be hard enough to produce a reflected wave of large amplitude, soft enough that the critical angle is larger than the incident angle defined by the proximities of sender and receiver to the boundary, and no ground wave is present. This results in a full (180°) phase shift at reflection. The frequency response is the mirror image of that shown in Figure 3.14: direct and reflected waves now interfere negatively for low frequencies, but show positive interference for those wavelengths equal to 1/2 and 3/2 of the path difference. (After [101, 102].)

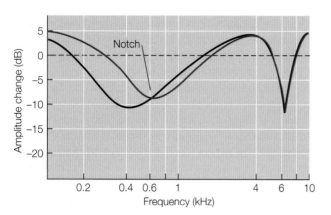

FIGURE 3.16 **Boundary interference for airborne sounds at low incident angles and over soft substrates** Frequency responses for propagation over ground covered with fresh snow (black) and soft earth inside a woodland (red). As in Figure 3.15 direct and reflected waves interfere negatively for low frequencies and certain higher frequencies. But here ground and surface waves restore the propagation of the lowest frequencies, leaving a "notch" at intermediate frequency values. Such soft boundaries also exhibit varying strengths of reflection and amount of phase shift as a function of frequency. This rounds and broadens the curves seen over hard surfaces. (After [101, 102].)

terrestrial animals, two processes can minimize these changes if the boundary is sufficiently soft and porous [8, 102, 279]. When sound from a distant source strikes a very hard surface such as asphalt, most of the sound energy is reflected toward the receiver from a single point on the surface. However, reflection from a softer and more porous boundary occurs in a more diffuse region on the surface. While that part of the sound wave reflected from a central point forms the reflected wave described above, incident sound reflected from other nearby points will differ from the reflected wave in both amplitude and relative phase. The ensemble of reflected contributions from these adjacent points is called a **ground wave** and constitutes a third path by which a signal might reach a receiver (**Figure 3.16**). Ground waves are detectable only at low heights above the boundary, and attenuate least when propagating low frequencies: whereas a 200 Hz ground wave must propagate 200 m before its amplitude has dropped by 3 dB, an 800 Hz ground wave will decrease by the same amount after propagating only 8 m. Over sufficiently soft terrain, ground waves can restore some of the lowest frequencies in a signal that might otherwise be filtered out at low incident angles. Propagation for frequencies that are too low to avoid negative interference and too high to be restored by ground waves cannot be improved. This results in a conspicuous **notch** in the frequency responses of most terrestrial habitats. Ground waves can also provide low levels of sound that persist in refractive sound shadow zones. If the ground is sufficiently porous, incident sound can also be trapped in a thin layer just above a boundary, and continue propagating as a **surface wave**. This fourth path linking sender and receiver attenuates less slowly than the ground wave, but is also limited to lower frequencies.

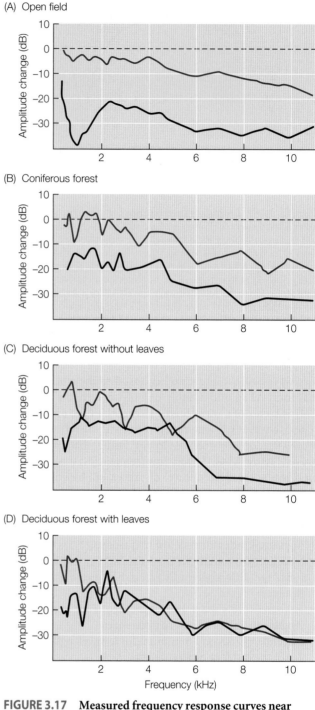

FIGURE 3.17 **Measured frequency response curves near ground in temperate terrestrial habitats** These graphs show the same four contexts of Figure 3.11. The frequency response for each habitat with both sender and receiver 10 m above the ground (red) is compared to response when both are at ground level (black). Ground level measurements show high attenuation at lower frequencies due to direct and reflected wave interference and some restoration of certain low frequencies due to ground and surface waves leaving a notch at intermediate values. All plots show complex patterns of frequency dependent scattering at intermediate frequencies; and a gradual drop in amplitudes at high frequencies due to heat and scattering losses. (After [218].)

In most terrestrial environments, scattering, boundary interference, ground waves, and surface waves all combine to produce an overall frequency response graph generating outputs with moderate-amplitude low frequencies, low amplitudes for the notch (200–700 Hz), and a gradual rise to a maximum near 1–2 kHz (**Figure 3.17**). At higher frequencies, boundary interference, ground waves, and surface waves have diminishing roles, and signal amplitudes decrease with increasing frequency due to heat and scattering losses. Ground-living animals that wish to communicate over long distances using sound must use either low frequencies (assuming they are large enough to do so efficiently) or high frequencies (at the cost of higher heat and scattering losses). The only alternative is to move sufficiently far above the ground that boundary reflections are negligible [84, 145, 230].

We now turn to the second type of boundary effect: that created when sender and receiver are both located between two parallel boundaries. For signals propagating in caves, water of intermediate or shallow depths, the SOFAR channel, or between the ground and a forest canopy, sound can be reflected (or refracted) back and forth between multiple parallel boundaries. Because sound is trapped between these boundaries, the propagating environment is called a **waveguide**. The direct wave and multiple reflections inside a waveguide interfere in complicated ways, just as multiple reflections interact inside a resonant structure [35, 106, 237]. As a result, certain frequencies (normal modes) are accentuated in waveguides, and others are filtered out (**Figure 3.18**). The lowest mode is called the **cutoff frequency** of the waveguide, since lower frequencies cannot propagate within it. As

a rule of thumb, the cutoff frequency has a wavelength about four times the thickness of the waveguide. Shallow bodies of water are a good example of a waveguide. The water's surface and the bottom form the two parallel reflecting surfaces. The depth of the sender in the water body, the acoustic properties of the bottom, whether the water's surface is choppy or smooth, and the temperature profile as a function of depth all play contributing roles in determining the cutoff frequency and the frequencies of normal modes. Cutoff frequencies for water bodies can be as high as 500–1000 Hz for a sender in a 1 m deep pool, and drop to 30–200 Hz for a 10 m deep lake or near-shore zone [307]. The low-frequency sounds exchanged by many shallow water fish are often below the ambient cutoff frequency; this limits their acoustic communication to very short ranges. Above the cutoff value, there is usually an **optimal frequency** whose amplitude is maximized during propagation; signal components with frequencies below and above this optimal value will tend to be filtered out with increasing propagation distances. The optimal frequency generally decreases as the depth of the sender is increased, and for a given location, there is usually an optimal depth and frequency combination that produces the maximal active space [155]. Male humpbacked whales (*Megaptera novaeangliae*) may take these factors into account in selecting both singing sites and singing depth during the mating season [237].

The third type of boundary effect occurs when both sender and receiver live in one medium but transmit and detect sound signals propagated across a boundary in another medium. For example, many small insects and hunting spiders communicate by exchanging sounds that propagate through their host plants [19, 22, 67, 68, 71, 152, 361]. Percussion, tremulation, tymbal buckling, or stridulation are used to generate vibrations that pass directly into the plant through the sender's appendages. In principle, such sounds could be propagated in a solid substrate such as a plant as longitudinal, transverse, Rayleigh, Love, or bending waves [217]. In practice, the signals used by these small arthropods appear to be propagated largely by bending waves [242]. Bending waves travel along the major axes of the plant, and unlike other substrate wave types, cause the entire cross section of

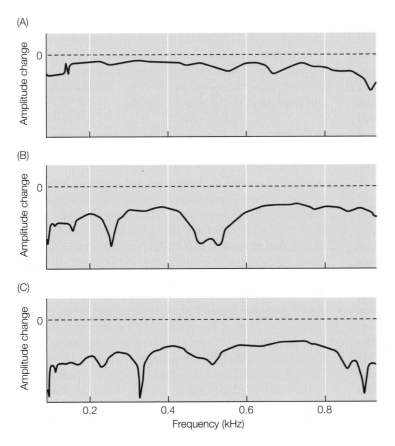

FIGURE 3.18 Predicted frequency response in a marine waveguide Low-frequency sounds propagating in a waveguide created by the ocean surface and bottom. The examples assume 60 m total water depth and positioning of the sender and receiver (e.g., humpbacked whales) at 15 m depth. Dashed lines indicate no change in signal amplitude; all points below this line are decrements in signal amplitude. Distances between sender and receiver: (A) 1 km; (B) 3 km; (C) 5 km. Note rise and fall of curves at increasing distance due to excitation of normal modes for this waveguide configuration. Cutoff frequency for this depth would be about 6 Hz and is thus barely visible in the graphs (After [237].)

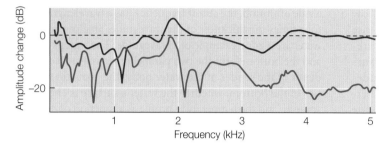

FIGURE 3.19   Frequency response of sound propagating plant stems Plots show frequency dependence of sound propagation over a 3 cm distance (red) and a 17 cm distance (blue) within the stems of the plant *Thesium bavarum*, a common host for substrate-signaling shield bugs (Cydnidae, Hemiptera). Interference between direct and multiple reflected waves inside the plant generates complex frequency response patterns that differ depending on where the receiver is located on plant. (After [242].)

a stem or leaf at a given location to oscillate in concert along a line perpendicular to the direction of propagation. Bending waves suffer little attenuation or damping during propagation, and are reflected at the root and branch tips back into the plant. Interference between direct and persistent reflected waves accentuates those frequencies that are compatible with the plant dimensions (normal modes) and filters out others. At normal-mode frequencies, bending waves cause some locations along the plant (acoustic internodes) to experience major oscillatory movements, while just a half-wavelength away (an acoustic node), there will be little motion at all. The locations of acoustic nodes and internodes along a plant vary with the frequency of the sound. Frequency-response graphs for plants show strong propagation of low frequencies (below 100 Hz), complex patterns of reduced propagation at intermediate frequencies (100–2000 Hz), and smoother variation at decreasing amplitudes for higher frequencies [20, 72, 164, 205, 242] The frequency response graph for plants thus shows the same kinds of peaks and valleys seen in other boundary contexts, but the better-propagated frequencies are largely compressed into the lower end of the bioacoustic spectrum (**Figure 3.19**).

Boundary propagation of animal signals also occurs at the water's surface (water striders and spiders), at the surface of the ground (fiddler crabs, snakes, rodents, and elephants), and on spider webs. Each of these two-dimensional boundaries has its own frequency-dependent properties. Propagation of ripple signals on the water's surface attenuates higher frequencies faster than it does low frequencies [20, 30, 112, 127, 191, 217]. Seismic signals in the ground can propagate in various ways, but those used by animals for communication are usually Rayleigh surface waves [3, 46, 257]. This is even true for burrowing animals, in which the relevant boundary is the surface of their tunnels [224]. As with ripple propagation on the water's surface, high-frequency Rayleigh waves tend to attenuate faster during propagation than do low-frequency waves [7, 46], although the degree of difference

varies with soil type [98]. And, as with waveguides, certain frequencies may be optimal for maximizing range and active space. Dry sand, for example, shows optimal propagation at 300–400 Hz, with lower values at lower and higher frequencies [3, 46]. In contrast with animals that signal via water and ground boundaries, spiders can design their webs to have favored acoustic properties. The multiple threads in a typical orb web can vibrate as longitudinal waves and as transverse waves (either parallel or perpendicular to the plane of the web). Optimal propagation in orb webs appears to occur via longitudinal vibrations in the radial threads of the web, and there is little frequency dependence up to 9–10 kHz [184, 228, 229]. Sheet webs of other spiders likely show some resonance and thus frequency dependence, but computations suggest that this resonance enhances propagation and normal modes are sufficiently close together to provide a nearly flat frequency response [253].

Temporal pattern distortion

As noted above, any change in the frequency composition of a signal will likely produce a corresponding change in the signal's waveform. Thus all of the processes outlined in the previous section can result in changes in the temporal pattern of a signal. Three additional propagation processes can modify temporal pattern directly: reverberation, added modulation, and dispersion. We take up each below.

REVERBERATION   We discussed earlier how a sound propagating in one medium can be scattered by intervening objects of appropriate size and acoustic impedance. Leaves and branches constitute the main source of scattering in forested habitats, and the swim bladders of schooling fish act similarly in aquatic environments. Depending upon the relative locations of the sender, receiver, and the scattering objects, one or more reflected versions of the signal can arrive at the receiver. These are called **reverberations** (or echoes).

Because scattering is frequency-dependent, a reverberation may emphasize different frequency components than does the original signal. It may also represent only a portion of the original signal. Depending on the duration of the original signal and the proximity of the scattering objects to the line connecting sender and receiver, reverberations may arrive after receipt of the direct signal and thus create a separate echo, or they may overlap and sum with the latter portion of the direct signal. An overlapped combination is invariably longer in duration than the original signal, and exhibits a "tail" that

(A)

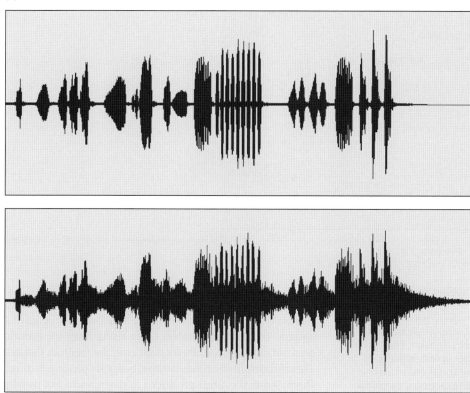

(B)

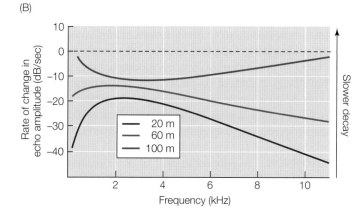

FIGURE 3.20 Reverberations in forests (A) Waveform of the song of a male lark sparrow (*Chondestes grammacus*) before (top) and after (below) propagation in a thick forest. This species normally lives in open habitats where reverberations are rare and thus has adopted none of the specializations used by forest birds to minimize song distortion due to reverberation. Note blurring of successive notes due to overlapping echoes and addition of long "tail" to song. (B) The length of added "tails" and blurring of successive notes are both greater when the rate of echo decay is low. This plot shows the rates of echo decay for different frequencies and different sender-receiver distances in a box-ironbark forest in Australia. At short distances, intermediate frequencies (1–3 kHz) reverberate longest and suffer the worst blurring of successive elements; at long distances, intermediate frequencies should experience the least blurring and minimal tail additions. (After [272].)

decays in amplitude with time (**Figure 3.20**). The duration of this reverberation "tail" has been measured in several terrestrial forests, and depends on the distance between sender and receiver, the component frequencies in the original signal, the density and sizes of scattering objects, and to a lesser degree, the height of the sender above the ground [150, 272, 300]. For distances less than 100 m, reverberation tails in forest are longest for frequencies with wavelengths similar to the dimensions of nearby leaves and branches, and shorter for either higher or lower frequencies. Longer distances result in longer tails. Reverberation tails are diminished if the sender moves higher above the ground; receiver height appears to have little effect on reverberation. Senders with more directional sound fields also produce less scatter and shorter tails [300].

Overlap of reverberation and a direct signal is bound to distort the temporal pattern that reaches the receiver. Longer or more rapidly repeated signals and longer tails are the most vulnerable combinations. Reverberations can fill in the initially silent intervals between successive syllables in a bird trill; a rapid and broad frequency sweep, such as that of a male cardinal, will be broadened in each time segment by overlapping the original segment with slightly lagging copies of it. Senders faced with "blurring" of a temporal pattern due to reverberations can select emission heights, component frequencies, and sound-field directionalities that generate minimal tails; limit the rates of amplitude or frequency modulations to values much shorter than the decay rates of tails; or avoid modulations and temporal patterns altogether.

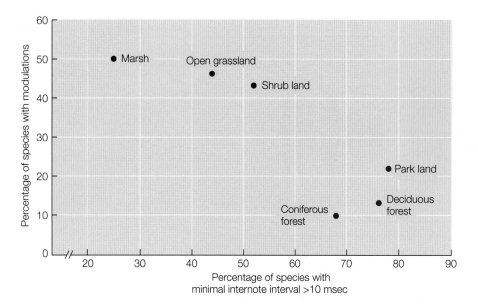

FIGURE 3.21 **Correlations between habitat and temporal patterns in birdsong** The horizontal axis in this graph indicates the percentage of 120 North American songbird (Oscine) species in each of 6 habitat types that space successive elements in their songs at intervals of 10 msec or more (abscissa). The vertical axis plots the fraction of those same species in each habitat that exhibit significant modulations in their songs (as evidenced by significant sidebands in spectrograms). As expected, if songbirds avoid song distortion due to reverberation, songs of forested species (deciduous forest, conifer forest, and parkland) are unlikely to show modulations and the shortest intervals between successive song elements are quite long. Songbirds of open country habitats (marsh, grassland, and shrubland) should have to worry less about reverberations, and they indeed show significantly higher fractions of species whose songs include amplitude and/or frequency modulations with very short intervals between successive song elements. These patterns remained statistically significant even after corrections for inclusion of bird families with many similar species. (After [372].)

The latter strategy can even turn reverberation into an advantage: echoes that overlap with a direct signal contain the same frequencies and can thus enhance the received signal against background noise [260, 330]. The risk of signal reverberations perhaps explains the widespread observation that forest birds include fewer and less dramatic modulations in their vocalizations than do birds that live in open habitats where reverberant scatter is minimal (**Figure 3.21**) [14, 27, 34, 43, 44, 129, 130, 156, 315, 332, 337, 355, 372].

ADDED MODULATION   In open terrestrial environments and under water, turbulence and eddies can make the medium quite heterogeneous. Large eddies typically spin off smaller eddies from their margins, these spin off even smaller ones, and so on, creating a series of different eddy sizes. As a sound signal propagates through such a medium, different segments encounter different densities and vortex patterns. The result is additional amplitude modulation of the signal waveform with greater distortion at higher frequencies and greater sender-receiver distances [279].

In terrestrial habitats, the major source of turbulence is wind which creates vortices that can modulate propagating sounds. In addition, bubbles of air rising from the ground on a hot day generate heterogeneities in the medium that can also add amplitude modulations to signals passing through them. Animal sounds produced at dawn and dusk usually have the least risk of added modulations, whereas midday winds and heat effects can produce significant alterations in the waveforms of propagating sounds. Richards and Wiley [300] reported up to 40 dB variations in amplitude imposed on propagating sounds in even mildly windy conditions. They also noted that this process modulated the envelope of waveforms at rates of 10–50 Hz over a wide range of conditions. Senders would have to modulate the amplitude of their emitted signals at rates higher than 50 Hz to avoid temporal distortion during propagation in these contexts.

DISPERSION   In bulk volumes of air or water, and in strings (e.g., spider webs), all sound frequencies propagate at the same speed. As a result, the alignment between a signal's frequency components is preserved during propagation even if their amplitudes are altered. This is not the case for bending waves in plants, waves on boundary surfaces (water ripples, Rayleigh and Love waves on solid substrates), or sound inside waveguides. In these situations, different frequencies propagate at different speeds, resulting in their temporal and spatial **dispersion**. Even if propagation does not change the relative amplitudes of a signal's frequency components, its waveform can become drastically altered if their alignment is changed due to dispersion. Propagation speeds increase with frequency in bending waves [217], but usually decrease with frequency for Rayleigh and Love waves [189]. In wave guides, propagation speed also decreases with frequency at least above the cutoff frequency [35]. The most complex situation is on the water's surface, where propagation velocity increases monotonically both below and above a minimum at 13 Hz [19, 29, 112]. In each of these contexts, a signal's waveform will be increasingly altered as it propagates farther away from the sender.

Web Topic 3.2 *Dispersive sound propagation*
Sound propagation with dispersion can result in major changes in a signal's waveform and in the speed with which the signal propagates as a whole. Here we provide more details on dispersive sound propagation for each of the contexts in which communicating animals are likely to encounter it.

Noise masking

Noise is any concurrent sound that obscures receipt of a sound signal by a receiver. At high intensities, noise can hinder receiver detection of a signal; at lower intensities, the receiver can detect the signal but noise hinders signal recognition (i.e., assignment of the received signal to an expected category). In many cases, noise, more than receiver sensitivity, sets the limits of active space size. The degree to which noise masks signals depends upon its amplitude, frequency, spatial distribution, and temporal pattern. Signals with a high amplitude relative to ambient noise—which create a high signal-to-noise ratio—are more likely to be detected and recognized by receivers. Signals with frequency compositions different from that of ambient noise will be favored in those species whose ears are tuned to a limited band of frequencies or can decompose sound signals into their frequency components (see later sections). A receiver may be able to ignore noise generated by a nearby directional sound source if its receiving organs are also sufficiently directional. The temporal properties of noise vary along a continuum ranging from fully continuous to sporadically intermittent. Continuous noise usually poses a greater problem for sound communication than intermittent noise. Below, we summarize the patterns of background noise in each type of environment, and outline the special challenges presented by noise from nearby conspecifics.

BACKGROUND NOISE   In both terrestrial (**Figure 3.22**) and aquatic environments, noise levels tend to be greatest

at the two ends of the biologically relevant spectrum [39, 49, 167, 168, 250, 307, 315, 331, 332, 335, 364, 365]. In air, the turbulence generated by wind creates low-frequency noise (100–1000 Hz). Intensities are highest in the lower end of this range, with values for moderate winds (1 m/sec) of 30–40 dB above the threshold of human hearing, (a reference called SPL; see Web Topic 2.3), and 60–70 dB (SPL) for strong winds (8 m/sec). Wind noise is generally greater in open habitats than in forested ones [250, 315]. Airborne noise can also be significant close to running streams and waterfalls and during heavy rainfall. While lower frequencies also

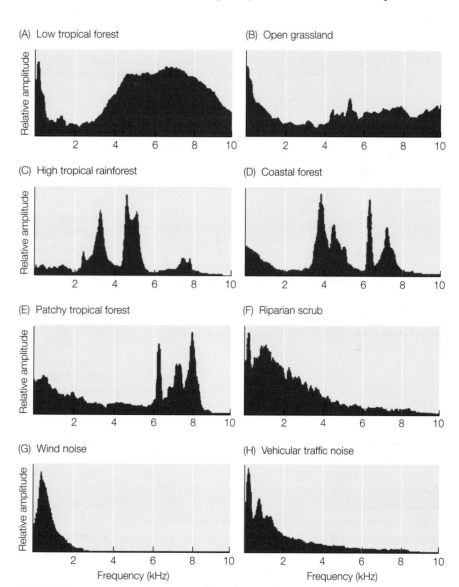

FIGURE 3.22 Frequency dependence of ambient noise in terrestrial habitats (A) Low tropical forest: low-frequency band due to wind; high-frequency band due to forest insects and birds. (B) Open grassland: high wind noise similar to (A) but reduced animal noise. (C) High tropical rainforest: very low wind noise; high-frequency noise due to cicadas and birds. (D) Coastal forest: similar to (C), but with wind noise and high-frequency bands shifted due to different cicada and bird species. (E) Patchy (gallery) tropical forest: similar to (D), but cicadas producing higher frequency noise. (F) Riparian scrub near a rushing stream. (G) Wind noise over a meadow. (H) Vehicular traffic noise. (A, B (Panama) after [315]; C–E (Cameroon), F (California Sierra Mountains), G (Colorado Rocky Mountains), and H (San Francisco, CA) after [335].)

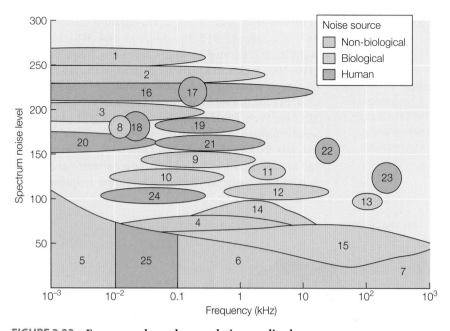

FIGURE 3.23   Frequency dependence, relative amplitude, and sources of ambient noise in the ocean   Natural nonbiological sources of noise: (1) undersea earthquakes; (2) sea floor volcanic eruptions; (3) lightning strikes; (4) heavy rain and storm winds on sea surface; (5) seismic noise; (6) calm surface wave noise; (7) thermal (molecular) noise. Biological sources of noise: (8) blue and fin whale calls; (9) humpback whale songs; (10) bowhead, right, and gray whale calls; (11) sperm whale echolocation clicks; (12) dolphin whistles and other social sounds; (13) dolphin echolocation clicks; (14) fish choruses; (15) snapping shrimp. Human sources of noise: (16) seismic airguns; (17) naval low-frequency active sonar; (18) ATOC and successor ocean temperature measurements using sound; (19) supertanker noise; (20) drilling rigs and dredgers; (21) large naval ships; (22) echo-sounding sonars; (23) side-scan sonars; (24) submarines; (25) general shipping noise. (After [65].)

tend to dominate these types of noise, noticeable amplitudes may persist up to 5–6 kHz [194, 335]. An increasing source of lower-frequency noise throughout the terrestrial world is vehicular traffic, which has a frequency spectrum and amplitudes not unlike those of wind [334, 335]. The major sources of ambient noise inside plants [57] or on the water's surface are wind and rain.

The major sources of continuous noise at the higher end of the terrestrial spectrum are advertising frogs, cicadas, crickets, and katydids [39, 335]. This band of noise begins to be significant around 2 kHz and can extend, particularly in the tropics, far into ultrasonic frequencies. All of these taxa can produce relatively continuous noise levels. Some cicadas alternate calling and silent periods, but often stimulate each other to call as a chorus, making for very intense bouts of noise. Frogs are more likely to call at night and can also chorus. The frequency zone between the upper margins of wind noise and the lower limits of frog and insect noise, (approximately 500–2000 kHz), is relatively quiet with respect to continuous noise. This is a band favored by many

birds and mammals for their vocalizations. However, few species of either group produce the continuous chorusing of frogs or insects, and their songs and calls constitute at most intermittent noise for other species. The major exceptions are breeding colonies [11] and the dawn chorus, which is a period of maximal avian and primate vocalization in many habitats [45, 58, 142].

We have seen that sound signals attenuate much more slowly in water than in air. Unfortunately for communicating animals, ambient noise also attenuates more slowly (**Figure 3.23**). This is especially true for low frequencies. In large bodies of water, very low-frequency noise (10–100 Hz) is generated by the Earth's seismic activity and lightning strikes. In polar regions, sea ice adds additional low-frequency sounds. Natural noise in the 100–5000 Hz range is mostly generated at the surface through wave action, breaking surf, and rainfall [235, 301]. Much of this surface noise is due to the entrapment of air bubbles that vibrate and then burst [87, 88, 190]. Wind is the major driving force for wave action in the open ocean, and noise levels in calm seas decrease with increasing frequency above 300 Hz. In stormy weather, surface wave noise can increase by 30 dB or more. Between 1 kHz and 100 kHz, snapping shrimp are the dominant sources of continuous marine noise [9, 28, 60, 289], although barnacles, sea urchins, and mussels contribute to the din [110]. Since these invertebrates live on the sea bottom, their noise is most intense in shallow coastal waters. This habitat also suffers high noise due to surf and breaking waves. Fish contribute only moderately to ambient noise levels. Most species use low-to-intermediate frequencies for communication (100–900 Hz) and call intermittently [232, 252]. A few species chorus, but usually during limited periods each day and in sufficiently shallow water that the sounds propagate only short distances. The calls and songs of baleen whales may also make a significant contribution to background noise in the ocean [83].

As with vehicular and airplane traffic in terrestrial environments, human activities are a major source of ocean noise [328]. Shipping produces noise in the 10–100 Hz range due to propeller cavitation; very large (supertanker) and very small but fast (jetski) vessels are of particular import. Modern mineral exploration uses reverberations from air guns above the sea floor to assess geological structure and content. Drilling, pile driving, and trench digging also produce intense local noise levels, and both military (<100 Hz) and fishing (>100 kHz) sonars contribute additional frequency bands. For each of the taxa of aquatic animals that use sound for communication, there is now a significant overlap between the frequencies used for communication and dominant frequencies of ambient anthropogenic noise [4, 32, 51, 78, 82, 85, 103, 116, 192, 198, 204, 215, 233, 285, 286, 308, 317, 321, 322, 328, 338, 339, 359, 363, 377, 379].

Web Topic 3.3 *Animal communication and anthropogenic noise*

To what degree does noise generated by human activities interfere with sound communication in animals? Recent studies in both terrestrial and marine environments suggest that there are increasing problems, but animals can sometimes adapt.

Communicating animals faced with ambient noise at frequencies overlapping their signals have limited options [49]. Where noise is not too intense, senders tend to try to maintain communication by increasing signal amplitude—the **Lombard Effect** [47, 48, 207, 288, 293]. They may also increase their signal durations [48] or signal repetition rates [191, 275, 285]. If the noise is sufficiently intermittent, senders can limit sound emission to the more quiescent periods [122]. Reciprocal acoustic exchanges between senders and receivers are used by many insects to confirm receipt of a signal despite noise [17]. Where noise is both intense and relatively continuous, the sole option is to shift signal frequencies to reduce overlap with the ambient noise. Such spectral shifts might occur over evolutionary time when noise spectra are sufficiently predictable, or within a generation, given sufficient flexibility in a species' sound-production mechanisms. A variety of animal species have evolved dominant signal frequencies that fall within a low-level "window" in the natural patterns of ambient noise [38, 40, 49, 81, 86, 201, 202, 315, 332, 365]. Short-term spectral shifts have also been reported in a number of avian and mammalian species living in urban areas with significant human noise (**Figure 3.24**) [160, 274, 294, 295, 333, 336, 362, 376].

CONSPECIFICS   Special problems can arise when a sender is surrounded by other sound-producing senders of the same species. In a breeding colony of penguins or other seabirds, calling is largely random and there is no overt attempt by neighbors to mask a given sender's calls. This is not the case in mating choruses of insects or anurans (frogs and toads), where males monitor each other's calling and attempt to call earlier, faster, or louder than nearby competitors. At high calling rates or with many nearby males calling concurrently, male calls overlap in time, and females have a difficult time assigning specific calls to specific caller locations. Faced with call overlap, females often invoke a **precedence effect** in which overlapping calls are assigned to the location of the first caller [386]. This sets up a competition in which males either jockey to call first or wait until a cluster of males has called and then follow with their own non-overlapping call. The typical outcome of this competition is either a high level of synchronous calling by neighboring male callers (e.g., full overlap), or completely alternate calling (e.g., no overlap) [26, 127, 128, 247]. In the first case, noise is maximized for

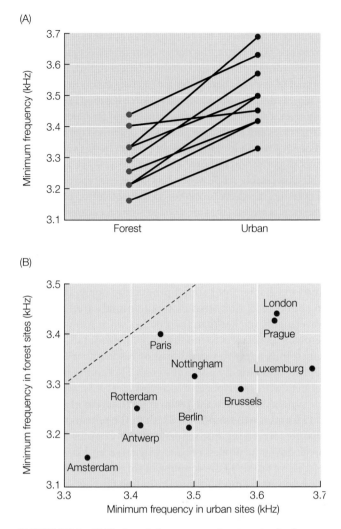

FIGURE 3.24 **Shifts in minimum song frequencies in the songs of great tits (*Parus major*) for urban versus nearby forest populations** For each of 10 European cities and towns, an average of 30 great tit songs were recorded in the urban environment, and another 30 songs were collected in a nearby forested site. (A) Average minimum frequencies for songs in each pair of sites (linked by lines). In every case, the average minimum frequencies are higher in the urban setting. (B) The same data plotted to show that there are no underlying geographical correlates to the minimal frequencies found in the songs. In each case, the urban population has shifted to a minimum frequency somewhat higher than that of the nearest forest population. If there had been no shift, points would fall on the dashed line in the graph. (After [336].)

the receiver females, and in the second it is minimized. We shall return to the evolutionary implications of these adjustments to conspecific competition in later chapters.

Sound Reception

Animals use their ears to detect, identify, evaluate, and localize ambient sound sources. As with other sensory modalities, communication is only one of several possible tasks mediated by hearing organs. The monitoring of parasites, predators, prey, and nearby obstacles (e.g., echolocation) can be more important functions for hearing than is conspecific signal exchange. As a result, the design of auditory organs is often a compromise determined by the relative importance to the animal of each of these potential hearing tasks. In addition, while an ideal ear would identify source locations accurately, and operate equally efficiently in near and far fields, over all frequencies and amplitudes, and with unlimited temporal and frequency resolution, the relevant physics makes such simultaneous perfection impossible. There are always trade-offs in which increasing scope or resolution for one type of acoustic data results in a decrease in the scope or resolution of another. The results are ears that are good at some measurements and poor at others, with different weightings in different taxa. In the remainder of this chapter, we examine the causes and consequences of a number of these compromises.

Fortunately, most of the principles and trade-offs outlined in Chapter 2 for sound production apply, in reverse, to sound reception. Just as small animals are limited to wavelengths smaller than their bodies when using horns or resonance structures to modify and radiate sounds, the same constraints apply to using horns or resonant structures to capture and amplify a sound signal. This ability to replace senders with receivers and vice versa, (at least within linear systems), is called the **reciprocity theorem of acoustics** [112], and it will simplify our task in the remainder of this chapter. Reversing our sequence from Chapter 2, we can break down the process of sound-signal reception into three successive tasks: (1) the coupling of sound in the medium to the receiver; (2) modification (including filtering, amplification, and impedance matching); (3) and analysis (characterization and interpretation of the perceived signal's temporal and frequency patterns).

Coupling between medium and the receiver

The auditory organs of animals have two major design constraints. First, all receptors used to detect sounds are derived from generalized mechanoreceptors. As a result, animals can detect a sound signal only if it moves some body part relative to the rest of the receiver's body. As we saw in Chapter 2, the energy in a propagating sound is divided between locally coordinated movements of molecules and local variations in pressure. In the near field close to a sound source, the contribution of molecular movement is dominant, and the propagating medium will exhibit a cohesive tidal surge back and forth. If a receiver can insert a flexible hair or filament into this tidal oscillation, and not have its entire body also move in concert, the differential movement of the hair and the body can be used to detect and even track the direction of the passing sound waves. In the far field, molecular movements are no greater than when no sound is present, and the major indication that a sound is passing is pressure variation. Far-field sound detection is thus more complicated, because a receiver will be able to detect passing sound waves only if the variations in pressure can be converted into relative movement of several body parts. Animals have evolved various devices to perform this conversion, and we humans rely on one such device for our own hearing.

The second and related problem is that a receiver's ability to capture a detectable amount of passing sound energy is dependent upon the relative acoustic impedances of the receiver and the propagating medium. If the two impedances are very different, as is the case for most terrestrial animals communicating through air, most of the incident sound energy at the receiver's ears will be reflected. Devices are thus needed to bridge the impedance differences and enhance sound energy capture. If the acoustic impedances of receiver and medium are very similar, as with crustaceans and fish in water, then the opposite problem arises: incident sound is fully absorbed into the receiver, causing its entire body to vibrate. Unless it can devise a way to make one part vibrate out of phase with the rest, it will be unable to detect much less track the incident sound waves. Below, we describe the common solutions to these coupling problems in four different acoustic situations.

NEAR-FIELD SOUNDS   Near-field sound detection poses somewhat different challenges in water and air. In water, the near field for a given frequency extends 4.4 times farther from the sender than it would in air. This makes near-field sound detection potentially more useful in water than in air. On the other hand, because the acoustic impedance of receivers and the medium are so similar, the animal must employ a coupling device that will vibrate in a sound field with a different phase or amplitude from that of the remainder of the body. The coupling device used by most aquatic animals is a fine hair, seta, or cilium that either projects from a sensory neuron or is connected to a sensory neuron. Such hairs are usually thin and stiff but have a flexible attachment to the receiver's body. Many display a fine feathering or other elaboration near the tip to increase friction with a moving medium. As the medium surges back and forth during the passage of a near-field sound, it drags the free tip of the hair with it. This rotates the base of the hair and stretches or compresses the attached sensory neurons. The neurons then send electrical signals to the brain for processing. Note that the hair tip must move either to a greater degree or with a different lag relative to any movements induced in its base and the whole animal by the sound. If all tissues move in concert, there will be no rotation of the hair and no electrical signals. Differential movement is guaranteed by making the mass and inertial properties of the hair significantly different from those of the animal. Since the animal is much larger than the hair, this is usually not difficult. Note also that the hair must be stiff: if it simply bends in the middle as the sound passes, there will be no rotation at the attachment point and no stimulation of the neuron. Aquatic arthropods and cephalopods (squids and octopuses) have many such receptor hairs on the surfaces of their bodies [37, 52–54, 89, 94, 131, 154, 284, 349, 360, 367]. Some are used to monitor the animal's movement and water currents, but others play important roles in the reception of near-field sounds (**Figure 3.25**).

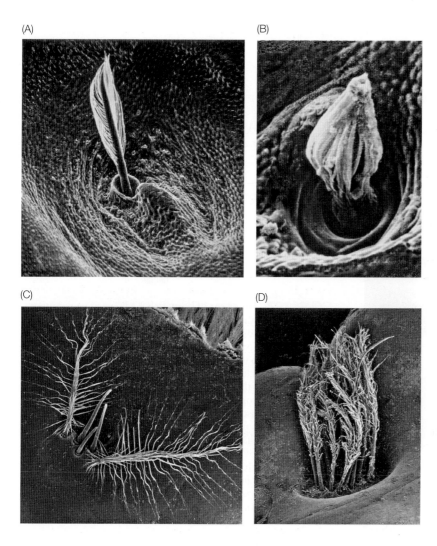

FIGURE 3.25 **Sensory hairs on exoskeletons of aquatic crustaceans** (A) Filiform hairs on carapace of rock lobster (*Palinurus vulgaris*). (B) Hair-fan receptor on carapace of European lobster (*Homarus gammarus*). (C) Hair-pit organs on claws of young crayfish (*Cherax destructor*), and (D) similar organs on adult crayfish of same species.

have interesting dynamic properties that make them useful for additional functions, as well. If the animal changes its velocity or direction, or if its tissues are set into oscillation by an incident near-field sound, the soft tissues respond at a more rapid rate than the statoliths, and as a result of the discrepancy, the sensory hairs are rotated. Whereas exoskeleton hairs described earlier continue to be rotated and produce nerve impulses as long as there is a steady flow of medium over them, statolith hairs remain rotated only until the statoliths can catch up with the rest of the body's new motion. Put another way, free hairs on the animal's surface measure the velocity of the nearby medium relative to the animal whereas statocysts monitor acceleration of the animal's tissues. Given their dynamic properties and the transparency of most aquatic animals to ambient sound, statocysts can easily serve as effective couplers and detectors of near-field sounds. It has been suggested that statocyst reception may account for the widespread startle response of cephalopods to sudden underwater sounds [54, 131]. Statocysts are also the likely receptor organs for conspecific sound signals in many marine crustaceans [53, 109, 143, 198, 199, 273, 284, 323].

Sense organs called **statocysts** are a variation on the same theme. These are typically fluid-filled capsules in which heavy masses called **statoliths** (sand, calcareous secretions, or a protein matrix) sit over the tips of an array of sensory hairs (**Figure 3.26**). Statocysts are widely employed by animals to distinguish up from down using gravity [50]. They

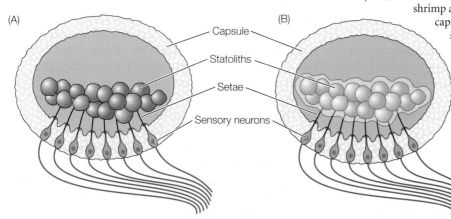

FIGURE 3.26 **Crustacean statocysts** These organs can be located at the base of the antennules (in crabs, lobsters, and crayfish), on the head (in amphipods), or on the tail (in mysid shrimp and isopods). They consist of open or closed capsules embedded in the exoskeleton. The capsules of crayfish and lobsters contain many small statoliths (A), whereas those of most other taxa host a single amalgamated mass (B). The statoliths sit on hairs (setae) that can rotate in their sockets just like the exoskeleton hairs shown in Figure 3.25. When the statoliths and the rest of the body move with different accelerations, the base of the hair moves relative to its socket, stimulating attached sensory neurons. The number of hairs per statocyst can range from 3 in isopods to over 400 in lobsters.

(A)

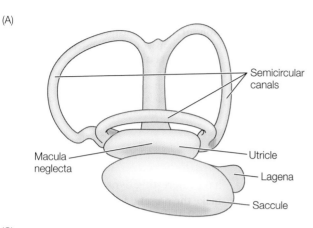

(B)

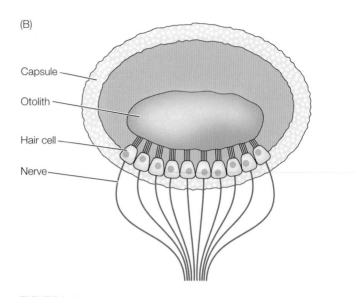

(C)

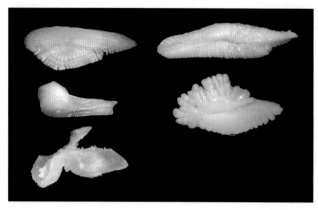

FIGURE 3.27 **Inner ear of bony fish** (A) One of paired inner ears from inside the head of a bony fish. The three orthogonal semicircular canals monitor movements in the three dimensions. Most fish rely in part on the saccule and lagena for hearing, and some species also use the utricle or macula neglecta. (B) The saccule, lagena, utricle, and macula neglecta each have a structure similar to that of the statocysts of Figure 3.26: a fluid-filled cavity is lined with sensory cells (called hair cells) and a single large otolith. Whereas crustacean statoliths rotate setae that in turn stretch or compress sensory cells, a fish otolith directly contacts the kinocilia and stereocilia of the hair cells. Hair cells do not have axons, but are innervated by auditory neurons. (C) Sample otoliths from different bony fish species (left column from top): New England hake (*Merluccius bilinearis*), morid cod (*Halargyreus johnsonii*), and opah (*Lampris guttatus*); (right column from top) white hake (*Urophycis tenuis*), and tilefish (*Lopholatilus chamaeleonticeps*).

Since statocysts are present in larvae of both invertebrates and fish, they may account for reports that pelagic larvae of many marine species find suitable settling sites by moving toward loud sound sources [248].

Most fish also use a statocyst design for the paired ears in their heads (**Figure 3.27**). In addition to three semicircular canals used to monitor a fish's movement in each of three directions, a typical fish ear contains three adjacent capsules (the **saccule**, the **lagena**, and the **utricle**), each of which has hair cells and a calcareous statolith (called an **otolith** in fish) with a density at least three times that of surrounding tissues. Sharks, rays, and some bony fish have a fourth capsule called the **macula neglecta**, in which a gelatinous mass replaces the calcareous otoliths of the other organs. In the majority of bony fish, the saccule and lagena appear to be the main acoustic sensors, and the utricle functions as a gravity monitor [104, 182, 200, 283, 378]. However, some catfish use the utricle for low-frequency hearing [280], and several herring species use it to detect the ultrasonic sounds of predatory cetaceans [287]. Sharks and rays rely on a combination of inputs from their enlarged macula neglecta and one or several of the other three organs [75, 76, 206].

In air, the near field for a given frequency ends very close to the sender, and the large impedance difference between the receiver and the medium seriously limits the transfer of energy into any sound-receiving organ. Despite these constraints, many terrestrial arthropods rely on surface hairs to detect near-field sounds (**Figure 3.28**). The body hairs of some caterpillars can detect the flight-generated airflow of hymenopteran predators at up to 70 cm [347]. The legs of spiders have arrays of specialized sensory hairs (**trichobothria**) that respond to nearby air currents. Away from a surface, detection ranges are limited to 15 cm; but close to surfaces, where near-field airflows are more channeled, ranges of 50–70 cm are possible [20]. Crickets and cockroaches have dense arrays of sensory hairs on their cerci that are highly sensitive to any type of air current [55, 136, 158, 326]. Female flies, male mosquitoes and midges, and worker honeybees have elaborate and often plumose antennae (**Figure 3.29**). Each antenna is attached to a mass of sensory cells (over 30,000 in male mosquitoes) called a **Johnston's organ** [23, 166, 234, 304]. These organs act as receivers of conspecific near-field sounds over distances of several body lengths, and usually have properties such that resonant modes of the antenna match species-specific sound frequencies.

Some terrestrial vertebrates also appear to communicate with near-field sounds. Elephants produce group coordination calls of very low frequency. Members of a typical herd are often well within the 20 m near-field limit of each others'

(A)

(B)

(C)

(D)

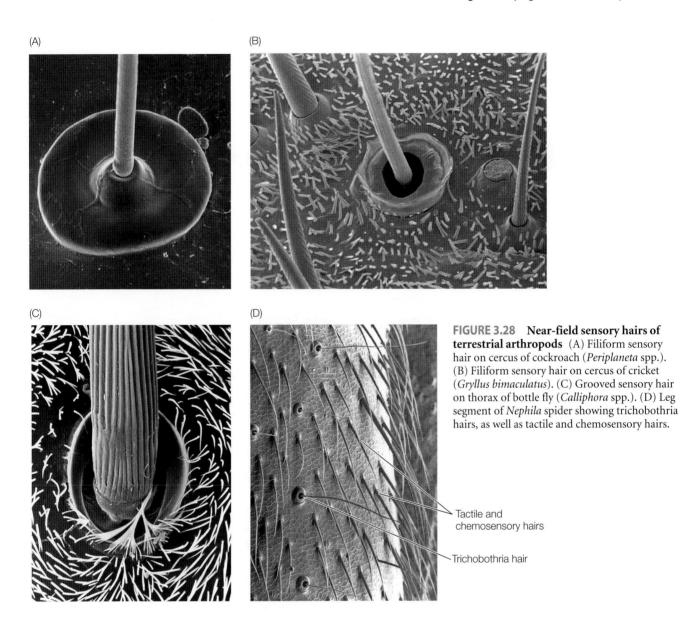

Tactile and
chemosensory hairs

Trichobothria hair

FIGURE 3.28 Near-field sensory hairs of terrestrial arthropods (A) Filiform sensory hair on cercus of cockroach (*Periplaneta* spp.). (B) Filiform sensory hair on cercus of cricket (*Gryllus bimaculatus*). (C) Grooved sensory hair on thorax of bottle fly (*Calliphora* spp.). (D) Leg segment of *Nephila* spider showing trichobothria hairs, as well as tactile and chemosensory hairs.

sounds. Although human ears do not respond to these low-frequency sounds in the far field, one can feel induced body vibrations when close enough to communicating elephants. It appears likely that elephants also use these near-field effects to exchange signals when sufficiently close together [275, 299].

FAR-FIELD SOUNDS IN AIR All terrestrial animals with ears use some form of thin membrane, called a **tympanum**, to couple airborne far-field sounds into their ears. There are two basic designs for tympana. In the first design, called a **pressure detector**, the tympanum is stretched over a closed cavity containing air at the same pressure as the silent ambient medium (**Figure 3.30**). Because the membrane is very thin and pliable, the combined acoustic impedance of the membrane and the air cavity is not very different from that of the medium; the similarity allows some incident sound energy to be absorbed. When incident sound waves raise the air pressure outside the

(A)

(B)

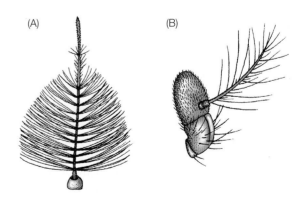

FIGURE 3.29 Near-field antennae of insects (A) Plumose antenna of male mosquito used to detect wingbeats of nearby female. (B) Antennal arista of female fruit fly used to detect near-field song created by courting male's wings.

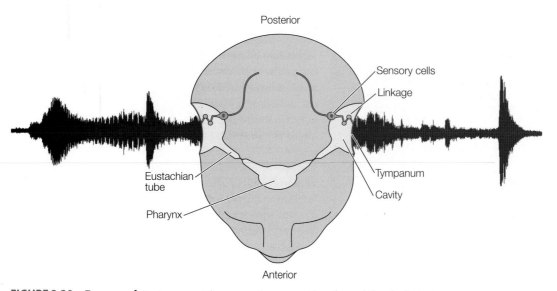

FIGURE 3.30 **Pressure detector ears** Diagrammatic cross section through head of a typical mammal with a pressure detector ear on each side of the head. Each ear is structurally and dynamically independent of the other. A thin tympanic membrane is stretched over a closed cavity in head. This membrane is bent inward when ambient sound pressure is higher than the resting pressure inside the cavity, and outward when outside pressure falls below inside level. A structural linkage (here, three articulated bones) connects movement of the tympanum to sensory cells in the inner ear. Each ear cavity connects by a Eustachian tube to the pharyngeal cavity. These tubes are normally closed during hearing but can be forced open to equilibrate cavities with ambient air pressure during silent periods.

membrane above that inside the cavity, the membrane is bent inward, and when the outside pressure falls below the cavity value, the membrane is bent outward. Incident sound waves thus create a synchronized inward and outward oscillation of the tympanum that can then stimulate attached motion sensors. Pressure detection works only if the tissues surrounding the cavity do not absorb enough sound energy to vibrate in concert with the tympanum: if they did, there would be no relative motion to stimulate the sensors. Given the large difference between the acoustic impedances of air and most animal tissues, this condition is easily met. Note that sensors attached to a tympanum change its impedance. As we shall discuss later, the sensors and attachments must minimize any increased reflection at the tympanum while transferring some sound energy to the sensors. Mammals typically have one pressure detector tympanum (eardrum) on each side of the head. Tympanal pressure detectors are also found in mantids, moths, lacewings, and beetles, in which they are largely used to perceive echolocating bats [16, 381, 382].

The second design also relies on differences in pressure on the two sides of the tympanum to bend it inward or outward. But here, the cavity behind the membrane is not closed: it is connected to a tube whose other end is exposed to the medium at another point on the body (**Figure 3.31**). Such a **pressure differential detector** allows the animal to sample the impinging sound field at two locations. When the pressure outside the tympanum is higher than that at the tube opening, the tympanum is bent inward, and when the reverse is true, it bends outward. Pressure differences on the

two sides of the tympanum can arise because one side experiences higher peak values than the other, or even if peak pressures are similar, because the two samples are out of phase. This design is widely found in insects, frogs, and birds. For example, locusts and grasshoppers have a tympanum on each side of the anterior part of their abdomen. Behind each tympanum is an air space bounded by a thin tissue wall. This wall abuts two air sacs in the center of the abdomen. The tissue walls for the cavity and air sacs are sufficiently thin and flexible that they propagate low- and intermediate-frequency sounds easily. Within this frequency range, each tympanum acts as the second source of pressure sampling for the other [246]. A functionally similar situation occurs in cicadas, which have a large central air sac acoustically coupled to the tympana on each side of the body [115]. Frogs, lizards, and birds have equivalent cavities connecting the air spaces behind the tympana on the two sides of their heads, and frogs may receive additional far-field input through the vocal sac and body wall [61, 63, 169, 176, 195, 255].

Katydids and crickets have two tympana on each foreleg. In crickets, the posterior membrane appears to be the major sound coupler [187], whereas in katydids, both membranes can respond to incident sound [244, 246]. The air space behind the tympana in each leg connects to a tracheal tube that winds through the upper leg and opens to the air through a spiracle on the same side of the thorax. In the katydid, each tympanum thus samples the sound field at two points. In crickets, however, the tracheal tubes from the left and right forelegs abut in the middle of the body before

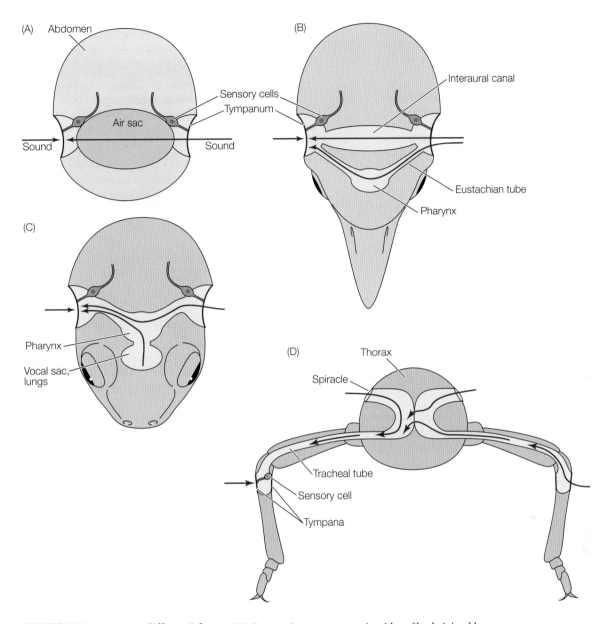

FIGURE 3.31 Pressure differential ears (A) Tympanic ears on opposite sides of body joined by one or more air sacs in center (grasshoppers, cicadas). Arrows show paths delivering sound to each side of left-hand tympanum. The same pattern occurs at the other ear. The tympanum moves when the pressures or phases of sounds on the two sides of its surface are different. (B) Tympanic ears on opposite sides of head connected by an interaural canal (most birds) or open Eustachian tubes and the pharynx (lizards and some birds). (C) Tympanic ears on the head connecting through large open Eustachian tubes and the pharynx with additional sound paths from the lungs, vocal sac, and body wall (frogs and toads). (D) A pair of tympana (one anterior and one posterior) on each foreleg of a cricket. The posterior tympanum is usually the responding structure. Sound pressure at its external surface is compared to that delivered by a tracheal tube to the inside surface. This internal pressure includes inputs from a pair of spiracles on each side of the thorax, and from the tympana of the opposite foreleg. Katydids have a similar system, except that the two tracheal tubes do not meet, and thus pressures from spiracles or tympana on one side of the body cannot be compared to those to the opposite side.

turning to the spiracle. The membrane separating the two tracheal tubes at this point is transparent to lower frequency sounds. In principle, a cricket tympanum could thus receive samples of ambient sound pressures from four locations: (1) just outside the tympanum; (2) just outside the spiracle on the same side of the body; (3) just outside the spiracle on the opposite side of the body, and (4) just outside the tympana on the opposite foreleg. In practice, the input from the

FIGURE 3.32 Tympanal ears of water boatman (Corixidae, Hemiptera) (A) A submerged water boatman with a trapped air bubble around its body (arrows) for breathing. The air bubble acts as a resonant impedance matching device that converts water-borne sounds into airborne sounds. Normal modes of bubble range from 1.4 kHz when the bubble is fresh to 3.4 kHz as the bubble shrinks due to the water boatman's breathing. (B) External view of one of the paired tympanal ears on the thorax of the boatman. The external surface of the ear is always covered by the air bubble. A clubbed structure sits on top of the tympanum. (C) Section through one tympanal ear, showing shape of club and air space inside the animal under the tympanum. Sounds captured by the air bubble cause the tympanum to vibrate. This vibration in turn induces the club arm to rock and feed back on the tympanum to create strong resonant behavior. The right and left ears have normal modes at slightly different frequencies, presumably to provide sensitive hearing over the full range defined by the changing resonances of the air bubble. (B,C after [290].)

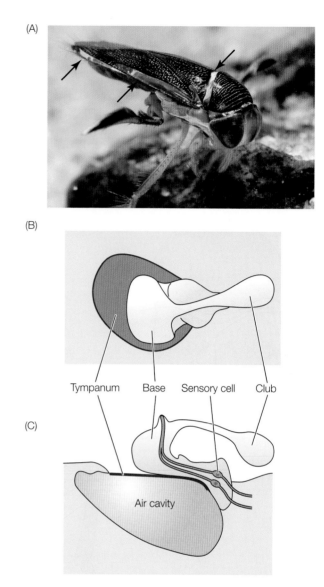

opposite tympana is negligible, but the other three sources contribute jointly to tympanum movement and direction [246]. Cockroaches do not have tympana, but may use tracheal tubes to convey airborne sounds to sensory organs inside their legs [324, 325]).

FAR-FIELD SOUNDS IN WATER   Particle velocities in an aquatic far field are usually too small to be detected directly by motion sensors. Receivers thus turn to large surfaces that can concentrate the sound pressure variations and convert them into localized tissue motion. The usual solution is to use an air bubble to capture waterborne sounds and convert their high pressure/low displacement motion into low pressure/high displacement motion that can be detected. Aquatic insects called water boatmen use air bubbles trapped around their bodies both for respiration and as sound-capturing mechanisms (**Figure 3.32**) [290, 291]. The more common strategy is to use an air bubble inside the animal. Because air is more compressible than water, sounds propagating through the wet tissues of the animal cause the bubble to compress and expand in synchrony with the rise and fall of pressure [133]. If a thin membrane around the bubble is connected to motion sensors in the ear, the aquatic animal will be able to hear far-field sounds. Since the far field for high frequencies begins much closer to a sound source than that for low frequencies, the linking of ears to air bubbles effectively increases the maximum frequency detectable by an aquatic receiver far from the sound source.

We have already discussed how bony fish use gas-filled swim bladders to maintain buoyancy and, in some species, produce communication sounds. Given the reciprocity principle for sound, it should not be surprising that some fish utilize the system in reverse by converting far-field pressure variation into synchronized expansions and compressions of the swim bladder. If this motion of the swim bladder membrane is coupled to the existing auditory organs, the differential motion induced in the hair cells and otoliths will generate electrical signals in the sensory cells as with near-field sounds.

A number of fish taxa have evolved linkages between their swim bladder and their ears (**Figure 3.33**). The effectiveness of these linkages can be measured by comparing acoustic sensitivity before and after deflation of the swim bladder or removal of the linkages [181, 385]. In some cod (Gadidae), squirrelfish (Holocentridae), and croakers (Sciaenidae), the swim bladder has a pair of tubular extensions that terminate very close to the saccule on each side of the head [182, 270, 283, 296]. In each of these fish, the linkages provide enhanced sensitivity to high-frequency and far-field sounds. Many members of the herring family (Clupeidae) exhibit similar extensions, but each of these terminates in one or a pair of air-filled bullae that abut the utricle in each ear. These structures provide sensitivity to very-high-frequency far-field sounds such as the echolocation pulses of foraging dolphins and porpoises [144, 214, 287]. Otophysans (including goldfish, minnows, zebrafish, piranhas, characins, neotropical electric fish, and catfish) have a chain of four small bones (called **Weberian ossicles**) on each side of the body that

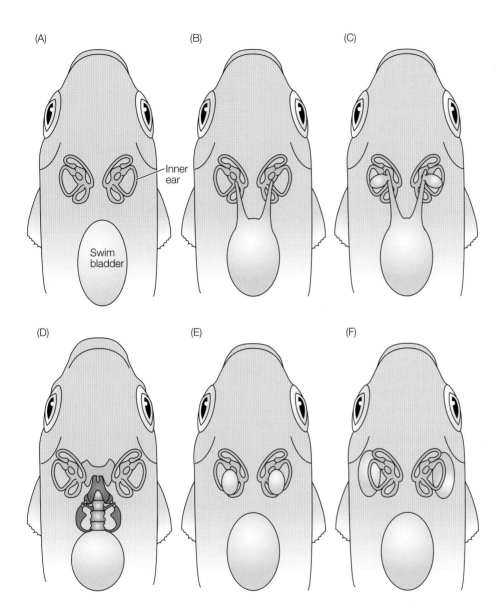

FIGURE 3.33   Role of swim bladder and other air bubbles in fish far-field hearing (A) Fish such as toadfish and sunfish lack any direct linkages between the swim bladder and the ears except intervening tissues. They are largely limited to near-field sounds. (B) Cod, squirrelfish, croakers, and drums have air-filled extensions of the swim bladder that terminate near the saccule in each ear. They can hear far-field sounds and much higher frequencies than toadfish. (C) Some herring and shad have swim bladder extensions similar to those in (B), but each of these terminates in one or more air-filled bullae that are linked to the nearby utricle. This combination makes them highly sensitive to the ultrasonic echolocation pulses of hunting dolphins. (D) Swim bladder of typical otophysan fish (goldfish, minnows, catfish) is connected to two parallel chains of 1–4 Weberian ossicles (blue). These chain ossicles use leverage off the backbone (grey) to transmit swim bladder vibrations to a perilymphatic sac (red). Latter in turn conveys vibrations to single endolymphatic sac whose fluids connect directly to each saccule. (E) African electric fish (Mormyridae) lack any special links to swim bladder, but each ear contains an air-filled sac just adjacent to the saccule. (F) Anabantoid fish such as gouramis also lack any links between their swim bladder and their ears. However, they are air-breathing and have special chambers for holding air bubbles for respiration laterally to each inner ear. The saccule is separated from the underlying air bubble by at most a thin epithelium or sheet of bone

links the swim bladder to the ear on the same side [283, 378]. This mechanism has resulted in some of the most sensitive and widest bandwidth auditory systems known in fish [200]. Eels lack both Weberian ossicles and extensions of the swim bladder, but some species still possess very sensitive far-field hearing. It has been suggested that the backbone may serve as the critical linkage in this group [320].

Other fish taxa use air bubbles that are separate from their swim bladders. Gouramis (Anabantoidei) are air-breathing fish that carry a small bubble of air in a suprabranchial cavity

FIGURE 3.34 Section through ear of toothed (Odontocete) whale Intermediate- and high-frequency sounds are captured by a thin "pan bone" section on each side of the lower jaw and conducted through special oil channels to the joints between the jaw and the skull. Sounds delivered by this route then propagate a short distance through soft tissues to the nearby ear. The ear consists of two loosely connected clam-shell like bones, the periotic bone on top and the tympanic bone below. These are suspended by ligaments in an air-filled cavity. Incident sounds cause a thin portion of the tympanic bone called the tympanic plate to vibrate near its junction with the periotic bone. A chain of middle-ear ossicles then modifies this movement and transfers it to the inner ear located in the periotic bone. Baleen (mysticete) whales have a generally similar structure except that the jaw does not appear to conduct sounds and the periotic-tympanic complex is less acoustically isolated from the skull. It is thought that the low-frequency sounds used by mysticetes are captured by the skeleton and conducted directly to the ears. (After [162, 265].)

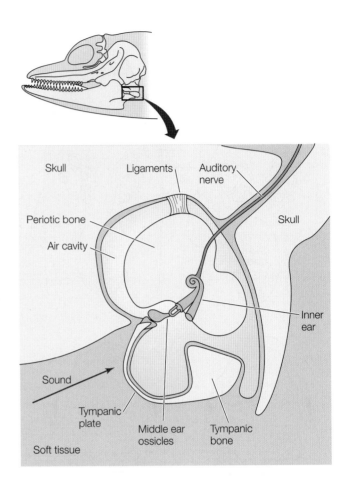

lateral to each ear. The vibrations induced in this bubble by ambient sound pressures are communicated through a very thin skeletal wall and into the saccule [178, 316]. African electric fish (Mormyridae) have a small air-filled bladder adjacent to the saccule of each ear [81, 111]. These air bubbles increase the sensitivity of the ear by 30 dB over the unaided range of these fishes' hearing (200–1200 Hz).

Among the fishes that produce sound signals, many, such as toadfish, damselfish, gobies, sculpins, and cichlids, lack elaborated coupling structures [180, 182]. Similarly, there are many fish with sensitive hearing that do not produce sound signals [179, 180, 182]. One explanation for the poor correlation between sound communication and hearing ability in fishes is that species with minimal adaptations do not need to communicate over long distances and thus can exchange sound signals with simpler ears. Elaborate ears in species with minimal sound communication may arise where hearing is important for finding prey or avoiding predators. In addition, several authors have noted that fish with the most sensitive ears tend to be found in acoustically quiet environments [5]. Thus the levels of ambient noise may be a better predictor of ear structure in fish than a reliance on acoustic communication.

The ears of whales (Cetacea) also require air spaces to mediate reception of underwater far-field sounds (**Figure 3.34**). Toothed whales and porpoises (Odontocetes) utilize high-frequency sound to echolocate their prey (see Chapter 14). Sounds focused and emitted through the head melon return as echoes and are captured by wide and very thin portions on each side of the lower jaw called "pan bones" [10, 262]. Sounds are then transferred between each pan bone and the corresponding ear by special fat- and wax-filled channels in the lower jaw [141, 171, 249, 262] The channels function as wave guides. One group of odontocetes, the beaked whales (Ziphiidae) may use both the jaw and a mass of fatty tissue in the throat to conduct sound to the ears [79, 80]. The odontocete ear is a clam-like structure, with a periotic bone on top and a tympanic bone below. It lies in a skull cavity

at the base of the lower jaw, surrounded by air-filled foam. The ear is not in direct bone contact with the skull, but is suspended from it by ligaments [162, 163]. This configuration prevents sound from entering the ear by bone conduction through the skull. The inner ear with its sensory cells is lodged in the periotic bone. The cup-shaped tympanic bone is loosely linked to the periotic bone on its top side, thick and dense on its bottom and internal sides, and very thin on the side opposite the end of the lower jaw. This thin portion of the tympanic bone is called the tympanic plate. Three articulated bones inside the air-filled tympanic cavity link the upper edge of the tympanic plate to the inner ear. They are positioned so that they rotate back and forth in a small arc around their center of gravity when the top of the tympanic plate moves in and out. Sound waves propagating from the jaw channels into nearby wet tissues exert oscillating pressures on the thin tympanic plate. Because the tympanic cavity is filled with air, it is compressible, and the top of the tympanic plate vibrates in synchrony with the varying sound pressures. The bottom of the thin wall is fixed to the heavy parts of the tympanic bone and acts as a hinge. These vibrations of the tympanic plate are then transferred to the inner ear by the chain of middle ear ossicles [138, 139, 265, 266].

Baleen whales (mysticetes) include the largest animals on the planet and are well known for their production of extremely low-frequency communication sounds [64, 97,

236, 350, 351, 366]. Coupling of ambient sounds to their ears is poorly understood. One suggestion is that parts of the skeleton are sufficiently similar to water in acoustic impedance that some incident sound energy is absorbed by the skeleton and conducted through the bones to the ears. Most mammals minimize bone conduction of absorbed or body-generated sounds (e.g., chewing) to their ears. One way to do this is to ensure that the primary point of articulation between the middle ear bones linking the tympanum to the inner ear is also the center of gravity of those bones [18]. Any vibrations conducted to the ear by the skeleton would then move all parts in synchrony and there could be no stimulation of the inner ear. Mysticete whales, true seals (Phocidae), burrowing mammals, and elephants all have very heavy middle-ear bones in which the articulation point is not the center of gravity [137, 140, 162, 163, 223, 264, 299]. The inner-ear capsule and the middle-ear bones might then move differentially in these animals, which allows the marine mammals to receive bone-conducted low-frequency vibrations from the water and the terrestrial animals to receive them from seismic vibrations.

SURFACE PROPAGATED SOUNDS   A variety of animals detect sounds by monitoring vibrations at the surface between two media. Insects and spiders living on plants are very sensitive to oscillations in the plant surface caused by internal bending waves [20, 68, 69, 71]. Many animals that live on or burrow in the ground can detect Rayleigh waves propagating on the surface of the substrate [46, 257, 258]. Water striders and spiders attend to ripples on the surface of the water, and orb spiders monitor both longitudinal and transverse vibrations of the strands in their webs [20, 29, 228, 368].

These species sense the motion of the surface and not the pressure variations inside either of the media. Waves on the water's surface attenuate rapidly and become distorted due to dispersion of the component frequencies. This limits the range of useful detection and characterization of water surface signals to under one meter [19, 368, 369]. Bending waves in plants are also dispersive, but attenuation is a bit less, and both spiders and insects living on the surface may be able to detect signals from up to several meters from the source [20, 164, 242]. Elephants appear able to detect very low-frequency Rayleigh waves on the Earth's surface at distances of kilometers [7]. In all of these contexts, low frequencies attenuate less rapidly than high frequencies, and most animals communicating at surfaces favor the lowest frequencies that their bodies can detect. Although sound speeds in water and solids are much higher than those in air, those on the surface of water or the ground are generally less than those in air [112]. For example, signal propagation on the water's surface is usually less than 1 m/sec, and Rayleigh waves in soil and bending waves in plants both propagate at speeds of 50–200 m/sec [46, 112, 242, 258]. As a result, signal propagation is slower on surfaces than within large volumes of air or water.

A receiver of surface signals must first couple the particle movements of the boundary into its body. Most species use their legs as the coupling devices [19, 68, 71], although whirligig beetles use their antennae [170], and some fossorial rodents lay their heads against the substrate or even bury them in the ground [256]. Legs have the advantage that they are jointed and held in position by adjustable muscle tension. This tensioned flexibility of the legs and the mass of the attached body make the entire animal a resonant device that can amplify the small particle movements of the boundary over successive cycles [112]. Spiders, fiddler crabs, and treehoppers can fine-tune the normal modes of this resonance by adjusting their posture and muscle tension [2, 3, 19, 66].

Once coupled into the body, two or more parts of the body must respond to these vibrations by moving out of phase so that auditory sensors can be stimulated (**Figure 3.35**). Spiders, scorpions, and their relatives have large numbers of **slit organs** embedded in their exoskeletons [20, 21]. These slits are V-shaped in cross section and particularly abundant near leg joints. Compression of the legs due to induced vibrations causes the openings of the slits to close, and this triggers motion sensitive cells located at the base of each slit. Spiders also have hairs that span leg joints and monitor leg flexure. Many insects have pit-like **campaniform sensilla** on their exoskeletons that function similarly to the slit sensors of spiders [302, 352]. In addition, internal **chordotonal organs** monitor the flexing of insect leg joints, and **subgenual organs** monitor differential rates of acceleration of the exoskeleton and the fluid hemolymph inside each leg [382]. All of these organs have been implicated in the detection of substrate vibrations [71, 90, 361]. In frogs, substrate vibrations absorbed by the body pass through the pectoral girdle and along a muscle that connects to the **operculum**—a moveable window—on the wall of the inner ear. Like fish, frogs have a saccule, and the vibrations transferred to it by the opercular movements allow the frog to monitor seismic sounds [258]. Like insects, most terrestrial mammals can detect some substrate vibration using existing flexure and touch receptors. Elephants and fossorial moles and rodents have evolved middle-ear bones that are much more massive than expected, given their body sizes. As noted earlier, these species appear to rely on differential inertia of the middle-ear bones and the rest of the body to stimulate the inner ear with vibrations absorbed from water or solid substrates and conducted by the skeleton to the ears [223, 225, 227, 263, 264, 299].

Modification of captured sound signals

The primary modification tasks undertaken by animal ears are amplification, filtering, and impedance matching. Note that these are similar to the same tasks required for sound radiation (see Chapter 2). Since these three processes are invariably linked, we outline below some of the known designs and discuss how they affect each of these tasks.

A number of terrestrial animals gather sound waves in the large open end of a horn and deliver it in an amplified form to one side of a tympanum. Horns also facilitate tympanic responses to incident sounds by providing a more gradual

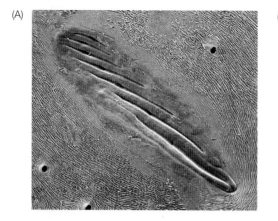

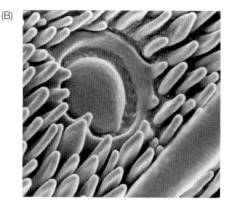

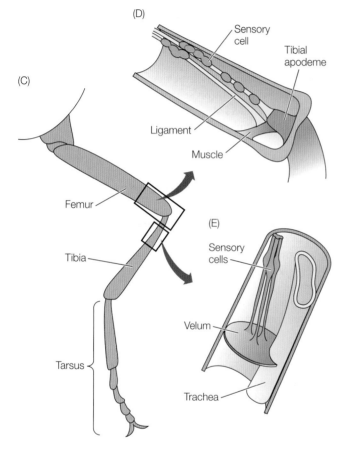

FIGURE 3.35 **Substrate vibration receptors of arthropods** (A) Lyriform organ consisting of multiple slit organs on the leg of a spider (*Cupiennius salei*). (B) Campaniform organ on antenna of silk moth (*Antheraea polyphemus*). (C) Diagram labeling segments of typical insect leg. (D) Section through lower femur of stink bug (*Nezara virdula*) showing femoral chordotonal organ. The organ includes two ligaments: one attached to the tip of the tibia at one end and the side of femur at the other, and the second to a muscle that flexes the tibia relative to the femur. The first ligament contains several sensory cells along its length and at its base on the femur wall. The second ligament has sensory cells only at its base. Both ligaments are stretched and relaxed as the tibial segment of the leg moves relative to the femur. (E) Section through the upper tibia of a green lacewing (*Chrysoperla carnea*), showing the subgenual organ. The membranous velum spread across the tibial cavity detects inertial movements of fluid hemolymph relative to the exoskeleton of the tibia. Sensory cells attached at their base to the tibial wall and by long dendritic filaments to the velum are stretched or relaxed as the velum moves differentially from the tibial wall. (D after [238]; E after [382].)

match between the low acoustic impedance of the open air and the higher impedance of the tympanum. The tracheal tubes used by katydids and crickets to sample sound fields act as horns with the large opening at the thoracic spiracle and the small opening just behind the tympana in each foreleg [241]. At the high frequencies typically used for conspecific communication, the trumpet-shaped tracheae of katydids can amplify the sound pressures delivered to the tympanum by nearly tenfold [246]. A V-shaped entry slit along the ventral midline of praying mantids provides a 1.8-fold amplification of incident sounds for the two tympana lining its walls [384]. While frogs and toads have tympana on the outsides

of their heads, reptiles, birds, and mammals all have their tympana at the end of a tapering canal inside their heads. This canal acts as a horn and can produce surprisingly significant amplification. Mammals extend the large opening of each canal with a horn-shaped **pinna** [312], and some birds achieve a similar effect by positioning special feathers around their canal opening [74, 176]. Note that all horns are resonant structures with a limited bandwidth: low cutoff frequencies are determined by the diameter of the large opening, and high cutoff frequencies depend on the length and shape of the horn [112]. The relative size, shape, mobility, and placement of the pinnae vary widely among mammals and can significantly affect their acoustic properties (**Figure 3.36**).

The ancestors of terrestrial vertebrates retained the fluid-filled inner ear of fish when they moved onto land. This created a major mismatch of acoustic impedances: how could the high displacement and low pressure movements of a tympanum exposed to airborne sounds be transformed into low displacement and high pressure fluid movements inside the inner ear? Early terrestrial vertebrates acquired several mechanisms that solved this problem [319]. A first step was the addition of two small membranous "windows" in the bony inner ear capsule. Pushing in on the **oval window** caused fluids to move in the cavity and bulge the **round window**

FIGURE 3.36 Alternative types of pinna design in terrestrial mammals (A) Simple ovate pinnae: brown greater galago (*Otolemur crassicaudatus*). (B) Triangular-shaped pinnae completely fixed at base, as found in canids and felids: red fox (*Vulpes vulpes*). (C) Immobile semicircular pinnae fixed to the side of the head, as found in monkeys and apes, mongooses, and some rodents: chimpanzee (*Pan troglodytes*). (D) Pinnae with tubular base allowing independent rotation, as found in many herbivores: black rhinoceros (*Diceros bicornis*). (E) Pinnae with the outer and upper quadrant deleted or notched, as found in some rodents and bats: Patagonian mara (*Dolichotis patagonum*). (F) Highly elongated pinnae, as found in aardvarks, rabbits, and some bats and marsupials: black-tailed jackrabbit (*Lepus californicus*).

Web Topic 3.4 *Levers and ears*
Different types of lever systems provide mechanical advantages and match acoustic impedances for improved hearing in animals. Here we outline the general classes of levers and indicate which are used in the middle ears of amphibians, reptiles, birds, and mammals.

outwards; pulling out on the oval window caused the fluids to move in the opposite direction and bend the round window inward. Another step was the evolution of a mechanical lever system connecting the tympanum to the oval window. This lever converted the large displacements of low pressure at the tympanum into small displacements of high pressure at the oval window. Finally, with the enlargement of the tympanum to a size much larger than the oval window, a large amount of sound energy could be captured by the tympanum and concentrated on the much smaller area of the oval window.

The physics of terrestrial (and secondarily aquatic) vertebrate middle ears has been studied in some detail [1, 59, 73, 135, 137, 148, 203, 208, 223, 226, 264, 266, 268, 312, 319]. As a rule, the relative size of tympanum and oval window is the dominant factor matching airborne to inner ear impedances. Tympana are reported to be 20–50 times as large as the corresponding oval windows for frogs, 10–13 times as large in lizards, 11–40 times as large in birds, and 10–32 times as large in mammals. Although amphibians, birds and reptiles, and mammals have evolved ossicular linkages independently [118, 211] and use somewhat different lever mechanisms (see Web Topic 3.4), the amount of pressure and velocity transformation achieved is fairly similar among the taxa. Typical

(A)

(B)

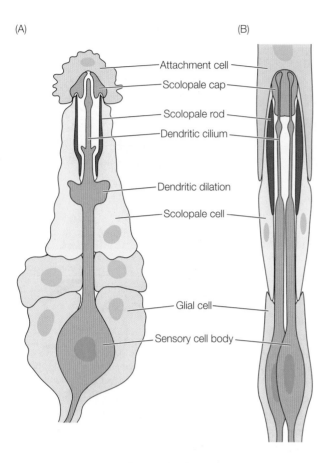

- Attachment cell
- Scolopale cap
- Scolopale rod
- Dendritic cilium
- Dendritic dilation
- Scolopale cell
- Glial cell
- Sensory cell body

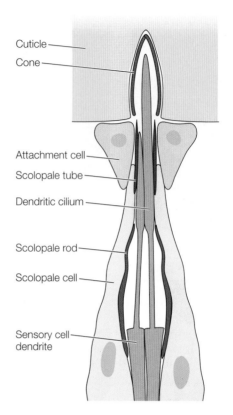

- Cuticle
- Cone
- Attachment cell
- Scolopale tube
- Dendritic cilium
- Scolopale rod
- Scolopale cell
- Sensory cell dendrite

FIGURE 3.37 Insect scolopidial mechanoreceptors In each example shown here, one to several sensory cells extend long ciliary dendrites into a sheath formed by scolopale cells. These contain stiff rods of actin and perhaps other motile proteins. Flexure of the dendrite stimulates the sensory cells which then send nerve impulses down their axons. Details of the structure of scolopidia vary with taxon and function. (A) Scolopidum from sensory organ in ear of a locust. (B) Scolopidum from femoral chordotonal organ of a lacewing. (C) Scolopidum sense organ on mouthparts of a beetle larva. (After [382].)

values for each mechanism show 10- to 50-fold changes in pressure and velocity. Since the effects of concurrent impedance matching mechanisms are multiplicative, acoustic impedances at the oval window can be increased several hundred times over that at the tympanum, and this can significantly improve the transfer of airborne sound waves into the fluids of the inner ear. Note however, that transforming oval window impedances beyond a good match for the inner ear fluids can be as ineffective as no transformation. There is an intermediate level of transformation that is optimal, and this may explain the relatively convergent values seen in the different vertebrate taxa.

Like a horn, the middle-ear complex of higher vertebrates has its own resonances and can be inefficient at frequencies significantly different from its normal modes [137, 264]. Many species have their middle-ear resonances matched to the sound frequencies of greatest importance. Elephants, mysticete whales, true seals, and burrowing rodents and insectivores that respond to low-frequency signals all have larger and denser ossicles than might be expected for their body sizes. Echolocating bats have their ears tuned to ultrasonic frequencies. Parallel adjustments must, of course, be made in the inner ear, and it is not always clear whether the middle ear or the inner ear plays the final role in limiting auditory bandwidths [314]. Other constraints, such as isolation of the inner ear from bone-conducted sound in most mammals [18] or the use of the eardrum as a sound radiator

in bullfrogs [292], may compromise the degree to which middle-ear resonances can be matched to behaviorally salient sound frequencies.

Detection and analysis of received sound signals

DETECTION   Once sound energy is extracted from the propagating medium and modified as necessary, stimuli are ready for detection. The sensory cells used for vibration detection are remarkably uniform within each of two major sound-signaling taxa (arthropods and vertebrates). Arthropods use bipolar neurons whose long ciliary dendrite is usually surrounded by a sheath of accessory cells [161]. The ciliary component contains nine pairs of parallel tubules arranged in a circle, but lacks the central pair of tubules of traditional cilia and flagella. In insects, several sensory cells may be clustered into a **scolopidium** (**Figure 3.37**). Scolopidial cells containing stiff scolopale rods form a sheath that encloses the sensory cell dendrites and bathes them in a specialized ionic fluid. The stiffness of the rods is adjustable by the scolopale cells

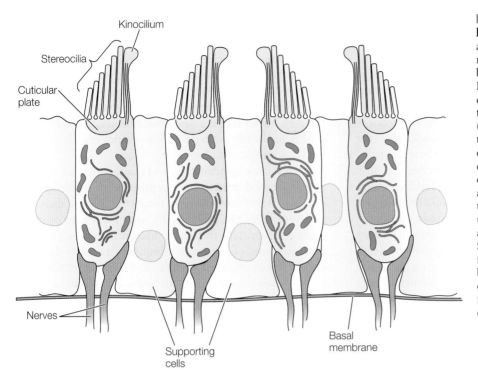

FIGURE 3.38 Vertebrate auditory hair cells Four sensory hair cells that are stimulated when either fluid movements, otoliths, or overlying membranes bend stereocilia toward the kinocilium. Both types of cilia are supported in each cell by a dense cuticular plate. Note that the two hair cells on left are oriented (polarized) differently from the two on the right. By having different hair cells oriented in different directions, the ensemble as a whole can monitor fluid, otolith, or membrane movements in any direction. Hair cells synapse with the dendrites of afferent nerves, which, unlike a typical neuron, are myelinated for rapid propagation like axons. Some hair cells may also receive input from efferent nerves that modulate their behavior and sensitivity. Supporting cells and a basal membrane provide a firm matrix in which the hair cells are distributed.

[382]. Scolopidia are the most common sensory structures within insect hearing organs.

The dendritic sheaths of arthropod mechanoreceptors may be attached to the base of a pivoting sensory hair (all arthropods), inserted into the hollow base of a sensory slit (spiders) or campaniform pit (insects), grouped under statoliths inside a statocyst (crustaceans), or clustered with similar cells into scolopidia to form chordotonal or subgenual organs [20, 382]. The majority of spider and insect receptors are stimulated by bending or compression; crustacean receptors tend to be stimulated by stretching. Arthropod mechanoreceptors can be extremely sensitive: at optimal frequencies, responses to displacements of 1 nm (10^{-7} cm) are common, and the hairs on cricket cerci may trigger at displacements of only 0.05 nm [20, 326, 353]. As a point of reference, the diameters of single atoms range between 0.03 and 0.3 nm!

Vertebrates also use ciliated sensory cells for sound detection, but lacking their own axons, these **hair cells** must synapse with other neurons [70]. Each hair cell hosts a single **kinocilium**, (showing the expected circular ring of nine pairs of tubules around a single pair), and an adjacent bundle of 20–100 **stereocilia** (more appropriately called **microvilli**), each containing thin filaments of the protein actin (**Figure 3.38**). The stereocilia are usually arranged according to height, with the taller ones closest to the marginally positioned kinocilium. The tips of the stereocilia are linked to each other such that motions induced by nearby fluid or tissue vibrations bend the entire bundle cohesively [175]. Bending the bundle toward the kinocilium depolarizes the hair cell, and bending it away hyperpolarizes the cell. Bending the bundle at other angles stimulates the cell in proportion to the fraction of the movement which parallels the kinocilium-midpoint axis. Vertebrate hair cells are thus inherently directional. Electrical changes in the hair cell modulate the impulse rates of the synapsed nerve cells. In many mammals and some birds, the kinocilium disappears in adults, but the directional polarization of the stereocilia responses remains fixed. The inner ears of most vertebrates contain two or more types of hair cell that differ in size, shape, and the number of nerves with which they have synapses. The functions of these different cell types differ in the different vertebrate taxa and will be taken up in the next sections. Vertebrate hair cells are as sensitive as those in arthropods; threshold displacements under 1 nm are common. Goldfish respond down to 0.1 nm, and squirrel monkey hair cells are reported to detect 0.01 nm displacements [283, 306].

FREQUENCY RANGE AND RESOLUTION   The bandwidths of auditory receptors can be maximized by adjusting their size, electrical and mechanical properties, and the ways in which they are linked to the medium or auxiliary structures. The orientation of directional hair cells in vertebrates relative to the passage of sound waves near a receptor organ, and the number and linkages between adjacent stereocilia can also be adjusted to diversify a cell's normal modes. Despite all these options, no single sensory cell is likely to have enough bandwidth to cover the frequency ranges provided by the coupling and modification organs. The solution is to have many cells, each tuned to a different normal mode (called its **characteristic frequency** in hearing organs), which collectively span the range of frequencies needed.

This solution solves another problem. Many animals must identify the frequencies in ambient sounds in order to discriminate between species-specific signals and noise, characterize the pattern in a signal, or use the differential propagation of different frequencies to estimate the location of a signal source. A single sensory cell may be able to depolarize and hyperpolarize in synchrony with the oscillating motion of the sound waves. The impulse rate of the attached nerve will then be equal to (or at least proportional to) the frequency of the sound (a phenomenon called **phase-locking**). If the brain can measure these impulse rates, it will have a reasonable estimate of the sound frequency. However, this only works up to the maximal rate at which nerves can fire (several kHz). In addition, a sensory cell can only track the peaks and valleys of the entire waveform. The variation in the amplitude of a complex sound containing many frequencies may correspond to only one or a few of those components (e.g., harmonics), or it may not correspond to any (e.g., beats). If multiple sensory cells are used, each tuned to a different normal mode, then each component frequency stimulating the sensory cells can be identified. Frequency resolution can be improved by adjusting the mechanical and electrical properties of the sensory cells to reduce their bandwidths. This may require more cells to cover the full frequency range needed, but it will provide improved frequency resolution.

Animal ears that decompose complex signals into frequency components are essentially performing a Fourier analysis of the sound. Like the spectrograms used by researchers, the ear provides the brain with the relative amplitude of each band of frequencies that it is able to extract. Also like spectrograms, ears tend to ignore relative phase information except when component frequencies are so similar that the ear cannot separate them. Then their relative phases may affect the apparent amplitude assigned to that band of components.

In some animals, the frequency spectrum of a sound is estimated by a single organ; in others, it is assembled in the brain using the inputs of many different organs. For example, each hair and bristle sensor, slit organ, or campaniform pit organ on the exoskeleton of an arthropod usually contains only one (insects) or up to four (spiders) sensory cells and thus has a limited frequency resolution for each sensor. However, by varying the physical properties of the attached hair, slit, or pit, the collective input from many sensors provides an animal with a wide frequency range and reasonable discrimination between different frequency components [20, 37, 53, 112, 347, 349]. For sensory hairs and bristles, length is a major property that can be adjusted in this way. In part, this is due to **boundary layers** of medium adjacent to the exoskeleton that are kept from moving in the sound near field due to friction against the exoskeleton surface (**Figure 3.39**). This inert boundary layer is thicker for low frequencies than it is for high ones, and is thicker in air than it is in water for a given frequency. In all media, a shorter hair may respond only to high frequencies, whereas a longer hair that sticks out past the large boundary layer may be the only sensor to detect low frequencies. Evolutionary fine-tuning can be achieved by varying the stiffness of the hair and its base attachment, how

far it can rotate, (often set by placing its base inside a walled socket of limited diameter), and the mechanical and electrical tuning of the attached sensory cells. In air, the cross-sectional area of the hair can also affect its normal modes. While such fine-tuning works well for frequencies below 1 kHz, it is difficult to move any hair or bristle at the velocities required for even higher frequencies. Sensory hairs, like slit and pit organs, are largely limited to low frequency sounds and near field conditions.

Insects use chordotonal organs to monitor flexing, stretching, and other proprioceptive tasks, and subgenual organs to detect movements of hemolymph inside their exoskeletons [108, 234, 382]. Again, multiple organs or organ regions can provide frequency discrimination by assigning separate zones of the overall frequency range to different organs. For example, Hemipteran insects that exchange substrate signals in plants use chordotonal organs in their legs to monitor low frequencies (50–100 Hz), and two subpopulations of subgenual organ scolopidia to separately monitor intermediate (200 Hz) and higher (700–1000 Hz) frequencies [71]. The subgenual scolopidia can detect differential bending of their dendrites as small at 0.1 nm at higher frequencies and are thus very sensitive.

Most terrestrial insects use tympana as the coupling organs for far-field hearing [147, 382, 383]. These may be located on any part of the body, including the thorax (tachinid flies and praying mantis), abdomen (cicadas and locusts), wings (lacewings), air-filled mouthparts (sphingid moths), or forelegs (crickets and katydids). In the majority of species, sensory scolopidia are directly attached to the inner surfaces of the tympana. This is true of grasshoppers, mole crickets, mantids, lacewings, tiger and scarabid beetles, tachinid and sarcophagid flies, moths, and some butterflies. In crickets and katydids, however, the scolopidia are attached along one side to the leg wall and along the other to the surface of the tracheal tube that forms part of their pressure differential ear system. Because this tube acts as a horn, the sound sample that it delivers to the ear is usually much greater than that received at the tympanum. Both the tympanum and the tracheal wall move relative to the leg wall, stimulating the scolopidia.

The number of scolopidia per tympanum in insects varies from 1–2 in noctuid moths up to several thousand in cicadas. The majority of species make do with 30–80 scolopidia. Frequency ranges and resolution seem better correlated with ear function than with numbers of scolopidia. Many insects such as moths, mantids, beetles, and lacewings use their ears to detect and avoid echolocating bats. Their hearing is typically limited to high frequencies (30–60 kHz). Conversely, female trachinid and sarcophagid flies that parasitize male crickets, katydids, or cicadas use their ears to find suitable victims. Their 100-scolopidia ears are mostly tuned to a narrow band of frequencies matching the male calling signals of their hosts, although ears of either sex may also detect bat echolocation sounds [147, 183, 297, 343].

Frequency resolution, but not necessarily frequency range, seems to be a priority for those insects that solicit

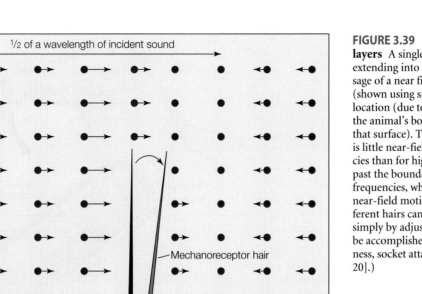

½ of a wavelength of incident sound

Mechanoreceptor hair

Boundary layer

Animal's body surface

FIGURE 3.39 Sensory hairs and boundary layers A single mechanoreceptor hair is shown extending into the adjacent medium during passage of a near field sound. Molecular velocities (shown using size of arrows) vary with horizontal location (due to wave action) and height above the animal's body surface (due to friction close to that surface). The boundary layer, in which there is little near-field flow, is thicker for low frequencies than for high frequencies. Long hairs extend past the boundary level for all but the lowest frequencies, whereas short hairs are exposed to near-field motion only for high frequencies. Different hairs can thus be made frequency-specific simply by adjusting their length. Finer tuning can be accomplished by varying hair thickness, stiffness, socket attachment design, etc. (After [348, 20].)

mates with acoustic advertisement or acoustic duetting. Strategies for achieving frequency resolution vary with the taxon (**Figure 3.40**). The tympana of grasshoppers and locusts are divided into regions of different thickness and acoustic properties. As a result, the normal modes of this surface are quite complex, with different regions being set into maximal vibration by different frequencies. The cluster of 60–80 scolopidia in each ear (called a **Muller's organ**) is connected to the tympanum by three legs of a tripod. Each leg contacts a different region of the tympanum, and thus different parts of the vibrating membrane stimulate different scolopidia in the Muller's organ. Complex sounds are thus broken down into component bands of 3.5–4.1 kHz, 5.5–6.5 kHz, and 16–19 kHz [36, 239, 342, 374]. A similar mechanism may operate in cicadas [345].

The acoustic scolopidia in katydid and cricket ears are arranged in linear rows along the axis of the tracheal tube to which they are attached [222, 244, 245, 341, 382, 383]. The scolopidia differ in shape, size, and electrical properties so that they form a graded series with units responding best to low frequencies at the proximal (close to the body) end, and those responding best to high frequencies at the distal (furthest from the body) end. Whereas some katydids show

a fairly even progression of characteristic frequency along the main axis of their acoustic organs, other katydids and most crickets show uneven progressions. Crickets, for example, have large numbers of scolopidia with characteristic frequencies close to the dominant harmonic of the male calling song. A second and more distal cluster favors the higher frequencies used for other types of calls in the male repertoire. In many species, the organ on its distal end terminates with a third broadly tuned cluster devoted to bat echolocation frequencies.

A sequential ordering of auditory units according to characteristic frequency within a hearing organ is called **tonotopy**. There are several reasons why tonotopy might be favored over more random dispersions of differently tuned sensory cells. Where characteristic frequencies are largely determined by mechanical properties of the sensory cells, adjacent cells with very different resonances may interfere with each other. When cells with similar resonance modes are clustered, this source of interference is reduced [211]. In fact, when adjacent cells have similar characteristic frequencies, there is often increased linkage between them, so that stimulation of one can trigger stimulation in similar cells nearby. In many tonotopic organs, a **tectorial membrane**

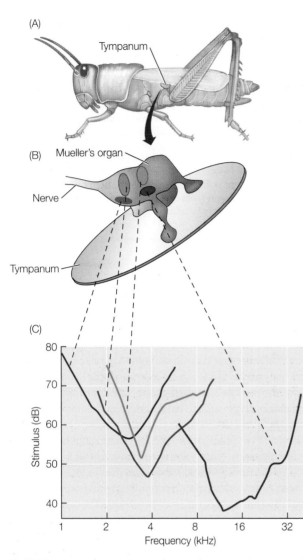

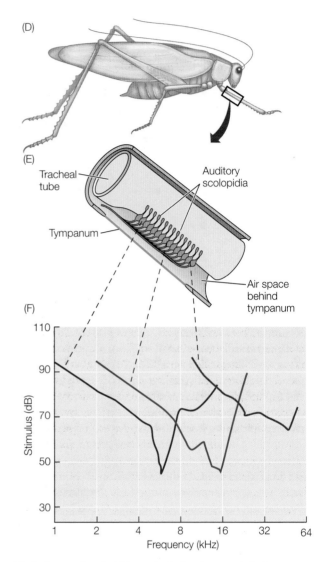

FIGURE 3.40 Frequency resolution devices in locusts and katydids (A) Side view of locust or grasshopper showing location of one of its two tympana just below the wing and on the first abdominal segment. (B) Expanded view from inside locust ear showing Mueller's organ and associated nerve attached to inside of tympanum. The approximate locations within this organ of the four key clusters of scolopidia are shown. (C) Plots of minimal stimulus intensity required to elicit a response versus stimulus frequency for single cells (*not* multiple cell averages) from each of the four key clusters of Mueller's organ scolopidia. The frequency with the lowest threshold intensity is the characteristic frequency of that unit. (D) Side view of katydid showing location of one of its two ears on the tibia of a foreleg. (E) Diagram of the inside of the tibia showing the linear array of auditory scolopidia (called the crista acustica) between the trachea and the membrane lining the space behind a tympanum. A subgenual organ located just above the crista is not shown here. (F) Sample threshold curves for three sensory cells in a katydid crista showing monotonic ordering of characteristic frequency according to location along the array. (B, C,E,F after [243].)

overlays the dendrites or stereocilia of the sensory cells. Because a membrane (unlike a solid statocyst) bends differently in different locations, local clusters of cells with similar frequencies can share stimulation. This makes the cluster of cells more sensitive to low-amplitude signals and allows for entrainment of all adjacent cells to a narrow frequency band [12]. Sound-induced oscillations of bare stereocilia or scolopidia would damp away quickly if exposed only to the ambient fluids [112]. A tectorial membrane provides a larger mass that "rings" long enough to trigger the relevant nerves. An even more effective mechanism for ensuring the sharing of local stimulation in tonotopic ears is overt movement by a sensory cell when stimulated by its characteristic frequencies. Mammalian outer hair cells do this by physically changing shape when stimulated, whereas those in other terrestrial vertebrates and the sensory cells in the Johnston's organs of mosquitoes and *Drosophila* wave their stereocilia or dendrites when excited [107, 123-126, 149, 211, 212]. This physical motion is then communicated to adjacent cells through direct contact or indirectly through a tectorial or other membrane. Such **auditory amplification** requires the expenditure of energy, and because it functions as a nonlinear feedback mechanism, can lead to dysfunctional behaviors such as the spontaneous production of sound by hearing organs. (This is one cause of human tinnitus, or "ringing in the ears" after loud sounds or head injuries). Despite some dysfunctional aspects, auditory amplification is apparently common in both arthropods and vertebrates, and provides a very effective means for improved detection of low-amplitude sounds and more refined frequency resolution.

Web Topic 3.5 *Auditory amplification*
Both arthropods and vertebrates use active motion of auditory receptors to amplify responses to low level sound stimuli. Here we examine some of the data demonstrating this mechanism in insects, mammals, and lizards.

All vertebrates use tuned hair cells for frequency resolution and amalgamations of many cells to insure broad frequency ranges. Most species use phase locking to measure lower sound frequencies. Depending on the species, fish detect and analyze acoustic signals in the saccule, utricle, lagena, and, if present, the macula neglecta. The saccules of most fish have two classes of hair cells with different electrical responses to sound stimulation and in some cases with different shapes and sizes [104, 283]. Each class favors a different portion of the species' frequency range even in toadfish, in which hearing is limited to frequencies less than 300 Hz, and goldfish, in which hearing can range from 50–5000 Hz. Most fish cannot detect sounds above 5 kHz, although as noted earlier, some herring and shad probably use special swim bladder coupling devices and their utricle to detect the ultrasonic (80–120 kHz) echolocation pulses of porpoises [144]. Frequency discrimination in fish without swim bladders or air bubble devices requires frequency differences of 10% or more, whereas in fish with such devices, only 3–5% differences are required. Although the goldfish shows some degree of tonotopic structure in its saccule, most fish do not exhibit tonotopic organizations in any of their hearing organs.

Amphibians retain the saccule, lagena, and utricle of their fish ancestors, and the saccule and lagena retain their auditory functions [195]. The saccule is the main receptor organ for low-frequency (<100 Hz) seismic signals transmitted from the pectoral girdle and through the operculum into the inner ear. In addition, frogs and toads have developed two other patches of hair cells in the inner ear: the **amphibian papilla** and the **basilar papilla**. Each of these patches is located inside its own tubular recess adjacent to the saccule, and each has an overlying tectorial membrane attached to the hair cells. The amphibian papilla is elongate and twisted in shape, with a tonotopic organization. It responds to moderate-frequency (80–1000 Hz) airborne sounds from a wide variety of sources. The basilar papilla is generally tuned to the higher-frequency components in the display calls of the males, and usually lacks tonotopic structure. Amphibian inner ears can function up to the frequency limits set by the mass and efficiency of their middle-ear structures (3–6 kHz),

and can discriminate between frequencies as long as they differ by more than 10–30% [104].

Lizards, birds, and mammals lack an amphibian papilla, but the basilar papilla of their amphibian ancestors has been elongated to accommodate a greater range of frequencies and, in some cases, to provide better frequency resolution. Whereas lizard papillae tend to be straight, those of birds must be curved or twisted to fit in the head, and those in mammals are so long that they are coiled like a seashell, and thus called a **cochlea**. In all three taxa, the papilla/cochlea contains three fluid-filled tubes (**Figure 3.41**). The central tube, the **scala media** is sandwiched between the other two and bathes the stereocilia of the hair cells. One outer tube, the **scala vestibuli**, connects to the fluid-filled cavity just inside the oval window, and the other, the **scala tympani**, connects to a cavity just inside the round window. Although the two tubes are linked at their opposite ends by a small duct called the **helicotrema**, fluid only passes between them during very slow pressure variations. The hair cells rest on a **basilar membrane** adjacent to the scala tympani, and extend their stereocilia into a tectorial membrane inside the scala media cavity. When sound waves are introduced through the oval window, fluid pressures rise and fall inside the scala vestibuli faster than fluid can move through the helicotrema. This causes the scala vestibuli to swell and contract. As it does so, it bends the scala media and its associated structures into and out of the scala tympani, causing the round window to move in and out to alleviate the resulting changes in scala tympani pressures. Because the basilar and tectorial membranes move differentially during this bending, the stereocilia are stimulated.

Web Topic 3.6 *Animations of vertebrate ears*
One movie can be worth a thousand still images. Here we provide links to a number of websites that provide free animations of middle and inner ears in operation.

Reptiles, birds, and mammals all have tonotopic ears. In birds and mammals, the hair cells tuned for the highest frequencies occur at the oval window end of the papilla and cochlea, respectively, and those responding to the lowest frequencies are found at the opposite end [121, 298]. In lizards, the low-frequency region may be at either end of their papilla; often, it is found in the middle between two regions of high-frequency hair cells [210, 213]. Lizards tend to have shorter papillae and fewer hair cells for their body size than birds and mammals, and their upper frequency limit is about 8 kHz. Birds have more refined middle-ear designs and longer papillae and can hear up to 10–12 kHz. The articulated middle ear ossicles and very long cochleae of mammals allow upper frequency limits of 150 kHz or higher.

Frequency resolution in reptiles, birds, and mammals is achieved through a combination of mechanisms. All of these taxa employ low-frequency phase locking, structural and electrical tuning of hair cells, and linking of adjacent cells through tectorial membranes. They also use local

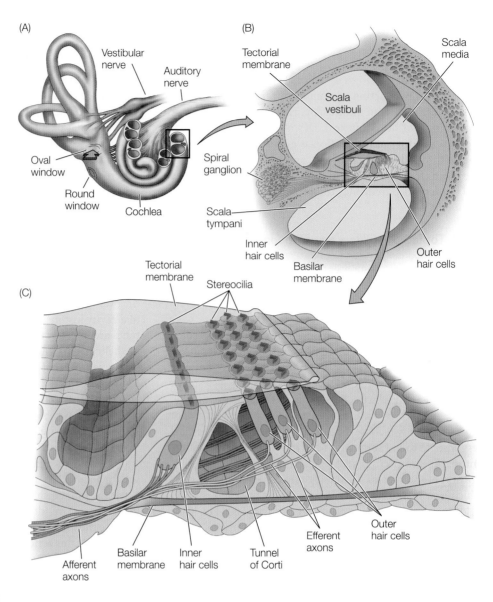

FIGURE 3.41 Structure of typical mammalian inner ear (A) Inner ear showing semicircular canals (on left), oval and round windows at base of cochlea, and coiled cochlea (right). (B) Cross section through cochlea at point indicated in (A). Sound enters the cochlea at the oval window and propagates down the scala vestibuli. This propagation causes the partition between the scala vestibuli and the scala tympani to bulge into and out of the latter. This in turn stimulates the hair cells in the partition. (C) Detail of mammalian hair cell distribution in partition. Hair cells are sandwiched between a tectorial membrane that lies on top of the hair-cell stereocilia and a basilar membrane the overlays the scala tympani. Bending of this sandwich causes the tectorial membrane to move differentially from the basilar membrane and produce a shearing force on the stereocilia. In most mammals, inner hair cells (which constitute at most 20–25% of the total hair cells in the cochlea) are arrayed in a single row parallel to the cochlear axis. These are the only hair cells that send auditory information to the brain. The more numerous outer hair cells are arranged in 3–5 parallel rows and respond to auditory stimulation by changing shape and amplifying the movements of the tectorial membrane. Bird papillae have a similar configuration except that the papilla is curved and not coiled. A cross section shows a continuous and graded series of hair cell types instead of two discrete classes, and the hair cells are distributed in a mosaic instead of in linear rows. A higher percentage (65–80%) of the hair cells in bird papillae send auditory information to the brain. Amplification is achieved by the movement of stereocilia of a subset of the hair cells rather than by the hair cells changing shape.

amplification by active hair-cell or stereocilia motion, at least for high-frequency resolution. In fact, each of the three taxa has independently evolved multiple hair cell types to divide up the tasks of detection and amplification [107, 121, 211, 212, 306]. Finally, birds and mammals taper their basilar membranes so that the high-frequency end is narrow, thick, and stiff, and the low frequency end is wide, thin, and flexible. When propagating sound waves, this heterogeneous membrane interacts with the fluids of the scala vestibuli and the scala tympani in complex ways [112]. Wave propagation along the papilla/cochlea is dispersive, and velocities vary with location and frequency. The result is the generation of stationary modes along the basilar membrane as a function of sound frequency: maximal bending occurs near the oval window end for high frequencies, and at successively greater distances along the membrane for lower frequencies. This is often called the **place principle** of bird and mammal hearing. The width and stiffness gradient of the basilar membrane is typically adjusted so that the location of maximal bending for a given frequency matches the characteristic frequencies of the nearest hair cells. The place principle thus augments the existing tonotopy of the organ. Because mammals have a much longer basilar membrane than birds, the place principle plays a much larger role in the fine-tuning of the cochlea than in the papilla (**Figure 3.42**).

Behavioral tests on various birds and mammals show minimally detectable differences in frequency of 2–4% for chickens and doves; 0.5–1% for owls, parrots and songbirds; 5–7% for chinchillas and gerbils; 1–5% for monkeys and elephants; and 0.1–0.5% for humans and echolocating bats [92, 196]. Behavioral responses are difficult to obtain on lizards, but neurobiological measures suggest that several species differ from most birds and mammals by not following Weber's Law (see Chapter 8) [12, 172]. Instead of showing a constant minimal percentage difference in discriminated frequencies, (the Weber's Law prediction), these lizards discriminate

FIGURE 3.42   Body size and vertebrate tonotopic inner ear design (A) Animals with larger body sizes usually have a larger inner ear (here measured by the length of their basilar membrane). However, mammals (black), which curl their cochlea, have longer basilar membranes than birds (red) or lizards (blue) of the same body size, and among the lizards, hearing specialists (4–6) have longer basilar membranes than nonspecialists (1–3). (B) Longer basilar membranes can host more hair cells, and this could be used to extend the frequency range of hearing (here measured as number of octaves (doublings) between the lowest and highest detectable frequencies). Although mammals tend to have larger frequency ranges than birds (with the exception of high-frequency

specialists (14 and 15)), and birds larger ranges than lizards, within each taxon there is little evidence that frequency range depends upon basilar membrane length. (C) An alternative use of the additional hair cells is that they increase frequency resolution by spreading each subset of the overall frequency range over a longer basilar membrane. This graph shows that mammals have the longest segments of membrane/octave, birds and hearing-specialist lizards show the next-longest segments, and lizard nonspecialists show the least. This pattern fits known levels of frequency resolution in these taxa. Note that the barn owl (10), an avian hearing specialist, has a segment length/octave value similar to that of mammals of the same weight. (D) Ranges of variation in numbers of hair cells/mm of basilar membrane in representative lizards, birds, and mammals. Birds and mammals of similar body size have similar total numbers of hair cells. Although birds tend to have shorter basilar membranes than mammals, they also have wider ones, allowing for more hair cells across that greater width. More hair cells across the membrane allow for refined temporal and amplitude resolution within a given frequency band. Species key: (1) European wall lizard (*Podarcis muralis*); (2) granite spiny lizard (*Sceloporus orcutti*); (3) alligator lizard (*Elgaria multicarinatus*); (4) tokay gekko (*Gekko gecko*): (5) bobtail skink (*Tiliqua rugosa*); (6) savannah monitor lizard (*Varanus exanthematicus*); (7) canary (*Serinus canaria*); (8) budgerigar (*Melopsittacus undulatus*); (9) starling (*Sturnus vulgaris*); (10) barn owl (*Tyto alba*); (11) pigeon (*Columba livia*); (12) chicken (*Gallus gallus*), (13) emu (*Dromaius novaehollandiae*); (14) mouse (*Mus musculus*); (15) horseshoe bat (*Rhinolophus ferrum-equinum*); (16) kangaroo rat (*Dipodomys merriami*); (17) gerbil (*Meriones unguiculatus*); (18) mole rat (*Spalax ehrenbergi*); (19) Norway rat (*Rattus norvegicus*), (20) chinchilla (*Chinchilla langer*), (21) rabbit (*Oryctolagus cuniculus*), (22) cat (*Felis catus*), (23) human (*Homo sapiens*), (24) bottlenose dolphin (*Tursiops truncatus*), (25) cow (*Bos taurus*), and (26) Asian elephant (*Elephas maximus*). (After [91, 96, 121, 172, 173, 177, 209-211, 311].)

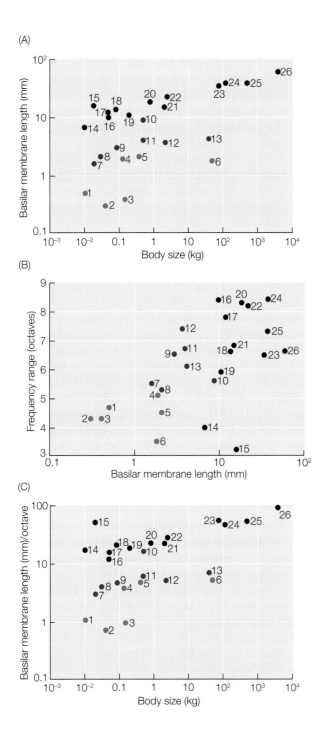

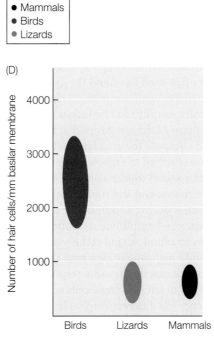

between two frequencies only if they differ by a fixed absolute difference in frequencies. Weber's Law is also partially violated for certain birds (such as parrots and owls) and mammals (echolocating bats) that can discriminate smaller than expected frequency differences for specific regions of their overall frequency ranges.

> **Web Topic 3.7** *Measuring auditory resolution*
> Many different measures have been developed to compute and compare the frequency, amplitude, and temporal resolutions of animal ears. Here we define a number of these measures and outline their differences and similarities.

DYNAMIC RANGE AND AMPLITUDE RESOLUTION   Single mechanoreceptors respond linearly to increasing signal amplitude only over a limited range. Each receptor has a minimum threshold amplitude that is required for it to respond at all, and at high enough amplitudes, its response saturates. As with frequency range, **dynamic range** is best extended by combining multiple sensory units with different thresholds and saturation points. The simplest example occurs in noctuid moths, whose tympanal ears contain only two scolopidia: both are tuned to the same characteristic frequency, but one is 20 dB more sensitive than the other. Bladder grasshoppers have a pair of auditory organs in each of six adjacent abdominal segments, each less sensitive than the one before it in the progression toward the posterior end [358]. Spiders often cluster up to 30 sensory slits into a compound **lyriform organ** [20]. Each slit in the cluster has a slightly different threshold and saturation level but covers an amplitude range of about 10 dB. By overlapping the ranges of different slits, the entire organ can provide nearly uniform coverage over a 40 dB dynamic range. Most animals with complex ears use a similar strategy of pooling the outputs of sensory cells or organs with different ranges. The overall dynamic range is usually limited by the lowest achievable threshold. Active amplification mechanisms and refined coupling and modification devices are most often used to extend the dynamic range to low levels.

Among other functions, amplitude resolution allows animals to compare reception of the same sound at two or more hearing organs and compute the angular position of the sound source, compare observed to expected amplitudes to estimate the distance to a sound source, compare display amplitudes of alternative mates, and use signal amplitude as a measure of aggressiveness during a contest. For a single mechanoreceptor, frequency and amplitude resolution will often be confounded: a low amplitude sound at the receptor's characteristic frequency and a high amplitude sound at some other frequency may produce exactly the same response in the receptor. One solution is to have multiple cells with the same characteristic frequency but different amplitude ranges (as in the noctuid moth ear). However, this creates another problem: small animals are limited in the number of sensory

cells they can accommodate in their ears, and each cell added to improve amplitude resolution is a cell that cannot be used to improve range or resolution of temporal, frequency, or location information. An alternative is to host dedicated brain cells that compare relative responses of each class of receptors to known tuning curves, identify likely confounds, and compute a corrected amplitude estimate. However achieved, wax moths and katydids achieve amplitude resolutions as low as 1–2 dB [15, 153], and typical mammals and birds show minimal resolutions of 0.8–4 dB [92, 329].

TEMPORAL RANGE AND RESOLUTION   A major constraint on both temporal range and resolution in animal ears is that some interval must elapse between two sounds or sound components in order for the receptor (or brain) to considers them as separate events. How long an interval is necessary? All acoustic instruments including ears must deal with a trade-off between temporal and frequency resolution: improving one invariably undermines the other (see Web Topic 2.4). Increasing the resonance of auditory receptors, whether electrically, mechanically, or through active amplifiers, will tune the receptors to increasingly narrow bands. The temporal cost is that it will also reduce damping, making the receptors "ring" when stimulated. The longer they ring, the less likely they are to discriminate between two closely successive events.

To a larger degree than for frequency or amplitude resolution, temporal resolution depends on processes that occur not in the ear, but in the brain of the receiver. There are thus many different measures of temporal resolution depending on what stage in the process is being tested. Responses at the coupling stage can be extremely fast: successive events need be separated by only 150–200 μsec to be discriminated as separate by noctuid moth tympana [244]. However, behaviorally determined temporal resolution tends to be significantly slower: minimal perceived gaps for test stimuli input to a single ear are 4–8 msec in insects and 1–4 msec in vertebrates [92, 119, 196]. Exceptions include the superior ability of some birds to discriminate between very fine temporal patterns in complex sounds [92, 93]; and, as we shall see in Chapter 14, echolocating bats and dolphins have evolved very special skills for measuring short time delays between successive acoustic events.

DIRECTIONALITY   Directionality refers to the ability of a receiver to determine the distance (range), horizontal angle (**azimuth**), and vertical angle (**altitude**) of a sound source. We briefly discussed amplitude and spectral contrast mechanisms that animals might use for ranging at the beginning of this chapter. How might animals determine the azimuth and altitude of a sound source?

Hearing organs that rely on near-field sounds in fluids can exploit the fact that the axis of molecular oscillation in the medium is significantly affected by the location of the sound source. The axis of movement points directly back toward a monopole source. Vibrational axes may not point

FIGURE 3.43 **Polarization patterns of hair cells in fish ears** All vertebrate hair cells are polarized: maximal stimulation requires that the cell's stereocilia be bent along an optimal axis. The separate hearing organs inside the inner ears of fish are usually subdivided into regions within which all hair cells have the same polarity. Pooled input from different regions enables the fish's brain to compute the location of a sound source. In each polarity map in this figure, the fish's head is to the left; upward is dorsal for the lagena and saccule maps, and toward the middle of the fish's side for the utricle maps. Arrows indicate the direction of bending that maximally stimulates hair cells in that region. The lagena and utricle usually have two distinct regions with opposing polarities. Polarization maps in the saccule vary among species. Hearing generalists (e.g., toadfish) divide the saccule into 4 regions with hair cell axes pointing in each of four perpendicular or opposite directions. Otophysans, which use ossicles to connect the swim bladder to the ear, have only two regions which are polarized vertically in opposite directions. A variety of other patterns are also shown. (After [282].)

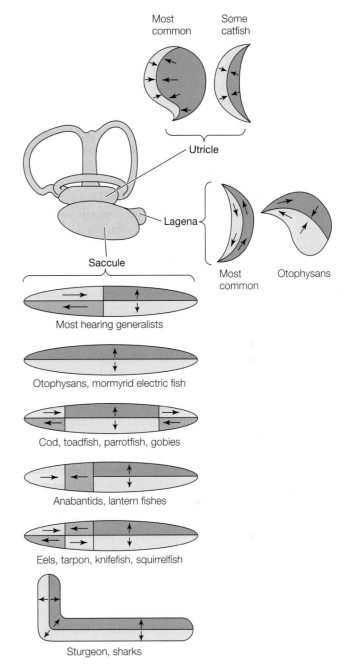

at the source when a receiver is sufficiently close to dipole or quadrupole vibrators, but there are still patterns in the near field which can be sampled and computations used to estimate source locations [157, 281]. Because each sensory hair on an arthropod's exoskeleton usually has its own preferred axis of movement, a receiver can identify or compute the direction of medium motion by sensing which hairs are most bent in the sound field [53]. The antennae and Johnston's organs of mosquitoes and *Drosophila* can also compare near-field fluid movements with their preferred axes of vibration [240, 305]. We have seen that vertebrate hair cells are maximally stimulated when their stereocilia are bent toward and away from the kinocilium. The saccules, utricles, and lagenas of fish are organized so that hair cells in each of several subregions are oriented in the same direction, but those in different subregions are oriented in different directions (**Figure 3.43**). Pooling the output from all these regions allows their brains to estimate the likely azimuth and altitude of near-field sound sources [283]. Fish hair cells are sufficiently sensitive that they may be able to use this mechanism at distances even beyond the usually defined transition between near and far fields [105].

Animals attending to far-field sounds, sounds propagated at boundaries, or sounds inside solid substrates estimate source azimuth and altitude by comparing a sound's times of arrival, amplitudes, and phases at 2–8 different hearing organs. Participating sensors include pairs of ears in crickets, katydids, grasshoppers, cicadas, lacewings, moths, and terrestrial vertebrates, and the subgenual, chordotonal, and touch/stretch receptors in the legs of elephants, hemipteran insects, and spiders. For an animal with a pair of ears, a sound will arrive simultaneously at the two ears only when it is located in the plane equidistant from them and perpendicular to the line joining them. The maximal delay in arrival at the two ears occurs when the sound source is located along an extension of the line joining the two ears. An animal can identify a source azimuth by rotating around the center of the line joining its ears until the sound arrives at both ears

simultaneously. The source will then either be in front of it or directly behind it. Alternatively, the times of arrival could be forwarded to the brain, which then computes the azimuth of the source without the head or body being turned. Note that if the line joining the two ears is parallel with the Earth's surface, there is no way this animal can detect the altitude of the sound source without invoking other mechanisms. If the line joining the ears is not parallel with the Earth's surface, the animal will have to rotate its body in both the vertical and horizontal planes before the sound will arrive simultaneously at the two ears. Such a movement could be used to estimate both azimuth and altitude. Alternatively, it could use other mechanisms to determine the sound azimuth, and forward the delay times to the brain to compute the altitude. Many owls have one ear set slightly lower on their heads than the other, and both cats and humans have asymmetries in the shape of their external ears that permit altitude estimation using time delays [41, 169]. Both azimuth and altitude can also be computed by a brain if the animal has more than two hearing organs and these are not all set at the same height in the body.

An animal's ability to estimate azimuth or altitude using interaural time delays is limited by its body size and the speed of propagation of the relevant sound. Body size is important because it limits the maximum distance that can exist between two or more auditory organs and thus the maximum delay between arrival times of a sound. Animals with sensors in their legs may have a certain advantage in this respect, because they can stretch out their legs to increase the interaural distance. Even so, the delay in a sound's arrival at multiple ears will be very small in small animals: in air, the maximum delay for a 1 cm insect would be only 30 μsec. In a human, the maximal delay averages 656 μsec, and in elephants, it can be over 3 msec. Medium also affects minimal resolution. The situation in water is more challenging than that in air: higher sound propagation speeds produce only a 7 μsec delay for a 1 cm snapping shrimp. On the other hand, bending waves in plants have significantly lower propagation speeds than sound in air: a 1 cm insect might experience up to a 100–200 μsec delay. This might make use of interaural time delays quite feasible for small plant-dwelling insects and spiders [20, 66, 71]. In addition to body size and sound speed, sound source resolution depends upon the minimal time delays that can be measured in the receiver's brain. Threshold values for measuring interaural time delays can be much smaller than those observed for discrimination of successive events at a single ear. Short delays can be measured by summing the outputs of the ears onto single cells in the brain. Different brain cells are preset to delay input from one ear relative to the other by fixed amounts. The brain cell that delays the input from the ear nearer the source by just the amount of delay in arrival times will show the strongest summed response. Knowing which cell responds then indicates which delay is present. In humans, this technique allows the detection of interaural delays down to 10 μsec.

Instead of comparing relative arrival times, an animal might compare the relative amplitudes of a sound at multiple ears. As with time delays, it could identify azimuth by rotating its head or body until both ears experience the same amplitude, or it could send differential amplitude information to the brain, where the azimuth could be computed. Most ears are too close together to exploit the slightly lower amplitude of a low frequency at a far ear due to greater spreading loss. However, if the wavelengths in the incident sound are smaller than the size of the animal's head or body separating the ears, diffraction and shadowing of the far ear can generate significant amplitude differences. Animals can make this work in ingenious ways. For instance, the ear-containing forelegs of katydid females are too small to produce diffraction even at ultrasonic wavelengths, yet the males emit

ultrasonic courtship calls. Females can still localize males because their bodies are large enough to produce significant diffraction differences at their spiracle openings. These differences are then conveyed to the tympana through the tracheal tubes [246].

As we saw in Chapter 2, diffraction when wavelengths are similar to the intervening object can be quite complicated. The amplitude of a diffracted sound at an animal's ear will depend on the sound wavelength, the size and shape of the animal, and the azimuth and angle of the sound source relative to the animal's body axis. Because different frequencies will be diffracted differently depending on the location of the sound source, an animal can use the perceived frequency spectrum of a known sound to estimate its location. The spectrum can provide both azimuth and altitude information [134]. Amplitude differences between ears can be magnified further using pinnae or other directional horns as auditory coupling organs. Pinnae are also very useful for resolving whether a sound source is in front or behind an animal, and many mammals can rotate their pinnae to provide single ear estimates of source azimuth. Some mammals have elaborate ridges and crenulations on their pinnae that alter the frequency spectrum of complex sounds at the ear in response to the altitude of the sound source. This is another way in which a single ear can provide elevation information. Note that all of these mechanisms require that pinnae and their elaborations be as large or larger than the relevant sound wavelengths. The small pinnae on small animals are effective only if their ears can detect and process sufficiently high frequencies [135].

What if animals are so small that they do not diffract sounds of interest and they cannot resolve the necessarily short arrival delays at multiple ears? The most common solution is to use a pressure differential coupling device [61, 169, 188, 195, 244, 246]. Grasshoppers, cicadas, frogs, lizards, and birds create a pressure differential system by linking the inner sides of their two tympana with air spaces or air-filled membranous sacs. Each tympanum then samples sound amplitudes at two locations. For example, cricket and katydid tympana compare the amplitude just outside their surface with that at openings of tracheal tubes on the animal's thorax. If the samples arrive on the opposite sides of the tympanum at the same amplitude and in phase, the tympanum will not move; any other set of relationships will cause the tympanal and tracheal membranes to vibrate in concert with the sound. For wavelengths larger than the distance between the two sample points, the samples on the two sides of the tympanum will be in phase when the sound source is located in the plane equidistant from the two sample points and perpendicular to the line joining them. At all other locations, the two samples will be out of phase and the tympanum will vibrate. Other factors being equal, the maximal amplitude of vibration will occur when the sound source is located along the extended line joining the two sample points. Although one pressure differential ear can thus provide some azimuth information, computations based on data from two or more ears provide even more accurate localization. Note that this system can break down when wavelengths are smaller than the distance between the sample points: if one sample is exactly one cycle behind the other, the tympanum will not move even though the two samples are no longer equidistant from the source. As sound frequency increases, the number of situations in which this ambiguity arises increases. At high enough frequencies, diffraction may finally be possible and the ears can then act as pressure detectors. Note that pressure differential ears become ineffective at very low wavelengths where the difference in phase between the two samples is small. Crickets manage to respond to lower frequencies than similarly sized insects by slowing down the delivery of the tracheal sample and thereby increasing the apparent phase differences [246].

Tachinid flies present another ingenious solution to the small animal problem [303, 305]. Their two tympana are located next to each other on the midline of the thorax. Each end of a semiflexible lever arm is connected to one of the tympana. The normal mode of this linked system is tuned to the dominant frequency in the calling songs of the male crickets that are parasitized by female flies. Oscillations of the tympanum nearest a calling cricket are transmitted to the more distant tympanum by the lever arm, suppressing the amplitude of the latter's oscillations due to the sound and delaying its response. This generates both an amplitude difference and a time delay between the two tympana that is large enough for the fly brain to use to estimate azimuth. A slightly different version of this solution is used by sarcophagid flies which locate and parasitize calling cicadas.

Most animals use some combination of mechanisms to achieve overall directionality, but different taxa use different combinations. How do the different combinations compare? Fish such as cod (Gadidae) use their heterogeneously polarized hearing organs to estimate azimuth and altitude of sound sources with accuracies within 12–20° [104]. Small terrestrial insects such as crickets and katydids, with pressure differential ears and higher frequency diffraction, manage similar azimuth resolutions of 15–20° but much less accurate altitude estimates (45° in crickets) [16, 119, 380]. Tachinid flies, with their unusual lever system, can achieve 1–3° azimuth accuracy [221]. Frogs, relying on pressure differentials, diffraction, and interaural time delays also manage a 15–30° azimuth accuracy and 23–45° altitude resolution [119]. Small birds, like frogs, rely on a combination of pressure differential, diffraction, and interaural delay mechanisms, and similarly achieve 20–30° azimuth accuracy [169]. Predatory birds, (in part because they are larger), rely more heavily than small birds on interaural time delays and achieve azimuth resolutions of 2–12° (hawks) and 1–7° (owls). Owls achieve similar resolutions for altitude angles and, like mammals, do so using frequency spectrum differences generated by complex feather structures around the ear opening. Mammals, with their extraordinarily high-frequency hearing and elaborate pinnae, tend to rely on diffraction and delay times for azimuth estimation, and frequency spectrum contrasts for sound source altitude [41, 42, 135]. Azimuth accuracy ranges from 1–5°

for dolphins, elephants, primates (including humans), cats, and many echolocating bats to 10–30° for most other non-fossorial mammals. Altitude resolution is usually more difficult than azimuth resolution in any given species; values range from 2–3° for dolphins and humans to 20–30° for other mammals.

Body Size and Sound

One of the more robust generalizations we can make about animal sound communication is that small animals usually communicate with high-frequency sounds and that larger animals tend to use low-frequency sounds. This pattern persists despite the large number of contributing factors and differences among taxa and habitats in the relative weighting of these factors. Several common threads, however, stand out as acting in concert to produce this pattern.

On the signal production side, we saw in Chapter 2 that it is difficult for an animal to produce an intense sound with a wavelength more than twice its body dimensions [24, 25]. Many animals *can* produce sounds with wavelengths longer than their bodies, but the cost is an increasingly rapid drop in the efficiency of sound radiation, and thus in sound intensity, with increasing wavelength. For animals communicating over a distance, that cost might be compensated for by the lower heat and scattering losses of slightly longer wavelengths. However, radiation efficiency drops off faster with increasing wavelength than propagation efficiency does. For wavelengths more than 2–3 times the size of the signaler, radiation losses will outstrip any propagation benefits. These effects are nicely exhibited in **Figure 3.44A**, which shows the dominant frequencies in the male display calls of insects. These are compared to the minimal frequencies (maximal wavelengths) that would still allow each insect to emit maximally intense sounds. As predicted, the frequencies favored by each species decrease with increasing body size. However, all observed frequencies fall below those predicted for maximal intensity radiation. With the exception of displaying male fruit flies, differences between observed and expected frequencies are small, or at most, moderate, in magnitude. This is likely due to a trade-off between lower radiation intensity and better propagation. Whereas all of these insects rely on far-field sounds to attract mates, fruit fly males perform their displays at close range, and females are within the near field of the male's sounds. Thus the flies can use much longer-wave sounds and still achieve the requisite active space. Such observations support the notions that there is a minimal frequency that an animal can use for long-distance communication, and that this limit is inversely correlated with the signaler's body size.

On the signal reception side, we get a similar effect, but for a different set of reasons. We saw in this chapter that the ability to identify the location of a sound source requires the presence of sufficiently high-frequency components in the propagated sound. For larger animals that have pressure detector ears, sufficient phase and intensity differences between ears used for localization exist only for wavelengths

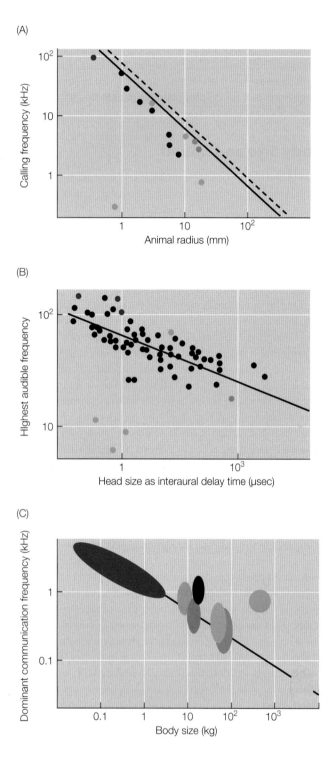

(A)

Calling frequency (kHz)

Animal radius (mm)

(B)

Highest audible frequency

Head size as interaural delay time (μsec)

(C)

Dominant communication frequency (kHz)

Body size (kg)

FIGURE 3.44 Three explanations for inverse correlations between favored sound frequencies and body size (A) Relationship between male advertising call frequency and body size in a sampling of terrestrial insects. Also shown are minimal frequencies for monopole (solid line) and dipole (dashed line) sources at each size that ensure maximal radiation efficiency and thus signal intensity at the source. Monopole sources: waxmoths (red), cicadas (green), mole cricket burrows (orange). Dipole sources: crickets and katydids (black) and fruit flies (blue). (B) Relationship between highest audible sound frequency and inter-aural distances for various mammals. Echolocating bats (red), burrowing rodents (blue), cats (green), humans (orange), and elephants (yellow). Others (black) are primates and rodents. Line is best fit to points excluding the burrowing rodents which use low frequency seismic signals.) (C) Fletcher model predicting optimal frequencies for sound communication after combining effects of sound production, propagation, and reception constraints (solid line) versus actual frequencies used by different birds and mammals (ellipses). Optimal frequency is here predicted to be proportional to body mass raised to the –0.4 power. Birds (red), cats (green), dogs (grey), monkeys (black), gorillas (purple), humans (orange), horses (blue), and elephants (yellow). (A after [25]; B after [135]; C after [114].)

smaller than the size of the receiver's head (including its pinnae) [134, 135]. Smaller animals that have pressure differential ears also require a sufficiently small wavelength before the phase or intensity differences at the two openings for an ear are detectable [246]. In both cases, localization requires the ability to hear sounds of sufficiently small wavelength and thus sufficiently high frequencies. The minimum frequency required for localization should thus decrease as the distance between the ears (pressure detector species) or sampling points (pressure differential species) increases. Note

that there is no penalty to localization if an animal can hear even higher frequencies than this limit, but in most species, higher frequencies suffer sufficiently higher propagation losses that it may not be worth attending to them. Thus we might expect the highest audible frequency that an animal can hear to decrease with increasing interaural distance. This is in fact what one finds for a wide sampling of mammals (**Figure 3.44B**). Again, there are exceptions to this correlation when sound communication is largely short-range or when the medium propagates only certain wavelengths: obvious examples include marine fish, burrowing rodents, and arthropods communicating on plant stems.

Ideally, one would like to combine all factors to see whether the negative correlation between body size and communication frequency for some individual factors is preserved. Fletcher [113, 114] has provided a general model for this correlation that integrates sound production, propagation, and reception components. He concludes that in general, the optimal frequencies for sound communication will always be inversely correlated with body size. He specifically predicts that the optimal frequency for a species will, on average, be proportional to its body mass raised to the –0.4 power (**Figure 3.44C**). Although available data appear to fit this prediction, there are still insufficient samples to exclude the possibility that optimal frequency is instead simply proportional to body length or width, and (since body mass is usually proportional to an animal's linear measure cubed), to body mass raised to the –0.33 power. Whatever the actual scaling, Fletcher's model and existing data support the idea that an animal's size seriously curtails the frequencies that it can use for sound communication. At times, this can be a handicap to the sender, the receiver, or both, but at other times it can be exploited in interesting ways. We shall return to this point in later chapters.

SUMMARY

1. Sound signals typically degrade during propagation to the receiver. Some species have developed acoustic adaptations that minimize degradation and maximize the **active space** of their signals. **Ranging** is the assessment of degradation in a signal; it allows a receiver or eavesdropper to estimate the distance to the sender.

2. The amplitudes of all frequencies in a sound signal decrease equally with distance from the sender due to spreading losses. Where the medium is stratified with respect to flow, temperature, or pressure, refraction may decrease sound amplitudes faster or slower than expected given spreading losses alone. Spreading losses are reduced by refraction when terrestrial senders call at dawn, are upwind, or call between the understory and the canopy in a dense forest; they are more severe on hot days, when the sender is downwind, or when the caller is above the forest canopy.

3. The frequency spectrum of a sound often changes as it propagates between sender and receiver. A plot of the change in amplitude for each possible frequency component is called the **frequency response** of that medium and situation. In most environments, amplitudes of propagating high-frequency sounds attenuate faster than those of low-frequency ones because of heat losses and scattering by objects in the sound path between sender and receiver. Heat losses and scatter are generally more severe in terrestrial environments than in aquatic ones.

4. A boundary between media can also alter the frequency response along the path between sender and receiver. It does so by reflecting waves toward the receiver. These waves can then interfere (positively or negatively) with the waves that have traveled directly between the two parties. In many cases, the reflected waves suffer a phase shift at the boundary, causing negative interference and cancellation of low frequencies. For sound in air propagating over soft earth, a **ground wave** can restore some of the lowest frequencies, at least over moderate distances. As a result, the frequency response graph for sound propagation in air but near the ground often has a **notch** in which intermediate frequencies are severely attenuated, but very low and high frequencies continue to propagate well. Boundary interference also results in high attenuation of low-frequency sounds propagating just under the surface in water. There are no restorative effects in water to minimize this filtering.

5. When sender and receiver are both located between the same two or more reflective boundaries, the medium for sound propagation becomes a **waveguide**. The complex reflections from the multiple surfaces result in certain frequencies (normal modes) being favored over others during propagation. Typically, there is a **cutoff frequency** below which propagation is negligible. Waveguides can seriously distort signal patterns over long distances.

6. A third type of boundary effect occurs when the sound propagates in one medium but the receiver is located in another adjacent medium. The receiver must then rely on whatever version of the sound is detectable on its side of the boundary. This is the case for insects exchanging signals on the water's surface, insects and spiders communicating through bending waves inside plants, and elephants and burrowing mammals communicating with seismic signals. All of these situations attenuate high frequencies faster than low frequencies, and the resulting sound waves, like waveguides, may have complex resonances and favored modes.

7. Temporal patterns in propagating sounds are most often altered by **reverberations** (echoes). Airborne sound signals propagating in forests acquire more reverberations than do similar sounds in open country. Echoes from fish swim bladders create reverberations under water. If the original signal has elaborate modulations, these will be degraded by added reverberations; if the signal consists of a long single frequency, it may be enhanced by becoming longer and louder to receivers. Most forest birds avoid modulations in their songs and calls. In open country, on the other hand, slow-amplitude modulations may be added to signals as they pass through wind vortices or heat bubbles. Open country birds thus favor rapid amplitude modulations for long distance signals, or rely more heavily on frequency modulation for pattern.

8. Sounds propagating on a water or solid substrate boundary, inside plants, or in air or water waveguides will suffer **dispersion**, in which different frequencies propagate at different speeds. This can create major distortions in the temporal and frequency patterns by the time a signal reaches a receiver.

9. Noise is a ubiquitous problem for animals communicating with sound signals. In most habitats, noise is more likely to mask low frequencies (due to wind in air and waves in water), and high frequencies (due to insects in air and snapping shrimp in water). Bending waves inside plants can be masked by rustling of the leaves and branches. Human-caused (anthropogenic) noise is increasingly a problem for sound-communicating animals in the wild. When faced with significant noise, animals may focus more on intermediate frequencies, limit signaling to periods of relative quiet, or increase signal amplitude to ensure effective long-distance communication.

10. All animal ears use differential motion between special mechanoreceptors and the rest of their body to detect sounds. As with sound radiation, coupling of ambient sounds into an ear is hindered if the wavelengths are significantly larger than the animal, and the acoustic impedance of the medium and the ear are sufficiently different. Animals have evolved special adaptations to enhance coupling, modification of captured sounds, and detection.

11. Small hairs on the exoskeleton and plumose antennae of arthropods move more easily in a near field than does the rest of the body. This differential movement provides the necessary coupling for near-field sounds. In other species, the internal sensory hairs are covered by heavy objects (**statoliths** in crustaceans and **otoliths** in fish). These masses accelerate more slowly in near fields than do the hairs and other soft tissues; and this differential movement bends the hairs and stimulates the sensory cells.

12. Terrestrial animals have thin membranes called **tympana** to couple far-field sounds in air into their ears. In mammals and moths, the tympana are stretched over closed cavities. The resulting **pressure detectors** compare external sound pressures to the reference pressure in the closed cavity. The tympana of grasshoppers, cicadas, crickets, katydids, frogs, lizards, and birds compare sounds sampled at two different locations in the far field and thus serve as **pressure differential detectors**. All use the exterior of a tympanum as one sample point. Crickets and katydids use tracheal tubes to convey a second sample taken outside their thorax to the other side of the tympanum. In grasshoppers, cicadas, and all terrestrial vertebrates except mammals, the insides of the two tympana are connected by an airspace or air-filled sacs. The second sample is thus taken outside the tympanum of the opposite ear.

13. In water, a number of fish use swim bladders or accessory air sacs the way terrestrial animals use tympana—to capture far-field sounds and convert them into oscillations of the cavity wall. Swim bladder movements are conveyed to the ears by small bones or intervening tissues. Toothed whales use fat-filled jawbones to convey far-field sounds to a thin bone in the ear that converts the pressure variations into vibrational motion. Baleen whales and true seals appear to capture sounds in various parts of their skeletons and convey them to ears that respond to bone-conducted sounds.

14. Boundary-propagated sounds are usually coupled into receiver bodies through their legs. Tension and posture can be varied to improve resonance of the body relative to frequencies of interest. Spiders use slits and insects use pits in the exoskeleton that are then compressed and expanded as a result of the coupled vibrations. These arthropods also use stretch receptors and blood movement detectors to monitor leg and body vibrations. Frogs absorb seismic signals through their bodies and convey them through a specialized muscle to their ears.

15. Terrestrial animals use horns, articulated chains of bones, or successive membranes of decreasing size to modify captured sounds and reduce impedance mismatches between the ambient medium and their bodies. While these devices can increase the sound energy delivered to inner ears, they usually have their own resonant properties that may constrain the animal's auditory range and resolution.

16. Arthropod and vertebrate auditory mechanoreceptors all have dendrites with a ciliary component. Vertebrate receptors accompany this **kinocilium** with multiple **stereocilia**. Bending, compressing, or stretching these mechanoreceptors constitutes the primary transduction step for hearing.

Both types of sensors can be extremely sensitive: threshold movements as small as or smaller than the diameters of single atoms are common in both groups.

17. Many insects use their ears largely to detect and avoid echolocating bats. They tend to have simple ears tuned to ultrasonic frequencies and few adaptations for frequency resolution. Taxa that use their ears for intraspecific communication usually have more sophisticated mechanisms for breaking complex sounds down into separate frequency bands and assessing the relative amplitude of each band. They thus perform some level of Fourier analysis. Broad frequency ranges and good frequency resolution can be achieved simultaneously by the presence of many sensory cells, each tuned to a different subset of the overall frequency range. In tonotopic ears, the sensory cells are arrayed in order of their resonant **characteristic frequencies**. Tonotopy reduces interference between stimulated cells with very different characteristic frequencies. In many tonotopic ears, the sensory cells are covered with a **tectorial membrane** that insures that all cells in a given band are stimulated simultaneously. In both arthropods and vertebrates, some sensory cells physically vibrate when stimulated; this can stimulate adjacent cells with similar characteristic frequencies leadung to **auditory amplification**. Frequency resolution can be further refined for lower frequencies through **phase locking**, in which nerve impulses being sent to the brain by the sensory neurons are synchronized with peaks in the waveform of the relevant sound frequency.

18. **Dynamic range** is largely set by the lowest amplitude signals that can be detected. Although arthropod and vertebrate mechanoreceptors are so sensitive that the potential dynamic range for animals should be enormous, ambient noise is often sufficiently high that it sets the floor for the effective dynamic range. Amplitude resolution, like frequency resolution, is largely limited by the number of sensory cells that are tuned to the same frequency but at different amplitude thresholds. In insects, amplitudes must differ by 1–2 dB at favored frequencies to be discriminated; equivalent thresholds for birds and mammals are 0.8–4 dB.

19. There is a trade-off in most auditory systems between frequency and temporal resolutions. Sensory cells that are narrowly tuned (large Q) will have worse temporal resolution than more broadly tuned and rapidly damped cells. In many cases, brain processes (e.g., phase-locking) can be invoked to improve temporal resolution. In practice, separate acoustic events will be discriminated by insects if they are separated by at least 4–8 msec; birds and mammals require intervening intervals of only 1–4 msec.

20. Animals can use any of four different kinds of information to identify the azimuth and altitude of a sound source. In near fields, hearing organs can often use the direction of molecular motion in the medium to estimate both azimuth and altitude. Time delays in the arrival of a far field sound at two or more hearing organs are often used either behaviorally (by rotating until the delay is zero) or computationally (by the brain) to identify the source azimuth. This cue

becomes decreasingly useful as receiver size gets smaller. Amplitude differences provide a third type of far-field cue, and azimuth can again be identified using turning or brain computation. Animals larger than relevant wavelengths rely on diffraction to create amplitude differences at two or more hearing organs; animals smaller than relevant wavelengths use one or more pressure differential organs that are inherently directional. The fourth cue, changes in far-field frequency spectra at the ear as a function of source angle, is used by many mammals and by some birds to estimate the altitude of a sound source.

21. Both sender constraints and receiver constraints contribute to the widely found inverse relationship between an animal's body size and the sound frequencies it uses for communication.

Further Reading

General reviews of environmental modification of propagating airborne sounds can be found in Wiley and Richards [371], Forrest [117], and Slabbekoorn [335]. Embleton [102] and Attenborough [8] provide more technical accounts of the relevant physics, and Brum and Slabbekoorn [49] discuss the sources and consequences of terrestrial noise. Substrate and boundary sound propagation is lucidly described by Markl [217]. Introductions to sound propagation in marine environments include Rogers and Cox [307] and Tyack [357]. Advanced marine treatments include Caruthers [56] and Tolstoy and Clay [354].

Fletcher [112] discusses the underlying physics of sound reception in both arthropod and vertebrate taxa. Michelsen and Larsen [244], Michelsen [246], and Greenfield [127] provide highly readable but more detailed reviews of sound propagation and hearing in insects. Other good reviews of arthropod hearing include Popper et al. [284] on crustaceans; Barth [20] on spiders; Wilcox [368] on water-surface dwellers; Čokl and Doberlet [71], Virant-Doberlet and Čokl [361], and Cocroft and Rodriquez [68] on insects communicating through plants; and McIver [234], Hoy and Robert [147], Keil [161], Robert and Hoy [303], Field and Matheson [108], Yager [383], Yack [382], and Robert [305] on the structure and function of insect mechanoreceptors in general and ears in particular.

Good reviews of vertebrate hearing strategies include Popper and Fay [283] on fishes; Lewis and Narins [195] on amphibians; Manley [210, 213] on lizards; Dooling et al. [92] and Dooling [93] on birds; Heffner and Heffner [134], Rosowski [312], Heffner [135], and Brown and May [42] on terrestrial mammals; and Ketten [163] and Hemilä et al. [138,141] on marine mammals. The evolutionary modifications required when the early terrestrial ancestors of whales and dolphins moved to the water are summarized by Nummela et al. [267, 269]. Recent discoveries about active amplification in hearing organs are described for insects by Göpfert and Robert [123, 124], and for vertebrates by Robles and Ruggero [306] and Fettiplace [107].

COMPANION WEBSITE
sites.sinauer.com/animalcommunication2e

Go to the companion website for Chapter Outlines, Chapter Summaries, and References for all works cited in the textbook. In addition, the following resources are available for this chapter:

Web Topic 3.1 *Transfer functions*
Here we discuss methods for measuring transfer functions of black boxes in general, and blocks of propagation medium specifically. What assumptions have to be met to make these measurements, and what happens if they are not met?

Web Topic 3.2 *Dispersive sound propagation*
Sound propagation with dispersion can result in major changes in a signal's waveform and in the speed with which the signal propagates as a whole. Here we provide more details on dispersive sound propagation for each of the contexts in which communicating animals are likely to encounter it.

Web Topic 3.3 *Animal communication and anthropogenic noise*
To what degree does noise generated by human activities interfere with sound communication in animals? Recent studies in both terrestrial and marine environments suggest that there are increasing problems, but animals can sometimes adapt.

Web Topic 3.4 *Levers and ears*
Different types of lever systems provide mechanical advantages and match acoustic impedances for improved hearing in animals. Here we outline the general classes of levers and indicate which are used in the middle ears of amphibians, reptiles, birds, and mammals.

Web Topic 3.5 *Auditory amplification*
Both arthropods and vertebrates use active motion of auditory receptors to amplify responses to low level sound stimuli. Here we examine some of the data demonstrating this mechanism in insects, mammals, and lizards.

Web Topic 3.6 *Animations of vertebrate ears*
One movie can be worth a thousand still images. Here we provide links to a number of websites that provide free animations of middle and inner ears in operation.

Web Topic 3.7 *Measuring auditory resolution*
Many different measures have been developed to compute and compare the frequency, amplitude, and temporal resolutions of animal ears. Here we define a number of these measures and outline their differences and similarities.

Chapter 4

Light and Visual Signal Production

Overview

Light is radiant energy. The sun is obviously the most important source of light on Earth. Sunlight not only warms the planet, but also provides the energy plants need to grow; plant growth in turn provides the primary source of food for other organisms in the ecosystem. Light is also the basis for vision. As long as there is some level of ambient light in the environment, an organism need do nothing to make its presence and location known. All organisms, plant or animal, can therefore generate visual signals passively as light reflects off their bodies and into the eyes of visual receivers. To understand the evolution of eyes and the constraints on visual communication, we need to examine the physical characteristics of light. In this chapter we will define light and describe its transmission properties and interactions with organic matter. We will then examine the ways that animals enhance their visibility to receivers with specific visual signals that employ color, movement, and even self-generated light.

Properties of Light

Our understanding of how light works has evolved over the past few centuries of scientific discovery. Eighteenth-century scientists thought that light was a stream of minute particles traveling in straight-line trajectories, or **rays**. Rays could be reflected by mirrors and refracted by prisms and lenses, and ray optics provided an explanation of how images were formed. The **wave theory** of light developed in the nineteenth century was able to explain the effects of interference and diffraction, while still taking into account the raylike behavior of light. The twentieth-century insights from **quantum theory** established the energy effects of light as it interacts with matter. Throughout these two chapters on light, we will make alternative use of the ray, wave, and quantum viewpoints. We now know that light is a type of electromagnetic radiation. The natural range of electromagnetic radiation is huge, and what we call visible light is only a small part of that range. In this section we describe the unique characteristics of visible light and explain why its properties are so useful for visual communication.

Characteristics of electromagnetic radiation

Electromagnetic radiation (**EMR**) is a rapidly moving form of energy. A stream of this radiation produces both electrical and magnetic forces as it passes any

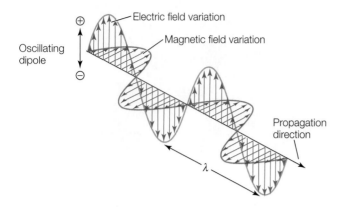

Oscillating dipole

Electric field variation

Magnetic field variation

Propagation direction

λ

FIGURE 4.1   An electromagnetic wave Electromagnetic waves transport energy through empty space. The waves are generated by the rapid oscillation of a negative charge relative to a positive one. They spread out from the source in all directions perpendicular to a line between the two charges. The electric force oscillates sinusoidally in the plane parallel to the motion of the charge vibration, and the magnetic force oscillates sinusoidally at right angles to the electric force. Both components have the same wavelength, indicated as λ.

given location. In Chapter 2 we defined a map of the direction and magnitude of a force at each point in space as a field. An **oscillating field** is one in which the direction or magnitude of the forces at each location varies with time in some cyclic way. Electromagnetic radiation consists of both an oscillating electric field and an oscillating magnetic field. It is generated when the motion of electrons in atoms and molecules is accelerated by an external energy source. The vibrating, negatively charged electron generates an electric field that in turn induces the magnetic field. The coupled electric and magnetic oscillations move outward from the vibrating charge at a critical speed, such that the two fields mutually induce each other indefinitely without gaining or losing energy. The planes of oscillation of the two fields are at right angles to each other, and both are perpendicular to the direction of travel (**Figure 4.1**). Electromagnetic radiation is therefore a **transverse wave**.

Because electromagnetic radiation possesses these wave properties, it shares with sound many of the same wave-related characteristics. As with sound, electromagnetic waves can be described by their **frequency** (f), which is inversely proportional to the **wavelength** (λ). Algebraically,

$$f = c/\lambda$$

where c is the speed of light. The traveling speed of electromagnetic radiation in a vacuum is approximately 3.0×10^8 m/sec for all frequencies. The frequency, or wavelength, of radiation depends on the composition of the excited material that produced it and the energy of that excitation. We perceive different frequencies of light as different colors. Light also varies in **intensity** and obeys the inverse square law: intensity decreases with distance from a point source as the inverse function of the distance squared. **Attenuation** of

light with distance is both wavelength specific and medium specific, so selective filtering of certain frequencies can occur. Light is directional and travels at different speeds in different media. It thus shares with sound the wave properties of **reflection**, **refraction**, and **diffraction** (see Chapter 2). Light waves differ from sound waves in that they require no medium for transmission, and in fact travel fastest in a vacuum.

In other respects light behaves like a rapid stream or **flux of massless particles**. The particles are viewed as packets of energy called **quanta**, or, for radiation in the visible range, **photons**. Quanta contain discrete amounts of energy; how much energy each quantum contains depends on the frequency of the radiation. Energy per quantum is given by:

$$E = hf$$

where E is energy in joules, h is Planck's constant, and f is frequency. The higher the frequency, and thus the smaller the wavelength, the greater the flux of energy. One way to envision the combined wave and particle characteristics of electromagnetic radiation is to view it as a rapid stream of minute elusive particles that creates a pulsating electrical disturbance in space.

The properties of any given light **beam** (a population of waves) depend upon the mix of frequencies that compose it (**Figure 4.2**). If all of the waves have the same frequency, the light is **monochromatic**, and if all of the waves are also in phase, the light is **coherent**, as in a laser beam. Natural light from the sun or a light bulb is spectrally complex and **incoherent** because it contains waves of many different frequencies and phases. When the electric (and therefore also the magnetic) vectors of the light are aligned, the light is **plane-polarized**.

Naturally occurring electromagnetic radiation spans a very wide range of frequencies, wavelengths, and energies. The tradition in EMR research is to express and graph this range along the wavelength axis, rather than along the frequency axis used in sound research. **Figure 4.3** summarizes the **electromagnetic radiation spectrum**, from the largest radio waves (10^9 to 10^{15} nm) through the intermediate categories of microwaves, infrared, ultraviolet and visible light to the tiny X-rays, gamma rays, and cosmic rays that are as small as or smaller than atoms (≤ 1 nm). Visible light comprises a narrow band of wavelengths around 10^3 nm. It is customary to refer to the wavelengths of visible light in units of nanometers (nm). One **nanometer** is equal to 1×10^{-9} meters. **Visible light** ranges from about 300 to 800 nm. In Figure 4.3, the expanded visible range depicts the human color classification associated with different wavelengths. White light as perceived by humans is an equal mixture of wavelengths between 400 and 700 nm.

The degree to which electromagnetic radiation acts like a wave or a particle depends on its wavelength and energy; long-wavelength radiation acts more like a wave, and short-wavelength radiation acts more like a stream of particles. These differences occur in part as a result of the relationship

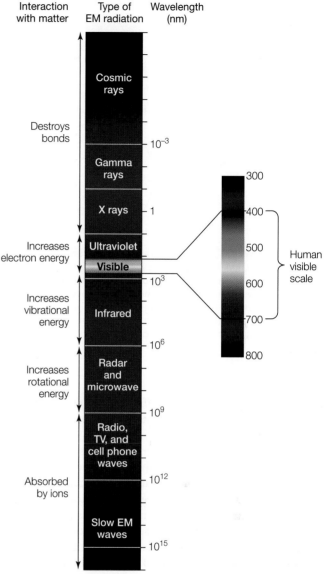

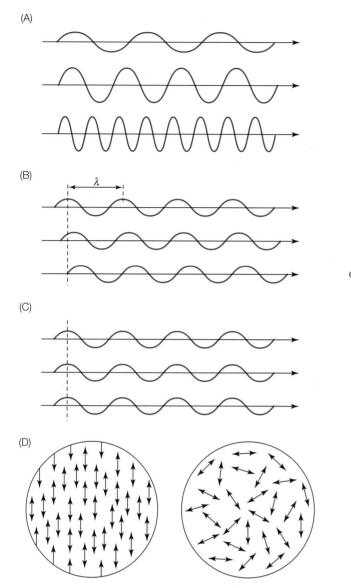

FIGURE 4.2   Complex, monochromatic, coherent, and polarized light   Parts (A–C) show the electric wave enveloped in beams of light traveling from left to right, and (D) shows the electric vectors of a beam traveling into the page. (A) Complex light contains waves of different frequencies and therefore different wavelengths. (B) Monochromatic light contains waves of the same frequency, but they are not necessarily in phase. (C) Coherent light contains waves with the same frequency and the same phase. (D) Plane-polarized light has parallel electric vectors, whereas unpolarized light has randomly oriented electric vectors.

FIGURE 4.3   The spectrum of electromagnetic radiation   Different wavelengths of electromagnetic radiation interact with matter in different ways and are therefore used for a wide variety of functions. On right, the visible light range is expanded. The boundaries between types of radiation are gradual. All wavelengths given in units of nanometers.

between the length of the wave relative to the size of the objects with which it interacts. Radio waves are meters to kilometers long and easily bend around most objects, as do long sound waves. X-rays are about the size of an atom, and they either pass right through most materials or are blocked and scattered by atomic nuclei. Gamma and cosmic rays are even smaller and contain enough energy to split atomic nuclei. Visible light is intermediate in wavelength between

radio and cosmic radiation and therefore shows characteristics of both waves and particles [53, 147].

How electromagnetic radiation interacts with molecules

Electromagnetic radiation affects molecules when it streams past them. The nature of the interaction depends on both the frequency of the radiation and the characteristics of the molecules. A brief description of this interaction will help

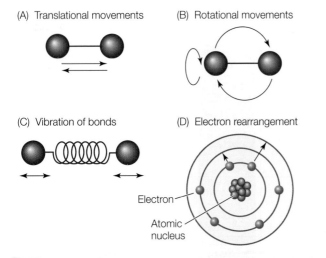

(A) Translational movements   (B) Rotational movements

(C) Vibration of bonds   (D) Electron rearrangement

Electron

Atomic
nucleus

FIGURE 4.4   Ways that molecules and atoms can oscillate (A) Translational oscillations, in which the entire molecule moves, as in Brownian motion and sound-wave transmission. (B) Rotational oscillations, in which the molecule rotates around one or more axes. (C) Bond oscillations, in which atoms within molecules vibrate relative to each other. (D) Electron orbital oscillations, in which electrons shift between orbitals.

us understand the suitability of different radiation frequencies for visual communication and the process of visual signal generation and perception.

Molecules can undergo several types of oscillations, as illustrated in **Figure 4.4** [11]. At the level of the whole molecule in gases and liquids, molecules move back and forth between collisions with each other (**translational oscillations**), and they may rotate around some axis (**rotational oscillations**). Within molecules, the component atoms vibrate toward and away from each other, with the chemical bonds between them acting like springs (**bond oscillations**). Within each atom, electrons rotate in defined orbits or shells around the nucleus. Electrons can shift from their lowest energy ground state orbital into higher energy level orbitals (**electron orbital oscillations**). For any given type of molecule, only certain frequencies of oscillations are allowed, and which frequency one finds in the molecule depends upon how much energy it possesses: the higher the energy, the higher the frequency for a particular type of oscillation. For electrons, higher energy corresponds to orbitals further from the nucleus. The energy difference between oscillatory states is lowest for translational oscillations, somewhat higher for rotational ones, intermediate for bond vibrations, and highest for electron shell transitions.

A molecule can absorb electromagnetic radiation only if the energy supplied by that wave is exactly equal to that needed to push this molecule into one of its higher oscillatory states. Because energy and frequency are proportional, this means that only certain frequencies of electromagnetic radiation can be absorbed by a given type of molecule. In fact, we can think of molecules as having their own vibrational

resonance frequencies (see Chapter 2) at which they absorb the corresponding frequency of radiation. Once energy is absorbed, it is trapped in the excited molecule until it can be lost through molecular collisions as heat (i.e., through translational oscillation) or coupled to a specific chemical reaction.

Because of the vast differences in wavelengths and energy, the different frequencies of electromagnetic radiation cause different types of oscillations and thus vary in their ability to be absorbed by natural products (summarized in Figure 4.3). Radio waves are so low in energy that they can be absorbed only by ions such as ionized gas particles and metals. Microwave radiation primarily affects rotational states. Small polar molecules such as water are especially prone to strong rotation under intense microwave radiation, the process by which moist food is heated in a microwave oven. Infrared radiation increases bond vibrations and causes essentially all materials to heat up. Visible and ultraviolet radiation affects electron states. Ultraviolet radiation is also strongly absorbed by ozone and water molecules. X rays, gamma rays, and cosmic rays contain enough energy to break bonds and destroy molecules by knocking electrons off and ionizing them [147].

Constraints on EMR wavelengths for visual communication

As we shall see in this section, most visual communication makes use of ambient light from the sun, which reflects off of most natural objects and travels very rapidly and in a straight line to the receiver. There are several reasons why the range of visible light is so narrow, and why both larger and smaller wavelengths don't work for animal communication. These reasons involve the availability of radiation from the sun, the absorption, reflection, and transmission properties of light (discussed above), and detectability constraints.

The availability of solar radiation on Earth is a fraction of that produced by the sun. The sun produces electromagnetic radiation primarily in the ultraviolet to infrared region as a result of its very high temperature. Low-energy radio waves and microwaves (which our sun does not generate in great abundance because it is too hot) are either reflected or absorbed by the ionosphere in the outer atmosphere. High-energy X-rays and cosmic radiation are strongly scattered and gradually lose energy as they collide with particles in the atmosphere. Most of the other very large and very small wavelengths are filtered out by the Earth's atmosphere. Some of the infrared radiation is absorbed by atmospheric water and carbon monoxide, but the remainder, mostly below 1000 nm, is able to penetrate and warm the Earth's surface. Most of the ultraviolet radiation below 300 nm is absorbed by the ozone layer of the atmosphere, but some reaches the Earth. Thus most of the electromagnetic wavelengths reaching the Earth's surface occur in the region of the spectrum between 300 and 1000 nm (**Figure 4.5**). About 83% of this radiation falls in the visible range. For animals living in aquatic environments, the spectrum of available light is even more

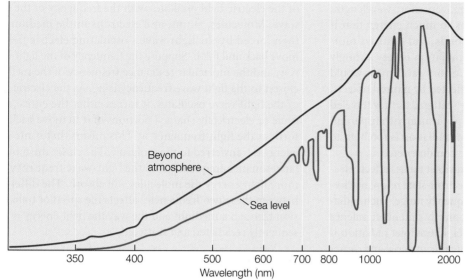

FIGURE 4.5   The frequency spectrum of irradiance from the sun The red curve shows the spectral distribution of sunlight just outside of the Earth's atmosphere. This curve is very close to that theoretically expected from black body radiation. The blue curve shows the spectral distribution of sunlight at sea level. Short wavelengths are both scattered by the atmosphere and absorbed by ozone. Long wavelengths are absorbed by oxygen, carbon dioxide, and water. The sharp dips are primarily the result of absorption by water. (After [246].)

restricted, because water strongly absorbs both infrared and ultraviolet wavelengths (**Figure 4.6**).

We turn now to the suitability of different wavelengths for visual communication and the question of detectability constraints. At the long-wavelength end of the spectrum, electromagnetic waves have such low frequencies and energies that they pass through and around nearly all nonmetallic and biological materials without being absorbed or reflected. Objects in the environment are transparent to this type of radiation. Thus radio and microwaves are neither available nor suitable as a medium for vision. Infrared radiation from the sun is far more available, and many types of molecules can absorb this frequency of radiation and become warmer in the process. Likewise, all objects in the environment with temperatures above absolute zero produce their own infrared radiation as a result of normal bond vibrations. This opens up the possibility of a type of "vision" based on the discrimination of objects in the environment that radiate and reflect different intensities of heat. Military technology uses infrared sensors to guide missiles and detect warm targets. But this modality has two limitations for communication. The first is that the thermal world is extremely noisy and variable. The second is that infrared radiation is quickly absorbed by any biological tissues. Since even the simplest receptor organ must have a cell membrane and cytoplasm between the source of the radiation and the responding molecules, the amount of energy reaching the latter will always be greatly reduced. Adding lenses and other structures to improve imaging would compound the problem further, as they would absorb more of the energy. In spite of these limitations, infrared radiation is useful to some animals. Pit vipers have evolved an infrared sensor with which they

detect the approximate location of their warm-blooded prey. The pit organ contains a thin membrane surrounded by air and is loaded with an array of neurons that respond to small changes in heat [57]. The low thermal mass of the membrane enables it to heat up quickly when the organ is aimed at an object warmer than its background. The heat receptors require five times more photons to generate a response than a visual light receptor. At best, a viper can detect a mouse at 30 cm and larger prey at somewhat longer distances [22, 134, 154]. Although vampire bats and a few insects have also evolved thermoreceptors as an aid in prey detection, an infrared detector does not come even close to an eye in its sensitivity and resolution [22].

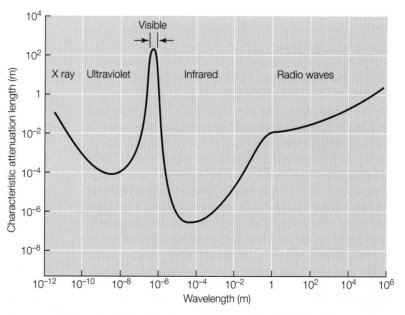

FIGURE 4.6   The transmission of different wavelengths of light by water Note the narrow but strong transmission peak in the visible range. (After [254].)

At the other end of the spectrum, very short-wavelength, high-frequency radiation contains so much energy that it ionizes atoms, breaks chemical bonds, and destroys molecules. Furthermore, this type of radiation is very strongly scattered, so little reaches the Earth. Short wavelengths would be largely scattered rather than reflected by natural objects, and reflected waves would again be scattered as they traveled to the receiver, so objects would appear nearly transparent. Moreover, the energy in this type of radiation is too high to be absorbed, so biological receivers couldn't detect it.

Radiation in the visible and ultraviolet ranges affects electron states. Many molecules have electron-state resonance frequencies corresponding to this frequency range. Such radiation is sufficiently powerful to temporarily shift outer-valence electrons up to another energy level. Ultraviolet radiation is especially strongly absorbed by all organic material and causes photoionization and burning of the skin. Luckily, most of this harmful radiation is absorbed by the Earth's atmosphere. The intermediate wavelengths of visible light can be absorbed without ionizing molecules, so they are potentially detectable without being destructive to biological tissues. These wavelengths also possesses ideal transmission properties for signals: they can reflect off objects and travel straight to the receiver to generate a fine-scale visual map of the external world. Finally, water and a few other materials *are* transparent to visible light. Since life and vision evolved in the sea, this fact alone could explain the restricted wavelengths for vision.

How visible light interacts with matter

Despite the implication in physics textbook illustrations that light bounces off objects as a ball bounces off a wall, this is not what happens at all. As mentioned in the previous section, visible light interacts with molecules by affecting electron states. Here we continue that discussion to understand the processes of reflection, refraction, and scattering.

MOLECULES UNDER THE INFLUENCE OF A LIGHT BEAM
Individual atoms usually consist of a balance of positive and negative charges and so are electrically neutral. However, if some of the atom's electrons are only loosely bound to the nucleus, an external electrical field may pull these electrons to one side of the atom. The atom is then said to be **polarized**. Similarly, when atoms join to form molecules, the combination may not have equal numbers of electrons at all locations. Such a molecule is said to be polar because one end may be more negative than another. If an electric field were applied to such a molecule, the molecule would tend to align itself with its positive side facing the negative side of the field, and vice versa. Alternatively, two atoms in a molecule might be pulled apart or pushed together under the influence of an external electric field. Whether it is electrons, atoms, or molecules that move, the application of an electrical field to a medium will generate some separation and alignment of the charges within it.

The electric field of a light wave is just such a polarizing force. However, rather than being a static force, the polarity of the electric field oscillates with the frequency of the light wave. Molecules, atoms, and electrons in the medium are then forced by the light wave's oscillating electric field to move back and forth. Suppose the frequency of the light wave is w, and the molecular resonance frequency of the medium closest to the light wave frequency is w_o. As the electric field of the light wave oscillates, it forces either the entire molecule or electrically charged portions of it to move back and forth at the light frequency w. This means that some light energy is transferred to the molecule. The closer the molecular resonance frequency is to the light-wave frequency, the more light energy the molecules will absorb. The difference between these two frequencies affects the speed of transmission through a medium and the way the light energy is subsequently reradiated and scattered.

SPEED OF LIGHT IN DIFFERENT MEDIA   The speed of an electromagnetic wave is maximal when traveling through a vacuum such as outer space. Even in a vacuum, however, it takes a finite time to establish the electric and magnetic fields, and thus the speed of light is not infinite (recall that $c = 3.0 \times 10^8$ m/sec). When light propagates through matter, the speed of propagation is generally reduced. This is because both the electrical and magnetic counterfields set up in the medium work against the external fields of the light wave, and thus hinder the process of inducing oscillations within the medium's molecules, atoms, or electrons.

The **speed of light transmission** through a transparent medium depends on the strength of the electrical and magnetic counterforces of the polarized molecules, the density of molecules, and the difference between the resonance frequency and the light-wave frequency. The density of atoms decreases the speed of propagation, because the more molecules or atoms there are available for polarization, the larger the counterfield that can be induced in a material by an external electric force. The closer the wave's frequency is to the resonance frequency of the medium, the slower the wave will travel. When the two are equal, the speed of propagation in the medium at this frequency approaches zero: all waves at this frequency are absorbed, and none are transmitted. If we denote the speed of transmission in a medium as v (where $v \leq c$), then a useful measure of relative speed is the ratio $n = c/v$, which is called the **index of refraction**. The more the medium slows down the propagation of electromagnetic waves, the larger the value of n. Hence n is small for air (1.00028) because of its low density of molecules, and is larger for liquids and solids such as water (1.33), glass (1.5) and diamond (2.4).

Metals are exceptions to these patterns. Metallic atoms are electron donors, and large arrays of these atoms form a solid medium in which the nuclei are densely packed in a matrix, but the valence electrons are free to move along the matrix. The resonance frequency of these unbound electrons is effectively zero. When the driving frequency of the electromagnetic wave is much greater than the resonance frequency, the motion of the unbound electrons in the presence of a wave is 180° out of phase with the driving wave. The opposing phase and complete freedom of the moving electrons leads to a counterfield in the metal that is exactly opposite to the external electric field. The result is that the electric field of an incident light wave is completely canceled by the counterfield inside the metal. Put another way, the speed of light propagation inside the metal is zero, but all of the incident light is reflected; there is no internal propagation, and the index of refraction is positive infinity.

SCATTERING   When molecules come under the influence of a light beam and absorb some amount of energy, they subsequently reradiate that energy. The reradiated waves are generally of the same frequency and wavelength as the incident wave. This reradiation process is called **scattering**. The nature of scattering in a given medium is multifaceted: it depends on the driving frequency of light; the resonance frequency of the molecules; and the size, density, and spatial pattern of the scattering units. These units may be atoms, molecules, particles, or larger structures, and we shall refer to them collectively as **scatterers** in subsequent discussions.

In Chapter 2, we learned that the scattering of a sound wave depends on the size of the encountered object relative to the size of the wave. The scattering of light waves conforms to similar basic principles. **Rayleigh scattering** occurs when the scatterers are much smaller than the wavelength of light, and Mie, or diffractive, scattering occurs when scatterers are similar to or larger than the wavelength. When a complex (white) light beam undergoes **Mie scattering**, all wavelengths are similarly scattered, and the result is incoherent white light. Familiar examples of matter that appears white due to Mie scattering include white milk, in which light is scattered by suspended fat droplets, and white clouds caused by scattering from water droplets.

Rayleigh scattering requires not only small scatterer size, but also a low density of scatterers, such that adjacent scatterers oscillate *independently* of each other. Under these conditions, the intensity of the reradiation from a single particle is strongly frequency-dependent according to the relationship:

$$I_s = \frac{\text{constant}}{\lambda^4}$$

Plugging in values for the wavelengths of red versus violet light, one can easily show that the intensity of scattering from a small particle is about four times as great for short blue and violet wavelengths as for long red wavelengths (**Figure 4.7**). This is ultimately a consequence of the higher energy of shorter wavelengths. The most familiar example of this frequency-dependent aspect of Rayleigh scattering is the blue color of the sky. Our atmosphere is composed of a mixture of small gas molecules at low density. Neighboring molecules are sufficiently far apart that they do not oscillate in synchrony under the influence of the same light wave. Sideways reradiations therefore are not likely to be canceled out with destructive interference as in a liquid or solid medium. Under these Rayleigh scattering conditions, overall scattering

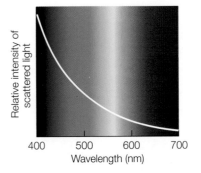

FIGURE 4.7   Frequency-dependent Rayleigh scattering Violet and blue wavelengths are scattered more than longer wavelengths.

is high, and shorter (violet/blue) wavelengths are scattered more than longer (red) wavelengths. Rayleigh scattering from molecules and small particles in the Earth's atmosphere is responsible for the blue daytime sky (**Figure 4.8**). Although water is also blue, the cause is not Rayleigh scattering, because water molecules are densely and uniformly packed and thus oscillate *nonindependently* under the influence of light waves. The blue hue, transparency, and reduced scattering of water are explained below.

TRANSPARENT VERSUS OPAQUE MATERIALS   Now we are in a position to answer the question of why some liquid and solid media easily transmit light while others do not. A key determinant of whether a medium is transparent to light or opaque is the difference between the light frequency and the medium's resonance frequency [249]. When the medium's resonance frequency is equal to the frequency of the light wave ($w_0 = w$), successive cycles of the light wave add up to produce very large amplitude oscillations in the molecules. As we saw in our discussion of sound (see Chapter 2), this buildup occurs only when the external oscillator has a frequency equal to a resonance frequency of the driven system. In the case of light as the external oscillator, the amplitude of the vibrations induced in the molecule build up to a critical point at which the molecule jumps into a higher energy state, absorbing all of the incident light energy completely. The excited molecule will momentarily vibrate violently, knocking into adjacent molecules to generate heat. Because the light energy is completely absorbed, there is no electromagnetic radiation to propagate on to the next molecule: absorption thus stops the propagation of a light wave and the medium is **opaque**. When the medium's resonance frequency is not close to the light frequency ($w_0 \neq w$), the molecules do not resonate when exposed to light, and the amplitude of any induced vibrations remains at a low level. Any energy transferred to the molecule from the radiation is then reradiated by the molecule at exactly the same frequency as that absorbed. The passage of a light wave through a **transparent** liquid or solid medium is thus the result of a rapidly propagated series of absorption and reradiation events through the medium's layers, and not the result of light waves merely

(A)

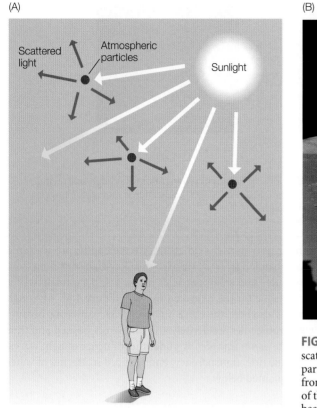

(B)

FIGURE 4.8 **Sky color** (A) Shorter wavelengths are more likely to be scattered multiple times by gas molecules and reach the receiver from other parts of the sky, whereas longer wavelengths are more likely to travel straight from the sun to the receiver and therefore appear slightly yellow. (B) A view of the moon's surface. Although the sun is shining brightly, the sky is black because there are no atmospheric particles to scatter the sunlight.

traveling through and around the medium's atoms, as often happens in a gas. A transparent medium is therefore composed of atoms and molecules whose resonance frequencies are quite different from the frequencies of visible light. Water, for example, has resonance frequencies in the ultraviolet and near-infrared regions but not in most of the visible range. It absorbs ultraviolet, infrared, and some red radiation, but transmits other visible light frequencies (see Figure 4.6). Water has a blue hue because it selectively absorbs longer red wavelengths within the visible range.

The second key to transparency is the uniformity of scatterer distribution and density. The reradiated light from each wave-stimulated scatterer is a secondary wave of the same frequency called a **wavelet**. The relative phases of the wavelets emanating from adjacent scatterers depend on whether the scatterers were struck by the same part of the electric wave cycle in the incident light wave or not. As mentioned above, in a gas the scatterers are sufficiently far apart relative to the wavelengths of ultraviolet, visible, and infrared light that reradiations have random phases and show no consistent pattern of interference. In liquids or solids, whole layers of scatterers are sufficiently close together and uniformly arrayed that they reradiate wavelets in phase with each other (coherently, as in Figure 4.2) and out of phase with the wavelets reradiated by more distant layers of molecules. This facilitates complex patterns of interference between the wavelets.

It turns out that reradiated wavelets traveling in the forward direction through a transparent liquid or solid (the direction of propagation of the external light wave) add up constructively and propagate the wave in the forward direction. All wavelets reradiating in other directions (sideways and backwards) will destructively interfere with those emanating from neighboring molecules and cancel each other out (**Figure 4.9**). This basic process, analogous to the propagation of a sound wave, explains the forward straight-line transmission of waves: only the forward-moving components are

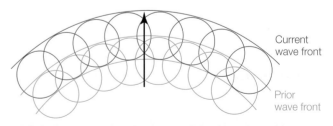

Current
wave front

Prior
wave front

FIGURE 4.9 **Huygens' principle of wave propagation** This theory explains why waves in a uniform medium travel in a straight line. Every point on a wave front can be thought of as a new point source for wavelets generated in the direction the wave is traveling (arrow). The wave front (arcs) at any instant conforms to the envelope of spherical wavelets (circles) emanating from every point on the wave front at the prior instant. Wavelets have the same speed of travel as the overall wave. Adjacent wavelets on the same front oscillating in phase with each other will undergo destructive interference, canceling out most sideways radiation, and wavelets coming from the next front will cancel out most backward radiation, so most of the reradiating energy occurs in the forward direction.

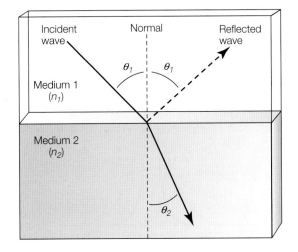

FIGURE 4.10   Reflection and refraction at a boundary A light beam traveling from a medium with a faster speed, such as air, into a medium with a slower one, such as water. The normal is a line perpendicular to the boundary between the two media. The angle of incidence, θ_1, is measured from the normal. In this case, the incident wave is bent toward the normal in the second medium θ_2. If the speed of the wave were faster in the second medium compared to the first medium, the wave would be bent away from the normal. Compare to Figure 2.14.

in phase. Thus in liquids and solids, incident electromagnetic radiation is generally propagated in straight lines through the material. Some light may be scattered to the sides in all materials during this propagation; the more ordered the scatterers in the medium and the closer together they are, relative to the wavelengths of the light, the lower the amount of sideways scattering.

REFLECTION AND REFRACTION AT BOUNDARIES   When a ray of light encounters a boundary between two media with different transmission speeds, a portion of it may be absorbed, a portion is reflected back into the first medium, and the rest is refracted and transmitted through the second medium (**Figure 4.10**). Reflection and refraction are actually just special cases of scattering by the boundary layers of molecules in the second medium, and the backward and forward components of the reradiating wavelets, respectively. At Web Topic 4.1 you can see an animated simulation of this process. For the idealized case of a perfectly smooth boundary, the angle of the reflected ray is equal to the angle of the incident ray. The tradition in light wave research is to measure these angles from "normal" incidence, represented as a line perpendicular to the surface.

The difference in the refractive indices of the two media at a boundary affects reflection and refraction in several ways. First, the larger the difference in the indices, the greater the proportion of light reflected. When the transmission speeds of the two media are not very different, most of the light will cross the boundary and enter the second medium. Second, the greater the difference in indices, the more strongly the

transmitted ray will be refracted (bent) at the boundary relative to the normal. Third, the direction in which the refracted ray is bent depends on which medium has the larger index. When the second medium has a larger index of refraction (slower transmission speed) than the first medium, the ray will be bent *toward* the normal (see Figure 4.10). When the first medium has the larger index, the ray will be bent *away from* the normal. Fourth, light reflected at a low-to-high index boundary undergoes a half-wavelength phase shift, but this shift does not occur at a high-to-low boundary. Finally, light reflected from a surface or scattered in the sky is partially or wholly polarized in the process. These details of reflection, refraction, and polarization do play a role in the production and reception of visual signals by some animals. Certain species can perceive the direction of polarized light and use it for foraging and navigation. For further explanations, video clips, and quantitative expressions for reflection, refraction, and polarization, see Web Topic 4.1.

Web Topic 4.1   *Light wave meets boundary*
View animated simulations of the way that scattered light is reflected and refracted at a boundary between two media in the form of reradiating wavelets. Obtain quantitative expressions for the effects of the indices of refraction in the two media on angles of reflection and refraction, proportion of reflection, and degree of polarization.

The intensity of reflection at a boundary also depends on the properties of the surface. If the surface is smooth, as depicted in Figure 4.10, incident light waves from a directional beam source are coherently reflected at certain viewing angles, giving the surface a bright, shiny, highlighted appearance. This is called **specular reflectance**. If the surface is uneven and bumpy, the incident light from a directional source is reflected in multiple directions and the reflection appears more uniform, matt, and lower in intensity. This is called **diffuse reflectance**. Moreover, for nonmetalic materials, light always penetrates the surface zone of a reflecting medium and interacts with several layers of molecules before being reradiated back out. The distance light travels into a medium is called the **skin depth**, and is approximately equal to one half-wavelength of the incident wave. Most biological tissue surfaces are semitransparent and heterogeneous within this light-penetrating zone. Internal boundaries can produce specular and diffuse reflectance, providing animals with ample opportunities to modify light waves into diverse visual signals as we shall see in the next section [199].

Light-Signal Generation

Visual signals are produced by the emission of light waves from the body's surface. Animals employ three very different mechanisms for generating light signals: pigments, structured surfaces, and bioluminescence. The first two mechanisms require the use of ambient **reflected light**, while the third involves **self-generated light**. For most light-signal generation mechanisms, coupling into the medium is part of the signal production process and is therefore somewhat simpler than it is for sound signal generation, but there are some specializations for enhancing reflected light signals. The key characteristic of light signals is their color, so before we begin to discuss production mechanisms, we need to define color.

Describing color

Regardless of whether the signal is produced by reflected or self-generated light, there are several quantitative properties we can use to describe all light signals.

The **brightness** of a light signal is its overall intensity, measured as the total photon flux produced by a sender in the units of radiance, photons s^{-1} sr^{-1} m^{-2} (where sr specifies the solid angle of the measuring instrument; see Web Topic 4.2). The brightness of a reflected light signal is a function of both the range of wavelengths reflected and the surface structure of the object or signaling organism. The wider the range of wavelengths reflected, the greater the overall intensity. A smooth surface will produce a higher intensity signal than a rough surface. If the visual receptor can discriminate between different light intensities, it can distinguish objects with subtly different reflectance properties. The brightness of a self-generated signal depends on the amount of energy used to produce the signal.

Two additional parameters are needed to describe the spectral composition of signals in addition to brightness: **hue**, which refers to the dominant wavelength or frequency of the light; and **saturation** (also called **chroma**), which refers to the purity of the dominant frequency (the presence of other frequencies adds gray to the dominant frequency and makes the color less intense). Both reflected and self-generated light signals may be characterized by their color. If the visual receptor can discriminate different wavelengths of light, it can enhance its ability to discriminate between objects with different reflectance properties beyond what is possible with the use of only intensity differences. As we saw in Figure 4.3, wavelength-specific color classes for humans range from 400 nm for violet to 700 nm for red. Some organisms can detect wavelengths as short as 300 nm in the ultraviolet region and as long as 800 nm in the near-infrared region.

A given color signal patch can be described graphically with a reflectance spectrum. **Figure 4.11** shows some plots of the reflectance spectra of a few colors. A patch that reflects all wavelengths equally appears white, and a patch that absorbs

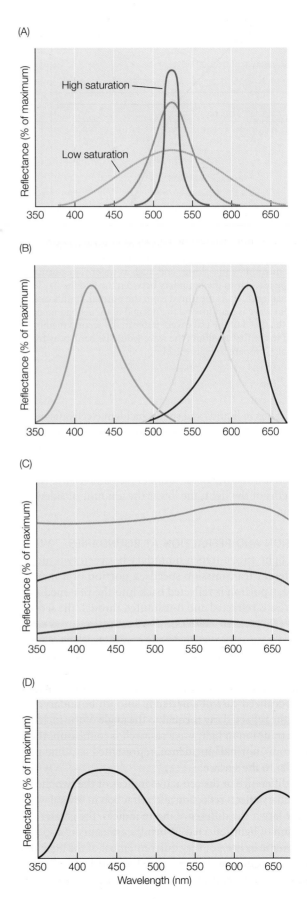

FIGURE 4.11 Idealized reflectance spectra for some colors The line colors on the graphs indicate the approximate hue represented by the spectral curve. (A) The color green at different levels of chroma or saturation. (B) Three different hues of similar saturation. (C) Three unsaturated colors of very different brightness. (D) Some colors, such as purple, are not monotonic but have two hue peaks. Natural colors do not necessarily have such smooth, symmetrical curves as those shown here.

all wavelengths appears black. Hue is defined by the wavelength of maximum reflectance for colors with a unimodal reflectance curve. Curves with a high peak of reflectance over a narrow wavelength range represent highly saturated colors, and broader (flatter) curves represent less saturated colors. The area under the curve is an approximate measure of brightness.

Humans see and categorize colors differently than most animals [87]. Several quantitative color-naming schemes have been developed, including the CIE chromaticity scale, the Munsell color system, and Frank Smithe's Naturalist's Color Guide [135]. However, all of these systems are limited to colors perceived by the human eye, which is insensitive to the ultraviolet range that many other animals can perceive. The generalized color classification system shown in **Figure 4.12** clearly differentiates the properties of hue, saturation, and brightness. The entire spectrum of hues that a species can perceive is ordered around the circle. Colors positioned around the edge of this circle are fully saturated, and the angle relative to some baseline hue (or the x, y coordinate in the horizontal plane) is a quantitative measure of the hue. Adding gray to these saturated colors leads to desaturation, as you see in the colors toward the center of the circle. The degree of saturation is therefore quantified as the horizontally measured length of the radius from the central vertical line to the color point. The vertical axis (z coordinate) represents brightness and contains shades of gray ranging from white at the top apex to black at the bottom apex. Any color can be represented within this three-dimensional space. To use such a system for a particular species, the range of hues in the circle would be restricted to those the species can perceive, and the spacing of the hues around the circle would reflect perceived hue differences (i.e., spacing would be irregular with respect to wavelength).

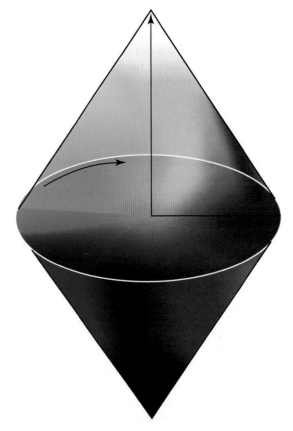

FIGURE 4.12   Three-dimensional double-cone representation of color space   The compass points of the horizontal circle represent the hues that the receiver can perceive. Points on the edge of the circle are spectrally saturated, and horizontal radii of different lengths from the central vertical line represent degrees of saturation. The vertical axis is the gray scale (brightness), with white at the top and black at the bottom.

Web Topic 4.2 *Quantifying light and color*
We have been able to measure the brightness of light for some time. Recent technological developments have made it much easier to quantitatively describe the hue and saturation of ambient light and animal color patches, including wavelengths that we cannot perceive.

Light signals can also be characterized by their **spatial characteristics**. These include the size, shape, surface features, color pattern, and contrast of color patches and body structures of the sender, as well as the posture, position and location of the sender relative to the receiver [199]. The visual acuity of the receiver, or its ability to make an accurate and fine-scale spatial map of all incident light rays entering its field, determines its ability to discern both small details of shape and pattern as well as to accurately locate a sender in space. Finally, **temporal variability** in intensity, color, and spatial characteristics can be used to generate a wide variety of signals. Complex patterns of colors and surfaces, color changes, flashing lights, changes in size and shape, and limb movement patterns are typical visual displays exhibited by animals. The last two features are more difficult to quantify than spectral features of signals. We shall deal with them in the final section of this chapter, on the modification of visual signals.

Pigments

Pigments are chemical compounds whose molecules undergo **selective absorption** of certain wavelengths within the visible light range via electron orbital shifts and scatter the remaining wavelengths. The receiver perceives the color of the scattered light. For example, a pigment that absorbs wavelengths in the violet and blue range will scatter green, yellow, and red and appear yellow, while a pigment that absorbs violet, blue, green, and yellow will scatter and appear red. All organic molecules absorb electromagnetic radiation in the short-wavelength (<300 nm), high-energy ultraviolet range, and if this is the only region in which they absorb, they appear colorless. It takes a special kind of molecular structure to absorb longer-wavelength, lower-energy radiation in the visible range.

A common feature of color-producing pigments is a long chain or network of **conjugated double bonds**. Conjugated double bonds consist of a series of carbon atoms joined by alternating single and double bonds. Such molecules have several resonance forms, because the electrons responsible for the double bonds can move from one side of the carbon atom to the other. These mobile electrons can more easily absorb light energy by shifting up to a higher energy level. When this shift occurs, the molecule is in an excited state. When these molecules collide with other molecules, the energy they have trapped is quickly lost as heat. The resonance energy required to shift an electron up to a higher orbital, and therefore the wavelengths of radiation absorbed, depend on the size of the molecule, the configuration of the conjugated double bonds, and the presence of oxygen-containing functional groups [146]. Five main chemical groups possess these color-producing characteristics: carotenoids, porphyrins, pteridines, purines, and melanins. The colors generated by such pigments are typically yellow, orange, red, black, and white. Natural true blue pigments are extremely rare. But there are some blue-green pigments, and as we shall see in a later section, animals produce violet, blue, and some green colors by means other than pigmentation.

In this section we briefly describe the chemical structure of each of the five pigment types, and its range of colors. We also mention the types of biological tissues in which these pigments are found, their biosynthetic pathways, and some costs and benefits of the pigments to the animals that bear them. In the figures we illustrate some of the **absorbance spectra** of common pigments and representative examples of animals bearing color patches with these pigments. In order for a color patch to intensely reflect the nonabsorbed colors, the pigment-containing layer on the surface of the animal may have to be coupled with a reflecting layer beneath it [214].

CAROTENOIDS   Carotenoid pigments are 40-carbon molecules that consist of a long symmetrical conjugated double-bond chain and a six-carbon ring at each end. The different forms of animal **carotenoids** have end rings that vary in their degree of conjugation and in the structure of their functional groups, and these structural differences affect their absorbance spectra. Colors typically reflected are yellow, orange, red, and pink. Red carotenoids, such as **astaxanthin** and **canthaxanthin**, generally possess one or two ketone functional groups, which shift the absorbance curve to longer wavelengths compared to yellow carotenoids such as **lutein** and **zeaxanthin** (**Figure 4.13A**). Carotenoids are widely used throughout the animal kingdom to generate color displays; they are found in most invertebrate phyla and all vertebrate classes. They are found in almost every type of integumentary tissue, including skin, scales, feathers, eggs, beaks, wattles, and eyes. Mammalian hair is the only tissue that has not been found to contain carotenoid pigments [15, 123]. **Figures 4.13B–E** show a few examples of known carotenoid signal color patches in a range of species. Colored tissues typically contain a mixture of carotenoid pigments, with differing proportions of red and yellow pigments determining the observed hue [75, 98, 200, 206, 226].

Carotenoids are lipophilic biochemicals that first evolved in archaea to reinforce cell membranes, and later developed into accessory light-harvesting pigments in photosynthesizing plants, algae, fungi, and bacteria. Animals lack the ability to synthesize these chemicals from precursors (with one unusual exception [136]), so they must obtain them from their diet by eating plants or algae directly or by ingesting herbivorous prey. The ingested carotenoids are transported through the blood directly to the integumentary tissue. The ancestral strategy was to sequester the unmodified pigment directly into the integument, but many animals have secondarily evolved biosynthetic processes that modify the compound to produce different colors. For example, yellow lutein, which is commonly available in the diet, can be converted to astaxanthin in some animals to produce a red hue [21, 123, 206, 207]. This manufacturing process occurs in the integument. Because carotenoids are lipid-soluble, they are often deposited within small oil droplets in skin, feather keratin, cuticle, scales, or other tissues. Mammals ingest β-carotene and split the molecule in half to make vitamin A, which is used for purposes other than color signals. Some crustaceans, especially shrimp, crayfish, and lobster, produce a dark violet-blue pigment called **crustacyanin** in their exoskeletons by binding carotenoids to large protein molecules in a **carotenoprotein complex** [20, 34, 133, 235, 248]. This is one of the few cases in which the absorbed wavelength is shifted to the long-wave end of the visible spectrum. The reason shrimp and lobster turn red when cooked is that the carotenoid-protein bond is broken, and the red color of the simple carotenoid (usually astaxanthin) is revealed.

Carotenoids play an important role in maintaining an animal's health. Because of their conjugated double-bond system, the molecules readily accept unpaired electrons from single oxygen ions, hydroxyl/peroxyl radicals, and other damaging metabolic by-products and thus protect cells from oxidative stress and ultraviolet damage [63, 193]. Carotenoids are also utilized during immunochallenges and embryonic growth. Because animals cannot manufacture the compounds and may have limited access to carotenoids in their diet, they may face a trade-off in allocating carotenoids to eggs, growth, and health versus color patch production. It should therefore not be surprising to learn that the intensity and hue of carotenoid-based color patches is often dependent on an animal's condition. Many studies of fish and birds have shown that poorly fed or parasitized individuals exhibit reduced color intensity [15, 68]. Carotenoid-based signals therefore have a strong potential to honestly reveal to receivers the health of the sender.

PORPHYRINS   **Porphyrins** are nitrogen-containing molecules with four pyrrole rings. Variants differ in whether the rings are laid out in an open chain or connected at the ends into a super-ring, and whether or not they are chelated with a metal ion. They possess the typical long series of conjugated

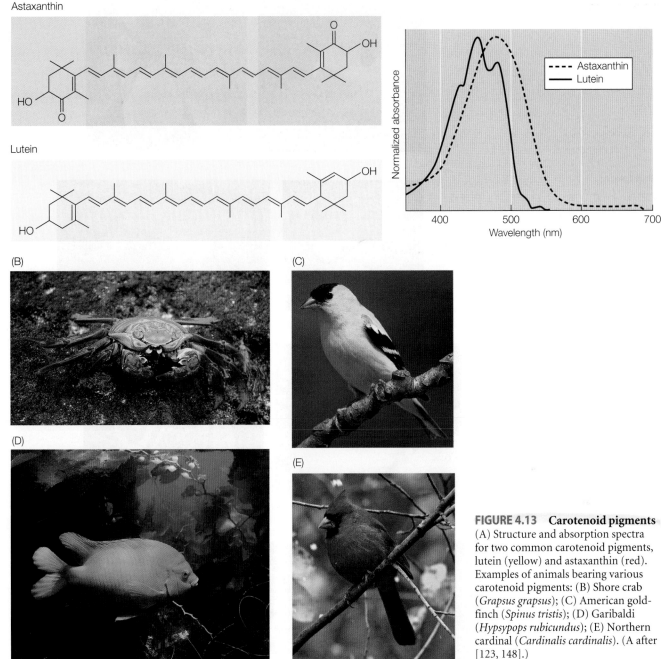

FIGURE 4.13   Carotenoid pigments
(A) Structure and absorption spectra for two common carotenoid pigments, lutein (yellow) and astaxanthin (red). Examples of animals bearing various carotenoid pigments: (B) Shore crab (*Grapsus grapsus*); (C) American goldfinch (*Spinus tristis*); (D) Garibaldi (*Hypsypops rubicundus*); (E) Northern cardinal (*Cardinalis cardinalis*). (A after [123, 148].)

double bonds that give them their pigment characteristics. As with carotenoids, the bond system can confer health benefits by absorbing damaging oxidative radicals. The pigments are manufactured by animals from a glycine precursor (an amino acid) [125].

Metalloporphyrins are super-ring molecules containing a zinc, magnesium, manganese, iron, or copper ion. Chlorophyll is a metalloporphyrin with a magnesium ion chelated in the center of the ring. Chlorophyll is of course the plant photosynthetic pigment that absorbs both blue and red light and reflects green. Hemoglobin is a metalloporphyrin with an iron ion at the center that produces a bright red color. Small capillaries near the surface of exposed skin patches produce red-hued signals in a variety of bird and mammal species [149]. Common examples of the use of blood to produce red colors can be seen in the wattles and bare skin facial areas of many birds (**Figure 4.14A**) and the estrous rump swelling of some female primates. In an unusual group of African birds called turacos (Musophagidae), copper-containing metalloporphyrin super-ring compounds known as **turacoverdin** and **turacin** are responsible for the color of green and red feathers, respectively (see Figure 4.14A). The copper is

(A) Porphyrin

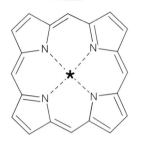

(B) Biliverdin

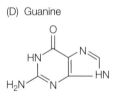

(C) Pteridines: pterin and flavin

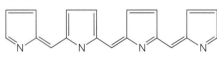

Pterin

Flavin

(D) Guanine

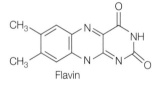

(E) Psittacofulvin

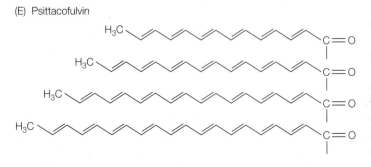

◀ FIGURE 4.14 Other pigments (A) Metalloporphyrins are super-ring porphyrin compounds with a central metal ion indicated by an asterisk. The unique red and green pigments of turacos (here, *Tauraco erythrolophus*) contain a copper (Cu) ion. Red skin, as in the wattle of this ground hornbill (*Bucorvus leadbeateri*), is colored by blood, with its hemoglobin metalloporphyrin pigment containing iron (Fe). The green pigment in plants, chlorophyll, is a metallopyrphyrin with a central magnesium (Mg) ion. (B) Biliverdins are chain porphyrins that produce the blue-green color of some bird eggs (here, American robin *Turdus migratorius*) and the blue-throated wrasse (*Notolabrus tetricus*). (C) Pteridines are crystalline pigments; pterins are responsible for the yellow, orange, and red colors of butterfly wings (here, *Colias philodice*), and riboflavin contributes to the yellow and orange skin color of these fire salamanders (*Salamandra* spp.). (D) Guanine produces the white specular reflectance of fish scales and the white iris of some birds (here, white-eyed vireo, *Vireo griseus*). (E) Psittacofulvins are unique pigments that generate the red, pink, orange, and yellow colors of parrots such as this scarlet macaw (left, *Ara macao*) and galah (right, *Eolophus roseicapilla*). (After [121, 125, 227].)

obtained from the birds' fruit diet. (Copper is a toxic ion common in fruit, so this unusual pigment may have evolved to detoxify the copper [86].) It is estimated that it takes about three months of feeding to accumulate sufficient copper for a complete feather moult [137]. The color of an individual turaco's plumage therefore has the potential to reveal important information about its health [125].

Natural super-ring porphyrin pigments (without a metal ion) are responsible for brown coloration in the spotting of some bird eggs, in the downy feathers of some nestling birds, and in the adult feather plumage of owls and bustards [125]. The pigments do not absorb infrared light, so they may function to reduce overheating of eggs and nestlings in open-cup nests. **Bilins** are open-chain porphyrins that generate blue-green colors. Biliverdin is believed to be a breakdown product of hemoglobin, which some birds use to produce blue-colored eggshells (**Figure 4.14B**). Although we would not normally think of eggs providing a color signal, in fact this idea has been proposed to explain the variation in blue color of the eggs of the pied flycatcher (*Ficedula hypoleuca*). Males are more likely to feed the nestlings in their nest if their mates produced bluer eggs. The female may be signaling her quality to the mate by depositing the valuable pigment in the eggshells [139, 140]. Blue-green biliverdins have also been found in butterflies and some reef fish [14, 46, 201].

PTERIDINES **Pteridines** are heterocyclic, nitrogen-containing by-products of the amino acids adenine and guanine, important components of DNA. The molecular backbone is a two-pyrimidine ring. These pigments are responsible for red, orange, yellow, and white colors (**Figure 4.14C**). Pteridines typically occur in crystalline form, as platelets or rod-shaped angular crystals. When densely packed, they play an important role as a coupling mechanism to reflect light. **Pterins** are one type of pteridine pigment found in the eyes and wings of insects, in the skin and eyes of lower vertebrates

(fish, amphibians, reptiles), in the eyes (iris) of birds, and the feathers of penguins [7, 14, 126, 159, 160, 202]. They usually produce yellow and orange colors, and often secondarily produce yellow fluorescence, a phenomenon described later in this chapter. Reptiles and amphibians often combine pterins and carotenoids to produce yellow and red skin coloration [224]. **Flavins** are a second type of pteridine pigment with an additional benzene ring. Riboflavin is a yellow pigment found in fish, toads, and salamanders [183]. It must be acquired from the diet (as vitamin B_2), and is critical for organismal growth and survival because of its important coenzyme role in the metabolism of carbohydrates, fats, and proteins [125]. **Purines** are a related class of pigment consisting of a pyrimidine ring fused to an imidazole ring. The most common one, **guanine**, is a silver or white crystal responsible for the specular reflectance of fish scales (**Figure 4.14D**) [26, 177].

PSITTACOFULVINS Even though carotenoids are readily available in the food and circulating blood of parrots, the psitticines do not utilize carotenoids as feather pigments like many other birds do [125]. Instead, parrots have evolved a unique set of pigments to generate their colorful red, pink, orange, and yellow plumage. **Psittacofulvins** are lipid-soluble linear polyene chains of seven to nine conjugated double bonds. Red parrot feathers appear to contain a mixture of these different-sized compounds (**Figure 4.14E**) [121, 227]. The molecules are manufactured by the birds during feather construction from basic building blocks, in contrast to diet-derived carotenoids. As a consequence, the expression of the color is relatively independent of the bird's condition [13, 113, 114]. Like carotenoids, psittacofulvins are good free-radical scavengers, but their absence in circulating blood may prevent them from conferring any health benefits [13, 111, 121]. The chemical structure of yellow psittacine pigments has not yet been determined. They contain no nitrogen, so they are not pterins. Some yellow parrot feathers fluoresce under ultraviolet light [3, 173, 182], ruling out carotenoids, which do not fluoresce. The yellow pigments are most likely to be psittacofulvin variants.

MELANINS **Melanin** is a generic term for a group of dark-colored pigments found in a wide variety of vertebrates, invertebrates, and plants [124]. Melanins are large polymer molecules composed of a core indole backbone. The indole unit contains a benzene plus a pyrrole ring with one nitrogen atom (**Figure 4.15A**). Many indole units are covalently linked together to form the whole pigment molecule. Melanins can differentially bind many different molecules and thus are often large and variable in structure from animal to animal. For this reason, the molecular weight of melanin has technically not been determined. There are two general categories of melanin that differ in composition, size, and absorbance characteristics. **Eumelanin** is the more common form, occurring throughout the animal kingdom. It is responsible for all black, gray, and most brown coloration found in animals (**Figure 4.15B**). The larger size of the eumelanin biopolymer

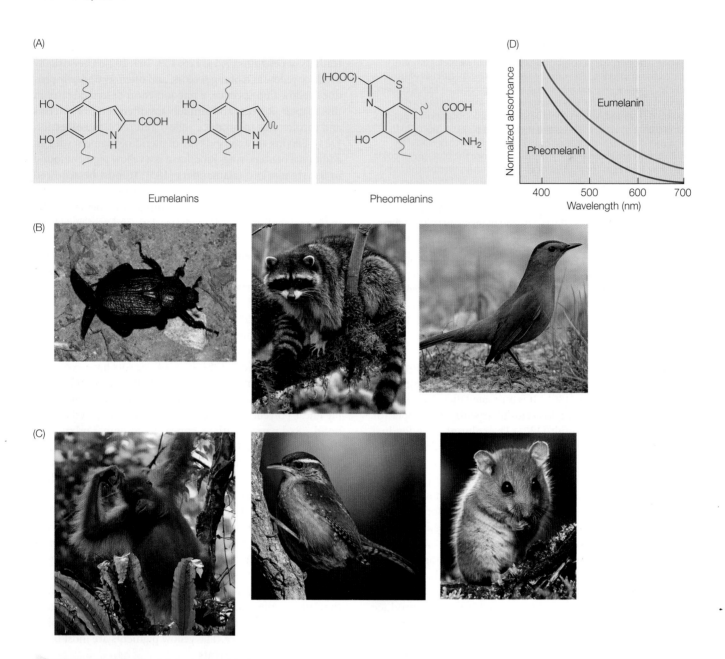

FIGURE 4.15 Melanin (A) Structure of the basic units of eumelanin (tyrosine) and pheomelanin (cysteine); curly red lines show sites of attachment of repeating units that form the polymer. (B) Row shows examples of black and brown eumelanin in a scarab beetle (*Osmoderma subplanata*), raccoon (*Procyon lotor*), and gray catbird (*Dumetella carolinensis*). (C) Row shows examples of reddish brown pheomelanin in orangutan (*Pongo pygmaeus*) and Carolina wren (*Thryothorus ludovicianus*); blonde, palomino, tawny, and golden fur, as in this dormouse (*Muscardinus avellanarius*), contains low levels of one or both melanins. (D) Absorbance curves for eumelanin and pheomelanin normalized relative to eumelanin. (After [124].)

and higher degree of conjugation are responsible for the near-uniform absorption of all visible and ultraviolet wavelengths. Eumelanin generally forms large, insoluble, rod-like granules. **Pheomelanin** is found only in the higher vertebrates (birds and mammals) and is apparently absent in poikilothermic

vertebrates (reptiles, amphibians and fish) and invertebrates. It is responsible for the generation of reddish brown and muted yellow colors (**Figure 4.15C**). The indole unit of pheomelanin contains a sulfur atom. The biopolymer is smaller and contains a lower degree of conjugation and fewer functional carbonyl groups, so absorbance at the long wavelength end of the spectrum is lower (**Figure 4.15D**). Pheomelanin forms smaller, globular granules of lower molecular weight than eumelanin. Eumelanin and pheomelanin may be combined to produce certain feather and fur hues [122].

Unlike carotenoids, melanins are endogenously manufactured by animals from basic amino-acid building blocks. Tyrosine is the key precursor of eumelanin, while the sulfur-containing amino acid cysteine is the source for pheomelanin. Vertebrates synthesize melanin in the peripheral tissues (skin) in structures called **melanocytes**. Melanocytes may be clustered near developing hair and feather follicles in birds and mammals, where melanin can be recruited into the keratinized structures. Poikilothermic vertebrates, on the other hand, synthesize melanin in specialized cells called melanophores [53]. We shall explore the signaling implications of these differences later. Because melanin can be synthesized by animals, production of melanin-based signaling patches is not as constrained by dietary access to chemicals as it is for carotenoid pigments. Freedom from this dietary constraint is probably responsible for the general finding that the size and darkness of signal patches are often independent of body condition [119] (although exceptions are accumulating [42, 60, 66, 219]). However, production of melanin-based signals is regulated by genetic and hormonal factors [33]. Males and females of many species differ in the expression of black signal patches. Several studies have shown that trait expression can be modified to that of the opposite sex by administering either androgen or estrogen blockers (depending on the species), and this evidence basically attests to genetic control of such sex differences [94]. More interesting is the association of within-sex variation in trait expression with variation in hormone levels. For example, the size of black patches in male house sparrows is positively correlated with levels of circulating testosterone [55, 56, 228]. Testosterone levels potentially could have many effects on aggressive behavior, reproductive success, and survival. Manipulation of nongonadal hormones such as pituitary-generated luteinizing hormone and thyroid-generated thyroxin can also affect melanin signal patches, and likewise could provide receivers with information about sender vigor, reproductive potential, and survival ability [33, 212]. More work remains to be done to determine whether the control mechanisms for melanin trait expression are configured to provide honest information about sender hormonal status, a topic we shall take up in Chapter 11 [16, 124, 127].

Melanin serves several other functions besides the production of color signals. Like the other pigments we have discussed that employ the conjugated double-bond system to absorb visible wavelengths, melanin is a powerful antioxidant that can scavenge free ions. It can accept singlet oxygen and hydroxyl/peroxyl radicals and thereby provide protection from oxidative stresses [120]. Because of the high density of carbonyl (COOH) units, melanin can even mop up the toxic positive cations of trace metals. As large polymers that cross-link with proteins, melanins also provide structural support and strength to tissues [195]. Accumulated evidence has demonstrated that melanin strengthens insect cuticle, bird feathers and beaks, and even plant seeds and fruit surfaces. Melanized hair, feathers, and cuticle are also believed to be more resistant to damage by microbial, mite, and flea parasites [255]. Melanin is well known as a photoprotectant against ultraviolet damage [85, 162]. Finally, melanin plays a role in animal thermoregulation. As melanized surfaces absorb a wide range of sunlight radiation, the photon energy is converted to heat. This heat can be used by animals to increase their body temperature when ambient temperatures are low.

Later in this chapter we shall see how melanin plays an important role in other aspects of visual signal production. It is used in conjunction with other pigments and structural colors to modify color hue. Melanin also serves to generate spatial pattern and color contrast with adjacent color patches, and is often the key player in the production of temporal variation in color signals.

OTHER PIGMENTS   Numerous additional pigments have been identified, some of which occur only in particular taxonomic groups and some of which have not been well described chemically. **Ommochromes** are a class of yellow, orange, red, and brown pigments used as colorants in several insect taxa, notably butterflies and dragonflies [156, 232]. The pigment is also found in the eyes of a variety of insects and crustaceans [14]. The compounds consist of polymers of the amino acid tryptophan. The larger, darker molecules are insoluble, sometimes solid, granules like melanin, and may be used in thermoregulation much as melanin is. **Papiliochromes** are another class of yellow to red pigments restricted to the papilionid butterflies [233]. The yellow and orange feather pigments of penguins and the downy feathers of precocial chicks in many avian species are not carotenoids but may be pterins [126, 128]. Some ducks possess green pigments that have not yet been characterized [125]. Finally, a true blue pigment was discovered in two species of dragonet fishes (*Synchiropus* spp.) but the chemical structure is as yet unknown [8, 54]. Manmade blue pigments contain metals such as cobalt, copper, tin, and aluminum. Blue pigments may therefore be rare in nature because they require such mineral elements, which are difficult for animals to acquire and metabolize.

FLUORESCENT PIGMENTS   Fluorescence occurs when a substance selectively absorbs some frequency of radiation and then emits a different frequency. Usually the absorbed frequency is high-energy, short-wavelength light, and the emitted frequency is lower-energy, longer wavelength light because the material dissipates some fraction of the initial excitation energy. **Figure 4.16A** illustrates the process by which a photon of light is absorbed, raising an electron to a higher energy state. Instead of slipping immediately back down to its ground state, the atom loses some energy to the crystal lattice or by collisional processes, then slips down to ground state while emitting a photon of lower energy. We are most familiar with the fluorescence of certain materials under UV (black light) radiation, because the excitation radiation is invisible to us while the emitted radiation yields vibrant colors in the visible range. Minerals and organic compounds that fluoresce under ultraviolet illumination have been put to many uses in medicine, microscopy, light bulbs, diagnostic tests, and the ecological tracking of plant and animal products. Yellow pterin pigments typically fluoresce, whereas carotenoids do not, so the presence of fluorescence

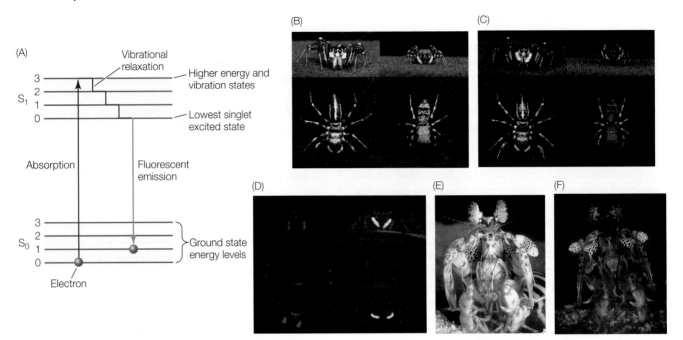

FIGURE 4.16 Fluorescent signals (A) The mechanism by which some pigments produce fluorescence under short-wavelength radiation. Diagram shows the ground state S_0 and a higher energy state S_1 for a single electron (blue) orbiting its nucleus. Within each energy state there are several vibrational states. An electron resting in its lowest energy ground state absorbs a photon of short-wave radiation and is elevated to a high vibrational energy level in the first excited state S_1 (purple arrow). It then loses some energy by first stair-stepping down to a lower vibrational level within S_1 in a process called vibrational relaxation (red). It subsequently slips back down to its ground state while releasing a photon of lower energy than the initial excitation photon (yellow arrow). (B–D) Ornate jumping spider (*Cosmophasis umbratica*). All images show male on the left, female on the right, front view above, top view below. (B) Under white-light illumination in human color-vision range. (C) Under UV illumination, white areas indicate UV reflectance, which is greater in the male. (D) Color image under UV illumination, showing bright fluorescent emission on the palps of the female. (E,F) Mantis shrimp (*Lysiosquillina glabriuscula*) in threat display mode showing yellow patches on the oval antenna scales and carapace under white light (E) and under blue incident light with a yellow camera lens filter (F). (A after [146].)

in a yellow-pigmented area indicates a noncarotenoid pigment. Only recently have a few fluorescent color signals been described for animals. The clearest case for a true signal occurs in the ornate jumping spider [106]. Males, but not females, reflect a striking pattern on the face and body in the ultraviolet range that is critical for female acceptance of a mating advance. Females, but not males, produce a bright green fluorescence on the palps that is critical for inducing male courtship (**Figure 4.16B–D**). Other potential fluorescent signals include the emission of yellow fluorescence from the yellow crown and cheek patches of the wild-type green budgerigar [3], and strong fluorescence of the yellow color patches on the oval antennal scales of a mantis shrimp (**Figure 4.16E,F**), which is estimated to enhance signal intensity by 10–30% at underwater depths between 20–40 m [118].

Red fluorescent color patches have been described in several species of marine fishes, and guanine crystals were implicated as the color-producing pigment [132]. These marine examples illustrate the fact that the excitation radiation for fluorescence is not restricted to UV, since only blue-green downwelling radiation is available at these depths. Fluorescence may also enhance color signals in other spiders, orchid bees, and some butterflies [2, 151, 243].

Structural colors

In contrast to the pigmentary mechanism of colored signal production, which employs *selective absorption* of certain wavelengths, the structural mechanism employs **selective reflection** to produce colors. The key property of this mechanism is the presence of **nanometer-scale structures** built into the integument (scales, feathers, skin, or cuticle) that generate specific patterns of scattering. This mechanism creates color displays not available from pigments. The majority of ultraviolet, violet, and blue colors produced by animals (and by some flowers as well) are structural in origin, and most of the green colors of animals are also structural. Yellow, red, and white colors can also be produced with structural mechanisms. Our understanding of the anatomy and physics of structural color production has expanded greatly during the past 50 years with the development of the electron microscope. In the figures below, we shall see both transmission electron micrographs (TEM, generated by beaming electrons at very thin slides of tissue with structures of heterogeneous densities which are differentially blocked) and scanning electron micrographs (SEM, generated by drying the specimen, coating it with a thin layer of gold, and recording the reflected electrons from the beam to yield a three-dimensional image). Major advances and reanalyses of structural color production by animals have been made in the past ten years [96, 178, 191, 211, 242].

(A)

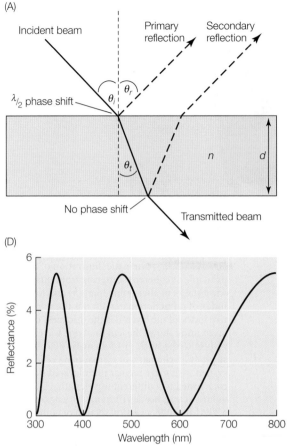

(B)

(C)

1000 nm

(D)

FIGURE 4.17 **Thin-layer interference** (A) Incident beam of light (solid line) reflects off both the top and bottom boundary of a thin layer of transparent material, producing both primary and secondary reflected waves (dashed lines). The layer of material has thickness d and refractive index n, which is greater than the index of the surrounding medium. The angle of incidence is θ_i, the angle of reflection is θ_r, and the angle of refraction is θ_t. (B) TEM of a distal barbule from an iridescent greenish feather of a European starling (*Sturnus vulgaris*), pictured in (C). (D) A thin-film optical prediction of the reflectance spectrum of a film + black layer system with each layer being 300 nm thick. (A after [171], D after [191].)

THIN-LAYER INTERFERENCE We begin this section with the structural mechanism that is simplest both to produce and to describe. The structure is a thin film of a transparent material with a moderate index of refraction [74]. When a ray of white light encounters the surface of the film, some of it will be reflected to produce a primary reflection, and some light will enter the layer and be refracted. This refracted ray will encounter the boundary at the bottom of the layer, and be partially reflected back into the layer, and refracted again as it exits the top surface boundary to produce a secondary reflection (**Figure 4.17A**). The primary and secondary reflections will be in phase for certain colors of light depending on the width of the layer, the refractive index of the material, and the angle of the incident light according to the Bragg equation

$$(m - \tfrac{1}{2})\lambda_{\max} = 2nd \cos \theta_t$$

where m is an integer, λ is the wavelength of maximum light reflection, n is the refractive index, d is the thickness of the layer, θ_t is the angle of refraction of the transmitted ray in the layer with respect to the normal. The ½ on the left side of the equation accounts for the fact that the primary reflection off of a higher-index boundary will undergo a 180° phase shift, which is equivalent to half a wavelength. When the Bragg condition is met, the primary and secondary waves are in phase and therefore coherent, so **constructive interference** of the reflected waves produces a colored beam that is intense and shimmery. A slight change in the angle of incidence or the viewing angle will cause the color to change slightly, but with greater changes in angle, the color will disappear. At this point, the primary and secondary reflections are out of phase and appear black as they cancel each other out through destructive interference. Colors that change with the angle of incidence are called **iridescent** [100, 164]. A familiar example of iridescence is the rainbow of colors viewed in a soap bubble or in an oil slick on a flat surface.

Black animals such as birds and insects sometimes have a layer of transparent keratin or cuticle over a solid layer of melanosomes. This layering generates an oily or iridescent sheen to their feathers or exoskeleton (**Figure 4.17B,C**). Keratin and cuticle have a refractive index of 1.5–1.6, and the thickness of the layer is often about 300 nm. This simple, two-layer laminar system actually produces several reflectance peaks that are integer multiples of the fundamental wave frequency (hence the m term in the equation), so the basically black animal has blue, green, yellow, or red highlights depending on the viewing angle (**Figure 4.17D**). Saturated colors usually cannot be produced with this form of **coherent scattering** [19, 30, 36, 191, 261].

MULTILAYERED REFLECTORS By stacking several layers of alternating high and low refractive index materials to form a

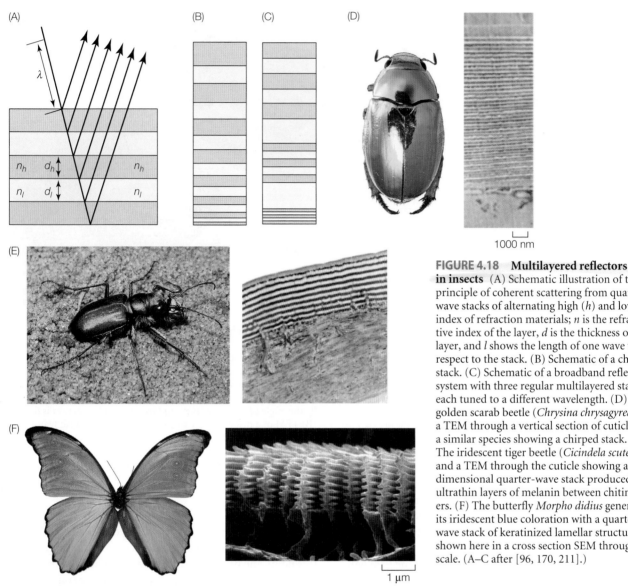

FIGURE 4.18 **Multilayered reflectors in insects** (A) Schematic illustration of the principle of coherent scattering from quarter-wave stacks of alternating high (*h*) and low (*l*) index of refraction materials; *n* is the refractive index of the layer, *d* is the thickness of the layer, and *l* shows the length of one wave with respect to the stack. (B) Schematic of a chirped stack. (C) Schematic of a broadband reflection system with three regular multilayered stacks, each tuned to a different wavelength. (D) The golden scarab beetle (*Chrysina chrysagyrea*) and a TEM through a vertical section of cuticle for a similar species showing a chirped stack. (E) The iridescent tiger beetle (*Cicindela scutellaris*) and a TEM through the cuticle showing a one-dimensional quarter-wave stack produced with ultrathin layers of melanin between chitin layers. (F) The butterfly *Morpho didius* generates its iridescent blue coloration with a quarter-wave stack of keratinized lamellar structures shown here in a cross section SEM through a scale. (A–C after [96, 170, 211].)

multilayered laminar array, very bright and saturated colors can be generated. To produce coherent reflections from all layers in the stack, the thickness and spacing of the more- and less-dense layers must be uniform and meet the Bragg condition. The most common structure contains 6–12 layers, and each layer has an optical thickness of one-quarter of a wavelength (allowing for the 180° phase shift at the denser boundary and the fact that each refracted wave travels twice through the denser layer before aligning with the reflected wave). Specifically, with incident light nearly perpendicular to the surface, the thickness of the layer times its refractive index is

$$\lambda/4 = n_l d_l = n_h d_h$$

where *h* specifies the higher index layer and *l* the lower index layer (**Figure 4.18**). This structure is called a **quarter-wave stack**. It is a one-dimensional color-producing array because periodicity occurs only along one axis (the vertical axis in

the figure; see also Figure 4.20). The hues produced with this mechanism may be violet, blue, green, or red, and their saturation and brightness are very high because of the constructive reinforcement of reflected waves from many layers. The saturation can be enhanced even further by adding a solid layer of melanin below the stack, which absorbs the wavelengths to either side of the peak wavelength (λ_{max}). The viewing angle over which the color can be seen is quite narrow, however, on the order of 20°. When one rotates a flat stack with respect to the angle of incident light or the viewing angle, the color turns to black because the multiple reflected waves are out of phase [164].

Numerous insects produce jewel-like iridescent body colors with such multilayered reflectors. Thin parallel layers of chitin with alternating layers of high and low refractive index materials are secreted by the epidermis during development and later harden during sclerotization (see Figure 4.18E) [96, 209, 211]. Butterflies such as the large blue *Morpho* species from South America produce their laminar arrays with the alternation of cuticle and air. Cuticle is composed of a protein plus a chitin polymer and has a refractive index of 1.56. The cuticular layers are nanostructural ridges of lamellae on the top side of the scales (see Figure 4.18F) [48, 95, 238]. There are also several variants on the multilayered stack that produce different color effects. In a **chirped stack** (see Figure 4.18B,D), the layers become progressively thinner from the exterior to the interior of the stack. This structure reflects different wavelengths from each depth zone, and results in a specular metallic gold reflectance, as observed in various golden beetles [152, 170]. Similarly, a double chirped stack, consisting of several series of regular stacks each tuned to a different frequency (see Figure 4.18C), produces the highly reflective silvery scales of the common silverfish isopod *Ctenolepisma* [101]. A "chaotic" stack, in which the layers vary randomly in thickness and spacing, produces a diffuse white reflectance [170, 209, 244]. Finally, some iridescent scarab beetles reflect colored circularly polarized light, generated by bowl-shaped helical stacks that reflect polarized light at different electric-vector angles [17, 65, 76, 192, 213].

In theory, stronger reflectance and more saturated colors should occur as the number of layers in the stack increases [40]. When the difference in refractive indices is not too large, reflectance can reach 100% with a ten-layer stack. As implied above, the uniformity of layer thickness and spacing also affects saturation. If the depth and regularity of the stack are at all dependent on an animal's condition, then we would expect the color properties of structural coloration to reveal honest information about the sender's health. In several butterflies, thermal stress and condition affect the intensity of coherently scattered ultraviolet wavelengths [88–91]. Leaf beetles (*Plateumaris*) exhibit extensive individual variation in color related to the spacing of cuticle layers of their elytra [99]. We shall discuss the functions of these traits with respect to communication in later chapters.

DIFFRACTION GRATINGS   A diffraction grating is a periodic structure of either slits or grooves that diffracts and splits light into several beams travelling in different directions. This process spreads white light into a spectrum of colors, and man-made gratings are used in various optical instruments. It was once thought that animals did not use diffraction gratings to produce color. However, a few animals can produce **reflective diffraction gratings** by making a series of parallel grooves or ridges on hard body surfaces. A reflective diffraction grating gives rise to coloration because certain wavelengths scattered from the tops of adjacent ridges will be in phase as a result of constructive interference (**Figure 4.19A**). The condition for colored diffraction is given by the diffraction equation:

$$w(\sin\theta_m - \sin\theta_i) = m\lambda$$

where m is an integer representing the order of the spectrum, w is the spacing between ridge peaks, θ_i is the angle of incidence, and θ_m is the angle of diffraction for a given spectrum

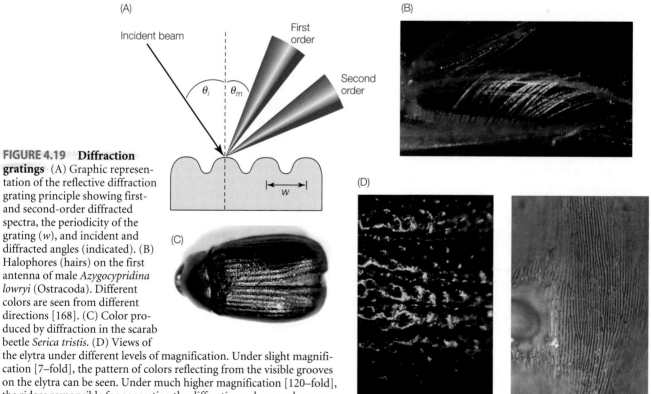

FIGURE 4.19 Diffraction gratings (A) Graphic representation of the reflective diffraction grating principle showing first- and second-order diffracted spectra, the periodicity of the grating (w), and incident and diffracted angles (indicated). (B) Halophores (hairs) on the first antenna of male *Azygocypridina lowryi* (Ostracoda). Different colors are seen from different directions [168]. (C) Color produced by diffraction in the scarab beetle *Serica tristis*. (D) Views of the elytra under different levels of magnification. Under slight magnification [7–fold], the pattern of colors reflecting from the visible grooves on the elytra can be seen. Under much higher magnification [120–fold], the ridges responsible for generating the diffraction colors can be seen, oriented perpendicular to the large grooves [222]. (A after [177].)

order. The first-order spectrum occurs at the closest angle to the normal and a second-order spectrum at a greater angle arises from coherent scattering from nonadjacent ridges. A bright and saturated first-order spectrum will be generated when the spacing of the ridges is on the order of a wavelength of visible light, and natural gratings have ridges spaced between 500–800 nm apart. Although the appearance of such a grating will appear iridescent, a rainbow of different colors will be seen at the same time and from different angles. Turning the animal will not result in a black zone as with the multilayered stacks described above. Such gratings have been found in numerous scarab beetles, the setae (hairs) of ostracod crustaceans, the wings of some flies, and in some Burgess shale fossils (**Figure 4.19B–D**) [70, 72, 169, 179, 222]. Diffraction colors likely serve an intraspecific communication function in the ostracods, as the reflective setae are exaggerated in males, but in other organisms the bright coloration is believed to warn or startle predators.

A diffraction grating consisting of parallel grooves or ridges is considered a one-dimensional array, because periodicity is present only in the one dimension perpendicular to the grooves and across the surface. Gratings can also be two-dimensional, if the surface is composed of an array of bumps. Moreover, if the spacing of the grooves or bumps is much smaller than the wavelengths of visible light, zero-order reflectance is produced [177]. Here, the reflectance angle spans 0° (i.e., normal to the surface or in the same direction as the incident wave). A zero-order, two-dimensional grating can produce a saturated color of a very short wavelength (ultraviolet or violet). This mechanism can also cause total transmission, or transparency, a phenomenon we will discuss later in the chapter.

MULTIDIMENSIONAL ARRAYS   Coherent scattering of a narrow band of wavelengths in multiple directions can be generated from two- or three-dimensional matrix arrays of scattering objects (**Figure 4.20**). Colors produced via this mechanism can appear either specular or diffuse, but are not usually iridescent because the same color is seen over a wide range of viewing angles. These arrays vary in how precisely

the scatterers are ordered, ranging from nearly perfect crystal-like arrays to quasi-ordered arrays, but the principal mechanism is the same. The scattering objects must be unimodal in size and interscatterer spacing, at least over short distances, so that reradiated waves of certain wavelengths are reinforced by waves from adjacent objects, while other wavelengths interfere destructively. Further details are provided in Web Topic 4.3.

Web Topic 4.3 *Dimensionality of structural color mechanisms*
Definitions of one-, two-, and three-dimensional arrays. Fourier analysis and other new techniques have recently been applied to the reflection patterns from periodic structural arrays to determine their dimensionality. We also review the photonic crystal approach to color production by periodic structures.

The most ordered arrays are found in insects. The examples illustrated in **Figure 4.21** are produced in different ways, but they result in a similar optical effect. The three-dimensional structures consist of crystal-like layers of closely and precisely arranged scatterers. The scatterers can be either hexagonally packed spheres surrounded by air, or the inverse, air spaces embedded in a denser lattice structure. The two examples shown in the figure both have sphere dimensions of about 250 nm, half the wavelength of light, and they produce bright opal green reflectance over a wide range of viewing angles. This biological color-producing structure is virtually identical to the crystalline structure of opal gemstones, which are composed of hexagonally packed solid silica spheres 150 to 300 nm in diameter. Such highly ordered structures are called **photonic crystals**, an industrial term for dielectric materials that prevent transmission of certain frequencies while allowing others. The allowed frequencies pass through the crystal while the disallowed frequencies are reflected to produce the color [47, 48, 77, 78, 129, 175, 176, 204, 242, 250].

Two-dimensional arrays of parallel collagen fibers in the skin of many birds and mammals are responsible for the blue and green coloration of facial skin, genitals, bills, and

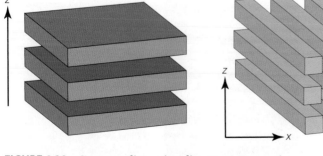
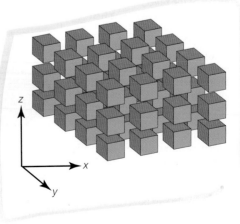

FIGURE 4.20   Scatterer dimensionality One-, two-, and three-dimensional array structures, respectively, composed of high- and low-refractive index materials. Arrows show the vector directions with periodic variability.

(A)

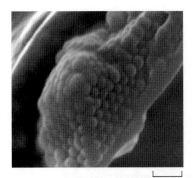

1000 nm

(B)

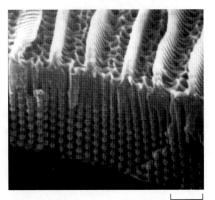

1000 nm

FIGURE 4.21   Photonic crystal lattice structures in insects
(A) The weevil *Pachyrhynchus argus* has a metallic turquoise coloration visible from any direction. An SEM of the interior of its scales reveals a three-dimensional lattice of closely packed transparent spheres about 250 nm in diameter (scale bar = 1000 nm) [174]. (B) The Kaiser-i-Hind swallowtail (*Teinopalpus imperialis*) possesses the typical ridges and netted cross ribs of papilionid butterfly scales, but the scale interior is filled with a diffraction lattice. The SEM cross section of a cover scale in the green region shows the cuticular lattice structure [51].

legs (**Figure 4.22**). It was once believed that this coloration was caused by Rayleigh scattering, but when full-spectrum radiometers were developed, no evidence for the increased scattering in the violet and ultraviolet range was found. Instead, reflectance spectra exhibited unimodal peaks within the visible range, and variation in hue was associated with variation in collagen fiber size and spacing. Collagen forms self-assembled, triple-helical fibers ($n = 1.42$) surrounded by a mucopolysaccharide matrix ($n = 1.35$) that maintains the regular spacing of the denser fibers. The fibers are arrayed perpendicular to the direction of the incoming light, and they coherently scatter a narrow range of wavelengths in multiple directions. The scattering structures are somewhat variable

FIGURE 4.22   Collagen arrays (A) Colorful skin bib of the pheasant *Tragopan caboti*, and cross section TEMs through dark blue (B), light blue (C), and orange (D) skin regions showing the highly regular packing of collagen fibers. The red color is produced by blood hemoglobin pigment. (From [190].)

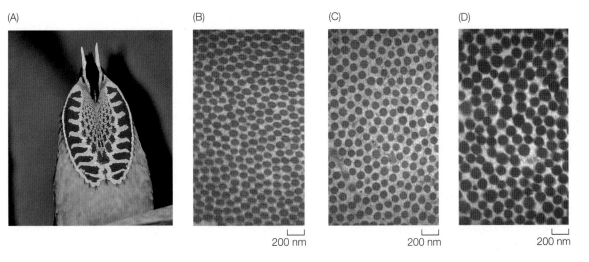

(A)   (B)   (C)   (D)

200 nm   200 nm   200 nm

(A) Feather structure

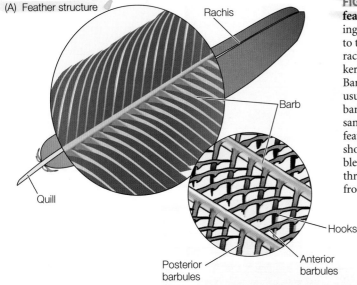

Rachis

Barb

Quill

Posterior barbules

Anterior barbules

Hooks

FIGURE 4.23 Non-iridescent avian colors produced in feather barbs (A) Anatomy of a typical wing or tail feather, showing rachis, barbs, and barbules. The main stem, or rachis, attaches to the skin via the quill. The barbs stick out perpendicular to the rachis. The barbs are moderately stiff shafts filled with a spongy keratin matrix, and the smaller barbules branch out from the barbs. Barbules from one side of a barb in the flight and tail feathers usually have small hooks on them that are attached to the nearest barbules on the next barb. This keeps all parts of the feather in the same plane. (B) TEM of the spongy barb matrix from the green feathers of the gold-whiskered barbet (*Megalaima chrysopogon*), showing very regular three-dimensional array of spherical air bubbles imbedded in keratin (30,000×). (C) SEM of the quasi-ordered three-dimensional array of scatterers in the spongy keratin matrix from blue-feather barb tissue of the eastern bluebird (*Sialia sialia*), showing complex networks of air and keratin channels (40,000×). (A after [41, 234].)

(B)

1000 nm

(C)

500 nm

in periodicity so the reflection is noniridescent. Because the difference in refractive indices is very small, the number of scatterers in colored skin must be increased by thickening the skin to produce a sufficient amount of reflectance [188, 190].

Birds have two fundamentally different mechanisms for generating structural feather colors, one based on a keratin and air matrix in the **barb** feather element, and another based on melanin and keratin in the **barbule** feather element. The anatomy of a bird feather is illustrated in **Figure 4.23A** to clarify these different feather parts. The barbs contain a spongy matrix of keratin and air spaces (refractive indices of 1.56–1.58 and 1.0, respectively [19]). The patterned distribution of keratin and air generates a three-dimensional tissue

structure that coherently scatters certain colors [218]. In some species the barb nanostructure consists of uniformly sized and spaced air bubbles connected by tiny channels and ordered in a regular array (**Figure 4.23B**). In other species the structure consists of a complex network of air and keratin channels, in which the air and keratin channels are approximately equal in shape and width (**Figure 4.23C**). A melanin layer often underlies the matrix to absorb other wavelengths. Non-iridescent blue and green colors can both be produced with these two structures. The periodicity of the scattering matrix is predictably larger for feathers with longer-wavelength hues. The more uniform and regular the array pattern, the greater is the intensity and saturation of the hue [187, 188].

The production of iridescent and metallic colors is based on very precise one- or two-dimensional arrays of melanin and keratin in feather barbules. Melanin has a high refractive index of approximately 2.0 [19, 215]. One-dimensional arrays are essentially quarter-wave stacks, as we have described above. The alternating layers consist of either stacks of flat melanosome platelets or rows of adjacent melanin rods [37]. The platelets or rods may be either solid or hollow air-filled structures [37, 58]. In the case of hollow structures, the primary reflecting surfaces are formed at the melanin/air boundaries, which differ more in refractive index ratio than the melanin/keratin boundaries. The addition of hollow air cavities inside the melanosomes increases the total reflectance per layer by an order of magnitude, so this strategy produces extremely brilliant iridescence and often 100% reflectance of incident light from the combined layers. These one-dimensional structures can be tuned to produce nearly any hue, including the blue, green, yellow, and red color patches of hummingbirds, sunbirds, and birds of paradise (**Figure 4.24A,B**). The intense color is often highly directional and used in brilliant flashing

FIGURE 4.24 Iridescent avian colors produced in feather barbules (A) TEM through the barbules of the blue throat feathers of the dusky starfrontlet hummingbird (*Coeligena orina*), showing a one-dimensional stack of air-filled platelets. (B) TEM of the one-dimensional multilayered laminar array from the green back feathers of the resplendent quetzal (*Pharomachrus mocinno*), showing rows of air-filled melanosome tubules. (C) TEM cross section of the hexagonally packed two-dimensional arrays of hollow melanosomes from the back feathers of the violet-backed starling (*Cinnyricinclus leucogaster*). (D) SEM cross section of the three-dimensional square arrays of solid melanosome rods (light-colored) in the peacock (*Parvo muticus*); the dark gray circles are air.

displays. Two-dimensional arrays are formed by a tight and orderly packing of melanin rods. **Figure 4.24C** shows the hexagonal packing structure of hollow rods responsible for the iridescent violet feathers of the striking violet-backed starling. The same mechanism produces the iridescent green plumage of some trogons and the iridescent golden chest feathers of turkeys [35, 37, 191]. Here again, the reflecting boundaries are created by the highly refractive melanin/air interface, but the two-dimensional structure means that the brilliant color may be viewed at a wider range of angles. A two-dimensional square array of melanin rods produces the metallic colors of the eyespots in the peacock's tail (**Figure 4.24D**). Solid melanin rods are connected in rows and columns by keratin. This square arrangement leaves round air tubes, which are also ordered in squares. Much like the collagen fiber arrays in colored skin (shown in Figure 4.22), smaller diameter rods and shorter spacing between them is characteristic of the short-wavelength violet and blue feather areas, and larger rod diameter and spacing is observed in green and golden brown feather areas [258, 263].

Evidence is accumulating that the hue and intensity of reflected light from these nano-structured feathers varies among individuals in several avian species, and that these color signals can provide information about sender condition and quality. For example, in several dichromatic species with blue males, the males with more blue or ultraviolet reflectance were older, larger, more successful in attracting mates, or in better condition [1, 10, 28, 29, 92, 93, 220]. A recent study on the satin bowerbird (*Ptilonorhynchus violaceus*), the males of which sport glossy blue-black plumage, has provided the first evidence that thickness of the keratin layer closely predicts the measured hue and ultraviolet chroma [30]. Stronger evidence for a fitness advantage associated with intense iridescence comes from peacocks, in which males with brighter tail eyespots have higher mating success [108]. The construction of precisely structured multidimensional barbule arrays is easily disrupted by poor condition or parasite loads in the bearers, making iridescent coloration well suited to function as condition-dependent signals [31].

INCOHERENT SCATTERING   Brilliant white coloration is generated by neither pigments nor periodic structures, but rather by structures that scatter all visible wavelengths. This mechanism is called **incoherent scattering**, which includes but is not limited to Rayleigh and Mie scattering. Any aperiodic structure with randomly oriented interfaces between low- and high-refractive index materials will produce white reflectance. In the case of white feathers, the spongy matrix of keratin and air spaces in the feather barbs contains air spaces that are larger and more variable in size than those shown in Figure 4.23. Such feathers have no melanin layer underneath the scattering structure [39, 191]. In white *Pieris* butterflies, the scales lack the sort of nanostructured laminar structures shown in Figure 4.18F, and instead are packed with small beads attached to the crossribs between scale ridges. The higher the density of beads, the greater is the reflected brightness. These beads are composed of a pterin pigment that absorbs ultraviolet wavelengths only and passes all other visible wavelengths [109, 138, 223, 253]. White beetles of the genus *Cyphochilus* are covered with elongated, thin flat scales whose interiors are composed of a random network of interconnecting cuticular filaments. The filaments have diameters of about 250 nm and a refractive index contrast with the surrounding medium of about 0.56. It is the aperiodic arrangement of these filaments that explains their brilliant white reflection, as other beetles with similar but regularly ordered filaments reflect specific colors [244].

Summarizing, there are four main categories of structural color production in animals: multilayered reflectors, diffraction gratings, multidimensional arrays, and aperiodic incoherent scattering. In a later section of this chapter, we shall point out some examples in which these mechanisms have been combined to make more complex colors. Because the structures are constructed from relatively hard materials, they fossilize quite well. The techniques we have discussed above for describing structural mechanisms in extant species have also been adapted to describe the coloration of extinct species. Multilayered reflectors have been documented in several ancient beetle species [174, 230], diffraction gratings were discovered in some Cambrian polychaetes and arthropods from 515 million years ago [169, 172], and the feathers of extinct birds and dinosaurs show the melanosome structure associated with black and rufous coloration today [105, 237, 262]. Structural mechanisms therefore represent an ancient strategy of color production in animals.

Bioluminescence

Some organisms from a wide variety of taxa can produce their own light via a biochemical process called **bioluminescence**. (Note that bioluminescence is not the same as fluorescence, which requires the input of ultraviolet radiation.) The majority of bioluminescent organisms are marine inhabitants; they include bacteria, dinoflagellates, ctenophores (comb-jellies),

cnidierians (jellyfish), squid (and other molluscs), annelid worms, shrimp (and other crustaceans), echinoderms, and many fish. Terrestrial examples include fungi, millipedes, centipedes, firefly and click beetles, and a few other insects [62]. Bioluminescence is absent in amphibians, reptiles, birds, and mammals. The process of light production, based on the reactions of a **luciferin** pigment (a generic term for several different light-producing compounds) and a **luciferase** enzyme, is similar for most of these species. The luciferin substrate is combined with molecular oxygen in the presence of the enzyme catalyst to produce an excited-state peroxy-luciferin intermediate, which quickly breaks down while releasing a photon of light. The spent luciferin by-product either reverts to active luciferin via internal resynthesis or is renewed from dietary sources [256]. The color of the emitted light is usually blue in marine species and yellow-green in terrestrial species. In bacteria, fungi, and a few other systems, light is emitted continuously, but in most cases luminescence occurs as flashes 0.1–1.0 seconds in duration. Flashing requires a rapid turning on and off of an enzymatic reaction, with reagents sequestered appropriately and subject to quick mobilization. Mechanisms for triggering the flash include a change in pH, the input of calcium or oxygen, and nervous control.

Although there are hundreds of types of luminous animals, and bioluminescence has evolved independently at least 40 times, there are surprisingly few luciferin molecular systems. In some cases this convergence can be explained by animals acquiring their luciferin through the food chain, but in other cases organisms have been shown to have evolved the ability to synthesize the same chemical independently. Bacterial luciferin is a reduced riboflavin phosphate that is oxidized in association with a long-chain aldehyde, oxygen, and a luciferase. The same compound also occurs in squid, a nematode, some fish, and tunicates. Dinoflagellate luciferin is thought to be derived from chlorophyll, and has a very similar structure. Vargulin is found in both the ostracod seed shrimp (*Vargula*) and the midshipman fish (*Porichthys*). Here there is a clear dietary link: the animals lose their ability to luminesce when deprived of luciferin-containing food. Coelenterazine is the most common of the marine luciferins, found in a variety of species from nine phyla. This molecule can occur in luciferin-luciferase systems, but is also found in photoprotein systems in which the luciferin, luciferase, and oxygen cofactor are bound together in a single unit. Firefly luciferin is used in a luciferin-luciferase system that requires ATP as the energy source, and either oxygen or nitric acid as the controlling agent [50, 61]. The process of converting chemical energy into light energy has a very high efficiency, around 90%, with very little heat generated as a by-product. However, it is an energetically expensive form of communication that is most useful for organisms active in dark environments [103, 110, 146].

The major driving force for the evolution of bioluminescence is not signaling to conspecifics, but signaling to predators and prey [62, 251, 252]. Anglerfish and some jellyfish use a bioluminescent dangling lure to attract unsuspecting

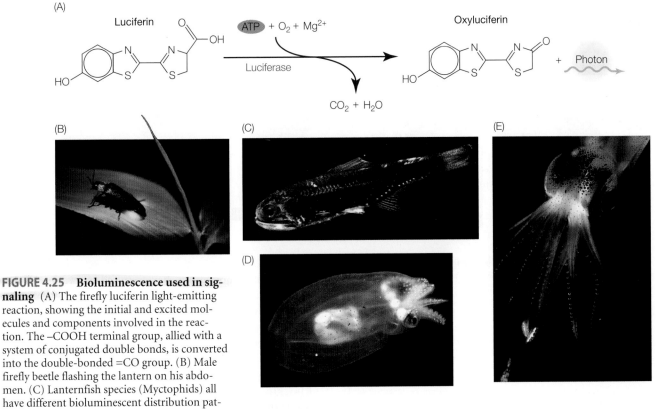

FIGURE 4.25   Bioluminescence used in signaling (A) The firefly luciferin light-emitting reaction, showing the initial and excited molecules and components involved in the reaction. The –COOH terminal group, allied with a system of conjugated double bonds, is converted into the double-bonded =CO group. (B) Male firefly beetle flashing the lantern on his abdomen. (C) Lanternfish species (Myctophids) all have different bioluminescent distribution patterns, which probably play a role in species recognition. (D) In the small octopus *Japetella*, females produce a ring of light around the mouth, believed to be a mating signal, only at certain times. (E) This small squid in the genus *Abraliopsis* has several different types of light organs. In addition to the bean-shaped ones at the tips of two central arms, it has small photophores covering the underside of its body.

prey [185]. Dragonfish (Malacosteidae) and flashlight fishes (Anomalopidae) apparently illuminate their immediate environment with a bioluminescent light to find prey [141]. Diverse antipredation strategies account for most cases of bioluminescence in the sea. Startling and distraction are two such strategies. Many organisms from bacteria to shrimp and fish flash when disturbed to warn predators or misdirect them to expendable body parts. Some even release a cloud of bioluminescent material as a smoke screen to blind or confuse predators in pursuit. Similarly, a newly-discovered deep-sea planktonic polychaete drops green luminescent "bombs" as decoys when it is disturbed [163]. Another strategy is the "burglar alarm," often by a captured prey, designed to attract secondary predators so that the prey may be released [43, 131]. Camouflage is the final major antipredation strategy. Diurnally active fish that live near the surface of the water are vulnerable to deeper dwelling predators with upward-directed eyes that can detect the prey's silhouette against the bright background. The prey species use bioluminescence to illuminate their bellies and reduce their contrast with the background [83]. Some transparent predators such as comb-jellies and jellyfish eat luminescent prey, which then become

visible inside the predator's body. To counteract the cost of this visibility, some predators have evolved pigmented gut linings or their own bioluminescent patterns to disguise their meals [62].

The use of bioluminescence for social signaling has evolved secondarily in a few groups (**Figure 4.25**). Familiar examples are the firefly beetles (Lampyridae), in which males emit repeated flashes of light to attract mates. The temporal pattern of long and short flashes is species-specific to ensure conspecific mating [18, 104, 107]. No less spectacular are the bioluminescent mate attraction displays described for ostracods (*Vargula*). Males eject puffs of luciferin into the water through nozzles near their mouth as they swim in a helical, and then upward spiraling path, while photically silent females observe and follow [24, 142, 143, 196, 197]. Female Bermuda fireworms (*Odontosyllis*) and pelagic octopods (*Japetella* and *Eledonella*) produce a distinctive luminescent display when receptive that appears to attract males [198, 251]. Ponyfish (*Leiognathus*) are small, laterally compressed, schooling fish that perform synchronized rhythmic flashing displays. Since males in this and other marine fish species have larger light organs, the display is likely to serve a reproductive function [67, 205, 257].

Modification and Coupling

The modification of light signals is so intimately related to strategies of **coupling** into the medium that we shall discuss these components together in this section. Animals modify

FIGURE 4.26   Color morphs of budgerigars (*Melopsittacus undulatus*) Above, schematic illustrations of the color-producing elements in feather barbs. (A) Green is produced by a combination of a blue-green coherent scattering layer with an air space/keratin matrix in the core, yellow pigment embedded in the keratin cortex above, and a melanin layer below. (B) Blue is produced as in (A), but with the yellow pigment removed. (C) For yellow, the melanin layer is removed. (D) White is similar to (C), but with the yellow pigment removed.

visual signals in four major ways: (1) by combining structural and pigmentary mechanisms; (2) by creating complex within-body color patterns; (3) by changing colors over a relatively short time scale; and (4) by employing postures and movements that maximally display colored body parts to receivers.

Combinatorial color-production mechanisms

Scattering and reflection of light from an animal's surface occurs passively, without any special adaptations, because of the difference in refractive indices between air or water and biological tissues. Some light also passes into the underlying tissue layers, where it is further scattered by the discontinuous densities of plasma membranes, organelles, and vesicles within cells. Producing colored reflectance requires either pigments or specialized nanostructures, as we have just seen. But some additional adaptations have evolved to further enhance the intensity, hue, saturation, and directionality of these color patches. Several of these adaptations involve combining pigments and structural colors, and others involve combining two structural mechanisms.

Color differences among the domestic breeds of the common parakeet, or budgerigar, provide a good introduction to many of the principles of combinatorial colors in animals (**Figure 4.26**). The wild-type green feather color is produced with three elements in the barbs: coherent scattering of blue-green frequencies from the spongy keratin/air matrix, the presence of yellow pigment embedded in the keratin, and a layer of melanin on the underside of the barb. Mutations that knock out or change the properties of one or more elements,

in different combinations, have resulted in a variety of colors. The yellow pigment present in green feathers absorbs the short-wavelength (blue) end of the spectrum reflected by the coherent scattering matrix, and the melanin layer absorbs longer (red) wavelengths. These two pigments combined with the structural scattering combine to produce a more saturated green color than the structural mechanism alone. Knocking out the yellow pigment results in a bluer hue. Knocking out the melanin layer while retaining the yellow pigment yields yellow. Knocking out both pigments, leaving only the scattering layer, results in a white feather. Changes in the periodicity of the scattering layer could potentially modify these hues [6, 38, 150, 188, 191].

Definitive studies on other species have corroborated the essential interactions between pigments and structural mechanisms that produce many, if not most, animal colors. Melanin is an essential underlayer for the production of saturated blue structural colors. In both an albino Stellar's jay (*Cyanocitta stelleri*) [216] and the white stripes of the otherwise blue butterfly (*Morpho cypris*) [260], the absence of the melanin produces a washed-out color, even though the scattering layer is properly tuned for coherent blue scattering. The melanin functions to absorb incoherently scattered white light and increases the color purity of the coherently reflecting layer. Similarly, carotenoid and other pigment-infused layers require some type of reflecting layer underneath to scatter back the nonabsorbed wavelengths. When the bright yellow feathers of the American goldfinch (*Spinus tristis*) are treated with a liquid of the same refractive index as keratin to fill the air spaces, they become transparent with a faint yellow cast, demonstrating the essential role of the white-reflecting spongy scattering layer [214]. Even more intense, brighter colors can be produced if the scattering layer is composed of pterin or guanine platelets [217]. The specular reflectance from the crystals increases the total amount of light reflected from pigmented tissue. Reflecting epidermal cells, called **iridophores**, that contain stacks of mirror-like platelets are found in the irises of many birds and the skin of reptiles, amphibians, and fish. A basal melanin layer is often included

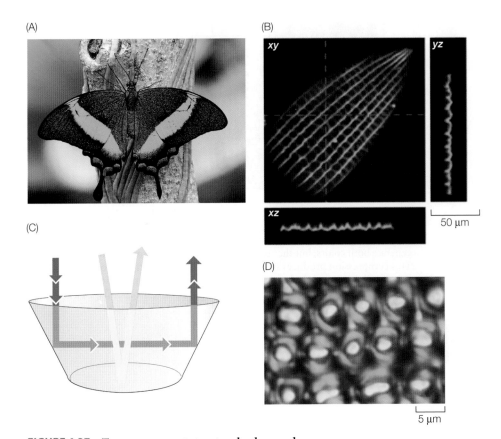

(A)

(B) xy   yz

xz   50 μm

(C)

(D)

5 μm

FIGURE 4.27   Two-component structural color production in a butterfly (A) The brilliant green color of the emerald swallowtail (*Papilio palinurus*) is generated by deep concavities on the scales that reflect yellow and blue in different ways. (B) Confocal micrographs (in florescence reflection mode) of a single scale showing pseudo-2D images of the top surface (*xy*) and cross sections in the *xz* and *yz* planes, indicated by dashed red lines. The scale surface effectively consists of rows of 4–10 μm diameter bowls. The bowls are lined with a multilayer stack of eleven alternating layers of air and chitin, each 75 nm thick. (C) Schematic of a bowl showing the mechanism of reflection of blue and yellow light under incident white light. Yellow light is reflected from light striking the bottom of the bowl at normal incidence; blue light is generated from light striking the sides of the bowls, which are angled at approximately 45° relative to the bottom surface. Short wavelengths striking the sides are selectively reflected across the concavity to the opposite side, where they are again reflected out of the surface. This doubly reflected light is both blue and strongly polarized. It creates a blue annulus around the yellow center point, as shown in the light microscope image in (D). At normal viewing distance, the pixilated yellow and blue are too small to resolve, so the colors combine to give a green appearance. (After [239, 241].)

to absorb nonreflected wavelengths and produce a more saturated color [6, 160].

A similar combinatorial strategy is employed by bioluminescent signalers to enhance the intensity of their chemically generated light. In fireflies, fish, and possibly other bioluminescent senders, the back lining of the photophore unit is filled with guanine platelets. This concave, mirror-like layer operates like the reflector of a headlamp to beam the light outward [172]. The most amazing example of this

strategy is found in deep-sea dragonfishes (Malacosteidae) that produce a beam of red light to illuminate their prey. The initially green bioluminescent light is converted to red light using a coupled fluorescent pigment and red filter [27]. Prey fish cannot see this red light, but the dragonfishes possess specially adapted visual pigments to see the reflections from their private flashlights [32].

Combinations of two structural mechanisms can also produce some amazing animal colors. The emerald swallowtail *Papilio palinurus* produces a remarkable color effect by combining a multilayered stack that produces yellow with deep concavities on the scale surface that produce blue (**Figure 4.27**). To humans at least, the blue and yellow reflections mix to produce green, but the blue component is strongly polarized and undoubtedly looks different to an insect receiver sensitive to the planes of polarized light [239]. Another example of structural color mixing has been described for the tiger beetle *Cicindela repanda*. The magenta background color of the elytra is dotted with blue-reflecting raised points that combine to make the beetle appear cryptic brown [210] (**Figure 4.28**).

The directionality of coherently scattered light from multilayered stacks can also be modified with secondary structural mechanisms. The viewing angle over which a receiver can see the color in a flat multilayered array is quite narrow, typically less than 40°. In some cases this creates a desirable flash effect when associated with a movement display (we will describe some examples later in the section on movement).

(A)

(B)

(C)

FIGURE 4.28 **Color mixing for crypticity in a beetle** (A) The tiger beetle *Cicindela repanda* appears cryptic brown, but a magnified view (B) of the elytra reveals a magenta-colored field covered with blue punctae. Multilayered stacks generate both colors, but the spacing between layers is compressed on the bumps. This compression produces shorter-wavelength blue reflection. Note the small rim of yellow around the bumps created by the gradual compression of the layers. (C) TEM of very fine scale pitted elytra surface showing the multilayered stack [210, 211].

(A)

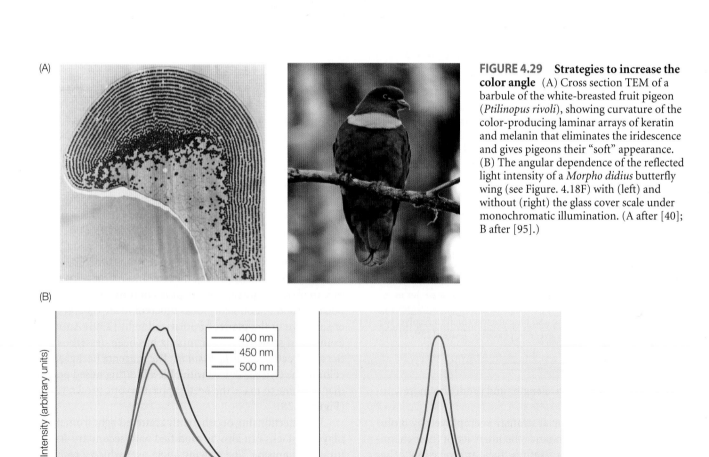

FIGURE 4.29 **Strategies to increase the color angle** (A) Cross section TEM of a barbule of the white-breasted fruit pigeon (*Ptilinopus rivoli*), showing curvature of the color-producing laminar arrays of keratin and melanin that eliminates the iridescence and gives pigeons their "soft" appearance. (B) The angular dependence of the reflected light intensity of a *Morpho didius* butterfly wing (see Figure. 4.18F) with (left) and without (right) the glass cover scale under monochromatic illumination. (A after [40]; B after [95].)

(B)

In other cases, the animal may be selected to broaden the angle over which the color can be observed. One mechanism for increasing the angle of color visibility is to give the laminar array a convex curvature [40]. In some birds that make use of multilayered arrays on their feather barbules, the barbule surface is rounded to give the laminar array a convex curvature (**Figure 4.29A**). This strategy eliminates the iridescent effect, so the color doesn't change with viewing angle or the angle of incident light. Another mechanism involves a transparent nanostructured layer that operates as an optical diffuser over the color-generating array. Some species of *Morpho* butterflies overlay the blue-producing wing scales with a second type of scale, called a cover or glass scale, that diffracts the blue reflected light over a wider viewing angle (**Figure 4.29B**) [238, 259].

Color patterns

From zebra stripes to leopard spots to the beautiful designs of butterfly wings, animals produce a stunning variety of complex **color patterns**. These patterns clearly evolved to serve several different adaptive functions, such as camouflaging the animal from its enemies, warning or startling predators that have detected it, and of course signaling important information to conspecific receivers. The details of these patterns are not specified in exhaustive detail by the genes. Instead, they are determined by developmental mechanisms with rules comprised of surprisingly few parameters. Over 50 years ago, mathematicians began to develop some theoretical models of pattern generation that were consistent with the observed variations among adult animals and body parts differing in size and shape [130, 145, 231]. At that time we did not know whether or not the models reflected real cellular and biochemical processes. More recently, developmental cell biologists have been able to uncover the developmental genetics and molecular signaling processes underlying color pattern determination [73, 166, 167, 186]. Their results indicate that the theoretical models did in fact capture the essence of pattern formation in animals [9, 97, 165].

A simple pattern such as alternating black and white stripes is generated, not by merely turning the black element on and off, but by two biosynthetic antagonists interacting in a cyclic way while diffusing over space. A generalized reaction-diffusion model is illustrated in **Figure 4.30**. The two (chemical) components are an activator and an inhibitor operating side by side. The activator generates more of itself by autocatalysis, and after some lag, it also activates the inhibitor. The inhibitor disrupts the autocatalytic production of the activator, hence the reaction part of the model. Meanwhile, the two substances diffuse outward over an area, but the inhibitor diffuses faster than the activator. Both the activator and inhibitor decay and the cycle can begin again. Depending on the lag times between initiation and inhibition, the relative diffusion rates, decay rates, and the distance over which the system flows, several cycles of the presence and absence of the activator responsible for pigmentation can be generated.

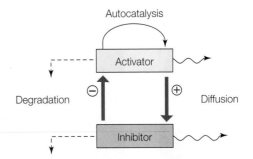

FIGURE 4.30 Reaction-diffusion model of color-pattern formation Two chemical antagonists, an activator (blue) at some spatial location that turns on some color-development process, and an inhibitor (pink) in an adjacent location that turns the process off. Both components diffuse over space (depicted on the right), with the inhibitor moving faster as indicated by the longer wavy line. Both components degrade with time (depicted on the left). If these antagonists control the production of a morphogen such as melanin synthesis in the neural crest of a fish or the melanocytes in a developing feather, a black and white bar pattern is generated. If the substances diffuse out in all directions from a point, a series of concentric circles is generated. (After [9, 189].)

Vertebrates provide the best fit to these models and melanin usually plays the starring role, although other colored pigments also may be involved. Melanin is manufactured in peripheral tissues and the synthesis of melanin granules is under the local control of chemical antagonists. Pigment cells originate in the neural crest of the early embryo, a tube running along the dorsal midline. These cells then disperse outward along the surface (ectoderm) of the embryo. An antagonistic process at this stage would generate longitudinal stripes on the animal. If an antagonistic process were established later in development along the length of a horizontal lateral line, vertical bars would be formed. Combining both of these processes and some variation in timing rules, a spot pattern could be produced [166, 167]. **Figure 4.31** shows a few examples of how these simple models can generate the patterns we observe. With recent advances in understanding the growth of avian feathers, slightly more complex reaction-diffusion models have been able to duplicate the intricate patterns found in birds [189]. Modeling the patterns of butterfly wings also requires some modifications, because the zones outlined by the wing veins are the repeating units of pattern. Eyespots on butterfly wings result when a reaction-diffusion process radiates out from a central point [157, 158].

Changing colors

Animals such as amphibians, reptiles, fish, and cephalopods, whose external surface consists of skin, a living and growing tissue, have options for short-term **color-changing mechanisms** that animals covered by the dead tissues of fur and feathers do not possess. (Note that the scales of reptiles and fishes are essentially very thick, hardened, keratinized skin,

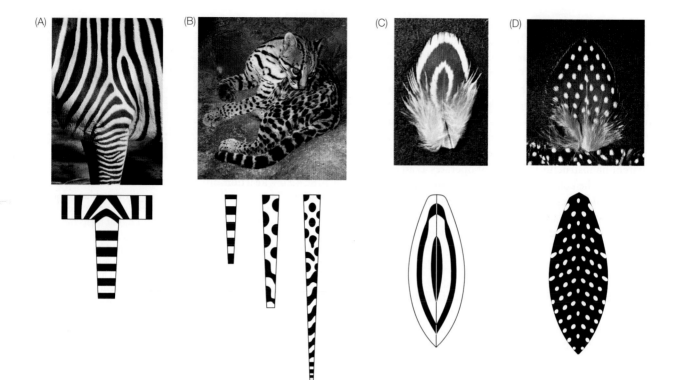

FIGURE 4.31   The fit between patterns predicted by reaction-diffusion models and real animal patterns (A) The scapular stripes of a zebra, where the leg meets the body, form a chevron pattern which is predicted by a model with this geometry. (B) Cylindrical tail models predict different patterns depending on their size and shape. Spots will appear on larger cylinders, bands on small cylinders, and transitions on tapering structures, as shown in the tail of the ocelot. (C) The concentric circle feather pattern, as seen in the flank feathers of the tinamou *Crypturellus tataupa*, requires a relatively simple 2-component model. (D) The array of offset dots pattern, as seen in the guineafowl *Guttera pucherani* feather, requires a 3-component model with two inhibitors. (A,B after [145]; C,D after [189]).

whereas the scales on butterfly wings are dead cuticle.) The color-producing cells lie beneath the exterior keratin layer and are generically called **chromatophores** [5, 6, 8, 102]. More specific names are given to cells that produce different colors: **melanophores** produce brown/black, **xanthophores** yellow/ochre, **erythrophores** red/orange, **cyanophores** blue, **iridophores** metallic violet/blue/green, and **leucophores** white. Gradual change in color with age or developmental stage, brought about by variation in the appearance or density of these chromatophores, is referred to as **morphological color change** [102]. More rapid color change, caused by variation in factors such as ambient temperature, stress, ambient light, background color, mood, and social contexts, is referred to as **physiological color change**. The pigment-bearing chromatophores (melanophores, xanthophores, erythrophores, and cyanophores, **Figure 4.32A**) are dendritic cells in which the pigment is packaged in many small cell

organelles or granules called **chromatosomes**. Changes in the dispersion of the granules within the cells are responsible for the color change. When the granules are dispersed throughout the cell, the animal appears colored, and when they are aggregated in the center of the cell, the animal appears pale. Iridophores and leucophores, on the other hand, are non-dendritic cells with color-producing organelles called **iridosomes** that contain transparent crystalline guanine platelets. The platelets are organized into parallel multilayered stacks interspersed by cytoplasm. These stacks generate coherent scattering of a narrow range of wavelengths via the structural mechanisms we discussed earlier. Hue changes are achieved by modifying the distance between the platelets: increasing the distance shifts the hue to longer wavelengths, and decreasing it shifts the hue to shorter wavelengths.

Granule dispersion and platelet orientation within chromatophores are ultimately controlled by either hormones or neurotransmitters. With hormonal control, color change may take an hour or two, whereas with neurotransmitter control, it can occur within seconds. Both pigment-based and structural chromatophores are embryonically derived from the neural crest and thus are neuron-like cells with G protein coupled receptors for pituitary hormones and cholinergic and adrenergic neurotransmitters on their membranes. Release of these stimulatory chemicals from adjacent cells initiates biochemical cascades that cause the translocation of platelets and granules within the cell. The movement of organelles occurs along the cell's cytoskeleton of microtubule tracks and actin filaments. **Motor proteins** such as tubulin, dynein, kinesin, and myosin contract or expand to drive this movement [4, 116].

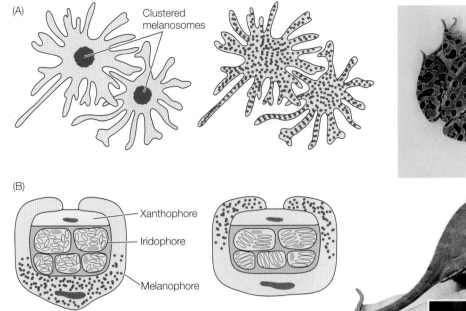

FIGURE 4.32 Chromatophores (A) Dermal melanophores in amphibians, reptiles, and fish consist of cells with long dendritic arms. Melanin granules (melanosomes) can migrate in and out of these arms. When the granules are clustered in the center of the cell, the animal appears light, and when they are dispersed, the animal appears dark. The dark leopard frog had been maintained in a tank with a dark background, the one on the left in a lighter environment. (B) The dermal chromatophore unit of these same vertebrate classes consists of three contiguous cell layers: xantho-phores containing yellow pigments, iridophores containing crys-talline platelets, and melanophores containing melanin granules. If the melanin granules are contracted to the deeper layer and the platelets are dispersed and randomly oriented as shown on the left, the skin appears yellow. If the platelets are organized in stacks as on the right, the skin appears green or blue, but simultane-ously dispersing the melanin granules over the top layer darkens the skin. The photos illustrate such color change in an individual female chameleon (*Furcifer minor*) taken 10 minutes apart; the upper photo shows the normal cryptic pattern, and the lower one shows the unreceptive female rejection display upon exposure to a male. (After [5, 6]).

Different-colored chromatophores can be combined in mosaic arrangements to create a wide variety of hues and color changes. For example, the dermal chromatophore unit of many poikilothermic vertebrates (fish, reptiles, and amphibians) consists of three contiguous cell layers (**Fig-ure 4.32B**). A top layer of xanthophores contains a yellow pigment (carotenoid or pteridine) that filters out (absorbs) short wavelengths. The middle layer of iridophores contains guanine platelets that can vary in their dispersion and orien-tation; if they are dispersed and randomly oriented, silvery white light is reflected, and if they are organized in stacks, blue wavelengths may be coherently reflected or scattered. The bottom layer of melanophores contains dispersable melanin granules, and the cells have dendritic extensions that cover the top layer. This dermal unit enables animals to

appear yellow, green, blue, or black depending on the disper-sion of melanin granules and guanine platelets [5, 59].

The most sophisticated example of color change for sig-naling purposes occurs in cephalopods (squid and octopus). The chromatophores in these animals are small multicellular organs that consist of a central pigmented compartment with brown, red, or yellow pigment connected to radial muscle fibers and motor neurons [23]. In the retracted state the sur-face of the chromatophore is highly folded and the color is inconspicuous. When the radial muscles are contracted, the structure is stretched out and the color becomes visible (**Fig-ure 4.33A**). The movement of the organ is under voluntary central nervous system control. By combining the expan-sion of its different-colored chromatophores, the animal can assume a variety of colors in rapid succession. These color signals are not only used to match the animal to its back-ground and startle prey, but they are also used in courtship, aggression, and group integration [25, 64, 115, 117, 144]. A few of the many patterned displays of the common reef squid are shown in **Figure 4.33B,C**.

Animals employ a variety of other strategies for short-term color change. Some birds and mammals have bare patches of skin designed to change color quickly as a conse-quence of changes in blood circulation. The intensity of red coloration is modified by dilating or constricting the blood

(A)

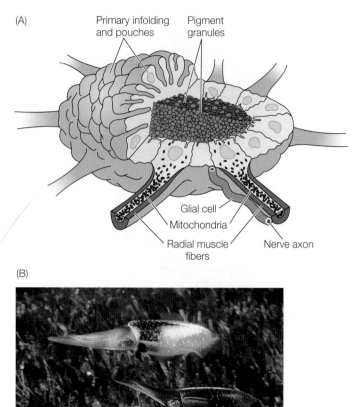

Primary infolding and pouches

Pigment granules

Glial cell

Mitochondria

Radial muscle fibers

Nerve axon

(B)

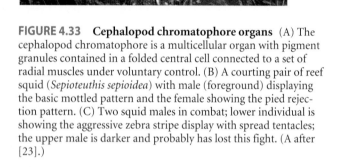

(C)

FIGURE 4.33 **Cephalopod chromatophore organs** (A) The cephalopod chromatophore is a multicellular organ with pigment granules contained in a folded central cell connected to a set of radial muscles under voluntary control. (B) A courting pair of reef squid (*Sepioteuthis sepioidea*) with male (foreground) displaying the basic mottled pattern and the female showing the pied rejection pattern. (C) Two squid males in combat; lower individual is showing the aggressive zebra stripe display with spread tentacles; the upper male is darker and probably has lost this fight. (A after [23].)

vessels under the skin, as seen in the blushing response of humans, the facial and head color of turkeys, the bare bottom of estrous female primates, and the red neck and legs of the breeding male ostrich (**Figure 4.34A**). Some beetles use an analogous fluid flow strategy to quickly change the color of their elytra. The cuticle contains a multilayered stack with porous low-density layers. Hydrating or dehydrating the cuticle changes the refractive index differences of the layers and alters the wavelengths reflected by the stack. Male Hercules beetles (*Dynastes hercules*) can change from green to black upon hydration [71, 194]. **Figure 4.34B** shows a Chrysomelid beetle that is gold in its resting state as a result of a hydrated chirped stack; expulsion of fluid causes the stack to become translucent, thereby revealing the red coloration of a pigmented layer beneath the stack [236]. Another strategy is to develop a small color patch on some part of the body that may be covered with another body part or with feathers and

thus unnoticed except when the animal flashes the color by uncovering it. Anole lizards, for instance, extend a brightly colored dewlap during push-up displays to conspecifics (**Figure 4.34C**). The roadrunner can pull aside the feathers near its eye to reveal a bright orange, white, and blue patch of skin (**Figure 4.34D**). Birds, butterflies, and other insects with folding wings can flash color patches by changing wing positions.

Postures and movements

Most species potentially have a very large number of different orientations, postures, gestures, and movements they can combine into a visual signal that symbolizes some type of information. These movements contribute temporal and spatial variation to visual signals [199]. Most animals of course move frequently as they go about their business of foraging, reproducing, and avoiding predators, so to produce a visual signal, a **movement display** or **posture** must be distinctive from these ordinary behaviors. Visual displays typically involve waving appendages in stereotyped ways, adopting stiff poses, or locomoting in unusual ways. Different movements may be performed sequentially to create complex displays. Quantifying the movement components of displays is very challenging, because one must specify movement pattern, speed, direction, amplitude, and sequence. A traditional way to examine displays is to parse complex sequences into smaller discrete categories by appearance and specify their frequency and sequence, as was done in a recent study of display components in birds of paradise [208]. This technique facilitates a comparative evolutionary approach to display form, but risks some inherent subjectivity with respect to category boundaries. Another strategy is to quantify multiple temporal, spatial, and morphological variable components and subject these to multivariate analysis. Composite features that capture aspects of three-dimensional movement

(A)

(B)

(C)

(D)

FIGURE 4.34   Other color-changing strategies (A) Birds and mammals can increase the redness of their skin by dilating surface blood vessels, as in this breeding male ostrich (*Struthio camelus*). (B) The Chrysomelid tortoise beetle (*Charidotella egregia*) changes from yellow to red in a time frame of about 1.5 minutes. (C) Anole lizards, such as this *Anolis roquet* male, can stretch open a patch of colored skin on their throats with a special hypoid bone. (D) The greater roadrunner (*Geococcyx californianus*) normally covers most of its postorbital skin with feathers, so that only a small amount of the white and blue coloration is visible. During aggressive interactions with rivals and during courtship, the feathers are pulled back to expose the bright orange region.

patterns and velocities can be defined. This approach was successfully applied to display components of the jacky dragon (*Amphibolurus muricatus*) [153, 184], and has the advantage of identifying traits of individual displayers potentially related to their display vigor and success in mate attraction or defense against rivals. Experimental approaches using models, robots, or videos can uncover the effectiveness and function of different display components [112, 155, 161, 180, 181, 221].

Movement displays are often enhanced with color patches on the moving body part. **Figure 4.35** shows several examples of visual displays that are augmented with color. Specular colors with a narrow viewing angle, such as structural colors produced with multilayered stacks, can generate brilliant flash displays when coupled with movements. For example, in

the aerial display of the male Anna's hummingbird (*Calypte anna*) to a perched female, he orients himself to simultaneously face the sun and the female at the bottom of the dive so that a brilliant flash of color from his throat feathers is delivered [164, 225]. Hummingbirds even seem to use muscles to affect feather orientation within the gorget, aligning them all in a flat plane. In some New World scarabid dung beetles, the head and large triangular prothoracic shield reflect a visible and ultraviolet iridescence that is best seen from a position directly facing the insect. The less iridescent male horn is silhouetted against this shield. Since horn size is indicative of male size, this directed color display may signal competitive ability to other males and mate quality to females [245]. A final example of flash color can be found in numerous butterfly species with iridescent specular reflectance on the wings. In

(A)

(B)

(C)

(D)

FIGURE 4.35   Posture, movement, and color (A) Male Wilson's bird of paradise (*Cicinnurus respublica*) in head-down perched posture displaying brilliantly colored feathers, blue skin, and purple curly tail plumes to female. Yellow neck ruff, red mantle, and green breast shield (not visible here) are expanded during different steps in this complex display. (B) Male vervet monkey (*Chlorocebus pygerythrus*) posturing to display colorful genitals. (C) Male fiddler crab (*Uca tetragonon*) displaying colored enlarged claw. (D) Male blue-footed booby (*Sula nebouxii*) performing courtship dance in which the blue feet are alternately raised.

some species the entire wing may flicker between alternating colors as the animal flies [229]. In other cases, colorful patches may wink on and off during flight [203, 240]. Movement or color patches may also be combined with signals in other modalities such as sound or scent. For example, the black and white tail of the ring-tailed lemur (see Figure 6.17C) is not only capable of creating a conspicuous visual display, but it is also sometimes anointed with an olfactory gland product and used to waft the scent to receivers. The movement associated with making an auditory signal may sometimes be exaggerated with a coupled visual display.

When visual displays involve a vigorously repeated movement, they are likely to entail a high energetic cost. The performance of the display may therefore reveal important information to receivers about the sender's condition, size, or experience. To view some active examples of movement displays, visit Web Topic 4.4. Complex movement displays are characteristic of courtship in animals, a topic we take up in Chapter 12.

Web Topic 4.4 *Movement displays*
See photos and video clips of visual displays in a variety of animal species.

Transparency

This section on coupling strategies to maximize color reflection would not be complete without a discussion of coupling strategies that *minimize* total reflection, rendering an animal partially invisible. **Transparency** is primarily a pelagic marine phenomenon, for a good reason. There is no place to hide in this environment, so becoming transparent is the most effective way to avoid being detected by predators. There are a few examples of transparency in freshwater and terrestrial organisms as well (**Figure 4.36**). It is virtually impossible for a multicellular animal to make its entire body completely transparent. Soft-bodied animals with thin tissues, such as

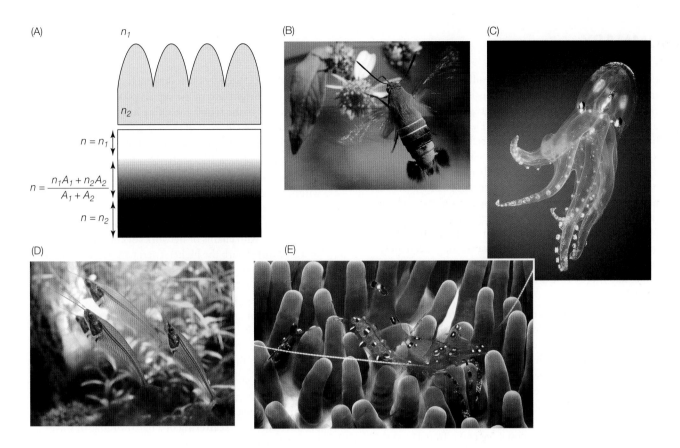

FIGURE 4.36 **Transparency: An antireflection coupling mechanism** (A) A nipple array of protrusions smaller than half a wavelength of light experiences a gradual change in refractive index rather than a sharp discontinuity (n_1 is the refractive index of the external medium, n_2 is the index of the surface of the organism). (B) The sphingid clearwing moth *Cephonodes hylas* uses a nipple array to make its wings completely transparent. Other examples of transparent organisms include (C) the octopus *Vitreledonella richardi*, (D) the glass catfish (*Kryptopterus bicirrhis*) and (E) anemone shrimp (*Periclimenes holthuisi*). (A after [81].)

comb-jellies and jellyfish, are most successful. Many partially transparent animals use strategies such as body flattening or elongation to disguise their shape. Eyes and guts, in particular, always absorb and scatter light, but elongation of these organs can minimize the shadows they cast. In some species they are surrounded with mirror-like reflectors [81, 82]. Large-bodied animals with endo- or exoskeletons cannot be completely transparent, because these dense tissues always scatter light.

Nevertheless, animals have evolved several mechanisms to greatly reduce scattering and increase transmission of light through the body. Nanostructural modifications to the external body layer can reduce surface scattering. One such mechanism, observed on the wings of some moths, is to cover the surface with a uniform array of submicroscopic protrusions (see Figure 4.36A). Because the protrusions are smaller than half a wavelength of light, the index of refraction at a given layer is the average of the relative projected areas of

the protrusions and the external medium. This surface structure effectively causes a gradual increase in refractive index from the outside to the inside of the animal, which eliminates scattering at the external-to-body boundary. However, further transmission of light inside the body would normally be scattered by cell membranes, cell organelles, lipids, and other structures that create boundaries of heterogeneous refractive indices. Theoretical studies suggest that for functional cells and muscle tissue to be transparent, the size, distribution, and refractive index of cell components would have to be constrained to particular ranges [79]. Close and uniform packing of similar-sized objects at scales over half a wavelength of light would achieve this objective [12]. Another possible strategy is to add a high-index substance to cells and tissue so that refractive index increases and boundaries are reduced. It will be quite interesting to discover the full range of cellular adaptations for transparency and the costs they may entail.

SUMMARY

1. Light is a form of **electromagnetic radiation** that, unlike sound, can travel in a vacuum. It consists of an oscillating electric field and a magnetic field at right angles to each other and to the direction of travel. Electromagnetic radiation exhibits the wave properties of reflection, refraction, diffraction, spreading loss, and attenuation. Its frequency is inversely proportional to its wavelength, but it can also be viewed as a rapid stream of tiny energy packets. The natural **electromagnetic radiation spectrum** spans a huge range of wavelengths. Low-frequency (long-wavelength) radiation in the radio range bends around most objects in the environment and acts like a **wave**. High-frequency (short-wavelength) radiation ionizes and scatters molecules and therefore acts more like a **flux of massless particles**.

2. **Visible light** is a very narrow range of electromagnetic radiation in the middle-frequency range. Only in this range can electromagnetic radiation interact constructively with organic matter to facilitate visual communication. Lower frequencies possess too little energy to be detected by organisms except as heat, higher frequency radiation possesses too much energy to be absorbed, and biological tissues are essentially transparent to both extremes. Nearly everything absorbs ultraviolet and infrared waves. The intermediate energy of visible light can be partially absorbed by organic molecules without damaging them, and this energy may then be either reflected or coupled to a detector. Due to atmospheric absorption of the higher and lower frequencies, most of the radiation from the sun that reaches the Earth's surface is in this narrow visible range.

3. The **speed of light transmission** is slowed to varying degrees when it travels through different media because the molecules pose an electric counterforce. Transmission through air is only slightly reduced because molecular density is low and irregular. **Rayleigh scattering** selectively scatters shorter wavelengths, which makes the sky appear blue. When light waves encounter a boundary with a liquid or solid medium, they interact with the surface layers of molecules and will undergo some combination of reflection, transmission, and absorption. If the medium molecules do not possess a **resonance frequency** close to the frequency of the light waves and are uniformly arrayed, the waves will be mostly transmitted through the medium layer in a series of reradiations, but travel speed will be slowed. Such a medium—like water or glass—is **transparent**. If the medium molecules have a resonance frequency in the visible light range, these frequencies will be absorbed, the energy will be dissipated as heat, and neither transmission nor reflection will occur. If the medium is very heterogeneous in density, light waves will be scattered and reflected at the discontinuous boundaries.

4. Most animals make use of **reflected light** from the sun to produce visual signals, while only a few produce their own light. Regardless of the source of light, all visual signals can be characterized by the following four properties: (1) intensity or **brightness** of the signal; (2) spectral composition or color of the signal, including **hue** and **saturation**; (3) spatial characteristics of the signal; and (4) temporal variability in intensity, color, and spatial properties.

5. **Pigments** are chemical compounds whose molecules undergo **selective absorption** of certain wavelengths within the visible light range via electron orbital shifts and scatter the remaining wavelengths. The receiver perceives the color of the scattered light. All pigments are organic compounds with chains or networks of **conjugated double bonds**. Most absorb the higher-energy short wavelengths in the visible range and therefore appear yellow, orange, or red. Pigments include **carotenoids**, **melanins**, **porphyrins**, **pteridines**, and **purines**. Some, like guanine, are crystalline in form and reflect and scatter all wavelengths, thus producing white. A few more complex pigment molecules produce green and blue-green colors. Melanin is a very large molecule that absorbs nearly all wavelengths and appears black or brown.

6. Ultraviolet, violet, and blue colors are usually produced by **nanometer-scale structures** in the integument that produce **coherent scattering** of a narrow range of wavelengths. Most green colors and some yellow-to-red colors are also generated with this mechanism. Coherent scattering can be achieved in several different ways, including the alternation of thin layers of high- and low-refractive-index materials (**quarter-wave stack**), diffraction gratings, and two- and three-dimensional ordered arrays of scatterers with diameters about half the size of a wavelength of light. Layered structures and gratings can generate very bright **iridescent** colors. Diffuse white color can be generated with a structure of larger, irregular scatterers that produces **incoherent scattering**.

7. **Bioluminescence** is the strategy of chemically self-generated light that is most effective in dark environments and most commonly found in the marine environment. There are only about half a dozen chemical mechanisms for producing light. The basic process involves a luciferin substrate that is combined with molecular oxygen in the presence of a luciferase catalyst to produce an excited-state peroxy-luciferin intermediate, which quickly breaks down while releasing a photon of light. Bioluminescence appears to first arise in response to nonsignaling selective pressures for avoiding or facilitating predation, and then may be co-opted for communication functions.

8. Visual signals often make use of several **coupling** mechanisms to generate brighter and more saturated color signals. Two or three color-producing mechanisms may be combined to achieve this objective. In particular, pigment layers on the surface need to have a structural reflecting or scattering layer underneath to direct the nonabsorbed wavelengths back out. Similarly, many of the structural color mechanisms, including thin layers, multilayer stacks,

diffraction gratings, and multidimensional arrays, often incorporate a layer of melanin pigment to absorb wavelengths outside of the tuned reflection structure. Some animals have evolved anticoupling mechanisms to render them transparent.

9. Animals modify color signals using temporal and spatial variation to produce more distinctive or conspicuous visual signals. One strategy is the development of **color patterns**, which often involves laying down geometric patterns of melanin and one or two other colors. Another strategy is to evolve **color-changing mechanisms**, such as covering and uncovering color patches with another body part, or evolving chromatophore cells or organs that can move pigment molecules and tissues. Finally, color patches are often coupled with **postures** and **movement displays** that feature the appendages or body parts bearing the color.

Further Reading

For a highly readable treatment of most of the topics in this chapter, including the properties of light and its emission, absorption, and scattering from a biologist's perspective, see the book by Johnsen [84]. Another outstanding source of information on all aspects of physics described in this chapter, including electromagnetic waves and optics, can be found at the HyperPhysics website by Nave [147]. Weiskopf [249] provides an excellent layperson's description of how light interacts with matter. Barrow [11] is the best source of information on the effects of radiation on molecular motion. Watterman [246, 247] offers a basic summary of the use of polarized light by animals, but see newer references in Web Topic 5.2. Fox [44, 45] and Nassau [146] are classic books on color. The book on bird coloration by Hill and McGraw [69, Vol. 1] has excellent chapters on color measurement, classification, pigments, and structural color-producing mechanisms. Good reviews of color in invertebrates include Ghiradella [49], Parker [171, 172, 177], Berthier [14], Kinoshita et al. [96], and Seago et al. [211]. For transparency, see Johnsen [79, 80], and for bioluminescence see the review by Haddock et al. [62] and the Bioluminescence Webpage [61]. Ball [9] reviews color pattern formation in the natural world. See Leclercq et al. [102], Mäthger et al. [116], and Aspengren et al. [4] for reviews of color change in vertebrates. Rosenthal [199] provides a good overview of spatiotemporal dimensions of visual signals.

COMPANION WEBSITE
sites.sinauer.com/animalcommunication2e

Go to the companion website for Chapter Outlines, Chapter Summaries, and References for all works cited in the textbook. In addition, the following resources are available for this chapter:

Web Topic 4.1 *Light wave meets boundary*
View animated simulations of the way that scattered light is reflected and refracted at a boundary between two media in the form of reradiating wavelets. Obtain quantitative expressions for the effects of the indices of refraction in the two media on angles of reflection and refraction, proportion of reflection, and degree of polarization.

Web Topic 4.2 *Quantifying light and color*
We have been able to measure the brightness of light for some time. Recent technological developments have made it much easier to quantitatively describe the hue and saturation of ambient light and animal color patches, including wavelengths that we cannot perceive.

Web Topic 4.3 *Dimensionality of structural color mechanisms*
Definitions of one-, two-, and three-dimensional arrays. Fourier analysis and other new techniques have recently been applied to the reflection patterns from periodic structural arrays to determine their dimensionality. We also review the photonic crystal approach to color production by periodic structures.

Web Topic 4.4 *Movement displays*
See photos and video clips of visual displays in a variety of animal species.

Chapter 5

Visual Signal Propagation and Reception

Overview

Almost all living organisms, plant and animal, are sensitive to light. Plants of course use the energy from the sun to grow, and they can turn their leaves in the direction of the sun to maximize their exposure. Even primitive one-celled organisms can orient with respect to the sun. The ability to detect light developed very early in the evolutionary history of life on Earth. The evolution of eyes that could form an accurate image of objects in the environment and a brain that could analyze the image required more evolutionary time. In this chapter we will describe the ecological context of visual reception, beginning with the way the environment affects the light arriving at the receiver. We will then discuss receiver strategies for collecting the light, perceiving it, and analyzing the spatial and spectral information. We cannot assume that animals see the world the way we do, because their visual systems have evolved in response to the amount of light available in their environment, as well as their needs for orientation, predator detection, and food finding. The capabilities of an animal's eyes certainly affect its use of vision for social communication.

Transmission of Visual Signals

The transmission of a long-distance visual signal from sender to receiver entails many of the same problems as long-distance sound transmission, such as attenuation, pattern loss, and background noise. However, the two modalities differ in several important respects. Most important, for reflected light signals, there is an additional step in the process, since the production of a signal is dependent on the availability and quality of the ambient light. Second, there is always background noise for visual signals, so the contrast between the signal and the background, as well as the perceptual ability of the receiver, takes on greater importance. Finally, light waves don't bend around objects as sound waves do, so visual signal transmission is much more strongly affected by the presence of opaque objects between sender and receiver than is sound transmission.

Overview of transmission steps

The major steps in the transmission of a colored reflected light signal are illustrated in the **ARTS diagram** of **Figure 5.1** [55, 126]. The letters of this easy-to-remember mnemonic stand for the four transmission steps, each of which is a process with an outcome that can be measured and depicted by an irradiance, reflectance, or absorbance spectrum. For reflected light signals, we begin with

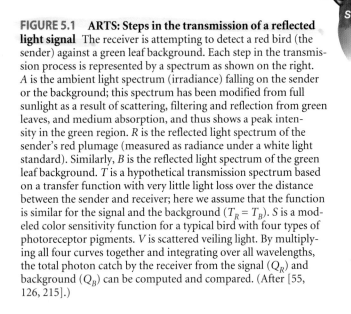

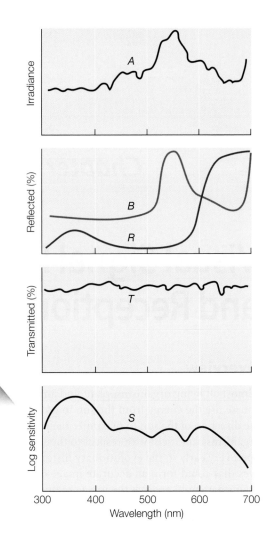

FIGURE 5.1   ARTS: Steps in the transmission of a reflected light signal The receiver is attempting to detect a red bird (the sender) against a green leaf background. Each step in the transmission process is represented by a spectrum as shown on the right. *A* is the ambient light spectrum (irradiance) falling on the sender or the background; this spectrum has been modified from full sunlight as a result of scattering, filtering and reflection from green leaves, and medium absorption, and thus shows a peak intensity in the green region. *R* is the reflected light spectrum of the sender's red plumage (measured as radiance under a white light standard). Similarly, *B* is the reflected light spectrum of the green leaf background. *T* is a hypothetical transmission spectrum based on a transfer function with very little light loss over the distance between the sender and receiver; here we assume that the function is similar for the signal and the background ($T_R = T_B$). *S* is a modeled color sensitivity function for a typical bird with four types of photoreceptor pigments. *V* is scattered veiling light. By multiplying all four curves together and integrating over all wavelengths, the total photon catch by the receiver from the signal (Q_R) and background (Q_B) can be computed and compared. (After [55, 126, 215].)

light from the sun (or moon and stars for nocturnal animals). Sunlight is scattered, absorbed, reflected, and attenuated, often in a frequency-dependent manner, depending on the medium, weather, time of day, and habitat. The **ambient light spectrum**, *A*, is the result. We measure this spectrum at the time and place where animals display, using a flat **irradiance** meter that accepts light over a 180° solid angle (2ϖ steradians). The unit of intensity is usually expressed as photon flux (measurements made in watts can be converted into photon flux; see Web Topic 4.2). This spectrum represents the absolute amount of light intensity at each wavelength available to senders for reflection.

The second step is reflection. As we described in detail in Chapter 4, the surface properties of the sender affect the wavelength-specific intensity of reflected light, and any process that reflects some wavelengths more than others results in a colored light signal. The **reflected light spectrum**, *R*, specifies the hue, chroma, and intensity of light emanating

from the sender. We measure this *R* spectrum with a narrow-tube **radiance** meter very close to the animal under standard lighting conditions to eliminate any scattered light input (see Web Topic 4.2), and express the spectrum as the fraction of white standard input light that is reflected at each wavelength interval. This measurement is essentially a transfer function, as described in Chapter 3 for sound transmission.

The third step is transmission from sender to receiver. Light propagation is attenuated and filtered by factors such as medium absorption, scatter, and total blockage from large opaque objects, and the degree of attenuation depends on the distance between sender and receiver. The **transmission spectrum**, *T*, is a transfer function that specifies the fraction of light lost at each wavelength interval during propagation due to these effects. At this point, we could multiple these three spectra together, $A \times R \times T$, to describe the color spectrum of light arriving at the receiver from the sender. Since the *R* and *T* spectra are expressed as fractions, they represent sequential loss of selective wavelengths at these steps.

The fourth step is perception by the receiver. This process depends critically on the **color sensitivity**, *S*, of the receiver. For simplicity, this step can also be represented as a spectrum. In practice, this spectrum is modeled based on knowledge of the wavelength absorption curves of the specific animal's photoreceptors [99, 215]; see Web Topic 5.1.

FIGURE 5.2 Colored ambient light in different habitats (A) Green forest light. (B) Purple light at low sun angles. (C) Blue light in the ocean. (D) Veiling light, the result of Mie scattering from the water droplets in fog.

We will learn more about color sensitivity in receivers later in the chapter.

The conspicuousness of the signal (or, for that matter, any visual target of importance to the receiver such as food, predators, and landmarks) is determined by the perceived contrast between the reflected light emanating from the sender and the reflected light emanating from the background. The **background spectrum**, B, can be quantified in the same manner as the signal, by measuring the light that reflects off objects or surfaces around the sender. Figure 5.1 shows this parallel series of steps for the background "signal." Assuming that the background is illuminated by the same ambient light spectrum as the target signal, and that propagation loss from the background to the receiver is similar, the color spectrum of light arriving at the receiver from the background can be computed as $A \times B \times T$.

Finally, photon input from veiling light, V, is an additional source of noise caused by glare and medium scattering that serves to reduce the contrast between signal and background;

its effects can be incorporated into the estimation of T. A similar process can also be undertaken to describe self-generated light signals, but the A spectrum would of course be omitted, the R spectrum would be measured in units of watts or photon flux, and the B spectrum in most cases would be black.

Armed with receiver sensitivity information, we can now compute the total photon (or quantum) catch of the receiver from the signal as $Q = A \times R \times T \times S$ and from the background as $Q = A \times B \times T \times S$, integrated over all wavelengths. The difference between these two spectra is the key determinant of signal conspicuousness.

In the following sections we will describe some of the effects of ambient light availability and propagation constraints on real animal signals, define the standard measures of brightness and hue contrast, and outline some of the strategies animals use to either increase their conspicuousness to conspecifics or decrease their conspicuousness to predators.

Availability and quality of ambient light

Both the amount and the spectral composition of available light can be very different in different environments (**Figure 5.2**). When the **ambient light** is itself colored, the apparent hue of a colored signal patch changes. A color patch that

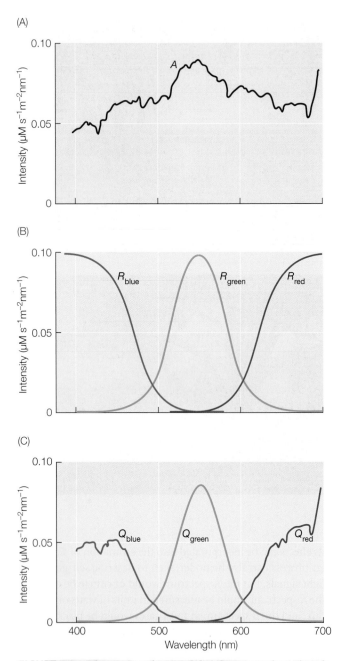

(A)

(B)

(C)

FIGURE 5.3   Change in color patch brightness under colored ambient light (A) Typical green irradiance spectrum from a forest habitat. (B) Reflectance spectra of hypothetical blue, green, and red color patches under white light illumination. (C) Change in the reflectance from the three patches when illuminated by the green ambient light in (A). The blue and red patches are greatly reduced in intensity, compared to the green patch. *T* and *S* are not considered here. (After [55].)

under white light conditions reflects strongly in a particular region of the spectrum will appear desaturated and less bright when viewed under an ambient light context that contains low levels of this spectrum region. In the extreme case where the reflected frequencies of a color patch are completely missing from the ambient light spectrum, the patch will appear

black. **Figure 5.3** shows an example of the change in reflectance for three patches—red, green, and blue—illuminated by a typical green forest ambient light. Although all three have equal brightness under white ambient light, the reflectance from the red and blue patches is significantly reduced relative to the green patch under ambient conditions. Thus animals can reflect colored signals only when those same frequencies are abundantly present in the ambient light.

In terrestrial habitats during the day, the three principal factors influencing irradiance spectra are the angle of the sun above the horizon, the weather conditions of the atmosphere, and the structure of vegetation. For irradiance measurements it is important to remember that light may arrive directly from the sun, indirectly from other parts of the sky, and indirectly from vegetation (**Figure 5.4A**). Scattered light from these components may be distinctly colored. In a forest with a continuous canopy, only filtered light arrives at the ground level, and this light has a green hue (**Figure 5.4B**). In woodland shade, where the canopy is not continuous, no direct sunlight is present, but there is much Rayleigh-scattered light from the blue sky, so the ambient light is bluish (**Figure 5.4C**). An open area or large forest gap will receive full white light illumination (**Figure 5.4D**). In a small gap in the forest, on the other hand, light arrives only directly from the sun and not from the blue sky, so long wavelengths predominate (**Figure 5.4E**). In any of these contexts, if the sky is cloudy a large amount of Mie-scattered white light will be present and all spectra tend to look less saturated, like that in **Figure 5.4F**. Total illumination can actually increase in shady sites when there is cloud cover. The angle of the sun determines the amount of atmosphere through which the light must travel before striking the Earth, with the shortest path occurring at noon and the longest path at sunrise and sunset. The longer the path, the greater is the scattering of blue wavelengths. This scattering partially eliminates blue wavelengths from direct sunlight, but scattered blue light arrives from the rest of the sky. In addition, selective absorption of middle (yellow) wavelengths by atmospheric ozone has its strongest effect at low sun angles, and the combined effects of this absorption and scattering produce purple ambient light (**Figure 5.4G**; see also Figure 5.2B) [17, 90, 173]. Finally, the presence of dried grass, leaf litter, tree trunks, and red leaves can alter irradiance spectra by reflecting yellow, orange, brown, and red wavelengths.

The terrestrial night sky exhibits variation in ambient light depending on the angle of the sun, the cloud cover, and the phase of the moon. **Figure 5.5** shows a series of spectra for twilight and evening sky colors on the same light intensity scale for comparison. Once again, we see the shift from white high-intensity illumination to purplish low-intensity illumination as the sun angle decreases. Light intensity under a full moon night sky is lower but basically white in color. On a clear moonless night, illumination is even lower and basically white, with a few spikes of light in the atmosphere from **airglow**. Airglow is caused by the recombination of oxygen (green and red peaks) and sodium (yellow peak) driven by UV

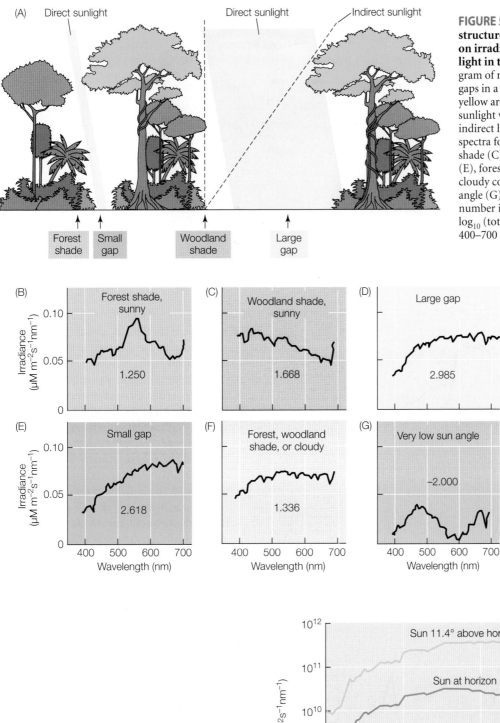

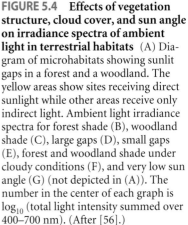

FIGURE 5.4 Effects of vegetation structure, cloud cover, and sun angle on irradiance spectra of ambient light in terrestrial habitats (A) Diagram of microhabitats showing sunlit gaps in a forest and a woodland. The yellow areas show sites receiving direct sunlight while other areas receive only indirect light. Ambient light irradiance spectra for forest shade (B), woodland shade (C), large gaps (D), small gaps (E), forest and woodland shade under cloudy conditions (F), and very low sun angle (G) (not depicted in (A)). The number in the center of each graph is $\log_{10}$ (total light intensity summed over 400–700 nm). (After [56].)

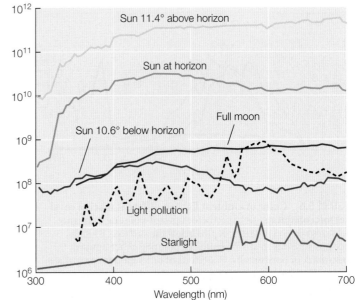

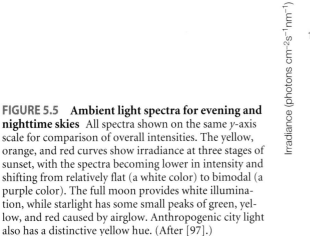

FIGURE 5.5 Ambient light spectra for evening and nighttime skies All spectra shown on the same *y*-axis scale for comparison of overall intensities. The yellow, orange, and red curves show irradiance at three stages of sunset, with the spectra becoming lower in intensity and shifting from relatively flat (a white color) to bimodal (a purple color). The full moon provides white illumination, while starlight has some small peaks of green, yellow, and red caused by airglow. Anthropogenic city light also has a distinctive yellow hue. (After [97].)

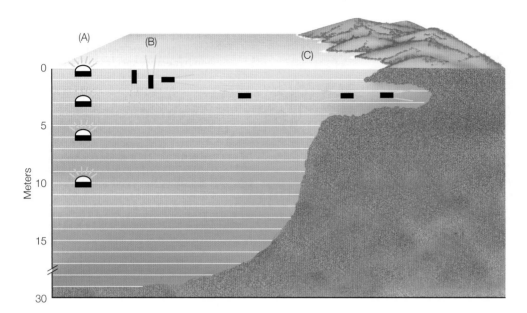

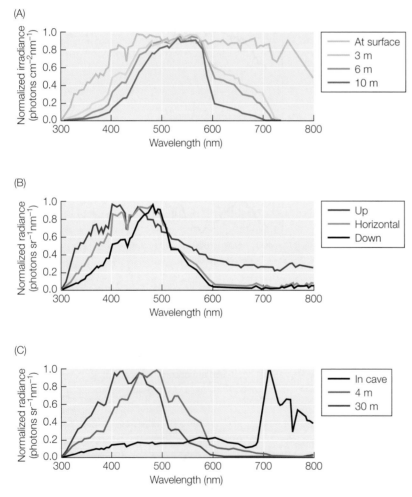

FIGURE 5.6 Color of ambient light in different marine coral reef microhabitats (A) Spectra for upward-pointing irradiance measures taken at different depths below the surface. (B) Radiance measures taken at 1 m depth but pointing either upward, downward, or horizontally. (C) Effects of different floor or surface features (horizontal radiance measures). All reflectance curves are normalized relative to their maximum value. (After [133].)

radiation from the sun, and at certain times of year becomes the spectacular aurora borealis light display in the arctic sky [17, 97].

Light available in aquatic environments depends on distance below the surface, angle of view, depth of the floor, substrate features, and particulate matter in the water [125, 126, 133]. **Figure 5.6** shows spectra from different kinds of measurements made in ocean reef habitats. Clear oceanic water becomes increasingly bluer with increasing depth below the surface. This color change occurs because the water strongly absorbs both ultraviolet and red-infrared radiation while transmitting blue and green wavelengths (see Figure 4.6). An upward-directed view obviously is brighter than either a horizontal or a downward view. The horizontal view is relatively green compared to the downward view, which is bluer. Similarly, shallow water is greener than deeper water. The green effect is caused by the presence of phytoplankton present in the sun-rich surface layer and in runoff from freshwater streams. Somewhat surprisingly, the water color in caves and under overhangs contains a noticeable amount of red and infrared. The red color is believed to result from encrusted red algae lining the cave, light filtered through coral, or fluorescence from algae and coral [133].

Similar principles affect light transmission in fresh water. A blue color is indicative of nutrient-poor water, low in phytoplankton and organic material. Freshwater lakes and ponds are often greenish as a result of the reflection of green and absorption of red and blue by phytoplankton and other plant material. Swamps and marshes often exhibit a yellowish hue from the dissolved products of decaying vegetable matter.

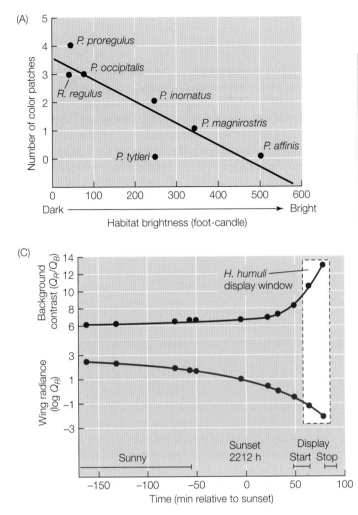

FIGURE 5.7 **Maximizing brightness contrast** (A) Old World *Phylloscopus* warblers possess yellow-green body plumage and from 0–5 white or pale yellow patches. They inhabit environments characterized by a range of available light from the brightest (open montane scrub) to intermediate (open forest) to dark (dense evergreen forest). Species in darker habitats have more patches of light color. (B) The contrasting patches, such as the rump patch on this *P. proregulus*, are featured in movement displays. (C, D) The male ghost swift moth (*Hepialus humuli*) times the beginning of its crepuscular display to coincide with the period when its silvery wings will contrast most with the grassy green background. The bottom curve shows decreasing wing quanta radiance as a function of time relative to sunset, and the top curve shows the computed contrast between the moth's wings and the background as a function of time. Display period is shown by the white boxed area. (A after [128]; C after [4].)

Streams collect the runoff of minerals that reflect in the red and infrared range, making the water appear reddish, brown, and sometimes black.

Contrast with background

The transmission efficiency of a visual signal depends critically on the type of **background** against which the animal is seen. Background noise is always present for any signal in this modality. Most backgrounds possess the same characteristics—brightness, hue, pattern, and movement—that correspond to the signal properties we identified early in the prior chapter. Senders can exploit any one of these, or a mixture, to develop a signal that presents a contrasting image against the backdrop. General strategies for these four types of contrast are described below, but it is important to realize that the degree to which these contrasts are perceived depends on the optical anatomy of the species. Each of these contrast-enhancing strategies can be reversed with a background-matching strategy to maximize camouflage and crypticity.

STRATEGIES FOR BRIGHTNESS CONTRAST   Animals can use **brightness contrast** (also called **achromatic contrast**) to enhance their conspicuousness. Brightness contrast can be exploited best in environments with backgrounds that are either extremely dark or extremely light. With a dark

background, such as occurs in forests, deep water, caves, or at night, white or bioluminescent colors contrast most; in bright, highly illuminated contexts, black or other dark colors will contrast most.

A good example of brightness contrast is provided by a study of eight leaf warbler species that occupy a range of habitats from low montane scrub to broadleaf forest. All eight species are primarily yellow-green but vary in the number of white patches on the wings, head, and rump (**Figure 5.7A,B**). The darker the habitat occupied by a species, the more patches of white it possesses [128]. A second example can be found in the ghost swift moth (*Hepialus humuli*), in which the silvery white males aggregate to display for a short period of time about an hour after sunset. The intensity of ambient light drops after sunset, but relative shortwave irradiance increases, resulting in maximum brightness contrast between the moth's wings and the grassy background at the time of display onset [4] (**Figure 5.7C,D**).

In the section above, we described how one would quantify the radiance arriving at the receiver from both the signal and from the background. For this analysis, Q_R and Q_B could be evaluated over a narrow range of wavelengths, or summed over all perceived wavelengths using the areas under the spectral curves. A useful measure of the brightness contrast between signal and background is the relative difference between the two:

$$C = \frac{Q_R - Q_B}{Q_R + Q_B}$$

C is zero when brightnesses are the same, negative when the signal is darker than the background, and positive when the signal is lighter than the background. Dividing by the sum of the signal and background radiance produces a symmetrical index that varies from −1 to +1, a range that is most appropriate for questions concerning contrast maximization against any type of background. A similar index with only background radiance in the denominator is more accurate for questions concerning the minimum or threshold contrast necessary for detection.

Some strategies of brightness contrast depend on the direction of the illuminating light. In media without a significant amount of Mie scattering, light comes from above. This means that the dorsal side of the animal is more strongly illuminated than the ventral side. **Countershading**, in which the dorsal side is more darkly pigmented than the ventral side, is thought to be a camouflage strategy that reduces brightness contrast for horizontal viewers and conceals the shadow cast by terrestrial animals close to the ground. In the marine environment, countershading reduces brightness contrast for overhead viewers, which see the subject against a dark background, as well as for viewers underneath the subject, which see it against a light background [182, 202]. In **reverse countershading**, a darker underside serves to increase the visibility of an animal that is illuminated from above (see Figure 5.9A). In open habitats at low sun angles, the illumination of the background depends on the position of the viewer relative to the sun. This effect is extreme in water surface environments. When the viewer faces toward the sun, it sees intensely bright, white reflectance; black signals are highly visible against this background. When the viewer faces away from the sun, it sees diffuse blue light from the sky, so colored and white signals are more conspicuous. Since senders cannot control their position relative to the sun and potential receivers, they cannot maximize conspicuousness in both contexts. Most seabirds are white, probably because this minimizes their conspicuousness against the bright sky to prey in the water below and reduces thermal stress in their open environment [76].

STRATEGIES FOR HUE CONTRAST   For species with good color vision, **hue contrast** (also called **chromatic contrast**) can be a very effective aid to discerning objects that do not contrast strongly with their backgrounds in terms of brightness. In the absence of information on the visual pigments in the eyes of the receiver, we can make an initial evaluation of the color contrast between the signal and background spectra arriving at the receiver. A simple method is to compute the difference between the intensity of the signal and the intensity of the background at each wavelength interval, and then determine the root mean square of these differences over all wavelengths that the species can distinguish as follows [55]:

$$D = \sqrt{\sum_{}^{all\lambda} \left[R(\lambda) - B(\lambda) \right]^2}$$

Any difference in hue, saturation or brightness will increase the value of D. In essence, D is the Euclidean distance of a line connecting the points of the signal and the background color on the three-dimensional color plot of Figure 4.12. Senders can maximize hue differences by using a saturated hue on the opposite side of the horizontal circle from the background hue, or they can maximize brightness differences along the vertical scale. If the color sensitivity of the receiver is known, then a more precise measure of hue contrast can be computed using the photon catch of each photoreceptor type from the signal and the background radiances. A similar Euclidean distance D value can then be plotted within the specific hue space for the species [99, 143, 215].

Recent studies of coloration strategies in a variety of animal taxa have tended to quantify both hue and brightness contrast, and recognize that most organisms face a trade-off between being cryptic to predators while being conspicuous to conspecifics (and pollinators in the case of plants) [69, 209, 210]. Cryptocity obviously entails a good match (low contrast) between basic body coloration and background for both hue and brightness. Conspicuousness can be achieved by maximizing either hue or brightness contrast of the whole body against the background or between adjacent patches within the body. Coral reef fish nicely illustrate these principles: the hues and patterns of these fish are spectacular, the color sensitivity of many species are known [123, 133, 134], and communication occurs in a variety of environments with different available light spectra (**Figure 5.8**). Species living in deeper (blue) water reflect blue body colors that closely match the background hue, which enhances their crypticity in an environment where there is no place to hide. The visual color sensitivities of these species are also restricted to the ultraviolet and blue end of the spectrum, implying that colorful signals are constrained by available light and the maximization of blue visual sensitivity for prey detection. Species living in shallower green water have expanded their color sensitivity to include longer wavelengths and often sport green or a combination of blue and yellow colors (see Figure 5.8A,B). The blue-yellow combination provides the strongest *hue contrast* in the marine environment where long

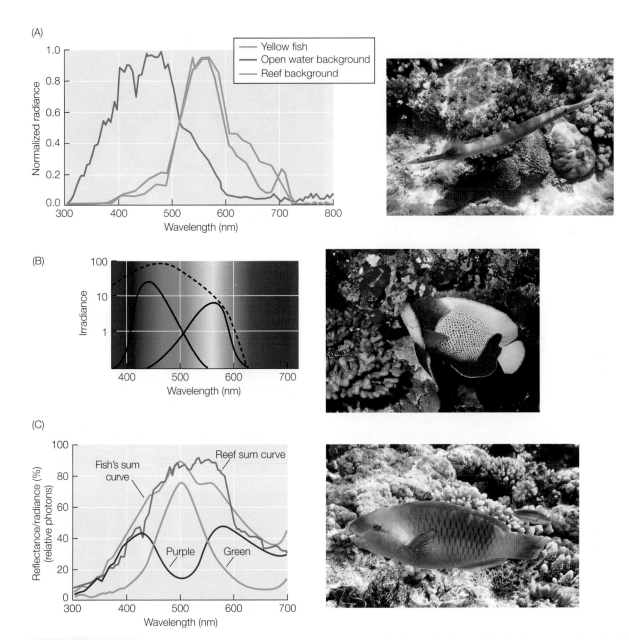

FIGURE 5.8 **Hue contrast and matching in reef fish** (A) The yellow color phase of the trumpet fish *Aulostomus chinensis* under calculated ambient light conditions (orange curve) matches an averaged reef background (green curve) very well, as seen in the photo, but contrasts strongly with the blue color of open water at the edge of the reef (blue curve). (B) The limited range of wavelengths at depths below 25 m (dashed curve) results in blue and red wavelengths are usually unavailable. Yellow-and-black or white-and-black striped patterns provide the strongest within-body *brightness contrast*. Fine-scale patterns of opposing hues can be quite conspicuous at close range but blend into a background-matching color at greater distances, as illustrated in Figure 5.8C.

yellow being the best within-body contrast colors (solid curves). The blue-girdled angelfish (*Pomacanthus navarchus*) shown here is a typical example. (C) The bullethead parrotfish (*Chlororus sordidus*) uses a fine-scaled pattern of contrasting blue-green, and purple coloration, which is visible at close distances, but which blends at farther distances (fish's sum curve) to match the background color (reef background sum curve) very well. (After [133].)

PATTERN CONTRAST   Terrestrial and shallow water habitats generally provide colored and patterned backgrounds that senders can exploit for either contrast or camouflage. As with hue and brightness contrast, within-body **pattern contrast** can either match or deviate from the background pattern [35, 57, 58]. Visibility-reducing strategies include matching patterns or textures, asymmetrical patterns, mimicry of unimportant environmental objects, outline disruption patterns, and countershading. Corresponding visibility-enhancing strategies include contrasting patterns or textures, symmetrical eye-catching patterns such as eyespots and repetitive bars,

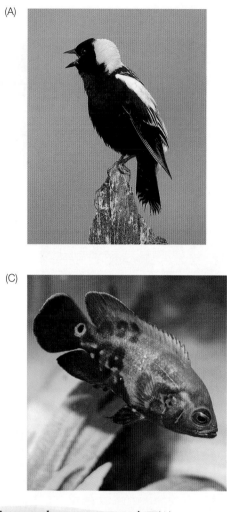

FIGURE 5.9 Conspicuous color pattern strategies (A) Reverse countershading in the breeding season plumage of the male bobolink (*Dolichonyx oryzivorus*), with dark belly and light back. (B) Outlining in the monarch butterfly (*Danaus plexippus*), a toxic species that uses its color pattern as an aposematic predator warning display. (C) Many fish, such as this oscar (*Astronotus ocellatus*), have eyespots on or near their tails, which may deflect predator attacks or serve as the focus of aggressive or courtship behavior. (D) The horizontal stripes on the rear legs of the okapi (*Okapia johnstoni*) contrast strongly with the vertical background pattern of its rainforest habitat; they are believed to help offspring follow their mothers in dense vegetation.

body outlining, and reverse countershading. A few examples are illustrated in **Figure 5.9**.

Researchers have started to test these hypotheses regarding pattern contrast with experiments and quantitative models derived from a deeper understanding of animal perceptual sensitivities [174]. We know that visual receivers can extract contours, outlines, and shapes of objects in a complex visual field and combine this information with color to detect and classify important targets (discussed in more detail later in the chapter). Patterns that facilitate visual construction of the target's outline tend to improve detection, and patterns that subvert this process tend to prevent it. **Outlining** the edge

of the body with contrasting colors clearly facilitates detection and recognition. The opposite strategy, called **disruptive coloration**, entails the use of bold contrasting colors that intercept the periphery and disrupt detection of the outline. Tests with avian predators show that artificial prey with disruptive patterns are less likely to be detected than prey with equivalent internal patches [42, 141, 183, 196]. **Symmetrical patterns** are more conspicuous than asymmetrical ones and reduce the effectiveness of disruptive coloration [44]. **Contrasting concentric circles** are believed to be highly conspicuous not only because they resemble eyes, but also because they are symmetrical patterns readily detectable by receiver visual systems [10, 195, 197-199]. Finally, **repetitive patterns** such as spots, vertical bars, and horizontal stripes can be highly stimulating to visual receivers in some circumstances, especially if they differ in pattern size or orientation from the pattern of the background [35, 101, 161]. For example, many fish species employ vertical bars as aggressive and mate attraction signals [13, 144, 162]. However, when viewed from a distance, fine-scale repeating patterns will blur, and if the average brightness is similar to the background brightness, the patterns may be cryptic in terms of hue contrast. Striped patterns involving black with red or yellow are often associated with toxic prey species—predators quickly learn to recognize and avoid such species—but contrast with the background pattern, brightness, or hue may also play an important role in this signaling context [6, 59, 98, 124, 155, 178]. We will discuss warning and other signals to predators in Chapter 14.

MOVEMENT CONTRAST   Both terrestrial and shallow water environments with vegetation usually present a moving background due to wind or wave action. The motion in both cases is relatively slow and sinusoidal with a frequency of ~1–2 Hz. Animals attempting to hide against such a background move with a similarly slow, sinusoidal pattern (e.g., vine snakes) [62]. The most conspicuous **movement contrast** pattern against such a backdrop is jerky, square-wave type motion of a higher frequency. For example, the push-up displays of *Anolis* lizards are jerky, rapidly accelerating up-and-down movements. The lizards' eyes are highly sensitive to such jerky movements [63, 158]. Males displaying in the field are responsive to the current level of background vegetation motion, and increase the speed and amplitude of their push-up movements as background motion increases [153].

Movement and striped color patterns often operate together in creating either crypticity or conspicuousness. Striped patterns parallel to the direction of motion, such as the longitudinal stripes of racing snakes and garter snakes that move rapidly in straight trajectories, serve to conceal the motion [21, 36]. Conversely, the movement of snakes with contrasting alternating rings is highly visible. The vertical body stripe pattern of zebras represents another example of a repetitive pattern that is perpendicular to the direction of motion. Numerous hypotheses have been proposed to explain the adaptive function of zebra stripes; it is possible that the stripes provide cryptic blending from a distance and under low light levels, or that they aid in thermoregulation, social functions, and tsetse fly avoidance [25, 181]. Another currently popular hypothesis invokes the concept of **motion dazzle**—that high-contrast patterns make it difficult for a predator to estimate the speed and trajectory of moving prey, especially when several prey individuals are moving in a group [201]. Computer prey-catching games have provided some evidence that high-contrast vertical band and zigzag patterns present more difficult moving targets for human gamers to capture than some other patterns, especially against patterned backgrounds [200]. Finally, **perpendicular display movement** relative to the major background pattern maximizes sender conspicuousness. Vertical jumping displays are often observed in habitats with horizontal backgrounds, such as grasslands [3, 26], and sideways movements

are especially conspicuous for animals displaying in vertically striped forest environments [121]. The claw-waving display of male fiddler crabs appears to be highly conspicuous regardless of the complexity of the background pattern [142].

Transmission from sender to receiver

The **transmission** of light waves from the object to the receiver is seldom maximal in natural conditions. Factors that reduce the intensity of the light waves reaching a distant object are similar to those that affect sound transmission: spreading loss or global attenuation, which is independent of frequency; and medium absorption, scattering and filtering, which are usually both wavelength and medium specific. When light spreads from a point source in all directions, its intensity falls off as the inverse square of the distance between source and receiver ($1/d^2$), just as sound intensity does. In principle, light reflecting off an infinitely large signal surface area experiences no fall-off in intensity over environmentally realistic distances. Real signal surfaces are somewhere between these two extremes of source sizes. We can only state that intensity declines approximately in proportion to $1/d$ and declines faster with smaller signal areas than with larger signal areas [52]. For light, the effects of global attenuation, medium absorption, scattering, and filtering are all combined into a single empirically determined parameter, the **beam attenuation coefficient**. The reciprocal of this value is **attenuation length**, the maximum distance at which a large and contrasting object can just be detected. Visual signal transmission distances are greater in clear air (about 6 km) than in dusty, humid, or foggy air (about 50 m). Maximum transmission distance in very clear ocean water is around 60 m, but is typically much less in freshwater environments where absorption from organic matter is strong [166]. Veiling light reduces transmission distances in both environments. Finally, **opaque environmental objects** located between sender and receiver obviously block signal transmission. Light waves cannot bend around objects as sound waves can. In a heavily vegetated environment, signal transmission can be improved only by reducing the distance between sender and receiver.

The hue of a colored signal patch can also change dramatically as the distance between sender and receiver increases. Figure 5.8C showed an example of how a fine color pattern at close distances changes to a blended color at a greater distance. Colored signal transmission in water is very much affected by selective wavelength absorption during transmission. Visually signaling aquatic animals like fish or squid generally view each other from a horizontal perspective. Even when the sender and receiver are relatively close to the surface, where longer red wavelengths are available to be reflected, the transmission of these wavelengths decreases with increasing distance between sender and receiver. A color patch that appears red at a close distance will appear black from 10 meters away (**Figure 5.10**). An ultraviolet signal patch would similarly change to black with increasing distance.

FIGURE 5.10 **A flash-illuminated photo of a school of red soldier fish** Closer individuals appear red, but more distant individuals appear dark as a result of strong absorption of longer wavelengths by water.

Signal detectability

Now that we have discussed the availability of ambient light, aspects of the contrast between a signal (or target object) and the background, and the principles of signal transmission, we can combine these components with the receiver's sensory abilities to assess the **detectability** of a signal under a range of environmental contexts. The proper technique for assessing detectability is to construct a **visual color model** based on measurements of ambient irradiance, object and background reflectance, viewing distance from the sender, and receiver sensitivity. Some models focus on a subset of these components. Details on constructing such models are explained in Web Topic 5.1.

Web Topic 5.1 *Constructing color models*
The only way to understand how animals with different visual systems than ours perceive colored objects is to build a visual model. The effects of hue and brightness contrast, receiver sensitivity, medium attenuation, and distance can be very precisely evaluated with this technique.

Let's concentrate first on the problem of detection from a distance. Detectability of course depends on the **apparent size** of the signal. As the distance between an object and a visual receiver increases, the size of the image the object projects onto the retina decreases. Apparent size is expressed as the solid angle subtended at the eye by the object's area in steradians, or the simple angle subtended by the object's length or diameter. This visual angle (in radians) is equal to the object's true length divided by the distance. At some point, a very small or distant object subtends an angle too small to be detected. The ability of the receiver to detect objects is therefore limited by the **resolving power** of its eye, which as we shall see later can also be expressed as an angle. When the

visual angle subtended by a distant object is less than the eye's angular resolution, the object will not be distinguishable from the background.

Given that the apparent size of a target object is larger than the receiver's minimum resolving angle, detectability is also dependent on the contrast (hue or brightness) between the object and the background. The lower the contrast, the less likely the object will be detected. The threshold of object-background contrast required for detection is affected by the amount of ambient light and the sensitivity of the receiver's eye. All eyes suffer certain unavoidable constraints caused by three types of **receptor noise** (details in Web Topic 5.1). These constraints are strongest under extremely low and extremely high levels of ambient light, and the lower the contrast between object and background, the greater are these constraints. These effects can all be combined into an estimation of the **contrast threshold** for detection of an object.

We can examine the consequences of viewing a particular target from increasing distances by measuring its intrinsic contrast at the source and then computing the degradation of the image caused by apparent size reduction, contrast reduction, and medium attenuation. A very simple expression for perceived contrast at distance d is

$$C(d) = C_0\, e^{-\alpha d}$$

where C_0 is the inherent contrast between signal and background at a distance of zero and α is the beam attenuation coefficient. The greater the distance between sender and receiver, the lower the absolute contrast compared to that which would be perceived very close to the sender [52, 55, 103, 125, 126]. Web Topic 5.1 presents a detailed example of a transmission model used to determine the **just-detectable distance** of fish employing different crypticity strategies [96].

Now we turn to recent studies that have examined hue contrast as a function of available light, background hue, and receiver color sensitivity. **Figure 5.11** illustrates a comparative study of color patch detectability in four species of manakins, small Neotropical forest birds that inhabit different levels of the canopy and therefore experience different light environments [83]. Females of all four species are cryptic green, whereas males are mostly black with one or two color patches on the crown, breast, or thigh. The species display at a different heights in the forest, and they differ strikingly in hue. Those displaying in the lower understory level sport color patches that maximize hue contrast; those at the level of the canopy maximize brightness instead. Several similar comparative

(A)

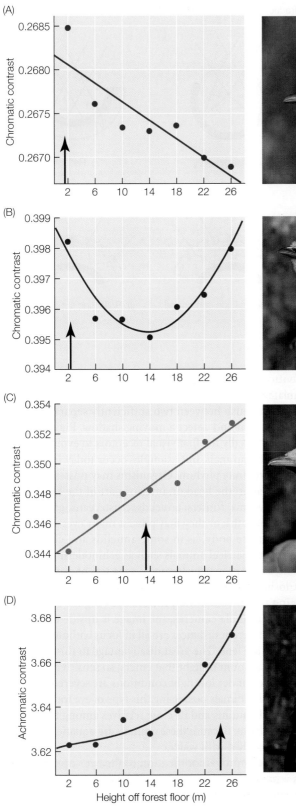

(B)

(C)

(D)

Height off forest floor (m)

Pipra coronata

Pipra filicauda

Pipra erythrocephala

Pipra pipra

FIGURE 5.11 Crown color strategies for male manakins displaying at different heights in the forest Graphs on the left show the calculated chromatic or achromatic contrast between the crown color and the background color at seven consecutive heights off the forest floor, taking into account ambient light and receiver spectral sensitivity. Arrows show typical display height of the species. The blue (A) and red (B) crowns of the two species inhabiting the lowest levels have the best hue contrast at the height where they display. The golden-headed manakin (C) would have a greater hue contrast if it displayed higher in the canopy. However, at the termination of its display in a shaded location, it moves into a small light gap where long-wavelength illumination is abundant, maximizing hue contrast. The highest canopy displayer (D) maximizes brightness contrast with its white crown. (After [83].)

Coupling from the Medium to the Receptor

To make a decent spatial image of the light patterns scattered off objects in the environment, a visual organ must first collect the light over as large an area as possible, bring it into the body through some transparent tissue, and concentrate and focus the light onto an array of photoreceptor cells. In this section we will discuss the evolution of eye design, revisit the mechanisms for producing transparent tissue that we introduced in the previous chapter, and examine the basic optical properties of focusing lenses.

The evolution of light-collecting strategies

The design of an animal's eyes and its resulting acuity reflect its visual needs. A sessile organism or a slow-moving herbivore does not need eyes to locate food or find its way about. However, such an organism most likely does need to perceive the direction of light so it can orient its body with respect to overhead sunlight, and it may need to detect the presence of a predator so it can take appropriate evasive actions. Eyes in such organisms are very simple and are designed to detect shadows and nearby movement. Once an animal becomes mobile, it needs eyes to navigate, so some type of crude image formation is required. Animals that move rapidly need a high level

studies on other avian species, chameleons, fish, and damselflies have also found that color patches of displaying males are optimally bright and contrasting given their particular light environment and visual sensitivity [49, 133, 134, 145, 188, 203].

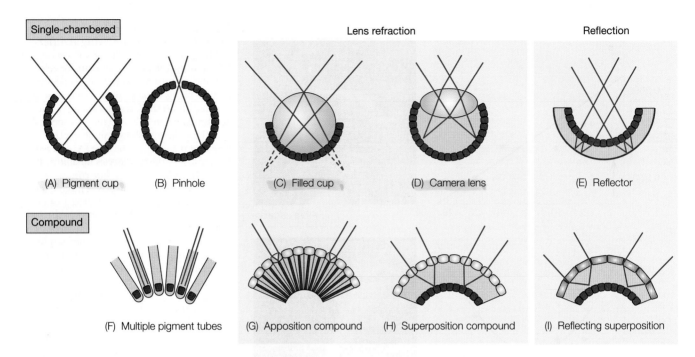

FIGURE 5.12 **Light-collecting strategies by animal eyes** The eye structures illustrated in A–I show the simpler and more evolutionarily primitive strategies on the left, and the parallel lens refraction and reflector strategies observed in single-chambered and compound eyes. Red areas are the retina, blue areas are lenses, purple lines are reflectors, and blue lines show light rays. (After [61, 110].)

of visual acuity to navigate accurately and hunt for food. Spatial vision first evolved in the Precambrian period, about 540 million years ago, and is believed to have fueled the Cambrian explosion of animal phyla [113]. Once visually-guided predators evolved, an arms race ensued that forced prey to develop armor and eyes to avoid being caught. These developments were followed by increased mobility, larger body size with skeletal support, and even better eyes. Intraspecific communication and social needs probably did not have a direct impact on the optical design of eyes in most taxa [29, 127, 133, 156, 157, 226]. The degree to which vision can be used for intraspecific communication is therefore dictated by the type of eye needed for foraging and predator detection.

The camera-lens eyes of vertebrates and cephalopods and the compound eyes of insects and many other arthropods are highly complex structures. Neither could have evolved in a single step; in fact, eyes are believed to have evolved independently at least 40 times [113]. The survey of animal eyes outlined below and illustrated in **Figure 5.12** provides some clues to the possible steps taken during the evolution of visual receptor organs [41, 61, 110, 111, 229].

The simplest type of eye is the **pigment cup eye** illustrated in **Figure 5.12A**. It consists of a pitlike depression that is lined by photoreceptor cells linked to nerves and isolated from the rest of the body by a light-absorbing pigment layer. The cupped shape permits detection of the direction of ambient light. Light shining directly over the eye will stimulate only

the bottom of the cup, while light entering at an angle will stimulate only one side. This eye cannot form an image, but it can distinguish between two sufficiently separated point sources of light and detect a moving shadow. Pigment cups were undoubtedly the first visual receptors to evolve in early multicellular animals, but examples occur today in virtually every invertebrate phylum. Organisms may possess two cup eyes located on the head or anterior part of the body, as in many flatworms, rotifers, annelids, crustaceans, gastropods, and chaetognaths; or they may possess many eyes spread out over the entire body, as in some annelids and chitons. In radially symmetrical animals such as echinoderms and cnidarians, eye cups may be located on the tips of arms.

Closing the cup to a small round opening creates a crude imaging organ called the **pinhole eye** (**Figure 5.12B**). As in the old pinhole cameras, the image is upside down, but objects at different distances are all in focus without the need for a focusing lens. The great disadvantage to this eye is the very small amount of light that enters, so sensitivity and the ability to see under low light conditions are severely restricted. The only animals possessing this type of eye today are the chambered nautilus and its close relatives among the cephalopod molluscs. Because these animals are aquatic, the body of the eye is filled with water.

Another simple modification of the eye cup is to fill it with a transparent medium and covering as illustrated in **Figure 5.12C**. If the cover and filling have a higher refractive index than the external medium and the outer covering is curved, a one-sided **lens** is created and light entering the eye is refracted. Such an eye will have greater light-collecting ability and some focusing power. However, the focal point of the lens lies behind the retina, so a clearly focused image is not achieved [225]. Eyes of this type are called **ocelli**

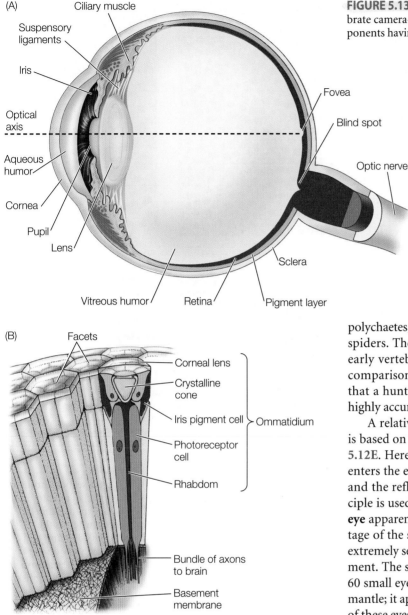

(A)

Ciliary muscle

Suspensory ligaments

Iris

Optical axis

Aqueous humor

Cornea

Pupil

Lens

Vitreous humor

Retina

Fovea

Blind spot

Optic nerve

Sclera

Pigment layer

(B)

Facets

Corneal lens

Crystalline cone

Iris pigment cell

Photoreceptor cell

Rhabdom

Ommatidium

Bundle of axons to brain

Basement membrane

FIGURE 5.13 **Eye structures** Key components of (A) a vertebrate camera-lens eye and (B) invertebrate compound eye. Components having a similar function are given the same color.

produce a bright inverted image that varies inversely in size to the distance of the object. Together with a high-density array of receptor cells and complex neuronal analysis, this eye can simultaneously analyze light entering the eye from a wide angle and produce the most accurate spatial map of objects in the environment that animals can achieve. **Figure 5.13A** shows a more detailed schematic anatomical drawing of the vertebrate camera-lens eye. This type of eye has evolved independently in two major taxa, vertebrates and cephalopod molluscs, and is also found in a few species from taxa with otherwise poor eyes, such as a family of pelagic marine polychaetes, a group of marine snails, and non-web-building spiders. These organisms, along with cephalopods and the early vertebrates, are all active pursuit-type carnivores in comparison with their less visual relatives, which suggests that a hunting lifestyle was the critical factor selecting for highly accurate visual imaging.

A relatively rare but interesting type of imaging system is based on parabolic mirror reflection, as shown in **Figure 5.12E**. Here, light passes through the retinal layer as it first enters the eye; it is reflected off of a curved mirrorlike cup, and the reflection is imaged on the retina. The same principle is used in mirror-type space telescopes. This **reflector eye** apparently forms a moderately good image. The advantage of the system is its large light-gathering area. It is also extremely sensitive to very small movements in the environment. The scallop, a sessile bivalve mollusc, possesses about 60 small eyes of this type distributed around the edge of its mantle; it appears to take advantage of the motion sensitivity of these eyes to rapidly shut its shell when nearby movement is detected [108]. The only other animals with this type of eye are the giant deep-sea ostracods (Crustacea), which possess two huge reflectors and appear to use the light-gathering potential of these eyes to see in the very dark environment of the oceanic depths.

The compound eye (**Figure 5.13B**) commonly found in arthropods (especially insects, crustaceans, and trilobites) evolved from the pigment-cup eye through a very different route. The intermediate eye is probably the **multiple pigment-tube eye** shown in **Figure 5.12F**, found today only in a few annelids, sea fans, and some starfish. This eye could have developed either from the clustering and narrowing of several eye cups into a unit or by the subdivision of a single cup with the addition of tubes. Each tube accepts light only from a unique point in space. Such an eye is highly sensitive to movement, and is therefore superior to the simple cup eye in being able to detect the presence of a predator before

(singular *ocellus*) and are found in terrestrial snails, some annelids, spiders, larval insects, and some nocturnal insects. These animals appear to be able to visually discern obstacles and detect conspecifics at very short distances (a few centimeters). Insects possess ocelli as secondary eyes that seem to be used for assessment of general light levels (e.g., for horizon detection and orientation via polarized light; see Web Topic 5.2).

It is but a small step from the filled eye cup to the true **camera-lens eye** (**Figure 5.12D**). All that is required is a space of low refractive index between the lens and the retina of about 1.5 lens radii. This space creates a two-sided lens with two refractive surfaces; it thus shortens the focal length of the lens and provides a sufficient gap behind the lens to bring an image into focus on the retina. Camera-lens eyes

it is close enough to cast a shadow. With the addition of a lens system covering each tube and tight hexagonal packing, a classical **apposition compound eye** is created (**Figure 5.12G**). Typical eyes contain a hundred to several thousand lens/tube systems, called **ommatidia**. Each ommatidium contains several radially arranged receptor cells with their photopigment regions projecting into the center of the ommatidium (see Figure 5.13B). The lens system narrows the acceptance angle of each ommatidium to a small solid angle that overlaps only slightly with that of neighboring ommatidia. Although each lens produces a small inverted image at the tip of the receptor cells, the entire ommatidium responds as a unit and averages the light intensity from the point in space that the unit perceives. The result is a crude mosaic image of the world. The apposition compound eye can see moderately distant objects with reasonable acuity and without the need to focus. This eye can also judge the distance to an object based on image size, and, like its primitive precursor, it is highly sensitive to slight movements [112]. A serious constraint is its poor sensitivity in low light conditions due to the narrow field of view of each ommatidium.

A dimly lit environment probably favored the evolution of the **superposition compound eye** (**Figure 5.12H**) in some nocturnal arthropods (fireflies, moths, and krill). The baffles between the ommatidia are replaced by a clear space between the lens system and the retina, and the lenses are modified so that they can accept light from a wider field. Light from a given point is therefore received by several adjacent lenses and is concentrated at a single spot on the retina. A bright, erect image is formed, and sensitivity is improved compared to the apposition eye with little loss in acuity [223]. A variant found in decapod shrimps and lobsters, called the **reflecting superposition eye**, uses small reflectors rather than lenses to collect light (**Figure 5.12I**). The compound eye in general is a marvelous lightweight solution for effective vision in a very small, highly mobile organism [109]. Although fine visual patterns are lost, general shape and rapid movement are easily resolved, and such an eye can be employed for visual intraspecific communication.

Transparent lenses

A lens must be composed of **transparent tissue** to pass exterior light into the photosensitive layer of the eye, and yet it also must have a higher refractive index than the outside medium to refract the light. Ocular lenses in both vertebrate and invertebrate eyes contain high densities of proteins generically called **crystallins**. These water-soluble, globular lens proteins have been co-opted from a variety of existing proteins and enzymes used for other functions [61]. The key to transparency is the ability of the densely packed molecules to maintain a locally uniform size and spacing pattern with an inter-scatterer distance less than half the wavelength of visible light [11, 206, 214]. Light waves radiate through the layers of molecules because of constructive reinforcement

(A)

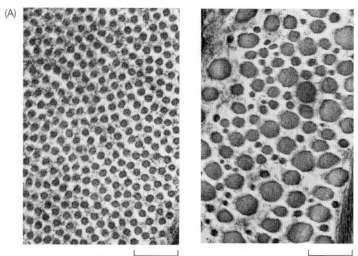

300 nm   300 nm

(B)

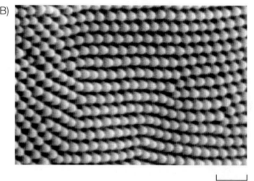

500 nm

FIGURE 5.14 Transparent lenses (A) TEM cross section through the transparent cornea (left) and opaque sclera (right) of a rabbit eye. The collagen fibers in the cornea are small and uniform in diameter (approximately 35–40 nm), and very evenly spaced, whereas the fibers in the sclera are highly variable in diameter (40–300 nm) and unevenly spaced. (B) SEM of the antireflective nipple array on the ommatidial surface of a lepidopteran (*Pieris napi*) eye. The uniform size and spacing of bumps at a periodicity less than the wavelengths of light creates a gradient refractive index that makes the surface transparent to light. (A after [139, 194].)

in the forward direction and destructive interference to the sides, as explained in Chapter 4.

The vertebrate lens is a capsule with a thin layer of epithelial cells on the front surface. These cells produce specialized elongated fiber cells that build up concentric layers in the lens [73, 207]. The fiber cells contain no nuclei or other organelles, only crystallin molecules. Because there is no blood supply to the lens cells, they are maintained by the epithelial cells and by nutrients in the **aqueous humor**. When lens cell maintenance breaks down in the ageing eye, the spatial pattern of crystallin becomes irregular and granular, and the lens becomes opaque, a condition we call **cataracts** [65, 208]. The vertebrate **cornea**, which is positioned in front of the lens, must also be transparent to light. The cornea is basically a layer of skin composed of long collagen protein fibers arrayed perpendicular to the incoming light. To achieve transparency, the fibers must be much smaller in diameter and more closely packed together than normal skin collagen fibers. **Figure 5.14A** juxtaposes a cross section through the epidermis in the transparent cornea region of the rabbit eye with the opaque white **sclera** region at the same magnification scale. The spacing between fiber centers in the transparent section is on the order of 65 nm, while the size and spacing of fibers in the opaque section ranges up to 285 nm and is highly variable, resulting in incoherent scattering of white light [30, 60, 139, 213].

Invertebrate eyes also contain lenses packed with high densities of crystallin protein molecules. In some species the proteins are very similar to those found in vertebrates, while in other species they are very different in chemical structure but maintain uniform packing and transparency in the same way [212]. For invertebrates with very good vision, such as the cephalopods with their camera-lens eyes, lenses are composed of specialized and elongated fiber cells similar to those found in vertebrates [92]. However, when the lenses are very small, as in the **crystalline cone** cells in the ommatidia of compound arthropod eyes or the lens capsules of small ocelli-type eyes, the lens cells secrete their own high densities of crystallin and are not so highly specialized. Ommatidium units often possess a cornea in addition to the lens. The arthropod cornea is made of thin layers of chitin. Chitin is a substance with a rather high refractive index, as we saw in the previous chapter, so adaptations are needed to reduce reflection at the corneal surface. In the compound eyes of Lepidoptera and at least some other arthropods, this problem is solved with the same strategy we described for transparent moth wings in Figure 4.36A, namely an array of tiny nipples or protrusions with a spacing of less than half a wavelength (**Figure 5.14B**) [194, 218, 231]. This structure mimics a layer of tissue with a graded refractive index, and allows light to be transmitted.

Focusing the light

Lenses must have a single-convex, double-convex, or spherical shape to refract and converge light to a focal point. To make a high quality image on the retina, most eyes must also have a dynamic mechanism to focus the image more precisely. This process is called **accommodation**. Accommodation is the ability to bring objects at variable distances from the eye into focus on the retina. The compound arthropod eye needs no mechanism for accommodation, but camera-lens eyes do require some focusing. The mechanism used for accommodation depends on whether the animal lives in the air or in the water. In water, there is little refraction at the water/cornea boundary because the cornea has a similar refractive index to water. The lens performs essentially all of the refraction and therefore must be very round and made of a very dense material with a high refractive index. Such a lens is very hard. To bring near and distant objects into focus on the retina, small muscles move the lens forward and

backward along the **optical axis** (**Figure 5.15**). This method of accommodation is found in cephalopods, fish, frogs, and salamanders. In terrestrial reptiles, birds, and mammals, the cornea provides the major refraction function, because the refractive indices of air and corneal tissue are different. The lens is needed only for fine focusing. The lens is therefore relatively soft, and accommodation is accomplished by changing the shape (curvature) of the lens. A rounder lens has a shorter focal length. For eyes in both terrestrial and aquatic environments, the relaxed state of the muscles connecting the lens to the eyeball is set at the distance most commonly needed by the animal.

Animals faced with very challenging visual environments have come up with some interesting and unique accommodations and adaptations [113]. Animals that must see well in both air and water, such as certain fish, most amphibians, aquatic turtles, diving birds, seals, and otters, have adopted various compromise solutions to the two alternatives outlined above. Compared to their wholly terrestrial relatives, seals and penguins possess a relatively flat cornea that has little refractive power in both air and water, and a lens that is rounder and performs most of the focusing. Diving birds, aquatic turtles, and sea snakes can also change the curvature of their corneas with special muscles when they transition between air and water. Rock-pool fish (*Dialommus* and *Mnierpes*), flying fish (*Cypselurus*), and four-eyed fish (*Anableps*) have either double corneas or double pupils with adjusted optics so they can see in both media simultaneously. Terrestrial ground-foraging animals such as land turtles and chickens have been shown to possess asymmetries in eye shape or lens shape so that they can view the ground in front of them with near-sighted vision while distant objects viewed at a higher angle are still in focus, a strategy called **lower-field myopia** [184]. A pinhole eye or camera-lens eye with a small round pupil has a greater depth of focus than a larger opening. Photographers use this principle to vary the depth of focus of objects in their photographs. We often squint to aid our resolution of a distant object for the same reason. Of course, the smaller hole reduces the total amount of light entering the lens of the eye or camera. Hunting spiders with camera-lens eyes employ a completely novel strategy—they move the retina to bring objects into focus. A final strategy is to increase the length of the receptor cells or depth of the retina, but this option is made at the cost of poorer resolution [113].

Controlling the amount of entering light

All but cave-dwelling and deep-sea animals experience daily variation in the levels of light illumination, so they need to find ways to reduce the light reaching the sensitive retina under bright light conditions but allow maximum light to enter under dim light conditions. In camera-lens eyes the **iris** serves this purpose. In humans and most diurnal vertebrates, the iris closes with a circular muscle so that the opening, or **pupil**, is round. A round pupil is ideal for diurnal animals with color vision because it cuts out light entering

(A) Terrestrial strategy:
Lens changes shape

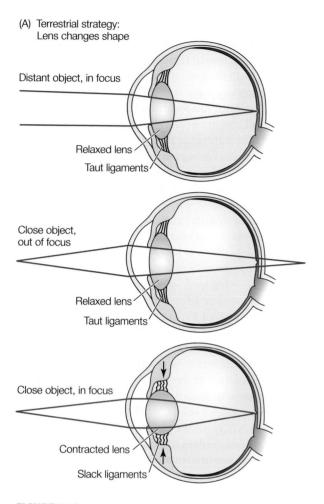

Distant object, in focus

Relaxed lens
Taut ligaments

Close object,
out of focus

Relaxed lens
Taut ligaments

Close object, in focus

Contracted lens

Slack ligaments

(B) Aquatic strategy:
Dense lens that moves

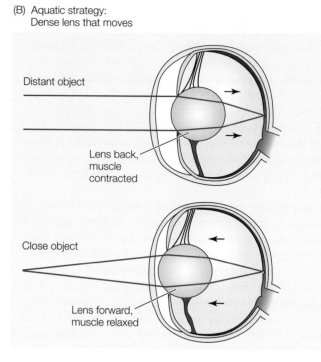

Distant object

Lens back,
muscle
contracted

Close object

Lens forward,
muscle relaxed

FIGURE 5.15   Accommodation strategies in air versus water (A) Terrestrial animals such as reptiles, birds, and mammals obtain most of their focusing power at the curved air–cornea boundary. A relatively flat lens focuses properly on distant objects, but close objects are out of focus until the fibers around the lens relax to make it rounder, which shortens the focal length and brings nearby objects into clear focus. (B) Aquatic animals do not obtain much focusing power at the cornea because the refractive indexes of water and cornea tissue are very similar. Aquatic animals have a dense, round lens that must be moved forward and back to focus on objects at different distances. For teleost fish illustrated here, the relaxed position is forward for close objects; muscles move the lens back to focus on distant objects. For elasmobranch fish and amphibians, the relaxed state has the lens in the back for viewing distant objects; the lens is moved forward to focus on close objects.

the perimeter of the lens, thereby increasing depth of focus and also reducing certain lens aberrations (discussed later in the chapter). However, a round pupil can only close so far. In crepuscular animals and nocturnal animals with some diurnal activity, the pupil must open completely to pass the maximum amount of light at night, and must close down to a tiny opening during the day. A slit pupil is far more effective in this case. The iris muscles must only pull the halves of the iris apart, like opening a curtain that is attached to the window at the top and bottom. The slit pupil can be closed to

a much smaller opening than a round pupil during the day. For example, the slit pupil of the cat can change the intensity of light on the retina 135-fold, compared to the 10-fold change possible in the round pupil in humans [89]. Whether the slit is vertical or horizontal depends on the visual field of the animal; forest dwellers that must look up and down have vertical slits, while open plains inhabitants usually have horizontal slits [89, 114, 127, 222]. **Figure 5.16** shows some examples of animal pupils.

An alternative method of reducing the light falling on the receptor cells is to use **masking pigments** in specialized cells close to the receptor cells. Pigment granules are moved to cover the receptors in bright light and retracted to expose the receptors during darkness. This strategy occurs in some vertebrates, notably teleost fishes, amphibians, and some birds, where the pigment epithelium serves this purpose [2, 48]. It is also an important strategy in the compound eyes of many flies and butterflies, where the pigment cells in each ommatidium can quickly block most of the light entering the rhabdom.

Reception of Visual Signals

We now move to the inside of the eye to discuss the photoreceptors that do the critical job of trapping light energy. We will examine the characteristics of the molecules that absorb light, the cells that house the receptor molecules, and the receptor cell array that performs the initial task of receiving the spatial image.

The visual pigment

As we learned in Chapter 4, some organic compounds can absorb photons of light in the visible range by elevating

FIGURE 5.16   Pupil shape in various eyes (A) Vertical slit pupil in a nocturnal arboreal snake (*Trimorphodon*). (B) Horizontal oval pupil in a pond frog (*Rana tarahumarae*). (C) The horizontally rectangular pupil is common in ungulates, such as horses, camels, and the domestic goat (*Capra aegagrus*), shown here. (D) The unusual W-shaped pupil in the European cuttlefish, *Sepia officinalis*. The function of this shape is not yet known, but the pupil can open and close very rapidly, as with the slit pupil [51].

electrons to higher energy states. These electrons then release energy in the form of heat as they slip back down to their original state. To use the energy in absorbed photons more effectively, living organisms had to devise a means of trapping the absorbed energy and quickly converting it to chemical energy before it was lost as heat. There are only a few types of pigment molecules found in nature that can accomplish this task. The key molecule that performs the initial photon capture is generically called a **chromophore**, to distinguish it from the other chemical components of a light-conversion system.

The first use of light energy by organisms was via **photosynthesis**—the process of converting photon energy into stored chemical bond energy in the form of **ATP** (adenosine triphosphate), which could then be used to fuel cellular metabolism. Most biologists are familiar with the green chromophore **chlorophyll** that absorbs red and blue wavelengths and performs the light-trapping job for green plants and algae (see Figure 4.14A). Chlorophyll traps photon energy with an associated electron transport chain (a biochemical cascade) that splits water into oxygen gas (O_2) and hydrogen ions (H^+) or **protons**. The protons are shuttled out of the cell or chloroplast, and the resulting electrochemical gradient across the cell membrane fuels the production of ATP when the proton reenters the cell or chloroplast at an ATPase enzyme site. Oxygenic photosynthesis first arose in cyanobacteria about 3000 million years ago (mya) and is responsible for Earth's oxygen-containing atmosphere as well as fixation of atmospheric carbon. Chlorophyll photosynthesis is believed to have evolved independently a second time, because the chlorophyll-binding proteins in all of the green prokaryotic bacteria are closely related to each other but are unrelated to the binding proteins in eukaryotes or to the chloroplast organelles of higher plants [115]. But there is another ancient photosynthetic pigment called **bacterial** (or **type 1**) **rhodopsin**, which occurs widely among extant archaea, bacteria, and eukaryotes and therefore must have arisen before their split approximately 3500 million years ago

[179]. The bacterial rhodopsin pigment is embedded in the cell membrane and operates as a proton pump, transporting H$^+$ out of the cell. As with the chlorophyll system, the electrochemical gradient and ATPase enzymes lead to the production of ATP [15, 95, 186].

Coincident with or very soon after the evolution of photosynthesis, light-dependent cells evolved photosensory systems so that they could detect light of different qualities. The ability to move toward areas of useful wavelengths and avoid harmful UV radiation would have been critical to the survival of the early light-sensitive cells [192]. There are only a few classes of pigment chemicals used for this sensory function. The early archaea appear to have adopted (duplicated and modified) their type 1 rhodopsin proton transport machinery for this new sensory function. This **sensory bacterial rhodopsin** forms complexes with a variety of transducers and secondary proteins, depending on the bacterial species. Instead of pumping protons, the trapped light energy cascades through these intermediary chemicals and, together with ATP energy from the transport rhodopsin system, triggers the flagellar motor system to move the cell. Some well-studied bacteria even have two sensory rhodopsin pigment systems tuned to different wavelengths to mediate movement toward longer wavelengths and away from UV wavelengths [191]. Besides rhodopsin, there are only four other classes of chemicals known to be used as photosensory pigments in single-celled bacteria and algae [82]. Rhodopsin-based systems have the fastest response times and therefore are the most widely used for highly mobile organisms.

A light-sensitive pigment for true vision must trap the light energy in a photon and convert it immediately into a nerve impulse. Rhodopsin appears to have been the best pigment for this function because of its ability to quickly amplify its effect on the cell membrane, as explained below. All multicellular animals with a visual organ connected by nerves to the brain and/or locomotory muscles use **type 2 rhodopsin**. We do not know exactly when or how type 2 rhodopsin first arose, but it appeared before the evolutionary split between the Protostomes (arthropods, molluscs, annelids, and flatworms) and the Deuterostomes (hemichordates, echinoderms, and vertebrates) during the Cambrian explosion 530 million years ago. These two animal groups show many similarities in their rhodopsin genes and amino acid sequences, and many differences with the bacterial type 1 rhodopsins.

Both types of rhodopsin consist of two components: **retinal**, a conjugated double-bond chromophore molecule derived from vitamin A$_1$ and **opsin**, a large polypeptide protein molecule (40,000 atomic mass units) that is anchored into the plasma membrane of a visual receptor cell. The structure of rhodopsin is shown in **Figure 5.17**. It consists of seven helices that span the membrane and are connected and held in a circular bundle. The retinal chromophore in its relaxed state as the 11-*cis* isomer forms a bent molecule that nestles horizontally inside the bundle. It is connected to the opsin via a lysine on helix 7 with a carbon-nitrogen

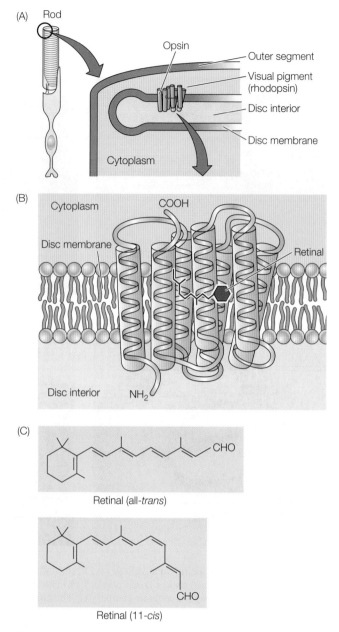

FIGURE 5.17   Chemical structure of rhodopsin (A) Rhodopsin is a two-component molecule anchored to the internal disc membrane of a photoreceptor cell, here a vertebrate rod cell. (B) The opsin component (green) consists of protein with seven transmembrane helices connected to each other in a bundle. The retinal component (red) is nestled within the bundle and attached to an amino acid residue within the seventh membrane-spanning region. (C) The chemical structures of the two isomers of retinal, all-*trans* and 11-*cis*. When the retinal is in its 11-*cis* isomer and attached to the opsin, it can absorb a photon of light. This transforms it to the all-*trans* isomer. Retinal reverts to the 11-*cis* form via one of three different processes. (After [229].)

double bond (called a **Schiff base**) and forms additional weak attachments with other helices. In the presence of light in the visual range, a single photon is absorbed by the retinal. This causes the retinal component to straighten into the all-*trans* retinal isomer. In the sensory type 1 and type 2 rhodopsins, the proton that is released from the Schiff base during isomerization is donated to the opsin instead of being pumped out [192]. The photoactivation of rhodopsin causes tilting and

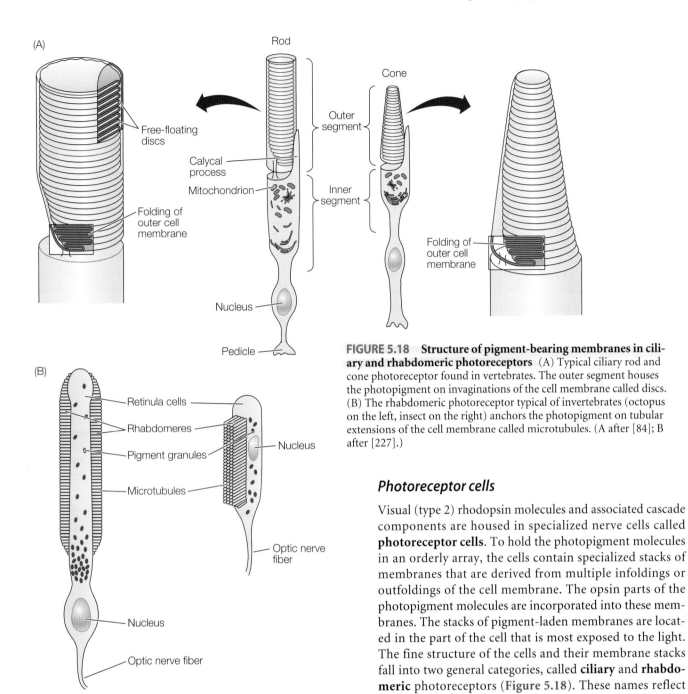

FIGURE 5.18 **Structure of pigment-bearing membranes in ciliary and rhabdomeric photoreceptors** (A) Typical ciliary rod and cone photoreceptor found in vertebrates. The outer segment houses the photopigment on invaginations of the cell membrane called discs. (B) The rhabdomeric photoreceptor typical of invertebrates (octopus on the left, insect on the right) anchors the photopigment on tubular extensions of the cell membrane called microtubules. (A after [84]; B after [227].)

Photoreceptor cells

Visual (type 2) rhodopsin molecules and associated cascade components are housed in specialized nerve cells called **photoreceptor cells**. To hold the photopigment molecules in an orderly array, the cells contain specialized stacks of membranes that are derived from multiple infoldings or outfoldings of the cell membrane. The opsin parts of the photopigment molecules are incorporated into these membranes. The stacks of pigment-laden membranes are located in the part of the cell that is most exposed to the light. The fine structure of the cells and their membrane stacks fall into two general categories, called **ciliary** and **rhabdomeric** photoreceptors (**Figure 5.18**). These names reflect the type of cell from which the photoreceptor evolved. Vertebrates have ciliary photoreceptors as their primary means of spatial vision, whereas most invertebrates with good eyes possess rhabdomeric photoreceptors derived from microvillous cells [41, 53, 228]. However, vertebrates and invertebrates retain the capacity to develop both types of sensory cells, which explains the exceptions to the pattern as well as the combined use of both photoreceptor types in some species [61].

The structure of photoreceptor cells is designed to maximize light absorption. The long tubular shape of both ciliary and rhabdomeric cells not only provides a tall stack of pigment-bearing membranes, but it also facilitates the trapping of light from a very narrow acceptance angle. Each cell acts as a **waveguide**, with internal reflection off the external cell

other movements of the helices, which initiates a biochemical cascade. In type 2 rhodopsins, the first step of the cascade involves an associated **G protein** located on the surface of the membrane adjacent to the rhodopsin (G protein is short for *guanine nucleotide-binding protein*, a family of proteins that exchange guanosine diphosphate (GDP) for guanosine triphosphate (GTP) in biochemical cascades). Subsequent cascade steps lead to massive amplification effects that ultimately open or close ion channels in the receptor cell's plasma membrane, thereby generating a nerve impulse [169]. The details of this amplification process, and the rhodopsin recovery process, differ for the two main types of receptor cells found in animal eyes [61].

membrane off propagating light waves down the length of the receptor. Waveguide properties operate best when the diameter of the photoreceptor is slightly more than one wavelength, or about 1 μm. In order for a photoreceptor to absorb incident light maximally, it must be aimed in the direction of the light entering the eye.

The two types of photoreceptors differ in several important ways:

- *Shape of pigment-bearing membranes* Advanced ciliary receptors have membranes shaped in flat round discs; the discs are stacked like pancakes, with their flat surfaces perpendicular to the direction of light. Rhabdomeric cells, on the other hand, pack their photopigments on rolled microtubular membranes. The tubules are stacked parallel to each other like the bristles of a brush and oriented perpendicular to the direction of light. The shape and orientation of rhabdomeric tubules and the strong anchoring of the rhodopsin molecules in the membranes make them inherently sensitive to the plane of polarized light.

- *Cascade amplification process* The two photoreceptor types use different families of opsins and G proteins and have different transduction processes. Ciliary receptors employ a transducin G protein. Their cyclic-GMP amplification process changes the membrane potential by causing the cell's sodium ion gates to close, thereby hyperpolarizing the cell (the inside becomes more negatively charged). Rhabdomeric receptors use a G_q protein and a phospholipase cascade that opens ion gates, thereby depolarizing the cell (the inside becomes less negative) [234]. The hyperpolarized or depolarized photoreceptor cell eventually synapses with a ganglion cell, which converts a change in membrane potential into a nerve impulse.

- *Rhodopsin recovery process* The spent *trans* retinal in ciliary cells reverts back to 11-*cis* by detaching from the opsin and diffusing out of the cell, where it is enzymatically converted to all-*trans* retinol, 11-*cis* retinol, and finally 11-*cis* retinal before reentering and reattaching to an opsin. A depleted cone cell requires 5–6 minutes, and a depleted rod cell 30–40 minutes, to complete this process, along with expenditure of ATP [119]. The spent chromophores in rhabdomeric cells can revert back to 11-*cis* by absorbing a second photon of light. Depleted rhabdomeric cells may therefore be able to recover more rapidly than ciliary cells.

- *Density of pigment molecules* The density of pigment molecules on the membrane surface is higher in ciliary cells than in rhabdomeric cells, meaning that more molecules are packed into a given length of photocell to yield a higher absorption coefficient per μm of length (0.035 μm^{-1} versus 0.007 μm^{-1}) [224]. Ciliary cells are therefore more sensitive to light per unit of cell length. However, rhabdomeric cells are two to four times longer and have a higher net photon capture efficiency of about 90%, compared to 20–80% for ciliary receptors [110].

Web Topic 5.2 *Perception of polarized light*
Rhabdomeric photocells are intrinsically more sensitive to the plane of light polarization than ciliary cells. Many invertebrates have specialized photoreceptors to make use of important environmental information provided by polarized light, and a few vertebrates are able to use it as well.

A typical receptor cell contains 10^8–10^{10} molecules of rhodopsin, depending on the species [122]. Because of the amplification cascade, only one of these molecules needs to absorb a photon of light for the cell to produce a response. Rhodopsin operates most efficiently with light wavelengths between 400 and 600 nm. Unlike the cut-off and multi-peaked types of absorption curves for typical carotenoid pigments (see Figure 4.13), the absorption curve for rhodopsin is a smooth bell-shaped curve with a peak in the middle of the visible spectrum around 500 nm. Rhodopsin is called **visual purple** because of its violet color in the 11-*cis* form when extracted from eyes. In a later section we will discuss the ways in which pigments with different absorption peaks are produced.

The retina

Photoreceptor cells are packed side by side to form a **retina**. The density of photoreceptors can reach 150,000 per square mm. The retina forms the photosensitive layer of all eyes, primitive and advanced. Each photoreceptor cell or group of photoreceptor cells forms a synapse with one or more nerve cells, and nerve cells form interconnections as they travel to the central nervous system. The ability of the retina to resolve fine details depends on the number of photoreceptor cells, the optical system of the eye, and the nature and complexity of the nerve connections to the brain.

Vertebrate photoreceptors are tightly packed and arrayed side by side with their active light-receptor end facing the back of the eyeball. The bottom end (inner segment) of the cell faces the incoming light and performs the waveguide light-trapping function. The nerve cells are attached to this end and extend in several layers toward the front of the eye, where the ganglion cell axons then bend and travel to the optic nerve, out of the eye, and to the brain. Light must travel through the nerve layer of the retina before it strikes the pigment-bearing end of the receptor cells. This arrangement is called a **reversed retina**. The retina is reversed in vertebrates because the eye develops as a vesicle that evaginates from the side of the brain or neural tube [41]. It is considered to be less efficient than the compound and camera-lens eyes of invertebrates, in which the active photoreceptors face the incoming light and nerves run straight back to the brain.

The vertebrate eye is also characterized by a **duplex retina** with two types of photoreceptor cells: **rods** and **cones**. These form two more or less independent visual systems, rods for vision in low-light conditions and cones for vision in bright-light conditions. This duplex system turns out to be

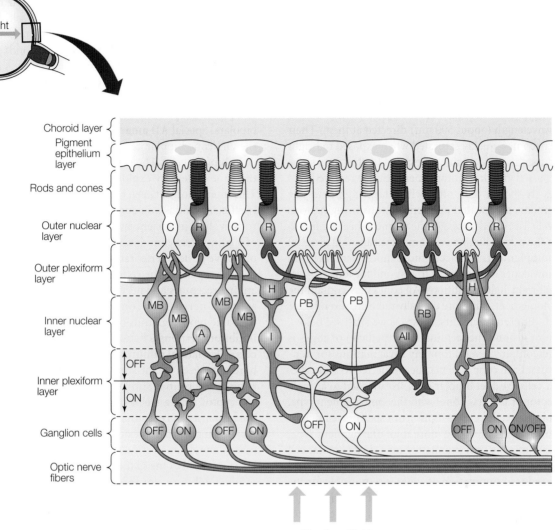

Light

Choroid layer
Pigment epithelium layer
Rods and cones
Outer nuclear layer
Outer plexiform layer
Inner nuclear layer
Inner plexiform layer
OFF
ON
Ganglion cells
Optic nerve fibers

Direction of light

FIGURE 5.19   Organization of the vertebrate retina The cross section of a vertebrate retina exhibits clear layers corresponding to functionally different cell types. The entire eyeball is encased in the choroid and fibrous sclera. The outer segments of the rods (R) and cones (C) face away from the direction of incoming light and toward the pigment epithelium cell layer, where isomerization of retinal takes place. Bases of the rods and cones are greatly enlarged to show the invaginations where synaptic contacts are made with retinal nerve cells. Two midget cone pathways are shown on the left in orange, with midget bipolar cells (MB) connecting to midget ganglion cells. A parasol cone pathway with parasol bipolar cells (PB) is shown in the middle in tan. For both cone pathways, ON systems make their bipolar-to-ganglion connections in a different sublayer of the inner plexiform layer from the OFF systems. A typical mammalian rod system is shown in red, with a rod bipolar cell (RB) and its associated specialized amacrine cell (AII) making connections to cone bipolar cells. An ON/OFF ganglion cell that responds to transient light changes is also shown. Axonless cell types are shown in green: horizontal cells (H) make lateral connections among cones or among rods, amacrine cells (A) make lateral connections within the inner plexiform layer, and interplexiform cells (I) make connections between horizontal cells, amacrine cells, and bipolar cells.

a highly efficient strategy for dealing with variable light levels, compared to rhabdomeric photoreceptors that lack this mechanism, because rods are energetically less costly to operate than cones [152]. Rods and cones can be distinguished by their physical shape and by the type of nerve connections they make with neighboring cells of the same type. The main types of nerve cells in the vertebrate eye are illustrated in **Figure 5.19**, and their functions are discussed below [137, 169]. These cell types form clear layers within the retina.

Rods are long, thin cells containing a stack of completely internal membrane discs that house the photoreceptor molecules. The base of each rod has a rounded spherule with a single invagination for synaptic connections with retinal nerve cells. One such neuron is the **rod bipolar cell**, which conducts the nerve response vertically into the retina. Between 15 and 45 neighboring rod cells connect to a single bipolar neuron, and their effects are additive. The consequence of this type of wiring is to increase visual sensitivity under low-light conditions: the light-gathering area on the retina for a single bipolar neuron is large, and a single photon absorbed by just one of the attached rods may cause a nerve response. However, since each rod is connected to an average of 2.5 different bipolar neurons, there is considerable overlap in the receptive field of each rod bipolar neuron and a reduction in the detail and sharpness of the eventual neural image. All of the rods usually contain the same photopigment. Rods are highly efficient receptors, absorbing 50 to 90% of the photons of optimal wavelength (about 500 nm) directed at them. Their response becomes quickly saturated when exposed to bright light, and they are then unable to respond further until some time has passed and the photopigments can be renewed.

Cones are shorter cells, often with a tapering point. The photopigments are located on membranous invaginations into the cell. The direct access of the pigment-bearing membranes to the extracellular environment means that renewal of the photopigments can occur more quickly and helps to explain how cones can continue to respond for long periods of time in bright-light conditions. All cones may contain the same photopigment, but more typically there are several cone types with different pigments; these, as we shall see below, permit color vision. The terminal base (pedicle) of cones is broad and contains several to many invaginations for neuron synapses. Each cone type may be connected to four types of bipolar cells. **Midget bipolar cells** are attached to a single cone receptor or to a few cones with the same pigment type. **Parasol bipolar cells** receive input from six to seven neighboring cone cells comprising more than one pigment type. The number of cone cells summating on a single bipolar cell is therefore less than the typical summation for bipolar rod cells. The reduced summation of cones is responsible for the higher resolution and finer-grained image of the bright light visual system. Midget and parasol bipolars can be further divided into two subvarieties, **ON bipolars** and **OFF bipolars**. ON bipolars *depolarize* (voltage becomes less negative) with an increase in the light stimulus to the photoreceptor, whereas OFF bipolars *hyperpolarize* (voltage becomes more negative) with an increase in the light stimulus and then activate with relative depolarization when the light is turned off. The bifurcation of each cone's response into separate on and off channels is maintained in subsequent neural connections up to the visual cortex and is responsible for the analysis of brightness and hue contrast [72, 169, 187]. We will examine color perception later in the visual processing section.

The responses from cone bipolars are transmitted directly to **ganglion cells**, which send long axons through the optic nerve and into the **lateral geniculate nucleus** of the brain, where the first level of visual processing takes place. The ganglion cell preserves the main response characteristics of the bipolar neuron; it is either a **midget ganglion cell** with a narrow field ON or OFF response or a **parasol ganglion cell** with a broader field ON or OFF response. Parasol ganglion cells have larger axons that enable the parasol system to respond faster than the midget system does. Ganglion cells respond with spike rate changes rather than the voltage changes seen in photoreceptors and bipolar cells. An ON ganglion

therefore *increases* its spike rate over its baseline resting rate when stimulated, whereas an OFF ganglion cell *decreases* its spike rate when the photoreceptor is stimulated. A third type of ganglion cell, called an ON/OFF cell, responds specifically to transient changes in light stimulus and receives input from both ON and OFF channels. Rod bipolars do not have their own ganglion cells. Instead, the primary connection of the rod photoreceptors to the optic nerve occurs via the cone bipolars. Special **AII amacrine cells** send excitatory input to the ON cone bipolar cells and inhibitory input to the OFF cone bipolar cells. The rod system therefore piggybacks on the cone system by exploiting the ON bipolars [137, 169]. This interaction between cone and rod systems may be highly efficient, and is also consistent with genetic evidence that rods evolved in ancient jawed vertebrates *after* the evolution of a completely cone-based color vision system in the more primitive jawless vertebrates [19, 33] (see Web Topic 5.3).

Other cells in the retina generate complex interactions among receptors and nerves [137]. **Horizontal cells** are flat, axonless nerves that connect adjacent photoreceptors of the same type (e.g., rods or cones). They appear to produce inhibitory responses between neighboring receptors, a process called **lateral inhibition**. Lateral inhibition enhances the border between a light-stimulated zone and an unstimulated zone in the retina. Similarly, stimulation of a single photoreceptor with a small spot of light causes a reduction in the responses of receptors in the surrounding region. **Interplexiform cells** span the outer and inner plexiform layers and make connections between bipolar cells, ganglion cells, and horizontal cells. One of their functions appears to be turning off the inhibitory effects of the horizontal cells to maximize sensitivity in crepuscular light conditions (i.e., both rods and cones contribute). **Amacrine cells** are also axonless nerves that make horizontal connections among bipolar cells, ganglion cells and other amacrine cells. There are many morphological subtypes of amacrine cells whose functions are unclear, but it is known that amacrine cells facilitate the initial phases of color analysis by making connections between midget bipolar neurons. They mediate the input into ON/OFF ganglion cells that detect changes in light intensity, and they may also be responsible for movement detection.

Antagonistic interactions, such as lateral inhibition, among nerve cells constitute the primary means for processing higher-level visual information as well, such as spatial, temporal, and color analysis. Animals differ in the degree to which such processing occurs in the peripheral visual organ (e.g., the eye and retina) versus in the brain. Vertebrates in general must significantly digest and reduce the information from the photoreceptors in the retina before sending it on to the brain, because of the constraints imposed by the reversed retina and eyeball-penetrating optic nerve [164, 211]. For example, the eye of the cat contains 130,000,000 receptors, but only about 170,000 fibers connect the retina to the brain. Invertebrates do not suffer this constraint and perform little retinal preprocessing. They do possess the equivalent of ON and OFF systems for contrast and hue analysis, but the bifurcation occurs at higher levels in the brain [140]. Such evidence for convergent mechanisms of visual signal processing between vertebrates and invertebrates indicates strong selective pressure for optimal receiver design.

Resolution and sensitivity

A good eye should have both *excellent acuity*, so that fine spatial patterns can be resolved, and *high sensitivity*, so that vision is possible under low-light conditions. It turns out to be rather difficult to maximize both features at once, because most adaptations to increase one feature generally decrease the other. For nocturnal animals, maximizing sensitivity is most important, so acuity tends to suffer. Diurnal animals do not need extreme sensitivity, so they tend to maximize resolving power. These **trade-offs** lead to different evolutionary adaptations for different ecological strategies. In addition, each individual animal shows efficient physiological adaptations, for instance by switching from a high-acuity cone system to an efficiently light-summing rod system as ambient light levels change. Each photoreceptor is also able to adapt to the prevailing ambient light by changing signal amplification, with the result that both cones and rods are able to function efficiently at light levels covering many orders of brightness.

MAXIMIZING SPATIAL RESOLUTION   **Resolution**, or resolving power, is the ability to perceive fine spatial patterns. This ability is determined by a combination of the quality of the image produced by the optical system of the eye, and the grain of the mosaic of photoreceptor cells in the retina. The degree of resolution is related to the precision with which an eye can split up light according to its direction of origin.

Three main factors potentially degrade **image quality** (i.e., the projected sharpness of the spatial details in a visual scene): scattering within the eye, focusing aberrations, and diffraction. All of these factors can lead to some blurring of the image. **Scattering** within the eye is caused by imperfections in the transparency of the tissues and eye media. Both compound and camera-lens eyes have this problem because multiple tissues and cells make up an eye, and there will always be some cell membranes and organelles present.

Focusing aberrations are more likely to occur in camera-lens eyes [113]. **Spherical aberration** is one source of aberration. No lens is perfect, and light passing through the edges of the lens cannot be focused at the same point as light passing through the center of the lens. The variable focal points cause blurring. Round lenses suffer most from this problem. The problem is minimized in a lens with a graded refractive index that is higher in the center compared to the edges (**Figure 5.20A,B**). **Chromatic aberration** is another source. This phenomenon arises from the fact that short wavelengths (blue end of spectrum) are refracted more than longer (red end of spectrum) wavelengths, so the focal point for blue objects is closer to the lens than the focal point for red objects (**Figure 5.20C**). You can verify for yourself how fuzzy the boundary is between adjacent red and blue stripes. Diurnal animals with

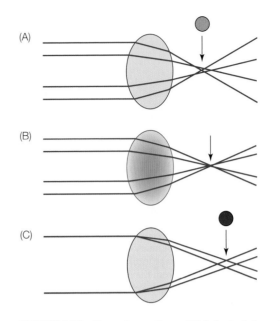

FIGURE 5.20 Lens aberrations (A) Spherical aberration in a lens with homogeneous index of fraction. A blurred image occurs at the focal point because refraction is greater at the edges of the lens where curvature is greater relative to rays entering the center of the lens. Gray circle indicates the size of the blur zone. (B) Spherical aberration can be corrected with a lens having a refractive-index gradient that is greater in the center and lower on the edges. This gradient reduces the degree of refraction at the edges of the lens relative to the middle, enabling the lens to create a sharp focal point and good resolution. (C) Chromatic aberration, with blue light focusing at a shorter distance than red light. This blurs the image of fine patterns comprised of different colors, especially an edge between red and blue. The size of the blurred zone at the focal point is indicated by the purple circle. (After [113].)

good color vision can limit the effects of chromatic aberration by blocking light rays entering the edges of the lens with a round pupil, and may add a mild yellow pigment to the lens or cornea to filter out UV wavelengths. Nocturnal animals cannot afford the reduced sensitivity created by these adaptations, and therefore employ a slit pupil coupled with a graded refractive-index lens [127]. **Diffraction** increases when the aperture of the eye is very small and approaches the length of a light wave. Compound eyes are more likely to reach this diffraction limit, dictating a lower limit on facet diameter of about 2 μm (or 2000 nm).

The effect of these image degraders can be estimated for a given eye by shining a pinpoint of light on it and measuring the **blur circle** falling on the retina. Given the constraints on image quality, it is uneconomical for an eye to have a retina grain that is finer than the size of this blur circle. A finer grain would result in unnecessary oversampling of the incoming light information. Ideally, the retina grain should match the image quality.

The **retina grain**, or sampling frequency of the retina, can be measured by finding the cutoff point where a fine spatial pattern, such as a grating of alternating dark and light bars, cannot be resolved. At the limit of the eye's acuity, adjacent

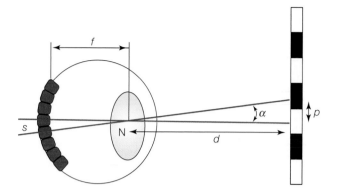

FIGURE 5.21 **Resolving power in a camera-lens eye** Schematic illustration of an eye viewing a grating on the right. N is the nodal point on the lens where the rays pass through without being refracted, because they meet the boundary of the lens at a 0° angle of incidence. In this schematic, f is the focal length of the lens, s is receptor separation, d is distance from the nodal point of the lens to the grating, and p is the distance between grating bars (light and dark bars equal in width). The angle of resolution in degrees or radians is α. When the grating size matches the receptor separation, $\alpha = s/f = p/d$. (After [113].)

stripes appear as one. We can make this measurement behaviorally (by training an animal to tell us when it can distinguish between two patterns of different dimensions) or anatomically (by measuring the spacing between photoreceptor cells in the retina). **Figure 5.21** illustrates the logic of matching between the grating pattern and retinal spacing. The best common currency for describing these two patterns is with angles in units of degrees or radians. The **minimum angular resolution** (MAR), $\Delta\phi$, can be estimated in two ways. The first way it can be estimated is $\Delta\phi = s/f$, the spacing between receptors (s) divided by the focal length of the lens (f). The second way it can be estimated is $\Delta\phi = p/d$, the distance between adjacent dark and light bars (p) divided by the distance between the grating and the nodal point of the lens (d). The smaller the value of $\Delta\phi$, the better the resolution. **Table 5.1** gives some values of MAR for different organisms. To get a sense of what these values mean, your little fingernail held at arm's length covers about 1°; this is the pixel size of the bee's world [113].

There are several ways to improve resolution, but each solution has constraints. The receptors could be *reduced in diameter* and/or packed more closely together. However, when receptor diameter approaches the wavelength of light, the diffraction limit is reached, and light entering the top of the cell will be partially scattered. Moreover, the waveguide effect of the long narrow photocell will be disrupted, causing light to leak into adjacent cells. Some type of baffle, such as a layer of pigment granules, needs to be positioned between adjacent photocells to prevent this leakage. Photocells in practice cannot be any less than about 2000 nm in diameter [85]. These effects set the primary limit on retina grain and resolving power. A second strategy for improving resolution is to *increase the focal length of the lens*. This is accomplished by making the lens curvature flatter. However, the eye must

be increased in size to accommodate the longer focal length. The size of the animal's head will ultimately limit how large the eyes can be. Both of these effects combined mean that within an animal group, larger body size is associated with higher visual acuity [104, 112, 180].

A widespread strategy for increasing resolution in vertebrate eyes with a mixture of rods and cones is to *increase the number of cones* relative to rods in the retina. Because of the reduced summation of cones on nerve cells, the effective receptor-separation factor s is lower for cones than for rods. Diurnal birds possess eyes with about 80% cones, compared to shorebirds, which are active day and night and have 50% cones, and truly nocturnal birds such as night herons, owls, nightjars, and oilbirds, which have 1% to 30% cones [138, 171, 172, 205]. However, diurnal birds pay the cost of reduced ability to see in dim light. A few mammals, such as squirrels and tree shews, have surprisingly high percentages of cones (50–95%) compared to most mammals [1]. Even primates, including humans, have only 5% cones averaged over the entire retina. Another compromise is to distribute a *high concentration of cones* in a small area of the retina at the focal point of the lens or in a horizontal streak and place a higher density of rods outside of this area. This cone-rich region is called an **area centralis**. With such a compromise, the animal can navigate well in dim light using its peripheral vision and can discriminate objects in line with the optical axis if some light is available. Most diurnal and some nocturnal vertebrates possess an area centralis. If even greater acuity is required, the

TABLE 5.1 *The resolution of selected animal eyes, measured as inter-receptor angle*

Species name	Degrees
Eagle (*Aquila*)	0.0036
Human (*Homo*) (fovea)	0.007
Octopus (*Octopus*)	0.011
Jumping spider (*Portia*)	0.04
Cat (*Felis*)	0.05
Goldfish (*Carassius*)	0.07
Dragonfly (*Aeschna*)	0.25
Hooded rat (*Rattus*)	0.5
Worker bee (*Apis*)	0.95
Shore crab (*Leptograpsus*)	1.5
Scallop (*Pecten*)	1.6
Wolf spider (*Lycosa*)	1.8
Sea snail (*Littorina*)	4.5
Fly (*Drosophila*)	5
Horseshoe crab (*Limulus*)	6
Cephalopod (*Nautilus*)	8
Deep-sea isopod (*Cirolana*)	15
Flatworm (*Planaria*)	35

Source: After [113].

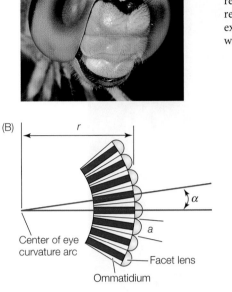

(A)

(B)

FIGURE 5.22 Resolving power in a compound eye (A) Dragonflies possess the best-resolving compound eye (see Table 5.1). (B) The interommatidial angle α determines the resolving power (MAR) of a compound eye. As in a camera-lens eye, this angle can also be expressed as a function of the facet aperture diameter a and the radius of eye curvature r, where $\alpha = a/r$. (After [113].)

Center of eye
curvature arc

Facet lens

Ommatidium

area centralis may also contain a **fovea**, a rimmed depression in the retina surface that bends the light outward to magnify the image. Foveas are found mostly in birds, shallow-water fish, diurnal reptiles, and higher primates; they are absent in the majority of fish, amphibians, and mammals. The primate fovea contains essentially 100% cones packed at very high density (324,000 cones mm^{-2} in humans, with a retina grain of 1750 nm spacing between cell centers) [1]. Many birds possess two foveas, a central fovea used for high-power lateral monocular vision and a temporal fovea for binocular vision in the forward direction [177, 222].

Resolution of the apposition compound eyes of invertebrates can also be expressed as MAR, as illustrated in **Figure 5.22**. In this case, the interommatidial angle can be estimated by the facet diameter $\Delta\phi$ (equivalent to s) divided by the radius of curvature of the eye surface r (equivalent to f). Compound eyes can also possess a region of higher acuity equivalent to that provided by a fovea. This is partly achieved by increasing the local radius of the eye, by flattening it to reduce the interommatidial angle. However, to obtain the desired increase in resolution, the acceptance angle of each ommatidium must also be reduced so that the fields of view of neighboring ommatidia do not overlap. Logically, one might think that this could be done by decreasing the facet diameter. Unfortunately, the facet size is usually already minimized at the diffraction limit, so a further

decrease in $\Delta\phi$ would not only reduce sensitivity but also blur the incoming light rays. So the aperture size of facets is actually increased in the fovea region, and the curvature of the facet lenses is decreased to narrow the acceptance angle of the ommatidia. Foveas often function to improve resolution in predatory insects; in some species they are present only in the male and are used to detect females [113].

MAXIMIZING SENSITIVITY   The **sensitivity** of an eye is defined as the number of photons caught per receptor per second while an animal is viewing a scene of standard radiance. Three features determine sensitivity: the diameter of the aperture or pupil (A); the angle over which each receptor accepts light (α), which as we saw above is determined by the receptor diameter and the lens focal length; and the proportion of photons entering the receptor that are absorbed (P). In equation form, $S = 0.62(A/f)^2\Delta\phi^2 P$, where the constant is equal to $\varpi/4$ to account for the round lens and receptor. The sensitivity of selected animal eyes is given in **Table 5.2**. One strategy to increase sensitivity is therefore to *increase the aperture diameter* of the cornea, pupil, and lens to let more light into the eye. A second strategy is to *increase the effective retina area covered by a photoreceptor*. This can be accomplished in a camera-lens eye with a duplex retina by increasing the ratio

TABLE 5.2 *The sensitivity of selected animal eyes, measured as photons captured per receptor*

Species name	Sensitivity	Light habitat
Marine isopod (*Cirolana*)	4200	Deep sea
Decapod shrimp (*Oplophorus*)	3300	Deep sea
Spider (*Dinopsis*)	101	Nocturnal
Horseshoe crab (*Limulus*)	83–317	Coastal, mainly nocturnal
Moth (*Ephestia*)	38	Nocturnal/crepuscular
Dung-beetle (*Onitis aygulus*)	31	Nocturnal/crepuscular
Hyperiid amphipod (*Phronima*)	38–120	Mid-water
Human (*Homo*)	18	Crepuscular
Scallop (*Pecten*)	4.0	Coastal sea-floor
Toad (*Bufo*)	4.0	Nocturnal/crepuscular
Shore crab (*Leptograpsus*)	0.5	Diurnal
Dung-beetle (*Onitis ion*)	0.35	Diurnal
Worker bee (*Apis*)	0.32	Diurnal
Jumping spider (*Phidippus*)	0.04	Diurnal
Human (*Homo*)	0.01	Diurnal

Source: After [113].

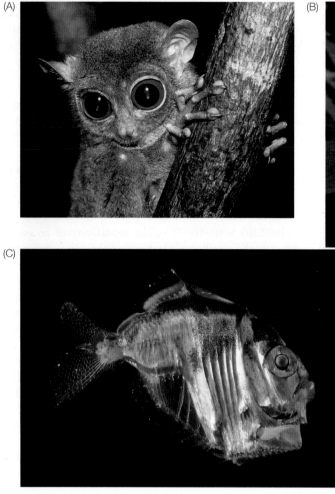

FIGURE 5.23 **Large nocturnal eyes** (A) A Philippine tarsier (*Tarsier syrichta*), (B) a western screech owl (*Megascops kennicottii*), and (C) a hatchetfish (*Sternoptyx*). In contrast to most nocturnal species, these species all have round rather than slit pupils. Owls and tarsiers have exceptionally mobile irises and well-developed eyelids to shut out excessive light, and mostly sleep during the day. Their nocturnal vision is largely achromatic, so they are not affected by chromatic lens aberrations. Deep-sea fishes have neither mobile pupils nor eyelids, as their environment is perpetually dark.

of rods to cones, since many adjacent rods sum their inputs. Many nocturnal animals, such as cats, rats, opossum, and seals have 95% or more rods, even in the area centralis, and a few species, such as deep-sea fish, have all-rod eyes [1, 32, 159, 221]. A third strategy is to *decrease the focal length of the lens* by making it rounder (increasing the curvature of the refracting surface). Then the retina must be brought forward by shortening the eye. The shorter focal length makes the image smaller but brighter. In conjunction with enlarging the lens, this creates a so-called **tubular eye** that can no longer be rotated in the head. Some owls and deep-sea fish have such an eye, with the consequence that they must turn their heads to look in different directions [113, 222]. A fourth strategy is to *increase the efficiency of photon capture* by making the photoreceptor longer or more densely packed with rhodopsin molecules. The wavelength of maximum absorption by rhodopsin can also be adjusted to match the peak region of ambient light [50, 91, 159]. Because several of these strategies will simultaneously reduce resolving power, a common strategy for nocturnal animals is simply to make the eyes extremely large to preserve some level of acuity (**Figure 5.23**).

The final strategy for increasing sensitivity is to *install a*

reflecting layer behind the photocells so that any light that passes through the retina and is not trapped by the receptor cells on the first pass is reflected back for a second pass at the receptors [222]. The reflecting structure is called a **tapetum**. It produces the eye shine characteristic of nocturnal animals. Like fish scales and the mirrored surface of the reflector-type eye, the tapetum is constructed of layers of guanine platelets. Most invertebrate eyes do not possess tapeta, but the house spider presents an interesting example. As we shall see below, spiders possess three to four pairs of eyes and each serves a different purpose. Tapeta often occur in the secondary eyes but not in the principal eye. Eyes with tapeta also have a reversed retina (i.e., the receptor cells are turned around, with their rhodopsin-laden membranes oriented away from the direction of incoming light) [110]. Presumably, the effectiveness of the tapetum is increased when the receptor cells are aimed at it. The common occurrence of tapeta in vertebrate eyes may be an adaptive factor favoring the reversal of the retina in this taxon [113].

TEMPORAL RESOLUTION **Temporal resolution** is measured by determining the flicker fusion rate. This is the rate at which a rapidly blinking light is perceived as continuous. Cones can respond to light and recover much more quickly than rods, so eyes with high cone ratios have better temporal resolution than eyes with high rod ratios. The human flicker fusion rate is about 50/second for cones and about 15/second for rods. Some animals with all-cone eyes have flicker fusion rates as high as 100–150/second. Invertebrates with rhabdomeric eyes have much higher flicker fusion rates than vertebrates. Invertebrate eyes, for example, can easily resolve the 60-cycle flicker of a fluorescent light and the rapid wing beat of a small flying mosquito [110].

Field of view

A final important aspect of visual reception involves the placement and orientation of the eyes on the receiver's head. For animals that use vision to detect and escape from predators, a very wide **field of view** is highly advantageous. This selective force leads to eyes placed on the sides of the head, or on eyestalks raised above the head, to give a nearly 360° field of view. In such cases, the eyes are mostly independent,

resulting in **monocular vision**, but there may be a small area of overlapping fields for **binocular vision**. Predatory animals, and especially those that actively pursue their prey, tend to have their eyes on the front of the head facing forward, so there is a large area of overlap and binocular vision, at the cost of a greatly reduced overall field of view. **Figure 5.24** illustrates a few classic examples of different eye positions and visual fields.

FIGURE 5.24   Fields of view and binocular vision The visual field of a single eye is approximately 170° in most animals; the greater the overlap of the fields of view of the two eyes for binocular vision, the smaller the total field of view. (A) The hawk and many other birds possess two foveas. The temporal fovea permits high-resolution binocular viewing in a narrow forward area ranging from 20–30°. Excellent monocular vision is enhanced by the central fovea. Monocular optical flow fields are indicated with dashed arrows. (B) In cats and dogs the visual angle between the two eyes is small so the visual fields overlap extensively. This gives excellent stereoscopic vision in the forward direction but a narrow overall field of view compared to that of the horse and hare. (C) The horse has a small area of stereoscopic binocular vision in front of its nose for foraging and must raise its head to view distant objects because the area directly in front of its head has a blind spot; the horse also has a broad horizontal band of monocular vision to each side. (D) Hares, rabbits, and dabbling ducks have unusually large monocular fields of view and upward-angled eyes that give them a band of good stereoscopic vision from front to back above their heads. (E) Jumping spiders have three main pairs of eyes, each of which views a different part of the visual field. The two anterior median (AM) eyes have a small retina with only a 10° field of view that can be moved with muscles to cover a total field of 58° (double-headed arrow). The anterior lateral (AL) eyes and the posterior median (PM) eyes have fixed but wider fields of view. This gives the animal excellent resolution in the forward direction as well as moderate vision to the side. (A after [222, 135]; B after [89]; C after [77]; D after [88]; E after [64].)

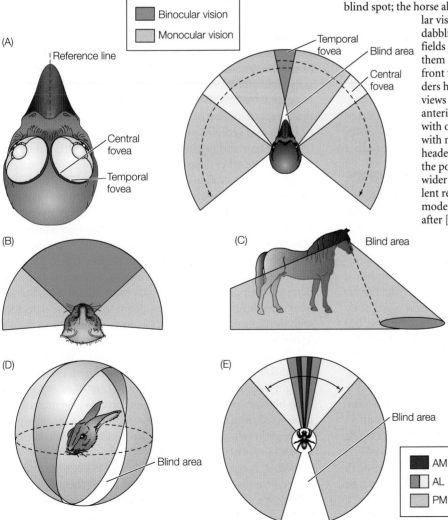

Visual Processing

Visual perception involves neurophysiological processes far beyond the scope of this book. In a nutshell, after substantial preprocessing in the vertebrate retina, nerve impulses are carried by the ganglion cells from both eyes, split, and are projected to the paired **lateral geniculate nuclei** (**LGN**) in the thalamus of the brain. These preliminary processing centers are composed of distinctive layers and contain retinotopic maps of the visual field (the spatial position of the neurons within the LGN layers preserves the spatial association of the ganglion cells within the retina). From there, messages are sent to the **visual cortex** of the brain, which is also layered. Spatial image analysis occurs here and takes some time to resolve. In this section we will examine three very important visual processing tasks from a nonphysiological perspective: color vision, feature detection, and distance perception.

Color vision

In Chapter 3 we learned that the detection and discrimination of different frequencies of sound is achieved by inherent frequency-dependent properties of the receptor cells, due either to their resonant frequency or to their location along a gradient. Similarly, the detection and discrimination of different light frequencies is achieved with photoreceptors that absorb different wavelengths of light. However, the similarity ends there. Because eyes must resolve spatial information, arrays of finely tuned frequency receptors would not work. Instead, color vision employs a small number of receptor types with overlapping wavelength sensitivities dispersed throughout the retina in a mosaic pattern, and the outputs of these receptors are combined and compared to produce the sensation of color differences.

Animals have found three ways to generate visual pigments with different absorption peaks [229]. One is to use a different variant of retinal. The common form of retinal found in rhodopsin is derived from vitamin A_1 and is called retinal$_1$. A second form, called retinal$_2$, is derived from vitamin A_2 and differs from retinal$_1$ only in possessing an additional double bond in the ring. When retinal$_2$ is combined with opsin, the pigment is called **porphyropsin**. The substitution of retinal$_2$ shifts the absorption peak of the pigment up about 20–25 nm when the rhodopsin peak lies at 500 nm, but the shift is higher, about 60 nm, when the rhodopsin peak lies at 560 nm [78]. Porphyropsin is found in many aquatic vertebrates such as teleost fish, amphibians, and aquatic reptiles.

The second and most widespread mechanism for altering the pigment absorption curve involves changes in the amino acid sequence of the opsin protein [18, 54, 232]. These affect the way the opsin component interacts with retinal. A single substitution can shift the peak up or down, and the effects of different substitutions are additive, so that pigments with absorption peaks as high as 620 and as low as 350 nm can be generated.

The third mechanism, found in birds, amphibians, lizards, snakes and turtles, is the addition of a colored oil droplet to the photoreceptor [43, 79, 81, 156, 217]. Carotenoid pigments are responsible for the color in the droplets, and specific colors are associated with specific photopigments. They act to narrow the absorption peak and shift it to longer wavelengths.

An animal living in dark, poorly illuminated environments must use all of its photoreceptors for spatial pattern analysis and brightness contrasts; it cannot afford the luxury of color vision. Such an animal is called a **monochromat**; it sees the world in shades of gray. Nocturnal, deep-sea, and cave-dwelling vertebrates typically have a very high proportion of rod receptors in their eyes. Species with some cones may be incapable of distinguishing colors if they possess only one type of cone photopigment. The cone system extends the range of dynamic vision to higher light levels than those permitted by rods alone and provides other advantages, such as temporal resolving power. Even when the cone pigment has a different spectral sensitivity from the rod pigment, wavelength comparisons are unlikely to occur because of the independent operation of the rod and cone systems. Animals known to have poor or no color vision include the rat, hamster, mouse, opossum, raccoon, genet, galago, owl monkey, most bats and deep-sea fish, dolphins, seals, whales, and cephalopods [94, 99].

Once a species has evolved two cone pigments with different spectral sensitivities, a form of color vision becomes possible. Such an animal is called a **dichromat**. The principal mechanism of color perception is the comparison of neural outputs from these two cone types; **Figure 5.25** shows a schematic representation of this process. A nerve cell in the retina or brain that receives positive input from the ON system of one cone type and negative or inhibitory input from the OFF system of the other cone type gives a unique response pattern to light stimuli of different wavelengths. Such neural units that receive opposing inputs from receptor cells with two pigment types are called **chromatically opponent cells**; they are the hue detectors of the visual system [46, 190]. At the same time, other cells receiving same-sign input from the two cone types effectively sum the output of both receptor types. They are insensitive to hue but highly sensitive to brightness, and yield divergent responses to black versus white stimulation. Subsequent combining of the outputs from the two hue detectors with the black-and-white detectors in the visual cortex results in the perception of hue, saturation, and brightness. It should now be clear why animals living in dark environments cannot afford color vision. Not only is the cone system less sensitive to light because of its reduced summation of receptor cells onto bipolar cells, but the opponent process also means that cone cells stimulated by light are sending antagonistic inputs to ganglion cells to enable color perception and enhance spatial and temporal visual analysis [175].

What does a dichromat actually see? We can ask animals to perform discrimination tasks that reveal their visual sensitivity to monochromatic light of different wavelengths and to mixtures of wavelengths. For the dichromat described above, all objects reflecting wavelengths between about

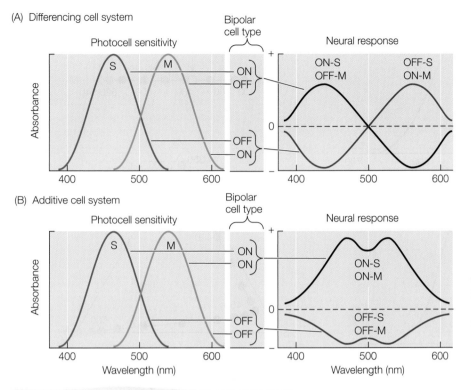

FIGURE 5.25 Logic of chromatically opponent cells Two cone types are required, one with a pigment absorbing maximally at a short (S) wavelength (475 nm) and another with a pigment absorbing at a medium (M) wavelength (525 nm) as shown in the two graphs on the left. The absorption curves overlap each other extensively. The middle panel shows the bipolar and ganglion cell types (ON or OFF) that combine their outputs at higher levels. The two graphs on the right illustrate the neural response at the level of the lateral geniculate nucleus of the brain. (A) In a differencing system, the ON bipolar cell output from one photocell type (here, a blue S cell) is combined with the OFF bipolar cell output of the other type (the green M photocell). The neural response is shown to the right in black. Other neural cells combine the OFF-S cell with the ON-M cell to produce the neural response shown in red. Any monochromatic light input will produce a unique set of responses from these two types of differencing cells. (B) In additive systems, output from the two ON cell types yields the neural response shown in black and output from the two OFF cells the response in red. These cells provide brightness information. Combinations of all four types of neural responses downstream lead to the representation of hue and saturation.

400 nm and 600 nm (primarily blue, green, and yellow for humans) would appear colored, and wavelengths smaller than 400 nm and larger than 600 nm would appear black (**Figure 5.26A**). This animal therefore would not perceive red colors. The ability of the dichromat to distinguish different wavelengths is maximal in the zone where the two photopigment curves overlap (i.e., in the green region). If a dichromat is asked to discriminate between monochromatic light of different wavelengths and an equiluminant gray light, there is a narrow region located between the wavelength peaks of the two photopigments in which the discrimination cannot be made. This is called the **neutral point**. However, dichromats treat this region as "green" and they perceive a continuous

range of chromatic colors from blue to green to yellow. Dichromatic color vision is characteristic of the majority of mammals, including squirrels, shrews, New World monkey males, cats, dogs, ungulates, many marine fish, spiders, crustaceans, and nocturnal insects such as ants and cockroaches [94, 99, 123, 176, 220]. Many amphibians are also dichromats, but their unique nocturnal color vision ability is facilitated by a color-opponent system involving two rod pigments [8, 70, 71, 75, 102, 120, 165].

Animals with three photopigments are called **trichromats**; these include apes and Old World monkeys, freshwater fish, some reptiles and amphibians, many insects, and some spiders. With the addition of a third pigment type, a second color-opponent system becomes possible, one between the S and M photocells, and another between the M and L photocells. The white/black system receives additive positive or negative input from two or all three pigment types. At least six types of nerve cell responses are predicted to occur in the initial brain nuclei for color encoding, and such cells have been documented in several species [7, 93, 190, 230]. Combinations of the outputs from these cells produce a rich range of hue and saturation perception such as that shown in **Figure 5.26B** for the honeybee. In trichromats there is never a neutral point or wavelength region that is indistinguishable from gray. The reason is that a wavelength that produces a neutral balance for one of the color-opponent systems will not be at the neutral balance point for the other color-opponent system.

Web Topic 5.3 *Evolution of primate color vision*
Diurnal primates reevolved trichromatic color vision from their dichromat ancestors. The peaks of the pigment absorption curves are not evenly spaced as they are in the honeybee. New World monkeys provide clues to how and why these patterns arose.

Butterflies, dragonflies, and jumping spiders have four or occasionally five photoreceptor types [20, 99, 106, 147, 157]. Possession of four cone pigments is the ancestral condition for vertebrates and persists in all birds, diurnal reptiles, and some fish and amphibians [34, 80]; see Web Topic 5.3. One function of an additional receptor type is certainly to extend the color perception range into the ultraviolet and near-infrared wavelengths. The second function is to improve the discrimination of small hue differences. Birds and turtles in particular add colored oil droplets with specific carotenoid

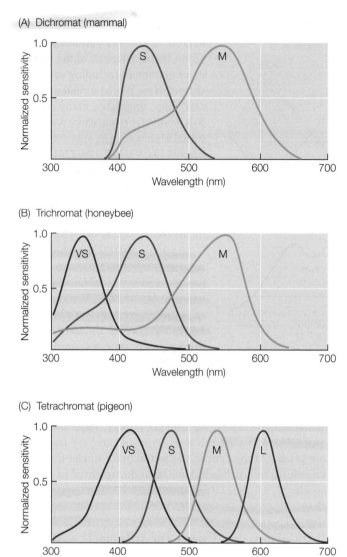

(A) Dichromat (mammal)

(B) Trichromat (honeybee)

(C) Tetrachromat (pigeon)

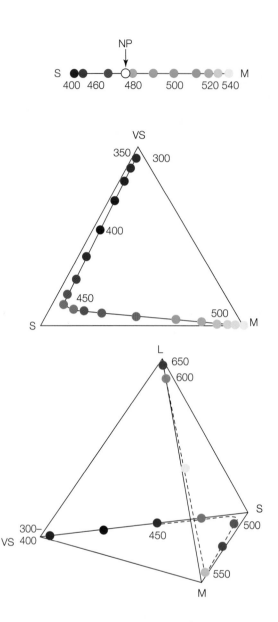

FIGURE 5.26 Photopigments and hue space Dichromat (A), trichromat (B), and tetrachromat (C), with $n = 2$, 3, and 4 receptor types, respectively. The graphs on the left show sensitivity curves for the photopigments, each of which absorbs at a characteristic peak wavelength: VS (very short), S (short), M (medium), or L (long). The graphs on the right show chromaticity diagrams of the hue space projected in n-1 dimensions. The loci of monochromatic hues are represented by the dashed line and the points, plotted every 10 nm in (A) and (B), and every 50 nm in (C). In (A), NP indicates the neutral point of a dichromat. (After [99].)

pigments to each type of cone cell. The oil droplets both narrow the spectral absorption curves and remove the second low absorption peak of rhodopsin in the UV range, as illustrated in **Figure 5.26C**, to increase wavelength discrimination [43]. Although the hue space of a trichromat can be represented in two dimensions, the hue space of **tetrachromats** must be represented as a three-dimensional tetrahedron. Birds have at least three chromatically opponent systems that combine input from spectrally adjacent pairs of receptor types [67]. Turtles have 12 chromatically opponent cell types, which are

believed to form five color-opponent channels based on combinatorial input from their four photoreceptor types [168].

In many taxa with three or more pigments, some pigments clearly serve different functions [99, 100]. The distribution of pigment-type cells in the retina varies with the species. Typically, some pigments are located in the upward-looking part of the eye while others are located in the lateral, forward, or downward-looking part. Some species appear to require different color sensitivity for different activities such as social communication versus food detection or orientation. Guppies, for example, use primarily green receptors to look upward during foraging, and red receptors to view conspecifics from the side [5]. Butterflies have those pigments necessary for food plant detection in the bottom half of the eye and pigments necessary for conspecific detection in the upper half [14, 106]. A specific function for the UV component of vision in birds has not been identified, but UV certainly plays a role in foraging, navigation, and plumage color signaling [12, 31, 43]. Juveniles of many species of fish exhibit UV sensitivity, which is often lost in adulthood. Sensitivity to UV wavelengths has been demonstrated to play a role in the foraging, shoaling, and short-distance visual signaling in some species [189]. There is evidence in goldfish that pigments operate under different illumination levels, suggesting that color perception needs vary at different times of day [146]. Finally, brightness perception and spatial resolution are often restricted to a subset of receptor types [99, 148]; in birds, for example, double cones that combine M- and L-sensitive receptors primarily function for achromatic vision [43, 67].

The pinnacle of photopigment complexity is found in stomatopod shrimp, which have 16 different photoreceptor types, 12 of which are specialized for color vision [37, 38, 105]. Sets

of photocell types occur in distinct regions and are specialized for hue perception, polarized light detection, and spatial resolution (**Figure 5.27**). Pairs of hue receptors have overlapping pigment-absorbance curves, but it not yet clear whether adjacent pigments interact in color-opponent systems or form separate wavelength band detectors. One theoretical advantage of a large number of very narrow-band pigment types is color constancy (see below), which may be particularly important in intertidal habitats where the ambient light spectrum changes greatly with depth [39, 154]. Color signals are clearly very important for these animals [28, 40, 131, 132].

Animal species obviously vary greatly in their ability to see color and in the quality of their color perception. Why has color vision evolved? There are some clear costs of color vision. Color vision compromises sensitivity, even for animals living

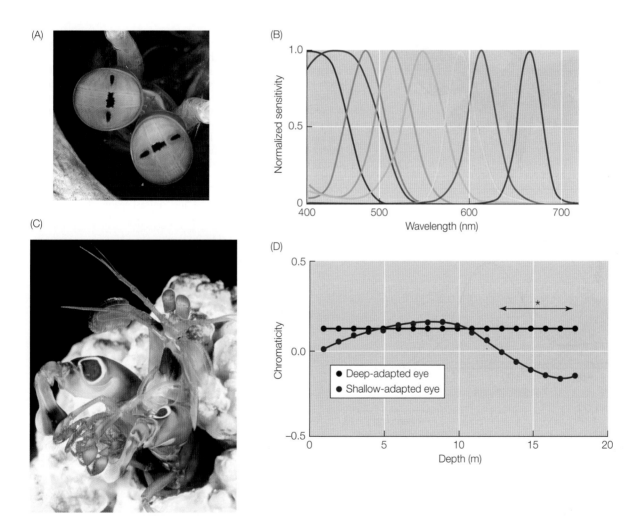

FIGURE 5.27   Color vision in mantis shrimp   (A) The apposition compound eye of a stomatopod (*Gonodactylus smithii*). The main color receptors are located in the first four rows of the prominent horizontal mid-band, and the two lower mid-band rows are specialized for detecting polarized light. Receptors in the peripheral, ventral, and dorsal hemispheres are responsible for brightness contrast and movement. (B) Calculated spectral sensitivities for shallow-water-adapted photoreceptors. Color of the curve approximates the wavelength sensitivity. (C) A shrimp in defensive posture defending its burrow, showing the prominent blue meral spots and the colorful maxillipeds and club-like raptorial appendages with which it strikes prey and rivals. (D) Computed chromaticity of the meral spot as viewed through a deep-adapted eye and a shallow-adapted eye, illustrating adaptive color constancy for the deep-adapted eye in particular (significant differences for depth range indicated by double-headed arrow). (After [27, 131].)

in well-illuminated environments, because of the antagonistic interactions described above [99]. The color system has lower spatial and temporal resolution than does the black-and-white system, so important visual tasks such as motion detection are "color-blind" [118, 185]. The border-enhancement effect characteristic of luminance borders does not have a color analog, so colors blend. Blue receptors in particular are so low in number that they do not contribute to the formation of chromatic borders and spatial resolution. These properties imply that color perception may make the distinction between slight brightness differences more difficult, and that reliance on color can cause two similarly-hued objects to meld or be confused. Nevertheless, these disadvantages may be small when compared to the great advantages of color vision [100]. **Object detection** against the background is without a doubt the most important

function of chromatic vision. **Object recognition**, and in particular the speed with which an object can be recognized, is greatly enhanced by color. The main target objects animals need to detect and recognize are of course food, predators, and conspecifics. In some cases, **navigation** and **orientation** may take advantage of chromatic information. As we shall see below, feature detectors and depth perception may combine color cues with other mechanisms. Possession of three or more photoreceptor types greatly enhances **color constancy**—the ability of the visual system to correct for variation in background and ambient light color so that target objects retain their characteristic hue [24, 149, 217]. Finally, **intraspecific communication** cannot be ruled out as a selective force in the refinement of color vision. **Figure 5.28** illustrates a particularly compelling case for the multiple advantages of color vision in a

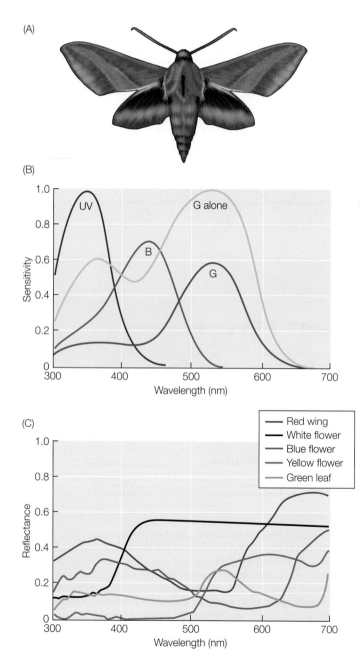

(A)

(B)

(C)

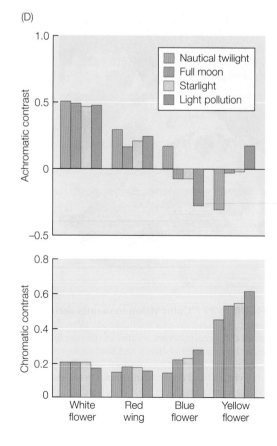

FIGURE 5.28   Chromatic and achromatic contrast for key target objects viewed by the crepuscular hawkmoth (A) *Deilephila elpenor*, showing red patches on wings. (B) Sensitivity curves for the three photoreceptors (UV, B, and G). The line for G alone shows the sensitivity of the green receptor used for the achromatic contrast calculations. (C) Reflectance spectra for the red patch, several flowers used by the moth, and the green leaf background. (D) Computed chromatic and achromatic contrast under different night sky illumination conditions (see Figure 5.4), taking into account the moth's spectral sensitivity and the color constancy adaptation of its photoreceptors. White flowers had the highest achromatic contrast under all illumination conditions, but yellow flowers had the strongest hue contrast against the background. (After [97].)

(D)

crepuscular hawkmoth, employing all of the principles we have discussed in this chapter.

Feature detectors

A narrow definition of a **feature detector** is that it is a single neuron that codes for a specific and complex stimulus. In the broader view, feature detection is part of the process of object recognition and classification. Evidence for feature detectors first arose in the 1950s and 1960s when neurobiologists refined the techniques for recording spike activity from single neurons in sensory organs and the brain. Studies on nerve responses in the ganglion cells and visual cortex of frogs, cats, and monkeys discovered cells that responded very selectively to different and specific types of stimulation patterns, such as edges, spots of a certain size, bars of specific widths and orientations, and objects moving in certain directions and velocities [9]. The idea was that these lower-level responses would combine and converge at each synaptic step at higher levels in the brain, with progressively higher-order cells coding for more abstracted functions and being more selective in their stimulus requirements. At some very high level in the brain, researchers expected to find cells whose responses were so specific that they would only fire when a particular image was viewed by the eye [136]. This concept was sometimes called the **grandmother cell theory** [74], based on the notion that a single cell would respond only to the image of a particular person's face (**Figure 5.29**, Model A).

Although appealing, this view is far too simplified to describe the process of visually analyzing and classifying objects in a complex visual landscape. The next generation of visual neurobiologists and cognitive psychologists showed that the receptive fields of these lower-level neurons became more distributed at higher levels (**Figure 5.29**, Model B). Distributed fields can represent the features of an object at a level of detail that is difficult to achieve with one or a few feature detector cells. According to this model of vision, the first step is detection of edges by retina cells. A set of lower-level neurons can collectively represent the contour elements of an object in detail, while higher-level neurons can code aspects of its outline, shape, texture, color, and motion [129, 130]. Moreover, a distributed representation would allow new objects to be represented as new patterns of activity across existing neurons. In fact, both of these processes probably occur together, along with complex lateral and feedback interactions among the layers of the visual cortex and LGN [159]. Computer vision models and MRI studies of awake animals and humans are leading to a better understanding of how real visual systems operate to solve complex object recognition and classification tasks [22]. Clever behavioral studies in birds, cephalopods, and bees have also begun to reveal the role of edge detection in navigation and object recognition [23, 86, 116, 117, 235, 236].

The one visual task for which the role of feature detectors is clearly understood is the rapid detection of important moving targets. For sit-and-wait predators such as frogs and lizards that strike at small fast-moving prey [219],

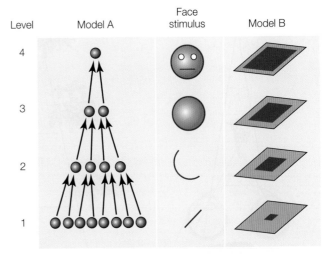

FIGURE 5.29 **Two extreme models of visual feature representation** Object detection and classification is believed to occur in a series of steps from the photoreceptor array to the brain, beginning with detection of edges, contour and shape construction, and incorporation of color information. The central column depicts the hypothetical representation of a face stimulus at different levels in the visual cortex, given input from retinal ganglion cells at the bottom and moving up to higher levels in the cortex. The levels correspond to specific layered areas in the visual cortex at increasing distances from the eye (1 = primary visual cortex area V1; 2 = area V2; 3 = area V4; and 4 = inferotemporal cortex area IT). Model A depicts the hierarchical model of a feature detector, where the neuron at the top level responds selectively to the face. Model B represents the diffuse receptive field model, showing the size of the receptive field increasing at higher levels. (After [170].)

and for prey animals that must make rapid evasive movements when predators approach, there is strong selection for quick visual recognition that immediately triggers a motor response. Recent studies on pigeons and locusts reveal functionally similar visual feature detector systems for detecting approaching (looming) objects [167]. The animal needs to know that the object is approaching on a collision course and how imminent a collision is. The relevant information is computed from the way the image of the object grows on the retina of one eye. Owing to the high spatial resolution of vertebrate eyes, the pigeon's detector can compute the time to contact while the object is still some distance away, and initate evasive movements in time to escape. The locust has a single large neuron on each side of the central nervous system about three or four synapses downstream from the photoreceptors. It is fed by a very compact neuronal microcircuit, which is tightly tuned to objects that are approaching on a collision course, and makes direct postsynaptic contact with motor neurons that activate the flight muscles. Such movement detectors may turn out to be fairly common in arthropod eyes and used for a variety of tasks [45, 66, 151].

Depth perception

Mobile animals need to have good **depth perception** so they can judge the distance to three important kinds of target

(A)

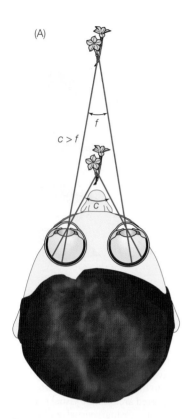

(B)

(C)

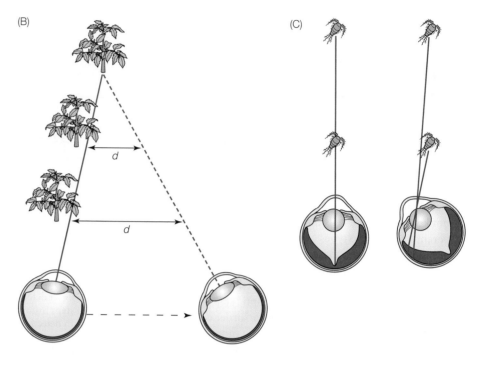

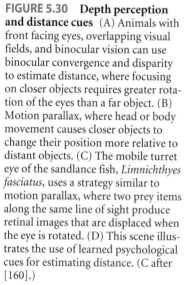

(D)

FIGURE 5.30 Depth perception and distance cues (A) Animals with front facing eyes, overlapping visual fields, and binocular vision can use binocular convergence and disparity to estimate distance, where focusing on closer objects requires greater rotation of the eyes than a far object. (B) Motion parallax, where head or body movement causes closer objects to change their position more relative to distant objects. (C) The mobile turret eye of the sandlance fish, *Limnichthyes fasciatus*, uses a strategy similar to motion parallax, where two prey items along the same line of sight produce retinal images that are displaced when the eye is rotated. (D) This scene illustrates the use of learned psychological cues for estimating distance. (C after [160].)

objects: environmental obstacles, prey, and predators. Species are expected to differ as to which visual targets are most important to them, and therefore which mechanisms they employ for depth perception. Navigating around objects without crashing into them places the greatest demands on animals that fly or run very fast. Estimating the distance to prey items is obviously most important for predators that pursue mobile prey. Perceiving the approach of predators is most critical for prey species that forage in exposed open habitat. Mechanisms for distance assessment can be divided into two main categories—physiological and psychological—and animals typically employ a mixture of these methods [163].

Physiological mechanisms for depth perception can be subdivided into binocular and monocular methods. Most mobile animals have at least some region of binocular vision, and if the output from the two eyes is properly cross-wired in the ocular regions of the brain [16, 47], they may have true **stereoscopic vision** [150]. Animals with this capability can integrate the two images coming into each eye and triangulate the location of objects in three-dimensional space using the disparity in image angles on the two retinas as illustrated in **Figure 5.30A**. However, the fact that an animal has overlapping fields of view does not necessarily imply that it possesses stereoscopic vision. For example, most birds have a fairly narrow region of binocular overlap; true stereoscopic vision has been demonstrated only in some owls. Overlap in other species may be primarily a consequence of the need for lateral optical flow fields that include a frontal region for navigation during flight (see Figure 5.24A) [135]. Animals with binocular overlap can also use **binocular eye convergence** to determine object proximity. This mechanism usually works in conjunction with **accommodation** and a fovea or area centralis, in which focusing on a close object stretches the extraocular muscles more than when focusing on a far object. As illustrated in Figure 5.30A, closer objects cause a greater retinal disparity or angular convergence than distant objects. Accommodation can also provide distance information in monocular vision, if the animal can sense the degree to which it must change the lens shape or position to bring an object into focus. **Motion parallax** is a powerful monocular mechanism for distance estimation. Moving the eye or the head causes closer objects to appear to shift position relative to more distant objects (**Figure 5.30B**). A variant of this mechanism is observed in the turret eyes of chameleons and the sandlance fish with their forward-positioned lenses and deep foveas. There is sufficient separation between the nodal point and the axis of rotation of the eye that retinal images are displaced when the eye is rotated (**Figure 5.30C**).

At least six **psychological cues** may be combined for distance estimation. These mechanisms operate with monocular as well as with binocular vision and generally must be learned to provide useful information.

1. **Object size**: Once the receiver has learned the typical size of relevant objects in its world, it can use this knowledge to evaluate distance, since closer objects appear larger and project a larger image onto the retina.
2. **Linear perspective**: The convergence of parallel lines or the gradual reduction of object size in a series of similar objects as distance increases provides another clue to relative distance.
3. **Aerial perspective:** More distant objects or scenes appear hazier and foggier.
4. **Color gradients:** For receivers with good color vision, distant objects are grayer in color due to Mie and Rayleigh scattering.
5 **Occlusion:** Closer objects appear to overlap and break up the outline of more distant objects behind them.
6. **Texture gradients:** Closer objects reveal more roughness and texture than distant objects.

The scene shown in **Figure 5.30D** illustrates all of these mechanisms.

SUMMARY

1. Transmission of a reflected light signal from sender to receiver occurs in a series of steps depicted in the **ARTS diagram**. (1) White light from the sun is modified by attenuation, scattering, and environmental filtering to produce the **ambient light spectrum** available to the sender. (2) The sender's surface selectively reflects certain wavelengths, described by a **reflected light spectrum.** (3) Reflected light is then transmitted a given distance through the environment to arrive at the receiver, and may be degraded by the combined effects of attenuation, scattering, and medium absorption as depicted in a **transmission spectrum.** (4) The **color sensitivity** of the receiver's eye determines the range of detectable wavelengths. The spectra in steps 2–4 are transfer functions which, when multiplied together with the ambient light spectrum in step 1, determine what is actually perceived by the receiver. A parallel pathway also describes the **background spectrum** arriving at the receiver's eyes at the same time.

2. **Ambient light** varies in hue and overall brightness in different habitats. The color of ambient light is often green in forests, but microhabitats may be bluish, yellowish, or purplish depending on the distribution of vegetation and the time of day. Ambient light in marine environments generally becomes bluer with increasing depth because of the strong absorption of long wavelengths. In shallow coastal and fresh water, bottom characteristics and particulate matter can result in other characteristic colors. Viewing

angle (up, down, horizontal) also affects available light in the three-dimensional aquatic world.

3. The **background** against which a sender's body or color patch is viewed has a very large effect on its conspicuousness. Strategies to increase conspicuousness involve maximizing **brightness, hue, pattern,** or **motion contrast**. Additional contrast strategies include **outlining, symmetrical patterns, contrasting concentric circles,** and **repetitive patterns**. It is to the advantage of prey animals to minimize these types of contrasts.

4. **Transmission** of a signal or target depends on the **distance** between sender and receiver and the combined effects of global attenuation, medium absorption, and scattering, which are expressed by the **beam attenuation coefficient**. Objects are visible from a greater distance in environments with low attenuation, such as clear air and ocean water. Fog, dust, and other particulate matter reduce transmission distance, along with large **opaque environmental objects** such as trees and rocks.

5. Aspects of receiver sensory ability affect the final **detectability** of a signal. The **just-detectable distance** of a signal or object depends on its **apparent size** and the **resolving power** of the receiver's eye. Large and contrasting objects are detectable from a greater distance as long as the visual angle subtended by the object is greater than the eye's angular resolution. If this constraint has been overcome,

the **contrast threshold** for detection depends on receiver color sensitivity. **Visual color models** are constructed to quantify the perceived hue and brightness contrast between a signal and its background.

6. Well-developed eyes are complex organs with sophisticated optical and cellular systems for collecting and concentrating light and with intricate nerve connections for spatial pattern analysis and resolution. The first eye took the form of a simple **pigment cup eye** lined with a retina that could detect the direction of light shining on it but that failed to form an image. Image formation became possible with the development of a focusing mechanism such as the **pinhole eye**, which suffered from low sensitivity. The advent of the refracting lens greatly improved sensitivity and eventually led to the **camera-lens eye**, which is both sensitive and capable of producing a well-resolved image. Another development involved multiple cups that were grouped and lengthened into tubes to restrict the angle of light acceptance to a narrow field. This design led to the **compound eyes** of insects and crustaceans.

7. A **lens** composed of **transparent tissue** is required to collect light over a large area and concentrate it for image formation. Transparent lenses contain specialized cells lacking organelles and other boundary structures that would cause scattering. The cells are densely and uniformly packed with proteins that give the tissue a high refractive index. The lens shape is usually doubly convex to refract light twice and focus the image on the light-sensitive retina. Strategies of **accommodation** (focusing ability) in camera-lens eyes depend critically on the environment. The cornea performs most of the refraction in terrestrial animals, while the soft lens can change shape for fine-tuning. Aquatic eyes possess a hard round lens that is moved forward and backward.

8. The energy of electromagnetic radiation in the visible range is captured by the visual pigment **rhodopsin**. It consists of a bent 11-*cis* carotenoid derivative (**retinal**) attached to a large protein (**opsin**). Upon absorption of a photon of light, the retinal is straightened to the all-*trans* isomer. This change causes a cascade of reactions that culminates in a nerve impulse. Rhodopsin molecules are embedded in the multiply folded membranes of long, thin photoreceptor cells. The **photoreceptor cells** are packed side by side in arrays to form a **retina** that retains the spatial arrangement of the incoming spatial stimulus patterns.

9. Evaluating the image falling on the retina requires a complex network of nerve connections. Most vertebrates have a **duplex retina**, with two parallel systems for use under high- and low-illumination situations. The low-light system uses **rod** receptor cells that are very long and highly sensitive to light. The nerve outputs from many neighboring rods sum together to increase the probability of photon capture and nerve response. While the rod system is highly sensitive in dark environments, its spatial resolution is poor due to the summation. The bright-light system uses **cone** receptors. Cone outputs are not extensively summated, so a fine-grain image is transmitted. They are connected to neighboring cones in complex additive and negative ways to enhance boundaries in the visual field and detect certain types of temporal and spatial patterns.

10. A good eye should have high acuity to resolve fine details and be sensitive to light under a wide range of ambient conditions. It is difficult to maximize both features because improvements in one tend to reduce the other. **Resolution** is maximized by increasing the lens focal length and by decreasing the diameter and length of receptors. **Sensitivity** is maximized by increasing the eye aperture, decreasing the lens focal length, increasing the diameter and length of receptors, and increasing photopigment density. Sensitivity and resolution therefore trade off against each other. Compromises include the rod/cone system of vertebrates and a central eye region of high resolution (**area centralis** or **fovea**) surrounded by a peripheral region of high sensitivity. Other modifications include reflecting **tapeta** and tubular eyes to improve sensitivity, and lenses that correct for **spherical** and **chromatic aberration**.

11. **Field of view** depends critically on the placement of eyes on the head. Lateral placement maximizes total field of view up to 360º but permits only a small region of binocular overlap. Front-facing eyes maximize binocular overlap. **Temporal resolution**, or flicker fusion rate, is much higher in cones than in rods, and is higher in rhabdomeric photoreceptors than in ciliary receptors.

12. Color perception requires at least two sets of photoreceptor cells with clearly different spectral sensitivities, and more complete chromatic differentiation requires three or four sets. **Monochromats** possess only one pigment type; they perceive intensity differences, but not color. **Dichromats** possess two visual pigments with different but partly overlapping absorbance curves. Color discrimination is based on **chromatically opponent cells**, ganglion cells that either add or subtract the neural output from neighboring photoreceptors of each color type. Ganglion cells receiving opposite-signed input from two cone types are responsible for hue discrimination, whereas ganglion cells receiving same-signed input encode brightness and saturation information. Dichromats can distinguish a wide range of colors but possess a **neutral point** in the middle of their spectral range. **Trichromats** have three photopigments and thus two chromatically opponent systems. They can perceive a wider range of colors and have no neutral point. Species with 4–5 receptor types frequently use certain wavelengths for special purposes.

13. The process of visual spatial analysis and **object recognition** proceeds through multiple layers of the visual organ and brain nuclei, beginning with edge detectors in the retina, evaluation of contours and outlines, and the addition of color and fine structure resolution. Recognition of objects appears to occur over a large and diffuse specialized region of the brain, rather than by a hierarchical process ending in a small number of cells. Specialized **feature detectors**, involving a small number of dedicated cells, seem to be limited to the detection of specific types of motion and are coupled to motor actions requiring fast responses, such as predator evasion or prey attack.

14. Depth perception is important for predators and fast-moving or flying animals. Such animals have evolved binocular vision, in which the visual fields of the two eyes overlap. The cost is a lower overall field of view; thus herbivores and prey species usually have eyes on the sides of their heads. These animals must use relative object size or parallax to judge distances.

15. Most of the properties of visual receptors are determined by the environment, lifestyle, and diet of the animal. Nocturnal animals must maximize sensitivity and therefore usually have reduced spatial and temporal resolution. Diurnal animals are adapted to high ambient light levels and can therefore maximize resolution and color vision. Predators and fast-moving animals require better depth perception and spatial and temporal resolution than prey species. These **trade-offs** largely determine the optics and wiring of the visual system, and therefore the opportunity to use the visual modality for communication.

Further Reading

The classic articles on transmission of light signals are by Lythgoe [125] and Endler [55], which should be augmented with more recent articles on color vision models such as Kelber [99], Vorobyev [216], and Montgomerie [143]. For perception of polarized light, see Horváth and Varjú [87]. The book by Land and Nielsson [113] provides an excellent review of adaptive structure and optics in all types of animal eyes. Fernald [61] reviews the contributions of genomics to the evolution of eyes and vision. Masland [137] and Solomon [190] provide readable reviews of the retinal and neurological mechanisms of color perception. For recent taxon-specific reviews of color vision see Briscoe and Chittka [20] for insects; Section II in Volume 2 of Ladich [107] for fish; Cuthill [43], Goldsmith [68] for birds; and Surridge [204], Yokoyama [233] and Ahnelt [1] for mammals. Kelber and Osorio [100] broadly review the functions of different color vision systems. Bruce [22] and Purves [163] are useful texts on cognitive perception.

COMPANION WEBSITE
sites.sinauer.com/animalcommunication2e

Go to the companion website for Chapter Outlines, Chapter Summaries, and References for all works cited in the textbook. In addition, the following resources are available for this chapter:

Web Topic 5.1 *Constructing color models*
The only way to understand how animals with different visual systems than ours perceive colored objects is to build a visual model. The effects of hue and brightness contrast, receiver sensitivity, medium attenuation, and distance can be very precisely evaluated with this technique.

Web Topic 5.2 *Perception of polarized light*
Rhabdomeric photocells are intrinsically more sensitive to the plane of light polarization than ciliary cells. Many invertebrates have specialized photoreceptors to make use of important environmental information provided by polarized light, and a few vertebrates are able to use it as well.

Web Topic 5.3 *Evolution of primate color vision*
Diurnal primates reevolved trichromatic color vision from their dichromat ancestors. The peaks of the pigment absorption curves are not evenly spaced as they are in the honeybee. New World monkeys provide clues to how and why these patterns arose.

Chapter 6

Chemical Signals

Overview

Chemical signaling is the oldest method of communication. From the earliest days of life in the Earth's oceans, single-celled organisms possessed the ability to detect and selectively take in different classes of chemicals needed for cellular metabolism. The detection of food was and still is the primary function of most chemical reception organs. At the same time that organisms are taking in food, they are eliminating metabolic waste products. Once such organisms evolved the ability to distinguish between the chemical compounds emanating from conspecifics and the chemical components of food, a primitive social organization and communication system existed. Early metazoans relied exclusively on chemical communication to synchronize gamete release and mediate fertilization between conspecific sperm and eggs, and some colonial species even used chemical signals to communicate alarm in the presence of a predator. More advanced organisms quickly evolved two types of chemical detection systems: smell and taste (in human terms). In this chapter we shall examine the types of chemicals and production methods used for chemical communication, the transmission properties of chemical odorants in different environments, and the evolution of chemoreception organs for the detection of chemical signals.

General Features of Chemical Communication

To transmit a chemical signal, individual molecules have to move the *entire distance* between sender and receiver. How can such movement be achieved? There are three basic ways.

1. Senders can use the **current flow** of air or water to carry the molecule to the receiver.
2. In the absence of a current, the molecule can only move by **diffusion**. Molecules naturally move along a concentration gradient, from a point of high concentration to a point of low concentration, but this requires a certain amount of time.
3. The receiver can move toward the signal and pick up the molecule by **direct contact**, so that the molecule doesn't have to move at all.

Contrasts between chemical, auditory, and visual signals

The propagation of olfactory signals differs in major ways from the propagation of auditory and visual signals:

DIRECTIONALITY   Unlike sound and light signals, which travel as orderly waves in a relatively straight path away from the source of emission, the diffusion of odorant molecules follows an irregular path from a point of high to low concentration. At any given moment, a diffusing molecule may move either away from or toward the source. A current-borne molecule will move in the general direction of current flow.

SPEED   Although chemical diffusion, like sound, requires molecular motion, the propagation of odors is much slower than that of sound. This is because in a sound far field, only the disturbance is propagated; in diffusion, the individual molecules must be propagated. Typical delay times between sender and receiver are milliseconds for sound, but seconds, minutes or even days for odors.

TEMPORAL PATTERN   Sound and light retain their temporal patterns as they propagate (although there may be some distortion over long distances); but neither diffusion nor current flow can sustain any initial pattern of modulation, because molecules do not move in synchrony. Any temporal pattern imposed on an olfactory signal during emission is lost within a short distance from the source [113].

SPECTRUM   The spectrum of olfactory signals (i.e., the different chemical compounds) cannot be arrayed in one continuous dimension, as can the frequency spectrum of sound and the wavelengths of light. This means we cannot use Fourier analysis to characterize olfactory signals, nor can we define a tuning or absorption curve for receivers. We therefore need a different method of categorizing and analyzing olfactory signals.

Forms of chemical communication

Because the selective detection and uptake of chemicals is a fundamental process of all living cells, chemical communication in a broad sense occurs at many biological levels. Chemicals that operate internally and facilitate communication between the brain and organs involved in growth, digestion and reproduction are called **hormones**. Chemicals that facilitate communication between conspecifics have traditionally been called **pheromones** and are, of course, the main focus of this chapter [52, 416]. Chemicals that are transmitted and detected between species are called **allelochemicals**, and are further subdivided on the basis of the costs and benefits to the sending and receiving species. When the receiver benefits, as when a predator detects chemical cues that help it to locate prey, it is called a **kairomone**. When the sender benefits, as when a predator produces a chemical that mimics the prey's pheromone to deceitfully lure it in, it is called an **allomone**. Chemicals that mediate mutualistic interactions benefiting both species are **synomones**. The inclusive term for all meaningful chemicals transmitted between organisms (within or between species) is **semiochemical**.

Semiochemicals are detected in two different ways: with **olfactory reception**, which involves the detection of airborne or waterborne chemicals from a distant source (e.g., smell), and with **contact reception**, which requires direct contact of the receptors with the chemical source (e.g., taste). Both olfactory and contact receptors may be used for the detection of food, conspecifics, and heterospecifics. As we shall see in the section on reception, many animals possess three separate chemical sensory systems, one for the detection of diffusing or current-borne chemicals, another for the identification of contacted food, and a third for the reception of social signals through contact.

Identifying chemical signals: Function before structure

Light, sound, pressure, and electric signals can be identified and described in the first investigative step, and the function can be determined later in a subsequent step. These modalities are forms of energy that we humans can easily detect, and signals in these modalities also are relatively easily distinguished from cues and nonsignaling behaviors. Chemical signals, on the other hand, are very difficult to identify without some knowledge of their signaling function. The first step in a study of chemical communication is typically the observation of a behavior that appears to be mediated or influenced by chemicals. This behavior then becomes the bioassay used in the search for the chemical compounds that elicit it. At all subsequent stages of collecting the potential chemical samples, purifying them, and identifying them, the researcher tests the product against the bioassay to be sure the critical signal compounds are still present (Web Topic 6.1). Moreover, the divergent transmission mechanisms for chemical signals, from deposited marks to long-distance current-borne emissions, are intimately tied to the basic function of the signal. While this chapter will focus on the principles of chemical signal production, transmission, and reception, we will need to include some mention of associated signal functions. By necessity, these functions are based on a receiver-response classification scheme (see Chapter 1).

An important distinction is made between the **primer** and **releaser** functions of pheromones. Primers produce physiological effects in receivers over time—for example, by stimulating the release of hormones that in turn affect the endocrine system of the receiver. Releasers produce immediate effects on the behavior of the receiver—approach, retreat, freezing, or other identifiable movements and postures. Releaser pheromones can be subdivided into a few basic categories, such as long-distance mate attractants, territory markers, trail markers, alarm signals, and group or individual recognition signals. We will make references to these functional categories in this chapter, as they relate to the design of chemical signals, but such functions will be discussed in greater detail in later chapters.

(A) (Z)–7–dodecenyl acetate

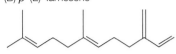

(B) β–(E)–farnesene

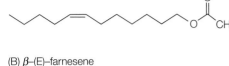

(C) Periplanone–B

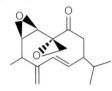

(D) Civetone

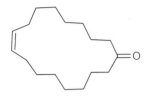

(E) 2-methylbut-2-enal

(F) 9–keto–2–decenoic acid

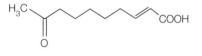

FIGURE 6.1   Some examples of airborne chemical odorants (A) The primary sex attractant component for the Asian elephant and about 140 noctuid moth species. (B) A common insect alarm substance and male mouse pheromone. (C) Cockroach (*Periplaneta americana*) sex attractant. (D) Civet (*Civettictis civetta*) sex attractant. (E) Rabbit (*Oryctolagus cuniculus*) mammary pheromone that attracts and guides pups to suckle. (F) One component of honeybee (*Apis mellifera*) queen substance.

Web Topic 6.1 *How are pheromones identified?*
To identify a pheromone, putative chemical secretions are collected, purified, and tested for biological activity with a good behavioral or physiological bioassay. The gas chromatograph plays a central role in identifying the chemical compounds.

Production of Chemical Signals

In this section we shall examine the range of chemicals used by animals to communicate with conspecifics, the sources of these chemical odorants, and the ways in which the chemicals are released into the medium by senders.

Types of chemicals used for intraspecific communication

The array of chemicals identified as pheromones is vast. All are organic compounds with a basic carbon skeleton. The major constraints on the chemical composition of pheromones are determined by the type of transmission (i.e., diffusion, current-borne, or contact) and by the medium (i.e., air versus water).

Airborne odorants must be **volatile** in air, that is, they must evaporate easily [406, 410, 411]. Volatility is primarily a function of molecular size and weight; larger, heavier molecules have lower volatility. The upper size limit for airborne pheromones is a molecular weight (MW) of about 300. Most airborne odorants contain between 5 and 20 carbon atoms. Molecules with more than 20 carbons are too large to diffuse effectively. Pheromones with less than 5 carbons may be rare because they are too volatile and possess too few options for species-specific structural variants. Within this size range, chemical odorants show a great deal of variation in shape and type of functional group. A few examples are shown in **Figure 6.1**. The majority contain a single functional group with one oxygen atom, such as an alcohol, aldehyde, or ketone, but some pheromones are acids or esters with two oxygen atoms (see Web Topic 6.2 for a summary of organic chemical classes). Compounds containing only carbon and hydrogen with no oxygen are called **hydrocarbons**; alkenes are unsaturated hydrocarbons with one or more C-C double bonds; and alkanes are saturated hydrocarbons with no double bonds. The shape of a molecule is an important determinant of its signal specificity. **Aliphatic** chemicals are composed of single or branching chains of carbon atoms (see Figure 6.1A,B,E,F). The position of any carbon double bonds and the presence of other atoms such as nitrogen or sulfur also affect the molecule's shape (several examples can be seen in the figure). If a long chain is connected at the ends, it becomes an **alicyclic** chemical, such as the civet sex attractant shown in Figure 6.1D. Chemicals with one or more 6-carbon (benzene) rings are called **aromatic** compounds (see Figure 6.1C). Such molecules are both more stable and more rigid than aliphatic compounds, so they can mediate highly specific receptor responses. **Stereoisomers** (two chemically identical compounds with different spatial arrangements) and **enantiomers** (two compounds that are mirror images of each other) are further examples of reception-specific shapes. Because the size, shape, and electric properties of an odorant molecule determine its ability to bind with a specific receptor site, semiochemicals are also called **ligands** (binders).

Some aquatic pheromones are low-molecular-weight compounds, much like airborne pheromones. However, the size restriction on airborne pheromones does not apply to waterborne pheromones. Organic compounds composed primarily of carbon (MW=12) and hydrogen (MW=1) can be less dense than water (composed of oxygen (MW=16) and hydrogen) and hence float regardless of their size. Large organic compounds such as polypeptides, proteins, and glycoproteins are often used as pheromones in water [67]. Regardless of size, waterborne odorants must be **water soluble** to disperse effectively and be detected by olfactory receptor organs. Solubility is determined by the molecule's **polarity**—the degree to which negative and positive charges are separated at different ends of the molecule. Polar substances ionize when they dissolve in water. Some examples of marine invertebrates whose pheromones have been described are illustrated in **Figure 6.2**.

Contact pheromones, on the other hand, are often **hydrophobic** (insoluble in water) compounds, such as lipids and other large hydrocarbons. Such chemical odorants are intended to last a long time. Many insects incorporate aliphatic hydrocarbons into the waxy layer of their cuticle; these pheromones can be detected only by contact with antennae [416]. Lipids are fatty, oily, nonpolar substances that are rubbed or deposited on a surface. Lipid secretions may also be useful binders or carriers for smaller, **lipophilic** (soluble in alcohol and ether) molecules that are the active pheromones [5]. Note that the suffix *-philic* means "to have an affinity for," whereas the suffix *-phobic* means "to lack an affinity for" or "to have an aversion for." Thus a water-soluble compound is hydrophilic and lipophobic.

> ### Web Topic 6.2 *Guide to organic compounds and biosynthetic pathways*
> Examples and a brief summary of organic compounds (lipids, steroids, terpenoids, peptides, proteins) and biosynthetic pathways mentioned in this chapter.

Some pheromones are unique for a given species. Sexual selection and the speciation process itself favor species-specificity of mate attraction pheromones in particular [325, 353]. Nevertheless, common biosynthetic pathways in organisms can lead to sharing of some pheromone components. Amusing convergences can occur, such as the fact that elephants and about 140 moth species possess the same primary sex attractant component, shown in Figure 6.1A [315]. Other pheromones, such as alarm substances, are often not species-specific, because there is little cost and perhaps a benefit to transmitting a danger signal that is detected by other species. There are two main strategies for achieving species-specificity for a chemical signal. One is to use a sufficiently large single molecule with a unique structure or chemical composition. The cockroach pheromone shown in Figure 6.1C is one example [73]. Peptide and protein pheromones can also be made species-specific with a change in just one or two amino acids in the sequence [375, 418]. The second strategy is to use a unique blend of compounds. Moth pheromones usually contain a main component, such as the one illustrated in Figure 6.1A, but there are several additional minor compounds required to obtain the full behavioral response [230]. These **pheromone blends** can contain just two, or several, components, and the ratio of components is usually specific for the blend. Other chemical signals are characterized by a large number of components, where the presence and proportion of each component can vary. These large blends often encode individual identity [5, 51, 78, 89, 221]. To distinguish such multicomponent chemical signals from small pheromone blends, the terms **pheromone mosaic** [193] and **odor signatures** [417] have been proposed.

Table 6.1 summarizes the results of a comparative study of mammalian sex attractants, territory markers, alarm or threat pheromones, and individual recognition odors [5]. Note that molecular weight is lowest for sex attractants, which need to be highly volatile. The number of components is greatest for recognition and range markers, which encode individual identity. The presence of aromatic rings in a compound increases chemical stability and durability, and is probably high for recognition and range markers because these pheromones need to last a long time and must withstand degradation from water and heat. Chemicals with carbonyl functional groups are more water soluble than chemicals with carboxyl or hydroxyl groups. This feature facilitates rapid volatilization and fade-out of alarm and threat pheromones. The chemical properties of pheromones are thus adaptively designed to meet the specific temporal and spatial requirements of each signaling function.

Production mechanisms

There are four main sources for the chemicals used as pheromones: (1) well-defined secretory glands that manufacture specific products to be released to the outside of an animal's body; (2) products released with excreted waste material; (3) products derived from plants, sometimes with minor modification; and (4) breakdown products from co-opted microorganisms.

TABLE 6.1 *Chemical characteristics of olfactory signals serving different social functions*

Characteristic	Sex attraction	Recognition	Threat/alarm	Range marks
Molecular weight (daltons)	91.0	140.1	189.9	208.1
Number of components (n)	8.4	10.5	8.3	16.1
Aromatic ring present (%)	11.9	21.0	14.5	16.1
Carbonyl group present (%)	0.0	18.3	41.9	28.0

Source: After [5].

(A)

(B)

(C)

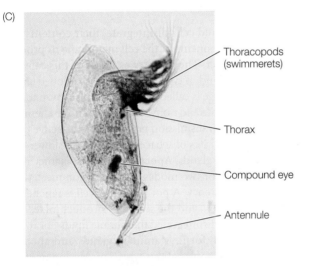

Thoracopods (swimmerets)

Thorax

Compound eye

Antennule

(D)

circles and releases a small volatile peptide sex attractant (cysteine-glutathione disulfide) that attracts multiple males. The males release a few sperm along with an egg-release pheromone blend (inosine plus glutamic acid). The female then discharges a cloud of eggs, and the males respond by releasing large clouds of sperm [163, 164, 313, 395, 426, 427]. (C) Mobile barnacle larva must settle adjacent to conspecifics in order to mate via a long penis as a sessile adult. Photo on the right shows several adult *Balanus amphitrite* (purple stripes) together with *B. eburneus* (ivory barnacle) and the bivalve mussel *Brachidontes exustus*. A large glycoprotein complex present in the cuticle of all larval stages and adults facilitates aggregation. Adults release a waterborne component that attracts free-swimming larvae from a distance. Larvae produce high concentrations of the protein in their feet (swimmerets), as shown by the purple stain in the photo of a *Balanus amphitrite* larva on the left. Larvae generate footprints as they move about on surfaces in search of a settlement location that are detected by other larvae [79, 106]. (D) The colonial sea anemone *Anthopleura elegantissima* releases a small volatile alarm pheromone from its tissue if wounded that elicits a rapid withdrawal response in adjacent individuals [181, 277].

FIGURE 6.2 Aquatic pheromones (A) *Aplysia* is a hermaphroditic sea slug that sometimes mates in chains. It attracts additional partners with a waterborne peptide pheromone comprised of 58 amino acid residues deposited in the eggs. This photo shows a mating group of 5 or 6 individuals adjacent to a cluster of eggs [292, 293]. (B)The polycheate ragworm *Nereis succinea*, shown here (male top, female bottom) in its swimming reproductive phase with developed gametes on the tail, uses two pheromones to synchronize external fertilization. The female first swims slowly in

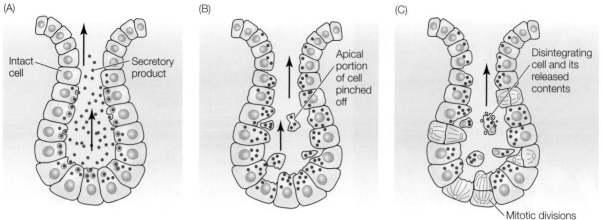

FIGURE 6.3   Exocrine gland secretion modes (A) Merocrine gland, showing vesicles containing the chemical product (red) as they connect to the cell membrane and release the chemical into the gland lumen. (B) Apocrine gland, with portion of cells pinched off and disintegrating along with the chemical product and lipid membrane. (C) Holocrine gland, showing entire cells moving into the gland lumen and disintegrating, while new cells are being produced in basal layer.

SECRETORY GLANDS   The bodies of both vertebrates and invertebrates contain numerous **secretory glands** composed of secretory cells that produce specific chemicals. All glands are formed by the ingrowth of an epithelial surface, from which develop one or more layers of columnar cells specialized for the production of pheromone components. **Endocrine glands** are internal to the body, and ductless; their products diffuse into interstitial cell spaces or directly into the blood stream for internal use as hormones, enzymes, and metabolites. **Exocrine glands**, on the other hand, may be internal, with ducts leading to anatomical surfaces, or external, on the integument itself. The functions of exocrine glands are to maintain the condition of the body covering, to release digestive metabolites, or to produce chemical communicatory signals.

Exocrine glands can be divided into three types based on their mode of secretion (**Figure 6.3**) [310]. **Merocrine** secretion is the most common type. The secretory product is collected in vesicles within the cells lining the gland. The vesicles move to the edge of the cell, where they coalesce with the cell membrane and discharge the secretory product into the gland lumen (a process called **exocytosis**). The product can consist of relatively pure pheromone. It is generally liquid in form, but can range from watery, with dissolved salts, to slimy, with protein or mucus. Secretion rates are typically fast, and release of the gland product may be under nervous control. Examples of merocrine glands include eccrine sweat glands, digestive glands, nasal mucosa glands, and glands that release volatile mate attraction and alarm substances. **Holocrine** glands produce their secretions via the death and breakdown of entire cells in the inner lining of the gland. The basal layer of the gland's epithelium continually produces new cells, which forces the old cells into the center of

the lumen. As the old cells disintegrate, their contents mix with the lipid components of the cell membrane to produce a thick, oily product. The primary example of this category is the sebaceous gland; it typically has a flask or lobed shape, and its fatty discharge is called **sebum** [8]. This sebum matrix contains both volatile and nonvolatile pheromone chemicals and modulates the transmission properties of the active pheromones. Secretion rates of sebaceous glands are slower than those of merocrine glands. **Apocrine** secretion is intermediate between the prior two modes in its secretion mechanism and product consistency. A portion of the cell is pinched off into the gland lumen, and the secretory product mixes with a small amount of disintegrated membrane lipids [145]. The product is milky in form. Mammary glands and apocrine sweat glands are examples of this type. The development and secretory activity of exocrine glands are often controlled by endogenous hormones [116].

Animal species vary in the number, size, and distribution of exocrine glands on the body. Ants certainly top the list for the number of discrete glands—the typical worker ant is a walking chemical factory, with around nine exocrine glands [177]. The combination of cooperative breeding in large colonies, dark underground nesting, and coordinated aboveground foraging by these small animals has favored a large repertoire of chemical signals. Each gland produces a different blend of chemicals and is strategically placed on the body to serve a specific function (**Figure 6.4**). For example, **poison glands** produce the venom that is used for subduing prey or defending the colony against enemies. The product may be simple formic acid, or a proteinaceous substance with neurotoxic or hemolytic properties. The secretions are stored in a large poison sac reservoir, which empties through a duct at the base of the sting. **Dufour's gland** opens into the poison gland duct but is separately controlled. It produces a mixture of volatile aliphatic hydrocarbons in most species that function to signal alarm or colony recruitment. The **anal gland** is often the source of the reproductive queen's mate-attraction pheromone [177]. **Sternal glands** are located on the underside of the abdomen and contain the main trail-marking pheromone in many species. Some species have additional

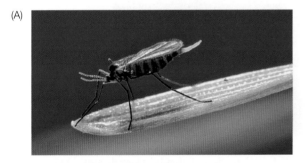

FIGURE 6.4 Ant glands Battery of glands in a worker of the Argentine ant (*Iridomyrmex humilis*): mandibular, maxillary, post-pharyngeal, thorax labial, metapleural, anal, Dufour's, sternal, and poison gland. (After [412].)

trail-marking glands on their legs or feet. **Mandibular glands** produce alcohols, aldehydes, and ketones which serve an alarm function in some species, and may be the source of the male pheromone in other species. In one group, this gland or its reservoir is very enlarged, such that grabbing the ant causes it to explode [60]. **Metapleural glands** are located in the thoracic region and contain antibiotic substances that protect the ant and the colony against pathogenic fungal spores, bacteria, and pollen grains [44]. The **postpharyngeal gland** is a reservoir for nonsoluble lipid hydrocarbons that are spread on the cuticle during self-grooming and shared among colony mates

during allogrooming and trophallaxis (food exchange). **Maxillary** and **labial glands** produce digestive enzymes.

Nonsocial insects typically have one primary gland that produces a volatile long-distance mate-attraction pheromone. Usually one sex undertakes this signaling task, termed **calling**, while the other sex possesses highly sensitive olfactory receptors tuned to the pheromone. The female is the calling sex in the majority of insects with a long-distance olfactory attractant (we discuss male-calling species below) [393]. Virgin females typically undergo a diurnal cycle of pheromone biosynthesis and release, calling for a few hours during a species-specific time period until they attract a male and mate [258, 335]. Pheromone production, release, and reception are controlled by a combination of hormones and neuroendocrine factors [97, 231, 314]. The pheromone in some species is a single distinctive volatile chemical, as in the Hessian fly (**Figure 6.5A,B**). Moth pheromones, by contrast, are more likely to be blends of several compounds. Closely

FIGURE 6.5 Calling females (A) The tiny female Hessian fly *Mayetiola destructor*, releasing pheromone from her ovipositor. (B) The key attractive chemical is a specific stereoisomer of a 14-carbon aliphatic chain acetate. (C) The pheromone blend components of the tobacco budworm moth *Heliothis virescens* and the congeneric *H. subflexa*. The two species share the same primary attractant aldehyde component (Z11-16:Ald). The minor components differ for the two species and serve to attract conspecific or repel heterospecific males. *H. subflexa* males require Z9-16:Ald and Z11-16:OH for attraction to conspecific females. *H. virescens* males require Z9-14:Ald for attraction to conspecifics and are repelled by the Z11-16:Ac *subflexa* component if it is added to the *virescens* main two-component blend. The Z number indicates the position of a C-C double bond. (D) Male *H. subflexa* on the left, male *H. virescens* on the right. (B after [130]; C after [386].)

(A)

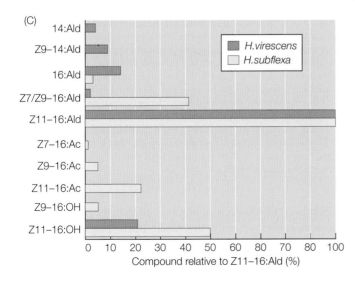

(B) Sex pheromone of *Mayetiola destructor*

OCOCH₃

(2S) – (*E*) • 10 • tridecen–2–yl acetate

(C)

14:Ald	
Z9–14:Ald	
16:Ald	
Z7/Z9–16:Ald	
Z11–16:Ald	
Z7–16:Ac	
Z9–16:Ac	
Z11–16:Ac	
Z9–16:OH	
Z11–16:OH	

H.virescens / *H.subflexa*

0 10 20 30 40 50 60 70 80 90 100

Compound relative to Z11–16:Ald (%)

(D)

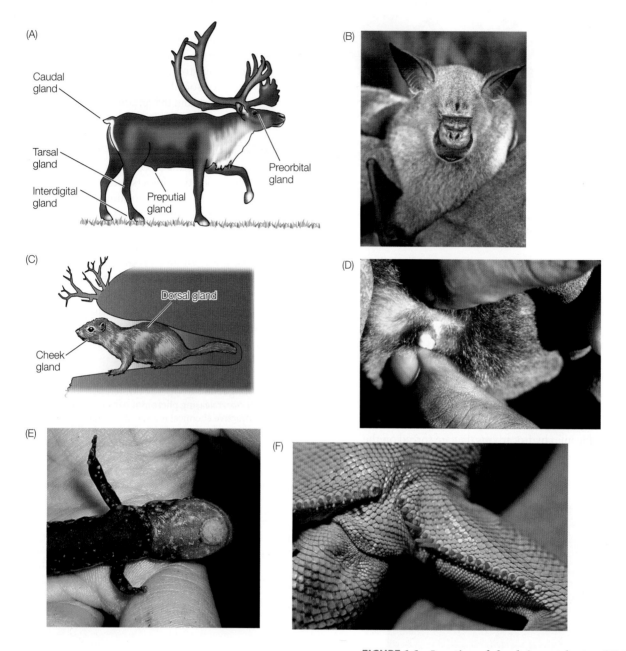

FIGURE 6.6 Location of glands in vertebrates (A) Reindeer (*Rangifer tarandus*) have five conspicuous glands: a caudal gland at the base of the tail; a tarsal gland on the inside joint of the hind leg; an interdigital gland between the hoof claws; a preputial gland on the penis; and a preorbital gland in front of the eye. (B) *Hiposiderous* bat with forehead gland. (C) Ground squirrels (*Spermophilus*) mark their burrows with the dorsal cutaneous gland and objects in their territories with the cheek gland. (D) Chest gland of *Phyllostomus hastatus* bat showing sebum carrier, spread on roost cavity and harem females. (E) Salamander males (*Aneides*) mark females during courtship with secretions from the mental gland on chin. (F) The desert iguana (*Dipsosaurus dorsalis*) and other iguanid lizards mark rocks on their territories with femoral gland secretions on thighs. (A after [358]; C after [56].)

related species may differ only in the functional group of the main component, or they may have the same set of compounds in different ratios [64]. The tobacco budworm moth *Heliothis virescens*, a crop pest in North America, and its close relative *Heliothis subflexa* in Mexico share three components of their blends but differ in other components (**Figure 6.5C,D**). The *H. subflexa* blend also contains three acetate components not found in *H. virescens*. These two species have overlapping ranges, and hybrid males are infertile, so selection should favor species isolation mechanisms. At least one of the acetate components appears to have evolved as an antagonist to repel heterospecific males [21, 386].

Terrestrial vertebrates often have one primary gland, strategically positioned on the body to serve the species' main olfactory signaling function [56]. **Figure 6.6** shows some examples. Mammals such as bats may possess a sebaceous

gland on the head, chin, or chest, so that the oily secretion can be rubbed on mates or oneself. Burrowing squirrels and rock hyraxes have a large sebaceous gland on the back that they use to mark the tunnels of their home burrow. Desert

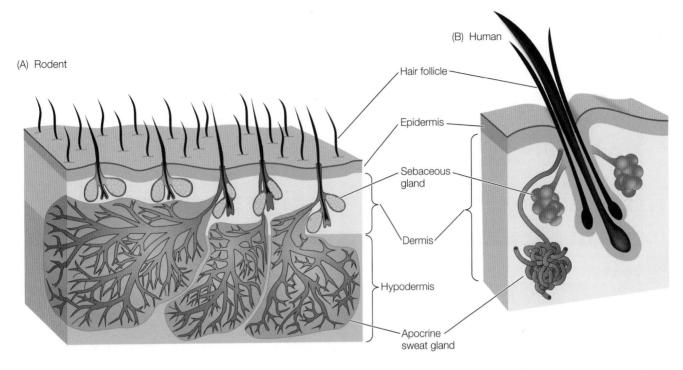

(A) Rodent

(B) Human

Hair follicle

Epidermis

Sebaceous gland

Dermis

Hypodermis

Apocrine sweat gland

FIGURE 6.7 Cutaneous glands in mammals (A) Mongolian gerbil (*Meriones unguiculatus*) cheek gland. (B) Human skin. Note the tubular form of apocrine glands and the globular form of holocrine sebaceous glands. (After [135].)

iguanas and some other lizards have femoral glands on the bottom of the thighs, again for marking their territories. Most birds have a single holocrine-type gland, called the **uropygial** or **preen gland**, located at the base of the tail. The oily secretion is rubbed on the feathers during preening. Although the main functions of the secretion are to waterproof feathers and kill bacteria, it may also serve to deter predators in some species. The possibility of olfactory communication in birds has barely been considered, but some evidence indicates a role in mate attraction in some species [23, 39, 153, 249, 327, 429]. Ungulates, in contrast, come closer to ants in having numerous (4–6) conspicuous glands that serve different signaling functions (see Figure 6.6A). The **caudal gland** located near the base of the tail produces volatile scents during the tail-raising alarm display. The **tarsal gland** marks the animal's resting sites and also releases volatile components. The **interdigital gland** leaves marks on the ground along the animal's path. The **preputial gland** contributes pheromones to the urine, and the **preorbital gland** is rubbed on upright twigs for territorial marking [148, 270, 272]. Males of many terrestrial mammalian species have especially large anal glands or preputial glands that add pheromonal substances to feces and urine, respectively [59, 328, 407, 428].

Some pheromones appear to be produced over the entire external surface of some animals. In the case of mammals, this is due to the presence of many small sebaceous and apocrine sweat glands associated with hair follicles (**Figure 6.7**). The sebaceous glands secrete sebum that maintains the condition of the skin and hair, while the sweat gland produces a watery secretion [135, 309]. Pheromones can be added to either secretion. Milk-producing mammary glands are in fact specialized apocrine sweat glands that also may secrete a volatile pheromone to guide young toward the teat [240].

Fish secrete a protective mucus layer on their surface from specialized cells in their skin rather than from glands. In one group of cichlid fish (*Symphysodon*) parents secrete mucus that is rich in proteins, which not only maintains contact with the brood of free-swimming fry but also provides food for the young. Volatile components, such as free amino acids, likely serve as the attractant [25, 75]. Some insects also have an array of epidermal gland cells that secrete pheromones through small ducts in the cuticle [311]. One interesting example is the green stink bug (*Nezara viridula*), in which mature males have highly specialized cup-shaped gland cells on their ventral side. A volatile sex attractant is released when the male is stimulated by female calling songs. The complex cell structure is believed to facilitate neural control of a final-stage chemical reaction leading to pheromone release [86]. Snakes do not have such skin glands, but they nevertheless emit critical sex pheromones from their body surface. Gravid female garter snakes (*Thamnophis sirtalis*) produce a blend of 13 long-chain (29- to 37-carbon) nonvolatile methyl ketone lipids via a holocrine-like mechanism in their skin that varies in composition to reveal their reproductive state [252, 253]. Reception occurs through tongue contact; males can distinguish receptive from nonreceptive females, and large from small females, on the basis of the ratio of lipid components [182, 347]. Under the influence of testosterone, male garter snakes add the lipid squalene to their skin lipid mixture to distinguish themselves from females [254].

The final example of diffuse surface pheromone production is the **cuticular hydrocarbon blend** on the surface of

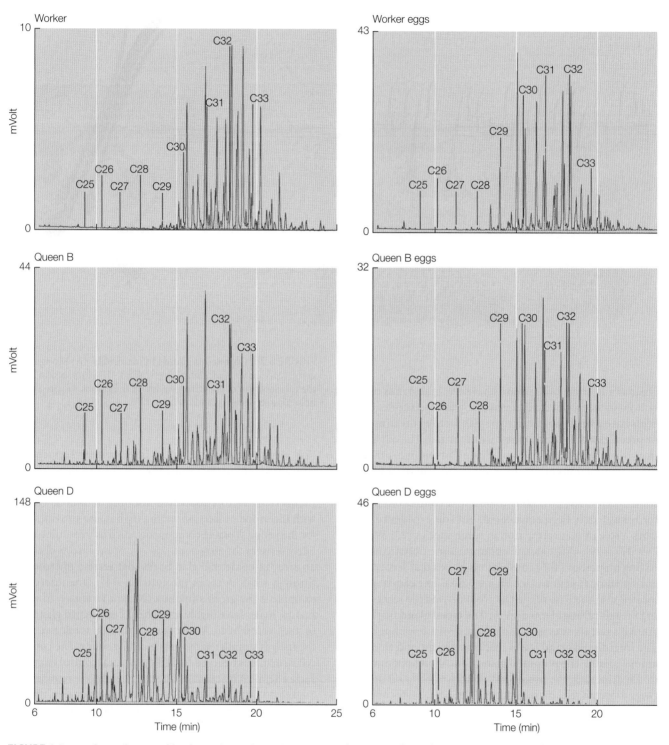

FIGURE 6.8 **Hydrocarbon profiles from the surface of workers, queens, and their respective eggs in the ant *Camponotus floridanus*** Graphs show gas chromatograms, where the relative proportion of each chemical component is reflected in its height on the *y*-axis and its molecular weight is approximately indicated by retention time on the *x*-axis; thin black lines with numbers show the retention times of *n*-alkanes with carbon chain lengths between 25 and 33 for better comparison. Queens and their eggs have more short-chain components, and the proportion of these components is greater for highly productive queens from large colonies (Queen D) compared to less productive queens from small colonies (Queen B). Workers can distinguish the eggs of workers and different queens, and will selectively eat worker eggs. Rubbing the hydrocarbon mix from a queen onto worker eggs prevents them from being eaten. (After [124, 125].)

essentially all insects. This pheromone is a mosaic or signature signal that serves a key role in individual, group, and species recognition [180]. Because the multicomponent hydrocarbon blend is species-specific, it may act as an allomone or kairomone in interspecific interactions. The insect cuticle is a matrix of chitin and protein that provides the exoskeleton to which muscles, trachea, and alimentary canal are attached. The **hemolymph** (blood) fills the interior space and exerts hydrostatic pressure against the stiff exterior. The cuticle is covered by a layer of lipids, mostly hydrocarbons (HC) that, along with waxy esters, waterproof the surface. The hydrocarbons are synthesized in **oenocytes**, cells in the abdominal integument that form a layer below the epidermal (chitin-producing) cells and are separated from the hemolymph by a basilar membrane. HC molecules are built de novo (from scratch) from fatty acid precursors in a sequence of carbon chain elongation steps. The HCs do not move directly into the cuticle, but are released into the hemolymph and then transported to target organs with the aid of large lipoprotein molecules called **lipophorin**. Target organs include the cuticle, the ovaries (where the HCs are used to cover and protect eggs), and exocrine storage glands (from which HCs are probably groomed onto the cuticular surface) [333]. Since HCs are nonvolatile, they can only be detected by contact receptors on the antennae or feet. In social hymenoptera, the HCs are shuttled to the postpharyngeal gland (or other head glands), where the blend is spread among colony members during allogrooming and trophallaxis to create a unique colony odor [239]. Recent studies have demonstrated that the reproductive status of females is superimposed on the colony signal with a subset of the hydrocarbon components [98]. Differences in HC signatures between workers and queens appear to be a direct consequence of ovarian activity on HC chain elongation [227]. The HC signature of queens is similar to the HC signature on their eggs (**Figure 6.8**), providing a mechanism by which workers can distinguish queen-laid from worker-laid eggs [124, 125]. The ability to recognize kin, nestmates, fertility status, individuals, and species via the HC signature signal has important consequences for mediating reproductive cooperation and conflict in insect societies, which we discuss in Chapter 13 [40, 150].

EXCRETED PRODUCTS   Animals eliminate a cornucopia of chemicals in their urine, feces, breath, and gills. While it is not at all surprising that a sensitive receiver can extract information about another individual's diet from **waste metabolites** in feces and urine [133], it may be less obvious how receivers might obtain information about gender, dominance status, fertility, body condition, health, and kin relatedness from excreted products alone [129-131, 408, 424]. Rodents are known for being particularly adept at signaling via their urine; but many other mammals, fish, and some invertebrates also use excreta for signaling purposes. It is relatively easy to demonstrate a relationship between excreta from a class of senders and a behavioral response or a priming effect on a class of receivers, but it is quite another matter to determine the active chemical and its source. The active chemical could be a true waste product, a pheromone produced in an internal organ and then transported to a specific exterior release site, or a secretory gland product released along with excreta.

The best example of informative waste products is the release of disabled **steroid hormone derivatives**. In vertebrates, the steroid hormones testosterone, estrogen, and progesterone responsible for gender differentiation during development, reproduction, and secondary sex characteristics, are produced in the testes, ovaries, and adrenal glands (see Web Topic 6.2). These hormones are derived from cholesterol, and in their active state they are largely hydrophobic (insoluble in water or blood). The molecules therefore must be transported to target organs in the body with specific binding proteins. Once they have been used, the hormones are not recycled, but are converted to hydrophilic derivatives in the liver via conjugation with glucuronide or sulphate ions, and then excreted into the urine by the kidneys. Conjugation requires enzyme catalysts and ATP, but this process must be less costly than breaking the molecules completely apart, because they retain their basic steroid shape. A receiver that evolves olfactory sensitivity to these **conjugated steroids** can distinguish male from female urine and detect the various reproductive stages in females. Extreme sensitivity of males to a sulfated progesterone released in the urine of females just prior to egg laying has been well described in the goldfish *Carassius auratus* (**Figure 6.9**) [349, 352]. Strictly speaking, this by-product chemical is a cue, not a signal, but for males it is a potent primer pheromone that stimulates sperm production [351]. One problem with using such a cue is that it is not very species-specific, since females of many fish species employ similar hormonal and metabolic pathways for developing and maturing their eggs and release similar steroid waste products. If females benefit by attracting only conspecific males, either species-specific blends or specially secreted steroid variants will evolve along with male sensitivity to them [353]. Several other aquatic species, including corals, have been found to use soluble variants of female-derived steroid hormones as pheromones to synchronize reproduction [110, 144, 288, 289, 303, 350, 354, 378]. Finally, excreted steroid-based pheromones can also have inhibitory effects on the reproductive physiology of same-sex individuals. Female zebrafish, *Danio rerio*, use waterborne pheromones to suppress reproduction in other females [144], and similar suppressive effects of the urine of dominant female urine on subordinate female cycling have been shown to occur in social cooperatively breeding mammals [30, 80, 169, 217, 242].

Some pheromones are species-specific chemicals produced in an internal organ and then transported with binding proteins to a watery external secretion such as urine, saliva, or sweat. The critical component of these pheromone transport systems is the binding protein, typically from the **lipocalin** family of proteins. Lipocalins are relatively small water-soluble proteins specifically structured to reversibly bind small

(A)

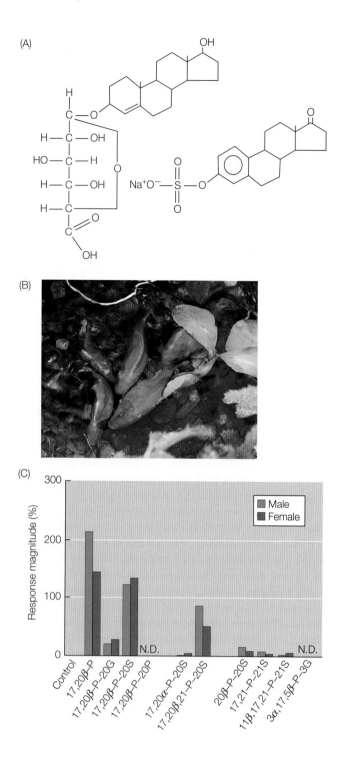

(B)

(C)

FIGURE 6.9 Conjugated steroids as pheromone cues (A) Examples of the conjugated steroids testosterone glucuronide (left) and estrone sulphate (right), the water-soluble waste elimination forms of the corresponding steroid hormones. (B) A spawning group of goldfish *Carassius auratus*; note the large gravid female in the center, who has attracted several males. (C) Olfactory response strength, measured as voltage transients from the olfactory epithelium surface, for male and female goldfish, to methanol control, unconjugated progester-one (17,20β-P), and nine conjugated 21-C gonadal steroids, indicated by the last letter G for glucuronide and S for sulphate conjugates. Not only is the receiver response strongest to the free and sulphated (17,20β-P-20S) forms of progesterone, but milt (sperm) production by the male is also highest after exposure to the mixture of these two components. N.D. indicates no response detected. (C after [352].)

Females exposed to the pheromone become primed for reproduction and receptive to the male's attempts to mount [41, 42, 295]. Another example is the female hamster mate attraction pheromone, called **aphrodisin**, which motivates males to mount. A key active component appears to be the lipocalin protein itself, which is secreted in the vaginal mucus and detected by contact. Several volatile ligands bind to this protein and may provide an airborne component to the pheromone [54, 55]. Recently, a lipocalin transporter was discovered in the sweat glands of the horse. In a search for possible volatile molecules that were able to bind to this protein, **oleamide** (a long-chain unsaturated fatty acid) turned out to be the best candidate [91]. A pheromonal function for this ligand has not yet been found in the horse, although oleamide is known to have hypnotic effects in humans and rats. This example represents an unusual case of finding the chemical first.

Urine and feces are excellent vehicles for the release of secretory gland products. As mentioned earlier, numerous vertebrates have specialized glands located near the anus, penis, or cloaca that release pheromone products during defecation or urination. Beavers (*Castor* spp.) build mud mounds and mark them with **castoreum**, a urine-based exudate from the castor sacs, and possibly with anal gland secretions as well. Receivers can distinguish the marks of different individuals. The general function of the signal appears to be territorial defense [337, 394]. The preputial gland of male rats (*Rattus norvegicus*) produces three chemical components released in urine markings; two components attract the opposite sex, and one component attracts the same sex [199]. Several species of terrestrial salamanders and lizards use scent-laden fecal pellets to mark their territories [141, 178, 190, 218]. A different twist on the mixing of gland and waste products is observed in the sac-winged bat, *Saccopteryx bilineata*. Males possess an unusual pair of sacs on the front wing membranes that have extensively folded skin but no secretory cells. Every day, the male cleans out the pocket and in it mixes secretions from the genital region and the gular (chin) gland with urine and saliva, creating a sweet-smelling perfume that he wafts at females (see Figure 6.15A) [389, 390].

The urine of rodents is especially packed with scents that impact most aspects of these mammals' social behavior,

hydrophobic molecules in an internal cavity (**Figure 6.10**). These proteins not only carry the insoluble ligand to target release sites, but they also mediate the release of the volatile pheromone [33, 136]. A good example of such a pheromone transport system is the courtship pheromone in the saliva of boars and hedgehogs. **Androstenol** and **androstenone** are similar 19-C testosterone derivatives produced in the testes that bind to the lipocalin and are delivered to the submaxilliary salivary glands. Male boars salivate copiously upon exposure to a female, and the water-soluble pheromone is thereby released.

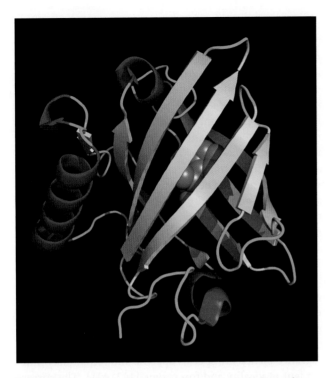

the process of presenting and "testing" cell components for their identity, unique sets of MHC peptides with 8–10 amino acids are produced that are later expelled through the kidneys and urine. These peptides are believed to form a nonvolatile chemical blend that a receiver can detect through contact. By temporarily binding with certain urinary metabolites, the MHC peptides also create a volatile blend of smaller compounds [320]. Mice can distinguish the urine of two inbred individuals differing at only one MHC locus using either the volatile or the nonvolatile fraction [356, 420]. Despite the potential use of this information for recognizing individuals, mice use MHC odors primarily for recognizing kin and selecting unrelated mates [186, 248, 298, 307, 419, 421]. Other vertebrates, including humans, also appear to use olfactory cues to select mates with maximally different MHC alleles; an obvious benefit of this preference is to assure MHC polymorphism in one's offspring [297].

Male mouse urine contains 3–4 times more protein than female urine. Most of it stems from the **major urinary proteins** (**MUP**) that are encoded from a different multi-gene family located on chromosome 4. At least six highly polymorphic loci are involved, and each individual mouse expresses 10–15 MUP types. These proteins are lipocalins produced in the liver that can pass through the kidney filter into the urine. They pick up volatile ligands in the urine and slowly release them over time [33]. As with the MHC olfactory products, MUP odors comprise both a volatile component (the ligands) that is detected at a distance, and a nonvolatile component (the proteins themselves) that can be detected only by contact. The MUP system appears to have at least three functions. One function is clearly individual recognition; males respond to a urine mark deposited on their territories by other males by approaching, investigating, and making their own marks around it [31, 74, 184, 187]. Marks by a brother with the same MUP type as the territory owner are not perceived as different. Females also leave scent marks and have their own MUP signatures. A second function is to provide current information about the status and health of scent markers. True waste products and dietary by-products can apparently bind to MUPs and are released in the volatile fraction. If a receiver can associate these volatile and more variable components of a urine mark with the stable genetically derived individual identity information in the nonvolatile component, it can obtain updated information about the condition of known scent markers [32, 185, 323]. The third function is to provide gender information. Males possess an additional class of MUPs that bind and release two adult male pheromones, thiozole and brevicomin [12, 301, 322] (see Figure 6.10). Exposure to these pheromones elicits aggression by other males and accelerates puberty and estrus in females [243, 269, 285].

The final source of odorant in male mouse urine arises from the preputial gland, which empties its product into the urethra. The active pheromone is a blend of two isomers of **farnesene**, a 15-C alkene (see Figure 6.1B). Levels of the pheromone are higher in dominant males, so it may function

FIGURE 6.10 Three-dimensional structure of a lipocalin protein Lipocalins are small soluble proteins consisting of five to ten strands that form a barrel (yellow) surrounding a cavity that is lined with hydrophobic residues that bind lipophilic molecules (purple). There are several classes of lipocalins serving different functions, including the male house mouse urinary protein (MUP), shown here with its ligand 2-(sec-butyl)-4,5-dihydrothiazole ('thiazole'). The lipocalin protein slowly releases the thiazole from scent marks the male deposits around his territory. Females detect this airborne component through olfactory receptors and learn to associate this odor with other individual-specific odor characteristics of the male. Other functions of lipocalin proteins include: retinol-binding proteins that transport vitamin A to the eyes; bilin- and carotenoid-binding proteins that produce pigment complexes; apolipid proteins that transport pheromones to sweat glands; olfactory proteins that participate in olfactory reception; and several others [34, 136, 322, 324].

from territory marking and dominance interactions to mate choice, reproduction, and parental care [52]. The house mouse, *Mus musculus*, has been studied best, and because its genome has been sequenced, we know quite a bit about the genetic basis underlying both signal production and reception. There are three main sources of odorants: major histocompatability complex odors, urinary protein odors, and secretions from the preputial gland, encompassing all of the excreted-product mechanisms mentioned above.

The **major histocompatability complex** (**MHC**) comprises about 50 polymorphic gene loci on chromosome 17 of the mouse, each one with about 50 alternative alleles. This hypervariable gene complex functions as a self–nonself recognition system in all vertebrates, so that an immune response can be mounted when foreign cells are detected. The loci encode an individually unique suite of transmembrane proteins that reside on all cell surfaces throughout the body. In

as a status indicator. It stimulates aggression in other dominant males and is suppressed in subordinate males, while being highly attractive to females [165]. Given the great complexity of olfactory information contained in urine from all of these sources, it is not surprising to find that when mice encounter marks from other individuals on their territories, they do not urinate on top of them as many other mammals would [194], but instead surround them with many small marks of their own so as not to mingle the scents [185].

PLANT-DERIVED COMPOUNDS   Plants possess different metabolic pathways from those found in animals, and therefore can synthesize some compounds that animals cannot (see Web Topic 6.2 for examples). As the base of the food chain, plants provide the basic dietary compounds required by animals. Plants interact chemically with animals in numerous other ways as well (see Chapter 16). For example, toxic or distasteful secondary compounds may be added to green leaves to reduce consumption by herbivores, and attractive scents along with nutritious fruit, pollen, and nectar evolve to facilitate symbiotic associations with animal pollinators and seed dispersers. Animals, including humans, exploit plant chemicals in many ways, one of which is to acquire, sequester, or modify certain **plant secondary compounds** as precursors for pheromones. This strategy has been best documented in **phytophagous** (plant-eating) insects—particularly some beetles, flies, moths, and butterflies—and in one group of bees.

The specialized association between Euglossine bees and orchids is both complex and fascinating. Orchid flowers are notorious for mimicking the female pheromone of some hymenopteran species to lure males that receive pollen while fruitlessly attempting to mate with a deceptive female-bee-like petal (see Figure 16.24). However, some orchid species lure insect pollinators with honest rewards. In the case of orchids specializing in the bright metallic Euglossine bees, the orchid offers the males floral scents that they exploit to produce a mate attractant (**Figure 6.11**). The males collect fragrant terpenoid and aromatic compounds by applying labial gland lipids to the petal surface, which acts as both a solvent and carrier. They scrape and pack the mixture into large pouches on their hind legs using stereotyped leg movements [62, 334]. A floral blend is made by gathering scents from a variety of natural sources, including other flower species, dead wood, and feces. The male bees defend small territories in the forest understory that contain perch stems for displaying and mating. They display with hovering flights or wing buzzes depending on the species. Fragrances are actively released during these displays in which the chemical contents of the pouch are transferred to tufts on the legs, followed by dispersal of the scent by the wing movements [29, 123]. The fragrance blends are species-specific and remarkably independent of locality and forest type [122, 431]. The current hypothesis for the evolution of this signaling behavior is that fragrances are difficult to acquire and may indicate some aspect of male quality to females [62, 121, 432].

A very different route to plant-derived pheromones is found in certain moths and butterflies that have subverted the plant's secondary compounds intended to deter herbivory [283]. Alkaloids are one such class of herbivore-deterring compounds commonly found in plants and fungi. They are hydrophobic, multiple-ring, nitrogen-containing compounds with a bitter taste and toxic pharmacological effects (morphine, quinine, strychnine, codeine, caffeine, cocaine, and nicotine are some examples). The caterpillar larvae of some Lepidopteran species have evolved special mechanisms to detoxify these plant compounds and thereby have become specialists on particular host plants. The moth *Utetheisa ornatrix* not only detoxifies the pyrollizidine alkaloids in its host plant *Crotalaria*, but the larvae sequester the toxins and use them in adulthood to protect themselves from predators. Adult

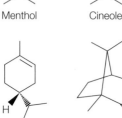

Myrcene   Menthol   Cineole

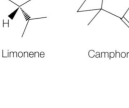

Pinene   Limonene   Camphor

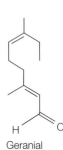

Geranial   α–Farnesene

FIGURE 6.11   Terpenoids and Euglossine bees A selection of the terpenoid essential oils collected by Euglossine bees from orchid petals and from other flowers, wood, and feces. (The term *essential* indicates that the oil is the fragrant essence of the plant from which it is extracted, not that it is indispensable.) The chemicals are scraped up using special brushes on the forelegs. The chemicals are transferred from there by rubbing the brushes against combs on the middle legs, and these combs are then pressed into grooves on the dorsal edge of the hind legs. The chemicals are squeezed past the waxy hairs, which block the opening of the groove, and into a sponge-like cavity inside the hind tibia, visible in this photo of a male *Euglossa imperialis*. (After [62].)

(A)

(B)

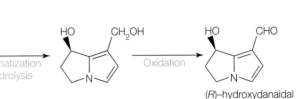

Monocrotaline

(R)–hydroxydanaidal

FIGURE 6.12 **Pyrollizidine alkaloids and *Utetheisa*** (A) *Utetheisa ornatrix* moth being cut out of a spider web in which it was caught because it is distasteful to the spider. (B) The larval host plant, *Crotalaria*, produces the pyrollizidine alkaloid monocrotaline, which deters feeding by most other insects except *Utetheisa*. The adult male *Utetheisa* converts a portion of his sequestered alkaloid (blue) into hydroxydanaidal, a courtship pheromone that females use to select mates. (After [118].)

female moths also divert the alkaloids to their eggs, and they acquire additional toxins for this purpose in spermatophores received from males during mating. Mating competition among males has led to the evolution of a male pheromone called **hydroxydanaidal** derived from the alkaloids (**Figure 6.12**). Female preference for males with more intense pheromonal signals provides a direct benefit in the form of more alkaloids for the eggs and better offspring survival [35, 96, 189].

There are numerous other examples of phytophagous insects that appear to use modified plant compounds for their mate attraction pheromones [37, 168, 215, 219, 282, 283, 317, 397]. Males commonly employ this strategy for both long-distance calling and short-range courtship. In contrast, when females are the calling sex, the main pheromone components do not resemble plant compounds and are clearly built de novo via basic metabolic pathways. In some cases, it is not clear whether male callers truly sequester and modify a host plant compound, or whether they have evolved a method of synthesizing a compound that mimics the host plant. For example, the male long-distance sex and aggregation pheromones of several bark beetle (Scolytid) species were initially believed to be acquired from plant terpenes, detoxified in the gut, and released as a waste product pheromone. But definitive studies have now shown that the beetles can synthesize the pheromone de novo in specialized gut-lining cells from basic acetate building blocks through an alternative metabolic pathway [146, 155, 342, 343] (see Web Topic 6.2). The interesting question remains: why do male callers, but not female callers, use or mimic the host plant volatiles? The most common reproductive strategy among phytophagous insects is for females to locate an appropriate host plant for oviposition (egg laying) using cues from plant volatiles, and then to call in males with a long-distance species-specific pheromone. When males emerge first or find isolated patches of host plants, they benefit by advertising their resource via the emission of a pheromone related to the plant volatiles to which females are already sensitive [219, 260, 302]. It appears that in such male-calling species, females usually require both the male pheromone and the plant odors, which guarantees that they will find both a mate and an oviposition site [219, 317]. Many short-distance courtship pheromones in males are also similar to plant compounds; they may have evolved in response to the female's previously evolved mechanism of priming pheromone and egg production with olfactory host-plant cues [20].

BACTERIA-DERIVED VOLATILES   Microorganisms such as bacteria and fungi provide another source of volatile chemical odorants in some animals. Bacteria can be used to break down large nonvolatile molecules into smaller volatile odorants. Such **bacterial breakdown products** may yield pheromone blends and signature odors. Species that co-opt bacteria for this purpose evolve small pockets or cavities of skin that retain moisture or urine to provide an optimal growth chamber for bacteria [7]. Bacteria also need to be provided with some type of nutritional substrate to break down. For example, **guaiacol** (**Figure 6.13A**) is a key component of an aggregation pheromone in desert locusts (*Schistocerca gregaria*) produced in the gut by bacteria. The bacteria occur in the lining of the hindgut where the fecal pellets are formed, where they break down a plant-derived precursor, vanillic acid, with a decarboxylation process [101]. In female mammals, volatile fatty acids produced by vaginal bacteria are believed to affect male sexual responsiveness. Female rats whose bacterial flora was killed by injections of an antibiotic were not attractive to males, compared to females with an intact flora [262]. Chimpanzee females produce several fatty acids in vaginal secretions (**Figure 6.13B**), which vary among individuals; but only one, isobutyric acid, increases during the ovulation period [257]. Microbes are known to reside in a variety of other mammalian glands, but the specific volatile products and their functions are not well known [273].

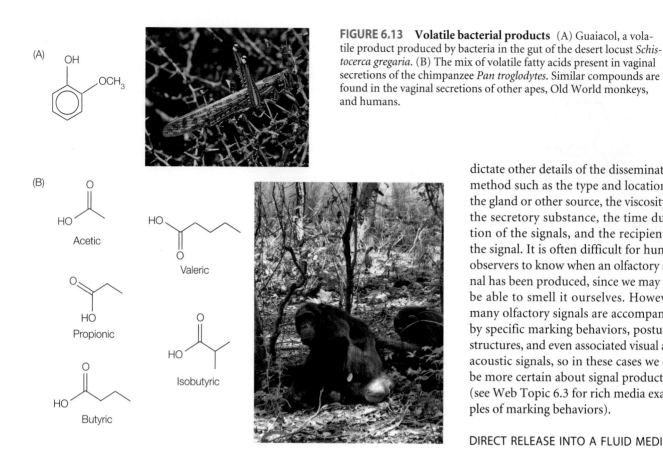

FIGURE 6.13 **Volatile bacterial products** (A) Guaiacol, a volatile product produced by bacteria in the gut of the desert locust *Schistocerca gregaria*. (B) The mix of volatile fatty acids present in vaginal secretions of the chimpanzee *Pan troglodytes*. Similar compounds are found in the vaginal secretions of other apes, Old World monkeys, and humans.

dictate other details of the dissemination method such as the type and location of the gland or other source, the viscosity of the secretory substance, the time duration of the signals, and the recipient of the signal. It is often difficult for human observers to know when an olfactory signal has been produced, since we may not be able to smell it ourselves. However, many olfactory signals are accompanied by specific marking behaviors, postures, structures, and even associated visual and acoustic signals, so in these cases we can be more certain about signal production (see Web Topic 6.3 for rich media examples of marking behaviors).

DIRECT RELEASE INTO A FLUID MEDIUM Directly releasing odorants into an air or water medium has the potential to provide fast delivery of chemicals, but the ability of the sender to target specific receivers and control the duration and distance of signal transmission varies greatly depending on how the chemical is produced and released. The simplest method of direct odorant release is to evert the gland or otherwise expose glandular tissue to the air or water medium, so that the chemical evaporates gradually from the surface and is carried away by ambient currents. This strategy is commonly found in calling insects such as moths, beetles, praying mantises, and hymenoptera. The sender often assumes a **calling posture**, in which the wings are spread or moved aside and the abdomen is raised, curled, or extended to expose the gland tissue (see Figure 6.5A) [27,

Dissemination methods

Animals use a variety of methods to release and couple odorous chemicals into the medium. These methods can be categorized along two axes: (1) where the chemicals are placed, and (2) how elaborately they are released (**Table 6.2**). The *where* category can be subdivided into direct release into a **fluid medium**, spreading odorants onto the **sender's own body** surface, and deposition of secretions onto **other solid surfaces**. The *how* category includes simple **passive exposure**, **piggy-backing** on another activity, and **specialized release behaviors**. The boundaries between these categories are fuzzy and graded. The function of the chemical signal will

TABLE 6.2 *Categorization of scent dissemination mechanisms based on where chemicals are placed and how elaborately they are released*

Complexity of release	Placement		
	Into fluid medium	On own body	On solid substrate
Simple exposure	Expose gland tissue	Ooze onto surface or hairs	Burrow/nest odor, footprints
Piggy-backed on another activity	Enurination	Grooming	Dung pile
Specialized activity	Spraying, self-generated current	Self-marking	Deposit onto substrate or conspecific

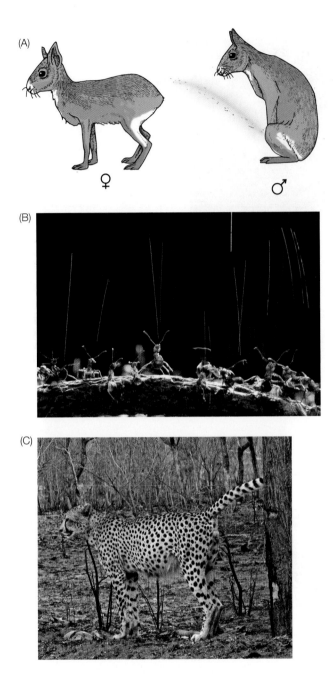

(A)

♀   ♂

(B)

(C)

FIGURE 6.14 Dissemination by spraying (A) Male mara *Dolichotis patagonum* enurinating on his female mate. (B) Wood ants (*Formica rufa*) spraying formic acid, which serves as both a defensive chemical and an alarm pheromone. (C) Male cheetah (*Acinonyx jubatus*) spraying urine against a tree.

receivers would be best categorized as a piggy-backing strategy, in which the urine stream is used as the delivery vehicle. Female fish release estrogen-containing urine in the vicinity of males, often pulsing the release to produce high-concentration bursts of odorant [10]. Female mammals also produce urine in the presence of males as a means of signaling their reproductive state. Both sexes of the mara (*Dolichotis patagonum*), a colonial and monogamous South American rodent, enurinate (spray urine) on conspecifics; females in particular deliver a rear-directed spray to repel the advances of unwanted males (**Figure 6.14A**) [290]. Other cases of spraying entail specialized glands and behaviors that can produce intense streams of liquid substances directed toward close targets—often predators or enemies. The bombardier beetle (Carabidae), for example, combines hydroquinones and hydrogen peroxide in a reaction chamber to make a fiery blast of steam [95]; the skunk (*Mephitis*) sprays a foul-smelling mixture of sulfuric compounds from its anal gland [415]; the termite soldier (*Nasutitermitinae*) exudes a string of toxic glue from its head gland [117, 308]; and harassed cephalopods release ink clouds into the water surrounding them [161]. These noxious chemical mixtures are known as defensive allomones. They can also be used to deliver chemical signals to conspecifics. For example, the formic acid spray of *Formica* ants is a highly volatile chemical that is used for both defense and the recruitment of colony mates to the source of danger (**Figure 6.14B**) [412].

Another specialized method of disseminating odorants directly into the medium is to release an odorant into a **self-generated current**. A variety of behaviors and movements enhance the transmission of chemicals by spreading the odorants or creating current flows. Winged animals that can hover, such as some bats (**Figure 6.15A**), the Euglossine bees, described in the previous section, and some moths have glands or scent-permeated hairs placed strategically so that the scent is wafted toward a target receiver. Aquatic animals use locomotory appendages (lobsters) or tails (newts and fishes) to waft pheromones toward receivers (**Figure 6.15B,C**) [15].

DEPOSITION ONTO SENDER'S OWN BODY Rather than emit chemicals directly into the medium, senders may first deposit or spread the chemicals onto their own bodies. Specialized surfaces or structures may be employed to maximize subsequent evaporization of the volatile components. One strategy is to use areas of **bare skin** as an evaporization site. Oozing secretions from glandular tissue onto adjacent skin surfaces is observed in mammals, such as from the temporal gland to facial skin in male elephants, from glands in the vagina to

143, 173, 179, 198, 213, 268, 299]. The function of calling is long-distance mate attraction, so chemical release is relatively slow and continues for several hours so that distant receivers can detect the scent and fly toward the source. In at least one moth, the sender actually produces a fine aerosol of liquid droplets and pumps the abdomen repeatedly, resulting in one of the highest known release rates for such a signal [214]. Some fish have fin glands that slowly release chemical odorants into ambient currents [279]. Senders obviously have little or no control over the directionality of current-borne chemical signals when they use environmentally available currents in this manner.

Much more control can be achieved in a short-distance signal by the sender forcefully **spraying** a liquid product. Cases of the release of odorant-laden urine toward specific

(A)

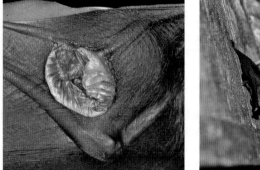

(B)

FIGURE 6.15 **Dissemination by current generation**
(A) *Saccopteryx bilineata* bat male (right) performs a hover display in front of female while opening the wing sac (left) and wafting scent. (B) During underwater courtship in the newt *Lissotriton vulgaris*, the male (left) wafts pheromone from a large abdominal gland toward the female by flicking the tip of his tail. (C) Female lobster *Homarus americanus* (right) blows her gill current containing pheromones into the male's shelter. The male (left) draws water through his shelter and fans it out with his pleopods to advertise his mating status. (C after [13].)

(C)

labial skin in primates and canids, and from lymph glands to the axillae (armpits) in humans. Fish, amphibians, and reptiles also have skin glands from which pheromones are slowly disseminated [8, 273, 316]. **Hair dispersal** is a second frequently used strategy for disseminating scents (**Figure 6.16**). Hairs increase the surface area for vaporization of both liquid and oily chemical secretions, and their structure is often modified to maximize their effectiveness. For example, hairs may have a scaly surface to hold more material, or they may have a hollow or tubular structure to help "wick" the pheromone down the length of the hair. These specialized hairs are called **osmetrichia**. Mammalian sebaceous glands are often associated with stiff, elongated, or dark-colored hairs [271]. Moth and butterfly species that have evolved male courtship pheromones possess osmetrichia or **hair-pencils** located on the abdomen, legs, or wings. The hairs are normally hidden

within a glandular pocket but are everted and fanned out during courtship; the male places himself in front of the female so that the pheromone reaches her sensory antennae.

Animals spread semiochemicals on their own bodies as part of normal maintenance grooming, an example of piggybacking scent dissemination onto another activity. Rodents have particularly large **Harderian glands** behind their eyeballs; these glands secrete lipid material into the lacrimal duct, and the secretion is then groomed over the entire body. The chemical composition of the secretion varies as a function of age, sex, and season [61, 208, 345, 371]. Male meadow voles *Microtus pennsylvanicus* engage in more grooming in the presence of females, and females are more attracted to males that groom more, suggesting a clear communication function for the grooming behavior [132]. Similar observations have been made for the hedgehog (*Erinaceus europaeus*)

FIGURE 6.16   Dissemination with hairs
(A) Crested auklets (*Aethia cristatella*) produce a tangerine-like odor from specialized wick feathers in the interscapular area, which is spread onto the nape feathers during ruff sniffing interactions between pairs and groups of birds. (B) With his winter coat, the male saiga antelope (*Saiga tatarica*) develops a hair tuft below the eye, which is believed to disseminate secretions from the preorbital gland. (C) SEM of hair structure showing cuticular scales from the tarsal gland hairs of black-tailed deer (*Odocoileus columbianus*, left) and a trough-shaped hair from the ventral gland of the Mongolian gerbil (*Meriones unguiculatus*, right) [271]. (D) Male moth (*Euploea core*) everting his bright yellow hairpencils during courtship.

[90]. We have already seen the important role of grooming hydrocarbon secretions from the postpharyngeal gland or its equivalent onto the exoskeleton in essentially all social insects as a colony identification marker. Although birds are not generally known for communicating chemically, they do groom uropygial gland secretions onto their feathers, and some species include semiochemicals in the secretion that have a communication function. For example, the crested auklet (*Aethia cristatella*), a monogamous seabird, possesses a distinctive tangerine-like scent closely associated with courtship and neck-sniffing behaviors [104, 152]. Several songbird species in the genus *Pitohui* native to New Guinea deposit a highly toxic alkaloid onto their skin and feathers as a means of predator deterrence [111]. In a similar vein, many birds and mammals groom toxic chemicals derived from other

species onto their body surfaces for the purpose of deterring predators, ectoparasites, or microbial pathogens. For example, monkeys spread benzoquinones from millipedes on their fur to repel mosquitoes; canids roll in decomposed material; hedgehogs chew plant material and spread it on their fur; and birds stand in ant colonies to fumigate their plumage. The strategy of acquiring the odors of other species (allomones) is called **self-anointing** [402].

There are many examples of more specialized self-marking behaviors by which senders transmit their chemical signals to conspecifics. Ring-tailed lemurs (*Lemur catta*, see Figure 6.17C) rub the secretions from their inner wrist glands onto their long tails, and then waft the tail in the direction of rivals to present a combined visual and olfactory threat [195]. Males of many mammalian species spread the same glandular secretions on their fur that they also deposit on their territories. In the hartebeest antelope (*Alcelaphus buselaphus*), males rub their preorbital glands onto their shoulders during the rut season [148]. In the closely related topi (*Damaliscus lunatus*), males urinate in a puddle and then kneel down to dredge their horns and face through the mud. Analogous self-marking behaviors have been described for other antelope species, several rodents, and the aardwolf (*Proteles cristatus*). The proposed reason for this behavior is that territorial intruders can learn the odor of the territory owner, so when they encounter him they can avoid costly fights [149].

DEPOSITION ONTO SOLID SUBSTRATES   Scent gland secretions can be deposited onto other solid surfaces such as environmental objects and conspecifics. In contrast to the spraying, current-generating, hair-evaporating, and self-marking methods of scent dissemination described above, where there is a close association in time between release of the odorant by the sender and its reception by the target, there may be a significant time delay between sending and receiving for **deposited marks**. These marks are intended to last a long time and the sender may be quite far away when the signal is detected. As with the other placement methods, substrate marks can be deposited via simple exposure, piggy-backed onto other activities, or applied with specialized marking behaviors.

Passive deposition techniques include leaving olfactory chemicals in footprints and marking burrows and nests by moving around and lying down in them. In some species, strategically placed glands ensure that marks are deposited during these routine activities. For example, the foreleg glands of impala (*Aepyceros melampus*), four-horned antelope (*Tetracerus quadricornis*), and pigs are located in positions that press against the ground when the animals lie down, and are likely to scent-mark the bedding site [148]. Burrow-dwelling *Spermophilus* ground squirrels have a dorsal field of about 60 apocrine glands extending down their backs, which passively mark their burrow as the animals move around inside (see Figure 6.6C) [159]. An interesting example of passively deposited group odor has been described in the desert isopod (*Hemilepistus reaumuri*). Each monogamous pair and their large brood of 50–100 offspring inhabit and defend a family burrow. The cuticular hydrocarbons from all family members are rubbed onto the burrow walls as they move through the tunnels, and each member consequently acquires the composite family hydrocarbon blend [232].

The range of animal species that leave **footprints** for conspecifics to follow is very diverse. Special glands, such as the interdigital glands on the feet of ungulates, have evolved expressly for this purpose (see Figure 6.6A). A major cost of leaving such trails is that the signaler can be followed by eavesdropping predators. In the case of ungulates, males follow female tracks during the rut (mating season), and females also use the olfactory tracks of other females to maintain a loose aggregation or to increase dispersion [221, 270, 414]. The galago (*Galago alleni*), a nocturnal arboreal primate, urinates on its feet to leave olfactory marks on the major branch routes of its territory [72]. Ant scouts deposit olfactory footprints from hind tarsal glands so they and their nestmates can find their way back to the colony after foraging trips [36, 176]. Snails leave mucus trails by default, but these can provide very useful information for both conspecifics and predators [403]. In a study of the predatory rosy wolfsnail *Euglandina rosea*, wolfsnails were found to readily track the

slime trails of both prey snails and conspecifics. The trails contain a nonvolatile glycoprotein component and a volatile component, each of which independently provides directional information. A left-right chemical asymmetry in the trail is believed to be responsible for this directional information [344] (discussed further in Figure 6.31B).

The most common example of mark deposition that is piggy-backed onto another activity is the addition of gland secretions to feces or urine to mark territorial boundaries. This method of territory defense is very common in mammals. The feces are typically deposited in consistent and visible places, such as dung piles and latrines, making them easier for receivers to locate, approach, and investigate [6, 46, 47, 273, 360, 416]. Male cats (Felidae) are well known for spray-marking on vertical structures in their territories (see Figure 16.14C). The urine contains strong-smelling sulfur and amine compounds, as well as lipids and waxy esters that fix and retain the scents for long periods of time [170, 304, 305]. Several terrestrial salamander species dot their territories with pheromone-laden fecal pellets as a highly efficient way to declare ownership [141, 178, 190]. Territory marking is rare in invertebrates and aquatic animals, but is well-developed in one group of ants, the African weaver ants (*Oedophylla longinoda*) [177]. Territory boundaries are marked with a matrix of colony-specific pheromones in fecal droplets. The ants defend their territory vigorously, and when fighting on an area marked with their own colony marks, they enjoy an owner advantage.

Direct deposition of gland products as territorial and dominance status markers is a very important and conspicuous behavior in terrestrial vertebrates, especially in mammals and lizards. In most cases, gooey glandular secretions are rubbed or pasted on conspicuous objects on the territory such as trees, stumps, twigs, and rocks; along boundaries and trails; and around dens and burrows [273]. **Figure 6.17** illustrates some of the special marking postures used. How do senders make marks that last, yet still provide useful olfactory information when a receiver eventually does approach it? We have already hinted at the concept of **keeper substances**, nonvolatile carrier materials that fix the more volatile active components to a surface and retard their evaporation or degradation. The lipids and waxy esters in the urine of cats and in the anal gland secretions of beavers appear to help retain the volatile components of these territorial scent marks. Lipocalin proteins that bind and transport pheromones to glands and urine also control the release of volatile pheromones and extend their longevity [183]. The chin gland secretion of rabbits (*Oryctolagus cuniculus*) used in territorial marking is also mostly composed of proteins. When a male becomes dominant, a new component, 2-phenoxyethanol, is added to his secretion. This chemical is not the active pheromone and is not even detected by the rabbits, but instead operates as a fixative that makes his marks more durable [166]. Mammalian sebaceous secretions contain primarily sebum, a nonvolatile mixture of lipids that serves as a carrier. Sebum interacts with the volatile pheromone in a

(A)

(B)

(C)

FIGURE 6.17   **Deposited marks with visual component**
(A) Male Thomson's gazelle (*Eudorcas thomsonii*) depositing dung and urine at a territorial dung pile. (B) Klipspringer *Oreotragus oreotragus* male (right) depositing a glob of scent from his preorbital gland onto a twig. (C) Ring-tailed lemur (*Lemur catta*) rubbing anal gland secretions on a vertical post. Glands on the inside of the wrist, visible in this photo, are used to spread scent on the tail, which is waved toward a rival in aggressive encounters.

remarkable way that optimizes pheromone evaporation. If the carrier and the pheromone are both polar substances, or if they are both nonpolar substances, then carrier and pheromone will bind strongly and almost no pheromone will be released. Sebum is actually weakly polar, so it releases polar pheromone molecules at an intermediate rate. In a humid environment, water molecules will compete for the hydrogen bonding sites on the surface layer of the sebum/pheromone mixture, and the pheromone will rapidly evaporate [318]. Therefore, if a receiver approaches such a mixture and sniffs, licks, or otherwise moisturizes the surface, a puff of volatile pheromone will be immediately released. Sebum thus retards the loss of pheromone from a deposited mark, keeping the detection zone very small, but facilitates the release of scent when a receiver is there to investigate.

Gland products may also be directly deposited onto conspecifics. Male insects and mammals are known to mark females as a chemical mate-guarding strategy. In insects, males transfer substances with the ejaculate that enhance

fecundity and inhibit remating of the female. In some cases the spermatophore acts as a plug, and in other cases peptides in the ejaculate serve this function [53, 82, 263, 348, 379]. In *Phyllostomus hastatus* bats, males mark both the defended roost site and the females in their harem with chest gland sebum (see Figure 6.6D); it has been suggested that these marks serve as a cue for synchronizing reproduction among the harem females [306]. Such marking behaviors by male bats may be responsible for the ability of individuals to distinguish colony mates from non-colony mates in other species [38]. Another context for conspecific marking is maternal labeling of offspring. This behavior is most important in species that combine their young in large crèches. Female Mexican free-tailed bats (*Tadarida brasiliensis*) live in huge colonies and reproduce synchronously, leaving their young pups in large clusters within the cave while they forage at night. Upon their return, each must find her own offspring in a mass of hungry begging young. These mothers use multiple cues to find and identify their own pups, including auditory cues, but it is believed that they also label their offspring with secretions from a muzzle gland [151, 238, 259].

Web Topic 6.3 *Marking behaviors and displays*
More examples of the behaviors involved in producing and disseminating scents.

Transmission of Chemical Signals

As mentioned earlier in the chapter, if chemical signals are to move the distance from sender to receiver, as opposed to the receiver approaching to contact a deposited mark, they can do so in two ways: via diffusion and via current flow. Both of these processes can occur at the same time. However, one or the other will generally predominate, depending on the environmental context. If there is no or very little current flow in the environment, then diffusion will be the prevailing cause of movement. If there is current flow, it will predominate and the effects of diffusion will be negligible. Flows can exhibit different patterns depending on their speed and the medium. The relative importance of diffusion versus different types of flow can be thought of as occurring along a gradient that depends on the velocity of the flow, the spatial scale, and the viscosity of the medium. This gradient can be expressed quantitatively by the dimensionless **Reynolds number**, *Re*:

$$Re = \frac{UL}{v} = \frac{\text{Inertial forces}}{\text{Viscous forces}}$$

where U is mean flow velocity, L is the characteristic length or distance, and v is kinematic fluid viscosity. The Reynolds number was developed in part to quantify the point where a flow changes from a **laminar** pattern to a **turbulent** pattern, so it expresses the ratio of inertial forces to viscous forces in the medium [87, 99, 114, 287, 369]. For more details on this derivation, see Web Topic 6.4. For the general discussion that follows, a heuristic understanding of this range of variation in *Re*, illustrated in **Figure 6.18**, will suffice. When *Re* is very small (<1) because the relevant distance is short and flow is negligible, molecules will move via the process of diffusion. When *Re* values are intermediate because the medium is viscous or the flow rate is low, laminar flow will occur. When *Re* values are high, the flow will be turbulent. These three conditions present very different transmission contexts for chemical signals in terms of the speed and distance of molecular movement and the spatial pattern of the molecules within the stimulus field. These transmission properties are described in this section. In the signal processing section later in the chapter, we will take up the strategies receivers use to navigate toward the sender under these different conditions.

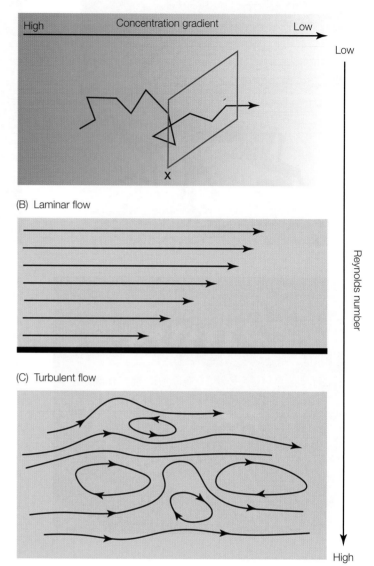

FIGURE 6.18 **Three types of molecular movement as a function of Reynolds number *Re*** (A) Diffusion occurs at very low *Re* in the absence of medium flow. Molecules move along a random path from regions of higher to lower concentration. The rate of diffusion is given by the number of molecules passing through a window (red) at point X along the gradient. (B) Laminar flow occurs at intermediate *Re*, where flow velocity is low relative to viscous forces. Characterized by smooth, constant current flow with layers moving parallel to each other. Velocity of flow is slower close to a boundary as indicated here by shorter arrows. (C) Turbulent flow occurs at high *Re* when flow velocity is high and inertial forces predominate; characterized by random eddies, vortices, and other flow fluctuations.

Web Topic 6.4 *Chemical transmission models*
Quantitative models for molecular movement under the conditions of diffusion, laminar flow, and turbulent flow attempt to describe the temporal and spatial characteristics of the active space for chemical signals and cues.

Diffusion

Molecules move down their concentration gradients. This is not because of some form of repulsion among similar molecules, but to unequal numbers of molecules moving in each direction. If we set up a sampling window perpendicular to the concentration gradient, more molecules will, on average, cross the window from the high to low concentration sides

FIGURE 6.19   Diffusion model of the spread of a single puff Concentration of odorant released in a single instantaneous puff at time t_0 and subsequent time periods is indicated by color density. The dashed circle shows the outer boundary of the active space that is above the receiver's detection threshold. As molecules diffuse away from the source point, the active space initially increases, reaches a maximum radius at t_2, then shrinks and is completely faded below threshold by time t_4 [43, 410, 411].

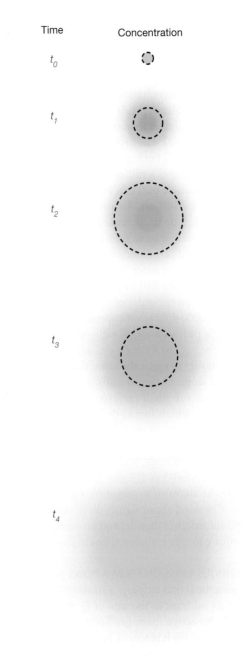

than the reverse, because there are more of them on the high concentration side available to do so. This produces a net movement in one direction across the window. This process is called diffusion (see Figure 6.18A). The rate of diffusion depends upon the steepness of the concentration gradient and the ease with which a particular type of molecule moves in a particular medium. The slope of the concentration gradient can be described as the change in concentration of a population of molecules over a given distance. The **diffusion coefficient**, D, is a measure of the ease of movement of a molecule in a medium; it depends on the size of the molecule, how the molecule interacts with the medium, and how the medium molecules interact with each other.

We can predict the concentration of diffusing molecules at any time and at any distance away from the source of the molecules for a variety of emission strategies and ecological conditions. If we know the amount of chemical released and the minimum threshold of detection by the receiver, the stimulus field can be described. The time course of chemical spread and fadeout, as well as the distribution of molecule density within the stimulus space, can also be precisely computed (see Web Topic 6.4). However, we can easily visualize what is taking place using graphic techniques.

SINGLE PUFF   Suppose a single **instantaneous puff** of odorant is released quickly. We can monitor the concentration of the odorant at various times after the release and at various distances or radii from the source. Q is the number of molecules released in the puff. If the puff is emitted away from any boundaries (such as the ground), it will diffuse out symmetrically in all directions. If the emitter is on the ground, the same number of molecules must diffuse into half as much space, so the concentration is then twice as high for any point in space or time. Given a receiver detection threshold of K molecules/cc, the zone of detection changes in time as shown in **Figure 6.19**. At t_0, all the odorant is concentrated at the source. Shortly thereafter, at t_1, some of the molecules have diffused outward. A sphere (if the animal is near no boundaries) or a hemisphere (if the animal is on the ground) defines the **active space** within which a receiver would be able to detect the signal. Within the active space there is a uniform concentration gradient increasing toward the center. By t_2, the radius of the active space has increased further. However, the same number of molecules are spreading over a larger and larger volume as diffusion progresses. Eventually molecules leave the current boundary of the active space faster

than they arrive from the release point. When this happens, the radius of the active space begins to decrease again. In the example, we see that the sphere is smaller by t_3 than it was at t_2. By t_4 the active space has shrunk, and the concentration of odorant is less than K everywhere; the puff has diffused away.

The maximum size of the active space, r_{max}, depends on the ratio of molecules released to the detection threshold concentration raised to the one-third power, or $(Q/K)^{1/3}$. The time to reach r_{max} depends on $(Q/K)^{2/3}$ and the reciprocal of D. Since both the maximum radius and the time to get there depend on Q/K to a power less than 1, it takes quite a large change in Q/K to make a big change in either the maximum radius or the time to reach that maximum. In fact, to double the radius of the active space, an animal must increase Q/K

by a factor of 8. Transmission of a single puff over a significant distance will require releases of very large numbers of molecules or a very high degree of receiver sensitivity. We can also calculate the time at which there is no longer any point in space with a concentration above K. This is called the **fade-out time**, and it also depends on $(Q/K)^{2/3}/D$. Fade-out time is always 2.7-fold greater than the time to r_{max}, regardless of the values of Q, K, or D. The higher the diffusion constant D, the faster the signal will both spread and fade [43, 114, 287, 366].

Pure diffusion is a rather slow process, and the distance over which a single puff can effectively diffuse in a reasonable time period is short. An attempt to increase the distance by increasing Q or to increase the spread time by decreasing D will quickly lead to spatial and temporal scales where current flow and turbulence take over. These constraints are even more severe in water than in air. Values of the diffusion constant range from 0.01 to 1 cm²/sec for volatile compounds in air, but are on the order of 0.00001 in water, which translates to a 10,000-fold slower speed. For example, the release of a volatile puff with a Q/K ratio of 1500 would take 13 sec to reach the r_{max} of 6 cm in air, whereas in water it would take 33.5 hours to reach the same size [99]! Diffusion transmission of single chemical puffs is a viable communication strategy only for small terrestrial organisms such as ants and termites, which live at the ground/air boundary layer where there is little or no wind. A highly volatile chemical puff released in this environment can diffuse to its maximum radius in a few seconds, and the signal fades in a minute or less. The transmission distance range of 1 to 10 cm is a useful scale for alarm or aggregation signals in small colonial insects. Chemical alarm signals do occur among close-living aquatic animals, but they depend on currents and turbulence, and require longer emission times to spread.

CONTINUOUS EMISSION FROM A POINT SOURCE   Now consider an animal that employs a strategy of **continuous emission** of an odorant at the rate of Q molecules/sec. As with the puff, the active space gradually expands away from the source. Eventually it reaches a point at which diffusion across the boundary of the active space exactly equals the input of molecules at the source. When this happens a steady state is achieved, and the active space will neither expand nor contract as long as the animal continues its output. The distribution of odorant within the active space will form a smooth concentration gradient from the center outward as long as currents do not disrupt it.

Diffusive spread of a continuously emitted odorant is unlikely to be a useful signaling strategy, especially for mate attraction, because of numerous constraints. First and foremost, diffusion conditions will not be sustainable over long time periods and large distances, because current flow and turbulence will quickly take over. It makes far more sense for a long-distance chemical sender to position itself where it can release pheromone into an optimal current. Secondly, small body size places constraints on olfactory mate attraction signals. An organism may be too small to

produce enough odorant to diffuse far enough to significantly increase encounter rates [115]. The estimated critical body size below which such a strategy becomes inefficient is actually quite small—under 1 mm; so bacteria, protozoa, and rotifers, animals too small to use diffusable pheromones, may instead employ high mobility to contact conspecifics. The final constraint is an aquatic environment, which as we saw above leads to extremely low chemical diffusion rates. For continuous emission in water, a compound with a given Q/K will generate an r_{max} 10,000 times as large as it would in air. However, the time required to reach this maximum distance will be 10^{15} times as long as that in air: in practice, never! That means that with continuous emission in water, the radius of the active space *could* be very large, but given the time available to most aquatic organisms, only a tiny fraction of this r_{max} can be achieved in any bout of emission. Once again, an obvious way around this problem is to make use of water currents. Another option is some type of movement by the sender while it continuously emits a pheromone. Copepods are small (1–3 mm) aquatic crustaceans that often swarm in particular locations for mating, but their vision is poor at best, and males use olfaction to find gravid conspecific females. A female strategy in some species is to hover in place while releasing pheromone. Females generate a small, constant feeding current with their maxillipeds, which spreads the pheromone in a cloud around them. This action effectively increases the chemical diffusion rate to approximately the values observed in air. Based on the trajectories of males that zigzag and swim in loops around the female while repeatedly approaching and retreating until she apparently accepts sperm transfer, the active space of the pheromone cloud is approximately 20 mm in diameter (**Figure 6.20**). Distribution of the pheromone within this space is predicted to be patchy but to form a rough gradient that enables males to locate the female [209].

CONTINUOUS EMISSION FROM A MOVING SOURCE   Animals may also release odorants continuously while moving substantial distances to form a chemical trail. For example, as ants forage, they often deposit a thin trail of pheromone by dragging their abdomen or stinger on the substrate. These trails may lead colony mates to foraging sites, or they may function to orient a scout back to the home nest [177]. Some animals, such as snakes and slugs, lay odor trails on the substrate during normal locomotion that can be followed by conspecifics seeking mates [83, 137, 286, 344]. More remarkable, mate attraction trails can even be produced in aquatic environments. Females of some other copepod species than the hoverers described above leave three-dimensional trails in open water that last 30 sec to 2 min, depending on species size. As we discuss later in the chapter, males can clearly track these trails in the appropriate direction [19].

Trail characteristics of **moving source emission** can be modeled as a series of small single puffs applied to successive locations along a line. Since they are close together, the active spaces generated by adjacent drops ("puffs") will overlap and

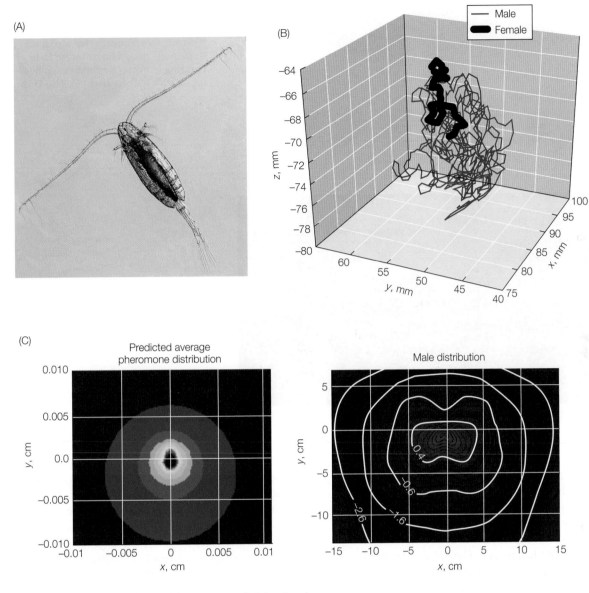

FIGURE 6.20 Pheromone cloud in a copepod (A) A female *Pseudocalanus elongates*, approximately 1.3 mm in length. During courtship display, she orients her head upward and rocks her body through a 45° angle while generating a feeding current with the maxillipeds and releasing a pheromone. The pheromone is believed to be an amino acid or small peptide. (B) Male and female trajectories in a 40-second video clip of courtship. Males speed around the female in zigzags and loops, occasionally touching her, for up to 5 minutes. (C) Predicted average concentration of pheromones around a diffusing female using a diffusivity of 10^{-2} $cm^2\ s^{-1}$ (left) and contour plot of log concentration of male positions relative to the female (right). The *x*- and *y*-axes represent the horizontal and vertical, respectively. (After [209].)

merge. The overlapping active spaces will expand out to some maximum radius to the sides and then eventually collapse back in as the odorant diffuses away. This will produce a long elliptical active space at any one instant. As the trail-layer moves on, the leading edge of the ellipse follows the signaler and the trailing edge finally diffuses away enough odorant that all points near it have concentrations less than *K*, as shown in **Figure 6.21**. The length of the trail's active space, *L*, depends on *Q/K* as well as on *D*. Not surprisingly, the length of the trail is also 2.7 times the location of the maximum radius of the active space. The maximum diameter of the active space at this location and the time it takes before the trail at any given location has totally dropped below *K* can also be calculated. The speed with which the ant walks does not affect the length of the trail, but it does affect the thickness of the active space. The thickness of the active space (like a single puff) is independent of *D*. This gives the trail-layer several degrees of freedom to control the width, length, and longevity of a trail independently [43].

Transmission by current flow

An animal that emits a continuous stream of odorant for the purpose of long-distance communication is highly likely to encounter the effects of some type of current flow in the

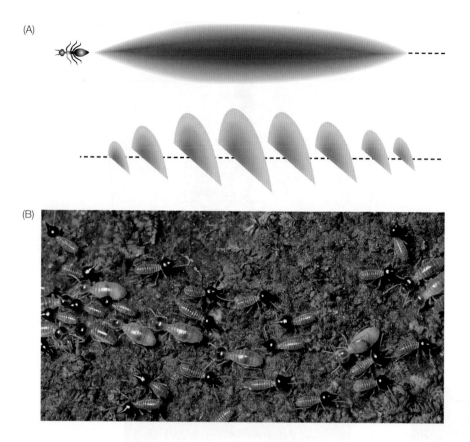

FIGURE 6.21 Pheromone trails (A) The active space of an ant or termite trail is essentially a sequential series of small single puffs along a linear transect. (B) Foraging termites (*Nasutitermes*) following an odor trail. (A after [43, 410, 411].)

environment. Intuitively, current flow increases both the speed and the maximum transmission distance of a chemical signal, but only in the down-current direction. Although diffusion processes may still be operating, they will be largely overwhelmed by the **advection** (horizontal transport, as opposed to the upward transport of convection) of the chemical molecules by the current. The diffusion properties of the chemical therefore become less important, and the flow dynamics of the medium dominate the transmission process. In principle, it should be possible to model the spread of molecules in a current flow if one knows the velocity of the flow. In practice, modeling this process is extremely difficult because most flows are turbulent.

When a medium flows in an orderly fashion, with the same velocity and direction of flow throughout the region of interest, it is called **laminar flow** (see Figure 6.18B). When a medium flows in an irregular fashion, it is called **turbulent flow** (see Figure 6.18C). Turbulence is created when the flow is hindered in one location relative to other adjacent locations. This hindrance generates torque on the medium, which then generates vortices (whirls), eddies (backflow and partial whirls) and waves. The pattern of vortices and whirls is complex and chaotic. Large vortices created initially produce smaller vortices, which in turn produce even smaller vortices. The energy contained in the larger vortices is continually transferred to the smaller ones, and small vortices are eventually dissipated as heat. The pattern is also continually changing in time and highly unpredictable. The most

important consequence of turbulence from our perspective is the rapid mixing and irregular transport of chemical odorants in nature.

A continuous input of force, or energy, is necessary to maintain a flow. In both the atmosphere and ocean, the main physical forces that cause flows are: (1) large scale temperature differences, which cause density differences in the medium and buoyancy differentials that result in vertical cycles of air and water as well as horizontal mass movement; and (2) Coriolis forces from the rotation of the Earth that result in air and ocean circulation patterns. Normal wind speeds (with the exception of major storms) range from 1 to 10 m/sec; normal current flows in the ocean are slightly lower. Three combined factors determine whether a flow will develop turbulence or remain laminar: the velocity of the flow, the density of the medium, and the distance scale over which one is observing the flow. The action of these factors is expressed by the Reynolds number mentioned earlier and described more fully in Web Topic 6.4. For our purposes, it is sufficient to understand that turbulent flow is more likely at higher velocities, in denser media (i.e., is greater in water compared to air), and over longer distances. Laminar flow is more likely at slow velocities, in less dense media such as air, and over shorter distances [99, 114].

LAMINAR FLOW   If the flow is laminar, then the spread of a continuously emitted chemical can be predicted fairly accurately. The most appropriate model combines diffusion and

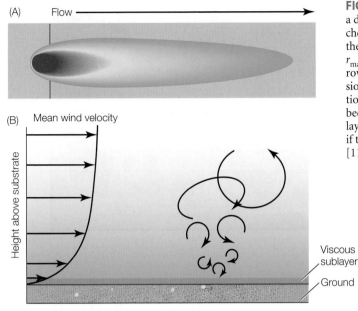

(A) Flow

(B) Mean wind velocity

Height above substrate

Viscous
sublayer

Ground

FIGURE 6.22   Laminar flow (A) The active space predicted by a diffusion model with advective flow. Color gradient depicts the chemical concentration gradient. The maximum distance at which the signal can be detected directly downstream is the same as the r_{max} without any flow. The main effects of current flow are to narrow the active space of the odorant and to increase the transmission speed. (B) Wind speed profile and turbulent eddies as a function of height above the substrate. Close to the substrate, eddies become smaller and wind speed is reduced, forming a viscous sublayer. The laminar characteristics of this sublayer are diminished if the wind speed is high or the substrate surface is rough. (A after [113]; B after [287, 401].)

advection: diffusion leads to the spherical spreading of the odorant, but the center of the active space moves with the flow. The active space has an ovoid shape as shown in **Figure 6.22A**. Along the downstream axis, the concentration falls off smoothly as $1/r$ and is independent of velocity. As velocity increases, transmission speed increases and the width of the active space narrows. Pure laminar flow conditions are unlikely to occur in natural environments, but quasi-laminar conditions sometimes may be found at the boundary layer between a flowing medium and a solid substrate. The velocity of the medium flow is reduced as it gets closer to the stationary "no-slip" substrate, and eddy size also decreases (**Figure 6.22B**) to create a viscous sublayer. Small animals living on this substrate can take advantage of the chemical gradients in this layer. For example, blue crabs (*Callinectes sapidus*) live on the benthic substrate and use the chemical gradient information from metabolites released from the siphons of clams to locate their prey [398]. A possible pheromonal example may occur in blennies (*Blennioidei*), small teleost fishes that also live on the benthic substrate. Males defend breeding territories and nest sites to which they attract gravid females. The large breeding males, but not the small sneaker males, have a large testicular gland that is believed to continually secrete a mate-attracting pheromone [279].

TURBULENT FLOW   Since most signaling situations do not meet the laminar flow condition, the above model does not have wide applicability. Plume models have been developed that replace the diffusion coefficient with a set of **dispersion coefficients** for the advection of molecules along the horizontal and vertical axes, and average the concentration of odorant at each location over small increments of time [43, 294, 366]. Field researchers make empirical measurements of

the typical rates of spread of smoke or ions in horizontal and vertical directions under different wind speeds and use these coefficient values to predict the plume structure. A graphic depiction of the active space that results from such turbulent flow models is shown in **Figure 6.23**. Current velocity affects the shape in nonintuitive ways. Low velocity leads to a long and narrow space stretched in the downwind direction. When velocity is high, it not only distorts the active space along one axis, but it also "whisks" molecules away faster from the surface of the active space than would simple diffusion alone. This dilution effect causes the active space boundary to shrink in closer to the source than it would in the absence of a current. Strong flows can thus decrease the effective range of a continuous emitter.

How well do these models estimate the active space of real biological odorants? Once again, we must let the animals show us what they perceive. In field trials, cages of male moths were set out at different distances from a source releasing the species-specific pheromone. Pheromone release rates approximated the typical values for female moths and were measured in each trial, along with wind speed. The male detection threshold was observed and measured in a controlled wind chamber; male moths flutter their wings and orient toward the source when they have detected the pheromone. The observed active space was far wider than any of the models predicted [119]. Moreover, the distribution of odorant inside the active space at any instant in time

FIGURE 6.23   Time-averaged turbulent flow model Example of a Gaussian model of the active space under turbulent flow, in which empirically measured dispersion coefficients are used instead of diffusion coefficients to estimate molecule transport in a current. (After [366].)

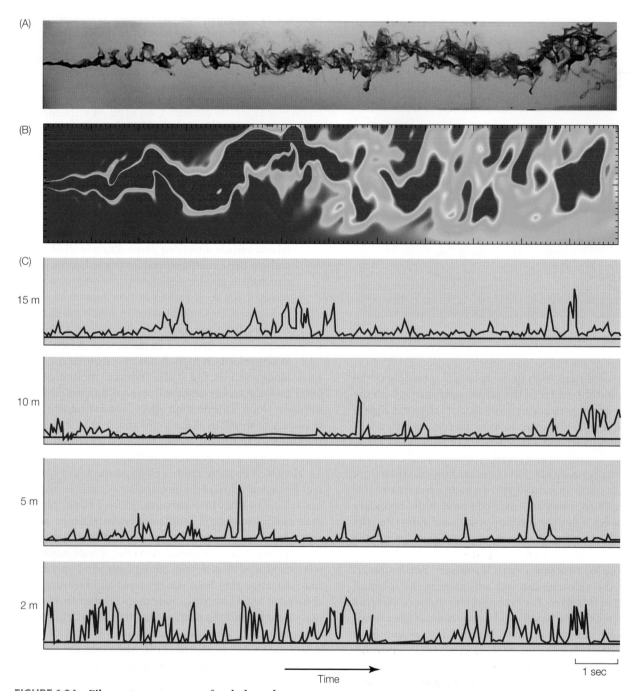

FIGURE 6.24 Filamentous structure of turbulent plumes
(A) Chemical dye released into an open-channel water flume in which a fully-developed turbulent flow was established [396]. (B) A numerical simulation model of turbulent flow [233]. (C) Fluctuations through time of an ion tracer released from a source point, measured simultaneously at different distances from the source. (C after [274].)

was extremely patchy and filamentous, as illustrated in **Figure 6.24**. Surface roughness and variability in wind direction have enormous effects on transmission patterns. The main plume filament can meander back and forth and loop up and down if vortices are large, and the maximum transmission distance can be very unpredictable. Time-averaged models

thus do not capture the realistic structure of active spaces for chemical stimuli released into turbulent currents. Given this nonuniform distribution of odorant, how can a receiver possibly detect an odor gradient and locate the position of a long-distance olfactory signaler? We take this up in the section on processing later in the chapter.

Other environmental effects on chemical signal transmission

In addition to ambient wind and currents, several other environmental factors affect the transmission and longevity of chemical signals. *Ambient temperature* affects the vapor pressure of volatile compounds in air. Volatility increases with higher temperatures, which causes chemicals to spread faster and fade sooner. It is therefore very important to know the vapor pressure of a pheromone compound for the temperatures under which it is used by the animal. Animals evolve signaling compounds with characteristics that operate best under the ambient temperatures they typically encounter [6]. High temperatures, especially in open habitat, also cause thermal convection and increased turbulence. Direct sun exposure, especially *ultraviolet radiation*, leads to chemical decomposition, so marks deposited in exposed locations need to be stable chemically to withstand ionization. *High humidity* and *rain* cause the most significant environmental challenge to deposited-mark longevity. As humidity rises, the evaporation of odor molecules from a surface increases as these molecules compete with water molecules for surface binding sites. It is common knowledge that dogs can track better when the ground is moist, because more odor molecules are released. Rodents can find seeds in the litter more efficiently under moist ground conditions for the same reason [381, 382]. Some animals stop depositing scent marks during rainy seasons because the odors don't persist long enough. Range marking by animals living in the hot and humid tropics requires special adaptations, such as protection by keeper substances, to retain the marks [5]. *Vegetation structure* of the habitat is another relevant environmental parameter. Odorous chemicals can adsorb to leaves and be released at a later time, affecting temporal strategies for signaling [300]. In aquatic environments, suspended clay particles can have the same effect [14]. Tree trunks and leaves in a forest mechanically disrupt current flow, thereby increasing turbulence and widening the plume spread. The temperature inversion in a forest on sunny days, with higher heat in the canopy and cooler temperatures in the shade below, reduces flow and traps odors, but gaps in the canopy create ventilation chimneys [18, 128]. Hedgerows and forest edges cause **odor shadows**, within which prey animals might avoid being detected by olfactory-orienting predators. *Topography* and *surface structures* also affect current flow patterns: breezes tend to move up the sides of valleys during the day, and down the slopes during the night [142].

One final point. We learned in earlier chapters that sound and light signals composed of mixtures of different frequencies usually undergo *dispersion* during transmission, whereby they split or segregate as a function of frequency. Chemical blends undergo the same process if diffusion plays a role in transmission. Blends typically contain components with different volatilities, so more volatile components will diffuse faster and farther than less volatile components. This dispersion principle leads to the prediction of concentric rings of active spaces for a multicomponent diffusing ant alarm signal [49, 71]. Similarly, the more- and less-volatile components of deposited marks have active spaces of different sizes and fade-out times of different durations [185]. Dispersion thus facilitates the decoding of scent-mark age in the urine deposits of mice [69]. Dispersion probably is not a factor for current-borne transmission of chemical blends, where diffusion does not play a role and mass transport keeps the blend components together in their original proportions [385].

Chemoreception

As mentioned early in this chapter, **chemoreception** was the earliest communication modality and is an essential feature of living cells. Both single-celled organisms and individual cells within the bodies of multicellular organisms must be able to recognize relevant extracellular chemicals and selectively transport them through the membrane. Biochemists use the term "signaling" to describe intercellular chemical exchanges in the same sense that organismal biologists use the term to describe interindividual chemical communication. While some of the chemical detection mechanisms employed between cells within a body are similar to those employed in chemical signaling between individuals, the chemosensory organs of advanced multicellular organisms have very different tasks to perform, because they must obtain access to the environmental "soup" of chemicals, recognize the important ones, and then couple the detection to a nerve impulse. The design of **chemosensory organs** differs for the two types of reception: olfactory and contact. These differences relate not only to the mechanisms for coupling environmental chemicals into the receptor, but also to the kinds of sensory cells and types of neural connections to the brain [376].

Coupling from the medium to the receptor

OLFACTORY RECEPTOR INTAKE   Olfactory receptors detect airborne or waterborne chemicals arriving from distant sources, which may include potential food items, predators, and chemical cues and signals from conspecifics. The receptor organ must therefore be designed to maximize exposure of the sensory cells to large volumes of the ambient medium. **Olfactory organs** are either internal cavities through which the medium is drawn, or hairlike projections exposed to ambient currents. In vertebrates and invertebrates, the organ is located at the most anterior projection of the head. Several mechanisms have evolved to maximize chemical exposure and capture.

In terrestrial air-breathing vertebrates (amphibians, reptiles, birds, and mammals) the nasal organ is an integral part of the respiratory system and the intake mechanism is **sniffing**. The inlet (nose or nares) is a duct that serves to clean and humidify the air before passing it to the **olfactory epithelium**, a tissue inside the nose that contains an array of sensory cells. The outlet is the lungs. During normal breathing, most of the entering air flows straight to the lungs, but a portion is diverted into a side pocket containing the olfactory epithelium (**Figure 6.25A**) [85]. The animal can increase the fraction of air diverted to the olfactory epithelium with more vigorous sniffs [205, 341, 362]. When sampling odors, mammals typically display brief bouts (1–2 seconds) of high-frequency sniffing (4–10 Hz) [85, 404, 405]. Elasmobranch fishes

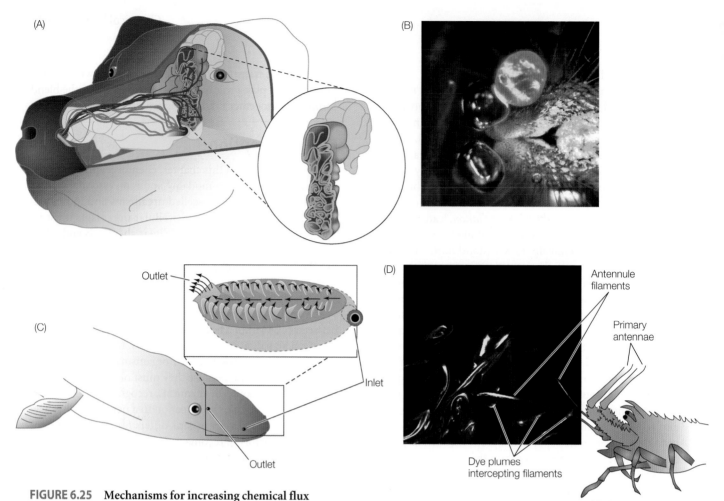

(A)

(B)

Outlet

(C)

Outlet

Inlet

(D)

Antennule
filaments

Primary
antennae

Dye plumes
intercepting filaments

Mechanical
lobster

FIGURE 6.25 **Mechanisms for increasing chemical flux to olfactory receptors** (A) Airflow through a dog's nose during a sniff, showing rapidly moving air (red lines) flowing along the top of the nasal cavity and back to the large and convoluted olfactory epithelium (green, inset). Slower moving respiratory air (blue lines) filters through non-sensory epithelium in the lower and lateral region of the cavity and into the lungs. (B) Two air bubbles of a diving water shrew (*Sorex palustris*); one is touching a wax object (green). (C) Separate olfactory organ in a fish (European eel, *Anguilla anguilla*), inset shows airflow (black arrows) over laminae of the olfactory epithelium. (D) The large primary antennae of the spiny lobster (*Panulirus argus*) function primarily for mechanoreception, while the more slender antennules, with several branching filaments, function primarily for olfaction. The filaments are covered with an array of fine sensory hairs that house the chemoreceptor cells. Rapid downward flicking of the olfactory filaments allows water to penetrate the chemosensory hair array. Here, real antennules attached to a mechanical lobster were flicked in turbulent flowing water containing a slowly released fluorescent dye. Higher dye concentrations are indicated by lighter colors. Note several dye plumes intercepting and penetrating the antennule filament. (A after [85]; C after [362]; D after [211].)

(sharks and rays) also have olfactory chambers connected to the mouth and use **gill action** to move respiratory water over the olfactory epithelium. Most aquatic mammals cannot smell underwater because they must hold their breath, but a few species have evolved special mechanisms to accomplish this task. Small diving mammals such as water shrews and moles employ a **bubble-touching** strategy, in which they blow a small air bubble through their nose while underwater, touch it to an object of interest, and then withdraw the

bubble back into the nose (**Figure 6.25B**) [68]. Amphibians exhibit a variety of strategies, depending on stage of metamorphosis and how aquatic the adult stage is, but many species can smell underwater; frogs trickle water through the nares into side chambers containing contact chemoreceptors [105]. In teleost fishes the olfactory organ is completely independent of the respiratory system. It consists of a chamber located in front of the eyes with a separate inlet and outlet (**Figure 6.25C**) [362]. The chamber contains lamellae covered by a thin sensory epithelium to generate a large, convoluted surface area for the reception of chemical odorants. Some fish species use jaw movements to create a **pumping** mechanism that increases the flow of water through the nose [17], whereas others use **ciliary action** to generate flow over the lamellae.

Arthropod olfactory detectors are located on paired antennae that extend forward from the head. The sensory cells are contained in small cuticle-enclosed hairs, called **sensilla** (singular *sensillum*) arrayed along the main antenna structure. The flow of environmental medium around the sensilla is restricted by the boundary layer effect described earlier (see Figure 6.22B). Special mechanisms are therefore needed to increase the access of chemicals to the sensillum surface. In flying insects such as moths, **wing flapping** produces pulses of rapid airflow that increase chemical penetration to the sensory

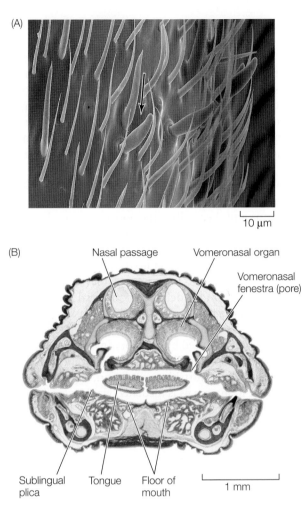

(A)

10 μm

(B)

Nasal passage

Vomeronasal organ

Vomeronasal fenestra (pore)

Sublingual plica

Tongue

Floor of mouth

1 mm

(C)

FIGURE 6.26 **Contact receptors** (A) Special sensilla (arrow) on the antennae of the carpenter ant *Camponotus japonicus* specialized for detection of non-nestmate cuticular hydrocarbons. By responding only to unfamiliar CH mixes, it contributes to a nestmate-versus-foreigner recognition system. (B) Transverse section through the snout of the gecko *Coleonyx variegates* at the level of the vomeronasal organ. The sublingual plicae are ridges that seal the mouth around the tongue when the floor is raised, forcing chemical-laden material into the vomeronasal organ. (C) Flehmen performed by a pronghorn (*Antilocapra americana*) male after tasting female urine.

hairs [237] (see Figure 6.33). Aquatic crustaceans such as lobsters, stomatopods, and crabs engage in **flicking** of their whip-like antennae as a mechanism equivalent to the sniff (**Figure 6.25D**) [212]. The fast downstroke significantly increases chemical penetration, and during the slower return stroke and stationary pause the chemicals remain trapped between the hairs, where they can be absorbed and analyzed [147, 211]. For each of these coupling mechanisms mentioned above, the pulsing of chemical-laden medium is the key to both increasing the flux of environmental chemicals into the receptor and preventing habituation of the sensory cells, so that repeated samples can be made [100].

CONTACT RECEPTOR INTAKE   Contact chemoreceptor organs, as the name implies, must come into direct contact with a chemical-laden surface in order to respond. Their design has as much to do with their location on the body as with their size, shape, and function. **Taste receptors** are one major functional class of contact chemoreceptor. In nearly all animals, the individual organs are small pores, pegs, buds, or bristles containing a few to several hundred sensory cells. Arrays of these small units occur in locations involved in feeding—around the edge of the mouth in fish; on the tongues of air-breathing vertebrates; and on the feet, mandibles, and antennae of arthropods. Taste receptors are sensitive to a few critical components of food, such as sugars, salts, amino acids, and bitter or sour compounds, and help

the organism evaluate whether a captured item is appropriate to consume or is toxic. They have little or nothing to do with social communication.

Contact receptors for conspecific communication tasks such as recognition of colony members and appropriate breeding mates, or for very specialized recognition of prey and host species, are another major functional class. Earlier we mentioned the cuticular hydrocarbon (CH) profiles in social insects and other arthropods that act as signature odors to distinguish colony and family membership and ovarian breeding status. Animals investigate each others' cuticle surfaces by **antennation**, touching the surface of another individual with the mobile antennae. Arthropod antennae contain a variety of sensilla structures with distinctive morphologies. Long thin bristles with multiple pores are associated with olfaction, and shorter bristles or pegs with a single pore at the tip serve a contact reception function [264, 284, 336]. A thicker, medium-length bristle type with multiple pores on the end responds specifically to non-nestmate hydrocarbons in an ant species (**Figure 6.26A**) [291]. Other hymenopteran species have since been found to specifically recognize any type of CH mix that *differs* from the familiar colony odor. This strategy of detecting foreigners, coupled to aggressive attack response behavior, may be the most efficient way to defend the colony against conspecific and heterospecific enemies [201, 380]. In contrast, the cockroach (*Periplaneta americana*) has antennal contact receptors that respond selectively to the blend of three CHs that characterize this species [330]. The tentacles of molluscs also contain distinctively structured chemosensory units for olfactory and contact reception [45, 329]. Contact receptors are especially important for detecting nonvolatile and insoluble compounds.

Terrestrial vertebrates, most notably amphibians, reptiles, and mammals, possess a large chemosensory organ called the **vomeronasal organ** (**VNO**). This is conspicuously absent in birds, Old World primates, apes, marine mammals and turtles, some bats, and crocodiles [127]. Fish do not have a separate VNO, but their main olfactory organ contains a mixture of the three neuronal cell types—ciliated, microvillous, and crypt cells—that became segregated into separate olfactory and VNO organs during the evolution of tetrapods [160] (see Web Topic 6.5). The VNO specializes in the detection of large nonvolatile chemicals such as urinary proteins and lipid secretions, but also responds to small soluble chemicals conveyed by water or mucus. The amphibian VNO is a side pocket of the main olfactory chamber accessible through narrow grooves or fissures in the nares. Material is drawn into the opening by **nose-tapping** or **trickling of water**. In reptiles, the hemispherically shaped organ is completely separate from the nasal cavity and lies between it and the mouth (**Figure 6.26B**). Access is only through the roof of the mouth using the **tongue**, which is flicked in the air or touched to a substrate and then pulled back into the mouth. The floor of the mouth is raised, pushing chemical-laden fluid into the VNO pores [134]. Snakes are particularly dependent on vomeronasal stimulation for both normal sexual behavior and prey tracking. The mammalian VNO consists of a pair of tubes situated above the roof of the mouth and may be accessed through the mouth, nose, or both. A variety of mechanisms help deliver nonvolatile chemicals to the organ. **Nose rubbing** transfers contact chemical signals directly between individuals. The receiver may lick or touch its nose to the urine or secretion or another individual. In some species the **nasopalatine duct** (a groove between the nose and mouth) serves as a channel to move nonvolatile chemicals into the VNO. Elephant males dip the tip of the trunk in female urine and then deliver the nonvolatiles to the vomeronasal opening [316]. After contacting such sources, many male mammals perform a behavior called **flehmen** in which the head is raised and the upper lip is curled (**Figure 6.26C**). This behavior appears to force nonvolatiles into the VNO. Another mechanism for delivery of material is a **vascular pumping action** caused by dilation of blood vessels and sinus tissue around the organ, which is initiated when a receiver's interest has been aroused by olfactory investigation [261]. Much of the intraspecific chemical communication involved in mate attraction, courtship, copulation, aggression, and parental care in mammals is mediated via the vomeronasal organ (see Web Topic 6.5). Stimulation of the vomeronasal system affects, and is strongly affected by, steroid and neuroendocrine hormones involved in reproduction [127, 251, 373, 375].

Chemosensory reception

RECEPTOR CELLS   **Chemoreceptor sensory cells** function similarly in vertebrates and invertebrates, despite their very different housing and evolutionary origins (**Figure 6.27**) [4, 174, 203]. All chemosensory cells other than vertebrate taste bud cells are **bipolar neurons**. The cell bodies form a sensory **epithelium** sheet; one dendrite extends to the surface where it meets the ambient medium, and the other dendrite forms a long axon directly into the brain. Chemosensory nerves in the

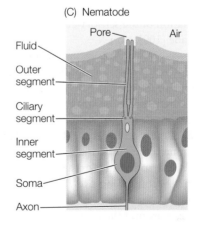

FIGURE 6.27   Chemosensory cells of vertebrates and invertebrates Main body of the olfactory sensory neuron shown in blue. (A) Olfactory epithelial cells in a rodent. Sensory cells are arrayed in a large columnar matrix sandwiched between the support cells; cilia extend into the mucus layer. (B) Typical insect sensillum showing a ciliary sensory cell inside a cuticular cover; pores allow access of chemicals into the lymph-filled lumen. (C) Nematode chemosensory cell. (After [3].)

majority of animals and organs are derived from **ciliary cells**. The vertebrate vomeronasal epithelium and taste buds, as well as a few invertebrate sensilla, are composed of **microvillous cells**. The vertebrate olfactory epithelium also contains a small proportion of microvillous cells [120]. The cilia or microvilli project out from the dendritic end of the cell that is exposed to the medium and are the sites of chemical reception. They are bathed in **lymph**, a watery mucus layer that protects the vulnerable nerve cells against toxins, regulates the access of chemical odorants to the sensory cell's membrane receptor sites, and enzymatically breaks down spent pheromone molecules [112, 196, 372, 388]. Support cells sandwiched between the sensory cells isolate the nerve cells electrically and secrete some of the lymph and mucus products. Unlike most other nerve cells elsewhere in the body, vertebrate olfactory cells have a short lifetime (30–90 days) and are continually regrown from stem cells at the base of the epithelium [88, 244, 365, 367]. At least one function for this ongoing **neurogenesis** in the olfactory system is to replace cells that have been damaged by toxic dust or infected by external pathogens [236].

Nerve impulses are generated as follows. Chemicals from the medium either diffuse through the mucus layer to the sensory cell membrane or are captured (bound) by specific binding proteins in the mucus and transported across. The chemical or loaded odorant-binding protein then binds to an **olfactory receptor protein** anchored in the membrane of the sensory cell. In most cases, this protein is a **G protein-coupled receptor** (GPCR) belonging to the same general protein class as the visual pigment rhodopsin. Like opsin, the chemosensory receptor proteins consist of seven transmembrane helices with a binding pocket in the center. Variations in animo acid sequences in the pocket region determine the nature of ligands that can bind to the receptor protein. Each cell possesses only one or a few different types of receptor proteins. When a ligand binds to a receptor, a G protein coupled to the receptor protein is activated. This activation triggers a biochemical cascade that opens ion channels in the membrane. The cell becomes depolarized and electrical action potentials are transmitted down the axon of the neuron to the brain. As mentioned in Chapter 5, G proteins are so called because they use guanosine triphosphate as a molecular switch. Vertebrate ciliary and microvillous cells have different protein receptors in their cell membranes, and the G proteins to which they are coupled lead to different second-messenger cascades: ciliary cells employ G_{olf} coupling proteins that use cyclic nucleotides (cAMP) as the second messenger in the cascade, whereas microvillous cells employ G_{aO} and G_{ai2} coupling proteins that use inositol triphosphate (IP_3) as the second messenger [109, 376, 425]. Insect olfactory neurons employ yet other GPCR systems as well as non-second-messenger, directly odor-gated ion channels [278]. These various signal transduction mechanisms differ in their speed and specificity of response and are likely adapted to serve the needs of species with different lifestyles and signaling functions [203, 364].

The genes that encode chemosensory receptor proteins have now largely been identified for vertebrate olfactory, vomeronasal, and taste organs; for fruit fly (*Drosophila*) sensilla; and for the nematode *Caenorhabditis elegans* [57, 70, 81, 107, 172, 226, 376, 377, 392]. These genes belong to several large **multigene families**. Vertebrate and invertebrate olfactory genes show no sequence convergences. The vertebrate contact organ genes (taste bud and vomeronasal) are similar to each other but completely different from the vertebrate olfactory genes, indicating that these genes have been recruited independently within each taxon and then expanded by gene duplication and divergence [281, 413]. A given species may have from one hundred to a few thousand such genes, comprising 1–4% of the genome and spread over several chromosomes [24]. They encode the different variants of the receptor proteins embedded in the receptor cell membranes. Since each receptor cell expresses only one or a few of these genes, chemosensory organs may contain a large number of different receptor cell types sensitive to different chemicals. Vertebrate olfactory neurons are strict in expressing only one gene and thus one type of protein receptor per cell [235, 340]. Humans possess about 350 functional receptor genes (and many nonfunctional genes), whereas mice have about 900 olfactory genes and 300 vomeronasal-plus-taste genes. *Drosophila* has 60 olfactory and 60 taste genes; each cell tends to express two gene types, one encoding a specialist receptor protein and the other encoding a generalist protein type. The nematode has over 1000 genes but only 32 olfactory neurons, because most cells express many gene types [24].

NEURONAL ORGANIZATION   In the olfactory systems of both vertebrates and *Drosophila*, the axons for *all* of the receptor cells expressing the same receptor gene converge at one or two consistent locations in the first brain processing center, the **main olfactory bulb** in vertebrates and the **antennal lobe** in the insect [4, 174, 203]. These sphere-shaped convergence sites are called **glomeruli** (**Figure 6.28**). Each glomerulus receives input only from one receptor type, so glomeruli perfectly preserve the information about receptor cell type. As many as 25,000 sensory cell axons may converge onto a single glomerulus in mammals, and up to 50,000 in male insects. Spatial maps of glomerulus positions in the bulb or lobe are very precise, bilaterally symmetrical between the two sides of the brain, and consistent between individuals of the same species. There is little evidence for a **chemotopic ordering** (systematic spatial arrangement) of glomeruli positions in the olfactory bulb related to chemical features, although some classes of chemicals do show coarse clustering patterns [188, 192, 256, 267, 355, 368]. From each glomerulus, dedicated second-order neurons (called **mitral cells** in vertebrates and **projection neurons** in invertebrates) carry the signal to higher processing centers in the brain. Although most excitatory integration between glomeruli occurs in the cortex, there are several types of **interneurons** that produce inhibitory interactions between different types of glomeruli. Like the olfactory sensory cells, these interneuron cells

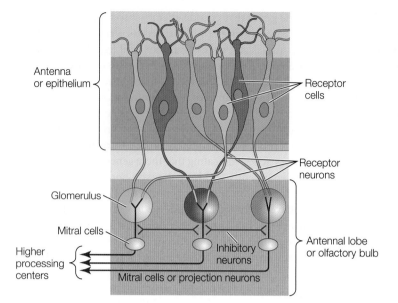

FIGURE 6.28 **Sensory cell organization** In vertebrates and many invertebrates, sensory cells of different types, shown here as colors, are dispersed throughout the sensory organ (olfactory epithelium or antenna). Receptor cells send their axons to spherical modules called glomeruli in the first processing center of the brain. A ubiquitous feature of all olfactory systems is the convergence of axons from the same receptor type to one or a few glomeruli. Second-order neurons, called mitral cells, (blue) then convey the glomerulus type information to higher centers. Several kinds of lateral interneurons (red) may cause inhibitory interactions among glomeruli. (After [58, 174, 265].)

undergo continual replacement by new cells [9, 66]. They play an important role in odor perception and discrimination in both vertebrates and invertebrates, as we shall see below in the section on olfactory coding.

Another striking convergence between some vertebrates and insects is the existence of **dual pathways** to the brain, one for general olfactory input and another for specific pheromonal input. These tasks require very different processing strategies [76, 140, 364] (**Figure 6.29**). Two olfactory pathways are found in the males of some insects, such as moths, in which females produce the long-distance mate attractant (see Figure 6.29B). The males' antennae have long, morphologically distinct sensilla sensitive only to the pheromone components [175, 200, 216, 222, 321, 339, 430]. Axons from these sensors converge in the **macroglomerular complex**, a specialized set of large glomeruli in the antennal lobes of the brain, with each glomerulus representing one of the blend components. Projections then extend to the inferior **lateral protocerebrum** of the brain. General olfactory input from host plants goes to glomeruli in the main antenna lobe in both males and females, and on to the **lateral horn of the protocerebrum**.

In fish, there is evidence for three pathways to different areas of the olfactory bulb: one for pheromones via the crypt cell receptors, one for food odor detection via the microvillus receptors, and a third for alarm substances via the ciliary receptors [160, 162]. This heterogeneity was the precursor for the separation into two very distinct chemosensory organs in tetrapods. In rodents and other tetrapods, the **main olfactory bulb** receives input from the nasal olfactory epithelium, and sends the information on to higher brain centers such as the **piriform cortex**, with some subsequent connections to the hypothalamus (Figure 6.29A). The **accessory olfactory bulb** receives input from the vomeronasal organ. There is more extensive cross-connectivity among glomeruli in the accessory olfactory bulb, implying greater peripheral preprocessing in the vomeronasal system compared to the olfactory system [109]. The accessory bulb then sends projections to subcortical regions of the brain including the amygdala and hypothalamus [373]. These two brain regions are part of the limbic system that regulates emotions and the autonomic nervous system. The amygdala processes emotional reactions of fear, pleasure, and aggression and the hypothalamus regulates reproduction and sexual arousal via neurohormones. Primer effects of pheromones operate through this pathway. A possible functional reason for the microvillous VNO cells and their different transduction mechanism mentioned earlier may relate to this primer role. VNO sensory cells are not only extremely sensitive to very low stimulus concentrations [223], but they also respond to constant stimulus input with long-duration trains of action potentials and little habituation. Together with the VNO's ability to take in large nonvolatile chemicals, the sustained sensory response facilitates these endocrinological priming effects [50, 167, 206, 251, 375].

Web Topic 6.5 *Vertebrate dual chemosensory system*
The vomeronasal organ does not function solely for pheromone reception, nor the nasal organ solely for environmental chemical reception. The two organs have overlapping roles. Vertebrates also possess several other chemoreception organs that are described in this online unit.

ODOR CODING   The olfactory systems of most animals have the ability to detect and discriminate a huge range of odorants. This ability arises both because of the very large number of different receptor cell types, and because of overlapping responsiveness of the cell types to a range of odorants that can combine at higher processing levels to create the perception of different odors. The olfactory receptor proteins on the sensory cell membrane do not present a lock-and-key, match or nonmatch binding site for whole odorant molecules. Instead, they are believed to be responsive to different odorant features, such as the functional group and molecule size,

(A) Rodents and other vertebrates

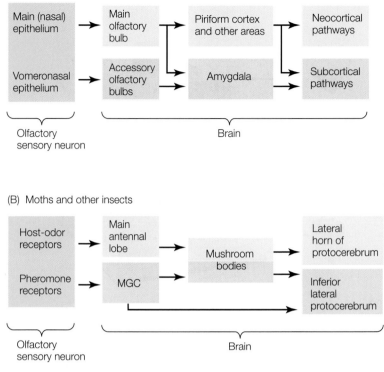

(B) Moths and other insects

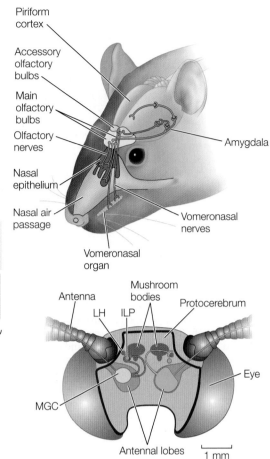

FIGURE 6.29 **Separate pathways for pheromone and general olfaction functions** General olfactory pathway shown in blue, pheromonal pathway in yellow. (A) Terrestrial vertebrates receive airborne odorants via the nasal epithelium, which projects to glomeruli in the main olfactory bulb and on to the piriform cortex and neocortical pathways. Species with a vomeronasal organ (VNO) have a second pathway leading to glomeruli in the accessory olfactory bulb, the amygdala, and subcortical pathways. (B) In moths and other insects, host-odor receptors send axons to glomeruli in the main antennal lobe and then send projections to the main calyces of the mushroom bodies in the protocerebrum and terminate in the lateral horn (LH), as illustrated on the right side. Males of some species have a second pathway, shown on the left side, receiving input from specialized pheromone sensilla on the antennae that send axons to large glomeruli in the macroglomerular complex (MGC) of the antennal lobe. Projections extend to the inferior lateral protocerebrum (ILP), mostly bypassing the mushroom body. (After [76, 339, 416].)

shape, and length [11, 191, 197, 202, 374]. A given receptor type therefore responds to a variety of different ligands, and the cell's excitatory output (rate, amplitude, and duration of neural spike trains) varies depending on how closely the ligand matches the cell type's specific feature requirement. Furthermore, a given ligand stimulates many different receptor cell types (17 on average), which causes a signature pattern of glomeruli to "light up" with activity in the olfactory bulb. **Figure 6.30** illustrates a simplified version of this **combinatorial code** concept.

In reality, the coding system is more complex than the figure suggests [255, 319, 357, 364]. It has been likened to a three-dimensional bar code, where the third dimension represents the strength of the response in the stimulated glomeruli [48, 157, 224, 267, 355, 363, 392]. Cell types vary in the

range of ligands they can accept. Some cells are very narrowly tuned, whereas others are broader. For example, the responsiveness of 35 *Drosophila* receptor proteins to a panel of 110 test chemicals was fully characterized [157]. A few responded strongly to only one test chemical and weakly to some other odorants, while others were activated by up to 30% of the test chemicals, including odorants with little structural similarity. The receptor cells have a certain level of basal activity, and some of the test odorants inhibited this activation, so some odors can act as antagonists. Similar studies in mice give equivalent results [246, 331]. Furthermore, the responses of glomeruli vary as a function of the concentration, with more glomeruli types lighting up at higher concentrations [255, 363, 364]. Different concentrations of the same odorant may therefore be perceived as qualitatively different stimuli. The versatile olfactory receptor organ can therefore respond to both specialized and generalized stimuli over a wide sensitivity range. Moreover, combinatorial activation and inhibition across different receptor types, mediated by interneurons and horizontal nerve connections at higher levels, facilitates the discrimination of a large number of different chemicals [1, 24, 346, 422]. A coding system that combines positive and inhibitory input of different receptor types is called an **across-fiber pattern strategy**, analogous to the coding of color by the visual system. Across-fiber patterning is the rule for olfactory detection of ambient environmental odors. In

Odorant features   Receptors   Odorants   Receptor arrays

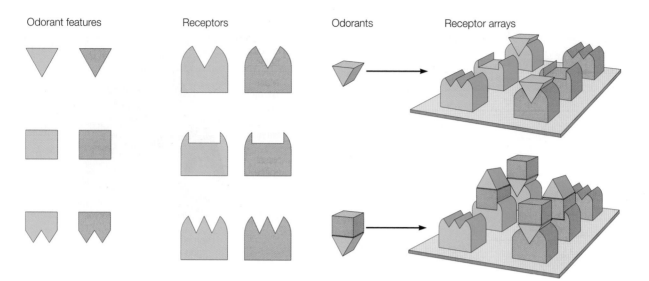

FIGURE 6.30 **The combinatorial code idea** The receptors shown in color on the right are those that recognize the odorant feature on the left. Each odorant has several features that can be recognized by several different receptors, and each receptor can recognize several different odorants. The identity of an odorant is distinguished by the combination of receptors that it stimulates.

conjunction with processing, learning, and memory at higher brain levels, this type of coding system enables receivers to detect and distinguish a large range of odors, including new ones; learn associations between odors and important events (food, predators); and habituate to unimportant odors. The continual generation of new interneurons and new inhibitory interactions among glomeruli plays a central role in the learning process [234, 247, 266, 423].

The alternative coding strategy, in which the input from one receptor type is preserved at higher brain levels without additive and inhibitory interactions among different types, is called a **labeled line strategy**. Frequency coding in the auditory cochlea is an example of this strategy. There is considerable debate about the extent of this strategy for chemoreception. Labeled line coding seems to be supported for the taste system [70] and promoted as the mechanism for single-component pheromone reception [109, 126], but is unlikely to operate in a pure form [77]. Insect and vertebrate pheromone receptors are often narrowly tuned and extremely sensitive, and may combine with a variable number of other receptors, depending on the sensory task. Arthropod sensilla can be easily tuned to one chemical odorant or a blend. One or a few receptor cells can be housed within a single cuticle-covered unit located on the antennae. The lymph can contain specific pheromone-binding proteins that transport predetermined chemical ligands to the sensory cell membranes, and the receptor proteins can also have narrow ligand specificities [216]. Vertebrate nasal olfactory cells can employ similar strategies to achieve narrow tuning, even though all

cells in the olfactory epithelium share the same mucus. The glands that secrete the nasal mucus also synthesize lipocalin odorant-binding proteins. While some of these proteins may function to remove toxic compounds, others are similar to the lipocalin proteins that carry lipophilic pheromones to release sites in senders [154, 296]. These same carriers may selectively transport the pheromones to sensory receptor sites in the receiver. All single molecules shown to trigger pheromone responses in mammals, such as androstenone in pigs, the rabbit mammary pheromone, and specific volatile components of male mouse urine, act via the main olfactory system in a labeled line-type manner [103, 229, 332]. Specialized "atypical" glomeruli have been implicated in these pathways. The vomeronasal organ also contains mucus with odorant-binding proteins, but the curious cross-connections among receptor types, glomeruli, and mitral cells means that direct representation of single chemicals may not always occur. In rodents at least, where MUPs and MHC-derived peptides play an important role in individual and strain recognition, the vomeronasal system may be designed in part to detect and compare pheromone mosaics and odor signatures [108, 109]. Thus chemosensory coding strategies even within the same organ may contain heterogeneous and dynamic coding strategies [94, 210, 220].

Finding the source

Animals live in an **odor landscape** that consists of overlapping plumes of countless chemicals released by other organisms. Individuals will be under strong selection to move toward the source of an odor emanating from prey, a potential mate, or a recruiting conspecific, and away from the source of an odor emanating from a predator or alarmed conspecific. How do they extract the critical directional information? In all four of the main signal modalities, an intensity gradient exists which is strongest at the source and declines with increasing distance from the source. The spatial distribution of signal intensity within the stimulus field therefore provides important information about the location of the sender. Olfactory signals differ from sound, light, and electric signals in that their stimulus fields lack directional components other than the intensity gradient; however, as we shall see, there are sometimes special features associated with olfactory signals that help to provide spatial information.

ORIENTATION IN A SMOOTH GRADIENT   In a smooth chemical gradient field generated by diffusion, organisms may employ two different mechanisms to guide themselves toward (or away from) the stimulus source. **Kinesis** is an indirect type of guiding mechanism in which an organism responds to changes in the intensity of the stimulus by altering its speed of motion or rate of turning. The behavioral responses are themselves undirected in the sense that all turns are oriented randomly with respect to the orientation of the stimulus field. The organism approaches (or avoids) the source by means of a biased random walk. For example, an animal attracted to a stimulus changes its direction of locomotion to an arbitrary direction more frequently if the stimulus level is decreasing, and reduces its turning rate to maintain its present course if the stimulus is improving. The organism never needs to obtain information concerning the direction of the goal, and a single receptor that can detect intensity differences or changes with some form of memory or sensory adaptation (e.g., habituation) is sufficient. **Taxis**, on the other hand, is a direct form of guiding in which the organism obtains information about the orientation of the stimulus field in relation to itself and is more likely to turn in the appropriate direction than in another one. This results in a more straight-line approach to (or retreat from) the stimulus source. There are two fundamentally different ways of obtaining the directional information. **Tropotaxis** involves *simultaneous sampling* by two or more receptors separated on different parts of the organism's surface; here, the organism directly measures the spatial gradient by comparing the intensity at the different body positions. **Klinotaxis** involves *sequential sampling* by a single receptor moved from one place to another; here the organism measures a temporal gradient, then infers the spatial gradient from information about how the receptor was moved [139, 204, 416]. These basic principles apply to intensity gradient orientation in all modalities [114].

Tropotaxis using true **stereo olfaction** requires that the paired nostrils or receptors be placed sufficiently far apart. Larger animals and those with wide heads can accomplish this easier than animals with small or narrow heads. Numerous adaptations, such as the wide head of the hammerhead shark (*Sphyrna*), have evolved to increase the distance between receptors or ensure that air or water is drawn from the side rather than from the front of the animal, [361, 391]. Placing receptors on long antennae provides even greater separation. The forked tongues of many snakes perform the same function [338, 359]. Mammals with well-developed olfactory systems, such as rats and dogs, have nostrils designed to draw intake air in laterally [85, 312, 409]. Also required for true stereo olfaction is isolation of the two nasal passages and parallel neural pathways that eventually converge for bilateral comparison. In normal circumstances of attraction to a stimulus, if the receptors are stimulated unequally, the animal turns in the direction of the one more strongly stimulated. Once both receptors are receiving equal stimulation, the animal moves in a straight line and thus directly approaches the source. One way to determine whether stereo olfaction is being used is to remove, deactivate or stimulate just one of the two receptors and place the animal in a gradient stimulus field. This should produce biased turning to one side or inability to accurately locate the source.

Klinotaxis requires either back-and-forth movement of the head or movement of the entire body. Wormlike organisms move their heads in an undulatory manner to either side of their direction of locomotion. A difference in stimulus intensity between the two sides will cause the animal to turn toward the side with the greater intensity. One way to experimentally test for the presence of this method is to pulse the signal at a rate similar to the rate of head movement, which should cause consistent turning in one direction. Comparison of two subsequent samples requires an accurate means of detecting changes in odorant concentration. Insects possess special neurological devices to accomplish this, consisting of interneurons in the olfactory bulb that flip-flop in their response whenever there is a change in the spike rate of a receptor cell [2, 207]. Given equal receptor sensitivity, klinotaxis requires a greater concentration gradient than tropotaxis, because sampling time is shorter. Both sequential and simultaneous sampling methods require good mechanisms for removing and flushing detected molecules away from receptor cells to facilitate rapid repeated sampling [99, 112, 196, 385, 388].

FOLLOWING TRAILS   Chemical trails are produced when mobile animals inadvertently or purposefully leave odorous molecules in their wake. Odor trails are usually deposited on a substrate, but a few organisms leave short-term trails in open water or air, like a jet tail. Receivers generally do not have great difficulty staying on a trail once they detect it; the challenge is tracking it in the right direction. Dogs are superb at this task, nearly always getting it right. While searching for the track they move quickly and sniff in short bouts. Once they have found the trail they slow down, make a few longer sniff bouts at several footprints along the trail, and within 3–5 seconds are able to decide which direction the trail maker was headed. They are able to detect the difference in concentration of diffusing scents in the air from a minimum of five consecutive footprints made 1 second apart and up to 20 minutes earlier [171, 370]. Copepod males use a similar strategy when they have encountered the diffusing odor trail of a gravid female, at first zigzagging back and forth to figure out the direction of the concentration gradient, then

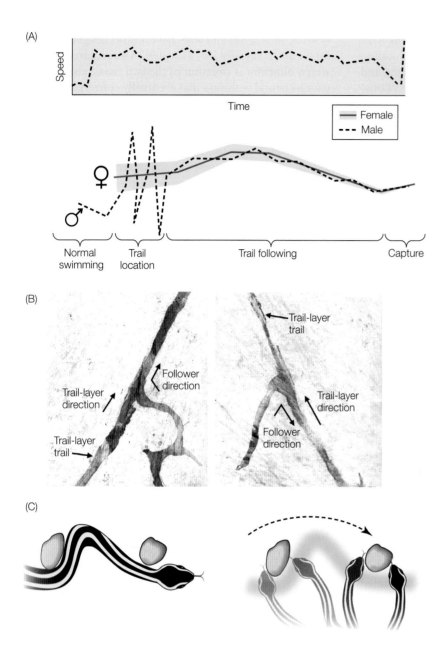

FIGURE 6.31 Trails with directional information (A) This summary of trail following in *Temora* copepods shows the path and diffusing trail of a female as she moves from left to right and a following male's trajectory. Temporal phases of the encounter and the male's corresponding swimming speed are also shown. (B) Mirror-image trail tracking by wolf snails (*Euglandina rosea*): (left) original trail laid from bottom to top being successfully tracked by follower; (right) mirror image of same trail made by pressing a transparency sheet over the top and peeling it back, being tracked in the wrong direction. (C) Female garter snake (*Thamnophis sirtalis*) deposits lipid pheromones (green) on the sides of objects against which she pushes; male tracker can determine direction of travel based on which side has more pheromone. (A after [102, 400]; B after [344]; C after [137].)

ORIENTATION IN A TURBULENT FLOW FIELD Orientation in a chemical plume carried by a current from a stationary source usually requires an altogether different strategy because of the turbulent and patchy nature of the stimulus [385]. Although it was once believed that animals could average successive samples and determine the approximate direction of the concentration gradient, we now know that the short-term variations in current flow and chemical concentration are too severe for this simple strategy to work. Since the signal is carried by a current, the source has to be upstream. If the animal can couple reception of the olfactory signal with detection of the direction of the current, then the direction of the source can be determined. **Anemotaxis** is the term used for orientation with respect to the wind in terrestrial habitats, and **rheotaxis** refers to orientation with respect to currents in water. Another potential source of directional information in a turbulent plume is the temporal and spatial structure of chemical patches. The intermittent signal varies at two scales: the large-scale meandering of large patches across the landscape at time intervals of seconds to minutes interspersed by large zones of no signal, and the fine filament structure and small eddies that occur within patches that vary over a time scale of milliseconds. At distances close to the source, the large patches will have higher concentration peaks, sharper edges, and short, rapid internal filament peaks, whereas at greater distances, patches will have lower peaks, fuzzier edges, less discernable filament structure, and larger no-signal zones (see Figure 6.24C). These distance effects will be more exaggerated when mean current flow and

tracking the trail by staying within its edges (**Figure 6.31A**). Snail trails appear to contain intrinsic directional information, presumably because the wide mucus deposition has a right–left asymmetry in chemical composition (**Figure 6.31B**). Some male snakes and lizards follow odor trails laid by conspecific females using their forked tongues, which can spread to twice the width of their heads and with simultaneous samples detect the trail edge [84]. Snakes have special directional information available to them, because the female deposits skin lipid scents selectively on the sides of objects against which she pushes while traveling (**Figure 6.31C**) [137]. Finally, an interesting example of polarized odor trails has been described for stingless bees, in which foragers deposit a decreasing density of odor marks on foliage along their flight route from the feeding site back to the nest [280].

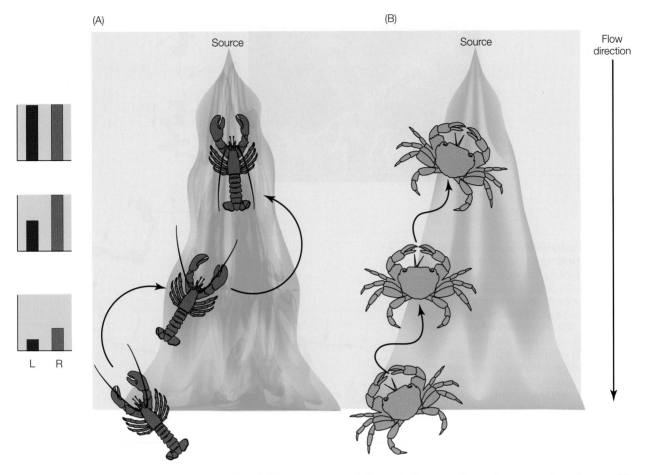

FIGURE 6.32 **Orientation in aquatic turbulent fields by substrate-walking animals** (A) Lobsters live in turbulent water and have long antennae, so they meander between the two edges of the plume using tropotaxis; bar graphs show outputs of left and right sensors at each position. They also make use of the increasing concentration gradient (klinotaxis) and possibly the within-patch fine structure as they walk up-current [16, 211]. (B) Crabs live in less turbulent estuarine environments where flow conditions are more laminar. The edge of the plume is relatively sharp under these conditions, so they walk up-current using slight turns to detect the presence of the edge. Crabs require a slow flow—their performance degrades significantly with either no flow or a fast turbulent flow. (After [399].)

overall turbulence are greater; and because water is more viscous than air, the spatial and temporal scales for the same current velocity will be smaller [276].

An animal walking on a substrate can easily detect the direction of a current by the forces it exerts on its body. It detects these forces through **mechanosensors**, specialized hairs on the body or antennae analogous to the particle detectors for near-field sound described in Chapter 4. Much of the work on substrate-walking animals has been done on crustaceans that combine detection of chemical plume structure with detection of water current direction, called **eddy rheochemotaxis**. Upon detection of an informative chemical, both lobsters and crabs begin a meandering up-current walk, although they use different mechanisms to follow the stimulus (**Figure 6.32**). A flying or swimming animal cannot make use of mechanosensors because its entire world is drifting. To distinguish the effects of their own movement

from that of the ambient medium flow, it must reference its up-current progress with respect to visual landmarks. This works well for terrestrial animals flying within view of earthbound structures, and is called **optomotor anemotaxis**. Male moths attempting to locate calling females use this sampling method. When the male detects the odorant, he flies upwind. If he loses the scent, he flies across the wind current in a zigzag pattern, called **casting**, until he catches the scent again and turns upwind [28, 65, 275]. A typical male flight trajectory is shown in **Figure 6.33**. Moths and crustaceans also make use of the within-patch filament structure to gauge how close they are to the source [63, 212, 245, 384, 387]. Swimmers in open water, such as sharks and salmon, have no access to visual landmarks. They appear to use an internal pattern of counterturning or zigzag swimming to try to detect the two edges of a plume and gradually move up the concentration gradient [401]. Recent models of

4 m

FIGURE 6.33 Orientation by moths flying in a windy turbulent field In a field trial, female gypsy moth (*Lymantria dispar*) pheromone was continuously emitted from the position marked with a star. Soap bubbles were simultaneously produced to monitor the position of plume patches and the wind direction (arrows). The tracing shows the trajectory of a male released from the lower left corner. When the male was within a plume patch (indicated by thick lines) he flew upwind with shallow zigzags, and when he was outside of any odor patches he made longer flights perpendicular to the wind direction until he found another patch. Inset shows a male gypsy moth with large antennae flapping his wings to increase reception of pheromone he has detected. (After [14, 92, 93].)

searching algorithms confirm that the most effective strategy is to spend some initial effort making wide exploratory movements to obtain more information on the likely source location before attempting to approach it. These algorithms are now being adapted for olfactory robots designed to search for chemical leaks, drugs, and explosives, a practical spin-off from the study of chemical signaling in animals [22, 225, 241, 250, 383].

SUMMARY

1. Olfaction is the oldest method of communication, having evolved from chemical mechanisms that primitive organisms used to identify food and locate mates. The sender's ability to control the transmission of chemical signals is limited, which explains the evolution of more sophisticated signaling modalities in more advanced animals. Individual signal molecules must move the entire distance from the sender to the receiver via **current flow** or **diffusion**, or the receiver must approach and make **direct contact**. In comparison to light and sound transmission, olfaction is considerably slower, its stimulus field contains no directional information, temporal patterning is not possible, and it possesses no linear array of variants equivalent to the spectra of sound and light frequency. Chemicals that facilitate communication between species are called

allelochemicals. Chemicals used for communication between conspecifics are called **pheromones**. Combinations of several chemicals in a specific ratio are called **pheromone blends**, and variable mixes of many chemicals are called **pheromone mosaics** or **odor signatures**.

2. Pheromones are organic compounds differing in size, shape, composition, and polarity. Airborne odorants must be sufficiently **volatile** to vaporize, and waterborne pheromones must be **water soluble**. Pheromones are produced in four ways. **Secretory glands** are well-defined structures near the body's surface that manufacture, store, and release highly specific chemical products. Glands are classified by secretion mechanism: **merocrine**, **holocrine**, or **apocrine**. The deposition of these gland products is usually associated with specific behaviors and social circumstances,

leaving little doubt as to the general function of the signal. **Waste metabolites** in urine, feces, saliva and sweat released from body orifices and organs associated with digestion and reproduction often provide useful information about gender, condition, reproductive stage, and dominance status. Internal gland products may be transported to these externally released fluids with special binding proteins. A few species derive pheromones from **plant secondary compounds** and from **bacterial breakdown products**.

3. Odors can be released into the medium in three basic ways: by **passive exposure** of glandular tissue, by **piggy-backing** on another activity, and by **specialized release behaviors**. Any of these mechanisms may be employed to deliver odors to three types of locations: direct release into a **fluid medium**, deposition onto the **sender's own body**, and deposition onto **other solid surfaces**. Some common dissemination methods include **spraying, hair dispersal, self anointing, self-generated current, footprints**, and **deposited marks**.

4. In the absence of any current flow, chemical odorants can be transmitted from sender to receiver via **diffusion**. If the volatility of the chemical is known, the spread of the odorant with time can be precisely modeled for a single **instantaneous puff**, **continuous emission**, and **moving source emission**. If the threshold concentration of detection by receivers is also known, the **active space** can be described. Diffusion is likely to operate only over short distances. It is probably most useful for small organisms, such as ants and termites, that live within the viscous boundary layer of a substrate and for some small aquatic organisms that live in relatively still water.

5. Long-distance transmission of chemical odorants must be coupled with environmental or sender-generated **current flow**. If the flow is slow and **laminar**, an elongated active space with a smooth concentration gradient is formed. However, most flows are **turbulent**, characterized by a complex, unpredictable, and variable pattern of eddies, whirls, and vortices. The stimulus field of an odorant in a turbulent flow is a patchwork of filaments and plumes that spreads in a wedge from the source. Odorant dispersal is also affected by vegetation, weather conditions, humidity, and topographical features of the landscape.

6. **Chemoreception** can be subdivided into two general categories: **olfactory reception** of airborne or waterborne chemicals from a distance source and **contact reception**. Olfactory reception begins by drawing odorant-laden medium into the chemoreception organ with mechanisms such as **sniffing** air into the nasal cavity, **pumping** water into nares, or **flicking** antennae. For contact reception, the receiver approaches the source and makes **direct contact** with the chemical.

7. **Chemoreceptor sensory cells** form a sheet of **epithelium** within **chemosensory organs**. External chemicals must pass through a mucus layer surrounding the sensory cells. The mucus layer contains binding proteins for transporting important odorants as well as compounds that defend the sensitive tissue and neurons against toxins and pathogens. The sensory cells are **bipolar neurons** derived from **ciliary** or **microvillous cells** that form projections into the mucus. Anchored on the membranes of these projections are **G protein–coupled receptors** that temporarily bind with certain odorant **ligands**. Binding activates a chemical cascade and nerve impulse. A limited range of odorant molecules will stimulate a given sensory cell. There are a large number of sensory receptor cell types, each sensitive to different chemical features and encoded by a different gene from **multigene families**.

8. At the first olfactory processing region in the brain, all of the axons from sensory cells of the same type converge on a few spherical **glomeruli**. Second-order neurons pass the information from each receptor cell type on to higher processing centers. A given chemical is recognized by a **combinatorial code** of stimulated glomeruli. Some animals have **dual pathways** to different parts of the brain, one primarily for analysis of the huge array of general environmental chemicals and one primarily for analysis of specific odorants, usually pheromones. Many terrestrial vertebrates possess two separate chemosensory organs, the **olfactory organ** for detection of airborne odorants and the **vomeronasal organ** for detection of contact and waterborne odorants.

9. The purpose of long-distance olfactory signals is to attract or repel receivers. If the signal is transmitted by diffusion or laminar flow, receivers can employ either **klinotaxis** (sequential sampling from two locations) or **tropotaxis** (simultaneous comparison from separated receptors) to determine the direction of the concentration gradient, and orient their movement accordingly. If the signal is transmitted in a turbulent flow field, receivers will couple detection of the odorant with assessment of the current flow direction to move up-current toward the source. Substrate-walking animals can use **mechanosensors** to detect current flow direction. Flying or swimming animals must assess their movement relative to visual landmarks and will initially make wide zigzagging exploratory movements before approaching the source.

Further Reading

Two recent books provide in-depth coverage of all aspects of chemical signaling described in this chapter, one by Müller-Schwarze [273] for vertebrates and another by Wyatt [416] broadly integrating all animal groups. The Nov. 16, 2006 issue of *Nature* (no. 7117) contains a series of excellent review articles on olfaction and taste. Recent information on the vomeronasal organ can be found in Evans [127], Brennan and Keverne [47], Halpern and Martínez-Marcos [158], Dulac and Wagner [109], and Baxi et al. [26]. Niimura and Nei [282], Liman [228] and Roelofs et al. [326] review current perspectives on the evolutionary dynamics of chemosensory genes. Dethier [100] and Kepecs et al. [205] summarize mechanisms of sniffing. Recent reviews of olfactory sensory mechanisms include Su et al. [364], Spehr and Munger [357],

Tirindelli et al. [373], Touhara and Vosshall [376], Imai [188], and Kaupp [203]. For odor coding see Malnic et al. [246], Hallem et al. [156], Saito et al. [331], and Soucy et al. [355]. Vickers [385], Weissburg [401], and Cardé and Willis [65] provide excellent explanations of turbulent flow and orientation in turbulent plumes.

COMPANION WEBSITE
sites.sinauer.com/animalcommunication2e

Go to the companion website for Chapter Outlines, Chapter Summaries, and References for all works cited in the textbook. In addition, the following resources are available for this chapter:

Web Topic 6.1 *How are pheromones identified?*
To identify a pheromone, putative chemical secretions are collected, purified, and tested for biological activity with a good behavioral or physiological bioassay. The gas chromatograph plays a central role in identifying the chemical compounds.

Web Topic 6.2 *Guide to organic compounds and biosynthetic pathways*
Examples and a brief summary of organic compounds (lipids, steroids, terpenoids, peptides, proteins) and biosynthetic pathways mentioned in this chapter.

Web Topic 6.3 *Marking behaviors and displays*
More examples of the behaviors involved in producing and disseminating scents.

Web Topic 6.4 *Chemical transmission models*
Quantitative models for molecular movement under the conditions of diffusion, laminar flow, and turbulent flow attempt to describe the temporal and spatial characteristics of the active space for chemical signals and cues.

Web Topic 6.5 *Vertebrate dual chemosensory system*
The vomeronasal organ does not function solely for pheromone reception, nor the nasal organ solely for environmental chemical reception. The two organs have overlapping roles. Vertebrates also possess several other chemoreception organs that are described in this online unit.

Chapter 7

Short Range Modalities

Overview

The chapters on audition, vision, and chemoreception described modalities that can be used to communicate over considerable distances. We conclude our review of sensory biology with an examination of three modalities that are effective only over very short ranges. **Touch stimuli** require direct physical contact between the receiver and another solid object. **Hydrodynamic stimuli** are generated by cohesive displacements of fluid medium (air or water) that can be detected when sufficiently close to the receiver. This general definition would include some near-field sounds discussed in Chapters 2 and 3, but would exclude the propagated pressure variations of far-field signals. **Electrical stimuli** result from electrical fields produced by other animals or distortions in the electrical fields produced by the receiver. As we shall see, these three modalities share a number of common properties and even a similar evolutionary ancestry.

Touch

Touch signal generation and propagation

Touch (also called tactile or thigmotactic reception) is probably the most ubiquitous of sensory modalities. Nearly all organisms have some mechanism to detect contact between themselves and other objects. This is necessary to avoid obstacles during locomotion, capture and manipulate food items, and coordinate access and occupation of a shelter.

TOUCH AND SOCIAL INTERACTIONS   Touch can also play a key role during social interactions between animals. While the early phases of animal conflicts are typically mediated with visual, auditory, olfactory, or electrical displays, touch is a common component of more escalated stages (**Figure 7.1**). Male hartebeest (*Alcephalus buselaphus*) groom each other's cheeks and necks when visual displays have not resolved a contest but the animals are not yet ready to engage in full combat [406]. Mountain zebras (*Equus zebra*) challenge each other by laying their heads on the opponent's shoulders and pressing down [120]. Once a conflict does escalate to physical combat, touch sensors are invariably recruited to coordinate a combatant's movements and obtain tactile information indicating the opponent's strength, size, and agility. After a contest is resolved, the loser may perform submissive and appeasement actions that involve touch. A subordinate African buffalo (*Syncerus caffer*) lowers its

(A)

(B)

(C)

FIGURE 7.1   The role of touch in conflict signaling (A) Many fish species perform ritualized mouth wrestling during contests. Here two red-spotted sandperch (*Parapercis schauinslandii*) on a Philippines coral reef engage in a lip-lock. This behavior includes both strategic and communication components. (B) Harem owner and intruder male plains zebras (*Equus quagga*) begin competitive interaction by placing their heads on each other's shoulders. Either may then press down on their opponent or escalate into more overt fighting. (C) A male mouflon sheep (*Ovis ammon*) that has just lost a contest is licking the back of the winner.

head and places it gently under the abdomen or between the rear legs of a dominant animal [406]. Subordinate mouflon sheep (*Ovis musimon*) lick the dominant on the neck, and chimpanzees (*Pan troglodytes*) appease a dominant by gently touching its genitals [390].

Mating contexts show a similar increase in the use of touch as an interaction escalates (**Figure 7.2**). Initial displays rely on longer-distance modalities because one or both sexes may be wary of close approach. At intermediate stages, one sex (usually the male) will groom, stroke, or rub against the other. In addition to tactile stimulation, this provides an opportunity to assess chemical signals on the potential mate or apply additional chemical signals. Many male mammals announce their intention to mount a female before doing so: male kangaroos grasp the tail of the female; gazelles stretch their front leg forward to stroke or tap the inside of the female's rear leg; and a wide variety of species such as elephants, rodents, antelopes, and rhinos place their head on the back of the female [390, 406]. Females can also use touch signals at early stages of courtship: unreceptive female

harvester ants (*Pogonomyrmex*) mounted by a male produce a stridulation that propagates directly through the points of body contact and causes the male to dismount [271]. Touch sensors usually play key roles in pair positioning and coordination during mating. It has even been suggested that females discriminate between different male consorts according to the level of tactile stimulation of the female reproductive organs during copulation [112]. It is not clear to what degree female stimulation might account for the wide variation in duration of copulation in mammals: for example, the average chimpanzee copulation lasts eight seconds, while wolves and some weasels copulate for hours [107].

Touch is widely used for social integration (**Figure 7.3**). Many animals greet the arrival of a known individual with physical contact. Ants, bees, wasps, and termites rhythmically tap the antennae or body of a new arrival with their own antennae; this provides them with identifying olfactory information about this individual, but also lets the latter know that it has been detected and assessed [179, 419]. Mammalian greetings include inserting the tip of the trunk in the

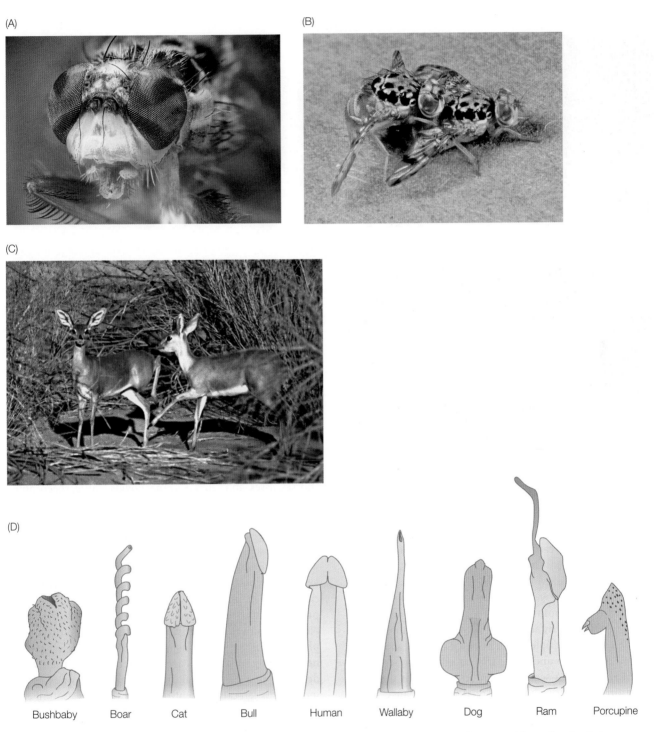

(A)

(B)

(C)

(D)

Bushbaby   Boar   Cat   Bull   Human   Wallaby   Dog   Ram   Porcupine

FIGURE 7.2   The role of touch in courtship (A) Both sexes of medflies (*Ceratitis capitata*) have elongated aristae on their antennae. These fine filaments are visible dangling over the fly's white face in this close-up. During courtship, the male faces the female and, using a rocking motion, uses his aristae to spread hers outward. (B) If she accepts these actions, that is the signal that he can move behind her to mount and mate [52]. (C) Male steenbok antelope (*Raphicerus campestris*) tapping the inside rear leg of a female to indicate his intention to approach and mount for mating [390, 406]. (D) Variation in penis structure of male mammals. Competition between males to provide tactile stimulation to females during copulation has been proposed as one explanation for the elaborate differences between species [107, 112, 194, 360]. Penis elaboration in some taxa such as damselflies is also designed to remove sperm of prior consorts during copulation. In mammals, sperm from prior matings has usually migrated from the vagina through the cervix into the uterus by the time another male mates with a female; thus, sperm removal is a less likely explanation than stimulation for the diversity of mammalian penis structures. (After [107, 112].)

(A)

(B)

(C)

FIGURE 7.3 **The role of touch in social integration** (A) Two weaver ants (*Oecophylla longinoda*) antennating each other as one returns to the nest. (B) A male chimp (*Pan troglodytes*) confirms his subordinate status by touching the scrotum of a dominant male. (C) Allogrooming (or in birds, allopreening) is widely used to confirm affiliative bonds or status within a group. Here, a mourning dove (*Zenaida macroura*) allopreens its mate.

new arrival's mouth (elephants), pressing noses or mouths together (ungulates, carnivores, and rodents), draping a hand over the back in a loose hug (chimpanzees), or shaking hands (some human cultures) [120, 390].

Perhaps the most widespread integrative behavior involving touch is **allogrooming**, the grooming of one animal by another. As with many activities involving touch, allogrooming often serves multiple functions. One obvious goal is the removal of parasites and debris from the body surface of the groomed animal. A second common function is the reaffirmation of affiliative bonds or relative dominance status between the groomer and groomed animal. Finally, allogrooming provides an opportunity for the groomer to examine olfactory cues on the groomed individual and perhaps add additional ones. Allogrooming is common in the social insects (ants, bees, wasps, and termites), but is apparently absent in social spiders (which self-groom extensively) [179, 419]. Allopreening by birds is widespread but spottily distributed within any given order or family. Colonially nesting seabirds, penguins, doves, caracaras, kites, kestrels, vultures, storks, egrets, rails, some geese and ducks, anis, owls, parrots, oilbirds, swifts, mousebirds, hornbills, toucans, ravens, crows, jays, long-tailed tits, babblers, bulbuls, white-eyes cactus wrens, estrildid finches, and cowbirds allopreen, whereas

many closely related species do not [55, 134, 169, 170, 222, 233, 313, 328, 335, 366]. Nearly all terrestrial mammalian mothers groom their offspring. Allogrooming among adult mammals is less ubiquitous but occurs in carnivores, equids (horses, asses, and zebras), swine, bovids (cattle, sheep, antelopes), deer, primates, and some rodents [110, 120, 122, 245, 293, 406].

Prolonged physical contact (huddling) is another mechanism for social integration that involves touch stimuli. As with greeting and allogrooming, huddling can lead to the exchange of both olfactory and tactile information. Huddling can also aid in heat exchange. Colonial corals, hydroids, entoprocts, ectoprocts (bryozoans), annelid worms, mollusks, barnacles, and tunicates live their entire lives in prolonged physical contact with conspecifics [57]. Linkages between neighboring individuals are sufficiently tight that tactile detection of aversive or alarm movements by one or a few individuals can cause an entire colony to react. A number of herbivorous insect nymphs (Hemiptera) and larvae (Lepidoptera, Hymnenoptera, and Coleoptera) form aggregations in which prolonged contact stimuli play a major role in group coordination [78, 88, 91, 128, 133, 148]. Certain moth, butterfly, sawfly, and weevil larvae form daily (or nightly) processions in which lines of individuals migrate between foraging and resting sites using a combination of tactile, auditory, and chemoreceptive stimuli (**Figure 7.4**) [89–91, 129–131, 133, 229]. Some insect larvae, including several of the processionary species above, adopt circular formations in which each individual moves into lateral contact with two others to form a rosette, a behavior called **cycloalexy** [91, 131, 210, 211, 392]. Whereas most fish and amphibians do

FIGURE 7.4 The role of touch in coordinating group movements in social insect larvae (A) Procession of weevil larvae (*Phelypera distigma*) using a combination of trail pheromones and continuous touch stimulation through contact. (B) Close-up of the same species shows contact between successive individuals. (C) Cycloalexic formation of the same species when threatened or not foraging.

not huddle, a variety of snakes and a few lizards aggregate in hibernacula where they are often found in contact with many other individuals [150]. Some communal dens may house multiple species of snakes, and some aggregations, such as those of the red-sided garter snake (*Thamnophis sirtalis parietalis*), may number in the thousands [39, 355, 356]. A variety of birds (in many cases, the same ones that allopreen) huddle when perched or at night roosts. Examples include partridges, bobwhites, chachalacas, owls, anis, parrots, black

vultures, Inca doves, swifts, nuthatches, wood swallows, martins, long-tailed tits, starlings, and ravens [313]. Huddling mammals include bats, rodents, carnivores, pinnipeds, elephants, hyraxes, primates, swine, hippos, and some of the more social marsupials [122, 406]. Pairs of horses and zebras will rest standing parallel to each other but facing in opposite directions so that the head of each can easily rest on the rump of the other [406]. Many toothed whales and porpoises (Odontocetes) rest or even swim with fins or flippers touching a conspecific [10, 82, 389, 413]. In many huddling organisms, the choice of partners, position in the huddle, and degree of prolonged contact can provide information to conspecifics about social bonds, relative status, or physical condition.

Touch can also be used to provide environmental information. Some ants recruit fellow workers to new food finds by tapping them with their antennae and forelegs. They then lead the recruits to the food, staying close enough together to maintain physical contact [179]. Honeybees perform dances inside their hives to direct nestmates to new food finds. While some of the information is transferred through sounds (see Chapter 14), recruits may also monitor the dancer directly with their antennae. Some stinging, toxic, or distasteful arthropods stridulate when captured by a predator; it has been suggested that these vibrations, transmitted directly through points of physical contact between prey and predator, may serve as warning signals to the predator (see Chapter 14) [58, 271, 277, 421].

CLASSIFICATION OF TOUCH STIMULI   Touch stimuli can be classified along several independent axes. One obvious axis is the intensity of the stimulus. At one extreme, a single courtship caress may be barely detectable by a receiver. Two male antelopes that have locked horns in a pushing contest both experience a persistent touch stimulus of intermediate intensity. A high-intensity contact, such as a strong bite or peck, may cause the receiver pain or even tissue damage.

A second axis for classifying touch stimuli is the temporal pattern of the stimulus. Useful parameters include the duration of the stimulus, whether it is repeated and with what frequency, and whether its amplitude or frequency is modulated over the duration of the stimulus or over a sequence of repeated stimuli. Unlike the situation in audition or vision, temporal pattern in a touch stimulus may be generated either by the sender or by the receiver. Arthropod antennation of a conspecific produces a distinct temporal sequence of touch stimuli on the recipient and is a sender-generated pattern. On the other hand, an animal that slides a sensitive finger (primates) or whisker (rodents) across a surface receives a temporal pattern of touch stimuli that depends both on the texture of the surface and the speed with which the sliding is performed. The pattern in this case depends in part on how the receiver samples the surface. Many animal social exchanges involving touch will have temporal patterns that depend on actions or properties of both the sender and the receiver.

A third axis is the type of touch receptor involved in processing a given stimulus. **Thermoreceptors** are widespread in animals and separate ones may be specialized for assessing higher-than-normal (hot) and lower-than-normal (cold) temperatures. Severely intense stimuli, whether due to temperature, pressure, or chemical exposure, are usually handled by specialized pain receptors (**nociceptors**). Nonpainful pressure on an animal's body surface will stimulate both **mechanoreceptors** just under the surface and deeper-tissue **proprioceptors** that monitor the tension of muscles or flexing of joints [333]. Mechanoreceptors are often subdivided into those that are **slow-adapting**—responding continuously to a persistent pressure or tension—versus those that are **fast-adapting**—responding to rapid changes in pressure or tension, but not to continued stimulation [258]. Slow-adapting receptors are often recruited to measure static displacements, whereas fast-adapting receptors are used to monitor the velocity or acceleration of touch stimuli. Mechanoreceptors can also be divided into those with a narrowly and sharply bounded zone of sensitivity in the skin (**focused**) versus those with a more gradual decrease in sensitivity at increasing distances from a single central hotspot (**diffuse**). When an animal touches a surface or is touched by another animal, that touch is likely to stimulate a combination of thermoreceptors, proprioceptors, and mechanoreceptors with different adaptation schemes and sensitivity zoning. Touching different types of surfaces with different intensities and temporal patterns will stimulate different combinations of touch receptors in different ways. The input to the brain of the receiver from a single touch stimulus can thus be quite complex.

FIGURE 7.5   Mechanisms by which ion channels in cell membranes can act as mechoreceptors (A) Here, a protein ion channel that is normally closed is tightly linked to the adjacent lipid bilayer that constitutes the cell membrane. When the membrane is stretched by osmotic swelling or by deformation due to collisions with external objects, the increased tension on the membrane causes the ion channel to open, allowing ions to move down their concentration gradient (either into or out of the cell), in turn triggering other cell responses. This is a likely mechanism for archaeal and bacterial systems. (B) Alternatively, the protein ion channel is attached to an external tissue matrix outside the cell on one side and to the cell's internal cytoskeleton on the other. Bending or collisions with the external matrix put tension on the channel, causing it to open. This appears to be the most common mechanism in most animal mechanoreceptors. (After [374].)

The fourth axis for the classification of touch stimuli is the function of the stimulus. It is useful to distinguish between the generation of touch stimuli that are solely incidental to some strategic function such as fighting, copulating, feeding, or huddling to keep warm, and the generation of stimuli for the purpose of transmitting touch signals to a receiver (e.g., the caress of a female during courtship by a male). In practice, many actions by animals will fall between these extremes and fulfill both strategic and informational functions. When two antelopes lock horns or two fish lock jaws during a fight, the prolonged contact between combatants allows each to gauge the other's skill and strength, as well as give each the opportunity to push the other off balance or otherwise make it more vulnerable to subsequent injury. We should also remember that even when contact between two animals is largely for the exchange of information, touch may be only one of the modalities providing that information: for example, the marking of one animal with a pheromone by another typically provides both olfactory and tactile information. Identifying the modalities involved in contact helps to estimate the relative role of tactile communication in actions that may have multiple functions.

Touch signal reception

Mechanoreception evolved very early in the history of living organisms. Many members of the Archaea and Eubacteria have mechanosensitive ion channels in their cell membranes [243, 274, 275, 314]. These channels respond to changes in the tension or thickness of the cell membrane's lipid bilayer by allowing otherwise excluded ions to cross the boundary (**Figure 7.5**). This flow can change the electrical field across the cell membrane and trigger further responses to that electrical change by the cell. It has been suggested that mechanosensitive channels originally evolved to detect and counter osmotic swelling or shrinkage of single prokaryotic cells; they may be the earliest of all sensory transducers [326].

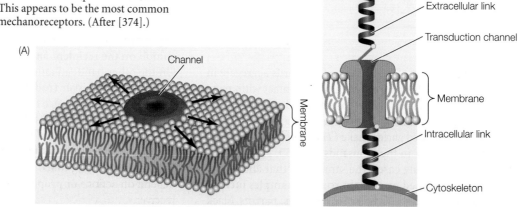

(A)

Channel

Membrane

(B)

Extracellular anchor

Extracellular link

Transduction channel

Membrane

Intracellular link

Cytoskeleton

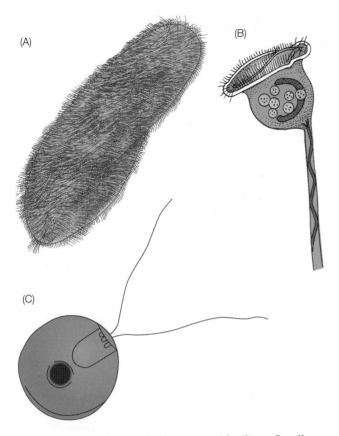

FIGURE 7.6 **Single-celled eukaryotes with cilia or flagella**
(A) *Paramecium* with numerous cilia over its entire cell surface.
(B) Stalked *Vorticella* with cilia surrounding the large open
umbrella it uses to trap prey. (C) *Chlamydomonas* with two large
flagella used for locomotion.

TOUCH IN SINGLE-CELLED EUKARYOTES   Although mechanosensitive ion channels may occur at any point on the membrane of single-celled organisms, they are often concentrated
in regions likely to encounter obstacles during locomotion or
foraging. The cell membrane of the single-celled ciliate *Paramecium* is covered with motile cilia that beat in synchrony
to propel the cell along a line parallel to its long axis (**Figure 7.6**). When the anterior part of the cell encounters an
obstacle, deformation of the cell membrane, either through
direct contact or indirectly by pressure on cilia in that region,
causes local mechanosensitive channels to open and admit
calcium ions [113, 114, 197, 260, 383]. The calcium influx
depolarizes the otherwise negative electrical charge inside this
region of the cell (relative to the outside), and this in turn
triggers a regenerative depolarization (action potential) that
sweeps over the entire organism's surface [32]. The action
potential in turn elicits an influx of calcium ions into the
central space of the cilia, and this causes the ciliary motors to
reverse direction. The paramecium thus backs away from the
obstacle. When the posterior end of a paramecium collides
with an obstacle, mechanosensitive channels in that region
trigger an outflow of potassium ions, resulting in hyperpolarization of the cell, faster beating of cilia in the normal

direction, and thus acceleration of the forward motion of the
organism [114].

   The membranes covering the cilia in *Paramecium* do not
appear to host mechanosensitive channels; instead, the channels are located in the cell membrane between the cilia [259].
A similar situation occurs in the stalked and sessile ciliate
Vorticella, which defensively retracts its large funnel-shaped
foraging apparatus when bumped by larger organisms [220,
357]. Here again, the sensitive channels are located not in the
ciliary membranes, but in that of the cell body [219]. Flagellated algae such as *Chlamydomonas* and *Spermatozopsis* also
change their direction of travel when encountering obstacles.
The ionic mechanisms appear to be similar to those identified
in ciliates, but the mechanosensitive channels of *Chlamydomonas* are located in the membrane encasing each flagellum
as well as that encasing the cell body [176, 425, 426]. Some
ciliates have evolved two kinds of cilia: those devoted to locomotion, and others that are nonmotile and function to alert
the cell to the presence of nearby objects and surfaces [237,
295].

INVERTEBRATE TOUCH RECEPTORS   We noted in prior
chapters that the auditory, visual, and olfactory receptor cells
of animals often include a modified cilium in the part of the
cell exposed to the stimulus. Not surprisingly, this also turns
out to be true for many invertebrate touch receptors. There
are several reasons why cilia might be so frequently recruited
for sensory transduction. First, as we have seen, many single-
celled organisms, including the likely ancestors of animals,
evolved linkages between stimulus-specific ionic channels in
their membranes and their locomotory cilia or flagella. Second, the cilium is one of the few cell organelles whose internal contents can be isolated from the rest of the cytoplasm by
a selectively permeable plate at its base. Finally, the cilium has
a very high surface-area-to-volume ratio, and this, plus the
limited internal volume, permits rapid changes in the ionic
and chemical composition of its internal spaces. All of these
features make ciliated cells preadapted for rapid detection of
ambient stimuli, the triggering of appropriate responses, and
rapid recovery back to prestimulation states.

Web Topic 7.1 *Cilia and sensory receptors*
Why are cilia so often associated with sensory cells in animals?
Here we expand on the reasons listed above for their role
in mechanoreceptors, and provide more details on the links
between ciliary structure and ciliary function in general.

   Unlike the motile cilia of *Paramecium*, animal mechanoreceptor cilia usually host abundant mechanosensitive
ion channels on their membrane surfaces. Stretching or
distortion of these channels produces a local depolarization of the cilium that spreads to the cell body of the receptor cell and triggers action potentials in attached axons or
synapsed nerve cells. Mechanoreception in most animals
triggers sensory nerve responses 10–100 times faster than

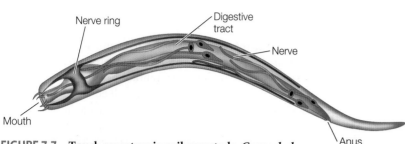

FIGURE 7.7   Touch receptors in soil nematode, *Caenorhabditis elegans* Mechanoreceptor and chemoreceptor dendrites extend from a nerve ring around the pharynx to six labial bumps around the mouth and a pair of saclike sensory organs positioned on opposite sides of the head. All of these mechanoreceptors have ciliated ends and respond to strong touch on the head due to collisions or during mating. The six red nerves (three for the anterior half of the worm and three for the posterior) extend long nonciliated dendrites anteriorly to monitor soft touches to half the external cuticle. These nerves help mediate locomotion, defecation, egg laying, and other activities in which monitoring subtle body contact is important [57, 374, 407]. (After [374].)

either photoreception or chemoreception [227, 374]. This is because touch stimuli produce receptor depolarizations directly, whereas vision and chemoreception require a chemical cascade linking the stimulus to nerve excitation. Ciliated touch receptors are known in hydroids (Cnidaria [180]); comb jellies (Ctenophora [294, 379]); flatworms (Platyhelminthes [127, 422]); scaly worms (Scalidophora [2]); round worms (Nematoda [311, 422]); segmented worms (Annelida [44]); moss animals (Bryozoa [358]); mollusks (Mollusca [64, 244, 338]); arthropods (Arthropoda [51, 61, 98, 224, 280, 327, 423]); sea urchins and crinoids (Echinodermata [5, 34, 132]); and tunicates (Urochordata [66]). In many of these invertebrate taxa, sensory cilia are directly bent by contact with other solid objects. In others, a layer of soft tissue overlies the ciliated dendrites, and pressure indirectly transmitted through the soft tissue distorts the mechanosensitive channels. The hard exoskeletons of arthropods do not support free cilia; instead, neurons with ciliary dendrites are buried in soft tissues below the chitinous outer wall and connected to the bases of hard spines, or setae, that rotate in their socket bases when displaced by external objects. Most invertebrates distribute their touch receptors unequally over their bodies: free-living flatworms, ribbon worms, snails, and arthropods concentrate touch sensors on their anterior ends; sessile filter feeders such as entoprocts (Entoprocta), moss animals (Bryozoa), bivalves (Mollusca), lamp shells (Brachiopoda), peanut worms (Sipuncula), and tunicates (Urochordata) host the highest densities of touch receptors in their extensible (and vulnerable) foraging organs (lophophores for bryozoans and lamp shells, tentacles for peanut worms, and siphons for clams and tunicates) [57].

Although most invertebrate touch receptors contain modified cilia [57], some taxa host both ciliated and nonciliated mechanoreceptors in the same animal [142, 374]. For example, male nematode worms use ciliated touch organs on

their anterior end to detect obstacles and coordinate mating, but use nonciliated and naked nerve endings located along the sides of their body for other functions (**Figure 7.7**). Similarly, stretch receptors along muscles and across appendage joints in a typical arthropod are not ciliated, whereas bristles, setae, and chordotonal organs invariably have ciliated dendrites. Many arthropods also have nonciliated thermoreceptors [256].

TYPES OF VERTEBRATE TOUCH RECEPTORS   In contrast with invertebrates, most terrestrial vertebrates do not have ciliated touch receptors [144, 161, 256, 258, 334, 404]. This is surprising, because the cells in the majority of vertebrate tissues host a cilium at some time in their development [77, 329, 361]. While cilia are retained as surface mechanoreceptors in some organs, such as the kidneys, where they monitor urine flow, they are lost in most other terrestrial vertebrate touch receptors.

Vertebrate touch receptors can be divided into reasonably distinct categories [208, 298, 339]. Thermoreceptor, nociceptor, and some genital-region neurons have highly branched and naked dendrites that spread out at the boundary between the epidermis (the external layer of skin) and the dermis (the next innermost layer containing blood vessels, nerves, and secretory glands). These are called **free nerve endings** because of their simple dendritic structure [242]. Vertebrates also have a variety of more complex mechanoreceptors (**Figure 7.8**). Clusters of oval **Merkel cells** are present in the inner margin of the epidermis in bare regions of the body, and around hair follicles in mammalian hairy regions [163]. They are not themselves nerve cells but have synaptic junctions with the dendrites of associated neurons. Their epidermal side has fine spikes that interdigitate with adjacent epidermal keratinocyte cells. Deformation of the skin surface or the movement of a hair will deform the membranes of the Merkel cells. These membranes are known to host mechanosensitive ion channels, and the cell cytoplasm contains neurotransmitter substances that can be released to excite the nearby nerve dendrites [123, 163, 285]. Merkel cells are generally slow-adapting and spatially focused mechanoreceptors. They thus continue to activate their associated nerves during a sustained displacement of attached tissues. Merkel cells are present in all vertebrates from fish through mammals. In mammals, they occur in highest densities in

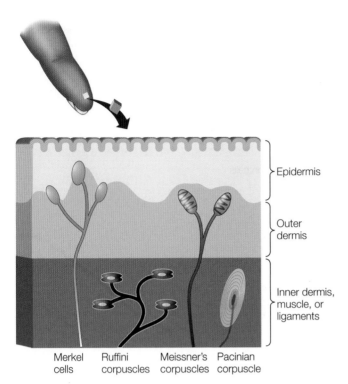

Merkel cells   Ruffini corpuscles   Meissner's corpuscles   Pacinian corpuscle

FIGURE 7.8 **The four most common types of skin touch receptors in vertebrates** Taxon-specific touch receptors are usually variants of one of these general types. The unencapsulated Merkel cells are slow-adapting receptors and can thus monitor persistent pressures. They are embedded in the lower epidermis and are only sensitive to stimuli just above them. Ruffini corpuscles are also slow-adapting, but their deeper location makes them sensitive to stimuli over a much wider area of skin. They are largely used to monitor tension and stretching in the dermis or in the roots of teeth, hairs, feathers, and scales. Meissner's corpuscles are fast-adapting receptors that ignore steady stimulation but are very accurate at tracking rapidly varying stimuli. Being close to the skin surface, they respond to stimuli just above their location. Pacinian corpuscles also respond to varying stimuli, but being deeper in the dermis and highly sensitive, they respond to stimuli over a much larger area of skin surface.

fingers, soles, lips, the noses of moles, around hair follicles, and throughout the mucosal linings of the mouth and anus [285]. **Grandry corpuscles** consist of many flattened Merkel cells alternating with bare nerve terminals (**Figure 7.9**). These are densely distributed inside the bill skin of birds [144] and in touch papillae of crocodiles [404]. The stacking of multiple layers in Grandry corpuscles converts the otherwise slow-adapting Merkel cells into a fast-adapting monitor of pressure variation.

The remaining types of vertebrate surface mechanoreceptors all consist of **encapsulated touch organs**: nerve endings sheathed by specialized Schwann cells [298]. Schwann cells are widely used by jawed vertebrates to wrap segmented myelin sheaths around axons and thereby accelerate the propagation of nerve impulses. The Schwann cells used in terrestrial vertebrate mechanoreceptors have a somewhat different geometry. Typically, a pair of flattened and parallel Schwann cells forms a "sandwich" with a single axon terminal between them. The tip or small lateral extensions of the axon may extend out of the sandwich and attach to other nearby tissues. Deformation of the ensemble causes differential movement of the sandwiched part of the axon and its tip or extensions. The resultant stretching of the membrane triggers mechanosensitive channels in the axon and action potentials in the nerve. Schwann cells are probably recruited into mechanoreceptors because they are uniquely able to attach to parts of bare axons and hold them in place when the axon tip or extensions are moved by touch stimuli.

The simplest axon-Schwann cell combination is the **Ruffini mechanoreceptor** [158, 160, 317, 342, 346,

352, 376, 405, 410]. This type of mechanoreceptor occurs in all vertebrates and tends to be located on the ligaments that connect adjacent bones, (including those holding teeth and tusks in their sockets), across joint capsules, around hair follicles, and in the bill skin of birds. A Ruffini axon typically divides into 8–32 small endings, each of which is sandwiched between two flattened Schwann cells (**Figure 7.10**). The tips of the nerve endings are surrounded by extensions of the adjacent Schwann cells that attach to collagen fibers in the nearby tissues. When those tissues are deformed or stretched, the Ruffini nerve ending is also stretched relative to the rest of the axon and thus stimulated. Ruffini organs are

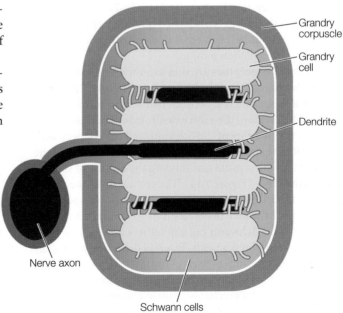

FIGURE 7.9 **Section through a Grandry corpuscle in the bill of a goose** Grandry cells respond to touch but are not themselves nerve cells. They are organized in layers and synapse with dendrites of the corpuscle nerve that lie sandwiched between them. Note fingers of Grandry cells that interweave with adjacent Grandry cells and with surrounding Schwann cells. Grandry corpuscles, unlike Merkel cells, are fast-adapting touch receptors. (After [144].)

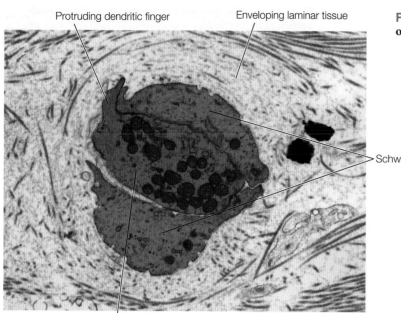

Protruding dendritic finger

Enveloping laminar tissue

Schwann cells

Single axon dendrite

FIGURE 7.10   Electron microscopic cross section of Ruffini corpuscle ending (After [376].)

usually slow-adapting and thus provide a measure of the current tension on the adjacent tissues. Because they tend to be positioned deeper beneath the skin surface than Merkel cells, a single deformation of the skin can trigger many Ruffini nerves. Their region of sensitivity is thus more spatially diffuse than that of Merkel cells. **Golgi tendon organs** are nearly identical in structure and response characteristics to Ruffini organs. However, they have one end attached to a muscle and the other to the muscle's associated tendons (which in turn link muscles to bone), and act as slow-adapting proprioceptors monitoring tendon tension.

Lanceolate receptors are slim axon terminals that form a palisade around hair follicles in mammals [161, 298, 377]. Each axon is sandwiched between two Schwann cells. Within its middle region, the axon extends many short dendrites between the margins of the Schwann cells to connect with the membrane surrounding the hair shaft on one side of the receptor and the capsule surrounding the outside of the follicle on the other (**Figure 7.11**). The conical tip of the receptor axon lacks these short axonal connections with adjacent membranes. Instead, it floats relatively freely in the intervening space and a Schwann cell cap tethers it loosely to both walls with long slim filaments. External deflection of the hair will cause compression of parts of the internal hair sheath on one side of the follicle and stretching on the other side. Either effect could trigger mechanosensitive channels in attached lanceolate receptors. Whereas continued hair deflection would produce persistent deformation of the middle parts of a lanceolate receptor, the tip is likely to move initially when the hair does, but then be pulled back to its central location by the tethers. Takahashi-Iwanaga [377] has argued that motion of the tip likely accounts for the well-known fast-adapting properties of lanceolate receptors, whereas

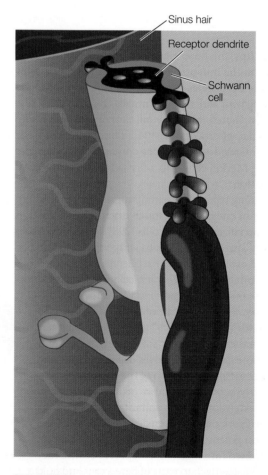

Sinus hair

Receptor dendrite

Schwann cell

FIGURE 7.11   Lanceolate receptor on sinus hair Tiny dendritic protrusions of the receptor axon (red) extend into spaces between surrounding Schwann cells (blue) to contact adjacent shaft of mammalian sinus hair (brown). Differential stretching then stimulates the dendritic processes. (After [377].)

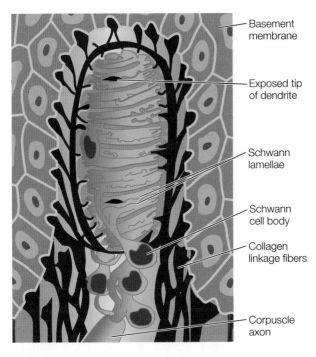

- Basement membrane
- Exposed tip of dendrite
- Schwann lamellae
- Schwann cell body
- Collagen linkage fibers
- Corpuscle axon

FIGURE 7.12 Fine structure of Meissner's corpuscle Nerve dendrite enters from axon at bottom and rises through center of corpuscle. At intervals, flat protrusions from the dendrite are sandwiched above and below by the tips of two Schwann cells flattened into lamellae (light blue). Because the Schwann cell lamellae usually enclose the dendrite protrusions completely, only a few protrusion tips are visible from the outside (black). The core of the corpuscle consists of a vertical stack of these sandwiched discs with the Schwann cell bodies and nuclei (dark blue) below and outside the stack. This core is enveloped by a matrix of linked collagen fibers (red) and the entire structure is surrounded by basement membrane cells (pink). (After [378].)

persistent hair deflection in the middle portions of the sensor may contribute to slow-adapting responses.

A **Meissner's corpuscle** consists of an elliptical capsule filled with a stack of discs (**Figure 7.12**) [45, 72, 298, 342, 378]. Each disc consists of two flattened Schwann cells. The capsule is oriented with its long axis perpendicular to the skin surface and is located in the dermis just beneath the epidermal boundary. One or two axons enter each capsule on their internal end and spiral between the Schwann discs up toward the skin side of the organ. At intervals, flattened swellings of the axon are sandwiched between the two Schwann cells making up a disc. Axons have fine projections that stick out between the Schwann cell margins to contact the outside capsule. Discs are connected to the capsule along their margins but are not connected to each other. Deformation of the overlying epidermis is transmitted to the capsule wall by surrounding tissues. Displacement of the capsule wall moves the margins of the discs and the fine axonal projections along with it, but the more indirectly connected disc centers and the associated axons do not move immediately. This displacement stretches the axon and stimulates mechanosensitive channels in its membrane. Soon afterward, the tension

between the disc margins and centers overcomes the inertia of the disc centers and the axons and they, too, move until all parts of the discs and axons return to their undeformed relative positions. Meissner's corpuscles are thus fast-adapting mechanoreceptors that detect changes in displacement of the skin but ignore steady displacements. Their nerves can track oscillating touch stimuli over a range of 2–100 Hz. Because they are just beneath the epidermis, Meissner's corpuscles are highly focused receptor organs.

The final class of axon-Schwann mechanoreceptors comprises the **lamellated corpuscles** [298]. There are several types. **Pacinian corpuscles** are large mechanoreceptors (up to 1 mm) that often occur in clusters like bunches of grapes [26, 152, 207, 208]. Each capsule contains a single bare axon that is encircled by 20–60 Schwann cell lamellae arranged in layers. The inner lamellae are not continuous around the axon, but instead cluster into two semicircular shells that abut along clefts on opposite sides of the axon. Small processes extend from the axon and through the clefts on each side toward the outer lamellae. The latter have no clefts and completely enclose the axon and inner lamellae. They also differ from the inner lamellae in having fluid-filled connective tissue layers between successive lamellae. The multiple layers of lamellae and the fluid-filled partitions attenuate slow deformations of the capsule wall but effectively transmit rapid transients to the axon. Pacinian corpuscles are thus fast-adapting receptors, and can track successive peaks of oscillating stimuli at rates of 50–1000 Hz. Because Pacinian corpuscles are extremely sensitive and tend to be located deep in the dermis, they have a very broad and diffuse zone of sensitivity; a single corpuscle may respond to a vibrating stimulus applied nearly anywhere on a human hand. **Herbst corpuscles** in birds and reptiles have a similar lamellated structure (**Figure 7.13**) [144, 404]. Like Pacinian corpuscles, they have an inner core of layered lamellae, but this is in turn surrounded by a matrix of collagen fibrils. Herbst corpuscles may or may not be surrounded by an encapsulating external sheath. These organs are also fast-adapting mechanoreceptors able to track oscillating stimuli over the range of 50–1000 Hz. The zones in which mammalian external skin blends into mucosal linings (e.g., the lips, eye conjunctiva, and anal and genital zones) contain a third kind of lamellated corpuscle called a **mucocutaneous end organ** [257]. These are usually encapsulated and contain a tortuous skein of small axon terminals. The latter are individually wrapped with several layers of Schwann lamellae [159, 209, 257, 298, 420]. Like the other lamellated receptors, they are fast-adapting. Finally, a variety of even smaller lamellated **simple corpuscles** has been reported in fish, amphibians, reptiles, birds, and mammals [11, 144, 298, 332, 348, 403, 404]; like the other lamellated receptors, these appear to be fast-adapting.

Vertebrate Ruffini and Golgi tendon organs provide proprioceptive monitoring of ligament and tendon tensions respectively. A third type of proprioceptor, the **muscle spindle organ**, complements these tension-sensitive organs by monitoring the current *length* of each of the body's striated

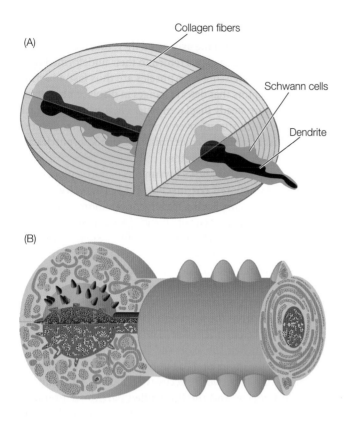

(A) Collagen fibers

Schwann cells

Dendrite

(B)

FIGURE 7.13   **Fine structure of Herbst corpuscle, an avian version of the mammalian Pacinian corpuscle** (A) Several cutaway views to show the internal structure of the elliptical corpuscle. The outer layers (yellow) consist of two-dimensional mats of collagen fibers. The core consists of onionlike layers of Schwann cells surrounding a single nerve dendrite. (B) Enlarged view of inner core of Herbst corpuscle showing Schwann cell layers (blue) and enclosed dendrite (red). The Herbst corpuscle inner core resembles that of Pacinian corpuscles except that the Herbst dendrite ends in a bulbous enlargement instead of remaining cylindrical as it does in Pacinian corpuscles. In both kinds of receptors, the dendrite sends out fine processes into the spaces between enveloping Schwann cells. Both Herbst and Pacinian corpuscles are fast-adapting touch receptors. (After [144].)

muscles [40, 164, 221, 228, 258, 283, 333, 334, 391]. Muscle spindles are encapsulated swellings containing 8–10 specialized (**intrafusal**) muscle fibers attached in parallel with a nearby muscle. Each intrafusal fiber is wrapped in a helical fashion by bare axon terminals. Stretching or contracting of the nearby muscle produces a similar stretching or compression of the intrafusal fibers and their attached axon wrappings. Muscle spindles differ from other mechanoreceptors in that they also receive efferent nervous input from the spinal cord and brain. This input can cause the intrafusal fibers to contract to varying degrees and thus modulate their sensitivity and response properties. The combination of efferent input and differences among the intrafusal fibers in size and shape allows muscle spindles to behave as both slow-adapting and fast-adapting sensors.

FUNCTIONS OF VERTEBRATE TOUCH RECEPTORS   Vertebrate mechanoreceptors, like those of invertebrates, are heterogeneously distributed over the animal's body surface. In addition to the assignment of particular receptors to particular body parts, different vertebrate body regions can host different mixes of the available mechanoreceptor types. The body surface of most mammals can be divided into large regions of **hairy skin** and smaller patches of **bare (glabrous) skin**. Bare skin typically has very high densities of mechanoreceptors with fine-scale spatial resolution. The slow-adapting Merkel cell receptors are present in bare skin of nearly all mammals, whereas fast-adapting Meissner corpuscles are limited to the sensitive finger pads of primates and rodents, the foot pads of cats and

elephants, the bare heads of mole rats, and bare patches on the tails of woolly and spider monkey tails [45, 138, 178, 196, 232, 378, 411]. Pacinian corpuscle densities vary from being largely absent (human facial skin), to relatively sparse (raccoon hands and primate fingers), to abundant (the foot pads of elephants where, in combination with Meissner corpuscles, they provide fast-adapting sensitivity to substrate-borne sounds, as discussed in Chapter 3) [207, 258, 316, 342, 388, 411].

Three types of hairy skin can be distinguished in mammals by the type of hair that is present [161, 253]. **Vibrissae**, or **sinus hairs**, are very long, thick hairs whose main function is mechanoreception (**Figure 7.14**). They are found in all mammals except humans. The most common form is as whiskers on the muzzle, but vibrissae can also be mixed in with smaller hairs in eyebrows and with the denser pelage on forelegs, abdomen, or feet. In naked mole rats and manatees, vibrissae form a sparse grid over the entire body [97, 337]. Vibrissae are also called sinus hairs because the middle shaft of the follicle below the skin's surface is enveloped by a blood-filled sinus. This allows vibrissae, unlike other mammalian hairs, to move inside the follicle when bent by external stimuli. By varying the lengths, resonant properties, and distribution of muzzle sinus hairs, mammals can often obtain information about the shapes, textures, and likely identities of touched objects [50, 171]. Not surprisingly, vibrissae host many more touch receptors around the internal hair shaft than other hair types do. Beginning just below the emergence of the hair from the skin and moving toward the base of a vibrissa, there is a succession of a few dozen free nerve endings, 1–5 small Pacinian corpuscles, up to 2000 Merkel cells surrounded by a palisade of 20–100 lanceolate receptors, and finally 1–10 Ruffini endings about midway down the follicle length. Additional Pacinian corpuscles may be found below the follicle, and clusters of Merkel cells may occur in the skin between vibrissae. **Guard hairs** are the outer and intermediate-size hairs seen on most furred mammals; they are represented in humans as eyelashes, scalp hair, and beards. Guard hairs host an ordering of touch receptor cell types similar to that of vibrissae, but with many fewer cells per follicle (e.g., 20 or fewer Merkel cells, 50 or fewer lanceolate cells, and few if any Pacinian corpuscles). **Velus hairs** create the fine

Tissue
parts:

Touch
receptors:

Epidermis

Hair shaft

Sebaceous
gland

Blood sinus

Hair follicle

Hair bulb

Free nerve
endings

Merkel
cells

Pacinian
corpuscle

Lanceolate
receptor

Ruffini
receptor

Sensory
nerve

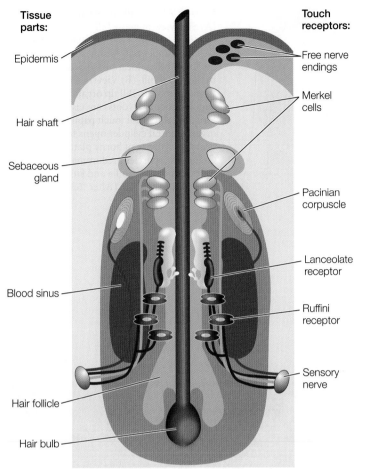

FIGURE 7.14 Touch receptors associated with a typical mammalian vibrissa (sinus hair) The hair shaft and its follicle are surrounded over part of their length by a blood-filled sinus. The hair is rooted at its interior end by a hair bulb. Depending upon whether they are slow- or fast-adapting receptors and have diffuse or focused sensitivity zones, different types of receptors occur at different depths below the skin surface. (After [161].)

pelage covering the body of terrestrial mammals. These hairs are innervated by 1–5 free nerve endings, small numbers of modified lanceolate cells, and few if any touch receptors of other types. However, it is worth noting that all hairs on a typical mammal have at least some touch sensitivity.

The erogenous skin of mammalian genitalia hosts few if any Ruffini, lanceolate, or Pacinian corpuscles, but can be densely packed with free nerve endings, Merkel cells, Meissner corpuscles, and mucocutaneous end organs [79, 80, 159, 209, 298, 382, 420]. In females, these are present in very high densities in both the glans of the clitoris and in the inner surface of its prepuce where it overlies the glans. A similar distribution occurs in the penis of most male mammals. Whereas mucocutaneous end organs tend to lie perpendicular to the skin in a flaccid mammalian penis, they become oriented parallel to the surface as penile erection stretches the skin. In this orientation, they are highly sensitive to the stroking motions of mammalian copulation [209]. In most male primates, the tip or glans (if present) of the penis is equipped with both free nerve endings and numerous encapsulated touch organs, whereas the overlying surface of the prepuce often has few touch receptors. Human males have the opposite configuration: the glans penis contains few receptors except for free

nerve endings, but the interior surface of the prepuce that overlies the glans of a flaccid penis contains high densities of Merkel cells, Meissner corpuscles, and mucocutaneous organs. In the erect human penis, this highly sensitive side of the prepuce experiences repetitive frictional stimulation during copulation. It is worth noting that human circumcision of the male prepuce typically removes a significant fraction of the sensitive mechanoreceptors initially present in intact organs [80].

Like mammals, birds use muscle spindles, Golgi tendon organs, and Ruffini endings as body proprioceptors, and free nerve endings as thermo- and nociceptors [121, 144]. Merkel cells tend to be more patchily distributed in birds than in mammals, with the highest densities in and around the bill, toes, legs, and parts of the wings [144, 151, 163]. The most ubiquitous touch receptors in birds are the fast-adapting Herbst corpuscles: these occur throughout the body skin, around feather follicles, in strings beside long bones and blood vessels, and in joints and muscles [144, 429]. Birds that dabble or probe for food with their bills tend to have very high densities of free nerve endings, Merkel cells, Herbst corpuscles, Grandry corpuscles, and Ruffini endings along the relevant bill edges or just inside their bills [144, 162, 163, 384]. In waterfowl, 20–200 papillae containing varying mixes of these mechanoreceptor types are present on both the upper and lower inner bill surfaces; these papillae form a **bill tip organ** (**Figure 7.15**) [33, 144]. It has been suggested that bill touch receptors are particularly important for avian foragers whose visual fields do not overlap in front of their bodies [273]. Chickens and parrots have touch papillae only inside their lower bills [139]. Wading shorebirds have a series of tiny pits packed with Herbst corpuscles on the outside of their long bills; these receptors help the bird to sense prey buried in the soft substrate [144, 299]. Seed-eating songbirds often have dense concentrations of Herbst corpuscles on that part of the bill that is used to hold seeds for hulling, and woodpeckers have dense concentrations of touch receptors in their extensible tongues [144]. Fish, amphibians, and reptiles also rely on heterogeneously distributed free nerve endings, muscle spindles, Merkel cells, Ruffini endings, and a variety of lamellated receptors for their sense of touch [11, 164, 221, 343, 365, 403, 404, 408, 424]. Many benthic fish, amphibian larvae, and some turtles and snakes have sensory barbels

(A)

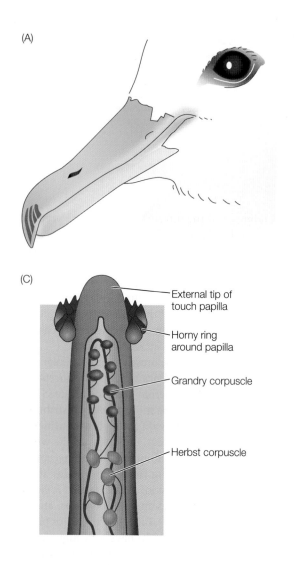

(B)

(C)

External tip of touch papilla

Horny ring around papilla

Grandry corpuscle

Herbst corpuscle

FIGURE 7.15 Bill tip organ of goose (A) Side view of goose bill showing internal locations (red) of 4 of the 100 touch papillae in the upper bill. Another 180 occur in the lower bill in locations opposite to those in the upper bill. (B) View of the upper bill from the mouth. Bill tip organ (red) and papillae openings (dots) are visible. (C) Structure of typical touch papilla in bill organ. Central cylinder opens to outside through a rosette of horny plates. Inside, Grandry corpuscles predominate in the outer half of the papilla and Herbst corpuscles in the inner half. (After [33, 144].)

(A)

(B)

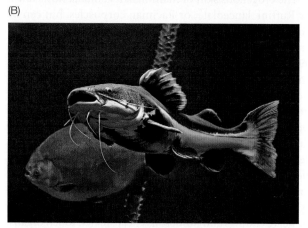

FIGURE 7.16 Sensory barbels on fish These filamentous sensory organs extending from the head host both touch and chemosensory receptors. (A) European sturgeon (*Acipenser sturio*); (B) Red-tailed catfish (*Phractocephalus hemioliopterus*).

or tentacles on their heads (**Figure 7.16**) [136]. Such organs usually contain chemoreceptors as well as touch receptors in the forms of free nerve endings and Merkel cells. The lips of moray eels have large numbers of lamellated touch receptors that are used to identify prey in dark crevices or when foraging at night [11].

All vertebrates send the neural information collected by their multiple touch receptors to special processing regions in their brains. As sound is represented in the brain in tonotopic maps and vision in retinotopic maps, touch is represented in vertebrate brains in one or more **somatotopic maps** [144, 226, 281, 330, 334, 415, 416]. Representations of touch stimuli are organized such that input from adjacent parts of the body are processed in adjacent parts of the brain map. However, the size of each region in the map depends not on the size of the represented body parts, but instead on the density and diversity of touch receptors in that region. A somatotopic map is usually quite a distorted representation of the animal's body (**Figure 7.17**). Emphasized body regions include the face and hands in primates [202, 203, 267]; the teeth and whiskers in rodents [74, 75]; the flat bill of the duck-billed platypus [241, 265]; the sensory rosette on the head of the

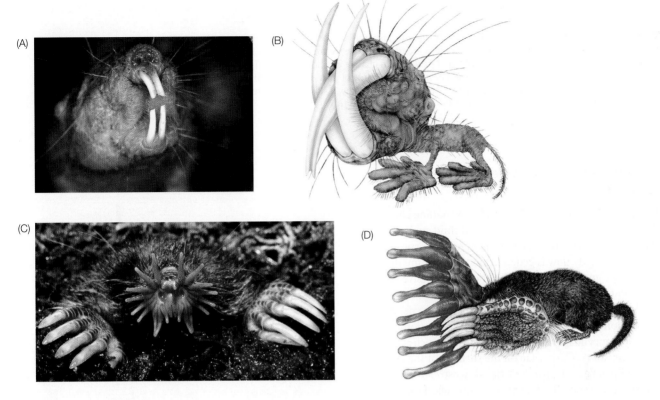

FIGURE 7.17 Specialized touch reception and brain processing in two burrowing mammals (A) Head of a naked mole rat (*Heterocephalus glaber*). This root-eating species burrows in dry soils and depends heavily on vibrissae and other receptors on its head and at the roots of its teeth for touch perception. (B) Somatotopic representation of a naked mole rat's touch receptors in its brain. (C) Head and foreclaws of a star-nosed mole

(*Condylura cristata*). This insectivore burrows in wet mud hunting worms, insects, and other invertebrate prey. Instead of head vibrissae, its main touch organs are 22 fleshy appendages arranged in a rosette around its nostrils. Papillae (Eimer's organs) on the appendages contain free nerve endings, Merkel cells, and lamellated corpuscles. (D) Somatotopic representation of star-nosed mole's touch receptors in its brain. (After [73, 75].)

star-nosed mole [75]; and the beak, tongue, and claws of parrots and owls [241, 265, 415, 416].

MECHANORECEPTOR ION CHANNELS   Researchers have made enormous strides in the identification of the transmembrane proteins that comprise the mechanosensitive ion channels in animals [142, 314, 361, 374]. In many cases, the genes that govern assembly and properties of these proteins are also known. All mechanosensitive ion channels identified to date belong to one of two families of proteins. The **transient receptor potential** (**TRP**) family includes mechanosensitive channels that admit depolarizing calcium ions when stretched [104, 142, 361, 374]. The **degenerin/epithelial sodium channel** (**DEG/EnaC**) proteins convert stretching into depolarization of the receptor by admitting sodium ions [142, 256, 374]. Most animal taxa studied to date host both kinds of protein channels. Regardless of the taxon, ciliated mechanoreceptors always appear to use TRP channels [77, 329, 361]. This is consistent with the role of calcium ion channels as governors of ciliary and flagellar activity in single-celled eukaryotes. Most, but not all, nonciliated mechanoreceptors rely on DEG/EnaC channels [142, 256]. Thus the ciliated touch receptors in the head of the nematode *Caenorhabitis elegans* use TRP channels, whereas the nonciliated

touch receptors on the side of the body rely on DEG/EnaC channels [311, 312, 361, 374]. Fruit flies (*Drosophila*) have TRP channels in all of their chordotonal organs, including the auditory Johnston's organ, but also have DEG/EnaC channels in free nerve ending receptors devoted to thermo- and nociception [1, 256, 361, 374]. In vertebrates, most of the encapsulated touch receptors use DEG/EnaC channels [256, 318, 374], whereas Merkel cells, despite being nonciliated, use TRP channels [104, 163, 256]. Free nerve-ending mechanoreceptors in vertebrates use either TRP or DEG/EnaC channels [256]. Why the two mechanosensory ion channel systems are distributed the way they are remains unclear.

Hydrodynamic Reception

Whereas touch receptors provide an animal with information about nearby solid objects, hydrodynamic receptors provide similar information about nearby fluids. One difference is that tactile receptors can monitor either steady pressures or movement of nearby objects, whereas hydrodynamic receptors only monitor relative fluid movement. We first define the types of stimuli that are salient to hydrodynamic receptors, and then examine the types of receptor designs required to respond to such stimuli.

FIGURE 7.18   Example of wake left by swimming fish Fish enters at bottom and swims directly upward and out of window at top. Particle velocities at successive times after fish first enters window are shown as small arrows: single dots mean no change in velocity above that prior to fish passage; larger arrows indicate direction and magnitude of particle velocity above initial state. Note that even 7.8 seconds after the fish enters the window, a significant perturbation of particle velocities remains adjacent to fish's trajectory. (After [167].)

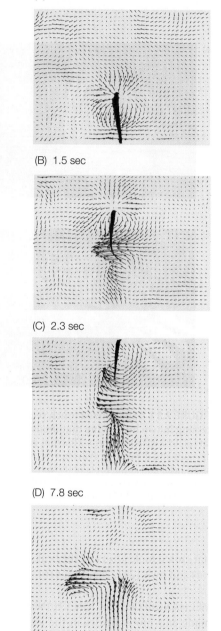

(A) 0.6 sec

(B) 1.5 sec

(C) 2.3 sec

(D) 7.8 sec

Hydrodynamic stimulus generation and propagation

WHAT ARE HYDRODYNAMIC STIMULI?   As outlined in Chapter 2, particle motion in an undisturbed fluid such as air or water is random in direction and has a characteristic average velocity. If a steady flow is generated in the medium, the particles acquire an additional and directionally cohesive velocity component. An animal such as a barnacle secured to the substrate but exposed to this current will experience a net drag in the direction of the flow. Since the animal's location is fixed, any flexible appendages exposed to the current, such as its cirri, and any sensory cilia or setae on the surfaces of the cirri, are likely to be deflected with their tips downstream relative to their bases, which are fixed upstream. This deflection is generated by the difference between the directional velocity of the medium and the zero velocity of the animal. All hydrodynamic stimuli result from a difference in velocities or accelerations between two spatially separated locations in a fluid medium. To detect such a stimulus, a receiver must thus sample at two or more locations simultaneously. In the case of the barnacle, sampling occurs at the point of fixation to the substrate and at the tips of its cirri extended into the current.

A variety of situations can generate hydrodynamic stimuli. While a benthic fish sitting on the ocean bottom, can, like a barnacle, experience a hydrodynamic stimulus when exposed to local water currents, a swimming fish in otherwise static water will also be exposed to a hydrodynamic stimulus, because its body is moving at a velocity that is different from that of the surrounding fluid. Even if an animal is fixed to a substrate in a static medium, any active movement of its body parts through the surrounding medium will create hydrodynamic stimuli. Activities of animals on the water's surface create surface waves that in turn generate subsurface hydrodynamic stimuli perceptible to nearby organisms, including foraging fish. Near-field sounds in water impinging on the otoliths in a fish's inner ear can also be considered a hydrodynamic stimulus since, by design, the dense toliths will move more slowly in the sound field than will the ambient water and the rest of the fish.

Like near-field sounds, the amplitudes of hydrodynamic stimuli fall off very rapidly as they propagate away from the source. Whereas near-field sound sensors such as the fish's inner ear can detect sounds at any point in the near field, hydrodynamic receptors require stimulus source proximities that are only a small fraction of the near-field limit [217]. The hydrodynamic active space for a typical fish is only 1–2 body lengths [85]. Also unlike near-field sounds, which disappear from any given location immediately after passage, hydrodynamic stimuli can persist at biologically detectable levels for a considerable time after generation. For example, a swimming animal typically generates a **wake** behind it (**Figure 7.18**). Depending on the animal's body size, undulation pattern, and swimming speed, this wake may consist of localized laminar eddies, a single long chain of swirling vortices, two parallel chains of vortices, or even more turbulent patterns [38, 167, 168, 296, 297, 393, 394]. Vortices can persist in water for several minutes after passage of even small fish, and enhanced velocities of fluid medium particles can last even longer. This creates the possibility that another animal could use hydrodynamic sensors to track the creator of the wake over tens of meters and eventually catch up to it [167, 324, 325]. The structure and size of wakes may even provide hydrodynamic trackers with information on the elapsed time since the wake was created, and the size, speed, shape, and type of wake creator [86]. Note that tracking wakes does not violate the 1–2 body-length limitation on detecting hydrodynamic stimuli: in this context, it is the wake that provides the immediate stimulus for the tracker, and not the creator of the wake, which may be tens of meters distant.

> ## Web Topic 7.2 *Hydrodynamic stimuli*
> The physics of hydrodynamic stimuli can be quite complex. We here outline some principles that can be used to categorize different types of phenomena.

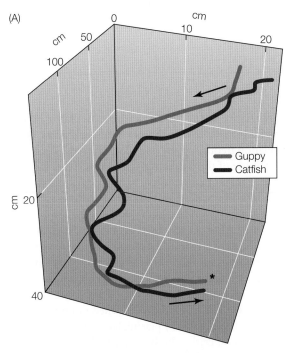

(A)

USES OF HYDRODYNAMIC SENSING Few animals can avoid creating currents, eddies, or vortices when they move in fluid media. Animals take advantage of this fact and use hydrodynamic sensors to detect the presence of nearby predators, prey, obstacles, or conspecifics. Silverfish, cockroaches, mantids, and crickets all have densely haired cerci at the ends of their bodies that alert them to air movements created by the wing beats of predatory wasps, flies, and bats or by the rapid tongue strike of a toad [47, 67, 115, 199, 385]. Fish nip off the extended siphons of buried clams after homing in on their exhalant water currents [85]. Other fish that feed just below the surface can use hydrodynamic sensors to detect the surface waves created by insects or spiders on the water's surface [284]. Blind cave fish overtly create turbulence as they move and use distortions in the reflected eddies and vortices to alert them to the proximity of obstacles [99, 172]. Some nonstridulating crickets create small air puffs with their wings as courtship and agonistic signals that are detected by conspecifics' cerci [175, 218]. A number of fish taxa use tail-beating or fin-fanning as hydrodynamic signals during agonistic and courtship interactions [35, 46, 53, 54, 206, 270, 341]. Schooling fish, squid, and cuttlefish rely on arrays of hydrodynamic sensors along their bodies to monitor movements of conspecific neighbors in the school and thus coordinate spacing and school cohesion [59, 62, 85, 147, 205]. Copepods, predatory fish, and seals can track hydrodynamic wakes and appear to use this ability to pursue moving prey (**Figure 7.19**) [36, 100, 102, 108, 324, 325, 347].

FIGURE 7.19 Hydrodynamic tracking by aquatic animals (A) Trajectory of swimming guppy as monitored in infrared light, which these fish cannot see. A catfish encounters the wake of the guppy and follows the guppy trajectory very closely until it catches and eats guppy (at asterisk). Catfish can track prey with hydrodynamic cues, olfactory cues, or a combination of the two. (B) To see if animals can track a wake using hydrodynamic cues only, Denhardt and colleagues trained a blindfolded harbor seal to follow the wake (solid yellow line) left by a small submarine. The example shows close following of submarine's track by the blindfolded seal. The dashed line shows the trajectory the seal would have followed if it had heard sounds from the submarine. Covering the whiskers of the seal as well as the eyes prevented the animal from successful tracking. (A after [324]; B after [102].)

(B)

| 00:00 sec | 02:84 sec | 05:76 sec |
| 07:32 sec | 10:00 sec | 12:84 sec |

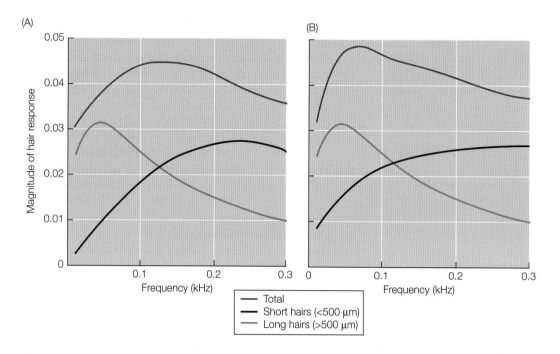

(A)

(B)

Total
Short hairs (<500 μm)
Long hairs (>500 μm)

FIGURE 7.20   Differential air current stimulation of sensory hairs on cercus of a cricket Each cercus is covered with sensory hairs of different lengths. Hair responses vary depending on hair length, frequency of stimulus oscillations, and stimulus amplitudes (wind currents in graph (B) have average velocities 100 times those in graph (A)). Short hairs require much higher frequencies (>0.2 kHz) for maximal stimulation than long hairs (0.03–0.04 kHz), due to the boundary layer. As stimulus amplitude increases, the minimal effective frequency to stimulate short hairs decreases, since the boundary layer becomes thinner. The frequency producing the maximal response for the overall cercus thus decreases from 0.14 kHz in (A) to 0.06 kHz in (B). (After [262].)

Hydrodynamic stimulus reception

As with touch, hydrodynamic sensing surely arose very early in the evolutionary history of animals. Sense organs that are stimulated when contacted by moving objects can also be stimulated by moving fluids. It is thus not surprising that hydrodynamic receptors are widespread among animal taxa.

INVERTEBRATE SURFACE RECEPTORS   As we saw in the prior section on touch, the vast majority of aquatic invertebrates host numerous mechanosensory cilia or setae on their external body surfaces. Many of these sense organs can be deflected by either solid objects or by water currents, eddies, and vortices, and thus do double duty as both touch and hydrodynamic sensors [60]. Their efficiency in the latter role is limited by the same **boundary layer** constraints that limit body surface near-field sound detectors (see Chapter 3). Because of frictional drag, the relative velocity of an animal and a surrounding fluid is zero at the animal's surface and rises asymptotically at greater distances. The boundary layer is that zone of fluid adjacent to the body surface within which relative velocities are insufficient to stimulate the hydrodynamic hairs or setae. This zone is thinner if the velocity of

the animal relative to the fluid as a whole oscillates: higher oscillating frequencies lead to thinner boundary layers. For a given oscillation frequency, the boundary layer also decreases as the amplitude of the oscillations increases. Thus hairs of a given length might be insensitive to hydrodynamic stimuli containing low-frequency variations, but quite sensitive to those of the same or even lesser amplitudes if the component frequencies are higher. At very high frequencies, even at high amplitudes, most invertebrate hairs cannot keep up with the rapid movements of the fluid. As a result, invertebrate hydrodynamic sensors tend to be most sensitive to intermediate frequencies of oscillation.

Many species of arthropods have turned boundary layer constraints into useful tools by placing hairs or setae of many different lengths on critical body surfaces (**Figure 7.20**). By monitoring which size classes of hairs are deflected and which are not, an animal can decompose a complex hydrodynamic stimulus into its different frequency components [17, 18, 195, 262, 353, 354, 368]. This is equivalent to the Fourier analysis of sounds performed by many animal ears (see Chapter 3). In addition, by having hairs or setae over all sides of its cerci (crickets, roaches, locusts, and mantids), legs (spiders), body (caterpillars), or the tail (crayfish and lobsters), the animal can sample the hydrodynamic stimulus field at many different locations [18, 48, 51, 109, 368, 380, 381]. This facilitates the spatial mapping of the hydrodynamic field. Because hairs and setae usually are polarized, (e.g., are maximally excited by deflection in one direction and inhibited by deflection in the opposite direction), an ensemble of sensors with diversely distributed polarities permits the animal to identify flows moving in any direction. As with touch sensors, the inputs from the multiple hairs or setae are usually integrated in the brain of the animal to form a somatotopic map (**Figure 7.21**) [199–201, 414]. The combined information on the direction

(A)

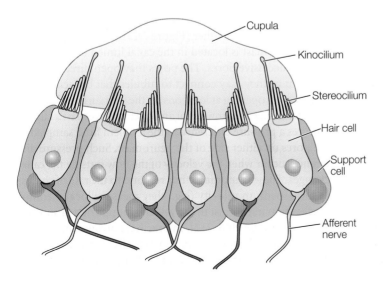

FIGURE 7.21 **Neural map of sensory hairs from cricket cerci** (A) Each of the two cerci on the posterior of a cricket hosts about 1000 sensory hairs that respond to air currents. Colored arrows around the cricket indicate possible directions of such currents: red currents come from the left side of the animal and blow to the right, blue-green currents blow in the opposite direction, yellow currents blow toward the head, and purple-black currents blow toward the tail. Different hairs respond to currents moving in different directions. (B) Each hair in each cercus sends a neuron to the last abdominal ganglion in the central nerve cord. This dorsal view of the ganglion uses the same color coding as in (A) to indicate which neurons in the ganglion are most stimulated by a current moving in a given direction. If one divides the ganglion into left and right halves, it is apparent that each half of the ganglion has its own somatotopic map of wind currents: currents blowing from the left to the right of the animal stimulate nerves on the left side of each half of the ganglion and those blowing from right to left stimulate the right side of each ganglion half. Neurons responding to head and tail currents are also represented, but here the line connecting them is perpendicular to the cricket's body axis. (After [201].)

(B)

of flow at each point, the shape of the hydrodynamic flow field, and the frequency composition of the stimulus can then be used to estimate the location, distance, size, and likely identity of the stimulus source.

LATERAL LINE SYSTEMS   A **lateral line** is a linear or curvilinear array of hydrodynamic sensors. The best-known examples occur in fish and aquatic amphibians [85, 103, 205, 246, 247, 276], but similar arrays are present in many cephalopods

(particularly squid and cuttlefish) [60, 62, 64]. Most of these species are relatively large, mobile, and aquatic. It is thought that this combination of traits has favored linearizing the disposition of their hydrodynamic sensors along their main axis of locomotion. As we shall see, this allows the animals to sample hydrodynamic stimuli systematically at multiple locations along their body, and to discriminate between those stimuli generated by their own motion through the water versus those generated by movements of other nearby objects or local currents.

The hydrodynamic sensor of fish and amphibians is the **neuromast** (**Figure 7.22**) [85]. It consists of a variable number of hair cells similar in structure to those present in the fish inner ear (see Chapter 3). Each hair cell hosts up to 150 stereocilia staggered in height with the tallest adjacent to a peripherally positioned kinocilium. Hair cells are polarized: deflection of the stereocilia toward the kinocilium stimulates the hair cell whereas deflection in the opposite direction inhibits it. Within a neuromast, the hair cells are organized such that one half are oriented in one direction and the other half are oriented in the opposite direction. Each neuromast has two sets of afferent nerve fibers that send information to the brain. One set of nerves increases its output when similarly oriented hair cells are stimulated, while the other set is inhibited by its oppositely oriented hair cells, and reduces its output. Additional efferent nerves allow the brain to modulate the sensitivity of the hair cells. The hair cells in a neuromast are nestled among nonciliated support cells and

Cupula

Kinocilium

Stereocilium

Hair cell

Support cell

Afferent nerve

FIGURE 7.22 **Neuromast organ from body surface of a fish** Sensory hair cells are embedded in support tissues and covered externally by a gelatinous cupula. The cupula touches kinocilia and stereocilia on each cell and moves relative to the fish's body in hydrodynamic currents. Hair cell kino- and stereocilia are polarized so that stimulation is maximal along one axis of the cell surface. Adjacent cells often have opposite polarities. Hair cells with similar polarities are innervated by afferent axons that tend to be bundled together. Each hair cell also synapses with an efferent neuron (not shown here) that allows the brain to modulate its sensitivity.

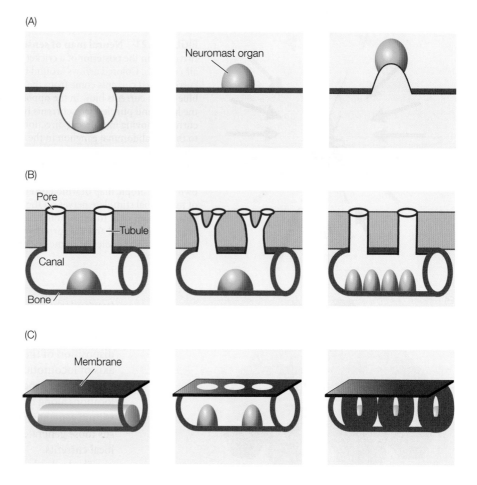

FIGURE 7.23 Alternative configurations of neuromasts in fish (A) Superficial neuromasts: these can be located in pits (left), mounted simply on the body's external surface (center), or elevated onto a fleshy papilla (right). (B) Narrow canals are fully enclosed by bone; they sample outside water currents through tubules and pores. Tubules may be simple (left); highly branched before reaching pores (center); or at such long intervals along the canal that many neuromasts may be positioned between two successive pores. (C) Wide canals are bony only in the interior half; the exterior half is closed by a membrane that is exposed to the ambient medium. Rays use wide canals containing one long neuromast along the entire length (left). Some bony fish have wide canals with small neuromasts at intervals and pores in the exposed membrane (center). Others have bony septa that divide a wide canal into successive segments (right). The neuromasts may nearly fill the cavities between septa or be placed so one nearly occludes a hole in the center of each septum. (After [83, 106].)

encased in a sheath of mantle cells. A gelatinous **cupula** rests on top of the stereocilia and kinocilia of the neuromast and is maintained in a vertical position by its connections to the hair cells and surrounding tissues. When exposed to a flow in the ambient fluid, the cupula is dragged away from its vertical position until the tensile force in the attached tissues balances the fluid-dragging force. The latter increases with the velocity of the fluid flow. As a result, the amount of deflection of the cupula and its attached stereocilia is directly related to the velocity of the hydrodynamic flow. The hair cells then indicate the amount of deflection of their stereocilia by modulating the impulse rates of synapsed afferent nerves. A neuromast thus provides the brain with a measurement of the flow velocity in its vicinity. The brain can then ascertain the direction of flow by comparing which hair cells are activated, inhibited, or unresponsive.

The skin of fishes consists of an outer epidermis, an underlying dermis, and depending on the species and part of the body examined, scales or scutes that are either embedded in the dermis or staggered in overlapping layers between the epidermis and dermis. Most fish have two types of lateral line systems [37, 83–86, 409]. **Superficial neuromasts** are located in the epidermis above any nearby scales. They can be flush with the surface, in pits with varying degrees of closure, or on pegs and papillae extending beyond the surface (**Figure 7.23A**). This variation in neuromast height affects the boundary layers that can be penetrated and thus

the neuromast's frequency range. It also affects the degree to which the neuromast is exposed to hydrodynamic stimuli generated by the animal's movement through the water. Superficial neuromasts can detect and measure steady and varying hydrodynamic flows up to a frequency of 70–80 Hz. The same afferent nerve may represent more than one superficial neuromast, making spatial resolution through superficial neuromast input fairly coarse.

Canal neuromasts are located inside fluid-filled canals buried in the dermis (sharks and rays) or in dermal bone (bony fishes) [83, 276]. **Narrow canals** connect to the ambient medium through regularly spaced pores, and adjacent canals may connect to each other (**Figure 7.23B**). In many fish, one canal neuromast is located in the canal lumen between each pair of successive pores. The pores may open directly to the medium or they may connect to tubules that branch several times before opening at locations on the skin surface. The flow of water inside a narrow canal and over a neuromast's cupula requires a pressure difference in the ambient fluid sampled at the pores on either side of the neuromast. Such pressure differences occur when the velocity of the flow outside the canal changes. These neuromasts thus respond to the rate of change in local medium velocity (e.g., they measure local medium acceleration in contrast to the velocity monitoring of superficial neuromasts) [106, 217]. In practice, narrow canal neuromasts do not detect steady flows or low frequencies below 30 Hz, but they accurately track rapid changes in medium flow up to 100 Hz [85, 86]. By acting as high-pass filters, they are particularly effective at detecting the rapidly varying stimuli created by nearby prey or predators while remaining insensitive to low-frequency noise created by the fish's own swimming, nearby surf, or rushing currents [117–119]. In contrast with superficial neuromasts, each narrow canal neuromast has its own afferent nerve providing the brain with a very fine spatial sampling of the hydrodynamic field.

As with superficial neuromasts, the configuration of canal systems can vary and the same individual may host multiple designs [83, 106, 204]. Whereas narrow canals are entirely enclosed by bone, **wide canals** contain bone only in the interior half (**Figure 7.23C**). They enclose their outer portion with a thin membrane that may or may not contain pores. This membrane is exposed to the ambient medium and acts as a transmitter of hydrodynamic stimuli into the canal. Wide canals may incorporate a single large neuromast along their main axis, multiple neuromasts at a regular spacing, or alternatively, be segmented by bony partitions, with openings almost filled with a neuromast. These various modifications turn wide canals into resonant structures that are tuned to certain frequencies of hydrodynamic stimuli but ignore lower or higher rates of oscillation.

The most conspicuous lateral line canal in fishes (**trunk canal**) runs along the side of the body from near the gills to the tail fin. Many fish have additional **head canals** encircling the eye and tracing the long axis of the lower jaw. The relative distribution of superficial, narrow-canal, or wide-canal systems varies markedly and is related to the specific hydrodynamic stimuli, the patterns of locomotion, and the habitat in which the fish lives [106, 204, 205, 289]. A typical bony fish has narrow canals and lines of superficial neuromasts on the head, narrow canals on the trunk line, and superficial neuromasts scattered over the body or at least surrounding the trunk canal (**Figure 7.24**). Reduced canal lines and enhanced abundances of superficial neuromasts are common attributes of slower-moving deep-sea fish, eels, and blind cavefish. These species have often retained only the wide membranous type of canal. Some potentially fast-swimming fish also have numerous wide canals on the head and body but glide or swim with special postures to reduce body turbulence when foraging. When both superficial and canal systems are present, they send information to the brain separately [37]. The localization of nearby prey and the tracking of wakes are complex tasks that are likely to require the melding of both kinds of information with or without augmentation by auditory, visual, or olfactory inputs [85, 291, 292]. As with other modalities, fish lateral line activity is collated in the brain as one or more somatotopic maps. These maps summarize and compare the amplitude and direction of water flow at many points on the fish's body [85, 143, 323]. For example, consider a fish facing the source of an oscillating hydrodynamic stimulus [106]. When the stimulus creates a sufficient flow toward the fish, this will tend to move the fish's head backward. Because fish are relatively stiff objects, the head's movement will impart the same direction and magnitude of acceleration to the entire fish. As the hydrodynamic stimulus moves past the fish's head and propagates farther away from the source, it will be attenuated by spreading and frictional losses. Since the entire fish is accelerated at the same rate and direction by the stimulus, but successively more distant regions of water are accelerated at decreasing rates toward the fish's tail, the *relative acceleration* of the water outside the pores of canal neuromasts varies dramatically along the length of the fish. At the fish's head, the stimulus accelerates the ambient water backward faster than it accelerates the fish. A neuromast near the head will thus record a *backward acceleration* of the nearby water (relative to that of the fish). At the

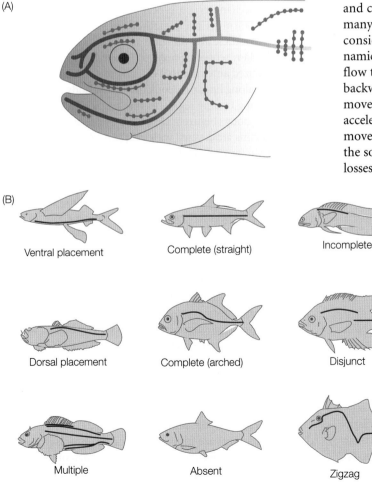

(A)

(B)

Ventral placement   Complete (straight)   Incomplete

Dorsal placement   Complete (arched)   Disjunct

Multiple   Absent   Zigzag

FIGURE 7.24 Distributions of lateral line systems in fish (A) Typical bony fish with canal systems in red and lines of superficial neuromasts in green. Red canals on head typically connect to trunk canal along entire side of animal. (B) Different locations and extents of trunk lateral line in different bony fishes. (After [83].)

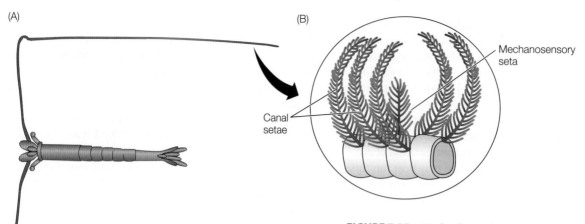

(A)

(B)

Mechanosensory
seta

Canal
setae

FIGURE 7.25 **Hydrodynamic sensors on antennae of pelagic shrimp (Penaeidae)** (A) Overall view of shrimp showing natural positions of extended and trailing segments of long antennae when the animal is swimming. (B) Magnified view of a trailing segment showing enclosed canal created by curved and overlapping setae with smaller mechanosensory setae inside the canal. These canals and their mechanosensory setae function like the lateral line canals and neuromasts in fish. (After [105].)

fish's tail, the fish is being accelerated backward faster than is the ambient water; a canal neuromast here will interpret this as a *forward acceleration* of the nearby water. At some point between the head and the tail, the acceleration of the water and the fish are equal; canal neuromasts at this point perceive no relative acceleration.

Because of the rapid attenuation of a hydrodynamic stimulus as it propagates away from its source, a fish can identify the likely angle between its body axis and the source location by noting which neuromasts in the linear arrays receive the strongest stimulation. The magnitude of the stimulus at these particular neuromasts provides one measure of the size of the source. A possible confusion between a large stimulus far away and a small stimulus close by can be resolved by noting how many adjacent neuromasts are stimulated: the front of a hydrodynamic stimulus expands as it propagates from the source and a more distant source will stimulate a larger patch of adjacent neuromasts along the lateral line. If the source creates a series of vortices in its wake, the size and direction of motion of each vortex can be estimated by the patterns of stimulation along the fish's body. These and similar comparisons between different regions in the somatotopic map allow the fish's brain to estimate the location of a stimulus source (direction and distance), its size, its direction of motion, its pattern of motion, and if a wake is being tracked, how long ago it was created and which way the source was moving [37, 86, 106, 137]. These are very remarkable feats that help explain why lateral line systems are nearly ubiquitous in both cartilaginous and bony fishes.

While all juvenile cephalopods have linear arrays of ciliated sensory nerve cells along their heads and arms, only the faster-moving squid and cuttlefish retain them as adults [62, 64, 249, 373]. Unlike the hair cells of vertebrates, each cephalopod sensory cell lacks stereocilia and instead hosts up to 200 similarly sized kinocilia. These sensory cells also lack a cupula, and their cilia are directly exposed to the ambient water. Cephalopods' lateral lines are as sensitive to water movements as those of

fish, and also use sensory cell polarization to record the direction in which their cilia are deflected [62, 63]. They respond to oscillating stimuli over a range from 20–600 Hz, with maximal sensitivities around 75–100 Hz [59, 236]. Similar arrays of ciliated mechanoreceptors occur along the bodies of predatory arrow worms (Chaetognatha) and several authors have proposed that these also function as lateral lines [57, 124–126, 193, 304]. Many schooling prawns (Penaeidae and Sergestidae) have specialized antennae that extend out from opposite sides of the head along lines perpendicular to the body axis and terminate in long trailing segments that are normally parallel to the prawn's body (**Figure 7.25**) [105]. The terminal segments of each antenna are covered with long curved setae that overlap each other to form a "canal" on one side of the segment. Mechanosensory setae are located at intervals inside this canal. It has been suggested that the terminal segment of each antenna acts as a separate lateral line, and that their length and spatial separation from the segment on the other antenna allow for very effective mapping of the local hydrodynamic field.

MAMMALIAN HYDRODYNAMIC SENSORS   The same sinus hairs used for touch in terrestrial mammals are used by some aquatic species to detect hydrodynamic stimuli (**Figure 7.26**). Like many predatory mammals, seals, sea lions, walruses, and otters all have pronounced facial vibrissae (whiskers). Recent work has demonstrated that harbor seals (*Phoca vitulina*) can use their vibrissae to track the hydrodynamic wakes of conspecifics [347] and even those of robotic submarines [100, 102]. It not yet known whether the highly elaborated vibrissae of walruses and bearded seals are also used for hydrodynamic as well as tactile sampling [250, 272]. Manatees and dugongs (Sirenia) are unusual among mammals in having well-innervated sinus hairs scattered over much of their body surface. It has been suggested that this facilitates the mapping

(A)

(B)

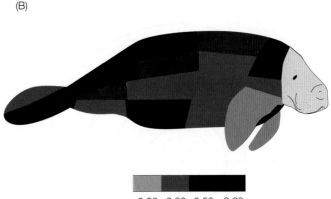

0.20 0.30 0.50 0.60
Hair density (# hairs/4 cm^2) on body

FIGURE 7.26 **Mammalian hydrodynamic sensory systems** (A) Highly sensitive vibrissae (whiskers) of harbor seal (*Phoca vitulina*) used to track wakes, as shown in Figure 7.19. (B) Map of varying densities of sensory vibrissae on body of male Florida manatee (*Trichechus manatus latirostris*). Vibrissae are also abundant on the head but are not shown in this figure. (After [337].)

of ambient hydrodynamic fields in a manner similar to that of lateral line systems [337]. All baleen whales (Mysticeti) have significant numbers of sinus hairs around their jaws, on their heads, within tubercles, and in some species, near the blow hole [250]. The associated follicles host large numbers of lamellated corpuscles and a high level of innervation. While these hairs are clearly mechanosensory, the degree to which they sense hydrodynamic stimuli remains unknown. Most toothed whales (Odontoceti) are hairless as adults, with the notable exception of river dolphins (Platanistidae) [250]. The latter exhibit stubby vibrissae above and below the mouth that are thought to have both tactile and hydrodynamic functions [248]. Muzzle vibrissae in other aquatic mammals such as water rats (*Hydromys chrysogaster*), water opossums (*Chironectes minimus*), and water shrews (Talpidae and Soricidae) may also be used for both tactile and hydrodynamic sampling [73, 101]. The largely aquatic duck-billed platypus (*Ornithorhynchus anatinus*) has no vibrissae on its body, but instead has over 60,000 specialized "push-rod" mechanoreceptors scattered randomly over its bill to detect hydrodynamic stimuli from nearby prey [269, 320]. Each sensor consists of a thin epidermal rod that can move freely within a surrounding sheath but is attached at its base to Merkel-like and lamellated receptors. These structures are reminiscent of the papillae in the bill tip organs of ducks and geese.

Electroreception

Properties of electrical stimuli

All atoms consist of a central nucleus containing uncharged neutrons and positively charged protons. This nuclear core is surrounded by one or more shells of negatively charged electrons. Depending on the kind of atom and the circumstances, electrons can be stripped away from the outer shells, leaving the source atom positively charged overall. If the lost electrons get added to the outer shells of other atoms, the latter become negatively charged. An atom or a molecule that contains unequal numbers of protons and electrons and is thus electrically charged is called an **ion**.

Positive ions exert a repelling electrical force on other positive ions; they exert an attractive electrical force on negative ions. Similarly, negative ions repel other negative ions and attract positive ones. A single symmetrical ion (**monopole**) radiates its electrical forces uniformly in all directions (**Figure 7.27**). As with other forces considered in this book, the amplitudes of these forces decrease as we sample them at increasing distances from the ion. A map of the direction and magnitude of the electrical forces at all points surrounding an ion is called its **electrical field**. If we place a positive ion and a negative ion close together but not touching, they jointly produce a **dipole electrical field**. Instead of radiating directly outward, the lines of force emanating from each ion are bent toward each other, creating a curvilinear electric field. Adding more ions to the ensemble can create even more complicated electrical field patterns.

The electric fields encountered by animals do not propagate like sound waves or the electromagnetic waves that constitute light [189, 191, 192]. When positive and negative charges become unequally distributed in a local region, the region is surrounded by an **electrostatic field**: any change in the electric field at the source is represented immediately everywhere around that source; there is no propagation delay. In the absence of propagation, electrostatic fields are not distorted by reflective interference, refraction, or diffraction. As with sound, light and chemical stimuli, and for similar reasons, the amplitude of an electrostatic field decreases with increasing distance from the source. For electrostatic monopoles, the magnitude of the field decreases with the square of the distance between the monopole and the point of sampling, and for dipoles, the rate of decrease depends on the cube of that distance [216]. Fields with more complicated geometries will consist of dipole, quadrupole, and even higher order components. Close to the source, all of these components contribute to the electric field. Because higher-order components attenuate faster with distance than lower-order ones, even complicated sources act like dipoles at distances greater than the size of the animal [191].

(A)

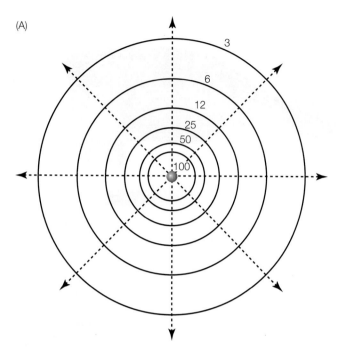

(B)

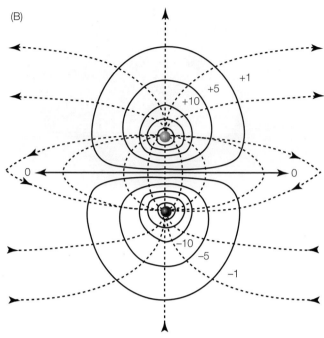

FIGURE 7.27 **Maps of some simple electric fields** (A) The electric field lines of force (dashed) around a monopole are straight and radiate out in all directions from the single charged source at the center. Solid lines connect all locations on the map that have the same electrical potential (here defined as the energy, measured in volts, needed to move a positive test particle from infinity to some point on the map). For a monopole, these are circular since the lines of the electric field are straight. Voltages decrease at successive distances from the charged source (see sample equipotential lines and their values). (B) A similar plot around a dipole. Here, the electric field lines of force are curved, and the corresponding lines connecting all points at the same electrical potential appear as ellipses flattened on the side between the two charges. Note that the electrical potential is zero on the line between the two charges and equidistant from them.

Mapping electrical fields around a charged body is often a complicated task. Instead, one typically maps a related measure called the **electrical potential**. This is the amount of energy expended (or released) were one to move a small positively charged particle from one location to another inside an electrical field. It is usually measured in **volts** and thus referred to as a **voltage**. Because the electric field amplitude decreases with distance from an electric stimulus, voltage also decreases with distance from the source. Electrical potential maps usually graph the amount of energy it would require to move a charged test particle from infinity, (where a given source's electrical field is essentially zero), to an intermediate point on the map. To make these maps easier to interpret, all locations that have the same voltage are connected by solid lines (see Figure 7.27). This is similar to outlining all points at the same altitude on a topographical map of terrain. A positively charged body will be surrounded by lines representing positive voltages with the higher-value lines closest to the body; a negatively charged body will be surrounded by negative voltages with more negative lines closest to the body. The amount of energy it would take to move a test particle between any two locations on

this map is just the numerical difference in mapped voltages between the start and end points. The rate at which voltage is changing at a given point, (e.g., the voltage gradient), is simply the amplitude of the electric field at that point.

Web Topic 7.3 *A primer on electric signals*
Here we provide a somewhat more quantitative introduction to the physics of electric fields, potentials, resistance, capacitance, impedance, and related topics.

There are a surprising number of electrical field sources in nature. In air, cloud processes build up large electrostatic potentials that eventually discharge as lightning. Lightning discharges permeate not only the air but also nearby bodies of water, creating electrical "noise" in both media [181]. In still waters, differential leaching of positive and negative ions into the water from exposed substrates can create electrochemical fields much as a man-made battery does. Ocean currents that flow perpendicular to the Earth's magnetic field lines can generate electric fields in the same way that an electrical generator creates power by moving a coil of wires past a magnet. Even living organisms that swim sufficiently rapidly through the Earth's magnetic field can generate weak electric fields around their bodies [216]. Finally, there are large-scale **telluric electrical fields** in the Earth's solid mantle and oceans that are generated by interactions between the Earth's magnetic field and the sun's magnetic and electromagnetic activities. These tend to be of relatively low magnitude in temperate and tropical oceans, but they can be considerably larger in the Earth's polar regions and during solar magnetic storms[225].

Biological sources of electric fields are also widespread. The most common mechanism for generating an electric field uses the same principle as the differential ion leaching

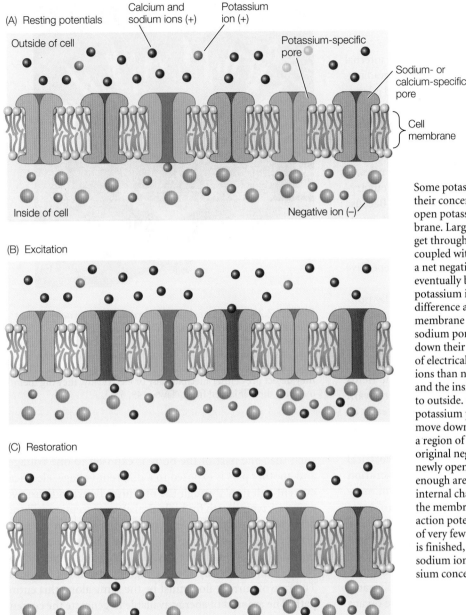

(A) Resting potentials

Outside of cell

Calcium and sodium ions (+)

Potassium ion (+)

Potassium-specific pore

Sodium- or calcium-specific pore

Cell membrane

Inside of cell

Negative ion (–)

(B) Excitation

(C) Restoration

FIGURE 7.28 **The generation of bioelectric potentials by living cells** (A) The cell actively transports positive potassium ions from outside the cell into the cytoplasm inside the cell, and moves positive sodium ions from inside to outside. Sodium pores in the membrane are closed, preventing sodium ions from leaking back in the other direction. Some potassium ions, however, can leak back down their concentration gradient to the outside through open potassium-specific pores in the cell membrane. Large negative ions inside the cell cannot get through any pores. The loss of potassium ions coupled with the retention of negative ions creates a net negative charge inside the cell. This charge eventually becomes strong enough to slow further potassium ion loss, resulting in a steady voltage difference across the membrane. (B) When the cell membrane is excited by an appropriate stimulus, sodium pores open and allow sodium ions to move down their concentration gradient toward a region of electrical negativity. Eventually, more positive ions than negative ones accumulate inside the cell, and the inside becomes positively charged relative to outside. (C) Sodium pores then close and more potassium pores open, allowing potassium ions to move down their concentration gradient toward a region of relative negativity. This restores the original negative charge inside the cell. Many of the newly opened potassium pores then close, and just enough are left open to maintain a steady negative internal charge. The rapid swing in potential across the membrane during (B) and (C) is called an action potential. It actually requires the movement of very few ions across the membrane, and after it is finished, the cell quickly removes any acquired sodium ions and restores the initial internal potassium concentration.

mentioned above. Many organisms import positively charged molecules such as potassium ions into their cells until the internal concentration is much higher than that of the external medium. Initially, this positive ion buildup inside the cell is electrically balanced by high concentrations of negative ions. Selective pores in the cell membrane then allow a small fraction of the potassium ions to move down their concentration gradient to the outside while preventing the balancing negative ions from leaving the cell. The result is a buildup of excess negative charges inside the cell until the growing electrical field halts the egress of further positive ions. The inside of the cell now has a steady negative electrical potential relative to the outside medium. When many adjacent cells in an organism maintain a similar electrical potential across their membranes, the entire tissue may act as a monopole source. Where different tissues in an animal's body hold different electrical charges, the electrical field can be more complex ranging from that of simple dipoles to more complex geometries [216, 225, 418]. Injured animals typically produce stronger electrical fields than intact ones as a result of the more direct connections between their tissues and the medium [418]. Aggregations of organisms, such as plankton blooms, may also generate large-scale zones with different ionic contents from those of unpopulated zones, and thus create large and complex electric fields between the zones [225].

Varying bioelectric fields are usually created by **excitable** tissues. A common precondition for excitable cells is a prior internal negative potential as described above. Excitation then involves the opening or closing of selective membrane pores in response to a relevant stimulus (**Figure 7.28**). The most commonly triggered pores allow positive ions, such as sodium or calcium, to move across the membrane. These ions are usually scarce inside the cell but abundant outside; opening the relevant pores allows them to move down their concentration gradient into the cell. This inward flow of positively charged ions is also accelerated by the initially negative charge inside the cell. The positive ion influx rapidly neutralizes the internal negative charge and, given a sufficient concentration gradient, may even make the inside of the cell positively charged relative to the outside. A subsequent egress of potassium ions restores the inside of the cell to its steady negative condition. Note that the movement of very few ions is needed to create major changes in the electrical fields around the cell. This sequence, sometimes with different ions playing the relevant roles, is the basis of **electrical excitability** in single-celled eukaryotes, motile plants, and animal nerves and muscles. Because adjacent nerve and muscle cells often act in concert, excitability can generate significant and often oscillating electrical fields around active animals [216, 418]. Oscillating fields are commonly generated by active locomotion and the repetitive mouth and gill movements of respiring fish.

The creation of steady or varying electric fields by non-biological sources or as a side effect of other biological functions is called **passive electrogeneration**. Sense organs that are dedicated to the detection and analysis of passively generated electric fields are said to perform **passive electroreception**. Because most passively generated electric fields are steady or only slowly varying in amplitude, the relevant electroreceptors are usually most responsive to steady or low frequency stimuli. In contrast, some species of fish have modified muscles or nerves, called **electric organs**, whose sole function is the generation of rapidly varying electrical fields around the fish. This is called **active electrogeneration** and it is used in three contexts that we shall discuss in more detail below: (1) to stun prey or deter predators; (2) to create an electric field whose distortion by nearby objects can be exploited for hunting in the dark or navigation in cluttered environments (**electrolocation**); and (3) to communicate with other animals using electrical signals (**electrocommunication**). Because the associated stimuli have rapidly varying amplitudes, **active electroreception** requires sense organs that are tuned to higher frequencies than those organs used for passive electroreception.

Passive electroreception

ELECTRORECEPTOR DESIGN   When mobile electrons or ions are placed in an electrical field, the electrical forces move them toward the part of the field of opposite polarity. Their movement constitutes an **electric current**. The magnitude of this current depends upon the voltage difference between their start and end points and the **electrical resistance** of the path along which they move. The resistance of a given path segment depends upon its length and its intrinsic **resistivity**: for the same path length, air has a much higher resistivity than fresh water, which in turn has higher resistivity than seawater. Higher resistance paths conduct smaller currents for a given potential difference; higher potentials will generate larger currents through a given resistance. In general, the current path will follow the lines of force of the electrical field. Thus the current paths between two ends of a dipole will be curved, just like the underlying electrical field lines.

FIGURE 7.29   Head of short-beaked echidna (*Tachyglossus aculeatus*) Note darkened and wet tip on end of the snout. This mucus-covered patch of skin contains both electroreceptors and mechanoreceptors. Echidnas normally forage by poking the snout into moist or sandy soil or into active termite or ant nests. When they detect prey either mechanically or electrically, they dig them out with their enlarged front claws [149, 331].

When an organism is inserted into an electrical field such that one part of the body is exposed to one voltage, and another part is exposed to a different voltage, electrical current will flow through the organism between these two points. A typical path for biological electroreception will traverse several successive resistances—the ambient medium on one side of the animal, the animal's skin, a portion of the internal tissues of the body, the skin on the other side of the animal, and finally another region of ambient medium. The rate of current flow must be the same along this entire path. The amount of energy available to generate the current is limited by the voltage difference between the start and end points. How that energy is expended along the path depends upon the relative resistances of the successive segments: segments of high resistance require more energy to maintain a given current flow, and thus "use up" a larger fraction of the overall voltage difference than segments of lower resistance. We can thus partition the overall voltage difference driving the current into a set of smaller voltage differences required to pass that current through each successive segment. The voltage difference seen across each resistance in the series is called that resistance's **voltage drop**.

Animal electroreceptors are stimulated when a minimal voltage difference of the right polarity is applied across them. The voltage drop across an electroreceptor is largely determined by the resistance of the tissues in which it is embedded: if those tissues constitute a large resistance relative to the ambient medium and other internal tissues, then the receptors will be exposed to a large fraction of any ambient voltage difference. If the medium constitutes a significantly higher resistance than the tissues, then there may be insufficient voltage change across the tissues and their embedded receptors

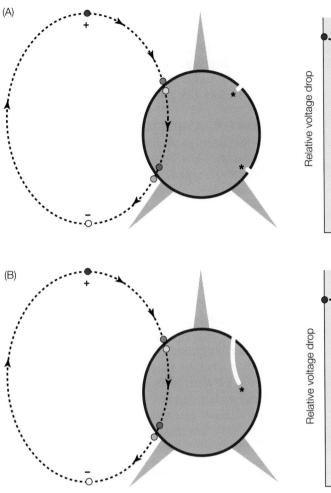

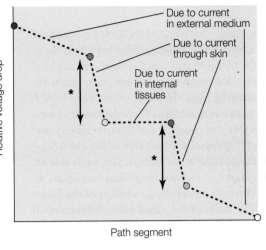

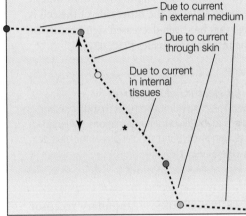

FIGURE 7.30 Strategies for detecting electric fields in different aquatic habitats Diagrams on the left show typical current flow (dashed line) from an external electric field through a fish (seen here in cross section with electroreceptors marked as asterisks); graphs on the right show successive drops in applied voltage on that current path. (A) Freshwater environment in which the resistivity of the medium is much higher than that of the internal tissues of fish. Voltage drop due to current through the medium thus "uses up" a large portion of potential voltage stimulus. For osmotic reasons, the skin of freshwater fish has very high resistivity, and a respectable voltage drop (line with arrows) is still possible where the current passes through the skin. The electroreceptors are positioned with one side open to a short low resistance canal (to sample the outside field) and the other side exposed to tissues just below the skin. (B) Marine environment in which resistivity of medium is only half of that of the inner tissues of fish. Some voltage drop occurs across the skin, but a larger drop is due to the location of the receptor cells inside the fish at some distance from the outside sampling point. A long but low resistance canal (white) connects one side of each receptor to the outside electric field. The other side of the receptor samples an internal voltage which is lower than that outside because the current has had to flow through the intervening tissues. Receptors in marine fish can be as far as 20 cm from the external opening of their associated canal.

to make electroreception worth attempting. Whereas air can maintain electrical fields well, it has a much higher resistivity than living tissues. This explains why electroreception is almost unknown in terrestrial animals. The only known exceptions are the several species of echidnas (Monotremata). These primitive mammals have electroreceptors on their long thin snouts, which they use to probe ant and termite nests when foraging [268, 321, 331, 418]. Echidnas may be the exception that proves the rule, since their snout tips are always covered with a wet mucus that minimizes the resistance between their electroreceptors and the electrical fields generated by their insect prey (**Figure 7.29**) [149].

Fresh water and seawater both have much lower electrical resistivity than does air, and it is not surprising that all electroreceptive animals other than the echidnas are aquatic. While living in water reduces the medium resistance considerably, it does not guarantee that most of the voltage gradient will be focused on electroreceptors. In fact, freshwater fish live in a medium that has a higher resistance than their internal saline tissues. However, that internal salinity also creates an osmotic challenge as water tends to move into their bodies, and salt ions tend to leak outward. To avoid the loss of internal salt ions and an influx of water, freshwater fish have evolved skins that minimize either flux. The result is an outer skin with an electrical resistance much higher than either the ambient water or the animal's internal tissues. This

is a perfect situation for electroreception, and freshwater animals routinely have their electroreceptive organs embedded in their skin right where the maximal voltage drop will occur (**Figure 7.30A**) [212, 427].

Marine fishes usually have less of an osmotic problem and in fact may benefit from allowing diffusion across their skins. However, the medium resistance is then only slightly less than that of their skin and tissues. Marine fishes deal with this problem by locating their electroreceptors deep inside their bodies. This places as much tissue between the receptors and the outside medium as possible, and thus maximizes the available resistance between the inner side of the electroreceptive cells and the outside of the animal. The outer side of the receptor cells plugs one end of a long canal that opens at the opposite end through a pore on the outside of the body (**Figure 7.30B**). The walls of the canal are constructed of highly resistant materials, while the interior is filled with a very low-resistance jelly. The outer ends of the receptor cells are thus exposed to voltages only slightly less than that at the opening of the pore, whereas the inner side of the cell is exposed to a voltage that has been decreased significantly by the resistance of the intervening tissues [212, 427].

Web Topic 7.4 *Bioelectric field resources*
Here we provide web links to a number of sites that show electroreceptive animals in the wild, further explanations of electroreception and electrogeneration, and simulations of the electric fields generated by teleost electrolocators.

AMPULLARY ELECTRORECEPTORS   Ampullary receptor organs are the main sensors used for passive electroreception. They respond only to steady or very slowly varying electric stimuli. Each contains a cavity filled with low-resistance jelly [56] that narrows into a canal and is exposed to the medium through a pore in the skin. As noted above, the canal linking the cavity and the pore can be short (freshwater species) or long (marine species). Electrosensory cells are located in the base of the organ where they are surrounded on two sides by support cells and expose their outer side into the cavity. Whereas freshwater species tend to have 1–20 sensory cells per ampullary organ, marine species may have thousands of cells per organ [212, 427]. Primitive fishes such as sharks, rays, skates, and sawfishes (Elasmobranchii), ratfishes (Holocephali), coelocanths (Coelocanthiformes), lungfishes (Dipnoi), sturgeon and paddlefish (Acipenseriformes), and bichirs and reedfish (Polypteriformes), as well as larval, and in some cases adult, salamanders (Urodela) and caecilians (Gymnophiona), all share a similar sensory cell design [41, 42, 65, 140, 141, 212, 305–309, 427, 428]. This consists of an externally facing surface hosting a single cilium and several to many microvilli, and an internal surface with synapses linking the cell to afferent (but no efferent) nerve fibers [212, 427]. Fossil evidence and the widespread occurrence of ampullary electroreceptors in primitive fishes suggest that these organs evolved very early in the history of vertebrates (**Figure 7.31**) [4]. Because they and lateral line neuromasts develop from similar embryonic tissues, show similar ciliated and microvillar structure, and project to adjacent areas in the brain, it has been suggested that electroreceptors evolved from lateral line neuromasts, or that both types of receptors evolved from some earlier lateral line cell with multiple functions [65, 76, 87, 140, 278, 306, 309, 310].

OTHER PASSIVE ELECTRORECEPTORS   While many primitive bony fishes (sturgeons, paddlefish, bichirs and reedfish) have retained the ancestral mixture of hydrodynamic and electroreceptive lateral line organs, the predecessors of advanced forms (Teleostei) lost their electroreceptors and retained only the lateral line hydrodynamic system [4, 309]. Two groups of teleost bony fishes then secondarily reacquired passive electroreceptors (see Figure 7.31). One group consists of the African elephantfishes (Mormyridae), frankfish (Gymnarchidae), and featherback fishes (Notopteridae). The catfishes (Siluriformes) and neotropical knifefishes (Gymnotiformes), which are related to each other but unrelated to the prior group, independently reacquired electroreceptive organs. Although each of these taxa hosts electroreceptive organs that resemble primitive ampullary organs in that they respond only to steady or slowly varying electric fields, all but one species use electrosensory cells with only microvilli on the exterior surface and no cilia. In addition, all of these low-frequency teleost receptors require electric signals of opposite polarity to those that stimulate the ampullary electroreceptors of primitive fishes and amphibians [212, 427, 428]. These differences reflect the prior loss of all ampullary receptors in early teleosts and the evolution of new designs without ancestral constraints.

Passive electroreception has also reappeared in the mammalian order Monotremata, a taxon far removed from its electroreceptive fish ancestors. The terrestrial echidnas mentioned above are one subset of this order. The other living member of this taxon is the aquatic duck-billed platypus (*Ornithorhynchus anatinus*) [263, 264, 266, 268, 320, 321, 345]. We noted in the prior section on hydrodynamic stimuli that the platypus has extensive hydrodynamic sensors scattered randomly over the surface of its ducklike bill. In addition, the bill hosts 40,000 modified mucous glands and associated free nerve endings that are arranged in rows parallel to the long axis of the bill and function as sensitive electroreceptors when the platypus is submerged and foraging for its invertebrate prey. No other mammals appear to be electroreceptive; although it was suggested at one time that star-nosed moles might host electroreceptors on their nasal star [145], this has been shown not to be the case [73].

PASSIVE ELECTRORECEPTOR PHYSIOLOGY   Passive electroreception systems can be highly sensitive. Examples of threshold sensitivities are compared to the strengths of ambient electric fields for marine species in **Figure 7.32**. Marine sharks, rays, and skates can detect electric fields, (measured as voltage gradients), as low as 0.005 µV/cm, and some species may be able to detect fields as small as 0.001 µV/cm [43, 157, 213, 214, 418, 427]. Similar sensitivities are found in electroreceptive marine catfish [43]. Many marine species

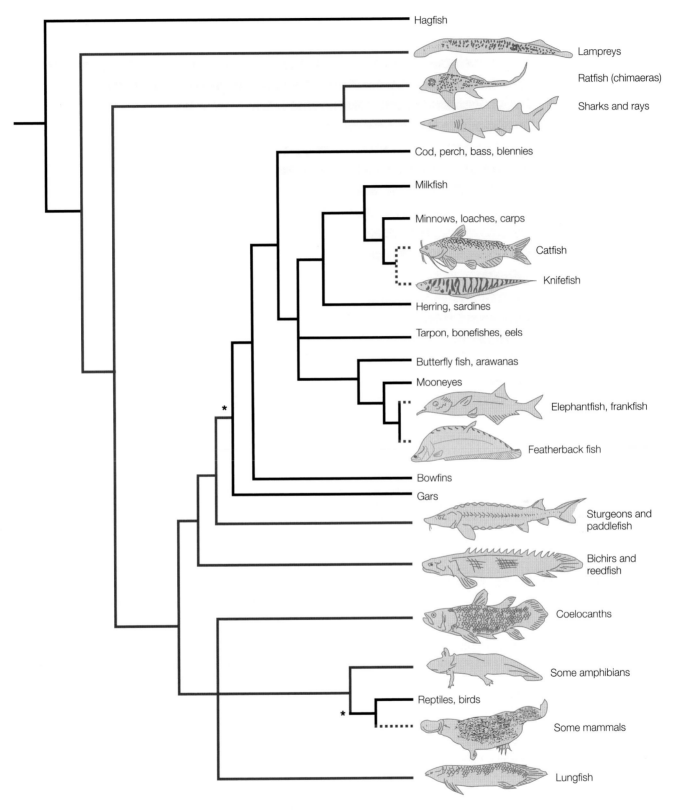

FIGURE 7.31 **The distribution of electroreception across the vertebrate phylogenetic tree** Red lines link all taxa that have true ampullary electroreceptors. Black lines link taxa whose ancestors lost their ampullary receptors. Asterisks indicate likely points at which ampullary receptors were lost. Dashed red lines indicate secondary reacquisition of electroreceptors after original organs were lost. Electroreception was initially a widespread sensory modality in vertebrates. (After [287].)

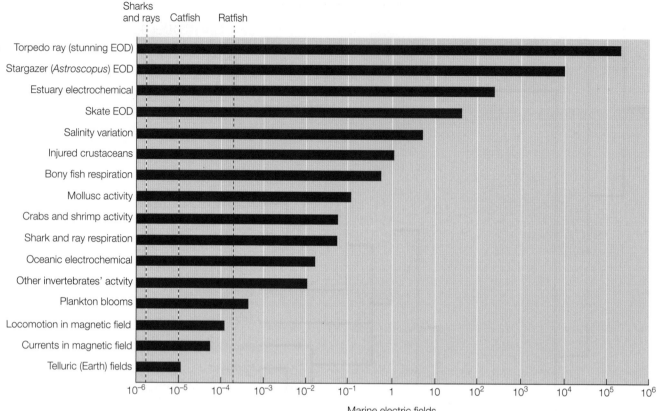

FIGURE 7.32 Common electrical stimuli and electroreceptor thresholds for marine habitats Various electrical field sources discussed in the text are listed along the left side of the graph and their average electrical field strengths are shown. Note that this is a logarithmic scale with a trillion-fold range of possible electrical field levels. Torpedo rays, stargazers, and skates all have special organs designed to produce electric stimuli. An EOD is a discharge of such an electric organ. Other biological sources of electrical fields listed here are incidental consequences of locomotion and respiration activity. Vertical dashed lines show threshold sensitivities for various kinds of marine animal electroreceptors. (Data from [43, 68, 157, 213, 216, 225, 303, 417, 418].)

exhibit specific adaptations that account for this extreme sensitivity to electrical stimuli. As we saw earlier, increasing the length of the canals linking the external sampling pores to the ampullary organs can significantly increase the amount of intervening tissue and thus the fraction of the overall voltage gradient that can be applied across the receptor cells. To facilitate canal lengthening, marine sharks, rays, skates, and coelocanths cluster their ampullary organs into a small number of locations with the relevant canals radiating out to distant pores (**Figure 7.33**). Because all organs in these clusters will be exposed to the same tissue voltage on their inner sides, any differences in the voltage drops that they record will be due to differences in the voltages at their respective pores and not to internal voltage variation [81, 215, 225, 418, 427]. As with the rod receptors in vertebrate eyes, as many as 100 electroreceptor cells in an ampullary organ may synapse with the same neuron; these neurons emit impulses continuously,

and stimulation of the receptors modulates this tonic rate with one polarity of stimulus causing increases and the other polarity decreases [427]. In addition to convergence of many receptor inputs onto the same nerve, and many nerves into a common nerve branch, the animal's brain pools the inputs of thousands of nerve branches. This cumulative convergence of ampullary organ input greatly increases the chances of detecting low-level stimuli [43]. The animal's brain further enhances sensitivity by subtracting the likely electrical noise generated by its own locomotion and ventilation from the raw information provided by the ampullary organs [43].

As a result of these combined adaptations, a foraging shark or stingray can detect the passive electric fields of prey buried in the sand or invisible in the dark at tens of centimeters and in some cases up to a meter away [387, 418]. Most sharks and skates encase their eggs and developing embryos inside leathery egg capsules. Predators can use passive electroreception to detect the tiny electric fields generated by the ventilation activity of embryos inside the capsules. However, the embryos develop their own passive electroreceptors at an early stage, and when they detect the passive electric fields of approaching predators, they cease ventilation and other motion to avoid being detected in turn [362–364, 386]. Hatched offspring also use electroreception to detect approaching predators in the dark. Adult stingrays use their passive electroreceptors not only to locate prey but to locate buried conspecifics for mating or social aggregation [363, 386, 387].

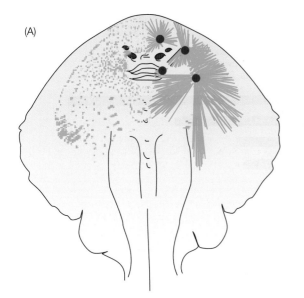

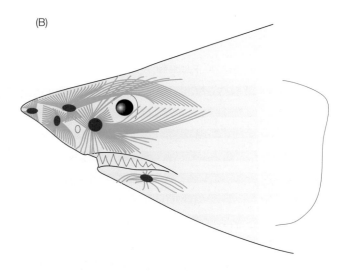

FIGURE 7.33 **Ampullary canals in rays and sharks** (A) Ventral view of thornback guitarfish (*Platyrhinoides triseriata*). The left side of the animal shows the distribution of ampullary canal openings (blue dots) on skin surface. The right side shows canals (blue lines) under the skin radiating out to openings from four regions (red dots) in which ampullary electroreceptors are clustered. All receptors in the same cluster experience the same voltage on their internal tissue side, but different voltages on their external side depending on the voltage near the opening of their canal. (B) Ampullary canals (blue lines) and ampullary electroreceptor clusters (red dots) in the head of the bull shark (*Carcharhinus leucas*). (A after [288]; B after [81].)

Freshwater species have their ampullary organs embedded in their skins, with very short canals. Because the skin and the freshwater medium constitute relatively high resistances compared to internal tissues, variations in tissue resistivity inside the animal have little effect on the voltage drops measured by the receptor cells. Freshwater species thus do not need to cluster their ampullary organs but instead can spread them out over the body or at least over those body regions most likely to encounter the electric fields of prey, predators, or conspecifics. As can be seen in **Figure 7.34**, receptors of freshwater species require stimulus amplitudes about 10–100 times larger than do those of related species in marine environments. At the same time, note that freshwater electric fields tend to be about 10–100 times greater in amplitude than marine fields from similar sources and measured at similar distances [43, 216, 319]. As noted earlier, the ampullary organs (or teleost equivalent) of freshwater species use only 1–20 sensory cells to achieve the necessary sensitivity.

Because most passive electrical fields are—or at least at a moderate distance, resemble—dipole sources, the electrical field lines will be curved. This presents a problem for an animal attempting to locate and approach the source of the field. One strategy is for the animal to compare the amplitude of the electrical field on the right and left sides of its body and then adjust the angle of their long axis until these stimuli are equal. This will place them parallel to the electric field lines at each successive location. Such a strategy will cause the animal

to trace the curved electric field lines back to the source, and this is in fact what is seen in nearly all passive electroreceptive species examined [191, 216].

Web Topic 7.5 *Adaptations for passive electroreception*
The early acquisition of passive electroreceptors by primitive fish was surely a key adaptation that facilitated this taxon's subsequent radiation and eventual dominance in aquatic habitats. In both marine and freshwater habitats, a number of strategies are employed to enhance passive electroreception.

Active electroreception

We have seen that passive electroreception is nearly ubiquitous among the cartilaginous (sharks, rays, skates, and ratfish) and primitive bony fishes (bichirs, reedfish, paddlefish, sturgeons, coelacanths, and lungfish), and that some species, such as stingrays, routinely use this sense to locate buried conspecifics. Given these preadaptations, it should not be surprising that certain fish taxa have gone a step further and modified nerves or muscles specifically to produce electric fields that can be used in social communication [4, 68]. Stingrays and skates belong to the same order of fishes (Rajiformes). Nearly all species of skates (Rajidei), but no stingrays, possess paired spindle-shaped **electric organs** in their tails that produce electric discharges at facultative intervals [198, 234]. These discharges are of too low a voltage to stun prey, but are more intense than most passive electric fields the skates would encounter (see Figure 7.32) [303]. The only exception is electrical noise generated by their own muscles; since their brains control this activity, it can be computationally removed from the overall electrical signal input [290]. Skates rarely discharge their electric organs when solitary, but do so frequently when in social groups [49, 362]. Whereas young skates tune their passive electroreceptor organs to frequencies typical of approaching predators, adults shift the tuning to match the frequency composition of conspecifics' electric discharges [362, 364]. These observations all suggest

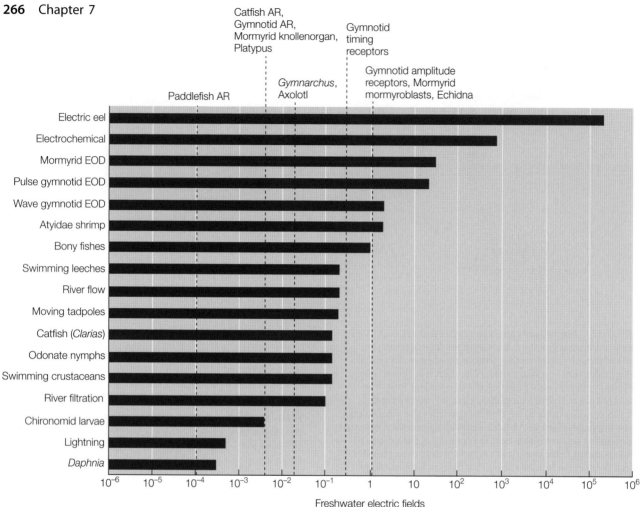

FIGURE 7.34   Common electrical stimuli and electroreceptor sensitivities in freshwater habitats Electrical field sources in freshwater are listed along the left side of the graph and their average electrical field strength is shown. The logarithmic scale covers a trillion-fold range of possible electrical field levels. Vertical dashed lines show threshold sensitivities for various kinds of freshwater animal electroreceptors measured either at single organs or at the brain level (AR, ampullary receptors). Note that the amplitudes of electrical fields in fresh water tend to be higher than for similar sources in seawater. Note also that corresponding thresholds for response, even for members of the same taxon (e.g., catfish), are correspondingly higher in freshwater contexts. Electric organ discharges (EODs) are used by the electric eel to stun prey. Gymnotid and Mormyrid fish produce smaller EODs for electrolocation. Other animals listed produce electrical fields as a side effect of their locomotion or respiration. (Data from [8, 16, 68, 116, 149, 177, 181, 225, 251, 320, 340, 369, 371, 375, 396, 402, 417, 418, 428].)

that the primary function of skate electric discharges is social communication.

The other order of electrogenic raylike cartilaginous fishes consists of the electric rays (Torpediformes). Members of this group possess a very large electric organ in each of their two body-wings [28]. These organs produce high voltage discharges that are used to stun prey [22, 255]. Electric rays in the genus *Narcine* also have a second smaller electric organ just posterior to each large organ [27]. These **accessory electric organs** produce discharges that are only 1–2% of the

voltage generated by the larger organs. While other electric rays such as *Torpedo* lack accessory organs as adults, embryos have the rudiments of such structures, which are lost as development proceeds [135]. It remains unknown whether these accessory organs, when present, mediate social interactions in rays as they do as in skates.

As noted earlier, most modern teleost fishes lack electroreceptors and thus engage in neither passive nor active electroreception. However, common ancestors of the worldwide catfish taxon (Siluriformes) and their neotropical knifefish relatives (Gymnotiformes) reacquired ampullary-like electroreceptors that allow their descendants to locate prey, avoid predators, and interact with conspecifics using electric signals. Independently, a cluster of African species consisting of elephantfishes (Mormyridae), frankfish (Gymnarchidae), and featherback fishes (Notopteridae) also acquired ampullary-like passive electroreceptors. Like the skates and rays, a subset of these electroreceptive teleost taxa have also evolved electric organs. Members of at least six families of catfish (Malapteruridae, Mochocidae, Clariidae, Bagridae, Siluridae, and Plotosidae) have electric organs [4, 431]. Whereas African electric catfish (Malapteruridae) use electric discharges to stun prey and thus resemble electric rays, the other catfish produce episodic low-amplitude electric discharges for social communication and thus resemble skates [13–16, 156,

Freshwater species   Marine species

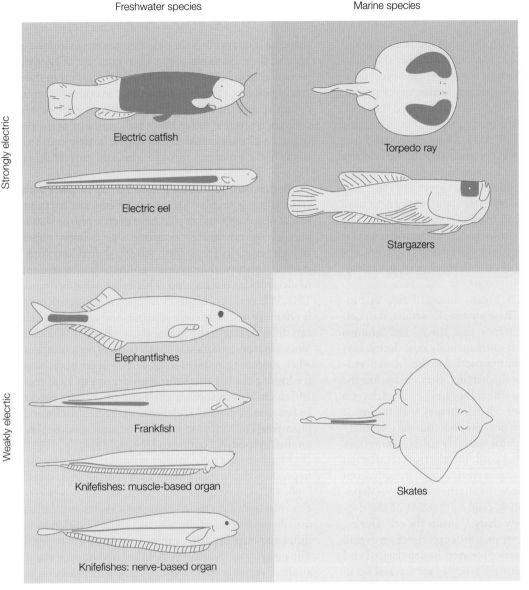

Strongly electric

Electric catfish

Torpedo ray

Electric eel

Stargazers

Weakly electric

Elephantfishes

Frankfish

Knifefishes: muscle-based organ

Skates

Knifefishes: nerve-based organ

FIGURE 7.35 Shape and location of electric organs in fish Electric organs are shown as dark patches within the outline of the animal. Examples are organized with regard to habitat (columns) and strength of electric organ discharges (rows). (After [31].)

336, 431]. In contrast to catfish, all knifefishes (except electric eels), elephantfishes, and frankfish have electric organs that are continuously discharging. A little more than half of the known species of knifefishes and the single known species of frankfish maintain steady and high discharge rates at all times and are called **wave fish**; the remaining species of knifefishes and all elephantfishes vary the rate of discharges over a wide range and are known as **pulse fish** [3, 19, 370]. The continuous electric discharges of knifefish, elephantfish, and frankfish electric organs reflect the fact that all of these species use these organs and associated receptors for electrolocation. In contrast to the prior teleosts with electric organs, two genera of marine stargazers (*Uranoscopus* and *Astroscopus*, Perciformes) reacquired electric organs without prior or concomitant acquisition of electroreceptors [12, 30, 65]. Stargazers hidden in the sand discharge their electric organs in bursts while engulfing passing prey with their large mouths, but the discharges are of too low a voltage to have any apparent effect on the prey [322]. The biological function of stargazer electric organs remains to be clarified [286].

ELECTRIC ORGANS   The electric organs of fish are usually modified muscle cells [4, 30]. In elephantfishes, frankfish, and knifefishes, larval fish use one set of muscles as electric organs, and these are either modified or replaced by other nearby muscles to form adult electric organs [19, 230, 231]. In skates and elephantfishes, the electric organs are modified muscles in the narrow section of the tail connecting the wider body to the caudal fins (**Figure 7.35**). Rays produce their electric organs from muscles located in the wings on each side of the body. The electric catfish *Malapterurus* uses trunk muscles for its electric organs. Frankfish and most knifefishes have also co-opted trunk muscles that extend from just behind their gills to the tip of the tail as their electric organs. Interestingly, the one family of knifefishes (Apternonotidae) with nerve-based electric organs begins embryonic development with the same modification and innervation of muscle cells that develop into the larval electric organ in other members of the order. However, these modified muscles are then lost during further development, and the motor nerves that would have controlled the larval electric organ undergo their own modification to become the adult electric organ [19, 231]. Some catfish and one stargazer genus (*Uranoscopus*) produce coordinated electric discharges in the sonic muscles used to produce drumming sounds with the swim bladder (see Chapter 2) [156, 286]. The other genus of stargazer, *Astroscopus*, appears to have developed its electric organ from eye muscles [412]. The widespread recruitment of muscle cells as electric organs is not surprising, given the focus of most passive electroreceptors on muscle-generated electric fields of prey, predators, and conspecifics [4, 65, 68].

All muscle-based electric organs consist of disk- or cup-shaped **electrocyte cells**. These are usually arranged in columns like stacks of coins. In most taxa, the stacked columns are arranged in parallel with the main body axis; electric rays and stargazers have the columns parallel to their dorsal-ventral axis [19, 30]. When not discharging, electrocytes, like the muscle or nerve cells from which they have evolved, have a resting potential across their cell membranes, with the inside of the cell 50–80 mV more negative than the outside. Electrocytes in all species except the stargazer *Astroscopus* receive nerve input from the spinal cord; the electrocytes derived from eye muscles in *Astroscopus* receive input directly from the brain. When a nerve fires, the attached electrocyte depolarizes on one face of its disk, causing that side of the electrocyte to become positively charged inside the cell. During peak discharge, an electrocyte may thus experience an overall 130–150 mV voltage difference between its depolarized and polarized faces. When many electrocytes are stacked up in a column and all depolarize the same side of their disks at the same time, the overall voltage difference between the two ends of the column is the sum of the voltages across the constituent cells. Thus the column of 6000 stacked electrocytes in an electric eel can generate a voltage difference of 700–900 volts.

The voltage difference between the two ends of a depolarized electrocyte column causes current to flow inside the fish from one end of the column to the other, out through the skin, back through the medium toward the opposite end of the fish, and again through the skin to the point where it started. Assuming that the fish benefits by maximizing the range over which its electric discharges can be detected, it should attempt to maximize the amount of energy in the discharge (measured as power) which is transferred into the medium. This transferred power is equal to the product of the voltage gradient generated outside the fish and the amount of current flowing in the circuit. The fish can increase the voltage gradient outside of its body by stacking up more electrocytes in each column. It can increase the current that flows through the circuit by increasing the number of parallel columns that discharge at the same time. Since there is a limit on how many electrocytes can fit inside a given animal, there is a trade-off between numbers of electrocytes per column and the number of parallel columns in the electric organ. For a fish of a given size, the optimal number of columns, and thus the optimal number of electrocytes per column, is one that makes the overall resistance of the electric organ and its associated tissues approximately equal to that of the outside medium through which the current flows [68, 191].

In practice, this means that fish living in water with low resistivity (e.g., marine habitats and freshwater habitats with lots of dissolved salts) should have electric organs with many parallel columns and few electrocytes per column. Fish living in high resistivity (most fresh waters) should have long thin electric organs with only a few columns but large numbers of electrocytes per column. This is just what is found [68, 191]. Whereas the resistivity of open marine environments is relatively homogeneous, estuaries and freshwater habitats can differ dramatically in resistivity values [174, 238, 367]. Since electroreceptors and electric organs both function best when adapted to ambient resistivities, it is not surprising that freshwater electric fish show high degrees of local receptor and electric organ specialization and consequent speciation. In fact, the freshwater knifefishes in the neotropics and elephantfishes in Africa are some of the most speciose fish taxa on their respective continents [3, 4, 92, 191].

Many species have further refined their electric organs to produce discharges with carefully customized waveforms [190, 372]. One way that waveforms can differ between species is in duration: the discharges of skates and catfish are usually long in duration, whereas those of different knifefishes and elephantfishes can be long or short [14, 19, 191]. The polarity of the electric field created outside the fish is another variable trait (**Figure 7.36**). It depends on which of the electrocyte faces is innervated and depolarized, or if both are activated successively, which face is activated before the other [19, 30]. The detailed shape of the waveform can also be fine-tuned [19, 68, 191, 239, 286, 371]. If only one face of each electrocyte depolarizes, the discharge waveform will be **monophasic** (i.e., show a voltage deflection in only one direction, either positive or negative, with a subsequent return to neutrality). If the two faces of the electrocytes depolarize successively, the waveform will be **diphasic** (i.e., first deflecting in one direction from neutrality, and then deflecting in the opposite direction before returning to the neutral point). Some diphasic electric organs result from separate innervation on the two faces of each electrocyte; in others, a depolarization triggered by nerve input to one face stimulates sufficient current in the electrocyte to depolarize the opposite face after a short delay. Apteronotid knifefish produce diphasic waveforms because the nerves that form the electric organ are bent into a J- or hairpin shape and polarity reverses as an action potential propagates from one end of the nerves to the other. Finally, some knifefishes and all elephantfishes innervate their electrocytes through an attached **stalk**. The stalks depolarize first, and this then triggers one or both faces

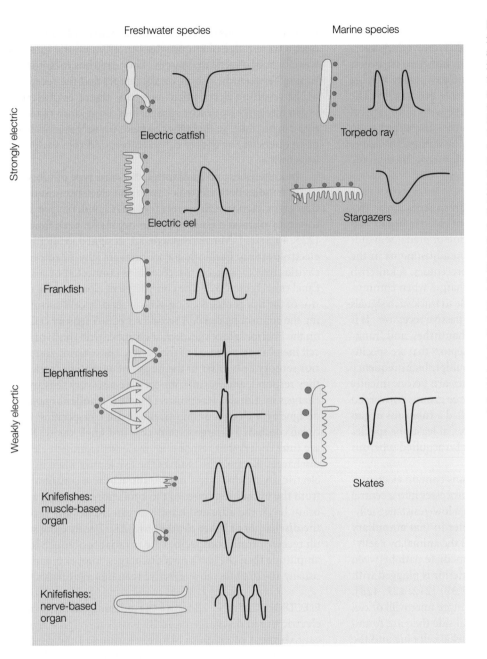

Freshwater species

Marine species

Strongly electric

Electric catfish

Torpedo ray

Electric eel

Stargazers

Weakly electric

Frankfish

Elephantfishes

Skates

Knifefishes: muscle-based organ

Knifefishes: nerve-based organ

FIGURE 7.36 Examples of different electrocyte innervations and resulting EOD waveforms in electric fish For each example, the figure on the left shows the shape of an electrocyte (light blue object) with its anterior side facing left and its dorsal side facing upward. The shape is a simplified cross section for electrocytes that are modified muscles (all but lower left example), and the disposition of the nerve in the knifefishes with nerve-based electrocytes. The location of the innervation to muscle-based electrocytes is shown as dark blue dots. On the right of each example is a typical waveform (red line) generated by that combination of electrocyte shape and innervation. Waveforms move upward when the head of the fish is positive relative to the tail, and downward when the reverse is true. Innervation directly on one side of the electrocyte tends to produce monophasic signals, the polarity depending on which side is innervated, whereas innervation of attached stalks often produces multiphasic signals. (After [19].)

to depolarize in turn. A stalked electrocyte can thus produce **triphasic** and even higher-order waveforms. The relative amplitude and order of the multiple phases can be adjusted by varying where the stalk is located and the size, geometry, sensitivity, and excitable properties of the stalk and faces of the electrocyte. Further diversity of waveforms is achieved in knifefishes and elephantfishes by the asynchronous depolarization of different parts of the electric organ or multiple electric organs, variation in overall shape and thickness of the electric organ along its length, and any differences in electrocyte geometries and thus the waveforms generated by successive segments of the electric organ [19, 30, 68].

What are the advantages of producing such diverse electric waveforms? One obvious benefit for the speciose

knifefishes and elephantfishes is that multiple species living in the same habitat can have their own signature waveforms [7, 187, 191]. Another factor is predator avoidance [370, 371]. Catfish worldwide use their passive electroreceptors to locate prey. Discharging an electric organ in a habitat where a catfish might be lurking would seem a very risky thing for a smaller fish to do. This risk can be reduced by modifying the discharge waveform so that it is less likely to be detectable by a catfish. As with sounds, the waveform of an electric discharge consists of the sum of many pure sine wave frequencies. Long-duration waveforms contain low-frequency components that are absent from short-duration waveforms. The passive receptors of catfish are most easily stimulated by steady or slowly varying (low-frequency) electric fields.

Thus a prey species might consider using a short-duration discharge to reduce detection by catfish. However, the average voltage of a monophasic waveform is non-zero regardless of duration, and thus contains a significant steady voltage component that would remain detectable by a catfish. The average voltage of a diphasic discharge with symmetrical positive and negative phases is zero: such a discharge has no steady (zero-frequency) component. The optimal discharge to avoid detection by catfish is thus symmetrically diphasic and of short duration.

TUBEROUS ELECTRORECEPTORS   Adjustments in the frequency composition and waveform of electric organ discharges to facilitate species discrimination or predator avoidance must be matched by concomitant adjustments in the frequency sensitivity of a fish's electroreceptors. A knifefish that uses short-duration diphasic discharges when communicating with conspecifics will be unable to track such signals with its low-frequency ampullary-like passive receptors. It is thus no surprise that knifefishes, elephantfishes, and frankfish all have a second class of electroreceptors that are specifically tuned to waveforms containing multiple high-frequency components. Although the knifefishes are taxonomically unrelated to the other two taxa, all three groups converged on a similar electroreceptor design called a **tuberous organ** [212, 427, 428]. And not to be outdone, at least one species of catfish that preys on knifefishes has also acquired tuberous organs [6].

Both ampullary and tuberous organs contain electroreceptive cells that extend their sensory processes into a central cavity. This cavity is usually filled with a low-resistance jelly. The two types of electroreceptors differ in that ampullary cavities are connected to the outside of the animal by a jelly-filled canal and pore, whereas the immediate path between the tuberous organ cavities and the exterior is plugged with a column of epithelial cells (**Figure 7.37**) [212, 427, 428]. Tuberous sensory cells also host many more microvilli or cell membrane infoldings on their external side than are found on ampullary receptor cells. The epithelial cell plug and the elaborated sensory cell surfaces of tuberous organs hinder the passage of steady and low-frequency currents through the receptor cells, but are electrically transparent to higher frequency currents. The excitable portion of a tuberous organ sensory cell is its interior side. Here the cell membranes are designed to be electrically resonant and to respond most strongly to stimulation by a particular frequency; lower and higher frequencies produce a much lower response in synapsed nerves [223, 225]. The resonant frequency is typically set to the dominant frequency in the waveform of a given species' electrical discharges [184, 186, 223, 428].

Knifefishes, elephantfishes, and frankfish all possess two functionally distinct classes of tuberous electroreceptors [65, 223, 427]. **Amplitude coders** provide accurate measures of the magnitude of detected electric stimuli; given the high-pass filter properties of tuberous receptors, relevant stimuli are usually discharges by the fish's own electric organs or

discharges by other nearby electric fish. **Time coders** respond over a wide range of amplitudes but monitor the rate and waveform of electric discharges. They can thus monitor the timing between the fish's own discharges and the reception of the resulting electric field as well as the rates of electric organ discharge by nearby conspecifics. Both kinds of receptors can be tuned so that they respond to a very narrow band of frequencies, or widely tuned to respond to a variety of waveforms.

In knifefishes and frankfishes, the same type of tuberous organ handles both electrolocation and electrocommunication tasks. Elephant fishes, on the other hand, have an anatomically distinct class of tuberous organs for each function [223, 402, 427]. **Mormyromasts** are designed explicitly for electrolocation. Each organ partitions its jelly-filled central cavities into two successive chambers connected by a small canal (see Figure 7.37). The compartment closest to the exterior of the fish contains a ring of 6–13 sensory cells surrounding the connecting canal. A second set of 3–5 sensory cells sits on the interior side of the interior chamber with much of their cell membranes exposed to the cavity. The exterior and interior sensory cells differ in the ranges of amplitudes to which they respond, and the interior cells are also very sensitive to changes in stimulus waveforms [24, 402]. **Knollenorgans** are designed for electrocommunication with conspecifics. Each organ contains 3–5 large sensory cells in a single cavity. These are time coders that respond in phase with ambient electric discharges. Because the brain of the fish knows when its own electric organs discharge, it can subtract these contributions from the overall responses of the knollenorgans. Thus at the brain level, the adjusted knollenorgan input only records the discharges of other nearby fish [23]. Because the stimuli received from conspecifics are invariably of much lower amplitude than the fish's own discharges, knollenorgans are usually 10–20 times more sensitive than mormyromasts [29].

ELECTROLOCATION   Because the current generated by an electric organ discharge typically passes through a significant segment of medium before returning to the generating fish, an opportunity exists to monitor any modification in that current created by objects other than the medium along the external current path. Such monitoring would provide a mechanism for detecting and even characterizing nearby objects at night or in murky waters where vision is limited. Knifefishes, elephantfishes, and frankfish have all taken advantage of this opportunity with a process called electrolocation.

There are three main modifications that an external object can make in the electrical field and associated current outside an electrically discharging fish [20, 69, 173, 174, 301, 402]. If the object can conduct the current but has a higher or lower resistivity than the medium, the amplitude of the current will be decreased or increased, respectively, depending on the size and proximity of the object. The amplitude coders in the tuberous receptors will monitor such an overall change. At the same time, the imposition of an object in the

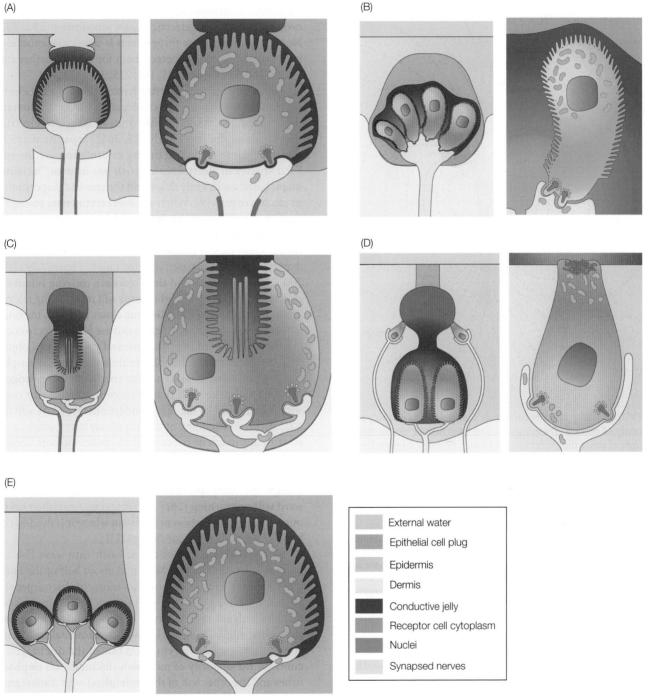

FIGURE 7.37   Tuberous electroreceptive organs in teleost fish (A) Catfish, (B) knifefish, (C) frankfish, (D) elephantfish mormyromast, and (E) elephantfish knollenorgan. Each example shows the overall structure of the organ on the left and a sample electroreceptive cell on the right. (After [212].)

Legend:
- External water
- Epithelial cell plug
- Epidermis
- Dermis
- Conductive jelly
- Receptor cell cytoplasm
- Nuclei
- Synapsed nerves

electrical field outside the fish can distort the distribution of current flow along the fish's body (**Figure 7.38**). An object with resistivity lower than the medium will cause the current lines between it and the fish to converge and create an electric "hotspot" on the fish's skin. An object with a higher resistivity will cause the current lines to spread out and create an electric "shadow" on the area of skin nearest the object. By comparing the stimulation levels of the many amplitude coders along its body, the fish can reconstruct where and how large the nearby object is. Finally, objects that do not conduct electricity may still become electrically polarized when placed in an electric field. Once fully polarized, they can then pass on the electric stimulus to the medium as if they were conducting. However, because it takes a while to build up the polarization, such an object slows down the passage of the waveform. The time that it takes to polarize a nonconducting

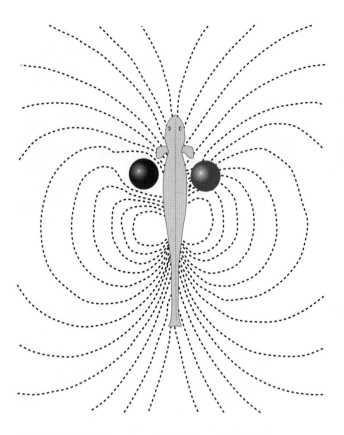

FIGURE 7.38 **Distortion of fish's electric field due to proximity of objects** Dashed lines show approximate electric field lines of active dipole field around fish. Current will follow these lines. The red object to left of the fish has higher resistivity than ambient medium. It thus deflects electric field lines so that the current flows around and not through the object. Result is an "electric shadow" on the fish's skin nearest the object where very little current flows and thus electroreceptors are minimally stimulated. The blue object to the right of the fish has a resistivity lower than that of the surrounding medium. This distorts the electric field lines so that more current flows through the object than would otherwise flow through this location. The result is an elevated concentration of current through the skin of the fish closest to the object and thus maximal stimulation of the electroreceptors in that region. (After [370].)

object depends upon its **capacitance** and varies with the object's size, shape, and composition [174, 395]. Thus a delay in the arrival of a discharge waveform at a fish's time coders can be used to indicate the proximity of a nonconducting object [397]. An object with significant capacitance will slow down the passage of some frequency components in an electric waveform more than others. As a result, a change in the waveform can also be used by a fish's time coders to detect the proximity of a nonconducting object, and the nature of the waveform changes may even provide information about the size, shape, and composition of the object [401]. In general, nonliving objects such as stones and dead wood have little capacitance, whereas living objects such as aquatic insects and other fish will have significant capacitance [400, 402]. The need to detect capacitive differences as well as resistive

ones during electrolocation has surely been one factor in the evolution of diverse electric organ waveforms [282, 367]. Note that each of these three effects is small, and most electrolocating fish cannot detect them for targets farther than one body-length away [191, 192, 225, 301, 431].

As they do for touch, hydrodynamic, and passive electrical stimuli, the brains of electrolocating fish create somatotopic maps of the spatial distribution of active electrical stimuli along the fish's body [25, 301]. Separate maps for processing amplitude and timing information are the rule. Each species of electrolocating fish has its own "baseline" maps, based on its body shape and the size and disposition of its electric organs [9]. When an object comes near enough to the fish, it will distort the electric field and be represented in the brain maps as an "image" or a "shadow." Concurrently, similar maps may be created for any passive electric or hydrodynamic stimuli associated with the target (**Figure 7.39**) [261, 300, 302]. If the fish and the object are moving relative to each other, the location of the image will move in the relevant brain maps, and as the two become closer together, the image will grow larger and become less diffuse [301]. Electrolocating fish do not confound the size and proximity of objects [350, 398, 399, 401]. The fish's brain appears to accomplish this separation by comparing the amplitude of the edge of the electric image with that at its center; this ratio varies with the distance to the object independently of its size [69, 402]. However, bending of the fish's body may be confounded with approach or movement of nearby objects because both change the ambient electric field around the fish. Some knifefishes avoid this problem by keeping their bodies rigidly straight and using specialized fins to move backward and forward without turning [261]. Other species bend their bodies but feed this information to the brain where it is used to correct the electric map accordingly [21].

Electrolocating species divide easily into wave fish and pulse fish. Wave fish (frankfish and about half of the knifefish species) produce continuous trains of discharges that are nonsinusoidal in waveform but very regular in emission rate. As with sounds, this type of periodic waveform consists of harmonics with the lowest (fundamental) frequency component equal to the emission rate of the discharges. In contrast, the majority of pulse fish (including all elephantfishes and the other half of the knifefishes) emit discharges of

FIGURE 7.39 **Simulated stimulation of receptors in the skin of a knifefish (*Apteronotus albifrons*)** The fish is initially swimming forward when it detects the prey (*Daphnia* sp.) 2.7 cm away (0 msec). Position of *Daphnia* is shown as a dot with vertical line indicating proximity of prey to closest patch of fish skin. The knifefish halts its forward movement and begins to swim backward (200 msec) until the prey is close to the fish's mouth (600 msec). Relative levels of stimulation of various receptors are shown for each region of fish body surface: (A) active (tuberous) electroreceptors; (B) passive electroreceptors: and (C) lateral line hydrodynamic receptors. Note concurrent movement and increased intensity of "hotspots" (red) for each type of stimulus along fish's body. (From [300].)

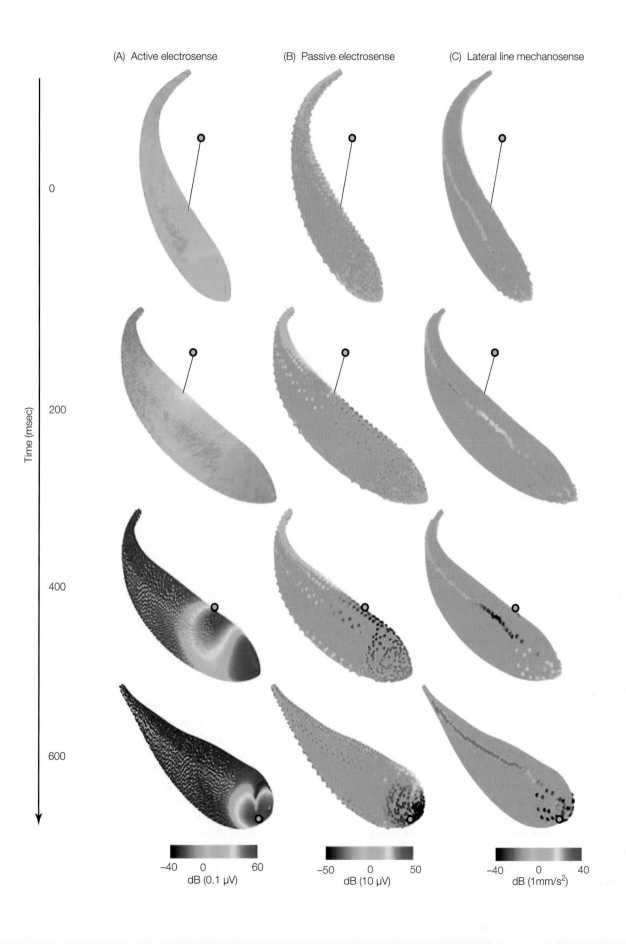

(A) Active electrosense   (B) Passive electrosense   (C) Lateral line mechanosense

short-duration multiphasic waveforms at irregular intervals. Pulse fish waveforms consist of a continuous and wide band of component frequencies. These differences in temporal and frequency pattern have important consequences for how these species use electrolocation.

Wave fish tend to live in deep or rapidly flowing rivers where they encounter few objects other than predators, prey, and conspecifics [92, 252]. Discharge rates can thus be adjusted so that component frequencies are those that are most affected by the capacitances of objects of interest. Resonant tuning of tuberous electroreceptors to match the frequencies contained in the adjusted discharges provides significant signal amplification while eliminating responsiveness to electrical noise. The proximity of objects with the appropriate capacitance will both delay the arrival of discharge signals at the fish's receptors and distort the waveforms; inappropriate objects will produce minimal lags or waveform changes and go undetected. The continuous high rate of discharge provides very fine-grained temporal information. This allows the electrolocating fish to track the movements of passing objects and localize them very accurately [186]. These are essential skills in deep rivers where prey objects moving with the current may appear and then disappear downstream very quickly.

Pulse fish favor shallow tributaries, streams, and floodplains [92, 252]. In contrast to deep-river species, these fish encounter large numbers of inanimate obstacles in addition to prey or predators. Many species have personal lairs to which they return each dawn after nocturnal foraging. Homing requires the ability to identify landmarks within their home ranges even in complete darkness. Electrolocation of inanimate objects may play a large role in this ability [351, 399–401]. Object discrimination by evaluating their size and capacitances requires discharge waveforms with many more component frequencies than suffice for wave fish. In these cluttered habitats, both the electrolocating fish and nearby objects move only at relatively slow speeds: there is thus no need for the fish to seek the fine-grained tracking information required by wave fish in fast flowing rivers. In short, pulse fish appear to have sacrificed rapid detection and detailed tracking of passing objects in exchange for exquisite object discrimination; wave fish have opted for the opposite trade-off. We shall see in Chapter 14 that similar trade-offs, with similar solutions, are faced by echolocating bats and cetaceans.

ELECTROCOMMUNICATION When a skate or catfish discharges its electric organ, nearby conspecifics can be assured that this is a social signal. Knifefishes, elephantfishes, and frankfish electrolocate incessantly and thus produce a continuous string of electric discharges. The only way these species can use electric discharges for communication is to modulate their usual pattern of emission and/or change its waveform. This is precisely what they do (**Figure 7.40**) [70, 239, 254, 344]. Wave fish have somewhat more limited options than pulse fish. Typical modulation patterns used for

communication include "chirps" (a sudden and brief increase in discharge rate), gradual accelerations or decelerations in discharge rate, "warbles" (cyclic increases and decreases in discharge rate), and brief discharge cessations. Pulse knifefishes maintain much more regular discharge rates than do elephantfishes (all of which are pulse species), and thus tend to show modulation patterns that are fairly similar to those of wave fish. Elephantfishes tend to have much more irregular discharge patterns. This gives them a much wider range of possible modulations. Typical examples include making their irregular pattern very regular for a period, accelerating and/or decelerating the discharge rate over time, emitting pulses in pairs or triads, emitting bursts of many pulses at either constant or stepwise-increased rates, (the latter often called "rasps"), or briefly ceasing emission. Modulations of the emission rate may be accompanied by concurrent modulations of the emission amplitude and/or waveform. For example, wave fish often reduce their emission amplitude during chirps. Each of these types of modulations can be strung together to create more complex signal combinations. For example, a wave fish chirp or a pulse fish burst is often preceded or followed by a brief cessation. Modulations of electric discharge rate and waveform are widely used by knifefishes, elephantfishes, and frankfish to mediate territorial defense, aggressive interactions, dominance, courtship, and coordination of foraging schools [70, 111, 153–155, 182, 183, 185, 188, 235, 239, 254, 344, 430–432]. In some taxa, electric signals are integrated with sound, touch, or hydrodynamic components to create even more complex signal repertoires [93, 95, 96, 349].

Electrocommunication signals must be sufficiently distinctive that they can be discriminated from other causes of electric discharge modulations [68, 254]. Discharge rates vary with ambient water temperature in most electrolocating fish [359, 372], and sudden changes in medium conductivity, for example after heavy rainfall, can also modify waveforms [238, 367]. Many species will increase their discharge rates to improve sampling when they first detect a nearby object or just prior to escaping a perceived threat. Both wave and pulse fish will modify their discharge rates to reduce overlap with the discharges of other nearby fish. This is called the **jamming avoidance response** [21, 174, 431].

Identity information is often important to electrolocating fish [93, 187, 189, 191, 239, 344]. Most nonsocial species are territorial and benefit by adjusting their aggressive behavior based upon the species, gender, and individual identity of intruders. In more social species, the ability to identify the age, sex, and dominance status of nearby conspecifics can be very useful. Knifefishes, elephantfishes, and frankfish encode identity information in their electric organ discharges by varying the discharge rates, waveforms, or both. For example, males of some pulse knifefishes (e.g., *Brachyhypopomus pinnicaudatus*) and elephantfishes (e.g., *Marcusenius macrolepidotus*) modify their electric organs as they mature to produce discharge waveforms that are much longer in duration than those of females or immature fish [240, 254]. The

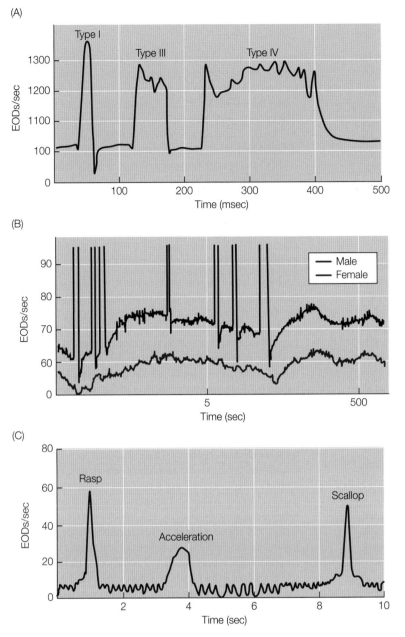

FIGURE 7.40 Modulations of electric organ discharge (EOD) rates by communicating teleost electric fish (A) Different types of "chirps" by the knifefish *Apteronotus leptorhynchus*, a wave species with a high baseline EOD rate. All chirps are transient increases in the EOD rate over time, but differ in the magnitude of the increase, the length of time the elevated rate is maintained, and the degree to which there is a concomitant decrease in EOD amplitude. Type I chirps are usually used by fish to advertise their presence on a territory. Type II (not shown here) chirps are briefer than the others and exhibit a much smaller increase in EOD rate. They are used by both sexes as an aggressive signal. Types III and IV are produced by males during courtship. (B) Coordinated variations in EOD rates of courting male and female knifefish (*Brachyhypopomus pinnicaudatus*). This is a pulse species. Males maintain an overall rate higher than that of females and periodically add "chirps" (shown here as sudden rises and falls over a large EOD rate range). (C) Three stereotyped variations of EOD rate in an elephantfish (*Brienomyrus brachyistius*). All elephantfish are pulse species. Rasps appear to function as courtship signals, accelerations as aggressive signals, and scallops as advertisement signals (usually by males). Although rasps and scallops appear similar, rasps have a fast initial rise and slower decay, whereas scallops show a more symmetrical rise and fall of EOD rate. (A after [111, 153, 235, 430–432]; B after [370]; C after [71].)

longer duration waveforms contain lower frequency components than the shorter female discharges. These long duration waveforms have the advantage that females can assess the sex of conspecifics at greater ranges using both their ampullary-like and tuberous receptors, but it has the disadvantage that predatory electric eels and catfish can better detect the males' discharges and attack them as prey [165, 370, 372].

Wave fish often have different baseline discharge rates in the two sexes: in the knifefish *Sternopygus macrurus*, males have discharge rates at least twice those of females, whereas males in the related *Eigenmannia virescens* have lower rates than females [254]. As noted earlier, the electric organ of most juvenile electric fish is different from that of adults and produces a distinctly different discharge waveform [4, 68, 371, 372]. Dominance status in social species is indicated by discharge rate in some wave fish and waveform details in pulse fish [153, 155, 372]. Most strikingly, individual identity is apparently encoded in subtle waveform differences by a number of species and then used by conspecifics in behavioral decisions [94, 146, 166, 279, 315, 344].

SUMMARY

1. Touch is the most widely used short-range sensory modality in living organisms. Animals rely on touch (also known as tactile stimulation) to resolve conflicts, mediate courtship and mate choice, reaffirm affiliative and dominance relationships, coordinate group movements, and exchange environmental information.

2. Touch stimuli can vary in amplitude, temporal pattern, informational versus strategic salience, and types of sensory receptors stimulated. The latter include detection and assessment of temperature (**thermoreceptors**), pain (**nociceptors**), positions and tensions of muscles and joints (**proprioceptors**), and both steady (**slow-adapting**

mechanoreceptors) and changing (**fast-adapting mechanoreceptors**) contacts with the receiver's body surface.

3. Prokaryotes (Eubacteria and Archaea) respond to cell membrane distortion by changing ion conductances in the region of perturbation. Cell membrane sensitivity to touch is also widespread in single-celled eukaryotes but is most often limited to the membranes enclosing or just adjacent to flagella and cilia. Most invertebrates use modified cilia as their primary tactile receptors, although some nematodes and arthropods detect tactile stimuli with both ciliated and nonciliated nerve endings. Touch receptors of invertebrates are often clustered on the head and appendages, and around the genitalia.

4. Vertebrates have a diverse set of touch receptors. **Free nerve endings** monitor temperature, pain, and sexual organ stimulation. **Merkel cells** and **Grandry corpuscles** occur in dense clusters on lips, bills, digits, and around hairs and feathers. They monitor steady pressures and are slow-adapting. **Encapsulated touch organs** consist of bare nerve endings surrounded by two or more onionlike layers of Schwann cells. **Ruffini mechanoreceptors** and **Golgi tendon organs** act as slow-adapting proprioceptors in muscles and joints. **Lanceolate, Meissner's, Herbst's, mucocutaneous end organs**, and **Pacinian corpuscles** are fast-adapting touch receptors that monitor movements of external objects along a receiver's body surface. These occur in high densities around the bases of bird feathers and mammalian vibrissae (also called sinus hairs), and in the bill tip organs of birds.

5. Two different sets of ion channels and associated genes are used by animal touch receptors. Those stimulated by sensory cilia admit calcium ions into the cells during stimulation, whereas nonciliary receptors favor entry of sodium ions. Both systems can be found in the same animal in nematodes, arthropods, and vertebrates.

6. **Hydrodynamic stimuli** are characterized by locally cohesive movements of fluid medium (e.g., water or air). Examples include eddies, vortices, wakes from moving animals, and short-range currents. Like near-field sounds, the amplitudes of hydrodynamic stimuli fall off quickly with distance from the source, and few animals can detect them at distances greater than 1–2 body lengths from their source. Unlike near field sounds, which persist at most for milliseconds, hydrodynamic stimuli can persist for many minutes after generation.

7. Biological uses of hydrodynamic stimuli include location of respiring but otherwise static prey, tracking of moving prey by following their wakes through the medium, detection of approaching predators, coordination of locomotion within schools of fish or squid, and mediation of aggressive interactions and courtship in insects.

8. Most invertebrates monitor hydrodynamic stimuli with cilia mounted on their body surfaces. The cilia of squid and prawns are organized into linear arrays that allow the animal to map the ambient hydrodynamic field. Fish and larval amphibians accomplish the same functions with ciliated receptors overlain with a gelatinous **cupula** like those in fish and amphibian ears. **Superficial neuromasts**, arrayed in lines along the surface of the body, monitor steady and low-frequency stimuli. **Canal neuromasts** are located inside small tubes with multiple openings just under the animal's surface, and specialize in assessing higher-frequency stimuli. The combined input from the two types of neuromasts provides the animal's brain with detailed spatial information about the amplitude, direction of flow, and spatial shape of ambient hydrodynamic fields.

9. The **vibrissae** (sinus hairs) of mammals can act as hydrodynamic sensors. Seals use the vibrissae on their snouts to track the underwater wakes of fish or conspecifics. The duck-billed platypus has hydrodynamic sensors on its bill that it can use to detect the eddies and vortices generated by the motions of aquatic invertebrate prey.

10. Active use of an animal's nerves and muscles, and ionic heterogeneities between anatomical regions generate both varying and static electric fields around its body. These weak fields are detectable at most a few body-lengths away from the animal, but within that range, temporal variations in the fields are propagated with no delay. The creation of these fields is called **passive electrogeneration**, because it is incidental to the primary function of the relevant organs.

11. Most primitive fishes (lampreys, sharks, rays, skates, paddlefish, reedfish, coelacanths, and lungfish), and some amphibians use ciliated receptor cells to detect passive electric fields (**passive electroreception**). These electrically sensitive receptor cells are grouped into **ampullary organs** spread over the animal's exterior surface but most often concentrated in the head region. Although the ancestors of advanced bony fishes lost their ampullary organs, some descendants such as catfish, knifefishes, elephantfishes, frankish, and featherback fishes, and the duck-billed platypus (a mammal) reacquired ampullary-like receptor organs. Most species use their passive electroreceptive organs to detect weak and slowly varying electrical fields produced by prey. Stingrays use passive electroreception to locate hidden conspecifics, and some catfish and sharks monitor nonbiological electrical fields to navigate at night or in murky waters.

12. Skates and some catfish have modified selected body muscles into **electric organs** that produce medium-intensity electric discharges as social signals. This is called **active electrogeneration**. Knifefishes, elephantfishes, and frankfish evolved similar electric organs, but use them both for **electrocommunication** and to probe their immediate environment for the presence of nearby objects (**electrolocation**). Both functions rely on modified electric receptors called **tuberous organs** that are sensitive to the high frequencies that make up the very-short-duration and rapidly repeated electric organ discharges. Objects of different sizes, shapes, compositions, and proximities distort the electric field around a fish in different ways. By combining the input from many tuberous receptors spread over its body, the fish's brain can characterize the location, motion, and many properties of nearby objects. Electrolocation allows species with tuberous organs to navigate around obstacles and find prey at night or in very turbid waters.

13. To avoid disrupting their navigational capabilities, electrolocating fish use modulations in the rate and waveforms of their electric organ discharges for social communication. Modulations include sudden bursts of discharges at high rates, variations in discharge amplitude, subtle changes in waveforms, and brief cessations in discharging. Differences in waveform pattern created by varying the geometries and structure of their electric organs facilitate species and gender recognition in species-rich communities. In some species, waveforms are individually distinctive, allowing for recurrent social interactions, maintenance of dominance hierarchies, and sexual selection with no or little reliance on other modalities.

Further Reading

Martinac [274] provides an overview of touch perception in prokaryotes and its early links to membrane permeabilities. Good introductions to the two gene- and ion-specific families of mechanoreceptors can be found in Gillespie [142] and Syntichaki [374]. The touch receptor organs in birds are reviewed by Gottschaldt [144], and a good introduction to mammalian touch receptors is provided by Munger [298]. Halata [161] discusses how different cutaneous receptor organs are distributed among different types of mammalian skin.

Budelmann [60] provides a good summary of hydrodynamic stimulus reception in aquatic invertebrates, and Barth [17] and Jacobs [199] review similar processes in terrestrial spiders and crickets, respectively. Coombs [86], Janssen [204], Janssen and Strickler [205], and Bleckmann [37] provide complementary introductions to the lateral line systems of fish. More detail on the receptors of lateral line organs can be found in Coombs et al [83].

Keller [225] provides an excellent overall review of electroreception in fish, and Pettigrew [320] summarizes the role of electroreception relative to other modalities in the duck-billed platypus. Passive electroreception is covered in more detail by Wilkens and Hoffmann [418], Collin and Whitehead [81], and Tricas and Sisneros [387]. Electrolocation is reviewed by Nelson [301] and von der Emde [402], electrocommunication by Hopkins [191] and Carlson [70], and both by Kramer [239]. More detail on electric organs can be found in Zupanc and Bullock [431] and Caputi et al. [68]; similar detail on electric receptors is provided by Zakon [427, 428] and Jørgensen [212].

COMPANION WEBSITE
sites.sinauer.com/animalcommunication2e

Go to the companion website for Chapter Outlines, Chapter Summaries, and References for all works cited in the textbook. In addition, the following resources are available for this chapter:

Web Topic 7.1 *Cilia and sensory receptors*
Why are cilia so often associated with sensory cells in animals? Here we expand on the reasons listed above for their role in mechanoreceptors, and provide more details on the links between ciliary structure and ciliary function in general.

Web Topic 7.2 *Hydrodynamic stimuli*
The physics of hydrodynamic stimuli can be quite complex. We here outline some principles that can be used to categorize different types of phenomena.

Web Topic 7.3 *A primer on electric signals*
Here we provide a somewhat more quantitative introduction to the physics of electric fields, potentials, resistance, capacitance, impedance, and related topics.

Web Topic 7.4 *Bioelectric field resources*
Here we provide web links to a number of sites that show electroreceptive animals in the wild, further explanations of electroreception and electrogeneration, and simulations of the electric fields generated by teleost electrolocators.

Web Topic 7.5 *Adaptations for passive electroreception*
The early acquisition of passive electroreceptors by primitive fish was surely a key adaptation that facilitated this taxon's subsequent radiation and eventual dominance in aquatic habitats. In both marine and freshwater habitats, a number of strategies are employed to enhance passive electroreception.

Chapter 8

Decisions, Signals, and Information

Overview

Communication would be useless if animals never made decisions. The signals typically emitted by senders provide information that receivers incorporate into making decisions that affect both parties. To understand communication, we thus need to have a thorough grounding in how animals make choices and decisions. We first tap into classical economics to identify the very best way that any decision could be made. We then look at how well animals (and people) achieve this ideal, and explore some of the shortcuts to and deviations from optimal decision making that have been observed. Signals can provide information to receivers only if the two parties share a code. We classify the different types of coding schemes that animals use, and provide some measures of their effectiveness.

Animal Decisions

A spider monkey hangs by her tail in the canopy of a Costa Rican forest (**Figure 8.1**). It is the dry season and the monkey is thirsty. Below her, she has spotted an enticing pool of water nestled in the flared roots of a nearby tree. She is tempted to descend to get a drink. Spider monkeys are not very agile on the ground, where they are particularly vulnerable to large snakes or cats [48]. The monkey is faced with a decision: descend now to get a drink despite possible danger, or remain in the canopy and hope that it will rain or that she will find water in a less dangerous spot later.

All organisms make such decisions every day. Decisions can be made consciously and rationally, as humans claim to make them, or they can be the result of relatively automatic neuronal or chemical processes [32]. Usually, the process is somewhere in between these extremes. Whatever the process, decision makers choose between available alternative actions. The alternative actions may be overtly physical activities—in the case of the spider monkey, climbing down the tree or swinging off to another canopy location—or they may be physiological changes, such as the initiation of breeding condition in a bird or changing sex as in some fish. The consequence to a decision maker of a particular action is called its payoff. Ideally, a decision maker should pick the alternative action that provides the best payoff. This presumption is a starting point for both human economic theory and evolutionary biology [76, 269, 333, 413].

If a decision maker knows exactly what the payoff will be for each alternative action, the decision is obvious. In this case, the decision maker has **perfect**

FIGURE 8.1 A spider monkey (*Ateles geoffroyi*) pauses in canopy to scan for predators on the forest floor

information. However, nature is rarely so straightforward. More often, a given action can lead to more than one possible outcome, and the animal cannot know for sure which will occur. In economics, this is called a **risky choice.** For example, if the spider monkey descends to drink, she will either encounter a predator and get one payoff (e.g., death) or she will encounter no predator and she will ascend the tree with her thirst satisfied. If she does not descend but, in moving to another location makes a predator on the ground aware of her presence in the canopy, the predator might track her subsequent movements and try to attack her when she finally does descend to drink. An alternative consequence of her decision not to descend might be that in moving to another location she does not attract the attention of a predator; the payoff in this case is that her level of safety from ground predators remains high—for the moment. In general, if a decision must be made between two alternative actions, and each action can result in either of two different outcomes, there can be up to four different payoffs that need to be considered. How can one choose the best action in this risky context?

The seventeenth-century mathematician and philosopher Blaise Pascal provided a useful answer. He warned against taking an action that could potentially produce a very high payoff when the probability of realizing that payoff is very small. An alternative action with a smaller payoff but with a very high probability of being realized might be the better choice. Pascal thus argued that decision makers should compare the **expected values** of alternative actions. Expected values are computed by **discounting** (multiplying) each payoff by the probability that it will occur and adding together all the discounted payoffs for a given action [83]. Expected values are also called **stochastic payoffs.**

Pascal's recommendation seems to impose a heavy computational task on decision makers. An optimal decision maker would have to estimate all possible payoffs after converting them to some common currency, estimate the relevant discounting probabilities for each payoff, combine these numbers into overall expected values for each action, and then pick the action with the best overall expected value. While our examples so far involve only two alternative actions and two alternative outcomes per action, many real decisions will involve more alternatives in both components. This will make the task even more challenging. As we shall see later, animals (and people) have evolved shortcuts that simplify the task but sometimes lead to errors. Whether actually invoked or not, Pascal's rule identifies the optimal strategy that any entity could adopt, and this allows us to compare this standard with what animals (and people) really do.

Not all decisions are equally challenging computationally. For example, our spider monkey needs to monitor the likelihood that there is a predator near the drinking tree. Since the alternative probability—that there is no predator—is simply 100% minus the first value, she needs to estimate only one number. Once she has estimated that probability, she can compute her overall expected value of descending to drink by adding the payoff of getting a drink now, discounted by the probability that there is no predator nearby, to the payoff of being eaten, discounted by the probability that a predator is present. She should next compute her overall expected value for staying in the canopy. This equals the payoff of not descending but being tracked, discounted by the probability that the predator is present, plus the payoff of moving without being tracked, discounted by the probability that a predator is absent. Because she only needs to estimate one probability (that a predator is present) to make all the computations, the task is not so daunting.

In more complex decisions, each alternative action might have its own independent discounting probabilities. Consider a choice between two raffles (**Figure 8.2**). Suppose the chances of winning are 5% in Raffle 1 and 10% in Raffle 2. The decision maker would need to know both probabilities to decide which raffle has the higher expected value. Tracking two sets of probabilities is harder than tracking one, and there appear to be fewer shortcuts to this task. However, the basic computations are similar to those in the single probability case. Note that what differs in these two types of decision making is not variability in the payoff values: in both cases, any decision is likely to generate four different payoff values that need to be considered; what differs is the number of probabilities that need to be estimated by the decision maker.

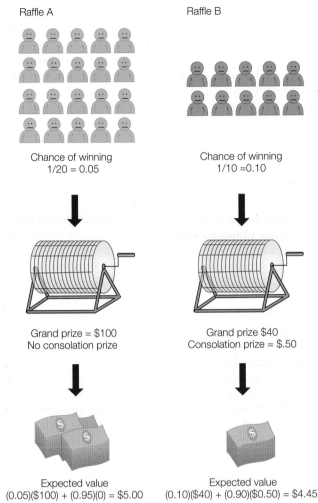

Raffle A

Chance of winning
1/20 = 0.05

Grand prize = $100
No consolation prize

Expected value
(0.05)($100) + (0.95)(0) = $5.00

Raffle B

Chance of winning
1/10 =0.10

Grand prize $40
Consolation prize = $.50

Expected value
(0.10)($40) + (0.90)($0.50) = $4.45

FIGURE 8.2 **Pascal's method for comparing expected values (average payoffs)** Consider Raffle A in which 20 people put their names into a raffle drum. Any one participant thus has a 1/20 = 0.05 probability of winning the raffle, and a 1.00 − 0.05 = 0.95 chance of losing. If a participant wins, they get $100; if they lose, they get nothing, as there is no consolation prize. The expected value for Raffle A can be computed to be $5. In Raffle B, only 10 people put their names into the raffle drum and thus any one person has a 1/10 = 0.10 probability of winning. If a participant wins, they get $40 and if they lose, they receive a consolation prize of $0.50. The expected value for Raffle B is $4.45. Thus, a participant who wants to maximize profit should enter Raffle A instead of Raffle B. Note that nobody in Raffle B actually receives $4.45: everyone receives either $40 or $0.50. However, someone comparing expected values in order to select a raffle will do as well or better, on average than anyone using any alternative rule.

Acquiring Information

Probability meters

As exemplified above, decision makers face uncertainty about which of several alternative **conditions** is currently true. For the spider monkey, the alternative conditions are that a predator is present or that it is not. Suppose the monkey

maintains a set of virtual meters in her head to track important probabilities. For her current problem, she could use two meters, one for the probability that a predator is present and one for the probability that one is not (**Figure 8.3**). However, since there are only two alternative conditions, and one has to be true, the probability that one is currently true is just 100% minus the other. The monkey can thus monitor all that she needs to know using just one meter. Let us assume that she uses the meter tracking her estimated probability that a predator is present. When the needle is to the left of center, the monkey believes that a nearby predator is unlikely; when it is to the right, her best estimate is that a predator is likely to be nearby. When the needle is straight up, she estimates that there is an equal (50:50) chance that a predator is or is not nearby (**Figure 8.4**).

Prior probabilities

If there are two equally likely possibilities (predator nearby versus predator not nearby), and the animal has absolutely no other prior information, then the needle should be set at 50%, since as far as she is concerned, the presence and absence of a nearby predator are equally probable. Most animals do not have to start with chance guesses, however. Natural selection on the monkey's ancestors and her own prior experience combine to provide a better-than-chance estimate that a predator is present. This baseline value can then be fine-tuned or overridden by recent experience. Perhaps over the last few months, predators were found to be present on 40% of the days. This prior information allows the monkey to set her meter needle at 40%.

Red lines and optimal decisions

Now that the monkey has a meter value, how might she use it to make a decision? Ideally, she might discount each of the four alternative payoffs with the meter probabilities, add up the discounted payoffs for each action, and pick the best overall average. However, there is a useful shortcut available. This involves drawing a red line on the meter: when the needle on the meter is below the red line, the monkey should

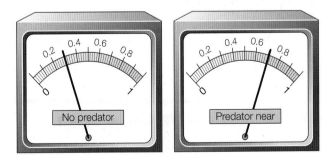

FIGURE 8.3 **Probability meters** Current estimate by a spider monkey that there is no predator at the foot of the tree (A) and the alternative probability that there *is* a nearby predator (B). Note that the monkey only needs to keep track of one meter since the meter on the right is 100% minus the value in the meter on the left.

(A)   (B)   (C)

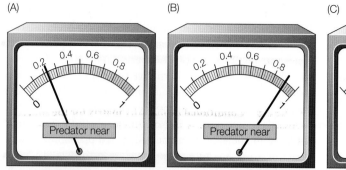

FIGURE 8.4   Possible probability readings on a monkey's predator proximity meter (A) The monkey's current estimated probability that a predator is near is very low. (B) The current estimate of the probability that a predator is nearby is very high. (C) The presence or absence of a predator is equally likely.

descend and drink; when it is above the red line, she should stay in the canopy (Figure 8.5). The placement of this red line depends only on the four possible payoffs. Even simpler, it depends on the *differences* between payoffs of "right" versus "wrong" decisions in each of the two conditions. When a predator is present (the first condition), staying in the tree is the right action and descending is the wrong one. When no predator is present (the second condition), the reverse is true. The optimal place for the monkey to place a red line on the meter depends on the ratio of the differences in payoffs between right and wrong actions for the two conditions. The red line is of course no guarantee: if the monkey estimates a low probability that a predator is present, thus putting her needle below the red line, and decides to descend and drink, a jaguar just might pop out and eat her. The red line is a best guess; that's all. However, it will yield the same success rate that she would have obtained had she used Pascal's calculations. In addition, the monkey only needs to store the

differences in payoffs between right versus wrong decisions to compute red lines; she does not need to store the absolute values of all possible payoffs. Red line metering is most useful when the probabilities of alternative conditions change more or change more often than do the alternative payoffs.

Just as with prior probabilities, the value of the optimal red line can be set by an instinctive wariness shaped through natural selection over prior generations, or it can be set through the monkey's own personal experience. In general, a red line set initially by heritable traits will need some fine-tuning. As the physiological state of the monkey changes, the payoffs of right and wrong decisions also change and influence the appropriate location of the red line: a very thirsty monkey may do best to set the red line higher than one that is not thirsty. In addition, the placement of the red line will vary with different classes of individuals; the payoffs of making a particular decision might vary for the two sexes, for older versus younger animals, or for those who are currently parents versus those who are not. Some adjustments will be needed over time to keep the red line positioned so that it accurately reflects current payoff values.

Web Topic 8.1 *Computing "red lines"*
If one knows the average consequences of "right" and "wrong" actions in a given situation, one can calculate what the "red line" probability used in decision making should be. If one kind of error is more costly than another, the red line will have to be adjusted to make this kind of error less likely.

Gaining additional information

Animals continually use their sense organs to sample what is going on around them. Our spider monkey has eyes, ears, and a nose that can be used to detect predators. Actually seeing a large cat near the drinking tree will resolve all uncertainty about whether a predator is nearby or not. Forest predators typically hide while stalking prey, and thus may not be visible. But the monkey might detect the rustle of understory plants as a jaguar tries to hide itself near the drinking tree. Rustles do not guarantee that a predator is present, but if something unseen is making a sound, then it might be a predator. Rustles are a **cue** (see Chapter 1): they are less

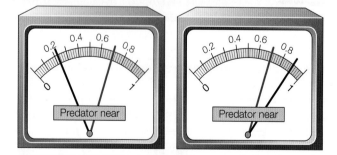

FIGURE 8.5   Adding red lines to probability meters Given the relative consequences of correct versus incorrect choices when predators are present and when they are not present, one can compute a threshold probability that on average provides the best decisions that a monkey could make. We can draw this as a "red line" on the monkey's "predator near" meter. When the needle indicating the monkey's currently estimated probability that a predator is near is below the red line (left meter), the optimal behavior, on average, is to descend and drink. When the needle rises above the red line (right meter), the optimal strategy is to stay in the canopy.

reliable than actually seeing the predator, but they occur sufficiently often when jaguars are hunting that it pays monkeys to listen for them.

Even with highly sophisticated sense organs, it is rare that animals can determine for sure whether a predator is nearby or not, whether a novel item is food or poison, whether a male seeking to mate with them will be a good or bad father, or whether their rival will attack or flee. This is because cues rarely provide perfect information that resolves all uncertainty. Instead, as with the rustle heard by the monkey, most cues are imperfectly correlated with conditions of interest to animals. A rustling could easily be generated by a breeze or by foraging birds. The more often jaguars are the source of rustling, the more it pays to attend to the sound and be cautious. However, even then, it might take an infinite number of rustlings to resolve all uncertainty about whether a jaguar is present or not. As we shall see in Chapter 9, such certainty is rarely worth seeking.

A decision maker may also acquire useful information from other animals. Suppose there is a second monkey in a nearby tree, and he actually sees a jaguar hiding itself near the drinking site. He produces a loud alarm call to alert other monkeys that he has spotted a predator [56]. If such calls are sufficiently correlated with predators being nearby, this resolves a lot of uncertainty and the female monkey may then decide to move off in the canopy. Or both monkeys may begin giving the alarm call until all the local spider monkeys have been alerted and join them in the chorus. The jaguar, realizing it has lost all chance of surprise, will slink off to stalk some less wary prey.

The alarm call used by these monkeys is a **signal** (see Chapter 1). Like a cue, the call is correlated with a condition of interest to the animals; here the condition is that a predator is nearby. Unlike a cue, which arises incidentally through some other activity, a signal is generated by a sender specifically to influence a receiver's decision. There are many possible reasons why a sender would spend the time or energy to send a signal to a receiver. In the case of the monkeys, the receiver might be the mate or a family relative of the sender. Alternatively, individual monkeys might have a reciprocal arrangement in which each alerts the other if they spot a predator first. Or it is possible that getting the entire monkey troop to call at the same time is an effective way to tell a predator that the game is up and it will do better by going elsewhere to hunt. Whatever the sender's motivation, the call helps the receiver make her decision. However, to make use of that call, the decision maker needs one additional piece of information.

Conditional probabilities and signal coding

Signals and cues help decision makers because both are correlated with one or more of the answers to a question that the decision maker would like resolved. Our monkey is faced with a question—is there a predator nearby?—that has two possible answers. We have seen that a predator may rustle the vegetation when hunting and thus provide a cue to the

Cue	Conditions	
	Predator nearby	Predator absent
Rustling bushes	70%	15%
No rustling bushes	30%	85%

FIGURE 8.6 Conditional probability matrix for the source of rustling sounds in a hypothetical spider monkey forest Rows list the two alternative cues: rustling sounds versus no rustling sounds. Columns list the two alternative conditions: a predator present versus predators absent. Cell values are the conditional probabilities that the relevant cue on the left will be experienced when a given condition is true. Note that each column should add up to 100%. Rustling does not provide perfect information because some predators manage to remain completely silent when stalking, and birds and wind can produce rustling even when no predator is present.

monkey about its presence. Given enough time and patience, we could record the fraction of time that relevant predators rustle the vegetation and compare this to the fraction of time that relevant predators make no sound. These fractions are estimates of the **conditional probabilities** that a rustling sound will or will not be produced when a predator is nearby. We could also record data to estimate the conditional probabilities that rustling does or doesn't occur when predators are absent. We can summarize all of these conditional probabilities in a **cue matrix** such as the one shown in **Figure 8.6**.

Web Topic 8.2 *Types of probabilities*

There are several types of probabilities. Knowing the different types and the relationships between them is necessary background if one wants to compute the amount of information in an animal signal, or wants to identify an optimal strategy when the consequences of different strategies occur probabilistically.

We can see in this cue matrix that rustling is not perfectly correlated with the presence of a predator, nor is silence perfectly correlated with the absence of predators. This is because some predators are very, very quiet, and both wind and harmless forest animals can generate rustling sounds. Rustling versus not rustling are thus useful cues for monkeys but neither will ever provide perfect information that completely resolves all uncertainty about the presence of a predator. Perfect information is available only when the figures in the upper left and lower right cells of the cue matrix are 100%.

We can compute a similar matrix of conditional probabilities for alarm signals (**Figure 8.7**). In this case, we want to summarize the conditional probabilities that a sender monkey will or will not call when a predator is present, and the corresponding probabilities that it will or will not call when predators are absent. A sender may fail to be 100% accurate in signal emission for several reasons: it may have heard a rustling that was not generated by the movement

Signal	Conditions	
	Predator nearby	Predator absent
Alarm call	85%	7%
No alarm call	15%	93%

FIGURE 8.7 Coding matrix for alarm call emission by a sender spider monkey Rows list the two alternative signals: alarm call versus no alarm call, and columns again list the two alternative conditions: predator nearby versus predators absent. Cell values are the conditional probabilities that the relevant signal on the left will be produced *when* a given condition is true. Note that the table does not have to be symmetrical: the value in the upper left cell does not have to equal that in the lower right (although it could in some cases). While this signal system is more accurate than the rustling cue, it also fails to provide perfect information.

of a predator; it may see or hear a predator but mistake it for something else; it may be young and inexperienced as to when to call; or it may not want to take the risk of calling and causing the jaguar to start tracking it instead. Whatever the reason, signaling systems, just like cues, are rarely 100% accurate; nearly all provide only incomplete information. The conditional probability summary for signals is called a **coding matrix.**

Updating

When a spider monkey detects a rustle (cue) or call (signal), she can combine this event with her prior probability estimates to **update** the position of the needle on her meter. Updating is possible only if the monkey has some prior knowledge of the conditional probabilities that a given cue or signal will occur when a given condition is true. She thus must have a version of the relevant cue or coding matrix somewhere in her head. As with other decision-making ingredients, this matrix could be instinctive and largely genetically inherited, learned through prior experience, or some combination of the two. However it is acquired, it is an essential component if cues or signals are to be incorporated into decisions.

Bayes' theorem specifies the optimal way to combine prior probabilities, cue or signal matrices, and current events to compute an updated probability. In terms of the likelihood of a predator in the understory, a Bayes' analysis combines the detection of a cue or signal, the relevant conditional probability matrix for different cues or signals, and the most recent prior probability. The resulting probability represents the fraction of the time that the condition of interest is likely to be true when the particular cue or signal has been received. Suppose our monkey is a Bayesian updater and initially sets her meter needle at 40% based on the average frequency of predator encounters over the last few months. When she hears another monkey call, she combines this prior probability of 40% with her knowledge of the coding matrix for alarm calls to produce a new Bayesian estimate and moves her needle to the new position. If calls are reasonably well correlated with predator presence, the needle will jump to a higher probability value, say to 61%. Bayesian updating predicts bigger jumps in the needle when the prior probability is closer to chance. So this first jump in the needle will be large (40% to 61%). Suppose the monkey now hears a second call. This time she uses 61% as her prior probability that a predator is present, (since that is now her best guess), and computes a new Bayesian estimate of 78%. The needle moves up again, but not as far as after the first call. If the monkey keeps hearing calls, her meter needle will move to higher and higher values, but each step will be smaller than the prior one. If instead there are periods with no calls, the monkey will use its Bayesian updater to record a "no call" period, and move her needle back down a notch.

Web Topic 8.3 ***Bayesian updating***
Here we outline the calculations that a computer or statistician would use to combine prior probabilities and a coding matrix to compute an updated probability. No animal or machine can produce a better estimate with only this information.

If after several calls, the monkey's needle crosses the red line, she should stay in the canopy. If the needle remains below the red line, then the optimal behavior for the monkey is to descend and drink. The fact that she may not change her behavior after hearing several calls does not mean that she did not receive information or make a decision. This exemplifies the observational problem of documenting decisions by animals when a signal or cue only reinforces what they were doing when the signal or cue was detected: a decision may have been made, but the absence of a change in behavior makes it hard for us to detect that decision. Despite these challenges, Bayesian updating has now been demonstrated in a variety of animal taxa (**Figure 8.8**) [19, 249–253, 407].

So far, we have assumed that receivers can easily classify the alternative signals and cues into discrete categories. What if this is not the case? For example, consider another species of monkey that uses high-pitched calls to warn about eagles, and low-pitched calls to warn about terrestrial predators. The correct response to an approaching eagle is not to sit in the top of the forest canopy, but to hide under the foliage. The best action if warned of a leopard's presence is to get into the trees. The task for the monkey is then to discriminate between high- and low-pitched warning calls. This is no problem if there is never any overlap between the pitch of calls given for eagles and those given for terrestrial predators. However, if alarm call pitch were sufficiently affected by caller fear level, body size, sex, or age, alarm call pitches might not fall into discrete ranges. While the average pitch might differ for eagles and leopards, some overlap might occur such that the same pitch could be sounded both when eagles were seen and when leopards were spotted. What should a receiver monkey do, given this uncertainty?

FIGURE 8.8   Western scrub jays (*Aphelocoma californica*) cache nuts during the summer for retrieval during the winter Although acorns are a more natural food, jays will happily take peanuts and other domestic seeds provided at bird feeders. Jays are very discriminating and will lift and shake multiple peanuts before choosing one for caching. Research has shown that Bayesian updating provides the best explanation for how scrub jays update their relative estimates of which peanut is the best one [252].

The optimal decision behavior in this case can be predicted using **signal detection theory**. This process also invokes prior probabilities (in this case, for eagle versus terrestrial predators); Bayesian updating given receipt of a signal; and the relative payoffs of right and wrong decisions which determine the location of a red line (**Figure 8.9**). However, instead of defining a threshold probability, this shortcut identifies a threshold call pitch below which the monkey should respond as if a terrestrial predator were present, and above which it should act as if an eagle were nearby. As with discrete signals, this method requires only that decision makers estimate and store the *differences* in payoffs between right and wrong decisions; the absolute values of each payoff are not needed. Different receivers will be likely to set their red lines at different pitches, because the payoffs of right versus wrong decisions or prior probabilities may differ with the age, sex, and recent experiences of the receiver. Recent studies suggest that mammals have two different decision-making regions in the brain: a relatively slow process that tracks cumulative probabilities and a faster signal detection system [400].

Dynamic Decision Making

We treated each decision making episode above as it if were independent of prior and later decisions. Some decisions are indeed made this way. However, animals often face a string of successive decisions that provide them with opportunities to sample both probabilities and payoffs, and thus to adjust red lines as the sequence proceeds. Successive sampling can be useful over both short and long time scales. Short-term sequences occur within aggressive contests or courtship. A participant may begin such an interaction with only minimal prior information about other interactants. After the first round of cue and signal exchanges, each party decides whether or not to continue the exchange. If the exchange is continued, each remaining participant updates its estimates of conditions of interest in the other parties, and these are used as the prior information as they enter the next round. This process continues until one or both parties make a decision that ends the interaction.

Longer-term sequences exploit an event in the present to help refine potential decisions in the future. For example, if there is some way to determine which condition was actually true after a decision has been made, (e.g., the jaguar steps out of the bushes when it sees that the monkey has remained in the canopy), our monkey can use such observations to adjust her long-term prior probability estimate that predators will be found nearby. Second, if signals or cues were detected, she can use these same observations to fine-tune the conditional

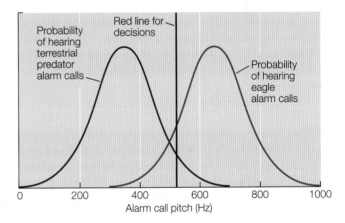

FIGURE 8.9   Decision making when two signals overlap in the critical property distinguishing between them Suppose hypothetical monkeys give different alarm calls for terrestrial predators and approaching eagles where the critical property distinguishing the two alarm calls is pitch (dominant call frequency). Because of variations in caller age, sex, size, and experience, the pitch of a given type of alarm call differs a bit among the monkeys. The graph above shows the distribution in pitch values for each kind of alarm call in this population: the higher the vertical axis value, the more common that pitch is for that type of call. Although calls for terrestrial predators tend to have lower pitch than eagle alarm calls, there is clearly some overlap at intermediate pitches. Receivers will thus have to define a threshold pitch (red line) below which they will respond as if it were a terrestrial predator alarm, and above which they will respond as if it were an eagle alarm. The optimal position of this red line can be predicted using signal detection theory, and will depend on the prior probabilities of encountering eagles and terrestrial predators and the differences in payoffs between right and wrong decisions. If eagles have been more common lately or the payoff difference between right and wrong decisions when eagles are present has become greater, the optimal red line will be further to the left. If the opposite is true, the optimal red line will be positioned further to the right. Receivers relying on the optimal red line in deciding how to respond will have higher fitness than receivers using any other criterion.

probability matrix in her head that keeps track of how well alternative cues or signals are correlated with alternative conditions. Last, she can evaluate how well the most recent decision matched the expected payoffs of her action and make appropriate adjustments in those expectations for future decisions. Experience with unexpected outcomes can be used to expand the list of alternatives in future decisions. In short, each of the critical components in decision making—prior probability estimates, coding matrices, and the relative consequences of right and wrong decisions—can often be fine-tuned by evaluating the situation after a decision.

The opportunity to learn from prior decisions may be one reason why animals sometimes appear to make "wrong" decisions. Unless they acquire their decision ingredients through genetic inheritance, an animal that always makes the right decision cannot know the payoffs of wrong decisions. This hinders any estimation of the expected values of alternative actions. Also, payoffs can change over time: what might have been an accurate estimate last month may no longer be appropriate. Unless it is too expensive, an animal may make a wrong decision just to get this information. We thus need to distinguish between animal decisions that are **exploitative** (undertaken primarily to achieve the optimal consequence) and those that are **exploratory** (undertaken to update information about probabilities or payoffs of decisions). The relative frequency of exploitative versus exploratory decisions will depend in part on the dangers and costs of wrong decisions: a monkey's decision to nibble an unknown species of fruit may be a low-cost "wrong" choice if most fruits in the habitat are nontoxic. And it may turn out to be a right decision if the monkey identifies a new and nutritional food source. On the other hand, descending to the forest floor despite hearing an alarm call may be a prohibitively expensive way to adjust probabilities and payoffs. Some contexts constrain the value of exploratory decisions because the consequences occur so long after the decision that conditions have changed and adjustments would be irrelevant. Other decisions occur so few times during the life of an animal that information gained by exploratory decisions is never called upon later. Whether an animal makes exploratory decisions will depend, again economically, on whether the immediate costs of a "mistake" are or are not outweighed by later benefits, and whether conditions change enough to warrant trying to track changes [250].

Biased Decision Making

In economics, decision making in which (1) expected payoffs of actions incorporate relevant discounting probabilities; (2) expected values of alternative actions are compared; and (3) the action that provides the optimal expected value is adopted is called **normative**. A decision maker that uses a normative process and usually achieves the optimal outcome is said to be **rational** [186, 212]. How often do animals make optimal decisions? Do they really set meters and thresholds in their heads, store adjustable probabilities, use

Bayesian updating, and otherwise make rational decisions in a normative way? Or are humans the only organisms that can make normative and rational decisions, and all other animals simply automatons responding reflexively or because of historical precedent to selected stimuli [342]?

We increasingly understand how both human and animal brains work, and have identified specific brain regions in many species that mediate the different stages of decision making. Where behavior decision processes can be examined quantitatively, updating and either some sort of metering or signal detection threshold best explains animals' choices [62, 136, 250, 252, 334, 336, 407]. Despite initial skepticism, behavioral ecologists have accumulated several decades' worth of data showing that most animals indeed make optimal decisions most of the time [223, 224, 304, 308, 385]. However, both animals and humans sometimes fail to make optimal decisions. How can we explain this mix of rational and irrational behaviors? The current consensus is that animals and humans largely try to make rational decisions, but there are economic and physiological constraints that limit execution of the full protocol. The results of these constraints are **biased decisions**. Much research has gone into identifying different kinds of decision biases and why they occur when they do. Because the primary function of animal communication is the facilitation of receiver decisions, biases and constraints on optimal decision making have surely affected and been affected by concomitant signal evolution. We cannot fully understand signal evolution without a good understanding of the evolution of decision biases.

Types of biases

A list of major decision biases is provided in **Table 8.1**. Because readers will be most familiar with their own decision making, we introduce these first using human examples. For intermediate values of probabilities and payoffs, people tend to make decisions that are largely rational. However, when one or more payoffs or probabilities approaches a potential extreme, biases regularly creep into decisions. One well-studied bias is **risk sensitivity** [213, 214]. There are two kinds of risk sensitivity. In the case of **risk aversion**, the decision maker favors a sure bet over a risky one even if the risky action has a better overall expected value. The opposite bias is **risk proneness**, in which the riskier choice is favored even though the more sure bet has a higher expected value. The distribution of risk sensitivity in humans is complicated. Given the same absolute amount that could be gained or lost, people tend to be risk prone when faced with gains of low probability or losses of high probability; they are risk averse when faced with gains of high probability or losses of low probability. These biases tend to be stronger for losses than for gains. The less certain human subjects are about their estimates of the probabilities in a decision, the stronger risk biases tend to be [403, 405].

Humans also show **context-dependent** biases during decision making. If people truly estimate expected values in an absolute sense, (the presumption in most economic

TABLE 8.1   *Types of biases reported in animal and human decision making*

Bias Type	Normative Pattern	Biased Pattern
RISK SENSITIVITY:	Expected values computed using actual payoff values and actual probabilities	Payoffs or probabilities or both are weighted before being combined into expected values
Payoff bias	Utility is independent of state of decision maker and linear function of available payoffs	Utility of given payoff is reduced if decision maker is in a good state and exaggerated if in a poor state
Probability bias	Probabilities not weighted before using to compute expected values	Very low or very high probabilities are weighted disproportionately in computing expected values
Loss aversion	The magnitudes of losses and gains are treated equally in computing expected values	The amount of bias due to payoff or probability biases is greater for losses than it is for gains
CONTEXT DEPENDENCE:	Only the relevant payoffs and probabilities are used in computing and comparing expected values	The specific identity of options or the presence of irrelevant options affects decision making
Intransitivity	If alternative A is preferred over B, and B is preferred over C, then A will be preferred over C	If alternative A is preferred over B, and B is preferred over C, there is no guarantee that A will be preferred over C
Irregularity	A preference for a given alternative cannot be increased by providing additional alternatives	A preference for an alternative can be increased by adding additional alternatives
Prior history	A current decision depends only on current probabilities and payoffs, and not on relative rewards obtained in prior decisions	Major changes in state of the decision maker during prior decisions are used as a criterion of choice in a current decision
Framing	Preferences between alternatives are not affected by the perception that outcomes will be gains versus losses	Preferences are altered depending on whether decision is perceived as a gain or loss relative to recent decisions
HEURISTICS:	Decision making uses full process of computing and comparing expected values	Shortcuts involving only part of the normative process or rules of thumb that generally give same results are used for decisions
Recognition	Familiarity with alternatives not a factor in choice	Decision maker is biased toward more familiar alternatives
Linear operator	Decision maker uses full Bayesian updating of relevant signals and cues before making choice	Decision maker uses weighted sum of prior probability estimates and detection of signal/cue instead of Bayesian updating
Best-of-N	Decision maker compares all available alternatives using optimal signal/cue before making a choice	Decision maker sets some number N, smaller than the total number of alternatives available and selects best alternative from the first N encountered
Threshold	Decision maker compares all available alternatives using optimal signal/cue before making a choice	Decision maker sets some threshold value of a relevant cue or signal, and accepts first alternative encountered that equals or exceeds that threshold
Elimination	Decision maker compares all available alternatives using all relevant signals/cues before making choice	Decision maker first uses one signal/cue to eliminate some alternatives before using other signals/cues to make final choice

models), an action A that has a higher expected value than an alternative action B should also be a better choice than action C if the expected value for B is higher than that for C. This principle is called **transitivity**. In practice, people sometimes make a series of decisions that turn out to be intransitive—even if they estimate A as better than B and B as better than A, they sometimes choose C when offered a choice between A and C [402]. Another consequence of absolute value estimation is that the availability of inferior choices should not affect decision making. This is the principle of **irrelevant alternatives**. For example, adding a third choice with a less desirable outcome to a prior two-alternative decision should not alter the decision. This predicted behavior is called **regularity**. Some people's decisions violate regularity. Similarly,

the history of prior decisions, other than through updating, should be irrelevant to a current one. In practice, people often bias their decisions based on current or prior alternatives that theoretically should be ignored [95, 241, 376, 404, 421]. How options are explained to people can significantly affect choice. This is called a **framing effect** [214]. For example, when the relevant probabilities and payoffs are identical, describing a choice in terms of gains will often result in different behavior than describing the same choice in terms of losses [225]. Finally, emotions such as fear, anger, and moral repugnance, among others, are well known to affect human decision making [142, 242].

The numerous biases discovered in human decision making have stimulated behavioral ecologists and psychologists

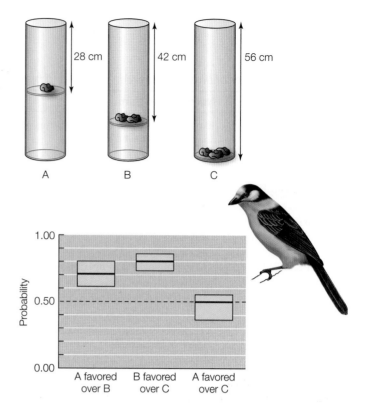

FIGURE 8.10   An animal decision that is not *transitive* Gray jays (*Perisoreus canadensis*), like scrub jays, collect food in the summer and cache it for the winter. A researcher habituated a number of wild gray jays to come to feeding stations when he whistled. Finding that they liked raisins but were wary about going very far into a screen tube to get them, he created several tube–raisin combinations, arranged them two at a time near a perch, and then let the jays pick which tube they preferred to enter to get the raisins. Tube A held only a single raisin placed 28 cm from the entrance. Tube B contained two raisins placed further inside the tube at 42 cm. Tube C contained 3 raisins placed 56 cm from the entrance. As shown in the graph, tube A was usually favored over tube B at a median probability (red line) of 70% to 30%, and tube B was favored over tube C with a median value of 80% to 20%. (Orange boxes show different levels of variation in choice around these medians). If choice depended only on the absolute values of the options, we would predict that tube A would be strongly favored over tube C. This is the principle of transitivity. It is clear from the graph that the jays showed no preference for tube A over tube C: the average choice fell right on the chance value of 50%. One explanation for this intransitivity is that jays focus on the difference between alternative payoffs, and do not have preferences fixed by the absolute value of payoffs. (After [414].)

to look more carefully at animal parallels [17, 136, 183, 212, 338, 374]. In fact, many of the human biases show up in animals. Sensitivity to risk was recognized early in the studies of animal foraging [50, 385]. Results are similar to those found in humans: animals tend to be risk averse where maximization of payoffs is favored (e.g., amounts of food obtained in animals and fiscal gain in humans), and risk prone where minimization of payoffs is favored (e.g., delay times in

receiving food in animals and fiscal losses in humans) [210]. The bias in risk prone/loss situations tends to be larger than that in risk averse/gain contexts [260].

Intransitivity (**Figure 8.10**) has been demonstrated in honeybees and jays [369, 415], and violations of regularity (**Figure 8.11**) when less desirable alternatives are added have been found in hummingbirds [15, 16, 187] and honeybees

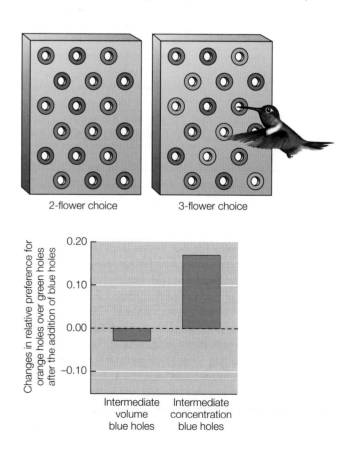

FIGURE 8.11   An animal decision that is not *regular* A decision process is regular if adding less desirable alternatives does not change the relative preferences for the initial alternatives. Researchers examined whether wild rufous hummingbirds (*Selasphorus rufus*) made regular choices when choosing between colored artificial "flowers"—holes drilled in plastic blocks and surrounded with a colored ring. In the initial (2-color) choice, orange-ringed holes were filled with 20 μl of 40% sugar solution, and green-ringed holes filled with 40 μl of 20% solution. After sampling both hole types, the birds showed a clear preference for the higher-concentration orange-ringed holes. In subsequent 3-color tests, the original two sugar concentrations were augmented with additional blue-ringed holes: these either contained 30 ml of 10% sugar solution (a volume intermediate between orange and green holes but a concentration less than either) or 10 ml of 30% sugar solution (a concentration intermediate between orange and green holes but a volume less than either). Adding blue holes with an intermediate sugar concentration increased the preference of the birds for the higher-concentration orange holes relative to the green ones; adding blue holes with intermediate volumes increased the preference of the birds for the higher-volume green holes relative to the orange ones. The birds did not prefer blue holes, but the addition of blue holes changed the relative preferences for orange holes over green holes. Switching the colors of the rings between trials and using different colors had no effect on these results. The likely explanation for this irregularity is that decisions are based not on absolute values of payoffs but on the differences between available alternatives. (After [16].)

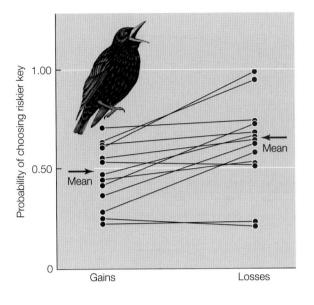

Probability of choosing riskier key

1.00

0.50

Mean

Mean

0

Gains   Losses

FIGURE 8.12   An animal example of a framing bias during decision making Researchers taught captive European starlings (*Sturnus vulgaris*) to peck at either of two adjacent keys in a test apparatus to gain access to food pellets. Birds were initially rewarded at both keys with the same fixed rewards: for some birds, the reward was small (1 pellet per peck) whereas for other it was large (7 pellets per peck). All birds then participated in a second exercise in which one key consistently provided 4 pellets, whereas the second adjacent key randomly provided either 2 pellets or 6 pellets. Both keys thus gave the same average number of pellets (4), but the second key was riskier than the first. This second exercise was a *gain* for birds trained on 1 pellet/peck, since the payoffs were always higher; it was a *loss* for the birds trained on 7 pellets/peck, since the payoffs were always lower. The process was run repeatedly so that each bird had a chance to do the second exercise at least once as a gain and at least once as a loss. When results were tallied, 12 of the 14 birds favored the riskier key in the second exercise when they experienced it as a loss; 8 of the 14 birds favored the non-risky key when they experienced the second exercise as a gain. Taken together, 11 of the 14 birds were more risk prone when faced with losses than when faced with gains. The graph shows differences in chances of choosing the riskier key for 13 of the birds in two training situations. Framing bias, in which decision makers are risk prone or risk averse depending on whether they perceive a decision as leading to losses or gains, respectively, is common in human beings as well. (After [260].)

and jays [370]. When jays, starlings, doves, horses, and locusts are offered a choice between two poorly known alternatives or alternatives with equal expected value, they will often bias their decision based on the history of prior rewards [63, 88, 261, 327, 328, 364, 416]. Framing effects (**Figure 8.12**) have also been demonstrated in starlings [260]. One common denominator in the demonstrated animal examples of intransitivity, irregularity, history effects, and framing is that the biases tend to show up when the initial alternatives are not overly different in payoff values. When there are big differences, the animals usually make unbiased and rational decisions. To show biases, researchers thus have to make the decision challenging by providing alternative expected values that are similar or even equal.

Explanations for biases: Curvilinear utility functions

One explanation for the mismatch between optimal decision theory and actual behavior is related to how animals (or people) translate a given payoff into biologically (or fiscally) useful units. The eighteenth-century mathematician Daniel Bernoulli noted that winning a lottery can mean a lot if you are poor but have little impact if you are already rich. He concluded that decision makers should not compare raw payoffs, but instead compare the degree to which each payoff would bring the decision maker closer to full satisfaction. He called the current state of satisfaction of an animal, given its recent access to relevant payoffs, its **utility**. There are several likely functions that could describe how access to a payoff might change the satisfaction state of a decision maker (**Figure 8.13**). If the function were a straight line (**linear**), then the initial state of the decision maker should have no effect on the magnitude or direction of the change in state created by a new payoff: a given value of payoff would always result in the same change in utility. If, however, that function is curved, then the initial state of the decision maker can significantly affect the

change that a given payoff produces in the animal's utility. A **concave** (decelerating) function will produce smaller changes in utility as the initial state of the decision maker increases. A **convex** (accelerating) function will produce larger changes in utility as the initial state of the decision maker increases. These predictions follow from a general mathematical rule known as **Jensen's Inequality** [379].

What do curved utility functions have to do with risk sensitivity? Consider a simple two-person gamble of $10 which is settled by the flip of a coin (**Figure 8.14**). The actual amount of money lost by one party will be exactly equal to that gained by the other. However, if one of the parties is poor and the other rich, the loss or gain of $10 can have quite different utilities. For poor people, utility generally increases in an accelerating manner with increasing wealth: the increase in utility associated with winning is much greater than the loss in utility associated with losing. The situation is reversed for the rich player: losing the bet decreases his utility more than winning the bet would increase it. The poor player should be risk prone whereas the rich player should be risk averse. The difference between utility gains and utility losses grows larger as the difference between alternative payoffs increases. For example, if the players raised the stakes to $50, the risk-averse rich player should be even *less* willing to play this game, whereas the risk-prone poor player will be even more eager. Risk-sensitive biases should become stronger as the variability in payoffs increases.

While replacing payoffs with the outputs of curvilinear utility functions helps explain some of the decision biases seen in animals and humans, it provides no explanation for others. Psychologists have responded to these discrepancies by proposing that both probabilities and payoffs are

(A) Linear

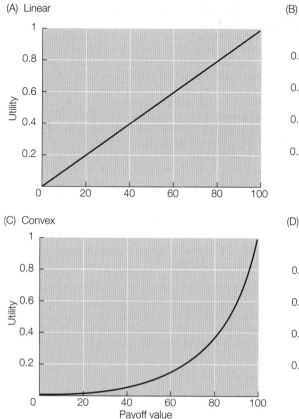

(B) Concave

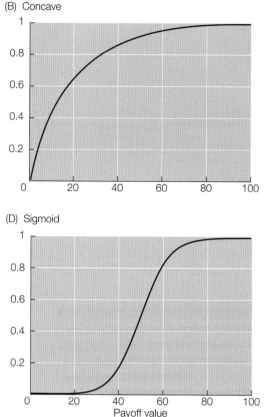

(C) Convex

(D) Sigmoid

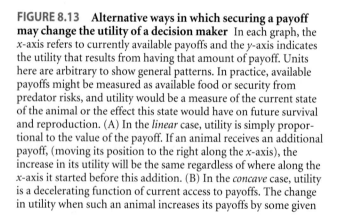

FIGURE 8.13 **Alternative ways in which securing a payoff may change the utility of a decision maker** In each graph, the *x*-axis refers to currently available payoffs and the *y*-axis indicates the utility that results from having that amount of payoff. Units here are arbitrary to show general patterns. In practice, available payoffs might be measured as available food or security from predator risks, and utility would be a measure of the current state of the animal or the effect this state would have on future survival and reproduction. (A) In the *linear* case, utility is simply proportional to the value of the payoff. If an animal receives an additional payoff, (moving its position to the right along the *x*-axis), the increase in its utility will be the same regardless of where along the *x*-axis it started before this addition. (B) In the *concave* case, utility is a decelerating function of current access to payoffs. The change in utility when such an animal increases its payoffs by some given amount will be smaller if the starting point on the *x*-axis is higher than if it is lower. Put another way, increasing access to payoffs has diminishing returns. (C) In the *convex* case, utility is an accelerating function of current payoffs: increasing access to current payoffs by a given amount will have only minimal benefits if the starting point is at the left side of the *x*-axis, but can have very large benefits if the starting point is far to the right. (D) Probably most resources such as food, water, or shelter actually have *sigmoid* utility functions: it takes a minimal access to any of these commodities before the animal's status starts to increase (an initially convex condition). If the animal has access to intermediate levels of resource, the utility function is roughly linear, and once the animal has a large amount of resource, adding more has diminishing benefits (a concave function). (After [182].)

transformed in one's brain using curvilinear functions before the outcomes of alternative actions are compared. This double-transformation model is called **prospect theory**, and it has had reasonably good success in predicting risk-sensitive biases in human decisions [213, 214, 402, 404, 405].

> ### Web Topic 8.5 *Prospect theory*
> Prospect theory seeks to explain known decision biases in humans by arguing that decisions are made not by direct comparisons of probabilities and payoffs, but instead by comparing psychological surrogates created using curvilinear transformations of the raw probabilities and payoffs.

Explanations for biases: Weber's law

We saw in the prior section that risk-sensitive biases due to curved utility functions should get stronger as the variability in payoffs of risky actions increases. Is this true? One widely used measure of variability within a set of values is the set's **variance**. This is the average of the squared differences between each value and the overall **mean** (average) of the set of values. A related index, the **standard deviation**, is the square root of the variance. Do decision makers become more risk sensitive as one increases either the variance or the standard deviation of an action's alternative payoff values?

In fact, neither the variance nor the standard deviation is a very good predictor of risk sensitivity in animals or people.

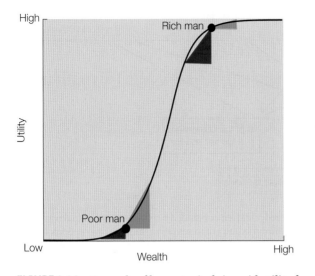

FIGURE 8.14 Example of how a typical sigmoid utility function can explain risk aversion or risk proneness Consider two individuals who differ in current wealth: a poor man and a rich man. They gamble by flipping a coin: whoever wins gets $10 from the other. Because the poor man's low wealth puts him in the convex region of the utility curve, the increase in utility if he wins the coin toss (blue) is greater than the decrease in utility if he loses (red). He should thus be risk prone and eager to gamble. The rich man's wealth puts him in the concave part of the utility function. His utility decreases more if he loses the coin toss than his utility would be increased if he won. His optimal behavior is to be risk averse and avoid the gamble. Note that the magnitude of the change in wealth (shown in the graph as the horizontal base of each triangle) is exactly the same for both players whether they win or lose; it is the curvature of the utility function that results in differential gains and losses (the vertical side of each triangle).

What *does* seem to predict risk sensitivity is another measure of variability, the **coefficient of variation**. This is the ratio between the standard deviation and the mean of a set of values and is usually given as a percentage. For example, if we know that the coefficient of variation of a set of values is 10%, then we know that the standard deviation will be 100 when the mean of these values is 1000, but will be 50 when the mean of the values is 500. The coefficient of variation is thus a measure of *relative variability* among the values. Current data suggest that animals and people consider two actions to be equally risky if the relevant payoff values have the same coefficient of variation; they consider one of two actions as more risky if it has a higher coefficient of variation in payoffs than the other action [371, 420].

Why risk sensitivity in animals and people should depend on relative variation among the possible payoffs for an action is not immediately obvious. One proposed answer is that this simply reflects how animals and people measure and store sensory information in their brains [129, 211, 339]. An amazing property of most animal and human sensory systems is that they can measure stimuli over enormous ranges. For example, humans can accurately identify the pitch of a sound over a thousandfold range of frequencies, and the loudness of a sound over a millionfold range of sound amplitudes. The apparent cost of this wide measurement range is that the error in measurement gets larger as the value being measured gets larger. In fact, for a given type of measurement, the ratio between the range of likely errors and the true value is roughly a constant. This finding is known as **Weber's Law**, and while there is continuing debate about the underlying physiological process [67, 86, 103, 200, 241, 387, 419], the rule seems to be quite general.

If decision makers obey Weber's Law, then larger payoffs will be stored with greater error, and thus variability, than smaller ones. This makes the variance and the standard deviation of the stored payoffs poor indices of risk. In contrast, the coefficient of variation corrects for the rescaling generated by Weber's Law and is thus a more reliable index of risk. Weber's Law may thus explain the otherwise curious observations about risk-sensitive decisions in animals and people.

Weber's Law may also explain why animals are often risk averse when payoffs of risky actions vary in the amount of a food reward, but risk prone when the payoffs differ only in the delay in providing that reward [14, 211, 339]. The rescaling of perceived values according to Weber's Law causes smaller payoffs to be weighted more heavily than larger payoffs in the computation of the expected value for a risky action; as a consequence, the perceived expected value of a risky choice will be lower than that for an otherwise similar sure bet. Where the variability is in the amount of food, animals will pick the larger expected value (the sure bet) and thus be risk averse; where variability is in the delay in access to food, animals will pick the smaller expected value (the risky choice) and thus be risk prone.

All of these applications of Weber's Law assume linear utility functions. Since we know that many animal and human utility functions are curved, but also that Weber's Law applies to the sensory systems of most animals, it seems likely that both of these processes can contribute to risk-sensitive biases. How they interact remains to be studied.

Web Topic 8.6 *Weber's Law and risk*
When animals scaling magnitudes according to Weber's Law try to estimate the most likely expected value of an action with several possible outcomes, that estimate is likely to be skewed toward the smallest payoff. If smaller payoffs are worse, this may lead them to be risk averse, and if smaller payoffs are better, to be risk prone.

Explanations for biases: Other considerations

DIFFERENTIAL PAYOFFS One explanation for the violations of transitivity and regularity seen in both animals and people is the possibility that decision makers compare **payoff differences** instead of individual absolute payoffs. We saw in our analysis of simple decisions that a decision maker could make optimal choices simply by comparing the differences in

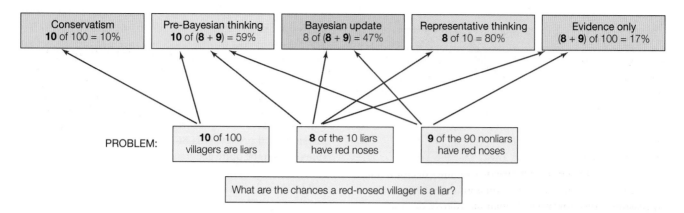

FIGURE 8.15 **Human shortcuts to Bayesian updating** Researchers Zhu and Gigerenzer gave simple Bayesian updating problems to human subjects. The task was to estimate the chance that one of two conditions was true given observation of a signal or cue that was correlated with one of the conditions. One sample problem described a visitor to a village who is told that 10 of the 100 villagers are liars, 8 of the liars have red noses, and 9 of the non-liars have red noses. The visitor encounters a villager with a red nose. What are the chances that he is a liar? All of the ingredients needed for a Bayesian update have been provided. While adults typically gave the correct Bayesian answer, children in grades 4–6 often used approximations. One approach (conservatism) was to ignore the relevance of the signal/cue completely and rely only on prior probabilities. In the example, 10% of the villagers are liars and 90% are not: the children thus reasoned that the red-nosed villager is unlikely to be a liar. Alternatively, children ignored the prior probabilities and focused instead on how often the signal/cue occurred in the entire population (evidence-only thinking). In the example, 8 + 9 = 17 villagers have red noses for an overall probability of 17%. Because red noses were rarely encountered, children concluded a red-nosed individual was likely a liar. A third method (representative thinking) was to look only at the conditional probabilities: if red noses are closely linked to liars, then seeing a red nose should be good evidence that the person is a liar. In the example, the conditional probability that a liar has a red nose is 80%, whereas only 10% of non-liars have red noses. A child using this reasoning would conclude that there was a good chance that the red-nosed person was a liar. A fourth approach uses the ratio between the relevant prior probability (here, that villagers are liars) and the overall probability of seeing the relevant signal/cue (here, that villagers have red noses). In our example, this ratio would be 10% divided by 17% = 59%. This is almost the same as the Bayesian estimate except that it ignores the conditional probability that a liar will have a red nose. The authors thus call this method a pre-Bayesian approach. Finally, many older children and adults used all the necessary information to compute the Bayesian estimate of 47%. While this example focuses on young humans, animals might use any of these methods as well to shortcut the Bayesian updating process. (After [437].)

payoffs of "right" versus "wrong" decisions in each condition to the current estimates of probabilities of those conditions. They do not need to store and manipulate the absolute values of each alternative payoff.

A more general case for decisions based on payoff differences, instead of absolute payoff values, has been made by Alisdair Houston [181]. His argument takes into account not only the immediate payoff for a decision but any effects that this decision has on future payoffs. Given that animals are likely to make some errors in decisions, Houston shows that optimal decision making *should* focus on differences between currently alternative payoffs, not absolute values, and that intransitive and irregular decisions are an unavoidable cost of this strategy. Recent animal studies have provided widespread support for the use of payoff differences in decision making as the explanation for irregular and intransitive decisions [15, 16, 187, 369, 415].

HEURISTICS Another proposed explanation for decision biases is the use of **heuristics** or "rules of thumb." Heuristics are "fast" and "frugal" shortcuts to the optimal decision-making process that work well much of the time, but may err during the remainder, creating some of the decision biases listed earlier [134, 268]. For example, an animal that has no time to do Bayesian updating may do just as well by tracking some form of running average. At any moment, it takes its current estimate of a probability or payoff, weights it in some way, and combines this with the detection of new cues or signals, also weighted, to generate a new estimate. If the weighting of the prior estimate is larger than that of the new observation, the animal will be relatively conservative, and its running average will change only slowly over time. If it weights new observations heavily, and discounts the past with a low weight, then its estimates will be highly responsive to apparent changes in the environment—but also susceptible to irrelevant noise and random variation. This type of weighted tracking (also referred to as using a **linear operator**), can produce reasonable if imperfect results as long as it uses a suitable weighting scheme [81, 136, 180, 209, 270].

Although people often use Bayesian updating, some, especially younger children, will use a variety of heuristics (**Figure 8.15**) instead [133, 175, 437]. These range from keeping track of frequencies (counts) instead of probabilities (which can yield the same answer as Bayesian estimates); invoking parts but not all of the Bayesian computation (yielding results of moderate accuracy); relying entirely on the prior estimate rate (a tracking with 100% weight on the past); or relying entirely on any new observation (a tracking with 0% weight on the past). Younger children in a variety of cultures initially use the simpler and less accurate heuristics, but by 11–12 years of age, at least half are using frequency counts or Bayesian updating [437].

A variety of other heuristics to simplify decision making have been proposed and to some degree documented in both animals and people [36, 127, 134, 135, 194, 249, 266, 322, 335, 424]. The **recognition heuristic** simplifies decision making when there is little information about any of the options by biasing choice toward a familiar alternative or away from unfamiliar ones. The widespread phenomenon of dietary neophobia (avoidance of novel foods) is an example [41, 259]. How animals classify objects and conditions depends on the mechanisms by which they generalize. Certain kinds of **generalization tasks** can lead to biases like overall neophobia or attraction to certain types of stimuli [128]. Where animals must choose from a large number of alternatives that may differ across many different aspects, evaluation of each of *N* alternatives with regard to some shared aspect can be nearly as effective as full Bayesian comparisons. This procedure is known as the **best-of-*N* rule** in behavioral ecology and the **one reason heuristic** in the human psychology literature. The optimal number of alternatives that should be sampled before making a choice will vary with context. Alternatively, the decision maker could set a threshold value of a common aspect and then accept the first sampled alternative that meets or exceeds the threshold. When there are no aspects sufficiently common to compare alternatives using threshold or best-of-*N* protocols, a default approach is elimination of the worse alternatives, until a small enough set is obtained for best-of-*N* or threshold comparisons. Whether best-of-*N* or threshold rules are effective proxies for Bayesian decision rules depends on both the economics of sampling and the degree to which focal cues and signals are correlated with conditions of interest [251].

As noted earlier, a reliance on past decisions, even when this is misleading for a current decision, has been recorded in animals as diverse as locusts, birds, horses, and people. In a variety of feeding experiments, animals faced with similar expected values chose the action that had resulted in the largest change in nutritional state during prior training sessions [63, 260, 261, 327, 328]. In contrast, a study on gray jays showed a bias toward the alternative with the least change in nutritional state during training, but at the same time the least risk [416]. Horses deprived of information about current food patch quality and abundance fall back on matching what they find with longer-term prior averages [88]. In these examples, either the current probabilities and payoffs were unknown (horses), or the expected values of the alternative actions were equal (birds and locusts). Either situation makes application of the standard decision process challenging, and resort to a **past history heuristic** seems to be a fast and frugal solution.

Web Topic 8.7 *Brains and decision making*
Much recent research has sought to identify the parts of the brain that are used to evaluate payoffs, track probabilities, perform updating, and invoke biases during human and animal decision making. Integration of the theories of decision making and the underlying neurobiology is now the focus of a fast-growing field called neuroeconomics.

Coding Strategies

Now that we have a detailed understanding of how animals make decisions, we can turn to how signals might be recruited into the process. The function of signals is to provide information that receivers can use in making decisions. A signal can only provide information if senders differentially produce it when they perceive a particular condition to be true. If senders were to produce the signal randomly, without regard to concurrent conditions, receivers would gain no new information by attending to that signal. When senders can emit any of several alternative signals, the same logic applies: alternative signals will provide information to attending receivers only if senders are sufficiently **consistent** in emitting one signal when a certain condition is true, and a different signal when different conditions are true. The ways in which senders associate signals with conditions constitute that species' **coding scheme**.

The overall coding scheme of a species can usually be partitioned into distinct **signal sets**. A signal set contains those signals that are associated with conditions that are alternatives to each other. Put another way, a signal set includes only those signals that a receiver might use to answer the same question. For example, if a species has different alarm calls for different types of predators, the relevant signal set consists of those alternative alarms plus silence (a condition of no predator present). Most of our discussion will focus on coding schemes for a given signal set. However, signal sets may not always be independent of each other; when they are not, we need to examine the ways in which they work in concert. The complete list of signal sets constitutes a species' **signal repertoire**.

Signal set design

The shaping and refining of signal sets can range from an evolutionary process that takes thousands of generations in most animals to the rapid learning of new words in humans. Whatever the time scale, set refinement is affected by many different factors. We have noted some of these factors in earlier chapters, and others will be presented in succeeding chapters. The processes by which signals evolve and become associated with particular conditions are covered in Chapters 9 and 10. In this chapter, we shall focus on coding schemes that have already evolved and examine how accurately and consistently they provide information to receivers. It will be useful first to identify several options in how signals can be associated with conditions in coding schemes.

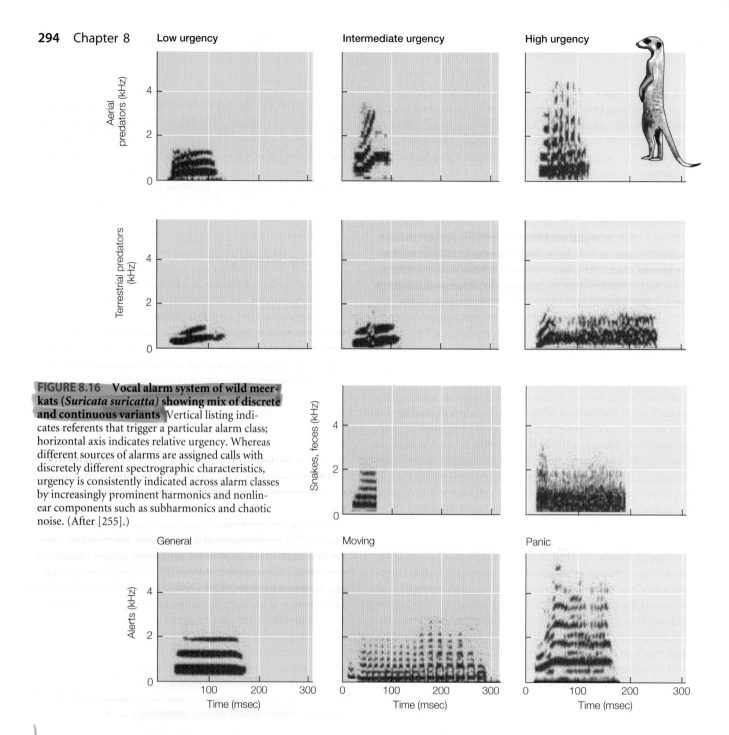

Low urgency   **Intermediate urgency**   **High urgency**

FIGURE 8.16 **Vocal alarm system of wild meerkats (*Suricata suricatta*) showing mix of discrete and continuous variants** Vertical listing indicates referents that trigger a particular alarm class; horizontal axis indicates relative urgency. Whereas different sources of alarms are assigned calls with discretely different spectrographic characteristics, urgency is consistently indicated across alarm classes by increasingly prominent harmonics and nonlinear components such as subharmonics and chaotic noise. (After [255].)

SIGNAL SET PLURALITY Plurality refers to the number of different signals that constitute a signal set. Within a signal set, alternatives for a given signal may be **discrete** in that they are individually distinct and without structural intermediates, or they may be **continuous**, with realizable intermediates between any two alternatives in the set. Most signal sets are either discrete or continuous, but some sets may consist of a mixture of the two (**Figure 8.16**). Discrete signals in the same set differ from each other by having disjunct values of one or more signal properties. These differences between alternatives must be sufficiently large and consistent (both between and within senders) that receivers can easily discriminate between them and assign each to an expected category in that coding scheme. Discrete coding schemes can be classified by the number of alternatives in the signal set

[155]. The simplest discrete signal set is an **implicit binary** one: when one condition is true, the sole signal in the set is turned on, made visible, or emitted; when the relevant condition is not true, the sole signal is turned off, hidden, or not emitted. Examples include signals used to announce an event such as the appearance of a predator, mark a site such as a territorial boundary, or declare the continued presence of the sender. **Explicit binary** sets contain two discrete alternatives, one of which must be "on": examples include the different coloration patterns or sounds that identify the sex of a sender. Discrete signal sets can include more than two alternatives: examples include different adult male, adult female, and juvenile plumages in many birds, and distinct alarm calls emitted for three or more types of predators in meerkats [256] and vervet monkeys [57].

The alternatives in a continuous signal set are potentially infinite in number: they differ in the values of one or more properties, and intermediates are possible between any pair of alternatives within a fixed range. An example would be a set that varies the amplitude or the frequency of an emitted sound depending on the probability that the sender is about to attack an intended receiver. In practice, only those variants in a continuous signal set that differ by more than the minimum resolution of a receiver's sense organs can be discriminated. In this sense, both discrete and continuous signal sets are perceptually finite. A major difference is that the values of allowable signal properties are fixed in discrete coding schemes, whereas properties of continuous signals can take any value within a given range. Note that continuous signals can usually be ranked according to one or more variable properties; continuous signals assigned to ordered conditions often preserve the order of the conditions in the order of the signal properties. In contrast, discrete signals assigned to adjacent ordered conditions often show maximal property differences [184].

Discrete signal sets are typically assigned to alternative conditions that are themselves discrete, such as species, sex, age class, and individual identity. Continuous sets are typically assigned to continuously variable conditions such as health, vigor, body size, hunger, level of aggressive intent, or predator proximity. However, as we shall see below, many coding schemes map discrete signal sets onto continuous conditions, signal propagation over a distance may convert initially discrete signals into more continuous sets, and receivers often classify continuously variable signals into discrete perceived categories. The plurality of a signal set may thus change multiple times between the initial conditions that triggered the signal and a receiver's recruitment of a signal into a decision.

SIGNAL PERSISTENCE Some signals, such as plumage patterns in birds that molt only once or twice a year, are fixed at creation and persist for long periods of time. Other signals, such as sounds or electric signals, are typically short in duration and last no longer than it takes them to propagate past the receiver. Hydrodynamic, tactile, and olfactory signals can have intermediate persistence times. It would seem that efficient signal design would match the persistence of signals with the persistence of the correlated conditions: if conditions are unlikely to change, why not create a long-lasting signal once and continue expressing it indefinitely? Similarly, if conditions are changing rapidly, it may be confusing and even dangerous to continue transmitting a prior signal that no longer reflects current conditions. In practice, signal persistence *is* often matched to the associated conditions. Species that do not change sex during their lives, such as birds, typically use persistent plumage patterns to indicate their sex. Birds in which dominance status is largely determined by body size and fighting ability and thus fixed for long periods may use an exposed chest patch to signal their status and thus head off any unnecessary fights. At the other end of the scale, the rapidly shifting logistics during overt conflicts are most often associated with transitory signals such as sounds or body movements.

While anatomically based signal sets, called **ornaments**, are often associated with long-term conditions, and action signals with short-term conditions, this is not always the case: for example, the red, white, and blue coloration on the head of adult male mandrills (*Mandrillus sphinx*) remains fixed and on display throughout a male's life, whereas the same set of colors on the head of an adult male turkey (*Meleagris gallopavo*) can be partly hidden when the bird is not displaying to females (**Figure 8.17**). Many songbirds stop learning new songs at maturity. The size and complexity of their adult

FIGURE 8.17 Fixed versus concealable ornaments in adult male animals (A) Male mandrill (*Mandrillus sphinx*) showing fixed red, blue, white, and gray facial coloration. (B) Male wild turkeys (*Melagris galloparvo*) infuse blood into head tissues to expand ornament tissues producing a similar red, white, blue, and gray pattern. However, unlike the mandrill, male turkeys can deflate many of the ornaments when not displaying to females or challenging competing males (C).

song repertoire can thus serve as persistent indicators of their condition before maturity [43, 297, 381, 382]. White fringes on wing feathers of female eider ducks (*Somateria mollissima*) provide persistent evidence of past immune challenges [156]. A sender can always convert a transitory signal such as a sound or electrical signal into a persistent one by continuous repetition. The angle of a continuously erectile structure like a crest, or the amplitude of a continuous growl can be modulated quickly to turn what is otherwise a persistent signal into a rapid series of transitory ones. Overall, it is striking how many evolutionary adaptations have arisen to extend or shorten signal persistence to match the duration of the relevant condition.

RELIABILITY GUARANTEES

As detailed in Chapter 10, signal sets can be characterized according to the mechanisms by which they guarantee **reliability**. Here we define reliability as the average probability that a receiver will correctly infer current conditions after receiving signals from a given set. Nearly all senders pay some costs or take some risks to generate signals. Where sender and receiver have identical interests, and no additional cost or penalty is imposed on a sender that emits a signal inappropriate for the current condition, we say the signal system is **cost free**. Where sender and receiver interests are not identical, senders may be tempted to cheat by giving a signal that does not represent the true conditions. A variety of mechanisms can be used to guarantee reliability in those cases (**Figure 8.18**). For example, **handicap signal sets** either impose an additional cost on senders that violate the accepted encoding scheme or provide an enhanced benefit to senders that honor that scheme [206, 265, 436]. Handicap signals are common during courtship, where females rely on male signals as indicators of male health, age, status, or genetic quality; when the sender is declaring a need that the receiver can help ameliorate; when prey signal to predators their likelihood of escaping if chased; and in aggressive interactions where opponents may be dissuaded by signals that are reliably correlated with aggressive intent, fighting ability, or experience. The honesty of **proximity cost signals** depends on the risk that a sender takes when it emits a threat signal close to an opponent. Such a sender must be ready to follow up that threat, or it could suffer serious injury. Senders drawing signals from a handicapped or proximity signal set clearly have some degree of choice, but sending an erroneous signal risks paying higher costs or receiving lower benefits than senders that honor the encoding scheme.

Index signal sets consist of alternative signals that are obligatorily (as opposed to just expensively) associated with specific alternative conditions. Usually, the alternative conditions concern some state of the sender such as health, body size, age, or reproductive condition. We have noted in earlier chapters that it is very difficult for a small animal to create a loud sound with a wavelength much larger than its body. While there are some anuran and bird species that manage to produce sound frequencies somewhat lower than expected for their body size [19, 108], low call frequencies usually provide a reliable index of caller body size [27, 107, 126]. Some authors have argued that index signals are actually just cases of handicap signals in which the costs of inappropriate signaling are extreme [368]. Whether index signals sets are seen as a category of their own or as extreme samples of handicap signal sets, we would expect that senders of signals from an index set have minimal freedom in selecting which signal to emit.

Conventional signals are those in which the coding scheme's assignment of alternative signals to alternative conditions is completely arbitrary. For example, matching of a song or call type recently emitted by another individual is often associated with escalated aggression in songbirds [410], whereas in other taxa, vocal matching is associated with appeasement and affiliative intentions [406]. The assignment of these signals to conditions is a *coding convention* that is shared by senders and receivers. Conventional signals can be cost free when sender and receiver have identical interests. When that is not the case, reliability guarantees are necessary. These are usually achieved by regular testing of sender honesty by receivers and some form of aggressive response or other punishment when a sender is discovered to have given an inappropriate conventional signal [265, 368, 409].

MULTIVARIATE SIGNALS

No signal modality can be fully characterized by measuring only one of its properties. This is because all modalities have multiple properties that can vary, and a value for each property must be specified to characterize any particular signal. For example, many animals produce sounds containing multiple frequencies that are harmonically related (see Chapter 2). A full description of even a single utterance must include: (1) its overall duration and average amplitude; (2) the frequency of its fundamental (lowest) component; (3) the relative amplitudes of all harmonics of that fundamental; and (4) a measure of the variations of the components in (1), (2), and (3) and the overall waveform amplitude during the course of the utterance. Characterization of a single uniform color patch on an animal's exterior requires specification of its shape, size, hue, brightness, and saturation. If the patch is iridescent, hue, brightness, and saturation must be specified for different viewing angles. Most olfactory signals contain multiple chemical components: which of a set of likely components are present in a given signal, the absolute concentrations of at least some of them, and the relative ratios of component concentrations must be specified to characterize an odorant at any given location and instant. Equivalent numbers of properties must be specified to provide full descriptions of touch, hydrodynamic, and electrical signals.

Any signal property that varies could potentially be used to create different alternatives in a signal set. When multiple properties within a modality vary independently, the many possible combinations provide a very large set of alternatives for coding. Variable properties within a modality are called **channels** or **components**, and thus signals relying on variation in multiple properties are called **multichannel** or

(A)

(B)

(C)

(D)

FIGURE 8.18   Examples of signal systems with reliability guarantees (A) *Handicap Signal:* Male tarbrush grasshoppers (*Ligurotettix planum*) compete for territories by countercalling with "shuck" calls. Call duration and calling rate appear to be costly, since males cannot increase both at the same time. Males with the highest combinations win disputes, and the combination appears to be a reliable indicator of male motivation and strength. (B) *Conventional Signal:* Blue-throated male morph of ornate tree lizard (*Urosaurus ornatus*). Blue versus non-blue is a conventional signal indicating status whose honesty can be tested by opponents' challenges. (C) *Index Signal:* Tigers stand on their back haunches and scratch the trunks of trees to mark their territories. The height at which the scratches occur is an unbluff-able index of the size of the tiger that made them. (D) *Proximity Cost Signal:* Senders must be prepared to follow through with threat signals given at close range. The risk at close proximity thus guarantees honest signaling [144, 171, 272, 395, 391].

multicomponent signals [6]. Many animals complicate the options further by recruiting more than one modality into a single display. Such displays are called **multimodal signals**, and they permit an even greater number of property combinations than a signal using only one modality. Whether relevant multiple properties belong to the same modality, different modalities, or some mixture, these ensembles can be classified together as **multivariate signals** [155].

In practice, not all variable properties are useful for coding: some properties suffer sufficient distortion or noise addition during propagation that receivers should ignore them even if senders were to use them to produce multiple alternatives. Other factors such as limitations on receiver sensory organs, concerns about predator eavesdropping, or energetic constraints may eliminate certain variable properties from

useful signaling. Even after these properties are set aside, there are invariably a large number of combinations of properties that can be assigned to one or more signal sets. How these combinations are divided up between multiple alternatives for a single signal set versus distributed over multiple signal sets, and thus multiple questions, has been the focus of much discussion [49, 117, 166, 188, 203–205, 278–280,

310, 312, 321, 326, 352, 353, 355, 361]. Although there are a number of competing classifications of multivariate signals, three main categories are regularly noted. We take these up in turn (**Table 8.2**, p. 300).

Where different signal properties in a multivariate combination are assigned to the same set of alternative conditions, we say that the coding scheme for that signal set is **redundant**. Because these properties are all assigned to the same set of alternative conditions, their values tend to be correlated with each other. Each property in a purely redundant set provides a receiver with the same information as any of the others, and when several of these properties are given in combination, the receiver response is identical to that elicited by any one of the properties alone. As we shall see below, purely redundant signal sets may be rare by this definition. Redundant coding is useful when suitable receivers occur over a wide range of noise environments or at variable distances from the sender. For example, male red-winged blackbirds (*Agelaius phoeniceus*) give a territorial display in which they bow forward, expose their red epaulets, and produce a wheezy song. In dense cattail marshes, the song can often be heard by dispersed conspecifics at greater distances than the red epaulets or bowing posture can be seen. However, in open and windy sites, the visual properties of the display may be detectable at much greater distances than the acoustic ones. The redundancy of having both visual and acoustic properties indicating the same condition thus enables these birds to use one display in a variety of habitats [38, 39]. Redundant signals also provide a check or "backup" when one property is misapplied by senders. For example, some male flat lizards (*Platysaurus broadleyi*) mimic female coloration to evade dominant males and gain proximity to females. While this works from a distance, they are unable to mimic the redundant pheromone signals that indicate sex, and at close range they are quickly identified and chased off by dominant males [426].

At the other extreme from redundant signals are those whose properties are assigned to different questions and thus can vary independently of each other. These are often called **multiple-message** signals. Many courtship and territorial signals indicate certain information (e.g., reproductive state, health, or aggressive intent) through one set of signal properties, but also provide species, sex, or individual identity information through other properties. As a result, properties of the plumage, posture, movements, and vocalizations displayed concurrently by birds such as the barn swallow (*Hirundo rustica*) tend to be uncorrelated and linked to different attributes of the displayer [118]. Similarly, songs of male rock hyraxes (*Procavia capensis*) contain different components that provide information about the singer's body size, condition, social status, and hormonal state [220]. In field crickets (*Gryllus campestris*), the fundamental frequency of the male's chirping song is an indicator of its body size, and thus of its nutritional condition when it was young, whereas its chirp rate and calling rate are both indicators of its current nutritional condition [192, 359, 360]. Similar

FIGURE 8.19 Multiple messages of pheromone mix in wing sacs of greater white-lined bats (*Saccopteryx bilineata*) The wing sacs contain no secretory glands. Instead, males clean them daily and refill them with a fresh mix of urine and secretions from their chin and genital glands. (A) Eight components of the sac mixture; a ninth component ($C_{15}H_{24}O_2$) has not yet been identified. (B) Multidimensional scaling (MDS) plot of similarities between wing sac pheromone mixtures of eight *Saccopteryx bilineata* and four *S. leptura* males. Greater within-species than between-species similarity (shown here by differential clustering of points by species) reflects species identity information. (C) MDS plot of wing sac pheromone mixtures for juvenile and adult male *S. bilineata* tend to be grouped in different parts of the plot and thus provide age information. (D) Correlation between concentrations of indole-3 carboxyaldehyde in pheromone mixture and the forearm length of males. The concentration of this compound in the pheromone mixture provides information about body size. Variations in the concentrations of the unidentified component and 2,6,10-trimethyl-3-oxo-6,10-dodecadenolide in the pheromone mix provide individual identity information. The composition of bacterial flora in sacs also varies between males and may reflect an individual's physiological condition. Males release the sac mixture into the faces of roosting females while hovering before them in flight or by shaking the opened sac at adjacent bats of either sex when roosting (see Figure 6.15). (After [53, 54, 411, 412].)

assignment of different acoustic properties to different questions is widespread in anuran calling [125, 126]. Different components in the femoral gland secretions of lizards provide simultaneous information about different aspects of the sender's current physiology [244, 245], and pheromone mixtures in the wing sacs of greater sac-winged bats (*Saccopteryx bilineata*) provide concurrent information about the species, age, body size, and physiological condition of scenting males (**Figure 8.19**) [53, 54, 411, 412]. The interval between flashes indicates species identity for many fireflies [239], and in at least one species (*Photinus ignitus*), the duration of each flash indicates the size of the nutritious spermatophore that a male is likely to contribute to a female [71].

There is a third class of multivariate signals in which the coding relies on specific combinations of signal properties. We shall call these **interactive coding schemes**. The simplest case occurs when redundant signal properties given in combination are more likely than any of the properties alone to elicit a receiver response. Such schemes may also tend to elicit a faster or stronger receiver response, or increase the memorability of signal alternatives [148, 149, 355]. This type of interactive coding is called **enhancement** [310, 312]. Enhancement effects may be quite heterogeneous within the same signal set: a dominant or primary property may show a much smaller enhancement when secondary properties are added than the converse [167]. In a variety of nocturnal frogs and numerous birds, vocal properties of calling displays act as the primary stimulus of receiver responses but can be enhanced if receivers are close enough to observe visual properties as well [243, 311, 393, 394]. Enhancement appears to be a common coding strategy for interspecific signaling; redundant acoustic, color, or olfactory properties of aposematic (warning) signals

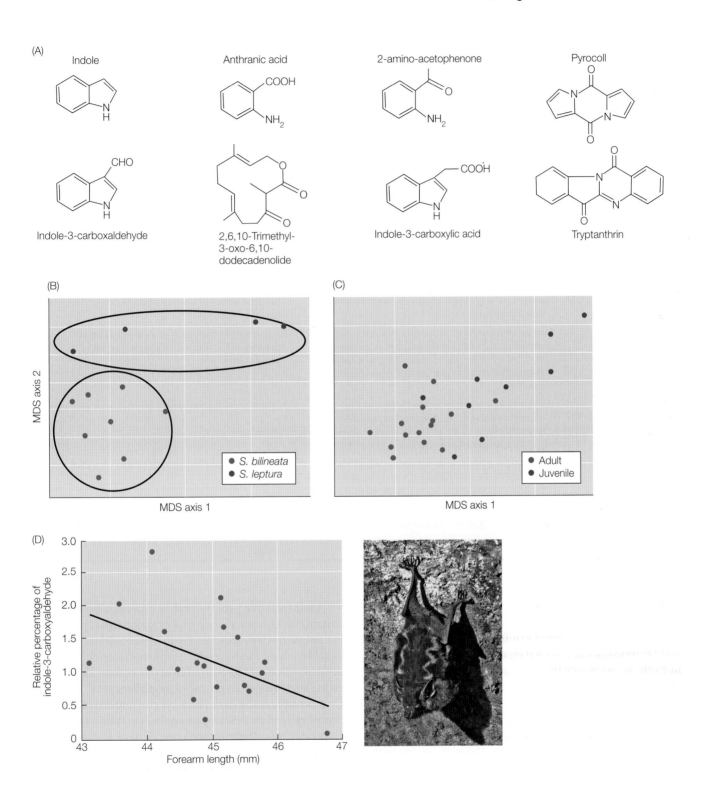

(A) Indole, Anthranic acid, 2-amino-acetophenone, Pyrocoll, Indole-3-carboxaldehyde, 2,6,10-Trimethyl-3-oxo-6,10-dodecadenolide, Indole-3-carboxylic acid, Tryptanthrin

(B) MDS axis 2 vs MDS axis 1; S. bilineata, S. leptura

(C) MDS axis 1; Adult, Juvenile

(D) Relative percentage of indole-3-carboxaldehyde vs Forearm length (mm)

enhance the memorability of toxic prey to potential predators [353, 354]. Bumblebees learn more quickly which flowers are more profitable when the flowers signal their identity through both color and olfactory properties [226].

A second interactive variant of redundant coding requires negative correlations between the values of the redundant properties: if a sender tries to make one signal property more extreme, it cannot help but reduce the values of some redundant property. This has been called **trade-off coding** or **multitasking** [166]. A variety of songbirds include trills in their songs that consist of a rapid repetition of a frequency-modulated note. Because trills require a rapid opening and closing of the beak to synchronize vocal tract resonances with syringeal sound production [174, 425], many songbirds show a negative correlation between how rapidly they can produce successive trill notes and the bandwidth exhibited

TABLE 8.2 *Alternative coding schemes for multivariate signals*

Coding Type	Definition	Property Correlations
Redundancy:	Multiple signal properties assigned to same set of alternative conditions; any property in redundant set provides same information as other properties in that set	In pure form, redundant signal properties relatively strongly correlated (can be either positive or negative correlations)
Multiple-Messages:	Different properties in the signal assigned to different questions and thus different sets of alternative conditions	In pure form, properties assigned to different questions are completely uncorrelated
Interactive Coding:	Combinations of signal properties provide different information or elicit different receiver responses than individual properties given singly	Variable levels of correlation between interacting properties
Enhancement	Suitable concurrent values of redundant properties elicit faster or stronger response in receiver than any of the properties exhibiting that value singly	Redundant properties relatively strongly correlated; correlation may be positive or negative
Trade-offs (**multitasking**)	It is difficult or expensive for sender to produce extreme values of two or more properties in the signal simultaneously	Relevant properties are strongly correlated so that extreme values of one tends to inhibit extreme values of other(s)
Amplifiers	One property or set of properties (the amplifier) is not assigned to any conditions, but its presence in signal enhances receiver assessment of another property or set of properties that is assigned to conditions	No correlation between values of amplifier and properties for which it enhances receiver assessment
Dominance	Two or more properties normally assigned to different questions and eliciting incompatible receiver responses are included in same signal but only dominant property elicits response	Properties are normally uncorrelated with each other
Modulation	This is a special case of dominance in which the presence of the subordinate property increases the chances that a receiver will respond to the dominant property	Properties are normally uncorrelated with each other
Emergence	Different combinations of different values in multiple properties are assigned to different alternative conditions within the same question	Properties are uncorrelated with each other, allowing for a large number of combinations of property values

Source: After [166, 310, 312].

by each note (Figure 8.20) [323]. Female swamp sparrows (*Melospiza georgiana*) favor males with either high trill rates or high bandwidths. However, their strongest preference is for males with combinations of trill rate and bandwidth that are near the upper limit set by the trade-off [10]. Paired male tree finches (*Camarhynchus parvulus*) are more likely to show trill rates and note bandwidths near their trade-off limit than are unpaired males [59], and brown skua (*Catharacta antarctica*) males that call near their trade-off limits have higher reproductive success [195]. Strut rate and a key component of the display sounds produced by lekking male sage grouse show a similar trade-off, and again males with combinations nearest the trade-off limit have the highest mating rates [132]. Similar trade-offs are present between the visual and acoustic components of male cowbirds (*Molothrus ater*) [66], call duration and calling rate in tarbrush grasshoppers [143], and the visual and olfactory signals of sagebrush lizards (*Sceloporus graciosus*) [396]. Trade-off coding could presumably function as a handicap to ensure reliability: only those senders in sufficiently good condition are likely to be able to perform near the trade-off limit.

An interactive process that is superficially similar to enhancement occurs when one or more signal properties are not themselves associated with any condition, but given suitable values, their presence improves receiver detection, discrimination, or memorization of one or more other properties. Such properties are called **amplifiers** [147, 159–162]. One example is the dark patch on the ventral side of male jumping spiders (*Plexippus paykulli*). While this patch increases proportionately with overall body size, it does not change with swelling of a male's abdomen, an indicator of that male's recent nutritional success. When displaying males show their abdomens to potential female mates, the black patch acts as a reference against which females can compare abdomen size and thus male condition [392]. It thus amplifies the fine differences in abdomen swelling made visible by the male displays. Another example is the pterin pigmentation in the wing scales of male sulphur butterflies (*Colias eurytheme*): these pigments absorb any diffuse reflectance of ultraviolet light from the wings, and thus amplify the ultraviolet iridescence that the males advertise during courtship flights near females [216, 356]. There appears to be much

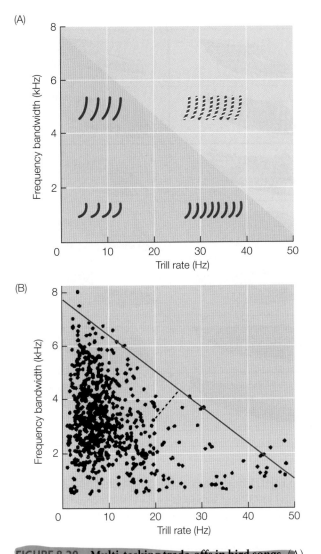

tails in fights with other males. The elaborate tail plumage appears to act as an amplifier to make female comparisons of male display movements easier. In other species, however, the physical properties of tail plumage appear to be the main display properties used by females to select mates. Tail size is noted by female widowbirds and the degree of iridescence on the peacock's tail by peahens. In these species, male display movements appear to facilitate, and in the case of the peacock, amplify the ornament properties being compared by females [247, 248, 320, 331].

When two or more signal properties that normally provide uncorrelated information if given singly are given in combination, several coding schemes are possible. If the combination results in no changes in receiver decisions and actions from presentation of these properties singly, this is simply another example of multiple-message coding. However, it is also possible that the appropriate receiver responses to these two signal properties' values are at least partially incompatible. One way to resolve this conflict, called **dominance** [310, 312], is for the receiver to weight one of the two properties more heavily, so that its consequent action is appropriate for the condition represented by the dominant property. Social play is widespread among vertebrate animals and often incorporates signals and actions that are also used separately in courtship or aggressive contexts [45, 89, 101]. In many species, the properties promoting play appear to dominate the usual implications of properties associated with courtship or aggression [23–25, 305, 306, 316]. Chimpanzees (*Pan troglodytes*) combine a variety of facial features with different calls to create a very large number of combined expressions. Different facial configurations and sounds are usually correlated with different conditions experienced by the sender. When conditions and appropriate expressions conflict, one of the conditions and its expressions or calls typically dominate the others in receiver responses [309]. In some cases of signal property dominance, the presence of the subordinate signal property, which by itself is normally associated with conditions not currently present or relevant, increases the probability that the dominant property will trigger a response appropriate to the dominant property's condition, or triggers a stronger response. This enhancement of a dominant property by a subordinate one is called **modulation** and was first described in social insects [176, 257]. One function of modulation may be to capture the attention of a receiver with one signal property and then provide the relevant information with another. This may be the explanation for joint visual and seismic components in the courtship displays of male wolf spiders [165].

Finally, variable signal properties that elicit no or incompatible responses when given singly may be combined to generate large numbers of different multivariate permutations, each of which is then assigned to an alternative condition. The assignment of different combinations of the same properties to different conditions is called **emergence** [310, 312]. For example, both acoustic and visual properties of male calling displays must be detectable to elicit competitive responses

FIGURE 8.20 Multi-tasking trade-offs in bird songs (A) Physiological predictions on limits of typical songbird to produce rapidly successive trills (horizontal axis) while simultaneously producing trill notes with wide bandwidths (vertical axis). Slow trill rates with large bandwidths (upper left), slow trill rates with small bandwidths (lower left), and high trill rates with small bandwidth (lower right) are all physiologically possible. Combinations outside the green triangle (e.g., upper right example) are not physiologically feasible. (B) Sample trill rates and bandwidths from 34 species of songbirds with the upper physiological limit to combinations estimated by the fitted red line. Relative performance of a given song can be measured by distance (dashed line) between the actual combination of trill rate and bandwidth and the nearest point on the red line: shorter distances constitute better song performance than longer distances. (After [323, 324].)

variability between species as to which properties of a display serve as amplifiers versus which provide the information of primary interest to receivers. In some species of lekking birds such as the greater sage grouse (*Centrocercus urophasianus*), the vigor of male displays appears to play a much larger role in female choice than does the size or ornamentation of their heads and tails [106, 131]. In fact, females have been observed to solicit matings from males who had lost their

FIGURE 8.21   Speculum plumage in ducks While males of many duck species have species-distinctive plumages, females tend to have less distinctive brown plumages, but share speculum color with the males. Multiple species often aggregate on the same ponds, and when startled by predators, manage to fly off in single-species groups. The speculum colors on the inner top half of their wings, which are usually covered when the ducks are on the water, are exposed during flight, providing a clear marker of species identity. Here are the speculum patterns for both sexes of seven common North American dabbling ducks (genus *Anas*). Note that the colored components of the speculum occur in standardized locations across the genus with different color combinations assigned to different species. (After [155].)

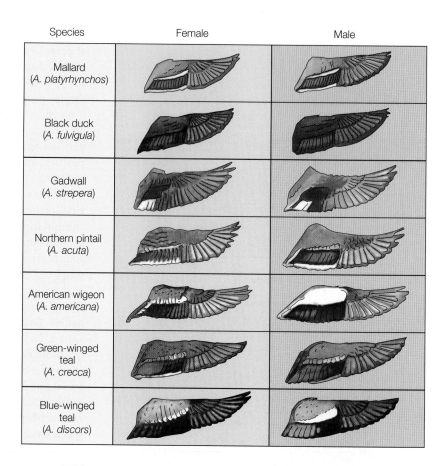

Species	Female	Male
Mallard (*A. platyrhynchos*)		
Black duck (*A. fulvigula*)		
Gadwall (*A. strepera*)		
Northern pintail (*A. acuta*)		
American wigeon (*A. americana*)		
Green-winged teal (*A. crecca*)		
Blue-winged teal (*A. discors*)		

in diurnal poison dart frogs (*Epipedobates femoralis*) [289]. The most common use of emergent multivariate signals is to provide identity information for a species, population, group, or an individual. Even though a given sender produces only the identification signal for its own species, each such signal must be sufficiently different that it is not confused with signals emitted by other sympatric species. This usually requires a signal combining many different variable properties. Thus, species identities of diurnal animals such as butterflies, fish, and birds are often signaled using composite color patterns. A particularly vivid example is the species-specific variation in the speculum feather combinations on the wings of dabbling ducks (**Figure 8.21**).

The provision of individual identity information poses problems similar to that for species identification: animals that live or breed in colonies, live in stable groups with differentiated statuses and roles, or need to discriminate between known territorial neighbors and intruders all benefit by having unique signals that differentiate them from a large number of others [398]. Again, emergent multivariate signals typically provide the diversity of signals needed to assign each individual its own unique marker. For example, colonial vertebrates such as penguins [4, 61, 208, 237, 367], bats [104], and seals [401] all rely on subtle frequency and modulation differences in contact calls to localize mates and offspring in large assemblies of conspecifics. Members of relatively stable groups may require individual markers if they perform

unique roles. Thus group-living long-tailed tits (*Aegithalos caudatus*) [373] and baboons [341] exhibit consistent individual differences in the frequency compositions of many of their social signals. Among paper wasps (*Polistes fuscatus*), individuals are distinguished by unique facial markings [397]. Dolphins (*Tursiops truncatus*) have individually distinct "signature whistles" that they use to declare their own presence. They are also able to solicit the attention of other group members by mimicking the latters' signature whistles [196, 197, 357, 380, 406]. The ability to discriminate between signals of territorial neighbors and those of intruders has been noted in lekking *Drosophila* [427] and a wide variety of vertebrates including fishes [233–235], amphibians [20, 21, 120, 193], reptiles [51, 244, 301, 408], birds [55, 113], and mammals [201, 286, 307, 349, 350]. These different taxa use different modalities and sets of properties to declare individual identity, but all rely on emergent combinatorial signals.

A trade-off exists between maximizing redundancy and maximizing the diversity of different messages that can be provided by signals. How might the conflict between these different coding goals be resolved? Theoretical models by Nihat Ay and colleagues [6, 7] suggest that an intermediate mix of redundancy and diversity is generally optimal. A cluster of properties that are strongly correlated with a given condition provides suitable redundancy, while multiple clusters, perhaps using different modalities, that are only weakly correlated with each other provide sufficient message

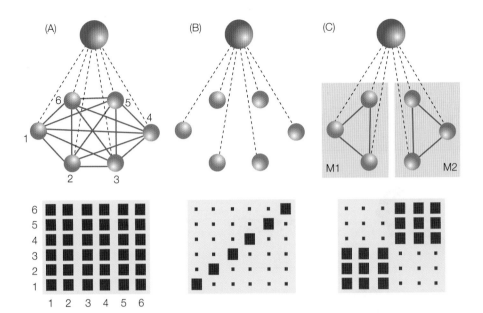

FIGURE 8.22 Alternative trade-offs between redundancy and multiple messages in coding schemes Top: three different ways (A–C) to distribute correlations among 6 different signal properties (red circles) that are then transmitted concurrently to a receiver (blue circle). Bottom: matrix summarizing degree of correlation between values of each pair of properties: large squares indicate high level of correlation between property specific to that row and property specific to that column; small squares indicate low level of inter-property correlation. (A) Completely redundant coding scheme in which every property is highly correlated with every other property. (B) Completely nonredundant coding scheme in which each property is only correlated with itself and each property provides different information to the receiver (multiple messages). (C) Intermediate (and optimal) coding scheme in which properties within a cluster (e.g., 1, 2, and 3 form one cluster, and 4, 5, and 6 form another) are highly correlated with each other, whereas there is only weak correlation between clusters. This scheme provides both some redundancy and some multiple messages. Note that it may often be easier to correlate properties within a modality (M1 and M2 in C) and thus use multiple modalities to generate different clusters of properties. (After [6].)

diversity (**Figure 8.22**). Even within clusters, properties need not, and in fact should not, be perfectly correlated with each other. There is growing support for these theoretical predictions. A number of authors have identified properties of multivariate signals that are partly redundant, but also partly sources of independent information. Examples include the courtship displays of the brush-legged wolf spider (*Schizocosa ocreata*) [130] and the overlapping roles of olfactory and bioluminescent signals in some fireflies [239]. In a variety of taxa, signal properties are used as indicators of aspects of a larger overall condition. For example, the chromatophores that produce coloration in fish, amphibians, and reptiles typically consist of three layers of pigments and structural color materials (see Chapter 4). Each of these layers can serve as an indicator of nutritional and physiological states of the animal; the overall combination produces a color that is an amalgam of multiple inputs on the animal's condition

[145]. Similarly, patches of color in the plumage and bills of many birds reflect aspects of the owner's health and vigor: carotenoid, melanin, and iridescent colors are each affected by different aspects of the bird's physiology [8, 44, 94, 168, 173, 375]. A similar case can be made that the composite properties of bird song are multiple reflections of the health and condition of the singer at a given moment [368]. In each of these cases, there are partial correlations between the physiological parameters being assessed and consequent partial correlations between the indicator signal properties. Each signal property is thus partially redundant in its representation of the overall condition of the sender, while providing a bit of nonduplicated information about a specific component of that condition.

SERIAL SIGNALS We saw in earlier chapters that each species has a limited range of signal properties that it can detect. For sound signals, the ranges of detectable frequencies, amplitudes, and modulation rates define a perceptual space into which each type of sound must be packed. As more sound signals are added to a repertoire, this perceptual space becomes more densely crowded, and the likelihood that a receiver will confound two different signals increases. There is thus a perceptual limit for any modality on the diversity of multivariate signals that can be discriminated by receivers with reasonable reliability. One solution to this problem is to limit the number of multivariate signals to a small set that is widely dispersed in the perceptual space so that individual signals are unlikely to be confused with each other, and then accommodate all further needs for signal diversity by producing sequential combinations of these highly discriminable elements. Such **serial coding** can generate an unlimited number of perceptually distinctive signals [292, 293]. Bird and whale songs are examples of serial coding: highly distinctive notes are strung together with very brief silences between notes, and the sequence is sung as a unit.

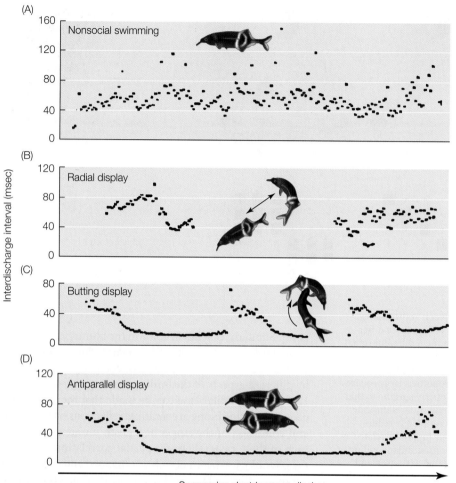

(A) Nonsocial swimming

Interdischarge interval (msec)

(B) Radial display

(C) Butting display

(D) Antiparallel display

Successive electric organ discharges

FIGURE 8.23 **Signaling and non-signaling variation in elephantfish (*Gnathonemus petersii*) electric organ discharge rates** Time intervals on vertical axes of graphs are inversely related to discharge rates: low interval values indicate high discharge rates and vice versa. (A) Normal variation in discharge rates when not signaling. (B) Smooth slowing and acceleration of rates during radial display. (C) Abrupt cessations followed by increasing discharge rates during butting display. (D) Steady high rates of discharge during antiparallel display. (After [281].)

There are a number of ways in which serial signals can be generated. The simplest approach is to repeat a single element of fixed (and usually short) duration at a regular interval. The choice of element and the rate of repetition can be assigned to different conditions and the different serial signals can then provide different information. This has been called **impulse rate coding** [155]. The rate of repetition can be fixed and species specific, as in the male advertisement calls of some frogs and toads [126], or it can vary with different repetition rates assigned to different conditions, such as the stages of aggression or courtship signaled by the electrical discharges of knifefishes (Gymnotidae) and elephantfishes (Mormyridae) (**Figure 8.23**) [178, 179, 221, 399]. In a second and related class of serial signals, called **alternation coding**

[155], a single element is repeated, but both the duration of successive elements and the intervals between them can be varied. Alternation coding is widely used by fireflies to provide species identity information (**Figure 8.24**) [239].

Serial signals can become quite complicated when more than one type of element is present. There is no a priori limit to how many successive elements can be combined into a serial signal; so serial signals of infinitely variable duration and composition are theoretically possible. Even for a small set of alternative elements and a small number of elements per signal, the number of possible combinations of elements, and if element order is important, the number of permutations within each combination, can become very large. For these reasons, serial signals with multiple element types are known as **combination signals**. Most species do not assemble elements at random into combination signals. Instead, there is normally a set of rules that defines which elements can occur in which parts of the signal, which elements can follow which other elements, whether repetition of the same element is allowed, and so forth. Any set of rules that limits the composition and ordering of serial signals is called a **syntax**. Although the term was originally associated with acoustic signals, the concept of rules limiting how successive elements can

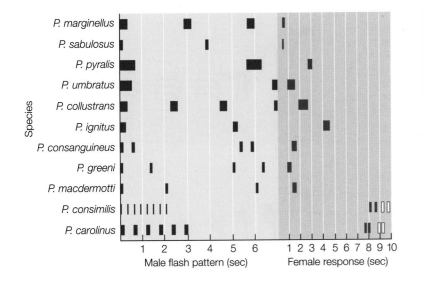

FIGURE 8.24 Species and sexual differences in flash duration and repetition rate among 11 North American species of the firefly genus *Photinus* Male flash patterns are shown on the left in blue and female responses on the right in red; open bars are optional female responses. The duration of flash is shown by the width of the bar, and the time interval between flashes by the spaces between flashes. (Note the difference in the time scales between male and female flashes.) (After [239].)

be strung together into serial signals can apply to any modality in which elements of relatively short duration are possible.

Two general types of syntax have been described for animal signals [258]. A **phonological syntax** defines the allowable combinations of elements that when emitted singly have no information value. This does not mean that different elements or sequences of elements in a phonologically serial signal cannot provide information about different conditions; as we shall see below, they can. However, such elements never provide that information if emitted singly. The syllables and phonemes comprising human words are recombined according to a given language's phonological syntax; most are not used singly with associated meanings. Most songbird species have serial songs with variable numbers of elements and somewhat variable composition; however, each species seems to have syntactic limits to these variations to ensure that their songs are easily discriminated from those of other species [55]. Since most of the elements in these songs do not have significance when given singly, the relevant syntax is phonological. Serial courtship songs that obey a phonological syntax are also found in mammals as diverse as bats [22, 34, 79, 80], mysticete whales [60, 271, 315], hyraxes [220], and rodents [177, 274]. Phonological syntax also characterizes serial calls produced by primates in a variety of social contexts. Examples include the territorial calls of gibbons and siamangs [121, 157, 277], and the obligately serial calls of titi monkeys (*Callicebus moloch*) [347].

In contrast, a **lexical syntax** defines allowable combinations of elements that can each provide information when given singly. Human sentences are an example of a serial signal defined by lexical syntax: each word (and in some languages, idiomatic phrases) can be used singly with communicative effect, and words and phrases can be combined to transmit composite information as long as the combination follows that language's lexical syntax rules. Combination signals based on a lexical syntax are uncommon in animals, but certain cases meet the minimal definition given above.

Capuchin monkeys (*Cebus olivaceus*) produce five basic types of calls: chirps, squawks, whistles, trills, and screams [348]; trills and whistles can be further divided into four subcategories each. Each of these call types is associated with a particular context, and can be given singly. However, 38% of calls are serial signals consisting of combinations of 2–4 different call types. Clear syntactical rules apply: squawks often follow chirps, but the converse is never heard. In fact, chirps rarely follow any other call types. Similar constraints apply to other possible combinations. Capuchins largely produce lexical combination signals when the situation is a mixture or intermediate between the contexts that typically elicit the component call types singly. Two combination call types of cotton-topped tamarins (*Saguinus oedipus*) also exhibit lexical syntax when the relevant conditions for the calls given singly are true at the same time [64]. Campbell's guenons (*Cercopithecus campbelli*) have six basic alarm call sounds that can be combined lexically to produce nine differently assigned combinations [302, 303]. Studies of wild chimpanzees (*Pan troglodytes*) have identified a variety of calls that are given both singly and in combination with each other and with tree drumming [72, 73]. Again, most combinations occur when several relevant conditions are concurrently true. Interestingly, the contextual specificity of chimpanzee serial signals is greater than that of the component single calls or drumming alone.

Each of the coding schemes outlined for multivariate signals can be applied to serial signals. *Redundant* serial signals are common. Continued repetition of the same signal is often employed to increase the likelihood that a receiver will detect and correctly identify the signal. King penguins (*Aptenodytes patagonicus*) [236] and chaffinches (*Fringilla coelebs*) [42] increase the number of repeated elements in their vocalizations as ambient noise levels increase. Western grebes (*Aechmophorus occidentalis*) increase the number of repeats in their calling for receivers at great or unknown distances [298].

Multi-message serial signals link different elements or multi-element phrases to different conditions. For example, the note-complex phase of white-crowned sparrow (*Zonotrichia leucophrys*) songs provides individual identity information, whereas the trill at the end of the song codes for species identity [291]. Baby sac-winged bats (*Saccopteryx bilineata*) produce serial isolation calls that are used by their mothers to find and identify which pup is theirs. The

FIGURE 8.25   Sample chickadee calls from the Carolina chickadee (*Poecile carolinensis*) This species has 6 common call elements in its chick-a-dee calls (A–E, and a hybrid element, D_h) [30]. Like the 4-element call of the black-capped chickadee (*P. atricapillus*) [152, 153], the Carolina chickadee call begins with introductory elements (here A or E), and then proceeds to a sequence of B–D notes with varying numbers of each element, or even the absence of an element, providing a variety of combinations. (After [115].)

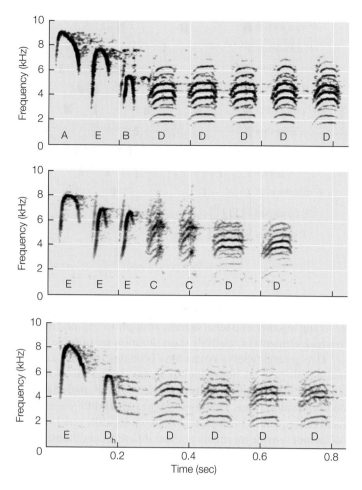

individual identity information resides in several complex syllables toward the end of the call [219]. North American tits (*Poecile* spp.) produce a complex "chick-a-dee" call that is associated with coordinating flock movements, maintaining contact, recruiting conspecifics to mob predators, and providing an "all clear" after a predator leaves. The chick-a-dees of most species consist of 4–6 basic call elements (labeled A–E) with some species having multiple variants of one of the types [30, 116, 153]. This serial call typically begins with introductory A or E notes, and then progresses in the order B–C–D (**Figure 8.25**). There may be up to 30 repetitions of a given element before the next type begins (e.g., AABBCCCD). Calls may omit one or more types (e.g., AAADD). Experimental playbacks that reverse element order (e.g., CACACA or DCDC) elicit much lower response levels than playbacks of calls that follow the normal syntax [65]. Different combinations are given in quite different contexts [105], and the fact that specific note types are replicated more often in the call when specific conditions are met (e.g., more C or D notes at new food discoveries or when mobbing predators), whereas others are concurrently reduced or dropped, suggests that each element type might provide a different type of information [9, 31, 105, 114, 154, 155, 254].

Interactive coding is also found in serial signals. Barking and growling are used independently as threats by timber wolves (*Canis lupus*); when used successively, the combination is perceived as even more threatening [358]. This is a clear example of *enhancement*. In several bird species with simple repetitive songs, the number of times that the constituent element is repeated within a single song, and hence the duration of the song, is varied to escalate or de-escalate territorial contests [232, 262, 267, 345]. This can also be considered a form of enhancement.

Trade-offs can be present in serial signals. In several bird species, males that repeat elements at high rates may slow down toward the end of the serial signal. This effect has been called "drift" and appears to reflect a trade-off between replicating the same elements rapidly early in a serial signal and maintaining that high rate through to the end of the series. Drift may thus be a measure of male stamina and health [230, 231]. In blue tits (*Parus caeruleus*), females mated to males that show minimal drift have larger clutches and produce more sons than those mated to drifting males [96, 325]. The "ballerina" courtship dance of Lawes' six-wired bird of paradise (*Parotia lawesii*) is a serial display that goes through five

very different phases and lasts for 20–45 seconds [363]. Males clearly differ in their performance of different phases of this display, and females are highly concordant in picking mates from among a wide range of males [330]. It is highly likely that there exist trade-offs in performance between different phases of this display.

Alerting elements fill a role in serial signals similar to that of amplifiers in multivariate signals. Alerts generally do not provide information by themselves, but they are invariably placed at the beginning of serial signals where they call a receiver's attention to the signal elements that follow. Examples include a loud note at the beginning of rufous-sided towhee songs (*Pipilo erythrophthalmus*)[344], and similar elements at the onset of Richardson's ground squirrel (*Spermophilus richardsonii*) alarm calls [389]. *Anolis* lizards typically begin a serial display of pushups with several accelerated or exaggerated renditions when visibility is limited or the nearest conspecifics are at some distance. These initially exaggerated elements thus appear to be alerts [111, 299]. In contrast, jacky dragon lizards (*Amphibolurus muricatus*) begin display with an alerting tail flick that is much slower, longer in duration, and wider in its arc than subsequent elements in the display [318, 319]. Male *Vargula annecohenae*, Caribbean ostracod crustaceans, display to females at sunset by squirting small packets of bioluminescent chemicals into the

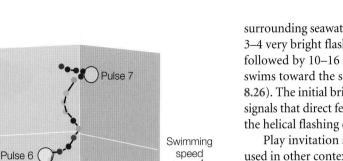

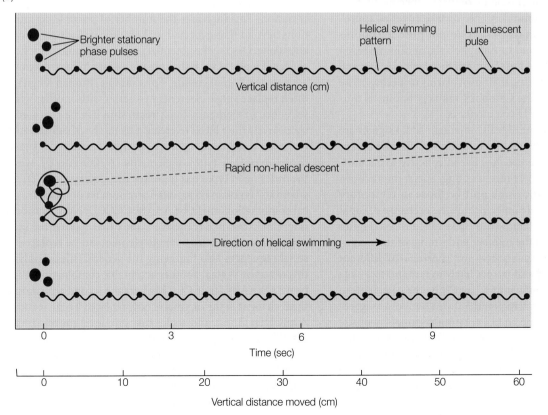

surrounding seawater. A typical male display begins with 3–4 very bright flashes emitted from a starting location, followed by 10–16 dimmer flashes emitted as the male swims toward the surface in a helical trajectory (**Figure 8.26**). The initial bright flashes appear to serve as alerting signals that direct female attention to the performance of the helical flashing display that follows [346].

Play invitation signals followed by signals normally used in other contexts can *dominate* the latter when presented serially, just as they do when present concurrently [25, 305]. As with multivariate dominance, dominance by one set of components may be quantitative and not absolute. For example, 51% of chimpanzee (*Pan troglodytes*) drums given in isolation are associated with travel, and 41% of uncombined pant-hoots are associated with food or responses to food. The serial combination of

FIGURE 8.26 Alerting components in light flashing display of male Caribbean ostracods (*Vargula annecohenae*) (A) The three-dimensional display shows successive positions of a displaying male (beads on trajectory), swimming velocity (color of bead as given by scale on right of plot), and points at which light pulses are emitted. Early in the display, the male stays at the same level in a looping, low-velocity trajectory and emits a series of alerting flashes. Then it begins the main display which consists of a spiraling trajectory toward the surface with a more rapidly emitted series of dimmer light flashes. (B) Temporal sequence of flashing in four displays by the same male. The size of the filled circles indicates the relative brightness of flashes in same display. (After [346].)

FIGURE 8.27   Coda patterns in sperm whales (*Physeter macrocephalus*) A sperm whale coda consists of 3–12 broadband clicks; different coda patterns differ in the number and timing of the clicks. The graph shows the patterns of the 23 different codas (A–W) shared by a population of about 400 sperm whales in the vicinity of the Galápagos Islands. Different populations differ in their sets of used codas and thus constitute vocal dialects. Codas appear to be exchanged between members of the same social group, and one whale may echo the coda pattern of another [343, 365, 422, 423]. (After [422].)

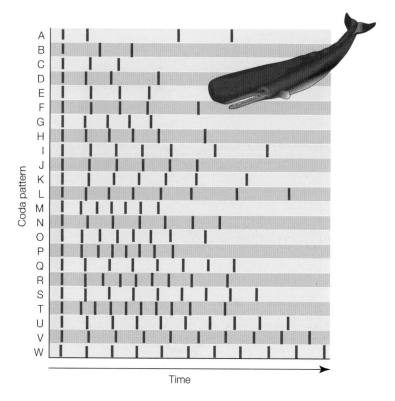

pant-hoot followed by drums occurs much more often during travel (50%) and much less often in association with food (18%). Thus drums dominate pant-hoots when given in combination. A clear example of serial *modulation* has been described in African forest monkeys [438]. Campbell's guenons (*Cercopithecus campbelli*) emit different alarm call types when encountering eagle versus leopard predators. Sighting of distant predators, treefalls, and other less threatening situations elicit several low-frequency booms followed by a series of eagle or leopard alarm calls. Whereas eagle or leopard alarms given alone elicit similar calling and flight by other troop members, the same calls elicit only general wariness when preceded by booms. Prior booms thus appear to modulate the salience of the two alarm calls. Interestingly, sympatric Diana guenons (*Cercopithecus diana*) form mixed troops with Campbell's guenons and respond to Campbell eagle and leopard calls by giving their own predator-specific calls. In both cases, the Diana monkeys flee or take cover as appropriate. Although Diana monkeys do not produce booms themselves, they respond to Campbell alarm calls preceded by booms exactly as do the Campbell monkeys (i.e., general wariness without emitting their own alarms or fleeing). The modulating effect of the Campbell monkey booms is thus fully recognized by the Diana monkeys.

As with multivariate signals, emergent serial signals are often used to provide *identity information*. Female sperm whales (*Physeter macrocephalus*) live in cohesive social units of 10–12 adults and their recent young. Social interactions between whales in the same unit or those in temporary aggregations of multiple units include the exchange of rapid strings of 3–40 broad bandwidth clicks called *codas* (**Figure 8.27**). Codas differ in the number and timing of their component clicks [423]. Each social unit shares 5–20 different codas, and individual animals are likely to respond to a coda by a member of the same unit by matching the coda pattern [365]. Units in the same region that form temporary aggregations also share some codas, creating regional "acoustic clans" [343]. Codas are an example of how repetition of a single element can be varied to produce a large number of emergent signals. Serial combination of many different vocal elements to provide information about individual and social group identity is also widespread in parrots [2, 13, 37, 69,

102, 172, 417, 418] and songbirds [40, 122, 123, 287, 288, 296, 388].

LIMITS ON MULTIVARIATE AND SERIAL SIGNALS   Multivariate and serial coding can generate very large numbers of alternative signals. Note that individual elements in serial signals can themselves be multivariate with multiple property variation. This increases the number of possible permutations beyond that expected with multivariate or serial variants alone. Despite the potential for very large and complicated signal repertoires, only humans appear to have taken full advantage of serial coding. For example, human languages often replace single elements in a series with another whole sequence of elements. Such **recursion** can be repeated at successively finer levels to create a hierarchy of substitutions. Consider a chickadee-like syntax that creates serial signals made up of four types of elements (A,B,C, and D), where the ordering of types has to follow the sequence ABCD, and where any type can be repeated any number of times. A typical combination might be AAABBCCCCD. Now consider a non-chickadee-like recursive syntax that follows the same general rules but allows single elements to be replaced by serial sequences also following the general rules. Applying this recursive syntax, another acceptable combination might be AAAB(AABCDD)CCCCD where the second repetition of the B element in the original is now replaced by the sequence AABCDD. One could make this one level more complicated by replacing the second A in the inserted sequence with AAAAABCD to give another allowable combination: AAAB(A(AAAABCD)BCDD)CCCCD. Clearly, recursion can generate an enormous number of serial signal permutations with only a few basic elements.

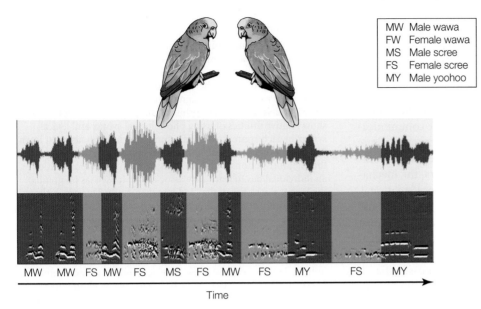

MW	Male wawa
FW	Female wawa
MS	Male scree
FS	Female scree
MY	Male yoohoo

MW MW FS MW FS MS FS MW FS MY FS MY

Time

FIGURE 8.28 **Duet by a mated pair of yellow-naped amazon parrots (*Amazona auropalliata*)** The top part shows the waveform, and lower part shows the corresponding spectrogram (frequency scale: 0–10 kHz) of the duet. The time scale for both is 0–7 seconds. Male contributions are in brown, and female contributions in green. Duets are serial combinations of three types of elements: "wawas," "screes," and "yoohoos" [431]. Each sex has its own wawa and scree variants, and these constitute most of each duet. Yoohoos are produced only by males and invariably end the duet. By varying the number, timing, ordering, and version of these elements, a single pair can generate a large repertoire of different duets while staying within basic phonological syntax rules. (Courtesy of Dr. C. Dahlin.)

In principle, recursion could be used in either a phonological or a lexical syntax to increase signal diversity, but evidence that animals use this technique in either way is so far lacking. Starlings (*Sturnus vulgaris*), nightingales (*Luscinia megarhynchos*), and sedge warblers (*Acrocephalus schoenobaenus*) all produce complex and variable songs based upon a phonological syntax in which successive phases of the song typically occur in the same order and each phase draws on a its own set of allowed syllable elements [55]. Phases are never inserted into other phases. The duets of mated pairs of yellow-naped amazon parrots (*Amazona auropalliata*) follow a similar type of syntax, with the two sexes contributing in alternation and each drawing on its own allowed set of variants for a particular phase (**Figure 8.28**) [75, 431]. Again, there is no evidence that whole phases ever replace elements in other phases. Northern mockingbirds (*Mimus polyglottus*) produce serial song bouts consisting of successive blocks in which each block consists of the same elements repeated 3–4 times and the same block is repeated only after a very long series of other blocks [87]. The order of different blocks may vary with different bouts, but again, there is no evidence of phonological recursion.

Not surprisingly given the rarity if not absence of phonological recursion, no example of lexical recursion has been described for any animal's natural syntax. Laboratory experiments suggesting that male starlings can learn to discriminate different songs according to lexical syntax rules [124] are controversial [68, 317], and there is no evidence that starlings exhibit lexical recursion in the wild. The absence of lexical recursion in animal signals is often cited as a major difference between animal communication and human language [109, 164, 191]. In part this may be due to a general rarity of any form of lexical syntax in animal signals. The few accepted examples consist of combinations of signals that can be used singly for different conditions but are combined when both conditions are true or when the current condition is an intermediate between the two others [64, 309, 347, 348]. The primary exceptions to this pattern involve several sound signal combinations in Campbell's guenons and wild chimpanzees [72, 73, 302, 303]. However, even here, no recursion is present.

Another instructive contrast between animal and human communication is that emission of most animal signals is tied to the current or very recent occurrence of the associated conditions whereas human language often refers to objects, events, states, or actions that are not currently present and may never have existed [28, 119, 190, 191]. Signals in human languages that can be used whether relevant conditions are present or not are often called **symbols**, and it is argued that they reflect persistent representations of objects, events, or conditions in the brains of those using them [85, 435]. While the degree to which animals incorporate abstract representations of objects and their relationships into their decision making remains poorly understood [58, 99, 169, 317], there is currently no persuasive evidence that animals use symbols (as defined above) in their communication. Animal examples that initially appear to be symbolic usually fail under closer scrutiny. Worker honeybees dance for hive-mates to indicate an earlier location of food or hive sites, but the dance for a given site occurs immediately after a foraging or scouting trip, not days later. As we have noted, current plumage in birds and call frequency in crickets represent physiological conditions at the time of the bearer's last moult and thus may not reflect their current condition. However, in both cases, the signals are locked into physical properties that do not change for long periods and thus do not require any symbolic brain representations. Some other examples of senders giving signals when the usual referent is absent are simply sender errors; as we shall see in later sections, effective communication does not require that senders never make mistakes. Truly deceitful signals by animals do occur, but they are uncommon, and as we shall discuss in later chapters, the mechanisms promoting them do not appear to be representational.

Most animal communication systems thus do not use symbols, lexical syntax, or recursion, key features of most human languages [29, 110, 164, 190, 191, 295]. Are there economic reasons why animals might adopt one coding strategy and humans another? Before presenting some possibilities, it will be useful to reformulate the differences slightly. Signals in both strategies provide receivers with information about past, current, or imminent "events." One can consider each event to be a combination of specific actors, actions, and in some cases, action magnitudes [295]. One strategy for representing events with signals is simply to create a list of all important event types and assign a different signal to each. If there were many relevant event types, multivariate or serial signals might be required to generate sufficient signal diversity. Senders and receivers would then share a learned or inherited "lookup" table indicating which signals were correlated with which events. Phonological syntax rules would suffice to filter out stimuli that were not allowable signals in that lookup table. Now, consider an alternative coding strategy in which senders and receivers would share a learned or inherited lookup table listing a set of signal elements that were each correlated with a specific actor, action, or magnitude. Events could then be represented by combining the signal elements corresponding to the relevant actors, actions, and magnitudes into a serial signal. Lexical syntax rules would specify where in a serial signal each type of signal element should fall, and how different combinations should be parsed and interpreted.

Each strategy has its benefits and its costs. Consider animals whose lookup table is limited to N entries. Those using the first "phonological" strategy could at most communicate about N events. Those using the second "lexical" strategy would distribute their N signal elements over likely actors, actions, and magnitudes. Even for small lexical tables, the number of potentially signaled events would be much greater than the number of assigned elements because of the large number of potential combinations. Thus the lexical strategy allows many more events to be represented with a given number of signal elements than the phonological strategy. The costs, however, of undertaking a lexical scheme are usually higher than those for a phonological one: the more complex parsing and interpretation processes require augmented neural machinery, there is a greater risk of error given the larger number of possible signal combinations, and the dynamics of acquiring signal-condition associations may be insufficient to ensure that senders and receivers all share

FIGURE 8.29   Suitability of phonological versus lexical syntax schemes as a function of numbers of events to be signaled Lefthand matrices summarize three contexts which differ in the number of events, here treated as combinations of objects and actions that need to be signaled (blue cells in matrices). Simulation models were used to produce the graphs on the right that show relative utilities of using phonological syntax and lexical syntax in each context. Models assume that individuals must learn to associate a given signal with a given event or event component through interactions with conspecifics; the more opportunities an individual has to learn new associations (abscissas on graphs), the higher the utility of that coding scheme. (A) This graph shows a small number of possible events (4 × 4 = 16), of which only 5 (31%) are biologically relevant. Phonological coding schemes always do better than lexical ones. (B) This graph shows an intermediate number of total possible events (36) of which 20 (56%) are important. Here the two coding schemes give very similar utilities with a slight advantage to lexical coding at lower numbers of learning occasions (left of dashed line), and to phonological schemes at higher numbers (right of dashed line). (C) High number of total possible events (100) of which 65 (65%) are important. Here lexical coding is favored for all numbers of learning occasions. Note that the minimal number of learning occasions required before utility is nonzero increases for both curves as the number of relevant events increases (After [295].)

the same lookup tables. A dynamic model assuming learned acquisition of lookup tables found that a threshold fraction of the possibly encoded events must be utilized before lexical schemes are better than phonological ones [294, 295]. This threshold decreases as the maximal lookup table size, N, increases; however, the minimal number of learning occasions required to acquire a lookup table also increases with N (**Figure 8.29**). These trends would make lexical coding more likely for animals with larger brains and longer lives. Where the total number of events that require separate signals is not large, the phonological strategy seen in most animals is not only sufficient, it is more favorable economically.

A quite different constraint on the use of lexical coding schemes is related to signal reliability guarantees. It is quite impractical to insure reliability of serial signals by imposing handicaps on the usage of each element in a serial combination: some combinations would become so expensive to send that senders would never send them [228]. Reliability must therefore be enforced by ignoring elements or combinations that prove to be deceitful. This creates a new problem because an individual receiver who ignores any elements used in a deceitful combination must then make adjustments in the coding matrix assignments of all other combinations that use those elements [229]. This will then require adjustments to other elements that are at times combined with the second set of changed elements, and so on until the entire matrix has been altered. Clearly, this individual will be unable to communicate with others retaining the original coding scheme. Lexical coding is thus only likely where there are no temptations to deceive, or where enough members of the population are willing to enforce the existing code by punishing deceitful signalers.

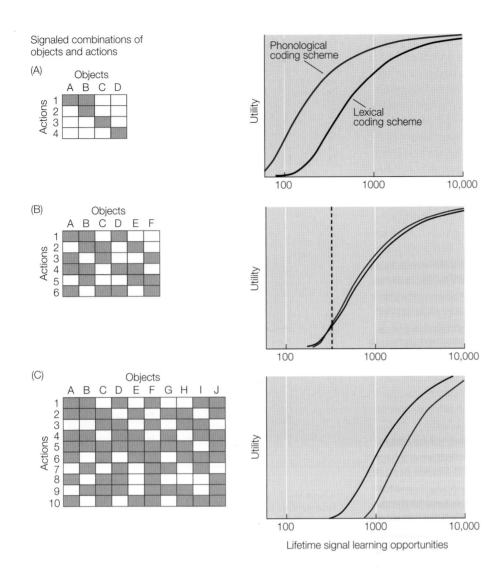

Signaled combinations of objects and actions

Mapping schemes

In a typical communication sequence, a sender emits one of a set of alternative signals based on its assessment of current conditions. During propagation, the signal may be altered in a variety of ways. The receiver then assigns the propagated signal to one of an expected set of alternatives, and uses its version of the coding scheme to perform an update. Each of these steps can be considered a **mapping** of one set onto another: the sender maps emitted signals onto conditions, propagation maps propagated versions onto emitted signals, and receivers map a table of expected signals and correlated conditions onto propagated signals. There is no guarantee that mappings will preserve plurality. For example, senders may map a set of discrete signal alternatives onto a set of continuously varying conditions. Even if the alternative conditions are discrete, the sender may map them onto a smaller number of discrete signals (a **reduction mapping**) or even a larger number (an **augmentation mapping**). Propagation of emitted signals over long distances is likely to change their plurality: reverberations can cause a discrete signal to arrive at a receiver in several successive and different versions. Noise

and distortion can also change a discrete set of emitted signals into a relatively continuous set by the time they arrive at the receiver. Finally, receivers may assign continuous signals (whether generated by senders or propagation or both) to discrete categories at any point in the chain between sensory reception and decision making. Even if senders emit discrete signals, receivers may assign the propagated versions to fewer or more discrete categories during processing.

An **isomorphic mapping** is one that preserves plurality. If conditions vary continuously, an isomorphic mapping by senders onto a continuous signal set and a similarly isomorphic mapping by receivers of received signals, with no propagation distortions between the two, would preserve the maximal amount of information. Yet fully isomorphic mapping seldom occurs in nature. Early ethologists were struck by the fact that a majority of animal displays are discrete, stereotyped, and performed at a "fixed intensity" even though the conditions they represented must surely be continuous [12, 74, 282, 285, 377]. Most birds have 5–6 discrete and quite different threat displays where one continuous signal would seem more informative [185]. Why would senders reduce the amount of information provided to receivers by mapping a finite set of discrete signals onto continuous conditions?

Similarly, studies of receiver responses and tests of signal discrimination make it clear that receivers often assign perceived signals to discrete categories even when senders emit continuous signal sets [163, 170]. They also may reduce plurality. A widespread example, called "cryptic binary coding" [155], occurs when a receiver uses a template to identify one out of a large number of alternative signals and classifies perceived signals into either fitting or not fitting that template. This is a common strategy when receivers are faced with identifying own versus other species, a mate or offspring in a large colony, or members of the same group that share a common odor or call versus nonmembers that have other odors or calls. Receiver categorization can occur in the first sensory stages, (a process called **categorical perception**), or instead be individually discriminated at the sensory level (**continuous perception**), but then assigned to a discrete set at a later stage in the decision process [158, 433]. Categorical perception has been identified in a number of animal taxa (**Figure 8.30**), and is an important component of human speech processing [18, 47, 90, 92, 93, 98, 158, 218, 240, 290, 329, 432]. If senders go to the trouble of producing continuous signal sets, why would receivers give up information by lumping received signals into discrete categories?

There are a number of conditions under which reductions in plurality may be adaptive. The encoding and decoding of continuous signal sets usually takes more neural machinery than dealing with a smaller number of discrete categories [82]. Resolving a continuous frequency range into fine threshold differences might require a longer cochlea than would broad categories. It may also take more time to map continuous signals onto continuous conditions, and continuous mappings will likely be more prone to assignment errors by either party. As we shall see in Chapter 9, the **value of information** (as measured by its utility; e.g., how that information increases or decreases the survival and reproduction of communicating individuals) does not always increase with the **amount of information** that signals provide. Where the provision or extraction of additional information imposes costs greater than the benefits, we would not expect either senders or receivers to invest in processing that additional information. Economic constraints on both parties thus limit the amounts of information that it pays to exchange.

There are other less obvious factors favoring discrete mappings and reduced plurality. If receivers cannot avoid errors in their assignment of propagated signals to expected alternatives, it may not pay for senders to encode finely separated signals; a few clearly distinct and discrete signals may suffice [202]. Categorical perception, or some other receiver scheme that reduces plurality, may also favor corresponding signal simplification by senders. Recent coevolutionary models confirm Darwin's proposition [78] that small discrete sets of alternative signals should evolve divergently such that they are not simply different but opposite ("antithetical") in property values [184]. This requirement limits

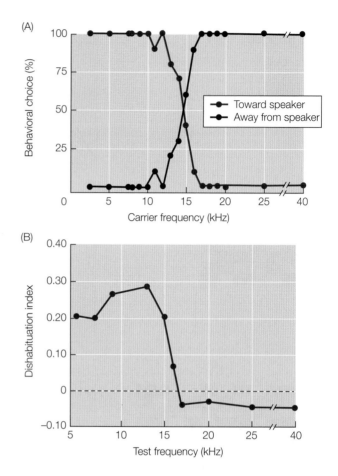

FIGURE 8.30 Categorical perception in Polynesian crickets (*Teleogryllus oceanicus*) Females are attracted to the display songs of males (4–5 kHz), but flee from the high-frequency sonar calls of insectivorous bats (25–80kHz). (A) Evidence for categorical labeling: tethered females received playbacks of synthesized male calling songs with different carrier frequencies. They then either turned toward or away from the playback speaker. Stimuli with frequencies below 13–16 kHz were treated as conspecific song, and higher frequencies as the sounds of a likely predator. (B) Evidence for categorical discrimination: tethered females received playbacks of batlike 20 kHz sound until they habituated (response to the fifth pulse was only 10–20% of the first). A test stimulus of a different frequency was played as the sixth pulse and the original 20 kHz stimulus then presented as the seventh pulse. The graph shows strength of the response to the seventh pulse depending upon the frequency of the sixth pulse. While any frequency tests below the 13–16 kHz threshold dishabituated the test stimulus, no frequency above that threshold did so. This shows that above the categorical threshold, crickets do not discriminate between stimuli of different frequencies. (After [432, 433].)

the number of suitable signal variants available to senders (**Figure 8.31**). Where there is a conflict of interest between senders and receivers, the presence of sufficient receiver error may stabilize the communication system without requiring overt signal handicaps [84, 100]. Finally, if senders have imperfect information about signaled conditions, signals are costly, and signal codes are at least partially heritable (leading

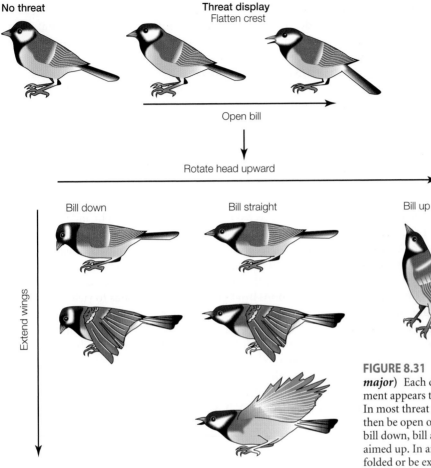

No threat

Threat display
Flatten crest

Open bill

Rotate head upward

Bill down   Bill straight   Bill up

Extend wings

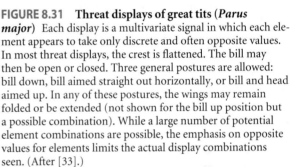

FIGURE 8.31   Threat displays of great tits (*Parus major*) Each display is a multivariate signal in which each element appears to take only discrete and often opposite values. In most threat displays, the crest is flattened. The bill may then be open or closed. Three general postures are allowed: bill down, bill aimed straight out horizontally, or bill and head aimed up. In any of these postures, the wings may remain folded or be extended (not shown for the bill up position but a possible combination). While a large number of potential element combinations are possible, the emphasis on opposite values for elements limits the actual display combinations seen. (After [33].)

to genetic correlations between sender and receiver coding schemes), selection can lead to the coexistence of multiple coding schemes in the same population [35]. This is in fact one mechanism for the generation of "personality" types in wild populations.

Measures of coding effectiveness

In an ideal world, senders would always know exactly which condition was true, always emit the same signal for a given condition (**consistency**), and always assign different signals to different conditions (**uniqueness**) [140]. Emitted signals would be completely unaltered during propagation, and receivers would always be able to identify which signal the sender emitted and, having perfect knowledge of the shared code, infer which condition elicited that signal. In real life, communication is rarely ideal. Senders may be inconsistent and produce more than one signal when a given condition is true, and/or coding may not be unique resulting in a given signal being elicited for more than one condition. Either or both deviations from perfect coding may arise simply because senders err in their identification of a current condition. Young, sick, or distracted animals may additionally make mistakes about which signal to emit for a current condition. Choosing which signal to emit is, like signal reception, a decision process: that means that senders may exhibit any of the

decision biases and shortcuts listed earlier for receivers. These decision biases will contribute to imperfect signaling. Finally, when senders and receivers do not have identical interests, senders may intentionally emit an inappropriate signal to benefit themselves at the expense of the receivers. Given certain circumstances, animal communication systems tolerate a persistent level of deceit while still being evolutionarily stable [1, 84, 390]. All of these factors can cause senders to deviate from the ideal case.

The degree to which an emitted signal is correctly identified by a receiver despite propagation distortion and noise is known as its **efficacy** [148]. Both senders and receivers can contribute to signal efficacy. Senders can limit long-distance communication to signals that are specifically designed to resist distortion in their particular environment. Receivers can invoke sensory and brain designs that maximize their ability to separate true signals from noise (**detection**), and correctly identify the alternative signals that senders may emit (**discrimination**). Where sender and receiver do not have identical interests, they may compete to impose the efficacy costs of communication on the other party [207].

Receiver errors reflect a trade-off between being liberal in assigning a newly perceived signal to a given category and condition—making a high number of **correct assignments** (also called **hits**) for that category but also many **false alarms**—versus being very conservative in making assignments, making many **correct rejections** for that category but also many **misses** that should have been recorded [253]. The total receiver error is the sum of the false alarms and the misses. It cannot be reduced for given levels of sender error, transmission efficacy, and receiver acuity. Instead, the contributing fractions of each type of receiver error are inversely related: adjustments that decrease one tend to increase the other [428, 429].

Reductions in effectiveness can occur at each of the successive stages—sending, propagating, and receiving. Most measures of signal set effectiveness compare how well a given input is mapped onto the relevant output. This can be undertaken for the entire communication sequence, or broken down into mappings at each stage in that sequence. At the sending stage, conditions are the input and emitted signals the output; at the propagation stage, emitted signals are the input, propagated signals the output; at the receiver stage, propagated signals are the input, and "interpretation" (assignment of input to alternative categories) is the output. **Forward measures** look at how well inputs are mapped onto outputs. **Backward measures** examine how well the true input can be inferred from receiving the outputs. Both approaches are used in animal communication studies. Below, we outline the logic behind some specific measures and show examples of how to use them.

FORWARD MEASURES OF SIGNAL EFFECTIVENESS   We shall first examine discrete coding schemes. The cell values of a coding matrix are the conditional probabilities that a given output (row) will be observed when a given input (column) is true. The cell values in a given column should sum to 1.0 (meaning there has to be some output when a given condition is true). A sender matrix summarizes how often an average sender assigns each signal to each condition; a propagation matrix maps the signals emerging from propagation onto those emitted by senders; and a receiver matrix summarizes how a set of expected categories are mapped onto propagated signals. An overall matrix can also be computed that combines the cumulative effects of the sender, propagation, and receiver matrices. In practice, one rarely if ever computes effectiveness measures for a species' entire signal repertoire. Instead, the repertoire is partitioned into its multiple signal sets, and measures are computed for each set.

There are three easily computable forward measures for discrete coding schemes. The first is a measure of consistency. A coding matrix would be perfectly consistent if a given input always elicited the same output. One begins by identifying the maximal cell value for each column in the relevant coding matrix. **Average consistency** for the matrix is the sum of the maximum cell values for each input column devalued by the prior probability that this input will occur. Average

consistency values can range from 100% for perfect coding down to the equal probabilities that a given output would be emitted by chance. If there were three outputs for a given input, consistency would be minimal if each output occurred $1/3 = 33\%$ of the time.

Average consistency only accounts for one kind of deviation from perfect coding: it ignores deviations from uniqueness in which a given output is produced for more than one input. A second forward measure, the **index of association** [139], is simultaneously sensitive to both kinds of deviations. In the absence of signal information, a receiver's best guess as to which of several alternative conditions is currently true will be based only on the relative frequencies with which the conditions have occurred in the past. The index of association measures the improvement in the rate of correct guesses when a receiver also has access to signals. This index varies from 100% for perfect coding down to 0% for a coding matrix with equal values in all cells. Because it examines all deviations from perfect coding, it tends to yield lower values than the average consistency measure for the same matrix. As with the average consistency, the index of association requires knowledge of the current prior probabilities for all inputs.

A third measure, the **determinant**, is computed from the coding matrix alone and is thus independent of prior probabilities [215]. It measures the heterogeneity of matrix cell values. A matrix for perfect coding will have the maximum heterogeneity in cell value distribution and its determinant will be 100%; a matrix in which all outputs are equally likely and thus have the same cell values will have a determinant of 0%. The more similar (e.g., homogeneous) any two columns or rows in a matrix, the smaller its determinant will be. Although determinants are classically defined only for matrices with equal numbers of inputs and outputs ("square" matrices), one can transform a matrix with unequal inputs and outputs to produce an equivalent measure on the same scale as for square matrices [434].

To see how these forward measures might be used, consider a species of monkey that gives one of three types of alarm calls when senders decide that a predator is nearby. One alarm call (S1) is most often given when leopards are suspected nearby; another (S2) is most common when eagles are detected; and a third (S3) is usually given when other potentially dangerous animals, such as snakes, approach. A fourth implicit signal, "no alarm call," is most often associated with the absence of any detected predators. The sender's coding matrix for these monkeys would thus list the alternative predator conditions as the input columns and the alarm call options as the output rows. Using some hypothetical conditional probabilities in the cells, the encoding matrix for alarm signals might appear as in **Figure 8.32A**.

The rows in this sender's matrix have been ordered so that the cells in the diagonal running from upper left to lower right (in bold) contain the maximal values for any column. The corresponding rows are the dominant outputs (signals) that might occur when that predator condition is true. The

(A) Sender coding matrix (%)

Signal	Condition			
	Leopard	Eagle	Other	No predator
Alarm S1	**78**	9	8	3
Alarm S2	6	**82**	5	1
Alarm S3	3	1	**62**	6
No call	13	8	25	**90**

(B) Receiver coding matrix (%)

Assigned signal	Sender emitted signal			
	Alarm S1	Alarm S2	Alarm S3	No call
Alarm S1	**96**	2	2	2
Alarm S2	1	**93**	2	3
Alarm S3	1	1	**90**	1
No call	2	4	6	**94**

(C) Overall coding matrix (%)

Assigned signal	Condition			
	Leopard	Eagle	Other	No predator
Alarm S1	**75**	10	9	5
Alarm S2	7	**77**	7	4
Alarm S3	4	2	**56**	6
No call	14	11	28	**85**

(D) Prior probabilities (%)

Condition			
Leopard	Eagle	Other	No predator
5	8	12	75

(E) Forward measures (%)

Matrix	Consistency	Index of association	Determinant
Sender	85	48	34
Receiver	94	80	75
Overall	80	38	26

FIGURE 8.32 Forward signal effectiveness measures for monkey alarm signals at close range (A) Coding matrix for sender, listing conditional probabilities that a sender will emit a given signal or be silent (rows) when a given condition (columns) is true. The maximal probability for each column is marked in blue. (B) Coding matrix for receiver, showing conditional probabilities that receivers will assign each emitted alarm signal or silence to one of the possible sender action categories. (C) Overall coding matrix combining both sender and receiver conditional probabilities. (D) Prior probabilities that each condition will occur. (E) Summary of forward measures for each stage and overall coding scheme when monkeys are near to each other. See text for computation details.

average consistency for the sender's coding matrix is computed by multiplying each column's maximal cell value by the prior probability (listed in **Figure 8.32D**) that this condition will occur, and adding up the products. If leopards are encountered 5% of the time, eagles 8%, other suspicious predators 12%, and no predator is present the remaining

75% of the time, the average consistency for this sender's matrix would be 85%, the index of association 48%, and the determinant 34%.

If receivers are so close to senders that propagation distortion is not an issue, and if they never make any errors in identifying which signal is sent, the forward measures computed above are sufficient. However, even if receivers are close enough that propagation distortion is minimal, they may misidentify which signal was emitted due to their own lack of attention or perceptual limitations. We thus may need to examine how accurately an average receiver monkey assigns a given perceived signal to each of the expected signal categories. These probabilities can be summarized in a receiver assignment matrix. To compute the average consistency of the receiver matrix, the dominant values are weighted by the relative frequencies with which senders send each type of signal. An example of a receiver assignment matrix is shown in **Figure 8.32B**. The corresponding forward measures are given in **Figure 8.32E**.

We can now combine the contributions made by both parties into an overall matrix that summarizes the assignment of perceived signals by receivers when a given condition is true. This is the matrix shown in **Figure 8.32C**. The values for each forward measure are summarized in Figure 8.32E. Note that in every case, the forward measures for the overall coding matrix are smaller than the values for either component matrix.

When sender and receiver are sufficiently far apart, propagation distortion and increased receiver errors will alter signal effectiveness. An example is shown in **Figure 8.33**. Here the sender coding matrix (see Figure 8.33A) and prior probabilities are the same as in Figure 8.32. While the sound is propagating between sender and receiver, distortions cause some alarm calls to acquire properties that make them more similar to others, and the presence of noise causes eagle calls to sound like some new kind of signal 11% of the time. These effects are summarized in a transmission matrix (see Figure 8.33B). The receiver has a harder time consistently assigning each emitted signal to an expected category, and must find some way to assign the new noisy sound to one or the other of the expected categories. The receiver assignment matrix is shown in Figure 8.33C. Again we can combine all stages into an overall coding matrix (see Figure 8.33D). The three forward measures for each component matrix and the overall matrix are summarized in Figure 8.33E. Note that all three measures have lower overall values than for communication using the same signals but at close range (see Figure 8.32).

BACKWARD MEASURES OF SIGNAL EFFECTIVENESS Backward measures evaluate how well signals help a receiver infer the current condition. There are three steps required to compute such a measure. These are best understood by examining the example shown in **Figure 8.34**. Suppose we begin with the same prior probabilities (see Figure 8.34A) and overall coding matrix (see Figure 8.34B) outlined in Figure 8.32.

(A) Sender coding matrix (%)

Signal	Condition			
	Leopard	Eagle	Other	No predator
Alarm S1	**78**	9	8	3
Alarm S2	6	**82**	5	1
Alarm S3	3	1	**62**	6
No call	13	8	25	**90**

(B) Transmission matrix (%)

Propagated signal	Sender emitted signal			
	Alarm S1	Alarm S2	Alarm S3	No call
S1-P	**85**	3	17	3
S2-P	6	**80**	12	1
S3-P	5	4	**67**	8
S4-P	0	11	0	5
No call-P	4	2	4	**83**

(C) Receiver coding matrix (%)

Assigned signal	Propagated signal				
	S1-P	S2-P	S3-P	S4-P	No call
Alarm S1	**64**	9	15	0	3
Alarm S2	9	**80**	12	64	1
Alarm S3	2	1	**55**	6	8
No call	25	10	28	30	**88**

(D) Overall coding matrix (%)

Assigned signal	Condition			
	Leopard	Eagle	Other	No predator
Alarm S1	**45**	13	19	8
Alarm S2	16	**61**	17	7
Alarm S3	6	5	26	13
No call	33	21	**38**	**72**

(E) Forward measures (%)

Coding Matrix	Consistency	Index of association	Determinant
Sender	85	48	34
Transmission	81	52	37
Receiver	79	43	0
Overall	65	10	3

FIGURE 8.33   Forward signal effectiveness measures for monkey alarm signals at long range   Same signal set as outlined in Figure 8.32 but with matrix values adjusted for fact that sender and receiver are far apart, resulting in significant degradation in signals during propagation. (A) Sender coding matrix (same values as in Figure 8.32). (B) Propagation transmission probabilities. (C) Receiver signal assignment matrix. (D) Overall coding matrix incorporating sender, propagation, and receiver contributions. (E) Forward measures of signal effectiveness for sender, receiver, and overall coding scheme at a distance.

Receivers would use this coding matrix, the prior probabilities, and their assignment of a stimulus to a given expected signal category to compute updated probabilities that each alternative condition is true. We can call these the "inferred" condition probabilities. Assembling the inferred condition

probabilities for all possible signal assignments, and assuming Bayesian updating, we get the a posteriori probability matrix shown in Figure 8.34C.

Note that there are many paths by which a receiver might assign a perceived stimulus to signal category S1: it may be that the sender correctly emitted S1 when it saw a leopard, and the receiver correctly identified this signal as an S1. Alternatively, the sender might have correctly emitted an S2 when it saw an eagle, but the receiver mistakenly classified what it heard as an S1. Not all paths leading the receiver to assign a signal to the S1 category begin with sighting of a leopard. We thus need to partition the matrix (see Figure 8.34C) into those paths which correctly link the current condition with the receiver's probability estimate for that condition and those which lead to incorrect associations. This is accomplished by combining the overall coding matrix (see Figure 8.34B) with the a posteriori probability matrix (see Figure 8.34C). We shall call the result a **reliability matrix** (see Figure 8.34D). Values in this matrix which lie along the diagonal from upper left to lower right are the probabilities that the receiver will correctly identify the current condition. Off-diagonal cell values contain the receiver's estimates that other conditions might be the current one. A perfect coding scheme will have 1.0 in each cell along the main diagonal of its reliability matrix, and 0 elsewhere. A coding scheme with the same value in all cells, which provides no new information, will just regenerate the prior probabilities along the main diagonal.

We can compute an **average reliability** by weighting the reliability for each input condition by the prior probability that it will occur, and adding these up. The average reliability for the coding scheme of Figure 8.34 when monkeys are close together is 73%. While this seems high, note that the reliabilities for the alarm signals (S1, S2, and S3) are only 31–48%; it is the reliability of "no signal" when no predators are present that inflates the overall average. We could of course compute an average reliability for only the signals, and this would be considerably lower. Average reliability, like the forward consistency measure, will be 100% for perfect coding, and equal to the sum of the squares of the prior probabilities for coding matrices that provide no information (all cell values equal).

Reliabilities measure the "endpoint" probabilities after receiving and processing signals; the endpoints for different coding schemes might be equal even though the receivers started with very different prior probabilities. One thus might want a **relative reliability** index that measures how much the signal system changes reliabilities from prior probabilities: we would expect a more effective coding scheme to produce a larger change. To put all such changes on the same scale of 0–100%, one can divide the observed change in probabilities by the maximum possible. This maximum will always be the

(A) Prior probabilities (%)

Condition			
Leopard	Eagle	Other	No predator
5	8	12	75

+

(B) Overall coding matrix (%)

Assigned signal	Condition			
	Leopard	Eagle	Other	No predator
Alarm S1	75	10	9	5
Alarm S2	7	77	7	4
Alarm S3	4	2	56	6
No call	14	11	28	85

(C) A posteriori probabilities (%)

Inferred condition	Assigned signal			
	Alarm S1	Alarm S2	Alarm S3	No call
Leopard	40	3	2	1
Eagle	9	60	1	1
Other	12	8	57	5
No predator	39	29	40	93

(D) Overall reliability (%)

Inferred condition	Actual condition			
	Leopard	Eagle	Other	No predator
Leopard	**30**	7	5	3
Eagle	11	**47**	6	4
Other	12	9	**35**	9
No predator	47	37	54	**84**

Average reliability: 73%   Relative reliability: 36%
Mutual Information: 0.428 bits   Relative information: 36%

FIGURE 8.34 Computation of reliability and mutual information measures for the coding system of monkeys communicating at close range The prior probabilities of the different predator situations are summarized in the vector (A), and the overall coding matrix based on both sender and receiver assignment matrices is shown in (B). These components can be combined using Bayes' theorem to generate the a posteriori probabilities (C) that a receiver would compute for each possible condition after signal reception. Combination of the overall coding matrix (B) and the a posteriori probabilities (C) generates an overall reliability matrix (D) that incorporates all errors by senders and receivers and the process of updating. The main diagonal of this matrix (**bold**) lists the probabilities the average receiver will estimate for each predator condition when that condition is in fact true. The off-diagonal values are the receiver's estimated probabilities that the current condition is not true. Average reliability uses the condition prior probabilities and the main diagonal of the reliability matrix to compute an overall average reliability. Relative reliability indicates the average change in the main diagonal values of (D) relative to the prior probability estimates in (A). Mutual information and relative information are basically the same as reliability and relative reliability, respectively, but rescaled based on binary questions (bits).

(A) Prior probabilities (%)

Condition			
Leopard	Eagle	Other	No predator
5	8	12	75

+

(B) Overall coding matrix (%)

Assigned signal	Condition			
	Leopard	Eagle	Other	No predator
Alarm S1	45	13	19	8
Alarm S2	16	61	17	7
Alarm S3	6	5	26	13
No call	33	21	38	72

(C) A posteriori probabilities (%)

Inferred condition	Assigned signal			
	Alarm S1	Alarm S2	Alarm S3	No call
Leopard	19	6	2	3
Eagle	9	36	3	3
Other	20	16	24	7
No predator	52	42	71	87

(D) Overall reliability (%)

Inferred condition	Actual condition			
	Leopard	Eagle	Other	No predator
Leopard	**11**	7	6	4
Eagle	11	**23**	10	6
Other	15	15	**15**	11
No predator	63	55	69	**79**

Average reliability: 64%   Relative reliability: 14%
Mutual Information: 0.156 bits   Relative information: 13%

FIGURE 8.35 Computation of reliability and mutual information measures for the coding system of monkeys communicating at a long distance The process of computation is the same as for Figure 8.34. Note lower values for all measures at this greater distance.

change generated by perfect coding. Again, we would want an average for this ratio across all conditions. The resulting average relative reliability for the example in Figure 8.34 is 36%. This represents the average change in probability estimates when a receiver attends to signals as opposed to ignoring them.

The reliability values for close-range communication (**Figure 8.34**) can be compared to those for long-range communication (**Figure 8.35**). In the latter case, the sender matrix and the prior probabilities are the same, but transmission effects and additional receiver assignment errors result in a considerably different overall coding matrix. The average reliability for signals at a distance is 64%, and the relative reliability is now down to 14%.

Web Topic 8.8 *Measures of discrete signal effectiveness*
Here we describe how to compute both forward and backward measures of discrete signal effectiveness using matrix algebra. Code for relevant Matlab routines is provided.

Web Topic 8.9 *Mutual information measures of signal effectiveness*
Here we provide an example of the computations needed to compute the average effectiveness of a signal set using mutual information.

Information theory provides a second widely used backward measure called **mutual information** [372]. Here, all probabilities and any changes in probabilities are converted into standard units called **bits**. If a **binary question** is one that has only two equally likely answers, a bit is the amount of information that is required to know which answer is true. Using this standard, it would require two binary questions to identify which of four equiprobable answers were true; the results would then provide two bits of information. Eight equiprobable answers would require three binary questions. Any problem can thus be rescaled according to how many binary questions would need to be answered to provide a complete answer (**Figure 8.36**). Bits need not be integers: it would take 1.58 bits to resolve which of three equiprobable answers was currently true. A quantity called **entropy** measures the uncertainty associated with any question by computing the number of binary questions required to completely resolve it. We are particularly interested in how receipt of a signal changes that uncertainty (reduces the entropy). One first computes the amount of information (in bits) before signaling that would be required to completely resolve the current question, and then how much uncertainty remains after receipt of the signal. The difference between the two values is a measure of the amount of information gained. Because this reduction in entropy is possible only if there are consistent correlations between conditions and signals, it is called mutual information. Note that entropy measures use the same basic data as reliability: a coding matrix, prior probabilities, and updated conditional probabilities computed using Bayes' theorem. However, they rescale the results into a standard set of units based on a logarithmic scale (bits). In the example of Figure 8.34, the mutual information is 0.428 bits, and that of Figure 8.35, 0.156 bits. Again, the measures show a decrease with distant communication.

As with reliability, the absolute number of bits provided by a set of signals is often more interpretable if it is compared to some maximum value. We usually want to ask how much the signals have contributed to resolving the initial uncertainty. We can thus provide an entropy measure equivalent to relative reliability by dividing the mutual information by the total number of bits that would have been required initially to remove all uncertainty. This ratio can be given as a percentage. If it is 0%, then little progress has been made by attending to signals; if it is 100%, then all initial uncertainty was resolved. We shall refer to this percentage as **relative mutual information**. In the example of Figure 8.34, relative mutual information is 36% and in that of Figure 8.35, it is only 13%. Note that each value is very similar to the relative reliability for the same example.

Each of the prior examples demonstrates an important principle that Jan Kåhre [215] has called the Law of Diminishing Information. This states that information is generally lost as it is passed through a succession of encoding and decoding steps. The information in the original encoding step can be preserved only if each successive recoding step is perfect (i.e., the reliability values are all 100%). As a result, the final overall matrix can never have higher forward or backward measures than the original one. The one exception is the determinant, which can approach 0 for one component matrix but be greater than 0 for the overall case. This is because the combined matrix has lower homogeneity than the component matrix with a near-zero determinant.

Web Topic 8.10 *Signal detection theory and signal effectiveness*
Signal detection theory provides a third backward measure of signal effectiveness that is widely used in psychology and neurobiological research.

MEASURES FOR CONTINUOUS SIGNALS   In theory, the forward and backward measures outlined for discrete signal sets should be adaptable to continuous signal sets. The one exception is relative mutual information: the maximal possible entropy for continuous variables is infinity. We thus cannot compare the mutual information provided with the maximum possible. The obvious approach for computing effectiveness measures when signals are continuous is to replace the sums used to compute averages with integrals [70]. Unfortunately, evaluating integrals requires knowing more about the distributions of input and output variables than finite samples can provide. One approximation is to segment the range of each variable into bins and thus create a discrete matrix whose cells contain the fractions of samples that fall in a particular row and column. One can then proceed as if the original data were discrete. This approach turns out to have several practical problems. One is that the fractions in the matrix cells may be only crude approximations of the true conditional probabilities that would be obtained if we had an infinitely large sample. Secondly, the numbers of samples that fall within particular rows or columns may be unequal, biasing the influence of some rows and columns over others in the total estimates. Finally, all measures are sensitive to how finely the input and output variables are segmented: too coarse a partitioning tends to hide patterns in the matrix, whereas too fine a division results in too few samples per bin to be representative. A variety of clever protocols have

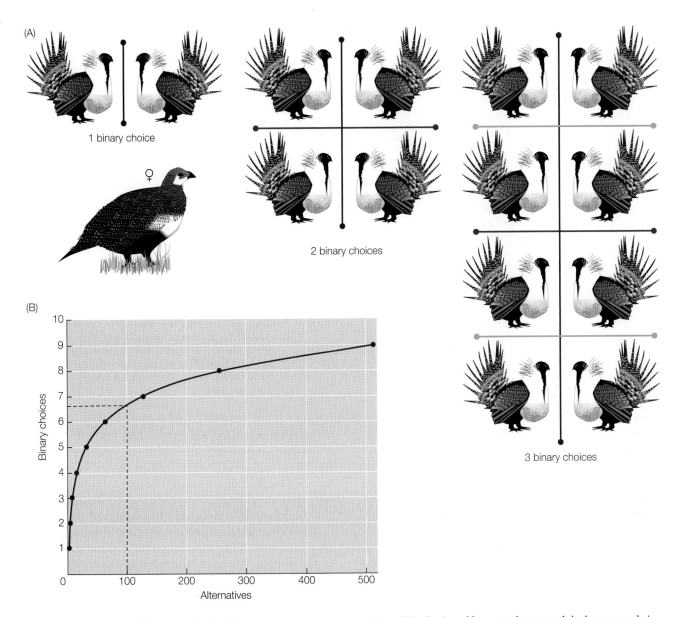

FIGURE 8.36 Binary choices (A) A female sage grouse must choose one mate from several similar males. If she has two males to choose from (left example), she needs to make only one binary choice (i.e., between the one on the left and the one on the right; red line indicates single choice). If she has to choose between four males (middle example), she needs two binary choices: between the left pair and the right pair (red line), and then between the individuals in the remaining pair (blue line). If she has to choose from eight males (right example), she can find her mate after three binary choices: between the left and the right group of four (red line); between the top and the bottom pair in the remaining group of four (blue line); and between the top and the bottom male in the remaining group of two (one of the two green lines). (B) A plot of the number of binary choices required to select one out of various numbers of equally likely alternatives. Although the graph was drawn using numbers of alternatives that require a whole number of binary choices (e.g., 2, 4, 8, 16, 32, etc.), one can easily see that any number of alternatives can be accommodated on this scale by allowing for fractional numbers of binary choices. Thus it would take 6.64 binary choices to select one out of 100 equally likely alternatives (dashed lines).

been devised to minimize these problems. Examples include using unequal bin sizes; defining continuous functions that describe the input and output variable distributions (kernel density estimation); extrapolating the measures to infinite sample sizes; and using the spacing between points in graphs of the output variable versus the input variable to compute entropies [46, 77, 217, 222, 300, 313, 378].

When the input and output variables are distributed in an approximately normal (Gaussian) manner and their relationship is roughly linear (or can be linearized with suitable transformations), there is another approach. First, we use the statistical method of **regression** to identify the best-fitting straight-line relationship between inputs and outputs (**Figure 8.37**). The scatter around this line is a measure of the

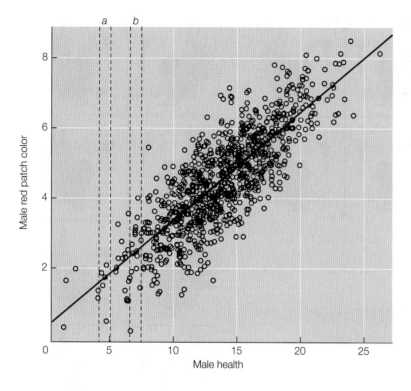

FIGURE 8.37 Hypothetical example of a continuous coding scheme Here, signal value (quantified as an index combining the size and intensity of a male's red belly patch on the vertical axis of the graph) is mapped onto some relevant condition (here, an index of male health along the horizontal axis). Sample points are shown as small circles. Overall, the color index appears to increase with the health index. A regression line (red) has been fitted to these data using standard statistical techniques. If one divides the horizontal axis into small segments (such as a and b), one can use the vertical scatter of points around the regression line within any segment as a measure of the coding scheme's consistency. The degree to which the color index predicted by the regression line is different for different segments, as indicated by the regression line's slope, is a measure of the uniqueness of this scheme. The square of the correlation coefficient for a regression (denoted by r^2) is an overall indicator of the effectiveness of this coding scheme, because it depends both on the regression slope and on the scatter of points around the regression line. The value of r^2 varies between 0 and 100%; in this example, it is 66%. Note that the computation of r^2 corrects for differences of scale: if we were to multiply all color index values by 3 and redo the regression, we would still get $r^2 = 66\%$ for these samples.

consistency of the coding scheme, and the relative slope of the line reflects the coding scheme's uniqueness. An overall forward measure of signal effectiveness that is sensitive to both deviations from perfect coding is the square of the **correlation coefficient** (usually denoted by r^2). This number can be interpreted as the fraction of the variation in the output variable that is explained by variation in the input variable. It will vary from 0% when there is no relative slope to the regression to 100% when there is a strong slope and no scatter. Under linear and normal conditions, mutual information (MI) in bits can be computed from the squared correlation coefficient using $MI = -0.5 \log_2 (1 - r^2)$ [77]. Note that the use of r^2 or MI where there are serious deviations from linearity or normality may underestimate the effectiveness of the corresponding coding scheme [215, 222].

CODING EFFECTIVENESS IN ANIMAL SIGNAL SYSTEMS   Interest in quantitative measures of signal consistency and reliability has varied historically. A burst of studies in the 1960s and 1970s computed mutual information values based on correlations between observed actions of senders and receivers [3, 26, 91, 151, 199, 246, 332, 430]. The problem with this approach is that actions by either party depend not only on the conditions encoded in the signals, but also on prior probability estimates, access to ambient cues, and the relative payoffs of alternative actions. Any given action may reflect not the most recent signal, but long prior ones. There is thus no guaranteed link between emission of any given signal and the next action by either party. This recognition undermined enthusiasm for such measurements. Research from 1990 to 2000 on reliability guarantees renewed interest in measuring

the effectiveness of animal signals [138, 198, 368]. Recent efforts focus on quantification of the correlations between conditions and signals by both senders and receivers. These are not easy tasks in any system. One typically begins with large samples to identify correlations between emitted signals and conditions of likely interest to receivers. Experimental manipulations are then performed to confirm associations between signals and conditions, and to characterize each party's mapping strategies. Neurophysiological monitoring is increasingly used to verify and refine receiver signal processing rules. Once discrete coding matrix tables or continuous signal regressions are available, these can be combined with the frequencies of alternative conditions obtained in nature to compute average measures of signal effectiveness.

To date, there are few studies that provide all of the requisite tests and frequency data for both senders and receivers. However, because overall measures are generally limited by the smallest contribution in the communication sequence, we can at least put an upper limit on the overall measures by looking at the most accessible phases. In general, this is the sender phase, where both relevant conditions and emitted signals can often be monitored. Recent efforts have focused on conditions external to the sender (e.g., alarm and food signals) or internal states of the sender that are easily measured or manipulated (e.g., sender body size, health, or hunger state). **Table 8.3** provides a sampling of such data. While it remains possible that some of the signal sets cited in this table are normally used only when additional redundant signals, concurrent signals in other sets, or ambient cues are present to provide additional information, they are representative of the general pattern that is emerging from animal

TABLE 8.3 *Measures of signal effectiveness for some animal signal sets[a]*

Input vs. output variables	Signal set type	Sender effectiveness (%)	Reference
Parasite resistance versus chest spot number and size in female barn owls	Continuous	16	[351]
Coccidial parasite load and beak color in partridges	Continuous	22	[284]
Male age and size versus proximity to bandwidth/trill rate trade-off limit in swamp sparrow songs	Continuous	32	[11]
Body size versus dominant frequency in male calls of 12 anuran species	Continuous	36 average (9–78 range)	[368]
Immunocompetence and color of head combs in male red grouse	Continuous	37	[283]
Body size versus formant spacing in male red deer roars	Continuous	39	[340]
Body condition versus male stickleback red courtship coloration	Continuous	44	[273]
Time since last feeding versus begging rate in swallow chicks	Continuous	65	[238]
Hawk/owl predator versus type of alarm call in Siberian jays	Discrete	72 (51; 55)	[146]
Terrestrial versus aerial stimuli and red squirrel alarm call types	Discrete	86 (69; 74)	[141]
Alarm and activity coordination calls in wild chimpanzees	Discrete	89 (59; 70)	[72]
Terrestrial versus aerial visual stimuli and chicken alarm call types	Discrete	95 (71; 80)	[150]
Protective alkaloids in spermatophore versus courtship pheromone in arctiid moths	Continuous	95	[97]

[a]Values are provided only for the relevant sender stage of communication. For discrete signal sets, the first value is the average consistency with the index of association and matrix determinant (in that order) in parentheses. Continuous signal sets are characterized by the r^2 values of regressions of signal on condition values.

signal studies: sender signal effectiveness is usually intermediate between random signaling and perfect coding, and thus we expect overall signal effectiveness to be intermediate at best [368]. The general conclusion is that intermediate signal effectiveness is often "good enough" to justify both parties participating in communication. We take up why this might be the case in the next chapter.

So, what is information?

There has been much discussion in recent years about the most useful definitions of *information* in biology [52, 76, 112, 137, 189, 215, 263, 264, 275, 276, 342, 362, 366, 374, 383, 384, 386]; (see also Web Topic 1.2). Much of this discussion has been sparked by the elucidation of the genetic code in DNA, but genes are a sufficiently special case that they do not provide an adequate definitional model for the use of cues and signals [189]. Part of the challenge in defining biological information is that one can identify no single entity, like energy in physics, that is transferred between source and recipient. Also, unlike an exchange of food or heat, the provision of information changes one or mores state in

the recipient of the information, but the transfer does not decrease or "use up" the ability of the source to provide the same information to other recipients [5, 227]. The provision of biological information is best described as a process that requires several steps. At the source, information is created by the emission of a stimulus whose form differs from that of alternative stimuli and whose emission is persistently correlated with the same subset of alternative conditions. A recipient then uses receipt of the stimulus to change some internal state. A useful change of state is possible only if the recipient has prior knowledge of the source correlations between stimuli and conditions. Thus cues and signals provide information only if both source and recipient invoke the same set of correlations between stimuli and conditions.

The complex nature of information provision has triggered considerable discussion over how or even whether one should try to measure the amount of information provided by signals [342]. The poor record of mutual information measures in accounting for sequential actions of senders and receivers led many workers to abandon all information measures in animal communication systems. Others have argued that the processes of stimulus classification, updating, and decision making are all too inaccessible or interwoven to warrant trying to measure their effectiveness singly, and thus the only useful measure of information provision is the degree to which signals affect sender and receiver utilities [76, 366]. However, many of the critical unanswered questions about animal signal evolution require separate measures of the amount of information a given signal set provides and the utility value of that information. As we saw earlier, most animals appear to follow optimal decision-making protocols, but there are many biases and shortcuts that result in imperfect decisions: how does imperfect decision affect the demands on signal effectiveness? Similarly, there are many reasons why signal effectiveness in animals appears to be intermediate even when the receivers seek optimal decisions. How effective does a signal set have to be for the value of information to justify the costs of engaging in communication? Questions like this can only be answered if we distinguish between and measure separately the amount of information and the utility that signals generate. Whereas receiver processing of signals was poorly understood a half-century ago, there has been enormous progress since in the design of conditioning paradigms [163], neurophysiological monitoring (see Web Topic 8.7), and the experimental manipulation of signal systems. It is increasingly possible to compute the maximal amount of information that could be obtained from a given communication set, and then parse observed reductions below this maximum among the successive phases of the communication process. The forward measures outlined above and relative reliability (or the mutual information equivalents) provide a good starting point for measuring the amount of information provided during the predecision phases of the communication process. In the next chapter, we take up the ways in which the amount of information contributes to the value of information and decision making.

SUMMARY

1. All animals use their sense organs to accumulate **information** about previous and current conditions around them. Cumulative information can be stored in the nervous system in various ways; a common mechanism is a list of alternative conditions and probability estimates for each alternative. Animals use their current state of information to make **decisions** about future actions. Optimal decision making adopts the action with the best **expected payoff**. The expected payoff of an action is the sum of the possible consequences for that action, each **discounted** (multiplied) by the currently estimated probability that the condition leading to that outcome exists.

2. **Cues** and **signals** are stimuli correlated with conditions of interest to receivers: cues are generated for reasons irrespective of the presence or responses of receivers, whereas senders emit signals because they and receivers generally benefit by the provision of this information. Receivers combine the receipt of a cue or signal with knowledge of the correlations between the stimuli and the conditions of interest to **update** their probability estimates that alternative conditions are true. Optimal receivers can then make a decision on future actions by computing and comparing expected payoffs, or by comparing the new probability estimates to threshold values (red lines) that are based on relative payoffs of alternative actions.

3. There is increasing evidence that many animals, including humans, track the current probabilities of ambient conditions, update these probabilities upon receipt of cues and signals, and make optimal decisions much of the time. However, both humans and animals sometimes deviate from optimal decision making. Causes include nonlinear functions relating payoffs to actual utilities (risk effects); nonlinear scalings of stimuli in sensory systems (Weber's Law); the time and effort saved by shortcuts to Bayesian updating; exploratory decisions to update condition lists and probabilities, and the need for fall-back strategies (heuristics) when alternative expected payoffs or probabilities have similar values.

4. A species' overall **signal repertoire** can usually be broken down into separate **signal sets**, each of which includes all alternative signals that might be given to provide information about the same question. The rules by which senders and receivers associate conditions and signals in a signal set constitute that set's **coding scheme**. Animal coding schemes are highly diverse. At a gross level, they can differ in **plurality** (i.e., whether they consist of continuous alternatives or discrete alternatives, and if the latter, in how many alternatives); **persistence** (how long an emitted signal continues to reflect a correlated condition); and the level of **reliability guarantees**. The latter range from **conventional signals** with no intrinsic guarantees for reliability, to **handicap signals** that create additional costs or reduced benefits when given unreliably, to **index signals** which are physically impossible or at least extremely difficult to produce unreliably.

5. **Multivariate signals** encode information in more than one concurrent signal property. This provides even greater diversity in coding schemes. The multiple properties may all provide the same information, as in **redundant signals**, or each property might provide totally different information, as in **multiple-message signals**. Multivariate signals of most animals provide a mix of redundant and multiple message information. Information provided by multivariate signals may also depend on interactions between

properties. **Interactive coding schemes** include: **enhancement** (in which redundant properties given concurrently produce stronger receiver response than if given singly); **trade-offs**, (in which performance of some properties at extreme values inhibits performance of other concurrent properties at extreme values); **amplifiers** (properties that never provide information themselves but facilitate information provision if invoked concurrently with others); **dominance** (in which multiple properties provide independent information when given singly, but one property dominates receiver responses when given concurrently); **modulation** (a special case of dominance in which the presence of the dominated properties enhances receiver responses to the dominant property information); and **emergence** (in which combinations of properties provide different information to receivers than any does when given singly).

6. **Serial signals** consist of multiple elements emitted in succession. All interactive coding schemes listed for multivariate signals can occur using successive elements in serial signals. Because the number of potential permutations of elements in serial signals can be enormous, most species limit allowable options using **phonological syntax rules** (in which individual elements given singly provide no information), or **lexical syntax rules** (in which individual elements can provide information). Human speech relies on phonological syntax to make words, and lexical syntax to make sentences; most animals only use phonological syntax.

7. Coding schemes can be treated as **mappings** of an output on an input: senders map emitted signals onto perceived conditions, propagation maps transmitted signals onto emitted signals, and receivers map expected categories on transmitted signals. Few of these mappings are **isomorphic**: that is, the plurality of the input is not preserved in the output. Senders may map discrete signals onto continuously varying conditions. Propagation often converts initially discrete signals into noisy continuous ones. Receivers often divide continuous propagated signals into discrete categories, either early in sensory processing (**categorical perception**) or at any later stage in the process. Changes in plurality at each stage invariably reduce the information transmitted.

8. Measures of **signal effectiveness** attempt to quantify the average amount of information provided by a signal set. **Forward measures** like **consistency**, the **index of association**, and matrix **determinants** quantify the correlations between alternative outputs (emitted signals) and relevant inputs (conditions). **Perfect information** is provided only when all dominant consistencies are 100% and the index of association and matrix determinant equal 1.0. **Backward measures** such as **reliability** quantify the fraction of time that a receiver relying on signals correctly infers the current condition. **Mutual information** is a measure that rescales reliability into units based on equivalent binary questions (**bits**). Existing data indicate that most animal signal sets have intermediate consistencies of 30–80%; perfect information appears to be very rare in animal communication.

Further Reading

The classic text by Raiffa [333] provides a very readable introduction to optimal decision making as originally envisioned by human economists. Good explanations of how these ideas may apply to animal decisions can be found in McNamara and Houston [269], Real [337], Giraldeau [136], and Dall et al. [76]. Deviation from rational decision making in animals is reviewed by Real [338] and Kacelnik [212], and in humans by Paulus [314]. Macmillan and Creelman [253] remains a standard reference on signal detection theory, and the application of these ideas to biology can be found in Wiley [428, 429]. Coding scheme strategies of animals and humans are reviewed by Hailman [155]; and Candolin [49], Partan and Marler [312], and Hebets and Papaj [166] provide classifications of coding schemes for multivariate signals. A challenging but thorough treatment of measures of information can be found in Kåhre [215], and Searcy and Nowicki [368] provide a recent review of reliability in animal signaling. An excellent review of the definitions of biological information is provided by Jablonka [189].

COMPANION WEBSITE
sites.sinauer.com/animalcommunication2e

Go to the companion website for Chapter Outlines, Chapter Summaries, and References for all works cited in the textbook. In addition, the following resources are available for this chapter:

Web Topic 8.1 *Computing "red lines"*
If one knows the average consequences of "right" and "wrong" actions in a given situation, one can calculate what the "red line" probability used in decision making should be. If one kind of error is more costly than another, the red line will have to be adjusted to make this kind of error less likely.

Web Topic 8.2 *Types of probabilities*
There are several types of probabilities. Knowing the different types and the relationships between them is necessary background if one wants to compute the amount of information in an animal signal, or wants to identify an optimal strategy when the consequences of different strategies occur probabilistically.

Web Topic 8.3 *Bayesian updating*
Here we outline the calculations that a computer or statistician would use to combine prior probabilities and a coding matrix to compute an updated probability. No animal or machine can produce a better estimate with only this information.

(continued on next page)

Web Topic 8.4 *Signal detection theory*

Here we examine in more detail the logic and mathematical basis for signal detection thresholds. The ways in which prior probabilities, relative payoffs, and costs of different kinds of errors combine to determine the optimal threshold are explained, and some additional properties of decision making that can be extracted from studying signal detection processes are presented.

Web Topic 8.5 *Prospect theory*

Prospect theory seeks to explain known decision biases in humans by arguing that decisions are made not by direct comparisons of probabilities and payoffs, but instead by comparing psychological surrogates created using curvilinear transformations of the raw probabilities and payoffs.

Web Topic 8.6 *Weber's Law and risk*

When animals scaling magnitudes according to Weber's Law try to estimate the most likely expected value of an action with several possible outcomes, that estimate is likely to be skewed toward the smallest payoff. If smaller payoffs are worse, this may lead them to be risk averse, and if smaller payoffs are better, to be risk prone.

Web Topic 8.7 *Brains and decision making*

Much recent research has sought to identify the parts of the brain that are used to evaluate payoffs, track probabilities, perform updating, and invoke biases during human and animal decision making. Integration of the theories of decision making and the underlying neurobiology is now the focus of a fast-growing field called neuroeconomics.

Web Topic 8.8 *Measures of discrete signal effectiveness*

Here we describe how to compute both forward and backward measures of discrete signal effectiveness using matrix algebra. Code for relevant Matlab routines is provided.

Web Topic 8.9 *Mutual information measures of signal effectiveness*

Here we provide an example of the computations needed to compute the average effectiveness of a signal set using mutual information.

Web Topic 8.10 *Signal detection theory and signal effectiveness*

Signal detection theory provides a third backward measure of signal effectiveness that is widely used in psychology and neurobiological research.

Chapter 9

The Economics of Communication

Overview

Communication is not free: participation imposes costs on both senders and receivers. Over evolutionary time, one expects that signal effectiveness will be fine-tuned so that the benefits of communicating exceed the costs for both parties. This evolutionary fine-tuning can be complicated, because any given signal exchange may benefit the two parties differently and in multiple ways; at the same time, the costs may be multiple and different for the two parties. A variety of economic tools have been developed to predict when and how communication is likely to be a viable strategy in animals.

Biological Economics

Merging two disciplines

A dictionary defines *economics* as "an analysis of the ways in which human societies allocate and manage resources." Biologists have extended this concept by relaxing the limitation to human societies and considering management and allocation of resources by any organism. The terms *allocate* and *manage* in the biologists' view are also interpreted more generally: while conscious decision making remains one option, biologists also apply the same economic analysis techniques to anatomies, physiologies, and behaviors that are largely heritable and thus not a function of conscious choice.

For any biological trait, including behavior, an economic analysis begins by quantifying the **benefits** and **costs** of that trait. Benefits and costs can be compared and their combined effects expressed either as a difference (benefit minus cost) or as a ratio (benefit/cost or cost/benefit). Where outcomes are uncertain, benefits and costs need to be weighted by their respective probabilities before they can be combined. The weighted combination of benefit and cost is called a **payoff.** If it is known or suspected that the subjects of an analysis could have adopted some alternative forms or values of the trait, biological economists typically measure or estimate the payoffs for each alternative to see whether the trait actually adopted by the animals did or did not have the highest payoff. The expectation of biological economics is that the trait form or value that we see in nature is the one with the best payoff among the realistically feasible alternatives. If it is not, biologists often spend a lot of research effort to find out why. Usually, it turns out that there was some important factor that was inadvertently omitted in computing the original payoffs. Many unsuspected factors contributing to biological benefits and costs have been discovered in this manner.

Biological economics can be applied to animal communication just as it can be applied to any organismal trait. However, the analysis is complicated by the fact that communication is a transaction between at least two individual animals. We thus need to compute the payoffs for the sender of a signal, and the corresponding payoffs for any receivers of that signal. If the sender could have sent any of several alternative signals in a given context, we must compare the payoffs to all parties of each of the alternative signals, as well as those of the signal sent. While all parties might agree on the optimal signal, it is also possible that the signal yielding the highest payoff for one party is not the same signal that would yield the highest payoff for one or more of the others. This is thus a much more complicated problem than studying the economics of some non-communication trait in a single individual. Luckily, a variety of economic tools and concepts have been developed that allow us to identify the optimal scenarios in nature. In fact, these approaches have uncovered principles of animal economics that appear to be just as general as those for human economics that have won Nobel Prizes. The goal of this chapter is to explain the basic logic of these economic tools and concepts, and lay the groundwork for their application in later chapters.

Individual versus evolutionary economics

In the last chapter, we examined the economics of individual decision makers and how access to reliably coded signals might help receivers make optimal decisions. In this chapter, we want to move beyond the level of individual transactions to describe what happens when different sender-receiver pairs in a population attack the same decision problem with different signals and different receiver behaviors. Where either senders or receivers have more than one option when selecting how and when to communicate, we say they have access to multiple **strategies**: here, each strategy is one way of entering into a communication transaction. Typically, senders have access to several alternative strategies for signaling, and receivers might recruit any of a number of strategies to respond to (or ignore) those signals. The payoff to a sender of adopting a given strategy usually depends, in part, on what response strategy is employed by the receiver (Figure 9.1). Similarly, the payoff to a receiver of entering into a communicative transaction will depend, in part, on the signaling strategy adopted by the sender.

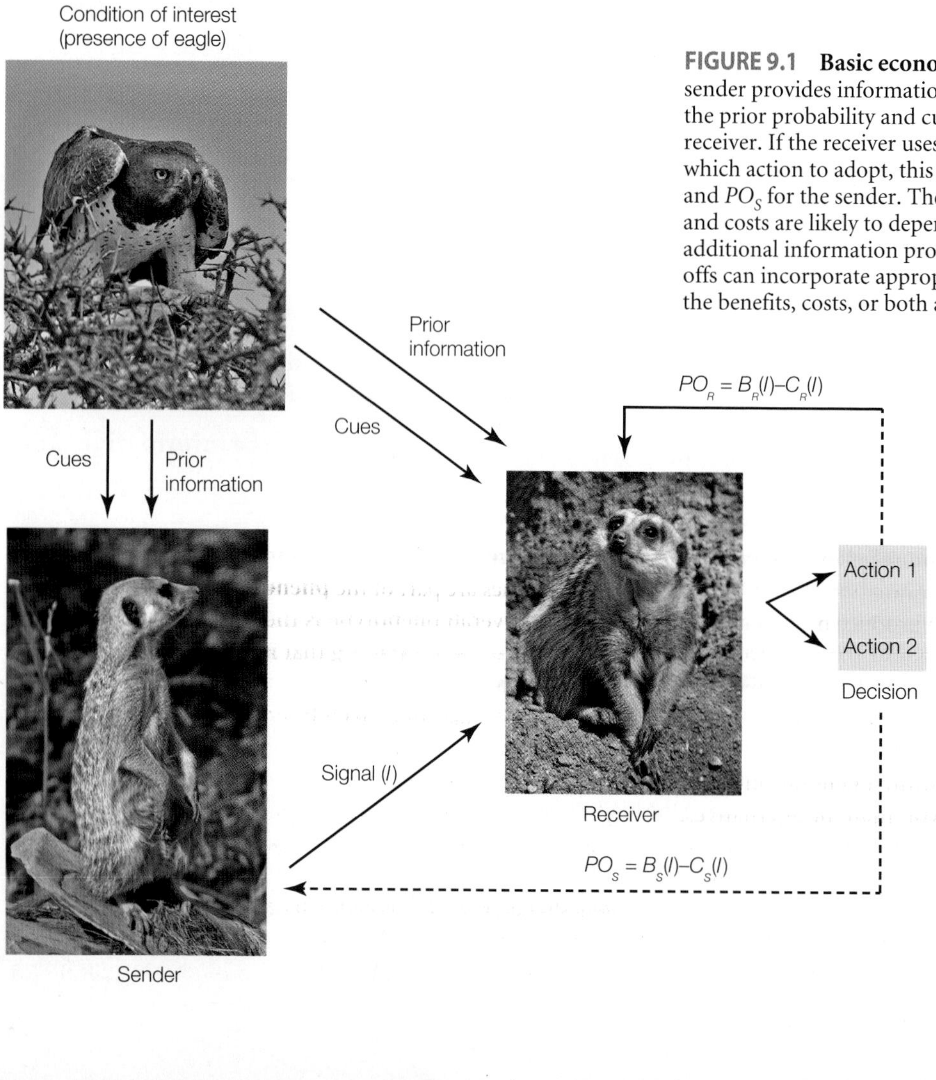

Condition of interest
(presence of eagle)

Prior information

Cues

Cues   Prior information

$PO_R = B_R(I) - C_R(I)$

Action 1

Action 2

Decision

Signal (I)

Receiver

$PO_S = B_S(I) - C_S(I)$

Sender

FIGURE 9.1 Basic economics of communication Here, the sender provides information (I) to the receiver above and beyond the prior probability and cue information already available to the receiver. If the receiver uses this additional information to decide which action to adopt, this leads to payoffs PO_R for the receiver and PO_S for the sender. These payoffs and their composite benefits and costs are likely to depend on the quality and quantity of the additional information provided by the sender. Note that the payoffs can incorporate appropriate probability discounting if any of the benefits, costs, or both are stochastic.

(A)

(B)

(C)

FIGURE 9.2   Many signal traits in animals are shaped by both sexual selection and natural selection (A) Male sharp-tailed grouse (*Tympanuchus phasianellus*) must display competitively in front of females if they want to be chosen for mating (sexual selection). However, these same displays attract eagles and other predators that exert a counterforce on display evolution (natural selection). (B) A similar trade-off occurs in male túngara frogs (*Engystomops pustulosus*), whose mating calls not only attract females, but also attract predatory bats (*Trachops cirrhosus*). (C) Female flies (*Ormia ochracea*) locate male crickets (*Gryllus rubens*) through their advertisement calls, and deposit larvae on the crickets that will eventually eat and kill them.

One goal in this and following chapters is to identify principles that can predict which combinations of sender and receiver communication strategies are most likely to be seen in nature. This will require a description of what happens over many successive generations when different pairs of senders and receivers in a population can adopt different combinations of strategies. For example, suppose that all animals in a population initially use the same combination of sender and receiver strategies. Imagine that some minority of the population begins to use a different combination of strategies for which both senders and receivers get a higher average payoff than the rest of the population. If this causes the minority senders and receivers to produce more offspring than those in the majority, and if these strategies are at all heritable, the next generation will consist of more individuals adopting the new combination of strategies, and over many generations, the new combination of strategies may replace the initial one. **Selection** occurs within a generation when one strategy has a higher average payoff than the alternatives. Darwin proposed a distinction between selection on strategies that enhance growth and survival (**natural selection**) and that on strategies that augment numbers of mates and/or numbers of effective matings (**sexual selection**) (**Figure 9.2**). The same strategy may experience natural selection in one

direction and sexual selection in another. A classic example is that of the peacock's tail; natural selection favors shorter tails, to facilitate escape from predators, whereas sexual selection favors the longer tails preferred by peahens. **Evolution** occurs when prior strategies are replaced by alternatives over successive generations. Whereas the payoffs of any given transaction characterize the economics of that particular sender and receiver, **evolutionary economics** focuses on many individual transactions, compares the average payoffs of different strategy combinations, and predicts how these affect the relative abundances of strategies in successive generations.

Strategies are part of the **phenotype** of an individual animal. The overall phenotype is the sum of all existing traits in an animal and everything that it does. One can also speak of a specific phenotype, by which is meant one of a set of alternative values for a given trait. As noted above, different individuals in a population may exhibit different phenotypes, including different communication strategies. Some of the phenotypic variation between individuals in a trait, such as the communication strategy each adopts, is due to differences in the genes that each animal carries. The entire ensemble of genes in each animal is called its overall **genotype**; different subsets of the genes in the overall genotype affect different traits. One important property of a population of animals, as

Astrapia nigra

Astrapia mayeri

Astrapia rothschildi

Astrapia stephaniae

Astrapia splendidissima

Hybrid zone

FIGURE 9.3 **Variation in adult male plumage in five species of the New Guinea bird-of-paradise genus *Astrapia*** The mountainous habitats and small ranges of these birds resulted in locally isolated populations (see separately colored ranges). Over many generations, natural selection, sexual selection, and drift have caused the populations to become different species with noticeably different male plumages. A small zone of recent overlap between *A. mayeri* and *A. stephaniae* hosts hybrids that were for a time thought to be a sixth species.

opposed to individuals, is the **genetic heritability** of a given trait. This is the fraction of a population's phenotypic variation in a trait that can be ascribed to simple genetic differences. The rest of the phenotypic variation for that trait is a result of nonadditive interactions between genetic factors and any differences in environmental influences that affect trait values, including learning and innovation.

The evolutionary change between generations in the average value of a continuously variable trait or strategy, or in the frequencies of alternative discrete traits and strategies, is called the **response to selection**. Its magnitude will depend both on the strength of selection, and on the degree to which strategies are heritable between generations. Strategies may be genetically heritable, learned culturally from other animals, or some mixture. The higher the heritability, the greater will be the response to a given strength of selection between generations. The majority of the strategies discussed in this book have a significant genetic heritability; they can thus exhibit **biological evolution**. Some traits, such as words in humans and vocal themes in certain whales and birds, are largely learned and thus reflect **cultural evolution**. It is not

uncommon for behavioral traits to show the results of both biological and cultural evolution. The dynamics of biological and cultural evolution may be different, but the basic processes of selection, replication in new individuals, and intergenerational change are essentially similar.

In real populations of animals, selection is only one factor that determines the relative abundance of alternative strategies in successive generations. Another is simple chance. While relevant payoffs of a communication strategy will affect its adopter's survival, reproduction, or both, random detrimental events—such as catching a disease or being caught by a predator—and random beneficial events—such as finding an unusually safe nest site or enjoying a very mild winter—can also play important roles. Any of these chance factors could cause an animal to survive and reproduce at a high rate despite using a communication strategy that usually yields a lower-than-average payoff. Changes in strategy frequencies across generations that are due to random events are called **drift**. Whereas selection is usually the dominant force in very large populations, both drift and selection play major roles in the evolution of strategies in small populations (**Figure 9.3**).

Selection and drift can vary the frequencies of strategies that are already present in the population, but they do not result in new strategies. With only selection and drift operating, evolution would stop once all members of a population used the same strategy. New variants are therefore a necessary condition for continued evolution. Where alternative strategies are at least partially heritable genetically, genetic

mutations can add a steady trickle of new alternatives into each generation. In most animals, genetic mutation rates are fairly low but persistent. Where alternative strategies are at least partially learned, **innovation** and learning errors can generate variation. For both genetically heritable and learned traits, **migration** between populations can constitute another source of new alternative strategies on which selection and drift can operate.

Evolutionary Models with Minimal Genetics

As we have noted, the within-generation economics of communication can become very complicated. Senders and receivers can each choose from among several alternative strategies, and their payoffs will depend upon the decisions of both parties. Senders are likely to adopt one set of strategies when they must compete to be chosen by a receiver (e.g., hungry nestlings begging to be fed by their parents or males competing to be chosen by a female) and another when they compete *not* to be chosen by a receiver (e.g., prey trying to convince a predator to chase somebody else). Senders will benefit by advertising themselves honestly in some situations, but in others, they may be better off sending dishonest signals. Receivers also may be a heterogeneous lot: even when senders try to direct signals to specific receivers, the world is full of eavesdroppers who often have different agendas than the intended receivers. If communication economics are this complicated within one generation, the evolution of alternative strategies across generations would seem even harder to predict.

Human economists face similarly complicated situations. Their typical solution is to devise an **economic model** in which the numbers of variables and alternative strategies are reduced while as much of the essence of the transactions as possible is retained. The simplification rests on making specific assumptions about what aspects of the transactions are relevant. Standard mathematics or thousands of computer simulations are then applied to the model to derive predicted behaviors and patterns. The economists then look to see how well the predictions and patterns generated by the model are realized in real human economies. Alternative models for the same phenomenon can be compared to see which one best predicts observed behaviors, and which assumptions lead to the best predictions.

Biological economists do the same thing. Complex biological processes are simplified into mathematical models that capture the context of a given process as much as possible. Standard evolutionary mathematics or simulations are then applied to each model to derive predictions and patterns, which are then compared to actual patterns seen in nature. In many cases, biologists can manipulate conditions in the field to test the predictions generated by alternative models. Evolutionary modeling has played a major role in our current understanding of animal communication. Because simplified models can have great generality, they have predicted patterns in unstudied taxa that have subsequently been verified. Evolutionary economic models will play a key role in subsequent chapters.

One major simplification used in biological economics is the assumption that the known complications of genetics at generational time scales tend to average out over longer evolutionary time scales. This simplification is known as the **phenotypic gambit** [337, 347, 497, 567, 569]. The three widely used modeling methods below typically assume some form of asexual reproduction in which heritability is present but genetic complications are absent; this allows them to focus on changes in phenotype frequencies across generations. These approaches can then be compared to more complex models that incorporate more genetic details to see if and when the phenotypic gambit is justified.

Simple optimality models

Simple optimality models assume that the benefits and costs to an animal adopting a given strategy are independent of what other parties are doing. One first determines whether benefits and costs should be combined additively or multiplicatively [311] and whether outcomes are sufficiently uncertain that components should be weighted by probabilities. One then computes the overall expected payoffs for each alternative strategy and identifies the combination with the highest expected payoff. The latter strategy is the one predicted to be dominant in nature. If it is not, researchers typically reexamine whether the right components were included in or excluded from the model; look to see whether there has been sufficient evolutionary time for the species to evolve to the state assumed by the model; or look for genetic, developmental, or phylogenetic constraints that prevent the population from having evolved to the predicted one [339, 344, 481, 568, 581, 639, 646, 665]. Because the payoff to an individual of nearly any communication strategy is, by definition, not independent of what other parties are doing, simple optimality models are rarely of much use in studies of animal communication. They have, however, been very useful in understanding strategy selection in nonsocial contexts such as how long to remain feeding in a given patch, which plant species to eat, when to molt, when to seek shelter from weather, or how much time to allocate to different activities.

Evolutionary game theoretic models

When the average payoff of adopting a given strategy depends upon the frequencies with which this and alternative strategies are currently being used in the population, a situation known as **frequency dependence**, **evolutionary game theory models** can be invoked to predict evolutionary outcomes [567, 645, 818]. The standard approach is to compute the average payoff to a "player" who adopts a particular strategy and then encounters other players adopting the same or different strategies in proportion to each strategy's current abundance in the population. Some care is required at this step to include any consequences that particular strategies may have on the values of payoffs and availability of

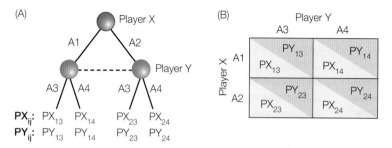

(A) Player X ... A1 ... A2 ... Player Y ... A3 ... A4 ... A3 ... A4

PX_{ij}: PX_{13} PX_{14} PX_{23} PX_{24}
PY_{ij}: PY_{13} PY_{14} PY_{23} PY_{24}

(B) Player Y ... A3 ... A4

Player X ... A1 ... PY_{13} PX_{13} ... PY_{14} PX_{14}

A2 ... PY_{23} PX_{23} ... PY_{24} PX_{24}

FIGURE 9.4 Alternative formulations of an evolutionary game In this simple example, there are two types of players (X and Y). Each has a choice of two alternative actions (A1 versus A2 for X, and A3 versus A4 for Y), and each party has to choose an action without knowing what its opponent has chosen. (A) Extensive form version of this game. The dashed line indicates that at the time Y chooses an action, it does not know what action X has chosen, and therefore at which node it is sitting. There are four possible combinations of actions, two types of players, and thus eight total possible payoff values: the row labeled PX_{ij} gives the payoffs for X and the row below (PY_{ij}), the concurrent payoffs for Y. Payoffs are labeled with subscripts indicating the combination of actions taken by X and Y respectively. (B) The same game collapsed into matrix (normal) form. There are four cells in the overall matrix corresponding to the four possible combinations of actions. Each cell is divided in half with the payoff to X for that combination in the lower left triangle, and the payoff to Y in the upper right triangle.

alternative options [406, 472]. One then looks for relative abundances of the alternative strategies that constitute an **equilibrium**—a distribution at which it does not pay for any player to switch strategies. A strategy that does better than mutants with different strategies when it is common and the mutants are rare is known as an **evolutionarily stable strategy** (**ESS**). An ESS may consist of only one dominant strategy, with the frequencies of alternative strategies declining to zero, or it can be a stable mixture of different strategies. As an example of the latter, ESS models have been invoked to explain the persistence of mixtures of different personality types in a population [128, 190, 341, 583, 753]. A given set of alternative strategies may have no, one, or many ESSs. When multiple ESSs are possible, the one that is seen in nature may simply depend on historical chance. There are two major approaches used to identify ESSs (**Figure 9.4**). The **matrix method**, also called **normal form**, summarizes the average payoffs of all possible combinations of strategies in a table [567]; the **extensive form** approach traces out all possible trajectories of interaction along branches of a tree and computes the average payoffs at the terminal points of each branch [181, 317, 415]. Extensive form games can often be collapsed into an appropriate matrix table; however, reconstituting an extensive game from the normal form table alone is usually not possible. Extensive form is also the most appropriate way to describe games with sequential moves.

ESS approaches are widely preferred over simple optimality models for studies of animal social interactions [223], and they are appropriate for nearly any case of communication. As an example, consider male katydids choosing between two alternative habitat types in which to sing and attract females for mating. One habitat may provide better acoustic propagation than the other, giving callers in that habitat larger active spaces and thus access to more females. However, as the density of callers in the preferred habitat increases, competition between males for the same set of females also increases. There will surely come a point at which calling from a branch in the lower-propagation but sparsely settled habitat is more advantageous than calling in the high-propagation but densely settled habitat. One could use an ESS model to look for the equilibrium distribution of males at which it would not pay for any individual to move to the other habitat. We shall discuss the application of ESS approaches to a wide variety of communication contexts throughout this book.

Adaptive dynamics models

Although ESS studies look for stable equilibria, not all frequency-dependent systems will evolve to an equilibrium. **Adaptive dynamics** approaches examine all evolutionary trajectories, whether they lead to equilibria or not. ESS analysis can thus be considered a subset of adaptive dynamics approaches. Dynamic analysis was largely developed to understand nonlinear systems in physics and engineering [775]. It turns out that physiological processes such as sound production, social interactions, and communication can behave in nonlinear ways and thus are logical subjects for dynamic analysis. An adaptive dynamics model begins by specifying one or more equations that compute the rate of change of each strategy's abundance per generation. One then assigns some initial frequencies for each strategy, applies the equations to predict relative abundances of each strategy in the next generation, and then uses these new frequencies as the initial conditions to predict a third generation. One can thus track the evolutionary trajectory of the population over any number of successive generations (**Figure 9.5**). Unlike ESS models that focus on stable equilibria, studies using adaptive dynamics models can uncover both stable and unstable equilibria. They may also identify **branching points** (where a trajectory can divide into multiple trajectories); **limit cycles** (recurrent cycling between alternative strategies); sudden **bifurcations** (in which the system makes a radical change in state after passing a **tipping point**); and chaotic trajectories that never retrace a prior path nor converge on a common end point [203, 206, 309, 310, 396, 408, 518, 590, 632, 735]. (See the introduction to nonlinear dynamics in Web Topic 2.8). A striking example of a strategy that requires this kind of model is found in side-blotched (*Uta stansburiana*) and viviparous (*Lacerta vivipara*) lizards [13, 755]. Each species has three different color morphs, each of which becomes dominant for a year or two before being displaced by one of the others. Each color morph has its own behavioral strategies for mating. The full cycle takes six years in the side-blotched lizard and four years in the viviparous lizard. Similar systems have been

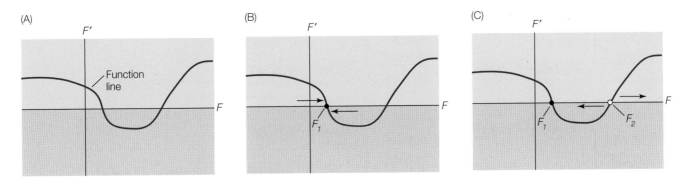

FIGURE 9.5   Graphic example of a simple adaptive dynamics model   Let F be the fraction of a population using a given signaling strategy and F' be the rate of change of F per generation (or other appropriate time interval). (A) This graph, called a *phase space plot*, shows possible values of F on the horizontal axis and possible values of F' on the vertical axis. For a specific adaptive dynamics model, only certain combinations of F and F' are likely to occur; in this example, these realizable combinations are indicated by the function line. Any evolution in this population has to move along the function line. For combinations of F and F' above the horizontal axis (pink zone), $F' > 0$: F will increase in the next time interval and the population will move to the right along the function line. For combinations below the horizontal axis (blue zone), $F' < 0$: F will decrease in the next time interval, and the population will move left along the function line. (B) Whenever the function line crosses the horizontal axis (i.e., when $F' = 0$), and, barring drift, the population will remain at the current value of F in the next time interval. This is called a *critical point*. The function in (B) has two critical points. Suppose the population has arrived at F_1. If drift causes the population to drop to a lower value of F, it will be in the pink zone favoring an increase in F and a return to F_1. If drifts leads to a higher F, the population will fall into the blue zone, favoring a reduction in F back to the critical point F_1. Thus, F_1 is a *stable critical point* or *equilibrium*: once there, the population tends to stick there. (C) Suppose another population is sitting at the second critical point, F_2. If drift causes F to fall below F_2, the population will be in the blue zone, and F will keep decreasing in subsequent intervals; if it drifts above F_2, it will keep increasing indefinitely until $F = 1.0$. F_2 is thus an *unstable critical point*. This demonstrates how the same dynamic system can lead to quite different outcomes depending on the initial frequency of the strategy. Functions more complicated than the one shown here can generate sudden jumps in frequency, cyclic patterns, and chaotic trajectories.

found in other lizards, bacteria, and sessile marine organisms [284, 454, 756].

While a one-time simultaneous exchange of signals by two animals might be modeled by matrix ESS analyses, the long sequences of exchanged signals and responses typical of mate choice, aggressive contests, or competitive begging by offspring would seem best modeled by extensive form games [414, 415]. Finding ESSs for such games is computationally challenging, and it is not obvious whether or not a population will follow a trajectory to a single ESS [181]. A subset of adaptive dynamics called **genetic algorithm modeling** may be useful in these contexts. Some minimal heritability, random mutation, and differential survival of alternative strategies as a function of ambient strategy frequencies are

built into the relevant equations of change [602]. Computer simulations then track a population with some initial composition (often randomly assigned frequencies of alternative strategies) over a sufficient number of generations that any asymptotic outcome becomes evident. This approach has been applied to some simple communication games in animals and the results suggest that, as with other adaptive dynamics approaches, more outcomes than convergence on a single ESS can occur [362, 363, 724].

Genetic Complications

Before we can turn to evolutionary models that include more detailed genetics, we need to examine briefly those aspects of modern genetics that are likely to affect evolutionary trajectories. This will also provide an opportunity to introduce some genetics-based phenomena that will be discussed at later points in the book.

Genetic concepts

GENES AND CHROMOSOMES   Animal genes are linked linearly along protein-bound **chromosomes**. Different versions of a gene that can appear at the same **locus** (location) on different copies of equivalent chromosomes are called **alleles**. The set of all genes in an organism or any discrete subset of the total is called a **genome**. Most animals divide their nuclear genomes into multiple linear chromosomes. **Homologous chromosomes** host the same loci, but not necessarily the same alleles. Depending upon the species and sex, a single cell nucleus may contain one copy of each chromosome (**haploidy**), a homologous pair (**diploidy**), or even multiple homologous chromosomes (**polyploidy**). Mitochondria in the cytoplasm also contain genetic material. Each mitochondrion of an animal cell contains 2–10 copies of a circular chromosome, and each cell contains a few to many mitochondria. During **meiosis** (the creation of gametes), all nuclear chromosomes are duplicated, and usually one copy of each chromosome ends up in a separate gamete (sperm for males, and eggs and polar bodies for females). While aligned spatially during meiosis, subsets of genes can be exchanged between homologous chromosomes so that alleles at one locus now find themselves linked to different alleles at another locus (**recombination**). The members of a given homologous pair

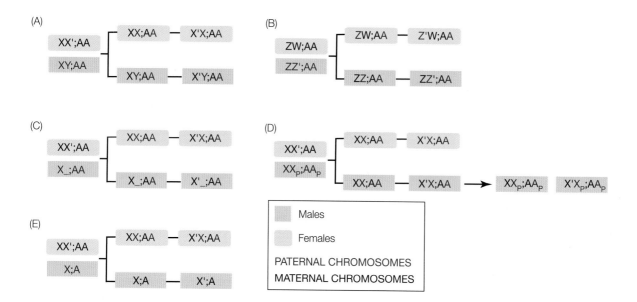

FIGURE 9.6 **Five alternative patterns of chromosomal sex determination in animals** For each pattern (A–E), the colored boxes represent a genotype, with the genotypes of two parents on the left and the genotypes of possible offspring on the right. Within each genotype, sex chromosomes are on the left of the semicolon and autosomes (non-sex chromosomes) are on the right. (A) A male heterogametic system, in which males have XY sex chromosomes and females have XX. Primes track individual chromosomes in the homogametic sex. This system is common in placental mammals, many flies, and some fish. (B) A female heterogametic system, in which males are ZZ and females are ZW. This system is present in birds, butterflies, many reptiles, and some fish. (C) An X0 system in which sperm with an X produce females and sperm without an X produce males; males *do* contribute active autosomes to both sexes of offspring. This system is found in some spiders, gnats, aphids and grasshoppers. (D) A paternal genome elimination system in which all chromosomes from the father are inactivated or eliminated during development (denoted by subscript P). Only active and retained chromosomes are passed on to offspring, and in male offspring, all chromosomes from the father are inactivated or eliminated during development. This occurs in some mites, beetles, scale insects, and midges. (E) A haplodiploid system in which all fertilized (diploid) eggs become females, and unfertilized (haploid) eggs become males. Put another way, only female offspring have fathers. This system is present in some rotifers, mites, beetles, and most Hymenoptera (wasps, bees, and ants). In most of these systems, external factors (temperature, relative body size, proximity of females, etc.) may fine-tune or even dominate the sex determination process.

(each with its own set of alleles at constituent loci) end up with members of other pairs in separate gametes (**random assortment**). Recombination and random assortment can thus create a very large diversity of gametes during meiosis. Under some circumstances, certain alleles at different loci co-occur in the same gametes at frequencies higher or lower than expected, given random assortment and reasonable levels of recombination. This tendency of certain alleles to be inherited together is called **linkage disequilibrium**.

Despite these simple components, the links between genetics and phenotypic traits and strategies can get quite complicated. The genetics of sex determination provides a good example [103, 183, 262, 265, 323, 358, 439, 550, 551, 630, 637, 729]. In asexual species or species with an asexual reproductive phase, there may be no meiosis, and offspring are produced as genetic clones of the parent. In typical sexual species, the nuclear genomes of a sperm and an egg contribute equally to an offspring nuclear genome with the adult number of chromosomes. However, offspring mitochondria, and thus most mitochondrial genes, usually derive only from the mother's egg; male sperm contribute minimally to an offspring's mitochondrial genome. A single nuclear genome produces males and females in hermaphrodites: simultaneous hermaphrodites such as terrestrial snails and sea slugs produce both sperm and eggs at the same time; sequential hermaphrodites such as slipper snails and some fishes (wrasses, parrotfish, gobies, and basses) begin life as one sex but then change to the other sex as they grow larger. The sex of many animals with separate males and females is determined by whether the members of paired **sex chromosomes** host functionally analogous loci (a **homogametic pair**) or predominantly loci with different functions (a **heterogametic pair**) (**Figure 9.6A,B**). In marsupial and placental mammals and many flies, males have the heterogametic pair (XY) and females the homogametic one (XX); in butterflies and moths, snakes, and all birds, the reverse is true (males are ZZ and females ZW). Fish, amphibians and turtles vary as to whether males or females are the heterogametic sex, and in some species of fish and frogs, both systems occur but in different populations [262, 549, 620].

Other taxa show more complicated mechanisms of sex determination (**Figure 9.6C–E**). For example, the fertilized diploid eggs of certain rotifers, nematode worms, spider mites (Tetranychidae), and insects such as thrips (Thysanoptera),

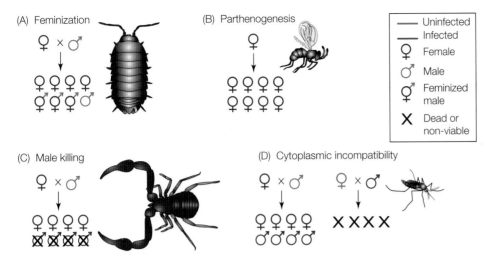

(A) Feminization

(B) Parthenogenesis

(C) Male killing

(D) Cytoplasmic incompatibility

—— Uninfected
—— Infected
♀ Female
♂ Male
⚥ Feminized male
X Dead or non-viable

FIGURE 9.7 **Alternative effects of infection by the bacteria *Wolbachia* spp. on sex determination in arthropods** The bacteria live in the host's cytoplasm and can be passed only from mothers to offspring. It is thus in the bacteria's interests to bias host reproduction in favor of female offspring. (A) Infected mothers produce normal daughters, but most sons are feminized by the bacteria to look and act like females. This occurs in some isopods, true bugs, and butterflies. (B) In certain mites, thrips, and wasps, infected females reproduce asexually via parthenogenesis, without needing or producing males. (C) In a variety of flies, moths, beetles, and pseudoscorpions, infected male offspring die without maturing or reproducing. (D) Most commonly, infected males produce no viable offspring unless their female mates are infected with the same strain of *Wolbachia*. Infected females reproduce successfully with uninfected males, and all offspring will be infected. This is found in mites, isopods, crickets, bugs, beetles, wasps, flies, moths, and butterflies. (After [839].)

some beetles (Micromalthidae and Scolytidae,) some scale insects (Coccidae, Aleyrodidae, and Margarodidae), and nearly all ants, bees, and wasps (Hymenoptera), become females, whereas unfertilized haploid eggs become males (**haplodiploidy**). In other mites (Ascoidea), beetles (Scolytidae), scale insects (Diaspodidae, Coccoidea), and midges (Cecidomyidae), both males and females begin as fertilized eggs but *all* of the chromosomes provided by sperm are inactivated or later eliminated. Some spiders, aphids, gnats, and grasshoppers begin as diploid offspring but then delete one of the sex chromosomes to create males while retaining both sex chromosomes for females [729]. Sex determination in a wide variety of arthropods (including up to 65% of all insect species) is complicated by parasitic bacteria (*Wolbachia* spp.) that are passed to succeeding generations via the cytoplasm of infected females' eggs [809]. The bacteria have evolved mechanisms to increase their own reproduction by reducing the production of male arthropod offspring relative to female offspring (**Figure 9.7**) [839]. Finally, in a variety of fish, amphibians, and reptiles, sex chromosomes are either absent or easily overridden and sex is determined by environmental factors such as ambient temperatures experienced by eggs or larvae [114, 182, 183, 517, 536, 620, 812].

GENOTYPES AND PHENOTYPES The ways in which different genes in a genome interact to influence specific phenotypic traits can also be complex. A typical diploid animal will have two alleles for any given locus. These alleles might be identical (**homozygous genotype**) or different (**heterozygous genotype**). When the animal is heterozygous for a trait, the phenotypic expression may be the result of the effects of the **dominant** allele masking any effects of the **recessive** allele. In other cases, a heterozygous genotype may show **incomplete dominance**—a phenotypic mixture of the two alleles' influences that favors one of them; **co-dominance**—an equal mixture of effects; or **overdominance**—in which the phenotypic expression is completely outside the range expected for either allele alone. When a single locus affects more than one trait, the effect is called **pleiotropy**. The value of a given trait often depends on the contributions of many loci: where a given gene's expression is independent of the activity at other loci, the pooled contributions are called **additive effects**; where the activity of a locus varies depending on which alleles are present at other loci, the complex interactions are called **epistasis** (**Figure 9.8**). The entire ensemble of allelic interactions mapping an animal's genotype onto its phenotype is called its **genetic architecture** [371].

Phenotype

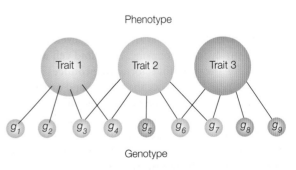

Genotype

FIGURE 9.8 **Epistasis and pleiotropy** A sample genetic architecture showing pleiotropic genes (g_3, g_4, g_6, g_7) and epistatic complexes (g_1–g_4 affecting trait 1, g_3–g_7 affecting trait 2, and g_6–g_9 affecting trait 3). Note that the same gene may be involved in both pleiotropic and epistatic processes. (After [371].)

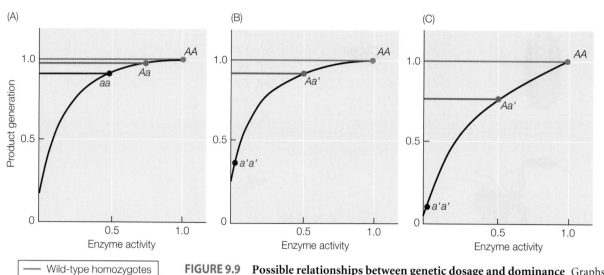

FIGURE 9.9 **Possible relationships between genetic dosage and dominance** Graphs show decelerating generation of a phenotypic product (vertical axis) given linear increases in the relevant enzyme as genetic dosage increases (horizontal axis). (A) Here, both the "wild-type" allele *A* and an uncommon mutant allele *a* produce the same enzyme, but each *A* allele produces twice as much enzyme as each *a* allele. Although heterozygotes (*Aa*) only produce 75% as much enzyme as wild-type homozygotes, the difference in phenotypic product is negligible because of the decelerating dynamics. Homozygous mutants (*aa*) produce half as much enzyme as homozygous wild types and the result is a noticeably different phenotype. The wild-type allele thus appears *dominant* to the mutant allele. (B) Here, a different mutant allele (*a′*) produces little if any enzyme. The heterozygote now produces only half as much enzyme as the wild-type homozygote; the heterozygote phenotype is somewhat different from that of the wild-type homozygote, but more similar to the wild type than to the mutant. This would likely be termed *incomplete dominance*. (C) Here the conditions are the same as in (B), except that the dynamics of product generation have a slower deceleration with increasing enzyme levels. The heterozygote generates a phenotype intermediate between the two homozygous phenotypes, a condition called *co-dominance*. (After [664].)

The properties of genetic architecture are in part a result of cellular biochemistry [6, 91, 142, 416, 443, 664, 869]. Genes provide blueprints for building protein enzymes that either run a cell's metabolism or regulate the genes that produce metabolic enzymes. The many enzymes in a cell interact to create complex branching or circular biochemical pathways: biochemical products of one enzyme become substrates for the next enzyme in the pathway. Most enzymatic pathways show a decelerating curve when product output is compared to amount of enzyme present (**Figure 9.9**). The amount of an enzyme produced by a cell depends on the **gene dosage**, the number and activity levels of alleles that typically produce that enzyme. A diploid cell which is homozygous for the gene that normally produces a given enzyme (genotype *AA*) has twice the dosage, and up to twice the enzyme production, of a cell in a different individual that is heterozygous with one allele producing the focal enzyme and another producing less of that enzyme or a different enzyme altogether (genotype *Aa*). Thus the effects of the normal allele will be considered *dominant* to those of a less effective allele; but because enzyme dynamics are decelerating, halving the dosage may produce only a small change in the amount of enzyme product. It has thus been argued that dominance among trait alternatives is a simple outcome of

enzymatic dynamics. Similarly, because nearly all biochemical products depend on a chain or loop of different enzymatic reactions, many phenotypic traits will arise through epistatic interactions between genes at different loci. Pleiotropy arises because the same enzymatic product may be a precursor to several subsequent pathways and thus several final phenotypic traits. It is not hard to see how the same sets of alleles might exhibit concurrent pleiotropy and epistasis [865]. Complex traits such as sexual behaviors or body ornaments likely involve multiple and intricate pleiotropic and epistatic effects [130].

The biochemical basis for allele interactions can have complex repercussions. As an example, we again turn to sex determination. Where one sex is determined by being heterogametic (XY males in mammals or ZW females in birds), mutations beneficial to that sex accumulate on the sex-specific chromosome (i.e., Y or W). Since many of these may be detrimental to the opposite sex, selection will favor reduced recombination between the sex-specific chromosome (Y or W) and the shared chromosome (X or Z) during meiosis [702]. Even if the sex-specific chromosome originally hosted many of the same genes as the shared chromosome, these are likely to be replaced over evolutionary time by other genes specific to that sex or even by noncoding genetic material.

The shared chromosome may exacerbate this divergence further by adding large numbers of genes specific to itself [67]. The result is that, compared to the homogametic sex, the heterogametic sex will have only half the dosage of those genes still on the shared chromosome. This creates a dosage problem since many of the genes on the shared sex chromosome have to interact with genes on the **autosomal** chromosomes (the non-sex-determining chromosomes) for normal biochemistry [231, 243, 541, 817]. It would be extremely complicated to change all of the relevant autosomally regulated biochemical pathways depending on whether a developing offspring was a male or a female. What happens instead in some species, such as *Drosophila*, is that the gene expression of the single X chromosome in XY males is doubled to match the normal expression in XX females. Mammals solve this problem by silencing one of the X chromosomes in every female cell during early embryonic development. In marsupials, it is the paternally inherited X that is silenced in all cells. In placental mammals, silencing is random, and females of placental mammals thus consist of a mosaic of cells, some with an active paternally inherited X, and some with a maternally inherited one. Birds and butterflies do not show whole chromosome dosage compensation, but instead vary expression of specific genes or sets of genes to achieve the same end [584, 817].

The differences between recombination rates and patterns of expression between the sex-specific and shared chromosomes in genetically sex-determined species create consequences that we shall encounter again in later chapters (**Figure 9.10**). Under many circumstances, the sex chromosomes appear to undergo faster and more extensive evolutionary turnover than the autosomes in the same animals [140]. For related reasons, species in which females are the heterogametic sex (birds and butterflies) are more likely to show more extreme sexual dimorphism than species in which males are the heterogametic sex (mammals) [8, 237, 551, 621, 696]. Because the shared and unshared chromosomes in species with heterogametic sexes have few genes in common, selfish mutant genes can arise on one of them that hinder the production of gametes not carrying a copy of the mutant gene's chromosome, and thus generate very unequal sex ratios of offspring (**meiotic drive**) [426, 785]. In mammals and a few arthropods, only

one allele at an autosomal locus (either from the father or the mother, depending on the locus) is allowed to be active (**genetic imprinting**). Which allele is active can affect subsequent evolution. As discussed further in Chapter 12, linkage disequilibrium effects that build up between genes coding for female mate preferences and male display traits can significantly modulate the process of sexual selection and subsequent evolution of sexual dimorphism and sexual signals

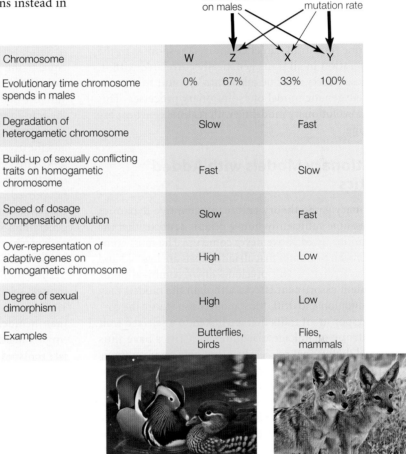

		Sexual selection on males		Male-based mutation rate
Chromosome	W	Z	X	Y
Evolutionary time chromosome spends in males	0%	67%	33%	100%
Degradation of heterogametic chromosome		Slow		Fast
Build-up of sexually conflicting traits on homogametic chromosome		Fast		Slow
Speed of dosage compensation evolution		Slow		Fast
Over-representation of adaptive genes on homogametic chromosome		High		Low
Degree of sexual dimorphism		High		Low
Examples		Butterflies, birds		Flies, mammals

FIGURE 9.10 Evolutionary consequences of different types of sex determination Whereas males are the homogametic sex (ZZ) in butterflies and birds (green column in table), they are the heterogametic sex (XY) in most flies and mammals (orange column). At the top of table, the relative influence on the four types of chromosomes of sexual selection on males and the male-based mutation rate are shown by the sizes of the arrows. These obvious asymmetries result in major differences in the evolutionary trajectories of the two sex determination systems. Since each male bird has two Z chromosomes, whereas each male mammal has only one Y, the same selective forces on males produce evolutionary changes faster in ZW systems than in XY systems. In both systems, the heterogametic chromosomes (W and Y, respectively) tend to degrade over evolutionary time and host fewer active genes; this occurs faster in XY systems than in ZW systems. Mutations that favor one sex but are deleterious in the other will tend to build up faster on the Z chromosome than the X chromosome. In contrast, mechanisms for dosage compensation tend to build up faster in XY than in ZW systems. The homogametic sex chromosomes tend to accumulate adaptive genes faster than autosomes: the difference is often greater in ZW systems than in XY systems. As a result of these many differences, strong sexual dimorphism and behavioral divergence between the sexes is significantly more common and greater in degree in ZW than in XY systems. (After [621].)

[461, 500]. Finally, faster and more extensive rates of evolution of genes on sex chromosomes may grant those chromosomes a priority role in species recognition and thus speciation processes [464, 666, 686].

It should be clear from these examples that a complete genetic model for the evolution of any particular trait or set of traits would likely be very complicated. Modes of inheritance differ for mitochondrial, autosomal, and sex chromosome genes. Species differ in whether an adult or a particular sex is haploid, diploid, or polyploid. A complete genetic model for a given case would require detailed specification of the patterns of dominance, epistasis, pleiotropy, and linkage disequilibria. When traits are genetically correlated through pleiotropy or linkage disequilibrium, selection on one focal trait can elicit indirect responses in the next generation in all traits genetically correlated with it. All of these effects must be taken into account to provide a complete genetic model of evolutionary processes. This has made evolutionary modeling with realistic genetics very challenging.

Evolutionary Models with Added Genetics

Evolutionary game theory and many adaptive dynamics models assume sufficiently simple genetic contexts that the complications listed above never come up. The most common approach presumes that all individuals are haploid and reproduce asexually. This precludes any recombination or independent assortment effects, although the models may permit mutation and drift. Most models also ignore the correlated responses of traits that are linked to a selected trait by pleiotropy or linkage disequilibria. Critics have thus expressed doubts that models focusing only on phenotypic selection and ignoring genetic constraints can make valid predictions about evolution [251, 333, 446, 528, 610, 741]. A common concern is that genetic constraints may interrupt evolutionary trajectories before they reach an ESS. The counterargument is that given enough time, there will usually be sufficient mutations to "kick" a population out of genetically "stuck" states and back onto a moving trajectory. Such a system will stop only when it reaches an equilibrium such as an ESS. An analogy is an evolutionary "streetcar" that makes periodic stops at which it sits until it acquires the right mutant passengers to get it moving again [246–248, 368, 369, 559]. This argument has not resolved everyone's concerns [520, 608, 834], and various approaches have been developed that mix the prior approaches with more realistic genetics. We describe four approaches below that will be mentioned in later chapters. All follow the lead of adaptive dynamics by formulating an equation that predicts the changes in trait or allele distributions between a parental and offspring generation. Where feasible, the methods then try to track where successive applications of this equation might lead an initial population over many generations.

> **Web Topic 9.1** *Equations of change and evolutionary models*
> Many evolutionary models begin with a basic equation that assigns an initial state, predicts intergenerational change, and then projects a population through successive generations. Here we introduce and compare some of the equations currently used in this type of analysis.

Quantitative genetics

The first approach, called **quantitative genetics**, invokes the **breeder's equation**. This equation sets the response to selection (measured as the change in the mean phenotypic value of a trait between the parental and the offspring generation) equal to the product of the trait heritability (measured as the ratio of the trait's additive genetic variance to its phenotypic variance in the population) and the **selection differential** on the trait in the parental generation (measured as the difference between the mean value of the trait for those that breed and the overall average in the parental population) [115, 253, 539]. This relationship was originally derived to guide the selective breeding of domesticated plants and animals, and has proved very effective in those contexts. The key insight is the partitioning of the causes of phenotypic change between generations into separate genetic and selective contributions, each of which can be measured independently and then combined multiplicatively to make predictions.

The traditional version of the breeder's equation ignored indirect selection on traits caused by pleiotropy. In the early 1980s, more general versions of the equation were derived that included such **polygenic** effects [153, 498–501]. The single valued response to selection in the original equation was replaced by a vector listing the responses to selection of each of a set of correlated traits. Heritability was replaced by a (matrix algebra) ratio of the additive genetic and phenotypic variance/covariance matrices (listing variances of each trait along the main diagonal and covariances between traits in off-diagonal cells), and the original single selection differential was replaced by a vector of selection differentials for the same set of traits. Each selection differential was now measured as a covariance between phenotypic values of a trait and fitness as measured by the relative reproductive successes of individuals expressing a range of trait values. Epistasis and dominance are not additive effects and thus are not included in the genetic variance/covariance matrix, but if present, can contribute to the phenotypic matrix. The equation can be further simplified by combining the phenotypic variance/covariance matrix and the selection differentials into a single vector relating phenotypic trait values and fitness. This vector can be portrayed as the slope at a particular location in an **adaptive landscape** where different combinations of phenotypic trait values define any given location in the landscape, and the height of the landscape indicates the corresponding fitness at that location (**Figure 9.11A**).

FIGURE 9.11 Evolutionary analysis using adaptive landscapes and quantitative genetics (A) Adaptive landscape for two phenotypic traits (antler size and the time spent calling by males during breeding season) in a hypothetical deer species. The two quantitative traits define the horizontal and vertical axes of the landscape. The fitness of males having a particular combination of traits is indicated by contours on the plot: darker greens indicate higher relative reproductive success than lighter greens. This example has a fitness peak in the upper left and an even higher fitness ridge in the lower right. The direction of selection at any point in the landscape is up the steepest nearby slope. (B) The yellow circle encompasses the additive genetic variation (G-matrix) in one possible deer population. The shape is circular because there is no genetic correlation between the two traits; variants are equally common in all directions around the mean (center) combination. Subsequent evolution will favor those variants higher up the slope toward the fitness ridge on the lower right, and thus move the population to this ridge over subsequent generations. (C) An alternative population has the same total amount of variation in trait combinations. However, in this population, antler size and the time spent calling are positively correlated, generating a thin ellipse instead of a circle. Selection is most likely to move the population along the longest axis of the ellipse. If it can cross the shallow valley, this population is more likely to end up on the fitness peak on the left. This example demonstrates how the combination of selective forces (landscape) and the available genetic variability (G-matrix ellipses) can predict evolutionary trajectories. (After [771].)

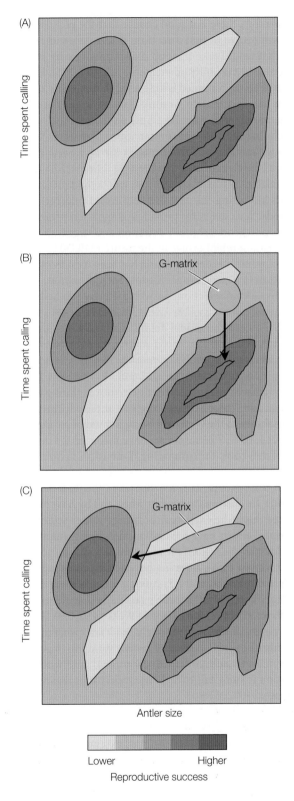

Antler size

Lower   Higher
Reproductive success

Adaptive landscape maps have a long tradition in genetics and ecology and provide a useful way to view evolutionary processes [258, 673, 754, 868]. Because a peak in this landscape represents a local maximum of fitness, a population whose mean trait values place it on top of that peak is likely to stay there. Where a landscape exhibits high ridges instead of or in addition to peaks, a population might move easily along the ridge over time, but not off of the ridge. What happens when a population with a given mean combination of traits is not currently on a peak or ridge? While average fitness will increase if the population mean moves up a slope, it cannot do so unless there are at least some variants in the population on the uphill side of the landscape. Drift is unlikely to move populations to peaks in most situations [177, 249]. This leaves current levels of genetic variation and future mutations as the main determinants of any such trajectory. Current genetic variation is summarized in the polygenic equation's genetic variance/covariance matrix, called the **G-matrix**. The G-matrix can be visualized by plotting each individual in the focal population on a graph using each trait's additive genetic values as coordinates (**Figure 9.11B,C**). Genetically uncorrelated traits will create a circular cloud of points, whereas genetically correlated traits will group into an elliptical cloud with its long axis at an angle relative to the graph axes. A circle or ellipse (for two correlated traits) or multidimensional ellipsoid (for more than two correlated traits) can then be drawn that encloses some given percentage (say 95%) of the points.

Projecting this ellipse or ellipsoid onto the adaptive landscape while preserving its shape and orientation allows one to see whether the long axis of the G-matrix, which provides the greatest potential for evolution, is pointed at a nearby adaptive peak or not. If this long axis points upslope toward a nearby peak, (and is thus parallel to the local selection gradient vector), there is a good chance that the population can and will evolve in that direction; if it is perpendicular to paths that would take the population to a nearby peak, that peak may be currently unattainable [25, 404, 771]. If the time interval being examined is long enough that many small mutations can accumulate, or mutations of large effect are common, one can plot a similar ellipse on the adaptive landscape enclosing the different combinations that are likely to arise through mutations. This **M-matrix** can be viewed in the same way as the G-matrix: if mutations at different loci tend to be correlated, the M-matrix ellipse will be flattened with a long axis indicating the path of likely evolution over long time periods. Such diagrams provide a graphical representation of the polygenic breeder's equation and suggest how a given population might evolve in the future [249, 502].

The polygenic breeder's equation has been used to model the evolution of a wide range of traits including male display signals, sexual selection, growth patterns, life history, and body part allometry [18, 129, 423, 462, 473, 486, 498, 500, 757]. To keep the model details manageable, most of these studies assumed that the G-matrix remains constant between generations or evolves at most slowly, there is no net effect of epistasis or dominance, and mutation rates are high enough to replace the genetic variation removed during each bout of selection. These assumptions have attracted criticism, with the stability of the G-matrix being a topic of particular concern [245, 657]. The responses have included the development of new tools for comparing G-matrices before and after selection [25, 127, 771], the measurement of G-matrix stability in real systems [212, 296, 655], and the undertaking of computer simulations to predict when and by how much G-matrices might be expected to change under selection [700]. Parallel efforts have focused on identifying **quantitative trait loci** (all the regions on an organism's chromosomes that affect a given trait), assessing how quickly these regions accumulate mutations, and quantifying the relative roles of regions that directly affect a trait versus those that are modifiers of direct effect regions (**Figure 9.12**) [450, 539, 711]. Finally, a variety of efforts have sought to identify what maintains the surprisingly high levels of genetic variation in populations even under substantial selection [434], and whether this variation is sufficient to facilitate trajectories that reach adaptive peaks [502, 516].

Extended quantitative genetics

A second approach, which we shall call **extended quantitative genetics**, also relies on the polygenic breeder's equation but drops one or more of the simplifying assumptions. One tack relaxes the presumption that epistasis is negligible over evolutionary time. The most common strategy is to include at least some epistatic interactions in evolutionary models [130, 154, 370]. A distinction is then made between positive epistasis (where a mutation enhancing the function at one locus enhances the functional performance at other loci in an epistatic complex), and negative epistasis (where a beneficial mutation at a given locus decreases functional effectiveness of

FIGURE 9.12   Quantitative trait locus (QTL) mapping of male song and female song preferences   Songs of two species of Hawaiian crickets, *Laupala paranigra* (*Lp*) and *L. kohalensis* (*Lk*), differ in pulse rates, and females prefer their own species' song. DNA from each species was fragmented to identify 90+ markers distributed throughout the genomes [612]. Hybridization between the species then provided a diversity of marker combinations from which linkages between markers could be computed [750]. The resulting linkage map units (usually equivalent to chromosomes) are shown at the bottom of the graphs for *Lp* (A) and *Lk* (B). Correlations across all the hybrid individuals between the presence of a particular marker and either a species' female preference or its male song were scaled relative to chance predictions using a logarithm of the odds (LOD) method [106]. LOD values of 3.2 (dashed line) or higher imply that genes closely linked to (but not necessarily including) the associated markers contribute significantly to that trait; these loci are indicated with asterisks. In both species, multiple QTL sites were significantly associated with male song, but only one site was significantly associated with female preferences (Lp1 and Lk1). Significantly, this female preference QTL (dark red zone) overlapped one of the male song zones (light red zone) in each species, and the relative locations of this overlap region were similar in the two species. The fact that at least one region of these crickets' genomes affects both male song and female preferences may help explain the rapid speciation of this genus in Hawaii.

other loci) (**Figure 9.13**). These models suggest that interplay between pleiotropy and positive epistasis can significantly change the G-matrix over successive generations, and jointly fuel evolutionary trajectories by providing new genetic variation [155, 371, 372, 648, 649, 825, 865]. In contrast, negative epistasis tends to stabilize the genetic architecture and make it resistant to selection, a process called **canalization**. A number of experimental systems have been used to test the predicted ability of epistasis to modulate evolutionary trajectories [469, 537, 757, 833].

A more ambitious form of extended quantitative genetics abandons all of the usual genetic simplifications. Properties such as epistasis, dominance, linkage disequilibria, changing G-matrices, and polyploidy are explicitly incorporated into evolutionary models. This can only be accomplished with careful tracking of genetic variances, covariances, and even the shapes of trait and allele frequency distributions across generations. Modern computers now make such tracking of statistical details feasible, and general protocols and equations for predicting intergenerational change are available [50, 119, 120, 160, 463, 808]. This approach has been applied to several physiological [460] and behavioral contexts including the evolution of mate choice and social cooperation [199, 301].

Extended adaptive dynamics

A third approach extends adaptive dynamics approaches by incorporating genetic correlations between multiple traits. Although most applications still assume asexual haploids, the relevant intergenerational equations now include an M-matrix summarizing the variances and covariances among likely mutations [22, 206, 229, 518, 521, 572]. While

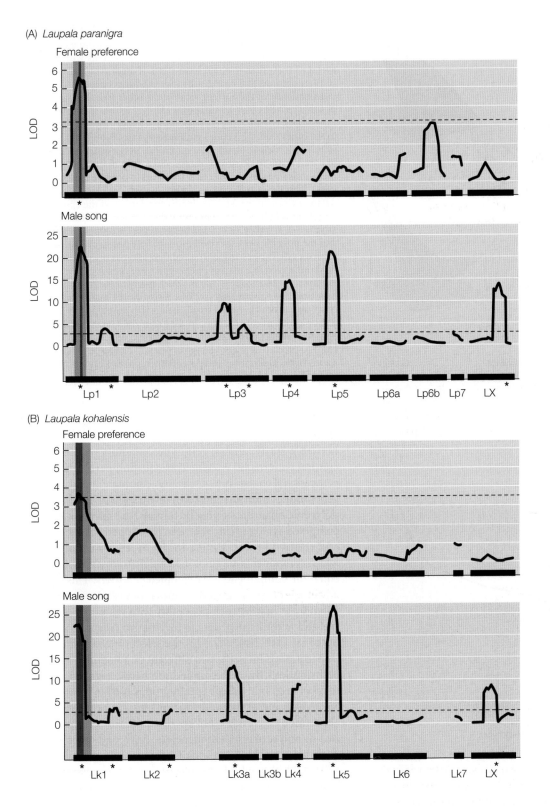

(A) *Laupala paranigra*

Female preference

Male song

Lp1   Lp2   Lp3   Lp4   Lp5   Lp6a   Lp6b   Lp7   LX

(B) *Laupala kohalensis*

Female preference

Male song

Lk1   Lk2   Lk3a   Lk3b   Lk4   Lk5   Lk6   Lk7   LX

quantitative genetics models only classify predicted outcomes as stable or unstable, adaptive dynamics models allow the discrimination of many more classes of outcomes and a wide variety of evolutionary trajectories [21, 309, 310]. Also unlike quantitative genetics models that focus on relative abundances of different alleles and traits, adaptive dynamics models explicitly track absolute population abundances and thus can incorporate ecological constraints. However, the inclusion of population ecology rests on the convenient assumption that mutations are largely of small effect and occur only at low rates. According to this assumption, ecological dynamics occur at much faster rates than evolutionary dynamics. A recurrent goal of extended adaptive dynamics is the identification of widely applicable dynamic measures that

(A)

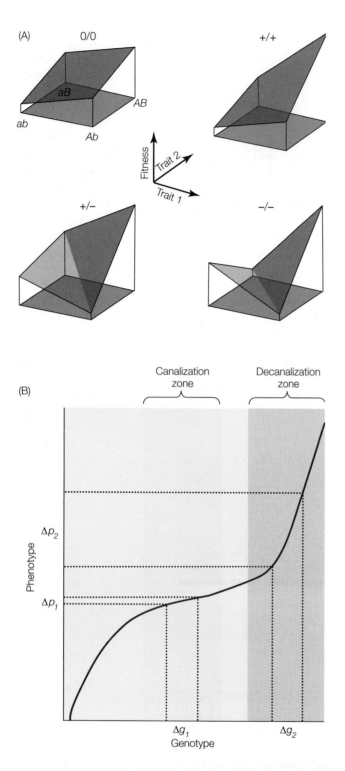

FIGURE 9.13 **Patterns and effects of epistasis** (A) Adaptive landscapes for two phenotypic traits: trait 1 varies from *a* to *A*, and trait 2 from *b* to *B*. The population currently sits at *ab* with a lower average fitness than it would have at *AB*. If there is no epistasis (0/0), mutations that change one trait produce changes in fitness independently of the current value of the other trait. In mutually positive epistasis (+/+), mutational changes in one trait toward the fitness peak enhance any corresponding changes toward the peak in the other trait. This combination accelerates evolutionary trajectories. It is also possible that mutations in the genes underlying one trait result in negative epistasis and a drop in fitness, whereas mutations in genes for the other trait lead to positive epistasis (+/−). This limits the possible evolutionary trajectories from *ab* to *AB* to one side of the adaptive landscape. Finally, if the epistatic interactions between the genes for the two traits are mutually negative (−/−), a "valley" is created between the current population and the possible peak that evolutionary trajectories are unlikely to cross. (B) The consequences of different types of epistasis can be summarized in a map of phenotype versus genotype. In this example, the blue zone is one in which many of the relevant genes have negative epistatic interactions: a significant change in genotype (Δg_1) produces only a small change in phenotype (Δp_1). This is thus a zone of canalization. In the green zone, many positive epistatic interactions cause the same amount of genotype change (Δg_2) to produce a major change in phenotype (Δp_2). This is thus a zone of decanalization and potentially rapid evolution. (A after [469]; B after [371].)

changes across generations in the relative frequencies of any heritable entity: alleles, genotypes, phenotypes, and even cultural patterns [279, 280]. This equation partitions cross-generational change into a selection component and a cross-generational transmission component [670–672, 681]. Like polygenic quantitative genetic models, the selection term is computed as a covariance between the property being tracked and relative fitness. Unlike in the breeder's equation, in which components are combined *multiplicatively*, in the Price equation, components are combined *additively*. This has the advantages that (1) uncertain outcomes can be accommodated by weighting each possibility with relevant probabilities and adding the products together to create an average outcome [340, 344, 347]; (2) the selection component can be partitioned into effects at different population levels (e.g., contributions through competition between groups versus that between individuals within groups [367, 671, 681, 824, 857, 859]); and (3) the additive transmission component can accommodate epistasis, linkage disequilibrium, nonrandom mating and sexual selection, meiotic drive, and other genetic complications that most quantitative genetics models avoid [280, 281]. When both the selection and transmission components are present, the analysis is exact and complete. However, many applications assume the transmission component to be negligible as part of a phenotypic gambit. This can lead to some of the same limitations noted earlier.

A disadvantage of the Price equation is that it only predicts information about the change in the mean value of the tracked entity and provides little information about variation in entity values in the offspring generation [279, 280, 301, 347]. One would need such details to track evolutionary

can predict the type of trajectory an evolving system will take. Applications of such measures have provided new insights into the processes of dietary selectivity, flowering seasonality, life history strategies, sex determination, sexual selection, altruistic behavior in viscous populations, polymorphism, and speciation [207, 208, 244, 517–521, 811].

The Price equation

The final approach to modeling the evolution of complex traits invokes the **Price equation**, which can be used to track trajectories over many generations. Quantitative genetics dodges this problem by assuming a constant G-matrix, and adaptive dynamics ignores recombination and other genetic complications as sources of distributional variation by assuming asexual reproduction. Only the extended quantitative genetics approach includes all contributions to trait variation in successive generations. Although usually applied only to single transitions between generations, the Price equation has proved extremely useful in defining the conditions under which various traits will or will not evolve. Applications include evolutionary models of sex ratio determination, differential dispersal of the two sexes, social interactions among relatives, dynamics of social networks, and competition between versus within groups [278–281, 302, 344–346, 349, 350, 680, 683, 786].

Comparing evolutionary modeling techniques

So: Which of the evolutionary approaches (summarized in Table 9.1) is best for studies of animal communication? Proponents of each of the approaches are quick to identify the shortcomings of competing approaches: ecologists and geneticists don't like the low mutation rate assumptions of adaptive dynamics [1–3, 51]; adaptive dynamicists argue that evolutionary game theorists and quantitative geneticists are obsessed with equilibria, are blind to alternative evolutionary trajectories [207, 208], and ignore relevant ecological feedbacks [591, 592]; extended quantitative geneticists disparage

assumptions of constant G-matrices and no epistasis [49, 371, 450, 463, 657]; and population geneticists consider the inability of the Price equation to track trajectories over many generations to be a serious limitation [301, 528, 633]. Most individual animals (and genotypes) are not locked into a single strategy their entire lives, but can vary strategies conditionally within a range of phenotypic plasticity called the **norm of reaction** [330, 405, 408, 736, 848]. Ignoring phenotypic plasticity makes evolutionary modeling simpler, but it can seriously compromise the validity of predictions.

There is no question that detailed genetic models that incorporate all complications are the ones most likely to make accurate predictions about evolution. However, many of the relevant parameters are difficult and even impossible to measure in field studies. On the other hand, one can usually identify relevant phenotypes, relate these to relative fitness, and get some sense of heritabilities in real populations. The question is then how often these more measurable parameters suffice for evolutionary studies. As we shall see in subsequent chapters, the answer is: reasonably often. It is also telling that the different accounting methods applied to the same problem often give surprisingly equivalent results: analyses of major topics affecting animal communication such as runaway sexual selection [18, 423, 461, 500, 521] and biased cooperation among genetic relatives [280, 281, 301, 344, 350, 364, 680–682] usually lead to similar conclusions regardless of which of the above methods is invoked. For some topics,

TABLE 9.1   *Summary comparison of evolutionary model methods*

Model	Advantages	Limitations
PHENOTYPIC GAMBIT	Simplifies models by assuming haploid genetics where "like begets like"	Cannot handle some evolutionary processes; largely focuses on single traits
Simple optimality	Payoffs assumed to be independent of what other individuals are doing	Limited to nonsocial phenomena and thus rarely relevant to communication
Evolutionary game theory	Allows payoffs to depend on what other individuals are doing; focused on finding stable equilibria	Has no explicit way to handle non-equilibrium evolutionary trajectories
Adaptive dynamics	Can track any evolutionary trajectory whether toward equilibrium or not	Requires low mutation rates and faster population dynamics than evolutionary change
AUGMENTED GENETICS	Greater genetic reality than phenotypic gambit	Increased complexity and interpretative difficulty
Quantitative genetics	Examines evolution of multiple traits at once, allowing for correlations between them	Assumes G-matrix constant over evolutionary time; ignores nonadditive interactions (epistasis)
Extended quantitative genetics	Allows G-matrix to change over time; allows for some nonadditive interactions	Computational complexity and interpretive difficulty
Extended adaptive dynamics	Tracks interacting trajectories for multiple traits; incorporates ecological parameters	Assumption of low mutation and slow evolutionary rates may be unrealistic for behavioral traits
Price equation	By using additive instead of multiplicative model, can incorporate any kind of genetic complication	Cannot predict variances in next generation, limiting predictions to one-generation transitions

the dynamics when not at an equilibrium can be relevant, and then the choice of method can matter. But in general, it is reassuring that multiple modeling approaches applied to the same topic often yield similar outcomes. We have presented the alternative methods here in some detail because each of them is cited at some later point in this book. Taken in combination, these multiple approaches have created a powerful set of predictions and cross-checks, and the result is a very solid theoretical background for economic studies of animal communication.

Evolutionary Currencies

All of the above models require an appropriate currency to compare the payoffs of alternative strategies. Whether one wants to track changes in allele, genotype, or phenotype frequencies across generations, success will always be limited by each individual's production of offspring. A suitable measure of the reproductive success of an individual is called its fitness. This term derives from Spencer's and later Darwin's references to "survival of the fittest" as the general outcome of natural selection and subsequent evolution [193]. This might be better phrased as "survival of the fitter" because selection can only sort through those alternatives that are present, and an absolutely "fittest" strategy that would beat all possible alternatives may not necessarily appear. Fitness is now perceived as a relative currency comparing the reproductive success of realizable alternatives [407]. Generally, we expect the version of a trait that confers the highest relative fitness to dominate in successive generations [670–672].

Unfortunately, there is no "one size fits all" measure of fitness. Natural populations of animals vary in size, density, structural stability, spatial distribution, life histories, and genetic composition. Each of these factors can affect the suitability of different measures of fitness. As we have seen, all evolutionary models invoke their own suite of limiting assumptions and simplifications. Because different treatments make different assumptions, appropriate measures of relative fitness may vary, depending upon the modeling strategy. In some cases, these give similar predictions because their respective assumptions are mutually satisfied. But this is not always the case, and it can be important to select that measure whose assumptions are most compatible with the model [72, 107, 108, 196].

Measures of relative individual fitness

The two most commonly cited measures of relative fitness are derived from demographics [141, 524, 768]. Where a focal population is relatively stable in size (namely, numbers of individuals), the total number of offspring that a given individual contributes to the next generation over its lifetime, R_0, can be a useful measure of fitness. In a stable population, the global average should be just that required to replace each individual. Because sexually reproducing diploid animals require two parents per offspring, it takes two offspring on average to replace each mated pair. If sex of offspring is not considered, the population average of R_0 should thus be 2.0. Since half of these offspring will be of the same sex as any one parent, one can alternatively count the number of offspring of the same sex produced by an average parent. In this case, the expected R_0 for a stable diploid population is 1.0. Male reproduction tends to be highly variable in most species; as a result, the number of daughters produced by the average mother is most often used to compute a population average. However tallied, animals with fitter strategies will have higher values than the population R_0 and those with less fit strategies will have lower values. **Relative fitness** for an individual is then the ratio between its value and the population average. Theoretically, R_0 should be computed by combining average annual survival and reproduction rates over the potential lifetimes of all individuals in the population. In practice, it is estimated by measuring the average **lifetime reproductive success** of a sample cohort for a given strategy [164, 622].

If the population is increasing or decreasing, the timing of offspring production can be as important to relative fitness as the total number produced in a lifetime: a female who begins breeding earlier than the average in an increasing population will produce reproducing offspring sooner than later breeders and thus have more total descendants. In a declining population, females who start breeding later than the average have an advantage. An appropriate measure of relative fitness is then the **finite rate of increase**, λ, which measures the exponential increase or decrease in population size assuming a constant per capita rate of change. It equals 1 in a stable population, is greater than 1 in a growing population, and is less than 1 but nonnegative for declining populations. One can also compute an individual finite rate of increase, λ_{ind}, that takes into account not only how many offspring one individual produces, but also when during its lifetime it produces them [108, 573]. As with R_0, the average λ_{ind} values for animals adopting a particular strategy can be compared to the population average or to the average values for alternative strategies to predict evolutionary trajectories. In some cases, R_0 and λ_{ind} will give similar predictions (**Figure 9.14**); in others, the greater sensitivity of λ_{ind} to age of sexual maturity as well as a different weighting of other life history parameters can generate different predictions [638].

While the operational dichotomy of using R_0 for relative fitness in stable populations and λ in changing ones seems simple, it is not always easy to decide whether observed fluctuations in population size reflect chance variation around a stable average, a recurrent cycle, or instead some long-term trend. Evolutionary predictions can differ between and within measures, depending on the age at which offspring are counted [108]. Other factors can make the proper choice of measure difficult [72, 107, 588]. For example, where competition mostly affects reproductive rates, R_0 may be the better measure; where competition reduces survival of all age groups equally, λ is likely to be the better measure [614]. One may thus have to assess *how* competition affects individuals before selecting a fitness measure. Measure selection becomes even more complicated when there is significant spatial or temporal

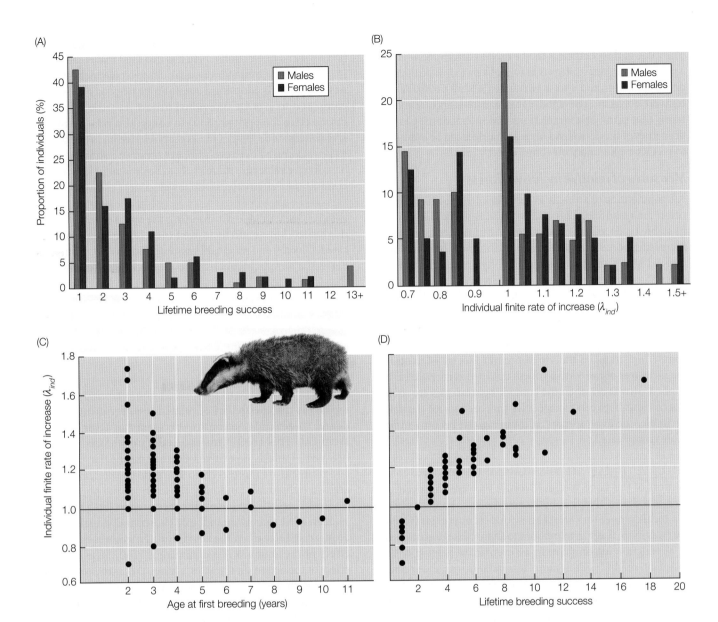

FIGURE 9.14 **Individual fitness measures in European badgers (*Meles meles*)** (A) Relative proportions of males and females producing specific numbers of offspring in their lifetimes (an approximation of R_0). Note the strong skew in lifetime breeding success around the value expected for a stable population (2.0, given that only half of the offspring will be of the same sex as the focal parent). Many individuals do not even replace themselves, but this deficit is more than made up by a few others that produce far more than needed for their own replacement. (B) The individual finite rates of increase (λ_{ind}) for the same individuals. Note the peaking of the λ_{ind} distribution at the stable population expectation (1.0), but again, there is a long tail in the distribution toward higher values. (C) The sensitivity of λ_{ind} to the age of first breeding: while variation is higher for early breeding, values of λ_{ind} tend to be higher the earlier an animal begins breeding. (D) The relationship between λ_{ind} and lifetime breeding success in this population. The plot would be a straight line if the two measures were equivalent; the slight curvature at higher values is due to fact that λ_{ind} weights offspring produced later in life less than those produced earlier, whereas lifetime breeding success counts all offspring equally. The curvature in this study is only slight, suggesting that either measure might be suitable for correlating alternative strategies with fitness. (After [225].)

heterogeneity in environmental conditions or when the relative fitness of a given strategy depends on the abundances of competing alternatives (frequency-dependence) [712].

Alternative fitness measures can be derived from the relevant dynamics. One approach extracts measures that predict the ability of a mutant strategy to invade a resident population whose members are using an alternative strategy. The relevant equations allow for nearly any kind of dynamics, including nonlinear processes, and although the extraction can be complicated, a measure equivalent to the finite rate of increase for the invading mutant can be computed. This measure is called the **invasion coefficient** in evolutionary biology and the dominant **Lyapunov exponent** in nonlinear dynamic analyses. The value of this coefficient can be used to determine whether a particular mutant strategy is or is not likely to invade the resident population, and if so, at what rate [589, 590, 690]. Where traits are polygenic, relevant dynamic equations include matrix terms that summarize transitions and interactions. Certain properties of the derivatives of these matrices can be examined to predict evolutionary outcomes directly [521].

Measures of relative inclusive fitness

All populations of animals experience intraspecific competition for resources. This competition is particularly acute in stable or decreasing populations, but is even present in increasing populations. Differences in fitness between animals adopting alternative strategies often reflect differential success in competition for resources. Given limited resources, the less competitive animals will have fewer descendants and lower fitness than the more competitive ones. One strategy's gain must lead to another's loss, and a loss by one strategy provides an opportunity for gain by another [509]. For competing strategies that do not require social interaction, the measures of individual fitness outlined earlier are easily applied. However, when the competing strategies are social interactions between conspecifics that are at least partially heritable, the evolutionary accounting can become much more complicated.

INCLUSIVE FITNESS   Consider two potential resource competitors of the same species. Because they are conspecifics, there is a chance that they both carry the same strategy genes. If a focal competitor (*actor*) knew that its opponent (*recipient*) was more likely than random to share strategy genes with it, a reduction in competition, or even active helping of the recipient, might result in more descendants with those genes than if the actor ignored this information. Similarly, if the actor knew that the recipient was less likely than random to share its strategy genes, aggressive and harmful behaviors towards the recipient might reduce the latter's survival and reproduction, releasing resources for the offspring of the actor and others carrying the same strategy genes as the actor. To track the evolutionary success or failure of genes promoting social actions, one thus needs to integrate the effects of the social action on the survival and reproduction of all relevant participants.

Several types of accounting have been proposed [364, 365, 678, 721, 790]. One approach, which sums the effects of all cases of a given social action on the fitness of a focal individual, measures what is known as **neighbor-modulated fitness**. An alternative measure, called **inclusive fitness**, sums the effects of a focal individual's performance of the social action on its own and all recipients' fitnesses. With either method, each effect is discounted by some measure of the probability that the affected individual shares the relevant genes before it is added to the total. In the second method, the discounted sum of the fitness effects of a particular social

action is added to what the actor's own fitness would be if it had no social interactions at all; it is this total that is called the inclusive fitness of that individual. The average inclusive fitness for animals adopting a common strategy or role is then compared to average individual fitness (i.e., the measures discussed in the prior section) to compute a **relative inclusive fitness** value [344]. One expects relative inclusive fitness to be maximized over evolutionary time, and strategies that have higher values are more likely to be present in future generations than alternatives with lower values [344, 364].

HAMILTON'S RULE   Computing the sums for neighbor-modulated fitness and inclusive fitness can be quite complicated. Given some reasonable assumptions, W. D. Hamilton derived a simple inequality, now called **Hamilton's rule**, that follows from both methods and predicts the direction of evolution from minimal data [364, 365]. Hamilton's rule states that genes that cause an actor to perform an action that causes a change −*C* in its own fitness while altering the recipient's fitness by an amount *B* will be favored in subsequent generations if

$$rB - C > 0$$

where *r* is the **coefficient of relatedness**, a measure of the degree of genetic relationship between the actor and the recipient. This coefficient is usually normalized to vary between −1 and 1. A positive value of *r* means that the actor and recipient are more likely to share the relevant allele than are two individuals randomly chosen from the population, whereas a negative value means they are less likely to share such an allele than two randomly chosen individuals. Hamilton's rule focuses on the actor, and *C* measures the **direct effect** of the action on its own fitness [110]. The **indirect effect** of the action on the actor's fitness, *rB*, is the equivalent change in the actor's fitness that would be needed to produce the same increase or decrease in the number of gene copies that is created by the change in the recipient's fitness, *B*. The sum of the direct and indirect components indicates the change in the inclusive fitness of the actor, given the action. Hamilton's rule can be derived using many of the modeling methods listed earlier with the same result [139, 145, 153, 279–281, 301, 344, 364, 365, 510, 561, 595, 677, 678, 680–682, 684, 721, 786, 788]. Note that *r* is a best guess that two animals share a given allele; the appropriate formula for its estimation can vary with the spatial structure of the population, whether the available data are allele or phenotype frequencies, and the appropriate measures of *B* and *C* [301, 720, 721, 786]. Two common approaches to measuring *r* are demonstrated in **Figures 9.15** and **9.16**. Hamilton's rule always gives a correct accounting if one knows the underlying genetics and measures *B* and *C* as effects of specific alleles or allele combinations on fitness. It may not always give the right answer if one invokes the phenotypic gambit and computes *B* and *C* as effects of alternative phenotypes and strategies [280, 301, 344, 680, 681]. In these cases, adjustments in the

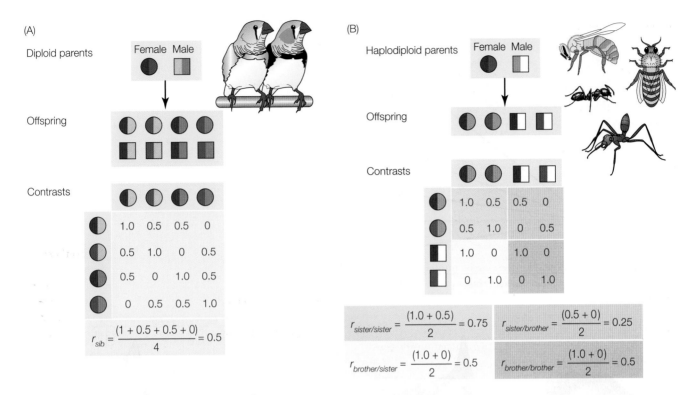

FIGURE 9.15 Genealogical (pedigree) computation of the coefficient of relatedness (r) Here, r estimates the average probability that two individuals share an identical allele inherited from a common recent ancestor. (A) Diploid female and male parents, typical of most vertebrates. Assume the two parents have no recent common ancestors, and that the male parent is indeed the father of all family offspring. There are four possible offspring genotypes, which occur equally in sons and daughters. The contrast table shows the probability that an autosomal allele present in the genotype of any one of the offspring listed in the rows (here females, but it could be either sex) is also present in a given sibling genotype listed in the columns (again, either sex possible). Assuming each type of offspring is equally likely, the average relatedness between the individual on the left and a randomly selected sibling is the sum of the corresponding row in the table divided by 4.

For all rows, $r_{sibling} = 0.5$. (B) Hymenoptera (wasps, ants, and bees) have a haplodiploid sex determination system: females are diploid, whereas unfertilized eggs become haploid males. If a female parent mates with only one male, there are two possible daughter genotypes and two possible son genotypes. Contrasts here are more complicated than in the diploid case: the average r between a daughter and any one of her sisters is 0.75; between a daughter and her brother, 0.25; between a son and his sister, 0.5; and between two brothers, 0.5. Note the asymmetry in brother/sister relationships depending upon which is the focal individual (row marker). Because they are also essentially "haplodiploid," genes on the X chromosome in mammals and the Z chromosome in birds will show relatedness patterns identical to those of autosomal genes in the Hymenoptera.

definitions of r, B, and C—and where fitness effects are not additive, the addition of a third interaction term—will usually maintain the predictive power of Hamilton's rule [338, 344, 561, 594, 679–681, 721, 745, 759, 789].

Despite its apparent simplicity, Hamilton's rule allows for a variety of alternative evolutionary trajectories [274, 298, 365, 366, 509, 512, 515, 721, 846, 847]. For example, any of B, C, and r can take either positive or negative values. There are four combinations that can lead to $rB - C > 0$. When all three terms are positive, ($B > 0$, $r > 0$, and $C > 0$), then the actor's actions decrease its own fitness (since $-C < 0$), but help increase the fitness of the recipient ($rB > 0$). If r or B is sufficiently large, $rB - C > 0$, and the action will be favored. This is called **altruism**; the actor reduces its own fitness to increase the fitness of another individual. If $B > 0$ and $C < 0$, then the actor experiences a "negative cost" which is equivalent to a

positive benefit, and thus both parties benefit from the action. This is called **mutualism** and is always favored for positive r, and sometimes even for negative r as long as r and B are sufficiently small. When both B and $C < 0$, the actor gains at the expense of the recipient's loss: the relevant action is then **selfish**. This is always favored when $r < 0$ and when r is small enough if $r > 0$. Finally, when the recipient is sufficiently dissimilar to the actor ($r << 0$), it may pay for the actor to perform an action at a cost to itself ($C > 0$) that harms the recipient and reduces its fitness ($B < 0$). This is called **spite**. A negative relatedness ($r < 0$) is not that unthinkable: if the average relatedness in a population is 0, and an individual is related to itself as $r = 1$, it follows that its average relatedness to any other member of the population must be negative. The magnitude of this negative relatedness is larger if the population size is smaller. Thus spite is predicted to be more likely

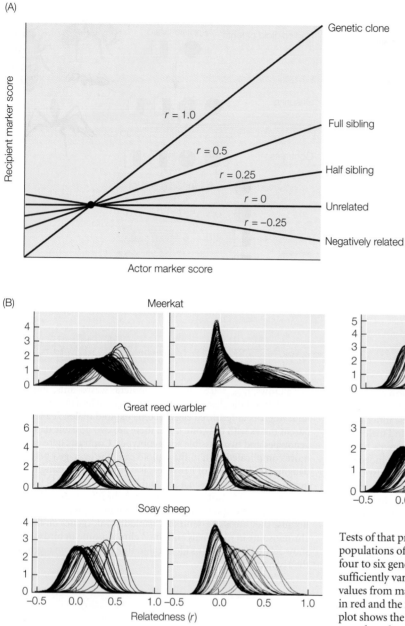

(A)

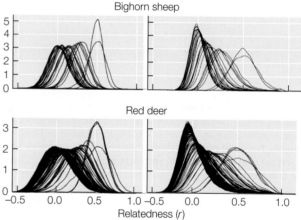

FIGURE 9.16   Regression computation of the coefficient of relatedness (r)   (A) Variable enzymes or DNA markers are scored in potential actors and recipients. Scores for a particular class of recipients (e.g., neighbors, cousins, or offspring) are then plotted against corresponding scores of a likely set or class of actors. The slope of the best-fit regression line that passes through the population mean is then an estimate of the average r between these recipients and actors [79, 540, 679, 745]. Expected lines for some classical pedigree relationships are shown. Full siblings have the same two parents; half siblings share only one parent. (B) If parents are truly unrelated, as in Figure 9.15, pedigrees and regression methods should produce the same r values for a given type of relationship.

Tests of that proposition are graphed here using five well-studied populations of birds and mammals whose pedigrees are known over four to six generations and for which 8–30 microsatellite loci were sufficiently variable for scoring [185]. Two methods of estimating r values from markers were used: the results of one [679] are shown in red and the results of the other [540] in green. Each curve in each plot shows the relative abundances of possible relatedness values in a random draw of 10,000 pairs of genotypes from a wild population with the measured allele frequencies. Curves for repeated random draws from this same population are overlain. Although separate peaks for unrelated (0), half-sib (0.25), and full-sib and parent-offspring (0.5) pedigree relationships are usually visible, there is considerable variation around the expected values. This variation indicates that pedigree assumptions of minimal prior drift, selection, or inbreeding may often be violated in the wild. Pooling the five species, between 91% and 99% of randomly drawn pairs were found to be unrelated ($r < 0.25$); only 0.2–7.5% of pairs had $0.25 \leq r < 0.5$; and a minimal 0.1–1.7% had $r \geq 0.5$. This suggests that close relatives of any focal animal constitute a very small minority in most bird and mammal populations. (A after [338]; B after [185].)

in small populations or in populations subdivided into small groups (**Figure 9.17**) [298, 338, 364, 652].

Each of the possible trajectories above relies on the presumption that at least one member of an interacting pair of

animals can estimate whether a potential competitor does or does not share copies of the relevant genes. Although we have largely focused on the actor adjusting its actions according to that information, the same arguments could apply to an actor

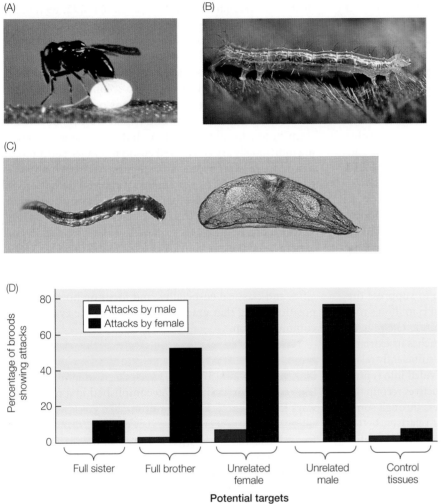

(A)

(B)

(C)

(D)

FIGURE 9.17 Helping and harming in the parasitoid wasp, _Copidosoma floridanum_ (A) A haplodiploid female wasp laying eggs on a moth egg (host). A single host egg may contain several eggs from the same mother wasp, or eggs from several different mothers. (B) Each wasp egg divides clonally into many larvae that begin eating the caterpillar host from the inside. In a host with only two wasp eggs from the same mother, but multiple eggs from many different mothers, the average relatedness of all larvae ($r = 0$) is intermediate between the relatedness of cloned larvae from the same egg ($r = 1$) and that for larvae from different eggs ($r < 0$). This sets up the context for spite. (C) Some larvae from each egg become soldiers (left) that never leave the host, while others (right) pupate and emerge from the host as adult wasps. (D) Female soldiers routinely attack and kill unrelated larvae to reduce competition; male soldiers are much less likely to attack any other larvae. In addition, female soldiers frequently attack males with whom they share a mother. This is because the average relatedness of wasp larvae inside a caterpillar is very high (especially if many females are clones), and the relatedness of a female to a full brother is often less than that average (due to the haplodiploid sex determination system). Female soldiers will even attack uncloned females from the same mother, but much less often than they attack brothers. All of these effects can be predicted using Hamilton's rule and the relevant values of r [299, 318–320]. (D after [319]).

always doing the same thing but a recipient who adjusted its response according to information about shared genes. In either case, at least one of the parties must estimate the relative likelihood of sharing the relevant genes when compared to two individuals randomly selected from the population. Does this happen in nature? In fact, animals can and do use any of three basic mechanisms to evaluate genetic similarity: greenbeards, kin cues and signals, and recognition of limited dispersal.

GREENBEARDS The classical **greenbeard** is a phenotypic marker generated by a pleiotropic gene or one of a small number of very tightly linked genes that advertise the presence of the gene or genes in a given individual [195, 364]. The same greenbeard genes also have the ability to recognize the marker in other individuals and promote actions that either help those also carrying the marker (and thus the genes), or harm those lacking the marker (and their genes). This definition has been extended to include genes that cause their carriers to settle preferentially in locations favored by other greenbeard carriers. The favored type of site then becomes the shared marker. Greenbeard actions can be facultative,

such that the actor adjusts its actions according to the presence or absence of the marker in a recipient, or they can be obligate, such that the actor directs the same action toward all encountered individuals. In the latter case, the recipients' responses vary according to whether they share the marker or not [303, 677].

There are several reasons why greenbeards might be uncommon in nature. First, a mutant "falsebeard" that showed the marker but failed to perform any actions associated with the greenbeard genes (thus avoiding the personal costs of those actions) could easily invade and eventually displace a population of greenbeards [196, 303]. Second, greenbeard genes might promote actions that favored their own propagation but decreased the propagation of other unlinked genes. Given the larger number and likely greater fitness effects of the latter, selection would then favor mutations in the nongreenbeard genes that suppressed the greenbeard gene actions or markers [344]. Third, some types of greenbeard actions such as obligate helping and any kind of harming require a minimal frequency of the relevant genes in the population before there is any fitness advantage to greenbeards [298, 303]. This would be hard to achieve in large well-mixed populations, but might be feasible in small populations or those subdivided into small groups. Finally, genes for facultative helping by greenbeards do not face a minimal threshold frequency, and are likely to go to fixation (100%) if they can dodge falsebeard invasions and suppression by other genes; at that point, everyone would have the marker and it would no longer pay to adjust actions [196, 303]. As a result of some or all of these obstacles, greenbeards occur only rarely in nature. Most known examples involve genomic conflicts inside an animal's body [123, 359] or occur in microbial species (see Chapter 16). However, **Figure 9.18** provides one verified example involving whole animals.

KIN CUES AND SIGNALS   Whereas greenbeard individuals share only a single pleiotropic gene or a few closely linked genes, individuals that are descendants of a recent common ancestor—**kin**—are likely to share alleles at many different loci [364–366]. Because any locus might exhibit sharing, there is no counterselection by other loci to suppress the helping of close kin or harming of non-kin if Hamilton's rule justifies the actions [196]. Economically justified helping of kin is often called **kin selection** [566]. It reaches its extreme in some arthropods and mammals in which sufficiently related members of the same social group are divided into reproductives and nonreproductives. Nonreproductives recoup the costs of their contributions to colony success through some combination of enhanced reproduction of relatives, better survival by not having to disperse, and the mutual benefits

of their coordinated activities [281, 364, 692, 694, 695]. Social groups whose members show (1) a generational overlap among members, (2) strongly differential reproduction throughout life, and (3) cooperative brood care and group maintenance have classically been said to be **eusocial** [593, 861, 862]. Thus, ant, bee, and wasp colonies with overlapping generations, one or a few queens performing most if not all of the reproduction, and female workers or soldiers contributing to colony functions but usually not reproducing individually are considered eusocial; similar partitions of reproductive effort but including both sexes are seen in eusocial termites [862]. As with many categorical definitions in behavioral ecology (those for "leks" and "information" come to mind) there are now many examples of "eusocial" species in which one of the classic criteria is violated or slightly less extreme, and the question arises what to call them [175, 179, 180, 222, 412, 490, 693]. We shall accept that there is often a gradient between the extremes of non-interacting conspecifics and classical eusociality, and focus on how a species' position along that gradient might affect the evolution of communication.

Whenever kin selection contributes to the relevant social economics, there may be a benefit to the discrimination of kin from non-kin [374, 694]. Such discrimination is called **kin recognition** and can be accomplished in a number of ways [564]. **Genetic kin recognition**, like greenbeard recognition, relies on genes that produce different markers in different kin lines, but does not require that the corresponding recognition and action-promoting genes be the same marker gene or be tightly linked to the marker gene. Recombination between marker and action genes can change the evolutionary dynamics significantly. Models of such systems predict eventual fixation of the marker genes and the loss of any reason to discriminate between kin and non-kin for many likely contexts; in some cases, limit cycles are possible [722]. The exceptions that preserve genetic kin recognition require either (1) high mutation rates coupled with low rates of recombination between loci and low rates of dispersal from natal locations or (2) independent selective pressures

(A)

(B) *GP-9* expression

Trait	Female genotype		
	BB	*Bb*	*(bb)*
Queen condition on emergence	Larger and fatter	Smaller and leaner	Usually die
Queen settlement	Founds new colonies	Joins exisiting colonies	Usually die
Worker preference for queen number	One per colony	Several per colony	Several per colony
Worker regicide	Accept one *BB queen*	Kill intruding queens lacking *b* allele	Kill intruding queens lacking *b* allele

FIGURE 9.18   A greenbeard gene in the fire ant *Solenopsis invicta* (A) Fire ant workers control whether new queen immigrants are allowed to stay and breed or are killed. (B) A single gene, *GP-9*, produces an odorant-binding protein that is expressed differently by the gene's *B* and *b* alleles. The large *BB* queens can found new colonies or join existing ones, smaller *Bb* queens only join existing colonies, and *bb* queens die. Ratios of *Bb* to *BB* workers in a colony vary with the genotypes of recent queens and their mates (which can be either *B* or *b*). Although colonies of 90% or more *BB* workers only permit their single queen to be *BB*, as few as 5–10% *Bb* workers can cause *all* workers to accept hundreds of *Bb* queens and kill *BB* queens. The *b* allele is thus a facultatively harmful greenbeard gene. Because it is lethal when homozygous in females but not lethal in males, it cannot go to fixation [448, 449, 719].

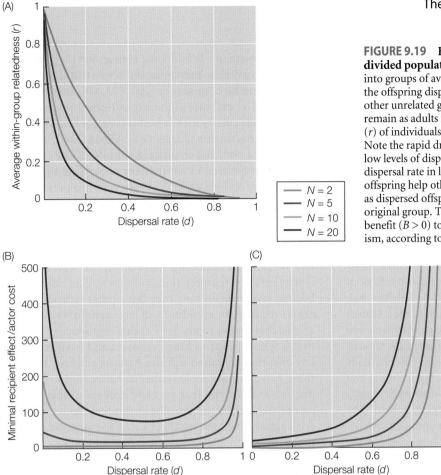

(A)

Average within-group relatedness (r)

Dispersal rate (d)

N = 2
N = 5
N = 10
N = 20

(B)

Minimal recipient effect/actor cost

Dispersal rate (d)

(C)

Dispersal rate (d)

FIGURE 9.19 **Helping and harming in viscous and sub-divided populations** Assume a stable population is divided into groups of average size N. Each generation, a fraction d of the offspring disperse from the parental group and settle in other unrelated groups, while the rest of the offspring (1 − d) remain as adults in the parental group. (A) Average relatedness (r) of individuals in groups as a function of dispersal rate (d). Note the rapid drop in within-group relatedness with even very low levels of dispersal, and the lower relatedness for a given dispersal rate in larger groups. (B) Suppose that nondispersing offspring help other group members at a personal cost, whereas dispersed offspring neither help nor harm members of that original group. The graph shows the minimal ratio of recipient benefit (B > 0) to helper cost (C > 0), that would justify altruism, according to Hamilton's rule, for different group sizes and dispersal rates. Helping is generally more likely (requires a lower ratio) for intermediate dispersal rates and for small group sizes. (C) Alternatively, suppose dispersers harm members of unrelated groups (B < 0) at personal cost (C > 0), whereas nondispersers neither harm nor help unrelated individuals. The graph shows the minimal ratio of recipient harm to actor cost that meets the Hamilton's rule criterion. Harmful behavior is favored for all group sizes at relatively low ratios when dispersal rates are low, but becomes increasingly unlikely at higher dispersal rates and for larger group sizes [232].

that favor retention of multiple markers in the population [28, 300, 428, 722, 755, 757].

All other mechanisms of kin recognition require learning distinctive cues. Certain contexts make it likely (but not guaranteed) that nearby conspecifics are kin, as with adult male group members in most birds, and adult female group members in many mammals. As long as those contexts are stable, the assumption of likely kinship is usually justified. Where such kin-related contexts are transitory, as with nest- and littermates, learning to recognize individual identity cues and signals while still in the kin group allows animals to continue to dispense help or harm selectively in other contexts [474]. Similarly, individuals raised together may acquire shared signals, such as olfactory secretions or specific calls, that persist sufficiently to identify them as kin in other situations [233, 734, 749]. Finally, individuals may be able to identify a number of shared physical or olfactory cues when kin are still living together that they can then use as templates for **phenotype matching** in future contexts [388, 734]. In the absence of suitable kin templates, an individual could also use itself for phenotype matching [360, 375, 376, 563]. We discuss these alternative mechanisms for kin recognition in greater detail in Chapter 13. It should be noted that all of these methods assume that animals raised together *are* kin; they need not be, and there may be other reasons why genetically diverse members of a given colony or social unit share common cues or signals [863].

LIMITED DISPERSAL A third mechanism for selective actions arises when there is limited dispersal of offspring away from parents [364]. Such populations are said to be **viscous**, and individuals living close to each other are usually more related to each other than they are to the general population. No separate kin markers or cues are required. By itself, limited dispersal would favor increased altruism by actors to nearby recipients. However, limited dispersal also increases resource and mate competition among relatives, which can change the relevant values of B and C sufficiently to make the Hamilton's rule criterion unfulfilled [281, 682, 786, 802, 844]. The competition is exacerbated if the population is subdivided into discrete and spatially separated groups [511, 515, 787, 811, 843, 844]. As we saw earlier, spite against nonrelatives is increasingly likely as group sizes get smaller [298, 509]. In situations where some offspring remain with their natal families and others disperse, the retained offspring are slightly inclined to be altruistic toward their neighbors, whereas the dispersing offspring are strongly inclined to be spiteful to members of the distant populations in which they settle (**Figure 9.19**) [232]. In birds and mammals, offspring of one sex tend to remain in the parental group and offspring of the other sex tend to leave and disperse; as predicted, the retained sex is often slightly helpful, whereas the dispersing sex is likely to be harmful to others with whom it settles [436]. Where local competition costs outweigh the discounted benefits of helping in subdivided populations, altruism within the group may be enforced through punishment of offending members [165, 282, 297, 432, 510, 514, 619, 691, 845, 866].

ALTERNATIVE ACCOUNTINGS   As we saw in the discussion on evolutionary modeling, different accounting methods make different simplifying assumptions and partition model components differently. Whereas inclusive fitness models divide up selective effects into costs to altruistic actors and indirect benefits through recipients that share similar genes, **group selection** models divide the same overall selective effects into costs to altruists and competitive benefits to the groups in which they live [761, 805, 824, 857, 858]. Both approaches can be derived from the same quantitative genetics or Price equation base, and both give similar predictions when predicated on similar assumptions [275, 279, 300, 302, 346, 384, 510, 512, 513, 680, 681, 685, 811, 838, 846]. Unfortunately, they are often based on different assumptions. For example, comparisons have been made between the robustness of group selection models that focus on genes (usually assuming asexual reproduction and no complications) and the limitations of inclusive fitness models that track changes in phenotypic frequencies [210, 633, 806, 814, 860]. The proper contrasts between group selection and inclusive fitness should focus on either the genes in both cases or the phenotypic effects in both cases, and in the latter case, identical assumptions about how genes are related to phenotypes should be invoked [681]. Different types of accounting *can* provide different perspectives on what is important during evolution. Rather than pitting methods against each other, it is much more productive to pool insights, contrast which assumptions affect which outcomes in each model, and compare outcomes while varying assumptions in the same ways across suites of models. This is why fluency in multiple modeling methods is an important skill.

Optimal Live History Economics

Trade-offs and optimization in evolutionary economics

Evolutionary economics invariably boil down to finding optimal trade-offs. Adoption of a given strategy will generate both benefits and costs to the adopter's fitness. If the strategy is a transaction between two or more animals, such as communication, both parties are likely to experience fitness benefits and costs of participation. Individuals adopting the strategy with the greatest excess of fitness benefits over costs will produce a greater fraction of the next generation. Over time, if this more productive strategy comes to dominate in the population, we can say that evolution will have *optimized* this trait within the relevant constraints. As we have seen, constraints include the range of alternatives that are initially available (a result in part of the prior history of the taxon), correlations between traits, and the species' genetic architecture. Chapters 2–7 in this book are largely devoted to listing

relevant constraints. While some authors have worried that constraints may be so numerous and severe that there is rarely much room for optimization [251, 333, 446, 528, 716, 730], the cumulative evidence of the last half-century suggests the contrary. Particularly with regards to animal behavior, both theoretical treatments [20, 25, 281, 339–344, 347, 348, 350, 567, 569, 632, 646] and empirical results [9, 224, 482] suggest that most traits seen in nature have undergone significant evolutionary optimization. We begin with this presumption in the following chapters, but examine relevant constraints where they appear to have been particularly limiting.

Life history trade-offs

One of the most important trade-offs in an organism's life is that between reproduction and survival: an animal that lives an exceptionally long time but neither reproduces nor assists relatives to reproduce will have zero fitness; another animal that invests so much in its first reproductive attempt that it ignores approaching predators and is killed before any offspring are raised will also have zero fitness. Between these two extremes are intermediate allocations of reproduction and survival that generate non-zero fitnesses, and among these, there will often be an optimal allocation [855]. For animals that have multiple successive opportunities to breed, the optimal allocation will depend upon its age and condition on each occasion. The optimal strategy would then be a program specifying the appropriate allocation at each successive opportunity. Communication often plays a crucial role in facilitating both an animal's survival and its successful reproduction. We thus expect communication strategies to be closely associated with lifetime allocation programs.

REPRODUCTIVE EFFORT   The allocation of resources to the reproduction component of fitness is called **reproductive effort** [266, 656, 768, 855]. One generally expects increased reproductive effort to produce more offspring. The consequent costs include reduced survival and reduced resources for future reproductive episodes. The impact of these costs will depend on the fraction of the total resource available that is committed to a given reproductive episode. For example, one would expect that producing a 1 kg baby would have a smaller impact on a 100 kg female than on a 10 kg female. By the same logic, a female that produces many small offspring might suffer the same impact as a similarly sized female that produces a few large ones as long as the total mass of the offspring was similar for the two females. This suggests that the ratio between the mass that an animal contributes to reproduction and its own mass might be a useful measure of reproductive effort. There are some interesting consequences of this proposition. In a stable population, the average female should just reproduce herself in her lifetime. Given the parameters with which metabolic costs typically scale with body size [853], one can calculate that individual replacement by any female, whatever her size, will require that reproductive expenditure over her lifetime which produces a net offspring body mass 1.4 times her own (**Figure 9.20**) [146–148].

(A)

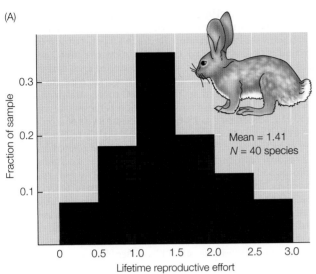

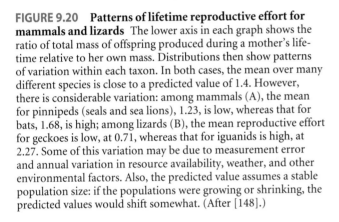

Mean = 1.41
N = 40 species

(B)

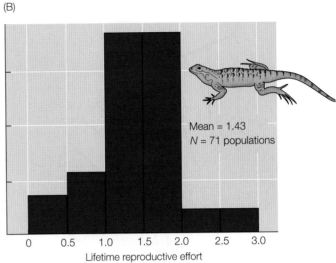

Mean = 1.43
N = 71 populations

FIGURE 9.20 **Patterns of lifetime reproductive effort for mammals and lizards** The lower axis in each graph shows the ratio of total mass of offspring produced during a mother's lifetime relative to her own mass. Distributions then show patterns of variation within each taxon. In both cases, the mean over many different species is close to a predicted value of 1.4. However, there is considerable variation: among mammals (A), the mean for pinnipeds (seals and sea lions), 1.23, is low, whereas that for bats, 1.68, is high; among lizards (B), the mean reproductive effort for geckoes is low, at 0.71, whereas that for iguanids is high, at 2.27. Some of this variation may be due to measurement error and annual variation in resource availability, weather, and other environmental factors. Also, the predicted value assumes a stable population size: if the populations were growing or shrinking, the predicted values would shift somewhat. (After [148].)

If the population is in a growth phase, the optimal lifetime reproductive effort will be somewhat higher, depending on the rate of growth and average lifespan of the animals, than if it is stable [149]. With or without this adjustment, each species should have an optimal lifetime reproductive effort (measured relative to body mass), and for animals living in stable populations, that optimum should be approximately the same for all species. This prediction appears to be generally true even for humans [118, 148].

REPRODUCTIVE VALUE   How should an animal divide this total reproductive effort among the various opportunities for breeding in its lifetime? It could expend the entire allotment in one breeding episode, or instead produce many successive reproductive episodes with smaller levels of reproductive effort/episode. A useful measure for predicting optimal allocation programs is **reproductive value** [266, 855]. This estimates an individual's cumulative reproduction from now forwards. For a stable population, reproductive value equals the sum over all future reproductive episodes of an individual's **fecundity**—the expected number of offspring produced in each episode—discounted by its **survivorship**—the probability that it will survive to that episode. If the population

is changing in size, a correction term is added that weights earlier versus later breeding episodes differently.

As with estimates of lifetime reproductive success (R_0), those occasions furthest into the future tend to add least to reproductive value because relevant survivorship values will be small. At birth, the reproductive value for an animal in a stable population is equal to R_0, and reproductive value then changes with successive stages in its life cycle. If a species has an early phase in life when there is no reproduction and mortality rates are much higher than later (as in the cases of vulnerable larvae or nestlings), the reproductive value of an individual starts off at a low value, and only rises once it has passed successfully through the period of acute mortality risk. Reproductive value in most animals thus rises in the early years, peaks at early maturity, and then declines as the animal approaches its expected maximal age.

OPTIMAL LIFE HISTORY PROGRAMS   An optimal life history is one in which an animal has adjusted its reproductive effort so that it maximizes its reproductive value at each age [855]. To see how this optimization might be determined, subtract current fecundity from current reproductive value and denote the remainder as the **residual reproductive value**. In general, there will be a trade-off between current fecundity and residual reproductive value: increasing current reproductive effort will increase fecundity but at the expense of reducing future survival or reproduction, thus reducing the residual component; decreasing reproductive effort will reduce fecundity but increase the residual component. The optimization task is then to find, for any given age, the reproductive effort that maximizes the sum of current fecundity and residual reproductive value. Where that optimum occurs depends on how each of these components changes with increasing reproductive effort: these relationships might be linear, accelerating, decelerating, or even sigmoid, depending on the animal's age, its ecology, the density of the population, and a variety of other factors [107, 733]. There are three

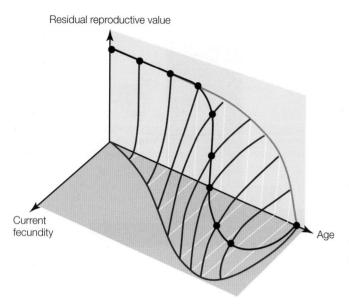

Residual reproductive value

Current fecundity

Age

FIGURE 9.21 Optimal life history trajectory for a typical animal with multiple seasons of potential reproduction Red lines show the likely curvature for trade-off between current fecundity (a direct result of current reproductive effort) and consequent residual reproductive value at selected ages. Early in life, even small levels of fecundity result in major decreases in future survival and reproduction. The trade-off curves straighten and eventually curve in the opposite direction as the animal matures and gains experience: at maturity, moderate levels of reproductive effort and consequent fecundity only slightly reduce residual reproductive value. The green line indicates the gradual decrease in residual reproductive value as an animal ages, accumulates injuries and disease, and becomes less competitive with younger breeders. The blue line indicates the maximal possible fecundity at each age, given the trade-off curves. The black line and dots indicate the optimal trajectory as an animal ages: the path shows gradually increasing reproductive effort and fecundity, usually at levels well below the current maximum but more closely approaching maximal values near the end of life. (After [656].)

qualitatively distinct options at each age [656]: the animal could refrain from breeding entirely, it could undertake an intermediate reproductive effort, or it could go all out and expend as much reproductive effort as possible. For short-lived species, there may be only one chance at reproduction; these animals should adopt the third (maximal effort) option the first chance they get [144, 167]. Such species are called **semelparous**. Species that are longer lived and have multiple occasions in which to breed are said to be **iteroparous**. For most iteroparous species, the three effort options become optimal at successive stages of life: young animals avoid reproduction, middle-aged animals expend intermediate effort, and old animals put all remaining resources into their final reproductive episodes (Figure 9.21).

At each age, animals can fine-tune their strategies according to their current condition, levels of habitat seasonality, and available phenotypic plasticity [330, 408, 768]. For example, mammalian reproduction involves three successive periods: pregnancy, nursing, and offspring dispersal. Seasonal breeders can usually time reproduction so that at least one of these periods coincides with the annual peak in food abundance, but not all three. Species with high adult survival rates outside the breeding season would be expected to schedule reproduction so that pregnancy occurs during food peaks: this reduces the risk to adult females even at the expense of reduced survival of any one year's offspring. Species with low adult survival outside of the breeding season should adopt timing that favors offspring survival even at the expense of further reductions in adult survival. In at least one group of bats, this is what is found [98].

Evolutionary models of life history predict that animals will exhibit close to optimal reproductive effort at each available episode [107, 108, 141, 149, 266, 326, 592, 614, 656, 710, 855]. In general, the data support these predictions [118, 148, 524, 768]. Different signaling strategies may be required at different life history stages to facilitate these different levels of reproductive effort [295, 425, 445, 452, 471, 597]. However, different signaling strategies face their own trade-offs and constraints. In the next section, we examine some of the major ways that signaling interfaces with life history strategies.

Optimal Signaling Economics

The evolutionary economics of specific traits, such as producing or attending to a given signal, are invariably complicated by trade-offs. Many of these trade-offs arise because some limited commodity is utilized by many different activities [408]. Animals require energy to forage, display, build refuges, migrate, and molt; an increased investment in one of these activities will often reduce the energy available for others. Time is a frequently limited commodity: diurnal animals have only so much daylight in which to pursue all their necessary activities. How much time should be spent displaying to a reluctant female versus searching for another? Many vertebrates draw on accumulated carotenoid reserves for both body coloration and responses to infection: what is the optimal allocation to these separate functions? Attention can be a limited commodity: it may be difficult for an animal to focus on feeding and watching for predators at the same time. In all of these cases, there are multiple ways to cut a limited commodity "pie": the economic task is to identify the optimal allocation. This allocation will depend critically on how the size of each pie slice affects fitness. Whenever fitness is positively correlated with slice size, increasing the size of one slice will increase fitness (a benefit) but it may concomitantly decrease the size of one or more other slices and thus decrease their contributions to fitness (costs).

What this means is that one cannot fully describe the evolutionary economics of any given trait by only listing its direct effects on fitness; one must also examine any indirect effects that investing in this trait has on other traits that rely

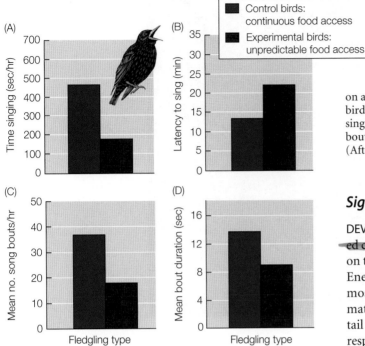

FIGURE 9.22   Effects of differential food access after fledging on singing by male starlings (*Sturnus vulgaris*) the following spring Control birds had continuous access to as much food as they wanted. Food for experimental birds was removed for 25% of daylight hours each day but at random times. Experimental birds made up for the shortages and attained heavier weights on average than controls. In the following spring, experimental birds spent less time singing than controls (A), took longer to start singing when exposed to a nearby female (B), sang fewer song bouts per hour sampled (C), and sang for shorter song bouts (D). (After [113].)

Legend: Control birds: continuous food access / Experimental birds: unpredictable food access

Signals and physiological reserves

DEVELOPMENT   In most species, energy reserves are a limited commodity. Every signal will generate some energy impact on the participants whether before, during, or after signaling. Energy surely plays a significant role in the development of most sender and receiver signaling organs. The larynx of a mature male hammer-headed bat (see Figure 2.41) and the tail of a peacock are a sufficiently large part of their owners' respective body masses that they must be expensive to generate. The pheromone mixes of cockroaches, the brightness of iridescent colors, the sizes of adult beetle horns, stridulatory organs in crickets, and song repertoires in songbirds all vary depending upon access to food during development [162, 214–216, 241, 390, 424, 453, 547, 634, 737, 738, 744]. Nutritional stress during fledging can even have long-term effects on later singing by male birds (Figure 9.22) [113]. Organs that play roles in both signaling and foraging may have different optimal configurations for the two tasks: development that favors a configuration optimal for one role will compromise the effectiveness of the organ in the other role. For example, the beaks of songbirds are used for both foraging and song emission: heavier bills in finch species provide better seed cracking but retard beak movements during singing [39, 159, 387, 411, 658, 659, 762, 849]. If signaling is favored over foraging, the trade-off can slow down the buildup of energy reserves required for many other traits; if foraging is favored, the range of modulations and the degree of effective coupling of a bird's calls to the medium can be severely constrained. The wings of birds have evolved specific shapes to make flight energy-efficient. Respiration and wing beats are often linked physiologically in birds [74, 84–87, 117, 238, 239, 288–290, 548, 800, 801]. The development of a particular wing shape commits most birds to a particular pattern of wing beating, and the links to respiration may then limit the allowable durations, amplitudes, and rates of calling both when the bird is in flight and when it is giving perched displays that combine wing movements with vocalizations [170].

Some signaling structures—such as larynges, beetle horns, and cricket stridulatory organs—are created once, early in an animal's development. But others—such as antlers and feathers—must be redeveloped multiple times in an animal's lifetime. In addition to providing insulation, plumage is widely used for signaling in birds. Coloration patterns provide

on shared and limited commodities. As we saw in the previous section, there is usually a fitness trade-off between fecundity and residual reproductive value. A common cause of this trade-off is one or more lower-level trade-offs between multiple traits that rely on a shared commodity. In many cases, increasing one trait will increase fecundity but because this reduces access to some shared commodity, another trait that normally helps to augment survival and future reproductive value will have to be curtailed. A complete accounting must thus examine the links between each trait and the components of fitness, and the links between traits through shared commodities.

The direct benefits of signaling are a major focus of the remaining chapters in the book. In the following section, we list many of the indirect effects of engaging in signaling. Most of the cases noted below will be considered costs, although a few, such as allogrooming in primates, may provide indirect benefits beyond the transfer of information. We focus here on three of the commodities that commonly generate trade-offs between traits: physiological reserves, physical integrity, and information storage (brains). Note that these commodities are not necessarily independent: an animal that has been physically damaged by a predator or fighting may be less efficient at foraging and thus have reduced energy reserves. A fourth commodity, time, is not treated separately but occurs as a recurrent thread within and between the others. Signaling can create different indirect effects at different stages in a life cycle. For each commodity below, we shall discriminate between impacts incurred during the acquisition of signaling prerequisites (development), those associated with preservation and repair of signaling apparatus (maintenance), and those experienced only during signaling (usage) [744].

FIGURE 9.23 **Bowers and decorations of Australian and New Guinea bowerbirds**
Male bowerbirds spend enormous amounts of time building display bowers out of sticks and
decorating interiors and foregrounds with colorful objects to display to visiting females. (A)
The avenue bower of the great bowerbird (*Chlamydera nuchalis*) decorated with stones, snail
shells, green fruits, and pieces of green glass. (B) The hut-shaped bower of the vogelkopf bowerbird (*Amblyornis inornatus*). Males carefully sort different decorations into separate piles.
This male has collected acorns, black fungi, and *Pandanus* flowers. (C) Another vogelkopf
bower with a vast number of acorns in the foreground and a small pile of orange fruits at the
entrance to the bower.

information about species, age, sex, and individual identities, and feather elevation is often used to mediate agonistic interactions. All birds molt at least once a year, usually after their breeding season. Those with separate breeding and nonbreeding plumages typically molt a second time prior to the next breeding season. Molting is energetically expensive, leaches calcium from the bones, uses up fat supplies, and suppresses the immune system [15, 313, 487]. Some species become flightless during the postbreeding molt, which may retard the replenishment of energy reserves [111, 276, 277]. All of these side effects make the timing of molt an important strategic decision [46, 321, 385, 653]. As we discuss in detail in Chapter 12, color production in avian plumage is sufficiently expensive that it is often dependent on the condition of the molting bird. Nuptial color production draws not only on general energy reserves, but also on the availability of specific nutrients such as carotenoids and minerals [31–34, 574, 611, 751]. While all feathers wear down and need periodic replacement, those with signal functions impose significantly higher costs on the process.

MAINTENANCE   Once developed, signals vary in the degree to which they impose maintenance costs on their bearers' energy reserves. The exoskeletons of arthropods are irreparable following the final molt (crickets) or metamorphosis (beetles): there is little that a mature arthropod can do to reverse subsequent wear and tear on any external signaling organs [12, 192]. Species such as reptiles, birds, and mammals that replace their exterior coverings at intervals are not as constrained, but despite the opportunity for periodic renewal, most birds and mammals spend considerable time grooming and maintaining their respective exteriors. Heavy signaling organs such as the enlarged claw in fiddler crabs, long tails in birds, and large larynges in mammals may increase the energetic costs of locomotion and reduce the agility of predators [11, 104, 113, 607]. Male birds-of-paradise and manakins that display on the ground spend significant time each day cleaning away litter and pruning intruding vegetation at their display courts [90, 174, 314, 530, 675, 676, 810]. Male bowerbirds spend an even larger daily effort expanding, elaborating, and restoring the elaborate bowers in which they display to females (Figure 9.23). In addition to regular maintenance, they often have to repair bowers that have been pillaged by competing males [89, 674]. A variety of animal species decorate or augment their nests to attract mates, and these ornaments may need regular replacement or reshuffling [732]. Pheromonal mixtures, such as those of the sac-winged bat (*Saccopteryx bilineata*), may also need to be replenished on a regular basis [131, 819, 820, 821].

Brains and associated nervous tissues are energetically expensive because they must maintain ionic gradients in the

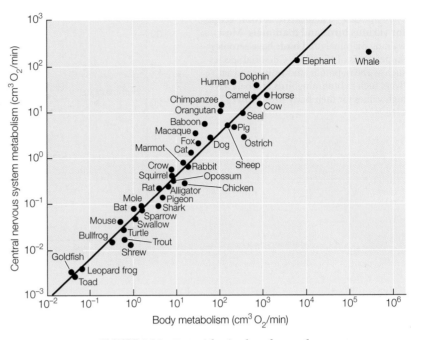

FIGURE 9.24   Logarithmic plot of central nervous system (brain) metabolism versus body metabolism in various vertebrates The red line shows the regression over the entire sample. The proportion of brain metabolism to body metabolism averages 4.8% for ectothermic vertebrates and 5.5% for endothermic ones. Primates, including humans, and dolphins have significantly higher proportions than expected for their body size; very large animals such as ostriches and whales have significantly lower proportions. (After [599].)

nerves in preparation for use [504, 626]. In addition, most animals maintain a steady base level of nerve impulse firing even when they are not moving about or communicating. The brains of most vertebrates constitute 1% or less of body mass; there are exceptions, such as songbirds, primates, and a few other mammal species, but even in these cases, the brain accounts for less than 3% of body mass [184]. Despite the small amount of nervous tissue involved, the energy used by a vertebrate brain even at rest is typically 2–8% of the overall energy budget [599]. This range is quite consistent, regardless of body size or thermoregulatory strategy (**Figure 9.24**). There are a few exceptions: very large mammals such as fin whales and elephants fall on the low end of this range (at 0.5% and 2% ,respectively), whereas nonhuman primate brains regularly contribute 11–13% to daily energy expenditures. Humans are extreme cases even for primates, using 20% of their energy budgets on nervous system activity (although their brains are only 2–3% of total mass). The other main exceptions to the 2–8% rule are electric fishes whose continuous electric organ discharges and large brains, particularly the massive cerebellum, can contribute up to 60% of these fishes' daily energy costs [623]. Within any group of mammals, resting metabolic rates tend to be higher in species with brains that are large relative to their body sizes [418, 419]; birds do not show a clear relationship between

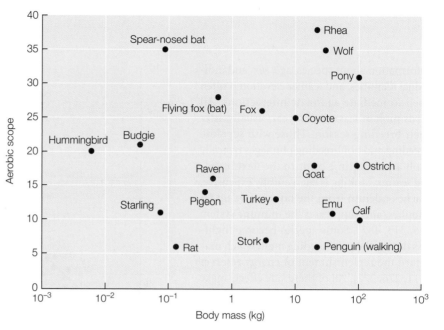

FIGURE 9.25 Facultative aerobic scope for various birds and mammals Most walking mammals and birds have aerobic scope values of 10 or less. Cursorial species such as rheas, ponies, and wolves and active flyers such as hummingbirds and bats have high values for their body sizes. (After [77, 116].)

relative brain size and resting metabolic rates, but they do show a negative correlation between relative brain size and flight muscle mass, suggesting some sort of energetic trade-off between brain activity and flight [417]. Taxa differ in the metabolic strategies they have evolved to meet the high energetic costs of nervous system maintenance [763]. While only a portion of any animal's brain is involved in communication, it surely contributes to the maintenance costs for both senders and receivers.

USAGE   The energetic impact of producing signals depends upon the energetic cost per signal emitted, the rate at which a given signal is emitted, and the current energy state of the sender. We first examine the costs per signal emitted [578]. The absolute energetic cost of any given activity increases with an animal's body size: it costs more to raise a leg or emit a call if you are larger. However, bigger animals have more energy to spend. The energetic impact of an activity such as signaling thus depends upon the cost relative to that animal's overall energy budget. Animals can expend energy in either of two ways. **Aerobic processes** burn available fuels with oxygen during the activity; as long as sufficient oxygen and fuel are available, they can sustain long bouts of the activity. **Anaerobic processes** do not require oxygen during the activity and can produce a large amount of energy quickly. However, they create by-products that must later be metabolized, and their accumulation leads to fatigue. Most terrestrial animals use aerobic muscles for steady locomotion and low-cost signaling, but force these into anaerobic states during short bursts of activity. The musculature of fish consists of a mosaic of aerobic (red) and anaerobic (white) muscles. Aerobic muscles constitute 6% of the total muscle mass in most sedentary and slow-swimming fish and 20% in fast swimmers; for fish and for fast-swimming marine invertebrates (e.g., squid), anaerobic processes play a larger role than they do for terrestrial animals [327, 715].

Two useful reference points against which any given activity's aerobic costs can be compared are (1) the resting

metabolic rate of that species, and (2) the maximal aerobic energy expenditure that the animal's circulatory and metabolic systems can sustain. Resting and maximal aerobic metabolic rates are usually correlated with each other, and often increase in parallel with increasing body size [70, 578–580]. The ratio between the aerobic energy costs of the activity and the resting level of the organism is called **aerobic scope**. The interpretation of aerobic scope values is complicated by large differences between species: resting metabolic rates for endothermic (warm-blooded) birds and mammals tend to be ten times as large as those for ectothermic (cold-blooded) arthropods, fish, amphibians, and most living reptiles. Even within a taxon all of whose members share the same thermoregulatory strategy, either the resting metabolic rate or the maximal energy expenditure that an animal can support can vary [69, 95, 489, 699, 701, 726].

Despite this diversity in energetic strategies, most animals have a **maximal aerobic scope** that ranges from 5 to 25 times resting energy levels (**Figure 9.25**): examples include insects when not in flight (2–23), amphibians (5–27), reptiles (6–12), birds (8–20), and terrestrial mammals (5–10) [70, 77, 94, 116, 137, 287, 577, 578, 625, 713, 782, 828]. In frogs and insects when not in flight, maximal scope values are seen in calling males [668]. In other taxa, maximal scopes are set by locomotory costs: high-end exceptions to the 5–25 rule include cursorial species such as rheas, canids, and horses (33–36) and insects during flight (30–200) [47, 116, 668, 713]. Not surprisingly, given the low percentages of aerobic muscle in their bodies, most fish show low maximal aerobic scope values (2–6) [163, 327, 458]. However, intermediate and larger species have significant ranges of "anaerobic scope" that they can call on for burst swimming.

Maximal aerobic scope values provide a reference against which one can compare the aerobic costs of animal signals. For example, typical values for arthropod sound production range from 2 to 20 times resting metabolic rates [36, 357, 395, 542, 668, 669, 698, 708, 852]. Different displays by the same species may vary in scope because different sets and numbers

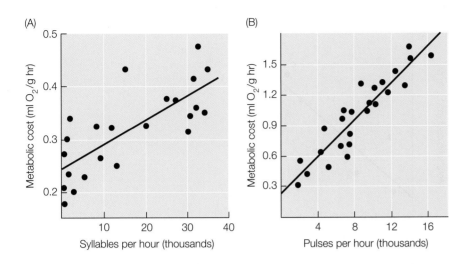

FIGURE 9.26   Energy costs of calling in ectothermic species that can vary the number of elements per call (A) Male *Ligurotettix coquilletti* katydids produce chirps consisting of variable numbers of syllables. The rate of syllable production is the product of the number of syllables per chirp and the chirp rate. The graph shows the monotonic increase in energy costs as the rate of syllable production increases. (B) Male *Hyla versicolor* frogs produce trills with variable numbers of pulses. Measured energetic costs of calling increase linearly with the overall rate of pulse production [783]. (A after [36]; B after [836].)

of muscles are recruited [166, 477]: for example, cricket courtship calls are twice as expensive as male advertisement calling [357, 395]. A surprising number of male arthropods have calling costs that are close to or even greater than the maxima for nonflight activities. Calling also appears to be expensive in many frogs: scopes of 10–24 times resting metabolic rates have been recorded, particularly in smaller species [336, 667, 668, 776, 836]. In both arthropods and frogs, call duration contributes much less to overall energy costs than does calling rate [36, 336, 783, 835, 836]. However, in species that can vary the number of elements per call, the product of elements per call and call rate is often a better predictor of overall energetic costs than call rate alone (**Figure 9.26**).

In contrast to the expensive calls of arthropods and frogs, individual vocalizations by perched or standing birds tend to be energetically cheap: the crow of a rooster has a scope of 1.15 or less, a lower value than for drinking, feeding, and even preening [96, 138, 402]. Songs and alarm calls of a variety of bird species are only 1.05–3.4 times resting levels [283, 441, 636, 830–832]. Begging by nestling birds was initially thought to be energetically expensive; data now show that typical scopes range from 1.02–4.09 [29, 441, 506, 570]. As with insects, avian vocal displays that include extensive movements or rigid muscle-dependent postures can cost more: the strut of a male sage grouse has an aerobic scope of between 14–17 [815]. Costs/vocalization do tend to increase with repertoire diversity in songbirds [294].

Where they have been measured, the per-signal costs for modalities other than sound show similar ranges of variation. Visual displays of male swordtail fish to females are only 1.02–2.0 times resting metabolic levels [186]. Metabolic rate for male bank voles (*Clethrionomys glareolus*) increases by a factor of 1.08 while they have enlarged preputial glands used for courtship signals [688]. Males of the gymnotid electric fish *Brachyhypopomus pinnicaudatus* increase the duration and amplitude of their electric organ discharges during the breeding season. Given that the fish discharge their organs continuously even when at "rest," the energetic cost of the increased power in each discharge has a scope of about 1.10 [727]. Fireflies expend 5.3

times resting metabolic levels per flash [867]. Aerial displays by birds ought to be very expensive although some species adopt special modes of flight to reduce these costs [382].

As noted above, the rate of signal emission is most often the major factor determining signaling energy costs. The highest rates of signal emission are usually associated with male advertisement, although again, there is considerable variation. At the low end are irregularly given calls such as the crowing of roosters: these have low per-signal costs and at usual emission rates consume only 5–6% of a rooster's daytime activity budget [138]. In contrast, males of the moth *Achroia grisella* advertise themselves with 30–50 ultrasonic pulses/second for up to 8 hours, and male cicadas snap their tymbals 50–500 times per second for hours at a time. The frog *Hyla microcephala* produces 6000–10,000 calls per hour during its four-hour nightly chorus, and another hylid, *Scinax rubra*, emits 2000 calls per hour over a seven-hour stint [836]. Male hammer-headed bats flap their partially opened wings while hanging from a favored branch, and call 3000–7000 times an hour for three to four hours a night throughout their two- to three-month breeding seasons [97]. Male sage grouse (*Centrocercus urophasianus*) can perform over 1000 struts each morning during their two- to four-month breeding seasons [312, 815], and male great snipe (*Gallinago media*) perform their drumming display at rates of 150–210 per hour for hours every evening over a five to seven week period [397, 522]. Almost uniformly, a sender's aerobic energetic costs increase (usually linearly) with signal emission rates (**Figure 9.27**) [29, 36, 329, 373, 395, 529, 698, 708, 776, 815, 831, 836]. The sonic muscles of most fishes are largely anaerobic, limiting drumming to short bursts at high rates [16, 264, 494, 603, 714]. An exception is the midshipman fish (*Porichthys notatus*) whose sonic muscles have special aerobic adaptations that allow it to drum at moderate rates for sustained periods [53].

The longer term impact of repeated signal emission is usually reported as a multiple of daily resting energy expenditures, (analogous to aerobic scope used for shorter periods), or as the fraction of the total daily energy expenditures

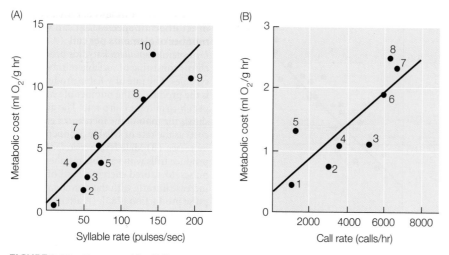

FIGURE 9.27 Interspecific differences in typical calling rates and associated metabolic costs Higher rates of calling in both male insects (A) and male frogs (B) lead to higher energetic costs of calling. Species in (A): 1, *Requena verticalis* (katydid); 2, *Scapteriscus borelli* (mole cricket); 3, *Oecanthus celerinictus* (tree cricket); 4, *Oecanthus quadripunctatus* (tree cricket); 5, *Anurogryllus arboreus* (cricket); 6, *Gryllotalpa australis* (mole cricket); 7, *Cystosoma saundersii* (cicada); 8, *Anurogryllus muticus* (cricket); 9, *Neoconocephalus robustus* (katydid); and 10, *Euconocephalus nasutus* (katydid). Species in (B): 1, *Engystomops pustulosus* (túngara frog); 2, *Hyla cinerea* (green tree frog); 3, *Hyperolius viridiflavus* (reed frog); 4, *Hyla gratiosa* (barking tree frog); 5, *Hyla versicolor* (gray tree frog); 6, *Pseudacris crucifer* (spring peeper); 7, *Hyla squirella* (squirrel tree frog); and 8, *Hyla microcephala* (yellow cricket tree frog). (After [668].)

attributable to these signals. The first measure can be converted to the second by noting that daily energy expenditures for most animals range from 2–7 times resting metabolic rates [30, 112, 703]. At the low end of the impact scale, begging by nestling birds constitutes only 0.02–0.22% of daily energy costs depending on nestling age [29]. While bird song usually has low per-signal costs, many species sing so often during the breeding season that significant costs can accrue. For example, male great reed warblers (*Acrocephalus arundinaceus*) that spend 50% or more of their time at their song perches average daily energy expenditures of 3.3 times their resting levels; in contrast, nonsingers expend only 2.2 times their resting values [373]. This implies that at least 33% of the daily energy budget of active singers is spent in singing. Electric organ activity accounts for 11–22% of the daily energy budget of male *Brachyhypopomus pinnicaudatus* (gymnotid electric fish), but only 3% of that of females; the difference is accounted for by the longer duration and larger amplitude of the male discharges [727]. The high per-signal costs and emission rates of displaying male sage grouse [815] and snipe [398] generate daily expenditures four times those of resting levels; nondisplaying male expenditures are only twice resting levels. This suggests that at least 50% of the daily energy budgets of active males of these two bird species is expended in display activity. Calling by male frogs can account for an estimated 26–56% of daily energy budgets [668].

The third factor affecting the energetic impact of signaling is the sender's energy reserves. A frog or sage grouse may expend a large fraction of its daily energy budget in signaling, but this is sustainable as long as the animal can replenish its energy reserves between display bouts and there are no concurrent demands for energy that are unmet because of this high rate of display. It is often useful to distinguish between **income spenders** whose daily energy costs are compensated by daily income even during periods of high expenditure, and **capital spenders** who accumulate energy during one period and then rely solely on the accumulated reserves during a subsequent period of high expenditure [88, 219, 438, 768]. While this distinction is most often noted with respect to breeding behaviors, it also applies to migration or molting activities.

At the capital-spending extreme are insects such as mayflies (Ephemeroptera), caddis flies (Trichoptera), some true flies (Diptera), and many moths (Lepidoptera), in which adults do not eat at all; all energy used for reproduction must be obtained as immatures [451, 598, 784, 850]. Both sexes of several baleen whales reduce or stop feeding during their annual migrations from feeding grounds to warm-water breeding sites. Seabirds and penguins often nest sufficiently far from feeding areas that one or both members of a mated pair go for long periods without feeding. Males of many ungulates, pinnipeds, and elephants reduce or even halt food consumption during the breeding season (**Figure 9.28**) [543, 615, 661, 662]. The primary advantage appears to be the release of foraging time for breeding efforts [616, 651, 856]. The cost of reduced feeding in any of these taxa is a steady loss of body mass as the breeding season progresses [169, 273, 617, 651, 698]. Loss of body mass can also occur in species that continue to feed throughout the year. Sage grouse breed five to six months after the last leaf production by their sagebrush food plants. Although males continue to forage throughout the breeding season, remaining leaves are of poor quality, energy expenditures exceed income, and they invariably lose weight [815]. Losses of body mass over the breeding season also occur in displaying great snipe and in calling toads [152, 398].

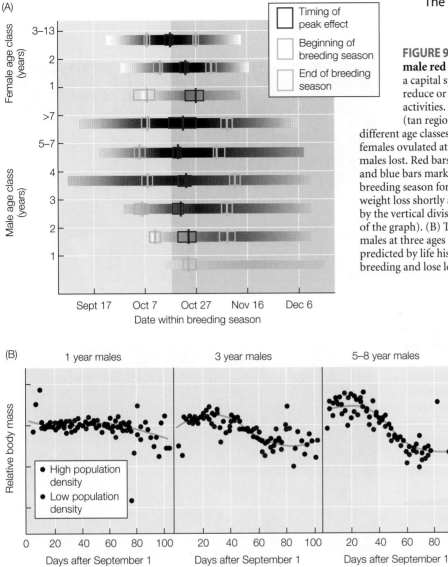

(A)

FIGURE 9.28 **Patterns of weight loss by rutting male red deer (*Cervus elaphus*)** The red deer is a capital spender during the breeding season; males reduce or even eliminate foraging in favor of breeding activities. (A) The relative timing of female ovulation (tan region) and male weight loss (blue region) for different age classes of each sex. The darker the bar, the more females ovulated at the corresponding date or the more weight males lost. Red bars indicate timing of peak effect and green and blue bars mark the effective beginnings and ends of the breeding season for each age/sex class. Mature males show peak weight loss shortly after peak ovulation by females (indicated by the vertical division between the light and dark portions of the graph). (B) The magnitude of the drop in body mass in males at three ages as a function of days after September 1. As predicted by life history theory, younger males invest less in breeding and lose less weight than older males. (After [617].)

Income spenders try to adjust total energy expenditures so that they match foraging income. Because many animals cannot display and forage concurrently, there is often a trade-off between these two activities. European robins, which sing and feed in the daytime, show a clear trade-off between the time they spend in singing and foraging [796, 797]. Nightingales, which sing but do not feed at night, adjust their song rates according to their energy reserves at dusk [795]. Other species of songbirds increase song rate or song performance if provided with supplemental food [44, 73]. Katydids trade off time and energy put into calling with the size and quality of the edible spermatophore that they pass to females during mating [37]. Anurans appear to trade off attendance at choruses with the energetic level of display when attending a chorus [133, 134, 613]. Begging nestling birds and juvenile male mice that scent mark tend to reduce their growth rates if they consistently signal at high levels [133, 134, 328, 459, 613, 709].

Most species fall along a gradient between perfectly balanced daily energy budgets and those making major energetic expenditures with no daily replenishment [585, 770]. Whether a species is a capital or income spender (or most likely,

something in between), the links between signaling costs and energy reserves provide a rich substrate for information about a sender's condition. The amount and character of a male vole's preputial gland secretions are dependent upon its access to food supplies prior to the breeding season [534]. The pheromone streaks used by male *Drosophila grimshawi* to court females reflect the male's current protein levels [221, 433], and the cuticular hydrocarbons used as signals by other *Drosophila* species also provide honest information about the sender's metabolic condition [80, 391]. The fasting by rutting male ungulates generates specific metabolites in their urine that females can use to evaluate male condition (**Figure 9.29**) [23, 562, 600, 601, 854]. We pick up these relationships again in later chapters.

Signals and physical integrity

DEVELOPMENT AND MAINTENANCE   An animal's body is another critical commodity whose successful nourishment and protection can significantly affect its fitness. Threats to the physical integrity of the body include structural damage or loss of body parts due to wear, accidents, social conflict,

FIGURE 9.29 Fasting and olfactory signals in male moose (*Alces alces*) Adult male moose begin fasting several weeks before mating begins [600]. (A) During this period, they scrape shallow pits in the earth into which they urinate and either splash urine back on their bodies with their hooves, or wallow in the pit. (B) Females are highly attracted to pits containing fresh urine and will also wallow in them [601]. Rutting males simply stand around during the time they normally would spend foraging; thus fasting is not a simple consequence of lacking sufficient time to feed and court. The shift from foraging to fasting results in major differences in metabolites present in rutting male urine. It has been suggested that this provides honest signals to females about the ability of a male to handle fasting stress and thus his relative physical condition [854].

or attacks by predators. In addition, the animal's physiological functions may be disrupted by parasites, disease, and adverse climatic conditions. The impact of any of these events can range from minor perturbations to death. Note that optimal life history strategies may not demand maximal body preservation: the best mix of current reproduction and residual reproductive value at a given life history stage might indeed call for serious physical risk. Different species and sexes within a species are likely to have different optimal body preservation schemes.

Development and maintenance of body integrity often overlap. Where tissues are replicated at intervals, such as the molting of arthropod exoskeletons or avian plumages, damaged parts can be replaced with new ones. Animals that undergo a significant metamorphosis during development, such as many marine invertebrates, holometabolous insects, and anuran amphibians, also have an opportunity to replace damaged larval body parts with undamaged adult ones. Although adult regeneration of lost appendages is widespread across animal phyla, it is irregularly distributed within any given taxon [68, 269, 545]. Arthropods normally replace lost appendages during one or more successive molts. Many crustaceans and hemimetabolous insects molt many times during their lives and can thus regenerate appendages until their final molt. Holometabolous insects do not molt again after metamorphosis and thus do not regenerate appendages lost as adults. Fish can regenerate lost fins, most amphibians can regenerate lost limbs, and many lizards can regenerate lost tails, but except for the annual replacement of antlers by deer, birds and mammals lack appendage regeneration [14, 59, 105, 455].

Many of the species that *can* regenerate, and even some that cannot (rodents), **autotomize** (cast off) appendages when attacked by a predator. In arthropods, autotomy of legs and antennae is most common, whereas in lizards and mice, tails are the main autotomized appendages [59, 440, 442, 545, 575, 748]. Autotomy leaves an attacking predator holding

the disassociated appendage or distracted by its continued movements while the attacked prey escapes [269, 389]. While autotomy may increase its practitioner's immediate survival, there can be major costs. These include reduced mobility or agility that constrains subsequent foraging success, predator escape, home range defense, mate searching, courtship display, and copulation orientation [56, 58, 59, 172, 173, 269, 315, 355, 538, 746]. Most lizards store fat reserves in their tails, and tail autotomy can significantly affect subsequent energy budgets [24, 59]. Autotomized crickets and lizards become more timid than intact animals, reducing their foraging rates, and the combination of restrictions on foraging and the energetic costs of regenerating the missing organ can cause lower body weights, reduced growth, and reduced reproduction in autotomized individuals [57, 59, 76, 109, 205, 440, 538, 545, 546, 587, 870]. Most arthropods have tactile, hydrodynamic, and chemosensory receptors in the tips of their appendages; loss of one or more of these appendages can impede subsequent monitoring of cues and

(A)

(B)

(C)

(D)

(E)

(F)

FIGURE 9.30 Accumulated skin scars on cetaceans Many species bear tooth scars from conspecific fights. Examples include (A) Curvier's beaked whales (*Ziphius curvirostris*), (B) Risso's dolphins (*Grampus griseus*), (C) bottlenose dolphins (*Tursiops truncatus*), and (D) sperm whales (*Physeter macrocephalus*). The small circular dots (E) on this Bryde's whale (*Balaenoptera brydei*) are scars from bites of cookie-cutter sharks (*Isistius brasiliensis*), and the raked scars and missing pieces in the tail flukes (F) of this humpback whale (*Megaptera novaeangliae*) reflect prior attacks by killer whales (*Orcinus orca*).

signals. The loss of chelipeds and their chemosensory organs in male crayfish inhibits subsequent location and assessments of females [66], and loss of sufficient numbers of legs used for both sensory and display functions in male wolf spiders reduces courtship duration and intensity [101]. Prior loss of legs also changes the outcomes of agonistic contests between male jumping spiders [791]. The trade-offs between autotomy and consequent costs likely explain why regeneration and autotomy are so spottily distributed within a taxon.

Tissues that are not replenished by metamorphosis, periodic molts, or regeneration wear out over time. As noted earlier, the stridulation apparatus of crickets gradually abrades after the last molt [192], and many mammals exhibit increasing scars and partial or complete loss of ears, tails, horns, fins, and digits with age (**Figure 9.30**) [27, 78, 220, 271, 292, 306,

307, 383, 393, 535, 543, 615, 618, 717, 743, 765]. There is then a premium on maintaining what is left. Self-grooming is widespread among arthropods, reptiles, birds, and mammals [52, 55, 171, 187, 230, 234, 259, 304, 394, 640, 647, 723, 739, 777, 871], and allogrooming is common in social arthropods, birds, and mammals [45, 52, 82, 93, 188, 252, 260, 456,

526, 552, 687, 766]. Demonstrated functions of grooming include the removal of ectoparasites (most taxa); cleaning of gills and antennae (Crustacea); conditioning of external coverings with secretions (birds and mammals); and spreading of social pheromones (social insects) [55, 82, 188, 234, 252, 304, 526, 687, 739]. As we shall see in the next chapter, many displays are often refined versions of preening or grooming motor patterns. Allogrooming can also take on secondary signal functions as outlined in Chapter 13. Most fish cannot self-groom and few allogroom conspecifics, but a number of species as well as a variety of crustaceans groom ectoparasites off of other species: the host fish become groomed, and the cleaners eat the removed organisms [176].

USAGE   A variety of different agents pose threats to the physical integrity of animals that engage in communication. Predators maim or kill their prey; they may be members of different species or cannibalistic members of the same species (spiders and mantids). Parasitoids are animals (usually wasps, flies, or staphylinid beetles) that specialize in finding and laying their eggs or larvae on the eggs, nymphs, larvae, or adults of specific host animals; the parasitoid larvae then consume and kill their hosts. Parasites also consume nutrients from either outside a host (ectoparasites) or inside (endoparasites); most do not kill their host but typically reduce its physical condition and energy reserves. Some species practice **dulosis**—making slaves of other animals. Dulosis usually does not damage slaves, but it terminates their fitness and thus is equivalent to death. Slavery primarily occurs in ants and victims may be either conspecifics or heterospecifics [189, 399, 410, 483, 851]. Combatants, whether members of the same or different species, can inflict injuries or death on each other. Physical conflict is common in competition for resources, mates, shelters, or dominance status; some species also engage in physical punishment when group members violate expected behaviors [165, 432, 866].

Physically damaging agents can impose costs on animal signalers in two ways: by eavesdropping on signals to detect and locate senders or receivers, or by mimicking a sender's

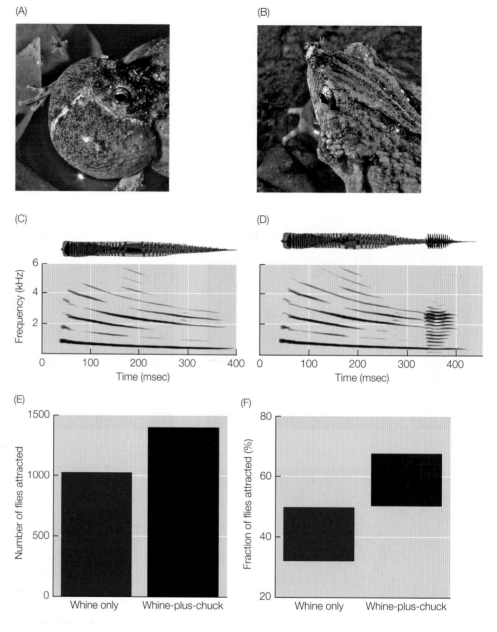

signal to attract and exploit receivers or other senders [378]. Examples of the first risk are widespread. Calling by male frogs can attract both giant water bug and bat predators [392, 643, 725]. Playbacks of a wood quail chorus attracted five different species of quail predators [361]. Hungry tree swallow nestlings that call more receive more food from their parents but are also more likely to attract grackle predators [523]. A wide range of parasitoids and blood-sucking parasites detect and locate their hosts using the hosts' auditory or olfactory signals (**Figure 9.31**) [12, 75, 257, 470, 508, 628, 689, 706, 826, 840, 874]. In both wild guppies and swordtail fish, predators home in on males using the same display traits favored by females in mate choice [242, 718]. Male pied flycatchers with brighter coloration are more often taken by hawks than are dull males, and while firefly flashes have minimal energy costs, they can suffer very high risks imposed by visual

◀ **FIGURE 9.31   Selectivity of blood-sucking flies (*Corethella* sp.) to calls of male túngara frogs (*Engystomops pustulosus*)** (A) Flies are attracted to frogs by the calls of displaying males. (B) Flies tend to aggregate on the snouts of nearby frogs (of either sex), where the flies pierce the skin and suck blood. (C) Waveform (top) and spectrogram (below) of a simple whine call of a male túngara frog. (D) More complex whine-plus-chuck of a male túngara frog. Female túngara frogs are more attracted to the whine-plus-chuck than to the whine alone. So are the flies, as shown by the greater numbers (E) and greater fractions (F) of flies attracted to a playback of a whine-plus-chucks versus a playback of whines alone. This preference by the flies is not due to easier localizability of the more complex call, but more likely because additional prey (females) are present when males emit the complex call. (C–F after [75].)

predators [527, 758, 867]. Threat signals during conflicts often require close proximity to opponents and thus entail a greater risk of injury [209, 663, 780, 781, 823]. The electrical signals of both gymnotid and mormyrid electric fish are widely exploited by predatory catfish with electrosensory capabilities (**Figure 9.32**) [772, 773]. In spiders, multimodal signals are more likely to attract predators than single-modality ones [707], and prey localization by predatory bats is more accurate when frogs emit more complex calls [643].

A variety of agents exploit other species by mimicking their signals. Bolas spiders (*Mastophora* spp.) attract and catch male moths by mimicking the sex pheromones of female moths on a sticky blob swung on the end of a silk thread [379]. A predatory katydid (*Chlorobalius leucoviridis*) mimics the "wing-flick" sounds of receptive female cicadas; this attracts male cicadas that are captured and eaten [560]. A similar trick is used by predatory *Photuris* fireflies that mimic flashes of female *Photinus* fireflies normally given to flashing male *Photinus*; males that approach the *Photuris* sufficiently closely become its prey [533, 816]. A host of different arthropods mimic relevant acoustic and pheromonal signals, allowing them to live and reproduce as "social parasites" inside the colonies of termites, ants, wasps, and bees [7, 41, 136, 235, 731]. Pheromones of female cockroaches (*Blatella germanica*) cause males to assume postures that allow females to feed on nutritious male secretions prior to copulation; immature nymphs emit a pheromone that tricks males into immobility long enough to steal secretions [236]. Slave-making ants use a variety of mechanisms to match, eliminate, or manipulate the colony-specific odors of groups that they will enslave [63, 124]. Many orchids release pheromones similar to those of female wasps and have flower structures that mimic the shape of female wasps, attracting male wasps that copulate with several successive flowers, thus cross-pollinating them [305]. Whydahs that lay their eggs in the nests of other bird species mimic the songs of their hosts, and several avian nest parasites produce eggs that match the color patterns of their hosts' eggs [132, 650]. Fish that groom ectoparasites off of other species are often mimicked in color and patterning by predatory species that approach hosts presenting for grooming and bite them; such aggressive mimicry is common among coral reef fish [604].

Communicating animals may increase their vulnerability to dangerous agents in additional ways. Persistent signals such as elongated tails in birds may reduce agility sufficiently that predator escape is hampered [104, 113, 378, 607]. Both senders and receivers may be distracted from predator surveillance during signal exchanges [161, 226]. Male song performance in passerines can be significantly affected by diseases and parasite levels in the singers [218, 293, 316]. We shall discuss the complex interactions between signaling and disease in later chapters.

Signals and brains

DEVELOPMENT AND MAINTENANCE   Communication relies on the functioning of both sender and receiver brains. Even when a signal is largely dependent on nonbrain anatomical features such as the color of a bird's plumage or the size of a fiddler crab's large claw, the fluffing or hiding of plumage and the manner in which the crab's claw is waved are mediated by the sender's brain. Receivers process sensory inputs, store codes, and make decisions using their brains. This raises the question of whether brains, like energy or physical integrity, are limiting commodities. While it was once thought that the numbers of brain neurons, and thus brain sizes, were fixed upon maturity, it is now recognized that adult brain composition may be significantly dynamic: large numbers of new neurons can be created in central brain areas and then migrate to the olfactory bulbs in mammals, the song control centers in songbirds, and the hippocampus in both taxa [17, 43, 331, 334, 525, 624, 804]. Adult neurogenesis occurs even more broadly in insects, fish, amphibians, and reptiles [447, 728, 760, 822, 875].

While some brain areas can be augmented, given sufficient need, there is evidence that augmentation carries trade-off costs. In both birds and mammals, the target areas for new brain neurons usually increase in volume during one season, and then shrink at a later one [42, 48, 102, 200, 202, 752, 804, 873]. In most cases, augmentation occurs prior to or during a period when the activities overseen by the enlarged area increase, and shrinkage when those activities wane (e.g., caching food for winter, singing in the spring, and defending larger territories during breeding periods) (**Figure 9.33**). Similarly, where certain activities, such as song learning in some species of songbirds, are limited to one sex, the relevant brain nuclei, initially present in both sexes, atrophy in the sex not performing that activity as the animal matures [38, 627]. Seasonally enlarged cerebral nuclei of the song system in songbirds expend much more energy than the same nuclei in their unaugmented state [4, 586, 644, 837]. All of these observations suggest that there are costs to maintaining an enlarged part of the brain when it is no longer needed. Similar conclusions arise from cross-species comparisons: while brain size scales with body size according to very standard rules in animals, certain species have larger brains than expected, given their body size, and this often correlates with some higher-than-average social or ecological function [83, 121, 198, 256, 324, 325, 420, 422, 475, 507, 605, 606, 641,

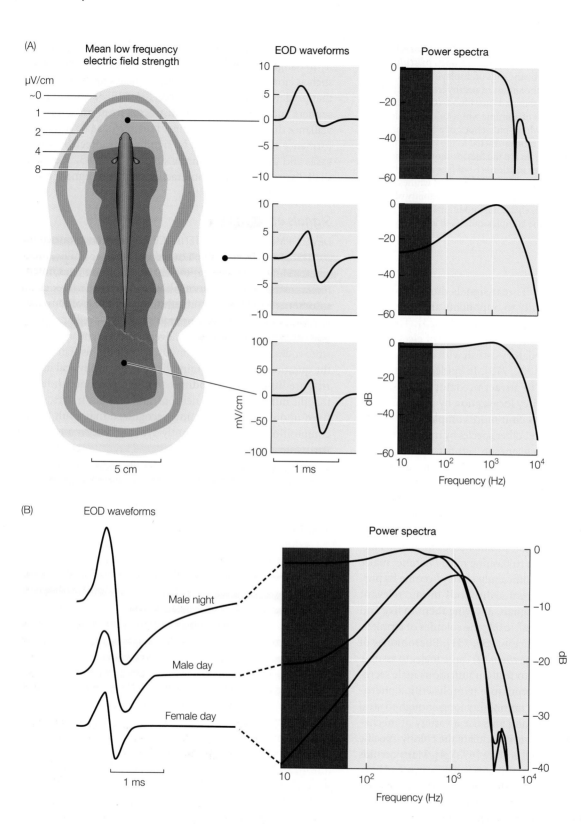

(A)

Mean low frequency electric field strength

EOD waveforms

Power spectra

(B) EOD waveforms

Power spectra

Male night

Male day

Female day

697, 728, 740, 760, 864]. In most cases, the areas that are enlarged are ones known to oversee the expanded activities. While concerns have been expressed about uncritical interpretations of brain size and function correlations [157, 380, 493], the weight of the evidence strongly suggests that brain

regions are enlarged only when the additional functional benefits outweigh the costs.

The size of a brain or any part of one depends not only on the number of neurons but also on their size. The diameter, conduction speed, and energetic costs of brain neurons are

◀ **FIGURE 9.32   Predation risks of electric fishes** Most neotropical gymnotiform electric fishes utilize both low- and high-frequency components of their electric organ discharges (EODs) for social interactions, courtship, and electrolocation. As discussed in Chapter 7, low frequencies are detected by the fishes' ampullary electroreceptors and higher frequencies by their tuberous electroreceptors. Sympatric predatory catfish have only ampullary electroreceptors and thus can detect only the low-frequency components in the EODs of gymnotiform prey. (A) One species of gymnotid, *Brachyhypopomus pinnicaudatus*, can produce slightly successive discharges at the head and tail ends of the body. At distances greater than 5 cm from the fish, negative interference between these two waveforms cancels out the lowest frequency components in the EOD, making it hard for predatory catfish to detect the signals (middle waveform and power spectrum; red zone indicates frequencies detectable by catfish). At closer distances, and especially near the head (top waveform and power spectrum) and tail (bottom waveform and power spectrum), one of the discharges is more intense than the other and the full spectrum of the EOD is available for sampling by nearby conspecifics. (B) While females always employ this strategy, males shift to different waveforms at night when they are actively courting females and harassing neighboring males. This strategy trades off a greater risk of predation for a greater range over which the males can interact socially with conspecifics. (After [774].)

acquire its code [291, 496]. At one extreme are cases in which the relevant brain circuits are fully laid down during development even if the animal is raised in isolation, and are little changed if such exposure is provided at any stage; at the other extreme are cases in which social learning is the *only* source of information about the code. There is considerable evidence that long-term learning, particularly social learning of signals, involves different processes of synaptic plasticity, neural energetics, and brain region coordination than does fixation during development [35, 40, 43, 156, 401, 457, 629, 660, 705, 872]. This does not mean that the two processes need be exclusive for any particular brain function. In fact, the same brain circuits used for signals that are not socially learned in one taxon may be elaborated to accommodate social learning in others [54, 255]. Few examples fall at the extremes noted above: instead, the acquisition of signal codes usually requires a mixture of intrinsic (often genetic) influences and social learning [660]. It is largely the relative weighting in this mix that varies between taxa, signal functions, and even between senders and receivers using the same signal set.

Neither senders nor receivers of most arthropods, fish, amphibians, and living reptiles require exposure to signaling

strongly linked: large-diameter axons are faster than small-diameter ones, especially if they are myelinated, but there are construction and maintenance costs for this speed [829]. Conduction speeds, and thus neuron sizes, are tightly linked to brain functions: where stimuli from multiple sensory sources converge on the same target neurons simultaneously, larger neurons are usually obligatory. Similar constraints apply to connections between the two sides of the brain, or between the dispersed substations overseeing complex processes such as bird song (**Figure 9.34**). Here, the timing and coordination of activity in different brain regions can be critical [421, 430, 697]. Nutrient supply may constrain both the density and frequency of firing for a given part of the brain, and thus limit its potential functions [841]. Regional parameters such as sensitivity, learning capabilities, and connectivity define the appropriate neuronal types, densities, and dynamics. All of these factors will impose trade-offs on the optimal size and structure of different brain regions [26, 635]. In addition to within-brain trade-offs, there may be significant trade-offs between the brain and other body organs in the allocation of energy and other resources [417–419, 626]. At least in parrots, increased cognitive abilities require not only larger brains but also larger genome sizes [19].

Communication hinges on senders and receivers acquiring similar codes for a given signal set. Code acquisition typically involves a mixture of genetics, innovation, practice, chance events, and any of several types of learning [61, 557, 558, 609]. An important consideration is the degree to which either party relies on **social learning**— the monitoring of other animals that are using the relevant signal set—to

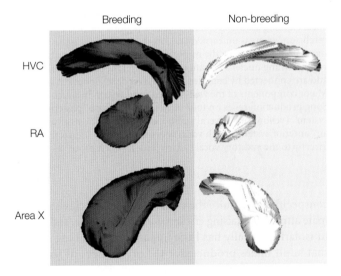

Breeding   Non-breeding

HVC

RA

Area X

FIGURE 9.33   Seasonal changes in volume of cerebral brain regions (nuclei) overseeing the song system of the three-spotted towhee (*Pipilo maculatus*) The HVC and RA nuclei oversee the production of learned songs, and the Area X nucleus participates in song learning. During the breeding season, the sizes of HVC and RA are 288% and 235% larger, respectively, than in the nonbreeding season, and Area X is 160% larger. As the annual breeding season ends, the enlarged nuclei shrink back down to nonbreeding sizes. Similar patterns of change, but with varying magnitudes, are seen in most songbirds. Whereas changes in HVC volume are due to changing rates of neuronal death versus new neuron incorporation, changes in RA volume appear due to variations in neuron size and the density of dendritic branching. These seasonal changes are largely, but not solely, triggered by varying levels of circulating testosterone [71, 102, 798, 804]. (After [102].)

(A) Songbird

(B) Parrot

(C) Hummingbird

(D) Human

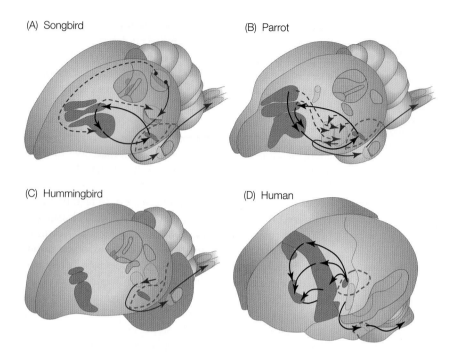

FIGURE 9.34 Brain allocations and patterns of connectivity in four vocal learners Vocal learning is the imitation of heard models in the acquisition of vocal signals. It is found in songbirds, parrots, and humans, and in some hummingbirds, cetaceans, bats, and possibly seals and elephants. Where examined, vocal learning always involves dedicated regions (nuclei) in the cerebrum, whereas unlearned and largely innate vocal signals are most often overseen by the midbrain and medulla. The multiple vocal learning nuclei are scattered over different parts of the brain and are connected by complex links and feedback loops. (A) Major components of the well-studied song system in songbirds. Song production relies on four nuclei constituting a "posterior system" (yellow) and several additional nuclei (red) constituting an "anterior system." Both vocal systems are linked indirectly and directly to the auditory vocal system (blue), which processes and categorizes heard songs. The posterior system stores currently learned song templates and when triggered, sends the relevant nerve signals to the syrinx and respiratory muscles to produce a song. The anterior system oversees learning new songs and modifies the fine structure of existing songs according to social context. Some of the known connections between nuclei in the same system and in different systems are shown as black, red, and dashed arrows. As shown in Figure 9.33, some nuclei in each system may vary in size seasonally. (B) Known nuclei in the vocal learning system of parrots and a few of the known connection linkages. (C) Hummingbird vocal learning has been little studied, but both anterior and posterior vocal systems appear to be present, as are relevant auditory processing centers. (D) Vocal learning nuclei in human brains with proposed color-coded equivalences to bird systems [255, 429, 431]. (After [431].)

conspecifics to acquire the signal codes for species-specific mate attraction. Raising crickets, katydids, cicadas, and frogs in isolation usually has little impact on the calling songs that adult males produce nor the preferences of females for a species-specific song pattern [308, 353]. Some senders may make minor strategy adjustments with continued social experience [197, 228], and females may become more discriminating with practice [381, 468]; but overall, social learning plays only a minor role in most species. In many of these ectothermic taxa, pulse rate and other parameters in male song change with ambient temperature, and interestingly, so do female preferences [64, 270, 308, 353]. This phenomenon prompted the suggestion that males and females might rely on a common pattern generator for both signal generation and species recognition [10]. This would account for the temperature effects, and because the same genes might oversee generator development in both parties, coevolution of the male signal and female preference would be assured. Whereas some early studies on hybrid crickets supported these notions [409], subsequent studies found different parts of insect brains overseeing signal generation and reception, and different genes affecting male signaling and female preferences [62, 81, 125, 126, 352, 354, 704]. While shared pattern generators remain unlikely, more recent genetic studies have found that many genes of small effect influence male signal and female preference traits in these taxa. Examples include the acoustic signaling of Hawaiian crickets (see Figure 9.12), olfactory signals in *Drosophila*, color patterns in *Heliconius* butterflies and Japanese killifish, and fin size and female preferences in zebrafish [92, 268, 286, 322, 335, 356, 484, 553, 750]. Instead of a single gene controlling both signal and receiver templates, there are various degrees of overlap in the suites of genes overseeing each party's behaviors.

At the opposite extreme are many vocal signals of songbirds, parrots, and humans for which social learning plays a major role in acquisition of both sender and receiver codes. In each of these taxa, both parties begin code acquisition

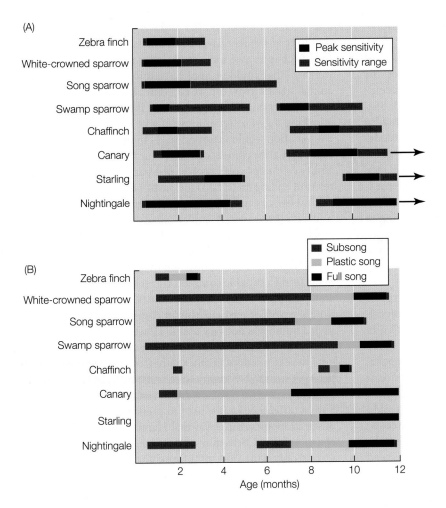

FIGURE 9.35 Interspecific variation among songbirds in timing of song learning (A) Sensitive phase in which young birds of various species are most likely to memorize songs of neighboring adult conspecifics. For each species, the black band indicates peak sensitivity and the red band shows the range over which there is any sensitivity. Note that some species may have two successive sensitive periods within a year and then halt further learning. Canaries, starlings, and nightingales have annually recurring sensitive periods when they can learn new songs (arrows). (B) Timing of the three primary steps in the motorsensory phase of song acquisition: subsong (soft, rambling, with little repeatable structure); plastic song (with recognizable repeats but significant variation), and crystallized full song (clear motifs that are distinct and repeatable). (After [413].)

with juvenile monitoring of adult vocal signal production (**Figure 9.35**); senders then progress into a motorsensory phase in which they produce a diverse medley of memorized and invented sounds, gradually organize these into repeatable motifs, and then winnow down the larger set into a final crystallized repertoire that is an appropriate match to what they heard as juveniles or are currently hearing as recent dispersers [135, 217, 413, 427, 532]. These processes are brain-intensive and are suspended during nightly sleep periods to consolidate progress made during the day [178, 194, 204, 261, 747, 827]. Each of these taxa imposes a number of constraints to ensure that juveniles only attend to suitable models. All show species-specific and largely genetic predispositions to favor model sounds that fall within some acceptable range of characteristics. Juvenile songbirds often focus on a key species-specific component in adult songs to identify conspecific models [554–556, 764]. Although captive parrots seem able to replicate nearly any sound, studies of species in the wild show much more conservative and species-specific ranges of calls [100]. Humans also have similar limits on the types of sounds that are acceptable speech components [488]. Variation is further constrained by limiting the memorization phase to a **critical period** when suitable models are most likely present

[413], and to tutors with whom the potential pupil has particular social or spatial relationships [5, 168, 254, 386, 437]. Female cowbirds and human parents provide feedbacks that steer a youngster's winnowing phase towards a suitable repertoire [488, 596, 842]; begging nestlings make adjustments after hearing nestmates' calls [403]; and nestlings of nest-parasitic cuckoos experiment until they find a begging call that elicits feeding by foster parents [503, 544].

In the prior examples, senders and receivers acquire their signal codes with similar levels of social learning. A common case in which one party relies more on social learning than the other involves identity signals. The signal components that provide individual identity information are usually generated with minimal social learning: examples include the unique markings on the heads of social wasps, individually distinct parent or offspring calls in communal breeders like seabirds, penguins, seals, bats, and ungulates, and the unique olfactory signatures of many mammals [799]. Diversity in individual identity signals arises from diversity in the senders' genetic backgrounds, body sizes, environmentally modulated growth patterns, innovation, diets, and accidental events [150, 233, 267, 272, 377, 411, 565, 571, 778]. The only way receivers can acquire these codes is by learning to associate a given signal with a given sender.

A similar process frequently underlies receiver acquisition of species identity signal codes [227, 285, 478, 767, 793]. Where species are highly sexually dimorphic and the costs of mating with the wrong species differ for the two sexes, the relative importance of social learning for receiver code acquisition may differ between males and females [60, 263, 351, 465–467, 479, 485, 495, 642, 742, 792, 794, 803]. As with bird song, receiver acquisition of species and gender identity codes may require exposure to suitable templates during a critical period and show little subsequent refinement; such code acquisition is called **imprinting**. A variation on this theme is the acquisition of codes for bird and mammal alarm calling [400]. In most species, senders need little social learning to produce species-specific alarm calls, but may require exposure to suitable examples before they can limit alarm calls to correct stimuli.

USAGE   Information stored in the brain, unlike energy or body parts, cannot be "used up": the same information can be signaled again and again with no requisite decrement in its content [491]. However, brain tissue is energetically costly to maintain, and preserving acquired information imposes a persistent cost on the individual. Another consequence of preserving information is that regions of the brain assigned to specific functions may "fill up." While it is possible to add neurons and synapses, we have seen that this is usually costly at multiple levels. For a given brain function, these constraints lead to performance trade-offs such as between speed and accuracy of decision making. For some animals, physical risk is sufficiently costly or their brains are sufficiently large that they rely on two parallel brain circuits: one that is "quick but sloppy" and another that is "slow but accurate" [122, 158, 240, 250, 505, 807]. The fact that one circuit does not suffice confirms that it is too expensive or physically difficult to build a single circuit that maximizes both speed and accuracy. Parts of the brain assigned to different tasks may also interfere with each other: the tools for individual identity recognition or mate choice may confound those used for species recognition [65, 480, 531, 654]. In addition to time, energy, and risk considerations, brain trade-offs surely contribute to the widespread reliance on imperfect signaling by animals. As noted in Chapter 8, biases and rules of thumb typically place smaller demands on brains than do perfect signals and completely Bayesian receiver protocols. The costs of these reduced brain investments are increased decision errors by receivers that can affect both parties.

The Value of Information

In Chapter 8, we examined a number of measures of the amount of information provided by a signal set. However, the amount of information is useful to a decision maker only if it is relevant to the current or subsequent decisions. We

thus need some way to filter and weight signaling information in terms of its relevance to fitness [151, 191, 211, 444]. In this chapter, we have reviewed a variety of "economic weights" that could be applied to information provided by animal signals. How can we combine all these economic considerations into an average expected fitness for an individual adopting a given signaling strategy? For receivers, such a combination would incorporate (1) the average probabilities that alternative contexts might occur, (2) the current probability estimates given receipt of a particular signal and the relevant coding matrix, (3) the payoffs of alternative responses, and (4) any indirect effects of engaging in communication with the sender. A similar calculation for senders would replace the payoffs to receivers with those for senders and the indirect receiver costs of communication with those experienced by senders. A combination of the effects of all components using a suitable common currency, which for this general example we can call **fitness units**, results in an average fitness payoff of participating in communication. By itself, such a number is not very useful. However, it becomes much more useful if we compare it to the same party's average fitness were it to adopt an alternative strategy. The difference between two such computations is the **value of information** that the first strategy provides over the second [332, 769]. Such a difference can be used to predict the likely direction of selection in the same way that Hamilton's rule or the Price equation is used. Selection will favor the first strategy over the second only if the value of information is positive.

The simplest application of the value of information is to ask is whether it pays for a receiver to attend to a given signal set as opposed to ignoring it. We saw in Chapter 8 that an animal without access to signals might optimally base its decisions on a comparison of the prior probabilities that a given condition is true with a "red line" value based on alternative payoffs. Although such a receiver always adopts the same action unless prior probabilities or payoffs change, this default strategy actually works quite well and is invariably better than choosing actions at random. Switching over to a strategy of relying on signals must provide at least as good a payoff as the default strategy, and a better one if the receiver is to compensate for any additional costs of communication. A key variable that determines whether the value of information for signaling versus nonsignaling is positive is the average reliability of the signal set [99, 476, 576]. This must be greater than a threshold value set by the default strategy fitness and the costs of signaling, and can lead to negative values of information if it is below that threshold [492, 813]. A signal set may have a positive value of information at close range, but a negative value at a distance where the signals are no longer sufficiently reliable (**Figures 9.36** and **9.37**).

If there are significant costs to the receiver of using signals, there will likely be an optimal reliability for receivers

(A) Condition priors (%)

Condition			
Leopard	Eagle	Other	No predator
5	8	12	75

(B) Overall coding matrix (%)

Assigned signal	Condition			
	Leopard	Eagle	Other	No predator
Alarm S1	75	10	9	5
Alarm S2	7	77	7	4
Alarm S3	4	2	56	6
No call	14	11	28	85

(C) Action/condition payoff matrix (%)

Action	Condition			
	Leopard	Eagle	Other	No predator
Climb tree	**700**	200	700	700
Hide in bushes	200	**800**	400	800
Stand alert	0	0	**800**	800
No change	0	0	600	**1000**

(D) Actions and conditions (%)

Selected action	Condition			
	Leopard	Eagle	Other	No predator
Climb tree	30	7	5	3
Hide in bushes	11	47	6	4
Stand alert	12	9	35	9
No change	47	37	54	84

Average payoff without signals: 822 fitness units
Average payoff with signals: 847 fitness units
Value of information: +25 fitness units

FIGURE 9.36 Computation of the value of information for signals given at close range Using the coding matrices outlined in Figure 8.34, we compare a receiver monkey who slavishly attends to alarm signals versus one who ignores all alarm signals. At close range, the relative reliability of the alarm signal system is 36%. We assume that the cost of listening for alarm signals is negligible and that average payoffs of each strategy depend only on the prior probabilities of each predator appearing, reliabilities of alarm signals for each condition, and payoffs of particular action–condition combinations. (A) Prior probabilities that a given predator condition will occur in any given observation period. (B) Overall coding matrix combining sender, transmission, and receiver effects. Rows are expected categories to which the receiver assigns propagated signals. (C) Hypothetical payoff matrix for a receiver adopting alternative actions (rows) given alternative conditions (columns). Cell values are in arbitrary "fitness units." All combinations result in lower fitness than the combination in which no predator is present and the monkeys make no change in their behavior (lower right cell of table). Zero values in the lower left corner of the matrix imply the death of the monkey taking that action in that condition. Average payoff of any action equals the summed cell values of that row, each discounted by that condition's prior probability; this calculation gives an average payoff of 660 fitness units for "Climb tree," 722 for "Hide in bushes," 696 for "Stand alert," and 822 for "No change." "No change" is thus the optimal default action if receivers ignore signals. (D) The fraction of time that a receiver paying attention to signals would perform the actions listed on the left when the conditions heading the columns were true. This is simply the matrix in Figure 8.34D with the appropriate actions replacing the perceived predator condition by the receiver. Bottom: average payoffs to a receiver ignoring signals, to a receiver attending to signals, and the value of the information (difference of two payoffs). The value of the information is positive and this argues that receivers should attend to these signals. See Web Topic 9.2 for computational details.

that is greater than the minimum threshold but less than that required for perfect information [99, 201, 476]. Reliability depends on investments by both senders and receivers. If sender expenditures are significant, senders will favor the minimum reliability that justifies receiver attention to signals. If the optimum for receivers is higher than the sender's minimum, a conflict can arise between the two parties over their relative levels of investment in the signal exchange [99, 435, 476].

The value of information combines all of the components that we have discussed in this and the prior chapter into a single economic metric that can be used for evolutionary modeling and prediction. In general, we can use fitness as the relevant currency for payoffs. However, there may be conditions in which payoffs are sufficiently cumulative that the gain or loss in fitness for the same decision increases with with time. In this case, some transformation, such as the logarithm of fitness, may be more appropriate in computing the value of information [213].

Our discussions above show why evolution rarely favors the provision of perfect information. As long as perfect information is not favored, both parties are bound to make mistakes and this need not be construed as cheating or dishonesty. The frequency of errors will depend on the probabilities that different conditions will occur, the relative payoffs of correct versus incorrect receiver decisions, and the costs of investing in signaling [476]. While the example above compared signaling to nonsignaling, the value of information can be just as useful when comparing the evolutionary economics of two alternative signaling strategies for the same decision process. Examples include the optimal mix of signals used in conflicts [779], and the relative suitability of combinatorial versus noncombinatorial syntax when senders may cheat [492].

Web Topic 9.2 *Computing the value of information*
Here we review several approaches to computing the value of information for animal signaling contexts.

FIGURE 9.37 **Computation of the value of information for signals given from a distance** Here are the same four matrices as in Figure 9.36, adjusted for distance effects. In this case, although the sender coding matrix is unchanged, the propagation distortion reduces the relative reliability of signals at a distance to 14% (see Figure 8.35). The value of the information is now negative: receivers should not attend to these signals at this distance.

(A) Condition priors (%)

Condition			
Leopard	Eagle	Other	No predator
5	8	12	75

(B) Overall coding matrix (%)

Assigned signal	Condition			
	Leopard	Eagle	Other	No predator
Alarm S1	45	13	19	8
Alarm S2	16	61	17	7
Alarm S3	6	5	26	13
No call	33	21	38	72

(C) Action/condition payoff matrix (%)

Action	Condition			
	Leopard	Eagle	Other	No predator
Climb tree	**700**	200	700	700
Hide in bushes	200	**800**	400	800
Stand alert	0	0	**800**	800
No change	0	0	600	**1000**

(D) Actions and conditions (%)

Selected action	Condition			
	Leopard	Eagle	Other	No predator
Climb tree	11	7	6	4
Hide in bushes	11	23	10	6
Stand alert	15	15	15	11
No change	63	55	69	79

Average payoff without signals: 822 fitness units
Average payoff with signals: 811 fitness units
Value of information: −11 fitness units

SUMMARY

1. **Biological economics** combine the **benefits** and **costs** to an individual of expressing a given anatomical or behavioral trait into a net **payoff**. The expressed value of a trait is part of an individual's **phenotype** and is a result of both heritable components—the individual's **genotype**—and environmental influences, including learning. **Evolutionary economics** shifts the focus from the individual to the average payoff of a given **strategy** when alternative strategies are also present in the population. **Selection** occurs within a generation when individuals adopting one strategy gain a higher payoff than do those adopting alternative strategies: the relevant payoffs for **natural selection** concern fecundity and survival, whereas those for **sexual selection** involve successful competition for matings.

2. The proportion of a population adopting a given strategy changes over successive generations. Some of these changes are due to selection in prior generations. The degree to which selection in one generation contributes to changes in strategy frequencies in the next depends upon the **heritability** of the strategy: higher heritability results in greater

responses to selection in successive generations. Other factors contributing to the changes in strategy frequencies over generations include **drift** (random variation in individual survival and reproduction), genetic and cultural **mutation**, and **migration** into and out of the population. Change in the proportions of alternative heritable strategies over multiple generations is called **evolution**.

3. It ought to be possible to predict the trajectory of subsequent evolution, taking into account the payoffs of alternative strategies. This effort can become complicated when, as with communication, the payoffs for one strategy depend on how often other strategies are currently being employed. This is called **frequency dependence**. One approach is to use **economic models** that simplify the evolutionary process to the minimum needed for useful prediction.

4. Evolutionary models based on the **phenotypic gambit** assume that the details of genetic heritability rarely affect the direction of evolutionary trajectories. **Simple optimality models** make this assumption and focus on

evolutionary trajectories in which the payoffs of a given strategy do not depend upon how many other individuals are adopting that or other strategies. Because signaling payoffs do depend on the actions of both senders and receivers, simple optimality models are rarely of use in the study of communication. In contrast, **evolutionary game theory** explicitly incorporates frequency dependence in the payoffs of alternative strategies. This method looks for equilibria at which it does not pay for any individual to switch to another strategy. An equilibrium strategy that cannot be invaded by rare alternatives when it is itself common is called an **evolutionarily stable strategy** (**ESS**). A third approach, **adaptive dynamics**, begins with an equation that describes changes in strategy frequencies between generations, including any necessary frequency dependence, and applies this equation recurrently to track evolutionary change over successive generations. Equilibrium is only one possible outcome of an adaptive dynamic model: other possible outcomes include **limit cycles** (repeating patterns), **bifurcations** (sudden instabilities and transitions), and chaos (completely unpredictable trajectories).

5. There are a number of genetic complications that might undermine the phenotypic gambit. The nuclei of different species and different sexes of the same species may contain different numbers of copies of each **chromosome**. Depending upon how finely the nuclear **genome** of an animal is divided into separate chromosomes, **recombination** and **random assortment** during **meiosis** and sexual reproduction may not fully randomize which genes are found together in offspring, resulting in persistent **linkage disequilibria**. In many animals, sex is determined at the chromosomal level: in **haplodiploid** species such as many arthropods, females are diploid, whereas males are haploid. Among birds and mammals one sex (females in mammals and males in birds) has two **homogametic** (similar) sex chromosomes; the other sex has a pair of quite different (**heterogametic**) sex chromosomes.

6. More subtle complications involve interactions between genes. One of the two **alleles** at a chromosomal **locus** in a diploid animal may be **dominant**, or the two alleles may combine their effects; which happens depends on the **dosage** levels of the enzymes they produce. A given allele may affect more than one trait (**pleiotropy**), and the expression of any one allele may depend on what alleles at other loci are doing (**epistasis**). Where sex is determined by chromosomal differences, special adjustments must be made in the dosage levels of active alleles to maintain normal biochemistry. This in turn can lead to quite different evolutionary forces on the two kinds of sex chromosomes, differences in their evolutionary rates, conflicts between the sexes, and complicated patterns of sexual selection and signal evolution.

7. Several evolutionary models incorporate at least some of these genetic complications. **Polygenic quantitative genetic models** allow for correlations between multiple traits caused by pleiotropy or linkage diequilibria. While this is less constraining than the phenotypic gambit, it still ignores epistasis and assumes that the additive genetic determinants of heritability (summarized in a **G-matrix**)

do not change significantly between generations. Evolutionary equilibria using this approach are conveniently visualized as peaks or ridges in **adaptive landscape** maps. **Extended quantitative genetics models** minimally allow for certain levels of epistasis that can cause the G-matrix to evolve between generations. More ambitious approaches relax nearly all constraints imposed by other models, but are computationally intensive and not easily interpreted. Like polygenic quantitative genetics, **extended adaptive dynamics models** allow for genetic correlations between multiple traits. Unlike quantitative genetics methods, these methods can track multiple kinds of trajectories besides equilibria, and seek indicators that will predict which trajectory is likely in a given case. **Price equation** modeling partitions the response to selection into additive components, in contrast to the multiplicative components used for quantitative genetics models. This allows it to handle many of the genetic complications avoided by other methods. The cost is that other methods must be invoked to provide the information needed to apply it repeatedly over many generations.

8. The evolutionary modeling methods differ markedly in their assumptions and methods, yet they often produce similar general predictions. One advantage of multiple approaches is the opportunity to refine a general result by invoking that model whose assumptions are most met in the specific case. The generality of the predictions made using one method can be tested by invoking another method with quite different assumptions.

9. The primary currency used to compare the payoffs of alternative strategies is **relative fitness**. For populations of stable size, fitness is measured as the average **lifetime reproductive success** of individuals adopting a given strategy. Where population size is changing, the **finite rate of increase** is usually a better predictor. However, competition, spatial patterns, and temporal heterogeneity can all shift propriety from one measure to the other. **Invasion coefficients** are analogues of finite rates of increase that can be extracted from adaptive dynamic models.

10. None of the usual fitness measures may suffice for social behaviors since strategy adoption by one individual might affect the fitness of another carrying the same alleles for that strategy. **Relative inclusive fitness** takes this possibility into account is by combining the effects of an animal's actions on itself and others, each discounted by a **coefficient of relatedness**. This coefficient compares the probability that the animal and a recipient of its actions share alleles with that computed for two randomly chosen animals in the population. A donor animal may adopt costly actions to help (**altruism**) or harm (**spite**) another animal, depending on whether their relatedness is above or below the population average. **Hamilton's rule** specifies the relative benefit, cost, and relatedness values that allow such behaviors to evolve. Suitable recipients might be recognized by donors because they exhibited markers associated with shared alleles (**genetic kin recognition** and **greenbeards**), or because both donor and recipients grew up in **viscous populations**. Spite becomes more

likely in small or sub-divided populations and in dispersing offspring than in retained offspring. **Group selection** accounting partitions selective effects in a different way from inclusive fitness, and while equivalent, can provide additional insights into evolutionary processes.

11. Evolutionary economics are most often used to identify optimal strategies. **Life history** strategies depend on trade-offs between **fecundity** and **survivorship** that determine fitness. **Reproductive value** at any age is the sum of an individual's current and future fecundities, each discounted by the probability that the individual will survive to the relevant age. There is usually a trade-off between current fecundity and the remaining contributions to reproductive value. **Reproductive effort** is the expenditure of resources, usually scaled to body size, during a given life stage. An optimal life history is one that adjusts reproductive effort so as to maximize reproductive value at each stage in life.

12. Signaling puts demands on commodities used for multiple functions and thus generates trade-offs between signal benefits and costs. The development and maintenance of signaling organs impose energetic costs on the organism. Per-signal energetic costs range from minimal expenditures for singing birds to near maximal **aerobic scope** values in calling frogs and insects. Even for low per-signal costs, significantly repetitive signal emission can constitute 20–50% of daily energy budgets. The impact of signaling energy costs depends on whether the participant is an **income spender** or a **capital spender**.

13. Risks to **physical integrity**, such as disease or damage, impose additional costs on communicating animals. Although the ability to regenerate lost body parts is widespread among animals, it is most often limited to immature life stages. Most species practice self- or allogrooming to maintain the condition of their external surfaces. Damage risks come from predators, parasites, parasitoids, slave-makers, and conflict adversaries. Some of these may gain proximity to communicating victims by mimicking or eavesdropping on their signals.

14. Brains are costly organs; enlargement of one region to augment a specific function imposes trade-off costs on other brain regions, organs, and functions. Variation in signal code acquisition strategies often reflects this trade-off. In many arthropods, fish, amphibians, and reptiles, mate-attraction signal codes are largely inherited; in contrast, oscine birds' song codes require extensive social learning. Many individual identity signals have a mixed strategy; the sender requires little if any learning to produce the signal, but the receiver must learn to identify and interpret the signal. Brain trade-offs likely contribute to the reasons that most animals use imperfect signaling rules of thumb for decision making.

15. The **value of information** is the difference in expected fitness between two alternative signaling strategies. Because receivers that ignore signals have default strategies for decision making that are still far better than chance, a switch to signals must do at least as well as the default strategy and even better if it is to make up the costs of participating in communication. The threshold at which a receiver should pay attention to a signal rather than ignoring it is determined in large part by the signal set reliability. Senders and receivers may disagree over the optimal level of reliability.

Further Reading

Kokko [472] provides a highly readable introduction to the logic and methods of evolutionary modelling. Overviews of optimization economics as applied to animal behavior can be found in Parker and Maynard Smith [646] and Krebs and McCleery [481]. A more quantitative treatment is provided by McNamara et al. [582]. Maynard Smith [567] provides an overview of evolutionary game theory, and Dugatkin and Reeve [223] review a variety of applications of this approach to animal behavior. Nowak [632] introduces the logic of dynamic modeling, and Dercole and Rinaldi [203] review adaptive dynamics and its applications. The excellent introduction to nonlinear systems by Strogatz [775] may help readers to understand adaptive dynamics modeling. A good source reference for the interactions between evolutionary processes and genetics is Charlesworth and Charlesworth [143]. Classical and polygenic quantitative genetics analyses including the study of quantitative trait loci is reviewed in a detailed textbook by Lynch and Walsh [539]. The use of adaptive landscapes to visualize quantitative genetics models is nicely outlined by Steppan et al. [771] and Arnold et al. [25]. Frank [279] and Grafen [347] explain the utility of the Price equation for evolutionary modeling and show how this differs from alternative approaches.

An overall review of different measures of fitness and their relation to life history contexts can be found in Brommer [107]; and both Grafen [337] and West et al. [846] explain the logic behind inclusive fitness, kin selection, and group selection partitioning of fitness. A very readable summary of the differences between altruism, spite, and greenbeards is provided by West and Gardner [847]. Stearns [768], Roff [710], and Brommer [108] provide reviews of current thinking about the evolution of life histories. The energetic and physical integrity costs of signaling are reviewed for multiple modalities and taxa by Searcy and Nowicki [744]. Trade-offs in brain design and elaboration are discussed by O'Reilly [635] and Wang [829]. Sherry [752] and Nottebohm [631] describe seasonal changes in the sizes of avian brain regions as relevant functions wax and wane. For a good introduction to the value of information, see Stephens [769]; and for a more detailed application of this metric to animal signaling, consult Koops [476].

COMPANION WEBSITE
sites.sinauer.com/animalcommunication2e

Go to the companion website for Chapter Outlines, Chapter Summaries, and References for all works cited in the textbook. In addition, the following resources are available for this chapter:

Web Topic 9.1 *Equations of change and evolutionary models*

Many evolutionary models begin with a basic equation that assigns an initial state, predicts intergenerational change, and then projects a population through successive generations. Here we introduce and compare some of the equations currently used in this type of analysis.

Web Topic 9.2 *Computing the value of information*

Here we review several approaches to computing the value of information for animal signaling contexts

Chapter *10*

Signal Evolution

Overview

How do communication signals originate and evolve? Signal exchanges have intrinsic properties that set their evolution apart from that of most other adaptations. We have seen that the communication process requires an interdigitating set of adaptations by both sender and receiver. Each party pays costs to participate, but recoups its losses through benefits provided by the other party. How can such an exchange get started? Unless both parties play their role from the start, won't an initiator have to pay costs that are not compensated? The accepted answer to this question is that one or the other party must first evolve precursors of their eventual signaling role for reasons other than communication. Scenarios for signal origin thus fall into two main categories: sender preadaptations and receiver preadaptations. Once such precursors are in place, the other party can then take advantage of them, and this will be sufficient to initiate subsequent coevolution. Further refinement and elaboration of the signaling exchange leads us to the question of how honest information is encoded in the signal, given the fact that senders may benefit from exaggeration and deceit and receivers may benefit from eavesdropping or ignoring signals altogether. We shall explore the models and evolutionary processes by which most communication signals remain reliable most of the time, and the conditions under which honest signaling occasionally does break down.

The Evolution of Behavior

Evolutionary change is a sequential process. The powerful force of natural selection acts on the phenotypic and genetic variation among individuals in a population, favoring individuals with certain trait characteristics over other individuals in a given environment. Over successive generations, complex traits arise from less complex traits in a series of small steps. Each new adaptation is layered on top of preceding ones in an incremental process called **cumulative selection** [111, 118, 362]. Prior evolutionary history is very important in constraining the new adaptations that may evolve in a given environmental setting. With enough time, new adaptations can arise, set the stage for novel innovations, and be modified or lost. Evolutionary biologists use morphological traits and DNA-based genetic traits to construct **phylogenetic trees** depicting clusters of closely related species and the most likely ordering of sequential steps taken during speciation and taxonomic radiation (see Web Topic 10.1). Phylogenetic trees based on DNA sequences and large numbers of carefully chosen

FIGURE 10.1 **Partial evolutionary tree for the emberizid sparrows** This phylogenetic tree is based on complete sequencing of two mitochondrial genes. Song repertoire size has been dichotomized into either a single primary song type or a repertoire of three or more song types, and these two states have been mapped onto the tree. The song systems of these species are representative of other members of their genus. Two scenarios for reconstructing the ancestral state are shown here. The upper tree shows the scenario with a single song type as the ancestral state; asterisks show the four transitions that would have to occur, one of which is a reversal back to single song type. The lower tree shows the scenario with a repertoire as the ancestral state; three transitions would have to occur, one of which is a reversal. The lower scenario with the ancestor having a repertoire is the more parsimonious one, since it has fewer evolutionary transitions. (After [107, 281, 400].)

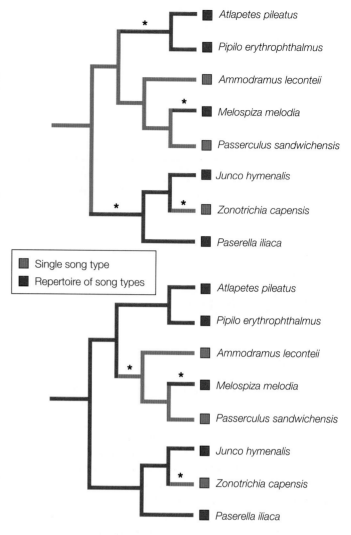

morphological traits tend to be more robust than those based solely on morphological traits. Behavioral traits evolve in the same way that morphological traits do, and in many cases, phylogenetic trees constructed with behavioral traits are highly concordant with trees constructed with morphological and genetic traits [46, 121, 491]. The forms of animal movement displays are particularly useful behaviors for phylogenetic analysis, because they are both evolutionarily fairly conservative and easy to identify and categorize. Once a robust phylogenetic tree has been estimated for a group of species, evolutionary biologists can then examine the occurrence of other traits in extant species and attempt to reconstruct the evolutionary scenarios for those traits. In tree-building and the reconstruction of ancestral traits, the logical principle of **parsimony** is used to select among alternative patterns. The most parsimonious scenario is the one with the fewest changes and the lowest complexity. However, sophisticated statistical methods sometimes select a less parsimonious solution as the most likely one. **Figure 10.1** shows an example of a reconstruction of the ancestral state of song repertoire size in emberizid sparrows.

> **Web Topic 10.1** *Estimating evolutionary trees*
> A guide to the process of reconstructing phylogenetic trees using multiple traits, and the measures used to describe their robustness and consistency. More examples of how to use phylogenetic trees to answer questions about signal evolution.

Once the evolutionary history of a trait has been described at the comparative species level, behavioral ecologists often attempt to understand the fitness costs and benefits of alternative strategies at the within-species level. Although for morphological traits a simple optimality approach may suffice, social behavioral traits such as mating strategies, aggressive behavior, parental behavior, and cooperative behavior require a different approach. In such cases, the fitness payoffs depend very much on what strategies other individuals in the population are using. As a result, a switch in strategy by one party may require a consequent switch in

the optimal strategies for other parties. Subsequent rounds of switch and counter-switch can lead to a complex coevolutionary trajectory. As explained in Chapter 9, these kinds of interactions are best envisioned as a game. Whether the game stabilizes at one or more equilibria can often be modeled and predicted with **evolutionary game theory**. This method specifically examines optimization when payoffs depend on the frequencies with which other organisms are using the available strategies. As we noted in Chapter 9, social interaction games are usually nonlinear systems, and thus may have no stable equilibria but instead exhibit recurrent cycling, sudden bifurcations, or chaos. The dynamics of social interaction games depend in part on the degree to which the two parties have conflicting interests. We might expect the dynamics of predator/prey interactions and contests over access to limited resources to have quite different dynamics and possible equilibria than mutualistic interactions in which both parties benefit from cooperative behaviors.

Communication signals are special traits that evolve for the purpose of influencing the actions of other individuals, so it is essential to view senders and receivers as two (or more) parties engaged in an evolutionary game. Senders are selected to produce signals that influence receivers to respond in ways that benefit the sender, and receivers are selected to attend to signals only if these help them make decisions that benefit them; otherwise, they should ignore the signals. In this latter case, senders will no longer benefit from producing the signal, and the communication exchange will be selected against. Therefore, for a signal to be maintained, it must contain enough useful information for receivers *on average* that they retain their receptive tuning and responsiveness to it. Here, again, the degree of conflict of interest is important. In cooperative contexts where the interests of the sender and receiver are aligned, such as in a beehive, selection on both parties favors efficient communication; dances by honeybee workers to communicate the location of food resources are highly reliable signals. However, in most communication exchanges there is some level of conflict, and the game theoretical perspective, with its emphasis on strategic interaction, can help us understand what conditions are needed to maintain reliable signals.

Models of Signal Evolution

As we begin the discussion of the role of sender preadaptations versus receiver preadaptations in signal evolution, it is important to keep in mind that selection acts on two components of signals: their efficacy and their strategic content [12, 136, 255]. The **efficacy component** ensures that the signal is detected and the message gets through to the receiver. As we learned in earlier chapters, signals are attenuated, degraded, and overlaid by irrelevant stimuli during transmission through noisy environments. By itself, this will favor conspicuous and distinctive signals that are easy to detect and discriminate. At the same time, such displays may be energetically costly to produce and may draw the unwanted attention of predators, parasites, parasitoids, and eavesdroppers. Optimal signal design from the efficacy perspective must take into account the cost and benefit trade-offs related to efficacy. The **strategic component** ensures the reliability of the signal. Receivers are under strong selection to respond to signals in ways that benefit them, so they will extract all the information they can from cues and subtle aspects of signals. By ignoring unreliable signals and cues, they select for signals that are constrained to provide reliable information. The strategic components of signal design often require senders to pay additional costs that guarantee honesty. Most signals contain elements of both design components, although efficacy is likely to be more important for long-distance signals, such as mate attraction signals, and strategic content will play a greater role in short-distance signals such as threat, courtship, and parent/offspring signals. In the two evolutionary scenarios described in the sections below, these two components come into play at different stages. The strategic component develops first in signals derived from **sender precursors**, and the efficacy component develops first in signals derived

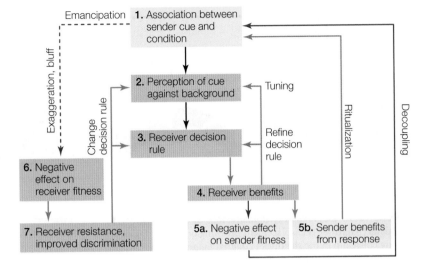

FIGURE 10.2   The sender-precursor model of signal evolution Blue boxes represent receiver steps, yellow boxes sender steps; green arrows indicate positive fitness effects, red arrows indicate negative effects. The first step (1) is the association between a cue and a condition by the potential sender. Receivers must be able to perceive or evolve receptors for the cue (2), and then incorporate the information into a decision rule and a response (3). If receivers benefit from their response (4), they will fine-tune their perceptual sensitivity, decision rule, and response. If senders do not benefit from the response (5a), they will attempt to decouple the cue/condition association. If senders do benefit from the response (5b), the cue will be modified via ritualization to maximize information transfer and will be transformed into a true signal. If the signal becomes emancipated from the condition, there is a possibility that senders could exaggerate or bluff, with a negative effect on the receiver (6). The receiver would then need to improve its discrimination ability and change or refine its decision rule (7).

from **receiver preadaptations**. The other component comes into play in subsequent evolutionary stages.

Signals derived from sender precursors

Figure 10.2 shows a flow diagram of the **sender-precursor model** of signal evolution. The first and critical step (step 1) is the existence of an association between a specific condition and the production of a motor pattern, structure, sound, or chemical cue by the sender. This association might arise because the cue is an inadvertent or unavoidable by-product of some activity performed in a specific context. The association may be initially imperfect, but it nevertheless may constitute a potential source of information to other animals as long as they can perceive it (step 2). Animals are generally designed to attend to such cues if the cues help them make better decisions [379]. In Chapter 8 we outlined how potential receivers might try to use these inadvertent cues in their decision making (step 3). If the information benefits the receiver (step 4), subsequent evolution may favor enhancement of the relevant links. For example, it may be worth the cost to modify the tuning of sensory organs to better detect and discriminate these cues. Cue reception is useful only to

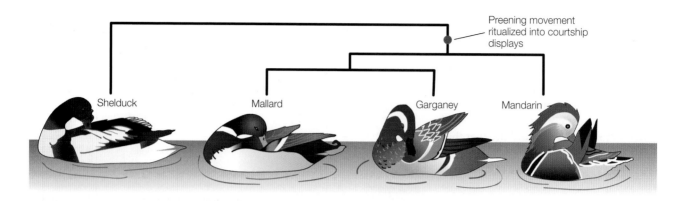

Preening movement
ritualized into courtship
displays

Shelduck   Mallard   Garganey   Mandarin

FIGURE 10.3   Degrees of ritualization of display preening in ducks (Anatinae) Displacement preening occurs sporadically during courtship in the shelduck (*Tadorna tadorna*) but is similar to normal preening and is therefore unritualized. The preening movement has become ritualized into courtship displays in more recently evolved species (at the red dot on the tree). In the mallard (*Anas platyrhynchos*), preening during courtship is partially ritualized in that the preening movements are restricted in range and directed at a patch of conspicuously colored feathers on the wing. In the garganey (*Anas querquedula*), ritualization has proceeded further, with movements that no longer serve any grooming function directed toward a light blue color patch. The mandarin drake (*Aix galericulata*) has the most highly ritualized preen, with the bill merely touching an enlarged conspicuous rust-red feather that becomes erect during courtship; the crest on the back of the head further emphasizes the preening movement. (After [299].)

the degree that the receiver has an accurate a priori estimate of which conditions elicit which cues (prior probabilities). Positive feedback from attending to cues may favor new and improved methods of code acquisition (learning, discounting). The value of cue information depends critically on how it is invoked to make decisions. Once attention to cues proves beneficial, selection may favor specific receiver preferences (biases) to regulate choices. Finally, the presence of this new information may favor the receiver adding new responses or refining existing ones (response refinement).

The bottleneck step is the next one: there will be no subsequent coevolution between the parties unless the change in decisions by receivers using the cues has some fitness consequences for senders (step 5). If sender fitness is increased, there will be positive feedback on the sender to improve the efficacy of the now incipient signal. **Ritualization** is the ethological term used to describe the refinement of an inadvertent cue into a true signal. This process involves one or more of the following four changes in cue features [509, 534]:

1. *Increased conspicuousness* makes the signal easier for the receiver to detect by enhancing the contrast between the signal and the background, through color, pattern, movement, or amplitude.
2. *Redundancy* allows the receiver to reconstruct the correct signal from an imperfectly received one [102]. The two main mechanisms for creating redundancy are repetition of the signal and the coupling of multiple parallel signal components (such as coupling a visual component with a sound).
3. *Stereotypy* minimizes variation in the display [353] and thus the number of categories into which the receiver must classify incoming signals. Display stereotypy may

also make it easier for the receiver to compare subtle differences in individual performance [548].
4. *Alerting components* attract the attention of the receiver. The addition of highly detectable but noninformational elements at the beginning of a display alerts the receiver that information is to follow [420].

Signals undergo varying degrees of ritualization. If the incipient signal is well associated with the context from the outset and provides all of the information the receiver needs, it may not be ritualized at all. A minimally ritualized signal may be difficult to distinguish from a cue. A classic example of the ritualization of preening into major courtship displays in ducks is illustrated in **Figure 10.3.**

Once selection favors sender ritualization, each change in the sender's signal design will result in corresponding selection on the receiver's perceptual tuning, code acquisition, decision preferences, and alternative responses. This will in turn have effects on the sender, which, if positive, will result in a new round of ritualization. The result is thus a cyclic loop of coevolution that can cause the final signal to be quite different from the original cue precursor. In many cases, inadvertent cues that served as precursors of signals remain within the behavioral repertoire of a species. With increasingly accurate phylogenetic trees in selected taxa, we can trace the sequence of changes that have occurred between cue and signal (**Figure 10.4**). Finally, a highly ritualized signal may become **emancipated** from the internal and external factors that originally triggered it. Emancipation removes the coupling between the initial cue and the condition and allows for the possibility of coupling the signal with new information. Strong selection from receivers, such as intense sexual selection by choosy mates, could result in a cue's development into a new signal that reveals strategic information about the

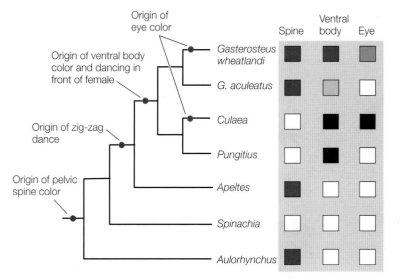

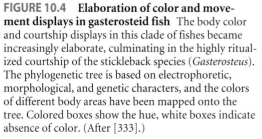

FIGURE 10.4 Elaboration of color and movement displays in gasterosteid fish The body color and courtship displays in this clade of fishes became increasingly elaborate, culminating in the highly ritualized courtship of the stickleback species (*Gasterosteus*). The phylogenetic tree is based on electrophoretic, morphological, and genetic characters, and the colors of different body areas have been mapped onto the tree. Colored boxes show the hue, white boxes indicate absence of color. (After [333].)

sender's condition (see the green loops on the right side of Figure 10.2). On the other hand, emancipation can also open the door for bluffing or deceitful manipulation of the receiver (see the dashed red line on the left side of Figure 10.2). The negative effect of unreliable information (step 6) will require the receiver to improve its ability to detect the truth, and may force it to refine its decision rule and responses (step 7).

Signals may originate from a wide variety of precursors with adaptive but noncommunicatory functions, including basic survival behaviors such as foraging, grooming, and fighting, and physiological processes involved in growth, maintenance, and reproduction. Because it is often possible to determine the nonsignaling source of a visual display, it is also possible to make an educated guess about the initial meaning of the cue as it evolved into a signal. In the sections below, we show how the precursor provides initial content and meaning for the signal.

LOCOMOTORY AND FEEDING APPENDAGES Animals recruit their moveable **locomotory and feeding appendages**—arms, legs, wings, mandibles, bills, and tails—to produce acoustic, visual, and tactile displays. Many examples of percussive and stridulatory sound production in arthropods and vertebrates were described in Chapter 2. One example not mentioned earlier that illustrates the inherent information principle is the flight sound of some insects. Flies, mosquitoes, bees, and moths produce a continuous tone with their wings while flying [484]. The fundamental frequency of the tone is directly related to the wingbeat frequency: smaller species tend to produce a higher-pitched sound than larger species, and within a species, wingbeat frequency and pitch increase with increasing ambient temperature and wing length. However, only a few species use their flight sounds for intraspecific communication. Male mosquitoes possess sound-sensitive hairs on their antennae that are tuned to the wingbeat frequency of females of their species; they use this ability to locate and identify prospective mates [80, 173, 394].

Some flies, wasps, and bees vibrate their wings in short-distance social contexts such as courtship and hive interactions, but it is not clear whether the sound or the tactile vibration is the more salient component [222].

The drumming of woodpeckers against hollow trees nicely illustrates the source and ritualization process of a percussive signal [330]. The chisel-like bill of woodpeckers not only enables these birds to dig into wood to obtain wood-boring insects, but it is also employed in the excavation of nest cavities in tree trunks. When a woodpecker forages or constructs a nest, it chops with a variable rhythm and thereby produces a sound with a variable temporal pattern and poor resonating quality. Paired birds give a ritualized tapping display during the nest construction stage as part of courtship. When the partner of an excavating bird approaches, the excavator moves to the outside of the cavity and taps rhythmically on the edge of the entrance hole. Most woodpecker species also produce a very loud drumming display that functions in territorial defense and mate advertisement. The birds usually drum on known hollow tree stumps or branches to improve the resonating quality and amplitude of the sound. The drumming in this case is more rapid and precisely patterned. This ritualization process of stereotyping the form and temporal pattern of the display not only distinguishes the social signal from its normal source but also serves to differentiate alternative mechanical signals [289, 471, 472, 541].

The movement of appendages is commonly incorporated into visual displays. Ritualization of these displays typically involves redundancy: enhancement of the movement with enlarging structures and bright colors, and stereotyped repetition of the movement. It is often the size of the structure, intensity of the color, and vigor of the movement that provides the receiver with useful information about the health, genetic quality, fighting ability, or motivation of the sender. Examples of such displays will be described in this and subsequent chapters.

INTENTION MOVEMENTS When an animal is about to embark on a mode of action such as attacking, fleeing, or copulating (to mention only a few), it must often prepare itself by assuming a certain posture, placing its limbs in certain positions, or exposing certain parts of the body that will be used in the act. Although the subsequent behavior

(A) Domestic dog (*Canis domestica*)

(B) Green heron (*Butorides virescens*)

(C) Black-headed gull (*Larus ridibundus*)

(D) Macaque (*Macaca* spp.)

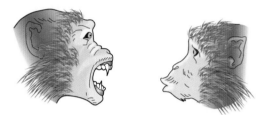

FIGURE 10.5 The principle of antithesis For each of the species illustrated here, the primary aggressive display is shown on the left and the submissive or friendly display on the right. Notice that display features such as body posture, orientation, head position, and degree of piloerection show opposite extremes in the two displays.

is always preceded by the preparatory action, the initial phases may be performed without completing the follow-up behavior. Such incomplete initial acts are called **intention movements** [106, 509]. Although an animal may honestly indicate its intentions, so-called intention movements may be ritualized signals indicating the motivational state that the sender is in, or even be used to mislead a receiver as to what the sender will do next. Whether ritualized or not, intention movements derived from a specific type of behavior cause perceptive receivers to anticipate specific subsequent actions by the sender.

The preparatory phases of attack behavior are intention movements that frequently become ritualized into threat signals. Some examples are illustrated by postures on the left in **Figure 10.5**. A directed stare at the rival, a tensed and forward body posture, the baring of teeth, bill, claws, horns or other weapons, and the protection or pulling back of sensitive body parts such as ears are common threat signals that are used in conflict situations. It is not difficult to see how such a signal can evolve. Before a dog can bite a rival or enemy, it must draw back its lips to expose its teeth. An individual that performs the preparatory movements but stops short of completing the action sends the message that it is more likely to bite now than before the signal was given. This information is likely to cause the rival to respond by retreating. Individuals that frequently prepare to bite but fail to follow through may be favored over individuals that only bare the teeth when they are

truly about to bite; such bluffers could win the contest without having to pay the cost of a fight. However, good experimental evidence that threat displays reliably predict subsequent attack by the sender has been found in birds [402, 464, 537] and ungulates [144, 526] for reasons we discuss later in this chapter. The reduced likelihood of becoming engaged in a costly fight will lead to exaggeration of the retreat-inducing intention movement, such as pulling the lips back more than is necessary. Ritualization often involves freezing the position, which gives the rival a moment to decide whether to retreat or stay and meet the challenge. This is a clear example of the ritualization process that occurs once the sender benefits from the receiver's response to its signal.

The displays that animals use to signal nonaggression or submission are often precisely *opposite in form* to those that signal threat or aggression. The right postures in Figure 10.5 illustrate the opposing submissive display for each of the threat displays on the left. Looking away from the opponent, turning the body away, closing the mouth, hiding weapons, and exposing sensitive body parts are common submissive gestures. Darwin [112] was the first to recognize the opposing nature of aggressive and submissive displays and termed the phenomenon **antithesis**. Aggressive and submissive displays may be antithetical in form because they arise independently from opposing actions (e.g., approach versus retreat). However, additional antithetical elements may be incorporated during ritualization to render them maximally divergent and reduce ambiguity [102]. A neural network model of two opposite-meaning signals [235] demonstrates this divergence process when senders benefit from making these two intentions clear to receivers. (See Web Topic 10.2 for a general description of neural network models.) The result is a coding scheme with strong reduction mapping of discrete, antithetical signals onto continuous conditions, as described in Chapter 8.

Web Topic 10.2 *Neural network models and feature detectors*
Neural net models provide a very useful way to explore the properties of recognition systems in animals. Feature detectors may emerge from such network training that mimics the ritualization process, or they may develop in noncommunicatory contexts and serve as sensory drivers for signal evolution.

Intention movements provide information to receivers in a variety of other contexts. Preparatory movements for flight are frequently ritualized in flocking birds to coordinate departures [13, 113]. When a bird on the ground is about to take off, it first crouches, raises its tail, and pulls its head in and back. It then extends its head and neck as it jumps into the air. The crouching portion of the behavior signals important information to other birds in a flock about what the individual is likely to do next. **Figure 10.6** illustrates a classic example of the evolution of courtship displays from flight intention movements in the Suliformes. Similarly, ducks in flocks use ritualized head

jerking as a signal to synchronize taking flight. Some of the displays observed during courtship have been interpreted as ritualized reproductive intention movements [14]. For example, in species with paternal care of offspring, male courtship displays often represent incomplete renditions of nest construction or parental care behaviors, and have the potential to reveal to the mate-searching female the male's skill or commitment to this critical egg-protecting behavior [261, 273, 373, 378, 385]. The presentation or manipulation of nesting material is observed in many birds, and courting male fish may perform ritualized nest-fanning displays. Copulation movements and display of sex organs form the basis of ritualized courtship in spiders and some mammals [146, 155]. See Web Topic 10.3 for examples of these displays.

Web Topic 10.3 *Rich media examples of ritualization*
Some classic examples of visual, acoustic, and chemical signals derived from sender precursors.

FIGURE 10.6 Evolution of courtship displays in Suliformes from flight intention movements (1) Gannets (*Morus* spp.) perform a slightly ritualized pre-take-off display before flying. (2) The sky-pointing courtship display posture of the boobies (*Sula* spp.) developed from the pre-takeoff display precursor by adding the upward-pointed bill and rotated wings. Both alternate wing waving (6) in the darters (*Anhinga* spp.) and slow wing waving (3) in the great cormorant (*Phalacrocorax carbo*) and close relatives evolved from the booby display. From state (3), rapid-flutter wing waving (4) evolved independently in Brandt's cormorant (*P. penicillatus*) and the pelagic and red-faced cormorant sister species (*P. pelagicus* and *urile*). The highly ritualized throwback display (5) of the European, king, and imperial shags (*P. aristotelis*, *carunculatus*, and *atriceps*) also developed from stage (3). These transitions are well supported by the phylogenetic relationships of these species. Thus the two most highly ritualized display types, (4) and (5), evolved from the flight intention precursor via at least two intermediate steps. (After [262].)

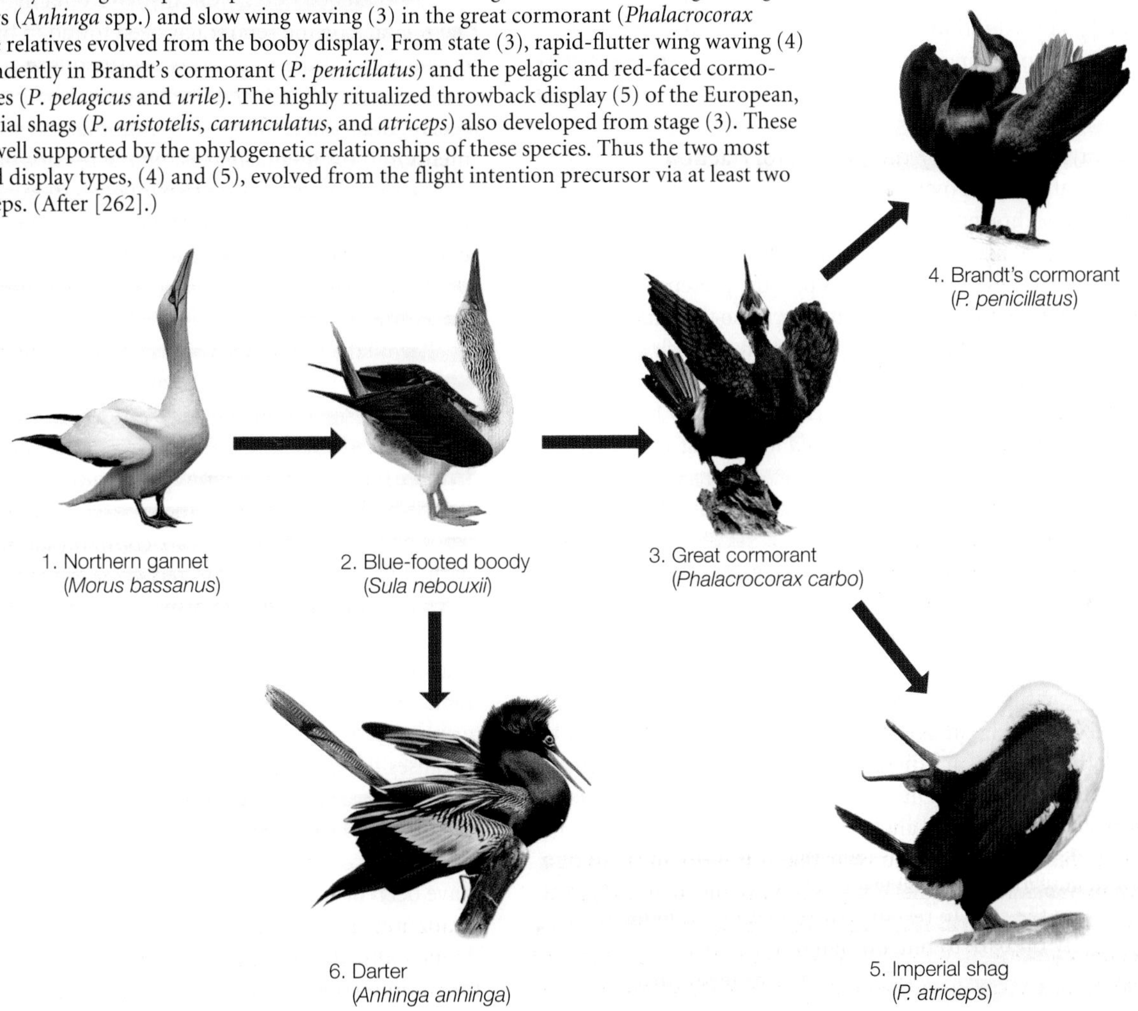

1. Northern gannet (*Morus bassanus*)
2. Blue-footed boody (*Sula nebouxii*)
3. Great cormorant (*Phalacrocorax carbo*)
4. Brandt's cormorant (*P. penicillatus*)
5. Imperial shag (*P. atriceps*)
6. Darter (*Anhinga anhinga*)

(A)

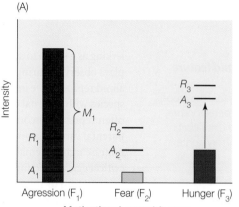

(B)

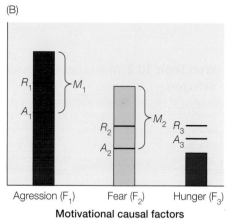

FIGURE 10.7 Motivational systems and conflict The activation of different motivational systems varies from moment to moment depending on the intensity of causal factors (F_i) for each system ($i = 1$, $2,\ldots$). Causal factors include components of an animal's internal state and external contexts relevant to the specific motivational system. Each system has associated with it a reactivity threshold (R_i) and an anticipatory attention threshold (A_i), which is always somewhat lower than R_i. The current motivational intensity for the system (M_i) is depicted as the difference between the total intensity of the causal factors minus the attention threshold A_i. In the two examples depicted, system 1 (F_1) represents aggression, system 2 (F_2) represents fear, and system 3 (F_3) represents hunger. In (A), aggression factors have increased past the attention threshold, due to the appearance of a rival, raising the A_i and R_i thresholds for other systems (e.g., F_3, arrow) so that the animal focuses entirely on chasing its rival. In (B), fear factors increase, perhaps because the rival decides to stand and fight, resulting in the activation of a second system. The strong fear stimulus partially inhibits aggressive motivation, and the animal is conflicted as to which behavior to perform. (After [16].)

MOTIVATIONAL CONFLICT   Conflicting social contexts provide fertile ground for incipient visual displays. Animals are viewed as possessing a complement of different **motivational systems** [44, 330, 510] (see Web Topic 10.4). Typical motivational systems include: hunger, thirst, aggression, fear, reproduction, thermoregulation, and grooming. Each system is regulated by its own set of physiological processes (proprioceptive, hormonal, and neurochemical mechanisms) and psychological mental states (desires, pleasures, and aversions). The state of these internal factors plus the presence of external factors are the *causal factors* that determine the current *stimulus intensity level* of the motivational system. Each system also has a current *attention threshold*, which causes the animal to attend to the relevant stimuli when the intensity level exceeds this value, and a *reactivity threshold* for taking action appropriate to the system. All motivational systems operate simultaneously, but the motivational level for some systems will be higher (or closer to its relevant threshold) than the motivational level for other systems at any given point in time (**Figure 10.7**). The system with the highest motivation level above the activation and reactivity thresholds is the one most likely to dictate the animal's current behavior [16, 199, 224, 225]. For example, if the animal has not fed for a while and is extremely hungry, it will be more highly motivated by hunger than by any other system and therefore will be more likely to engage in foraging behavior. Once it has fed, the hunger motivational intensity will drop, and another motivational system will command the animal's attention. In situations where two motivational systems have both exceeded their activation thresholds, the animal is said to be in **motivational conflict**. The most common motivational conflict situation is the simultaneous activation of aggression and fear systems that typically occurs

in encounters between well-matched rivals such as two territorial males at their joint boundary. Similarly, a three-way conflict between reproduction, fear, and aggression is believed to take place during courtship in many species. Three types of behaviors that occur in such contexts—ambivalence, displacement, and redirected behaviors—are thought to be the sources of some visual displays.

Web Topic 10.4 *Emotion, drive, and motivation*
Past and current views on these controversial issues.

Ambivalence behavior supposedly reflects the animal's indecision over which of two opposing motivational systems to attend to, and results in either (1) the **alternation** of intention movements characteristic of the two motivational systems; or (2) the **blending** of antithetical intention movements characteristic of the two systems into an intermediate form. The classic example of a ritualized display incorporating alternation between two systems is the zigzag dance of the male stickleback during courtship. The male is simultaneously motivated to attack the female who has just entered his territory and to lead her to the nest, and he performs a series of alternating to-and-fro movements between the female and the nest (see Web Topic 10.3 and Figure 12.28). An example of the blending of antithetical intention movements is the **broadside threat display** seen in encounters between rival males of many species. This posture is exactly intermediate between a highly aggressive opponent-facing posture and the turning away or rear-end presentation characteristic of a fearful or retreating animal. Ritualization frequently involves the "freezing" of this position into a stiffly held posture. The broadside threat not only reflects motivational conflict, but it also presents the largest possible surface area to the opponent. Since larger individuals tend to win fights against smaller ones, exaggeration of size is likely to benefit the sender if it causes the opponent to retreat. Ritualization therefore also involves exaggeration of body size with structures that enlarge the profile of the animals (**Figure 10.8**). More complex blended systems of the fear and aggressive motivational systems have been described in some species. Here there is a series of gradations between the pure fear and pure aggression states in which the level of each motivational system is encoded by different display components (**Figure 10.9**).

(A)

(B)

FIGURE 10.8 **Examples of ritualized broadside displays** (A) A sailfin blenny (*Emblemaria pandionis*) flaring its dorsal and ventral fins as a threat while defending its burrow. (B) Shaggy fur enhances the front and broadside profile of the American bison (*Bison bison*). When performing the broadside threat to a rival, the male arches his back, bellows, and keeps his tail stiff; if he moves, he does so in slow, short, stiff steps to maintain his broadside orientation to the rival [303]. (C) Enlarged head and throat structures increase the broadside profile of the helmeted basilisk (*Corytophanes cristatus*) [404].

(C)

Displacement behaviors are defined as acts that are apparently irrelevant to the motivational system in which they appear [17, 21, 277, 509]. They are recognizably similar to or derived from motor patterns normal to the species, but are usually short, incomplete, and nonfunctional in their usual sense. Displacement acts commonly appear during motivational conflict situations, where they seem to have nothing to do with either one of the conflicting motivational systems and are interspersed between relevant acts. Examples include mock feeding, drinking, preening, or sleeping during conflicts with rivals and during courtship (**Figure 10.10**). In primates, displacement acts such as scratching, self-grooming, body shaking, and yawning also appear in contexts of conflicting emotions. These contexts include social anxiety (such as being a subordinate in a group or being close to a dominant or unfriendly individual); cognitive stress (such as having to solve a difficult discrimination task); and in thwarting or frustrating situations (such as being prevented from achieving an expected goal) [77, 78, 206, 283, 291, 305, 309, 516, 523].

As with other types of behaviors we have discussed, displacement acts vary in the degree to which they have been ritualized and converted into signals. In the zebra finch

Increasing aggressiveness

Increasing fear

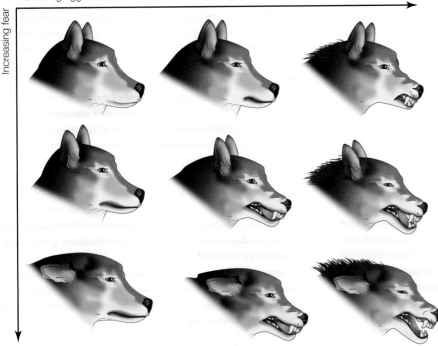

FIGURE 10.9 **Blended fear and aggression in the facial displays of the wolf** The upper left panel shows the relaxed posture, while the lower right panel shows the mixture of extreme aggression and fear. The wolf (*Canus lupis*) uses flattened ears to express degrees of fear and baring of the teeth plus piloerection to express aggression. (After [455].)

(*Taeniopygia guttata*), beak wiping is frequently interjected during courtship by the male (see Figure 10.10C). Beak wiping normally occurs during feeding and functions to clean the bill of food particles. In the related spice finch (*Lonchura punctulata*) and striated finch (*L. striata*), a low bow has become an integral part of the courtship display. The body position during the bow is very similar to the position taken during beak wiping, suggesting that the bow is a ritualized form of this beak cleaning behavior [354]. If displacement acts reliably occur in certain emotionally conflicting conditions, they could provide useful information to receivers about the motivational state of the sender. For example, in territorial disputes, the territory owner is strongly aggressive toward any intruder it meets in the center of its territory and fearful during encounters made outside the territory. At a point near the boundary, the two tendencies balance or come into conflict, and the performance of ambivalent or displacement behaviors is a good indication of how far the resident is prepared to go in defense of his territory. However, senders could benefit by using these irrelevant acts as tactical strategies to de-escalate a conflict, throw a rival off its guard, or otherwise manipulate the receiver.

The third type of behavior that is sometimes observed during motivational conflict situations is **redirected behavior**. These are behaviors in which the form of the act is appropriate to the context but it is directed toward an irrelevant stimulus. Examples include the attacking of a third-party bystander or an inanimate object during a fight; drinking or feeding movements directed at smooth surfaces or shiny pebbles by a thirsty or hungry animal; or a male copulating with an inanimate object after being rejected by a female. Redirected behaviors can become ritualized, as in the case of grass pulling in fighting herring gulls—they appear to seize and pull on clumps of grass as if they were the wings of their opponents. As with ambivalent and displacement behaviors, redirected acts could convey information about the conflicted motivational state of the performer. However, they could also be interpreted as intention movements, tactical acts by victims to deflect aggression, or mechanisms to release built-up over-stimulated drive tension [25, 84, 96, 401, 556].

AUTONOMIC NERVOUS SYSTEM   The **autonomic nervous system** of vertebrates is an involuntary system of nerves that controls the internal organs, such as the heart, blood vessels, lungs, intestines, eyes, and certain glands. The function of the autonomic system is to provide a rapid physiological response to a variety of stressful situations. For example, when an individual must exert itself to run or fly, the autonomic system acts to accelerate the heartbeat rate, dilate air

FIGURE 10.10   Examples of displacement behaviors (A) The threat display of the three-spined stickleback (*Gasterosteus aculeatus*) ritualized from displacement sand-digging, a nesting behavior. (B) The choking display in the herring gull (*Larus argentatus*) observed during boundary disputes is similar to the display between mated pairs during nest building. (C) Displacement beak wiping during courtship in the zebra finch (*Taeniopygia guttata*). This display is ritualized into the bowing display in related finches. (D) Displacement feeding in domestic cocks (*Gallus domesticus*) during a fight. (E) Displacement sleep in the European avocet (*Recurvirostra avosetta*) during a fight. Although displacement acts during agonistic encounters seem maladaptive, in the sense of placing the actor in a very vulnerable position, they are in fact brief and rudimentary, and the actor keeps a wary eye on its opponent. (A,B,D,E after [508]; C after [354].)

passages to the lungs, increase the blood supply to the muscles, and reduce intestinal activity. When ambient temperature increases or decreases, the autonomic system responds by increasing or decreasing blood flow to the skin and by fluffing or sleeking feathers or fur to regulate internal body temperature; and when ambient light changes, the autonomic system dilates or constricts the pupil of the eye to regulate the amount of light entering the eye. Any autonomic response that produces some visible change may provide the source for a ritualized visual display. Because the autonomic system is often involved in responses to stressful or fearful stimuli, visible manifestations of the response are often signals of fear. Examples include feather sleeking in birds and pallor and sweating in humans. When aspects of an autonomic response become ritualized, however, they may come under some degree of voluntary control and the normal function may even be reversed.

Piloerection, or the fluffing of feathers or hair, is a common thermoregulatory response with a highly visible manifestation that is frequently ritualized into visual signals in birds and mammals. The ritualization process usually involves restricting the area of erected feathers or hair to a small part of the body, elongating or coloring the feathers or fur to exaggerate the display, and the evolution of voluntary control of the muscles that cause piloerection [352]. Feather/hair posture regulates the insulation properties of the pelage. A sleeked posture minimizes the thickness of the insulation layer and helps to reduce the body temperature of a hot animal. A fluffed posture, in which hairs or feathers are partially erected without losing contact between adjacent ones, traps a thicker layer of insulating air and helps to increase the body temperature of a cold animal. A ruffled posture, involving extreme erection and separation of fur or feathers, breaks the insulation layer and leads to heat dissipation. It also gives the animal a ragged and enlarged outline. In birds, the movement of general body plumage is only slightly ritualized and follows the pattern that we would expect from thermoregulatory responses. Sleeked body feathers are associated with aggressive tendencies, a context in which individuals are preparing for action and need to minimize feather insulation to reduce body temperature. Fluffed body feathers are associated with submissive tendencies and reflect the thermoregulatory needs of an inactive individual. Highly ritualized feather erection displays in birds, on the other hand, involve feather ruffling of specialized regions of the body and are emancipated from thermoregulatory consequences (**Figure 10.11**). For example, in the Steller's jay (*Cyanocitta stelleri*), a fully raised crest

indicates maximum aggressive arousal and a fully depressed crest indicates submission [64]. In mammals, ritualized erected fur displays are associated with fearful, stressful, and submissive contexts. Raised fur along the neck, back, and tail is an important component of defensive threat displays in many species, as in the familiar domestic cat. A fluffed tail in some squirrels and New World primates is indicative of fear, alarm, and stressful contexts, and a raised head crest in the mandrill appears to have a conciliatory function [15, 207, 287, 357].

Several other notable physiological processes have become ritualized into visual displays. **Pupil dilation** has been ritualized into a visual display in many parrots (**Figure 10.12A**). Recall from Chapter 5 that the pigment epithelium layer of

FIGURE 10.11 Highly ritualized piloerection (A) The *Parotia* birds of paradise have several specialized plumage ornaments used to perform elaborate displays to females: wirelike feathers with spatula tips on the head are extended out to the sides; extremely elongated flank feathers are spread in a skirt during the ballarina display; a breast shield with iridescent feathers is raised and lowered to produce a flashing pattern [458]. (B) The elongated fur on the head and back of wild pigs, such as this javelina (*Pecari tajacu*) normally rests flat against the body, but is raised when the animal is excited or aggressively motivated [119, 306]. (C) The head crest and topknot of Gambel's quail (*Callipepla gambelii*) are fully erected in dominant and aggressive birds but are lowered and flattened to different degrees in submissive and fearful birds [201].

FIGURE 10.12 **Co-opting autonomic system structures with visual manifestations for ritualized signaling** (A) Parrots have colored irises and voluntary control over pupil dilation, which they contract repeatedly when excited. (B) The male magnificent frigate bird (*Fregata magnificens*) has inflatable throat pouches used for mate attraction display. (C) The Siamese fighting fish (*Betta splendens*) gives a frontal display with extended gill covers during aggressive interactions. (D) Yawn-threat in the crab-eating macaque (*Macaca fascicularis*); note additional ritualization of closed eyes with the white eyelids.

signal of embarrassment in humans that employs this autonomic system, as well as the red skin flushing in the combs and throats of birds and the sexual swellings of estrous primates [266, 268, 363, 370, 384, 466, 467]. **Sweating** occurs in humans, horses, and rats in stressful and fearful contexts, and is the source of informative olfactory cues or signals [1, 2, 81, 82, 105]. Similarly, **defecation and urination** are involuntary responses to frightening stimuli and may be the source of some defensive olfactory threat signals [15].

the retina performs most of the light regulation function in birds, freeing the iris for display. Aggressive or highly excited individuals display by rapidly expanding and contracting the pupil. Parrots that use pupil dilation as a signal invariably have a brightly colored iris and a ring of skin or feathers of a contrasting color around the eye to amplify the movement. **Respiratory structures** may also be ritualized for visual display. The inflated external vocal sacs of male prairie chickens and frigate birds are brightly colored and visually conspicuous courtship display components (**Figure 10.12B**). The proboscises of bull elephant seals, the trunks of elephants, and flaring nostrils in primates are all used in visual signaling. In some fishes the gill covers are enlarged and held out in frontal visual displays (**Figure 10.12C**) [61, 62]. **Yawning** is a physiological response that increases arousal when the environment provides little stimulation [28]. An open-mouth yawn also exposes the teeth, and in primates yawning has become ritualized and emancipated into a threat display (**Figure 10.12D**). **Vasodilation** of bare skin areas is a mechanism for cooling that can lead to a flush of red coloration. Blushing is a visual

RESPIRATORY SYSTEM   As we saw in Chapter 2, terrestrial air-breathing vertebrates have co-opted their **respiratory system** to produce vocal signals. Air is forced from the lungs or air sacs down a tube and through a narrow orifice that generates vibrations. Vocal cords and associated muscles, nerves, sacs, tracheal structures, tongues, and mouth/beak shapes have evolved to produce sounds, so essentially all vocal utterances are ritualized signals. By controlling the tension on the membranes, the size and shape of the tube, and the pattern of air pressure and flow, most terrestrial vertebrates can produce a variety of different sounds. High tension produces a high-frequency sound and low tension produces a low-frequency sound. Moreover, high-tension vibrating membranes are more likely to generate a nearly sinusoidal waveform and a tonal, whistlelike sound, whereas loosely vibrating membranes generate a nonsinusoidal waveform with more harmonics and a harsher, atonal sound [185]. Low-frequency sounds also have more closely spaced harmonics, which creates a broadband appearance on a spectrogram and facilitates the perception of formants arising from vocal tract filtering [149]. These

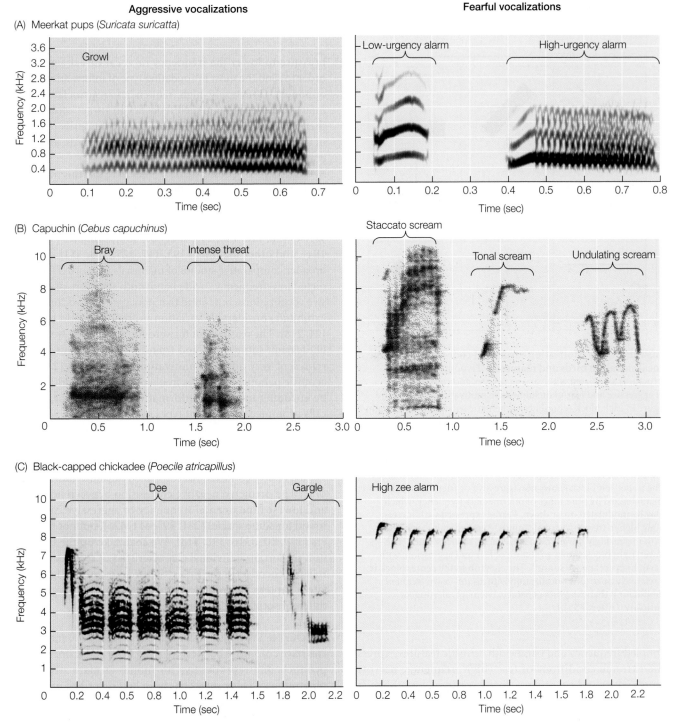

FIGURE 10.13 Antithesis in aggressive and fearful vocal signals within species Vocalizations given by aggressive and fearful individuals, shown on the same time and frequency scale for each species. Aggressive sounds are low in frequency and very broadband (growls), whereas fearful sounds are higher in frequency and more tonal (whistles and squeals). Fearful sounds are more variable because they often reflect different levels of urgency and may combine aggressive and fearful features.

acoustic features of vocalized sounds are often associated with individual differences as well as with the motivational state of the sender [501]. Aggressive senders usually utter sounds that are low in fundamental or dominant frequency and broadband and atonal in quality, such as growls, roars, and harsh cries (**Figure 10.13**, left side). Fearful senders, on the other hand, often utter sounds that are high in fundamental frequency and contain few harmonics; examples include high-pitched whimpers and tonal whistles (see Figure 10.13, right side). These two forms of vocal signals form an antithetical signaling system equivalent to the visual displays shown in Figure 10.5.

In the 1970s, Eugene Morton [355] proposed that frequency and bandwidth variants could constitute a graded or blended system of vocal displays for the expression of different ratios of aggression and fear, analogous to the visual system depicted in Figure 10.9. His **motivation-structural code** scheme for the close-range vocalizations of birds and mammals is illustrated in **Figure 10.14**. As suggested earlier,

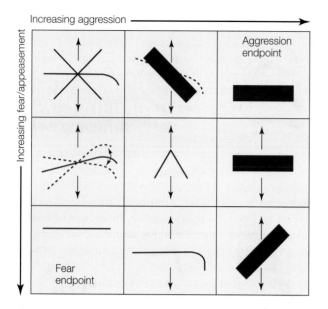

FIGURE 10.14 Motivation-structural code for short-range vocal signals in birds and mammals Hypothesized sound structures associated with varying degrees and combinations of aggressive and fearful or submissive motivational states. The figure in each block represents a schematic sonogram. The height of the note above the baseline (bottom of the block) indicates the frequency of the sound, and the arrows suggest the directions in which the frequency may vary. The thickness of the note represents the tonality and bandwidth of the sound. Dashed lines indicate that the note's frequency-modulated slope may change. The upper-left block shows nonaggressive or friendly vocalizations characterized by pure tones of intermediate frequency, which may be modulated up or down. Fear is expressed with increasingly higher-frequency tonal sounds. Aggression is expressed with increasingly lower-frequency, harsh, atonal sounds. The lower-right block shows a mixture of fear and aggression, with harsh high-frequency sounds. The chevron in the center block is a relatively pure tone vocalization with an up–down frequency modulation that expresses general excitement, alerting, or curiosity. (After [355].)

coupled frequency and bandwidth acoustic features express the extreme conditions of fear (high frequency, narrow bandwidth) and aggression (low frequency, broad bandwidth). Intermediate frequencies and bandwidths indicate conflicting motivational levels. In addition, the type of frequency modulation can provide further information on the motivation state of the sender within these broad categories. A downward-modulated frequency sweep tends to signal slightly greater aggressiveness, while an upward-modulated sweep signals less aggressiveness. In the middle of this scheme is the chirp or bark, a vocalization that is chevron-shaped in a spectrogram, which is proposed to signal ambivalence, indecision, curiosity, or mild alert. Morton also suggested a possible reason for the association between low frequencies for aggression and high frequencies for nonaggression. In all vocal vertebrate species, larger individuals can potentially produce lower-frequency sounds than smaller individuals. Since individuals with larger body size often win aggressive encounters with smaller individuals, there has been selective

pressure to lower the frequency of vocal threat signals in order to send the most intimidating type of information to an opponent. Conversely, high frequencies are associated with infant vocalizations and their tendency to attract adults. Furthermore, a fearful animal is likely to involuntarily tense all of the muscles in its body, including its vocal cords, and thus emit high-frequency sounds.

In the ensuing years, many studies have put this model to the test. The predicted coupling between frequency and tonal quality for aggressive versus nonaggressive vocalizations was supported in several corvid birds [228, 315], numerous canid species [56, 110, 148, 208, 422, 423, 454], primates [45, 179, 194], and other mammals [93, 147, 227, 290, 334]. Several broader comparative studies of mammals supported the predicted low-frequency / wide bandwidth structure of aggressive vocalizations, but found that nonaggressive vocalizations exhibited considerable variation in frequency and tonality [24, 218]. The reasons put forward for this finding were that: (1) fearful, submissive, and friendly vocalization contexts represent very different motivational states; and (2) other acoustic parameters besides frequency and bandwidth could be encoding important information. In particular, alarm calls prompted by different types of predators have very different acoustic structures [218, 260, 468]. Evidence for the predicted association between down-sweeping frequencies in more aggressive contexts and up-sweeping frequencies in less aggressive contexts was confirmed in some studies [228, 477]. The predicted negative correlation between body size and lowest fundamental frequency of aggressive calls was found for some species [231, 397, 411, 424], but not for others [149, 209, 316, 392, 413]. We shall take up the topic of how important individual characteristics of senders may be reliably encoded in the fine structure of acoustic signals in later chapters. For more examples of vocalizations given in different social contexts, see Web Topic 10.3.

ENDOCRINE SYSTEM   As we learned in Chapter 6, some species have co-opted chemicals derived from endocrine system products and reproductive hormones as mate attraction pheromones. This process has been best described in teleost fish, and Peter Sorensen and colleagues have elucidated the goldfish pheromone system as a classic model of signal evolution from sender precursors [482, 483]. Female goldfish release an array of steroid and prostaglandin hormones and their metabolites into the water in the stages leading up to egg release. During the preovulatory stage, a surge of gonadotropin (a protein hormone from the pituitary gland in the brain) stimulates the ovary to produce a mixture of steroid hormones, including pregnenes and androstenes that facilitate egg maturation. These hormones and their conjugated derivatives are then expelled into the water via urine (see Figure 6.9). After ovulation, when the eggs are in the oviduct, levels of prostaglandin hormone increase in the blood to regulate mating behavior and egg release. In the first evolutionary stage, these physiological regulators leaked out via urine and the gills as by-products, but no benefit would have accrued, because male receivers could not detect the chemicals. Females, however, had internal membrane G protein receptors for these chemicals to detect and regulate the cyclical hormonal changes. In the second evolutionary stage, male mutants co-opt these same G protein receptors into their olfactory cell membranes and develop the ability to detect the chemical cues released by gravid females. Only a subset of all the chemicals released by females may be detected by the males, because male receivers benefit most by developing extreme sensitivity to the best predictive chemical cues. The third stage of evolution, specialization and ritualization of the signal by females, may or may not occur, depending on whether females benefit from further elaboration. By this multiple-step process, a true chemical communication transaction could evolve, creating a specific pheromone blend and specialized receptors in the receiver.

The testosterone-derived pheromone in the male boar represents a much more ritualized signal, probably because males were selected to evolve an olfactory signal that both stimulated and benefited the female. In this case, a specialized androgen is produced in the testes and carried to the saliva glands by lipocalin-binding proteins. When the male is sexually aroused, he produces copious amounts of viscous, frothy drool, and the volatile chemical is released with his breath. The boar pheromone is highly attractive to female pigs, accelerates puberty in young females, induces estrous, and causes receptive females to assume the mating posture [50, 126, 393].

ANTIPREDATION DEFENSES   Defensive adaptations used against predators can easily become the source of alarm signals and other signals directed at conspecifics. Examples of signals derived from **antipredation defense tactics** vary in their degree of ritualization. At the most unritualized end of the spectrum are chemical substances released from injured conspecifics that cause strong fright or avoidance reactions [85, 123, 157, 542, 543]. Since these chemicals are only produced when an animal has been attacked and injured by a predator, they provide reliable public information about a nearby predation event that benefits receivers. Researchers appropriately refer to these substances as cues rather than signals, because it has been difficult to demonstrate a benefit to senders that is directly related to receiver response. Some members of the Ostariophysan fishes (schooling minnows, catfish, and characins) produce an alarm substance called **schreckstoff** in specialized secretory club cells embedded in the middle of the epidermis (see Chapter 14). Club cells are not close to blood vessels where they could release hormones into the blood, nor do they possess pores through which to release the substance to the exterior; they can release their secretions only when the skin is damaged. The cells are metabolically expensive, and allocation of resources to them is facultative. There is some evidence for antiparasite and antibacterial protection by these chemicals, a nonsignaling benefit [7, 479, 480]. A novel signaling benefit has been proposed in the European minnow (*Phoxinus phoxinus*). Breeding males develop their bright nuptial coloration under the control of

testosterone, and during this period, reduce their expression of club cells. If the absence of club cell products is detectable by females, it could serve as an indicator of male reproductive status and innate immunocompetence [240, 395].

A clearer case for chemical alarm signals to conspecifics derived from defensive chemicals can be made for the Hymenoptera [226, 536]. We learned about the formic acid defensive spray in formicine ants (see Figure 6.14B), which is sufficiently volatile to also serve as a chemical alarm recruitment signal. In other cases, the defensive chemical is a venom or sticky glue substance with a high molecular weight. Because such a substance is not volatile enough to diffuse rapidly, a smaller and more volatile chemical is secreted at the same time to signal alarm. For example, honeybee (*Apis mellifera*) venom consists of a complex mixture of proteins, histamine, serotonin, and acetylcholine, which together cause the inflammation experienced by the victim of a sting. Associated with the sting gland is a separate set of secretory cells that produce the alarm pheromone. The pheromone contains a mixture of volatile acetate and alcohol compounds that release different components of alarm response behaviors [525]. Similarly, the glands that produce the sticky substance used by termites in defense also produce volatile substances that cause the searching and recruitment alarm reaction in colony mates [99, 416]. This evidence makes a strong case for a ritualized alarm pheromone.

Sudden bursts of sound or strong vibrations are used by a wide variety of animals when they are threatened or grabbed by a predator. The predator's startled hesitation following such a display is sometimes sufficient to allow the victim to escape. Examples of sound-production mechanisms that have evolved in otherwise silent animals for the sole purpose of warning or escaping from enemies include the click and jump of click beetles, the stridulation buzz in lobsters and numerous insects, the hiss of some salamanders and lizards, and the tail rattle of rattlesnakes [10, 52, 165, 184, 222, 296, 317, 391, 456]. Such sounds have the potential to evolve into intraspecific communication signals. In some insects, similar sounds are given as both antipredator deterrents and social calls, but which function served as the precursor is not known. For example, cicada males use their tymbal to produce their mate attraction call and also to create an alarm call when grasped [295]. The female reduviid bug *Rhodnius prolixus* stridulates to reject copulatory attempts by males and also when she is grasped by a predator [307]. The clearest case for the predator-deterrent function evolving first has been made for anti-bat moth sounds. Various taxa of moths have evolved tymbals, castanets, or stridulatory organs to produce ultrasonic clicks that either warn bats of the moth's toxicity or jam the echolocation pulses of the bats [97, 528]. Ultrasonic courtship signals have subsequently evolved in a few arctiid, noctuid, and pyralid moths [94, 95, 427].

A final example of a defensive strategy that has been highly ritualized and emancipated from its original function is the flashing bioluminescent mate attraction signal in firefly beetles of the family Lampyridae. A phylogenetic analysis shows that the origin

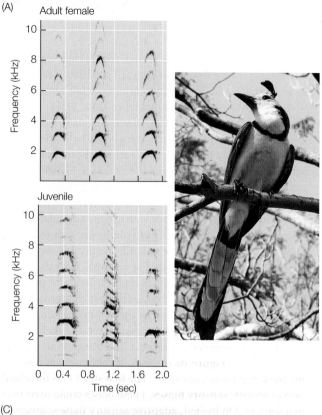

(A) Adult female

(A) Juvenile

(C)

of bioluminescence predates the origin of the Lampyridae [57, 58]. In the ancestral cantharoid beetles, the ability to produce and emit photic signals appears in the larvae and functions as an aposematic warning display to predators. The larvae produce a continuous glow, and are toxic and distasteful. Nocturnal predators such as mice quickly learn to use such light signals as a warning to avoid bitter food [517]. The elaboration of the photic organ in adults, and the ability to produce a flashing sexual signal, arose at least three times in the Lampyridae.

OTHER DISPLAYS   New displays can be co-opted from **existing displays**. Once an informative signal that elicits a specific type of response has evolved, senders can use this signal in another context in which the same type of response is adaptive. Although such signals may appear manipulative, both the receiver and the sender must benefit from this response in the new context. For example, both infantile and sexual behaviors may be used to deflect aggression. The

FIGURE 10.15   Displays developed from other displays (A) Female magpie-jays (*Calocitta formosa*) have co-opted the fledgling begging call for use as a fertility advertisement call. The structure of the call is the same, but it is louder and uttered for hours at a time by laying females. Adjacent territorial males and floater males are attracted and may feed the female, as well as mate with her [133]. (B) The male Nilgiri tahr (*Hemitragus hylocrius*) performing a courtship side-twist display to a female by turning his head just anterior to her hind leg (top photo); the movement is similar to the side-bunt gesture of a dependent kid desiring to suckle (bottom photo) [418]. (C) The width of the black chest stripe in the great tit (*Parus major*) is correlated with dominance status in males and is displayed with a head-up posture during aggressive interactions as shown here by the male on the right. Stripe width is also used by females to assess mates, with wider stripes being preferred. The bird on the left is a female, as indicated by the very thin breast stripe [364].

classic example is social presenting in many primates (see Figure 16.26). Estrous females present their rear end, often enhanced with colored sexual skin, to solicit copulation from a preferred male partner. The same posture is used by subordinates of both sexes toward aggressive dominants of the same sex as a greeting or appeasement gesture. The dominant individual often acknowledges the gesture with a brief ritual mock mounting, often leading to a reduction in social tension and reconciliation [122, 144, 220, 308, 481, 532]. A second example is the use of nestling food-begging behavior by adult females of many bird species during courtship (**Figure 10.15A**). The male often responds by courtship feeding the female. Courtship feeding has potential benefits to both parties, including improved nutritional condition for the laying female, a potential cue of the male's parental ability, and

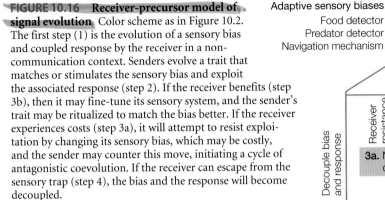

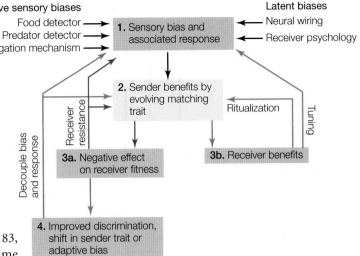

FIGURE 10.16 Receiver-precursor model of signal evolution Color scheme as in Figure 10.2. The first step (1) is the evolution of a sensory bias and coupled response by the receiver in a non-communication context. Senders evolve a trait that matches or stimulates the sensory bias and exploit the associated response (step 2). If the receiver benefits (step 3b), then it may fine-tune its sensory system, and the sender's trait may be ritualized to match the bias better. If the receiver experiences costs (step 3a), it will attempt to resist exploitation by changing its sensory bias, which may be costly, and the sender may counter this move, initiating a cycle of antagonistic coevolution. If the receiver can escape from the sensory trap (step 4), the bias and the response will become decoupled.

access to extra-pair mating opportunities by the male [183, 244, 511]. In a reversed-sex-role example along the same line, the "side and rear bunt" gestures of young goats (tribe Caprini) directed towards their mother's udder has been co-opted as a male sexual courtship display to females (**Figure 10.15B**) [418]. A third major class of co-opted displays is the use of aggressive signals between males by females to assess male dominance or quality. Females may initially eavesdrop on male contests and use male-to-male signals indicating fighting ability or dominance as cues, but selective pressure from female mate choice may lead to further elaboration of the male traits (**Figure 10.15C**) [41, 51, 364].

Signals derived from receiver precursors

In the sender-precursor model described earlier in this chapter, receiver sensory capabilities play a critical role in the second step—perception of a cue (see Figure 10.2). An informative cue that is more readily detected by receivers is more likely to become established as a signal than a cue that is difficult to detect. Thus signal modality is generally associated with the main modality (or modalities) employed by a species for noncommunicatory functions. For example, among spiders, visual courtship signals are found in taxa that hunt visually, such as jumping spiders; drumming and vibrational signals in ground-dwelling taxa that identify prey by surface vibrations, such as wolf spiders; and vibratory signals in species that capture prey in webs [544]. Furthermore, the design of long-distance signals in particular is strongly affected by environmental conditions and background noise; signal elaborations that maximize signal detection (i.e., the efficacy component) in a given species' signaling microhabitat by that species' receptor organ tend to predominate. Environmental characteristics and receiver sensory capabilities affect sender and receiver coevolution under any signal evolution model. The fine-tuning of signal features with respect to receiver sensitivity and environmental background is referred to as **sensory drive** [135, 136].

The **receiver-precursor model** of signal evolution, on the other hand, posits that specific sensory biases and associated responses arise first as a result of natural selection for noncommunicatory functions, and the detection-response system is subsequently co-opted by senders that evolve traits

matching these sensory system characteristics [23, 86, 161, 444–446, 448]. **Figure 10.16** shows a flow diagram illustrating the key evolutionary steps. The first step (1) is the evolution of specific **sensory biases**. These biases could arise from two sources. On the left, **adaptive sensory biases**, which are often associated with specific responses, evolve via natural selection to detect important environmental stimuli. Possible adaptations include a system that detects important food cues and evokes approach toward the stimulus, a system that detects predators and evokes fleeing or freezing, or a system that detects navigational cues. In the second step (2), senders evolve traits that match these specialized detectors and may take advantage of the associated response. An hypothetical example of how this might work: a species that feeds on red berries evolves a visual feature detector for round red objects, and foragers are attracted to such objects. Mutant male senders with red spots that mimic the berries are more likely to be detected and approached by females, and such mutants might therefore have greater mating success than males without the spots. Another source of sensory bias, called **hidden** or **latent biases**, could also instigate the first step in this model (shown on the right in Figure 10.16). In theory, such biases are incidental and selectively neutral consequences of how organisms are built. For example, artificial neural network models have shown that networks trained to recognize a certain type of stimulus sometimes produce sensory biases for novel stimuli as a by-product [20, 139, 140] (see Web Topic 10.2). Similarly, higher brain processes affecting receiver psychology may also produce biases for certain traits [172, 196, 197, 440, 441]. Through this pathway, the traits would appear to be arbitrary in design, in the sense that they are not associated with an adaptive feature-detection system. Moreover, they would not likely be associated with a specific response, but would be highly detectable because of a symmetrical shape, repetitive pattern, rapid movement, or pungent odor.

Regardless of whether the sender's signal trait is matching or arbitrary, the fitness consequence of trait evolution in step 2 could be positive, negative, or neutral, as shown in step 3. If

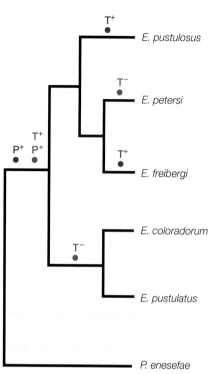

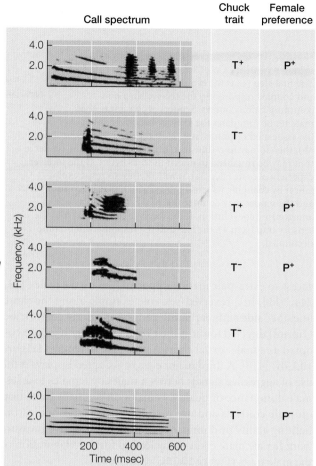

Call spectrum | Chuck trait | Female preference

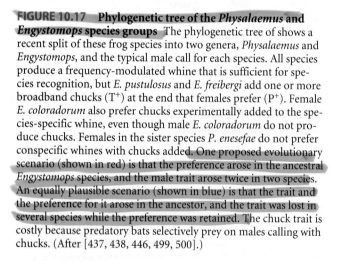

FIGURE 10.17 Phylogenetic tree of the *Physalaemus* and *Engystomops* species groups The phylogenetic tree of shows a recent split of these frog species into two genera, *Physalaemus* and *Engystomops*, and the typical male call for each species. All species produce a frequency-modulated whine that is sufficient for species recognition, but *E. pustulosus* and *E. freibergi* add one or more broadband chucks (T+) at the end that females prefer (P+). Female *E. coloradorum* also prefer chucks experimentally added to the species-specific whine, even though male *E. coloradorum* do not produce chucks. Females in the sister species *P. enesefae* do not prefer conspecific whines with chucks added. One proposed evolutionary scenario (shown in red) is that the preference arose in the ancestral *Engystomops* species, and the male trait arose twice in two species. An equally plausible scenario (shown in blue) is that the trait and the preference for it arose in the ancestor, and the trait was lost in several species while the preference was retained. The chuck trait is costly because predatory bats selectively prey on males calling with chucks. (After [437, 438, 446, 499, 500].)

it is neutral, no further coevolution will occur. If it is positive (3b), e.g., a female can find a male mate with such a trait much faster than a male without the trait, then a round of positive coevolution would follow that makes the sender's trait even more detectable. If the effect on the receiver is negative (3a), e.g., a female may be distracted from foraging or stimulated to breed before the optimal time, then we have a clear case of **sensory exploitation**, or **sensory trap** if the receiver's response is also exploited. A round of antagonistic coevolution between sender and receiver will ensue. The receiver should evolve ways to resist this exploitation. Eventually the receiver may evolve finer discrimination between the important environmental stimulus and the mimicking sender, and possibly separate response rules for the two stimuli (step 4).

HISTORICAL EVIDENCE FOR RECEIVER PRECURSORS Comparative studies based on phylogenetic tree reconstructions have been used to identify receiver sensory specializations that predate the evolution of the sender's signal trait. To support the interpretation of a sensory specialization as a precursor, basal taxa should have the receiver bias but not the sender trait, while terminal taxa should exhibit both receiver bias and the trait. Figure 10.17 illustrates the classic example of a preexisting acoustic bias for the broadband "chuck" call component of the male túngara frog and its relatives (*Physalaemus* and *Engysomops* spp.). In some members of this group, males add one or more chuck notes to the end of the frequency-modulated whine, and females prefer males that give these more complex calls. In one species, *E. coloradorum*, males lack the chuck, yet females prefer to associate with conspecific males to whose calls experimental chucks have been added. The chucks are detected by the inner ear's basilar papilla, which is tuned to the higher chuck frequencies. It has been hypothesized that a preference for chucks predated the appearance of the trait itself [445–447, 449].

Such a hypothesis is difficult to test for several reasons [314, 469]. First, the reconstruction of a phylogenetic tree always involves some uncertainty. Tree reconstructions use morphological and genetic characters from extant species to estimate the positions of the interior nodes. The accuracy of the node positions and branching patterns depends on the nature and quality of the character data as well as which species are included. Second, evolutionary reversals do occur,

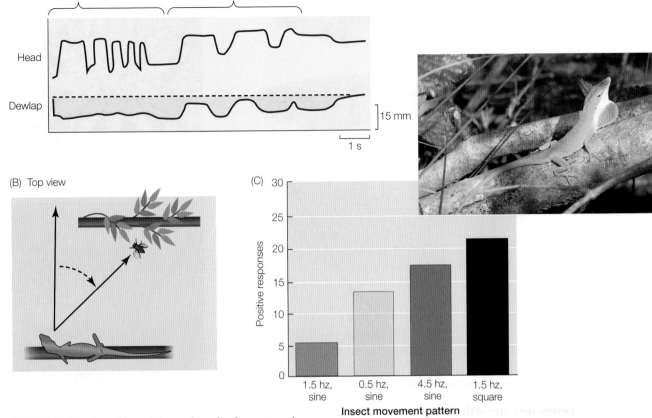

FIGURE 10.18 **Matching of the pushup display pattern in** *Anolis* **lizards with visual prey detectors** (A) The pushup display of anolis lizards combines a jerky square-wave head movement with repeated extensions of the colored dewlap. These types of display movements have exploited the species' visual adaptations for prey detection. (B) The detectability of a small insect positioned in front of a moving branch was tested by monitoring the response of the lizard, which changes the orientation of its head when it spots a prey object; the background vegetation was moved up and down in a 1.5 hz sine-wave pattern, and the insect was moved with a range of different patterns. (C) Responses were higher when the insect was moved with a sine-wave pattern that was fast or slow relative to the background movement, and highest with a square-wave movement pattern. (After [154].)

which can complicate the tree reconstruction and the drawing of evolutionary inferences. When the traits involved are sexually selected mate attraction signals, there is always the possibility that a trait initially evolved into highly elaborate signal in an ancestral species, along with the preference for it. During subsequent speciation, the trait could have been lost due to its high cost to the sender, while the receiver retained the detector for it [300, 410]. Third, if selection is strong on the sender and receiver, rapid coevolution may have occurred. The separate origins of the two adaptations therefore may not be preserved in the character mapping of extant species on the phylogenetic tree. The preexisting bias story in the *Physalaemus* frog group is beset by these difficulties. An equally plausible scenario is shown in Figure 10.17 suggesting that the trait and preference for it arose in the ancestor of the (new) genus *Engystomops*, followed by several losses of the male trait but not the preference for it. Studies on the sister group species *P. enesefae* demonstrated that females did not prefer

added chucks, yet the tuning of the basilar papilla is similar in both genera. No adaptive argument for a special sensitivity to "chuck"-like sounds was offered. The high cost for chuck-producing males from selective bat predation makes joint trait-preference evolution followed by male trait loss a much more biologically plausible scenario [51, 469]. A second proposed case of a preexisting bias for sworded tails in *Xiphophorus* fish species faces similar problems [32–35, 51, 336, 439].

EVIDENCE FOR MATCHING TRAITS   There are several clear examples of social signals that are designed to stimulate a receiver's preexisting feature detector, and in some of these cases, the selective force favoring the feature detector is also known. *Anolis* lizards are sit-and-wait predators that feed on moving insects. The eyes of lizards have a large peripheral retinal region with a moderate density of receptor cells sensitive to velocity-specific motion. These cells habituate rapidly to background motion such as the 1–2 Hz sinusoidal movement of leaves in wind. Experiments in which a prey object is moved in different ways in front of typical moving vegetation show that either higher-frequency or jerky (square-wave) prey movements are the most readily detected (**Figure 10.18**). The territorial pushup displays of male lizards involve square-wave motion, high amplitude, and rapid acceleration that are maximally different from the background [154]. The signal display pattern appears to have exploited the prey-tuned motion detectors in the receptor. Another example is the orange coloration on male guppies (*Poecilia reticulata*), which mimics fruit highly favored as food by the fish. What makes this case so compelling is that

(A)

(B)

Clusia

(C)

Sloanea

FIGURE 10.19   Matching fruit color in male guppies (A) These two male guppies (*Poecilia reticulata*) bear orange patches differing in area and chroma. Populations of guppies in freshwater streams of Trinidad differ in the average chroma of orange spots in males and in the strength of female preference for orange chroma. (B) Two examples of orange-colored fruit (*Sloanea* and *Clusia* spp.) that occasionally fall into the streams from the forest canopy. Guppies voraciously consume these fruits, which are one source of carotenoids that are the main pigment for the male color patches. (C) Guppies of both sexes were found to be especially attracted to orange painted discs placed in the water. The degree of attraction to orange correlated with the strength of the female preference for orange (carotenoid) coloration in males at the population level (each symbol represents the mean of a different population). The results suggest that the preference for orange coloration in males may have initially evolved as a sensory bias for detecting rare but rich food sources [191–193, 426].

among the populations studied, the strength of the female preference for orange males in different populations is proportional to the interest of both sexes in orange-colored discs in that population (**Figure 10.19**). A third example, involving the tactile modality, is the water mite (*Neumania papillator*), which locates its copepod prey by detecting their water vibrations. When males detect a female, they move their legs to create vibrations that mimic those caused by prey; in response, the female approaches and clutches the male; mating may follow if the female is receptive. Experimentally food-deprived females are more likely than satiated females to clutch males [406–409]. For some olfactory examples, we showed in Chapter 6 how male insects often produce a courtship pheromone that chemically mimics the products of the host plants to which females are attracted for oviposition,

and how male Euglossine bees collect fragrant floral oils and produce a blend to attract females.

Although most of the good matching-trait examples involve the exploitation of a food detection system, there are a few interesting nonfood examples. Most fiddler crab (*Uca*) species live in a mudflat environment. The crabs feed on the surface and rest and reproduce in burrows dug into the sand. Foraging individuals that have strayed from their burrows are vulnerable to rapidly approaching predators such as birds, and run to holes or any nearby vertical object to hide. In some species, males build sand hoods behind the entrance of their burrow, and females preferentially approach these visible vertical structures because they mimic temporary hiding places (**Figure 10.20A**). Hood-building males benefit by attracting potential mates, and females benefit by reducing their risk of predation [87, 88]. In a group of small freshwater fish called darters (*Etheostoma* spp.), breeding males of some species develop yellowish knobby structures on the spines of their dorsal, pelvic, or pectoral fins that mimic the appearance of eggs (**Figure 10.20B**). The breeding system of darter species with such knobs is based on male parental care—males defend a rocky overhang and attract one or more females, who glue their eggs on the underside of the overhang. Males then guard the eggs until they hatch. In many fish species with male egg guarding, females often prefer to oviposit in nests that already contain eggs, which ensures that the male is unlikely to abandon the nest. Female darters are strongly attracted to large males with large knobs. The knobs are located on the fin that is closest to the oviposition site, depending on the species' courtship behavior. It has been proposed that the structures give the illusion of

(A)

(B)

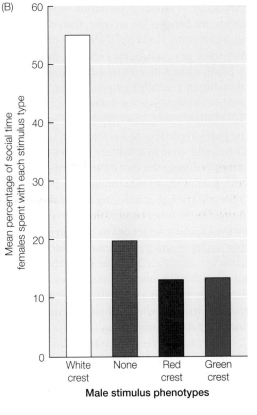

(B)

FIGURE 10.20   Sensory traps (A) Male fiddler crab (*Uca terpsichores*) waving in front of the sand hood he has constructed behind his burrow. Females frightened by a predator will quickly run to such vertical structures, including artificial pillars [88, 270, 271]. (B) During the breeding season, male guardian darters (*Etheostoma oophylax*) develop egg-mimicking knobby yellow structures on the tips of their dorsal fin spines that attract females to oviposit in their defended rocky overhang nest site. This male was collected and placed next to his overturned nest rock with maturing eggs glued to the underside. The dorsal fin placement of the false eggs may give females the impression of more eggs present in the nest, as the male courts in a broadside position under the overhang. Knobby structures have evolved four times in this genus on different fins. In species with egg structures on the pelvic or pectoral fins, males court females in an upside-down or frontal position, so that the ornaments are positioned close to the real eggs [31, 381–383, 403].

EVIDENCE FOR LATENT BIASES   Signaling consequences for latent biases are more difficult to document. Several experimental studies have documented female birds' preferences for novel artificial traits in males, such as colored leg bands, dyed feathers, added head tufts, and certain acoustic flourishes (Figure 10.21) [67, 247, 386, 398, 447, 457, 461]. While such experiments do reveal the existence of latent sensory biases, they are weak evidence for the evolution of sender traits via receiver precursors because the traits in fact have not evolved [23]. Thus receiver biases do not inexorably lead to the evolution of matching sender traits [162].

In most cases, the novel traits in these experiments represent natural stimuli that have been placed in a different

more eggs in the nest than in reality [31, 351, 381, 383, 403, 490]. As a final example, burrowing owls usurp ground squirrel burrows as nesting sites and defend the burrow against the squirrels using a long, loud, hissing vocalization. Squirrel pups are vulnerable to rattlesnake predation and adult squirrels learn to associate danger with the warning rattle of an encountered snake. The defensive hiss of the burrowing owls mimics the snake rattle and co-opts the squirrels' cautious retreat response to this sound [98, 380, 442]. The vocal threat and alarm signals of many animals resemble the rattle of a rattlesnake, and it is tempting to suggest that these signals may have arisen because of an innate or learned retreat response to this kind of sound.

(A)

FIGURE 10.21   Latent preferences (A) A male long-tailed finch (*Poephila acuticauda*) on the left, and a zebra finch (*Taeniopygia guttata*) on the right, sporting artificial white feather crests. (B) Females in both species prefer conspecific males with white crests over normal males without a crest. Females do not approach males with artificial red or green crests. The preference for this novel trait in two species has been proposed as evidence for nonadaptive neurophysiological processes with a phylogenetic basis. The crest preference could reflect a general avian predilection for elaborate facial ornamentation. (After [67].)

(B) Mean percentage of social time females spent with each stimulus type

Male stimulus phenotypes: White crest, None, Red crest, Green crest

position on the body, intensified in shape or color, or acquired from a closely related species. Thus responses may be the result of **receiver generalization** from familiar stimuli, a phenomenon called **supernormal stimulation** in the ethological literature and **peak shift** in the psychological literature. The ability to respond to novel stimuli that are intensifications of familiar stimuli is the consequence of built-in neurological recognition mechanisms for coping with variability [172]. For example, if animals obtain a positive benefit from responding to the larger, brighter, or faster stimulus, they are likely to respond even more strongly to an exaggerated stimulus outside the normal range [139, 141, 172, 196, 197, 441]. Moreover, visual pattern displays that are more symmetrical, more repetitive, or rounder may be more readily detected [140, 263, 487]. On the other side of this coin, extremely spectacular, unpredictable, and rapidly changing stimuli may be confusing, startling, or aversive to a receiver. Certain kinds of sounds, such as loud sudden-onset bursts and broadband pulses or rattles, can have a similar psychological impact. Stimuli with these characteristics may evolve into highly effective threat, alarm, or predator-deterrent signals [141]. An older term for such unpredictable and startling acts is **protean display**, after the Greek god Proteus, who was noted for his ability to rapidly change his appearance. The evolution of some animal display strategies and predator evasion tactics have been explained from this perspective [127, 205, 232, 233, 421].

ESCAPING THE SENSORY TRAP   As illustrated in Figure 10.16, when senders evolve a trait that matches a sensory bias in the receiver because they derive a benefit from the receiver's response, three possible evolutionary scenarios may occur. If sender exploitation of receiver biases neither harms nor benefits the receiver, then obviously there will be no coevolution. However, if senders continue to benefit, they may track any evolutionary changes in the receiver's sensory biases caused by natural selection unrelated to signaling to maintain a matching signal. If sender exploitation has a positive consequence for the receiver, then cycles of positive coevolution would ensue, leading to increased sensitivity on the part of receivers and elaboration of the trait in senders. This could arise in a solitary species that must search for mates, reducing the cost of searching and thus benefitting both parties. Examples include the hood-building fiddler crabs and the egg-mimicking fins of the darters mentioned above. On the other hand, if the receiver is truly manipulated by the exploitative sender to the detriment of the receiver, then cycles of antagonistic coevolution will ensue. This scenario is most likely to occur when the bias has evolved to meet an adaptive, noncommunicatory function such as food detection; in such a case, deceptive senders may distract receivers from foraging and possibly induce them to mate in nonoptimal times and places.

How can receivers escape from this sensory trap? Receivers may evolve resistance to the sender trait, but this could be costly in itself if it reduces sensitivity to the natural adaptive stimulus. If receivers evolve increased ability to distinguish between the sender's signal and the natural stimulus, senders may counter by evolving better mimicry. Eventually, receivers should evolve different recognition and decision rules for the two stimuli [428]. A convincing case for such coevolution has been described for splitfin fishes (Goodeinae), where males have evolved a yellow tail band that mimics worms and dragonfly larvae, the main prey of these freshwater fishes (**Figure 10.22**) [304]. In this case, the food detector became the social signal detector, and alternative sensory mechanisms now handle food detection.

There are a few documented cases of receivers still caught in a sensory trap. One example has been described in the garter snake (*Thamnophis sirtalis*) [470]. When captured by a predator, the female has a hypoxic stress response that involves cloacal gaping, which repels the predator by extruding feces and musk. Males exploit this response during

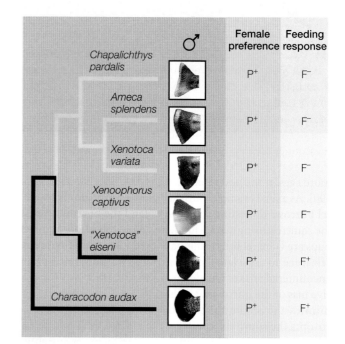

FIGURE 10.22   Antagonistic coevolution of a tail fin pattern in splitfin (subfamily Goodeinae) fishes These internal-fertilizing, live-bearing fish are endemic to Mexico and inhabit shallow freshwater habitats. The phylogenetic tree of a subset of six species is shown here. The males of several species have a terminal yellow band (TYB) that resembles a quivering worm or damselfly larva when the tail undulates. The TYB is absent in the basal taxon *Characodon* and faint in *X. captivus*; yellow lines show lineages with a TYB and the bicolored line shows uncertainty about TYB origin or loss. Females in all six species spent more time visiting males with experimentally painted TYBs relative to natural control males (P+). Males and females of the two species without a TYB (*C. audax* and *X. eiseni*) directed feeding responses (bites, F+) toward the yellow tails of heterospecific fish with a TYB, whereas males and females of the four species with a TYB did not (F−), implying that the latter have evolved resistance to the simple food stimulus. Both sexes in species with a TYB appear to have escaped a sensory trap by evolving alternative food recognition mechanisms, and the TYB has become a potent mate attraction signal. (After [304].)

courtship to achieve copulations—they crawl up on top of females and exhibit rhythmic pulsating waves of muscular contraction that interferes with female respiration. The female's cloacal gaping response permits easier intromission by the male. Females sometimes die of suffocation during these interactions, a significant cost indeed. Numerous other examples of sexual sensory traps that overwhelm female resistance—insemination by injection of sperm, hormonal components in mammalian ejaculate and insect spermatophores that speed up female egg laying, and male pheromones that tranquilize females—clearly prevent females from reproducing at optimal times and making optimal mate choices [22].

A final interesting situation of coevolution following exploitation of receivers by senders has been proposed for acoustic courtship signals in moths. As mentioned earlier, several moth taxa evolved mechanisms to produce ultrasound clicks that deter insectivorous bats, which then served as precursors for ultrasonic courtship signals. Such acoustic courtship signals have also exploited the ears that evolved in moths expressly to detect the ultrasonic echolocation calls of bats [95]. Ultrasound-sensitive ears are widespread in moths and precede the evolution of sound production. The response of moths to these sounds is an immediate evasive dive and hiding or freezing on a substrate. For males to exploit this receiver-response system of females with an acoustic courtship signal, the initial avoidance response of females to bat ultrasounds must be reversed to one of attraction [8, 9]. How might this happen? One proposed scenario is an intermediate stage in which the male produces a batlike sound that causes the female to dive and hide, and he then follows her to the hiding site and proceeds to court using olfactory signals [188, 474, 530]. The acoustic signal then diverges in pitch and pulse rate, and females modify their auditory tuning curves and response rules [160, 493]. Other scenarios do not require the olfactory intermediate phase and propose instead that male ultrasonic courtship signals exploit the motionless response of females to bat calls (the sensory trap scenario) but are softer and sufficiently distinctive from bat calls [108, 360, 361]. Regardless of the scenario, both males and females currently benefit from this signaling exchange, but there is lingering evidence of antagonistic interactions and risks to females that must have occurred during the evolutionary transition. In moth species that respond to both bat and conspecific acoustic signals, males are selected to make their ultrasonic clicks distinguishable from those of bats, and females evolve precise context-specific thresholds for responding with approach versus evasion [189, 190, 259, 427].

In summary, sensory preadaptations can serve as important sources of signals, especially mate attraction signals, and may be more widespread than previously thought [23]. The resulting signals may match obvious sensory adaptations or appear to be arbitrary in form. If the preexisting response is beneficial to receivers as well as to senders, the signal is highly efficacious from the start and there may be little need for coevolution and receiver fine-tuning. However, it is more likely that sender and receiver interests are not the same, and

some amount of tuning, code refinement, and ritualization will take place to hone the communication interaction into one that benefits both parties.

The Evolution of Reliable Signals

In the discussion above, we learned that both the receiver-precursor and sender-precursor evolutionary models could lead to a positive feedback loop in which both parties benefit and signals become elaborated. However, we also saw that senders can readily evolve deceitful or manipulative signals that benefit them to the detriment of receivers, leading to negative feedback loops and possible destabilization of the communication exchange. To go beyond these flow diagrams and evaluate the coevolutionary processes occurring during signal evolution, we need to examine the interactive strategies of senders and receivers. This strategic perspective of signal evolution will help us understand how signal properties can become shaped to provide reliable information that receivers benefit from heeding. Evolutionary game theory, with its emphasis on strategic interaction, is a powerful tool for modeling this aspect of signal evolution.

Conflicts of interest and the problem of honesty

In some signaling contexts, sender and receiver have no conflict of interest, because both parties rank the payoffs of alternative receiver responses in the same order. In such cases, both parties benefit from the accurate provision of information, and a cooperative signaling exchange can be maintained. As mentioned earlier, the classic example of this cooperation is the dance language of honeybees, in which a forager returning to the hive gives a signal that conveys the location of a food resource to her sister workers. The forager-sender benefits by recruiting other foragers to the food site, and the worker-receiver benefits from obtaining accurate information about where to find food. Even in other contexts, such as senders that exploit receiver sensory biases to make themselves more conspicuous to potential mates, the receiver may obtain a net benefit by finding appropriate mates more quickly.

In many communication contexts, however, sender and receiver have conflicting interests, because they rank the payoffs of alternative receiver responses differently. Under these conditions, animal senders will be tempted to provide misleading information so that the receiver performs the act that most benefits the sender. The strength of the selective pressure to deceive or manipulate depends upon the signaling context and the degree of conflict between the parties. Conflict of interest is greatest when two more or less equal competitors both desire the same nonshareable resource. Each would like the other to back down without a fight, and each would benefit from persuading the other that it is the better fighter by any means possible, including bluff. In the mate attraction context, both male and female benefit from mating with the correct species and therefore agree about the accurate transmission of species information. But females may only want to mate with a high-quality male, whereas males typically strive to acquire as many mates as possible, putting pressure on low-quality males to hide or exaggerate their quality. Similarly, an offspring in a brood of siblings may exaggerate its need for food to the parent in order to garner a larger share of the food for itself. Even in cooperative groups, where all members benefit from shared predator surveillance and alarm signaling, some individuals may try to avoid their share of sentinel duty to gain more foraging time; this puts everyone at risk or forces others to do more than their fair share.

Deception is the provision of inaccurate information by the sender, such that the sender benefits from the interaction but the receiver pays the cost of a wrong decision. Types of deceit include *exaggeration* or *bluff* (using a signal whose rank among ordered alternatives is different from that for the corresponding condition values), *lies* (use of the wrong signal among an unordered set of alternatives), and *withholding information* (not giving a signal when one would be appropriate). **Honesty** is the provision of reasonably accurate information by the sender. Honest signals, whether visual, vocal, olfactory, or electrical, are those in which some characteristic of the display is reliably associated with some attribute of the sender or its environment that is of interest to the receivers. Identifying this attribute specifies the kind of information transmitted by the signal. To fully demonstrate an honest signaling system, we would need to show that receivers not only attend to the signal and its variants, but also benefit from incorporating this information into decisions. As we discussed in Chapters 8 and 9, no signal is likely to be perfectly accurate. Both senders and receivers will make errors, and it may not pay for either party to give or seek perfect information. As we also saw in Chapter 8, signal reliability can be quantified, and animal signals are surprisingly reliable, even when there are strong conflicts of interest. Why is this so? Why don't more animals cheat? To answer this puzzle, we need to cast the interactions between senders and receivers as an evolutionary game. To understand the relevant models, we first need to review some basic precepts of evolutionary game theory.

Evolutionary game theory

As discussed in Chapter 9, evolutionary game theory is the modeling method of choice when the relative fitness of a strategy is frequency dependent. In an evolutionary game, the average payoff for an individual playing a given strategy is the sum of the discounted payoffs it would receive in pairwise contests with each type of strategy in the population. The discounting is simply the relative abundance of the opponent strategy and thus the probability that the focal animal would encounter it. If the average payoff for a given strategy is higher than the global average, that strategy will become more common in the next generation; if it is less than the global average, it will become less common [223, 503]. Such intergenerational change for a strategy ceases when either the strategy frequency goes to zero, (i.e., the strategy goes extinct), or the average payoff for the strategy becomes equal

to that of the global average. The population reaches an evolutionary equilibrium when all strategies cease changing. This can happen if either one strategy takes over the population (making the average payoff of playing against itself equal to the global average payoff), or differences in payoffs become balanced by differences in frequencies so that the average payoff of any one strategy equals the overall average of the remainder (making it again equal to the global average). Such an equilibrium is an evolutionarily stable strategy (ESS) [320, 324, 389]. A strategy that displaces all others once it is sufficiently common is a **pure ESS**, and a stable blend of strategies is called a **mixed ESS**.

A wide variety of evolutionary games have been studied. On one axis, such games can be categorized depending upon whether their alternative strategies are **discrete** or **continuous**. Discrete games can be summarized in two-dimensional matrices, called **normal form analysis**, and simple rules can be applied to find ESSs; continuous strategy games require more complex mathematical analysis (see Web Topic 10.5). On an independent axis, one can contrast games in which the payoffs in the normal form matrix are fixed (called **contests**), with those in which the payoffs in the matrix can vary depending on the relative frequencies with which alternative strategies are currently being played (called **scrambles** or **playing against the field**). Many behavioral games involve a series of cumulative bouts before a payoff is received. The ESS in such **dynamic games** is not a single action but a sequence of successive actions. Note that in all of these categories, the system may have no, one, or many possible ESSs.

Games can also be categorized according to whether relevant players are equal or not. In a **symmetric game**, all players have access to the same strategy set, are equally likely to win playing strategy A against an opponent playing strategy B, and get the same payoffs when playing strategy A against strategy B. In an **asymmetric game**, one or more of these conditions is violated. For instance, senders and receivers are unlikely to have access to the same strategy sets, and dominants and subordinates are unlikely to have the same chances of winning, even if each plays the same strategy against an opponent strategy. In asymmetric games, a player in one role is not playing against the other role; instead, a player competes with other players in the same role to post the best response to whatever the players of the other role are doing. Put another way from a communication perspective, there are several games being played out in an asymmetric game—one among senders to find the best strategy given what receivers might do, and one among receivers to find the best strategy given what senders might do. These games are still frequency dependent, but in more complicated ways [204].

Because of the appeal of normal form analysis, researchers initially partitioned many scramble, dynamic, and continuous games so that they could be approximated by simpler discrete games. However, much recent work has gone into devising methods for analyzing continuous games without forcing them into artificial categories [223, 367]. Discrete games are now often treated as special cases of a more general continuous game case. In addition, dynamic and sequential games are better analyzed not by collapsing payoffs into a matrix, but instead by plotting all successive steps on a tree and evaluating payoffs at each branch terminus. This is called **extensive form analysis** [100]. Each of the methods for finding ESSs has its own virtues and shortcomings [203, 331, 367], and we shall invoke all of them in subsequent discussions.

ESSs were originally defined as strategies that when common are: (1) the best response to both themselves and any other rare mutant strategy, or (2) if equal in payoff to another strategy, are a better response to the other strategy than the latter is to itself [320, 324]. While these criteria have survived various challenges, they do not specify the dynamics that might get a population to an ESS. Should one expect many natural populations to end up at an ESS? How might the trajectory be interrupted? Are all equilibria defined by those conditions alone stable? Several subsequent approaches have shed some light on these questions. As reviewed in Chapter 9, adaptive dynamics models track evolutionary trajectories of populations, whether they lead to stable equilibria or not [124, 335]. At any point in time, a population is characterized by the relative frequencies of the alternative strategies for some trait (**Figure 10.23**). If the average payoff for any of the strategies is above the global average, it will increase in frequency; other strategies will decrease, and the population will move to a new location on the adaptive landscape. When the population reaches a point where no strategy has above-average payoffs, it stops moving. This is an equilibrium. However, in nonlinear models, such as those describing frequency-dependent selection, there can be different kinds of equilibria [168, 169]. For a **stable equilibrium**, all likely trajectories move toward the equilibrium point, never away from it. If a population at a stable equilibrium point is perturbed to a slightly different location on the adaptive landscape, it will follow these trajectories back to the equilibrium point. All the trajectories surrounding an **unstable equilibrium** leave the equilibrium point; a population that arrives here will not stay but will move away over time. Some locations are **saddle points**; trajectories along one axis all enter the point, whereas those on a perpendicular axis all exit it. A population can move to a saddle point along one axis and leave it along another. Some equilibria are only circled by the allowed trajectories around them; the population can never reach such an equilibrium point. This generates limit cycles, a phenomenon that, though so far rare in behavioral contexts, appears to exist in some lizards [475, 476]. Given these diverse possibilities, how have so many biological traits, and particularly animal behaviors, located and settled at predicted ESSs [129]?

One important property that may solve this puzzle is **convergence stability** [142, 143]. An equilibrium that is convergent stable is also an attractor, in that mutant strategies that appear between the equilibrium point and the current population are favored over the current population. Convergence stability thus channels a population's evolutionary trajectories toward an ESS, at least in the vicinity of that ESS. It has been argued that most biological examples in which

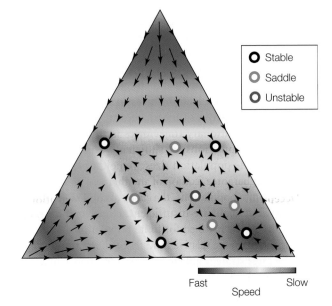

FIGURE 10.23   Adaptive dynamics of an evolutionary game The triangle delimits possible frequency combinations of three alternative strategies in a population. At the corners, only one strategy is present; along the edges, only two strategies are present; and inside the triangle, all three strategies are present, with each having a higher frequency the closer it is to the corner where it is the only strategy present. At the center of the triangle, each strategy constitutes a third of the population. Arrowheads show the directions that the population is predicted to move next if it arrives at a given point, and the background colors indicate how rapidly the population will move. This game models confrontations between teams of four players, each of which could use any one of the three strategies. The game has four stable equilibrium points, one unstable equilibrium point, and four saddle points. (After [177].)

the observed strategies appear to have reached a predicted ESS must also have been convergence stable [331]. Convergence stability is a bit harder to achieve if an ESS strategy is multivariate, but the challenges are reduced if the separate components are genetically correlated [294]. This may help explain the evolution of the many multivariate signals that we have discussed in previous chapters. Interestingly, an equilibrium that is convergent stable but not an ESS can lead to **evolutionary branching**; once the population reaches the equilibrium combination, subsequent evolution causes it to diverge into two or more trajectories [125, 168, 169]. This phenomenon has been linked to speciation; its relevance to signal evolution within a species remains to be explored. As an alternative, a strategy may be an effective **neighborhood invader**, in that it can displace any strategy that is nearby. This would allow the strategy to invade a population and, even when initially rare, move gradually to the ESS frequency distributions [18, 19]. Since neighborhood invaders are inherently convergence stable, this property alone could favor a population reaching an ESS.

The original ESS definitions assumed infinite population size. The strength of selection in this case was not relevant; thus payoffs could be equated with strategy fitnesses.

Recent studies have examined what happens when the evolving populations are finite [365, 367, 502]. This is relevant because drift becomes increasingly important as population size is reduced; even a convergently stable ESS strategy may disappear from a small population by chance. In addition, the relative strengths of selection (measured as the impact of a given payoff on a player's fitness), become important factors shaping the evolutionary trajectories in finite populations [515]. When populations are sufficiently small, the criteria used for infinite population sizes are neither sufficient nor necessary to guarantee fixation of an otherwise predicted ESS. For larger finite populations, the classical conditions become necessary, but again are insufficient. However, alternative conditions can be invoked for finite populations. These criteria are based on the relative fitnesses of alternative strategies were the rarer strategy able to achieve a frequency of a third of the population [49, 159, 365, 372]. Meeting these criteria largely depends on the distribution of payoffs for different pairwise combinations of strategies.

The criteria for finite populations grade into those for the classical case as the product of population size and selection intensity increases [512–514]. The relative time scales for selection and drift may also contribute to how far a small population deviates from classical ESS predictions [515, 554]. The relevance of finite population models to evolution of behaviors like signals depends on how often communication systems evolve in restricted populations and limited time scales. The fact that many classical ESS predictions appear to be realized in nature suggests that population sizes and time scales have not been that restrictive. However, there is one context in which local effects can be quite limiting: evolution within viscous networks. We take up these issues in Chapter 15.

Web Topic 10.5 *Evolutionary game theory*
Here we provide some more background on this modeling technique, including methods for finding ESSs in various games.

Classification of signals based on type of cost

In the early days of ethology, many signals were found to have evolved through the ritualization of behaviors that were or had been functionally appropriate to the contexts in which the signals are now given. Signals were believed to be honest indicators of underlying motivations because they were derived from physiologically or anatomically linked sources. With the rise of evolutionary game theory in the 1970s, this notion of signal honesty was questioned. Why should a sender give an honest signal? In aggressive contexts, for example, a threat signal that accurately predicted the sender's intention to attack would be readily overtaken by a mutant sender that always delivered a strong threat regardless of the sender's true intentions [321, 323]. Similarly, if senders wear their emotions on their sleeves, what prevents a clever receiver from using that information to exploit the sender? These ideas led to the view that senders were best characterized as "deceitful manipulators" trying to mask their true intentions and trick receivers into actions benefiting senders. Receivers in turn were characterized as "mind readers" trying to discount false signals, anticipate the true intent of the sender, and thus identify their own best countermove [117]. This process was envisioned as a never-ending arms race with increasing deceit and concealment of true intentions by senders, parried by increasing discrimination and exploitation by receivers. Except during temporary phases when the receivers have drawn ahead of senders, or where sender and receiver have common interests, signals would be largely deceitful and uninformative (discussed further in Web Topic 1.1).

Amotz Zahavi challenged this pessimistic view of signal honesty. He asserted that receivers have the upper hand, and ought not to respond to signals unless they carry some guarantee of honesty. Possible guarantees were briefly introduced in Chapter 8. One option is to require that the signal impose a production cost such that deceitful senders cannot afford to give an exaggerated signal, or produce it only in an ineffective way. By attending only to signals that are costly to produce, receivers in most contexts can obtain reliable information about sender quality. Zahavi referred to these costs as "handicaps," using the analogy of the weights added to the saddles of racehorses to reduce but not eliminate disparities in performance; the size of the handicapping weight is an indicator of a horse's quality [546, 547, 553]. Although Zahavi's idea was viewed skeptically at first [40, 114, 272, 322], subsequent game theoretic models demonstrated the evolutionary feasibility of handicap signaling, and the **handicap principle** is now widely accepted as an important mechanism for the evolution of reliable signals [181, 182, 241, 254, 255, 328, 399, 460]. Even signals that initially arose from sender exploitation of receiver biases can escape the sensory trap and evolve into honest handicap signals [304].

Since the mid-1980s, dozens of evolutionary game models of biological communication have been developed, each depicting different signaling contexts and comprising different game structures and sets of assumptions [217, 239, 258, 328, 460]. The common feature in most of the models that found at least some conditions for a signaling ESS was the assumption of a cost imposed on dishonest senders. Without such costs, senders could become dishonest, receivers would ignore the signals, and no evolutionarily stable state with informative communication signals would be attained. These costs ranged from signal production costs to receiver retaliation costs, reputation costs, and various types of physical and physiological constraints. The realization that the type of cost affects both the form of a signal and the specific kind of information it can encode has led to a useful classification of signals based on the type of cost (mentioned already in Chapters 1 and 8). Independent attempts to classify signals in this way have largely converged and provide a very powerful framework for understanding the evolution and diversity of animal communication signals [115, 217, 239, 327, 328, 520]. We shall distinguish between five broad signal classes—handicaps, indices, proximity signals, conventional

TABLE 10.1 *Classification of signals based on cost that guarantees honesty*

Category	Cost	Signal design	Type of information
HANDICAP SIGNALS			
Quality handicap	Differentially costly to produce or bear	Graded display, intensity correlated with sender quality	Condition, health, stamina, fighting ability
Signal of need	Differential benefits per investment cost	Graded display, intensity correlated with sender need	Motivation, need, resource valuation
INDEX SIGNALS			
Quality index	Physical/physiological constraint	Discrete or graded display, form linked to sender attribute	Body size, strength, age, natal area
Pointing	Informational constraint	Linked to target of sender's attention	Identify object of sender's attention
PROXIMITY SIGNALS			
Vulnerability handicap	Injury risk	Vulnerable posture	Willingness to take risks, confidence
Tactical threat	Injury risk	Attack preparation posture	Willingness to take risks, intentions
Defensive threat	Injury risk	Protection of vulnerable body parts from attack and injury	Nonaggressive intentions
CONVENTIONAL SIGNALS			
Cost-free signal	No or minimal cost, no conflict of interest	Low-amplitude signal derived from cue	Coordination of cooperative activities
Conflict conventional signal	Socially-imposed cost	Arbitrary form, antithetical discrete or graded signal	Status, motivation, willingness to escalate, fighting ability
MODIFIERS			
Amplifier	Constrained by sender attribute	Improve assessment of sender attribute	Reveal body size, display vigor, conspicuousness
Attenuator	Constrained by sender attribute	Hinder assessment of sender attribute	Hide body size, display vigor, conspicuousness

signals, and modifiers—and within each class we describe some of the more common subcategories as identified by Hasson [217], Hurd and Enquist [239], and Számadó [496, 497]. Hasson's classification scheme combines honesty guarantees with informational topics (e.g., signals about the sender versus about third party individuals or objects), and provides mathematical expressions for how the sender's fitness depends on the signal given and the receiver's response. Számadó's classification deals only with agonistic context signals. Table 10.1 summarizes our classification scheme based on how each type of signal ensures or affects signal reliability, and shows the relationship between the type of cost, the form and design of the signal, and the kinds of information the signal might provide.

HANDICAP SIGNALS The reliability of **handicap signals** is maintained by signal production costs, which distinguishes them from index and conventional signals, as we shall see below. Handicap signals are strategic, in the sense that all senders can in principle make all signal variants, but it costs more to make a bigger or more intense signal. Senders are assumed to vary in their ability to bear these production costs, or in their ability to reap benefits from a given amount of display investment. The different benefit/cost trade-off optima for individuals of different abilities leads to a positive association between display intensity and sender characteristics, which can be highly informative to receivers. The term "handicap" is somewhat misleading, because the process can include complex multiplicative effects of costs and benefits in addition to additive effects [171, 234] (see Web Topic 10.6). Nevertheless, the term is well ensconced in the literature and is understood to refer to signals made reliable by cost/benefit trade-offs. Handicaps can be further subdivided into two signal classes based on game models with different costs and assumptions.

The **differential-costs model** is the classic handicap model proposed by Zahavi and Grafen [182, 546, 547, 553]. The key assumption in this model is that senders differ in their ability to bear the cost of signaling. Furthermore, the type of cost should be relevant to the particular sender attribute about which the receiver wants information [274, 300,

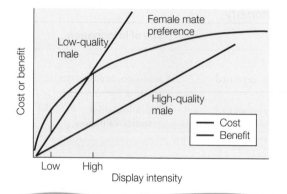

FIGURE 10.24 **Quality handicap model for mate choice**
The following assumptions are made in this model: (1) display-ing is costly, reducing survival of displayer; (2) low-quality males pay a higher marginal cost (steeper cost slope) than high-quality males (differential cost); (3) females are more likely to mate with a high-intensity displayer; (4) female preference for a given display level does not depend on male quality (i.e., there is a single prefer-ence curve for all males). In this model, the cost and benefit fitness components are additive, which means they must be measured in the same fitness currency. At the ESS, where each male's net ben-efit is maximized, high-quality males display at a higher intensity than low-quality males, so display intensity is an indicator of male quality or condition. (After [255].)

553]. Receivers are selected to pay attention only to those types of signals that impose costs linked to the type of infor-mation beneficial to receivers. Directional selection pressure from receivers favoring the most costly signal variants often results in extreme elaboration and exaggeration of the display character. Senders of poor quality or condition tend not to produce the more intense variants because they cannot bear the high signaling cost. Thus the form of the signal is associ-ated with its information content, typically a continuously varying signal parameter correlated with some sender attri-bute. Figure 10.24 illustrates the additive **quality handicap model**, in which poor quality senders pay a higher cost for a given intensity of display. Their best option is to display at a lower intensity than a high-quality individual, so signal inten-sity is reliably correlated with sender quality. Such signals are also called **quality indicators** and **condition-dependent sig-nals**. Handicapping costs can include production costs paid at the time of display, risk of predation or parasitism from highly conspicuous signals, and developmental costs paid earlier to grow the structures and organs needed for display-ing. Some examples below show how such costs can be linked to useful information for receivers.

Signals with energetically expensive production costs can inform receivers about the health, stamina, or foraging abili-ties of senders. For example, vocal and visual mate attraction signals must be repeated again and again. Female preference for males that not only produce high-quality displays but also repeat them at a high rate will increase the selection pressure on males to perform at the highest energetic level they can sustain. Unhealthy or poor-quality males cannot maintain such a costly display rate, and this fact will be detectable to

females. By selecting vigorous mates, choosy females obtain better genetic fathers, better parental providers, good forag-ers, or owners of food-rich territories. **Figure 10.25** illus-trates a well-studied example of a quality-indicator signal, in the form of an acoustic drumming display by a wolf spider. Males that display at high rates not only bear the energetic cost of display better, but they also have stronger immune systems, greater mobility, and more effective escape mecha-nisms against predators [6, 278–280, 298, 3106]. Similarly, repetitive countercalling contests between rival males can facilitate the assessment of relative condition, fighting ability, or motivation without the rivals having to engage in a physi-cal fight, as demonstrated in red deer and grasshoppers [90, 186, 187]. As a final example of a performance cost handicap signal, energetic jumping displays (called stotting) are given by gazelles to approaching predators (see Figure 14.5). Only individuals in good condition can afford to perform these actions well and still have energy reserves left over to run from the predator. Stotting vigor thus provides honest infor-mation to the predators, which subsequently avoid chasing the more able displayers [75, 76, 152, 371].

Developmental costs are borne before the onset of dis-play and can reveal either nutritional condition during development or more intrinsic aspects of genetic quality. The development of exaggerated display organs can also result in high maintenance costs for the bearer. A well-doc-umented example is the elongated tail feathers of the barn swallow (*Hirundo rustica*). Females prefer males with longer tails, pairing very quickly with males that have artificially enlarged tails compared to males with shortened tails. Long tails are a handicap for males—barn swallows are aerial for-agers that capture flying insects on the wing. Artificial tail elongation increases the drag on the tail and reduces agility and foraging efficiency. Males with naturally long tails are stronger, healthier, and more parasite-resistant individuals who can not only grow long tails and cope with the foraging handicap, but who also pass on their parasite resistance to their genetic offspring. Females thus obtain better-quality offspring by selecting long-tailed males [101, 342, 343, 345-348, 451]. Another good example of a quality handicap sig-nal with a development cost is red carotenoid coloration in birds and fish [151, 158, 221, 230, 275]. Since animals must acquire and extract carotenoid pigments from their diets, color patch intensity is believed to indicate foraging ability and nutritional condition [332, 374]. Bearers of such color patches may also sustain the maintenance cost of increased conspicuousness to predators. Males in many songbird spe-cies have repertoires of song types that they learn in the first months of life; females have been shown to prefer larger rep-ertoires and more complex songs [79]. One explanation for this preference is the *developmental stress hypothesis*, which argues that young birds need good nutrition to develop the brain nuclei overseeing song learning and production [65, 368, 369, 485]. Such developmental constraints may be severe enough that the resulting signal variants should be considered index signals rather than handicaps.

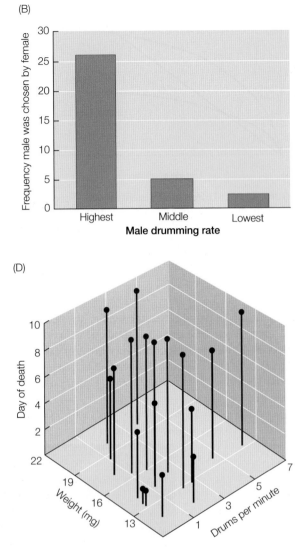

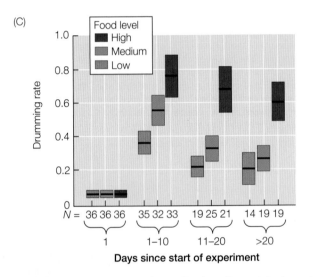

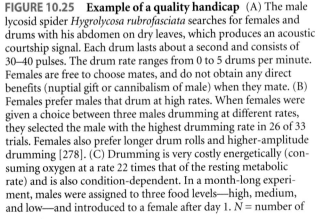

FIGURE 10.25 **Example of a quality handicap** (A) The male lycosid spider *Hygrolycosa rubrofasciata* searches for females and drums with his abdomen on dry leaves, which produces an acoustic courtship signal. Each drum lasts about a second and consists of 30–40 pulses. The drum rate ranges from 0 to 5 drums per minute. Females are free to choose mates, and do not obtain any direct benefits (nuptial gift or cannibalism of male) when they mate. (B) Females prefer males that drum at high rates. When females were given a choice between three males drumming at different rates, they selected the male with the highest drumming rate in 26 of 33 trials. Females also prefer longer drum rolls and higher-amplitude drumming [278]. (C) Drumming is very costly energetically (consuming oxygen at a rate 22 times that of the resting metabolic rate) and is also condition-dependent. In a month-long experiment, males were assigned to three food levels—high, medium, and low—and introduced to a female after day 1. *N* = number of

males remaining during each time period (of the 36 in each group at the start of the experiment). Males increased their drumming rate dramatically after the female was introduced, and males with higher food resources drummed at higher rates. Males kept in the presence of females had lower survival rates than males kept without females [279]. (D) For males forced to drum at high rates in the presence of females, those with higher drumming rates survived better than males with lower drumming rates. Individuals shown in red survived through the entire period. Body weight was also correlated with survivorship. Males with high natural drumming rates were more mobile, more immune to pathogens, and more effective at escaping from avian predators, so females obtain better-quality mates by preferring those that advertise with high drumming rates [5, 6, 298]. (B after [390]; C after [280]; D after [310].)

In the **differential-benefits model**, senders obtain different benefits for the same level of display (**Figure 10.26**). There still must be a signal production cost that increases with intensity of display, but all senders can experience a similar cost function. One version is the *Sir Philip Sidney begging*

game, in which senders—such as brood-mates—vary in their nutritional need and thus in the benefit they stand to gain by giving a begging signal to a potential donor—such as a parent—for food or water [250]. The cost may be an energetic cost, a risk of dying if not fed, or a predator attraction

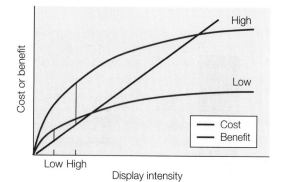

FIGURE 10.26 Differential benefits model Senders vary in the benefit they obtain for a given level of display intensity, and all senders experience a similar increase in cost of display. As in Figure 9.24, this is an additive model. At the ESS, senders that stand to gain a greater benefit will display at a higher intensity than senders with a lower benefit. (After [250, 255].)

risk. This model requires that the sender (the needy beggar) and the receiver (the donor of food) be genetic relatives, especially if the cost function is very low. An example that appears to fit this model well is begging by canary nestlings (*Serinus canarius*) (**Figure 10.27**) [269]. Hungry nestlings beg

more vigorously, and benefit more from being fed, than their better-fed siblings [267]. Although the absolute energy cost of begging does not appear to be very high, it does increase

FIGURE 10.27 Benefit and cost of begging in canary nestlings (A) Nestling birds gape, call, shake their bodies, and extend their necks upward to varying degrees depending on how hungry they are (i.e., with increasing time since their last feeding). Some finches, such as canaries and this house finch parent, regurgitate seeds to their young, doling out food to several or all nestlings during a single visit. (B) In an experiment with canary broods of three chicks ranked in size as big, medium, and small, all chicks were removed from the nest, food-deprived for a period of time, and then fed either 0, 0.25, or 0.5 ml of feeding mix, before being returned to the nest. Parental feeding rates were then monitored. Regardless of whether the starved chick was the big, medium, or small one, it begged more vigorously and was fed more than its siblings. (C) In another experiment, two size-matched chicks were fed by the experimenter. One chick was always fed after 10 sec of begging (the "low-beg" treatment), while the other chick was forced to beg for 60 sec (the "high-beg" treatment). Experimental feeding continued for one day, with both chicks receiving the same amount of food. Each point on the graph represents a single matched pair. The high-beg chicks gained less weight than the low-beg chicks. The marginal difference in mass loss between the two chicks due to energetic expenditure (high-beg minus low-beg) was positively correlated with the difference in postural begging intensity between the two chicks. (B after [265]; C after [269].)

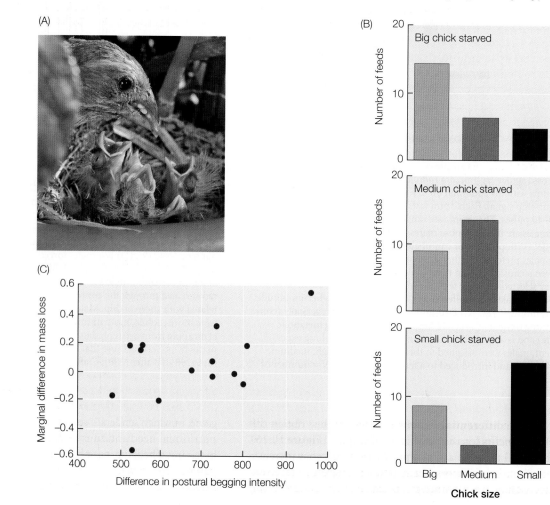

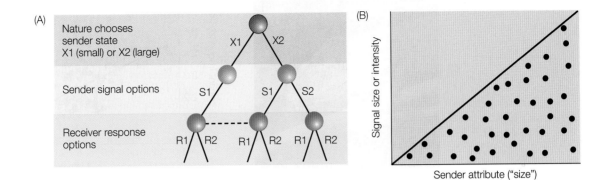

FIGURE 10.28   An extensive form game for an index signal (A) Senders in state X1 are constrained to use S1 whereas senders in state X2 can choose either S1 or S2. A receiver knows that the sender is X2 when S2 is given, but does not know the sender state when S1 is given; this uncertainty is indicated by the dashed line. (B) For a continuous index signal, an analogous game model would result in a triangular plot of signal size as a function of sender size, since large senders can give either large or small signals but small senders can give only small signals. The black line indicates a performance constraint. (A after [239].)

with increasing display vigor and duration [27, 292, 329, 429]. Another cost of begging is the attraction of predators [60, 210], which is greatest for open-nesting bird species and is a cost experienced equally by all of the nestlings in a nest, as modeled in this game. Differential-benefits models with additive costs and benefits such as the Sir Philip Sidney game can be used to explore the conditions for any **signal of need** or motivation.

As we saw in Chapter 9, one can often partition overall fitness into either additive or multiplicative components. Whereas most of the above models use additive economics, some authors have examined handicap evolution with multiplicative partitioning. Not surprisingly, these models also predict exaggerated sexual displays such as those described above. Here, senders differ in their ability to convert a given amount of display effort into more versus less attractive signals to receivers of the opposite sex [170, 171]. For many modeled scenarios like mate attraction, where costs are measured in terms of survival and benefits are measured in terms of offspring production, multiplicative partitioning of components is more convenient. Even in a differential-costs model with multiplicative fitness components, a variety of cost and benefit curve shapes and slopes can lead to an ESS in which display intensity is positively correlated with sender quality [171, 371]. Finally, it should be noted that there is no reason why only differential costs or differential benefits are affected by sender choice of signals; it is quite possible that both could combine to favor graded-intensity signals that reflect sender quality or motivation [302].

Web Topic 10.6 *Handicap controversies*
Zahavi's perspective, and the arguments of those who disagree with him.

INDEX SIGNALS   In the case of handicap signals, all senders can potentially produce any of the alternative signal variants, but senders of **index signals** have, at best, limited choices. An individual sender's index signal options are constrained by its physiology, anatomy, physics, experience, or awareness to those honest indicators of attributes of interest to receivers. Such signals are not strategic, because some senders simply

cannot produce some signal variants [239]. This category has been defined broadly by some authors to include **modifiers** (described below) [217, 328]. We feel modifiers should have their own category since index signals can provide information to receivers, whereas modifiers given alone cannot. Hurd and Enquist [239] call indices "performance-based signals."

There are two common categories of index signals. **Quality indices** provide reliable information to receivers about sender attributes such as body size, strength, age, and aggressiveness, because they are functionally and incorruptibly linked to these sender attributes. For example, tigers mark their territories by scratching as high as they can on tree trunks [505]; scratch height is an unfakable index of body size (see Figure 8.18). Some quality indices may not be so precise. If only low-quality individuals are constrained from making high-intensity signals whereas high-quality individuals make both low- and high-intensity signals, receivers will be uncertain about sender quality when they detect a low-intensity signal [239]. **Figure 10.28A** illustrates this point in an extensive-form game, and **Figure 10.28B** shows how a continuous quality index signal with a performance constraint might lead to a triangular distribution of signal "size" versus sender "size." Good examples of such triangular distributions can be seen in the relationship between chela size and chela performance strength in crustaceans, where some males grow claws that are very large but have low muscle mass and poor closing force [72, 540] (see Figure 11.21). Some examples of index signals are shown in **Figure 10.29**. Body size can be reliably signaled by low-frequency acoustic signals, because size places constraints on the maximum wavelengths that can be emitted (Figure 10.29B; see Figure 11.12C,D). Because fish cannot stand on a surface and accurately assess

(A)

(B)

(C)

(D)

FIGURE 10.29   Examples of index signals (A) The height of the casque (helmet) in the dwarf chameleon (*Bradypodion pumilum*) is correlated with jaw musculature and bite strength [492]. (B) The dominant frequency of acoustic signals is negatively correlated with body size in many species of frogs (here, *Uperoleria rugosa*), because smaller individuals are physically unable to produce sound waves of a viable amplitude larger than their body dimension. (C) Melanic male mosquitofish (*Gambusia holbrooki*) are more aggressive and have higher mating success than silver males [229]. (D) An alerted prey animal, such as this gerenuk (*Litocranius walleri*), stares in the direction of the potential threat. Markings on the face and forward-rotated ears reveal the direction of gaze and signal to a predator that it has been detected. (B after [424].)

another individual's relative height with parallel broadside displays (see Figure 10.8), they instead use signals such as tail-beating (see Figure 11.5B), indicating their size through the force of the pressure wave directed at a rival. Many fish have contrasting lines demarcating their body margins; long-bodied fish usually have horizontal lines, whereas deep-bodied fish have vertical lines [30]. It has been suggested that this demarcation unambiguously reveals the size of the fish [550, 553]. Similarly, push-pull and mouth-wrestling forms of ritualized fighting are reliable indices of weight and strength [29, 242]. Bite force in lizards is determined by the size of jaw adductor muscles and signaled with various head indices (see Figures 10.29A and 11.19). Agility may be revealed with certain types of moves during courtship dances or aggressive sparring bouts. A good example has been described in the fruit fly (*Drosophila subobscura*), where the courting male faces the female head-to-head and must quickly move from side to side following the female's lead. The male movement arc is greater than the female's arc, which imposes a severe performance test; males that lag behind are rejected as mates [319]. Indices of age evolve from signals that are linked with aging processes, such as song repertoire size in birds with open-ended learning, and changes in plumage, feather, or cuticle coloration (see Figure 12.24) [166, 176, 356, 518]. Age is associated with several components of quality, including breeding experience, immunocompetence, and ability to survive [63]. In songbird species with age-restricted (pre-dispersal) song learning, song-type sharing between newly settled males and their neighbors is inversely correlated with

dispersal distance, and could serve as an index of fighting ability if good fighters are more likely to gain territories close to their tutors [489, 538]. Finally, degree of melanization of general body plumage, skin, and pelage in some vertebrates has been proposed to act as an index signal of aggressiveness, sexual activity, and immunocompetence. The linkage between dark coloration and these other traits arises from pleiotropic and cross-reactive effects between the closely allied genes of the melanocortin system that regulate synthesis of brown and black melanin, and those affecting metabolic rate, immune response, stress response, and sexual steroid hormone production (see Figure 10.29C) [128, 229].

A variety of other sender attributes may be signaled by constrained index signals. Territory ownership can be signaled by the presence and confidence of an owner in its familiar area. Olfactory scent marks are especially reliable signals of ownership, since the owner's long-lasting signature marks fill the territory and the boundaries, and can be matched up with the scent of an encountered individual. Olfactory signals can also provide unbluffable indices of sex and reproductive state. Chemical pheromones that are derived from reproductive hormones and by-products reveal reproductive status in both males and females [483]. Although still controversial, alarm substances used by many schooling fish to disperse or cluster during predator attacks appear to be antibacterial agents sequestered in the vacuoles of skin cells that are released upon injury, and may therefore qualify as constrained index signals [542, 543]. Flashing white rump patches in mammals and white outer tail feathers in birds also provide reliable information from frightened and fleeing animals. Note that acoustic alarm signals do not possess this honesty guarantee [341].

The second common type of index signal is based upon **informational constraint**. In this case, the signal can be used only by a signaler that has access to particular information [239]. A stalked prey, perceiving a hidden predator, may stare at the predator, signaling to the predator both its own alerted state and the futility of continuing to stalk (see Figure 10.29D). Such a signal can be performed only by a sender that knows the location of the hidden predator [214]. Acoustic signals that operate like passwords also fall into this class of information-constrained index signals. For example, matching or mimicking the song or call of another individual is a potent mechanism for getting the target individual's attention. Female northern cardinals, who sing many of the same song types as the males in a population, match the just-sung song of their mate to solicit his assistance at the nest [202]. Neighboring male songbirds in many species with partially shared song repertoires address singing rivals with a direct song match or with another song type in the neighbor's repertoire, which indicates that they know the composition of the neighbor's repertoire [39]. Dolphins and parrots also match the signature calls of target individuals that could only be known by fellow group members [243, 337, 522, 527].

In practice, the assignment of signals to handicap or index categories has proved problematic. Because indices cannot be faked and thus do not need to impose differential costs or benefits on senders to be honest [328], some authors have taken this to mean that indices should impose *no* significant costs on survival components of fitness [519]. Since nearly any performed display imposes some energetic, risk, or time costs on the sender, such an interpretation would exclude any performance display from being an index. Similarly, structurally static signals such as plumage color that impose no costs while being signaled would have to be defined as indices. As other authors have pointed out, differences in structural signals could be handicaps reflecting differential costs during prior ontogeny [460]. These confounds can be avoided by reemphasizing that handicaps enforce honesty on senders by imposing costs or benefits of a magnitude that reveals some sender property of interest to receivers; it is presumed that any sender could perform or develop any given level of display, but the costs would differ depending on the property of interest to receivers [182]. Indices differ in that only some senders could produce a signal at some extreme level. That effort might be costly, but it is not the cost that ensures honesty.

There is a larger problem with the handicap-index distinction and it has to do with the obsession to assign every phenomenon to a single discrete category. It seems very likely that the same performance signal might function as a handicap in one population or species, but as an index in another. Any given display whose level can be varied will surely have upper and lower bounds set by physics, chemistry, or physiology. Many of these limits were noted in Chapters 2–7 of this book. The actual value of a limit might vary among senders of differing body sizes, ages, prior nutritional histories, or genetic differences. Within the limits for any given individual, the kinds of signal costs outlined in Chapter 9 increase as that individual adopts signal levels closer to at least one of its limits, if not both the upper and lower limits. The rate at which signaling costs increase with increasing signaling level is also likely to vary among individuals depending on their physical condition, prior nutritional and injury histories, and random genetic differences. For example, every animal possesses a mix of different kinds of muscles, each of which provides a different trade-off between speed, strength, and endurance. The mix varies subtly between individuals, and this can affect both a given animal's maximal signal performance ability and the rate at which costs increase as that maximum is approached. If average signaling levels in a population are near most senders' upper limits, honesty is most likely to be guaranteed by indices; if the average level is well below physical limits, then differential display may reflect handicap trade-offs. Without detailed knowledge of the physics and physiology, it may be impossible to know whether one sage grouse struts more slowly than another because its muscles simply cannot go any faster, or whether recent foraging failures have made it necessary to save energy for other functions.

This likely overlap between index and handicap constraints in the same signal can be exacerbated by continuing sender/receiver coevolution of signals [282]. Consider the pitch of loud calls in terrestrial vertebrates [328]. The fundamental frequency of the vibrating vocal cords is a function of cord mass, so low frequencies could serve as indices of body size. However, strategic investment in a little more tissue mass can lower the pitch to produce a signal that is then only reliable if it also incurs handicap costs. This appears to have occurred in some frogs [38]. But there is a way around this situation as well. The length of the vocal tract and the filter weighting of formants contribute to the perception of pitch in many vertebrates. Formant dispersion is also likely to be correlated with body size in such species, and unlike vocal cord mass, is less easily modified [149]. Thus, formant dispersion becomes a constrained index signal. Some species may be selected to dodge this constraint by elongating the vocal tract, either by a coiled tube in the throat or by extension of the larynx down toward the sternum, allowing deeper sounds to be emitted. However, eventually the signal pitch hits a body size constraint because respiration is inhibited by friction in the excessively long tubes. Thus subsequent rounds of evolutionary elaboration may change a signal from a handicap to an index and back again.

PROXIMITY SIGNALS   The primary cost of engaging in competitive agonistic interactions over access to an unshareable resource is the risk of attack by the opponent and the likelihood of injuries received from fighting. Agonistic-context signals evolve for the purpose of enabling opponents to assess which one of them would likely win such a fight, thereby resolving the conflict without having to resort to costly escalations. However, any type of threat signal is highly prone to bluffing and exaggeration unless it is backed up by some proof of true willingness to fight. The only fail-safe guarantee of an honest threat signal is a close approach toward the opponent. The honesty-guaranteeing cost for a true threat signal has therefore been called a **proximity risk**, and signals given close to rivals are called **proximity signals**. Hurd and Enquist [239] refer to this type of signal as an interaction handicap, because it involves a cost imposed by the receiver's response during an interaction. A graphic summary for the tactical threat model is illustrated in **Figure 10.30**. Although this model results in aggressive and submissive signals of arbitrary form, in fact the proximity cost is likely to select for tactical signal forms that further benefit strong and weak senders. At least three subclasses of proximity signals can be distinguished. One type is the **vulnerability handicap** [4, 520]. In this case, the sender approaches the opponent closely and displays or exposes a vulnerable part of its body. The sender not only assumes a proximity risk, but further handicaps itself by assuming a very vulnerable position. Zahavi described some signals he thought fell into this category, such as exposing the neck, belly, or flank, or assuming a relaxed posture [550, 553]. Such a pose would be very dangerous in a close encounter with a rival and would only be risked by high-quality individuals that were extremely confident in their ability to win a fight. Any broadside display performed close to a rival could be considered a vulnerability handicap. The parallel walk performed by rival male deer may contain an element of vulnerability [245]. A second type of proximity signal is the **tactical threat**. In this case, the sender approaches and assumes a prefight preparatory position or an attack-intention display, such as baring teeth, opening the mouth or mandibles, pointing the bill, and placing horns or other weapons in the ready-to-fight position. This subcategory, which has also been called a **negative handicap**, differs from the previous one in that the sender obtains a *tactical benefit* from its posture [496, 497]. Tactical threat signals have been described for aggressive behaviors in penguins [524], fulmars [138], some songbirds [237, 402, 537], and ungulates [496, 526]. A possible third signal form in this proximity-risk cost category could be **defensive threat** signals. Like tactical threats, signals in this category would include preparatory postures for defending oneself against an

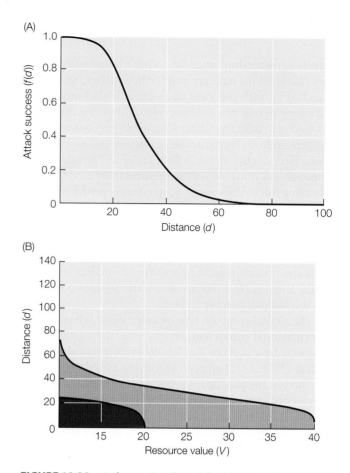

FIGURE 10.30   A threat signal model with proximity risk (A) The model assumes that each opponent knows its own fighting ability but not that of its opponent. The game has two steps. In the first step, each player can choose between two signals, S1 or S2, which have no inherent production cost. In the second step, each player can give up, attack unconditionally, or attack if the opponent does not withdraw. Weak and strong players pay different fighting costs depending on whether their opponent is weak or strong. The probability of a successful attack is a monotonically decreasing function of the distance d between contestants. There is an ESS for the following policy: When a player is strong, he signals S1; if the opponent signals S1, then the player attacks unconditionally, but if the opponent signals S2, the player waits for the opponent to flee. When the player is weak, he signals S2; if the opponent signals S1, the player withdraws, but if the opponent signals S2, he attacks unconditionally. This ESS policy is stable only if the distance d between opponents is less than a threshold value, called the honest striking distance. The region of stability is shown as the red region in (B) for different resource values (V). There is a second distance threshold, called the dishonest striking distance, above which players should always bluff (signal S1 regardless of strength) and opponents should always avoid responding. Between these two thresholds is a region, shown in green, in which the proportion of bluffers gradually increases. The positions of the two threshold curves are affected by the shape of the curve that relates attack probability to distance, which is in turn determined by the species-specific weaponry and fighting technique. The longer the reach of the weapon and the more mobile the attacking technique, the greater the honest striking distance. Mammals that fight with horns from a standing position will have a steeper attack probability curve, a shorter honest striking distance, and an almost nonexistent bluffing region (i.e., a sharp transition between honest signaling and no signal). Birds that chase and dive at their opponent have a more gradual attack probability curve and a larger distance region in which a mixture of honest and dishonest signals can be given. (After [497].)

(A)

(B)

(C)

FIGURE 10.31 **Examples of signals with a proximity risk**
(A) Possible vulnerability handicap example in great-tailed grack-les (*Quiscalus mexicanus*): the bill-up threat performed at close distance to rival [248]. (B) Tactical threat in slender crayfish (*Cherax dispar*), with claw weapons spread and held in attack-ready position. This display, given in close proximity, allows the rivals to assess each other's size and strength and gives each indi-vidual a tactical advantage, but may also lead to fighting and injury [71, 459, 539]. (C) Defensive threat in the cat, with ears pulled back protectively, teeth bared in a snarl (not visible in this photo), and tail fluffed and lowered. The cat also crouches low, so that it can quickly flip onto its back in full defensive posture, where teeth and claws of all four feet are available for fighting and the back is protected from a neck bite [55, 297].

imminent attack, such as rolling onto the back with all legs, claws, and teeth poised to defend, or turning to kick with a rear leg. Examples of vulnerability handicaps, tactical threats, and defensive threats are shown in **Figure 10.31**. We shall discuss these types of signals further in Chapter 11.

CONVENTIONAL SIGNALS   Some communication signals do not impose strategic costs on senders, nor are they physi-cally linked with sender attributes, yet they provide reliable information to receivers. Without strategic costs or index constraints, the form of these signals can be arbitrary, and the alternative signal variants can be correlated with alterna-tive conditions by any convention agreed upon by all par-ties; hence the name **conventional signals** [198, 239, 552]. Conventional signals remain honest if: (1) there is no incen-tive for either party to be dishonest; or (2) receivers punish dishonest senders. We treat each case in turn.

Cost-free signals with negligible strategic costs and only efficacy costs of communication can evolve when sender and receiver have **common interests**; that is, they rank the pos-sible outcomes of the interaction in the same order of pref-erence. Even if they differ slightly in the value they place on different outcomes, signaling will still be evolutionarily stable without the need for any form of constraint or receiver retali-ation [326]. Many examples of such cost-free signals seem to fit this context. During the final stages of courtship, after

mate choice has occurred, accurate signals are required to coordinate mating for the benefit of both parties. In exter-nally fertilizing fish, for instance, both male and female may give short-range signals to synchronous release of sperm and eggs [116]. For insect species in which females mate only once in their life, the courtship advances of a male toward a mated female are a waste of time and effort for both of them. Females in this situation give a distinctive signal; for example, the female *Drosophila subobscura* extrudes her ovipositor, which causes the male to immediately cease courtship. Food and alarm calls often seem altruistic and rife with conflict because the sender pays a cost while receivers benefit. But if senders of food calls [130–132] and alarm calls [47, 178, 555] benefit from foraging and defending in a group, then inexpensive signals accurately associated with these environ-mental contexts can evolve.

Another factor that increases common interest is high levels of relatedness between sender and receiver [415]. We learned earlier that the Sir Philip Sidney begging model pre-dicts stable signaling of need with low-cost signals if the par-ties are more closely related [43, 250, 325]. Honeybees give 18 different olfactory and vibratory signals to colony mates inside the dark hive. The signals attract, orient, and activate other individuals, regulate aspects of reproduction, solicit feeding and grooming, recruit foragers to food resource loca-tions, and coordinate colony fission/swarming [462]. Most of these signals, but probably not all, benefit senders and receivers alike, because of the common interest in productive

FIGURE 10.32   A conventional signaling aggressive interaction game in extensive form   Two contestants, Ego and Opponent, compete for an indivisible resource. Each may be weak or strong, both can give two possible signals (S1 or S2), and both perform one of three alternative fighting behaviors: give-up (G), attack (A), or pause-attack (P). There are four subgames, one for each combination of strong or weak Opponent and S1 or S2 by Ego. Only one subgame is shown here, in which Opponent is strong and Ego gives S1. Payoffs (not shown) are computed for each combination of state and fighting behavior for both parties to determine an ESS for each subgame (the * in this hypothetical example). Attacking always entails a cost, the magnitude of which depends on the relative strength of the contestants. This cost can be avoided with pause-attack if the opponent gives up, but there is a very large cost for pause-attack if the opponent attacks. Two equal opponents that both attack have a 50% chance of winning, and if both give up, the resource is randomly awarded. An ESS occurs if Ego signals first knowing his own state but not Opponent's state, and then obtains information on Opponent's state before deciding which behavior to use. At the ESS, Ego reliably signals his state with one of the two signals and Opponent believes him, and both use the same decision rule: attack an opponent that gives a signal of similar size to one's own signal, pause-attack an opponent giving a smaller signal, and give up against an opponent with a larger signal. (After [236].)

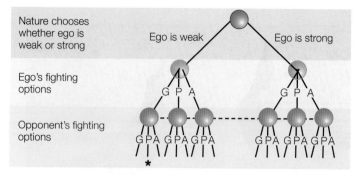

colony functioning shared by the closely related hive members. Similarly complex repertoires occur in the closely related members of wasp, ant, and termite colonies [536].

A final model for reliable signaling of need without strategic costs is the **pooling equilibrium** model, which is based on a simplified signaling system with a few discrete signal categories rather than a continuously varying signal of need. The model posits that senders are divided into two or more discrete categories of need, and all senders in the same category "pool" their need levels by giving the same signal [43, 284, 285]. At the ESS of the simplest two-category scenario, chicks beg only if they are above a threshold level of need. The receivers (parents, in this case) respond to the average need of signalers and feed all the chicks that beg. The model relies on relatedness among chicks and thus a certain level of common interest in not cheating, as well as the occurrence of satiated chicks that do not beg [59]. While this model works mathematically, in reality it seems quite easy for some variation in begging intensity to occur, and for parents to feed more vigorously begging chicks more food, which would destabilize the signaling system [328]. Existing evidence that chicks vary their begging signal intensity in a linear way with increasing hunger level does not support a pooling model [267, 301, 414, 529], but there could be other signaling contexts to which this model could be applied [59, 285].

When there is conflict of interest between sender and receiver, bluffing or exaggeration of inexpensive conventional signals would invade, so some type of **socially imposed cost** is required to maintain honesty. The obvious cost is **receiver retaliation**, which comes in the form of either: (1) receiver skepticism, (2) a retaliation rule in which receivers approach and check the sincerity of senders, (3) punishment of known

liars, or (4) future reprisals. Receivers have much more control over the reliability of conventional signals than of costly handicap signals because they can usually test and verify the sender's true condition. In principle, inexpensive conventional signals should have an advantage over costly handicap signals whenever the information receivers seek can be verified directly, such as information about sender fighting ability, aggressive motivation, or dominance. Signals with socially imposed costs are believed to be more resistant to breakdown over evolutionary time [42, 43, 285].

To adequately model conventional signaling between individuals with conflicting interests, a more complex game with a sequential structure and an exchange of information is required. The extensive-form version of this game is illustrated in **Figure 10.32** [137, 236]. The context is an aggressive interaction over a nonshareable resource that is worth fighting for. The two contestants are either weak or strong; they can give one of two possible signals to inform the opponent of their strength; and each has three fighting strategies: give up, attack, or pause-attack (which allows a giving-up opponent to flee before an attack can occur). An ESS is sought in which there is a reliable association between the sender's strength and the signal given, and a behavioral rule for each possible combination of contestant states. Suppose the sender knows his own state, but not his opponent's state, before he signals, and then obtains some information from the opponent that resolves this ambiguity. At the ESS, one signal is reliably associated with a weak state and the other with a strong state (it doesn't matter which signal signifies which condition, because both signals are arbitrary), and the stable **receiver retaliation rule** is to attack when both deliver the signal of strength, pause-attack when the opponent delivers a signal of weakness, and give up when the opponent delivers a signal of superior strength. This time-structured, extensive-form, interactional model provides a prime example of how a retaliation rule can maintain a stable conventional signaling system.

Some conventional signals are discrete, with antithetical signals for the messages of strength and weakness, whereas others are graded along a continuous parameter. Prominent color patches in some birds and lizards are classic examples of graded conventional signals. The size or hue of the patch is correlated with the dominance rank of the individual, hence their designation as **badges of status** (see Figures 11.24

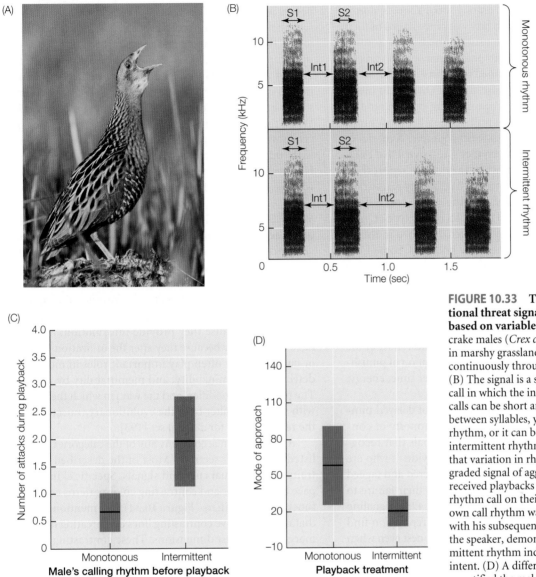

FIGURE 10.33 **The case for a conventional threat signal in the corncrake based on variable call rhythm** (A) Corncrake males (*Crex crex*) defend territories in marshy grasslands by calling nearly continuously throughout most of the night. (B) The signal is a simple, loud, two-syllable call in which the interval between sequential calls can be short and similar to the interval between syllables, yielding a monotonous rhythm, or it can be longer, yielding an intermittent rhythm. (C) To demonstrate that variation in rhythm constitutes a graded signal of aggressive intentions, males received playbacks of an intermediate-rhythm call on their territories. The male's own call rhythm was positively correlated with his subsequent aggressive approach to the speaker, demonstrating that the intermittent rhythm indicates stronger aggressive intent. (D) A different set of experiments quantified the males' responses to monotonous versus intermittent playback patterns. Males approached the playback speaker faster and more closely (lower mode of approach score) during the intermittent treatments compared to the monotonous treatments, suggesting vigorous aggressive retaliation against the stronger threat signal. The more threatening signal is, if anything, less costly to produce than the less aggressive signal. This simple graded threat signal is based on an arbitrary, Morse-code-like convention. (After [417].)

and 11.25). Large badge size deters aggressive challenges by small-badged individuals, and large-badged individuals usually win fights. The cost of possessing a large badge is aggressive retaliation from other large-badged individuals [339, 340, 375, 377, 430, 432–435, 465, 506, 533]. The evolution of status badges must be accompanied by frequent testing of the honesty of individuals with a badge size similar to one's own, while avoiding or ignoring individuals with larger or smaller badges. This behavioral rule is the key factor that makes it costly to deceptively exaggerate one's badge size. The results of experimental manipulations are consistent with these predictions. Male house sparrows (*Passer domesticus*) with experimentally enlarged black chest patches were involved in more aggressive encounters than controls, and their attackers had larger-than-average patches [338]. Naturally large-bibbed male Harris sparrows (*Zonotrichia querula*) with experimentally reduced patches had to fight very hard against flockmates for food, but eventually prevailed [431,

436]. A mismatch between the dominance signal and aggressive behavior in a wasp, *Polistes dominulus*, elicits costly social punishment [507]. Some transient threat signals, such as song-type matching and song-type switching in songbirds, call rhythm in corncrakes (*Crex crex*), vertical bar darkening in swordtail fishes (*Xiphophorus*), and head bobbing displays in a lizard (*Anolis carolinensis*) are also believed to be conventional signals of aggressive intentions with receiver retaliation costs [238, 349, 350, 417, 521]. **Figure 10.33** presents the experimental evidence for this assertion in the corncrake.

Punishment is widespread among social animals: dominant reproductives may punish lazy workers; victims of theft or parasitism may punish the usurper; males may punish females that refuse to mate, and parents may punish greedy young [92]. Here we are concerned with cases in which senders are punished for giving false signals. One could interpret the aggressive challenges against large-badged senders described above as a form of punishment, rather than mutual testing of sincerity in aggressive encounters. A model in which senders are punished when they give a signal that exaggerates their quality, but not when they under-report their quality, predicts that if assessment of true quality is accurate, senders will signal accurately, and the signal cost will be minimal at the ESS [285]. Real examples of this type of punishment are rare. The best one has been described in rhesus macaques, which often call when they find food. Sometimes food finders will fail to call, a case of lying by withholding information. If an individual is discovered feeding on a favorite item without having called, it is treated with considerable aggression [219]. While individuals may sometimes manage to get away with such lying, the effect of immediate punishment may be sufficient to keep the signal honest most of the time. The difficulty is how to explain the evolution of punishment, when it presumably costs the punisher time, energy, and risk [92, 167, 246, 293, 359, 412, 545].

Future reprisal cost constitutes a form of delayed punishment that can potentially stabilize the honesty of conventional signals among group-living animals that recognize each other and interact repeatedly. Individuals who are observed to behave deceptively are remembered and tagged with a poor reputation, which may result in their failure to obtain certain benefits in the future [54, 366, 425]. Signaling models with repeated interactions based on reputation find that cheap conventional signals can remain honest even when receivers have conflicting interests. Primates show the best evidence for such signaling interactions [180, 246, 286]. For example, macaques and baboons use soft grunts as signals of reconciliation or friendly intentions, which could be used deceptively to get close to another individual and attack it or, in the case of females, to grab an infant. Using a simple action-response game structure, the conditions under which this signal could be honestly associated with benign behavior were explored [473]. Actor strategies included: being truthful (signaling only when truly friendly); always signal (whether friendly or not); and never signal. Receivers' acts were: always flee; always stay and interact regardless of the signal; and believe the signal (stay only if signal given, otherwise flee). If the individuals only interacted once, truthful signaling and believing comprised a stable ESS only if the cost of lying was higher than being honest, or if both individuals ranked interacting the highest (common interest). However, if repeated interaction among known individuals was allowed, in which receivers remember whether each actor had previously lied or not and fleeing was always the response to a known liar, then honest signaling was stable even with no signaling cost and a conflict of interest. This ESS occurs because the future

benefit to the actor of being believed is greater than the short-term benefit the actor receives from lying. Field studies of interacting female rhesus macaques showed that approaching grunters almost never behaved aggressively (whereas 29% of nongrunters did); they were more likely to handle infants gently; and recipients were less likely to give submissive or anxious gestures toward approaching grunters [473]. While these results are consistent with the model, we don't yet know whether macaques remember and avoid interacting with individuals who have signaled deceptively in the past. However, chimpanzees do retaliate later against individuals who have stolen food from them [246].

MODIFIERS   **Modifiers** are traits that augment (**amplifiers**) or hinder (**attenuators**) the utilization of information provided to a receiver by cues or signals. By itself, a modifier provides no information [211, 213]: the coding matrix for one contains the same random probability in all cells. In this sense, modifiers differ dramatically from handicap, index, proximity, and conventional signals, which are all favored by selection because they provide information. Modifiers have evolved only because they alter the utilization of signals or cues, and they often play important roles in modulating detection, discriminability, and memorability by receivers. The presence of modifiers and the ways in which they interact with signals can alter both the evolutionary trajectories and the relevant ESSs for a signal set [195].

A modifier can accompany any of the categories of signals listed in the prior sections. Most of the described modifier traits involve visual cues and signals. Specific skin or carapace backgrounds and contrasting color patterns commonly function as amplifiers (**Figure 10.34**). We mentioned earlier that many fish have contrasting lines demarcating their body margins or longest dimensions. These contrasting bars and patterns could be conceived of as amplifiers of body size cues or enhancement of species recognition signals. At the other extreme, animal camouflage tends to break up the outline of the animal, make it the same color or texture of its background, or add structures that resemble nearby thorns, twigs, and leaves. Such traits thus attenuate the cues that predators or prey might use to detect the camouflaged animal [36, 103, 453, 488]. Traits that amplify handicap signals include the elongated plumage and wing markings that exaggerate the movements of displaying male birds [48, 153, 478] and flank markings that exaggerate the stotting display of antelopes [74] (see Figures 10.34B and 14.5). A variety of index signals show amplification. The collared lizard has contrasting white patches at the corners of its mouth that make the size of its jaw muscles, an index of bite force, more easily assessed by receivers during the gape threat display (see Figure 11.19). The markings around the eyes and ears of mammals amplify the direct stares at predators, an informational index signal (see Figures 10.29D and 14.3A) [214, 217]. A contrasting triangular marking on the abdomens of jumping spiders makes assessment of nutritional condition easier (see Figure 10.34C) [504]. Nutritional condition in spiders is generally

(A)

(B)

(C)

(D)

Wide
light region

Well-fed spider

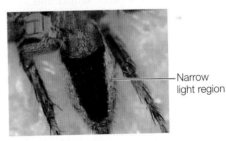

Narrow
light region

Poorly fed spider

FIGURE 10.34   Examples of amplifiers (A) Group-living species, such as herding ungulates and the flock-forming sparrows (*Aimophila ruficauda*) shown here, often have striped head or facial patterns that exaggerate the direction of gazing, amplifying cues that aid in coordinated group movement. (B) The exceptionally long head plumes of Gunnison sage grouse (*Centrocercus minimus*) exaggerate its head movements during the strut display, which contains more pops and head throws than the display of the closely related *C. urophasianus*. (C) Fish often sport body-spanning stripes in long-bodied species (zebrafish, *Danio rerio*) and vertical stripes in deep-bodied species (sailfin tang, *Zebrasoma veliferum*). Such stripes can provide cues of body size, but if the stripes extend into the fins, as in the case of the zebrafish's tail here, they could attenuate the estimation of body size. (D) The jumping spider (*Plexippus paykulli*) has a dark triangular marking on its abdomen that improves the assessment of condition. The expanded abdomen in a well-fed spider (top) produces a wider light region adjacent to the dark triangle compared to the narrow light region of a poorly fed spider (bottom). The abdomen pattern is revealed during the hunch display in contests between males [504].

considered a cue for fighting ability and stamina. Patches of bare skin may be colored in ways to amplify the visibility of parasite infections. This idea was proposed to explain the yellowish color of the air sacs in male sage grouse (*Centrocercus*

urophasianus), as it makes the hematomas induced by mite louse infections highly conspicuous [53, 249]. Many species use conventional signals to indicate social status. The cheek patches of great tits have been interpreted as a "canvas" that makes large numbers of plumage attacks on subordinates highly visible (see Figure 12.18C) [164].

Modifiers clearly occur in other modalities, but they have been little studied. Certain stereotyped notes in bird songs and mammal calls can be shown to provide no information by themselves, but instead alert receivers to the imminent vocalizations that do provide information [376, 396, 420, 494]. It is likely that some components in pheromone mixes serve as alerting stimuli without themselves providing other information. One function of multicomponent and multimodal signals may be the alerting of receivers through one component while providing relevant information through other components.

When signals consist of many components, it is not always easy to determine which components are signals or cues and which are modifiers. For example, the displays of male lekking birds, such as sage grouse, combine elongated plumages, sounds, and stereotyped movements. Significant correlations between male display rates and mating success, but no demonstrable correlations with plumage variation, suggest that the movements and possibly the sounds constitute the relevant signals in these displays, and that the elongated plumage serves largely as an amplifier [174, 175]. Similar partitioning of functions in other signaling systems has not been studied extensively, so it remains unknown how often plumage is an amplifier, as opposed to an informative signal, in birds. This is definitely an area that deserves more attention.

Causes of unreliable signaling

Most signals are believed to be honest most of the time, either because conflicts of interest are minimal or because appropriate costs are imposed on cheaters. But dishonest signaling is expected to occur at least some of the time [89, 120, 252, 495]. Receivers will tolerate a certain amount of deception as long as they obtain a positive value of information (see Chapter 9). There are several contexts in which senders may be able to get away with some dishonesty: an out-of-equilibrium outcome of dynamic coevolution between sender and receiver; a high level of perceptual error by receivers; multiple classes of senders with different costs and benefits of signaling; multiple classes of receivers whose responses provide different payoffs to the sender; and tactical deception. We take up these scenarios in turn and provide some examples of each one.

OUT-OF-EQUILIBRIUM COEVOLUTION   Apparently stable equilibria mask what are, in reality, always dynamic systems. Existing ESSs are continually being challenged by mutant strategies. In many cases, the mutants cannot compete against the current equilibrium strategies, and they disappear from the population. However, in some cases they can invade and this will destabilize the equilibrium. **Figure 10.35** illustrates an interesting model exemplifying the dynamic nature of communication systems [282]. Sensory processing by female receivers was modeled as a neural network, and the form of male courtship signals was modeled with a second neural network. Males and females interact, breed, and produce offspring, and payoffs are higher for female receivers when they mate with high quality males. At some equilibria, male signals were relatively honest and females thus benefitted from attending to their signals. However, one property of neural nets is the accidental generation of latent preferences. On occasion, mutant males produced signals that stimulated these latent preferences independently of the male's

(A) Costless signals

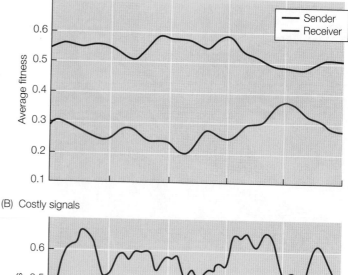

(B) Costly signals

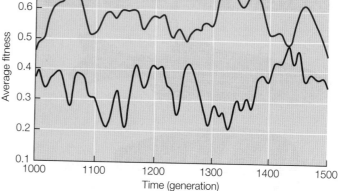

FIGURE 10.35   A dual neural net model of coevolving senders and receivers In this simulation model of the evolution of male mate choice signals, separate neural networks track the changes in the senders' signal features and the receivers' sensitivities to those features. Senders vary in qualities that receivers wish to assess. The sender network generates multi-dimensional signals indicating quality (honestly or dishonestly). The receiver network evolves perception for high-quality senders; more accurate receivers have higher fitness. When signals are costless (A), sender fitness is high, but receivers ignore signals. When signals are costly (B), receiver fitness is mostly high, but the fitnesses of both senders and receivers vary over time in opposite directions, as senders find ways to temporarily deceive receivers and receivers evolve better mechanisms for detecting honest senders. (After [282].)

quality. Such mutants could then invade and spread, and for a while, the signaling system was dishonest. Eventually female mutants with altered neural nets were able to discriminate between honest and dishonest senders, and the system converged back to an equilibrium in which signals that were accurate representations of male quality were attended to and favored. This model thus predicts periods of relatively stable equilibria punctuated by intervals of dishonest signaling. As we saw in Chapter 9, signaling systems are usually nonlinear and their evolutionary trajectories need not lead to equilibria at all; alternative outcomes include limit cycles, sudden bifurcations, and chaos [168, 223, 335, 387, 388, 476].

(A)

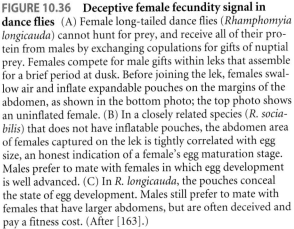

(B)

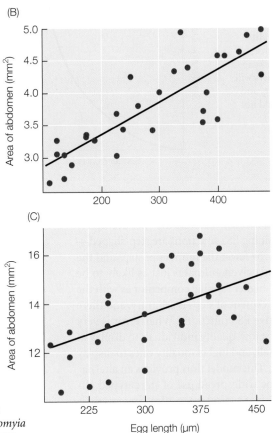

(C)

FIGURE 10.36 Deceptive female fecundity signal in dance flies (A) Female long-tailed dance flies (*Rhamphomyia longicauda*) cannot hunt for prey, and receive all of their protein from males by exchanging copulations for gifts of nuptial prey. Females compete for male gifts within leks that assemble for a brief period at dusk. Before joining the lek, females swallow air and inflate expandable pouches on the margins of the abdomen, as shown in the bottom photo; the top photo shows an uninflated female. (B) In a closely related species (*R. sociabilis*) that does not have inflatable pouches, the abdomen area of females captured on the lek is tightly correlated with egg size, an honest indication of a female's egg maturation stage. Males prefer to mate with females in which egg development is well advanced. (C) In *R. longicauda*, the pouches conceal the state of egg development. Males still prefer to mate with females that have larger abdomens, but are often deceived and pay a fitness cost. (After [163].)

How widespread are **out-of-equilibrium systems**? Deception is difficult to detect, but numerous examples of sexual conflict in which one party appears to be deceived at some cost to it have been described [22, 23, 419, 463]. A striking example can be found in some dance flies, where females deceive males about their egg development by gulping air and inflating abdominal sacs to obtain nuptial gifts (**Figure 10.36**) [163]. Limit cycles (see Figure 11.30) have been described for the colored visual signals of several species of lizards, and may occur more widely than was initially realized because few researchers have thought to look for them [156, 264, 476].

IMPERFECT INFORMATION   We outlined in Chapters 8 and 9 multiple reasons why it usually does not pay for senders to provide or receivers to seek perfect information through

signals. Evolving sensory systems face a trade-off between the range of detectable stimuli and fine levels of similar signal discrimination. The resulting intermediate optima often lead to inherent decision biases and invariably preserve some sender and receiver error. High levels of signal discrimination require high levels of risk, energetic expenditures, lost time, or nervous system allocation that are often not economically justified. As discussed in Chapter 8, many species have thus evolved *rules of thumb* that are cheaper and still do pretty well, but cannot avoid some residual error. Where significant efficacy costs must be paid, conflicts between senders and receivers as to who should pay these costs can also result in error-prone compromises [253]. Given that most animal signaling systems will exhibit residual errors by one or both parties, how does this affect the existence or stability of ESS combinations?

Several theoretical models have explored this issue [120, 251, 253, 257]. The general finding is that **perceptual error** does not change the ESS outcomes of signaling games qualitatively, but it may lead to quantitative changes in signaling strategies. In a numerical simulation version of the Grafen–Zahavi quality handicap model with different levels of error, receivers vary in their ability to accurately discriminate differences in display investment [253]. In error-free situations, reliability is high and the signaling ESS is a smooth monotonic function relating display investment to sender quality (**Figure 10.37**). When receiver error is higher, it may no longer

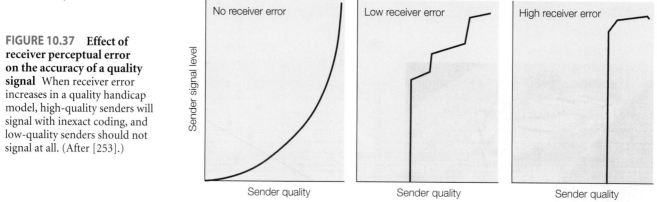

FIGURE 10.37 Effect of receiver perceptual error on the accuracy of a quality signal When receiver error increases in a quality handicap model, high-quality senders will signal with inexact coding, and low-quality senders should not signal at all. (After [253].)

pay senders to invest the highest possible levels of energy into displaying their quality, because receivers may not accurately detect it. The corresponding ESS functions are step-shaped—the more receivers err, the fewer the number of steps and the wider each step. When a given animal is just as likely to be confused with the next-higher ranking competitor as with the next-lower one, the placement of steps is arbitrary, and there are thus many alternative ESS functions. When error is very high, the ESS is for most low-quality individuals to display at one low level or not at all, and higher-quality ones to display at another higher level. This model thus provides an alternative explanation for the widespread use of stereotyped all-or-nothing signals given at typical intensities. Stereotypy is simply the honest ESS when receiver perceptual error is high.

In a model that combines the Sir Philip Sidney begging game with signal detection theory, a random error effect is added to the receiver-donor's perception of the sender-beneficiary's display investment cost [257]. As perceptual error increases, donors become more likely to inappropriately transfer resources to healthy senders and fail to transfer resources to needy senders, which harms both sender and receiver fitness. To compensate, senders must increase their investment in signaling to maintain a signaling ESS. This model demonstrates that the strategic handicapping cost required to maintain honesty has to be augmented with an additional efficacy cost of signaling to improve receiver discrimination accuracy. A somewhat similar model makes the assumption that receiver error is greater for higher-quality senders (receivers are more likely to mistake a high-quality signal for a low-quality one than vice versa), which forces them to differentially increase their signal investment cost over that of low-quality senders. This effect augments the handicapping cost, and could be achieved with the addition of an amplifier to improve detection of high-quality senders [120]. Examples of augmented signals given in noisy contexts include increased singing amplitude by finches when background noise is increased, additional white patches of warblers inhabiting darker environments, and multimodal begging signals in nestling birds [104, 266, 311].

MULTIPLE CLASSES OF SENDERS   All of the models we have discussed so far assume a single class of senders, so the ESS has been a single rule for mapping information onto signal form or display effort. What if there are **multiple classes of senders** with different stakes and potential costs? This situation has been examined with a variant of the Sir Philip Sidney game in which there are two classes of begging senders: one that begs honestly and one that always begs regardless of need [252]. The latter are thus deceptive some of the time. Receivers cannot tell the two types apart. The model allows for the two classes of beggars to differ in their cost of signaling and in their relatedness to the receiver. There are several conditions under which a signaling ESS can be maintained, with receivers responding to all begging signals: (1) the honest beggars are more closely related to receivers than the constant beggars; (2) the honest beggars pay a higher cost for begging that do constant beggars; or (3) both. It must also be the case that honest beggars are sufficiently common relative to constant beggars. Such variation in sender economics is likely to be common and so should be the existence of multiple sender classes. If receivers were able to distinguish the different sender classes, they would use different response rules for signals from each class.

Bluffing threats in mantis shrimp (*Neogonodactylus bredini*) may be an example of this phenomenon (**Figure 10.38**). Both sexes defend individual burrows from intruders with a claw-spreading threat display. Recently molted individuals who are soft and unable to deliver the powerful claw strike may nevertheless give the threat display in the hope that an intruder will not press an attack. Individuals frequently follow a threat display with a strike during intermolt, but never strike during the molt. About 15% of a reef's shrimp residents are in molt at any one time, and it appears that intruders cannot distinguish molting from intermolt residents. The probability that a recently molted resident retains its burrow is significantly greater if it threatens than if it does nothing, so the bluff is effective [3, 4, 486].

Another potentially common source for multiple sender classes is the life-history strategy of terminal investment in iteroparous species that can expect to reproduce over a series of episodes. Throughout most of life an individual should moderate its investment in reproduction to ensure its survival to the next episode. However, an individual perceiving itself to be in its final breeding episode (because it is old, sick, in poor condition, or faced with predators) should increase its reproductive investment to the maximum level [91, 535]. For males that employ a costly handicap signal to attract mates, terminal individuals may give a very strong

(A)

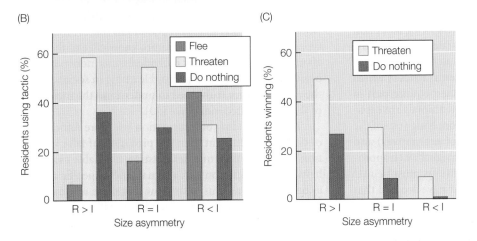

FIGURE 10.38 Bluff in a threat display (A) A mantis shrimp (*Neogonodactylus bredini*) at the entrance to its burrow performing the "meral spread" threat display, with claws extended forward in attack position. Each individual lives in and defends a burrow by delivering a snapping strike at invaders with its raptorial appendage. (B) Intermolt resident (R) tactics depend on whether they are larger or smaller than the invader (I). When residents are larger (R > I) or the same size (R = I) they usually threaten, and when smaller (R < I) they usually flee. (C) Molting residents never strike because they are more likely to injure themselves, but if they threaten anyway (bluff) they are more likely to retain their burrow than if they do not display, in which case they are either ousted from their burrow or flee. (After [3, 486].)

(B)

(C)

display for their condition compared to the rest of the male population [276]. If females use this signal to select high-quality mates and they cannot distinguish terminal from non-terminal males, some males would be providing dishonest signals. As long as such males are a sufficiently small fraction of the male population, the male signal may still be honest on average and worth the receiver's attention. Intriguing examples of dishonest signaling by terminal males have been described for the size of the red color patch in stickleback males in natural and experimentally low nutritional condition (**Figure 10.39**) [73], and for pheromone production in experimentally immune-challenged male mealworm beetles (*Tenebrio molitor*) [450].

MULTIPLE CLASSES OF RECEIVERS   What is the optimal strategy for senders when there are **multiple classes of receivers**, each of which can generate quite different payoffs

to senders through their responses? A model of this situation predicts that senders best benefit by using incomplete honesty to elicit different responses from different receiver types [443]. For example, consider a territory-holding male signaling to attract females but also attempting to deter rival

FIGURE 10.39 Terminal investment in the stickleback (A) Male three-spined sticklebacks (*Gasterosteus aculeatus*) vary in the size of the orange or red patch on their jaw. The color is produced by a carotenoid pigment, which is argued to be a costly handicap signal. Females prefer males with larger patches, because such males are better able to guard a nest of eggs. (B) The size of the patch relative to overall body size as a function of male condition (measured as fraction of body weight that is lipid) shows a U-shaped distribution: males in very poor condition can exhibit very large patches. Males maintained on a low-nutrition diet also developed large patches. Although this terminal investment strategy produces low-quality deceptive males, patch size is still an honest signal on average because the proportion of deceptive males in the population is small. (After [73].)

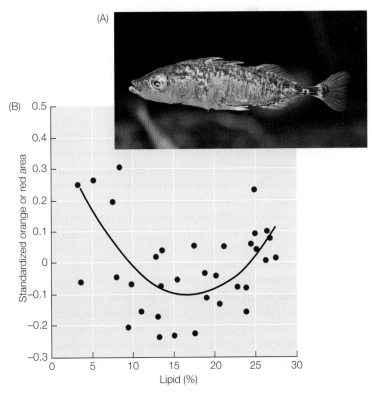

males. The receivers are females whom the sender would like to attract and satellite males who want to sneak matings with females attracted to the signal. This sender might attack any individual entering his territory, making the signal deceitful from the females' perspective, but possibly protecting the sender against the loss of paternity to an inferior satellite. Because this behavior also benefits the female that mates with the sender, it pays females to continue to be attracted to signals even though they are sometimes attacked.

TACTICAL DECEPTION   In contrast to interspecific deception and established intraspecific alternative strategies, such as the mimicking of females by sneaker males to obtain fertilizations, **tactical deception** involves short-term tactics in which the deception uses elements from an honest counterpart in the species' repertoire. As defined earlier, animals are capable of giving two types of false information: lying (use of the wrong signal among an unordered set of alternatives) and withholding information (failure to give a signal when appropriate). For lying to work, several conditions must be met [68, 150, 463, 531]. First, a species must have a signal that is tightly correlated with a particular context—for example, an alarm call associated with the presence of a predator on a significant proportion of occasions. Second, when individuals detect such signals, they must respond in a relatively stereotyped or consistent way, and must do so on a statistically significant number of occasions. Thus when animals hear an alarm call, they must consistently flee. Third, individuals must have the flexibility to manipulate the behavior of other group members by producing a species-typical signal in a novel context. Finally, if deception is truly occurring, the receiver should incur a cost.

A good example of an outright lie has been described in birds foraging in flocks, where one individual may give a false alarm call to scare competitors away from a rich food find [318, 341, 358, 452]. As the frightened birds flee to cover, the alarm sender flies in to feed before the others return. Although a certain amount of alarm unreliability is caused by errors (e.g., false alarms by young individuals who haven't learned which heterospecifics are truly dangerous [37]), senders have been shown to give intentional false alarms from which they clearly benefit. Great tits, for example, are likely to give false alarm calls when other birds are feeding at concentrated food sources, but less likely to do so when the food sources are dispersed. Moreover, dominant birds are likely to give a false alarm when the bird at the food source is another dominant bird, but not when the bird is a subordinate individual who can be easily displaced [341]. Initially it was believed that senders would not "cry wolf" too often, because receivers would learn to ignore the signal, and it would cease to be effective in true alarm contexts. However, several studies have shown that the incidence of alarm calls in the absence of a predator can be as high as 55–63%. The reason such a high level of dishonesty can be sustained is clearly the high cost of ignoring an honest signal (i.e., death) relative to the cost of heeding a dishonest signal (losing a bite of food) [200]. The

value of information thus remains positive. False alarm calls are also given by some male birds and squirrels during the fertile period of female mates to interrupt extra-pair copulations with competitors [344, 498]. Anecdotal accounts of other forms of seemingly deceptive behaviors are more prevalent in certain primate species than in others, especially in baboons and chimpanzees [69]. For instance, a young individual desiring a food item controlled by an adult may utter a scream normally given when attacked or threatened in order to recruit its mother, who chases off the offspring's "assailant" while the youngster grabs the food item [68]. Although this type of behavior appears to require knowledge of the whereabouts and status of nearby individuals, it is more parsimoniously explained as a case of rapid learning in social contexts; the actor does not truly need to understand the mental state of the manipulated individual [70].

The withholding of information is a more difficult form of deceit to document than the provision of false information. Not only must one demonstrate a reliable association between a specific context, a signal, and a specific response as before, but one must also show that there is some behavioral flexibility on the part of senders to either give or fail to give the signal in a way that benefits the sender in a given circumstance [150]. Modifying one's behavior based on the presence of possible eavesdroppers has been termed the **audience effect** of signaling [83, 312, 313]. Food-caching corvids, for instance, pay close attention to who may have seen them cache food, make false caches of nonexistent or trivial items when under observation, and delay approach to a cache or search at false sites when competitors are nearby [66, 109, 134]. Male roosters give food calls and the mock-feeding "tidbitting display" to attract mates, sometimes when no food is present, but food calls are always withheld in the presence of rival males [145]. Group-living primates typically give food calls when they encounter rich food sources, but individuals who discover a resource when alone often fail to call, or return alone to the food site later [11, 288]. Again, such behaviors imply learning through social experience, but they do not necessarily imply understanding of the deceptive meaning of these acts. Chimpanzees, however, do appear to withhold information with deceptive intent. A dominant individual may inhibit its intention movements and visual attention toward a desirable object in the presence of others, then leave the immediate area, hide behind a tree, and subsequently peek out, waiting for an opportunity to retrieve the object without a contest [69, 70]. Such behaviors imply a **theory of mind**—an understanding of the mental states of other individuals [405].

In conclusion, low levels of dishonesty may persist in many signaling systems, but signals must be reliable enough on average to justify receiver response. Otherwise, receivers will ignore signals, senders will no longer benefit from giving dishonest information, and the signals will disappear from the species' repertoire. A variety of costs and constraints maintain signal honesty, and signals must be costly in a way that explains why they provide reliable information.

SUMMARY

1. Evolution is a process of gradual change, as new adaptations become layered on top of preceding ones. Evolutionary biologists use morphological traits and DNA-based genetic traits to construct **phylogenetic trees** depicting clusters of closely related species. Robust trees can be employed to map the sequence of adaptations arising during speciation and taxonomic radiation. We can also assess the evolutionary progression of behavioral traits, including signal traits and receiver response traits, with these phylogenetic trees. But to understand the adaptive function of signals, we need to analyze the coevolutionary interactions and the fitness costs and benefits to senders and receivers. A signal cannot evolve and be maintained unless senders and receivers both benefit from the information provided.

2. Some new signals evolve from **sender precursors**, preexisting behavioral, physiological, or morphological traits that already provide informative cues to receivers. If the sender benefits from the decisions and responses made by receivers as a result of these cues, then the cue can be modified into a signal via the process of **ritualization**. Ritualization helps get the message across by making the cue easier to detect in a noisy environment, easier to discriminate from other behaviors, and, if learning is relevant, more memorable. If the ritualized signal becomes emancipated from the original cue association, it can become highly elaborated to provide new information to receivers, or it might be used to deceive receivers. Well-described sources of sender-derived signals include: **locomotory and feeding appendages; intention movements**; ambivalent, displacement, and redirected behaviors emerging from conditions of **motivational conflict**; visual and olfactory manifestations of **autonomic nervous system** functions; acoustic manifestations of the **respiratory system; endocrine system products**; and **anti-predation defense tactics**. New displays can also be co-opted from other **existing displays**. Signals derived from sender precursors provide information about the sender's intentions, mood, and physiological state, and may also be ritualized into mate attraction signals.

3. **Receiver preadaptations** also play an important role in the evolution of signals. The ritualization process leads to highly efficacious signal designs given the receiver's sensory capabilities and the signaling environment. The fine-tuning of signal features with respect to receiver sensitivity and environmental background is called **sensory drive**. A more specific source of new signals, however, is the exploitation by senders of preexisting **sensory biases** in the receiver that evolved to detect important environmental stimuli such as food or predators. A signal that mimics the features of these environmental stimuli, and may also exploit the associated receiver response, such as approach to food or flight from a predator, is called a **sensory trap**. If the receiver benefits from the sender's matching trait, then the trait may become elaborated further. More likely, the receiver pays a cost, and a round of antagonistic sender/receiver coevolution may ensue. To escape the trap, the receiver must improve its ability to distinguish between the sender's mimicking trait and the important natural stimulus, and evolve different response rules for each one. Signals derived from receiver precursors are usually mate attraction signals.

4. **Honesty** is the provision of accurate information by the sender, and **deception** is the provision of unreliable information. Honest communication occurs when both parties benefit on average from the signaling exchange. If sender and receiver have a conflict of interest because they rank the payoffs of alternative receiver responses differently, senders will be selected to exaggerate, bluff, lie, or withhold information. Receivers will counter by either improving their ability to detect such cheaters, or they will cease paying attention to the signal.

5. **Evolutionary game theory** is a powerful and essential tool for modeling the coevolutionary interactions, where the relative fitness for an individual playing a given strategy depends on the frequencies of strategies being played by others in the population. Strategies with higher fitness can invade, and those with lower fitness may be lost, leading to a balance in payoffs for one or more strategies called an evolutionary stable strategy. At this equilibrium point, no party can do better by shifting to another strategy. If a population is at a stable equilibrium, small deviations from this equilibrium lead to trajectories favoring a return to it, a property called **convergence stability**. Trajectories surrounding an **unstable equilibrium** move the population away from the point. Drift in small finite populations can greatly affect evolutionary trajectories. Game models can take many forms: alternative strategies may be **discrete** or **continuous**; interactions can be modeled in matrix-based **normal form analysis** as **contest games** with fixed payoffs, or **scrambles** with frequency-dependent payoffs; **dynamic games** with a sequence of successive actions require **extensive form analysis**; and games may be **symmetric**, where all players have access to the same strategy set, or **asymmetric**, with different strategy sets or payoffs for players with different roles.

6. Communication games are usually asymmetric, because senders and receivers have different roles and payoffs. Models that find conditions in which signaling and paying attention are an ESS generally include some payoff cost to senders that is greater when they cheat. Classifying signals on the basis of the type of cost that guarantees signal reliability provides important insights into the design of signals and the kinds of information that can be encoded in the signal. **Handicap signals** are those with an associated production cost, such as energy expenditure, predation risk, developmental investment, or vulnerability to attack. For the classic quality handicap signal, receiver selection pressure favors signal forms that impose the type of cost that reveals those aspects of the sender's qualifications of importance to the receiver. Higher-quality individuals can

bear the cost of the display better, resulting in high-quality senders delivering more intense or vigorous signals. A variant of the model shows that if senders vary in the benefit they obtain from some level of display, display vigor can reliably indicate sender need or motivation.

7. **Index signals** do not entail differential costs, but are reliable because their performance is functionally and incorruptibly linked to specific sender attributes. **Quality indices** are typically constrained by body size, strength, age, or fighting experience; senders of insufficient size, maturity, or strength simply cannot produce the high-quality signal. **Informational constraint** creates another type of index signal, in which signal production is linked to the sender's knowledge.

8. **Proximity signals** are threat signals given in conflict contexts, where the cost that guarantees honesty is a close approach to the rival. Senders must approach close enough to risk attack and injury for the display to be effective. There are three subcategories: **tactical threat** signal, in which the sender assumes a posture that enables it to immediately launch an attack; **vulnerability handicap** signal, in which the sender exposes vulnerable body parts to the rival; and **defensive threat** signal, in which the sender attempts to protect its most vulnerable body regions and places weapons in a position to defend itself.

9. **Conventional signals** are neither constrained nor costly to produce, and may be arbitrary in form. Their meaning is established by an agreed-upon convention. If there is no conflict of interest between sender and receiver, because they rank the receiver response alternatives similarly, then **cost-free signals** can evolve. If sender and receiver have conflicting interests, then a **socially imposed cost** is required to maintain reliability. Receivers test the conviction or call the bluff of a sender giving a strong signal by approaching closely, and may retaliate or punish if a mismatch between the signal and the sender's true state is detected.

10. **Modifiers** are traits that provide no information directly, but make it easier (**amplifiers**) or harder (**attenuators**) for receivers to assess certain aspects of a sender's character. Modifiers can work in conjunction with any of the signal classes mentioned above.

11. Most signals are believed to be honest most of the time because of the cost controls, but **dishonest signaling** is expected to occur some of the time, and receivers tolerate a certain amount of deception as long as there is an average net gain by attending to the signal. Causes of unreliable signaling include: **out-of-equilibrium systems**; **perceptual error** by receivers; **multiple classes of senders** with different benefits and costs that receivers cannot distinguish; **multiple classes of receivers**; and **tactical deception**.

Further Reading

The book on animal signals by Maynard Smith and Harper [328] covers much of the same ground as this chapter and presents the ideas on evolution of signals and classification of signals based on the type of cost. Maynard Smith [324] and Parker [389] provide good introductions to the principles of evolutionary game theory. Nowak [367] provides a lucid introduction to adaptive dynamics and shows how ESSs are only one of a set of possible outcomes. A more general overview of adaptive dynamics models can be found in McGill and Brown [331]. The edited volume by Dugatkin and Reeve [129] then shows how these principles can be applied to various evolutionary contexts. Searcy and Nowicki [460] present a highly readable review of signal reliability and explain some of the detailed examples we will cover in subsequent chapters. Readers interested in the original views of Zahavi [546–553] and Hasson [211–213, 215–217] should of course read their key papers on handicaps and amplifiers, respectively. Hurd and Enquist's [239] taxonomy of signals explains the logic of extensive-form games for evaluating types of signals. Arnqvist [22, 23] makes the case for receiver-precursor models of signal evolution.

COMPANION WEBSITE

sites.sinauer.com/animalcommunication2e

Go to the companion website for Chapter Outlines, Chapter Summaries, and References for all works cited in the textbook. In addition, the following resources are available for this chapter:

Web Topic 10.1 *Estimating evolutionary trees*
A guide to the process of reconstructing phylogenetic trees using multiple traits, and the measures used to describe their robustness and consistency. More examples of how to use phylogenetic trees to answer questions about signal evolution.

Web Topic 10.2 *Neural network models and feature detectors*
Neural net models provide a very useful way to explore the properties of recognition systems in animals. Feature detectors may emerge from such network training that mimics the ritualization process, or they may develop in noncommunicatory contexts and serve as sensory drivers for signal evolution.

Web Topic 10.3 *Rich media examples of ritualization*
Some classic examples of visual, acoustic, and chemical signals derived from sender precursors.

Web Topic 10.4 *Emotion, drive, and motivation*
Past and current views on these controversial issues.

Web Topic 10.5 *Evolutionary game theory*
Here we provide some more background on this modeling technique, including methods for finding ESSs in various games.

Web Topic 10.6 *Handicap controversies*
Zahavi's perspective, and the arguments of those who disagree with him.

Chapter *11*

Conflict Resolution

Overview

In disagreements between two animals over limited resources such as food, mates, or breeding sites, the interests of sender and receiver conflict to the maximum degree. Physical fighting sometimes results, but in many cases conflicts are settled with displays and ritualized methods of combat. How can communication resolve such disagreements? A contest is likely to be won by the individual that is larger, stronger, more experienced, or more motivated, and we would expect the opponents to signal this type of information. However, given the conflict of interest, we might also expect senders to bluff or exaggerate in order to convince the opponent to back down. Resolving conflicts is a multiple-stage process with assessment, signaling, escalation, and decision-making by both parties. We begin by examining the information an individual has available to it at the outset of a competitive encounter, particularly about its own fighting ability and motivation. We then review the theoretical and empirical literature on fighting strategies in animals and the nature of assessment and decision-making during the contest. Finally, we describe the variety of aggressive and submissive signals employed by animals in conflict situations. Most species have a repertoire of agonistic displays and acts that serve distinct functions, signaling dominance status, threats, preparation for attack, appeasement, and retreat. Our goal is to determine how the forms of these signals evolve to provide the information needed at various stages throughout a contest, and how they can remain reliable despite the strong conflicts of interest between competing parties.

The Process of Resolving Conflicts

Whenever two individuals simultaneously attempt to gain access to the same valuable resource, and that resource cannot be shared without a loss in fitness, a conflict situation arises. Examples of conflict situations include two males attempting to acquire the same territory or female mate, two hungry individuals desiring the same nonshareable piece of food, or two females or pairs interested in the same nest site. Both would prefer to win the resource without paying the cost of an escalated fight, so each would like the other to back down and retreat. Only one of them will win, the other will lose. In this sense, the interests of the competitors are diametrically opposed. However, both would benefit if the contest could be settled without a fight, and in this sense they agree.

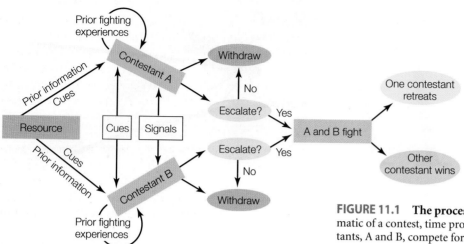

FIGURE 11.1   The process of conflict resolution In this schematic of a contest, time proceeds from left to right. Two contestants, A and B, compete for access to a resource. Both contestants may obtain information about the value of the resource from cues, and both may have prior information about their own previous fighting experiences. They exchange cues and signals, while potentially coming closer to each other. Either one may withdraw at any point. Either one may decide to escalate, and if the other contestant also escalates rather than withdraws, a fight ensues. Whichever contestant decides to quit first ends the fight and loses, while the other contestant wins, and the contest ends.

The fact that many aggressive interactions do end without a fight implies that communication has been successful in resolving the conflict.

Contest stages and information acquisition

While the two opponents in a conflict situation are likely to be of the same gender and to require the same types of resources to increase their fitness, they are rarely equal in **fighting ability** or **resource-holding potential** (**RHP**) (the probability of successfully defending a resource against a challenger), **motivation** (a function of need or resource valuation), and **experience** (dominance status and feedback effects from prior interactions). Each contestant would like to convey to its competitor that it is the superior or more motivated fighter. Selection pressure to bluff or exaggerate one's fighting ability, motivation, or experience ought to be strong. At the same time, each is trying to decide whether it is likely to lose, in which case it is better to cut one's losses and retreat. It is important to note that each player must assess its fighting ability, motivation, and experience *in relation to its opponent's*. Both individuals are therefore simultaneously senders and receivers, and an exchange of signals and cues takes place between them. Conflict resolution proceeds in stages and leads to the accumulation of increasingly more accurate information about the likelihood of winning. Throughout the contest, the competitors must make decisions about what signals and tactics to use and whether or not to continue. The flow diagram in **Figure 11.1** illustrates the key components of the conflict resolution process explained below.

As we have pointed out in Chapters 1 and 8, there are several sources of information that animals can use to make these decisions. Before a contest ever begins, animals will have amassed several types of *prior information* that help them decide how to react when they encounter a potential rival. Most important is the individual's knowledge of its own fighting ability. Fighting strategy models almost always assume that an individual knows its fighting ability, but they rarely consider how contestants acquire this information. The

answer is: from the outcomes of previous aggressive encounters. Over time, animals learn about their fighting prowess relative to others in their population and acquire a subjective estimate of their dominance status. This learning process has such important implications for the establishment of dominance hierarchies and nonviolent conflict resolution that we describe it in greater detail below. Another type of prior information relevant to potential aggressive conflicts is the expected frequency of encountering rivals. If rivals are very common, an animal may need to be careful about picking its battles. An animal may have prior information about a particular rival because it has previously fought with that individual or observed it in a third-party fight. Such information aids in the assessment of its chances of winning against that opponent. A final type of prior information is knowledge about one's need for a limited resource, which affects an animal's motivation to fight for it. With regard to a food resource, this knowledge is private information about hunger level, determined by internal physiology and time since the last meal. With regard to a territory, shelter, or mate, prior knowledge about the quality of the resource provides useful private information that may affect the animal's willingness to fight to defend it.

Most information about the value of the resource—for example, the size of a food item, the fecundity of a contested female, or the quality of a nest or burrow site—is acquired from *cues*. Although some information about a rival may come from cues such as body size and scars or injuries (evidence of the rival's fighting experience), signs of exhaustion, and chemical cues derived from winning or losing, most information about the rival will come from *signals*. All of the behaviors performed during aggressive interactions, ranging from threatening to submissive signals and tactics, are collectively referred to as **agonistic behaviors**. Agonistic signals include body-size amplifiers, badges of status, and behaviors that communicate aggression, threat, defensiveness, submission, appeasement, retreat, and victory. Rivals assess each other's strategic choice of signals and the performance of these signals to make decisions about quitting or continuing a contest. Contests generally begin with the least costly signals, and if these signals cannot resolve which contestant is most likely to win, the contestants escalate to a more intense and costly stage of engagement.

If cues and signals cannot resolve a conflict, the contest will escalate to a final combat stage involving physical contact. Exhaustion, injury, and even death may result, but a winner will usually emerge. The winner gains the resource—food or access to a mate, a burrow, or a territory—and enjoys the resulting fitness benefits. The winner may declare its victory with a victory signal, which, as we describe later, may ensure future victories.

Assessing one's own fighting ability

Evolutionary game models for the evolution of aggressive signals, such as the two models described in Figures 10.30 and 10.32, begin with the assumption that an animal knows its own fighting ability, but not that of its opponent. The models seek the conditions for an ESS in which senders honestly give one type of signal when they are strong and a different signal when they are weak [111, 199, 432]. Only knowledgeable animals can select the most appropriate signals to give. In theory, when animals know their own fighting ability with reasonable accuracy, the level of escalated fighting in the population is reduced [310]. Moreover, animals are better able to optimize their decision-making and net gain in potential conflict situations if they can assess their probability of winning against a particular opponent, persisting aggressively if the opponent appears to be weaker and retreating quickly if it appears to be stronger [302, 391]. How do animals determine their fighting ability?

If body size is an important determinant of winning, and body size varies greatly among individuals in a population and is readily assessable at the outset of a contest, then body-size-related cues and signals can provide sufficient information to make this assessment. But in many cases, body size may be difficult to assess or insufficiently variable, and many individuals are likely to cluster in the middle range of a normal distribution. In such cases, animals can accumulate information about fighting ability through the experience of fighting. Numerous studies on a variety of vertebrate and invertebrate species have demonstrated that a record of previous wins or losses influences the outcome of a current contest [88, 188, 189, 191, 221, 339, 356, 391]. When an animal has won a previous fight, it is more likely to win a subsequent fight; and conversely, if it has lost, it is more likely to lose again, even against a different opponent. This phenomenon is called the **winner-loser effect** [72, 99, 187]. Behaviorally, prior winners are more likely to initiate

confrontations and escalate faster in subsequent agonistic encounters, while prior losers are more cautious, passive, and likely to retreat when challenged. The mechanism by which these short-term responses to prior experiences affect subsequent aggressive behavior can be mediated either by learning or by physiological changes in hormone titers or neuroendocrine receptors [123, 135, 136, 189, 346, 350, 351, 447]. Depending on the species, levels of corticosterone, androgens, serotonin, catecholeamines, neuropeptides, or juvenile hormone in insects, have been shown to change following winning or losing. Thus the winner-loser effect serves as an important *rule of thumb* that animals with little cognitive ability can employ to estimate their rank and adjust their effort in costly contests.

An evolutionary game model examined the adaptive evolution and maintenance of such prior-experience effects and their ability to provide individuals with a mechanism to assess their relative fighting ability [314]. In this model, individuals are randomly assigned a true RHP (fighting ability), which they do not know. Animals then compete in round-robin contests among triads of individuals drawn randomly from the population. They revise their subjective RHP assessment in light of their experiences, raising their estimation of their RHP upward after a win and lowering it after a loss. Animals with a higher self-estimated RHP fight harder. Thus the greater the difference between an individual's and its opponent's RHP self-estimate, the more likely it is to win a contest, but if an animal overestimates its RHP, the cost of fighting increases. A contestant assesses neither its own absolute RHP nor the absolute RHP of its opponent. Thus, the strength of the winner-loser effect is affected not by the RHPs of particular individuals, but by the distribution of RHP in the population. An interesting outcome of this model is that the loser effect can exist alone, or it can coexist with a winner effect, but a winner effect cannot exist without a loser effect. This prediction is well-supported by the empirical literature; some studies find only loser effects, and there are no cases of winner effects without loser effects [189, 221, 339]. Loser effects may be stronger either because the initial magnitude of their effects on behavior are stronger or because their effects last longer. Losing likely incurs greater injury and fitness costs, so ratcheting aggression downward after a loss rather than ratcheting up after a win may be a conservative adaptive strategy.

The winner-loser effect plays a crucial role in structuring **linear dominance hierarchies** in animal societies [14, 36, 70, 71, 96, 101, 181–183, 264, 457]. The outcome of this simple, individual-based rule of thumb played out over many pairwise contests within a group often results in a highly linear hierarchy. This is an example of an **emergent property** that is not predictable as a linear sum of the events that went into creating it. We discuss emergent properties in more detail in Chapter 15. In general, group size must be relatively small, and individuals must be able to recognize each other to maintain a stable hierarchy [100]. In experiments using small groups of equal-weight cichlid fish (*Metriaclima zebra*), different rank

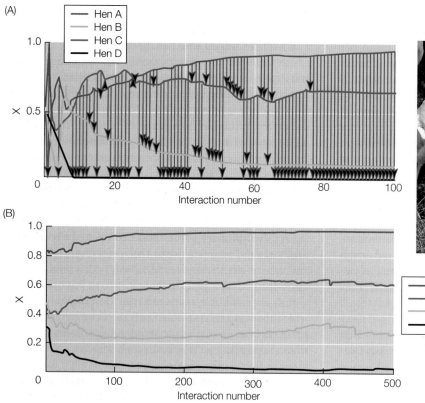

FIGURE 11.2 **Hierarchy formation in flocks of Leghorn hens** Groups of four hens unknown to each other are placed together in an enclosure and allowed to establish a dominance hierarchy. Each time an agonistic interaction occurs between two hens, the identities of the winner and the loser are noted. (A) A 100-interaction period for a group of four hens under continuous observation. The series begins with the initial introduction of the four individuals. X is a dominance index computed as the cumulative fraction of interactions won relative to total interactions. Each interaction is shown by an arrow. The origin and color of the arrow indicates the identity of the initiator of each interaction and points to the receiver (e.g., a red arrow pointing at the blue line means that hen A initiated an interaction with hen C). Note that in all cases the initiator has a higher dominance rank than the recipient (all arrows point down). (B) The emergence of dominance ranks averaged over 14 different groups for the first 500 interactions. This graph shows that the dominant individual emerges early, but the emergence of the remaining ranks occurs on a more variable time scale. (After [276].)

orders emerged when the same set of individuals was allowed to establish a dominance hierarchy on separate occasions, but the hierarchy was linear in both cases. This study showed that intrinsic differences among individuals played some role in determining rank, but that interactions and winner-loser effects were very important for linearizing the hierarchy [73]. Moreover, eavesdropping on pairwise interactions by third party bystanders likely also affects the hierarchy structure [97–99, 276]. **Figure 11.2** illustrates the establishment of a linear hierarchy in small flocks of domestic hens.

Contestants can also take advantage of winner-loser effects to improve their overall assessment of the probability of winning against a particular opponent by assessing prior contest effects in the opponent [189, 391]. This source of information has been called **social cues**. For example, an animal may be able to determine whether its opponent has lost previous fights by detecting signs of exhaustion or injury, as demonstrated in a study of jumping spiders [441]. Some species emit chemical cues that arise as by-products of winning and losing, which may enable contestants to assess their opponents' estimated RHP [23, 32, 43, 197, 222, 338, 488]. If winners gain an advantage from advertising their prowess, we would expect the evolution of victory signals [39]. Opponents could also learn about each other's previous fighting experiences by observing interactions in which they are not taking part. Such eavesdropping effects have been found in several species and will be covered in a later chapter [161, 218, 342, 343].

Types of resources and their value

Animals fight to gain access to limited resources. There are many types of contestable resources, but we can simplify by grouping them into four main categories: food, mates, shelters, and territories. These resources differ in several respects: the magnitude of the value of a single resource with respect to the lifetime fitness of an individual, the degree to which a resource's value can be assessed, and the degree to which it can be defended.

Food is a resource that is often sufficiently abundant that animals do not have to fight over it—the fruit of a mast-fruiting tree is a good example. Moreover, food is often shareable, so that a food morsel can be split into two pieces to resolve a potential conflict. If food is scarce or difficult to find and animals are hungry, a food item obviously increases in value. Finally, if food is clumped in space and has some durability, it becomes easier to defend for some period of time, compared to scattered food items.

Mates, or receptive members of the opposite sex, are another commonly contested resource. Typically, gravid or estrous females are a valuable and limited resource for males, who may guard them before, during, and after copulation to ensure that their sperm are the ones to fertilize the female. The value of a female may increase with her body size or age if larger or older females are more fecund. Female value also depends on the temporal distribution of female receptivity in a population. When female receptivity is asynchronous, the operational sex ratio (sexually competing males : receptive females) is very high. Each temporarily receptive female is extremely valuable, and male competition is intense. By contrast, when females are synchronous, most males are able to acquire a mate, and competition is lower.

Shelters such as nests, burrows, and webs are valuable because they involve a significant amount of investment by the animal that constructed it. If the structure also contains one's offspring, it is of course extremely valuable.

Territories are a highly valued resource for many species. A territory is a fixed area from which intruders are excluded by some combination of advertisement, threat, and attack. The territory usually contains food, and possibly shelters, nests, mates, and offspring. Acquisition of a territory usually involves fighting for it, followed by establishing the boundaries with adjacent territory owners and defending it against intruders. The loss of a territory is likely to cause the failure of a season or a lifetime of reproduction.

A general measure of the **resource value** of a contested item is the difference between the fitness gained by winning the contest and the fitness lost by losing the contest [155]. From this perspective, value depends an animal's current condition, need, or investment, and on how easy it is for it to obtain another similar resource item if it loses the present one. Resource value varies within and between individuals for a number of reasons. Properties of a resource item such as its size and quality determine the *objective* aspects of resource value. Other aspects of the value of a resource are more *subjective*. Value may depend on the abundance of similar resource items in the environment, with scarcer resources being more valuable. Value may also vary as a function of the physiological needs of the individual contestant. For example, a hungry individual places a higher value on a morsel of food than a sated individual. Finally, information about a resource's value might not be available to both contestants. For the food resource example, a focal animal is likely to know its own need, but not that of its rival. Likewise, the defender of a territory or an opposite-sex mate may know the quality of its resource, but an intruder may not be privy to this information [18].

These four general types of resources—food, mates, shelters, and territories—represent increasing levels of

ownership. Food and mates can be temporarily guarded, but shelters and territories must be defended by a committed owner. Naturalists have long noted that owners almost always win in contests against intruders, even if the owner is slightly smaller than the intruder. This common observation implies that ownership confers some additional advantage above and beyond fighting ability. The advantage of ownership has been ascribed to an increase in **motivation to fight** arising from prior investment in construction, defense, and reproduction [83, 103, 248, 358, 435, 450, 455]. For example, a male butterfly that has had a brief encounter with a female is more motivated to defend his display territory than a territorial male that has not yet interacted with a female [33]. In the case of larger territories, familiarity with the area and access to the resources on the territory could also improve aspects of fighting ability [86, 210, 216, 230, 373].

Fighting Strategies

The types of signals and behavioral acts used in aggressive encounters are closely linked to a species' **fighting strategy**. By fighting strategy we mean the displays, tactics, and the bases of the decision to continue versus quit that are used during extended conflicts over an unshareable resource. Some species fight by engaging in a continuous or repeated behavior such as grappling, wrestling, or chasing until one individual quits or flees. Other species proceed through a predictable series of stages involving increasingly costly mutual displays. Still others employ a mixture of unilateral and mutual displays and tactics in an unpredictable order while escalating or de-escalating during the encounter. In this section we will describe the major categories of fighting strategies, focused around three current conflict resolution models. These models differ in their assumptions about how contestants gather information during an interaction and how they make the decision to end it. The different models also make implicit assumptions about the kinds of signals and behaviors employed during a contest. Web Topic 11.1 provides more detailed descriptions of these models. Once we have laid the groundwork on fighting strategies, we will examine signals employed during contests.

Web Topic 11.1 *A detailed description of three conflict resolution models*
The sequential assessment model and its variants, the energetic war of attrition model, and the cumulative assessment model. The unit also discusses the strategies and challenges of distinguishing among these models with empirical data.

Assumptions and predictions of fighting strategy models

In this section we briefly describe three alternative fighting strategy models whose predictions have been examined and tested in a variety of animal species: the sequential assessment

TABLE 11.1   *Summary of the predictions and display characteristics of fighting strategy models*

	Sequential assessment (SAM)	Energetic war of attrition (E-WOA)	Cumulative assessment (CAM)
Decision based on	Information about opponent relative to self	Costs resulting from own actions	Costs resulting from own actions and inflicted by opponent
Requires matching behaviors	No	Yes	No
Assessment of opponent	Yes	No	No
Escalation	Not within a phase, but in sequential phases	Escalation or de-escalation possible within phases	Escalation or de-escalation possible within phases
Contest duration most strongly correlated with	RHP asymmetry between opponents	Loser RHP	Loser RHP(+) and winner RHP(−)
Contest duration increases with increasing mean opponent RHP	No	Yes	Yes
Display characteristics	Nondangerous index signals or ritualized fighting tactics	Energetically costly grappling or handicap signals with enforced matching	Dangerous displays

Sources: After [49, 355, 424].

model, the energetic war of attrition model, and the cumulative assessment model. A very important difference among these models is whether a contestant makes the decision to quit based on an assessment of its fighting ability relative to that of its current opponent (**mutual assessment**) or based on its prior or internal evaluation of its own fighting ability (**self-assessment**). Table 11.1 summarizes the assumptions and predictions of the three models.

SEQUENTIAL ASSESSMENT MODEL   The **sequential assessment model** (**SAM**) of conflict resolution is based on mutual assessment. It assumes that two competitors for a nondivisible resource initially know very little about each other's relative fighting ability, but acquire this information while interacting. A process of repeated rounds of displaying enables them to update their estimates of each other's relative fighting abilities [110, 115]. Fights consist of the repetition of one type of interaction or behavior that reveals some information about fighting ability to each opponent, albeit with some error. At each point in time, contestants have a current estimate of their probability of winning and an uncertainty term (standard deviation) for that estimate. The first few repetitions of the signal provide a great deal of information and reduce the error component. As the contest proceeds, the estimate becomes more accurate and the uncertainty decreases in a manner similar to statistical sampling. The ESS for such a dynamic game is an evolutionarily stable policy called the **giving-up line**. When one contestant's current estimate crosses this line, it is sufficiently certain of its lower fighting ability that it makes the decision to quit (**Figure 11.3A**).

The sequential assessment model assumes that the displays are index signals or ritualized fighting tactics that reveal the sender's fighting ability (RHP). A variant of the model allows for several signals to be used that vary in their costs and accuracy of encoding aspects of fighting ability. In fish,

for example, lateral display may provide only partial information on body size, tail beating may improve the size estimation, and mouth wrestling or head butting may provide information on strength. Opponents begin the contest with one type of display or behavior from the agonistic signal repertoire. This behavior is repeated and, as in the model with one behavior, each contestant's assessment of its chances of winning improves with time. If there is still uncertainty and the giving-up line has not been crossed, the next stage of the contest begins. They use (or add) a second behavior that provides new information. The contest escalates through several stages if neither individual estimates that it will lose (**Figure 11.3B**).

The sequential assessment model predicts that contests will be longer and more variable in duration the more similar the two contestants are in fighting ability. This happens because two very closely matched contestants need to sample each other's displays for a longer time before the error term is small enough for them to distinguish the small difference in their abilities, but sometimes the high error at the beginning of the contest leads one individual to quit early. This model also predicts longer contests over higher-valued resources. The multistage model similarly predicts that more closely matched contestants will proceed through more stages. Contestants will begin the encounter with the lowest-cost display in their repertoire, and in subsequent stages will use or add more costly displays and behaviors that provide more information. Although the model does not explicitly require contestants to match behaviors, it is likely that they will do so because they rank the costs of alternative displays in the same order. Similarly, the sequence of behavioral acts used should be independent of a contestant's relative fighting ability. Finally, there should be no escalation within a stage as contestants refine their estimates, and the duration (or number of acts) of completed stages should be independent of fighting ability.

(A)

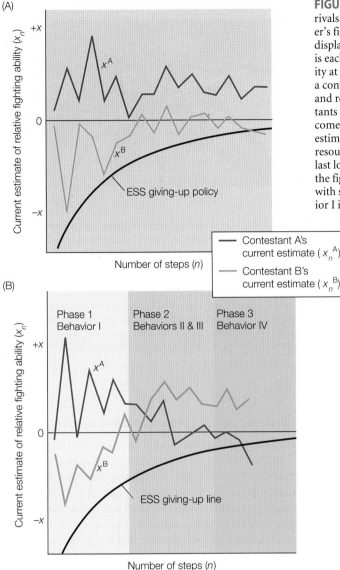

FIGURE 11.3 **The sequential assessment model of fighting** Two rivals, A and B, begin the contest with little information about each other's fighting ability, but acquire this information by sampling repeated displays. The x-axis represents time or the step number, n. The y-axis is each contestant's current estimate of its own relative fighting ability at a given step, x_n. The curved line is the ESS giving-up line; when a contestant's estimate crosses this line, it is selected to end the contest and retreat. The curve is low at the outset of the contest because contestants lack sufficient information to decide whether to quit, but the curve comes closer to $x = 0$ with time because repeated sampling makes the estimate of relative fighting ability more accurate. When the value of the resource increases, this curve shifts down, and contests are predicted to last longer. (A) The game with one behavior, showing the trajectories of the fighting ability assessments for the two contestants. (B) The game with several behaviors, which are used in phases. The least costly behavior I is used repetitively in the first phase (yellow), and if differences cannot be resolved, the second phase (green) begins with the more costly behaviors II and III. The third phase (blue) involves the use of behavior IV. Contests with more closely matched competitors will last longer and escalate through more phases, whereas contests with disparate competitors will be shorter. (A after [110]; B after [116].)

energy reserves, or upon reaching the maximum attainable cumulative display expenditure [313, 354]. Like the other WOA models, the form of the display or competitive behavior must force the two contestants to perform matching signals, but for a different reason. In the classic WOA models, a contestant must not reveal its intentions or bid time, which would enable the opponent to select a slightly longer persistence time and win. In the energetic war of attrition, the contestants must expend energy at the same rate; otherwise one individual could cheat by delaying or reducing its effort to save energy. On the other hand, competitors may jointly vary the intensity of displays or the rate of display during the contest. Because the contest behavior is essentially a costly handicap signal, the display is expected to honestly reveal an individual's stamina. Stamina is unlikely to be related to body size or any other visible cue and more likely to be related to nutritional condition, fat reserves, or health.

One key prediction for the energetic war of attrition model is that the winner should exhibit better nutritional condition or energy reserves than the loser. In addition, contest duration should be positively correlated with the condition or energy reserves of the loser. This relationship arises because better-condition losers are able to fight longer before they quit. Similarly, since competitors assess only themselves, contests between two closely matched high-quality rivals should be longer than those between two closely matched low-quality individuals. This prediction distinguishes the energetic war of attrition from the sequential assessment model, which predicates contest duration on the relative difference between the competitors rather than on their absolute fighting abilities. The types of conflict behaviors envisioned by the energetic war of attrition model include noninjurious matched interactions such as aerial chases, spiral flights, and possibly countercalling contests.

ENERGETIC WAR OF ATTRITION   The **energetic war of attrition model** (**E-WOA**) only superficially resembles the classic symmetric and asymmetric war of attrition models, in the sense that two contestants persist in a lengthy interaction of matched behaviors until one of them reaches its quit time. In the classic WOA models, the quit time is a "sealed bid" determined before the contest starts and drawn at random from a specific probability distribution function (see Web Topic 11.1). The energetic war of attrition model dispenses with the biologically unrealistic notion of a sealed bid, and instead assumes that each contestant sets its persistence time based on an assessment of its own current ability to continue in a costly interaction. This model is therefore a *self-assessment model*, without any assessment of the rival or of relative RHP. Opponents engage in a contest of endurance using repeated costly displays or a continuous energetically intense activity. Contests end when one of the contestants reaches a threshold level of costs that it can bear, when it runs out of

(A)

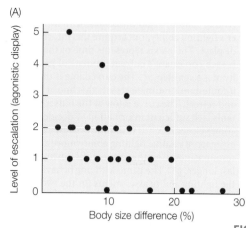

(B)

FIGURE 11.4 **Stages of escalation in jumping spider contests** (A) The *y*-axis shows the level of escalation through the five typical agonistic displays of the jumping spider (*Hypoblemum albovittatum*): (1) hunch display, (2) stave display with abdomen raised to the side and mutual flicking of the third leg, (3) head butt with legs erected and pushing against each other, (4) cheliceral lock, and (5) grappling. The more similar the males are in body size (weight), the higher the stage of escalation reached during contests. (B) Two males performing the head butt display, with legs extended and head pushing against the opponent's head. (A after [211, 474].)

CUMULATIVE ASSESSMENT MODEL The **cumulative assessment model** (**CAM**) describes a fighting strategy in which a contestant's decision about whether to continue or quit an extended agonistic interaction is based on a cumulative *sum of the costs imposed by the opponent's actions*. It primarily applies to species that engage in potentially injurious behaviors. Contestants do not evaluate the opponent's RHP relative to their own, but instead keep track of the damage inflicted on them by the opponent, and quit when they reach their own damage tolerance threshold [355]. This *self-assessment model* was developed as an alternative to the sequential assessment and war of attrition models, and can be applied to cases in which contestants do not match behaviors during an encounter and can escalate and de-escalate independently of each other. Contestants also cannot assess the effects of the damage they inflict on the opponent, except to the degree that the opponent is still fighting or has quit. The payoffs to contestants depend on aspects of individual quality such as defensive skill, energy efficiency, and ability to tolerate costs. The model can be generalized to contests involving noninjurious displays, but in this case there must be some component of external cost beyond the control of each individual that increases with time, such as physiological stress, predation risk, lost foraging time, or lost mating opportunities.

This fighting model predicts that contest duration should be positively correlated with loser RHP, as in the energetic war of attrition model above, but it should also be negatively correlated with winner RHP. This occurs because a high-quality winner will inflict damage on its lower-quality competitor at a higher rate, causing the lower-quality individual to reach its quitting threshold even faster. Contestants are likely to escalate at different rates. Specifically: (1) in short contests, the interaction is so brief that high-quality individuals perform at high levels at the outset; (2) in long contests, high-quality individuals escalate their display rate and intensity at a faster rate than low-quality individuals; and (3) in intermediate-length contests, low-quality individuals begin at a lower intensity than high-quality individuals but escalate at a greater rate because they cannot tolerate a long contest.

Evidence supporting the alternative models

In principle, these models should be readily distinguishable based on their different predictions relating contest duration to contestant RHP: self-assessment models predict a positive correlation between contest duration and *loser RHP*, whereas mutual assessment models predict a negative correlation between contest duration and *RHP asymmetry*. However, a recent simulation analysis showed that if contest duration is closely correlated with loser RHP as a result of self-assessment, a spurious negative correlation between contest duration and RHP asymmetry will also occur [138, 440]. This happens because very small competitors are likely to give up very quickly, and they are more likely to have substantially larger rivals, so greater RHP asymmetry becomes associated with shorter contests as a by-product. Moreover, the cumulative assessment model predicts a positive correlation between duration and loser RHP and a negative correlation between duration and winner RHP, which generates an even stronger spurious negative correlation between RHP asymmetry and duration [49, 331, 440]. Earlier studies that examined contest duration only as a function of RHP asymmetry, and concluded that there was assessment of the rival and a good fit to the sequential assessment model, therefore may have missed possible cases of self-assessment because they did not look at the strength of the simple correlation between absolute loser RHP and duration. More recent tests that examined all possible correlations have found evidence for all three models in different species.

All species that escalate contests in stages, with both contestants performing the same behaviors repeatedly in a

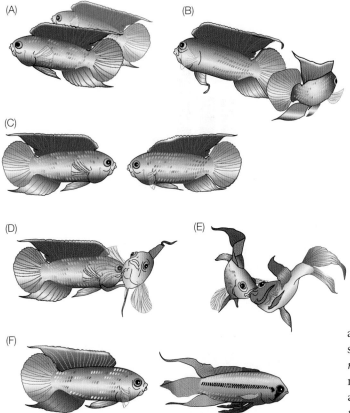

FIGURE 11.5 **Contest escalation stages in *Nannacara ano-mala*** (A) Broadside display, (B) tail beating, (C) frontal orientation, (D) biting, (E) mouth wrestling, and (F) termination, in which the loser on the right lowers its fins and displays midline darkening. (After [212].)

display, (2) the stave display with mutual flicking of the third leg, (3) head butt, (4) cheliceral lock, and (5) grappling [474]. Each stage involves a greater risk of injury. The more similar the body size of the two contestants, the higher the stage of escalation as predicted by the sequential assessment model (**Figure 11.4**).

The more specific predictions of the sequential assessment model were tested in the cichlid fish, *Nannacara anomala* [113, 116, 212]. Contests proceed in clear stages, beginning with a broadside display and tail beating and escalating to biting, mouth wrestling, and finally to circling with mutual biting (**Figure 11.5**). Tail beating stops once mouth wrestling and biting begin, and contestants flash distinctive color signals to coordinate stage shifts [198]. The predictions were supported in that the larger fish nearly always won and contest duration and the variation in duration were negatively correlated with the difference in contestant size (**Figure 11.6**). As also predicted, the duration and number of acts for completed phases were not correlated with weight asymmetry, and contest duration did not increase with the mean absolute size of contestants. Contests in the convict cichlid (*Cichlosoma nigrofasciatum*) proceed in a similar fashion and also meet many of the model's predictions [237]. A rigorous test of alternative fighting strategy models with a fiddler crab (*Uca pugilator*) concluded that the sequential assessment model provided the best fit [367]. In natural contests between males for breeding burrows, contestants used one or more agonistic elements that varied in intensity from a no-contact extension of the enlarged claw to flipping over the opponent. Contest

predictable order of escalation, fit the general spirit of the sequential assessment model. Red deer stags (*Cervus elaphus*) are a classic example [75, 76]. Males first engage in a roaring countercalling exchange from a distance; the frequency of the roar immediately distinguishes young males (higher frequency) from mature males (lower frequency), but among mature males there are no differences in frequency. Instead, the rate of roaring is evaluated for a period of time. Roaring uses the same thoracic muscles employed in fighting and the ability to keep up in a countercalling contest is a good indicator of a male's condition. If the roaring rate of two males is similar, they approach and perform a parallel walk, a broadside display with the head held high, providing visual information on relative body height and condition (number of antler points is correlated with age and condition). If this exchange doesn't resolve the contest, they then engage in a pushing contest with locked antlers that finally determines which male is the stronger one. This ritualized form of fighting provides ultimate information on the relative weight and strength of the two opponents. Another good example was described in the jumping spider *Hypoblemum albovittatum*, where contests always proceed through five predictable stages of mutual escalating behaviors: (1) the facing-off hunch

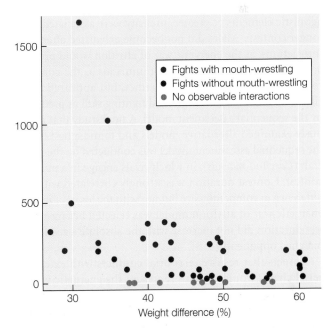

FIGURE 11.6 **Fight duration as a function of the weight difference between opponents in *Nannacara anomala*** Both the mean duration and the variance in duration increase as the contestants become more similar in size. (After [113].)

(A)

(C)

(B)

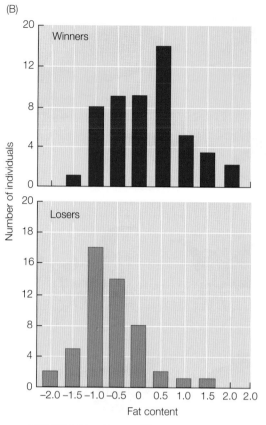

FIGURE 11.7 **Damselfly contests** (A) Male ebony jewelwing damselfly, *Calopteryx maculata*. (B) Comparison of the fat content of losers versus winners of fights, showing higher average fat content for winners. Fat content is measured as standardized residuals from a regression of fat on lean dry body mass. (C) Strong positive relationship between the loser's fat content and contest duration, suggesting that the loser's quitting time is determined by self-assessment of fat reserves as predicted by the energetic war of attrition. (B after [292]; C after [313].)

duration increased as males became more similar in size, and agonistic elements of greater intensity were associated with longer contests. Males did not become exhausted after long interactions, as the energetic war of attrition would predict, and the behaviors used were not injurious as the cumulative assessment model would predict, but appeared to be ritualized displays of strength and fighting skill as predicted by the sequential assessment model. A final study that rigorously examined alternative models and found a best fit for the sequential assessment model was conducted on the shore crab (*Carcinus maenas*), in which rivals engage in a pushing contest. Contest duration was strongly correlated with the difference in competitor size but not with the loser's size. The energetic war of attrition model was rejected because contest duration did not increase with the absolute size of size-matched opponents [415].

Insects that resolve territorial conflicts with extended aerial chases would seem to represent the energetic war of attrition model. Such chases are noninjurious but energetically expensive, and they enforce the matching of behaviors, because the two contestants encircle each other, coordinating their movements. Dragonfly and damselfly males fight to defend oviposition sites in patches of vegetation on ponds. Contest dynamics support many of the specific predictions

of the E-WOA [78, 291, 292, 360, 361]. In *Calopteryx* damselflies, escalated contests involve a series of bouts in which males initially face off by hovering in front of each other for 3–5 seconds using rapid wingbeats, and then begin a prolonged chase that lasts up to five minutes, involving a complex series of ascents, descents, spiral turns, and rapid changes in direction. As predicted by the energetic war of attrition model, the fatter male (measured via post-contest extraction of thorax and abdominal lipid) nearly always wins, and contest duration is positively correlated with the fat content of the loser (**Figure 11.7**). Contest duration was also correlated with the mean combined fat of the two contestants, as predicted. In a study of another damselfly species, territorial males that chased a tethered rival for 20 minutes had significantly lower fat after the chase compared to the control territory owners, indicating that fat reserves are rapidly depleted during long fights. However, it is very difficult to completely rule out mutual assessment during these chases. The fat content of males is tightly correlated with muscle power output, suggesting that the insects possess a mechanism for adjusting muscle contractile performance to match the rate at which energy can be mobilized from stored fat reserves [293]. If higher muscle performance by fatter males translates into higher top flight speed, then males could be assessing each other's speed during a chase. On the other hand, if higher muscle performance translates into greater flight efficiency, then stamina is the critical feature characterizing male quality, and the energetic war of attrition could in fact represent the contest dynamics well. Botfly males (*Cuterebra austeni*), which defend mating perches by way of high-speed chases, may also fit the energetic war of attrition model, as fatter males show a competitive advantage [229].

Males of many butterfly species engage in circling flights in contests over perches, ending with one male fleeing in a horizontal flight, but in these cases, it is more difficult to evaluate why particular males win than it is in the case of damselflies [225]. Residents or perch owners generally win in contests with invaders, and older males are also more likely to win [226–228, 231, 232]. In the speckled wood butterfly (*Pararge aegeria*) that defends sunspots on the temperate forest floor, males with higher body temperatures win, suggesting that agility is an important component of these aerial flights [426]. But no measures of body size, wing loading, or flight muscle mass have been related to win/lose status or contest duration in any species [225]. Comparative studies have shed some light on this issue. Species in which adult energy stores derive only from the larval or the prereproductive stage and are gradually depleted during the reproductive period exhibit fat-constrained fighting behavior consistent with the energetic war of attrition self-assessment model. An energy-consuming fighting strategy with self-assessment of stamina may provide the best honesty-guaranteeing mechanism of conflict resolution in such cases. In contrast, species that feed as adults engage in aerial contests that appear to involve mutual assessment of fighting ability (agility) and motivation (resource valuation) [229].

Several studies that set out to compare the predictions of all three fighting strategy models concluded that the cumulative assessment model fit best. This conclusion was based primarily on finding a greater increase in contest duration as a function of loser RHP compared to relative RHP, and a negative correlation between duration and winner RHP. Several studies also reported longer contests for large size-matched opponents than for small size-matched opponents. Examples include: the jumping spider *Plexippus paykulli*, which escalates a contest with behaviors similar to those described earlier for *Hypoblemum albovittatum* [439]; the orb-weaving spider *Metellina mengei* [45]; the fiddler crab *Uca mjoebergi* [331]; the amphipod *Gammarus pulex* [91, 368]; the tree weta *Hemideina crassidens*, a nocturnal orthopteran insect native to New Zealand [224]; Cape dwarf chameleons *Bradypodion pumilum* [424]; and male fallow deer *Dama dama* [215]. In all of these species, contests escalate and include physical fighting, so the damage cost assumed by the cumulative assessment model likely occurs.

A particularly compelling case for the cumulative assessment model was made for the house cricket (*Acheta domesticus*) [48, 162, 163]. House crickets have an agonistic display repertoire of 13 behaviors, including displays without physical contact, tactics with light contact, and tactics with hard contact. The functions of the displays range from various levels of aggressive threat to submissive signals, defensive signals, and fighting tactics (**Table 11.2**). Many displays are performed unilaterally rather than mutually, so there is no mechanism to enforce matching of behaviors or intensities by the opponents except for mandible sparring and wrestling. Contests begin with the least costly displays (stridulation, antenna lashing, mandible flare) and escalate to more costly acts involving greater risk of injury (e.g., head charge, mandible lunge, wrestling). Eventual winners perform more of the costly behaviors, escalate faster, and exhibit overall higher oxygen consumption and glucose mobilization than losers. Body weight is the single most important determinant of winning, which is not surprising, given that the winner of an escalated wrestling contest may throw the loser. Weight asymmetry is negatively correlated with contest duration, but the simple correlations between contest duration and loser's weight (positive) and winner's weight (negative) are stronger (**Figure 11.8**). Together, this evidence suggests that the winner imposes increasing costs on the loser, who ends the conflict when he has reached a cost accrual threshold, as predicted by the cumulative assessment model.

Although it is clear that the cricket and other species use self-assessment in deciding whether to escalate or back down from a conflict, many of these species use obvious signals and displays in the early stages of contests which sometimes resolve conflicts before physical contact commences. For example, dwarf chameleons engage in contests that vary greatly in duration and degree of escalation from simple lateral displays to open mouth threats, headshaking, biting, mouth wrestling, and jaw locking. Contest outcome is influenced by the height of the ornamental casque (see Figure 10.29A), the relative size of the pink patch in the center of the flank, and by previous fighting experience [424]. Rejection of some form of mutual assessment seems unwarranted in this case, given the important role of the casque, the color patch, and experience. Several recent studies have described fighting strategies in which mutual assessment is important in the precontact stages of contests, and self-assessment in the contact stage. For example, in the jumping spider *Phidippus clarus*, males guard females against rivals. Contests begin with foreleg waving (an index signal of body size) and abdominal substrate vibration (likely associated with motivation). The next stage of escalation involves foreleg fencing. The duration of this stage and the probability that it escalates to wrestling is related to the difference in body weight between two rivals, and relative body weight is a better indicator of the probability of winning than size per se because animals vary in condition. This result suggests that foreleg fencing enables contestants to assess relative weight via a sequential assessment mechanism. The duration of the final escalation stage with grappling is

(A)

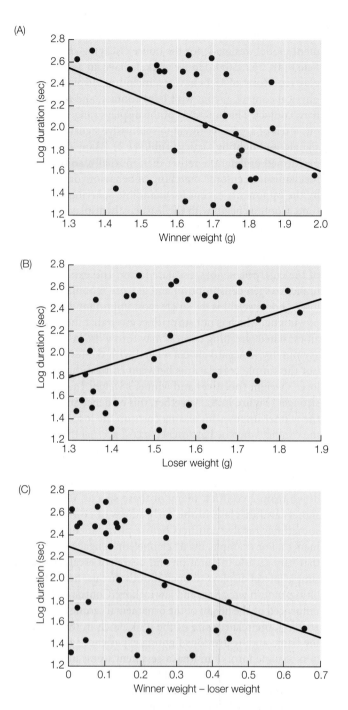

(B)

(C)

(D)

Levels of aggression

0. Mutual avoidance

1. Pre-established dominance

2. Antennal fencing

3. Mandible spreading (unilateral)

4. Mandible spreading (bilateral)

5. Mandible engagement

6. Wrestling

Decision

Shaking

Rival song

Winner   Loser

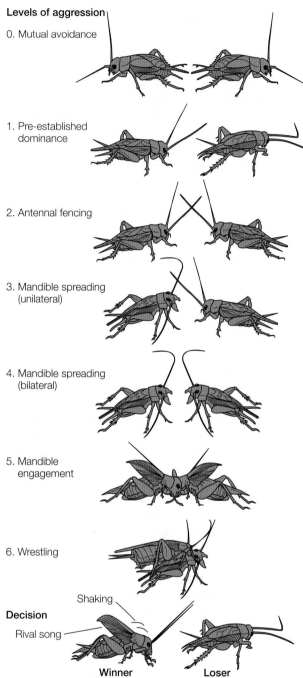

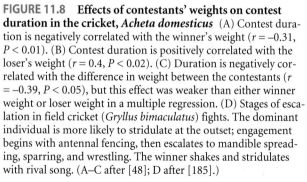

FIGURE 11.8   Effects of contestants' weights on contest duration in the cricket, *Acheta domesticus* (A) Contest duration is negatively correlated with the winner's weight ($r = -0.31$, $P < 0.01$). (B) Contest duration is positively correlated with the loser's weight ($r = 0.4$, $P < 0.02$). (C) Duration is negatively correlated with the difference in weight between the contestants ($r = -0.39$, $P < 0.05$), but this effect was weaker than either winner weight or loser weight in a multiple regression. (D) Stages of escalation in field cricket (*Gryllus bimaculatus*) fights. The dominant individual is more likely to stridulate at the outset; engagement begins with antennal fencing, then escalates to mandible spreading, sparring, and wrestling. The winner shakes and stridulates with rival song. (A–C after [48]; D after [185].)

more highly correlated with the loser's body weight than with relative weight. The study concludes that spiders switch to a different decision rule at this stage—self-assessment of opponent-imposed costs [105]. Similarly, in killifish (*Kryptolebias marmoratus*) and fiddler crabs (*Uca pugilator*), rivals adopt mutual assessment at earlier stages when deciding whether to escalate the contest from mutual display to physical contact, and once a contest has escalated they switch to self-assessment to decide how long to escalate [190, 367].

TABLE 11.2   *The agonistic display repertoire of the house cricket* Acheta domesticus

Display category[a]	Cost[b]	Function and information
SIGNALS		
Raising of cerci	0.02*	Defensive act by loser or subordinate
Prebout stridulation	0.02	Owner or dominance signal
During stridulation	0.02	Mild aggressive threat
Postbout stridulation	0.02	Post-fight victory signal by winner
Flaring of mandibles	0.08	Aggressive threat, intention to bite, reveals weapon size
Shaking	0.49	Aggressive motivation signal
LIGHT CONTACT		
Head butt	0.37	Ritualized fighting tactic
Foreleg punch	0.37*	Ritualized fighting tactic
Mandible spar	0.45	Mutual ritualized fighting tactic
Antenna lash	0.68	Aggressive motivation signal
Stridulation with lash	0.70	Multimodal motivation signal
HARD CONTACT		
Rear kick	0.37	Defensive act by loser or female guarder
Head charge	0.61	Overt attack, often by burrow intruder
Mandible lunge	0.61	Bite
Wrestling	0.83	Mutual mandible lock with head twisting

Sources: After [162, 163, 185].

[a] Displays are grouped into three categories: signals, tactics with light contact, and tactics with hard contact, which represent increasing levels of risk of injury.

[b] Cost was measured in a metabolic chamber and values are given in units of ml O_2/g/sec; values indicated by * are estimates.

It seems likely that contests in many species entail some assessment of rivals followed by reliance on self-assessment during the final escalation and physical contact phases of a fight. In addition to examining the factors affecting each contest phase and the nature of the behaviors and escalation patterns, it is important to determine which opponent characteristics, such as body size, weight, weapon size, or energetic status, determine the duration of different stages, and how well these characteristics are correlated with the probability of winning. With this information, we should be able to evaluate the true function of agonistic behaviors, and specifically whether they evolve to: (1) reveal something about the sender's body size or other aspects of fighting ability; (2) test and wear down the stamina of the opponent; or (3) inflict sufficient costs on the opponent that they quit. These strategic and tactical signaling issues will be addressed in a later section of this chapter.

Role of resource value and other asymmetries

Up to this point, we have focused on fighting ability as the primary determinant of the outcome of an escalated contest, and on the degree to which opponents assess each other's fighting abilities. Nevertheless, all of the fighting strategy models include a term, *V*, for the value of the resource that the winner obtains. A general expression for the payoff of winning is $PO = pV - C$, where *V* is the value of the resource, *C* is the cost of fighting, and *p* is the probability of winning. Animals can afford to pay a greater cost to obtain a more valuable resource, either by persisting longer or by using riskier behaviors, and performance of higher-cost behaviors might improve the chances of winning the resource [106]. In this section we review the way that different fighting models conceptualize the role of **resource value asymmetry**.

A variant of the war-of-attrition model investigates the case in which contestants know their own *V* but not that of their opponent [34]. Contestants are assumed to have equal fighting ability but to differ in their current *V* according to a random distribution. At the ESS, the contestant with the higher *V* persists for longer and wins the contest (see Web Topic 11.2). The context envisioned for this model is conflict over food resources, where animals in the population differ in their current hunger state. Opponents are not able to assess each other's hunger, and in fact, benefit by keeping this information private; a contestant that is not hungry should not reveal this fact to the opponent, because it is always possible that the opponent is not hungry either, in which case

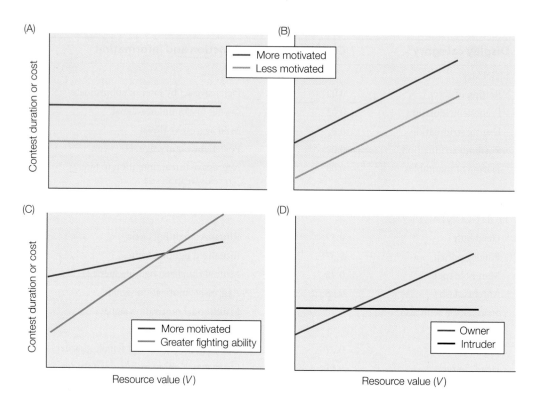

FIGURE 11.9 **Contest duration as a function of resource value for different forms of resource assessment** The *x*-axis represents objective resource value, the *y*-axis represents persistence time or cost each contestant is willing to invest in the conflict. Red line indicates contestant with the greater motivation in all cases. (A) In this case, there is no assessment of resource value. Contestants differ in their internal motivation (subjective resource value): the more motivated individual should persist longer and win (assuming no difference in fighting ability). (B) In this case, the same resource is assessed by both contestants. Both evaluate objective resource value, and increase their persistence accordingly; contest duration should increase with increasing resource value, and the more motivated individual should always win. (C) In this case, there is mutual assessment of own and opponent's resource value. Resource value and fighting ability may differ for the contestants. Here, one individual has the higher motivation, and the other has a slightly greater fighting ability. (D) In this case, there is asymmetric information about resource value in ownership context. The owner has precise information about the value of the resource it controls; the intruder assumes the average population value of the resource. Owner likely to win when resource value is higher. (After [18, 114].)

the food may be won cheaply [34]. A graph of the simple predictions for this type of subjective resource assessment is shown in **Figure 11.9A**. Empirical tests of this model are often based on staged contests between individuals that have been food-deprived for varying lengths of time. The result is nearly always that the hungrier individual fights harder for the food and wins. Contest duration did not increase as a function of resource quality in any of these food deprivation experiments, as predicted by the model, because duration is determined by the quitting time of the less hungry contestant [18, 114].

Web Topic 11.2 *Resource value and ownership asymmetries in fighting strategy models*
Further details on the models discussed in this section, where contestants may differ in resource value or ownership roles. This unit also examines the evidence for resource and ownership assessment, and the implications of whether neither, one, or both contestants have access to this information.

The sequential assessment model has a great deal of flexibility for exploring alternative forms of resource assessment. In the simplest case, the value of the resource is an inherent property of the resource, such as the size of a food morsel or the reproductive status of a female, that is perceived by both contestants at the outset of the contest. Objective resource value is therefore the same for both contestants and known at the onset of the interaction [110]. Fighting ability is unknown at the onset and experienced during the contest, as described before. The higher the resource value, the lower the giving-up

line in Figure 11.3A,B (see also Web Topic 11.2) Contest duration and contest cost are both predicted to increase as *V* increases, because both parties are highly motivated to win the resource (**Figures 11.9B** and **11.10**).

In the case where both fighting ability and resource valuation vary between contestants and are not correlated with each other, i.e., are drawn from separate distributions, the ESS solution is more complex. Contestants are assumed to gather information about each other's resource value and relative fighting ability during the contest. An individual's own *V* does not change as it learns more about its opponent's *V*. Two giving-up curves are generated, where the individual

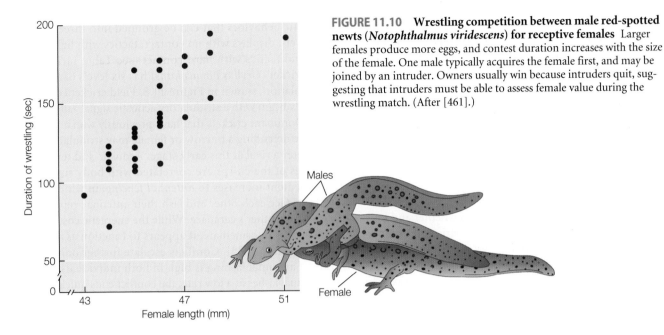

FIGURE 11.10 Wrestling competition between male red-spotted newts (*Notophthalmus viridescens*) for receptive females Larger females produce more eggs, and contest duration increases with the size of the female. One male typically acquires the female first, and may be joined by an intruder. Owners usually win because intruders quit, suggesting that intruders must be able to assess female value during the wrestling match. (After [461].)

with the higher *V* has the lower curve and therefore the longer persistence time, other things being equal. The duration and outcome of the contest depend on whether the individual with the greater fighting ability has the greater or smaller resource value [114, first model]. A slightly smaller competitor could potentially win against a larger competitor if it placed a higher value on the resource. The model predicts that contest duration will increase if either contestant's *V* increases, but it also predicts different slopes of duration versus resource value for the winner and the loser (**Figure 11.9C**). The longest contests will occur when one animal has the higher *V* and the other has the higher fighting ability. Increasing the range of population variation in resource value makes the outcome of fights more strongly determined by asymmetry in resource value, and increasing the range of population variation in fighting ability makes the outcome more strongly determined by asymmetry in fighting ability.

Another level of complexity arises when one contestant has good information about resource value but the other does not. Such **asymmetric information** is most likely to occur in ownership situations, where the owner has an accurate estimate of the value of a female, shelter, or territory resource as a result of information gathered before the contest, but the intruder can only estimate a mean value for the particular resource based on population averages. Owners in control of a high-quality resource are predicted to persist longer than owners with a low-quality resource, so contest duration will increase as a function of *V* for owners but remain relatively constant for intruders (**Figure 11.9D**). The consequence of these relationships is that intruders will tend to win primarily low-quality resources, whereas owners will retain control of high-quality resources. For a detailed discussion of when and how animals assess resource value, see Web Topic 11.2.

Contests between owners and intruders could be resolved quickly and at low cost, without any assessment of fighting ability or resource value, if the **owner/intruder role** asymmetry itself were used to make the decision to persist or quit. Maynard Smith called this concept, in which the intruder respects the owner's prior residency without considering relative fighting ability or resource valuation, the **bourgeois strategy**. The bourgeois strategy set specifies that owners escalate immediately, while intruders always retreat [303]. Since the owner/intruder role distinction is usually very clear and readily perceived by both parties, the bourgeois strategy suggests one answer to why owners nearly always win in contests against intruders. The first models, using a war of attrition framework with assessment of the role asymmetry, found some conditions for a stable bourgeois ESS, but it tended to occur only when the resource value was small [166, 299]. Modelers also discovered that there were two ESS outcomes under these conditions, a *common-sense* strategy in which the owner persists longer while the intruder retreats, and a *paradoxical* strategy in which the intruder persists longer while the owner retreats. Another serious problem with these early models is that they fail to account for the density-dependent effect of territories becoming rare and thus extremely valuable as suitable area becomes filled with bourgeois owners, which could favor a highly aggressive "desperado" strategy by nonowners [155]. Both of these issues were tackled in a recent model with feedbacks between individual behavior and population dynamics, which found that *partial respect for ownership* could evolve from an initial population of no respect using only the owner/intruder asymmetry [236] (see Web Topic 11.2).

A survey of field observations and experiments led to the conclusion that there is generally some correlation between owner/intruder roles and either resource value, fighting ability, or both [170]. Territory owners have achieved their status because they successfully won a contest to acquire the territory, so good fighters are more likely to accumulate among the population of owners. In addition, they have invested in defending the boundaries, developing a truce with the adjacent neighbors, mating, and establishing a nest with off-spring, so their valuation of the territory is likely to be high. Models that incorporate such correlated payoff effects along with the ownership role asymmetry find improved stability for the partial ownership-respecting strategy [236, 269]. Thus recent theory and empirical observations lead to the conclusion that animals should assess both fighting ability and resource value before engaging in an escalated fight, rather than rely on the payoff-irrelevant bourgeois convention.

Agonistic Signal Repertoires

So far we have addressed the question of whether and how animals *assess* fighting ability and motivation in order to make optimal decisions in conflict situations. Now we need to address the question of how animals *encode* this information. In order to resolve conflicts, animals need at least two **agonistic signals**: an aggressive threat signal, and a submissive or de-escalation signal. However, most species possess more than these two in their agonistic signal repertoire, and typically have more aggressive than submissive signals. Tinbergen [449] noted that agonistic signals comprised the largest percentage of the total signal repertoires of gulls, and proposed that different types of signals were needed for communicating in different contexts, such as with more versus less threatening opponents, at far versus close distances between opponents, and in contexts requiring different tactical approaches. Moynihan [333] emphasized the use of diversions and surprise tactics during conflicts, which require a diversity of signals and tactics. Andersson [12] proposed that threat signals derived from attack intention movements eventually become ritualized, emancipated, and invoked as bluffs, thereby becoming ineffective as threats and requiring the evolution of new threat signals. The fighting strategy models discussed above—sequential assessment, energetic war of attrition, and cumulative assessment—shed a different light on the question of why species have so many agonistic signals by focusing on cumulative costs and decision-making during contests. Signals and tactics with different costs and benefits are needed to fine-tune the investment in a contest and facilitate the resolution of a conflict with the lowest possible risk. Furthermore, the models imply that contestants may sometimes benefit from signaling information about fighting ability, resource value (motivation), and prior experience (dominance). We therefore begin this discussion of agonistic signals by describing the agonistic signal repertoires of two species—the house cricket and the little blue penguin—in which the risk and effectiveness of the different displays and tactics have been well described. Analyses of signals in both species have also attempted to assign different functions to the displays and evaluate the types of information encoded.

Two examples of agonistic signal repertoires

Male crickets fight over ownership of burrows and access to females. The house cricket (*Acheta domesticus*) has 13

agonistic behaviors that can be grouped into three intensity levels: displays with no contact, tactics with light contact, and tactics with hard contact (see Table 11.2). The field cricket (*Gryllus bimaculatus*) has six levels of agonistic escalation, shown in Figure 11.8. Field cricket conflicts typically begin with a stridulatory acoustic signal called *rival song*. Dominant crickets that have previously won a fight or that are defending a burrow or female may stridulate more and deter a rival at this early stage. Acoustic and temporal features of the chirps are correlated with body size [52]. Engagement increases to *antennal fencing*, in which contestants face each other and lash their antennae repeatedly against the other's carapace. While the energetic cost of this behavior is intermediate, it appears to function as a signal of motivation [185]. Conflicts escalate further only if the rate of antennal lashing is high in both individuals; if one individual lashes at a low rate, the conflict ends quickly with the retreat of this individual. *Shaking* (also called *juddering*), a quick rocking of the whole body forward and back several times, is a second, and more energetically costly, motivation signal that is performed more often by the eventual winner. The next stage of escalation is a unilateral *mandible flare*, which may or may not be countered by the rival. This display appears to provide information about weapon (mandible) size, an indicator of fighting ability. With intact crickets, the conflict may terminate at this stage, but with blinded crickets, which are not able to use this visual rival assessment signal to settle the contest, the contestants escalate quickly to physical fighting, and such fights are longer. Escalated fights involve *mandible locking* and *wrestling*. Fights may sometimes break off and be interspersed with *unilateral biting*, *head butting*, and *head charging*, whereupon wrestling may continue. One individual may *throw* the other, which causes the other to flee or be chased by the thrower and terminates the fighting bout. There are two defensive acts: a *cerci raise*, which often precedes the defensive *rear kick* employed by a cricket defending a female or burrow. Finally, the winner of a fight performs a *victory display* consisting of body shaking in conjunction with the stridulation call.

The little blue penguin (*Eudyptula minor*) is the smallest penguin species, but it has an extraordinary agonistic display repertoire and a high frequency of aggressive interaction (colony average of 0.34 interactions per minute). Breeding colonies form in the spring along the coasts of New Zealand and southern Australia. The birds prefer to nest in covered areas such as in caves, burrows, or under dense vegetation. Cave colonies are the largest, and pairs defend nesting hollows along the cave walls. Cave-dwelling populations have agonistic repertoires consisting of 4 vocalizations (*bark*, *yell*, *growl*, and *bray*) and 18 postural displays and tactics listed in **Table 11.3**. Several acoustic features of the high-intensity bray call that males use to advertise for females and interact with other males at a distance are correlated with body size and mass, and therefore provide an initial estimate of fighting ability. The postural displays are employed to communicate short-term changes in motivation and intentions. They can be ranked according to their risk of injury, which is based on the typical distance between opponents at which they are displayed and whether they involve movement toward or away from the opponent. An aggressive interaction consists of a sequence of 2 to 28 acts from this agonistic repertoire. There are 5 offensive displays that indicate increasing levels of aggressive motivation: *direct look, directed flipper spread, point, zig-zag approach*, and *flipper spread approach*. An opponent can back down at an early stage of interaction using two submissive displays, *face away* and *indirect look*. If the contestants have gotten close and one decides to quit at that point, it uses one of the distance-increasing displays, *low walk* or *submissive hunch*. Contests escalate with 5 different contact tactics, which incur increasing risks of injury. *Bill-lock fights* are the most intense and riskiest form of fighting, as the eyes can be pecked. Vocalizations, particularly growl, bark, and

yell, often accompany these postural displays, especially during the fighting phase. Most contests end with a clear winner, who gives the *bowed flipper spread* display as a victory signal [462–466].

The need for honesty guarantees

Aggressive signals bear the greatest need for honesty guarantees because of the high degree of conflicting interest between sender and receiver. As revealed in the two examples above, species possess repertoires of agonistic signals that provide different types of information about the sender, including aspects of fighting ability, aggressive motivation, intentions to attack and retreat, dominance, and victory. How can this information be reliably encoded, given the conflict of interest and temptation to bluff or exaggerate? Because of the reciprocal signal exchange and the generally aggressive context, receivers have at their disposal more

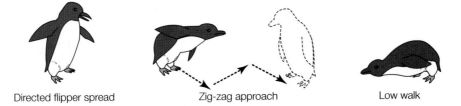

Directed flipper spread   Zig-zag approach   Low walk

TABLE 11.3 *The agonistic display repertoire of cave-dwelling populations of little blue penguin* Eudyptula minor

Category	Display or tactic	Risk level[a]	Distance[b]	Movement[c]
Defensive, distance-increasing	Low walk	1	<1	Away
	Submissive hunch	1	<1	Away
Defensive, stationary	Face away	2	<1	Stay
	Indirect look	2	<1	Stay
Offensive, stationary	Direct look	3	>3	Stay
	Directed flipper spread	3	2–3	Stay
	Point	3	1–2	Stay
	Bowed flipper spread[d]	3	1–3+	Stay
Offensive, distance-reducing	Zig-zag approach	4	2	Toward
	Flipper spread approach	4	1	Toward
Contact	Bill to back	5	0	Toward
	Breast butt	5	0	Toward
	Bill to bill	5	0	Toward
	Bill slap	5	0	Toward
	Bill lock/Twist	5	0	Toward
Overt aggression	Attack	6	0	----------
	Bite	6	0	----------
	Fight	6	0	----------

Sources: After [462, 463].

[a]Risk level is an estimate of the probability that the actor will be pecked by its opponent; displays are listed in order of increasing risk.

[b]Distance indicates the typical physical distance in meters between sender and opponent at which each display is given.

[c]Movement shows the direction the sender moves with respect to the opponent during the execution of the display.

[d]The bowed flipper spread is a victory signal by the winner.

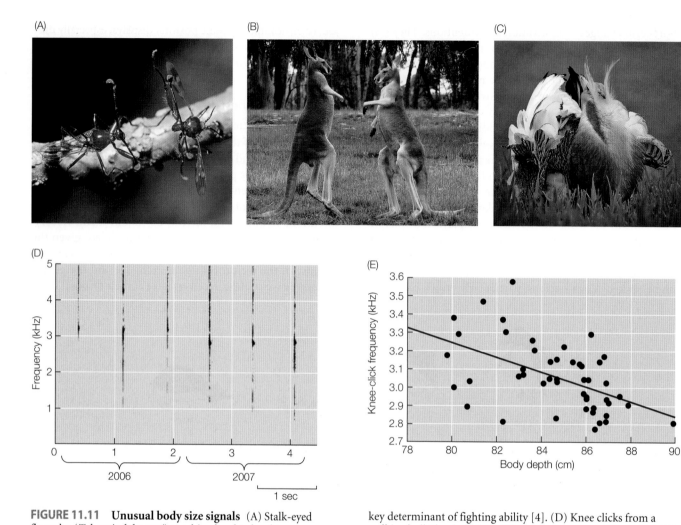

FIGURE 11.11 Unusual body size signals (A) Stalk-eyed fly males (*Teleopsis dalmanni*) perching head to head to assess each other's size based on eyespan. (B) Kangaroos (here, *Macropus rufus*) fight by standing up and boxing with forelegs, which enables assessment of relative size. (C) The length of the white whiskers prominently displayed by the lek-breeding male great bustard (*Otis tarda*) is positively correlated with body weight, a key determinant of fighting ability [4]. (D) Knee clicks from a walking bull Eland antelope (*Taurotragus oryx*; see Figure 12.24D) were recorded in two consecutive years. Note the consistency of frequency within years and the lower frequency in the second year. (E) There is a negative correlation between frequency of knee clicks and body size of bull Elands. (D,E after [51].)

mechanisms for imposing costs for dishonesty on senders than they do in other functional signal categories. Receiver retaliation, such as approaching to test or punish senders, is a powerful mechanism receivers can use to verify the reliability of conventional signals. The proximity risk of performing displays at close distances to the opponent can guarantee the honesty of true threat signals. Physiologically constrained index signals can reveal unbluffable aspects of sender attributes. Vigorous and energetically costly displays that use up energy stores can most profitably be borne by contestants in good condition to reliably indicate stamina. Signals with different forms and costs clearly provide different kinds of information, and may vary in their reliability. In the remainder of this chapter, we shall describe examples of conflict resolution signals that provide different kinds of information as a consequence of their design and

honesty-guaranteeing cost. We shall also point out examples of bluffing and deception and relate them to the sources of dishonesty outlined in Chapter 10.

Fighting Ability Signals

Sender characteristics that determine fighting ability include body size and weight, weapon size and strength, physiological condition, and stamina. **Fighting ability signals** can provide reliable information about each of these features, as described below. Fighting experience, skill, and age can also affect fighting ability. We shall defer the discussion of prior fighting experience to the section on dominance signals. While age could be a proxy for experience, age per se is a more important feature of mate attraction signals, so we shall defer the discussion of age signals to the next chapter.

Body size indicators

Body size is a very important determinant of winning an escalated fight in a great majority of animal species, particularly in those with a high degree of variability in body size. Variation is high in species with indeterminate growth such as fish, amphibians, lizards, and many invertebrates, and in species with variable larval nutritional states such as insects. Body size is sometimes important in mammals, and is less important for birds. When body size is a primary determinant of winning, this information is signaled and assessed as early as possible in an aggressive encounter so that a clearly smaller contestant can quickly concede and retreat.

If body size can be assessed visually, i.e., where there is sufficient light, visual acuity is adequate, and rivals are sufficiently close, then there may be no need for specific body size signals, and body size remains a cue. However, it is more likely that some form of ritualization converts a body size cue into an index signal. A common visual mechanism for achieving this conversion is the evolution of a display posture that most advantageously reveals body size. Typically this is a broadside display (see Figure 10.8), but standing very tall or stretching out appendages can serve the same function (**Figure 11.11**). These displays invariably become subject to the addition of apparent-size enhancing structures, such as raised fins, extended gill covers, manes, head casques, and eye stalks. While one might think that such bluffing structures would reduce the accuracy of body size assessment, selection drives the size of these ornaments to their physical or physiological limit. Larger animals are often able to produce or bear relatively larger ornaments, whereupon they become handicap signals of body size or condition [235].

Another visual mechanism is the addition of **amplifiers**—contrasting body-spanning color marks and stripes—that facilitate the assessment of body size. We discussed the principle of amplifiers in Chapters 8 and 10—markings that span, outline, or encircle relevant body parts, and make it easier for receivers to detect and assess the size or orientation of the sender's body. Amplifiers are believed to be honest by design because they are constrained by the underlying cue or signal they amplify. The designs of some animal markings, such as the alternating black and white bars on the tibia of some spiders (**Figure 11.12A**) and the horizontal

or vertical stripes on many fishes (see Figure 10.34C), have been proposed as amplifier signals of body size [332, 483]. No study has yet demonstrated that animal receivers are able to assess another's body size more accurately when amplifier marks are present, but human receivers are able to use the striped pattern on the body sides of pipefish (**Figure 11.12B**) to accurately judge the fish's body size [31]. An alternative explanation for the function of striped and barred patterns is that they enhance the **illusion** of length or size. Humans are familiar with the slimming effect of vertically striped clothing and the fattening effect of horizontal stripes. Similarly, a segmented line is perceived as longer than a solid line of the same length, the so-called Oppel-Kundt illusion (also called the filled-space illusion, **Figure 11.12C**). In humans, the illusionary effect reaches a peak at an intermediate number of

(A)

(B)

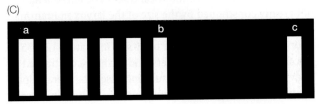

(C)

FIGURE 11.12   Size amplifiers   (A) The threat display of the Mediterranean tarantula (*Lycosa tarantula*) with raised legs bearing alternating light and dark patterns. The markings are believed to aid in assessment of tibia length, which is correlated with body size [332]. (B) The sex role-reversed pipefish *Syngnathus typhle*. Males brood the eggs in a pouch. They are smaller than females and prefer mates that are larger and more ornamented with black markings. Here, two ornamented females (foreground) compete for a male (behind). Humans can more accurately estimate fish length when the markings are present [31]. (C) The Oppel-Kundt illusion. Which segment is longer, a–b or b–c? A shape filled with a regular pattern (as in the a–b segment here) appears longer, but in fact, the two segments are equal [57].

segments [57, 282]. Rats, chickens, pigeons, and fish are also subject to this illusion [94], so we should not dismiss the possibility that optimally-configured striped and barred patterns in many animals *amplify* the perception of body dimensions in the true sense of this word.

As discussed in Chapters 2 and 3, the fundamental frequency of acoustic signals is highly constrained by body size—animals cannot produce and radiate sounds with wavelengths much larger than their body size. This body size constraint results in a negative correlation between lowest possible fundamental frequency and body size, both in inter-specific comparisons (see Figure 3.44) and in intraspecific comparisons (see Figure 10.29B). The correlation coefficient between body size and dominant call frequency ranges from −0.7 to −0.9 in many frogs and toads, so frequency is potentially a fairly reliable index signal of body size [26, 27, 59, 85, 385, 392, 469]. Vocal fold length, which determines the fundamental frequency of the call, appears to be strongly constrained by body size, which maintains the honesty of this key acoustic call component in most anuran species [129]. Exceptions are found in some low-density species where the strength of inter- and intrasexual selection for large body is weak [20, 427]. In more competitive chorusing species, rivals have been shown to assess body size via call frequency parameters and use this information in aggressive interactions [25, 85, 385]. However, males in several species have been shown to lower the fundamental frequency of their calls after being challenged by another male, in what has been interpreted as a deceptive signal [26, 59, 144, 153, 467, 469, 470]. The correlation between postchallenge call frequency and body size is only slightly less strong than the correlation with prechallenge calls in green frogs (*Rana clamitans*), and is stronger in bullfrogs (*Bufo americanus*), implying that the shift in call frequency is still constrained and relatively reliable [26, 186]. In cricket frogs (*Acris crepitans*), the magnitude of the frequency shift is associated with the strength of subsequent aggressive response, so in this case initial call frequency signals body size information while the *change* in frequency signals motivation level [467, 470].

In birds and mammals, the syringeal membranes and vocal folds, respectively, produce the fundamental frequency and any related harmonics of vocalized sounds, while the tube-like vocal tract serves as a filter, selectively amplifying certain frequencies called formants (see Chapters 2 and 3). Both anatomical structures ought to be constrained by body size, and therefore result in frequency components that reliably indicate the body size of the sender. But what is to prevent selective pressures from increasing both the mass of the vibrating source and the length of the vocal tract? Vocal fold length is not greatly constrained by the size of the larynx in mammals or the syrinx in birds, making the vibrating voice source a poor determinant of a reliable body size indicator [129, 353]. Vocal tract length, on the other hand, is more constrained by body size. As the length of the tube increases, the average spacing between successive formants increases. There is a strong

correlation between body size, vocal tract length and formant dispersion in rhesus and *Colobus* monkeys, domestic dogs, and red deer stags [128, 171, 376, 382, 437]. In the red deer, the stag retracts the mobile larynx down toward the sternum during the call to effectively lengthen the filtering tube, so it is specifically the formant dispersion at this point in the call that provides the body size information (**Figure 11.13**). In response to playback of resynthesized calls that simulate rivals of different body sizes, territorial stags are more attentive, reply with more roars, and extend their own vocal tracts farther to a larger rival stimulus [377]. Rhesus monkeys also respond differentially to calls of different formant structure [142].

Acoustic features of nonvocalized sounds can also signal body size information. We return to the house cricket example described earlier, in which males typically initiate aggressive interactions with stridulatory rival song. Unlike the patterned chirping of the male cricket's mate attraction song, rival song consists of sharper, lower-frequency single chirps that may be repeated at rapid intervals [3]. In a careful analysis that examined correlations among a variety of frequency and temporal chirp parameters as a function of body size, body condition, and winner-loser status, two acoustic parameters were consistently related to body size: the interval between the pulses within chirps, and the number of pulses per chirp—larger males produced chirps with more rapid pulses and more pulses total. Winning in crickets is affected by a combination of body size, weight, and motivation, but winners in this study did have smaller pulse intervals and also more rapid chirp rates [52]. In fish, which are indeterminate growers, body size is a critical determinant of winning. As we have shown in several prior examples, broadside displays are often performed at the start of an agonistic interaction. In croaking gouramis (*Trichopsis vittata*), grunting sounds are also produced. The dominant frequency and the amplitude of grunts are correlated with body size, as well as with winning [256]. Since the air bladder that resonates the sound is closely liked with body size, these acoustic parameters are physically constrained to provide reliable body size information. Interestingly, related fish species that do not produce sounds and rely more on visual signals during contests are generally more likely to escalate to fighting than the gouramis, suggesting that acoustic signals may provide more reliable body size information than visual signals.

Body size information can be encoded in other modalities as well. Various weakly electric fish species emit chirp-like electric organ discharges during agonistic interactions. Depending on the species, larger individuals produce higher amplitude and/or higher frequency discharges. This relationship arises because larger individuals have a larger electric organ [102, 453, 484]. Chemical signals have in some cases been linked to body size. In the rock lizard *Lacerta monticola*, the femoral gland secretions of larger individuals contain a higher fraction of cholesterol. Males respond aggressively to cholesterol presented on a cotton swab, and if one contestant in staged encounters between unfamiliar size-matched males

(A)

(B)

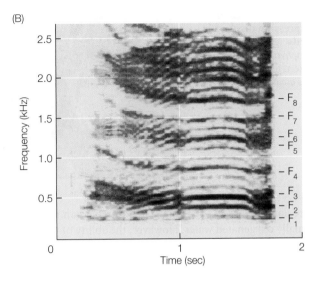

FIGURE 11.13 **Spectrogram of a red deer roar** During the approximately 2-second roar, the fundamental frequency rises to a plateau comprising most of the second half of the call. The evenly spaced frequency bands are integer multiples of the fundamental frequency and form the harmonic stack. This broadband sound is then filtered by the vocal tract, producing formants (labeled F_1–F_8) that sweep from higher to lower frequencies as the stag lengthens his vocal tract. The formants are not perfectly evenly spaced, suggesting that the vocal tract is not a perfect tube. The minimum values of the formant frequencies reached during the second half of the roar, which are determined by the length of the fully extended vocal tract, are better correlated with body size than the fundamental frequency. (After [376].)

was treated with extra cholesterol, it usually won [295]. Several other studies have shown that males distinguish among scent marks arising from senders of different body sizes and are more intimidated by scents from larger individuals [65, 255, 283]. However, such chemical signals are unlikely to be constrained by body size per se. The production of lipids in femoral gland secretions is dependent on testosterone, and testosterone is probably elevated in larger individuals because of their dominance status and prior history of winning. We shall discuss chemical threat signals later, in the section on dominance signals.

Stamina indicators

Stamina, or more generally, *performance*, is an important component of fighting ability because it determines how long a contestant can persist in an energetically costly escalated fight. Several recent studies have shown that various measures of whole-organism performance, such as sprint speed, treadmill endurance, push/pull strength, and bite or claw force, are higher for contest winners than for losers [165, 206, 260, 261, 431]. Stamina is affected by body condition (weight with the effect of linear body size statistically removed), fat and glycogen stores, the ability to mobilize energy stores, aerobic and anaerobic metabolic rates, parasite load, and immune response strength. Because these factors

are all internal, an animal should know its own performance ability at the outset of a contest, while providing minimal or no information via cues to the opponent. We would expect animals to signal information about their stamina only when such signals deter rivals from persisting in an aggressive interaction. Relative stamina information may become available once the contestants escalate the contest. There are a few clear examples of pre-escalation signals that provide stamina information relevant to aggressive interactions, and numerous others that provide general information about condition. These signals are variously classified as amplifiers, constrained index signals, or costly handicap signals.

An excellent example of a stamina signal is the lateral threat posture of many iguanid lizards. It is a broadside display in which the body is elevated with leg extension, the thorax is laterally compressed, and the throat flap is extended (**Figure 11.14**). The posture is held for extended time periods, and may be accompanied by foreleg pushups. Detailed studies of the side-blotched lizard (*Uta stansburiana*) show that the thoracic compression component is energetically costly because it interferes with respiration, and the lizard accumulates an oxygen debt and lactate buildup [40]. Only males in good condition can afford to perform the display frequently and for long durations, and this ability—as well as the probability of winning an escalated fight—is correlated with their endurance on a treadmill. This is one of the clearest examples of a handicap signal with a cost directly linked to the key quality feature—energy needed to fight. A similar argument has been made for the erection display of opercula (gill coverts) given by many fish during frontal engagement (see Figure 10.12C). Opercular display duration and frequency are significantly higher for winners than for losers. While its opercula are extended, the fish cannot pass water over its gill tissues, so oxygen uptake is restricted. In Siamese fighting fish (*Betta splendens*), oxygen consumption increases with the amount of time spent displaying, and males subjected to oxygen-reduced water produce less frequent, shorter

FIGURE 11.14 Lateral compression displays in lizards (A) Collared lizard *Crotaphytus collaris*. (B) Western fence lizard *Sceloporus occidentalis*.

opercular displays to their image in a mirror than males in normally oxygenated water. However, the ability of males to display in low-oxygen environments is not affected by their body condition per se. These fish possess an air-breathing organ that allows them to gulp air at the surface. Larger males can store more oxygen and have a higher capacity to sustain an oxygen debt, and therefore are able to display more vigorously without interruption [1, 68, 253].

Recall in our earlier discussion of contests in dragonflies that fat content affects contest duration, but the conclusion about whether such contests were based on mutual assessment or were conducted as energetic wars of attrition without opponent assessment was uncertain. The existence of a cue or signal associated with fat stores provides the potential for mutual assessment. The male rubyspot damselfly (*Hetaerina americana*) sports a brilliant red spot on the base of its wings.

(A)

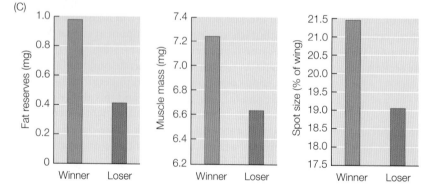

(B)

(C)

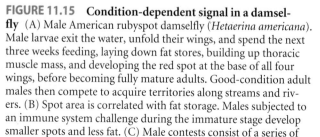

FIGURE 11.15 Condition-dependent signal in a damselfly (A) Male American rubyspot damselfly (*Hetaerina americana*). Male larvae exit the water, unfold their wings, and spend the next three weeks feeding, laying down fat stores, building up thoracic muscle mass, and developing the red spot at the base of all four wings, before becoming fully mature adults. Good-condition adult males then compete to acquire territories along streams and rivers. (B) Spot area is correlated with fat storage. Males subjected to an immune system challenge during the immature stage develop smaller spots and less fat. (C) Male contests consist of a series of ascending and descending flights of circular trajectory in which both males repeatedly face each other. They occasionally pause and perch, folding the wings as shown in the photo, allowing other males to assess their spot area. Following fights in which winners and losers could be identified, both males were captured and measured. The series of bar graphs show significant differences between winners and losers for fat reserves, muscle mass, and spot size; spot brightness, chroma, and body size did not differ between winners and losers. (After [78–80, 82, 372].)

(A)

(B)

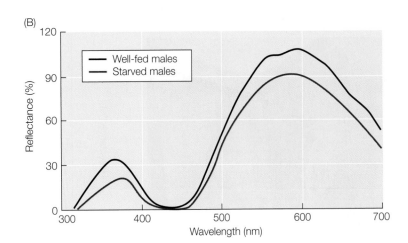

(C)

FIGURE 11.16 **Amplifier signals of condition in spiders** (A) Ventral view of *Lycosa tarantula* showing the abdomen patch. The length measurement (APL) changes more than the width measurement (APW) as the abdomen distends following feeding, and the overall area of the patch increases greatly. This is the same species shown in Figure 11.12A. (B) Reflectance from the dorsal abdomen of the jumping spider *Cosmophasis umbratica* for well-fed and starved males. Both the UV and yellow reflectance peaks are increased for better-fed individuals. (C) A front dorsal view of a male *C. umbratica* in the aggressive threat posture, with forelegs in the hunched position and abdomen raised to display the bright UV stripe to the rival. (B after [274].)

The size of the spot, but not its hue, chroma, or brightness, is correlated with the male's fat reserves and thoracic muscle mass at the time of his sexual maturation. Only males with sufficient fat reserves and large spots are able to gain territories and mate successfully. The winners of naturally occurring territorial contests have larger spots than losers but are not necessarily larger in other body dimensions (**Figure 11.15**). Experimental enlargement of spot area increases the probability of territory acquisition, demonstrating that competing males attend to the color patch as an assessment signal. However, as males age, their condition deteriorates but their patch area remains unchanged, so this signal does not provide current information about fighting ability for older territorial males [78–80, 159]. Studies of other dragonfly species provide mixed evidence for the signaling of body fat reserves with color ornaments [81, 131, 394].

Body condition greatly affects the outcome of wrestling fights in spiders, as we learned earlier in jumping spider *Phidippus clarus* contests. Males of many species signal their condition early in aggressive encounters with an abdomen display that is sometimes amplified with a contrasting color patch. One such condition-dependent amplifier signal was illustrated in Figure 10.31A for the jumping spider *Plexippus paykulli*, in which a V-shaped color patch emphasizes the

roundness of the abdomen [438]. A second example is shown in **Figure 11.16A** for the Mediterranean tarantula *Lycosa tarantula*. The abdomen-spanning black stripe is positioned at the point of maximal distension for a fully-fed individual. The length dimension increases more than the width measurement, so the patch becomes thicker after feeding, and the entire patch area is greatly increased [332]. A third example has been described for the ornate jumping spider *Cosmophasis umbratica* [273, 274]. Here, it is the intensity of UV reflectance that varies with condition. Reflectance is higher for recently fed spiders (**Figure 11.16B,C**). Abdomen stretching alters the UV spectral characteristics by changing the pattern of packing and overlapping of the iridescent cuticular scales.

Condition-dependence of plumage color in birds has been studied extensively. The current view is that carotenoid pigment colors and structural colors are most strongly affected by an animal's condition, while melanin pigment color patches are less condition-dependent and more affected by hormones and social interactions [93, 160, 177, 178, 213, 306]. Many of these color signal studies involve signals used in mate attraction, where male condition is often very important, and we shall discuss those examples in the next chapter. Melanin signals are commonly used to mediate male–male interactions and are regulated by aggressive interactions, dominance status, and testosterone; we shall review some of those studies in the section below on dominance signals. A field study of the eastern bluebird (*Sialia sialis*) illustrates the use of condition-dependent color as an indicator of fighting ability. Males have brilliant ultraviolet/blue plumage on the

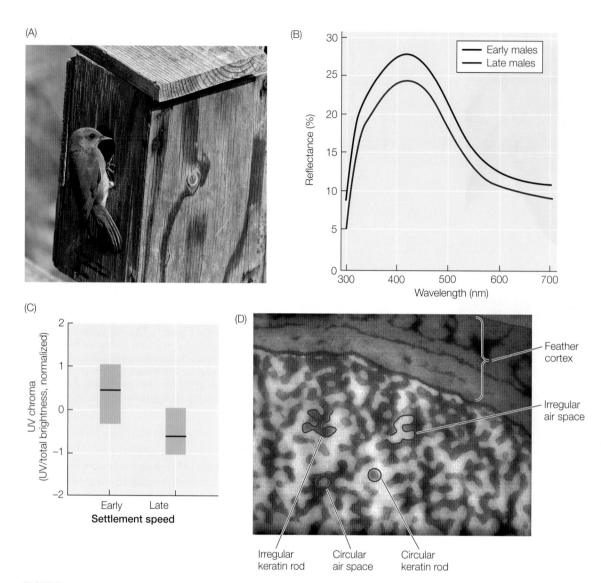

(A)

(B)

(C)

(D)

Feather cortex

Irregular air space

Irregular keratin rod

Circular air space

Circular keratin rod

FIGURE 11.17 **Condition-dependent indicator of competitive ability in bluebirds** (A) Male eastern bluebird (*Sialia sialis*) at a nest box. (B) Reflectance spectrum of blue plumage for the males who acquired the limited early nest boxes versus males who acquired later boxes. (C) UV plumage chroma for early versus late nest box males. (D) Transmission electron microscope image of a blue feather barb. (B,C after [409]; D after [408].)

head, back, wings, and tail (**Figure 11.17**). Brightness, hue, and chroma are significantly correlated with body condition, fat score, age, and reproductive success. The birds breed in cavities but cannot excavate their own holes, so they compete strongly for access to a limited supply of natural cavities or nest boxes. In an experiment to assess male competitive ability, a limited number of nest boxes was put out in early spring, and after birds had established residency in those nestboxes, more nest boxes were put out to attract the next round of competitors. The early group of nest box owners had significantly greater UV brightness, hue, condition, and fat scores than the later group of owners; this effect was not caused by a predominance of older males in the early group.

The UV/blue hue is produced by a semiordered matrix of keratin rods and air vacuoles in the feather barbs. An investigation of nanostructure variation showed that the number of circular rods determined the feather's UV/blue chroma, while the uniformity of circular rod diameter determined spectral saturation (see Figure 11.17D). The study concluded that plumage color reflects male condition and affects competitiveness in male–male interactions [408–410]

Acoustic signals can also indicate the sender's condition. In common loons (*Gavia immer*), pairs defend large territories encompassing all or part of fresh water lakes in North America. Males give territorial yodel calls in response to intruders. The dominant frequency of the call is negatively correlated with body mass and condition, but not with linear body size measures. Moreover, changes in male body mass and condition between years are correlated with changes in dominant frequency (reduction in condition associated with an increase in dominant frequency). An acoustic playback experiment to territory owners found that male receivers vocalized sooner and more often in response to lower-frequency yodels. Researchers

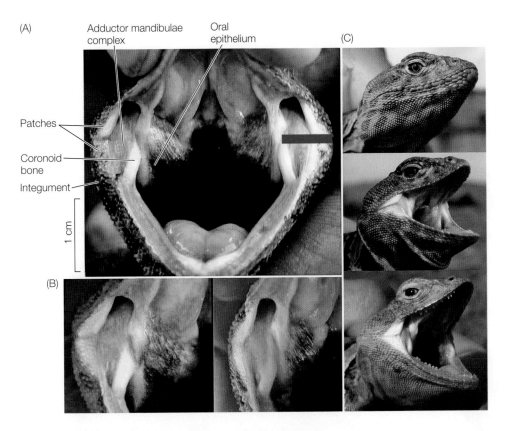

(A)

Adductor mandibulae
complex

Oral
epithelium

(C)

Patches

Coronoid
bone

Integument

1 cm

(B)

FIGURE 11.18 **Amplified gape display in the collared lizard**
Crotaphytus collaris (A) Frontal view of a gaping adult male.
The muscles of the adductor mandibulae complex (AMC) are
delineated medially by the coronoid bone, on which components
of the jaw-adductor muscles insert, and laterally by the mouth-
corner patches. The coronoid bones contrast with the melanic
oral epithelium, and the patches contrast with the adjacent integument. The horizontal red bar indicates width of the AMC [268].
(B) Comparison of the AMC and associated patch for two males
of similar body size. The male on the left has a relatively high bite
force while the male on the right has a relatively low bite force.
The broader muscle complex of the stronger biter is amplified by
its prominent mouth-corner patch. (C) The mouth-corner patches of collared lizards each consist of a folded pocket of skin with
reduced scales that unfolds as the mouth is opened.

suggested that the mass of the syringeal membranes must be
affected more by body condition than by linear size to produce
such a result [290]. Acoustic features of the loud, two-syllable
"wa-hoo" calls of male baboons have also been found to vary
as a function of male rank and competitive ability. Compared
to low-ranking males, high-ranking males call more often,
at faster rates, and for longer bouts, and also produce calls
with higher fundamental frequencies and longer "hoo" syllable durations. As males fall in rank and age, hoo syllables
shorten, fundamental frequencies decrease, and formant dispersion decreases. Because wa-hoos are often given while males
are running or leaping through trees, these subtle variations
in acoustic parameters may indicate a male's stamina [126,
127, 233]. Similar evidence for condition-dependent acoustic
parameters has been found in other bird and mammal species
[46, 120, 137, 140, 169, 238, 239, 296, 443, 458], but was not
found in the roars of lions [359].

Weapons

Weapons affect the strength of a contestant and its ability to
inflict injury on the opponent. In escalated fights with conspecifics, most animal species make use of either antipredator defensive tactics, such as a hindleg kick, or feeding structures, such as teeth, mandibles, bills, and claws. These are the
primary weapons for major groups like fishes, reptiles, and
birds. Because these structures are designed and optimized
for feeding, their size and effectiveness during a fight will
be strongly linked to body size. If the weapon is normally
not visible, it may be revealed in pre-escalation threat displays. For example, the canine teeth of mammals may be
displayed by pulling back the skin around the mouth, and
the mandibles of insects may be flared, as we saw in the
cricket example. Similarly, fish and lizards may perform a
mouth-gaping display. These displays reveal jaw size and are
thus considered index signals. In lizards, bite force is a major
determinant of fighting ability, and this feature is strongly correlated with head size [207, 261]. A few species have
amplified the signals so that true bite force can be assessed.
Male collared lizards (*Crotaphytus collaris*), for example, give
a gaping display that exposes the major jaw-adductor muscle
complex, and they have added white patches at the corners
of the mouth that amplify the signal (**Figure 11.18**) [268].
In a similar vein, we previously discussed fighting strategies
in the Cape dwarf chameleon, where casque height was one
of the key correlates of winning. The width and height of the
head determines bite force in lizards via the constraint of
jaw musculature support [176]. Although the casque is ornamental, its size is still constrained by head size and provides
a reliable index of fighting ability during the pre-escalation
broadside display [424, 425].

(A)

(B)

(C)

(D)

FIGURE 11.19 Some examples of highly elaborated weapons (A) European earwig *Forficula auricularia* with forceps on the back end of the abdomen. Larger males develop larger forceps; they are employed in both male competition and courtship [374, 452, 471]. (B) In narwhal (*Monodon monoceros*) males, the left tooth erupts at the end of the first year and develops into a spiraled tusk. It is used in display and fighting, and there is a high incidence of broken tusks and scars [411]. (C) A male dung beetle (*Onthophagus taurus*) uses its shield and horns to aggressively defend the entrance to its burrow containing a female and offspring. Only males above a threshold body size develop the horns; small horn-less males employ an alternative, sneaking strategy to obtain matings [318, 319]. (D) White rhinoceros, *Ceratotherium simum*, showing prominent horns made of keratin; males use their horns during conflicts and threaten with a head-low posture. The two individuals here are not threatening seriously, as their ears are not pulled back [117].

Some animal taxa have evolved specialized weapons for fighting conspecifics. Notable examples include the spines and chelipeds of crustaceans; spines, pincers, mandibles, and horns in some arachnids and insects; spines and tusks in a few frogs, and the horns and antlers of ungulate mammals (**Figure 11.19**) [109]. These weapons typically develop only in males, and particularly in species where males must defend either resources or females to achieve reproductive success. Many examples of large weapons occur in species that defend burrows, such as fiddler crabs, lobsters, shrimp, scarab beetles, and tusked wasps and frogs. The entrances

of burrows appear to be highly amenable to being guarded with frontally placed weapons. Similarly, some food resources and egg-laying sites are clustered or confined in ways that permit them to be guarded, as observed in many beetles. In other cases, males guard females directly. The most elaborate horns and antlers of ungulates are found in open-habitat large-bodied species that live in large herds, where males have the opportunity to monopolize many females [50]. For a few clades of weapon bearers, it has been possible to reconstruct the historical sequence of weapon evolution. The ancestral condition is generally the possession of small but very dangerous weapons, such as short sharp horns, fangs, or tusks. More derived species tend to have much larger and more complex weapons designed to facilitate ritualized fighting. In parallel with this trajectory, weapons shift from being serious armaments capable of killing an opponent to instruments of strength assessment and testing [109]. The diversity of these derived weapons is spectacular and generally related to the fighting strategy employed by a species, as illustrated in **Table 11.4** for bovid ungulates.

TABLE 11.4 *Fighting tactics, horn shapes, and associated threat displays for representative bovid species*

Representative species		Fighting tactic	Threat display
Common duiker	Klipspringer	**STAB:** Deliver short thrusts to flank, neck, or belly Horns short, straight, and smooth	Horning vegetation; low horn presentation
Oryx	Sable	**FENCE:** From standing or reared position, strike front of horns with oblique blow; over the shoulder stab from parallel stance Horns with long reach	High horn presentation; symbolic stabbing displays from a broadside stance (shown here for the oryx)
Kudu	Grant's gazelle	**WRESTLE:** From head-down, front-facing position, lock horns and push or twist opponent Horns twisted with relatively long catching area	Medial horn presentation with erect postures and head-flagging displays
African buffalo	Mountain sheep	**RAM:** Run toward each other with lowered head and hit with hard blows Horns very thick at base, recurved, downward facing tips	Low horn presentation
Hartebeest	Wildebeest	**KNEEL:** Drop down on front knees, deliver jabs or hook opponent's horns to throw off balance Horns medium length, hooked, points angled toward posterior	Kneeling posture, often preceded by mutual head-turning away before rushing approach toward each other; ground pawing and horning the ground

Sources: After [67, 139, 275, 286].

Enlarged weapons serve a *tactical function* during fighting by enabling contestants to hold, ram, block, grasp, or flip the opponent. A larger weapon obviously improves an animal's ability to defeat its opponent, up to the point where it becomes cumbersome and unwieldy. Where it has been examined, the contestant with the larger weapon usually wins, in some cases even after controlling for body size [24, 108, 184, 224, 319, 364, 418]. How this advantage is achieved is not always clear. A particularly insightful study of the dung beetle *Euoniticellus intermedius* illustrates the link between horn size and fighting ability [259]. Horn length and body size both determine contest outcome, but horn length is a more important factor in contests between two large males [364]. Horn length is positively correlated with performance in two types of laboratory tests: the force required to pull the beetle out of a tunnel, and the distance the beetle was able to run on a circular runway before exhaustion. The effects persist after controlling for body size. Horn length is also correlated with immune response strength [363]. The work on this species has led to the conclusion that horn length and physiological capacity are dependent on the same energetic resource pool during development. Weapon size thus also

(A)

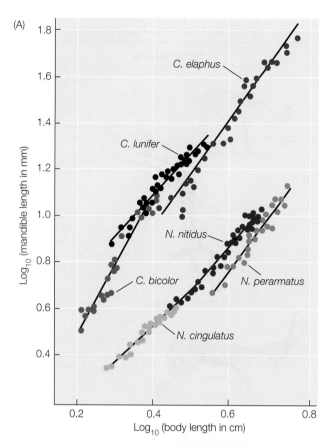

(B)

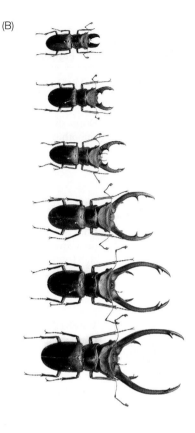

FIGURE 11.20 **Allometric relationship between weapon size and body size** (A) Log-log plots of male body length and mandible length in three representative species each of the stag beetle genera *Neolucanus* and *Cyclommatus* (Lucanidae). (B) Examples of the range of male body sizes and horn sizes in *Cyclommatus elaphus*. (After [223].)

serves as a reliable *index signal* of strength and fighting ability. The variation in weapon size among individuals is often greater than variation in other morphological traits, because weapon development is extremely sensitive to nutritional condition, parasite load, and body size [121, 132, 250, 251, 254, 416, 452]. Males in better condition are strongly selected to invest proportionally more in their weapons. This proportional investment can be quantified with **allometric analysis**, which measures the size or growth of a part of the body relative to the whole body using a log-log regression to obtain the exponent, or slope of the line (see Web Topic 11.3). A survey of the allometric relationship between weapon size and body size in stag beetles, earwigs, and fiddler crabs shows that the exponent is almost always positive and usually in the range between 1.5 and 2.5 [235]. Larger males can afford to allocate more resources to weapon development and benefit more from this investment than small males [37, 364]. This means that weapon size is not only a highly reliable *handicap signal* of male body size and condition, but that differences between competitors may be amplified in some cases to make relative assessment easier (**Figure 11.20**).

Theoretical models suggest that it is not worthwhile for small males to exaggerate or bluff their body size with a dishonestly large weapon (see Web Topic 11.3) [37, 156, 235, 337, 485]. Despite the compelling evidence summarized above for physiological linkages that maintain signal honesty, there are a few cases in which weapon size has become decoupled from fighting ability and provides unreliable information. In fiddler crabs, the size of the large claw in males is normally strongly correlated with body size and fighting ability, but males who have lost their claw regenerate a new one that is deceptively weak. The regenerated claw reaches the same size as the original claw and apparently is perceived by the same rules, but contains less muscle and has a lower closing force and lower pull-resisting force (**Figure 11.21**). A burrow-defending resident male with a regenerated claw suffers a competitive disadvantage in escalated contests against challengers. However, claw type does not affect the ability of nonresidents to win fights. This discrepancy is explained by the fact that nonresidents selectively challenge smaller residents, whereas defenders are forced to fight any male that challenges them [21, 63, 263]. This situation is similar to the mantis shrimp example described in Figure 10.38, where there are two types of senders with different coding rules that cannot be distinguished by receivers. Examples of unreliable weapon signals have also been described in other crustacean species [195, 479]. Cheaters may benefit by producing a more intimidating threat signal, but they pay the cost of reduced fighting ability when pressed and reduced escape speed [480].

(A)

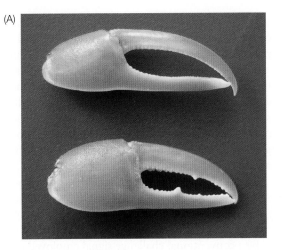

(B)

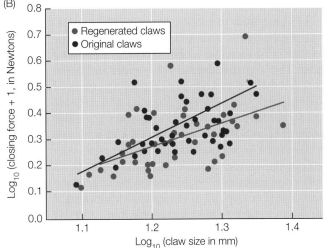

FIGURE 11.21 Deceptive claws in fiddler crabs (A) A regenerated claw (top) and an original claw (bottom) for two male *Uca mjoebergi* of the same body size. The regenerated claw is 20% lighter and more slender, and also has reduced interior tubercles. (B) Regenerated claws have a lower closing force than original claws. Also note the greater variance in force for large claw sizes compared to small claw sizes, as hypothesized for some types of index signals in Figure 10.28B. (After [262].)

Web Topic 11.3 *Positive allometry of weapons and ornaments*

A summary of several theoretical models developed to understand why the size of sexually selected weapons and ornaments is often proportionally larger in large individuals compared to small individuals.

Aggressive Motivation Signals

An animal's decision to persist, escalate, or back down in an agonistic encounter over a resource item depends on three motivating factors: (1) internal or physiological factors, such as hunger or reproductive stage, that influence need for a resource; (2) the quality and lifetime importance of the resource item; and (3) the relative fighting ability and motivation of the opponent. Information about the need factor will be obtained from self-assessment, and information about the resource will mainly be obtained from cues; these two factors together determine the benefit of gaining the resource. Information about the opponent's fighting ability and motivation will be obtained from some combination of cues and signals; this factor determines the potential cost of a contest and the likelihood of winning. Senders are selected to transmit the strongest possible evidence of their fighting ability to encourage the receiver to back down and quit. But should senders transmit accurate information about their motivation, resource valuation, and intentions? It was once thought that animals should suppress such information, so that the receiver is clueless about what the sender is likely to do next

and thus unprepared [87, 249, 300]. In reality, animals often *do* signal their motivation and intentions [173]. Both the penguin and the cricket agonistic repertoires described earlier contain aggressive motivation signals: the cricket uses antenna lashing, sometimes in conjunction with stridulation, as well as a mandible flare threat, and the penguin has a series of discrete offensive threat displays given at various distances from the opponent. But whether or not it is advantageous to evolve aggressive motivation signals might depend on the circumstances. On the one hand, if both contestants honestly were to signal their need and resource valuation, they could negotiate and assess which one of them is more motivated without having to escalate to a fight. Honesty guarantees would be required to reduce the temptation to exaggerate one's motivation. On the other hand, if an owner knows that its resource is of high value and it reveals this information to an otherwise unknowledgeable opponent, the opponent might be inspired to fight harder in order to take over the resource. In this case, one should signal ownership but no information about value.

It is useful to divide **aggressive motivation signals** into three subcategories: (1) **challenge signals**, which are usually performed at some distance from the opponent and represent directed low-level threats; (2) **general aggressive motivation signals**, which reflect need for the resource; and (3) **offensive threat signals**, which are performed close to the opponent and are often followed by an attack. The boundaries between these subcategories are not firm, but they correspond to signals with different designs and honesty-guaranteeing mechanisms. Note that motivation is a rapidly changing state rather than a quality that can be handicapped, so the differential cost quality handicap mechanism cannot serve as an honesty-guaranteeing mechanism for signals conveying this type of information about the sender [172].

Challenge signals

The first stage in an agonistic encounter is typically a **challenge signal** directed at a targeted rival individual from some distance. Because of the distance between sender and receiver, the signal needs to contain clear directional information that singles out the rival and grabs its attention. The signal's design must therefore evolve to contain a pointing component. The large distance between sender and target also means that threat signal honesty cannot be guaranteed by a proximity risk (see Figure 10.30, zone of unreliable signaling), so challenge signals need to possess some other type of stabilizing cost. If the pointing signal is to serve as a threat, senders must be prepared to back up their signal with a more escalated response *if need be*, i.e., if the target stands its ground or also escalates [179]. Moreover, the target individual must signal or respond in some way, withdrawing if weaker or not motivated to fight and approaching if strong and highly motivated. This type of signal would best be classified as a *conventional* challenge signal indicating aggressive intentions. These low-level threats are unlikely to lead to immediate escalation by the sender, but operate as a probe to test the aggressive responsiveness of the target. The cost of bluffing is the possibility of a retaliatory approach by the target, which could eventually escalate into an attack. Where low-intensity conventional threat signals exist in a species' repertoire, they are accompanied by a series of stronger escalating threat signals and weaker de-escalating signals.

A good example of this type of signal is the *direct look* stationary threat signal of the little blue penguin—the most commonly performed agonistic display in the penguin's repertoire. (see Figure 11.12). The sender stands tall, facing the rival and pointing its bill, and may enhance the visual display with a growl vocalization. The signal is likely to elicit a more aggressive response by the receiver, such as *directed flipper spread* or *point*, whereupon the sender is very likely to de-escalate with *face away* or *indirect look*, unless it truly intends to persist and escalate the interchange [464]. These signal and response patterns fit the conventional signaling model well. Among animals that interact frequently, maintaining a reputation for signaling honestly may also serve to preserve signal accuracy by senders [466].

Song-type and frequency matching, synchronized switching, and overlapping in songbirds are examples of acoustic challenge signals. Unlike visual signals, acoustic signals are relatively omnidirectional and can be beamed toward a narrow target only with specialized adaptations [84]. However, temporally synchronizing one's songs or calls with respect to another individual's vocalizations provides an unambiguous way to direct an acoustic signal to a target. **Song-type matching** occurs when a bird replies to a rival song with the same song type from its repertoire; **frequency matching** occurs in species such as chickadees that can shift the frequency of their song up or down; and **synchronized switching** occurs in bout singers that repeat a song type several times before switching to another one. Bird song researchers have long recognized that these strategies enable a bird to get the attention of a target rival and initiate a countersinging interaction [44]. Numerous studies have demonstrated that matching, switching, and overlapping often predict the sender's

subsequent approach toward the rival, so these conventional signals do function as aggressive challenges with some level of reliability [8, 60, 77, 89, 130, 247, 398, 459, 460]. Documenting a receiver retaliation response to such acoustic signals is more difficult, because one must employ interactive playback techniques. Among the small number of interactive studies that investigated song matching, a stronger receiver approach response to the matching treatment compared to the non-matching treatment was found only when neighbor songs were broadcast from the appropriate territory boundary [29, 60, 325, 336, 459]. Experiments conducted with stranger songs from the center of the territory resulted in equal and strong aggressive responses to both treatments, suggesting a ceiling effect [10, 308, 312, 348]. Song matching therefore appears to be a conventional challenge signal employed primarily in a territory defense context by neighbors who know each other's song type repertoires, a topic we shall take up later and in Web Topic 11.4.

General aggressive motivation signals

Once a target opponent has been notified, senders may next indicate their willingness to persist by demonstrating their ability to expend energy. Signals of need are stabilized by at least some expenditure of energy coupled with a high potential gain from winning, as proposed by the differential-benefits handicap model.

It has been noted in many studies that the level of aggressive motivation can be inferred from the **display rate** of a competitor [17, 47, 53, 68, 90, 107, 162, 205, 258, 265, 278, 456]. In field cricket agonistic interactions, males first approach each other, make antennal contact, and then begin to lash each other's carapaces with repeated whips of their antennae. When the frequency of antenna movement is high and similar for both males, the interaction escalates to physical fighting, but when the frequency differs greatly for the two males, the contest is settled immediately. (If the antennae of both males are removed, the males commence courtship behavior with each other.) Body weight asymmetry between contestants plays some role in determining the winner in crickets, but motivational asymmetry (ownership of a female or burrow) is also very important. Antennal fencing therefore appears to be used to assess relative motivation early in the interaction [185]. A second example involving song rate in song sparrows is illustrated in **Figure 11.22**. In the few minutes before a physical fight or chase, the challenger male that is attempting to expand his territory boundary elevates his song rate to maximal levels; the defender increases his song rate to half that of the challenger. Song rate does not predict who wins the encounter, but it identifies the male that is in the process of attempting to increase his territory area [39]. Weakly electric fish also employ increases in pulse or wave frequency to indicate aggressive motivation. In the brown ghost knifefish *Apteronotus leptorhynchus*, the frequency of both rapid bursts (chirps) and gradual frequency rises (GFRs) increase during aggressive encounters, but more so for the eventual winner, and chirps in particular precede bouts of physical contact [454].

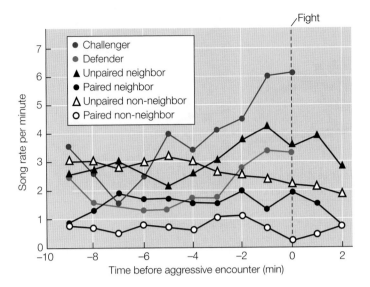

FIGURE 11.22 Song rates of territorial male song sparrows in the minutes before a fight A microphone array was used to simultaneously record an entire neighborhood of territorial males and locate their positions in a field. Challengers (red) were new males attempting to occupy an interstitial space between established males and expand the territorial boundaries; defenders (blue) were the particular males that they fought at the point of the dashed line. Challengers greatly increased their song rate; defenders slightly increased their rate, but not as much as challengers. Note that unpaired males had higher song rates than paired males. (After [39].)

In these and other examples, a higher display rate is associated with higher need or motivation and is routinely interpreted as a signal indicating a higher arousal state. Regardless of the modality, mutually signaling rivals can easily compare each other's display rate. Singing and antenna lashing are not energetically costly to perform, so it does not seem likely that the elevation of display rate during a short interaction would constitute a significant energetic constraint, as proposed by the quality handicap model. But there certainly is some generally increasing cost of displaying at a higher rate. Like begging in nestling birds, this type of signal appears to fit the criteria of the differential benefits handicap model.

If defended resources such as territories, nest sites, or females vary greatly in quality, and owners vary their display rate to reflect this quality, then intruders may use this information to decide how hard they should fight to acquire the resource. In the web-defending *Agelenopsis* spider, an invading loser's giving up time is positively correlated with web quality if the owner is defending its own web, but this correlation is lost if the defender has been experimentally relocated to a new and unfamiliar web. The number and diversity of signals increases in contests over higher-quality webs and both defender and intruder increase their contest costs, implying that intruders obtain web quality information from the defender [379–381]. Several other studies have attempted to get at this issue by manipulating the value of the resource from the owner's perspective in a way that is not directly perceived by an intruder. In a study of the bowl and doily spider (*Frontinella pyramitela*), the male defending a female knew if the female was of high value (a virgin) or of low value (because he had just finished copulating with her) when an intruder was introduced. In contests lost by the intruder, contest duration was far shorter when the female was of low value than when she was of high value; intruders may have thus gained some information about female quality from the defending male's behavior [270]. Sand goby (*Pomatoschistus minutus*) males excavate nests under stones and mussel shells lying on a sandy substrate. Potential nest sites were manipulated to appear larger or smaller than the actual excavatable area. Takeover patterns suggested that owners provided cues to invaders about true nest quality [278, 279]. In this deceitful experimental setup, males may have paid a cost by honestly conveying their motivation, but under natural conditions honesty is likely to be more beneficial in terms of attracting females and repelling at least some male rivals.

Offensive threat signals

Offensive threats are intention-to-attack signals designed to deter an opponent by threat of force. To qualify as a true threat, the signal must be associated with a significant probability of subsequent attack if the opponent does not retreat. The signal should be addressed to one well-defined receiver, and thus contain some clear directional information indicating the target opponent. For this reason, threats are usually discrete visual displays or postures. In order for the signal to be credible to the receiver, the display must be performed within close proximity of the receiver-opponent, so that signal reliability is guaranteed by a proximity risk. In other words, senders must demonstrate their honest aggressive intentions by being willing to accept the cost of a possible retaliatory attack by the opponent. Finally, receivers must do something in response, such as counterattack or retreat; a nonresponse is not acceptable, and should lead to the follow-through attack by the threatening sender [433].

Offensive threat displays are usually preattack postures or movements, so their form depends on the fighting tactic employed by a given species. We have already described several such intention-to-attack signals: a teeth-bared ears-back facial expression in carnivores, primates, and many other mammals (see Figures 10.5A,D and 10.9); a gape display in lizards and fish (see Figure 11.18); bill-forward postures in birds (see Figures 8.22 and 10.5B,C; see Table 11.5); the mandible flare in insects (see Figure 11.8); and cheliped spread or waving in crustaceans (see Figures 10.31B and 10.38A). In birds that fight with their wings, such as gulls and penguins, raising the carpal joint is usually part of the threat posture

LHF

HHF

WF

TABLE 11.5 *Risk and effectiveness of aggressive displays in American goldfinch observed during interactions at winter feeders*

Sender display[a]	Sender's next act (%)						Receiver response (%)				
	LHF	HHF	WF	Attack	Retreat	Win	LHF	HHF	WF	Attack	Retreat
LHF	0	12.1	67.2	5.2	15.5	0	15.5	24.1	60.3	0	0
HHF	0	2.5	62.0	13.9	10.1	11.4	3.8	7.6	74.7	2.5	11.4
WF	0	0	11.1	19.7	40.3	28.9	0	0.2	48.8	31.1	19.9

Increasing effectiveness

Increasing risk

Sources: After [366].

[a]LHF is the least intense threat, HHF is an intermediate-intensity threat, and WF, with bill opened and pointed toward opponent and wings extended ready to lunge, is the most intense threat.

(see Figure 10.5C and Table 11.5). Table 11.4 illustrates how the different attack intention displays of ungulate species are related to their fighting tactics: medial horn presentation in wrestlers, rearing up in rammers, kneeling with horns forward in species that fight on kneeled forelegs, and lateral horn or head waving in fencers and stabbers.

Many species have a repertoire of several discrete threat displays that can be arrayed along a gradient of increasing effectiveness in deterring an opponent. More effective displays are characterized by: (1) a close resemblance to an attack-ready posture; (2) a performance close to the opponent, demonstrating greater risk-taking by the sender; and (3) follow-through, with the sender attacking. A sender indicates its motivation by its choice of display [111]. A well-known difficulty in accurately quantifying threat effectiveness by measuring the probability of subsequent attack is the **successful-threat problem**: if a sender's threat succeeds in causing the receiver to leave, no subsequent attack will occur [179, 200, 417, 423]. To solve this problem, one uses a *trifactorial analysis of contingency*, which takes into account the receiver's response to the display, using analyses of the three-step sequence of the sender's display, the receiver's response, and the sender's subsequent act. This method was successfully employed in the study of little blue penguins to show that less intense threats often caused opponents to escalate, whereas more intense threats caused them to de-escalate [464]. During agonistic interactions, the penguins begin with the less intense threats and escalate in steps until the opponent backs down, in order to settle the conflict at the lowest possible cost. The five offensive threat displays, along with

their typical performance distance, probability of subsequent sender attack, and probability of opponent de-escalation, are listed in Table 11.3. The smaller the distance, the more likely the sender is to attack rather than back down, and the more likely the receiver is to de-escalate, a general pattern for serial threat displays in many species [200].

Table 11.5 demonstrates the trade-off of signal effectiveness versus risk of counterattack to the sender for contests by American goldfinches (*Carduelis tristis*) over small food items at winter feeders [366]. The most effective threats also bear the greatest risks. Numerous other studies on a wide variety of animal species, including other songbirds, seabirds, lizards, and crustaceans, have found similar effects [2, 11, 42, 43, 90, 112, 167, 201, 334, 400, 464, 466]. However, two studies on the great tit *Parus major* did not find support for the positive effectiveness versus risk relationship among displays, possibly because dominance hierarchies within flocks play an important role in this species [265, 477].

Acoustic and olfactory signals would generally not be expected to serve as intention-to-attack signals, because they do not provide good directional information. However, acoustic and olfactory elements often accompany visual signals, and may enhance the effectiveness of the postural display and provide individual signature information [43, 464, 466]. But one type of acoustic signal might qualify as an offensive threat display: **low-amplitude growls** in some mammals and **soft song** in some songbirds. Aggressive growls in food-defending canids elicit immediate retreat by nearby receivers [122]. Experimental studies on both song sparrows and swamp sparrows show that the signal that most

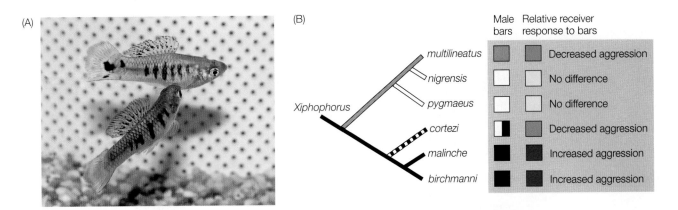

(A)

(B)

Male bars	Relative receiver response to bars	
multilineatus		Decreased aggression
nigrensis		No difference
pygmaeus		No difference
cortezi		Decreased aggression
malinche		Increased aggression
birchmanni		Increased aggression

Xiphophorus

FIGURE 11.23 **Evolutionary reversal of responses to a conventional attack intention signal** (A) *Xiphophorus cortezi* males expressing vertical bars shortly before an aggressive biting attack. (B) A phylogenetic tree of the northern swordtail clade (genus *Xiphophorus*), showing the presence of bar expression and the relative receiver aggressive responses to mirror images of themselves with versus without bars (removed by freeze-branding). Swordtail ancestors evolved the ability to rapidly express black bars, but in a different context. The common ancestor of the northern clade evolved bar expression as an honest signal of attack intentions (black). *X. malinche* and *birchmanni* respond more aggressively to bars (red), as expected for a conventional signal. In *X. multilineatus* (gray), the bars are absent in small males but develop once males grow larger, so the signal has shifted to one that honestly indicates body size; receivers respond less aggressively to barred compared to barless images (blue). *X. nigrensis* and *pygmaeus* have completely lost the bars (white), and receivers respond similarly to barred and barless images (yellow). *X. cortezi* is polymorphic (hatched); about half the adult male population lacks bars and these individuals are more persistent and usually win fights against barred males. Receiver response is stronger to the barless image (blue). This species could be in the process of losing its bars because of an invasion by a cheating "Trojan sparrow" strategy (defined in the section on status signals) [349]. (After [330].)

reliably predicts whether a territorial male will attack a taxodermic mount is the number of soft songs delivered; other acoustic signals such as normal (loud) song rate, song type switching rate, matching rate, or other calls, were not associated with attack, although both species also perform a wing-waving display associated with aggressive behavior [9, 22]. Soft song requires close proximity to the receiver in order to be heard. While the researchers argued against a proximity risk cost for this signal, because it might be confused with regular songs sung from a greater distance, in fact the structure of soft songs is sufficiently different from regular songs to be clearly differentiated. Wing-waving would confirm the location of the sender.

Some offensive threat signals are believed to be conventional signals stabilized by receiver retaliation. One example is vertical bar darkening in northern swordtail fishes [329, 330]. In two species, *Xiphophorus birchmanni* and *X. malinche*, rapid bar darkening occurs reliably and immediately before the fish bite at their image in a mirror, and the aggressive response to an image with bars is stronger than the response to an image without bars. However, in two other species *X. cortezi* and *X. multilineatus*, males reliably darken before biting, but they respond less aggressively to barred than to barless images. These two species are polymorphic for the male bar trait, and two other closely related species have lost the bars altogether (**Figure 11.23**). One advantage of doing a comparative experimental study of sender traits and receiver responses on a series of related species is that one can see how the same original signal might evolve along multiple paths, including ending up as a conventional signal, a body size indicator, or a polymorphic trait. One would never see this dynamic history by looking at just a single species. A second example of a proposed conventional threat signal is the headbob displays of *Anolis* lizards [201]. Bob types C, B, and A are associated with increasing probability of receiver attack. However, they are also associated with closer distances to the target. While the type C bob (see Figure 10.18) could be considered a low-intensity conventional pointing display, in which push-ups and lateral compression are combined to improve detection at a distance, type B and A displays are delivered at closer distances. Type A bobs lack the push-up

and compression components and seem to better prepare the sender tactically for an attack. This signal may be better classified as a threat constrained by a proximity risk. Crest-raising in birds and ears back in many mammals are similar conventional element signals that strongly predict attack but are given at such close range that they entail a retaliation risk and should be classified as proximity threats.

Dominance Signals

Dominance signals are signals of fighting ability and motivation with an added component reflecting prior experience. As discussed earlier in the chapter, prior fighting experience and winner/loser effects are critical processes for learning about one's own relative fighting ability and for the emergence of linear dominance hierarchies in social animal groups. Body size and condition differences cannot explain the linearity of hierarchies so commonly observed, because most individuals are clustered in the middle of these normally distributed variables. The experience of winning gives an individual important feedback about itself and makes it more confident in the

(A)

(B)

(C)

(D)

next aggressive encounter; similarly, losing makes an individual more likely to lose in a subsequent interaction. These experiential effects can become incorporated into signals. We have divided dominance signals into three subcategories: status signals, territory ownership signals, and victory signals.

Status indicators

Animals living in small stable social groups typically possess mechanisms for individual recognition and establish a dominance hierarchy based on repeated and remembered interactions with known group members. As groups become larger or more fluid in membership, individual recognition of dominance status becomes more difficult, a situation potentially favoring the evolution of a status signal. If a signal could honestly reflect an individual's dominance status or its accumulated confidence from prior aggressive interactions, it would enable rivals who had never encountered each other before to evaluate their relative aggressiveness and settle disputes without the need to escalate. Status signals were first described in several seed-eating avian species that form winter foraging flocks in which birds frequently engage in disputes over scarce small food finds [320, 386, 388, 401]. The signals are

FIGURE 11.24 Status badges in birds In each of these examples, the size or brightness of the patch is correlated with dominance status. Experimental manipulation of badge features modifies the responses of receivers, indicating that the variation is an informative signal. (A) Black head patch of male European siskin (*Carduelis spinus*): badge size is bimodally distributed, and older males have larger head patches on average [400–402]. (B) Black chest patch of male house sparrow (*Passer domesticus*): size is correlated with dominance status, testosterone, and aggressiveness [55, 56, 147, 148, 305, 320, 322, 323]. (C) White throat patch in both male and female eagle owls (*Bubo bubo*): patch brightness is correlated with body size in females and with reproductive success in males [357]. (D) White nest decorations in the black kite (*Milvus migrans*): the number of decorations brought to the nest is correlated with the breeding pair's age, territory quality, and dominance over other pairs [405].

called **badges of status** because they often take the form of a black melanin-based plumage patch on a white background located on the chest, throat, or head, and the size of the patch is correlated with dominance status (**Figure 11.24**) [404]. More colorful badges have been described in several lizard species, where the size or chroma of color patches on the throat, head, sides, or belly vary as a function of status (**Figure 11.25**) [475]. As described below, even olfactory badges

(A)

(B)

(C)

FIGURE 11.25 Colorful lizard status badges In each lizard species shown here, the size and saturation of the color patch is correlated with dominance and likelihood of winning a fight, and experimental manipulations of patch color affected the responses of receivers. (A) UV reflection of the throat patch in Augrabies flat lizard *Platysaurus broadleyi* [421, 476]. (B) Green flank patches in the Sand lizard *Lacerta agilis* [7, 347]. (C) Orange head color in the Mediterranean lizard *Psammodromus algirus* [295].

of status have been proposed for some insects, fish, lizards, and mammals. In nearly all of these examples, experimental enlargement of the badge leads to a temporary increase in status and intimidation of rivals by the bearer. Acoustic signals are less likely to serve as status badges, but may be correlated with dominance status through their ability to encode body size, motivation, and stamina.

The question immediately arises: What prevents a sender from cheating by sporting a badge that exaggerates its prior contest success or willingness to fight if challenged? The signals seem to be inexpensive to make and not physically or physiologically constrained. While experimental field studies with badge enlargement usually found that exaggerators rose in status, they were also subjected to far more aggression than before, especially from other large-badged individuals [321, 387]. Badge size reductions forced dominant individuals to fight much harder to win a contest [387]. The first evolutionary game model of status badges concluded that a mixture of aggressive individuals with large badges and nonaggressive individuals with small badges could be stable if the cost of escalated fights relative to the benefit of winning increased rapidly with increasing badge size, and if animals with large badges were often challenged by others with large badges [301]. Higher rates of aggressive interaction between pairs of individuals with similar-sized badges compared to pairs of individuals with different-sized badges have been documented in numerous species [404]. These observations fit the ESS predictions for a conventional aggressive signal: attack an

opponent with a similar badge, pause-attack if the opponent has a smaller badge, and retreat if the opponent has a larger badge. This prediction assumes that each contestant knows its own badge size (strength). The consequence of this ESS policy—greater aggression between more similar rivals—is the same as the prediction for signals of fighting ability and motivation discussed earlier. However, cheating is easier with a conventional signal, compared to a body size or performance signal. The cost of bluffing, by sporting an enlarged badge while lacking the corresponding motivation, strength, or experience, is believed to be high.

Subsequent game models identified another cheating strategy that can easily invade this signaling system—the **Trojan sparrow** or **modest aggressor** strategy, in which an animal displays the subordinate signal but behaves very aggressively [220, 349]. The models show that for a population to resist invasion by this strategy, there must be a second, contest-independent, cost to generally aggressive behavior. Aggressive behavior is facilitated by androgen hormones such as testosterone in vertebrates, juvenile hormones in invertebrates, and several neurohormones in the brain (see below). These endocrinological pathways generate a variety of physiological consequences that can impose costs on highly aggressive individuals. Testosterone increases metabolic rate, and the higher activity level of more aggressive individuals can result in greater exposure to risks, predators, pathogens, and injury [54, 288, 294, 403, 487]. A well-known proposal for the negative consequences of aggression is the so-called **immunocompetence handicap hypothesis,** which links elevated testosterone levels with suppression of the immune system [599–601]. The argument here is that the suppression of the immune system causes the animal to be more susceptible to disease, and that only high-quality individuals can bear the immunosuppression consequences of high testosterone levels. While testosterone has been shown to suppress immune function in some animals, notably reptiles and a few bird species, this effect is not widespread. If anything, the causal relationship is reversed—immune activation depresses testosterone [38, 158, 384]. Stress and other consequences of high levels of testosterone and fighting can impose other types of physiological costs on aggressive individuals [5, 41]. Thus for a variety of reasons, dominant individuals invariably pay costs to maintain their status but reap the benefits of priority of access to food, nest sites, and often females; subordinates have lower benefits, but also avoid the fighting and physiological costs and may have better survival. This trade-off generates frequency-dependent selection on subordinates and dominants to maintain the mixed strategy, and favors the honest linkage between testosterone, aggressiveness, and badge size or chroma [383]. There is ample evidence of correlations between testosterone and the magnitude of expression of badges and other sexually selected traits, as well as experimental evidence of a causal relationship [16, 35, 54–56, 118, 277, 306, 362, 375, 384]. Melanin-based traits are particularly likely to be linked reliably to testosterone and aggression because of the pleiotropic effects of the genes regulating the synthesis of eumelanin on these traits [95].

The final piece of this signal evolution story is the mechanism by which fighting experience feeds back to generate variation in badge size or chroma. In vertebrates, social experiences and sensory input activate the hypothalamus–pituitary–gonadal/adrenal axes to modulate steroid hormone production and plasma titers [15, 152, 317, 344, 390, 393, 481]. Circulating hormones eventually feed back to the amygdala and limbic systems of the brain that regulate neurohormones and motivational systems (see Web Topic 10.4). The aggressive milieu, especially rival challenges, agonistic interactions, and contest outcomes, affects hormone titers in predictable ways. As discussed earlier in this chapter, winning a fight leads to a short-term spike in plasma testosterone or other androgens in vertebrates, and losing results in either a reduction in androgen or an increase in the stress hormone corticosterone or catecholamines such as epinephrine and dopamine [135, 136, 189, 350, 447]. These changes have cascading effects on brain monoamine (serotonin) metabolism and neuropeptide expression that change the density of hormone receptors to produce longer-term effects in selected brain regions—cumulatively, the **win/lose scorecard.** Invertebrates have similar behavior-hormone feedback mechanisms, but use biogenic amines such as dopamine, serotonin, and octopamine directly [104, 192, 194, 245, 422]. Both vertebrates and invertebrates thus possess mechanisms to link these neurochemical changes to the production of informative status signals, if there is a selective advantage. Visual color signals and olfactory signals have the greatest ability to reliably link prior experience effects to signal form.

Many studies have demonstrated a correlation between signal variants and dominance, but only a few notable studies provide experimental evidence for prior experience effects on signals, and in some cases evidence for links with key hormones or neurotransmitters. In birds, melanin is deposited in the feathers at the time of moult in late summer or early fall. An experimental study on house sparrows varied the aggressive environment of groups of three males housed together during the moulting period. Groups with more similar pre-moult badges sizes had higher rates of aggression and grew larger badges than less aggressive triads, and dominant males grew larger badges than subordinates [305]. Colored skin structures change in relation to recent social experience in several vertebrates: the size and chroma of the comb in jungle fowl *Gallus gallus* [272, 352], the area and redness of facial skin in male mandrills *Mandrillus sphinx* [406, 407], the eye sclera color of juvenile salmon *Salmo salar* [430], and dorsal darkening in tree lizards *Urosaurus ornatus* [486]. *Anolis* lizards display darkened eyespots during aggressive encounters, which appear earlier in the eventual winner and reflect a suite of hormonal, catecholaminergic, and serotonergic differences between dominant and subordinate individuals (**Figure 11.26**) [240–242, 428, 429, 482]. In electric fishes, the electric pulse waveform lengthens and testosterone increases in dominant individuals [66].

Good olfactory examples include changes in the proportion of chemical components of the sex pheromone and cuticular hydrocarbons as a function of the actual number of wins or losses in the cockroach *Nauphoeta cinerea* [74 625, 119, 326, 328, 389], and status-associated increases in the proportions of hexadecanol and octadecanol in femoral gland secretions of rock lizards *Lacerta monticola* [293]. Urine pulsing by dominant Mozambique tilapia (*Oreochromis mossambicus*) undoubtedly reveals differences in sender plasma androgen levels to receivers [23, 146, 180, 340, 341]. Mammalian olfactory status badges have been described in hyenas, elephants, and sifakas and are likely to be widespread in this highly olfactory taxon [58, 271, 365]. The best-studied system, from the perspective of behavioral effects and neural and neuroendocrine mechanisms, comes from work on decapod crustaceans. All species examined show behavioral winner and loser effects. Following fights, winners show an increase in serotonin while losers show an increase in octopamine. Injecting these amines into the animals during encounters causes them to display aggressive or submissive postures, respectively. These neurotransmitters affect specific neurons in the brain, and they produce metabolites that appear to be released in the urine. Urine is excreted in pulses through

FIGURE 11.26 Eyespot darkening in Anolis lizards (A) *Anolis carolinensis* male in aggressive posture showing dark spot behind the eye. (B) In aggressive interactions, dominant males are faster to exhibit the dark spot and some subordinate males never show the spot. Addition of the dark spot causes receivers to retreat, and covering the dark spot with green paint causes receivers to attack. (C) Changes in serotonin levels in the hippocampus over time for dominant and subordinate males, showing an initial rapid increase, followed by a decline in the dominant male; serotonin is slower to peak and remains elevated for longer in the subordinate. (B,C after [241, 429].)

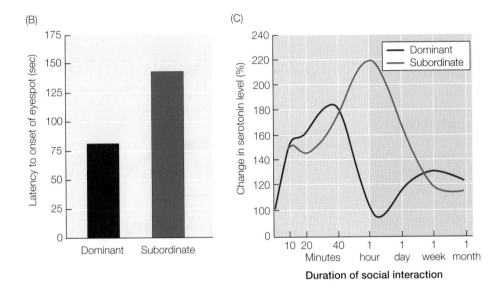

nephropores located near the base of the antennae and can be directed toward the opponent during contests with the gill current. Receivers can distinguish between the urine of unfamiliar dominant and subordinate senders. As expected, dominant individuals respond very aggressively to dominant urine odor, but subordinate individuals quickly retreat from this same stimulus [143, 193, 222, 327, 419].

Territory ownership signals

Territoriality can be viewed as site-based dominance [197, 448]. An animal defending a territory is dominant and behaves very aggressively in the center of its defended area, but becomes more cautious as it approaches its boundary. It behaves submissively or flees if it crosses the boundary and encounters a neighboring territorial owner. Territorial neighborhoods contain a mixture of **owners**, each dominant within its own more or less exclusive and contiguous area, and **floaters**, nonterritorial members of the population that are generally subordinate to the owners and either skulk unobtrusively on defended areas, exist in the interstices between territories, or occupy subprime habitat. Owners produce signals on a regular basis to announce their presence and good condition. A key design feature of **territory ownership signals** is some form of individual signature. An owner identifies itself via the signal in order to negotiate an agreement with neighbors as to the location of their common boundaries and distinguish neighbors from floaters [309]. All modalities can effectively encode individual signature information, as we discuss in Chapter 13. An example of signature information in a territorial signal is illustrated in **Figure 11.27**. The truce that often characterizes neighbor-neighbor interactions is known as the **dear enemy phenomenon** [442]. A neighbor is another dominant individual that has won its territory by fighting for it. While neighbors may fight during initial territory establishment, once the boundary location has been settled, both individuals benefit by respecting this boundary. Ritualized signaling and negotiation characterize subsequent interactions, and neighboring trespassers generally retreat if they are detected crossing the boundary unless they intend to expand their territory. Floaters, on the other hand, should be attacked and chased off immediately upon detection. Floaters may be subordinates or younger individuals, but they are potentially more dangerous adversaries if their intent is to take over the territory.

The functions of a territorial signal are to declare the presence of the owner, define the defended area, and repel potential intruders. All four of the main modalities can serve as territory defense signals, in the sense of declaring owner presence and encoding individual signature information, but they differ in their ability to provide the spatial and strategic

FIGURE 11.27   Coding of individual, population, and species information in the territorial song of the white-crowned sparrow (*Zonotrichia leucophrys*) Songs for this species always contain an introductory whistle, note complex, buzz, and terminal trill. The trill is strongly indicative of regional dialect, whereas the note complex, shown in red boxes, differs among individuals within a population and is highly consistent in a given individual. Playback of a neighbor's song with the note complex exchanged for another male's note complex results in the stimulus being treated as a strange male. (After [335].)

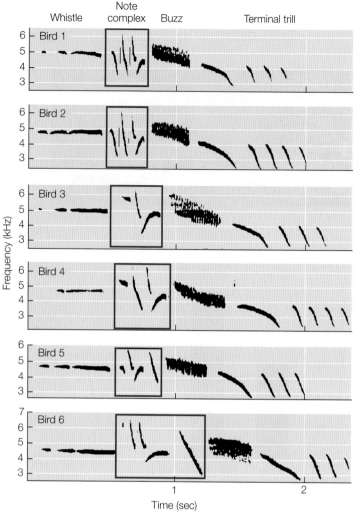

information. The optimal modality for encoding the spatial information depends on the size of the territory relative to the animal, the type of habitat (open or vegetated), time of activity (diurnal or nocturnal), the medium (water or air), and the method of locomotion (swimming, crawling, running, or flight). Acoustic, visual, olfactory, and electric signals allow members of a territorial neighborhood to interact in unique ways. But before we delve into these modality-specific characteristics of territorial signals, we need to consider the broader issue of how we determine the value of territorial signal components and their variants.

Field experiments designed to determine the essential elements of territorial signals are controversial. Theoretically, to test the repellent effect of a signal, one should remove the owner from its territory and replace it with an array of variant stimuli to assess how well each variant can keep intruders out [246].However, because these experiments are difficult to execute, a more common practice is to present the variant stimuli to owners on their territory. Thus, what is tested is the response of territory owners to different qualities of intruders rather than the response of intruders to different qualities of territory owners [77, 397]. Interpreting the responses of owners is problematic. On the one hand, we expect them to defend their territory vigorously against all intruders, and most vigorously against the intruders perceived as the greatest threat. This view is consistent with the results of many neighbor–stranger discrimination experiments, where the aggressive response to a stranger stimulus is much stronger than the response to a neighbor stimulus presented at the

boundary with that neighbor. On the other hand, a stimulus representing a very strong intruder should repel even a territory owner. Researchers have been inconsistent in their interpretation of a weaker response by an owner—is it a consequence of the owner being less threatened, or greatly intimidated? **Figure 11.28** illustrates the hypothetical peaked-curve relationship between response strength and threat signal

FIGURE 11.28   The hypothetical peaked-curve relationship between aggressive signal intensity and aggressive response intensity The receiver is assumed to be of average quality, and to respond most strongly to a signal of similar (average) intensity. Curve peak would shift to the right for a high-quality receiver. If the stimuli range from low to average intensity (region A), as expected for some conventional signals, then the stronger response will be given to the higher-intensity signals because the receiver is less threatened by the lower-intensity signals. If the stimuli range from average to high intensity (region B), as expected for index, proximity, and handicap signals of fighting ability and motivation, then the stronger response will be given to the lower-intensity signals, because the receiver is intimidated by the higher-intensity signals. (After [89].)

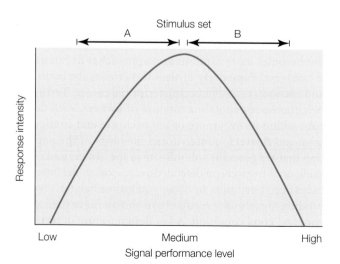

(A)

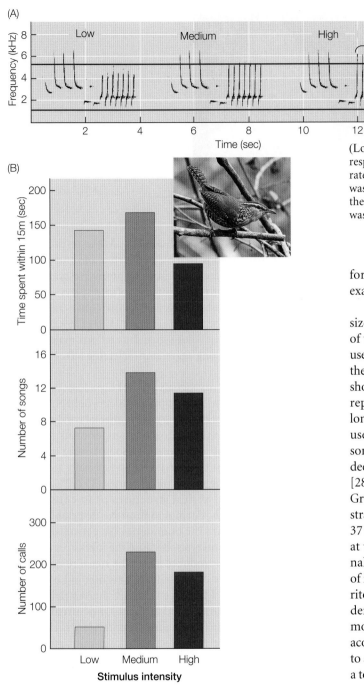

FIGURE 11.29 **Behavioral responses of territorial male banded wrens (*Thryothorus pleurostictus*) to playback of song stimuli differing in performance level** The band-width of trill notes is a key determinant of song performance quality in this species. (A) Example of a set of playback song stimuli in which the performance level of an average trill (Medium, bandwidth demarcated by horizontal red lines) has been modified by reducing (Low) or increasing (High) the upper frequency range. (B) Males responded most strongly, in terms of close approach, high song rate, and high call rate, to the medium stimulus. The low stimulus was approached aggressively with little singing or calling, whereas the high stimulus produced high levels of singing and calling but was less likely to be approached closely. (After [89].)

for relatively high intensity signals. **Figure 11.29** shows an example of evidence for this peaked curve response.

Acoustic signals are ideal in the contexts of large territory size, vegetated habitat, and nocturnal activity. The location of the signaling owner can be assessed if receivers are able to use amplitude, phase, and sound degradation cues to estimate the distance to the sender (see Chapter 3). Signals are typically short calls and songs with species-specific patterning that are repeated for some period of time, but a few species deliver long vocalizations a few times per day. Taxa that commonly use acoustic territorial signals include most birds [69] and some mammals (most notably canids and primates but also deer, pikas, sea lions, hyraxes, grasshopper mice, and koalas) [289], anurans [141], fishes [257], and arthropods [141, 157]. Group-living birds and mammals call together in a chorus, a strategy that permits assessment of group size [164, 234, 304, 371, 399, 478]. Territorial signaling is usually most intense at the beginning of the daily activity period: dawn for diurnal animals, dusk for nocturnal animals. The **dawn chorus** of many avian species seems to be an important ritual for territory-owning males, in which they sing very vigorously and demonstrate their presence (survival) and condition every morning [6, 61, 133, 252, 420]. Like visual territorial signals, acoustic territorial signals allow both owners and nonowners to simultaneously assess multiple signalers in a neighborhood, a topic we take up later in Chapter 15. And like electric signals, acoustic signals allow two individuals to countercall and modulate the signals on a short time scale in order to negotiate relative dominance and reveal differences in motivation. In territorial songbirds that learn song type variants from adjacent adult tutors, **countersinging** is a complex vocal exchange for resolving boundary disputes. Depending on the species, males may compare each other's song rates, song-type switching rates, matching rates, overlapping, or aspects of song and note fine structure [280, 451]. Each species uses a subset of these singing strategies in a graded series of escalated signals [62, 130, 266, 460]. Web Topic 11.4 describes the singing patterns and escalation rules of several well-studied species. Some frogs and mammals also use call variants for aggressive acoustic negotiation [28, 30, 145, 444, 468, 469].

intensity, which should be true for any aggressive signal and is consistent with all of the aggressive signaling models presented in this and the previous chapters. For a receiver of average quality, aggressive response should be strongest to signals representing a similar, average-quality sender, and weaker to signals representing very low and very high-quality senders. The peak of the curve should shift to the left if the receiver is a low-quality individual, and to the right if the receiver is high quality. Furthermore, the slope of receiver response strength as a function of signal intensity will depend on the intensity range of the stimuli presented and the type of signal—positive for relatively low intensity signals, negative

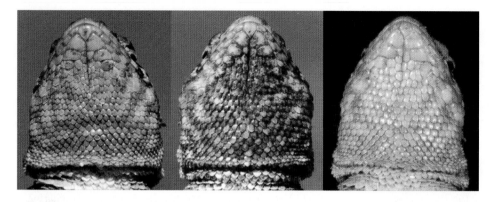

FIGURE 11.30 **Throat color morphs in the side-blotched lizard** *Uta stansburiana* Orange-throated males (left) defend large territories with many females; blue-throated males (center) defend smaller territories and intensively mate-guard a few females; and yellow-throated males (right) patrol a large home range and sneak copulations rather than defend territories or females. These photos show homozygous individuals for the three main color morphs. The color morphs are determined by a simple genetic polymorphism, so some heterozygous intermediates exist. The polymorphism is maintained by a cyclic paper-rock-scissors type of game in which each morph, when it is rare, can invade another morph, but when it is common, is invadable by the third morph [412–414].

Web Topic 11.4 *Songbird territorial negotiation*
Singing strategies and escalation rules in selected songbird species with song-type repertoires.

Visual territory signals are most effective for diurnal species defending small to medium-sized territories in open unobstructed habitat. The signals typically consist of a color patch, posture, or display movement. Species taking advantage of the continuous "on" time of visual territorial signals include many open-country birds and mammals and shallow-water fish. A potential disadvantage of visual signals arises when owners need to move off of their territories, or when nonowners need to coexist on or between territories without being harassed by owners. Several solutions have evolved to cope with these problems. One solution is to design colored badges that can be hidden or temporarily removed [168, 209, 316, 486]. A second solution is to evolve discrete badge classes of different hues associated with alternative territorial strategies, as observed in several bird, lizard, and fish species (**Figure 11.30**) [92, 175, 243, 244, 267, 414, 446]. Color patches can also vary in hue and size over a continuous range to encode information about sender condition or status. A study of the red-collared widowbird (*Euplectes ardens*) employed the framework outlined in Figure 11.28 to evaluate the deterrent effect of color variants. Males are mainly black with a red or orange collar and a long tail (**Figure 11.31A**). Collar area and hue (redness) are both correlated with body size and condition, and territory holders have larger or redder

collars and better body condition [13]. When territory owners were challenged with male taxidermic mounts sporting collars of different areas and hues, aggressive responses were lower to mounts with more intense collar displays, but owners who themselves had larger or redder collars were more aggressive toward the mounts (**Figure 11.31**) [369, 370]. The condition-dependent color signal thus reflects male fighting ability and has a strong aggression-deterring effect in male–male interactions.

Olfactory territory signals are not broadcast as volatile diffusable or current-borne clouds, which would be too subject to the vagaries of changing environmental conditions and currents to define the location of a sender or its territory. Instead, many small, low-volatility, long-lasting marks are deposited throughout the territory and especially around the perimeter and along major travel routes [151, 217]. Marks may be placed on vertical structures or in conspicuous scrapes, middens, or latrines to improve detectability with a visual component. The boundaries are patrolled and the marks renewed at intervals. Because the marks remain long after the owner has produced them, the signal remains "on" even in the absence of the sender. This method of territorial signaling is ideal for large territories, terrestrial habitats (including subterranean and arboreal), nocturnal species, and quadrupeds such as mammals and lizards. However, scent marks are not very effective as keep-out signals. Territorial neighbors often invade and deposit their own marks, requiring the owner to countermark in response. In some mammalian urine-marking species, owners completely **overmark** an invader's mark in order to mask it, whereas in other species the marks overlap slightly or are placed adjacent to the invader's to preserve the individual signatures [124, 219]. These **scent wars**, analogous to acoustic countercalling, probably explain why marks are much more dense at the boundaries between adjacent territories (**Figure 11.32**). When the marks contain both a genetically fixed, long-lasting individual signature component and a more volatile component that reveals information on the status of the sender and the age of mark, the pattern of scent marks in an area provides a public record of past interactions among the competitors [124, 196, 204, 445]. In mice, owners are under strong selection to frequently place fresh marks adjacent to any invader marks

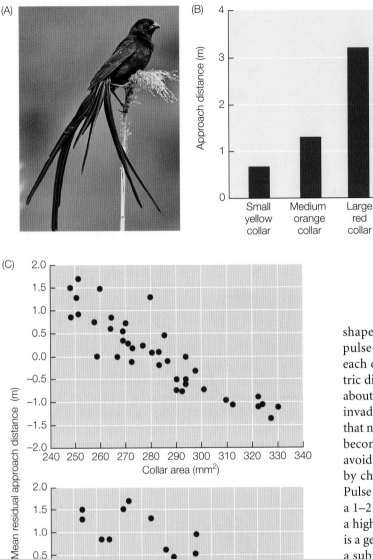

(A)

(B)

(C)

FIGURE 11.31 Deterrent effect of badge hue and size in a territorial bird (A) Male red-collared widowbird (*Euplectes ardens*). (B) Approach distance of territorial male to mounts with different collar signal intensities, illustrating reluctance to approach the large red collar. (C) Approach distance of males to the mounts as a function of their own collar area (top) and hue (bottom), illustrating closer approach by males with better collars. In staged contests over food between brown nonbreeding males with painted-on collars, red collars dominated orange collars, and orange dominated brown controls. In field manipulations of established territory owners, all males with enhanced-area collars maintained their territories, but males whose collars were reduced in size or hue either lost their territories, or suffered a reduction in territory size and an increase in intrusion rate. (After [369].)

shape is individually distinctive and consistent in several pulse-type species, so neighbors have a way to recognize each other [134, 307, 395]. Although the continuous electric discharge required for navigation has an active space of about one meter around a fish, the territories are small, so an invader is quickly detected and chased off. However, a fish that needs to move through another individual's territory can become electrically invisible by turning off its discharge to avoid detection. Electric fishes challenge and threaten rivals by changing their EOD wave frequency or pulse pattern. Pulse fishes (see Chapter 7) typically signal a challenge with a 1–2 second pause in their electrical discharge, followed by a high-frequency burst or chirp. A gradual rise in frequency is a general aggressive motivation signal in some species, and a submissive signal in others [454]. A completely subordinated fish will often go silent for long periods of time [324, 395]. A typical mutual aggressive behavior in both pulse and wave electric fishes is an **anti-parallel display**, in which the head of each fish is positioned opposite to the rival's tail. The electric organ is thus immediately adjacent to the rival's sensitive receptor. In the knifefish *Apteronotus leptorhynchus*, a wave species, the lower-frequency individual may sometimes increase its frequency to match the opponent's frequency [436]. Recall from Chapter 7 that this overlapping results in jamming, so that both fish are unable to electrolocate. Thus even electrical signals can be designed to effectively intimidate rivals.

Victory signals

Victory displays are postcontest signals given by the winner. Both of the agonistic repertoire examples described earlier in this chapter possess a victory signal: the *bowed flipper display* in the little blue penguin and *postbout stridulation* in the cricket. Another well-known example is the conspicuous visual and vocal triumph ceremony observed in geese and swan [284] (see Figure 13.12). But there are surprisingly

as a way of demonstrating their competitive superiority over neighboring owners to potential mates and other subordinate males [203, 378]. Owners immediately attack any male invaders they encounter, and invaders that recognize an owner by matching its odor with the scent of the predominant territorial marks will quickly retreat [149, 150, 287].

Many weakly electric fish in both the New World Gymnotiformes and the African Mormyriformes aggressively defend territories using **electric organ discharge** (**EOD**) signals [324]. Territory owners patrol their boundaries and maintain exclusive access to their area. The EOD waveform

FIGURE 11.32 Scent wars in mice (A) Two adjacent dominant male owners (represented by blue and red) regularly place small urine-based scent marks throughout their respective territories, especially at boundaries. Marks are regularly replenished; fresh marks are indicated with a darker color, aged marks with a paler color. The volatile components of the marks fade in about a day, but the nonvolatile components retain the signature information of the owner (see Chapter 6 and Web Topic 6.5 for chemical mechanisms). (B) One male (blue) invades the neighbor's territory and leaves scent marks. (C) Invaded neighbor (red) rapidly deposits countermarks near the intruder's marks. (D) Invaded neighbor returns frequently to place fresh marks on the invader marks. (After [203].)

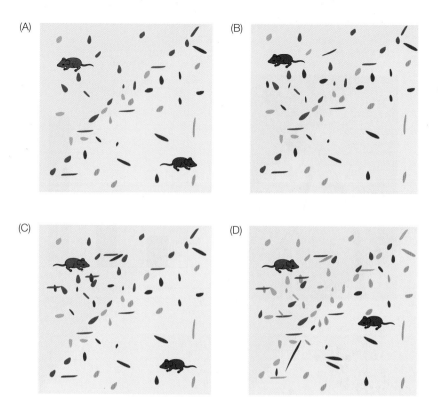

few additional reports of victory displays, either because few investigators have looked for them or because they are similar to other displays in the repertoire [39]. A major issue is whether the display functions within the contest dyad or is directed to third-party receivers. In the former case, the presumed function of the display is to **browbeat** the loser and discourage it from initiating a new contest. To achieve this end, the display should be designed to reveal the winner's confidence or condition. The penguin's bowed flipper display consists of the winner bowing low to the ground and spreading his flippers. One interpretation of this display is that the sender places itself in a vulnerable position for attack by the opponent and qualifies as a *vulnerability handicap signal* that could only be risked by a highly confident individual [39]. The postcontest stridulation of winning crickets and tree wetas, splashing displays in frogs, and vigorous singing by songbirds could signal the good condition of the winner to the loser [39, 125, 154, 162, 281, 473]. However, displays with a loud acoustic component are more likely to function as a **victory advertisement** to other same-sex rivals or potential mates. Models of victory displays suggest that both the browbeating and advertisement scenarios can lead to stable signaling [315]. The advertisement function of victory displays is an eavesdropping issue that we shall take up in a later chapter. We shall also consider postcontest signals and behaviors given by losers that may disguise their perception to others as weak subordinates.

De-escalation Signals

Any continuous dominance or aggressive motivation signal, such as a badge of status signal, crest raising signal, or variable display rate signal, spans the range from high to low status,

and the low end indicates submissive or nonaggressive intentions. But **de-escalation signals** are discrete, and often take a form that is antithetical to the aggressive signal for the species (**Figure 11.33A**; see also Figures 10.5 and 10.13). Both the cricket and penguin agonistic repertoires contain at least one discrete de-escalation signal. The penguin has several de-escalation signals, with different signals given at different stages

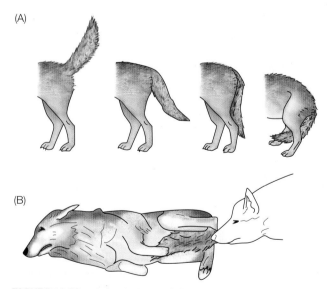

FIGURE 11.33 Submissive and de-escalation displays in wolves and dogs (A) Wolf packs form linear dominance hierarchies based largely on age and gender. Tail position forms a graded display of dominance and subordinance, with dominants displaying tail up, and subordinates displaying tail down. In extreme submission, the tail is curled under the belly. (B) In complete de-escalation, the animal rolls over onto its side or back and presents its belly to the dominant. Front paws may be folded across the chest, hind quarters lifted, and tail tucked under. (After [64, 311, 396].)

of a contest and in particular, at different distances from the opponent. If a penguin decides to back down at an early contest stage while it is some distance away, it performs a *face away* or *indirect look* display. But if the contestants have come very close or have engaged in physical contact, the quitting individual performs a *low walk* or *submissive hunch*. These two displays are designed to move the sender away from the rival. Other quitting displays appear more like vulnerability signals, in the sense that the quitter assumes a posture from which it is impossible to launch an attack (**Figure 11.33B**). For example, American bison (*Bison bison*) bulls fight by ramming their heads together and pivoting around each other to jab a hooking horn into the rival's flank. The de-escalation signal is a turn of the head to the side with the muzzle low and forward, in grazing position. This display fully exposes the neck and flank, but even a charging rival will come to a screeching halt [285]. The head-high broadside display is an aggressive challenge display in many ungulates, but an analysis of the context of lateral presentation in fallow deer has clearly demonstrated

that it is a signal given by losers to de-escalate a contest [214]. As in the bison example, such a display seems to place the submitter in a very vulnerable position. Why do quitters give such signals, rather than just flee, and why do winners accept them? These questions are addressed in an evolutionary game model, in which contestants play the hawk and dove game if one of them doesn't signal retreat upon learning that it is the inferior competitor. De-escalation signals can evolve under the following conditions: (1) when the value of the contested resource is not too high relative to the cost of injury; (2) when the extra benefit of winning an escalated contest (by dominating the loser in the future) is small; (3) when the the losing animal has poor options for safely running away; and (4) when the estimation of the difference in the resource-holding potential between the combatants is accurate but not perfect [298]. De-escalation signals, together with submissive and appeasement signals, are the aggression-reducing glue that holds social groups together, and we will discuss them in more detail in Chapter 13.

SUMMARY

1. Two individuals competing for the same nonshareable resource item represent the maximum degree of sender-receiver conflict, with both preferring that the other party back down. But both benefit if a physical fight can be avoided. Two competitors for the same resource are rarely equal in **fighting ability**, **motivation**, or **experience**, each of which can affect the likelihood of winning a fight. Animals make use of prior information, cues, and signals to assess their likelihood of winning a fight against a particular opponent. If one party estimates that it is likely to lose, it makes the decision to quit. If communicating cannot resolve the likely winner, the contest will escalate to costly physical fighting. The signals and tactics employed during contests are collectively referred to as **agonistic behaviors**, which include both aggressive and submissive acts.

2. Most evolutionary game models of conflict resolution assume that individuals possess some knowledge of their own fighting ability. Engaging in contests and remembering the outcome is the best way to acquire this information. Winning increases one's estimate, while losing decreases it. This **winner-loser effect** has a physiological basis that affects subsequent aggressive behavior. A cognitive sense of self is not required, only simple rules of thumb that adjust effort invested in a contest based on the outcome of previous contests. Formation of **linear dominance hierarchies** in group-living species is an **emergent property** of such winner-loser effects.

3. Animals fight to gain access to limited resources such as **food**, **mates**, **shelters**, and **territories**. The value of the resource item, the contestants' need for the resource, and the availability of alternative resource items determine how motivated a contestant is to fight for the contested

item. **Resource value** can be measured as the difference between the fitness gain if the contest is won and the fitness loss if it is lost. Resources such as territories, shelters, and sometimes mates may be defended for some length of time, leading to a level of **ownership**. Owners are likely to value the resource more than intruders because of their investment in defending the resource, and they may have better knowledge of its specific value. A high resource valuation is assumed to lead to an increased **motivation to fight**.

4. Species employ different **fighting strategies**—the signals, tactics, and combat methods used during extended conflicts. Fighting strategy models differ in their assumptions about the way contestants gather information during the interaction, which types of signals and tactics are used, and how they make the decision to end it. The **sequential assessment model** assumes that animals have little information about their rival's relative fighting ability at the beginning of a contest, but acquire this information in a **mutual assessment** process by performing repeated mutual displays. Increasingly costly signals and tactics are used in stages if cheaper signals do not lead to the resolution of the likely loser. When one contestant is relatively certain it will lose, it makes the decision to quit.

5. Two other fighting models are based on the notion that a contestant makes the decision to quit based only on **self-assessment** of its own fighting ability, rather than by mutual assessment. The **energetic war of attrition model** assumes that contestants engage in a matched and costly chasing or grappling contest, A contestant's decision about whether to continue or quit an extended agonistic interaction is based on its assessment of its current ability to continue in this activity. The first individual to quit loses the

contest. The **cumulative assessment model** assumes that that contestants inflict significant stress and injury upon each other during the conflict, and a contestant's decision about whether to continue or quit a fight is based on the sum of the costs it has sustained.

6. The three models differ in many of their predictions. Empirical tests of the alternative models have provided support for each one in different species. However, the energetic war of attrition model applies only to a few species that fight with extended chases. The cumulative assessment model is likely to characterize the escalated stage of fighting in some species. Most species undergo at least some initial signaling and mutual assessment, as described by the sequential assessment model, before escalating to a physical fight.

7. Other types of **asymmetric information** may be used to help settle conflicts. **Resource value asymmetry**—in which the contestants place different values on the resource—is expressed in the assessment phase of the contest and involves signals that indicate motivation to fight as well as fighting ability. For conflicts involving owned resources, an **owner/intruder role asymmetry** may be used to settle the contest. The **bourgeois strategy** is a solution in which contestants assess which one of them is the owner, and the invader then retreats. Although owners usually win against invaders, recent models and empirical results indicate that this simple rule is not stable, and that owners usually win because they have the better proven fighting ability and value the owned resource more.

8. Most species have a rich repertoire of **agonistic signals**, including signals that indicate body size, aggressive motivation, offensive and defensive threat, dominance status, victory, submission, appeasement, and retreat. **Fighting ability signals** include index signals and amplifiers of **body size**, and handicap signals indicating **stamina**. Stamina is usually dependent on condition, and informative signals include performance of energetically costly displays, as well as color and acoustic signals affected by condition. **Weapons** serve a tactical function during fights by enabling contestants to jab, wrestle, block, or flip the opponent. Extremely enlarged weapons occur in a few taxa where males defend burrows or compete intensely for females. Fighting is highly ritualized, and the size of weapon primarily serves as an assessment signal of body size and condition.

9. Because they are used only in conditions where there is a strong conflict of interest, signaling costs are needed to guarantee some level of signal reliability. **Honesty-guaranteeing costs** include receiver retaliation, proximity risk, physical/physiological constraints, and energetic costs, and thus may be classified as conventional, proximity threat, index, and handicap signals, respectively.

10. **Aggressive motivation signals** are important for species in which resource value and need vary significantly between contestants. The signals indicate a sender's willingness to fight for a resource item, and can be subdivided into three categories. **Challenge signals** are targeted toward a specific rival from a distance. Conventional visual challenge signals include directed looking postures, and acoustic signals include song matching and frequency matching. **General aggressive motivation signals** indicate level of arousal (desire for the resource and willingness to fight for it) and often take the form of energetically costly repeated displays. **Offensive threats** are given close to the opponent, where there is a significant proximity risk of injury, and are strongly correlated with subsequent attack. Examples include intention-to-attack postures and low-amplitude acoustic signals.

11. **Dominance signals** reflect an animal's prior fighting experience. Visual and olfactory **badges of status** are dominance signals that appear to be conventional in form but require a physiological link to aggressive hormones and neurotransmitters, in addition to an immediate retaliation cost, to remain honest. **Territory ownership signals** are site-based dominance signals that declare the presence of the owner, define the defended area, and repel intruders. Ownership signals usually encode individual identity as well as the owner's motivation and condition. Boundaries between neighbors are often dynamic and negotiated with acoustic (e.g., countersinging) or olfactory (e.g., overmarking) interactions. **Victory** displays are postcontest signals given by the winner, and may serve to browbeat the loser, demonstrate confidence, or advertise the victory to third-party receivers. **De-escalation signals** are antithetical in form to aggressive threat signals and are designed either to move the quitting individual quickly away from the dominant one, or to indicate the intention not to attack.

Further Reading

Two older books provide good introductions to animal aggression: Huntingford and Turner [197] review fighting tactics in a variety of species and describe the ecological and evolutionary context of conflict; Archer [15] stresses the hormonal determinants of aggression. Mason and Mendoza [297] focus on conflict in primates. For a thorough review of self-assessment and winner-loser effects, see Hsu [189] and Lindquist and Chase [276]. For a recent review of fighting strategy models, see Arnott and Elwood [18, 19], as well as the material in Web Topics 11.1 and 11.2. Useful reviews of agonistic signals by topic include: acoustic signals [120, 174, 437] olfactory signals [124, 151, 203, 327]; performance signals [208], weapons [67, 109], status badges [95, 404, 475] songbird territorial interactions [398], neuroendocrine approaches [194, 245, 345], and agonistic signaling models [2, 199, 200, 202, 220, 298, 315, 433, 434].

Go to the companion website for Chapter Outlines, Chapter Summaries, and References for all works cited in the textbook. In addition, the following resource is available for this chapter:

Web Topic 11.1 *A detailed description of three conflict resolution models*
The sequential assessment model and its variants, the energetic war of attrition model, and the cumulative assessment model. The unit also discusses the strategies and challenges of distinguishing among these models with empirical data.

Web Topic 11.2 *Resource value and ownership asymmetries in fighting strategy models*
Further details on the models discussed in this section, where contestants may differ in resource value or ownership roles. This unit also examines the evidence for resource and ownership assessment, and the implications of whether neither, one, or both contestants have access to this information.

Web Topic 11.3 *Positive allometry of weapons and ornaments*
A summary of several theoretical models developed to understand why the size of sexually selected weapons and ornaments is often proportionally larger in large individuals compared to small individuals.

Web Topic 11.4 *Songbird territorial negotiation*
Singing strategies and escalation rules in selected songbird species with song-type repertoires.

Chapter *12*

Mate Attraction and Courtship

Overview

At the level of species survival, the communication system involved in the meeting and mating of males and females could be viewed as a mutualistic endeavor for achieving successful reproduction. Both sexes presumably want accurate signals to identify the species and sex of potential mates and to exchange honest information on receptivity so that fertilization can proceed efficiently. It is certainly true that each sex is dependent on the other for successful reproduction and both benefit from mating. However, at the level of individual selection, a mixture of cooperation and conflict characterizes the interactions between males and females during periods of mating. The conflict is a result of fundamental differences in male and female strategies that arose very early in the evolution of eukaryotic diploid organisms with haploid gametes. Sexual strategies determine the basic social systems of all higher organisms and most aspects of the signaling behavior involved in mating. In this chapter, we survey the evolution of mate attraction signals, focusing on the mechanisms that insure reliable encoding of species identity and mate quality information. First, we identify the differences in male and female strategies that set the scene for conflict. We then discuss the important concept of sexual selection and briefly review the coevolutionary models of male-female interaction with an eye to what these models tell us about reproductive communication signals. The last section examines the signals and tactics employed during courtship and fertilization.

Male and Female Reproductive Strategies

Most plant and animal species reproduce sexually. Male and female reproductives are diploid (possessing two copies of each chromosome) and produce haploid gametes that fuse only with those of the opposite sex to make diploid zygotes. The advantage of sexual, versus asexual, reproduction lies in the processes of meiosis and genetic recombination, which yield offspring of individually variable genetic diversity. This diversity provides the raw material for adaptations to fluctuating and changing environments, including selective pressures from competitors, predators, and parasites [701]. Sexual reproduction presents adults with the primary challenge of locating conspecific members of the opposite sex, and communication plays the key role here. As illustrated in **Figure 12.1**, reproduction in sexual species can be broken down into a series of phases: (1) attraction and searching, (2) courtship, (3) mating, and (4) parental care. These activities entail time and energy costs, so in order to repeat the

FIGURE 12.1 **The reproductive cycle of continuous-breeding species** In the first stage, individuals engage in attracting or searching for members of the opposite sex. Once an individual has encountered a potential mate, it must decide whether to actually mate. After deciding to mate, the two individuals must achieve successful fertilization. Having produced zygotes, some level of parental care may be undertaken to ensure offspring survival. Before repeating the cycle, continuous breeders must recover from the cumulative costs of prior phases and develop new gametes.

cycle, individuals must recuperate and replenish their energy reserves before beginning a new one. Relative investment in these phases often differs for males and females. As the difference in investment by the sexes increases, conflicts of interest are more likely to occur.

The two sexes are defined on the basis of the size and mobility of their gametes. **Females** are individuals that produce large relatively immobile gametes, called **ova** (singular *ovum*), which are provisioned with a store of nutrients that the fertilized zygote will use during its early development. **Males** are individuals that produce small highly mobile and often flagellated gametes, called **sperm**, which usually contribute only DNA to the zygote. This difference in size and mobility, called **anisogamy**, simultaneously facilitates the meeting and fusion of conspecific gametes and sets the stage for conflict between the sexes.

Anisogamy is presumed to have arisen from an initial state of **isogamy**, in which all gametes are similar in size. Even in unicellular organisms with isogamous gametes, there are usually two mating types, and fertilization can only occur between opposite types. This dichotomy may have arisen to prevent cytoplasmic gene mixing and a conflict between nuclear and cytoplasmic genes—one type resigns itself to contributing no cytoplasmic (e.g., mitochondrial) genes to the next generation, so that only nuclear genes from the parents fuse and recombine [305–307, 309]. Once this asymmetry is established, various scenarios may lead to the cytoplasmic parental type becoming larger and the other

type becoming smaller [597]. Some models focus on the survival advantage of a large gamete that is subsequently exploited by a small one, while others focus on the evolution of a fast and mobile gamete that is attracted to a larger, less mobile and chemically attractive gamete to maximize gamete encounter rates. In either case, parental individuals that produce gametes of one of the extreme types fare better in conjunction with the opposing type than parental individuals with the ancestral intermediate type [91, 127, 160, 276–278, 309, 322, 529].

Anisogamy has several important consequences (**Table 12.1**). Because sperm are considerably smaller, male gametic investment per reproductive event is usually less than female gametic investment. Males therefore require less time and energy to replenish their sperm supply after a mating; females need a longer time, which varies with the size of their eggs and clutches. During this renewal time, animals are unreceptive. The greater the duration of the female's reproductive cycle relative to the male's cycle, the more the ratio of receptive males to receptive females is skewed in favor of males. The ratio of receptive males to receptive females is called the **operational sex ratio** [164, 376, 707]. When it is strongly skewed toward males, a male's reproductive success is limited by his access to females, and males will compete intensely among themselves to gain access to females [41, 528, 531, 680, 786]. Furthermore, the small size of sperm relative to

TABLE 12.1 *Summary of sexual strategies*

	Male	**Female**
ANISOGAMY	Small mobile sperm are cheap to produce in large numbers	Large eggs with yolk stores expensive to produce
	Sperm do not contribute to offspring survival	Larger egg increases offspring survival
	Shorter reproductive cycle	Longer reproductive cycle
OPERATIONAL SEX RATIO	OSR skewed in favor of males	Many available mates
	Limited by access to females	Limited by access to resources
	Compete for mates	Compete only for resources
REPRODUCTIVE SEX ROLES	Selection favors traits that increase number of females inseminated	Selection favors traits that maximize resource acquisition and mate quality
	Less choosy about mate quality	Choosy, mating with wrong species is costly
	Active, courting sex	Coy

eggs means that individual males can afford to produce large numbers of gametes, so they have the potential to fertilize many females. Male sperm competition leads to strategies that optimize the efficiency of locating and entering eggs, as well as the various strategies for mating described below. Females, on the other hand, do not benefit as much from maximizing their number of mates. Instead, they increase their fitness by adopting strategies that maximize their food intake for the provisioning of their gametes and zygotes. In short, females are limited by access to food, whereas males are limited by access to females.

Another important outcome of different levels of reproductive investment by the two sexes is a difference in choosiness of mates. Because of their typically higher investment in gametes and reproduction, females are choosier and have a higher threshold for acceptable mate quality than males do. This difference often results in a conflict of interest over mating, as illustrated in the model shown in **Figure 12.2**. When a male and female have encountered each other, they must decide whether a mating should take place. If the union is likely to produce high-quality offspring, both will benefit from mating. However, if the union would produce intermediate or low-quality offspring, the male may still benefit from the mating while the female's best option is to reject the male. This sexual difference is the basis for the traditional **reproductive sex roles**: competitive actively courting males and coy choosy females [530, 731, 787]. As the male's relative investment in reproduction increases and approaches the level of female investment, the zone of disagreement becomes smaller, and males become choosier. Nevertheless, females are still usually the choosier sex [382, 555]. In some species, males invest significantly in parental care and sex roles become reversed: males are then the limiting sex, females compete for them, and males become very selective of high-quality females. We shall discuss this interesting but relatively rare phenomenon later in the chapter, and focus for the most part on traditional sex roles.

Male–male competition leads to a variety of fitness-maximizing strategies. In sessile aquatic broadcast spawners, males have little option but to maximize sperm number and mobility; their gamete production costs are likely to be very high, and the dilution of sperm released into large volumes of medium may also make sperm limiting to females [410, 795]. If the organisms are mobile, however, several different **male mating strategies** become available [531, 676]. Males can move rapidly in search of females and sequester or guard them from other males to increase their probability of fertilizing all of a female's ova (female defense strategy). They can defend the resources females require as a way to attract and obtain preferential access to females (resource defense strategy). If neither females nor resources are readily defendable, males can position themselves in good locations for encountering as many females as possible and advertise themselves (self-advertisement strategy). Males can also engage in postmating activities that serve to increase the survival of zygotes, such as provisioning or guarding them. **Table 12.2** describes

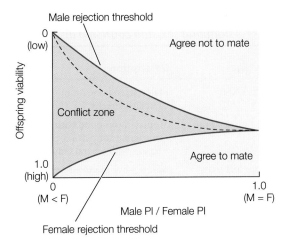

FIGURE 12.2   Mating conflict game This graph is the solution to an evolutionary game model between the sexes over whether or not to mate as a function of the viability of the offspring that would result as a function of the relative parental investment (PI) of the sexes, where PI = 0 represents no investment on the part of the male and PI = 1.0 represents an investment equal to that of the female. Each sex has the option to accept or reject the mating, taking into account the cost of finding another mate if the mating does not occur. Males and females have different mating threshold curves; the mating is accepted if parameter values fall above this curve. Both will agree to mate when offspring viability is high and males provide some investment, and agree to mate when viability is low and male investment is low. As the cost of finding another mate increases, the curves shift upward on the right side, such that mating is accepted with lower offspring quality values. This is a battleground model that delineates the conflict zone but not how it will be resolved. The dashed curve within this zone separates the region where selection is stronger on females to reject (above the curve) from the region where selection is stronger on the male to persevere. (After [530].)

the main alternative male mating strategies and outlines their association with which sex searches, which sex performs the primary mate attraction signal, and the nature of the courtship display.

These male mating strategies have major effects on mate attraction and courtship signals by determining which sex *searches* and which sex *signals* during the search and attraction phase of reproduction. They also affect the extent to which males must persuasively court females with elaborate displays. The question of how much effort males and females should invest in searching for mates has been investigated using an evolutionary game model in which both sexes have a continuous range of movement options from 0 to 100% of the receptive period. The ESS solution is for one sex to stay put and the other to search; both sexes searching some fraction of the time is unstable [244]. Although either sex could end up being the searching sex, the typical skew in operational sex ratio favors the roaming male ESS over the roaming female ESS (**Figure 12.3**). The sex that remains stationary therefore becomes the signaling sex. However, male mating strategies significantly impact this outcome. In female defense systems, males become the searching sex and

TABLE 12.2 *Male mating strategies and their association with searching sex, signaling sex, and courtship display intensity*

Mating strategy	Description	Searching sex	Signaling sex	Courtship
FEMALE DEFENSE	Males defend one or more females directly	M	F	Minimal display
Long-term association	Monogamous bond with one female or permanent harem with several females	M	F	Minimal display
Dominance hierarchy	Animals live in mixed-sex groups, dominant males have priority of access to females	M	F	Short-term consortship bond
Scramble competition	Females generally solitary or loosely clumped, males roam widely in search of females	M	F	Brief association for mating, can be coercive
RESOURCE DEFENSE	Males defend resources females need, such as food or nest sites	F	M	Moderate display
SELF-ADVERTISEMENT	Males position themselves in locations visited by many females	F	M	Highly elaborate courtship display

Sources: After [164, 244, 676].

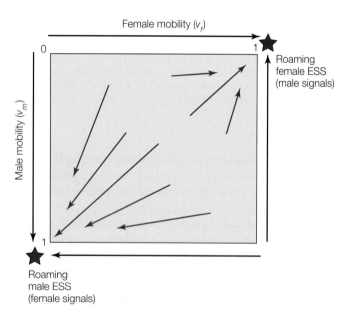

Female mobility (v_f)

Male mobility (v_m)

Roaming female ESS (male signals)

Roaming male ESS (female signals)

FIGURE 12.3 The mobility game In this continuous-strategy, asymmetric scramble game, each sex can vary its mobility while receptive from 0 (the animal sits and waits until a member of the opposite sex finds it) to 1 (animal moves during every time interval until it finds a mate). The male's velocity options are given by v_m and the female's by v_f. The solution is a double ESS denoted by the red stars for either maximally mobile females and stationary males (the roaming female ESS), or for maximally mobile males and stationary females (the roaming male ESS). The domain of attraction is stronger for the roaming male ESS because of the operational sex ratio favoring males. The sex that stays becomes the signaling sex. Species with roaming males include most mammals, reptiles, and insects. Species with roaming females include birds, frogs, and orthopterans. (After [244].)

females the signaling sex, as expected. But when males defend resources such as a food patch, nest site, or territory, they are selected to stay put and advertise the resource while females roam in search of good resources. Similarly, when males opt for the self-advertisement strategy, they establish display territories near locations with high female traffic. This system often results in aggregations of competitively displaying males called **leks**. When females are the searching sex, they are free to encounter, assess, and choose among a number of potential mates, and males may again evolve conspicuous and costly displays to attract them.

Signaling males often employ the visual and acoustic modalities, and mate attraction and courtship displays may be highly elaborate, especially in the case of lekking systems. By contrast, when males are the searching sex, females often employ the less costly olfactory modality for their attraction signals, and male courtship tends to be minimal or characterized by forceful control over females [676, 805].

Sexual Selection

Sexual selection is the evolutionary process that arises from competition among members of one sex (usually male) for access to the limiting sex (usually female) [134]. It typically leads to **sexual dimorphism**, the different appearance of the sexes, via the evolution of specific traits in the competing sex that improve individual mating success. Because females are most often the limiting sex and males the competing sex, we shall use female and male in the next sections as shorthand terms for the limiting and competing sexes, respectively. As previously mentioned, there are important exceptions in which females are the competing sex and males the limiting one, and even cases where both sexes compete for mates, which we take up later in the chapter.

Competition for mates can take two basic forms that affect the types of traits that evolve. **Intrasexual selection** involves overt competition among males to control or monopolize mating access to females. Male mating success here depends on an individual's relative weaponry, body size, muscular strength, aggressiveness, speed, endurance, and experience. As we saw in Chapter 11, many of these traits are hard to estimate directly during a contest, and this has favored the evolution of various displays that are correlated with the traits and used to mediate contests without overt escalation. **Intersexual selection** arises when females can choose which available males will serve as mates. Benefits of being choosy include access to a preferred male's territory, higher levels of paternal care, effective provisioning by the mate, minimal exposure to disease and parasites, and the production of male offspring that will themselves be favored as mates. Again, the relative benefits of mating with a particular male are often hard to evaluate directly during courtship, and intersexual selection typically leads to elaborate male structures, displays, vocalizations, and odors that provide females with information about the individuals as potential mates. In most species, both intrasexual and intersexual selection play roles in generating nonrandom mating success among males; the relative influence of the two processes will vary among species and contexts.

Although Darwin believed that sexual selection generally operated in concert with natural selection to improve the adaptedness of individuals to their environment, he recognized that the traits that evolve via sexual selection can have a negative effect on the survival of their bearers and thus work in opposition to natural selection. Therefore a more inclusive view is to consider these processes as strategies for optimizing the different components of fitness. Sexual selection leads to adaptations that maximize mating success or mate quality, while natural selection leads to adaptations that maximize fecundity and survivorship (**Figure 12.4**). Recognition of this trade-off is key to understanding the observed patterns of signaling associated with animal mating.

Intersexual selection models

Intrasexual selection, which affects the same traits and signals in competing senders and receivers, essentially follows the evolutionary processes outlined for contest signals in Chapter 11 and will not receive further theoretical treatment here. Intersexual selection jointly affects the sexually selected display traits in males and the preferences for particular values of those display traits in females [11]. Evolutionary models of intersexual selection thus must take into account complex coevolutionary interactions between the sexes. A diversity of models for intersexual selection has been proposed, ranging from verbal appeals to intuition to more formal game-theoretic and genetic treatments. Models differ in assumptions, and as discussed in Chapter 9, in the methods invoked to predict evolutionary trajectories. One model might focus on **direct selection** of female preferences for certain male traits that improve female reproductive success or survival,

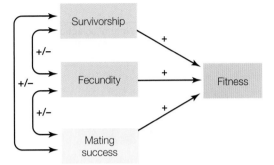

FIGURE 12.4   The components of fitness Fitness can be divided into three components: survivorship, fecundity, and mating success. Sexual selection describes variance in mating success (yellow), which includes successful fertilization and acquisition of high-quality mates, while natural selection generally describes variance in survivorship and fecundity. Curved double-headed arrows show the potential for positive or negative correlations among these components. (After [21].)

whereas another model might examine **indirect selection** for female preferences caused by a genetic correlation with another trait under selection. Some models find that likely initial conditions lead to an equilibrium solution of male trait value and level of female preference, while others result in nonequilibrium or cyclical solutions.

As we also discussed in Chapter 9, genetic models of evolutionary trajectories can track the consequences of different patterns of sex determination, recombination rates, mutation rates, and the strength and direction of selection on a trait. Quantitative genetic models have the additional advantages of specifying the response to selection for a polygenic phenotypic trait, and can better incorporate pleiotropic and correlated relationships among coevolving traits. Furthermore, they are based on parameters that we can usually measure in the field and laboratory. A very useful synthesis of intersexual selection models can be achieved with a **multivariate quantitative genetics framework**, which allows the models to be compared using a common language and viewed as subsets of a complete accounting of evolutionary change [201, 376, 395, 459]. All models in this framework consider the simultaneous coevolution of three traits: a **male display trait** or mating tactic (t), a **female preference** for the male display trait (p), and a **viability** or residual fitness trait (v) present although not necessarily of equal value in both sexes. Residual viability includes the fitness components of fecundity and survivability left over after excluding the fitness effects of mate number and quality.

Recall from Chapter 9 that the change in the mean value of a single quantitative trait after a generation of selection is approximately equal to the product of its heritability and the selection differential on it. In the case of multiple traits evolving in tandem, things are a bit more complicated. First, we need to incorporate any genetic linkages between the traits. As discussed in Chapter 9, this is usually accomplished by

(A) Initial conditions

	Sexual selection gradient			Natural selection gradient			Additive genetic variance			Additive genetic covariance		
	t	p	v	t	p	v	t	p	v	tp	vt	pv
Fisher process	+	0		^	0		+	+		+	0	0
Good genes	^	0	^	^	^	^	(+)	+	+	(+)	(+)	+
Direct benefits	^	+		^	^			+				
Sensory bias	^	0		^	+			+				
Sexual conflict	+	+			^			+				

(B) Equilibrium conditions

	Sexual selection gradient			Natural selection gradient			Additive genetic variance			Additive genetic covariance		
	t	p	v	t	p	v	t	p	v	tp	vt	pv
Fisher process	+	0		−	0		+	+		+	0	0
Good genes	+	0	+	−	−	+	+	+	+	+	+	+
Direct benefits	+	+		−	−		(+)	+				
Sensory bias	+	0		−	^			+				
Sexual conflict	+	+		−	−		(+)					

FIGURE 12.5   Conditions associated with alternative intersexual selection processes The three traits under potential selection are a male display trait t, a female preference p, and residual viability in either sex v. The tables show the predicted relationships between each trait and the two components of fitness (sexual and natural selection gradients; left-hand columns) and the genetic architecture (additive genetic variance and covariance; right-hand columns) for each intersexual selection process. (A) shows the initial conditions required to initiate that process, and (B) shows conditions expected at the equilibrium. In both tables, selection gradients can be positive (+), negative (−), stabilizing for intermediate values of the trait (^), or flat (0). Variance and covariance components can be zero (0) or greater than zero (+). Values in parentheses may or may not be important. Because these models all focus on the evolution of female preferences, all require additive genetic variance of p to get started (eighth column from left in (A); similarly, all equilibria in (B) require that the sexual and natural selection gradients on the male trait (first and fourth columns) have opposite signs. (After [201].)

combining the additive genetic covariances between the traits and the additive variances of each trait into a single G-matrix. The selection differentials on the three traits can be summarized in a vector that is usually combined with the overall phenotypic variation in the traits to produce a selection differential. Thus for multiple coevolving traits, the changes in mean values of the traits across a generation is approximately equal to the product of the G-matrix and the selection gradient. It will be additionally useful to subdivide the selection gradient into two components—a sexual selection component favoring increased number or quality of mates, and a natural selection component favoring fecundity and survival.

Five different models of sexual selection have been proposed to explain observed patterns of sexual dimorphism and display behaviors: the Fisherian runaway, good genes, direct benefits, sensory bias, and sexual conflict models. All invoke the same quantitative genetic process for intergenerational change, but they emphasize different components of that process. Note that the different subsets of the process invoked by these models are not exclusive—most can occur

simultaneously with others. **Figure 12.5A** summarizes the conditions that are most important for *initiating* each model process, and **Figure 12.5B** summarizes the likely conditions *at equilibrium*. For a more formal analysis of this quantitative genetic approach to sexual selection, see Web Topic 12.1.

Web Topic 12.1 *Quantitative genetic models of sexual selection*
An overview of the multivariate quantitative genetic approach to the evolution of female preferences and sexual conflict models.

FISHERIAN RUNAWAY MODEL   The **Fisherian runaway model** largely focuses on the G-matrix component of the quantitative genetics equation. R. A. Fisher was the first to propose a genetic process for the evolution of female preferences that could explain Darwin's observation of exaggerated, costly traits employed by males to attract females [183, 569]. The model begins with the accumulation of females in a population who share a mating preference for males with a particular cue or display trait. In this early stage, the preference need have no natural selection benefits and imposes no costs on the females expressing it. The accumulation could simply be a result of mutation and drift. However the preference is acquired, females that use it in selecting mates will produce offspring that carry both the preference genes from their mother and the display trait genes from their father. Once the population hosts a minimal number of females expressing the same preference, a positive covariance between the preference and trait genes builds up. If the covariance is large enough, a positive feedback loop is set in motion, and the resulting "runaway" process can produce a rapid elaboration of the male display trait along with shifts in average female choosiness to match these more exaggerated male trait values. Males with an extreme trait benefit because more females prefer them, and females benefit because their sons are preferred as mates. While there is *direct sexual selection* on the male trait, there is only *indirect selection* on the female preference via the

(A)

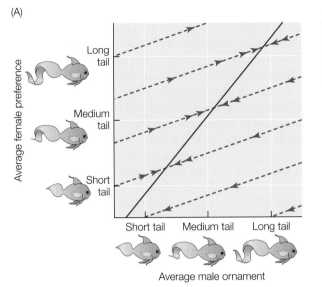

(B)

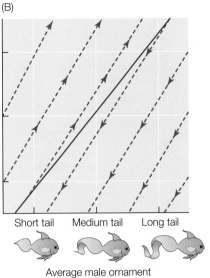

FIGURE 12.6   The Fisherian runaway model Both graphs depict the evolutionary equilibrium values for the average value of the male trait preferred by females (tail size, *y*-axis) as a function of the male trait (tail size, *x*-axis). If there is no cost of female choice, the outcome can occur anywhere along the red line of equilibria. The blue dashed lines with arrows show evolutionary trajectories. (A) If the covariance between female preference and male trait is small relative to the additive genetic variance, the population will slowly approach the equilibrium line and remain relatively stable. (B) If the covariance is large, the population will "run away" from an unstable equilibrium line, either increasing or decreasing. (After [394, 459].)

positive genetic covariance between the male trait and female mating bias for the trait (see Figure 12.5).

The Fisherian process with its runaway stage has also been called the "sexy sons" model because females only benefit by producing male offspring that have high mating success. As the runaway process continues, the extremely elaborated male trait is likely to become a hindrance due to developmental constraints, production costs, or predation risk. Display elaboration halts when the naturally-selected viability cost balances the sexually-selected mating advantage. Thus for a purely Fisherian trait, attractive males with extreme traits should suffer correspondingly higher mortality. Fisherian traits are sometimes referred to as "arbitrary" in form because their selection during the runaway process is related only to female attraction, not to male fitness [260, 363]. This unfortunate terminology overlooks the fact that even if preferences begin due to random processes, they often end up becoming informative indicators of male condition [373]. In addition, the Fisherian process is likely to accompany *any* of the other intersexual selection processes since all of them begin with expression of female preferences for particular male display traits. This has led to unnecessary disputes over whether a given example is due to Fisherian runaway or some other model: in most cases, both are likely to be contributing factors [373, 374, 376].

Quantitative genetic analysis of a pure Fisherian model predicts lines of equilibria with increasing values of both female preference and male trait expression (**Figure 12.6**), but only if there is no cost to females for being choosy [362, 394]. If females eventually experience a cost of mate choice, such as more time spent searching for high-quality males or increased effort rejecting low-quality males, the line collapses to a single point where the female's cost and benefit of choosing are balanced [568, 570]. The Fisherian process by itself is notoriously unstable and easily knocked into a nonequilibrium state, producing rapid changes and frequent differences in male and female trait values among populations. It has been argued that where a Fisherian process plays a major role, it could lead to rapid and high levels of speciation, such as are found in lekking birds-of-paradise (see Figure 9.3) [394, 526].

GOOD GENES MODEL   **Good genes models** assume that females can produce fitter offspring by preferring mates with evidence of heritable viability and health; the relevant evidence is usually an exaggerated male signal trait that reveals meaningful aspects of male residual viability. Because of the temptation for males to cheat, it is also usually assumed that the display is a handicap and only males in good condition can afford the most costly versions of the signal. The process given this combination of assumptions is often called the **indicator model** (**Figure 12.7**).

Good genes models start with the presence of some male display trait correlated with the male's heritable viability. This can happen when display intensity is condition dependent or indicates some other aspect of male viability such as age, body size, or immunocompetence. Assortative mating between females with the preference and males with the trait leads to indirect selection on the female preference via the positive genetic covariance between female mating bias and male viability (see Figure 12.5A). There is no direct sexual

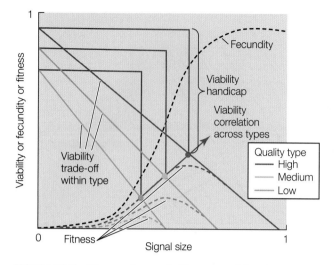

FIGURE 12.7 **The good genes indicator model** Relative male viability, fecundity, and net fitness (*y*-axis) as a function of their signal size (*x*-axis) for males of three different quality types: high (blue), medium (green) and low (orange). Mating success (fecundity) increases with increasing signal size (black dashed line). Higher-quality males experience a lower marginal cost (less steep slope of solid colored viability lines) when producing signals of different intensities, and also have higher viability in general (higher *y*-intercept). Dashed colored lines show the net fitness (= viability × fecundity), and colored dots show the optimal signal size and viability (survival rate) for each type of male. Evaluated across individuals at the equilibrium (red arrow), higher-quality males have higher net fitness, higher mating success, and higher survival. The vertical drop in fitness (viability handicap) is the same for all three male types in this case, but could differ for different fecundity curve shapes. In theory, reliable signaling could occur even if high-quality males experience higher marginal costs (steeper slope of viability line), as long as their viability line *y*-intercept was very high and the fecundity curve increased more gradually. (After [215, 216, 514].)

selection for the preference. Female preference for the indicator trait will result in females producing more viable sons and daughters. This indirect benefit of producing high-quality offspring will likely lead to females becoming increasingly more selective and consequently experiencing some costs of choosing good mates. The male display trait itself does not need to be genetically determined (e.g., it could be age-based), but if it is, a genetic covariance between the trait and female preference will be established that can augment the good-genes process with a Fisherian one [294]. The male trait and female preference can also undergo cycles of elaboration and diminution in a good genes framework. Thus a number of selective forces come into play during the good genes process and they combine to impose direct and indirect selection on all three traits [201, 315, 316, 567].

Because the good genes process contains the conditions for the Fisherian process, it is misleading to conceive of these two processes as alternative hypotheses. Instead, most theoreticians suggest that the Fisherian process is inevitable, and the issue is the degree to which good-genes effects are also present [166, 373, 459]. Both processes can

be combined into a single conceptual model of indirect benefits, sometimes referred to as the **Fisher-Zahavi model** of sexual selection. Numerous factors may determine the relative importance of the two processes; for instance, the Fisherian process is stronger in species with highly polygynous or promiscuous mating systems where a few males can succeed in attracting many females, while good-genes effects are stronger in monogamous systems. Indirect selection effects in general are believed to be weak compared to direct effects, because they depend upon a longer causal pathway from preference to male trait to total fitness (viability). A weak link anywhere in this pathway means that the entire pathway will be weak [364]. Key weak links are the potentially low additive genetic variance of total fitness and of the male trait itself, brought about by persistent female choice for certain male traits that quickly drives the preferred alleles to fixation and erodes genetic variance. This problem is called the **lek paradox**, a reference to the conundrum of how lek-mating females can continue to derive a good-genes benefit from their strongly skewed mate choice, generation after generation, for highly ornamented males that provide no resources [363]. In fact, a variety of factors are known to maintain genetic variation, including high mutation rates, heterozygote advantage, selection in variable environments, frequency-dependent selection, sexually antagonistic selection, flexible female preferences, and preference for condition-dependent traits [45, 104, 124, 241, 294, 571, 596, 638, 728]. These surely account for the finding that the additive genetic variation of total fitness in real populations is never zero but usually accounts for 1–10% of overall phenotypic variation [93]. Thus indirect selection on female preference remains a viable option [31].

DIRECT BENEFITS MODEL **Direct benefits models** represent the least controversial case of intersexual selection [363, 376]. Females that have a genetic predisposition to choose mates that provide them with material resources obtain the immediate benefit of an increase in their fecundity, survivorship, or offspring survival rate. No genetic correlations are required. The key component is the presence of some male cue or signal associated with the males' ability to provide direct benefits, which results in positive direct sexual selection for the preference trait (see Figure 12.5A). The direct benefits process does not require heritability of the male trait for its initiation—benefits to females can be determined entirely by environmentally generated differences among males, as long as some reliable cue or signal of male ability to provide the benefit exists. However, it is likely to become heritable as a consequence of the female selection pressure. This pressure will then lead to the evolution and elaboration of a male trait that indicates these benefits via the process outlined in Figure 10.2. Most models assume that a condition-dependent handicapping mechanism will produce a signal that honestly reveals male ability to provide direct benefits [263, 279, 580]. At equilibrium, both males and females will experience some naturally-selected fitness costs of this selection (see Figure

12.5B). The costs for females may include lost foraging time, predation risk, or delays in reproduction while searching for acceptable males, and males pay a cost of producing and displaying the trait. Net fitness should be positive for both sexes [580, 659].

SENSORY BIAS MODEL   No formal **sensory bias model** for female preferences exists. We have at best a verbal model proposing that female mating preferences evolve as a byproduct of natural selection on sensory systems used in nonmating contexts such as foraging or predator detection [37, 38, 363, 644–646]. As we described in Chapter 10, males that develop a display that mimics this feature may gain a mating advantage because females more readily detect them. In the quantitative genetics framework, the female preference is initially a preexisting naturally-selected sensory bias. If males then develop a display that mimics some aspect of this feature, and there is additive genetic variance for this display, a sexually-selected mating benefit will accrue to males possessing better matches to the feature. Female mating preferences thus emerge as a correlated response to natural selection on other behaviors. In other words, there is indirect selection for female mating preferences arising from pleiotropic genes that affect both mating and other behaviors [201, 202].

This sharing of the sensory system for two different functions is unstable and could be resolved in several ways. In one equilibrium solution, the female pays no cost for the male mimicry trait and her bias remains at a local, naturally-selected equilibrium point. Male display is maintained at a balance between natural selection that seeks to return the display to the original optimum and sexual selection that favors further exaggeration (see Figure 12.5B) [201]. A second possibility is that a genetic correlation develops between the male display trait and the female bias as a result of assortative mating. Once this covariance has been established, both the display trait and the bias for it now have the potential to be exaggerated via the Fisherian process, and there will be indirect positive sexual selection for the female preference. A third scenario is that the male display trait increases in size or intensity to the point where it becomes very costly; for example, it attracts predators or is energetically expensive to produce. It could then become a condition-dependent indicator of male viability, and female preference for the display could be indirectly selected under the good genes process. Finally, the male display trait could exploit the preexisting bias in a way that is manipulative and costly to females, for instance by causing them to mate more frequently than is optimal. Females would then evolve to resist, and males would counter by further increasing their expression of the manipulative trait, as described

in the chase-away model below. Sensory bias is unlikely to be an equilibrium solution by itself, but in conjunction with other coevolutionary processes may be an important mechanism for generating new signal systems, initiating speciation in different environmental contexts, and explaining some of the species-specific diversity in mate attraction signals [72].

SEXUAL CONFLICT MODEL   Female mate choice (intersexual selection) always occurs against a backdrop of male–male competition (intrasexual selection), and in some cases this competition may limit a female's ability to choose a preferred mate or reproduce at the optimal rate. In the 1980s and 1990s, observations of overt physical conflict over forced mating in water striders and cryptic conflict during internal fertilization in *Drosophila* flies generated new insights and a view of males and females engaged in a **sexual arms race** [21, 105, 112, 283, 533, 534, 605–607, 637]. Competition among males to circumvent female preferences leads to increasing male persistence parried by increasing resistance by females, as illustrated in **Figure 12.8**. The resulting **sexual conflict** (or **chase-away**) **model** depends on the existence of a female preference for male traits, previously evolved via one of the processes described above, to create the opportunity for sexual conflict [21, 211, 283, 299, 781]. Initial conditions for this model include positive sexual selection (direct benefits) on the female preference; positive sexual selection on the male trait; and stabilizing natural selection on the female preference (see Figure 12.5A). Once females evolve a preference for a male trait that provides direct or indirect benefits, males with the preferred trait are then under selection to further increase their mating advantage by directly manipulating female fitness, such as increasing the mating rate above that

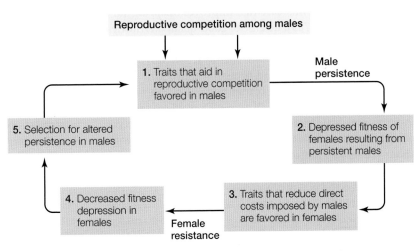

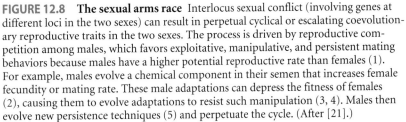

FIGURE 12.8   The sexual arms race Interlocus sexual conflict (involving genes at different loci in the two sexes) can result in perpetual cyclical or escalating coevolutionary reproductive traits in the two sexes. The process is driven by reproductive competition among males, which favors exploitative, manipulative, and persistent mating behaviors because males have a higher potential reproductive rate than females (1). For example, males evolve a chemical component in their semen that increases female fecundity or mating rate. These male adaptations can depress the fitness of females (2), causing them to evolve adaptations to resist such manipulation (3, 4). Males then evolve new persistence techniques (5) and perpetuate the cycle. (After [21].)

which is optimal for females. Sexual conflict is characterized by direct antagonistic selection between the sexes, where traits that enhance the fitness of one sex reduce the fitness of the other. The consequence is a negative natural selection gradient that balances the positive sexual selection gradient for both sexes (see Figure 12.5B). Even though females pay a direct cost by mating with manipulative males, they benefit by producing sons that are good manipulators. Other outcomes are also possible. Females may evolve insensitivity to the male trait, or evolve resistance against it, as a way to avoid such manipulation [639]. Nonequilibrium, cyclical solutions can occur [459, 741]. Finally, males are likely to incur significant costs to manipulate females, which can limit the extent of their manipulation.

Evidence for alternative sexual selection models

To rigorously discriminate among these processes, one would have to measure a number of the selection gradient and variance/covariance components listed in the columns of Figure 12.5. Several of the component values are common to all of the processes and therefore will not aid in discriminating among them. For instance, all require additive genetic variance for female preference as an initiating condition, and at equilibrium all predict positive sexual selection and negative natural selection for the male trait [82]. Even some of the remaining components make overlapping predictions for different processes, so demonstrating a key prediction for one process does not necessarily eliminate the others.

In principle, direct benefits are relatively easy to document. The resources that males can potentially provide to females include: a good supply of sperm to fertilize a female's clutch of eggs [275, 439, 494]; nuptial gifts and courtship feeding that enhance female fecundity [86, 95, 148, 226, 235, 503, 674]; male parental care that improves offspring survival [240, 576, 705, 783]; defense of good-quality territories that ensure a food supply or good nest site for the female and her offspring [99, 145, 146, 252, 594, 599, 601]; male defense against predators and harassing conspecific males [67, 122, 359]; and avoidance of transmittable disease [1, 624, 792]. What kinds of male displays might evolve to indicate these potential benefits to females? Numerous studies have described male signals that are significantly correlated with enhanced female reproductive success, which we describe in a later section of this chapter [477]. One concern with such studies is the potentially confounding observation that females mated to attractive males sometimes strategically increase their own investment in eggs [92, 368, 788]. This led to the **differential allocation hypothesis** which argues that females might be willing to pay a cost for such augmented investment, as long as they can reap the benefit of more or higher-quality offspring [673]. Thus a major reason why females might increase their investment when mated to an attractive male is to obtain indirect benefits. To accurately assess the magnitude of direct material benefits provided by males, careful experimental studies, such as cross-fostering of offspring with different parents where paternal care is

the issue, are required to sort out the role of maternal versus paternal effects, as well as environmental versus genetic sources of the reproductive benefits.

Over the past three decades, major research effort has been expended to find evidence of the good genes process. This requires measuring the fitness of males and females over at least two generations and looking for evidence of male trait heritability and the correlations between the male trait, female preference, and viability (especially the positive preference–viability covariance). A metadata review examined the association between male traits and their survival and found that most correlations were in fact positive, supporting the viability requirement of the good genes process [330]. Another metadata analysis of 22 studies examined the strength of the correlation between the male trait and the survival of the male's offspring, and found a small but significant positive average correlation that accounted for 1.5% of the variance in total offspring fitness [475]. Assuming maternal effects have been controlled for, this result can only occur if attractive males pass on their superior viability to their offspring. More recent studies have obtained similar results, and have also corroborated the heritability of the male trait and survival benefits for both male and female offspring [83, 149, 193, 327, 331, 728, 776]. Nevertheless, one recent review concludes that indirect effects may be a less compelling driving force for female mating preferences than direct effects and nonrandom mating caused by sensory exploitation, male–male competition, and female minimization of search and harassment costs [380].

The Fisherian process is typically documented with evidence that females mated to attractive males benefit by producing sons that are also attractive as mates [82, 168, 256, 339, 769]. However, only a few studies have argued for a pure Fisherian process by measuring the fitness components that might exclude good genes and direct benefits processes. One example is a study of the lek-breeding sandfly (*Lutzomyia longipalpis*), in which females approach small swarms of males located close to a prey host where the female flies obtain a blood meal before mating. Each female freely chooses a mate, presumably based on a pheromone signal. A laboratory study established small leks of males comprised of either preferred or nonpreferred individuals, and monitored the fitness consequences of females mating in these two contexts in terms of direct costs and benefits and the subsequent success of their sons and daughters (**Figure 12.9**). No direct benefit or good genes effects were found, but females mating with preferred male types did produce highly successful sons [339]. Similarly, a series of studies of *Drosophila simulans* has shown that the primary fitness benefit of mate choice to females occurs via the mating success of sons [288, 715–717].

Some procedures used by researchers to distinguish the roles of pure Fisherian from good genes effects in a given system have proven to be theoretically misguided [374]. For example, only the Fisherian process requires the existence of a positive trait–preference genetic covariance. To document this covariance, researchers conduct artificial selection

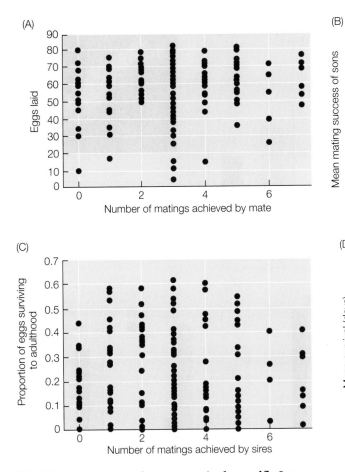

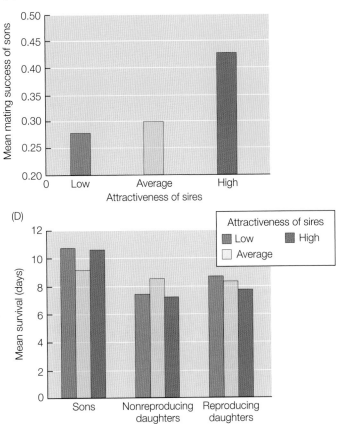

FIGURE 12.9 **Evidence for sexy sons in the sandfly** *Lutzomyia longipalpis* Males aggregate and display in swarms or leks, and females choose mates based on a male pheromone signal. Some males are consistently preferred by females over other males. In this laboratory study, males were pretested for attractiveness and sorted into small groups of low, average, and high attractiveness. Virgin females were released into one of the treatment groups and allowed to mate once with the male of their choice (known paternity), after which they were isolated and allowed to lay eggs. These offspring were then monitored for survival, fecundity (daughters), and mating success (sons). (A) The number of eggs laid by a female was not correlated with the attractiveness of her mate to other females (measured as the number of matings he achieved in the pretest), indicating no direct benefits of mate choice to females. (B) The sons of attractive sires were themselves more attractive. (C) There was no correlation of the fecundity of daughters (measured as the proportion of their eggs surviving to adulthood) with the attractiveness of their sire (i.e., the grandfather of the eggs). (D) Adult survival of sons, nonreproducing daughters, and reproducing daughters was not affected by the attractiveness of the sire. (After [338, 339].)

experiments on the male trait, making it either larger or smaller, to determine if the mean trait value preferred by females changes in the same direction. Such an effect has been demonstrated in stalk-eyed flies and guppies [81, 293, 785]. But this correlation is also very likely to occur in the good genes process and should not be used to distinguish between them. In a similar vein, one cannot distinguish between Fisherian and good genes processes by comparing the relationships between trait magnitude and male survivorship. Although we expect the "size" of Fisherian male traits to show a negative relationship with survivorship and good-genes traits to show a positive one, good-genes traits can have a slightly negative slope with survival, as long as this cost is more than compensated by mating success and offspring performance benefits [372]. Finally, the good genes process is expected to lead to female preferences for condition-dependent viability-indicating male display traits. But as we have already mentioned, all of the other processes also are likely to reach an equilibrium where trait elaboration is balanced by costs, and the outcome is often a condition-dependent trait.

In addition to these overlapping theoretical predictions, Fisherian, good-genes, and direct-benefit effects are not exclusive and can operate together, as suggested earlier. One such example is illustrated in **Figure 12.10**. The arctiid moth *Utetheisa ornatrix* has been the subject of a long-term investigation. In Figure 6.12, we described the male pheromone hydroxydanaidal, which is derived from a toxic pyrrolizidine alkaloid contained in the food plant of this species. During mating, males provide females with a spermatophore laden with the alkaloid. Females prefer males with a high concentration of hydroxydanaidal pheromone, which

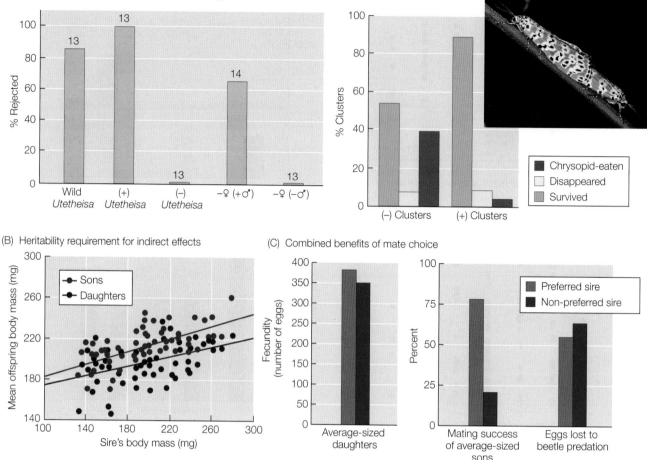

(A) Direct benefit of pheromone to female and eggs

(B) Heritability requirement for indirect effects

(C) Combined benefits of mate choice

FIGURE 12.10 **Multiple mate choice benefits in the arctiid moth *Utetheisa ornatrix*** Males provide females with a spermatophore laden with toxic alkaloids obtained from the food plant *Crotalaria*. Females base their mate choice on a pheromone signal that contains hydroxydanaidal derived from the plant alkaloids (see Figure 6.12). Larger males have more alkaloids and are preferred by females on the basis of information contained in the pheromone. Females obtain direct and indirect benefits from their choice of mates. (A) Wild adult moths, adults raised on alkaloid-containing plants (+) and alkaloid-free females (−) mated with alkaloid-containing males (+) are rejected as prey by spiders and other predators, compared to adults raised on nonalkaloid plants (−) or alkaloid-free females (−) mated to alkaloid-free males (−). Egg clusters laid by females given alkaloids (+) are more likely to survive than egg clusters not containing alkaloids (−). (B) Both male and female offspring inherit their father's body size phenotype. (C) Male offspring of preferred sires are themselves more preferred as mates; female offspring of preferred sires produce larger clutches compared to the female offspring of nonpreferred fathers, and their eggs are less likely to be eaten by beetles. (After [162, 318–321].)

is correlated with the amount of alkaloid they receive. The sons of preferred males are also preferred as mates, implying a Fisherian "sexy son" benefit. Larger males have higher levels of the toxin and the pheromone, and they are preferred as mates. Body mass is heritable by both male and female offspring, implying potential good-genes benefits. Finally, adult females and their eggs benefit directly from obtaining this

toxic compound, because they are more likely to be rejected as prey by spiders and other predators.

As discussed in Chapter 10, the sensory bias process under some conditions can lead to male traits that resemble important cues to which the female sensory system is tuned by natural selection. Several examples were described in earlier chapters, including mimicry of floral scents by some Euglossine bees; modifications of fins and tails in fish to resemble worm-like food prey items or eggs; and the construction of pillars behind burrows in fiddler crabs to which females run when startled by a predator. The best evidence for pleiotropic associations between behaviors sharing a common sensory system was described in Figure 10.19 for guppies, where strong among-population correlations were found between foraging preferences for orange objects and mating preferences for orange males [616]. A neural network simulation study testing the pleiotropy prediction of the sensory bias process found that selection for foraging on colored objects could lead to modest increases in mating preferences for the same colors as a correlated response [202].

The sexual conflict process leads to fundamentally different kinds of male and female traits that occur in the courtship phase rather than in the mate attraction phase of reproduction [21, 105]. Many of these traits we would not even classify as communication signals. When the conflict occurs prior to

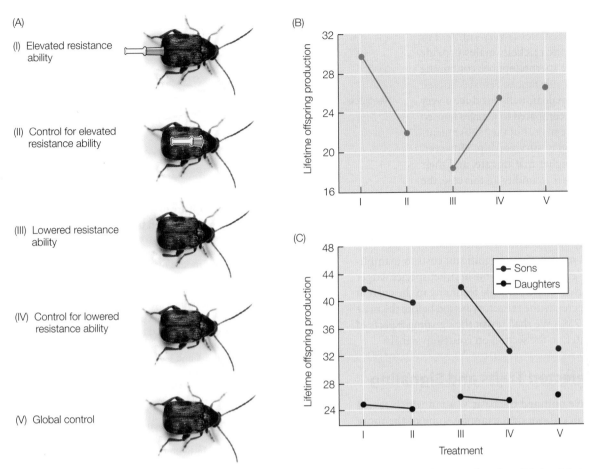

FIGURE 12.11 **Direct fitness effects of female preference behavior in a seed beetle with forceful male mating tactics** In contrast to most studies of sexual selection that manipulate the male trait to estimate fitness costs and benefits, this study manipulated female resistance behavior. Male Adzuki seed beetles (*Callosobruchus chinensis*) initiate mating by grasping females from behind and protruding their long penis toward the female's genital opening. Females vary in their tendency to move away from and vigorously kick courting males. (A) Treatments: (I) the resistance ability was elevated by gluing a plastic prong to the female's rear abdominal plate, which made it more difficult for males to achieve a genital grasp; (II) elevated resistance control treatment for treatment I was a prong glued to the elytra; (III) the resistance ability was lowered by ablating the female's lower rear legs; (IV) lowered resistance control treatment for III was ablation of a foreleg on one side and a middle leg on the other; (V) the global control was no manipulation. Groups of four virgin females with the same treatment were housed with four males and lifetime reproductive success monitored. (B) Offspring production was significantly higher for the elevated-choice treatment (I), and significantly lower for the reduced-choice treatment (III), compared to controls. (C) Lifetime offspring production of sons and daughters of the experimental females was not significantly affected by the treatment their mother had received. The trend for sons of lowered-resistance females to have higher fitness than controls goes in the opposite direction from the direct effects shown in (B), but implies an indirect benefit to females mating with manipulative males in the form of more successful manipulative sons. Direct fitness effects were far stronger than indirect fitness effects. (After [434].)

mating, the male traits can include structures, tactics, and chemical signals that enable males to copulate even if the female is unwilling. Biting and holding is a common tactic in elasmobranchs, seals, some beetles, lizards, salamanders, mammals, and even a few birds. Insects and amphibians can have special graspers that hold females even if the latter struggle to escape. In fish and hermaphroditic slugs, intromission organs may be elongated or turned into injecting darts. In some arachnids, males release a toxin that renders females temporarily unconscious. When the conflict occurs after mating, males typically evolve seminal fluid products that assist the sperm to swim in the female reproductive tract or stimulate the female gonads to increase egg production rates.

Female "preference" traits are expressed as levels of physical, chemical, or tactical resistance.

Support for the sexual conflict process is obtained by finding significant naturally selected costs of mate choice for females and evidence of traits that increase fitness in one sex while decreasing it in the other. Costs to females include lower survival and above-optimal mating rates arising from forceful or deceitful male mating tactics. **Figure 12.11** describes a study of a seed beetle with forceful male mating tactics and strong resistance by females. In this study, the female resistance trait was experimentally modified to assess the fitness consequences. Females benefitted when their resistance efficiency was enhanced, and lost fitness when their resistance ability was decreased. Sometimes the effect of males on females is very subtle. For instance, in domestic crickets, males provide females with a large spermatophore containing nutrients and other substances. Females housed with an attractive male initially produce a very large clutch of eggs compared to females housed with an unattractive male, but they subsequently suffer a reduction in survivorship. However, their sons have a higher mating rate, and their daughters have higher fecundity. In this case, the indirect benefits for females outweigh the direct costs [256]. Other studies find that the fitness of daughters is negatively correlated with the fitness of sons, clear evidence of sexually antagonistic selection [80, 119, 172, 189, 521, 558]. Males may also incur significant costs to manipulate females. For example, among bushcricket species, larger spermatophore mass increases male fecundity in the current clutch but also increases the refractory period before the male can mate again [738]. We shall revisit the signaling aspects of these conflicts later, in the section on courtship signals.

Sexually Selected Traits and Signaling

Now that we have outlined the basic ways in which sexual selection can proceed, we next examine some of the more widespread signals that have been the targets of sexual selection and ask what kinds of information these signals might provide to receivers. Females, who are usually the relevant receivers, use male signals to select mates. As discussed in the prior section, mate choice may benefit females directly by affecting their fecundity or survival, or indirectly by providing good genes and other advantages to their offspring. Several reviews have identified a wide range of variable male traits that different species have recruited for making mate choice decisions. Some remain cues, whereas others have been elaborated by sexual selection into signals; the same trait may act as a cue in one species and become a signal in another. Widely used criteria for mate choice include body size; morphological features such as color patches, weapons, and other body structures generally called ornaments; chemical signals; behaviors that function as tactile, acoustic, electric, visual, and hydrodynamic signals; externally constructed structures such as nests, bowers, and pillars; and material resources such as good territories and nuptial gifts [11, 335]. In this section we review these traits based on the presumed type of information they provide. We shall attempt to describe some of the mechanisms by which those traits serving as signals provide reliable information needed by choosy mates to make beneficial decisions.

Condition and health

Many models of intersexual selection predict that the size or intensity of sexually selected signals should reflect the overall physiological condition, health, and ultimately the general viability of potential mates, because only the better-quality

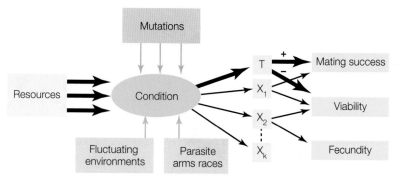

FIGURE 12.12 Genic capture of genetic variance by condition-dependent sexually selected traits An individual's condition is determined by the resources it acquires and makes available for allocation to fitness-enhancing traits. The thickness of the arrows reflects the amount of resources allocated. T is a sexually-selected male trait and X_1–X_k are other traits related to reproduction, foraging, and survival. Before any trait elaboration, natural selection would result in a low allocation of resources to T and greater allocation to traits X_1–X_k; but with increased sexual selection, more resources are diverted to T (thicker arrow), as shown here. Elaboration of T increases mating success at a cost to viability (indicated by the + and – on the thick arrows). Condition is affected by a large number of genetic loci and is therefore a large target for mutations (blue). As the male trait increases in cost and relative amount of resources allocated to it, its expression becomes increasingly dependent on condition and reflective of overall genetic quality. Environmental contexts that fluctuate in time and space, and coevolutionary arms races with parasites (orange) interact with an individual's genetic composition to determine its condition. Because condition is influenced by a large number of genes subject to mutation as well as selection, an ongoing mutation-selection balance maintains additive genetic variance for condition-dependent traits. (After [638, 727].)

individuals can afford to pay the cost of expressing the trait. As sexually selected traits become more costly to produce, their expression becomes more dependent on the sender's condition. Condition is presumably affected by a large number of loci, and it is therefore very susceptible to mutational disruptions and environmental fluctuations via many pathways. **Condition-dependent trait expression** should thus reflect much of the genetic quality of a male, a concept called the **genic capture hypothesis**. This concept is illustrated in **Figure 12.12** [316, 571, 638, 727]. Genic capture is believed to maintain genetic variance and constitutes a major resolution to the lek paradox of eroded genetic variance discussed earlier. The heightened condition-dependence of sexually selected traits should also amplify the phenotypic differences among males, making it easier for females to detect and compare differences in overall male genetic quality and avoid males with large mutation loads. Finally, female preference for the most costly, condition-dependent, and variable traits may weed out deleterious mutations from the population and improve the adaptedness of individuals to their environment [782]. Females that mate with more intensely signaling males thereby gain some combination of direct and indirect

(A)

(B)

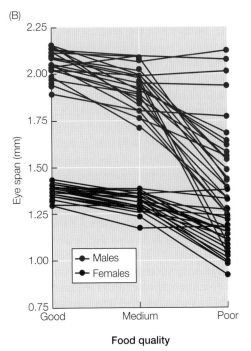

FIGURE 12.13 Condition-dependent signal of good genes in the stalk-eyed fly (*Crytodiopsis dalmanni*) (A) Males (upper left and lower right) have longer eyestalks than females (center) due to a female preference for males with longer stalks. (B) Siblings from different family lineages were reared on good, medium, and poor quality food. Graph shows eyespan lengths for the resulting adults, separated into males and females, with lines connecting mean values for members of the same lineage reared on different food. Increasing the level of food stress for developing larvae resulted in a greater decrease in eyestalk length for males compared to females. Body size also decreased with poorer food; after controlling eyestalk length for body size by computing relative eyespan, the female relationships with food quality became flat, while male relationships continued to show both an overall decrease and greater variation in eyespan length with poorer food conditions. Thus, the male sexually selected trait shows heightened condition dependence. Furthermore, males in lineages that performed well in good environments also tended to perform well in poor environments. This type of genotype by environment interaction implies that eyespan length is a good indicator of both current condition and good genes for resistance to environmental stress. (After [136].)

benefits. Selection will favor the evolution of female preferences that specifically target those male characteristics that provide the greatest benefits in the current ecological context.

A plethora of studies on wild populations have found correlations between trait expression and condition, but such correlations do not prove that the trait indicates genetic quality. Evidence that a sexually selected trait shows *more heightened* condition dependence than non-sexually selected traits of a similar nature adds some weight to correlational studies [123, 365]. However, such evidence is not easily obtained.

Experimental studies that subject animals to various types of stress, such as limited diets, brood size manipulations, high rearing densities, temperature shocks, or parasite infections, and then quantify the effects on trait expression, are often used to identify a cause-and-effect relationship between condition and the trait. Such experiments have several pitfalls. First, the stressors are often extreme and not always relevant

to the subject species' natural environment, so results must be interpreted cautiously. For example, zebra finch males subjected to severe food restriction as nestlings were found to have 50% smaller HVC brain nuclei and to sing shorter songs, but a second study that manipulated brood size, rather than limiting food, to modify nestling condition to the same degree found no effects of treatment on song duration or HVC size [89, 221, 695]. Second, the timing of the stress treatment with respect to the species' life history and the type of trait is critical. As noted in Chapter 8, some traits reflect condition over long time scales and are established at the time of juvenile development or, in the case of feather or horn growth, at the time of ornament development. Other traits, such as display rate or some types of skin/bill color, are more dynamic and respond to short-term changes in adult condition [270, 657]. Short-term and long-term condition-dependent traits reveal different kinds of information about the sender [100]. Third, unless the study is designed to test for genetic effects, a positive result cannot exclude a purely environmental effect of condition on the trait. Females that select males based on environmentally determined trait values can reap only direct benefits from their choice, not indirect genetic effects. The most persuasive studies of good-genes benefits are those that both manipulate the environment and evaluate genotype by environment interaction effects. **Figure 12.13** illustrates the results of such a study on a stalk-eyed fly. Some genetic lines maintained longer male eyestalks under all environmental conditions, whereas others grew shorter stalks in more stressful environments [135, 136]. These results strongly support the indicator model for good genes and the differential cost requirement of the quality handicap model.

Some traits are better indicators of condition and genetic quality than others. To understand the mechanisms that produce more reliable signals, we turn to the field of evolutionary developmental biology, or **evo-devo** [103]. Any trait is affected by two types of selection factors: extrinsic and intrinsic [204]. Extrinsic factors include environmental conditions such as predators, food, parasites, weather, season, habitat structure, and presence of conspecific individuals such as competitors, social cooperators, and mating partners. Intrinsic factors include the multitude of biochemical and physiological pathways that enable the organism to develop, grow, and function. These intrinsic pathways must be internally cohesive if the organism is to develop properly, thrive, and reproduce within the variable range of external environmental contexts. Developmental pathways also determine the form of sexually selected signal traits, whether structural, behavioral, or chemical. In order for a signal trait to correlate reliably with health or condition, the pathways required for its development need to be well integrated into many other organismal processes and thus regulated by many gene loci. Otherwise, the signal will not reflect sender condition. Some kinds of traits may be intrinsically more integrated than others. To be reliable, the signal trait should also be costly or constrained, as we have previously discussed. When choosy females start to exert selection pressure on a male signal trait by favoring those males with more extreme values, mutant males that have less costly ways to produce the exaggerated trait will be favored in subsequent generations. The mechanism for achieving this cost reduction is to modularize the pathways needed for trait development. **Modularity** means isolating the pathways and loci for a given trait so that it is less dependent on other developmental pathways; ideally, the trait becomes a stand-alone structure, behavior, or chemical. This process makes the trait less informative for choosy females. If complete modularity is achieved, the trait might become highly elaborated but only informative about a very narrow subset of the male's physiology. One then expects females to counter-respond by attending to additional trait components that restore a more overall assessment of differences in male condition. One strategy might be to focus on a new component that enforces a trade-off with another component. If the environment is variable, different signal components might be favored in different contexts. By this process, a dynamic, multicomponent signal that reflects different organismal processes can evolve [24, 25, 228]. A conceptual illustration of the role of integration and modularity in the evolutionary process is shown in **Figure 12.14**. Examination of specific types of traits below will help illustrate these processes.

COLOR PATCHES   Color patches are a common target of sexual selection in many birds, lizards, some amphibians, fish, insects such as butterflies, and various other arthropods. The size, hue, chroma, and symmetry of a patch can vary independently and provide information about different aspects of condition. **Carotenoid-based color patches**

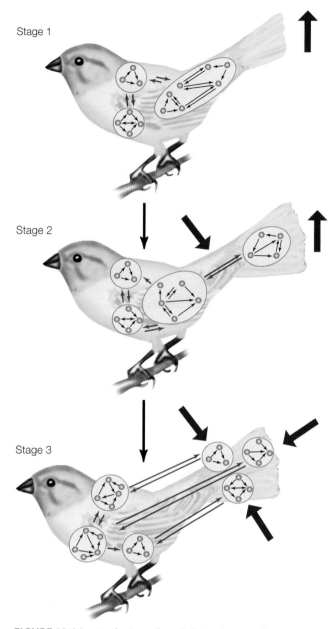

FIGURE 12.14   Evolution of modularity in sexual traits Small circles and thin arrows within the circles indicate individual physiological processes and their interactions. Large ellipses indicate modules of these physiological processes. Large brown arrows show the direction of external sexual selection forces. Stage 1: Female sexual selection favors the exaggeration of a detectable, well-integrated, condition-dependent trait that varies among males and indicates their quality; in this case, there is strong external selection for a longer tail as indicated by the upward brown arrow. Stage 2: As the tail becomes larger, its functions become less integrated with organismal functions, due to selection on males to reduce the cost of producing the long tail indicated by the second, oblique brown arrow. As a consequence, tail length becomes progressively less informative about male condition, except for those processes directly involved in tail production. Stage 3: Selection favors both further exaggeration of the tail trait, which is increasingly produced by trait-specific pathways as indicated by the three inward-pointing brown arrows, and greater integration of trait-specific pathways into organismal functions, as indicated by the long internal red arrows. Different components of the sexual trait therefore reflect different organismal processes, and in different environmental contexts may reveal different aspects of male quality. The result is a composite and dynamic sexual signal. (After [25].)

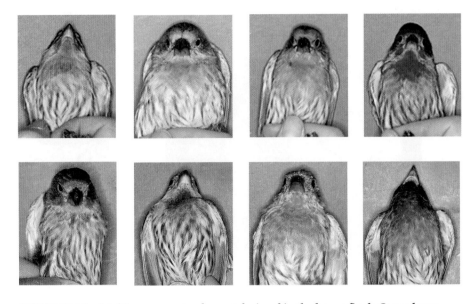

FIGURE 12.15 **Multicomponent color-patch signal in the house finch** *Carpodacus mexicanus* Birds in the top row show variation in hue from yellow to deep red. Birds in the bottom row show variation in patch size and saturation of the more typical red patch hue.

are particularly good multicomponent signals, because their expression is dependent on a mixture of external and internal processes. Carotenoid precursors must be obtained from the diet and therefore reflect the availability of specific foods and an animal's foraging ability. They must be ingested, selectively absorbed by the intestinal epithelium through endocytosis, converted to appropriate pigment forms, stored, and transported via protein carriers to target tissues, processes that require energy and many enzymatic reactions. The red chest patch of the house finch (*Carpodacus mexicanus*) is a well-studied example (**Figure 12.15**). Laboratory experiments that independently varied access to different types of pigments and overall food availability found that males on restricted diets were less able to produce chest patches with full red hue and intensity, while non-sexually selected plumage traits were not affected by the treatments [268]. Carotenoid color patches are also very sensitive to intestinal parasite infections that impede digestion [73, 267]. These studies found that carotenoid color patch production is not only energetically costly, but also more condition dependent than many other traits [592]. Color patch area, area symmetry, hue, and hue symmetry each vary to different degrees and are only partially correlated with each other. Wild males with redder patches and greater hue symmetry were more successful reproductively, while males with larger and more symmetrical patches had better survival [23, 26]. In the house finch and the guppy (*Poecilia reticulata*), carotenoid coloration is also heritable, permitting females to obtain indirect genetic benefits by choosing more colorful mates [82, 83, 265, 266]. Carotenoid color patches are under strong sexual selection by female choice in other avian, lizard, and fish species and show evidence of similar multicomponent signaling of different aspects of condition [81, 170, 177, 292, 366, 420, 426, 441, 461, 668, 766].

Melanin-based color patches are often found to function in intrasexual contexts, as described in Chapter 11, but they can also be the focus of female mate choice. Melanin, unlike carotenoids, is synthesized by animals from basic amino acid building blocks (mainly tyrosine). The biochemical pathway for melanin synthesis is long, complex, and tightly integrated, involving at least 80 gene loci; numerous enzymes; several hormones; and metallic cofactors such as zinc, iron, copper, and calcium (which must be ingested and therefore could be limiting). Moreover, the family of melanocortin regulatory genes is involved in a variety of physiological processes including the stress response, immune response, energy expenditure, sexual activity, and aggressiveness, as well as surface pigmentation. Melanin ornament expression is therefore likely to reflect the behavior and physiological condition of the sender [229, 328, 429, 455, 573, 578, 630, 687]. A survey of wild vertebrate species in which melanin-based coloration was studied in relationship to other traits revealed that darker individuals were more aggressive, more sexually active, and more resistant to stress [158, 233, 618]. In barn owls, females with darker breast markings are more successful reproductively and are preferred as mates by males (**Figure 12.16A**). Although dietary restriction usually does not affect melanin plumage coloration, an increase in stress via implantation of the stress hormone corticosterone does lead to paler plumage in nestlings [634]. Insects use melanin as a component of their immune response to encapsulate and kill parasites. Melanin precursors could be limiting in the specialized diets of some species and pose a trade-off between ornament production and immune function. As a consequence, melanin ornament trait expression has been found to be condition-dependent in some insect species (**Figure 12.16B**) [588, 685, 702].

FIGURE 12.16 Sexually selected melanin signals (A) Breast spottiness in barn owls (*Tyto alba*); male on left, female on right. Females with more and larger dark eumelanin spots have stronger antibody responses, better regulation of the corticosterone stress response, better ability to absorb calcium, earlier breeding age, higher reproductive success, greater adult survivorship, and offspring with better resistance to stress and ectoparasites. Spottiness is not condition dependent, but is highly heritable, and therefore qualifies as an index signal of immunocompetence. These correlates of eumelanin are pleiotropic effects of the melanocortin system. The within- and between-sex variation in coloration appears to be maintained as a consequence of sexually antagonistic selection, as spot size is positively selected in females but negatively selected in males [5, 156, 501, 627–629, 631–636]. (B) Dimorphic melanin coloration in the ambush bug (*Phymata americana*). Males have more melanin than females, and the amount of black in males is positively correlated with their mating success. The advantage of being darker is thermoregulatory: darker males heat up faster and can therefore spend more time searching for mates [588–590].

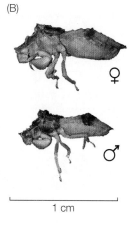

(A)

(B)

1 cm

Structural color patches, especially those based on iridescent reflectance, are some of the most spectacular visual displays in the animal kingdom. Because color patches are usually limited to males, we expect them to be under strong sexual selection by female choice. There is extensive correlational evidence that females of some bird species prefer brighter iridescent and UV-reflecting males [16, 46, 57, 304, 423, 520, 791]. This preference is also found in some species of fish [208, 367, 609, 688] and butterflies [354, 356, 527, 709]. The production of intense and saturated structural coloration requires the construction of precise laminar or lattice nanostructures in surface tissues. The development of these matrix structures is a multistage process [217, 582]. A disruption at any stage could easily result in a less-perfect matrix, so the appearance of structural colors ought to be highly dependent on an animal's condition as well as its overall genetic quality [186]. That structural integrity of reflecting surfaces is correlated with condition, mate preference, or components of fitness has been verified in both birds and butterflies [150, 152, 354, 356, 405, 432]. Experimental evidence that condition determines, and is not only correlated with, structural color properties has been demonstrated in jumping spiders (see Figure 11.16), eastern bluebirds (see Figure 11.17), the golden breast feathers of male turkeys [269], and UV plumage of nestling blue tits [324]. An outstanding example of a condition-dependent sexually selected iridescent signal, complete with genetic analysis, is described in **Figure 12.17** for the orange sulfur butterfly, *Colias eurytheme* [353, 355].

Condition-dependence of sexually selected **white color patches** and ornaments has been studied primarily in birds. White feathers were once believed to be inexpensive to produce because they do not require pigment synthesis or a specialized nanostructure, only random variation in air-keratin

FIGURE 12.17 Condition-dependent structural and pigment color signals in the orange sulfur butterfly *Colias eurytheme* Photo of the dorsal surface of a male under (A) normal white light and (B) UV light (with the source and viewing angles set to maximize UV reflection into the camera). Females base their mate choice primarily on the brightness of the UV reflection (purple shaded area in (C)). (C) Reflectance spectra from the yellow-orange area of the wing from the highest (solid line) and lowest (dashed line) UV-reflecting individuals. Graph also shows hue measurements: UV hue measured at the reflection peak in the 300–400 nm region (purple), and orange hue measured as the wavelength at the midpoint between the maximum and minimum reflectance values in the 450–700 nm region (orange). (D) Rearing larvae on high-quality versus low-quality diets affected aspects of both the visible and UV color components, but the effects were stronger for the UV component. A low-quality diet greatly reduced the UV brightness and the angular breadth over which the UV reflectance occurred, but not the UV hue. A low-quality diet reduced the hue and chroma of the yellow-orange reflectance. The horizontal lines in the boxes show the mean values; significant differences indicated by an asterisk. (E) The UV color is produced by a laminar-ridge nanostructure on the wing scales. Front and side view cross-sections through the cuticular structure on the scale surface shows three adjacent ridges, each with approximately four lamina. The orange-yellow color is produced by UV absorbing pterin pigment crystals situated below the laminar nanostructure. The pigment depresses the peak UV reflection slightly, but strongly depresses a diffuse UV reflection that emanates from the scales. This latter effect enhances the directionality and spectral purity of the UV iridescence, and amplifies the contrast between the UV reflectance and the scale background color as the male flaps his wings during flight. Individual male variation in UV reflectance caused by nutritional stress can be attributed to specific changes in the laminar ridge nanostructure. Brighter reflection occurs when the spacing between ridges (R) is smaller as shown in (F), and when there are more lamina. A wider angular breadth results from larger laminar termination distance (TD) as shown in (G), which is associated with longer lamina and shallower angles of insertion. Dietary restriction reduces the overall density of cuticle-based reflecting surfaces. UV reflection indicates phenotypic, but not genetic, quality of males. Females receive a nutritious spermatophore from the male during mating and benefit from mating with mates in better condition using UV reflectance as a mate quality cue. (After [353, 355, 643].)

boundaries to generate incoherent scattering (Chapter 4 [581]). Nevertheless, the size and/or brightness of white color patches is correlated with condition, dominance, or breeding success in several species [94, 151, 176, 206, 225, 234, 247, 451, 545, 671, 729, 747, 791]. Two experimental studies, one manipulating nutrition in juncos and the other imposing an immune challenge on female eiders, resulted in reductions in the size and brightness of the white patches in subsequently grown feathers (**Figure 12.18A**) [248, 452]. Good diet and condition increase the density of keratin and the number of reflecting surfaces in white feathers. However, white color patches may indicate mate quality through other mechanisms as well. The **revealing handicap mechanism** posits that some signals more readily reveal parasite infection [243]. Melanin and other pigments such as carotenoids and psittacofulvins strengthen the integument against abrasion and wear, and also protect against feather-degrading bacteria,

mites, lice, fleas, and other parasites [66, 233, 429, 454]. Depigmented feathers and skin are therefore more vulnerable to damage. Female barn swallows *Hirundo rustica* prefer males with both longer tails (see below) and larger white tail spots (see Figure 12.19A). *Mallophaga* lice chew holes in the white areas of feathers in preference to the pigmented areas, both creating visible damage and increasing the probability of feather breakage. Contrary to the expectation that barn swallow males with longer tails and larger white spots would have higher infestation levels, they actually had lower levels of infestation, implying that these males possessed better immune systems and could afford to display larger areas of white plumage [377, 378]. The frequent occurrence of white tail feather tips in many avian species (**Figure 12.18B**), which is not associated with tail length or shape as would be expected for a simple shape amplifier, suggests that white tips may function as a revealing feather-quality handicap signal [185].

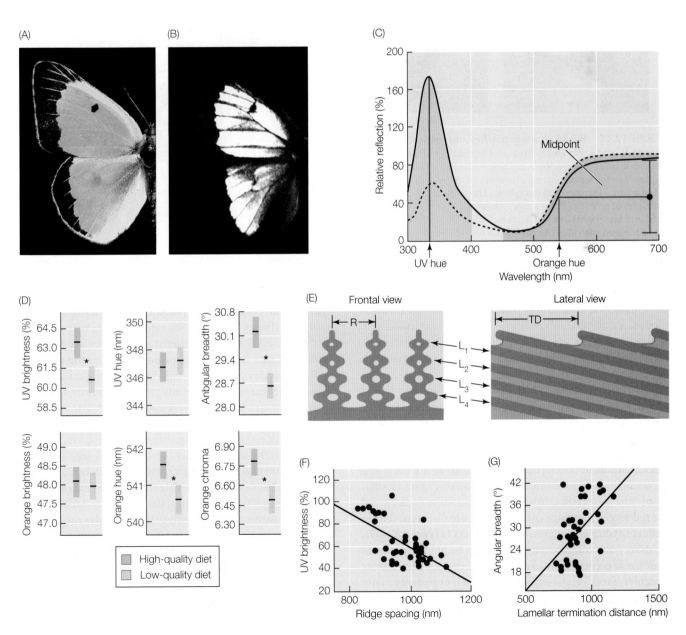

FIGURE 12.18 White plumage patches and patterns (A) A male dark-eyed junco (*Junco hyemalis*) flashing his white outer tail feathers in an aggressive posture. These white tail feathers are also displayed to the female during courtship. (B) White tips on the tail of a white-throated hummingbird (*Leucochloris albicollis*). (C) Examples of great tit (*Parus major*) white cheek patches with high (left) and low (right) immaculateness score. (D) Flank feathers of the red-legged partridge (*Alectoris rufa*) form alternating vertical rows of white, black, and brown stripes.

Contrasting white and dark adjacent patches create complex color patterns in many birds, as well as in some lizards, fish, and butterflies [655]. Variation among individuals in the sharpness, regularity, or continuity of a standardized species pattern may indicate differences in individual quality. For example, male and female shelducks (*Tadorna tadorna*) mate assortatively with respect to the **immaculateness** (sharp, continuous margins) of their white chest markings, and immaculate pairs are the most successful breeders [175]. Experimental disruption of the smoothness of the white/black margins of cheek patches in male great tits leads to a reduction in status and reproductive success (**Figure 12.18C**) [176]. The bold vertical black, brown, and white flank bars of the red-legged partridge (*Alectoris rufa*) are displayed during both courtship and aggressive encounters (**Figure 12.18D**). Birds in poor condition appear not to be able to pigment their feathers completely to the edge and therefore have more irregular bars [70]. Complex color patterns provide many more opportunities to advertise individual quality than simple hue patches alone [254].

ELONGATED TAILS   **Tail elaboration** is a sexually selected trait in some birds and fishes. Because the tail serves a critical function in locomotion, the potential cost of a larger-than-optimal tail is reduced foraging efficiency or ability to escape predators. In the well-studied European barn swallow, males possess outer tail feather streamers that are approximately 12 mm longer than optimum for flight maneuverability. Females prefer long-tailed males. Long-tailed males have higher extra-pair paternity rates and greater survivorship, and their offspring have greater resistance to parasites. Experimental tail elongation increases male attractiveness to females, but reduces foraging maneuverability and captured prey size in this aerial forager [464, 465, 467, 468, 471, 473, 478]. Tail elongation also decreases survival, but naturally long-tailed males are better able to bear the cost of experimental tail elongation than naturally short-tailed males [472]. However, recent studies show that the meaningful variation in male tail length is actually determined by naturally selected individual differences in optimal tail length, and the additional 12 mm elongation provides no further information to females about male quality [78, 491, 640]. The elongated tail of the male barn swallow (**Figure 12.19A**) may be an example of the emancipation and fixation of a sexually

(A)

(B)

(C)

FIGURE 12.19 **Famous tails** (A) Barn swallow (*Hirundo rustica erythrogaster*) in flight, showing both elongated outer tail streamers and white patches. Male tail length is a stronger focus of female mate preference and fitness in European populations compared to North American populations [498, 649]. (B) Jackson's widowbird (*Euplectes jacksoni*) males prominently advertise their tails in leaping aerial displays above their grassland habitat, and in tail-quivering courtship displays around the central tuft of their circular display court [12, 13, 15]. (C) The extraordinarily exaggerated tail display of the male peacock (*Pavo cristatus*). The features of the tail and male display behavior that females prefer appear to vary significantly among populations [424].

selected trait that now operates as an amplifier of a naturally selected trait. This example also shows us that not all sexually selected traits become condition-dependent indicators of sender quality at equilibrium [337, 650].

Tail length is clearly a target of female choice in several species of long-tailed widowbirds (*Euplectes* spp.) (**Figure 12.19B**). The primarily black-plumaged males perform conspicuous aerial courtship displays that showcase their tail plumes. Experimental lengthening of the tail increases male mating success but also results in a steeper decrease in body condition during the breeding season, compared to experimental tail shortening. Natural tail length is positively correlated with male condition in several species, but it is not yet known what benefits females might receive from their choice [10, 14, 583, 586]. Tail length among widowbird species is related to body size with strong positive allometry, suggesting that sexual selection drives tail size to the maximum that the birds can bear [128]. The classic long-tailed bird, and the inspiration for Darwin's treatise on sexual selection and current sexual selection models, is of course the peacock (*Pavo cristatus*) (**Figure 12.19C**). Studies by multiple research groups have generated conflicting conclusions about the main targets of female choice. Tail length seems to

have reached a physical limit, varies little among fully mature males, and does not seem to be directly selected at this evolutionary stage, although it presumably was under strong sexual selection in the past. Instead, the density, size, and reflective properties of the iridescent eye spots, and other components of the male display such as tail rattling, orientation with respect to the sun, and display vocalizations, appear to be the features currently affecting mate choice in different populations and species [133, 423, 548, 711, 794]. A controlled breeding experiment in one population provided evidence of good-genes effects, with both male and female offspring of sires with larger eyespots exhibiting faster growth and better survivorship [549, 550]. Avian tails, whether elongated or not, are excellent vehicles for color patches and other markings that reveal feather quality of the bearer, and thus can serve as revealing indices of male quality [185].

Fish tails are another matter entirely. All else being equal, females prefer longer-tailed males in guppies (*Poecilia reticulata*) and various swordtail species (*Xiphophorus*). However, careful experiments indicate that it is larger body size that females prefer, and longer tails merely give the illusion of larger total size [343, 625]. Females obtain good-genes benefits by mating with larger males, in the form of faster growth and better survival of offspring [604, 763]. In both fish taxa, some males deceptively invest in tail growth at the expense of their general body growth, reducing the opportunity for females to benefit from the large-size perceptual bias [39, 344].

BEHAVIORAL DISPLAY   Several studies and comparative analyses have examined the relative importance of **behavioral displays** such as acoustic signals and visual movement displays compared to morphological traits such as color patches, structures, and body size as targets of female mate choice. Behavioral traits tend to predominate, although there are certainly examples of female preferences for morphological traits [11, 184, 335, 706]. These reviews also point out that morphological traits are more important determinants of male–male competitive ability or dominance than behavioral traits. In lek mating systems, where males provide neither resources to females nor parental care to offspring, females exert extremely skewed mating preferences for certain males based on their behavioral display traits [14, 29, 141, 184, 218, 280, 422, 508, 647, 664, 700].

There are several reasons why behavioral traits may be better targets of female choice. First, behavioral traits are far more variable than morphological traits, providing females with a wider basis for comparing males [508]. Second, males typically repeat displays at high rates for several hours each day, so even though short-term display costs as measured in the lab may not be especially high, continuous displaying for days on end is very energetically costly. Coupled with the additional cost of reduced time available for foraging, repetitive displaying should be a reliable indicator of current male condition [253, 706, 746, 779]. Third, the **performance quality** of complex displays, including their fine structure, choreography, consistency, and rate, requires the integration of many morphological and physiological developmental pathways and involves more developmental genes, so behavioral display traits are not as easily converted into a modular ornament-specific pathway that costs less to produce. [24, 392]. Likewise, physically challenging display maneuvers push the musculoskeletal system to its limits. When female choice exerts strong selective pressure on males to display at their maximum performance level, we expect to observe variation among individuals that is closely tied to variation in physiological condition, health, morphology, and ultimately fitness [308, 311]. Finally, behavioral traits may be indicative of locomotor performance abilities associated with survival behaviors. Consistent with all of these arguments, a meta-analysis of the within-species correlation between trait size and survival of the signaler across forty species of vertebrates and invertebrates found a much stronger positive association for behavioral traits than for morphological traits [330].

Female preference for males that display more vigorously is widespread. Whether display effort is measured as the rate of repeated songs, calls, or movement displays, or as the duration of displays or display bouts, greater effort entails an increase in energetic cost, which can be a substantial proportion of daily energy expenditure (see Chapter 9). Numerous studies have demonstrated that males in better nutritional condition or health produce more vigorous and costly mate

attraction displays [123]. We have already described one such example, the condition-dependent drumming in a lycosid jumping spider (see Figure 10.26). Females in several cricket species prefer males that spend more time calling and chirp at higher rates within bouts. The amount and quality of food provided to adult crickets significantly affect these costly components of the acoustic signal, but do not affect carrier frequency or pulse duration [47, 257, 274, 286, 342, 657, 753]. The primary benefits of mate choice based on these costly signal traits for females appear to be direct: male crickets transfer sperm and material resources to females in their ejaculate as well as in the spermatophore, which is consumed by the female. Females that chose males in better condition have higher egg fertilization rates, larger clutches, and sometimes better survivorship [95, 754]. Furthermore, they may also obtain some indirect benefits—attractive sons and a higher rate of offspring survival [171, 256, 314, 393, 617, 683]. Anurans are another group of acoustic signalers in which calling is energetically expensive and females prefer males that excel in producing the most costly call parameters [779]. **Figure 12.20** illustrates the results of studies on the calls of the gray tree frog (*Hyla versicolor*), in which there is an energetic trade-off between call rate and call duration. In this case, females primarily focus on call duration, which is a heritable male trait. Sires with long calls produce more-fit offspring [775, 776]. Finally, song rate is an important mate choice criterion for many avian species [662]. Condition-dependence and mate choice consequences have been studied best in the zebra finch (*Taeniopygia guttata*). Males in better condition have both redder beak coloration and higher song rates, and are preferred as mates. Song rate is heritable, but also more dependent on short-term changes in resources and condition than beak color. Offspring of attractive sires produce heavier offspring and sons that also deliver higher song rates [54, 295].

Beyond basic display rate differences, the details of performance quality, such as the fine structure and choreography of acoustic and visual displays, are also targets of female choice. These signal components are often affected by conditions during larval or juvenile development rather than by immediate resources available to adults. For example, the dominant frequency of chirps in crickets is affected by nymphal nutrition, because poorly fed nymphs emerge as smaller adults with smaller stridulatory harps that produce higher-frequency chirps. Chirp rate and bout duration, as mentioned earlier, depend on immediate resources and adult condition [656].

Nestling songbirds raised on restricted diets or challenged with parasites grow up to sing shorter or less complex songs, develop smaller song repertoires, or copy their tutors less accurately [55, 87, 88, 285, 427, 511, 692, 695, 696, 798]. Neuroanatomical studies suggest that stress or dietary restriction during development reduces the size of the higher vocal center (HVC) nucleus in the brain, which is responsible for song learning and vocal production [89, 427, 551, 695]. The song control nuclei in the brain develop later than the rest

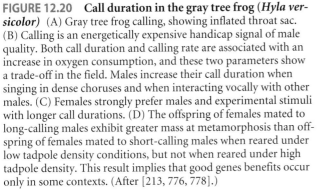

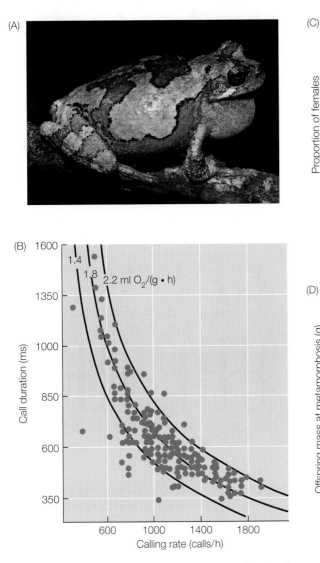

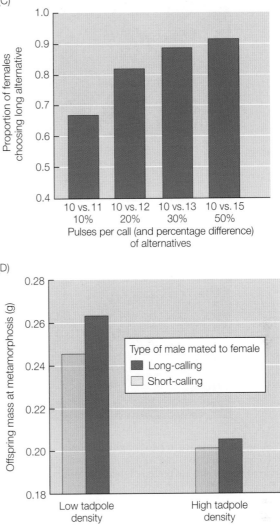

FIGURE 12.20 Call duration in the gray tree frog (*Hyla versicolor*) (A) Gray tree frog calling, showing inflated throat sac. (B) Calling is an energetically expensive handicap signal of male quality. Both call duration and calling rate are associated with an increase in oxygen consumption, and these two parameters show a trade-off in the field. Males increase their call duration when singing in dense choruses and when interacting vocally with other males. (C) Females strongly prefer males and experimental stimuli with longer call durations. (D) The offspring of females mated to long-calling males exhibit greater mass at metamorphosis than offspring of females mated to short-calling males when reared under low tadpole density conditions, but not when reared under high tadpole density. This result implies that good genes benefits occur only in some contexts. (After [213, 776, 778].)

of the nervous system, specifically during the nestling and juvenile stages in conjunction with auditory input, and therefore are vulnerable to environmental conditions and parental provisioning ability. Female birds show clear preferences for subtle details of song structure potentially affected by rearing conditions, such as longer and more complex songs, high-performance trills, higher-amplitude singing, more consistent songs and syllables, and local song dialects [32, 194, 222, 284, 511–513, 535, 561, 601, 661, 663, 697, 739]. **Figure 12.21** illustrates some examples of these high-quality songs. These song features are difficult to sing well because they require highly coordinated motor patterns of syringeal and respiratory muscles (see Chapter 2). High-quality singing thus reflects good nutrition during development, as well as good genes for sensory and motor coordination. This **developmental stress hypothesis** argues that signal reliability is guaranteed by the physiological constraint of juvenile brain development for song learning, not by differential production costs. By selecting males that sing high-quality songs, females gain direct benefits of more physically robust, stress-resistant mates who can acquire and defend a good quality territory and provide paternal care to their offspring; they may also gain indirect genetic benefits [33, 55, 326, 428, 551, 601]. A preliminary population genetic model of this hypothesis finds that a gene causing males to bias their song learning toward challenging song types, and causing females to prefer such song types, can spread when only males that are both high quality and reared under good conditions are able to produce the challenging song types [610].

(A)

(B)

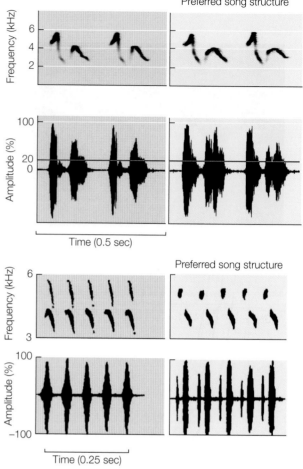

FIGURE 12.21 Song performance in two songbirds Both examples show the spectrogram (top) and waveform (bottom) for a song that is preferred (right) and not preferred (left) by females. (A) Songs of the dusky warbler (*Phylloscopus fuscatus*) contain frequency-modulated notes. Preferred males can perform a larger proportion of each note at a higher amplitude above the gray 20% amplitude threshold line. (B) Canary (*Serinus canaria*) songs consist largely of different trills. Preferred males can perform rapid double-note trills, whereas nonpreferred males perform single-note trills; the difference is evident in the waveform graphs. (A after [194]; B after [739].)

The basic principle of this hypothesis can be generalized to any behavioral display trait that requires a high level of neural and muscular coordination, multitasking, and possibly practice to perform well, such as the audiovisual display of the sage grouse, the leaping and aerial displays of some birds, and the courtship dances of some *Drosophila* flies [308, 444, 539]. A few recent studies have attempted to discover precisely how display performance links to physiological ability, survival behaviors, and genetic quality. In guppies, details of the performance of the sigmoid courtship display preferred by females are correlated with fast start kinematics, escape speed, and success in avoiding predators [515]. In a cricket, however, female-preferred high calling rate was negatively correlated with jumping ability. Both studies suggest that there are complex trade-offs among display traits, physiological capacity, and survival behaviors that make it difficult to identify a single indicator trait of genetic quality [393].

CHEMICAL SIGNALS Although chemical signals were traditionally viewed as providing primarily species and sex identity and sexual receptivity information, recent studies show that chemical signals also vary on an individual basis and can provide information about condition, health, and other aspects of mate quality [332]. Most of the studies investigating chemical mate assessment have been conducted on insects, lizards, and mammals. Chemical signals involved in mate choice are all multicomponent blends, where the relative proportions of key components as well as the presence or absence of some components are affected by a variety of physiological processes. This means that selection can operate in complex ways, sometimes in opposing directions on the different components, to yield individually variable blends that reflect the characteristics of the sender.

Chemical blends often vary with body size, a proxy for condition. Certainly in insects, adult body size reflects nymphal or larval rearing conditions; in vertebrates, size can reflect both rearing conditions and age. The female *Utetheisa* moth shown in Figure 12.10 uses the amount of pheromone released by the male to judge male size, which in turn indicates the larval diet quality of the male and the amount of protective alkaloids he will provide the female during mating [115]. In the cockroach *Nauphoeta cinerea*, both the amount and the ratio of three main chemical components of the male pheromone vary with condition, and females use the ratio information to select good-condition mates [107]. The same

(A)

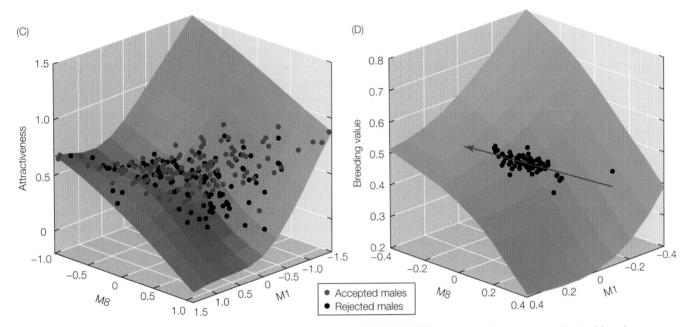

(B)   Main components of pheromone blend

Cuticular hydrocarbon	Heritability h^2	Correlation with body size	Selection gradient β
Z,Z-5,9-$C_{25:2}$	0.089*	0.240*	0.041
Z,Z-5,9-$C_{27:2}$	0.405*	0.043	−0.081
Z,Z-5,9-$C_{29:2}$	0.271*	−0.106	0.069
2-Me-C_{26}	0.124*	0.212*	−0.035
2-Me-C_{28}	0.050	0.247*	0.132
2-Me-C_{30}	0.062*	0.107	−0.045

(C)

(D)

- ● Accepted males
- ● Rejected males

FIGURE 12.22   Adaptive landscape analysis of female preferences for male cuticular hydrocarbon pheromones in *Drosophila serrata* (A) A mating pair of flies, male on left. Females in this species accept or reject courting males based on the cuticular hydrocarbons on the male's body. (B) Although production of each of the individual hydrocarbons is heritable, and three are significantly correlated with male body size (asterisks), the sexual selection gradients for individual hydrocarbons are low (right-hand column). (C) An adaptive landscape analysis (see Figure 9.11 for method) shows that fitness (measured here as male attractiveness and plotted as a blue surface) is strongly affected by the values of the additive genetic components in two independent sub-mixtures (M1 and M8) of male hydrocarbons. Points mark actual locations in this landscape of a large sample of accepted and rejected males. If an ellipse were to enclose these points to represent the G-matrix for this sample (see Figure 9.11), note that the longest axis of this ellipse, along which any future evolution is most likely to move, would be nearly perpendicular to the slope (selection gradient) of the adaptive surface. (D) A similar plot for a subset of the males for whom relative fitness could be measured by seeing whose pheromonal mixtures were most common in the next generation (male breeding values). The vector of greatest variance (red arrow) is 75° away from that expected were the G-matrix and selection gradient to be pointing in the same direction. Prior strong selection by females has "used up" most of the additive genetic variation among males along the axis of their preference. Subsequent selection will produce little change in underlying allele frequencies, since there is little variation along the axis where choice is exercised. (After [61, 273, 282].)

pheromones also signal social status of the male, but females prefer a different ratio of components from the one indicating male dominance, to avoid mating with manipulative males. Females mated to preferred males obtain both direct and indirect benefits [480, 481, 483, 486, 487]. Mate choice experiments in lizards, salamanders, and mammals find that both males and females prefer larger or better-condition mates and can make this discrimination using olfactory signals alone [174, 417, 438, 553].

Studies of Australian *Drosophila serrata* fruit flies have made impressive progress in understanding the underlying genetics of condition-dependent chemical mate choice signals. The flies aggregate around fruiting trees in the tropical forest and use blends of 8–10 cuticular hydrocarbon compounds to signal species identity, sex, and condition (**Figure 12.22**). Females prefer larger males and assess body size using two or three of the components that are positively correlated with size. These particular components, all methylalkanes, are believed to be costly to produce and thus are reliable indicators of male condition. The male hydrocarbon profile is heritable, and females gain some benefits in terms of offspring survival by mating with preferred males. However, males with extreme hydrocarbon blends sire lower-quality offspring, resulting in stabilizing selection on the male trait. Very intriguing is the finding that most of the genetic variance in the male trait is not focused on the component ratio preferred by females, but is orthogonal to the direction of sexual selection [60, 272, 273]. This means that there is little genetic variance for selection to act upon, so females may not obtain much genetic benefit from their choice. This research has called into question the idea that selection for multiple-component condition-dependent traits always captures and maintains genetic variance [742].

Chemical traits also possess the ability to indicate immunocompetence and current levels of infection. In mice, rats, lizards, and grain beetles, females can distinguish an infected from an uninfected male using olfactory cues alone, regardless of whether the infection is due to a bacterium, a virus, or a protozoan parasite [346, 417–419, 421, 441, 542, 598, 797]. These effects could be caused by several mechanisms: (1) infection might change the composition of commensal microbes that shape individual odors; (2) infection might trigger the major histocompatibility complex (MHC)-based immune response, which then alters urinary odor; or (3) activation of the immune system might alter excretion of other metabolic byproducts such as the stress hormone corticosterone, or reduce androgen levels [541, 612].

FLUCTUATING ASYMMETRY   **Fluctuating asymmetry** refers to subtle random differences between the two sides of a bilateral trait, as opposed to the functionally deliberate asymmetry of traits such as claw size in the fiddler crab. The degree of fluctuating asymmetry is thought to be caused by developmental instability as a result of genetic and/or environmental stress. It has been proposed that such asymmetries might therefore serve as cues about a potential mate's phenotypic or genotypic quality, and could become ritualized into a signal of mate quality [470, 479]. There are some examples of correlations between symmetry of bilateral traits and male mating success, but these correlations might be side-effects of mate choice for other sexually selected traits correlated with symmetry—for instance, locomotory or female-controlling appendages that enable males to compete better against other males [118, 303, 379].

Several experimental studies have demonstrated female preferences for males with symmetrical mate attraction signals. Examples of such traits include tail streamers in the barn swallow [466, 469]; conspicuous foreleg tufts in a wolf spider [733]; the vertical black side bars revealed during alternating lateral displays in several fish species [488, 489, 658]; pelvic spines in the stickleback [447]; chest plumage pattern in the zebra finch [709]; facial features in humans [224, 681]; and even one acoustic trait, stridulatory calls produced by different numbers of pegs on the left and right legs in a grasshopper [476]. However, there is no evidence that such mate preferences in any species provide females with a genetic benefit, as the degree of fluctuating asymmetry is almost universally found to be not heritable [400, 564]. Moreover, asymmetry

is generally uncorrelated with condition or with the absolute size of the trait, which would be expected if asymmetry indicated developmental instability and overall adaptedness to the environment [564]. The emerging consensus is that asymmetries in sexual characters may reveal trait-specific developmental instability caused by single-gene mutations, but that such asymmetries do not reveal overall genetic quality. Fluctuating asymmetry is therefore unlikely to be a major indicator of mate quality.

Genetic compatibility

So far in this chapter we have referred to a potential mate's genetic quality only in terms of "good genes." Mate choice based on elaborate condition-dependent male traits provides a heritable genetic benefit to all females. Each "good gene" is an allele that increases offspring fitness by some increment, regardless of the architecture of the rest of the female's genome. We expect all females in a population to show relatively congruent choice for the same elaborated male traits, and populations will thus undergo directional selection.

There is a second component of genetic quality which has to do with **gene compatibility**. A compatible gene is an allele that increases fitness only when situated in a specific genome. For example, a pairing of unlike alleles at a locus may result in **heterosis**—improved fitness in the heterozygotic offspring due to genetic dominance or overdominance. Or a particular allele may result in **epistasis**—a positive or negative fitness effect due to an interaction with an allele at another locus. Such fitness effects lead to **nonadditive genetic variance** [200, 312, 407, 496, 497, 591]. This type of genetic variance is *not heritable*. Nevertheless, females can benefit by selecting males with compatible genes [445, 495, 800]. In particular, the survivorship of their offspring may be enhanced via heterosis. The consequence of mate choice for compatible genes is that females in a population are expected to prefer males whose genotypes are different from or compatible with their own. Furthermore, populations will not respond to selection on genetically nonadditive male traits, but female preference for acquiring compatible genes *can* evolve. Good genes and compatible genes together comprise **total genetic quality** [495]. If there is a trade-off between a good gene and a compatible gene, the female mate choice criterion can fall anywhere along a gradient favoring one or the other, or females may attempt to optimize both components of genetic quality [591]. Models of good versus compatible genes also suggest that populations could cycle in evolutionary time between mate choice for good genes and compatible genes, because good-genes sexual selection erodes additive genetic variation. In time, reduced variation boosts natural selection for compatible genes, which subsequently rebuilds additive genetic variation [114, 298, 495].

Nonadditive genetic effects can explain a significant amount of phenotypic variance in offspring survivorship in controlled breeding experiments, and evidence that females base their mate choice on genetic dissimilarity from their own genotype is accumulating for both invertebrates and

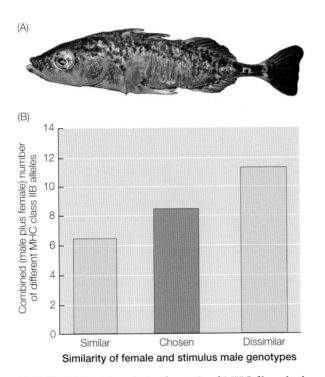

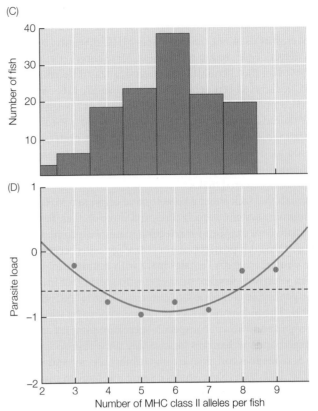

FIGURE 12.23 Mate choice for optimal MHC diversity in the three-spined stickleback (*Gasterosteus aculeatus*) (A) Healthy male in full nuptial color. (B) Results of choice tests by females presented with water from male tanks, where males differed in their MHC allele similarity with the female. The chosen male was the one that resulted in an intermediate combined male-plus-female allele diversity. Rejected males were those that would result in either a lower or a greater combined number of different alleles. When females had low allele diversity they preferred males dissimilar to themselves; when they had high diversity, they selected males more similar to themselves. (C) Frequency distribution of allele number in a natural population of sticklebacks; the mean is 5.8 alleles. (D) Fish with an intermediate number of alleles had lower parasite loads than fish with either small or high numbers of alleles. (After [2, 387, 463].)

vertebrates [358, 446, 463, 495, 544, 767, 768]. The primary benefits appear to be: (1) maintenance of sufficient **heterozygosity** at the MHC loci in vertebrates, which improves immunocompetence and offspring viability; and (2) sufficient outbreeding. Inbreeding exposes deleterious recessive alleles and causes significant fitness loss (called *inbreeding depression*) in all sexually reproducing organisms [129, 348]. Outbreeding reduces this risk. On the other hand, extreme outbreeding, for example, choosing a mate from a different population, can disrupt locally adaptive gene complexes. Thus **optimal outbreeding** occurs at some intermediate value. As we shall see below, there is also an optimal diversity for the MHC. To achieve the nonadditive genetic benefits of optimal outbreeding and optimal MHC variability, an animal must be able to detect the genotype of a potential mate and assess its similarity to its own genotype. How might they do this?

Vertebrates have co-opted the MHC-based adaptive antigen recognition system for this purpose. As mentioned in Chapter 6, the hypervariable MHC loci produce allele-specific glycoproteins that reside on all cell membranes, pick up foreign peptides in the cell, and present them to T-cells (lymphocytes in the blood), which then mount an attack to kill the pathogenic cells. MHC molecules have a groove on the outer surface that binds with a limited set of peptides. The ideal immune system has an optimally intermediate variety of peptide-binding types, enough to pick up foreign peptides from the current dominant parasites in the population, but not so many types that a large fraction of self-produced peptides are bound, overwhelming the ability of the T-cell system to fine tune its self versus nonself selectivity [510]. Females seeking sires that will give their offspring good immune systems should select mates that not only demonstrate good health, but also possess an array of MHC alleles that will complement their own [463].

Vertebrate species with a vomeronasal organ can make this potential mate compatibility assessment prior to mating using chemical signals. **Figure 12.23** demonstrates the ability of female three-spined sticklebacks (*Gasterosteus aculeatus*) to make this discrimination and select mates that optimize the MHC diversity of their offspring. Animals release MHC glycoprotein and peptide compounds in urine and sweat. The vomeronasal organ has a cluster of sensory receptor cells, called V2R neurons, that co-express specific MHC alleles [415], and a few olfactory neurons also express MHC alleles. The presence of MHC glycoprotein molecules apparently makes these sensory neurons sensitive to the same peptides picked up by the MHC glycoproteins operating in the immune system [62, 313, 686, 801]. This sensory system provides the self-referent template and the means for comparing it to the glycoprotein secretions of other individuals. Moreover, it is used not only to assess MHC diversity, but also to detect degree of kin relatedness. Mate choice preferences based on the chemical perception of allelic differences at MHC loci have been demonstrated in fish [2, 116, 396, 462, 463], amphibians [210], lizards [519], rodents [543, 544, 613, 793], and humans [614, 765]. Several of these studies show that females permitted to mate with their preferred male produce offspring with greater or optimal MHC heterozygosity and higher offspring survival, compared to females mated to nonpreferred males.

Animals that lack either a vomeronasal organ or a MHC-based immune system, namely birds and invertebrates, have an alternative mechanism for selecting mates with compatible genes: this is a post-copulatory strategy called **cryptic female choice**, whereby females mate with multiple males and then selectively fertilize their eggs with sperm that meet specific genotypic compatibility criteria [53, 161, 230, 682, 721, 799]. The biochemical signaling interactions by which eggs recognize and accept conspecific sperm are complex and multistaged, and mechanisms for assessing the within-species degree of relatedness or genetic compatibility are not known [108, 734, 750, 802]. Nevertheless, some mechanism must exist, because there is ample evidence for nonrandom genotype-specific fertilization that leads to enhanced offspring immunocompetence or optimal outbreeding [56, 65, 76, 106, 179, 182, 188, 196–198, 333, 347, 358, 560, 642, 723].

Female-specific preferences for compatible genes via both inbreeding avoidance and MHC diversity should result in optimally heterozygous offspring. As with compatible genes, this nonadditive genetic benefit is not heritable. However, recent models show that female preference for more outbred or heterozygous mates can evolve, especially in small inbred populations [200, 496]. A nonadditive genetic benefit thus becomes converted into an additive genetic benefit. Assuming that heterozygosity becomes associated with a costly ornamental male trait, a directional mating preference for a male-quality trait could evolve and provide females with indirect genetic benefits, while at the same time maintaining heritable variation. This idea implies that heterozygosity itself imparts good-genes benefits [85, 358, 381]. There are several examples in which more heterozygous males express higher-quality mate attraction signals and are preferred by females; the associated traits include repertoire size in song sparrows [602, 603], scent marks in house mice [310, 719], and plumage crown color in blue tits [209].

Age indicators

Females may base their mate choice on **male age** if the age of a potential mate affects their direct or indirect fitness benefits and if they have some means of assessing male age. A popular prevailing view is that old age is an indicator of high genetic quality. Older males demonstrate their superior quality merely by surviving, since lower-quality individuals carrying deleterious mutations are eliminated from the

population by mortality selection at younger ages [84, 437, 731]. Females that select older males should therefore obtain more viable mates that will pass on good genes to their offspring. A dynamic simulation model designed to explore the conditions under which male age could be an indicator of good genes supports this concept [42]. In this model, survival is determined by genetic variation in several viability-related traits. Females are assumed to have perfect knowledge of male age, and the model allows the strength of female preference for males of each specific age to evolve. Under a wide range of conditions, females evolve stronger preferences for middle-aged or older males compared to young males. However, this model does not address the types of cues or signals females use to assess male age. Another model has attempted to incorporate signaling effort and life-history trade-offs into the evolutionary process [369, 371]. In this case, there is between-male genetic variation in condition (i.e., resource acquisition). Each male allocates its limited resources to either costly advertisement display or survival, and the ratio of investment in display versus survival can be varied at each age to maximize lifetime reproductive success. Depending on the shapes of the advertisement-survival trade-off curves for males of different condition, advertisement and longevity can be positively correlated across males, and the optimal male strategy is to increase display effort with age, meaning that the most vigorous displayers are also the best survivors. Females can therefore use display vigor as a mate choice cue to obtain mates that are both older and of higher genetic quality and condition. However, under some model conditions honesty is not preserved, and an optimally behaving low-quality male may advertise more than a high-quality male of the same age. The low survivorship of such cheaters keeps their proportion in the population low, so females will still benefit *on average* by trusting display intensity as an indicator of viability.

These models provide predictions about the types of signals females use to assess age (cues or index signals as suggested by the first model, costly handicap signals as suggested by the second set of models) and the types of benefits (direct or indirect) females obtain by choosing older males. The empirical evidence provides many examples of male-age-based female preferences, and often a plausible male trait that females can use to assess age (**Figure 12.24**). Many of these studies involve extra-pair mating in birds, where investigators can compare the growth, immunocompetence, and survival of maternal half-sibs with sires of different ages and secondary sexual characteristics. A metadata analysis shows that older age is the most consistent characteristic of extra-pair sires [3]. The brightness of UV/blue plumage is one signal trait that increases with age in several avian species and is preferred by females [140, 334, 357, 398, 672, 677]. Acoustic signals such as song repertoire size, performance of rapid broadband trills, consistency of repeated notes, and early onset of dawn singing increase with male age in a variety of avian species and are correlated with female mating preferences [32, 33, 71, 120, 139, 207, 219, 222, 408, 563, 611]. These song

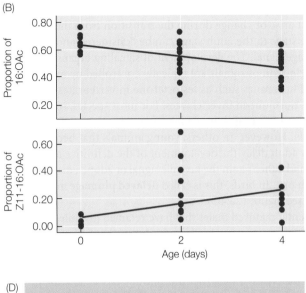

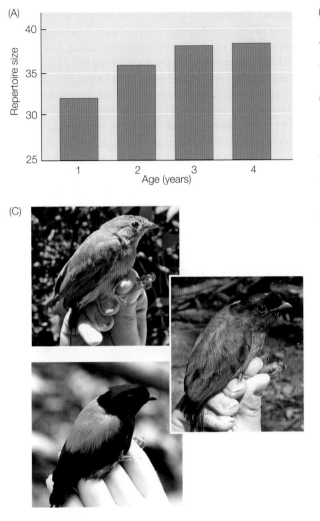

FIGURE 12.24 Signals of age (A) Repertoire size increases with age for the first few years in Swedish populations of the great reed warbler (*Acrocephalus arundinaceus*). Females prefer males with larger repertoires for both social and extra-pair mates [195, 251, 252]. (B) In the corn borer moth (*Ostrinia nubilalis*) the proportions of two components of the male chemical pheromone blend change with male age: the proportion of Z-11 hexadecenyl acetate (Z11-16:OAc) increases, while the proportion of hexadecanyl acetate (16:OAc) decreases. Females of this species prefer older males [399]. (C) Male long-tailed manakins (*Chiroxiphia linearis*) require five years to reach the definitive male plumage color (bottom), progressing from the first-year pattern of olive green with some red head feathers (top) through the black-faced pattern at three years (middle) [153, 449]. (D) Dewlap size in the eland antelope (*Taurotragus oryx*) bull increases with age. (A after [250]; B after [399].)

features are reliably correlated with age because they require learning, practice, and good condition. Such signals could be indices of age, costly handicaps, or a combination. A good-genes effect was found in the great reed warbler (*Acrocephalus arundinaceus*), in which repertoire size increases over the first few years of life (see Figure 12.24A), females prefer large repertoires in both social and extra-pair mates, and offspring of large-repertoire males have higher survival and offspring recruitment rates. Female warblers also obtain direct benefits with this preference, because older males acquire better territories [251, 252].

Other types of age-related cues and signals include antler size in deer [745], dewlap size in eland (see Figure 12.24D) [79]; crest height in newts [30]; gular pouch size in frigate birds [431], decreasing brightness of UV structural colors due to wear in butterflies [352, 527]; calling performance in some frogs and crickets [173, 193, 325, 749]; and chemical signals in mice, lizards, and some insects [399, 417, 440, 484, 523]. Genetic benefits of mating with older males have been documented in only a few species, because it is difficult to rule out direct effects [84, 193, 258, 357, 483]. Among insects in which males provide females with direct benefits in the form of protein-rich food supplements, females often prefer younger or middle-aged males, presumably because older males have depleted their resources, but the age cues are not known [22, 340, 341, 413, 579, 770].

All of the age signals described above constitute variation among adult age classes. In most species, mate attraction signals develop around the time of sexual maturation. Signals that arise or change during the transition from the last juvenile stage to the adult stage can be considered another type of age signal. The development of signaling organs and the production of signals are typically controlled by maturational hormones such as testosterone in vertebrates and juvenile hormone in invertebrates. For most species, sexual maturity and assumption of adult sexual signals occur concurrently. However, in others, young animals that become sexually adult delay the development of the definitive adult signals and either retain the juvenile signal or adopt a distinct subadult one. In birds, this is called **delayed plumage maturation** (see Figure 12.24C).

Recently matured males that have retained juvenile patterning, which in many birds is similar to that of females, may benefit by being able to dodge defending males and get close enough to females to obtain matings. However, this occurs sufficiently rarely that it cannot be a general reason for delayed maturation. A more general explanation for this delaying strategy is the **status-signaling hypothesis** [425, 667]. Delayed plumage maturation in birds occurs most often in sexually dichromatic species in which either female choice or male competition has favored costly signals of male quality via plumage color. It also occurs in species where competition is acute, such as on leks, or when access to a mate or scarce nest sites depends on experience and extensive knowledge of options. In each of these cases, a delay in the onset of breeding can be compensated by good future breeding opportunities. The duration of the delay period may vary among males depending on their condition, and both distinct subadults and pseudo-juveniles sometimes succeed in attracting mates and breeding. Subadult males signal their younger age and lower competitive ability, and presumably benefit from reduced aggression from other males and the absence of full signal production costs [132, 345, 449, 490, 621, 622, 704, 764]. For example, in the lazuli bunting (*Passerina amoena*), older and bluer males prefer and facilitate the establishment of adjacent territories by young males, perhaps because the mates of these younger males seek out the older males for extra-pair copulations [227]. The advantage of gradual and flexible development of sexual maturity and signaling may also explain age-based changes in acoustic and olfactory signals in many other species.

Parental ability and other direct benefits

As mentioned earlier in the chapter, females may chose males on the basis of the material resources they provide, such as good territories, food, parental care, or protection from predators and male harassment. Females benefit from these resources in terms of enhanced fecundity, survivorship, or offspring survival, and the magnitude of these direct benefits can be considerably greater than those of genetic benefits [364]. If females can assess the resource quality directly, for example by observing the size of prey gift items, sampling the territory quality, or monitoring male foraging ability directly, then elaborate signals with costly guarantees of reliability are

not needed. However, if females have no way to verify the quality of male contributions directly, then honesty guarantees are required to prevent dishonest signaling. For direct benefits like paternal care, there is a time lag between mating and subsequent care, so the signal is a "promise" of good care in the future. There are two ways for a male to cheat: (1) a low-quality male could give a deceptively high-quality signal but be unable to deliver the benefit; or (2) a high-quality male could give a corresponding high-quality signal to solicit and achieve the mating, but then withhold the benefit and use it to attract another female. Female preference for a condition-dependent handicap signal linked with **parental ability** should be sufficient to prevent the first form of cheating, although there is the conundrum that costly investment in the signal could reduce the final benefit that the male can deliver. To prevent the second form of cheating, females must evolve some type of conditional reproductive tactic that penalizes cheating males without compromising their own fitness [757]. One strategy is for females to limit sperm transfer to the time it takes them to consume a nuptial gift, so that larger gifts allow a greater male benefit, as shown in **Figure 12.25** [651, 720]. Another strategy is to mate multiply and avoid using the sperm of low-investing males. Alternatively, females could reduce their investment in the size of the clutch or the quality of the offspring of a low-investing male and remate quickly with another male [756]. Any female strategy that links the resource benefit she receives from the male to the reproductive fitness he receives from the mating will largely reduce sexual conflicts of interest and favor honest signaling that benefits both parties. Finally, an important determinant of the level of male investment in offspring and reliable signaling of investment ability is the male's opportunity for additional matings, i.e., the operational sex ratio. If opportunities are limited, male desertion is not a profitable strategy, and males benefit more by staying and contributing to offspring care. In this case, male advertisement is expected to be modest and honestly reflect their genetic quality and condition-dependent parental investment simultaneously. If additional mating opportunities are common, which is likely to occur if females themselves mate multiply to obtain indirect genetic benefits, then males will invest strongly in costly advertisement displays and their level of paternal care will be low [370].

Males share parental care with mates in the majority of avian species, and are the sole providers of parental care in numerous fish species and a few amphibians, insects, and birds. We would expect females to select for reliable signals of a male's motivation and ability to care for offspring in these species. For birds, the strength of the correlation between the intensity of sexual display and male parental effort depends on the level of extra-pair mating opportunities, as mentioned above [474]. Species with low levels of extra-pair paternity show strong positive correlations between primary mate attraction signal intensity and the male's investment in provisioning offspring, whereas species with higher levels of extra-pair paternity show either no correlation or a negative

(A)

(B)

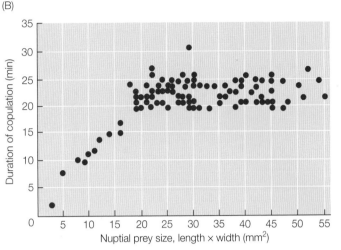

FIGURE 12.25   Assessment of direct benefits from nuptial gifts (A) A pair of mating hangingflies (*Hylobittacus apicalis*) hangs from an overhead leaf support while the female (left) feeds on the insect gift brought by the male. The female can terminate copulation by pulling away the tip of her abdomen. (B) The size of the nuptial prey determines the duration of copulation. When the insect is small the female terminates copulation quickly. Copulations of less than 20 minutes result in no sperm transferred and no eggs laid. (After [720].)

correlation between display intensity and male investment in care (**Figure 12.26A**). The negative correlation in some species implies that females prefer more attractive males, despite the fact that these males allocate more effort to attracting additional mates than to investing in the social mate's brood [157]. A second comparative study on a larger sample of species and signal types showed that the most reliable predictors of male parental effort are display behaviors closely linked to parental care, such as courtship feeding [477]. The rate of displays such as songs and flight maneuvers is moderately correlated with paternal effort, whereas plumage color and repertoire size are less useful predictors of paternal effort, and tail length and other plumage characters are strongly negatively correlated with paternal effort (**Figure 12.26B**). The latter male traits, while clearly condition-dependent as discussed earlier, appear to advertise good-genes benefits rather than direct benefits to females. In fish and other ectotherm species in which solitary males guard the eggs, most signals and cues used by females provide very reliable information about the male's ability to guard, measured as hatching success (**Figure 12.26C**). Body size and display rate in particular reflect male condition and ability to fan, guard, and remain vigilant at the nest throughout egg development [6, 36, 240, 477, 705, 710].

The reliability of signals for other types of direct benefits is currently mixed. Among the more reliable are signals indicating male fertility (sperm number). Body size and signals correlated with body size were the most frequently found traits positively related to the percentage of eggs fertilized [125, 406, 435, 477]. Enhancement of female fecundity, or clutch size, is another potential direct benefit a female could obtain via mate choice; but among all types of male traits examined, the correlations between male trait magnitude and female fecundity were very low; even for insects, the mean correlation strength did not differ for nuptial gift traits compared to other male phenotypic traits. While there are some excellent examples of females preferring males that honestly signal with beneficial gifts [86, 754, 755], these cases are offset by a number of studies of insects and spiders that show

detrimental effects of nuptial gifts on female fecundity or survival [735, 737]. We shall pursue this discussion of sexual conflict in the courtship section of this chapter. Finally, the abundance of food on a male's territory is an important criterion for mate choice in many avian species, and females often can extract some information about territory quality from the vigor and duration of costly male advertisement displays. Males on good territories can meet their energy needs efficiently and therefore can spend more time displaying. In experimental studies, augmentation of the amount of food on a male's territory increased his display rate and attractiveness to females [4, 48, 595].

Dominance

Females often mate with more **dominant males**. Does this occur because females specifically prefer dominant individuals who have proven their competitive superiority, or because dominant males exclude subordinate males and prevent females from exercising their choice of mates? This question underlies alternative views on how the two sexual selection processes—intersexual and intrasexual—interact. One view is that the two processes are complementary and reinforcing. Dominant males are here regarded as high-quality sires because of their demonstrated competitive ability against subordinates, so females should benefit by mating with strong, aggressive males [49, 68, 784]. An alternative view is that male strategies evolve to maximize the male's own fitness and may sometimes conflict with the optimal female strategy. When dominant males control access to females, they

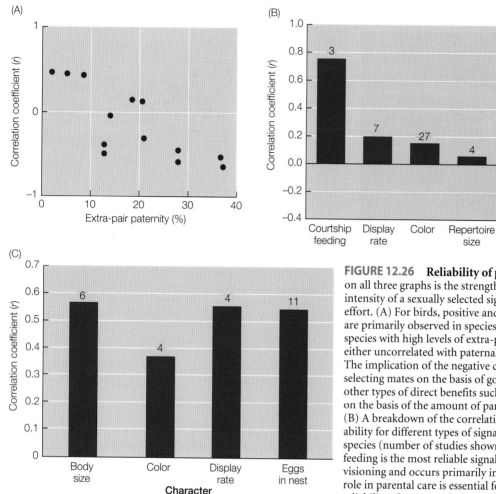

FIGURE 12.26 **Reliability of paternal care signals** The *y*-axis on all three graphs is the strength of correlation between the intensity of a sexually selected signal (or cue) and male parental effort. (A) For birds, positive and reliable paternal effort signals are primarily observed in species with low extra-pair paternity; species with high levels of extra-pair mating have signals that are either uncorrelated with paternal effort or negatively correlated. The implication of the negative correlations is that females are selecting mates on the basis of good genes, indirect benefits, or other types of direct benefits such as territory quality, rather than on the basis of the amount of parental care the male will provide. (B) A breakdown of the correlations of paternal care signal reliability for different types of signals in 48 studies of different bird species (number of studies shown for each category). Courtship feeding is the most reliable signal of subsequent male nestling provisioning and occurs primarily in seabird species where the male's role in parental care is essential for successful reproduction. The reliability of song and flight display rate as a predictor of paternal effort is strongly positive in some species but uninformative in others. Plumage color is marginally correlated with parental effort, repertoire size is uninformative, and the length of tails and other feather plumes is significantly negatively correlated with male parental effort. (C) A breakdown of the correlations between hatching success and various male signals and cues in 25 studies of fish and other ectotherms with sole male guarding of eggs (number of studies shown for each category). Body size and display rate both indicate male condition, which strongly affects a male's ability to guard eggs for the duration of development. Male color is a less reliable signal. The number of eggs already in the nest is a good predictor of hatching success, because males with more eggs are more motivated to stay and guard the nest. (A after [474]; B,C after [477].)

prevent females from making their preferred choice and may even impose costs on females [485, 486, 593]. A compromise approach is to accept that some form of male competition essentially always exists, and that it can have both positive and negative implications for adaptive female choice. The challenges are to understand the trade-offs and net fitness benefits, and to identify the circumstances in which male competition facilitates or constrains female ability to mate with preferred males [790].

Females obtain numerous benefits from mating with the winners of male–male contests. The direct benefits include enhanced resource provisioning [559, 577], parental care [97, 98], territory quality [17, 99], vigilance [559], and reduced harassment from other males [67, 122, 205, 554]. Indirect genetic benefits include sons that become dominant and successful in mating [375, 482]. When females prefer dominant males, they can adopt several communication strategies to facilitate their mate choice. First, females may adopt as mate-choice signals the same signals that males have evolved to indicate dominance status and fighting ability to each other [49, 69] (for an example, see Figure 10.15C). Badges, weapons, and indicators of body size and stamina are reliable signals of male dominance status, fighting ability, and condition

because rival males have put them to the test during competitive interactions. Female preference for male dominance signals may subsequently exert additional selection pressures favoring their elaboration. A review of species in which the same sexually-selected signal is used by male receivers for intrasexual conflict and female receivers for mate choice found that body size was by far the most common male trait, especially in invertebrates, fish, amphibians, reptiles, and mammals; but color patches, body ornaments, and acoustic signals are also sometimes jointly used [302]. Inter- and intrasexual selection operated in the same direction in 85%

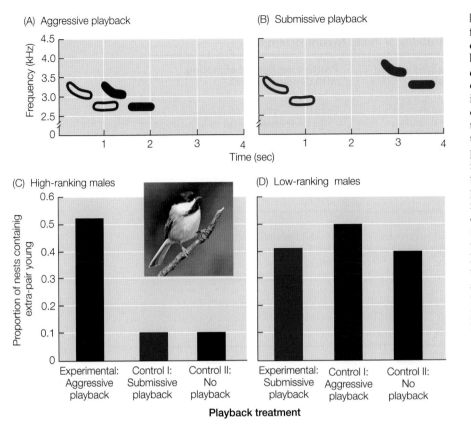

FIGURE 12.27   Evidence for female eavesdropping on male countersinging contests in the black-capped chickadee (*Poecile atricapillus*) Sound spectrograms of vocal interactions recorded during interactive playback experiments, where open notes indicate the song notes of the male subject and solid notes show the playback pattern. (A) Aggressive playback trials in which the simulated intruder matches the pitch and overlaps the songs of the subject. (B) Submissive playback trials in which the simulated intruder avoids matching the pitch and overlapping the songs of the subject. (C) High-ranking males that received the aggressive playback treatment (red) lost paternity significantly more often than high-ranking males that received submissive (blue) or control (black) treatments. (D) Low-ranking males lost paternity at similar rates in all three types of treatments. (After [460].)

of these studies, and in opposite directions in the remainder. A nice example of how competition can impose honesty on a trait preferred by females has been described in the three-spined stickleback [97, 98]. In Figure 10.39 we illustrated the cheating strategy of some poor-condition terminally ill stickleback males who increase their red coloration to equal that of good-condition males. When such males are placed in a competitive situation with other males, the social costs cause them to reduce their red coloration to avoid fighting with superior males. Competition thus increases the reliability of the red color signal as an indicator of health, condition, and parental ability.

A second female strategy is to **eavesdrop** on male contests. In this way females can not only observe which male wins an encounter, but can also attend to the signaling interactions. For example, female black-capped chickadees eavesdrop on male territorial countersinging interactions, and a female mated to a dominant male that "loses" a simulated contest with a highly aggressive playback subsequently cuckolds her mate in favor of neighboring dominant males (**Figure 12.27**) [460]. Similar evidence for female acoustic, visual, and olfactory eavesdropping has been described in other avian species, frogs, fish, and mammals [9, 155, 386, 404, 442, 524, 608, 740]. A third female strategy is to **incite** or encourage competitive encounters between males, so that females can improve their chances of mating with a dominant male. Females either give loud calls, specific displays, or behave in ways that attract or bring males into contact so that

they compete more intensely. This phenomenon is observed in group-breeding species such as colonial and lekking birds, pinnipeds, primates, and fiddler crabs [126, 281, 296, 516, 572, 574, 648, 665, 678, 803]. Eavesdropping and inciting have been argued to provide a more accurate and less costly means for females to identify competitively superior males than relying solely on direct assessment of single males [784].

As we have already mentioned, males can impose costs on females [21, 790]. Dominant males may provide less parental care than subordinates, and may also cause injuries, reduced fertilization success, reduced female lifespan, and increased energetic costs [191, 199, 402, 487, 522, 594, 700, 759]. Females in some species may actively avoid mating with dominant males if they can [593]. Numerous studies have compared female mating preferences under experimental conditions where male competition is prevented (e.g., by tethering the males) to the mating outcome when males are allowed to compete [790]. There is often a difference in the characteristics of the successful males, usually ascribed to male–male interactions interfering with courtship. For example, female tiger salamanders prefer males with long tails, but long tails get in the way during male fights; larger, shorter-tailed males hamper the courtship of preferred long-tailed suitors [297]. Similar results have been found in other amphibians, fish, insects, and numerous mammals [111, 181, 450, 485, 502, 547, 600, 670, 679, 789]. Female birds in general are more able to exercise their choice, but even here, male–male competition for territories restricts the subset of males

TABLE 12.3 *Female strategies for coping with male dominance as a function of the benefit received from mating with dominant males and the ability of males to control females*

	Negative net benefit of mating with dominant males	Positive net benefit of mating with dominant males
Weak male control over female access to preferred males	Females use male signals as cues to determine male dominance status and avoid mating with dominants. E.g., cockroach [487], baboon [690], water strider [20]	Females co-opt male dominance signals or eavesdrop on male-male contests. E.g., wrasse [758], mandrill [669], junco [451], rock sparrow [232], yellowthroat [714]
Strong male control over female access to preferred males	Strong sexual conflict with males forcefully manipulating females at a net cost; evolution of cryptic female choice. E.g., gartersnake [675], bean weevil [131], water strider [679], bushcricket [736], dung fly [58]	Females passively accept mating by local dominant male or incite aggressive interactions among males to insure dominant individual mates with them. E.g., wallaby [678], grackle [572], bearded tit [281], elephant seal [126]

among which females can choose [584, 585]. **Table 12.3** summarizes the discussion above in terms of two orthogonal aspects of male dominance from the female's perspective: the net benefit of mating with dominants, and the ability of dominant males to control female access to preferred mates. Although the two axes are depicted as dichotomous, they are in fact continuous, with gradations between the indicated extremes. As we shall see below, the female's ability to exercise mate choice depends on the species' male mating strategy and other aspects of ecology and reproductive biology.

Courtship

In prior sections, we have focused on particular signals as if they did not occur as part of a sequence. In fact, a solitary signal exchange rarely leads to copulation. Instead a variety of different signals are woven into a sequence called **courtship**. Courtship commences after a conspecific male and female have encountered each other, and entails all of the behaviors leading up to and including successful mating. Because males are usually less choosy and more eager to mate than females, courtship has traditionally been characterized as **persuasion** and **stimulation** on the part of the male and **coyness** on the part of the female. Most species have several courtship signals and behaviors, given by both sexes in a specific sequence (**Figure 12.28**). Signals are typically comprised of a mixture of modalities, with olfactory and tactile signals often playing a prominent role. Different signals provide different kinds of information at each stage of the courtship process. For example, the long-distance mate attraction signaler may need verification that the respondent is indeed a conspecific of the opposite sex, and not a predator or same-sex competitor. The

approaching individual may require additional information about mate suitability. Both parties may then want verification of sexual intentions and receptivity in the partner. In this section we shall first discuss some general principles that determine the intensity and nature of courtship. Then we examine the design of signals that provide various types of information, such as species and sex identification, receptivity, and synchronization of mating. We end on the theme of sexual conflict and discuss two more recent and somewhat opposing perspectives on courtship: the evolution of stimulating and manipulative tactics by males, and the importance of female signals, cooperation, and negotiation during courtship.

General principles of courtship intensity and character

In some species, courtship is characterized by elaborate ornamentation and display by males, while in other species there is a minimal exchange of signals before mating occurs. In addition, courtship may appear relatively cooperative or instead include aggressive and coercive interactions between the sexes. Some examples of such contrasts within the same taxon are provided in **Table 12.4**. In the following sections, we examine the factors responsible for this variation.

ROLE OF MALE MATING STRATEGY   Earlier in this chapter we mentioned that an important factor affecting the nature of courtship behavior is the outcome of the mate-searching game and the associated male mating strategy (see Table 12.2). Male mating strategies are the ecologically determined tactics males employ to compete against other males

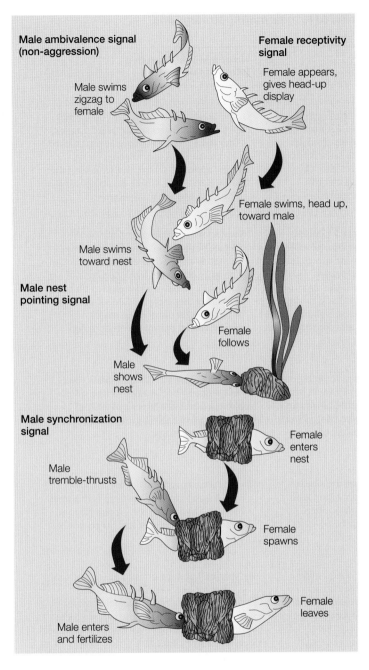

Male ambivalence signal
(non-aggression)

Male swims
zigzag to
female

Female receptivity
signal

Female appears,
gives head-up
display

Female swims, head up,
toward male

Male swims
toward nest

**Male nest
pointing signal**

Male
shows
nest

Female
follows

**Male synchronization
signal**

Male
tremble-thrusts

Female
enters
nest

Female
spawns

Male enters
and fertilizes

Female
leaves

FIGURE 12.28 Courtship sequence in the stickleback A
female approaches a red territorial male and gives the head-up display to indicate her gravid, receptive state. The male then performs
a zigzag display, while the female assesses its vigor and acquires
olfactory cues to the male's genetic diversity and health. The male
then swims down to the nest he has constructed and points toward
it. If she is interested, the female follows and enters the nest. The
male prods her back to stimulate her to spawn. Once her eggs are
laid, the female leaves; the male immediately enters the nest and
fertilizes the eggs. (After [461, 462, 502, 725].)

for access to females. In mating systems where males defend
resources and advertise their prowess and females are the
mobile searching sex, females have a choice of mates. Courtship in such species consists of male displays designed to

entice and persuade females to approach and mate. These
displays include colorful plumage, elaborate dances, circling
around the female, soft vocalizations, and gentle tactile signals such as stroking or licking. The male is never aggressive toward the female, lest she leave the territory. General
examples of this courtship style include most birds, resource-
defending ungulates, lekking and chorusing species, fish
with paternal care, and insects and spiders with male long-
distance mate attraction displays. In female defense mating
systems, males are the mobile sex. They attempt to control
one or more females on a long-term basis, consort briefly
with a series of fertile females, or engage in patrolling and
scramble competition for females. Because males must
engage in extensive contests with other males to gain access
to females, the consequent intrasexual selection leads them
to be larger and stronger than females. This places them in a
better position to physically dominate females. Particularly
when females are clustered in space, a single male may be able
to defend them against other males in a permanent harem.
Females have little option for choosing mates, so they typically do not resist and passively accept the harem male. They
signal their receptivity to the male, who copulates without
any significant display. Courtship in some cases may contain aggressive and threatening signal elements. In scramble
competition systems with serial consortship, and in species in
which the female gives a long-distance mate attraction signal,
females attract multiple males and potentially have a choice
of mates, but must succeed in rejecting unwanted suitors if
they are to exercise their preferences. Maximum conflict can
occur in these situations, and the outcome depends on the
ability of females to escape unwanted male advances.

ROLE OF FEMALE BEHAVIORAL FREEDOM   The **behavioral
freedom** of females to choose mates depends on their ability
to escape from coercive males. When females have high escape
potential, they possess the leverage to make males court extensively before agreeing to mate. In this situation, females are
likely to evolve preferences for attractive male displays, rather
than dominance displays. On the other hand, when female
escape potential is limited, males have more leverage in the
courtship, and are likely to use their larger body size, weapons, grasping appendages, and intromittent organs (penises)
to dominate their chosen mate. Aspects of the species' habitat,
body shape, mode of locomotion, and reproductive biology all
affect female behavioral freedom and the character of courtship [575]. Female ability to escape males is greater in species that live and move in three-dimensional habitats—air,
water, and forest canopies—than in species that live in two-
dimensional habitats where males can more easily pin females
down. This habitat effect is evident in the examples in Table
12.4. Another factor that affects female freedom is weaponry.
If females are larger and more aggressive than males or possess
formidable weapons, males must approach and court females
with caution. Male courtship in this case consists of affiliative or submissive displays, infantile behaviors, food gifts,
and patient assessment of the female's receptivity. Examples

TABLE 12.4 *Taxa with contrasting patterns of courtship related to male mating strategy and female behavioral freedom*

Taxon	Elaborate courtship (strong female freedom)	No or minimal courtship (strong male control)
FIDDLER CRABS These crabs live in large communities on muddy or sandy shores of rivers, estuaries, and bays. Both sexes dig and defend individual burrows. Males have a single enlarged claw that they wave in a species-specific pattern prior to mating [144].	In species in which mating and egg-brooding occur in the male's burrow, females are the searching sex. Females have complete freedom to select and enter the burrow of their choice. Male claw waving is elaborate, with a vigorous vertical or horizontal arc movement.	In species in which females brood in their own burrow, males are the searching sex. They display with a lethargic vertical claw movement from their own territory. Males approach a receptive female on the surface, seize her, stroke or tap her carapace, and attempt to copulate.
BIRDS With the exception of a few flightless species, volant birds live in a three-dimensional world, and wings are ill-equipped for grasping a female. Females are usually free to choose mates without coercion, and males rely on ornamentation and display to attract females [77, 223, 448, 456].	In perching species, males lack penises. Females can easily escape forceful males and must cooperate fully for successful copulation. The female expresses receptivity by flexing her legs, supporting the male mounting on her back, raising her tail, and opening her cloaca. Elaborate male mate attraction displays and courtship intention displays.	In waterfowl, mating takes place on the surface of the water, a two-dimensional habitat. Males have an intromittent organ. The male grasps the female on back of her neck with his bill while mounted on back, preventing her escape. Males of monogamous species must guard mates against forced mating by other males.
TURTLES Their armored shells and short legs make it difficult for males to seize females [51].	Fully aquatic species, such as sea and pond turtles, live in a three-dimensional habitat. Females can escape unwanted males and have full freedom of choice. Males court females with elaborate visual, olfactory, and tactile displays.	Terrestrial species and semi-aquatic "bottom walkers" live in a two-dimensional habitat. Males are larger and can pin females down against the surface, preventing their escape.
PRIMATES Female freedom is greater in arboreal than in terrestrial species [101, 102, 117, 691].	In arboreal species, the sexes are similar in size. Females solicit copulation with olfactory signals and displays, often mutual exchange of displays.	In terrestrial species, males are usually larger than females. Courtship is minimal; females signal receptivity, males respond forcefully. A dominant male is generally successful.
PINNIPEDS The body plan, designed for swimming, makes it difficult for males to grasp females [64, 403, 743, 744]	In species that mate in the water, males defend their underwater territories and perform complex vocalizations and diving displays to attract mobile foraging females.	In species that mate on land or ice, animals are less mobile. Females cluster, and males defend harems of females. Males are larger than females and can pin them down with a flipper over the back; mating is sometimes forced.
INSECTS Adults are volant in most species, but where females are constrained to a surface, males can pin them down. [21, 637, 679].	Flying species live in a three-dimensional habitat. Females generally have freedom to choose mates and can escape from forceful males. Males court females with visual and vibratory displays and olfactory signals.	Water striders live and mate on the surface of the water, a two-dimensional habitat. Males grasp and ride females; males of many species have elongated graspers to prevent females from removing them. Fig wasps are another example of surface-constrained females.

include most spiders, mantises, some chameleons, birds of prey, and spiny and carnivorous mammals [52, 169, 246, 323, 412, 689, 703].

OPTIMAL FEMALE COYNESS   In species where females do have a choice of mates, how much time and effort should

they put into assessing male quality? The more effort females devote to carefully evaluating males, the more likely they are to select a good mate and the stronger the selection pressure on males to engage in competitive courtship display. However, if females take too long to decide, they may pay a variety of opportunity and time loss costs. So we expect there to

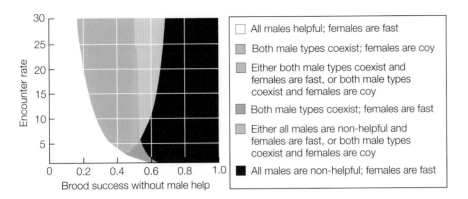

FIGURE 12.29 **Optimal female coyness in the "battle of the sexes" model** Females can be either fast (mate immediately) or coy (demand a delay for courtship and information gathering), and males can be either helpful (stay and help the female raise offspring after mating) or nonhelpful (leave after mating). The graph shows regions of stability for different male and female strategy combinations as a function of the relative brood success without male help (*x*-axis; 1.0 means male help contributes nothing beyond what the female alone can achieve) and the probability of encountering another mate if the male leaves or the female rejects the mating (*y*-axis). Coy females acquire information about an encountered male's propensity to help versus not help; they increase their chances of accepting a helpful male but pay a reproductive time cost for the delay. Female coyness and variation in male behavior can both be maintained under a wide variety of conditions (colored areas). Favorable conditions include a relatively balanced sex ratio; otherwise one or the other sex experiences severe difficulty in encountering another mate after rejection. If most males are unhelpful, then their searching competition is extreme and they would do better upon encountering a female to stay and help her. An intermediate benefit from male help also favors stable coyness; if male help is extremely beneficial, males will stay and help regardless of the female strategy; but if male help doesn't provide any additional benefit beyond what a single female parent can accomplish, it never pays for males to help. An intermediate rate of information transfer is also favorable—if it is too slow, the time cost becomes too high to make coyness and assessment viable; if it is very fast, the male type is immediately known, so no further inspection is needed. (After [458].)

be an optimal amount of time taken by females to assess a male before deciding to either mate with him or reject him and keep searching. Males, on the other hand, would prefer that females simply accept them upon encounter. Dawkins [138] proposed a simple game theoretic model to depict this **battle of the sexes** over mating. Females have two strategies: be *coy* (refuse to mate with an encountered male until some time has passed) or be *fast* (mate immediately). Males also have two strategies: be a *philanderer* (refuse to engage in a long courtship, and if a female does agree to the mating, leave immediately after) or be *faithful* (prepared to court a female and after mating stay and help rear offspring). The outcome of this game is unstable, with perpetual cycling among strategies [660]. To illustrate, say we begin with a population of mostly fast females and philandering males, and a context that yields benefits from some paternal help. If coyness is

effective in preventing nonhelpful philandering males from mating, then coy females and faithful males will increase in the population and the proportion of philandering males approaches zero. At this point, it no longer pays females to waste time in courtship display, so fast females start to invade. Once fast females are common, philandering males invade, and we are back to the beginning of another cycle. This model has two shortcomings. First, it is modeled as a contest between the encountering pair, so male and female fitness depends only on the current decision and ignores future reproductive success, which is affected by the probability of finding another mate and the evolving behavioral strategies of other individuals in the population [493]. Second, it does not capture the essence of information transfer and assessment by the female, which involves a trade-off between the gain in information and the loss from delaying reproduction with increased assessment time [752]. A better model that incorporates both of these features makes different predictions [458]. Over a wide range of reasonable parameters, populations with primarily coy females and faithful/helpful males can be stable (**Figure 12.29**). Neither male helpfulness nor female coyness can evolve if the other strategy is absent, so they mutually stabilize each other. This model can be generalized to any type of benefit females might obtain from more carefully assessing male quality.

Courtship signals

SPECIES AND SEX IDENTITY SIGNALS   Sexually reproducing species must be able to identify conspecifics and distinguish males from females. A single species-specific multivariate signal can provide both types of information, and examples in all modalities abound. Sexually dichromatic species, including many birds, reef fish, and butterflies, rely on visual signals for **species** and **sex identification**. The males usually possess some brightly colored surface pattern that clearly distinguishes them from other species, and females are less ornamented, typically more drab and cryptic in appearance, but still possess species-specific characteristics. Movement displays are an important distinguishing signal in non-web-building spiders [323]. Fireflies provide another visual signal example, where males flash in distinctive species-specific patterns and females answer with a shorter flash pattern following a specific temporal delay (see Figure 8.24). Acoustic signalers show similar patterns, with males broadcasting louder and more complex patterns that differ from those of other closely related species and females answering with their own calls. Sex-specific antiphonal courtship calling, with both sexes approaching each other, is found in numerous insect and frog species [27, 147, 163, 190, 249, 261, 262, 499, 698, 751]. **Figure 12.30** shows some examples of duets in several orthopteran species. Electric fish likewise have species-specific electric organ discharge waveforms, and females in some species produce a variant of the male waveform [40, 159, 178, 730, 796]. By far the most commonly employed signal modality for species and sex identification is chemical signaling. In a wide variety of vertebrate and invertebrate species, either a pheromone blend, in which

the presence or absence of some components specifies sex, or a more complex multicomponent chemical mixture with sex-specific proportions, provides identity information [317, 332, 500, 619]. For example, in moth species with female long-distance and male short-distance pheromones, the sexes share some of the same chemical production pathways and end products, but the male blend contains several additional compounds [399].

RECEPTIVITY SIGNALS   Males of most species remain sexually receptive throughout the duration of the breeding season, whereas females are receptive for short periods of time—specifically, when they have developed a clutch of eggs and are ready to oviposit or have finished rearing a brood and are ready to bear the next one. Females therefore benefit by

Ancistrura nigrovittata (Phaneropterinae)

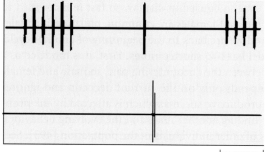

100 ms

Poecilimon affinis (Phaneropterinae)

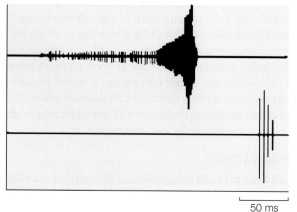

50 ms

Amblycorypha parvipennis (Phaneropterinae)

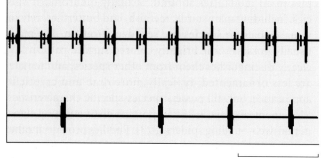

500 ms

Steropleurus stali (Ephippigerinae)

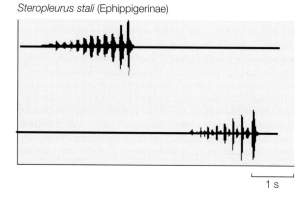

1 s

Bullacris membracioides (Acrididae)

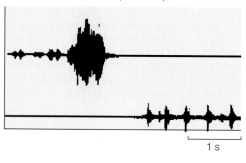

1 s

FIGURE 12.30   Male and female calls in several orthopteran species For each species, the oscillogram for the male call is above and the female call below; the timing between male and female calls is preserved. In general, the frequency spectrum is the same for male and female sound pulses, but the duration and pattern is usually simpler for the female's call, and the time delay of the female's call after the male call is critical for species recognition. The male bushcricket *Ancistrura nigrovittata* produces a short trigger call between his pulse series to elicit the female call. The bushcricket *Amblycorpha parvipennis* calls continuously, and the female intersperses her calls precisely after some of them. The female bushcricket *Steropleurus stali* replies with a call similar to the male's call. The female bladder grasshopper *Bullacris membracioides* also produces a longer call resembling the male's call, but lacking the loud terminal pulse. (After [27].)

(A)

FIGURE 12.31   Courtship displays derived from mating intention movements (A) A male jumping spider (*Habronattus americanus*) (right) raises his colorful tufted forelegs in a symbolic gesture to indicate that he has detected the olfactory mate attraction signal of a female (olfactory receptors are located on the leg tips); he also displays his red palps (sperm storage organs), which are visible in this photo [187, 430]. (B) The male greater kudu antelope (*Tragelaphus strepsiceros*) courts by following a receptive female and resting his chin on the female's shoulders. The male rests his chin on the female's back during mounting [167]. (C) Presentation or manipulation of nest material is a common element of courtship display in birds, as shown by this pair of great egrets (*Ardea alba*).

(B)

(C)

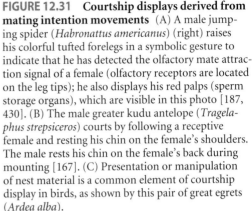

signaling to potential mates when they are receptive or fertile, and by turning off the signal when they are not, since they do not gain from being harassed by males when they are unreceptive. Nor should males waste their time courting unreceptive females, so there is little conflict of interest over honest signaling of **sexual receptivity** by females. Olfactory signals are by far the most commonly employed for this function, since they can be reliably linked to hormonal state [332]. Female mammals and fishes release steroid hormone products in their urine, and males are thus able to distinguish estrous from nonestrous females [59, 300, 301, 517, 556, 565, 699]. In crustaceans such as lobsters, crabs, and copepods, reproduction is coupled to the moult cycle of females; urinary chemical products possibly linked with moulting or ovigenesis signal receptivity to males [18, 35, 361, 684]. Acoustic signals are employed to signal receptivity in some animals. The female Barbary macaque (*Macaca sylvanus*) mates with multiple males throughout her approximately three-week estrous period and gives repeated calls during copulation. The fine structure of the calls, including duration, dominant frequency, and repetition rate, changes during this period. The primary function of the calls is to incite competition among males. Males can discriminate between calls given at different stages and respond most strongly to those given around the time when conception is most likely

to occur [552, 666]. Females in a few other mammal species (brown rat *Rattus norvegicus* and grey mouse lemur *Microcebus murinus*) and at least one bird (alpine accentor *Prunella collaris*) give specific vocalizations only during the days that they are fertile [90, 397, 443]. Visual throat color patch signals are correlated with reproductive state and clutch size in some female lizards, peaking at the time of ovulation [774, 804]. In several other species, the color intensity peaks after ovulation and mating, and appears to signal *nonreceptivity*; females employ the signal to deter the advances of harassing males [239, 518, 761].

SYNCHRONIZATION SIGNALS   **Synchronization signals** are used to coordinate gamete release and subsequent fertilization. They include mating solicitation signals by females, intention and timing signals by males, and mutual signals by both sexes. Examples of some male signals are illustrated in **Figure 12.31** and female signals in **Figure 12.32**. In nearly all birds, females must cooperate for successful fertilization. Males typically approach females closely and give a copulation intention signal, often an upright chest-out posture, sometimes with a prey item, nesting material, or leaf petal in the beak, and wait for the female to assume the mating solicitation display posture. This posture consists of flexing the legs, tilting the body forward or horizontally, spreading the

(A)

(B)

(C)

FIGURE 12.32 Mating acceptance signals by females (A) Courtship in birds such as these least terns (*Sternula antillarum*) often involves the male approaching with a food item. If the female is receptive, she crouches, spreads her wings, and moves her tail to the side so the male can stand on her back. (B) In rodents and many other terrestrial mammals, females solicit mating with the lordosis posture, shown here in a copulating pair of European hares (*Lepus europaeus*). The raised hind legs and arched back tilts the pelvis forward and facilitates male intromission. (C) A copulating pair of superb jewelwing damselflies (*Calopteryx amata*). The male (top) grasps the female with the claspers on the tip of his abdomen, and the female (bottom) must twist her abdomen around so that her genitalia contact the male's intromittent organ. The sperm-transferring structure is located on the underside of his abdomen.

wings slightly, and raising the tail up and to the side, which enables the male to hop on the female's back and position his tail under the female for the two birds' cloacas to contact in a **cloacal kiss**. Some birds engage in a mutual display of dancing, crest-raising, or head bobbing prior to copulating. Mammals perform a similar ritual, with the male giving an intention-to-mount display and the female arching the back and raising the tail up and to the side in a position called **lordosis**, which facilitates male penis intromission. In the typical frog species, females approach a preferred calling male and touch or butt his vocal sac. The two may engage in some further reciprocal touching, and then the male either leads the female to an oviposition site, or hops on her back while she moves to an oviposition site. With the male firmly hugging the female in a position called **amplexis**, the female lays her clutch of eggs and the male immediately fertilizes them externally [525, 777]. In contrast, newts and salamanders have a

form of internal fertilization, in which the female uses her cloaca to pick up a spermatophore that the male has deposited on the surface. The male must therefore maneuver the female to walk over the spermatophore, which may involve fanning or directly rubbing pheromones into her vomeronasal organ so that she follows him and stops at the right moment [19, 242, 264, 694].

Precise synchronization is especially critical in aquatic animals with joint broadcast release of sperm and eggs into the water column. In fish, male and female often engage in visual and tactile displays that increase in tempo as gamete release approaches. A nice example of courtship has been described in the blueheaded wrasse (*Thalassoma bifasciatum*), in which the male pales his normally intense coloration in response to the approach of an interested female, an action likely to minimize detection by competitor males. Dark spots appear on the tips of the male's pectoral fins, and he fans them at an increasing rate. When the female is ready to spawn, she signals by swimming rapidly upward, and the male follows immediately behind her while both release gametes [137]. Some fish use sounds to coordinate spawning [414]. In the croaking gourami (*Trichopsis vittata*), both sexes make loud sounds during aggressive encounters, but females emit short, low-intensity purring sounds to initiate and coordinate spawning [391]. Hermaphroditic fish with egg trading have an even more intricate courtship, since they must signal which sexual role they will assume. The sea bass (*Serranus subligarius*) rapidly changes color patterns to regulate this process. Individuals normally wear the male role pattern with vertical bars. Upon encountering another individual, one of them must adopt the female role, signaled by no bars and a solid dark tail. Once the pair has settled on opposite roles, the female individual flashes a V-shaped coloration pattern to synchronize gamete release [142, 143]. In marine invertebrates such as brittle stars and polychaete worms, females produce an egg-release pheromone just before spawning that induces nearby males to release their sperm (see Figure 6.2B) [620, 693, 762].

Web Topic 12.2 *More examples and multimedia clips of courtship sequences*

Here we describe some of the more unique forms of courtship, such as the use of display objects in bowerbirds, mutual displays in monogamous species, cooperative male displays, and the fascinating displays of spiders.

STIMULATION AND MANIPULATION   Males have evolved a huge array of strategies to stimulate or prime females into a receptive state, induce females to mate with them, and facilitate selective investment in their sperm and offspring. Given the relatively recent recognition of sexually antagonistic coevolution in sexual selection, we need to reconsider how often female responses to male courtship signals are beneficial to both partners, or are instead manipulative male strategies that benefit them at a cost to the female. There is sometimes a fine line between stimulation and **manipulation** and we often lack sufficient data to make the distinction.

Some male courtship signals have a priming function with long-term effects on female reproductive physiology. For example, song stimulation in birds enhances the reproductive effort of females. In mockingbirds and canaries, playback of conspecific male songs, especially more complex high-quality singing, causes females to initiate nest-building faster and to lay more and larger eggs [220, 384, 409, 416, 713]. The neuroendocrine pathways from auditory input to reproductive and behavioral responses are becoming better understood [436]. Female birds also respond to plumage color and leg band color manipulations of their mates with changes in their investment in eggs [360, 411]. This differential investment in reproduction by females in response to experimentally modified male signals is clearly an adaptive response [673]. Pheromone signals in male mammals also operate as primers to accelerate puberty and induce estrus in females [74, 212, 509]. These pheromones are detected through a combination of olfactory bulb and vomeronasal organ input, and the neurological mechanisms by which these chemicals affect behavior and reproductive physiology are being elucidated [349, 350, 433]. Much has been written about the pregnancy-blocking (Bruce) effect of the odor of a strange male in mice, which can cause a recently inseminated female to fail to implant the embryo. This phenomenon is often viewed as a manipulative male strategy to bring a newly encountered female into a receptive state [75]. However, in natural settings females can easily avoid contact with new male scent and can limit this costly effect to those circumstances in which it is adaptive for them to abort and initiate a new reproductive cycle [44].

Other male courtship signals have immediate effects on female behavior and therefore function as releasers. In Chapter 6 we mentioned the androgen-derived pheromone in the saliva of male boars, which causes females to stand quietly and adopt the lordosis posture for mating. Another compelling example of a true male courtship pheromone has been described in *Plethodon* salamanders. Courting males deliver a blend of proteinaceous pheromones to reluctant females by rubbing their mental gland secretions (produced in the skin under the chin, see Figure 6.6E) directly onto the nares of the female, where it enters the vomeronasal organ (**Figure 12.33**). The effect of the pheromone is to stimulate female receptivity and increase the likelihood of successful insemination [214, 290, 291, 390, 623]. As mentioned above, the male must guide the female over the spermatophore he has deposited on the surface so she can pick it up in her cloaca. This male chemical signal is believed to be stimulatory but not manipulative, as the female must raise her head to receive the signal, and if she is not prepared to mate she can freely leave. In some salamander species, however, males have evolved an injection method of pheromone delivery, in which they scratch the back of the female with modified teeth and rub the gland secretion directly into the female's circulatory system. This strategy certainly removes some control by the female. In yet other species, males forego the pheromone and grasp the female, depositing their spermatophore directly into her cloaca [19, 289]. Moths and butterflies also deposit a male pheromone released from specialized hairs on the abdomen or wings onto the antennae of the female during courtship (**Figure 12.34**). If the female is flying, the deposition of the pheromone causes her to land and sit quietly, and raise her abdomen to accept mating by the male. Females control mating and modulate their response depending on the species identity and male quality information in the signal [121, 180, 271, 500, 504, 557, 615, 771]. Finally, some male arachnids produce chemical signals or toxins that temporarily tranquilize females so they can mate with less risk of being cannibalized [43, 566].

Nuptial gifts of prey items or male-produced secretions represent another type of male courtship signal that can have both short- and long-term effects on females. When the gift is a true prey item, it can obviously have some nutritional benefits for egg-producing females. Among birds, courtship

FIGURE 12.33 **Alternate courtship tactics in salamanders**
(A) In *Lissotriton* newts, which mate underwater, the male (rear)
displays by wafting a pheromone produced in a large abdominal gland toward a female (front) with flicking motions of his
large crested tail. The male must coax the female to walk over a
spermatophore packet he has deposited on the substrate. (B) In
Plethodon shermani, the male (left) attempts to rub the secretions
of his mental gland directly on to the nares of the female, who lifts
her head to receive the chemical signal. The pheromone is a blend
that stimulates female receptivity. (C) A few species, such as the
mountain dusky salamander *Desmognathus ocoee*, have evolved
an injection method of delivering the mental gland pheromone,
in which the male rasps the female's back with his teeth and rubs
the pheromone directly into her bloodstream. (D) A fourth strategy, in which the male dispenses entirely with the pheromone,
is observed in species such as *Calotriton asper*. The male (darker
individual) forcefully grabs the female by biting her tail and wrap-
ping his tail around her body. In this position, the male's cloaca is
pressed against the female's cloaca and transfer of simplified spermatophores takes place [19, 289, 291].

feeding is most prominent in seabird and raptor species.
These species are monogamous, and feeding demonstrates
the provisioning skill of potential mates, provides a substantial amount of nutrition to the laying female, and improves
reproductive success for both partners [148, 226, 503, 674,
748]. In shrikes (*Lanius*), males impale prey on thorns in
conspicuous places for both their mates and potential extra-
pair mates, and the probability of mating with both increases
with prey item size [732]. In insects and spiders, the primary
functions of nuptial gifts are to increase the probability of
enticing a female to mate and prolonging copulation duration, which increases the amount of sperm transferred [651,
736, 737]. While this clearly improves the male's mating success, there may or may not be a significant nutritional benefit

to the female. In hangingflies and scorpionflies (see Figure
12.25), females prefer males that offer larger prey, but a male
may sometimes first feed from the prey item themselves
before seeking a female, or snatch the prey item away from
her after a successful copulation and use it to attract a second female. In the spider *Pisaura mirabilis* and in dance flies,
males offer females a prey item wrapped in a silk balloon.
Males sometimes cheat and offer balloons that are empty or
contain nonfood items, which females nevertheless accept.

Consumable secretions produced by males and offered
to females can attract females and provide nutritional benefits but may serve to accelerate the female's egg production
rate at a cost to her future fecundity. For instance, the dorsal
secretion of the male cockroach (*Blattella germanica*) is a tasty
but nutritionless substance that tricks females into crawling
onto the male's back, where copulation occurs [385, 507].
Similarly, the gelatinous spermatophylax added to the spermatophore by males of some cricket species contains mainly
glycine and free amino acids, and takes hours to consume. It
stimulates the taste receptors and is strongly attractive even
to females of species in which males do not add a spermatophylax to their spermatophore. A prevailing view of these
male-produced gifts, the **candymaker hypothesis**, suggests
that the secretions are a pheromonal mate attraction signal
that evolved via the sensory trap mechanism [652, 653, 760].
However, costly male gifts sometimes do provide females
with a nutritional benefit, and females may have some power
to counteract the males' temptation to cheat. The idea is that
females may penalize deceptive males by evolving conditional postcopulatory investment mechanisms that selectively

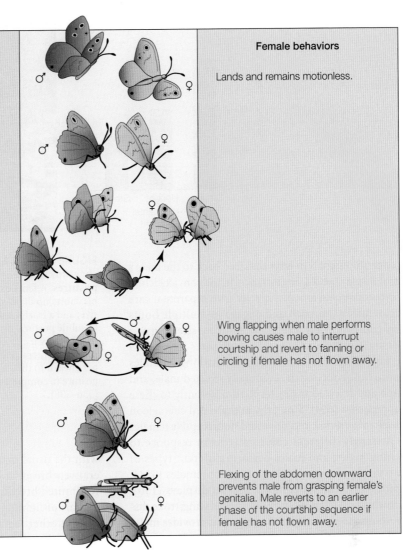

Male behaviors

Aerial pursuit
A perched male detects female on basis of color, size, and movement, and chases after her.

Fanning
Male lands behind female and rhythmically opens and closes his forewings, which reveals his large ocellus spot.

Circling
Male walks toward the female in a semicircle while fluttering his wings, placing himself in front of her.

Bowing/antenna orientation
Male bows on knees and tilts body forward, opening and closing forewings around female's antennae. Alternates with male rotating his antennae.

Copulation attempt
Male moves around behind and to side of female so both face same direction, he puts left forewing under female's right forewing.

Clasping
Male attempts copulation by flexing abdomen toward female to make genital contact. If female flexes abdomen toward male, successful copulation ensues.

Female behaviors

Lands and remains motionless.

Wing flapping when male performs bowing causes male to interrupt courtship and revert to fanning or circling if female has not flown away.

Flexing of the abdomen downward prevents male from grasping female's genitalia. Male reverts to an earlier phase of the courtship sequence if female has not flown away.

FIGURE 12.34 Courtship sequence in grayling butterflies (*Hipparchia statilinus*) Courtship steps in this species always proceed in the same order, illustrated here from top to bottom. Females accept the mating by remaining motionless throughout the sequence, or they can terminate the courtship at any point by flying away. Females can also interrupt the male's sequence with refusal behaviors indicated in red on the right, which cause the male to revert back to an earlier step and repeat the sequence from that point. Persistent males may have to repeat these steps many times before the female accepts the mating. Males possess specialized scales on the dorsal side of their forewings that produce a male pheromone. Females assess males on the basis of both the visual pattern of the male's dorsal ocelli spots, which are revealed during fanning and circling, and on the quality of male's pheromone blend, which is revealed during bowing. (After [557, 724, 726].)

reduce a cheater male's fitness [757]. This could be viewed as a retaliation strategy to ensure signal honesty. Whether the nuptial gifts of insects are a product of sexual conflict thus remains contentious [238, 546, 772].

NEGOTIATION   Although in some species females appear to passively observe and evaluate the male's courtship signals, in other species there is a significant amount of interactive signaling between the male and female. This interaction is best examined as a **negotiation** [626]. Behavioral ecologists have again borrowed theoretical constructs from economists to examine decision-making and interactive signaling in a market context [96, 245, 492, 505, 506, 540]. Negotiation in

economic markets involves a series of steps—partner choice, information gathering, haggling, and resolution (deal or no deal)—and the optimal decision at each stage depends on each player's outside (alternative) options, bargaining position, and the outcome of the prior stage. During the haggling stage, the seller sets an initial asking price and then decides how much to adjust the price based on each counteroffer by the buyer. Animal-relevant models of negotiation envision two players with different **reaction norms** that characterize how strongly they respond to a given "offer" by the partner. Evolutionarily stable interaction rules can indeed be found in which players are selected to negotiate the final outcome [457, 718]. Several studies have shown that females respond

(A)

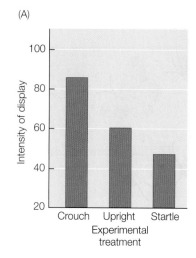

(B)

FIGURE 12.35 Changes in male satin bowerbird (*Ptilono-rhynchus violaceus*) display intensity in response to female postures A female enters a male's constructed bower to observe his courtship display. The male display can be intensely aggressive, and a fearful female may flee. (A) Successful males respond to subtle postural displays of females and reduce the intensity of their display if a female shows a startle response (B). Displays were produced interactively with a robotic female that can quickly raise her chest into the startle posture or lower her chest into the crouch posture (a component of the copulation solicitation signal). (After [536–538].)

to male courtship signals in ways that feed back to the male's subsequent display tactics. For instance, in the convict cichlid (*Archocentrus nigrofasciatus*), a species with biparental care of young, females approach and engage in multiple bouts of mutual courtship with several males before selecting one for spawning. Courtship displays for both sexes include tail beating, quivering, and mutual lateral brushing. Females spend more time courting the eventually selected male, and the chosen and rejected males respond differently to these female approaches [654]. Similarly, in the veiled chameleon (*Chamaeleo calyptratus*), receptive and nonreceptive females display strikingly different color patterns in response to approaching males, and males perform different types of courtship displays toward the two types of females [351]. Some spiders also exhibit complex mutual displays [130]. Female cowbirds (*Molothrus ater*) give a brief wingstroke display in response to preferred male songs that provides males with feedback about their attractiveness; males subsequently increase their rate of singing these preferred song types [780]. Studies with robotic females show very dramatically how males respond to different female signals during courtship. Robotic female whitethroats (*Sylvia communis*) performing different combinations of their jump displays and calls elicited different courtship display behaviors in males [34]. Reproductively successful male sage grouse (*Centrocercus urophasianus*) are able to adjust their display tactics as a function of female proximity, while unsuccessful males are less responsive to female distance [539]. **Figure 12.35** illustrates a final example of negotiation in satin bowerbirds (*Ptilono-rhynchus violaceus*); the courting male must be sensitive to the female's startle response lest he frighten her away.

Sex role reversal

Most of the emphasis in this chapter has been placed on *male* mate attraction and courtship signals, because intersexual selection is usually stronger on males as they compete for the attention of limited females [336]. However, females can be subject to both inter- and intrasexual selection in the same way as males, leading to **sex role reversal** [110]. If resources are limited, they may have to compete against and fight other females for resource access even in species

with the usual male-biased operational sex ratio. In cooperatively breeding vertebrates, females may have to compete for limited breeding roles as intensely as males [113, 255]. Signals of fighting ability and motivation directed specifically at other females will evolve with the same principles described in the previous chapter. For an example, see our cover photo of the *Eclectus* parrots, showing the bright red coloration of the female who competes intensely with other females for nesting cavities [259, 401]. If males invest even moderate time and effort in courtship and are restricted in the number of females they can mate, they will use any cues or signals available to identify the most valuable females. Thus males may also be choosy about mates. The greater the male investment, and the greater the variance in female quality, the stronger the selection pressure on males to pick the best females. Social systems in which males are choosy but still competitive and limited by access to females are called *partially role reversed*. In this relatively common situation, males will evolve increased sensitivity to cues of female quality. In insects, male mate choice of females is based primarily on fecundity and virginity indicators, using cues such as body size and evidence of recent mating [63]. In vertebrates, body size is a useful index of fecundity in species with indeterminate growth; age, experience, social rank, and condition can be correlated with female fecundity and thus male mating preferences in determinate growers [11, 335]. Mutual mate choice can result in both sexes expressing a preference for more partner ornamentation as a signal of quality [7, 8, 28, 231, 247, 287, 329, 383, 397, 628, 641, 773].

(A)

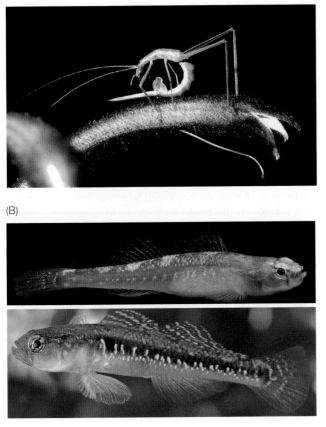

(B)

(C)

FIGURE 12.36 **Examples of partial and complete sex role-reversal** (A) A pollen bushcricket (*Kawanaphila nartee*) female is consuming spermatophylax provided by a male. The sex under the stronger sexual selection pressure varies in space and time. The operational sex ratio (OSR) may be male- or female-biased in different environments. When food is scarce, the OSR favors females and females vary more in quality, resulting in males becoming more choosy and preferring larger females [236, 237, 389]. (B) Male and female two-spotted gobies (*Gobiusculus flavescens*) are both highly ornamented, but in very different ways. Gravid females (top) sport bright orange belly coloration, caused by carotenoid pigmentation in the eggs, and transparent skin with some additional carotenoid chromatophores. Males (bottom) have colorful enlarged dorsal fins and flank markings that are displayed during aggressive encounters with other males and during courtship. Males defend a breeding territory, build a nest, and perform all egg-guarding. The sex roles change dramatically during a season. Initially males and females are equally abundant and males compete fiercely for territories and actively court females. Later in the season, male numbers decline, females begin to compete among each other for access to territorial males, and males cease intrasexual aggression and courtship activities [8, 192, 708]. (C) The wattled jacana (*Jacana jacana*) is a completely role-reversed species. Females (left) are 48% larger than males (right) and have a larger frontal shield and prominent wing spurs. Males perform all incubation and most chick guarding, while dominant females defend large territories encompassing the smaller territories of several males. The OSR is strongly skewed in favor of females. In this simultaneous polyandry system, females copulate and lay eggs for several of their males at the same time, so males often care for offspring not their own. Their nests are floating platforms in freshwater marshes [165].

In contrast to cases of partial role reversal, social systems in which males are choosy and females compete for males are called *completely role reversed*. This happens when male investment in reproduction is so large that it limits female fecundity and the operational sex ratio favors females. In this relatively rare situation, females will evolve not only highly conspicuous signals of reproductive quality and fecundity so that they can attract the best males, but also signals of fighting ability to compete with other females. Polyandrous mating systems, and reproductive systems with sole paternal care, are the primary examples of complete role reversal. In some species with a high level of paternal investment, the more reproductively limited and competitive sex switches seasonally from one sex to the other [192, 388]. **Figure 12.36** shows some examples of sex role-reversed species (see also Figures 11.12 and 13.12).

SUMMARY

1. Sexual reproduction requires conspecific males and females to attract and find each other and then to successfully court and mate. Because females by definition invest more in gametes than males do (**anisogamy**), the sexes have different optimal reproductive strategies. Males produce large numbers of small, mobile sperm at relatively low cost, and therefore have a shorter reproductive cycle than females, which produce fewer, large, costly ova at a slower rate. The greater the difference in gametic investment, the more strongly skewed the **operational sex ratio** of receptive adults becomes in favor of males. Male fitness is thus limited by the availability of females. Males compete among themselves for access to mates by employing various **male mating strategies**. Females usually have a wide choice of males, and benefit by selecting high-quality mates. These differences are the basis for the traditional **reproductive sex roles**, with males being the competitive and actively courting sex and females being coy and

choosy. Which sex searches while the other gives the long-distance attraction signal depends strongly on a species' male mating strategy.

2. **Sexual selection** is the evolutionary process that arises from competition among members of one sex for access to mates of the opposite sex. In contrast to natural selection, which maximizes fecundity and survivorship, sexual selection maximizes mating success. It typically leads to **sexual dimorphism**, differences between male and female phenotypes within a species. **Intersexual selection** leads to the coevolution of male mate attraction signals, female preferences for those traits, and viability consequences of both male traits and female preferences. This coevolution is best described using a **multivariate quantitative genetics process**. Different models of intersexual selection emphasize interactions between different components of that process. All models require that there be heritable variance in female preference behavior, so that mating biases can evolve. These processes are not exclusive.

3. The **Fisherian runaway model** is driven by **indirect selection** on female preferences caused by a positive genetic covariance between the male trait and the female mating bias for the trait. Females indirectly benefit from preferring males with highly attractive heritable traits, because their sons may inherit these traits and in turn be highly attractive to females—the "sexy sons" effect. Some evidence for the Fisherian model in natural systems exists, but it is often accompanied by effects of one or more of the other models.

4. The **good genes model** proposes that females produce fitter offspring by preferring males with traits that indicate their heritable viability or health. There is indirect selection on the female preference caused by a positive genetic covariance between female mating bias and heritable male viability. The male trait must be a costly, condition-dependent display signal that reliably indicates his viability. Male viability must be heritable, and the process will be accelerated if the display trait is also heritable. There is growing evidence in support of this model in natural systems.

5. The **direct benefits model** describes the situation in which females develop preferences for male characteristics that provide them with material benefits. There is **direct selection** on female preferences because of the immediate benefits. No covariance terms are required, and the male trait does not need to be heritable. Environmentally determined variation in males' ability to provide benefits is sufficient, as long as females have some cue or signal of this variation. Empirical support is most abundant for this model; however that may reflect the fact that it is the easiest of the alternatives to test.

6. The **sensory bias model** proposes that males evolve traits that exploit pre-existing sensory biases in females for relevant environmental cues involved in detection of food or predators. In this case, there is indirect selection for female mating preferences arising from pleiotropic genes that affect mating biases and other behaviors. A few good examples in support of this model exist, but it is believed to serve primarily as a mechanism to initiate male display traits and is likely to be displaced by one or more of the other model processes.

7. The **sexual conflict**, or **chase-away**, **model** envisions a **sexual arms race** in which males evolve traits that manipulate females to their detriment and females then evolve counter-adaptations. Sexual conflict is characterized by direct antagonistic selection between the sexes, such that traits that enhance the fitness of one sex reduce the fitness of the other. This model can lead to a stable solution, in which the costs and benefits to the sexes are in balance, or to cycles of male escalation and female counteradaptation. Evidence for this model is increasing as more studies look for it.

8. Persistent female preference for more extreme male trait values causes trait expression to be costly and thus strongly dependent on male condition. Condition is affected by many loci and thus presents a large target for mutations. According to the **genic capture hypothesis**, trait expression should reveal a male's overall genetic quality and viability. Many kinds of mate attraction signals show heightened **condition-dependent trait expression** and provide direct and indirect benefits to females that base their mate choice on such traits. Selection on males then favors increased isolation of the different physiological pathways needed for trait development, a process that results in developmental **modularity**.

9. All modalities have been exploited for sexually selected signals. **Color patches** are a common target of sexual selection because their expression is often condition dependent. **Carotenoid-based color patches** are generally more dependent on health and environmental conditions than **melanin-based color patches**, which tend to reflect stress resistance, immunocompetence, and aggressiveness. The brightness of **structural color patches**, including iridescent and white patches, is readily disrupted by poor condition. The **immaculateness** of the borders between contrasting color patterns can also reflect overall quality. Whereas **tail elaboration** is usually a reliable indicator of health and quality in birds, it is too often exploited to exaggerate body size in fish to be a useful mate criterion. The **performance quality** of vigorous **behavioral displays** using various combinations of modalities provides one of the best male trait targets for female preference because is it strongly tied to physiological condition, health, and viability, and is difficult to modularize. **Chemical blends** can directly reveal immunocompetence and current levels of infection. **Fluctuating asymmetry** is not a good indicator of overall quality.

10. Good-genes traits provide heritable additive genetic benefits to all females, but choosy females can also select males that have **compatible genes** with respect to their own genotype, a process based on nonheritable **nonadditive genetic variation**. A widespread benefit is optimization of **heterozygosity** at the MHC loci, which improves immunocompetence and offspring survival. Identification of suitably dissimilar mates is achieved via olfactory cues or

via **cryptic female choice** after copulation, which is accomplished by differential sperm selection. Similar mechanisms can also be used to avoid inbreeding and achieve **optimal outbreeding**.

11. Females may benefit by attending to signals and cues for additional criteria. One is **male age**; older males have proven their ability to survive, and show their age in various cues or signals. Another criterion is **parental ability**; males that provide females with food or other resources demonstrate their ability to provide for offspring. A third criterion is dominance; females may prefer **dominant males** and respond to signals that have evolved for the purpose of resolving male–male conflicts and status in selecting dominant mates. They may also **eavesdrop** on the outcomes of male–male contests or **incite** male competition.

12. **Courtship** facilitates successful copulation and consists of an exchange of signals between the sexes. If sex roles are traditional, male displays primarily serve to **persuade** and **stimulate** females, while females remain **coy** to further assess male suitability. Male displays are associated with female **behavioral freedom** to choose. Male courtship display is more elaborate in species in which the females search for mates while males defend resources and in arena display mating systems. Sufficiently realistic models depicting the **battle of the sexes** over mating find that a broad range of conditions favor coyness by females and honest signals of fidelity and helpfulness by males. Specific displays may provide information on **species** and **sex identification** and **sexual receptivity**. **Synchronization signals** coordinate the immediate act of gamete release and fertilization. In some species, males evolve mechanisms to stimulate and **manipulate** (or coerce) females to mate. In other species, males must cautiously interact with females using **negotiation** strategies. Partial and complete **sex role reversal** occurs in species with very high male reproductive investment and strong competition among females for resources, breeding opportunities, or mates. Appropriate mating signals are suitably reversed as well.

Further Reading

Sexual selection and the evolution of the relevant signals are extremely active fields of study with a huge literature base. Mating systems and strategies were outlined in two classical articles by Emlen and Oring [164] and Thornhill and Alcock [722] and more recently reclassified by Shuster and Wade [676]. Sexual selection models from a quantitative genetics perspective have been reviewed by Mead and Arnold [459], Fuller et al. [201], and Kokko et al. [376]. See Arnqvist and Rowe [21] and Parker [532] for perspectives on sexual conflict. The importance of condition dependence for the maintenance of reliable signal traits is well described by Tomkins [727], Cotton [123], and Badyaev [24, 25]. Reviews of specific condition-dependent traits include: color patches [154, 228, 455], tails [185], performance [392, 561, 562, 779], chemical signals [332], and fluctuating asymmetry [400, 564]. The role of genetic compatibility has been examined by Neff [495, 496], Fromhage [200], Mays [446], and Puurtinen [591]. Brooks [84] summarizes the advantages of selecting older mates; Møller [477] reviews direct benefits; and Wong [790] evaluates the consequences of male dominance on female choice. Recent reviews of courtship include those by Vahed [735, 737]; Ptacek [587]; Talyn [712]; Sullivan [706]; Fusani [203]; and Houck [291]. Sex-role reversal is examined by Bonduriansky [63], Berglund [50], and Clutton-Brock [109, 110].

COMPANION WEBSITE
sites.sinauer.com/animalcommunication2e

Go to the companion website for Chapter Outlines, Chapter Summaries, and References for all works cited in the textbook. In addition, the following resources are available for this chapter:

Web Topic 12.1 *Quantitative genetic models of sexual selection*
An overview of the multivariate quantitative genetic approach to the evolution of female preferences and sexual conflict models.

Web Topic 12.2 *More examples and multimedia clips of courtship sequences*
Here we describe some of the more unique forms of courtship, such as the use of display objects in bowerbirds, mutual displays in monogamous species, cooperative male displays, and the fascinating displays of spiders.

Chapter 13

Social Integration

Overview

Cooperative behavior develops when two or more individuals integrate their activities to achieve a common goal that increases the inclusive fitness of both. Examples include the coordination between a male and female attempting to mate and care for offspring; communication between parents and dependent offspring to assure that offspring needs are met; and efforts by the members of group-living species to maintain group cohesion, coordinate movement, and organize communal activities. Although the evolution of indiscriminate altruism is a theoretical possibility, a risk of cheating pervades nearly all cooperative actions in real animals. Donors in cooperative interactions must have evidence that their costs will be compensated by sufficient direct or indirect benefits; recipients must ensure that they will not be exploited by a potential donor. Animals have evolved three complimentary tools to minimize the chances of cheating and exploitation during cooperation. First, they limit cooperative interactions to classes of individuals that are relatively known and trustworthy. This requires suitable cues and signals to allow for such discrimination and recognition. Second, even for individuals that are potentially trustworthy, some honest evidence of need by recipients and capability to provide by donors may be demanded before the interaction proceeds. Again, this evidence is usually mediated by signals. Finally, a third type of signal may be required to coordinate the actual interaction. In this chapter, we first review the theoretical conditions for cooperation, and then examine the general principles of recognition. In subsequent sections we describe the roles of the three classes of cooperation signals (identity signals, honesty guarantees, and coordination) in each of three social contexts: male–female cooperation, parent–offspring interactions, and group cohesion.

Evolution of Cooperation

As discussed in Chapter 9, the origin and maintenance of cooperative behavior has been the focus of much theoretical research. Existing economic models of cooperation can be subdivided into two main classes, as illustrated in **Figure 13.1** [351, 537, 687]. The first class applies to cooperating interactants that are sufficiently similar genetically that any inequities in direct benefits and costs can be compensated through *indirect genetic effects*. We have already described the relevant greenbeard, kin selection, and limited dispersal economics in Chapter 9.

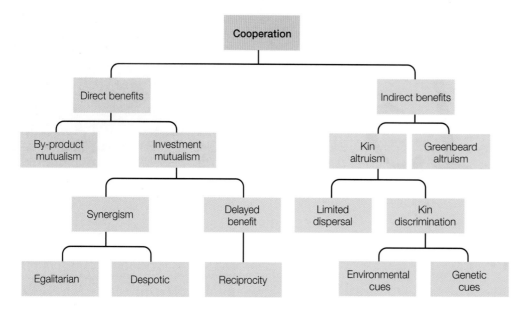

FIGURE 13.1 **A classification of the explanations for the evolution of cooperation** Direct benefits explain mutually beneficial forms of cooperation, and can be further subdivided into those that do not require additional investment (by-product mutualism) versus those that do require investment. If the investment involves synchronous cooperation or assured future benefits, it is classified as synergism, and this category can be further broken down into a continuum from egalitarian (low skew) to despotic (high skew) distributions of costs and benefits among group members. If investment results in a delayed benefit that occurs with some probability, it is classified as reciprocity. Indirect benefits explain altruistic forms of cooperation, which can be further subdivided into kin and greenbeard altruism. Kin cooperation can occur in structured or viscous populations with limited dispersal without specific recognition of kin, or in heterogeneous populations with some mechanism (genetic or environmental) for recognizing and targeting close kin. (After [44, 486, 687].)

The second class of economic models ignores genetic similarities but assumes that all parties experience a net *direct benefit* by cooperating. **Mutualism** is a general term for all direct benefit cooperation models. At first glance, it might seem that mutualistic cooperation would evolve more readily than indirect-benefit cooperation. In fact, it has turned out to be rather complicated. Consider first a situation in which a task with a fixed investment cost must be undertaken, and once the task is completed, all members benefit equally. The benefits are thus also fixed. Incubation of a joint clutch of eggs by communally nesting birds would be an example. The question is then whether the cost of undertaking the task will be divided equitably among all group members or not. There will always be a temptation to cheat and do less of the work, since both actors and slackers will gain the same benefit. This type of situation is called **by-product mutualism**, and usually leads to a mixed ESS in which a fraction of the group participates in the task and a remaining fraction declines to help [154, 244, 537, 589]. The optimal mix of helping versus slacking will vary with the relative costs, benefits, and time constraints.

In a second subclass of mutualism, the total benefit of some communal action increases synergistically, at least up to an asymptote, as the number of group members contributing and paying the investment costs of doing so increases [324, 486]. Given the asymptotic total benefit, there will be some intermediate number of participating members at which the net benefit per member will be maximized. Construction of a communal refuge, collaborative defense of a group territory, and group hunting of large prey are obvious examples [537]. This form of mutualism, called **synergism**, has proved to be highly vulnerable to unequal distributions of costs and benefits among group members. Attempts to understand the trade-off between cooperative and selfish efforts within groups have led to a spate of evolutionary game models, together called **skew theory**, that investigate just how inequitable the benefits and costs can be while still providing a marginally net benefit to all parties [91, 290, 302, 351, 509, 662].

Reciprocity is a third type of mutualism in which donors expend costs now to the benefit of recipients, but then must wait until the roles are reversed before they receive the benefits that compensate for those costs. This type of transaction is favored only if benefits usually exceed costs and the rotation of roles is sufficiently balanced [17]. As an example, vampire bats that have fed successfully on a particular night may regurgitate some of their blood meal to a colony mate that has been unsuccessful; on other occasions the recipient may be the successful forager, whereupon it feeds the unsuccessful first donor. The bats do not need to be kin to engage in these exchanges [692].

Evolutionary game models of any of these forms of mutualism rarely predict a pure ESS for cooperation. If such a pure state could evolve, all animals would be cooperators, and donors could contribute indiscriminately to recipients without worrying about insufficient compensation. Instead, the relevant ESSs are either a stable mixture of cooperators and noncooperators, or unstable systems that cycle through various weightings of the two strategies [437]. The instabilities are in part due to the cumulative economics required when a donor has to consider not only the current interaction but also past and subsequent ones, as in the **tit-for-tat rule** of cooperating with a particular partner only if he cooperated on the previous occasion. As the number of events that need to be tallied to justify cooperation increases, so does the number of alternative rules that could be invoked for choreographing those events. We examine the cyclical consequences of these alternative rules, along with the effects of population size and spatial structure, in Web Topic 13.1 and Chapter 15.

The one thread that recurs in all evolutionary models that predict persistent cooperation is that individuals should be highly selective in their choice of cooperation partners. Recognition of appropriate partners is therefore critical. We next examine the process of recognition and the evolution of identity signals, before turning to the types of identity, need, and coordination signals that animals have recruited in several common cooperative contexts.

Web Topic 13.1 *Direct benefit models of cooperation*
Here we review a number of current models used to explain the evolution of cooperative behavior without requiring kin selection or greenbeard compensations. We begin with the classic Prisoner's Dilemma model, and build up to iterated choreographies.

General Principles of Recognition

Recognition is a broad biological phenomenon ranging from identification of prey and predators to recognition of conspecifics, gender, neighbors, group members, friends, rivals, mates, kin, and self [508, 590]. The mechanisms employed by senders to encode the identity information, and the difficulty of the receiver's discrimination task, differ for these different levels of recognition. As we outlined for the detection of signals in Chapter 8, the greater the number of classes that must be distinguished, the more complex the signal must be to encode the variants and the more difficult the discrimination task. In this chapter we shall focus on the recognition of conspecific individuals for purposes of social integration and cooperation, but the basic principles discussed below are common to all levels of recognition.

The process of recognition

Recognition is the discrimination and identification of a target individual or group among a field of similar nontarget individuals or groups. It implies that a receiver has an innate or learned memory of characteristics unique to the target and has found a good match to that standard [35]. The process of recognition in an aid-giving context comprises three components as illustrated

in **Figure 13.2**. In the **production** component, the potential recipient of aid provides some information about its identity by producing a cue, signal, or label. In the **perception** component, the donor perceives the signal and assesses its similarity to a template of the target's unique signature stored in memory. In the **action** component, the donor decides whether the potential recipient's signal meets the criterion for a match. This decision is based on the prior odds that the recipient sender truly *is* the target gleaned from the cues, the evidence provided by the signals, and the costs and benefits of correct and erroneous assessment. The donor then takes an appropriate action toward the signaling recipient, such as to *attack* if the sender is perceived as a nondesirable target or *feed* if it is perceived as a desirable target. A definitive recognition response will not occur if any one of these steps fails, such as inadequate signature information on the part of the sender, or, on the part of the receiver, an inability to perceive signature differences, a decision not to discriminate, or a lack of an appropriate or distinctive bias in behavior toward targets.

Recognition mechanisms

The mechanisms animals use for recognition fall into four categories: spatial location, familiarity, phenotype matching, and recognition tags [392]. These mechanisms are based

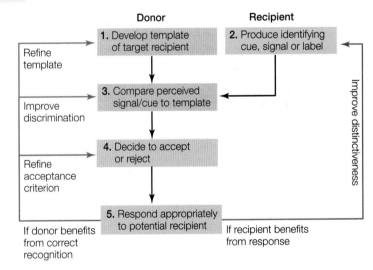

FIGURE 13.2   The process of recognition in the aid-giving context Recognition consists of three components: *the perception component* (blue), the *production component* (pink), and the *action component* (purple). (1) The donor develops a template of the target recipient either inately or via learning. (2) Potential recipients produce a cue, signature signal, or acquired label of their identity. (3) The donor then compares the perceived cue/signal against its stored template and assesses their similarity. (4) The donor decides whether or not the difference between cue/signal and template is sufficiently small to meet the acceptance criterion. (5) The donor makes the appropriate action toward the potential recipient, either aiding, cooperating, or joining if the decision is to accept, or witholding aid, retreating, or attacking if the decision is to reject. If the response benefits the recipient, the recipient will improve its signal distinctiveness, and if the response benefits the donor, the donor will refine its template, discrimination ability, and acceptance criterion.

upon different strategies for signal/cue production, perception, and decision making. Each varies in its accuracy, vulnerability to cheaters, and usefulness for recognition at different social levels.

SPATIAL LOCATION   Associating target individuals with a specific **spatial location** is the simplest mechanism of recognition. If the probability is high that only target individuals or groups will be encountered at a particular place, then the prior odds ratio for target identification at this site is high, and location is a good cue for recognition. There is no selection on recipients to give signature signals, and the action rule is to vary agonistic, cooperative or aid-giving behavior as a function of distance from this site. For example, an animal's nest usually contains only its own offspring, so directing all parental care to this site ensures that offspring receive the parent's attention. We have already seen in Chapter 11 how location is used for neighbor recognition in territorial systems. Location can be used as a mechanism for nondescendant kin recognition if family members remain clustered in an area. Location can even facilitate species recognition in animals such as swarming insects that form mating aggregations at specific times and sites. Using spatial location as a cue for recognition is a very simple rule of thumb that requires a small amount of learning and spatial memory but is easily subverted by cheaters. Both conspecific and nonconspecific parasites can insert themselves or their offspring into the target location and receive the benefits of aid-giving behavior bestowed there. The best evidence for the operation of this mechanism is immediate appropriate behavior directed toward nontarget individuals placed into the target site.

FAMILIARITY   **Familiarity**, or **associative learning**, is a recognition mechanism that requires prior experience from direct association with target individuals or groups, followed by the learning and memorization of their specific characteristics. An important feature of the familiarity mechanism is that the classes among which distinctions are made are arbitrary [212]. As the number of classes increases, the prior odds for correctly assigning a stimulus to a target class become smaller, so there is strong selection for signature signals. Individual-level recognition is always based on associative learning of specific individual characteristics and entails the largest number of classes and the greatest need for complex signature signals. Familiarity can also be used to identify kin, but target relatives must be unambiguously associated with receivers during the critical learning period. Group and colony recognition can be achieved with familiarity if all group members possess or acquire a common badge, such as the odor of a communally used food source or an acoustic signal dialect, which distinguishes them from other groups. Finally, familiarity can be involved in population and species recognition in birds and perhaps other vertebrates, where exposure to the songs and the appearance of parents during a critical early learning period causes young to prefer similar individuals as mates later in life. Familiarity is more difficult

to subvert than spatial location, but subversion does occur. Memory saturation in individual recognition systems may cause receivers to sometimes misidentify target individuals, as occasional nontarget individuals by chance may bear fairly similar characteristics to targets. If nontarget individuals are inserted into a family before learning has been completed, offspring and sibling recognition can be subverted. In colonial insects, parasites can sometimes gradually acquire the colony odor by interacting with peripheral colony members and may eventually be accepted into the main colony. For researchers to demonstrate that receivers are basing recognition on signature cues, it is important to remove any spatial location cues.

PHENOTYPE MATCHING   The **phenotype matching** mechanism functions to identify genetic similarity. In contrast to the familiarity mechanism, recognition via phenotype matching is the ability to assign stimuli to classes of relatedness relative to the receiver or known parent [212, 342]. When phenotype similarity is well correlated with genetic similarity, receivers can distinguish kin from non-kin in the absence of any prior experience with the target individual or group. The receiver must compare the phenotype of an encountered individual with a template acquired from a referent. The referent can be a known familiar relative or the self. If the referent is the family, then via familiarity and learning, the receiver acquires a template based on average genetically-based family characteristics, which may be visual, auditory, or chemical traits. The template may even be based on environmentally acquired cues that are correlated with kinship, such as diet-based colony odors [350]. In this case, the recognition system allows for recognition of unfamiliar kin. If the referent is oneself, then the receiver may be able to distinguish fairly fine degrees of relatedness, such as half versus full siblings. No learning is required if the sensory system possesses a built-in self-template [243, 653]. As we learned in Chapter 12, the vomeronasal organ of many vertebrate species contains receptor cells that express an individual's MHC alleles and mirrors the self-recognition process of the immune system. Chemical cues or signals containing MHC-derived glycoprotein waste products from others could easily be compared to the self-template to assess the degree of genetic similarity between sender and receiver. Plants and some invertebrates have an even simpler mechanism for self-recognition in the form of a single hypervariable locus with many rare alleles [299]. In the colonial marine ascidian *Botryllus schlosseri*, for example, larvae settle and fuse only with other individuals that share at least one allele at this locus, and are thus very close relatives [220, 221]. To demonstrate phenotype matching clearly, it is important to control for spatial location and familiarity cues by showing that unfamiliar siblings are distinguished from unfamiliar nonrelatives or that full versus half siblings within a litter are distinguished. Phenotype matching is relatively difficult to subvert because similarity is based on a unique allele or a set of polygenic characters that is unlikely to be matched by an unrelated individual.

(A)

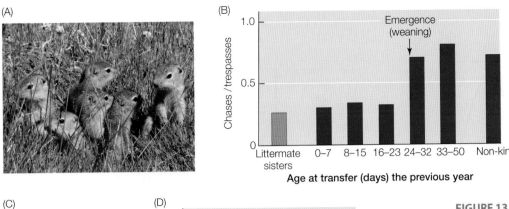

(B)

(C)   (D)

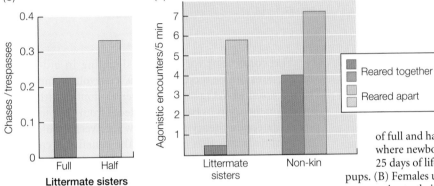

FIGURE 13.3   Multiple recognition mechanisms dependent on context in Belding's ground squirrels (*Spermophilus beldingi*) (A) Ground squirrels are colonial breeders. The females usually remain in their natal colony and therefore are often related, while males disperse from the natal colony to breed. Estrous females mate with several males, so litters may be composed of a mixture of full and half siblings. Each female digs her own burrow, where newborn young are placed and remain for the first 25 days of life. This photo shows a litter of recently emerged pups. (B) Females use only spatial location cues to recognize their young prior to their emergence from the burrow. Just prior to emergence, pups develop olfactory signature signals, and mothers and siblings become familiar with each other. Foreign pups (blue) introduced into a burrow prior to emergence are accepted by the mother and siblings; however, pups introduced at the time of emergence are subsequently treated aggressively. Differences in aggression toward introduced pups are still present a year later. (C) Yearling littermate sisters behave more aggressively toward littermate half sisters than toward full sisters, consistent with a recognition mechanism based on phenotype matching with a self-referent template. (D) Familiarity has the greater effect on aggressive interactions, but true kinship has a detectable effect. Yearling littermate sisters reared apart (in separate burrows) treat each other less aggressively than non-kin reared apart. (After [261–263, 392, 393].)

RECOGNITION TAGS   The **recognition tag** mechanism is associated with greenbeard models of cooperation. The production of a tag, perception of the tag, and cooperative behavior toward other individuals bearing the tag, are all encoded by a single gene or a set of tightly linked genes. The recognition template is not learned. Furthermore, cooperating individuals do not share genes at other loci and thus are otherwise unrelated, distinguishing this mechanism from kin selection [190]. Initially this model of cooperation was believed to be unsustainable because of the easy invasion of mutant cheaters bearing the tag but not the cooperative behavior, and because the greenbeard gene would be subject to conflict and suppression by other genes in the genome. Alternatively, if the cooperative behavior is advantageous, the tag and the behavior could spread to fixation so that all members of a family, population, or species possess it, making the tag system difficult to identify. The discovery of a greenbeard-like gene *Gp-9* in the red fire ant (*Solenopsis invicta*) illuminated the subtleties of such tags (see Chapter 9 and Figure 9.18). Other single greenbeard genes have since been discovered that code for cell-adhesion molecules. These molecules recognize and adhere to the same molecule in other individuals, and are responsible for beneficial cooperative reproductive aggregations in slime molds and toxin-protecting biofilms in bacteria and yeast [181, 338, 488, 606, 701]. Tag-associated altruism has been described in one vertebrate, the side-blotched lizard, whose territorial badge system is illustrated in Figure 11.32. Genetically similar but unrelated blue-throated males settle on adjacent territories and cooperate against the more aggressive orange-throated males. At least four distinct loci scattered throughout the

genome are required for cooperation, only one of which causes the throat-color polymorphism. Linkage disequilibrium among these loci is likely favored by the reproductive advantages arising from cooperation [599-601].

These mechanisms are not exclusive. Several different mechanisms could operate simultaneously in the same species to achieve different levels of recognition. For example, Belding's ground squirrels (*Spermophilus beldingi*) use pre-emergence familiarity to identify offspring and littermates, and phenotype matching to recognize half sibs from other litters [392] (**Figure 13.3**). A single mechanism can be used to simultaneously discriminate different levels. For instance, animals can learn by familiarity which individuals are kin if there is a reliable period of kin association. Any recognition system that is based on genetically determined phenotypic cues will also permit discrimination of levels of genetic similarity because relatives are always more similar than non-relatives. Thus a phenotype-matching species recognition

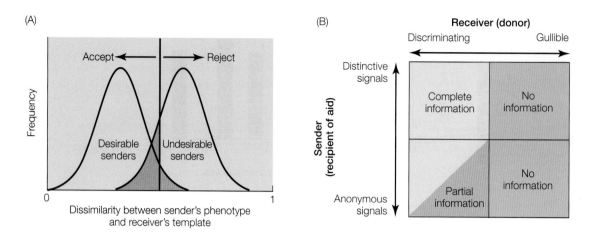

FIGURE 13.4 **Game context of recognition** (A) The frequency distributions of perceived dissimilarity between the phenotypes of desirable (left) and undesirable (right) senders. The red line shows the acceptance response threshold of the receiver. If the perceived dissimilarity falls to the left of the line, the sender is accepted. Because the two curves overlap, some undesirable senders will be accepted (false alarms, blue area) and some desirable senders will be rejected (missed detections, purple area). The position of the acceptance threshold line evolves to minimize the net cost of false alarms and misses. For instance, it will shift to the right if misses are more costly than false alarms and to the left if false alarms are more costly. The curve for desirable senders can also evolve and will move to the left to reduce the overlap between the curves when desirable senders produce more distinctive signals. (B) Alternative ESS outcomes for the recognition game. In the yellow zone, the sender produces a very distinctive signal that the receiver can easily recognize, and discrimination is always perfect. In the green zone, the sender partially reveals its identity, but some receiver errors occur. In the orange zone, the sender conceals its identity and the receiver is unable to discriminate, so it always responds favorably to the sender, regardless of desirability. (After [288].)

mechanism will also permit some discrimination of close and distant kin [212], and an individual recognition system based on perception of genetically determined signature signals will also encode family membership [32, 408]. It is therefore important to consider alternative hypotheses for recognition and to determine the primary evolutionary function of an observed discrimination ability.

Evolution of identity signals

When does it pay to reveal identity? Envision a mother bat or penguin returning to the communal crèche with a load of food. All of the young are begging in the hope of being fed, and she must find her own offspring among them. In a population of undistinguished offspring and indiscriminate parents, offspring would randomly receive food from all parents with young in the crèche. In a population of offspring that produce distinctive signals, parents with the ability to distinguish their own offspring's signals will feed them preferentially. If the benefits of recognition (more food from true parents) outweigh the losses (less food from nonparents), **identity signals** should be favored.

The conditions for identity signaling in aid-giving interactions can be modeled best as a coevolutionary game between sender-recipients and receiver-donors with signal detection theory as its basis [288]. Recall from Chapter 8 that signal detection theory tells us how receivers should make optimal decisions that maximize the benefits of correctly assigning signals to contexts while minimizing errors from false alarms and missed detections. **Figure 13.4A** illustrates this receiver decision problem for the **recognition game**, where there are two classes of senders—desirable and undesirable recipients of aid—whose phenotypes deviate from the receiver's target phenotype template to different degrees. Given that the phenotype distributions for the two types of recipients overlap to some degree, the receiver evolves a **response threshold** for accepting or rejecting the sender's request for cooperation. Now we allow the senders to modify their signaling phenotype. If senders evolve more individually distinctive signals, their frequency distribution curve shifts to the left and the difference between desirable and undesirable recipients' signals increases. After receivers have readjusted their threshold acceptance line, the probability of errors from false alarms and missed detections decreases. If desirable recipient's signals are highly distinctive and receivers develop very specific templates for them, the desirable sender's curve may shift completely to the left to become centered on zero phenotype dissimilarity [508]. One of the three possible outcomes in this model is *high signal distinctiveness* in senders combined with *fastidious discrimination* by receivers (**Figure 13.4B**). This outcome is favored when the cost to donors of aiding or accepting undesirable recipients is high (e.g., they are brood parasites or free-loading foreigners), or when the number of undesirable recipients is large, which would dissipate donor resources on non-fitness-enhancing individuals. A second outcome is *partially distinctive signals* with an overlap of desirable and undesirable sender phenotypes. In this case, receivers can discriminate among senders better than randomly but still make some errors, to the benefit of senders. The third outcome is complete selection against senders making distinctive signals, or *adaptive anonymity*, coupled with *adaptive gullibility* in receivers. Senders will conceal their identity, or fail to give any identity signals, only when receiver-donors are inclined to accept recipients in the absence of any identifying information. This occurs when the cost of accepting undesirable recipients is low, or when there are few undesirable recipients present. If receivers are inclined to reject recipients in the absence of identity information, senders will be forced to evolve distinctive signals.

Signals designed to encode identity will employ the coding strategy of **emergent multivariate signaling** described in Chapter 8. That is, signals are comprised of multiple components, usually in the same modality, and values of the different components are uncorrelated with each other, so it is possible to generate many unique combinations. Furthermore, identity signals are likely to be genetically determined to maintain signal persistence, and cheap to produce as there is no need for them to be condition dependent [138, 642]. All of the main modalities can be enlisted for identity signals. Visual signals typically consist of contrasting color patches varying in hue, shape, and pattern (**Figure 13.5**, see also Figure 13.25). Chemical signals are multicomponent blends with variation in both the mixture and the relative concentrations

(A) Red-billed queleas (*Quelea quelea*)

(B) Tawny-flanked prinia eggs (*Prinia subflava*)

(C) Wild dogs (*Lycaon pictus*)

FIGURE 13.5 **Visual identity signals** Examples of species with conspicuous and individually variable color markings. (A) Queleas nest in dense colonies, within which mated pairs must frequently locate each other. Mask color and shape and crown color vary independently to create many distinctive variants [137]. (B) The prinia has evolved extreme individual variation in egg color and spot and line patterns as a strategy for females to distinguish their own eggs from those of a parasitic cuckoo finch (*Anomalospiza imberbis*) [618]. (C) Wild dogs live in large packs and hunt cooperatively. Distinguishing among individuals at a distance aids in hunt coordination and maintenance of a dominance hierarchy [642].

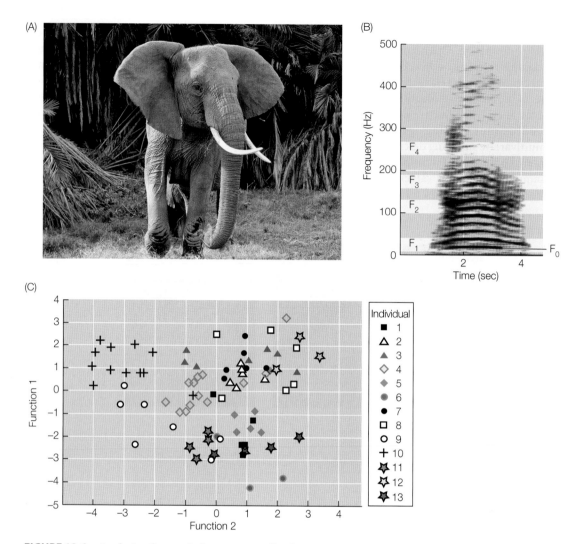

FIGURE 13.6   Analysis of acoustic features contributing to individual distinctiveness in the contact calls of female African elephants   (A) A female elephant (*Loxodonta africana*) giving a contact call, illustrating the forward and back motion of the ears that accompanies calling. (B) The call is harmonically rich and contains features related to the laryngeal source (fundamental frequency, F_0) and features related to the vocal tract filter, including the trunk (formants F_1–F_4, demarcated in yellow). (C) Nine source and filter variables were measured on calls from 13 individuals and subjected to a discriminant function analysis, which extracts orthogonal combinatorial discriminant functions. The scatterplot shows clear clustering of individual calls using the first two functions; each symbol represents a different individual. (After [400].)

of components (see Figure 6.8). Auditory signals can combine multiple orthogonal acoustic features, such as variable note shape, frequency modulation, harmonic structure, and biphonation to create individually distinctive signature calls (**Figure 13.6**; see also Figures 11.27, 13.7, 13.9, and 13.15; Web Topic 13.2 contains some rich media examples of individual signature calls). Even electrical pulses can vary in wave shape to provide individual signatures in electric fish [404, 451]. Identity signals usually serve one or more additional

functions, such as territory defense, group cohesion, gender identification, or status signaling [329].

Web Topic 13.2   *Examples of social integration signals*
Here we provide rich media examples of identity signals, pair integration signals, parent–offspring signals, and group integration signals.

Male–Female Integration

Mate recognition

Mate recognition is a term that applies to several social levels. At the most basic level, it refers to the mechanism by which conspecific members of the opposite sex are recognized, more precisely called **species recognition**. At the next level, it refers to the mechanism by which good quality mates are identified, called **mate-quality recognition**. We discussed this topic at length in the previous chapter. There are currently differing opinions over whether species recognition and mate quality assessment are separate or combined processes, so the discussion below will consider the evidence for these points of view. The third level of mate recognition, **partner recognition**, applies to species in which males and females form pair bonds for the purpose of reproductive cooperation. There we shall examine how paired individuals recognize and find each other, especially when they live in large mobile social groups where they may become separated.

SPECIES RECOGNITION VERSUS MATE-QUALITY ASSESSMENT   An important function of mate attraction signals is the identification of conspecific mates, so that individuals can avoid the cost and wasted effort of courting and mating with heterospecifics. These signals provide premating information on species identification and play a central role in the reproductive isolation of species, along with postmating mechanisms [134]. During speciation, when populations diverge genetically to become separate species, mating signals also diverge. According to **ecological speciation models**, when expanding populations encounter new ecological conditions or exploit new resources, population members undergo natural selection pressures that eventually lead to substantial genetic incompatibility and partial reproductive isolation from members of the original population. Divergent sexually-selected mating preferences for locally adapted mates may then amplify differences in mate attraction signals [533, 556, 659]. **Figure 13.7** illustrates an example of disruptive ecological selection in a species of Darwin's finch leading to divergent beak sizes, which affects the structure of male songs; visual and acoustic signal components likely contribute to assortative mating and increased genetic isolation of the two morphs. In **behavioral speciation models**, sexual selection for good genes or sexy sons, sexual conflict arms races, or sensory drive causes a reduction in cross attraction between members of different morphs or populations and eventually leads to reproductive isolation and speciation. In this case, rapid divergence in mating signals results from coupled evolutionary changes between male and female traits [64, 191, 343, 344, 455, 516, 689]. Notable taxa that have likely undergone profuse rapid speciation driven by sexual selection for mating signals include: body color patterns in African cichlids [566–568] and birds of paradise (see Figure 9.3); multimodal courtship behaviors in *Drosophila* [52, 133, 216, 298]; acoustic signals in Hawaiian crickets (*Laupala*) [448, 581, 583, 584], and electric wave signals in mormyrid fishes [8]. For both the ecological and behavioral speciation mechanisms, but especially the latter, it is likely that preferences for well-adapted or high-quality mates are responsible for subsequent species differences in mating signal form.

In Chapter 12 we made the point that mate attraction signals are multivariate signals with several components in the same or different modalities. A controversy exists concerning the presence of separate signal components functioning for species recognition versus mate quality assessment. One view proposes that there are separate signal components for these two functions and that receivers employ them in a **hierarchical rule** for mate choice, with the species-specific component setting primary limits on species recognition

and the mate-quality component facilitating comparisons among potential conspecific mates [7, 474, 475, 549, 556]. This perspective is more compatible with the ecological speciation model, and implies that species recognition signal components arise first and lead to reproductive isolation, followed by the evolution of mate quality signal components that modulate mating preferences. The other view proposes that multiple signal components are combined into a **one-dimensional preference function**, and mating decisions are based on a threshold value or relative comparison along this single scale [51, 100, 281, 317, 477, 534]. This perspective is more compatible with the sexual selection speciation model, and implies that mate attraction signals indirectly do the job of species recognition by identifying good-quality mates. So what is the evidence for species recognition signals?

There are three main arguments for the evolution of distinct signals or signal components that function primarily in species recognition. One is **reproductive character displacement**, in which populations of two closely related species differ *more* in phenotypic traits where they occur together (in sympatry) than where each species occurs alone (in allopatry). When two very similar species occur in sympatry, they interfere with each other's ability to identify conspecific mates. Mismating between heterospecific individuals results in hybrid offspring of low fitness, and selects for divergence of mating signals and receiver preferences that reduce the likelihood of heterospecific mating, a process called **reinforcement**. The typical indicator of reproductive character displacement is strong rejection of heterozygote mating by individuals from the sympatric population, and a greater likelihood of mating when heterospecifics from the allopatric populations are brought together. Numerous examples of this process have now been described [134, 218, 364, 446, 476, 515, 575]. A second line of evidence comes from genetic studies of the genes responsible for closely related species' differences in courtship behavior traits, compared to the genes responsible for intraspecific variation in courtship traits. If some of the same genes are involved in both inter- and intraspecific comparisons, then one could argue that mate choice processes influence species recognition; on the other hand, if different genes are involved, then different evolutionary processes regulate the two functions. A few studies have investigated this question using **quantitative trait locus** (QTL) mapping. As explained in Chapter 9, a QTL is a region of DNA that contributes to variation in a phenotypic trait and can ultimately be related to candidate genes. Studies of *Drosophila* species find that the loci controlling interspecific trait differences are usually *different* from the loci controlling intraspecific variation in courtship, suggesting that different genetic processes underlie microevolutionary changes in courtship traits used for intraspecific mate quality assessment and evolutionary changes in courtship traits arising during speciation [6]. However, some cases of the same loci involved at both levels may yet be found. For instance, novel signaling traits arising from a single mutation, a hidden ancestral gene, or polyploidy (whole genome duplication) are sometimes responsible for

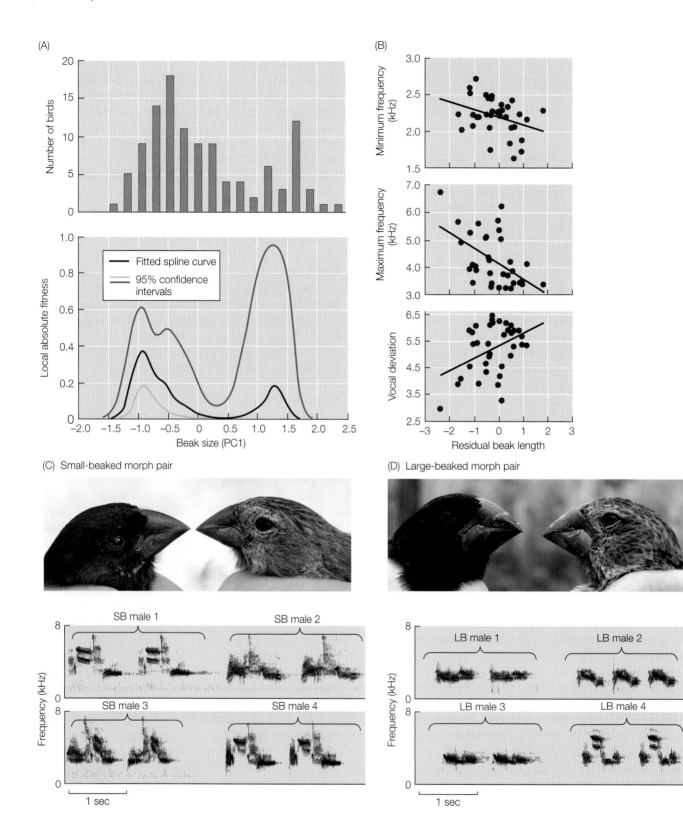

(A)

(B)

(C) Small-beaked morph pair

(D) Large-beaked morph pair

SB male 1   SB male 2

LB male 1   LB male 2

SB male 3   SB male 4

LB male 3   LB male 4

1 sec

1 sec

rapid species divergence [90, 215, 278, 372, 520, 557, 611, 656]. More taxa need to be examined from this genetic perspective to adequately evaluate this hypothesis [164, 666, 698].

Behavioral studies of mate choice preferences for manipulated courtship signals provide a third line of evidence for separate species and mate-quality signals. The components of multivariate mate attraction signals can often be separated into two categories, static and dynamic [192]. **Static components** are determined by the physical constraints of the signal production apparatus, show little variation within a species, and differ quantitatively between closely related species. They are subject to **stabilizing selection** by female preference for

◀ **FIGURE 13.7** **Partial ecological speciation and morphology-linked song divergence in the medium ground finch (*Geospiza fortis*)** (A) A bimodal distribution of beak size has developed in this Galápagos finch on Santa Cruz Island. There is clear evidence of disruptive selection, as survival is lower for individuals with intermediate-sized beaks than for birds with larger or smaller beaks. Gene flow between the two morphs is also restricted, indicating that this population of finches has undergone partial sympatric speciation. The divergence is likely caused by the distribution of seeds in this habitat, since beak size and dietary seed size are positively correlated across species of seed-eating Darwin's finches. (B) Beak size constrains song performance. Birds with longer, deeper, and wider beaks produce songs with lower minimum and maximum frequencies and frequency bandwidths, and greater deviation from the performance limit. (C) A small-beaked morph male and his mate, and (D) a large-beaked morph male and his mate; males are black, females are brown. The birds mate assortatively with respect to bill size. Spectrograms of songs from small-beaked (SB) and large-beaked (LB) males are illustrated below, clearly showing the lower frequencies and bandwidth of the larger-beaked males. The positive assortative mating may be facilitated by the propensity of female Darwin's finches to imprint on the songs of their fathers and use this information when making mate-choice decisions. (After [213, 214, 250, 267, 268].)

mean trait values, are more heavily weighted in preference decisions than other components, and presumably provide species identity information. By contrast, **dynamic components** are more variable within species and often correlated with sender condition. These components are subject to strong **directional selection** by female preference for more extreme values, often above the normal species range, and they most likely provide mate quality information. For example, many species of anurans and orthopterans sing in bouts of repeated pulsed calls. Dominant call frequency and pulse rate are usually static call features that vary little within and between males and differ significantly between species. Females prefer calls at or near the mean for the population. Call rate and call duration, on the other hand, are dynamic features that vary more within and between males, and females prefer higher

FIGURE 13.8 **Static and dynamic call features in the gray treefrog, *Hyla versicolor*** (A) Waveform of two consecutive calls, showing the five acoustic components; the coefficient of variation between males in the population for each component is shown in parentheses. (B) Results of female choice tests between pairs of calls differing in one parameter. (C) Frequency histograms of males exhibiting different call trait values. Static call features such as dominant frequency and pulse rate vary little among males (see (A), low coefficients of variation), and females show strong stabilizing preferences for the modal male trait values. These components are the main species identifiers. Dynamic features such as call rate, call duration, and pulse number vary more within and between males (high coefficients of variation), and females show strong directional preferences for the highest values of these components (here, duration). Females can distinguish calls differing by only one pulse and prefer the higher pulse number. Long call duration in *H. versicolor* is a reliable indicator of heritable male genetic quality (see Figure 12.20). (After [192–196].)

values. **Figure 13.8** shows the static and dynamic signal components of an acoustic mate attraction display in a treefrog. Similar distinctions have been found in several other anuran and orthoperan species [18, 41, 76, 79, 89, 98, 162, 185, 318, 431, 582]. Likewise, males in the closely related fruit fly species of the *Drosophila melanogaster* subgroup vigorously court and sing to females with some combination of sine wave and pulsed songs produced by wing vibrations. The inter-pulse interval (IPI) differs between species, and females exhibit a strong preference for the mean values of the IPI of their species. Pulse duty cycle and the vigor of display are more variable, and females show linearly increasing preferences for higher display effort [132, 216, 535, 631].

Given the existence of multiple signal components, some showing a stabilizing preference curve associated with species identity and others showing directional preference associated with mate quality, how should females combine the information to make an accept-or-reject mating decision? Recent models have explored alternative receiver cognitive processes for jointly assessing multiple signal components [99, 101, 317, 477]. Females could: (1) evaluate the components in a sequential hierarchical fashion, (2) simultaneously combine them as independent additive components, or (3) integrate them in a multiplicative manner. These alternatives can be distinguished by testing female preferences for different combinations of manipulated male signals. For example,

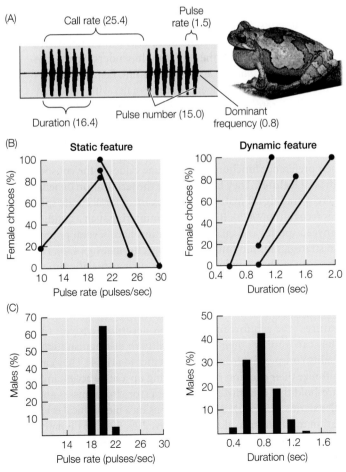

in the túngara frog (see Figures 2.6 and 10.17), the whine component of the male courtship call is an essential species identifier that differs in form from that of other related species, and the pulsed chucks that are variably added to the end of the whine increase the male's attractiveness. If the whine is experimentally modified by blending it with the whine of another species, it will not be chosen by a female, but if the whine is only slightly modified outside of the species-typical range and chucks are added, this "semi-heterospecific" call is preferred over the simple conspecific whine without chucks. Preference tests with a variety of call combinations support a model in which the species-specific whine component and mate-quality chuck component are combined multiplicatively into a single ordinal scale without feature weighting or hierarchical evaluation [477]. Similar results were obtained for a treefrog that combines call rate (mate quality) with pulse rate and fundamental frequency (species identity) in a multiplicative manner [100]. If the species identity component is characterized by a bell-shaped preference curve, the multiplicative strategy effectively results in a mate quality signal that is amplified when the species identity component is closest to the preferred population mean [101]. See Web Topic 13.3 for further details on these models.

When high-quality conspecifics resemble a sympatric heterospecific, a conflict exists between mate recognition and mate quality assessment, and selection may alter the cognitive strategy for weighting different signal components. To envision this conflict, review the overlapping curves in Figure 13.4A and consider the situation where the most desirable senders are located in the right-hand distribution tail where they overlap most with the undesirable senders. Spadefoot toads (*Spea multiplicata*) are a case in point. Females in populations allopatric to congeners prefer males with extremely high calling rates, and benefit with a 3.5% higher fertilization rate when mating with preferred males. In populations sympatric with a high-calling-rate congener, *S. bombifrons*, spadefoot females risk heterospecific mating and have shifted their preference function to the average values of the conspecific call features. They fail to receive the fertilization benefit, thereby trading off mate-quality assessment for accurate species recognition [475]. Receivers encountering such a conflict between species and quality recognition have several potential strategies for minimizing heterospecific mating [474]. (1) They can evolve nonphenotypic cues such as specific times, locations, and habitats where conspecific mates can be found. (2) They can develop preferences for alternative high-quality mate features that also differ from the heterospecific signal features (the reinforcement process described above). (3) They can improve their sensory ability to distinguish high-quality conspecifics from heterospecifics [444]. (4) They can optimize their decision rules by setting up a mate acceptance threshold that minimizes the cost of errors from misses and false alarms (see Figure 13.4A) or by using a hierarchical evaluation mechanism as described above [99, 508]. (5) Finally, receivers can refine their template of a high-quality conspecific by attending to additional signal components. The

addition of olfactory cues for species recognition along with visual and auditory signals is a particularly common strategy [47, 144, 238, 283, 334, 355, 426].

The discussion above has assumed that mate attraction signals and mate choice preferences involved in species recognition are *genetic* traits—in other words, that species recognition is hard-wired and innate. But *learning* can also be involved in species recognition under some circumstances [276, 484, 576, 665]. Early learning of parental phenotypes via imprinting (see Chapter 9) can facilitate assortative mating and lead to speciation and reinforcement. Population genetic models show that paternal imprinting in particular, in which young females imprint on the phenotype of their father and select mates with similar traits, is a more efficient mechanism for species divergence and maintenance than genetically hard-wired preferences [576, 646]. Paternal imprinting is very likely responsible for the assortative mating of the finches described in Figure 13.7. A role for sexual imprinting on either the male or the female parent has been described in many other birds and mammals, and in some fishes, insects, and spiders [158, 175, 246, 304, 462, 635, 665]. Learned avoidance of heterospecifics in sympatric zones, learning of habitat preferences, and even learning of mate attraction traits, such as passerine bird song, are other ways that cultural traits can reinforce premating isolation and speciation [19, 150, 159, 333, 380, 630]. The resolution of the controversy raised earlier about the existence of species and mate-quality recognition processes is that species differ in the number and types of components in their mate attraction signals depending on their biogeographic speciation history, their signal production mechanisms, the presence of sympatric congeners, and receiver cognitive abilities.

Web Topic 13.3 *Cognitive models of mate choice*
Here we outline computational mechanisms females employ to integrate several sources of information from mate attraction signals. More examples of multimodal mate attraction signals and characteristics of species recognition components in various taxa are also provided.

PARTNER RECOGNITION   For animals that form male–female pair bonds or same-sex bonds for the purpose of extended cooperative behavior such as sharing of parental care duties or joint resource defense, distinctive signals are needed for rapid **partner recognition**. The best-documented examples are the acoustic contact calls of birds and primates [329]. The majority of bird species are monogamous and form extended pair bonds because the male is selected to remain with his mate and contribute to parental care by provisioning nestlings and guarding mobile young. For the typical monogamous territorial species, the recognition task is not a particularly difficult one because the pair is relatively isolated on its territory and there are only a small number of neighbors and possible intruders that need to be

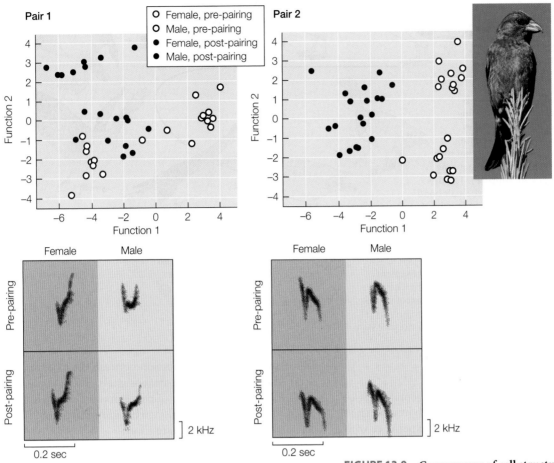

FIGURE 13.9 **Convergence of call structure within pairs of red crossbills (*Loxia curvirostra*)** In the field, the contact calls of paired red crossbills are similar to each other, but differ from the calls of other pairs. Captivity studies show that the convergence occurs during pair formation, as prospective mates initially differ in call structure. This figure shows spectrograms of the calls of two pairs that differ in their level of convergence. Scatterplots (top) show the first two factors of a discriminant function analysis of call features. Spectrograms (bottom) show male and female call structure before and after pairing. Pair 1 converged partially but remained statistically distinct, whereas pair 2 converged to statistically indistinguishable shared calls. Pairs with greater convergence showed higher levels of affiliation. Crossbills exhibit distinct ecotypes with different bill shapes, diets, foraging techniques, and vocal dialects, and birds pair assortatively with mates from their own ecotype, another example of sympatric ecological speciation in progress [161, 222, 301, 608]. (After [577].)

distinguished. Males and females quickly learn the individual characteristics of their mate's songs or calls and can distinguish mate from nonmate on this basis alone [396, 441]. Cooperative breeders similarly show the ability to recognize different group members on the basis of individualistic call structures [579, 702]. For species that are monogamous but also live in large mobile flocks or colonies, however, the recognition task is much more daunting, and in these instances there is strong selection for more complex calls with several independent components that can encode a large number of individual variants. Distinctive calls and discrimination of mates have been demonstrated in several parrot species, silvereyes (*Zosterops lateralis*), zebra finches (*Taeniopygia guttata*), goldfinches (*Spinus tristis*), and several other flocking cardueline finches [42, 87, 430, 517, 667, 668, 678, 679]. In the budgerigar, finches, crossbills, and the monogamous pygmy marmoset (*Cebuella pygmaea*), male and female call structure actually converges gradually following pair formation, which gives the pair a unique family signature (**Figure 13.9**) [222, 253, 430, 577, 610].

Signals in other modalities can also be designed to encode individual variants. Individually distinctive visual signals, employing variable patches of color, are used as a mechanism to distinguish mates and important companions in group-living wasps, *Quelea* finches, and African wild dogs (*Lycaon pictus*) [137, 640, 642]. Olfactory signals are used in many

mammal species and some seabirds for partner recognition [85, 197, 287, 297, 386, 605]. The Bruce effect in rodents, in which a female may abort a litter if exposed to a male who is not the sire of her current litter, clearly depends on olfactory discrimination of the original sire from any newcomer males [75]. The monogamous burying beetle (*Nicrophorus orbicollis*) presents an interesting case in which mate recognition is based on a simpler process than strictly individual identity signals. Heterosexual pairs cooperate to defend a carcass they have buried to rear their offspring. They do not attack each other, but they do attack any outsiders that attempt

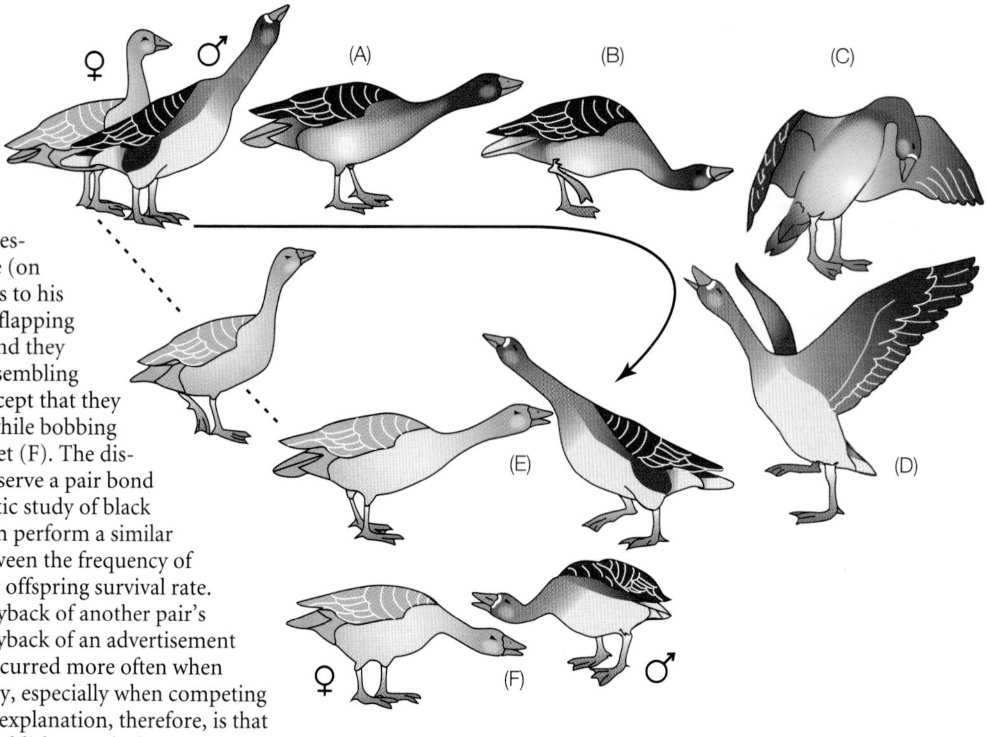

FIGURE 13.10 The function of the triumph ceremony in geese and swans In the graylag goose (*Anser anser*) triumph ceremony, the male approaches (A) and aggressively challenges (B) another male (on right, not shown). He then returns to his female while vocalizing and wing flapping (C,D). She approaches him (E), and they both adopt aggressive postures resembling those the male used in leaving, except that they orient obliquely past each other while bobbing their heads and vocalizing in a duet (F). The display was initially hypothesized to serve a pair bond maintenance function. A systematic study of black swan (*Cygnus atratus*) pairs, which perform a similar display, found no correlation between the frequency of display and pair bond duration or offspring survival rate. The display was stimulated by playback of another pair's triumph ceremony but not by playback of an advertisement call of an unmated male, and it occurred more often when many pairs were in close proximity, especially when competing for food. The most parsimonious explanation, therefore, is that the display serves as a threat to establish a pair's dominance status over other pairs. (After [176, 284, 335].)

to usurp the resource. The olfactory cue a female uses to distinguish her mate is a cuticular hydrocarbon associated with the male's recent breeding status and experience with a carcass, a **breeder's badge**, which an invader would typically lack. Experimental substitution of a different breeding male for the true mate results in acceptance by the female. In nature, carcasses are rare and a breeding pair is very unlikely to encounter another breeder. This breeding badge mechanism is therefore a low-cost but effective rule of thumb for distinguishing mates from dangerous intruders [427].

Pair cooperation

Monogamous pair bonding (sometimes referred to as **male–female partnerships**) generally occurs in species where the male performs a critical role in parental care. Males are selected to remain with the female after mating because they benefit more from increasing the survival of their offspring than from maximizing the number of females fertilized. Since females by definition also invest in offspring survival, the reproductive interests of monogamously mated individuals overlap strongly. Monogamy is the rule in birds and is found occasionally in some mammals, fish, insects, and a few other arthropods [116]. The bond formed between males and females may last only for the duration of one breeding cycle, or it may last for several cycles, sometimes for many years [49]. The exceptionally long-term partnerships in seabirds, geese, swans, parrots, and some songbirds appear to be favored because reproductive success increases with bond duration. The reason for this advantage arises from the improved ability of experienced breeders to obtain and defend high-quality territories, begin reproduction early in the season, and coordinate incubation and nestling care [48, 616, 658]. These activities

all require good communication between the partners and learned responsiveness to partner signals. In this section we shall examine three aspects of pair communication: mutual displays, hormonal integration of breeding stages, and negotiation of relative parental care effort.

MUTUAL DISPLAYS   Mated pairs typically perform joint displays together, often highly synchronized and sometimes quite elaborate [381, 674]. The timing and context of these displays vary, suggesting different functions for these signals beyond the vague notion of **pair bond maintenance** (**Figure 13.10**). When the joint displays occur after pairing and up to actual mating, they are likely to have a courtship function, as discussed in Chapter 12. When the displays occur during incubation exchanges or after a period of separation between the mates, they serve recognition and greeting functions. Displays that occur during conflicts with other conspecifics over space or resources obviously serve a joint-defense function. Other mutual displays may function primarily to advertise the mated status of the pair to other conspecifics. Evidence for and examples of these functional categories are described below. More examples are provided in Web Topic 13.2.

Some species engage in a long courtship period and use mutual displays as a means of **continued mate evaluation**. For example, great crested grebes (*Podiceps cristatus*) form pairs in mid-winter but don't start breeding until spring many months later. During this prebreeding period, they engage in several types of mutual courtship displays initiated by either sex. The most complex display is the weed ceremony (**Figure 13.11A**). Weeds are a nest construction component, so the display clearly indicates intention to reproduce, but the synchrony of the sequential display stages is remarkable and the display may be terminated if precise synchrony is not maintained [271]. This suggests that the birds are using

(A)

(B)

FIGURE 13.11   Mutual displays in mated birds   (A) The weed ceremony of the crested grebe (*Podiceps cristatus*). (B) The mutual sky-pointing display by male and female blue-footed boobies (*Sula nebouxii*).

the display to evaluate each other's commitment or quality. The continued mate evaluation explanation is very clear in mutual displays of the blue-footed booby (*Sula nebouxii*) (**Figure 13.11B**). The color of the foot integument varies with the nutritional condition of the birds, and if the color of the male's feet in particular is manipulated to make the bird appear in poor condition, the female reduces her rate of courtship and copulation with the mate. Interestingly, the equivalent manipulation of the female's feet does not alter male behavior toward her [644]. The female crested tit (*Parus cristatus*) uses a different mechanism to evaluate mate quality during courtship—she solicits copulation at exceptionally high rates. The poorer a male's condition, the more likely he is to refuse to copulate, whereas males in good condition never refuse. Females mated to poor-condition males engage in more extra-pair copulations [354]. There are numerous examples of females reducing their reproductive effort, seeking extra-pair copulations, or divorcing their mates to remate with a better male, implying that females continuously assess their current mate relative to other options [104, 115, 145, 155, 156, 248, 422, 449, 617, 625, 657].

Mutual displays can serve a **recognition** and **greeting** function in cases where mates spend substantial periods of time apart. For example, many birds with long bouts of alternating incubation and foraging, such as herons and seabirds, perform elaborate displays when the forager returns to the nest to relieve the incubator [674]. In the common heron *Ardea cinerea*, the arriving bird stands erect and raises its neck to the fullest extent. The neck feathers and crest are erected, the head is often pointed up, and wings are flapped while a raucous vocalization is given repeatedly. The bird on the nest does the same display, sometimes without standing [272]. Such greeting displays confirm the status of the approaching bird as the mate willing to take over incubation, provide multiple cues for partner identification by both

birds, and could potentially allow mutual assessment of partner condition. A similar greeting, recognition, and information exchange function has also been ascribed to the mutual displays of seahorses, in which the male broods the eggs in a specialized pouch and is visited daily by the female. In *Hippocampus whitei*, females have a larger home range that includes the smaller territory of a male, but the pair spend most of the day apart. Each morning they engage in a greeting ceremony (**Figure 13.12**) that allows the female to assess whether the

FIGURE 13.12   Greeting ceremony in the seahorse (*Hippocampus whitei*)   Each morning the female approaches the male and they perform a mutual greeting ceremony, which lasts about six minutes. They both develop a dark stripe on the outer margin of the dorsal fin, as seen in this photo, and lighten in color. The male in particular (right) changes to a bright yellow color. Side by side they both grasp a plant shoot, and with the male on the outside, circle around the shoot in a "maypole" dance. They then release the shoot and swim slowly parallel to each other with the male grasping the female's tail. They swim back to the shoot, repeat the maypole dance, and they slowly parallel swim until one of them, often the female, darkens and swims away. If the male has released a brood, the mutual display is prolonged and egg deposition and fertilization follow [670, 671].

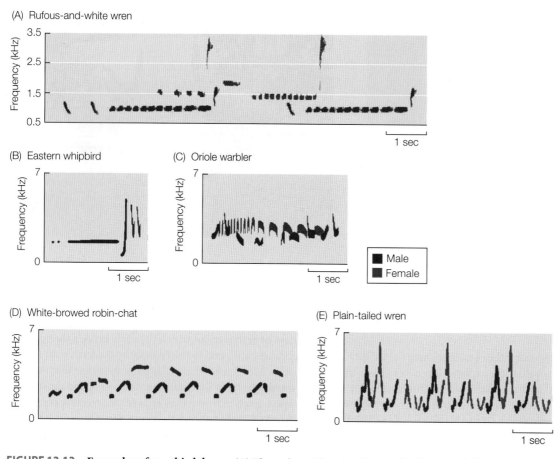

FIGURE 13.13 **Examples of songbird duets** (A) The male and female rufous-and-white wren (*Thryothorus rufalbus*) overlap their songs without precise timing; either sex may initiate and each may give one to several songs. Female songs are always of a higher frequency than male songs. (B) The female eastern whipbird (*Psophodes olivaceus*) gives characteristic notes at the end of the male's dramatic whip. (C) The song of the female oriole warbler (*Hypergerus atriceps*) overlaps the male song, but not with high precision. (D) The male white-browed robin-chat (*Cossypha heuglini*) initiates a duet and the female fits her notes precisely within the gaps of his song. (E) Male and female plain-tailed wrens (*Thryothorus euophrys*) fit their parts together precisely and cycle through a rapid succession of different phrases; several males and several females may chorus together, each sex singing closely in unison. (After [83, 410].)

male has hatched his brood. If so, she immediately gives him another clutch. The time required for the male to brood is about equal to the time needed by the female to produce a clutch of eggs, so the operational sex ratio is equal and there is no courtship role reversal [671].

Cooperative territory defense is another context favoring mutual display. In some territorial monogamous species, especially those that retain the same territory throughout the year or in subsequent breeding cycles, the pair defends their territory as a team. The male and female jointly approach a territorial intruder and present a visual or vocal display. In some species, the pair display in a highly coordinated duet; in others, the members of the pair display independently (**Figure 13.13**) [236]. The intensively studied Australian magpie-lark (*Grallina cyanoleuca*) provides key evidence supporting a cooperative defense function for duetting [232–235]. Male and female notes are precisely alternated during duets. Either

sex can initiate a duet, and both are equally likely to add their voice to the solo song of the mate to create the duet. The observed combinations of male and female song types are nonrandom, indicating that pairs develop specific answering rules to each other. Playback of a duet is more threatening, and elicits a stronger response, than playback of solo song. Moreover, pairs increase the precision of their duet with increasing duration of their pair bond, and more precise duets are more threatening stimuli than less precise duets. Duetting requires a high level of attentiveness to the mate, and the degree of coordination reveals this attentiveness to both the partner and to rival conspecifics [236].

Mutual displays can serve a **mate guarding** function by revealing the mated status of each partner to same-sex rivals. This strategy reflects a conflict of interest between the sexes when the solo signaler is attempting to attract additional mating opportunities. A mate-guarding function has been ascribed to duetting in a few species. Females duet to defend their mates against other females and demonstrate their male partner's mated status in eastern whipbirds (*Psophodes olivaceus*) and monogamous primates (gibbons, marmosets, and tamarins) [319, 521]. Evidence supporting this conclusion includes a female-biased sex ratio (high female competition for mates) and a high level of female aggression toward female solo songs and the female member of rival pairs. The reverse pattern, male mate-guarding of the female, appears to be an important function of duetting in the subdesert mestite (*Monias benschi*) and the tropical boubou (*Laniarius aethiopicus*) [210, 211, 561, 562]. Duetting can also operate as a mechanism for mates to locate and keep track of each other's whereabouts, a strategy that is favored in nocturnal and dense forest species [396]. This location function for duetting has been documented in neotropical forest wrens, where one member of the pair may sing to elicit a response by the mate, and subsequently approach the partner if they are far apart [368, 411]. Structural convergence of calls, which facilitates mate recognition, can also serve as an acoustic strategy for advertising pair status. Pair advertisement is not limited to birds and mammals. Monogamous pairs of neotropical cichlid fish pairs remain in close contact on their small territories and adopt a color pattern that advertises their paired, territorial status; the male aggressively defends the boundary of the territory while the female guards their brood [25, 654]. An insightful experiment with zebra finches (*Taeniopygia guttata*), a monogamous flock-dwelling species, illustrates the importance of advertising paired status. A male's responsiveness to the playback of his mate's call depends on his social audience. If the audience is a nearby mated pair, he readily responds and answers his mate's call, but if the audience is an unmated male and female he does not respond to his mate's call [667]. Paired birds may have a competitive advantage in colonial species.

Tactile interactions such as allogrooming, huddling, hugging, kissing, and tail-twining are common forms of mutual display between monogamous pairs of birds and primates [674]. While these actions may serve some practical functions, such as plumage hygiene and thermoregulation, they are often highly ritualized and closely connected with other ceremonial displays and post-aggression contexts. Pair bond maintenance and reduction of aggression are the classical explanations for these behaviors. In two avian species with long-term pair bonds, cockatiels (*Nymphicus hollandicus*) and guillemots (*Uria aalge*), a higher allopreening rate is positively correlated with reproductive success and mate retention, and negatively associated with aggression levels [361, 616]. Allopreening rate and willingness to reciprocate may indicate pair compatability or be used as a test to evaluate mate quality and commitment [617, 623, 703].

HORMONAL INTEGRATION OF REPRODUCTION   Males and females possess very different reproductive plumbing and hormonal cycles. As we saw in Chapter 12, males are typically in a continuously receptive state, with stored sperm available at all times, whereas females are not receptive until they have developed a clutch of eggs or are prepared to ovulate or gestate internal embryos. Female reproduction is a complex physiological process regulated by the serial release of pituitary and steroid hormones. In most species, the female requires a species-specific stimulus from a male at some point in her reproductive cycle. Examples include the induction of ovulation with vigorous copulation in some mammals, the enhanced stimulation of oviduct development in many birds with highly versatile or vigorous male song, and the tranquilizing effect of the male pheromone in pigs, salamanders, and butterflies [1]. These are mechanisms by which females can assure themselves of a conspecific and possibly also a high-quality mate. Classic studies of doves, rats, and other species illustrate the mutually stimulating interactions between male and female signals, internal hormones, and external stimuli that **synchronize hormonal cycles** and coordinate the entire reproductive episode [352]. Hormones can affect behavior in three ways: (1) by influencing the development of special structures involved in the performance of a behavior such as male claspers or display structures, or by stimulating oviduct enlargement or lactation gland development; (2) by influencing the peripheral nervous system that controls sensory input to the brain; and (3) by triggering behavioral mechanisms in the brain directly. In the presence of appropriate external stimuli, olfactory, auditory, and visual displays by one sex affect hormone production in the other sex and modify the latter's behavior, which in turn influences the hormonal state and behaviors of the first sex. **Figure 13.14** summarizes these interactions in the ring dove *Streptopelia risoria*, a monogamous species in which both sexes brood the eggs and young and feed them by regurgitating a milky substance from a specialized crop sac.

NEGOTIATION OF PARENTAL CARE   How much effort should each parent invest in parental care? Early models approached this question from the perspective of an evolutionary trade-off between investing time and energy in the current brood versus surviving to invest in future broods or abandoning the brood to seek additional reproductive opportunities [266, 395, 650]. Since males and females face different reproductive constraints, parental care was treated as a game between the sexes in which, after mating, each sex could either care or leave. If there is little improvement in offspring survival with parental care, both sexes are predicted to leave. If offspring survival is significantly improved with care by one parent, but having two parents adds little further benefit, then one parent is predicted to leave, and a conflict exists over which one stays. If two parents are significantly more successful than one, then both parents should care. Empirical studies to test the predictions of these models often focused on biparental species and typically removed or handicapped one of the parents to determine whether the other parent could compensate for the absence or reduced effort of the mate. The expectation was that the remaining (or unhampered) parent would increase its effort but not fully compensate. The reasoning is that if

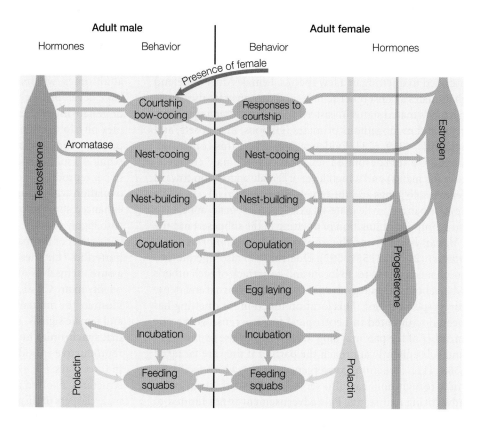

FIGURE 13.14 Reciprocal stimulation of reproductive behaviors in the ring dove (*Streptopelia risoria*) Temporal sequence (vertical axis) of male (left) and female (right) hormone profiles and behaviors. Males in breeding condition direct bow-cooing vocalizations to nearby females, who may be attracted or chased by the male. If the female is interested in the male, both initiate nest-cooing vocalizations under the influence of estrogen (in the male, converted from testosterone via aromatase in his brain). The female's nest-cooing self-stimulates the release of luteinizing hormone and follicular development, which leads to increases in progesterone. Nest building, copulation, and egg laying proceed. The sight of the eggs in the nest induces incubation behavior and suppresses courtship behavior in both sexes, testosterone in males, and progesterone in females. Incubation is maintained with the onset of prolactin release, which also causes the development of the crop sac and brood patch. When the young hatch, the parents feed them initially by regurgitating crop milk and later by bringing them grain. When the squabs are a few weeks old, prolactin decreases, testosterone in the male and estrogen in the female increase, and the cycle is repeated. (After [114, 187, 433].)

one parent can fully compensate, monoparental care would have been favored. A metadata analysis of mate removal and handicapping studies indicates that partial compensation is common, but cases of no compensation and complete compensation also occur [241]. Also recall from the battle of the sexes courtship model in Figure 12.29 that partial compensation favors faithful helpful males and coy female assessment of male parental investment.

Missing from this approach to parental care allocation between the sexes in biparental species is the fact that parents adjust their individual provisioning effort based on a variety of factors, including age and number of offspring, parental condition, mate's effort, and ecological conditions [696]. Recent models at the behavioral time-scale explore the consequences of short-term decision making on provisioning effort, and take into account communication between parents and offspring and between the parents themselves. For example, one model compares how two communicating and perfectly coordinated partners, versus two independently-acting partners, would respond to different kinds of experimental manipulations [2]. Another model finds that evolutionarily stable **negotiation rules** can evolve that enable two parents to interactively arrive at a final allocation of effort [406]. A third model, building on this negotiation model, incorporates uncertainty regarding brood need. When each parent has only partial information about offspring hunger

derived from nestling begging signals, parents rely on signals or cues from the mate to improve their estimate of offspring need [291]. A novel experiment using the biparental great tit tested the predictions of this model by manipulating the perception of offspring need for one of the parents. Intense nestling begging sounds were selectively broadcast at the nest only when this focal parent came to provision. Not surprisingly, the focal parent increased its provisioning rate, but so did the other parent [254]. This result clearly indicates that parents not only adjust their provisioning effort in response to offspring signals, but also that they communicate and pay attention to each other. A few studies have documented subtle vocal communication between nesting pairs of birds, but more studies are needed [209, 231]. The varied responses to different types of manipulation experiments observed in different avian species suggest that provisioning behavior lies on a *negotiation continuum*, which describes the extent to which parents respond to the actions of other family members and the quality of the information they receive [255].

Negotiation and coordination occur at other stages of breeding as well. Nest placement is a joint decision in many avian species. Males may propose sites, but females must agree. A study of this decision-making process in blue-footed boobies nicely illustrates the communication mechanism involved in this negotiation. Males perform low rates of a nest-pointing display at several potential sites to "feel out" the female's opinion. If the female performs the pointing display at any of these sites, the male's rate of display increases. If the female disagrees with the male's selections, they expand the number of sites investigated. Both sexes have absolute veto power over potential nest sites, in that a site is never accepted if one of the two partners fails to point at it. Maximum display rates by both partners occur at the site that is finally chosen [620]. Such collaborative tactics likely exist in many other species. Incubation is another stage requiring coordination if both sexes tend the eggs, since for most species only one bird can sit at a time, and leaving the eggs uncovered increases hatching failure. A study of cockatiels showed that pairs that were more attuned to each other—more synchronized in their activities, more responsive reciprocators of allogrooming, and less aggressive in general—were also better able to coordinate incubation bouts so that only one parent attended the eggs at a time, which significantly improved hatching success [616]. A study of male–female collaboration during incubation in Kentish plovers (*Charadrius alexandrinus*) also demonstrated each mate's ability to compensate for manipulated reductions in the partner's effort, so that one parent always attended the nest [330]. Once males are committed to staying and helping to rear the offspring, selection seems to favor the use of signals and cues between mated pairs to maximize efficiency and reproductive success.

Parent–Offspring Integration

Previously we discussed parental investment from the perspective of how the male and female parents allocate and coordinate parental care duties such as incubation, guarding, and provisioning of offspring. In this section we shall examine the relationship between parents and offspring. An inclusive definition of **parental investment** is: the time and energy devoted to offspring that increases the offspring's survival while decreasing the parent's ability to survive and invest in future offspring [650]. Parental investment therefore includes gametic investment (in eggs and sperm) along with zygotic investment (all post-fertilization effort). Viewed from this perspective, there is a potential conflict of interest between parents and offspring, because each individual offspring may selfishly prefer to receive more investment than the parent is selected to give in order to optimize its lifetime reproductive success [651]. Here we shall examine the extent of this conflict and the communication signals employed to regulate the amount of parental care given to offspring. But first, we review the mechanisms used by parents and offspring to recognize each other, because it is important that these costly behaviors be directed toward true offspring.

Offspring and parent recognition

Recall the recognition principles and mechanisms outlined earlier in this chapter. The sophistication of **parent–offspring recognition mechanisms** depends on the difficulty of the recognition task, which is affected by factors such as the development pattern of the young (altricial versus precocial), the existence of a fixed nest or roost site, family size, gregariousness of young from different families, and the duration of time young are left unattended while parents forage. The simplest effective recognition method will be used—typically location if there is a fixed nest site, and associative learning

of individual signal characteristics if offspring are mobile. Recognition is unidirectional in some species, with either parental recognition of offspring or offspring recognition of parents. Mutual recognition occurs when the task is particularly challenging. Even phenotype matching is adaptive in some circumstances.

USE OF LOCATION CUES   For territorial birds and mammals that have a small number of immobile young placed in a nest well-separated from adjacent broods, the **nest location** is the focus of parental recognition of offspring. Neither signature signals by offspring nor offspring recognition of parents is required. For example, the California towhee (*Pipilo crissalis*) is a typical monogamous territorial songbird. Parents do not respond differentially to cluck distress calls of their own and nonoffspring chicks, but do respond more strongly to distress calls played back close to the nest site than to calls played farther from the nest [39]. Similarly, solitary territorial swallow species (barn swallows, *Hirundo rustica*, and rough-winged swallows, *Stelgidopteryx serripennis*) do not recognize their own young or discriminate against heterospecific foster young. Fledglings of these noncolonial species produce begging calls that are less complex, less individually distinctive, and more difficult to discriminate than the calls of fledglings of closely related colonial species (**Figure 13.15**) [34, 356, 367, 407, 409]. Phocid seal mothers and pups likewise exhibit nondistinctive calls and do not rely on vocalizations to recognize each other, because they remain together at a fixed location throughout the lactation period (in contrast to Otariid seals, described below) [273, 282]. Although location is an efficient recognition cue in these cases, it leaves the parent vulnerable to brood parasites that dump nontarget offspring into the nest and to stray conspecific nonoffspring young. Avian host species that are heavily parasitized by cuckoos, cowbirds or conspecifics evolve the ability to learn to visually recognize their own eggs and eject foreign eggs [139, 140, 375, 526, 618]. However, recognition and selective rejection of foreign *nestlings* is rare in birds, primarily because the success of egg discrimination and parasite mimicry makes the incidence of successful parasitism too rare to select for the cognitive skill of vocal/visual nestling discrimination [217, 345, 370, 468]. Mammalian mothers are vulnerable to conspecific freeloaders but readily use olfactory cues at short range to identify their own pups and limit the loss of milk to nonoffspring pups [360].

UNIDIRECTIONAL RECOGNITION   In species with **unidirectional recognition** systems, which party produces the distinctive signal that the other uses for recognition depends upon the ecological context. Offspring bear the burden of recognizing the parent in most precocial birds, where families are relatively isolated, chicks feed themselves, and the parent leads them to good foraging areas and defends them against predators using directive vocal signals. Chicks rapidly *learn by association* the visual and vocal characteristics of the parent and respond appropriately to parent signals while avoiding

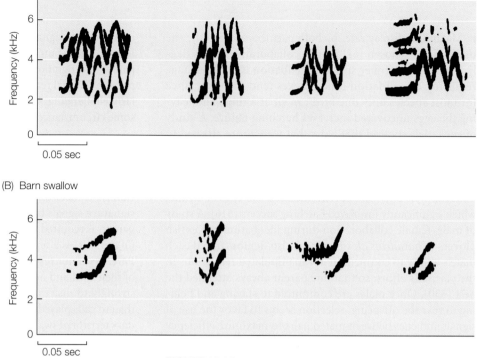

(A) Cliff swallow

0.05 sec

(B) Barn swallow

0.05 sec

FIGURE 13.15   Comparison of chick begging calls in colonial and solitary swallows   (A) Calls of four individual cliff swallow chicks, a colonial-breeding species, and (B) calls of four individual barn swallow chicks, a solitary-breeding species. Chicks of both species produce calls with two nonharmonically related voices. The cliff swallow calls are longer and contain multiple frequency modulations. Barn swallow calls are poorly defined at the beginning, with a tonal frequency sweep emerging from a burst of noise. Acoustic analysis of the variance in quantitative call features (duration, frequency peak, and range and period of the frequency modulation) indicates that cliff swallow calls contain approximately four bits more information than barn swallow calls (9.01 versus 5.15). Laboratory studies show that both cliff and barn swallow parents can more easily learn to discriminate among the begging calls of different cliff swallows than among the calls of different barn swallows. Furthermore, cliff swallow parents achieve discrimination criteria faster than barn swallow parents. The cliff swallow signaling system is positioned in the yellow zone of Figure 13.4B, with calls designed to facilitate individual recognition and receivers possessing good discrimination ability. The barn swallow system is positioned in the green zone of Figure 13.4B in lab studies, and in the field does not discriminate between calls from its own and heterospecific young (orange zone of Figure 13.4B). (After [367, 409].)

nonparent signals [131, 286]. Since there are often many chicks per brood and parental effort per chick is low, there is little pressure on parents to individually recognize their offspring [353]. Similarly, in fish that guard mobile fry, the fry initially learn to recognize their parents and parents eventually recognize their offspring using brood-specific olfactory cues [405]. In seagulls such as the laughing gull (*Larus atricilla*), which have breeding colonies with moderate spacing between nests and mobile chicks that require feeding (semiprecocial), parents arrive on the nesting territory with food and call to the offspring. Parental voices are individually distinctive, and chicks learn to recognize and respond selectively to them. However, parent gulls do not differentiate their young chicks from foreign chicks on the basis of vocalizations [36, 37]. Only later do parents learn to distinguish their chicks visually. Finally, ungulate mothers that hide their young fawns in dense vegetation and return at intervals to call them out and suckle also have a unidirectional offspring-recognition-of-parent system; this makes adaptive sense because hidden fawns calling frequently to identify themselves would risk revealing their location to predators [645].

By contrast, in some group-living species, offspring develop the signature signals and parents bear the primary discrimination burden. For example, nursery colonies of the Mexican free-tailed bat (*Tadarida brasiliensis*) may contain thousands of pups. A mother remembers the approximate location of the site where she last left her pup and thereby reduces the number of individuals she must examine by a large fraction. She then uses olfactory and auditory cues or signals to distinguish her own pup. Pups call and beg to every passing female. The decision rule results in a substantial number of errors; in a series of studies, 17% of mother–baby pairings were found to be unrelated [22, 401-403]. Other bat species show similar development of complex offspring-specific vocal signature signals and very accurate discrimination abilities by mothers [53, 148, 321, 340, 694]. Mothers, on the other hand, give less distinguishable acoustic signals, and even if pups are capable of some degree of recognition, they may benefit from responding to all female approaches in the hope of obtaining a meal from another mother. Group-living primate females similarly recognize the separation cries of their infants [513]. Unidirectional olfactory recognition of offspring is common in mammals, lower vertebrates, and invertebrates [180, 189, 360, 392]. In mice, it is specifically the highly polymorphic nonvolatile major urinary proteins (MUPs) that provide the stable source for the individual signature signal [637].

(A)

(B)

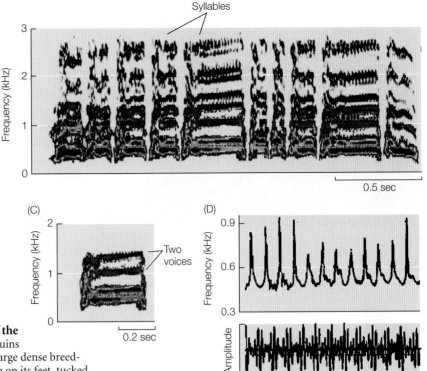

FIGURE 13.16   Two-voice vocal system of the *Aptenodytes* penguins   (A) Adult king penguins (*A. patagonicus*) huddling close together in a large dense breeding colony. Each bird is incubating a single egg on its feet, tucked under a fold of skin, and is awaiting the return of its mate from foraging at sea. Mates can only recognize and find each other vocally, not visually. (B) Spectrogram of an adult emperor penguin (*A. forsteri*) call showing multiple syllables and two simultaneous series of harmonically related bands of slightly differing frequency arising from the two-voice vocal system. (C) Close-up of a single syllable clearly illustrating the two voices. (D) Frequency (top) and temporal (bottom) domain illustration of a single syllable, showing beat formation generated by the two frequencies. The periodic frequency discontinuities coincide with the periodic amplitude fluctuations, allowing quantification of the beat period, here 10 msec. The beat pattern is highly resistant to degradation when transmitted through a colony. (After [10, 11].)

MUTUAL RECOGNITION   For most colonial species in which young from many families aggregate and intermingle while parents leave on foraging trips, a **mutual recognition** mechanism is favored. The mere existence of such crèches implies a highly effective parent–offspring recognition mechanism. In swallows, many seabirds, and ground squirrels, young are initially restricted to a nest or burrow site. Parents use nest location as the recognition cue at this stage, and experimentally inserted foreign young are not distinguished. Shortly before fledging, burrow emergence, or crèche formation, the offspring develop complex and distinctive signature signals, and parents learn to recognize their young as individuals [31, 37, 84, 142, 168, 295, 296, 320, 359, 393, 397, 409, 564, 591, 622]. Where offspring discrimination of parents has been studied, they usually are found to respond selectively to parental signals, and individual parent signals are as statistically distinguishable as the offsprings' signals [33, 397, 592]. Even if offspring are less discriminating as receivers than parents, their signals can act as a filter to aid the parents' search.

The most sophisticated mutual recognition systems appear to be favored when there are no locational cues

available. For example, seabirds with very large and dense nesting colonies, such as terns and most penguins, possess a mutual recognition system with signature vocal signals by both parents and offspring. Penguin species that incubate their eggs in a nest and thereby obtain some additional locational cues from the fixed meeting site have less complex signal structures in comparison to non-nesting species, such as king and emperor penguins (*Aptenodytes patagonicus* and *A. forsteri*), which incubate the single egg on the parent's feet [296]. Not only must male and female parents find each other during incubation exchanges, but the chick must also learn its parent's signature signals to facilitate successful reunions after foraging trips. These two penguin species possess a particularly complex 2-voice vocal production system that generates a beat pattern from the interference between two slightly different frequencies. Individuals' calls differ in beat period, frequency and amplitude modulation pattern, and frequency spectrum composition (**Figure 13.16**). Removing one of the voice components completely eliminates individual recognition capability [10, 11]. Seals are another taxon that nicely illustrates the factors affecting the sophistication of parent–offspring recognition systems. As mentioned above, true seals (phocids) haul out on shore or ice floes for the duration of the birth and lactation period. These seals are not very mobile on land, and females remain in constant contact with their pups. Some species show no discrimination of pup vocal signals and rely on olfactory cues to limit their nursing of other females' pups, while others show unidirectional female recognition of pup vocal signals. Eared seals (otariids), on the other hand, are more mobile

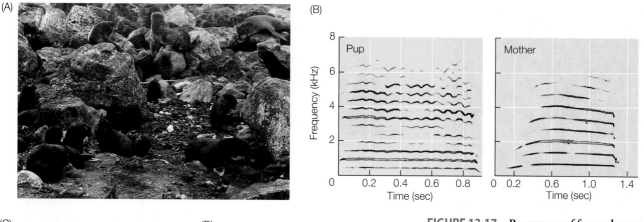

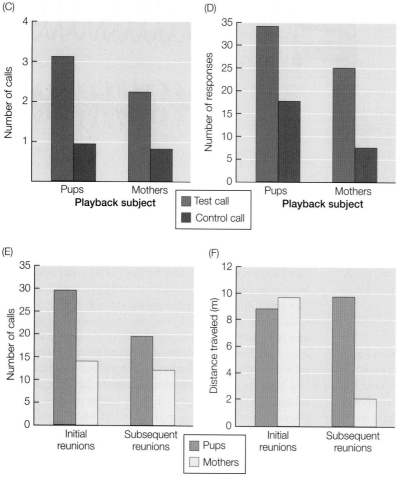

FIGURE 13.17 Responses of fur seal pups and mothers to each other's vocalizations (A) Subantarctic fur seal (*Arctocephalus tropicalis*) pups waiting for their mothers to return from foraging at sea. (B) Spectrograms of pup and mother calls. (C,D) Responses to playbacks. (C) Both pups and mothers gave more calls to playback of the other's calls compared to control calls from other individuals of the same age, but pups responded more vigorously than mothers. (D) Pups made more errors by responding to control playbacks than mothers did, both in an absolute sense, and after adjusting for overall response rate. This implies that pups have a more liberal threshold line for acceptance (farther to the right in Figure 13.4A) and respond to more false alarms, while mothers have a more stringent threshold and sometimes miss a genuine alarm. (E,F) Behaviors during natural reunions. Pups called more (E) and traveled a greater distance (F) to achieve a successful reunion. Pups pay a greater cost for a failed reunion and therefore invest more, a result consistent with parent–offspring conflict theory. Playback experiments using modified calls, by removing upper- or lower-frequency components and by changing the call shape, show that pups and mothers pay attention to different call features when discriminating each other's vocalizations. Pups primarily use spectral information (pitch) and the characteristics of the ascending frequency modulation during the first part of the mother's call, whereas mothers use the temporal pattern of frequency modulation in the pup's call. (After [107, 108, 274].)

on land. Females leave their pups at the rookery for one to three weeks at a time to forage. Upon returning, a female must reunite with her pup, which has probably moved some distance during that time. Pup vocal signals are far more stereotyped within individuals and more distinctive between individuals in Otariids compared to Phocids [273]. Mutual recognition based on vocal signals has been documented in at least four species of fur seals and sea lions [274]. Pups play an active role in the reunion process by approaching the mother when they detect her call, but signal acceptance thresholds as well as receiver discrimination cues differ for mother and pup (**Figure 13.17**).

Large gregarious mammals living in open habitats, such as ungulates and macropods, cannot hide their young in vegetation. The young are mobile and move with the herd shortly after birth. Recognition must be learned during the first few hours with a rapid imprinting process, and both mother and calf must participate. The mother quickly learns the characteristic scent of her single offspring and may subsequently label the calf with her own olfactory marks [223, 237,

524]. Later the mother learns the voice and appearance of her offspring for long-distance recognition, and the calf learns the voice, appearance, and odor of its mother [165, 166, 559, 560]. Dolphins and murres (Alcidae) that raise their young in the open sea face a similar challenge, but as they cannot employ chemical signals, they must rely primarily on vocal signals [275, 292, 349, 545].

FAMILY BADGE   When brood size is very large, individual recognition of each offspring is not feasible, and a **family identity signal** is favored. The desert isopod *Hemilepistus reaumuri* is an excellent example of a learned family-specific olfactory badge [363]. Monogamous pairs raise broods of 50–100 in an underground burrow that is vigorously defended against invasion by non-family members. If a family member is temporarily removed and rubbed against the body of a non-family member, its own family will subsequently reject it. The family odor is a meld of the hydrocarbons produced by each family member. The chemicals are exuded on the cuticular exoskeleton, wiped onto the tunnel walls as the animals move around within the burrow, and redeposited onto the backs of all other members. This system relies upon genetically based individual differences in hydrocarbon production to generate differences among families. The two (presumably unrelated) parents are different from each other and their offspring represent a genetic mix of parental genes affecting odor, so the family badge must blend all members' odors. This is an example of phenotype matching in which the referent is an average of all family members.

Parent–offspring conflict

Parents provide food and protection to their offspring, and offspring fitness and survival increase as a result. Although both parties share an interest in the successful rearing of the young, parents and offspring disagree over the exact length and extent of parental care. Parents are selected to maximize their lifetime production of young, and given that several broods may be raised in a lifetime, an excessive investment in one brood or offspring may reduce an individual's ability to invest in future offspring. Parents should thus allocate resources fairly equally to all offspring. Individual offspring, on the other hand, are selected to obtain a disproportionate share of parental care for themselves. However, offspring selfishness is not unbounded; it is limited by the fact that siblings represent an important component of an individual offspring's inclusive fitness (see Chapter 9). We can express the extent of conflict between parents and offspring from the "gene's-eye view" using Hamilton's rule. Parental inclusive fitness is maximized when the benefit B of increased aid to an offspring is equal to or greater than the cost C of lost future reproductive success, or $B/C \geq 1$. An offspring's fitness is maximized when this B/C ratio is equal to or greater than r, the coefficient of relatedness between the offspring and its siblings. When $1 > B/C > r$, there is a conflict between the two parties in which the offspring prefers more investment than is optimal for the parent [651]. The coefficient of relatedness

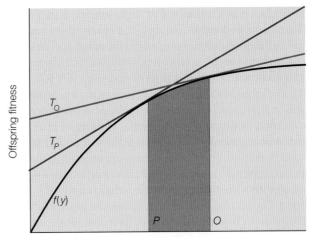

FIGURE 13.18   Model of parent–offspring conflict   Offspring fitness $f(y)$ is a function of the resources it receives from the parent (y). The benefit of investment decelerates at higher levels of investment. The parent suffers a linear reduction in future fitness as it invests more in current offspring. The optimum investment from the parent's point of view, P, occurs when the marginal benefits of feeding the current young equals the marginal costs of lost future reproductive success. This occurs at the point where a line with slope T_P is just tangent to the offspring fitness curve. The optimal investment from the offspring's point of view, O, occurs where the line with slope T_O is tangent to the offspring fitness curve. The parental line slope is always higher than the offspring line slope and the difference between P and O is a measure of the extent of parent–offspring conflict. (After [202].)

is 0.5 for full siblings and 0.25 for half siblings, so the degree of conflict increases when relatedness decreases. **Figure 13.18** illustrates this zone of conflict in a battleground model of **parent–offspring conflict.**

What is the evidence that such conflict actually exists? Numerous examples of behavioral disputes during weaning have been described. Some have argued that these disputes are likely a proximal outcome of maturational processes rather than ultimate genetic conflicts [29, 417]. For instance, cases of mammalian mothers that sometimes extend the suckling period for an offspring at the cost of foregoing the next reproductive bout are not driven by offspring demands but by maternal decisions in the face of variable and unpredictable environmental conditions [205, 647]. However, genetic conflicts can shape maturational processes. To demonstrate parent–offspring conflict, one would need to show that offspring exert real power and that the dispute has direct antagonistic fitness consequences for parents and offspring [312]. Currently there are only two such examples. One is **brood sex ratio conflict** between the queen and her worker offspring in the social Hymenoptera. The queen is equally related to her sons and daughters ($r = 0.5$) and prefers a 1:1 sex ratio, whereas the workers, her daughters, are more closely related to their sisters ($r = 0.75$), than to their brothers ($r = 0.25$), and therefore prefer a 3:1 ratio. Workers have a great deal of power in this conflict because they control which eggs are reared.

The conflict is often resolved at some intermediate point. In bumblebees (*Bombus terrestris*) the observed sex ratio is closer to the queen's optimum, and in the wood ant (*Formica truncorum*) the brood is female-biased close to the worker's optimum by selective elimination of male eggs [68, 505, 628, 652]. The second example occurs in mammals, where there is significant post-zygotic investment in fetal development by the mother, mediated by the placenta. The fetus and its placenta contain genes derived from both the mother and father. Paternally derived alleles in the fetus could potentially conflict with maternally derived alleles and cause a greater transfer of resources from the mother to the fetus than is optimal for the mother. The supression of maternal alleles by paternal alleles, called **genomic imprinting**, has been proposed for the actions at several gene loci [225-229, 420].

Parent–offspring conflict in the two examples above may be more evident because they are ongoing arms races between two relatively similar competitors—rival adults in the Hymenopteran example and powerful physiological agents in the mammalian example. Other potential cases of parent–offspring conflict may not be evident because they involve asymmetric partners and have reached an evolutionarily stable resolution. Offspring may be the winners in some parasitoid wasps, where mothers are powerless to challenge the siblicidal tendencies of their offspring because they do not attend the brood [347]. But in most species, offspring are at a disadvantage because of their small size and weak physical power, so parents have the upper hand and are better able to achieve their optimum [3, 312, 418, 651]. **Phenotypic resolution models** that examine the behavioral interactions between begging offspring and food-provisioning parents, with birds and insects in mind, have had some success in demonstrating the process of parent–offspring conflict.

Offspring begging models sort into two categories: honest signaling models and sibling scramble competition models [204, 532]. **Honest signaling models** view begging as an informative signal of offspring need to parents. Parents actively control the allocation of food to individual offspring in the brood by using the signal characteristics to decide whom to feed and how quickly to return with more food. These models predict that begging intensity should reflect offspring need or hunger, that parents should provision in relation to signal intensity, and that begging should be costly to ensure offspring honesty (see Figure 10.26) [201, 203]. A variant of the honest signaling model proposes that costs can be minimal if the signal is a less-informative dichotomous signal [45]. **Scramble competition models**, on the other hand, assume that offspring control food allocation by jostling for the best position when parents arrive with food, and that parents are passive providers of food to the closest or most active offspring. In these models, the signal reflects the competitive strength of offspring: stronger competitors are more likely to be fed, and competing must be costly [377, 457, 458]. Begging costs in both types of models may include the energetic performance cost of display to the individual, the brood-wide cost of attracting predators, and the loss in

inclusive fitness due to siblings that perish from underfeeding. These two classes of models can be considered extremes at the ends of a continuum of relative parental power from high to low. Scramble models predict overall higher levels and costs of begging compared to honest signaling models, but otherwise the two classes of models share many of the same predictions. More recent models have explored the combined effects of both signaling of need and sibling competition [289, 459, 518, 519]. For more details on these begging models, see Web Topic 13.4. It seems likely that the position of a species along the power continuum will depend on the context of parental feeding. Honest signaling is more likely to occur under uniparental care, when offspring are produced singly, and when the ability of parents to provide resources to offspring is high and the costs of provisioning are low (i.e., food is abundant). If offspring are produced in large broods and two parents care for them, offspring may have more control, and scramble competition becomes more likely. Parents can increase the spread of offspring competitive abilities by initiating early incubation in birds, which increases hatching asynchrony and the age and size distribution of nestlings. Finally, the power balance is likely to shift during ontogeny, with parents in better control when offspring are young and scramble competition increasing among older offspring [532]. A third modeling strategy, using **tug-of-war models** that let the degree of competitive asymmetry vary continuously, may hold more promise for evaluating the consequences of power asymmetries between relatives [91, 94, 95, 510].

> **Web Topic 13.4** *An overview of begging models*
> Here we describe honest signaling models, scramble competition models, tug-of-war models, and sibling negotiation models.

Within-family signaling interactions

Within families, signals and potential information flow can occur from offspring to parents, between siblings, and from parents to offspring. We take up these signaling interactions in turn. Rich media examples can be found in Web Topic 13.2.

SIGNALS BY OFFSPRING TO PARENTS   Begging behavior and tests of the begging models have been studied most extensively in altricial birds, whose nests and young are readily accessible to observation and experimental manipulation. **Begging signals** are remarkably similar among species, consisting of *multimodal* displays with vigorous postural movements, brightly colored gape (mouth), and repetitive vocalizations [309]. While some components of begging displays may be correlated and provide partially redundant information, growing evidence suggests that components provide different kinds of information about the nestlings, so that begging qualifies as a *multiple-message* display. Parents require information about nestling hunger level or need in order to decide how often to bring food to the nest and how

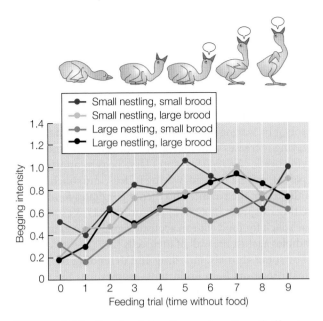

FIGURE 13.19 Begging intensity as a function of offspring hunger Nestling barn swallows between three and six days old were removed from their nests and subjected to a series of mock parental stimulus trials at 10-minute intervals that induced them to beg vigorously. Before the start of the experiment, nestlings were fed until satiated, and during the subsequent experimental trials they were not fed, so they became hungrier with time. All nestlings increased their begging vigor during the course of the trials. To examine whether long-term need also affected begging rates, nestlings were selected that differed in size relative to their siblings and in the size of the brood from which they were taken. Long-term need was predicted to be greatest for small nestlings from large broods, lowest for large nestlings from small broods, and intermediate for small nestlings from small broods and large nestlings from large broods. These expectations were partially met, in that smaller nestlings tended to beg more vigorously than larger nestlings. (After [371].)

of parents to fully feed the entire brood is constrained. The vigor of postural displays appears to provide information about both quality and need. Virtually every study that has examined parental food allocation as a function of begging display vigor has found a positive association [309, 525, 536, 669]. The body extension and shaking behavior of high-intensity begging potentially makes this display component energetically expensive. Although direct measurements of the magnitude of this cost offer mixed evidence, extended begging reduces growth rate in young nestlings and could provide honest information about need and quality at this stage [106, 311]. Experimental manipulation studies show that begging intensity increases with increasing short-term need [357, 371, 539, 669] (**Figure 13.19**). But nestlings in better condition, especially older and larger nestlings, can extend their necks higher and beg more vigorously at a lower energetic cost than smaller nestlings, so begging vigor can also reflect competitive ability [129, 447]. If parents benefit by extracting information on true need controlling for quality, they will have to rely on additional subtle cues and signals. Although begging vocalizations are often correlated with postural intensity, specific acoustic components such as time delay of call onset, call rate, call bout duration, and call pitch are better correlated with hunger in some species [199, 358, 388, 389]. The overall call rate of the brood is often a key determinant of parental provisioning effort, as shown in several experiments that augmented acoustic signals at the nest with playback and thereby increased parental visit rate [40, 428, 450, 483]. One study found a decreased visit rate when nestlings were muted [200].

Color signals can provide independent information about aspects of nestling quality such as health, immunocompetence, and general condition. Nestlings in many species have red-colored mouths surrounded by a yellow-colored fleshy flange (**Figure 13.20**). The chroma of these colored skin areas is determined by carotenoid pigments, and plasma levels of these pigments are correlated with various measures of condition and immunocompetence. Several experimental studies have shown that parents preferentially feed nestlings with

to apportion food to the neediest individuals. But they may benefit from obtaining information about offspring quality so they can selectively promote the survival of those young with the best chances of surviving, i.e., with the best reproductive value, especially if resources are limited and ability

Cavity nest   **Open cup nest**

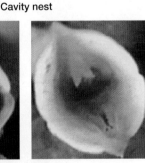

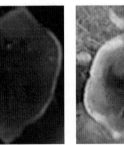

Jackdaw
Corvus monedula

Starling
Sturnus vulgaris

Dunnock
Prunella modularis

Blackbird
Turdus merula

FIGURE 13.20 Colored mouths of nestling birds Nestlings signal hunger with widely opened mouths. Nestlings of species that nest in cavities have wider, lighter yellow flanges that contrast more with the inner mouth color compared to species that nest

in open cups. The contrasting flange likely improves detectability of the young to the parents in the dark nest environment. Interior mouth color is not related to nesting habitat or amount of available light in the nest.

FIGURE 13.21   Parental care in burying beetles An adult *Nicrophorus vespilloides* beetle regurgitating food to a begging larva. Chemical stimuli from the parents trigger begging by the larvae. Experimental studies with *N. vespilloides* show that hungry larvae are more likely to beg and be fed, but that adults do not preferentially feed very hungry beggars over less hungry beggars. This evidence indicates that begging is a discrete signal rather than a graded signal of need. However, large begging larvae in asynchronous broods were more likely to be fed than small begging larvae, indicating a parental preference for investing more in offspring of higher value [602–604].

redder mouths or yellower flanges [143, 157, 167, 369, 538, 540]. In finches that feed nestlings by regurgitating seeds, there is a red flush at the onset of begging that becomes redder with increasing food deprivation (see Figure 10.27A). Parents in these species feed more than one nestling during a feeding bout and compared to single-load feeders, take longer to carefully decide which ones to feed, which may have selected for a finely graded signal of need [306, 308]. In some species, it is the intensity of UV radiance from skin or plumage that is associated with nestling quality or parental feeding rate [143, 294, 633]. Nestlings of hole-nestling species often have yellow mouths or contrasting yellow flanges (see Figure 13.20). This coloration probably improves detectability of the young against the dark background, but does not indicate the relative quality of individual nestlings [247, 269, 308, 310].

The begging signals of the young of non-avian species with parental provisioning are similarly related to need or quality. Begging vocalizations are important in mammals such as meerkats, where call rate influences feeding rate [383]. Call pitch and amplitude are reliably correlated with offspring hunger and parental feeding investment in piglets and human infants [681, 682, 685]. Insects employ tactile, vibrational, or chemical signals to solicit feeding from adults [390]. For example, the female burrowing bug (*Sehirus cinctus*) lays her eggs in a cluster. When they hatch, she both guards the nymphs and brings them nutlets from mint plants. The nymphs release a solicitation pheromone that indicates their hunger level. Addition of volatile extracts from underfed hungry nymphs caused females with normal broods to bring more nutlets. High levels of two key blend components, alpha-pinene and camphene, appear to differentiate hungry

from well-fed nymphs [325, 326]. Similarly, common earwig (*Forficula auricularia*) larvae differ in their cuticular profiles depending on their nutritional condition, but better-condition offspring are selectively provisioned, suggesting that the chemical cue indicates offspring quality [391]. Ant larvae solicit by waving their heads and mandibles, and the rate of waving increases with increasing hunger level [256]. The burying beetle (*Nicrophorus* spp.) larva begs by approaching the parent, raising its head, waving its legs, and touching the mouthparts of the parent, who then regurgitates food to the soliciting offspring [506, 602] (**Figure 13.21**). Vespid wasp larvae solicit for food with an acoustic signal by scraping their mandibles against the wall of their cells [277].

For some animals, guarding and defending the young against predators is an important—sometimes the primary—parental care duty. Offspring are selected to give **distress or alarm signals** to elicit parental approach and protective behaviors [390]. A well-documented example is found in the thornbug treehopper (*Umbonia crassicornis*), in which a brood of nymphs forms an extended cylindrical aggregation along the length of the host-plant stem (**Figure 13.22**). Attacks by predatory wasps are more likely to occur at the ends of the aggregation, and disturbed nymphs produce substrate-borne vibrational signals that cause the mother to approach the predator, fan her wings, and kick with club-shaped hind legs. The signal is a brief series of pulses lasting 30–40 msec. After one nymph is attacked and signals, adjacent nymphs repeat the signal, so that a rapid wave of signaling travels through the aggregation. Coordination among signaling nymphs is necessary to evoke the female's response and enable her to determine which end of the aggregation was attacked [117-119, 496]. Auditory distress and alarm signals are common in many species with mobile offspring that can become separated from the parent. These signals will be discussed in more detail in Chapter 14.

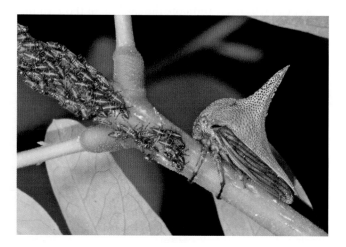

FIGURE 13.22   Substrate alarm signal by treehopper nymphs A female thornbug treehopper (*Umbonia crassicornis*) guarding her offspring. They will signal her as a group if a predator approaches. Thornbugs can communicate using vibrations they "hear" through sensors throughout their bodies [496].

SIBLING COLLABORATION   Although siblings within a brood or litter may compete with each other for parental care, they are nevertheless close kin and may benefit by communicating with each other to ameliorate some of the costs of competition. **Sibling negotiation** is one proposed strategy, in which siblings communicate among themselves in the absence of the parent to assess each other's relative hunger, and then modify their competitive behavior so that the hungriest sib is fed. Evidence for such negotiation has been described in barn owls (*Tyto alba*). Parents forage most of the night and bring a large prey item (e.g., a mouse), which is given to one chick, approximately once every hour. Chicks give begging calls regularly in the parents' absence, and both the call rate and call duration vary as a function of hunger level. They increase their begging intensity significantly when a parent arrives, but a well-fed chick reduces its begging rate in the presence of a hungrier nestmate. The chick with the longer call duration is usually the one fed by the parent [527–531]. Similarly, spotless starling (*Sturnus unicolor*) chicks engage in substantial bouts of begging in the absence of parents and seem to reduce their competitive levels of begging once a parent arrives [88]. A sibling negotiation model based on these types of interactions is described in Web Topic 13.4.

Collaborative begging occurs in a few species. In black-headed gulls (*Larus ridibundus*), chicks in increasingly larger broods (up to three) beg more synchronously and thereby both decrease their individual begging effort and increase the rate of parental provisioning [394]. Banded mongooses (*Mungos mungo*) produce communal litters of pups from multiple females; pups that beg at low rates nevertheless obtain more food as litter size increases, suggesting a spillover benefit (by-product mutualism) from additional littermates [38]. The pre-hatching vocalizations of precocial embryonic birds and crocodilians could also be considered a form of cooperative sib–sib signaling [307]. The primary function of these acoustic signals that increase shortly before hatching seems to be synchronization of hatching, with older eggs delaying and younger eggs speeding up their final development so that the brood hatches in a shorter period of time. These signals may also prepare the parent for the change in parental care duties, such as digging up the buried eggs in the case of crocodilians, and increased incubation, egg turning, and food provisioning in the case of birds [60, 82, 445, 664]. In an unusual insect example, larvae of a parasitic blister beetle aggregate to chemically and visually mimic the female sexual signal of their host species [542].

More closely related kin are expected to exhibit more cooperative, or at least less aggressive, behavior toward one another. This **kin nepotism** principle also applies to broods of siblings, which can vary in their relatedness from half to full sibs, depending on whether one or several sires fathered the brood [418]. An interspecific comparison of passerine birds found that the loudness of nestling begging calls increases as the relatedness among the members of a brood declines [78]. Ground squirrels can detect fine differences in relatedness based on olfactory cues and direct more

aggression toward half sibs compared to full sibs, even after controlling for familiarity [261] (see Figure 13.3). Olfactory kin recognition prevents cannibalism of siblings in tiger salamanders *Ambystoma trigrinum* [470, 472], spadefoot toads [471], a parasitic wasp [198], and a parasitic beetle [366].

SIGNALS BY PARENTS TO OFFSPRING   We have already described a major class of calling-out signals employed by parents to find offspring from which they have become separated. Parents broadcast individually-specific signature signals to elicit approach or reciprocal signaling from offspring that have hidden themselves, strayed away from expected locations, or were deposited in a crèche. In other contexts, parents give **directive signals** to offspring to elicit specific responses. For instance, parents usually possess a better knowledge of immediate environmental conditions, such as the presence and location of food, cover, and predators, than offspring do. Signals by parents encoding this environmental information can increase the probability that the young avoid predators, locate areas with good food resources, and remain in a cohesive group. Even if offspring are immobile, as in altricial nestling birds, noisy begging can attract predators, and parents of many avian species give an alarm call if they detect a nearby predator. The species-specific alarm signal causes the offspring to immediately stop begging [379, 479]. Either an "all-clear" signal or the arrival of the parent at the nest releases the vocal inhibition. The silencing response is largely innate, as it is too risky for nestlings to learn this behavior—they must get it right the first time they hear the signal [141]. If the young are mobile, parents carry the additional burden of directing the movements of the offspring to appropriate locations. In this situation, a repertoire of signals with specific meanings and responses is needed. The vocal repertoire of the western sandpiper (*Calidris mauri*) provides a nice example of different parental signals and their functions [286]. The typical brood of four downy precocial chicks leaves the nest shortly after hatching. Chicks are programmed to quickly imprint on and follow any nearby large moving body, which under normal circumstances is a parent. The family may travel several hundred meters in one day. Parents give four types of calls: a brooding call, a gathering call, an alarm call, and a freeze call. Each has a distinctive acoustic structure, occurs in specific circumstances, and elicits specific behaviors by the chicks, as described in **Figure 13.23**. Ungulate and other mobile mammalian offspring are similarly programmed to follow their mothers and learn their identifying characteristics, and mothers give vocal, visual, olfactory, and tactile signals to solicit approach and suckling. For instance, the mother caribou performs a head-bobbing display to encourage the approach of her young calf. Suckling initiation vocalizations have been described in a few ungulates, but it is more common for the mother to orient her body into the reverse parallel position for nursing and nudge or lick the calf to encourage it to search for the udder [677]. An olfactory example of a directive signal is the nipple pheromone in rabbits, mentioned in Chapter 6, that stimulates searching and suckling behavior by the pups. A

(A)

(B)

(C)

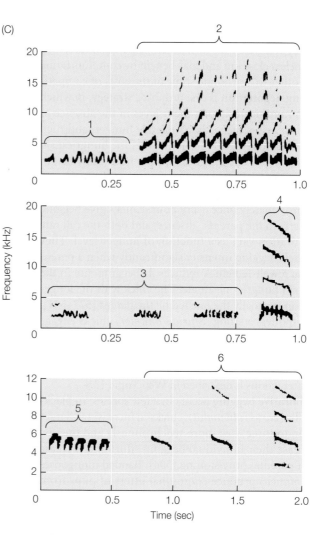

FIGURE 13.23 **Vocal repertoire of breeding shorebirds** (A) An adult western sandpiper (*Calidris mauri*) and (B) a brood of newly hatched chicks. (C) Vocal repertoire: (1) parental gathering call; (2) parental alarm call; (3) three examples of parental brooding call; (4) parental freeze call; (5) chick contact call; and (6) chick alarm call. (After [286].)

vibrational example is the abdomen wagging of *Polistes* wasp queens that induces larvae to salivate and transfer amino acids back to the queen and other colony members [74, 77, 188, 543]. Crocodilians are unusual among reptiles in possessing both extended maternal care and a repertoire of vocal signals between mother and hatchlings, including maternal growls to maintain brood cohesion, hisses for defense of the young, and various alarm and distress calls by the young [664]. Auditory, visual, and chemical alarm signals constitute a major class of directive signals given by parents to offspring, but the same signals are usually given to other nearby conspecifics as well, so we shall defer the discussion of these environmental signals to Chapter 14.

Parents can affect the development of their offspring in profound ways using strategies that sometimes qualify as communication signals. One strategy involves **prenatal chemical signals**. Many organisms have the ability to develop along alternative pathways depending on environmental conditions, a phenomenon called **phenotypic plasticity**. Parents experiencing variable environmental conditions can translate or

communicate these conditions to the young and modify their morphology to one that is adaptive in that particular context. For example, in deteriorating environmental conditions, ovipositing female insects can produce diapausing offspring, ready for dormancy, or winged offspring, ready for dispersal. Furthermore, choice of oviposition or host site can affect offspring size, gender, and level of offspring competition [425]. In vertebrates, maternal deposition of androgen hormones in eggs has a significant effect on offspring begging behavior, competitiveness, development rate, survival, and adult breeding success [429]. Maternal nutritional state can significantly affect metabolic rate, size, health, and offspring survival in mammals and other plants and animals, including humans [30].

A second strategy by which parents can influence offspring development involves **teaching**. A broad operational definition of teaching requires that an individual, A, modifies its behavior only in the presence of a naive observer, B,

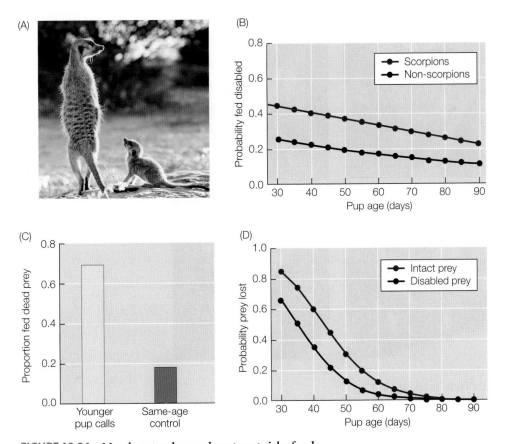

FIGURE 13.24 Meerkats teach pups how to eat risky foods
(A) Meerkats (*Suricata suricatta*) in the Kalahari Desert of southern Africa live in cooperative family groups composed of a dominant breeding pair and numerous related helpers (older than 90 days) who provision the current litter of pups. They feed on a range of vertebrate and invertebrate prey, many of which are difficult to handle and potentially dangerous, such as scorpions. Helpers devote much time and effort to teaching pups how to handle tricky food items by bringing them disabled prey and remaining to encourage and monitor their success. (B) Helpers adjust the items brought to pups as a function of their age. Younger pups are brought dead items or disabled prey. (Scorpions, in particular, tend to be presented in a disabled form.) As they learn to deal with prey, more and more live items are introduced. (C) The acoustic properties of the pup's begging calls change with age, and helpers use these cues to adjust their provisioning strategy. In playback experiments to groups with older pups, broadcasting the begging calls of younger pups caused helpers to bring more dead prey, compared to control trials broadcasting the begging calls of pups the same age as those currently in the group. (D) Pups improved their prey handling ability as they aged and lost fewer items, especially live intact prey. Experimental training of pups for three days on either disabled or dead scorpions resulted in greater handling success by those pups given the disabled scorpions. (After [638].)

documented examples. The ant *Temnothorax albipennis* lives in small colonies under rocks and employs tandem running to teach naïve colony mates the location of good food resources. The leader (teacher) slows her speed of running only when continually tapped by the follower's (pupil's) antennae. The pupil subsequently is able to make more rapid and direct trips to the food source and may become a teacher to other colony mates [183]. Pied babbler (*Turdoides bicolor*) adults often give a "purr" vocalization when they deliver food to nestlings, and nestlings learn to associate this call with food. This call not only stimulates nestlings to beg, but it is also used in a nonfeeding context to lure fledglings away from predators [494, 495]. Finally, in meerkats and solitary foraging carnivores, adults gradually introduce pups to live mobile prey in accordance with the pup's age and ability (**Figure 13.24**). Although there is a cost in the sense that the prey may escape, pups improve their prey handling skill as a result of these opportunities to practice [96, 638]. Teaching should evolve only when the costs to teachers of facilitating learning are outweighed by the subsequent fitness benefits they accrue once pupils have learned, and these benefits are a function of the difficulty of the task for young that do not receive teaching [639]. These conditions are most likely to be met for parents and offspring (or other very close kin) where there is a large discrepancy in knowledge and skills, an extended period of time for progressive teaching and learning, and a higher benefit-to-cost ratio threshold because of the high relatedness.

at a potential cost to A; and as a result of A's behavior, B acquires knowledge or skills more rapidly or efficiently than it would otherwise [96, 639]. This definition differentiates teaching from other forms of **social learning**, in which naïve animals acquire information by observing other individuals engaging in their usual behavior. Complex cognitive abilities are not needed, as can be seen in the following three recently

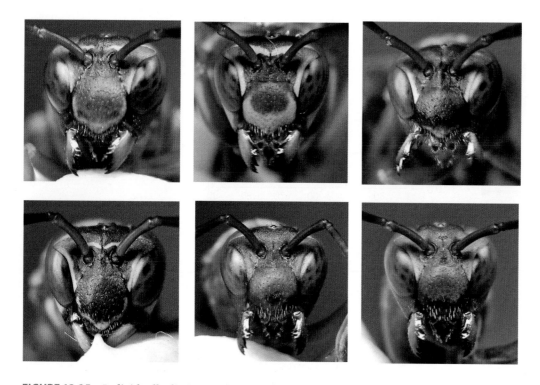

FIGURE 13.25 Individually distinctive facial markings in brown paper wasps (*Polistes fuscatus*) There is striking variability in the presence or absence of facial markings, and further variation in the width and length of all yellow markings, from small dots to long, bold stripes. Some wasps have eyebrows, outer eye stripes, and various patterns on the clypeus shield. Colonies of this species are founded by multiple queens, who fight to establish a dominance hierarchy. Several queens typically reproduce, and as workers emerge they integrate into the dominance hierarchy. Individuals whose facial patterns are altered with yellow paint receive increased aggression when initially returned to their colony, but eventually this aggression declines as colony mates become familiar with the altered individual.

Group Integration

Group living serves a variety of adaptive functions: warning and defense against predators, location of food, territory defense, migration, and cooperative reproduction, among others. The type of cooperative endeavor affects the size and stability of groups, the composition of groups, and the signals group members need to coordinate activities. In this section we shall examine the mechanisms used to recognize group members and describe some of the signals used to maintain group cohesion. Signals used to inform group members of environmental information such as the presence of predators or the location of food will be discussed in the next chapter.

Group recognition

The mechanisms used for **group-mate recognition** depend on the size and stability of groups, whether or not group members are tied to a fixed site such as a nest or territory, and the importance of recognizing specific individuals or classes within groups. Many forms of cooperative and altruistic behavior can evolve only when restricted to close kin, so the recognition mechanism plays a key role in determining the kinds of interactions that take place among group members.

When groups are relatively small and stable in composition, group-mate recognition can be based on associative learning of individual signature traits. Individual recognition facilitates the development of stable dominance hierarchies and reciprocal alliances within groups because each individual can remember the outcome of prior encounters with other members [26, 300, 649]. Development of a distinctive identity signal has clear advantages. As we learned in Chapter 6, house mice produce urine marks with unique chemical compositions. Subordinate group members are treated with aggression if they fail to maintain their olfactory marks, because a nonsignaler is perceived to have dispersed out of the group [270]. Similarly, co-foundress female paper wasps express highly variable facial markings (**Figure 13.25**). Experimental modification of the visual pattern turns a group mate into a stranger and results in increased aggression [640]. Paper wasp species bearing such distinctive facial patterns are characterized by a reproductive strategy in which egg-laying and foraging are shared unequally in relation to dominance rank. Cheating is prevented by monitoring and policing, which require a reliable individual identification system [585, 586, 641]. Individual recognition is also important in group-living species that direct greater levels of cooperation toward closer kin. Primates are a classic example. In Old and New World monkeys, females remain in their natal group while males disperse. Groups are composed of multiple matrilines arranged in a stable linear dominance hierarchy in which all female members of one matriline outrank or are outranked by all female members of another [597]. Affiliative and aggressive interactions are thus strongly affected by both matriline identity and dominance status. Clever playback studies using various types of calls show that responses are affected by the identity of the source caller, as well as by the context [109, 110, 112, 512]. For example, baboon females respond more strongly to a sequence of threatening and submissive calls that mimics the sounds of a dominance rank reversal *between* matrilines than to a sequence that mimics the sounds produced in a rank reversal *within* a matriline, indicating that the baboon females classify others on the basis of both individual rank and kinship [43]. As we describe in Chapter 14, many group-living bird and mammal species give loud calls with individual signature information that function to alert or recruit group members [97, 153, 265, 293, 329, 673, 702].

Fission-fusion societies are those that exhibit flexible changes in group size by means of the merging and splitting of subunits, generally caused by spatial and temporal variation in the distribution of resources or by variation in group members' needs and activities. For example, some primate species gather in large groups at safe locations for sleeping, but spread out in smaller foraging groups during the day. In other species, smaller cohesive family units sometimes merge to form larger temporary associations. At least some fission-fusion societies maintain an individual signature system so that individuals can keep track of previous associates. Because of the large number of individuals that must be distinguished, signals must be designed to encode a large number of variants, and receivers must have well-developed long-term memory capability. Elephants, dolphins, and parrots are classic examples, and their signals are invariably acoustic. For elephants, one can describe concentric circles of female association levels that include increasingly larger numbers of individuals of lower degrees of relatedness. Playback studies show that females can recognize up to 100 close and distant kin on the basis of their "rumble" contact calls (see Figure 13.6), and this discrimination is possible over distances up to 3 km [399, 400, 423, 613, 614]. Dolphins emit individually specific whistle calls that enable kin to find each other in large milling crowds and over long distances [279, 544, 545]. Several species of parrots have now been shown to possess loud contact calls with well-developed individual signature structures [87, 128, 678–680].

Some species embed their individual signature system within a group signature system. **Group labels** are usually acquired signal features, either learned acoustic signals or blended olfactory signals. Killer whales (*Orcinus orca*) provide one of the best acoustic examples [174, 179, 414, 683, 684]. The smallest social unit is a pod containing 5–20 closely related females. Each pod produces an average of 10 types of discrete calls, which are given repetitively in vocal interactions among pod members. Calves learn these calls during the first few weeks after birth, when the adult females

significantly increase their rate of calling. Call types differ among pods, but some pods share a portion of their calls with other pods, presumed by researchers to be descended from the same matriline. These associated pods form a larger social network called a clan; there is no sharing of call types between clans. There is a tendency for pods sharing call types to associate more often than those that do not share call types, although pods from different clans sometimes travel together. Such vocal traditions generally reflect genetic relatedness and in principle permit phenotype matching based on a learned family vocal signature template.

Other mammalian and avian group-living species show evidence of learned group-specific acoustic labels that promote defense of a common roost or foraging area against other adjacent groups [61-63, 72, 80, 81, 382, 439, 440, 492, 514, 579, 580, 655]. European badgers (*Meles meles*) offer a good example of an olfactory group label. Badgers live in groups of up to 27 individuals that defend a communal territory. All group members contribute subcaudal gland scent marks to the latrines that border the territory. The odor of the secretions is partly generated by bacterial flora in the subcaudal pouch and produces an individual-specific signature. Badger groupmates also mark each other, sometimes mutually, by pressing their backsides, and hence the openings of their subcaudal pouches, together. This behavior produces a shared group component to the odor that contributes to group cohesion [85, 86].

When group members are obligatorily associated with a communal nest or roost site and some unique feature of this site such as its odor is imparted to the occupants, a group label can be acquired passively. Group members can thus be unambiguously distinguished from nonmembers. This familiarity mechanism can function equally well for groups of related or unrelated individuals and is most effective for distinguishing among groups if the olfactory cues are derived from dietary sources in a heterogeneous environment [342, 350, 473]. However, cheaters that sneak into the periphery of the nest and acquire the nest odor can invade groups using such a mechanism. In eusocial insects, with their huge altruistic workforces and valuable stored resources, exclusion of nest parasites and non-kin conspecific usurpers is crucial, so a true kin recognition mechanism based on genetically-derived olfactory labels is favored [189]. In temperate zone wasps, whose nestmate recognition systems have been well studied, the queen rubs abdominal Dufour's gland secretions (and possibly other gland products) on the nest comb, where it is absorbed by the epicuticle of newly emerged adults. The new adults learn the colony odor in a process similar to imprinting. They establish this specific hydrocarbon profile as the colony template against which other individuals are compared by phenotype matching, and they remember the CHC profile for a long time even if isolated from the nest. The response of a receiver encountering another individual is all or nothing: the individual is either accepted and treated tolerantly or rejected and attacked—there is no graded aggressive response. The threshold of receiver rejection is lower for queens than for workers (cut-off line further to the left in Figure 13.4A), probably because the queen has more to lose by accepting a usurper than a worker does. Despite this nearly fail-safe system, queens of a few congeneric parasitic species have evolved the ability to enter host nests and mimic the host queen's recognition pheromone, then kill the host queen and dupe the host workforce into raising the parasite's offspring [103, 189].

If groups are composed of a mixture of individuals with different degrees of relatedness, kin selection theory predicts that individuals should direct more aid to close relatives than to distant relatives, as we noted above for vertebrates. In Figure 13.3 we illustrated this same effect in ground squirrels, where phenotype matching against a self-referent or learned family template results in reduced aggression toward close relatives. In the social insects, colony mates will consist of a mixture of full, half, and sometimes non-siblings if there is a single queen who mates polyandrously or if there are multiple egg-laying queens. Interestingly, despite an enormous research effort, no strong evidence of within-colony nepotism through promotion of full-sister reproduction has been found in bees [4, 73, 332, 502, 634], wasps [207, 487, 612, 624], ants [50, 149, 264], or termites [9]. The question arises: Is the lack of within-colony nepotism due to a constraint on reliable recognition cues for discrimination, or to high costs of discrimination relative to the benefits of cooperation? The consensus is that the high inefficiency cost that would result from selfish workers attempting to seek out, discriminate, and selectively rear closely related sister queens while neglecting less-close relatives reduces colony fitness in competition with other colonies, and favors strategies and behaviors that eliminate within-colony relatedness information [207, 303, 509, 634, 686]. This could be achieved through: (1) the muting of kin recognition cues by senders, (2) the raising of threshold tolerances by receivers, or (3) the active scrambling of recognition cues by those individuals that benefit from kin ambiguity [509]. In species with highly polyandrous queens, so many patrilines are generated within the colony that the incentive for workers to behave selfishly is greatly reduced. A larva surrounded by workers of whom 90% are from a different patriline would not benefit from advertising its kinship. This diminishes selection for distinctive subfamily discrimination cues and favors liberal nest-mate acceptance thresholds in workers [503]. In the case of multiple queens, collaborating reproductives are often closely related, which could make it more difficult and less beneficial for workers to distinguish among their larvae, but even where queens are unrelated, nepotism does not appear to be favored [149, 624]. The strategy of homogenizing and blending the colony odor through frequent contact, grooming, and trophyllaxis among colony members not only scrambles subfamily distinctions, but may also improve the discrimination of colony mates versus non-colony intruders [504, 590]. When it pays to differentiate levels of kinship, such as in conflicts over male rearing and brood sex ratio, the discrimination mechanisms readily evolve [505].

When groups are very large and unstable in composition, a group label is meaningless, and groups are more appropriately called aggregations. Examples include some wintering and migratory flocks of birds and herds of antelopes. Individual recognition is difficult to maintain because of the temporary nature of associations. However, if there is competition for resources within such aggregations and individuals vary in fighting ability, then an ability to recognize relative dominance status is favored. Visual **status badges** evolve under these circumstances [522, 574]. In Chapter 11 we noted that the conditions required for the evolution of reliable status badges include frequent testing of true fighting ability by individuals with similar badges and a significant cost associated with cheating. Reliable badges reduce the overall level of fighting for all individuals in the aggregation because they settle contests quickly without escalation. Similarly, **age badges** reduce aggression of older individuals toward younger individuals.

Appeasement signals

Animals living in stable social groups invariably experience conflicts of interest and disputes over resources, mates, and space that increase personal costs and diminish the benefits of group cooperation. High levels of aggression among group members might induce losers to leave the group, forfeiting the advantages of group living and reducing the benefits to the winners that stay. Even if no individuals leave the group, unfriendly interactions may jeopardize the willingness of some group members to cooperate in the future. It is therefore to the advantage of all group members to resolve disputes nonviolently with signals, prevent aggressive escalation, and reconcile quickly if a conflict does occur [13]. Behavioral tactics and signals that maintain peace in groups are called **appeasement signals**. Peacekeeping interactions can vary along two broad orthogonal axes. One axis reflects the symmetry of the interactants, which can range from relatively symmetrical interactions involving individuals of similar status giving mutual signals, to asymmetrical interactions involving individuals of different ranks or roles giving unilateral signals. The second axis reflects the duration of the consequences of the signal, which may be short- or long-lasting. **Figure 13.26** shows the repertoire of appeasement signals of a typical primate species, the Barbary macaque, and illustrates the range of contexts and partner symmetries for these primarily unilateral displays.

CONFLICT AVOIDANCE   A linear dominance hierarchy among group members who all know and recognize each other as individuals is the most basic mechanism for minimizing conflict in stable social groups. As we discussed in Chapter 11, dominance hierarchies are an emergent property of winner-loser rules of thumb, whereby individuals learn their relative rank and fighting ability and adjust their expectations for winning based on the outcome of prior conflicts. Most potential conflicts over a non-shareable item are thus resolved, and physical fighting avoided, once the two

FIGURE 13.26 **Visual and tactile appeasement signals in Barbary macaques (*Macaca sylvanus*)** This species has a repertoire of 37 gestures, including facial expressions, postures, movements, manual gestures, and tactile signals. Some of the appeasement gestures are shown here (see also the friendly lip-smacking facial expression in Figure 10.5). (A) A female giving the bared teeth display or submissive grin, further accentuated in this case with submissive belly slapping. (B) Hugging and teeth chattering in a reconciliation event among three individuals. (C) A friendly embrace between two juvenile individuals. (D) Extreme teeth chattering, an open-mouth gesture in which the lips and cheeks are contracted, the mandible is opened and closed rapidly, and the teeth are clacked. This gesture is used in a wide variety of circumstances to signal submission, appeasement, affiliation, or reassurance. (E) A subordinate male (left) presenting his hind-quarters to a threatening dominant male (right). The presenter is backing up toward the dominant while looking back and grinning or teeth chattering. (F) In response to presenting, the recipient male may accept the subordinate gesture and reassure the presenter that he will not attack by performing either a hip touch or a mock mount. In the hip touch, shown here, the recipient grasps the presenter on the waist and performs teeth chattering or lip smacking while the presenter reaches back to touch his own or the recipient's genitals. In the mock mount, the mounter climbs on the back of the presenter, clinging with hands around the waist and feet on the presenter's hind legs, and gives ritual pelvic thrusts; the mountee looks back and reaches back to touch the leg of the mounter, and both males perform teeth chattering [252].

individuals perceive their relative rank and the more subordinate individual backs down. The subordinate individual usually gives a submissive signal that indicates its intention to retreat and inhibits an attack by the more dominant individual. Submissive signals therefore fall into the category of unilateral asymmetric signals with immediate consequences. Submissive signals are typically discrete displays antithetical in form to the main threat displays of a species, such as looking away, hiding weapons, or uttering high-frequency vocalizations, as illustrated in Figures 10.5 and 10.13. In social insects, subordinates cower by lowering their heads and antennae while they are antennated by the dominant [688]. Additional examples were mentioned in the final section of Chapter 11, including exposing vulnerable body parts and adopting sexual or juvenile postures to thwart an attack. Primate submissive displays typically include presenting the rear end, grinning with bared teeth, pouting, teeth chattering, and lip smacking (see Figure 13.26). Dominant individuals may also give specific **conflict-avoidance displays** if they wish to approach a subordinate with nonaggressive or benign intentions. In baboons, dominants that approach subordinates while uttering "grunts" are significantly less likely to attack and more likely to show affiliative behaviors such as grooming than dominants that approach without grunting [111, 454]. Similarly, macaque females that vocalize while approaching a female with an infant are more likely to be tolerated and allowed to handle the infant than females that don't [593, 595, 691], and juvenile white-faced capuchin monkeys (*Cebus capucinus*) that give the trill vocalization while approaching others are more likely to interact affiliatively [219]. Dominants may also give **reassurance displays**, such as touching genitals and mock mounting, to young or fearful subordinates.

POSTCONFLICT RESOLUTION   Once a conflict has escalated to physical aggression, we would expect the loser to subsequently retreat and distance itself from the winner so as to avoid any further attacks. In some group-living species, however, such conflicts are often followed by **reconciliation**, or friendly postconflict reunion. In the minutes following the conflict, both former opponents exhibit anxiety behaviors such as scratching, self-grooming, and other displacement acts. The winner in particular may then approach and embrace, kiss, or groom the loser. Reconciliation acts that involve close contact, and require both parties to perform, signal the readiness of the former opponents to de-escalate despite the risk of renewed conflict. These exchanges subsequently reduce the occurrence of anxiety behaviors to baseline levels. Reconciliation is therefore believed to repair the damage to the relationship between the two parties caused by the conflict in the short term, and to maintain future affiliative and cooperative interactions in the long term [13, 14, 331, 482, 594]. Two individuals are more likely to reconcile following a conflict if their partnership is mutually highly valued, for instance if they are closely related or if they are frequent reciprocal grooming buddies or preferred sexual partners. Chimpanzees show the highest levels of reconciliation behavior, but such behavior is also common in New and Old World monkeys [463, 466, 547, 636], lemurs [453], hyenas [675], domestic goats [550], and dogs [125]. A smaller number of species also exhibit **third-party consolation** after conflicts, defined as the intervention of an individual not involved in the conflict to console the loser. In chimpanzees, consoling occurs if the two agonistic opponents have not succeeded in reconciling; for instance, they may not have valued the relationship sufficiently. The consoling individual is usually a highly valued partner or friend of the loser. Consolation behaviors include embracing, grooming, close sitting, and other types of touching, such as licking, gentle mouth wrestling, genital rubbing, and copulation. As with reconciliation, consoling also reduces anxiety behaviors in the victim [184]. Consolation is observed in chimpanzees [184, 596], bonobos [452], gorillas [126], domestic dogs [125], and possibly rooks (*Corvus frugilegus*) [565], but is rare in monkeys [93, 102, 523]. In dogs, a third-party individual that may not have even witnessed the conflict may be attracted by the whimpering sounds of the loser and approach to lick, sniff, or play. In rooks, a monogamous colonial species, conflicts occur only between nonmated individuals, and they never engage in reconciliation. Instead, both combatants seek out or are approached by their mates, and engage in a special bill-twining behavior observed only in this postconflict context. An alternative explanation for the function of this display in rooks is the advertisement of the pair bond alliance. True consolation implies that the third party appears to show empathy toward the loser of a fight and approaches in order to comfort or reassure.

ALLOGROOMING   **Allogrooming** is the grooming or preening of another individual. It is nearly ubiquitous in primates [555] and also occurs in some ungulates [242, 313, 421], a few rodents [623], carnivores [341], bats [693], cooperatively breeding birds and mammals [493], and social insects [419, 619]. **Figures 13.27** and **13.28A** illustrate allogrooming techniques in some of these species. Allogrooming is a classic **multiple function act**. First, it obviously serves a *hygienic function*. This claim is supported indirectly by the fact that allogrooming is directed toward parts of the body that the self-groomer cannot reach, such as the back, neck, and head [464, 704], and directly by the observation that allogroomers pick up and eat lice eggs [632]. Moreover, ectoparasite loads are lower in primate troops and insect colonies with greater allogrooming rates [469, 541, 676]. A comparative study of ungulates demonstrated that adult allogrooming was concentrated in lineages inhabiting closed woodland or forest habitat associated with increased tick exposure [421]. This hygienic service is clearly very beneficial to the recipient and entails some cost to the groomer in terms of time (primates spent up to 20% of their day engaged in allogrooming) and possibly reduced vigilance against predators and harassers [127, 378]. Why should an animal perform this seemingly altruistic behavior?

FIGURE 13.27 Allogrooming (A) Green wood hoopoe (*Phoeniculus purpureus*) pair. One individual preens the neck feathers of the other, which stretches its neck. (B) Social grooming in primates is unilateral. This photo shows grooming between two female Lowe's Mona monkeys (*Cercopithecus lowei*). (C) Mutual grooming is typical in horses, in this case Przewalski's horse (*Equus ferus*).

Allogrooming has a *calming, tension-reducing effect* on the recipient, releasing beta-endorphin in monkeys and lowering heart rate in horses [173, 305, 619]. It is a major stress-reduction strategy in primates and rats [113, 251]. In meerkats, subordinates groom the dominant breeding female to placate her [341]. Therefore, one immediate benefit to the groomer may be a reduction in aggression received from other group members. For female monkeys, the stress-reducing effects of being part of a large and stable grooming network translate into significantly higher longevity and offspring survival [598, 697].

Allogrooming also functions as a *commodity that can be traded in a biological market for other goods* [435, 436]. The benefits obtained in exchange for grooming may include reciprocal grooming; access to resources controlled by the recipient; a mating opportunity; tolerance or reduced aggression if the recipient is a dominant; or agonistic support, in which the recipient assists the groomer during an aggressive conflict. According to the **biological market model**, individuals do not compete over access to trading partners in an agonistic manner, bur rather outcompete each other by offering a valuable commodity, such as grooming. Individuals attempt to establish partnerships with those group members

that can offer the most valuable traded commodities given prevailing conditions. Many recent studies of grooming relationships in primates support the biological market model [27, 28, 105, 182, 224, 384, 481, 551]. Grooming is most frequent between related group members, within female matrilines in New and Old World monkeys and among males in chimpanzees, but does also extend to nonrelatives. Depending on the species, between 5% and 33% of grooming bouts are immediately reciprocated. Many partners tend to spend roughly equal amounts of time grooming each other, a behavior called *time-matching* (**Figure 13.28B**). Reciprocated bouts typically involve individuals of similar dominance status, and these partners engage in many repeated grooming bouts. A metadata analysis of 22 primate species provided robust evidence that females preferentially groom those females who groom them the most, even after controlling for possible effects of kinship [553]. But there can be large differences in grooming effort between initiator and reciprocator, and many grooming bouts are completely unreciprocated. These unequal bouts usually involve individuals of different rank, and the initiators, who invest the greater amount of time, are typically subordinates. The difference in grooming

FIGURE 13.28 Grooming reciprocity in chacma baboons (A) A family grooming bout, with the alpha male on the right, the alpha female in the middle, her oldest daughter on the left, and younger offspring in the front. (B) Time matching. The grooming time invested by each individual in a reciprocal bout is positively correlated with the grooming time invested by the partner. (C) Favoritism toward dominants. The disparity in grooming time between two reciprocal groomers is positively correlated with the difference in rank between them, with the lower-ranking individual contributing more grooming (a higher-ranking individual has a smaller rank score). These types of effects are strongest in large troops experiencing a high degree of resource competition. (After [27].)

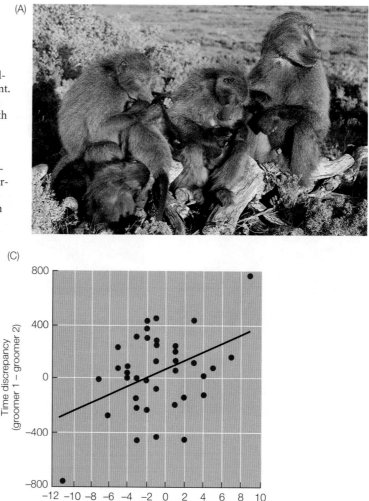

duration is positively correlated with the difference in rank between partners (**Figure 13.28C**). Dominants are less likely to groom subordinates, but they may provide other types of benefits, such as tolerance and reduced aggression, food, and agonistic support. Another metadata analysis of 14 primate species found a significant positive association between grooming and agonistic support received, although the correlation was relatively low [552]. Naturalistic experiments show that individuals are significantly more likely to tolerate, share food with, and provide agonistic support for an individual that has recently groomed them [147, 249, 578]. However, in species and populations with a high level of agonistic behavior, where the aid of dominants during conflicts is a high-value commodity, subordinates are likely to seek out dominants as grooming partners and groom them without receiving grooming reciprocation.

Finally, allogrooming serves a *signaling function*. The mere act of offering to groom another individual communicates the message that the target recipient is a valuable partner [643, 660]. In a biological market with a choice of partners, the value of a particular partner can change over time, depending on the need for services, changes in group composition affecting the number of partners who can offer particular services, and changes in partner quality, such as condition

and experience. Even well-established partners need to assess each other's relative power and social status. Rates of allogrooming invitations, refusals to be groomed, and relative time spent in reciprocal grooming provide information about the status of the partnership. For example, reciprocal agonistic support in male chimpanzees is a high-risk collaboration. Male alliances are fickle; one ally can be dropped overnight in favor of another. The equality of grooming exchanges can therefore be used to confirm the partners' willingness to support each other [245, 416, 434]. Another example is the common vampire bat, *Desmodus rotundus*, which lives in primarily female roost groups comprised of a mixture of related and unrelated individuals. Individuals that have been unsuccessful foragers for two nights in a row risk starvation and beg for food from roostmates. Food sharing by regurgitation of blood occurs between certain pairs of individuals, who may or may not be related. Food-reciprocating pairs also groom each other more often than expected, suggesting that allogrooming may facilitate the monitoring of the partner's potential to give or receive food [692, 693]. A final example of the signaling function of allogrooming occurs in social insects, where grooming serves to homogenize the colony odor among colony members and facilitates colony recognition [65, 66, 615]. Interestingly, in honeybees, certain

patrilines appear to be specialized for this allogrooming task [186, 327].

GREETING DISPLAYS Species that live in fission-fusion societies require a sophisticated "social brain" to negotiate separations, reestablish relationships, and reduce tensions at reunions [5, 16]. Fission-fusion societies include—in addition to the elephants, dolphins, killer whales, and parrots mentioned earlier—chimpanzees, bonobos, hamadryas baboons, spider monkeys, sperm whales, spotted hyenas, and humans. Signals are clearly important for managing interactions in these communal species, the most prominent being **greeting ceremonies** during fusions. Greeting ceremonies typically combine recognition signals, signals of approach with benign intentions, and expressions of mutual affiliation, reassurance, or assessment. Females of many primate species approach and embrace familiar individuals upon meeting. This gesture involves clasping the partner with one or both arms, a behavior derived from infant clasping, and it may be accompanied by kissing, cheek rubbing, olfactory sniffing, and mutual genital rubbing in the case of the remarkably

FIGURE 13.29 Greeting ceremonies (A) Two adult female bonobos (*Pan paniscus*) greeting with genito-genital rubbing; one female is clinging like an infant to the other, who carries the weight of both. (B) Female elephants (*Loxodonta africana*) greeting by entwining their trunks around each other. (C) Male hamadryas baboons (*Papio hamadryas*) greeting with genital touching. (D) Two adult female spotted hyenas (*Crocuta crocuta*) in greeting ceremony stance, each with a rear leg raised. Note the slight tear in the pseudopenis, indicating that this female has recently given birth for the first time.

sexual bonobo (**Figure 13.29A**) [15, 146, 206, 443, 456, 547]. Other mammalian species have distinctive tactile greeting signals, such as head butting in some cats, chest and flank nudging in horses and dolphins, and mouthing in canids [135, 362, 376, 461]. Probably the most spectacular ceremony to behold is the meeting of two closely related female elephant groups that have been separated for several days. They detect each other from a distance on the basis of "rumble" contact call signals. They then commence a full running charge toward each other while flapping ears, trumpeting, rumbling, and screaming. Upon contact, the groups mingle and engage in excited urinating and defecating, tusk clicking, trunk entwining, touching of the mouth and head, and backing into each other; these obviously emotional interactions may last up to 10 minutes [423, 424, 480] (**Figure 13.29B**).

Greeting ceremonies among male monkeys stand in contrast to the friendly greetings of females, even in the same species. They are more ritualized, more dependent on the dominance ranks of the individuals involved, and typically include genital display or touching. For example, in the hamadryas baboon, males first approach each other while maintaining eye contact and performing submissive lip smacking and ear flattening. When close, they adopt a face-to-face, rump-to-face, or side-by side body orientation, and lip smacking and ear flattening reach a peak. One male presents his rear and the partner reaches out to touch the first male's genitalia. The males then retreat without further aggression (**Figure 13.29C**). In the howler monkey (*Alouatta palliata*), males approach and emit throat rumbling and clucking sounds specific to the greeting context. Then while standing in a face-to-face posture and still vocalizing, each male grabs the other's shoulder with one hand and sniffs the partner's opposite armpit. Next they adopt a mutual rump-to-face position and sniff each other's genitals. Both males then retreat [152]. Quantitative studies of these male greeting events find that either dominants or subordinates may initiate greetings; that one participant may sometimes terminate the interaction; and that symmetrical and completed ceremonies are more likely to occur between males that are similar in status and fighting ability and familiar with each other [120, 152, 607, 690]. The offering of genitals to be touched by another male is interpreted as a highly risky behavior to test the partner's willingness to make concessions without threatening him. The spotted hyena (*Crocuta crocuta*) is another fission-fusion species with a similar greeting ceremony involving partners standing head-to-tail, lifting the rear leg, and mutually sniffing and licking the other's genitals. In this species, however, females are dominant over males and possess a large pseudopenis, which is erected during these greetings (**Figure 13.29D**). Hyena greetings are more likely to occur between females of similar dominance status than between males and females or between males [160]. These observations further support the notion that in greetings among the more competitive sex, two individuals of similarly high status must mutually subordinate themselves to reduce tensions and negotiate a truce during reunions [121].

Parrots, dolphins, and killer whales use complex vocal signal exchanges to mediate mutual recognition, assessment, and affiliation during reunions. We mentioned earlier that members of killer whale pod groups possess a repertoire of around 10 shared call types. As pods travel and forage, group members engage in vocal exchanges and are likely to immediately match the call type of another group member [414]. Similarly, bottlenose dolphins (*Tursiops truncatus*) sometimes respond to the signature whistle of a conspecific group member by emitting the same whistle [280], and spectacled parrotlets (*Forpus conspicillatus*) use specific call types when interacting with different familiar family members [680].

Vocal matching is therefore a mechanism for finding and addressing a particular social companion in a crowd. Orange-fronted conures (*Aratinga canicularis*) and galahs (*Eolophus roseicapillus*) take vocal matching one step further: during interactions with conspecifics, the birds morph their own signature contact call so that it gradually converges with the call of the other bird [23, 546, 663]. Call convergence seems to be associated with affiliative approach and flock joining. Conversely, morphing the call so that it diverges from that of another bird's call is associated with aggressive behavior and retreat. This ability of parrots to subtly modify their call "on the fly" in relation to the call of another individual in order to mediate flock fissioning and fusioning may explain their ability to imitate the speech of human caregivers [70].

Group movement coordination and decision making

Group-living animals must routinely make decisions about when and where to move and what activities to undertake. These decisions need to be made jointly if the group is to remain together; otherwise it is likely to split apart and the benefits of grouping may be lost. The decision-making process for groups is different from the process for individuals because it requires reaching **consensus**. If the interests of group members conflict, then some group members will pay a **consensus cost** if they abide by the group decision. Individuals must therefore weigh the benefits of grouping against this cost. The communication mechanisms employed to reach consensus depend on the degree of conflict for the particular type of decision; the size of the group; and whether decisions are made by a large fraction of group members (democratic), a small fraction of members (oligarchic), or a single leader (despotic) [123].

DEMOCRATIC DECISIONS AND QUORUM SENSING By pooling the knowledge and information of a larger number of individuals, more accurate decisions can often be made by **democratic** groups than by individuals. This concept has been called the **wisdom of the crowd** [337, 629]. To illustrate, if single individuals acting independently have a 0.6 probability of making the correct binary decision, averaging the responses of 100 independent individuals with the same goal leads to a nearly 1.0 probability of making a correct decision [314, 365]. When groups are composed of heterogeneous individuals, consensus decisions also produce less extreme decisions unless some systematic bias is present [122]. For example, homing pigeons released in a flock showed less circling around the start zone, no resting episodes, straighter beeline routes, and shorter homing times back to the loft than birds released individually [151]. In other decision-making contexts, such as deciding which of several alternative food patches to visit or when to travel, such averaging cannot be used, yet a consensus must nevertheless be reached. From a communication perspective, all that is needed to decide on a joint action is a mechanism of **quorum sensing**, whereby animals increase their probability of exhibiting

(A)

(B)

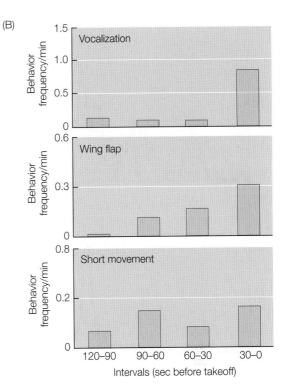

FIGURE 13.30   Coordination of group flight   (A) A flock of Canada geese (*Anser canadensis*) in the process of taking flight. Note the unanimous orientation and stretched neck posture of the geese still on the ground. (B) Rates of three behaviors of domestic geese (*A. domesticus*) in the two minutes before takeoff, subdivided into 30 sec intervals. Vocalizations increase sharply just before takeoff, while wing flapping and short movements increase somewhat earlier. (After [499].)

a behavior as a sharply increasing function of the number of other animals around them already performing this behavior [627]. Depending on the species, the quorum rule for taking action may be based on a simple majority, a supermajority, or a submajority. Good evidence of such majority rule decisions has been described for the initiation of group movement from a resting state in primate troops, ungulate herds, and waterfowl flocks. Animals may cast their vote by standing up, orienting their body or gazing in a consistent direction, or calling [122]. In geese flocks, for instance, the birds begin to stand up and wingflap about two minutes before takeoff. They make short walking movements in the preferred direction of travel and jerk their heads upward. These activities increase the arousal state of nearby geese, which are recruited into performing these behaviors. In the last 30 seconds before departure, the birds gradually increase their rate of calling, and just before takeoff most birds are vocalizing and the pitch of the calls increases [498] (**Figure 13.30**). Similar consensus-building **departure signals** have been documented in gorillas [621], capuchin and sifaka monkeys [56, 348, 413, 467, 648], and elephants [480, 613]. In many monkey and ungulate species, any animal may be the first to initiate the movement—there is no consistent leader—but this individual pauses and looks around to make sure others are following, and if there are not sufficient followers, it will return to the group [58, 67, 412, 467, 485, 499, 648]. Likewise in migrating waterfowl flocks, if a subset of individuals takes flight but the flock size is small (presumably too small to achieve optimal flight aerodynamics), they may turn around, land, and make a second liftoff attempt later when they can recruit a larger group of birds [20, 478].

OLIGARCHIC DECISIONS   An intermediate solution between total group majority rule and despotic decisions is to place the decision-making responsibility on a subset of the most knowledgeable individuals, an **oligarchic** strategy. The remaining individuals must acquiesce to following these leaders in the interest of maintaining group cohesion [124, 316]. What characterizes these leaders? *High motivation*, such as a strong need for a particular resource, is one important factor that causes some individuals to emerge as leaders. For example, the hungriest individuals are most likely to initiate and lead the march from a resting site to a resource patch. In mammals, lactating females have a stronger influence on group movements toward food patches and watering holes than nonlactating females and males [24, 163, 177, 500]. Similarly, food-deprived individuals take the front positions in fish schools, so that they may lead the group in their preferred direction [336]. *Bold temperament* is a second characteristic of leaders. Personality differences between individuals are often assumed to represent nonadaptive variation around an adaptive population average, but such differences can be associated with leader–follower roles. An experiment with pairs of stickleback fish showed that bolder individuals tended to lead excursions to a food patch while shy fish emerged as followers. Social feedback enhanced these roles, with bold fish inspiring faithful followership and shy followers facilitating effective leadership [239, 432]. *Knowledge* is a third key determinant of leadership. In schooling fish and roosting raven colonies, individuals with precise knowledge of food patch locations can guide others to the resource [507, 700]. Older individuals fly at the front of 80% of migrating broad-winged hawk (*Buteo platypterus*) flocks [385], and older female elephants tend to lead the movements of their family

(A)

(B)

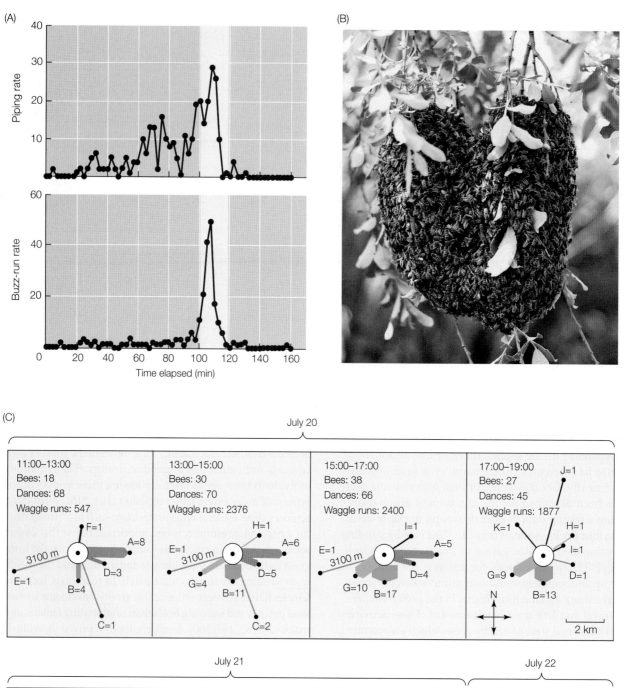

(C)

July 20

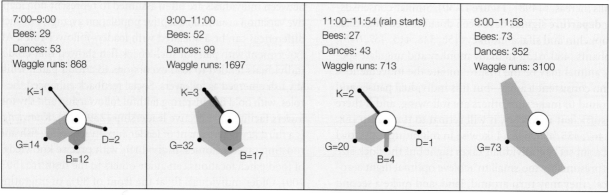

◀ **FIGURE 13.31   How honeybees move to a new home** When a colony has grown to a sufficient size, approximately half of the bees along with the queen will leave the home hive in a swarm and search for a new home, leaving the remaining bees and several young queens behind to continue reproducing. An oligarchy of about 200 scout bees is responsible for making the decision about when to leave and where to go. (A) About an hour before the swarm leaves (yellow stripe on graphs), scout bees begin to perform a high-pitched piping signal. The pipers burrow among the quiet bees while vibrating their flight muscles, which stimulates the quiescent bees to warm their own flight muscles. About 10 minutes before departure, the scouts perform the buzz-run display, buzzing their wings against other bees, to signal that liftoff is imminent. (B) The swarm bivouacs on a tree branch. The scouts then investigate possible new home sites. (C) Scouts that have discovered a potential site return to the swarm and perform waggle dances, indicating the angle and distance to the site in the same manner that foragers advertise food sites (see Chapter 14). Over a period of several days, other scouts investigate the advertised sites and perform dances for the sites (labeled A–K) they judge to be acceptable. The sequence of panels shows the direction to each advertised site (arrow angle); the distance to the site (arrow length); and the number of bees dancing (arrow width and the number shown next to the arrow). The duration of a dance is correlated with the perceived quality of the site. Within a few days, all dances point to the same site. A quorum has been reached. The scout bees then repeat the piping preparation and buzz-running displays, and the swarm lifts off. The scout bees lead the swarm flight to the new home. (After [501, 558, 571–573, 672].)

groups in times of drought because of their long-term memory of water holes [178]. In eusocial insects, when the colony is ready to fission and disperse, foraging scouts comprising only 5% of the colony are the only individuals with sufficient information to lead the colony to a new nest site [672]. In honeybees, scouts both make the decision about which site to adopt and lead the way during the final mass movement to the new home. Extensive signaling is involved. Each scout advertises the site she has found, recruiting other scouts to examine her site. The scouts winnow down the selection to a few sites and then a single site, using a quorum-sensing mechanism. Once agreement has been reached, they prepare the rest of the colony for flight and lead the way to the agreed-upon site. **Figure 13.31** describes the signals and process used by honeybees to decide on a new site and prepare for liftoff.

DESPOTIC DECISIONS   Models of group decision making predict that despotism, with one individual controlling when and where a group moves, should be uncommon. The leader, typically the dominant group member, is expected to make decisions that maximize its own fitness, which may often conflict with the interests of other members. If other members experience a significant consensus cost, they may not abide by the leader's decision and the group could split apart. The benefit of maintaining group cohesion must be extremely high to overcome these costs. Conditions under which it might pay to follow a single **despotic** leader include: (A) the potential for significant errors in assessing the best timing and direction of travel, (B) a large difference in

experience or knowledge between the leader and all other group members, so that the leader's average decision error is lower than the average error of other members, and (C) small group size [122]. Known examples of despotic groups are indeed relatively rare, as predicted, and do meet these types of conditions. For instance, chacma and hamadryas baboons (*Papio ursinus* and *P. hamadryas*), which inhabit harsh environments (dry mountains and deserts, respectively), have fission-fusion social systems in which the basic "fission unit" is a single male plus several females [92, 339]. Many units coalesce at night roosting sites and spread out during the day, driven by the low density of food and strong foraging competition. The large and aggressive males provide important protection for females and their young against conspecific harassers and predators, and also control and herd straying females. Females are not strongly bonded to each other but instead direct their attention and grooming toward the harem male. The male initiates group travel with intention movements, postures, and setting off in his preferred direction with a purposeful gait. He does not pause or look around to see who is following; the other group members usually do follow, sometimes reluctantly. If a female tries to initiate group movement, the others will not follow her unless the male also strikes out in the same direction. The experimental provision of food supplements that could be monopolized to varying degrees by the dominant male revealed that females usually abided by the male's decision to visit a distant provisioning site, even when it meant paying a substantial consensus cost [315]. Bottlenose dolphins provide a second example of single-individual decision making, in this case by temporary leaders who have superior information. Traveling bouts to new foraging patches are often initiated by a "side flop" display performed by a single male (**Figure 13.32**). Displaying

FIGURE 13.32   Leadership signal in bottlenose dolphin (***Tursiops aduncus***) The side flop display given by a single male often initiates traveling bouts. The performer jumps clear of the water and lands on its side. The display probably has both visual and acoustic components.

(A)

(B)

(C)

FIGURE 13.35 **Cooperative nest construction in weaver ants** (***Oecophyllas maragdina***) (A) The arboreal colonies of the Asian weaver ants consist of a single queen and up to 500,000 workers, which are subdivided into two castes. The more numerous "major" workers, shown here, perform most of the exterior construction of the leaf and silk nests. (B) Leaves are bent and pulled together by chains of workers grasping each other by the waist. Once the leaf edges have been pulled close together, they are held in position by a phalanx of workers while other workers retrieve larvae from nearby subnests. (C) Holding a larva in its mandibles, the worker touches the larva's head to the leaf surface and taps the larva with its antennae, which induces the larva to exude silk. The worker then waves the larva back and forth across the leaf stems like a weaver's shuttle, gluing the leaves together with drawn-out threads of silk. The larvae

thus operate as an auxiliary caste, and altruistically give up the silk they would normally use to make their own cocoons. The smaller "minor" workers remain inside the nests tending the brood once a nest is finished. This highly coordinated nest-weaving behavior is claimed to be the highest level of cooperation observed in ants [136, 257, 258, 260, 695].

products and dead bodies; foraging for food; and protecting the colony against predators and parasites. How do workers decide exactly when and where to engage in a particular task? The queen spreads a pheromone throughout the colony to notify workers of her presence and identify her eggs, which indirectly suppresses laying by the workers, but she does not direct workflow in the colony through central command signals. Rather, workers spend a substantial amount of time patrolling the nest, gathering information, and making use of a great many cues and a few signals to find jobs that need doing [208, 285, 570]. However, to avoid anarchy and wasteful searching and repetition, there is a basic organization with **division of labor** among the workers [46]. Task allocation is determined primarily by age, with newly emerged workers

performing "inside" tasks such as brood care and food processing, and older workers performing the more dangerous "outside" tasks such as foraging and colony defense. Some ant species have evolved physical castes, differing in body size and other morphological features, suited to performing specialized tasks more effectively [258]. Specific inside tasks are allocated based on perception of cues and a **self-organizing system**, which we shall describe in detail in Chapter 15. Outside tasks are communicated more by signals, which will be described in Chapter 14. **Figure 13.35** describes one of the more amazing examples of cooperation in eusocial insects, the construction of nests in *Oecophylla* weaver ants. See Web Topic 13.2 for additional examples of group integration signals.

SUMMARY

1. Social integration entails cooperating, giving aid, and synchronizing activities with other individuals. Because these behaviors usually involve a cost to the donor, they will not evolve unless there is a compensatory benefit. Models of cooperation are based on either indirect genetic benefits via kin selection or greenbeard altruism, or on

direct benefits arising from **mutualism**, **synergism**, or **reciprocity**.

2. Integration signals typically involve requests for aid by the sender and donation of aid by the receiver. Under most cooperation models, the receiver must bestow the aid on a known target individual to benefit, so **recognition** of the

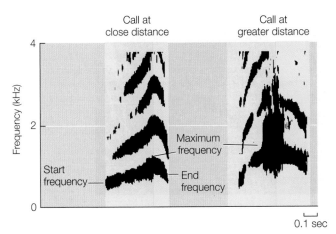

FIGURE 13.34 Changes in contact calls with increasing separation distance The coo contact call of the Japanese macaque (*Macaca fuscata*) changes gradually with increasing sender/receiver distance. Key acoustic features associated with distance are higher start, end, and maximum frequencies, longer call duration, longer duration to maximum frequency (yellow zone), and a noisy component at the maximum frequency, which gives the call a harsh sound. These effects may be caused by the higher motivation of more isolated senders to communicate and reunite with group mates. When initial coo calls are not answered, senders increase call duration, amplitude, and frequency modulation [322, 548]. (After [626].)

are likely to be answered and reunited with the group more quickly than subordinates [153]. Similarly, spider monkeys (*Ateles geoffroyi*) frequently answer the "whinny" call of a lost group member but only approach the call of a close associate [497]. This selective response to valuable group mates reveals the ever-present conflict of interest even in cooperative groups.

The wheeling flocks of starlings and pigeons and streaming schools of small silvery fish seem to have a collective mind of their own. In fact they are the result of highly synchronized following behavior and the ephemeral leadership of a few individuals [130, 337, 589]. The view of such swarms as an emergent property of relatively simple individual decision rules for cueing and orienting with respect to neighbors requires a social network approach. We shall therefore defer our discussion of swarms until Chapter 15.

Cooperative breeding

Cooperative breeding is a social system in which individuals care for offspring that are not their own. Cooperative systems vary along a continuum of reproductive sharing from **egalitarian** mutual care of several females' offspring in a joint brood to highly **skewed** care of a single female's offspring by nonreproductive helpers. Because several to many individuals contribute to the care of a brood, special signals beyond those discussed earlier for monogamous pairs are often required to coordinate caregiving duties.

Most vertebrate cooperative breeders (mammals, birds, and fish), whether in an egalitarian or a skewed caregiving system, defend a group territory that contains the food resources

needed by all adults and offspring. Defense itself is a group affair, and larger units tend to prevail over smaller ones, so territorial signals such as acoustic chorusing evolve to provide information about group size, and recruitment signals evolve to alert dispersed group members of a territorial intrusion [21, 71, 81, 230, 240, 398, 438, 490, 563]. Special signals for coordinating brood care are primarily found in joint-nesting birds. For example, ani (*Crotophaga* spp.) groups consist of two to four monogamous pairs. Pairs take turns working on the nest during the construction phase and signal their approach to the nest by calling together in flight so that all group members know the whereabouts of other group members. All adults contribute to incubation but only one sits on the nest at a time. When a bird intends to incubate, it flies to the nest with a green leaf in its bill, which induces the current incubator to leave. During nestling provisioning, a bird makes individual trips to the nest with food but subsequently tends to sit at the top of the nest tree as a sentry until the next bird arrives. Single dispersing birds searching for a vacancy give a loud, repetitive joining call [69, 489]. None of these signals occurs in single-pair species. The Taiwan yuhina (*Yuhina brunniceps*) is a small passerine bird with a very similar social system, but in this species all group members remain in close proximity most of the day and perform most breeding activities except incubation together. During nestling provisioning, the dominant male gives a flight call that synchronizes feeding trips to the nest, which has the effect of ensuring a more even distribution of food to all nestlings [587]. In contrast to this cooperation at the brood care stage, joint-nesting females in both species compete strongly during the egg-laying stage to increase the proportion of their eggs in the clutch—in anis by tossing out eggs laid by other females prior to their own first egg, and in yuhinas by physically tussling to block each others' access to the nest [588, 661]. Note that these females are unrelated to each other.

The colonies of advanced eusocial insects, such as army and leaf-cutter ants, honeybees, yellow-jacket wasps, and fungus-growing termites, have been likened to **superorganisms** because the activities of colony members are so well integrated and show little evidence of internal conflict [259, 569]. The number of individuals in a colony can reach 20,000 or more. These societies evolve when relatedness is high, resources are patchy, and intergroup competition is strong [511]. The majority of colony members are nonreproductive workers who raise the offspring of one or a small number of reproductive queens (or a monogamous royal pair in the case of termites). Workers are themselves offspring or close relatives of the queen, so colonies can be viewed as families that have undergone strong selection to cooperate internally and suppress conflict in competition with other families [511]. We mentioned earlier the lack of kin subgroup discrimination within colonies, which prevents the sort of wasteful competitive behaviors described above in joint-nesting birds. The tasks that must be accomplished by a colony include constructing a complex nest structure; provisioning the brood and workers; cleaning the nest of waste

(A)

(B)

(C)

FIGURE 13.35   Cooperative nest construction in weaver ants (*Oecophyllas maragdina*)   (A) The arboreal colonies of the Asian weaver ants consist of a single queen and up to 500,000 workers, which are subdivided into two castes. The more numerous "major" workers, shown here, perform most of the exterior construction of the leaf and silk nests. (B) Leaves are bent and pulled together by chains of workers grasping each other by the waist. Once the leaf edges have been pulled close together, they are held in position by a phalanx of workers while other workers retrieve larvae from nearby subnests. (C) Holding a larva in its mandibles, the worker touches the larva's head to the leaf surface and taps the larva with its antennae, which induces the larva to exude silk. The worker then waves the larva back and forth across the leaf stems like a weaver's shuttle, gluing the leaves together with drawn-out threads of silk. The larvae thus operate as an auxiliary caste, and altruistically give up the silk they would normally use to make their own cocoons. The smaller "minor" workers remain inside the nests tending the brood once a nest is finished. This highly coordinated nest-weaving behavior is claimed to be the highest level of cooperation observed in ants [136, 257, 258, 260, 695].

products and dead bodies; foraging for food; and protecting the colony against predators and parasites. How do workers decide exactly when and where to engage in a particular task? The queen spreads a pheromone throughout the colony to notify workers of her presence and identify her eggs, which indirectly suppresses laying by the workers, but she does not direct workflow in the colony through central command signals. Rather, workers spend a substantial amount of time patrolling the nest, gathering information, and making use of a great many cues and a few signals to find jobs that need doing [208, 285, 570]. However, to avoid anarchy and wasteful searching and repetition, there is a basic organization with **division of labor** among the workers [46]. Task allocation is determined primarily by age, with newly emerged workers performing "inside" tasks such as brood care and food processing, and older workers performing the more dangerous "outside" tasks such as foraging and colony defense. Some ant species have evolved physical castes, differing in body size and other morphological features, suited to performing specialized tasks more effectively [258]. Specific inside tasks are allocated based on perception of cues and a **self-organizing system**, which we shall describe in detail in Chapter 15. Outside tasks are communicated more by signals, which will be described in Chapter 14. **Figure 13.35** describes one of the more amazing examples of cooperation in eusocial insects, the construction of nests in *Oecophylla* weaver ants. See Web Topic 13.2 for additional examples of group integration signals.

SUMMARY

1. Social integration entails cooperating, giving aid, and synchronizing activities with other individuals. Because these behaviors usually involve a cost to the donor, they will not evolve unless there is a compensatory benefit. Models of cooperation are based on either indirect genetic benefits via kin selection or greenbeard altruism, or on direct benefits arising from **mutualism**, **synergism**, or **reciprocity**.

2. Integration signals typically involve requests for aid by the sender and donation of aid by the receiver. Under most cooperation models, the receiver must bestow the aid on a known target individual to benefit, so **recognition** of the

◀ **FIGURE 13.31**   **How honeybees move to a new home**   When a colony has grown to a sufficient size, approximately half of the bees along with the queen will leave the home hive in a swarm and search for a new home, leaving the remaining bees and several young queens behind to continue reproducing. An oligarchy of about 200 scout bees is responsible for making the decision about when to leave and where to go. (A) About an hour before the swarm leaves (yellow stripe on graphs), scout bees begin to perform a high-pitched piping signal. The pipers burrow among the quiet bees while vibrating their flight muscles, which stimulates the quiescent bees to warm their own flight muscles. About 10 minutes before departure, the scouts perform the buzz-run display, buzzing their wings against other bees, to signal that liftoff is imminent. (B) The swarm bivouacs on a tree branch. The scouts then investigate possible new home sites. (C) Scouts that have discovered a potential site return to the swarm and perform waggle dances, indicating the angle and distance to the site in the same manner that foragers advertise food sites (see Chapter 14). Over a period of several days, other scouts investigate the advertised sites and perform dances for the sites (labeled A–K) they judge to be acceptable. The sequence of panels shows the direction to each advertised site (arrow angle); the distance to the site (arrow length); and the number of bees dancing (arrow width and the number shown next to the arrow). The duration of a dance is correlated with the perceived quality of the site. Within a few days, all dances point to the same site. A quorum has been reached. The scout bees then repeat the piping preparation and buzz-running displays, and the swarm lifts off. The scout bees lead the swarm flight to the new home. (After [501, 558, 571-573, 672].)

groups in times of drought because of their long-term memory of water holes [178]. In eusocial insects, when the colony is ready to fission and disperse, foraging scouts comprising only 5% of the colony are the only individuals with sufficient information to lead the colony to a new nest site [672]. In honeybees, scouts both make the decision about which site to adopt and lead the way during the final mass movement to the new home. Extensive signaling is involved. Each scout advertises the site she has found, recruiting other scouts to examine her site. The scouts winnow down the selection to a few sites and then a single site, using a quorum-sensing mechanism. Once agreement has been reached, they prepare the rest of the colony for flight and lead the way to the agreed-upon site. **Figure 13.31** describes the signals and process used by honeybees to decide on a new site and prepare for liftoff.

DESPOTIC DECISIONS   Models of group decision making predict that despotism, with one individual controlling when and where a group moves, should be uncommon. The leader, typically the dominant group member, is expected to make decisions that maximize its own fitness, which may often conflict with the interests of other members. If other members experience a significant consensus cost, they may not abide by the leader's decision and the group could split apart. The benefit of maintaining group cohesion must be extremely high to overcome these costs. Conditions under which it might pay to follow a single **despotic** leader include: (A) the potential for significant errors in assessing the best timing and direction of travel, (B) a large difference in

experience or knowledge between the leader and all other group members, so that the leader's average decision error is lower than the average error of other members, and (C) small group size [122]. Known examples of despotic groups are indeed relatively rare, as predicted, and do meet these types of conditions. For instance, chacma and hamadryas baboons (*Papio ursinus* and *P. hamadryas*), which inhabit harsh environments (dry mountains and deserts, respectively), have fission-fusion social systems in which the basic "fission unit" is a single male plus several females [92, 339]. Many units coalesce at night roosting sites and spread out during the day, driven by the low density of food and strong foraging competition. The large and aggressive males provide important protection for females and their young against conspecific harassers and predators, and also control and herd straying females. Females are not strongly bonded to each other but instead direct their attention and grooming toward the harem male. The male initiates group travel with intention movements, postures, and setting off in his preferred direction with a purposeful gait. He does not pause or look around to see who is following; the other group members usually do follow, sometimes reluctantly. If a female tries to initiate group movement, the others will not follow her unless the male also strikes out in the same direction. The experimental provision of food supplements that could be monopolized to varying degrees by the dominant male revealed that females usually abided by the male's decision to visit a distant provisioning site, even when it meant paying a substantial consensus cost [315]. Bottlenose dolphins provide a second example of single-individual decision making, in this case by temporary leaders who have superior information. Traveling bouts to new foraging patches are often initiated by a "side flop" display performed by a single male (**Figure 13.32**). Displaying

FIGURE 13.32   Leadership signal in bottlenose dolphin (***Tursiops aduncus***)   The side flop display given by a single male often initiates traveling bouts. The performer jumps clear of the water and lands on its side. The display probably has both visual and acoustic components.

males were found to be *social brokers* who spent time with individuals from many different social units in the population and thus had a better knowledge of past and current activities of other groups and the location of undepleted foraging patches [373, 374]. Some other examples of group coordination signals given by dominant leaders in small groups include pack travel initiation in gray wolves (*Canis lupus*) [465] and foraging site movement vocalizations in green woodhoopoes [491].

GROUP COHESION   Once a group is on the move, members need to remain in close contact. When the habitat is open and group members are sufficiently close, visual cues can maintain group cohesion. In more densely vegetated habitat, visual contact is quickly lost and **acoustic contact signals** are required. Flocks of birds foraging through foliage keep up a constant twitter with repeated soft peep, tick, or chip vocalizations that reveal the location of the flock. When birds get out of earshot, their calling becomes louder and more frequent, often eliciting countercalling, which enables stragglers to reunite with the group [387]. Numerous avian species also call repeatedly while migrating at night (**Figure 13.33**). Quantitative data on contact call function has been much easier to obtain on foraging primate troops, whose call rate increases when the distance to the nearest neighbor increases and when the habitat is more densely vegetated [54, 323, 415, 442]. In some species the acoustic features of the calls—duration, frequency bandwidth, and modulation structure—change with separation distance, possibly reflecting fear and distress (**Figure 13.34**). If an individual becomes truly separated from its group, the vocalization may change qualitatively to a separation call, which is louder and longer than the contact call [54, 55, 57, 609]. Separation calling, especially by offspring, elicits approach by the parent or close associate [513]. When adults become separated and call, responses by the group may depend on the status or identity of the caller. For instance, in white-faced capuchins (*Cebus capucinus*), lost and calling *dominant* males and females

FIGURE 13.33 Nocturnal flight calls of migrating warblers Many species of passerine birds, cuckoos, and woodpeckers utter distinctive flight calls while migrating at night. Spectrograms of calls from six New World warbler species are shown here. There is a significant phylogenetic effect on syllable structure, while the diurnal habitat of each species affects frequency characteristics in ways similar to song (lower frequency, less modulation, and narrower bandwidth in denser vegetation). In some species, these same calls are also observed during the day in family groups with fledglings and in foraging flocks. Individual calling rate at night can reach 20 calls per minute, and increases with greater cloud cover. Calls are often reciprocated with answering calls. Young inexperienced birds call at higher rates than older birds. The calls are believed to stimulate migratory restlessness, maintain group cohesion, and reduce the variation and error of navigational headings. (After [169–172, 346].)

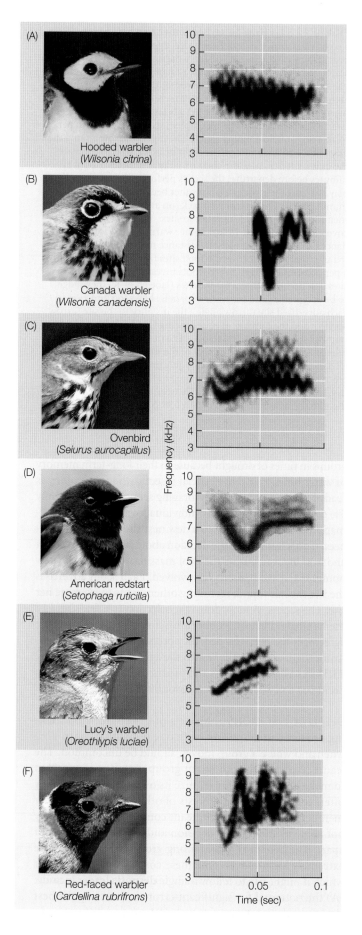

sender is essential. The process of recognition has three components: **production** of an identifying signal, cue, or label; **perception** of the signal, which entails comparison of the received signal against a previously acquired template of the target recipient; and some **action** by the donor based on a threshold decision rule to accept or reject the target. Donors acquire a template of the target through association with a **spatial location**, **familiarity** (associative learning of the target individual's characteristics), **phenotype matching**, or genetically-linked **recognition tags**.

3. **Identity signals** evolve when a potential recipient of aid or cooperation benefits from making its true identity known to donors. In response, donors evolve fastidious discrimination abilities to identify desirable target recipients when the cost of aiding random individuals is high. Signals in all modalities can be designed to encode individually distinctive signature information by increasing their complexity and the number of variable orthogonal components. Identity signals serve functions in territory defense, parent–offspring communication, pair coordination, family and group cohesion, and status signaling.

4. **Mate recognition** spans three levels. **Species recognition** entails distinguishing conspecific from nonconspecific members of the opposite sex. Species-distinctive signals arise as populations diverge and adapt to local ecological conditions. Sexual selection can also drive speciation and lead to species-specific signals. **Mate-quality recognition** is the mechanism by which good-quality mates are identified. Many species appear to possess multivariate mate attraction signals, with some less variable **static components** under **stabilizing selection** providing species identity information and other more variable **dynamic components** under **directional selection** providing mate quality information. Whether receivers use these two components with a **hierarchical rule** to first select conspecifics and then evaluate quality, or combine them into a **one-dimensional preference function** and apply an acceptance threshold, remains controversial and probably varies among species. **Partner recognition** is found in species that form male–female bonds for the purpose of cooperative reproduction. Partners develop distinctive signals that help them to find each other in a crowd of other conspecifics.

5. Extended **male–female partnerships** form when males are selected to help females care for offspring. Highly synchronized visual and acoustic **mutual displays** evolve to facilitate **continued mate evaluation**; **recognition** and **greeting** after a separation; **cooperative territory defense**; and **mate guarding** against same-sex rivals. **Tactile interactions** such as allogrooming and huddling not only serve a hygienic function but also may reduce aggression between pair mates. Sequential displays between breeding pairs help to **synchronize** male and female **hormonal cycles** and coordinate reproductive episodes. Pairs may improve the efficiency of their joint parental care efforts by employing **negotiation rules** to arrive at a final allocation of effort or by communicating and coordinating their trips to the brood with food.

6. Species with parental care require **parent–offspring recognition mechanisms** to identify their own offspring and to avoid bestowing costly care on nonoffspring. The mechanism employed depends on the rate of offspring development, type of nest or roost site, family size, degree of sociality, and foraging strategy. When immobile young are raised in a nest, **nest location** cues are used to identify offspring, but once the offspring become mobile and mingle with those of other parents, one or both parties typically evolve distinctive individual signature signals and refined discrimination abilities. **Mutual recognition** between parents and offspring evolves when the offspring of many parents are placed in a crèche. When broods are very large and isolated in a nest or burrow, a shared **family identity signal** is favored over individual identity information.

7. Parents and offspring disagree to a certain extent over the amount of parental investment to be provided, with offspring preferring more than the parent is optimally selected to give. The best examples of **parent–offspring conflict** are **brood sex ratio conflict** between queens and workers in hymenoptera and **genomic imprinting** in mammals. Two categories of **offspring begging models** offer potential resolutions to the conflict over offspring provisioning. **Honest signaling models** view begging as an informative signal of offspring need, which parents use to allocate food delivery. Sibling **scramble competition models** assume that begging is a strategy of jostling for food brought to the nest by the parent and view begging vigor as an indicator of offspring quality. These two processes mark the endpoints of a continuum from high to low parental power; which one prevails depends on the number of parents, the size of the brood, the style of food delivery, the availability of food, and the age of the offspring.

8. Multimodal **begging signals** by dependent offspring to parents usually increase in vigor as a function of offspring hunger. In response to begging for food, parents typically increase their rate of food provisioning. But food is not always selectively delivered to the hungriest offspring in the brood. Instead, it may be given to the largest or strongest sibling. Siblings may negotiate among themselves or collaborate to reduce competitive costs and increase parental care. Parents also signal to their offspring. **Directive signals** are given to mobile broods to maintain cohesion, identify appropriate food, and warn against predators. Parents in a few species **teach** their offspring about food handling and environmental hazards.

9. Mechanisms of **group-mate recognition** in animals that live in cooperative societies depend on group size and stability. When groups are relatively small and stable in composition, group-mate recognition can be based on associative learning of individual signature traits. Some long-lived species with **fission-fusion societies** can still manage group interactions based on individual recognition and long-term memory. When groups are large and associated with a nest, as in social insect colonies, recognition is based on an acquired **group label** or badge. Within social insect colonies with multiple queens or sires, there is an absence of subfamily discrimination, caused by selection

for a high tolerance threshold, muting of identity cues, or active scrambling of recognition cues. Large fluid flocks and herds do not require group identification and instead may use **status** and **age badges** to mediate associations and aggressive interactions.

10. **Appeasement signals** reduce costly aggression within stable groups. Subordinates give submissive signals to suppress attacks by dominants, and dominants give **conflict-avoidance displays** indicating benign intentions as well as **reassurance displays** to subordinates. After a conflict, competitors often engage in **reconciliation**, or other individuals may offer **third-party consolation**. **Allogrooming** serves to reduce tension and reaffirm bonds and future cooperation between valued reciprocating group members, in addition to its obvious hygienic function. **Greeting ceremonies** are common during reunions in fission-fusion societies. Greeting in female primates often entails hugging and kissing, whereas in male primates and dominant female hyenas, the displays are more ritualized and often involve genital touching.

11. Members of stable mobile groups must reach a **consensus** in deciding when and where to move next. Such decisions may be made by a large fraction of the group (**democratic**) using a **quorum-sensing** mechanism to assess when a majority threshold has been reached. Alternatively, a small fraction of the group (**oligarchic**) or a single leader (**despotic**) may make the decisions, which other group members must accept despite potential conflicts of interest. Once groups are on the move, they employ visual cues or **acoustic contact signals** to maintain group cohesion.

12. **Cooperative breeding systems** range from egalitarian mutual care of a joint brood to skewed rearing of a single individual's offspring by nonreproductive helpers. **Egalitarian** systems show evidence of competition during egg laying, but thereafter employ a variety of signals to coordinate territory and nest defense, parental care, and provisioning of the young. In **skewed** systems, helpers are closely related to the breeder, and selection favors suppression of conflict and highly efficient organization. Large colonies of eusocial insects employ **division of labor** based on age to allocate worker effort to different tasks. Worker activities are controlled not by central commands but by a **self-organizing system** of simple rules and individual decision making based on a large number of cues plus a few signals.

Further Reading

For a succinct review of evolutionary explanations for cooperation, see West [687]. The chapter by Sherman et al. [590] outlines the production, perception, and action components of recognition. A series of more recent reviews of these components can be found in a special 2004 issue of *Annales Zoologici Fennici* (vol. 41, no. 6). Individual recognition systems are reviewed by Tibbetts and Dale [642]; kin and group recognition are reviewed by Beecher [32], Komedeur [328],

Gamboa [189] and Mateo [392]. Kondo and Watanabe [329] discuss the identity information and social function encoded in the acoustic contact calls of birds and mammals.

Coyne and Orr [134] is the most recent reference book on speciation. Insightful articles on the controversy surrounding species recognition and sexual selection have been contributed by Castellano [99, 101], Arbuthnott [6], Phelps et al. [477], Pfennig [474], Price [484], Servedio and Noor [575], Pfennig and Pfennig [476], and Sobel et al. [611]. Overviews of cooperative behaviors between mated pairs can be found in Black [48], Wachtmeister [674] and Hall [236].

Important reviews of parent–offspring conflict include Trivers [651]; Mock and Forbes [417]; Bateson [29]; Wright and Leonard [699]; and Kilner and Hinde [312]. Relevant models are described in Royle et al. [532] and Parker et al. [460].

Aureli et al. [13] present a useful framework for classifying appeasement signals. Two useful books on peacemaking in primates are those by de Waal [146] and Aureli and de Waal [12]. Additional reviews of cooperation and reciprocal altruism in primates can be found in Schino and Aureli [554] and Silk [597]. Boinski and Garber [59], Conradt and Roper [122, 123], and King [316] provide useful overviews of group movement, decision-making, and leadership strategies.

COMPANION WEBSITE
sites.sinauer.com/animalcommunication2e

Go to the companion website for Chapter Outlines, Chapter Summaries, and References for all works cited in the textbook. In addition, the following resources are available for this chapter:

Web Topic 13.1 *Direct benefit models of cooperation*
Here we review a number of current models used to explain the evolution of cooperative behavior without requiring kin selection or greenbeard compensations. We begin with the classic Prisoner's Dilemma model, and build up to iterated choreographies

Web Topic 13.2 *Examples of social integration signals*
Here we provide rich media examples of identity signals, pair integration signals, parent–offspring signals, and group integration signals.

Web Topic 13.3 *Cognitive models of mate choice*
Here we outline computational mechanisms females employ to integrate several sources of information from mate attraction signals. More examples of multimodal mate attraction signals and characteristics of species recognition components in various taxa are also provided.

Web Topic 13.4 *An overview of begging models*
Here we describe honest signaling models, scramble competition models, tug-of-war models, and sibling negotiation models.

Chapter 14

Environmental Signals

Overview

This chapter reviews signals that provide information about environmental conditions. The primary focus is on signals to predators, signals to fellow foragers about predators, and signals to fellow foragers about food finds. We also examine a special case of environmental signals in which sender and receiver are the same individual: autocommunication. A widespread function of autocommunication is the detection and localization of food items. Environmental signals span a wide range of relevant economics and a corresponding diversity in the amount of information provided to receivers.

The Diversity of Environmental Signals

Environmental signals constitute a very heterogeneous assemblage. With some interesting exceptions, most environmental signals have to do with *food*: either finding and sharing food, or alternatively, trying to avoid becoming food. Most, but again not all, environmental signals provide ancillary locational information. However, outside these general threads, environmental signals exhibit a much greater diversity than any context that we have considered in prior chapters; this is true whether the signals are classified according to informational focus, basic economics, or honesty guarantees.

We shall consider four broad classes of environmental signals. **Predator deterrent signals** are given by a potential prey animal to discourage a predator from attacking it. The primary receiver is thus the predator, and the informational focus (topic) is whether the signaler is or is not suitable prey. The economics of these signals are in some ways similar to those of conflict resolution (see Chapter 11), in that senders provide information that could help both parties avoid an unnecessary escalation. They also have aspects in common with the economics of mate attraction signals (see Chapter 12), in that senders try to influence a receiver's choice; in the case of predator deterrent signals, however, the goal is to avoid being chosen. Because of inherent conflicts of interest between predators and prey, honesty guarantees will be required for such signals. **Alarm signals** are given by a sender to warn other individuals that it has detected a predator. In contrast to predator deterrent signals, the primary receivers are other potential prey in the neighborhood, and the informational focus is the predator.

Food signals advertise the detection and in some cases, the location and quality, of food discoveries. The primary receivers are other foragers. The economics of alarm and food signals are similar, and usually involve senders paying a cost to provide a benefit to receivers. As we discussed in Chapters 9 and 10, such ostensibly altruistic signaling requires some compensatory direct or indirect benefit to the sender to justify its actions. The sender cost is usually sufficient to guarantee honesty; however, there are exceptions, and we shall take these up in later sections.

Finally, **autocommunication signals** are emitted and received by the same individual. The ways in which the signal is altered en route provides information about the local environment. Because sender and receiver are identical, there is no conflict of interest, no need for costly honesty guarantees, and the relevant economics are focused on maximizing signal efficacy. We take up each of these four categories of environmental signals in turn below.

Predator Deterrent Signals

Predators and prey begin any interaction with an inherent conflict of interest. However, if a prey can provide a relatively honest reason why the predator should not attack it, it may be in the interests of both parties to share that information [127, 386, 728, 970]. Because of the conflict of interest, there is a strong requirement for honesty guarantees before any predator should attend to such signals. The relevant guarantees are usually handicaps, indices, or a mixture of the two. A complete predator/prey interaction progresses through five basic steps: (1) detection of the prey by the predator; (2) classification and evaluation of prey suitability; (3) approach by the predator; (4) physical attack and prey subjugation; and (5) consumption of the prey [236, 552]. Prey may produce signals in an attempt to interrupt this sequence at any of these stages; both parties benefit if a justified interruption occurs earlier rather than later in the sequence. If unsuccessful at interrupting the sequence at one stage, a prey may produce a different signal at a later stage. Below, we review possible predator deterrent signals roughly in the order in which they are likely to occur in a typical predator/prey interaction.

Detection and camouflage

The simplest and most common antipredator strategy is to avoid being detected by predators. Many species of animals have evolved color patterns, shapes, sounds, smells, or behaviors that minimize their chances of being detected by a predator. Such adaptations are called **camouflage**. Because many of these adaptations are attenuators of cues that would otherwise make the prey more detectable, some authors have included camouflage as a special case of communication [387, 388]. As pointed out by critics, the specializations of camouflaged "senders" do not lead to coevolved specializations by intended "receivers" and thus should not be considered communication [568, 761, 845]. We include camouflage in our discussion for two reasons. First, many animal signals have evolved to be easily detected at close range by primary receivers, but to be camouflaged and easily missed at a distance by predators. Thus camouflage at certain distances

or in certain circumstances may be a key property of many animal signals. Second, many of the adaptations facilitating camouflage are the exact opposites of those promoting signal efficacy: understanding one helps gain perspective on the other. There is also some ambiguity as to whether camouflage should be assigned to the detection or the classification and evaluation stage of predator/prey encounters [810, 856]. Consider a prey camouflaged as a leaf. If a predator ignores this prey because its current search image does not discriminate between leaves and other background, detection does not occur; if the predator perceives the leaf as a distinct object but rejects it as inedible, its evaluation is negative. Assigning a particular rejection to one or the other stage would require knowing intricate details of the neurobiology of the predator; but for our discussion, the distinction is really not that critical. Camouflage works at either stage one or stage two to deter the predator from stage three, approaching the prey. Given this perspective, we shall consider three basic categories of camouflage: **crypsis** (blending into the background); **masquerade** (mimicking inedible objects); and the creation of **decoys** (objects that distract predators from prey)[856]. Although our focus is on camouflage by prey, it is worth noting that many predators also adopt camouflage to get closer to unsuspecting prey. Many of the principles are the same for either party.

The basic goal of crypsis is to blend the cues generated by a prey animal into those of its background (**Figure 14.1**). This can be challenging, since different predators of the prey may rely on different sensory modalities or, within a modality, focus on different spectral or chemical ranges of stimuli. Whether prey that move between different backgrounds during their daily or life cycles do better by adopting a compromise crypsis or by specializing in only one of the backgrounds depends upon the distributions of predators, backgrounds, and similar prey [423, 577, 728]. Crypsis may also have to be compromised to facilitate other activities such as mate attraction, foraging, and thermoregulation [132, 852]. Although most studies of crypsis have focused on visual stimuli, crypsis is a potential strategy in any modality [730].

Visual crypsis depends both on the color and the pattern of prey markings (see Chapter 5 for sensory basis). Disruptive patterning obfuscates the body outline of the animal, and is most effective when evidence of bilateral symmetry is minimized [184, 185, 850–852, 855]. Color matching enhances crypsis further [88], and more complex backgrounds appear to facilitate more effective crypsis [208, 578]. Most bird predators appear to be deceived by both disruptive patterning and color matching of insect prey [858]. As prey themselves, birds occupying different strata in a forest may have colorations that ensure crypsis in their particular layers [318]. Much of the colored patterning on diurnal mammals appears to be designed for crypsis [127, 128, 861]. Countershading, in which the upper half of the body is darkly colored and the lower half lightly colored, is widely seen in both terrestrial and aquatic environments. One explanation is that animals colored and patterned to look like a background

(A)

(B)

(C)

(D)

FIGURE 14.1   A sampling of crypsis in various animal species Many animals mimic the visual patterns of their habitats to make themselves cryptic to both their prey and their predators. (A) This cuttlefish (*Sepia officinalis*) and its cephalopod relatives have complex skin pigment cells that can be adjusted to match nearly any ambient pattern. (B) A horned frog (*Megophrys nasuta*) is nearly invisible in the leaves and ground litter in a Borneo forest. (C) Tassled anglerfish (*Rhyncherus filamentosus*) host elaborate skin extensions that make them look like rocks covered with seaweeds and bryozoans. (D) Two individuals of the crab spider (*Thomisus onustus*) that have adjusted their body colors to match the flowers on which they wait for insect prey. The top individual usually lurks in the magenta center of its flower; it was temporarily on the surrounding white petals when photographed.

cast a shadow on their ventral sides when they are close to a substrate; this makes their ventral side conspicuous against the background and the remainder of their bodies. A lighter shade on their ventral side counters this shadow effect. Similarly, a dark-bellied fish would be clearly visible against the

sky-lit surface for a predator looking upward; coloring its ventral surface white makes it less conspicuous. While these explanations appear to account for some cases of countershading, they fail to account for others, so there must be other reasons for countershading that have not yet been considered [463, 723–725, 728, 729, 851].

Many fish, shrimp, and other aquatic invertebrates achieve crypsis by being optically transparent [133, 444]. Other marine invertebrates, particularly small crabs, decorate themselves with algae to blend into backgrounds [65, 428]. Cephalopods and chameleons are famous for their ability to adjust their body pigmentation rapidly to match that of their current background [375, 866]. As noted in Chapter 4, both taxa have also co-opted their ability to vary color patterning for social signals. The challenge of being cryptic to multiple target species has been studied in several systems. Crab spiders hide on flowers; their coloration is a compromise between that which would minimize detection by their hymenopteran prey and that which would make them least conspicuous to avian predators [888, 889]. Chameleons must also optimize cryptic body coloration, given risks of both snake and bird predation [865].

Masquerade camouflage is very common in terrestrial arthropods, many of which have evolved shapes and colors that emulate sticks (phasmid insects and caterpillars), leaves (mantids and phasmids), thorns (treehoppers), dewdrops (treehoppers), or feces (spiders and caterpillars) [726, 810]. Fish such as sea dragons (Sygnathidae) are disguised as seaweeds, octopus as sponges and worm tubes, and some nocturnal birds, such as potoos and frogmouths (Caprimulgiformes), emulate tree stumps while asleep during the day [374, 376, 418, 735, 857, 880]. **Element masquerades** copy a specific component within a larger ensemble of objects (e.g., a single stick in a bush) and are effective only within that larger context; **object masquerades** emulate an entire object (e.g., bird feces) and can be effective when present in a wide variety of backgrounds (**Figure 14.2**) [367]. However, even object masquerade effectiveness may vary with the type of background on which it is located. **Decoys** are objects placed near potential prey to distract predators. The best documented examples are the decorations that some species of spiders place in their webs. These may function to distract predacious wasps from the spiders [610, 611, 890, 901].

Notifying predators about prey suitability

Once a predator has detected possible prey, it often delays briefly to evaluate whether the prey is suitable, and if there are multiple suitable prey, which one to attack. This provides an opportunity for prey to signal their relative unsuitability to the predator and perhaps dissuade its further approach. There must, of course, be actual reasons why the signaling prey might be relatively unsuitable, and the signals produced must provide the predator with some sort of honesty guarantees if they are to be heeded. There are three basic types of predator notifications: predator detection signals, prey condition signals, and aposematic signals.

PREDATOR DETECTION SIGNALS   Where stealth and surprise are important in determining a predator's success, prey that detect the predator before it launches an attack may be less worth pursuing than those that remain unaware. It thus may pay for prey to signal to a stealth predator that it has been detected and for the predator to give up chasing those particular prey when so signaled. While this may seem an obviously good strategy for both parties, there are, as usual, trade-offs. A prey that signals may increase its risk if the predator has not in fact detected its presence; the prey also makes its presence known to any other predators, parasites, or parasitoids that may have not yet detected it. These costs are particularly acute for a prey that erroneously signals to a nonpredator stimulus and inadvertently notifies undetected predators of

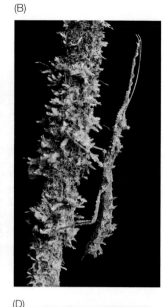

FIGURE 14.2 Animal masquerades (A) A praying mantis (*Hymenopus coronatus*) masquerades as a tropical orchid. (B) A stick insect (*Phenacephorus* sp.) adopts an element masquerade to copy one of the many lichen-covered branches in the forest. (C) A spider (*Phrynarachne decipiens*) masquerades as bird feces. (D) A tropical short-horned grasshopper (*Chorotypus* sp.) exhibits an element masquerade as another leaf in the forest.

its presence. Similarly, predators may be better judges than prey of whether or not they are in a good enough position to catch a fleeing prey. These separate trade-offs for the two parties can be modeled as an asymmetric game. Not surprisingly, an ESS favoring prey signaling and predator attention to the signals exists only for certain combinations of predator density, increased prey risks upon signaling, decreased chances of being caught if predators are detected early, predator costs of pursuing prey, and the availability of alternative prey when a predator's current stalk is disrupted [64, 298]. Note that coursing predators that pursue a prey until they catch up with it or outlast its flight (e.g., larger canids) gain no benefit from predator detection signals and are likely to ignore them. We would thus not expect prey to give predator detection signals to this type of predator.

Web Topic 14.1 *Models for environmental signaling*
A number of theoretical models are cited in this chapter. Here we provide a quick synopsis of the basic logic and methods used in many of these.

Despite the limiting conditions, **predator detection** signals are relatively common in vertebrates (**Figure 14.3**). Although the same signal can both notify predators and warn fellow foragers (see later sections), many solitary prey also give such signals, and their function can only be direct communication with an approaching predator. In addition, many species give predator detection signals only when they are sufficiently close to refuges or the approach of the predator is sufficiently slow that the signals are indeed honest: there

is no way the predator could catch an alerted prey before it becomes inaccessible. There are many clear examples of predator detection signals. Some species of fish flick their fins, anoles do additional pushup displays, whiptail lizards wave their forelegs, and other lizard species wave their tails at approaching predators [106, 163–165, 230, 499, 509]. Among birds, moorhens tail flick and motmots wag their long pendular tails when predators are nearby [604, 605, 695]. Both birds and mammals signal predator detection with alert postures while staring at the detected predator; some species have special head markings around their eyes or on their ears that amplify the stare and make it clear to the predator that it is the subject of their focused attention [127–131, 386, 861]. Some birds beam what were initially thought to be alarm calls at adjacent predators [986], and many ungulates precede

(A)

(B)

(C)

(D)

FIGURE 14.3 Predator detection signals (A) Pronghorn antelope (*Antilocapra americana*) stare fixedly at a predator. Note how the facial markings amplify the direction of stare. (B) Turquoise-browed motmots (*Eumomota superciliosa*) swing their pendulum-shaped tails from side to side when predators are detected. (C) In addition to antelope-like stares, red kangaroos (*Macropus rufus*)

and other macropods can rest their body weight on one leg and their tail while they thump the ground with the other leg to notify predators that they have been spotted. (D) A white-tailed deer (*Odocoileus virginianus*) flashes its fluffy white tail at a predator to indicate that it has seen the predator and is already fleeing.

(A)

(B)

(C)

FIGURE 14.4 Predator inspection (A) A herd of wildebeests (*Connochaetes taurinus*) inspects and trails a spotted hyena (*Crocuta crocuta*). (B) Similarly, a group of topis (*Damaliscus korrigun*) conspicuously monitors a male lion (*Panthera leo*). (C) A school of big-eye trevally jacks (*Caranx sexfasciatus*) inspects one of their major predators, a silvertip shark (*Carcharhinus albimarginatus*). While trailing sharks, jacks may also steal bits of prey after a shark kill, and rub their bodies against the rough skin of the shark to remove ectoparasites.

stares with specialized barks, whistles, hisses, foot stamps, or snorts [121, 127, 131, 666, 703]. Mongooses and squirrels raise their tails as predator notifications, and nocturnal rodents and larger macropods, such as kangaroos, stamp the ground with their feet to signal predator detection [61, 505, 647, 689, 718]. Many monkeys give alarm calls that function in part as predator notifications [16, 998, 999, 1005]. If predators continue to approach or appear sufficiently close, most prey will flee and many exhibit conspicuous white markings on their posteriors that act in part as predator detection signals [68, 126, 127, 131, 359, 403, 493, 494, 825, 837, 839].

Another form of predator notification, **predator inspection**, involves coordinated behavior by multiple prey (**Figure 14.4**). Again, it is largely directed at predators hunting by stealth. Participating prey assume alert postures, face the detected predator, and move as a group toward it [127, 131]. If the predator changes location, the inspecting prey follow it at a distance while continuing to orient their bodies toward it. Members of the inspecting group may concurrently produce any of the predator detection signals noted previously. The consequence is that all nearby prey are alerted to the

predator's presence and current location by this conspicuous activity. This clearly undermines any hopes of stealthy approach by the predator, and both savannah cheetahs and forest leopards will abandon subsequent hunting in the immediate area after being inspected [272, 999]. Inspection behavior is very common among fish, small passerine and colonial nesting birds, and savannah ungulates. Often, certain members of the inspecting group tend to approach more closely or track the predator more assiduously than others. In part, this is expected from game theoretic models of predator inspection strategies [215]. In addition, differences in inspection risk are correlated with dominance status in some fish [319], species identity in mixed bird flocks [630], relative youth in gazelles [272], and persistent personality differences in a variety of taxa [180, 217, 660, 930].

PREY CONDITION SIGNALS   Prey detection signals are likely to be ignored by coursing predators. However, if individual prey could demonstrate their reduced suitability relative to alternative prey, it might pay such predators to use this information before deciding whom to chase. Suitability might depend on health (sicker animals would be easier to catch); agility (less agile animals would be less likely to dodge final attacks); or energy reserves (lower levels would mean reduced endurance in a long chase) [11]. Again, the signals must be either indices or handicaps to guarantee ESS stability [386, 637, 917]. As is often the case with handicaps and

(A)

(B)

(C)

(D)

FIGURE 14.5   Stotting by various mammals Stotting (also called pronking) is widespread among mammals that live in relatively open habitats and are prey to coursing predators. Examples shown are (A) African springbok (*Antidorcas marsupialis*), (B) European fallow deer (*Dama dama*), (C) European roe deer (*Capreolus capreolus*), and (D) a large South American rodent, the mara (*Dolichotis patagonum*).

indices, honest signaling of condition puts more suitable prey at a disadvantage. However, once the signaling strategy is adopted by less suitable individuals, prey that fail to display become more conspicuous, and it will often pay for predators to choose them over the displaying animals. Hence it may be better to display even if this reveals the signaler's relatively lower health, agility, or endurance. Some salticid spiders mimic the shape and behavior of ants to discourage bird predators. However, other salticids preferentially feed on ants, and the ant mimics have thus evolved a separate set of postures and displays to persuade salticid predators that they are not really ants [613]. Anoles often perform pushup displays when predators appear, and the vigor and rate of these displays provides honest information to the predator about the speed and endurance of the signaler [510]. The durations and spectrographic compositions of predator calls from short-toed larks (*Calandrella rufescens*) are significantly correlated with caller health; birds weakened by infections have noticeably distinct calls that should be assessible by predators [496, 497]. Rabbits (*Oryctolagus cuniculus*) in good condition have more white on their tails than

less healthy ones; eagle-owl predators (*Bubo bubo*) use this trait to attack preferentially those in poor condition [661]. Wagtails (*Motacilla alba*) and various squirrels (Sciuridae) flag their tails when in predator alert states and tend to slow or stop flagging when feeding or grooming: predators may thus target those that are not flagging as being more vulnerable to capture [401, 693]. The stripes on zebras may serve as amplifiers to make it easier for predators to detect which individuals are running most slowly [530]. African savannah antelopes, North American deer, and the mara (*Dolichotis patagonum*), a large South American rodent, may "stot," "pronk," or "alarm walk" as they flee canid predators: this is a stiff-legged form of locomotion in which vertical motions can be nearly as pronounced as horizontal ones (**Figure 14.5**) [127, 214, 840]. The general consensus is that these displays are prey condition signals [123–127, 131, 271, 840]. Differential predator inspection may also be a prey condition signal: those prey in the best condition can afford to inspect the predator at closer range than those in worse condition and predators may take this into account in selecting whom to chase[125, 127, 131, 917].

Whether prey condition signals are indices or handicaps poses many of the same problems discussed in Chapter 10 [127, 568, 762]. Do gazelles stot differently because some can never jump as high or vigorously as others (index), or because some cannot afford to use up energy now that might be needed to flee at high speed later (handicap)? Similarly, while a direct stare at the predator is an index because only those prey who know where the predator is located can aim their stare in the correct direction, the length of time that the prey spends staring at a predator may depend upon how long it can defer additional foraging and thus be a handicapped indicator of its current energy reserves and escape potential. It is likely that many of these predator notification signals constitute a mix of both types of honesty guarantees.

APOSEMATIC SIGNALS   Aposematic signals, first introduced in Chapter 5, also signal relative suitability to potential predators. Unlike the signal types discussed above, which are emitted facultatively, aposematic signals are often transmitted continuously. And unlike crypsis, they consist of stimuli that make the animals *more* conspicuous against the background—e.g., red and orange colors against the green of vegetation or blue of water, loud and sudden sounds, or unusual and pungent odors. Aposematic signals advertise that the sender is defended by chemicals, dangerous weapons, spines, or stinging organs. Chemical defenses range from offensive tastes or smells that predators encounter immediately on attack to toxic compounds that sicken and even kill predators some time later. Many defensive chemicals of animals are secondary metabolites generated by their food sources or symbionts [518, 532, 626, 642, 655]. Typical chemicals include alkaloids, terpenoids, glycosides, phenols, and phenazines, and likely food sources include all of the other kingdoms: archaea, bacteria, protists, fungi, and plants. Just as some animals sequester and concentrate toxins in nonanimal foods, other animals further up the food chain acquire and store toxins present in their animal prey. Many toxins thus move up the food chain sequentially.

Whether a species adopts aposematic coloration when it acquires its defenses from another organism depends, in part, on the reliability of continued access to those defenses. Like other sessile taxa, sponges are highly vulnerable to predation. Most species contain a variety of toxic and distasteful compounds, some of which are generated by symbiotic bacteria [655]. Many dorid nudibranchs (gastropod molluscs without shells) specialize in eating sponges and are able to sequester many of the sponge toxins in their own tissues. Because the nudibranchs have a reliable supply of chemical defenses, nearly all dorids exhibit conspicuous coloration and patterning to warn off fish and other predators (**Figure 14.6**). Aeolid

nudibranchs specialize in eating cnidarians such as hydroids or sea anemones. They are able to ingest their prey without discharging the cnidarian stinging cells (nematocytes), and store these on their backs where they can sting attacking predators [4, 334]. Given their specialized diets, this is a reliable defense and these nudibranchs also exhibit bright and conspicuous coloration.

Two of the most toxic chemicals used in animal defense are the alkaloid nerve poisons tetrodotoxin and saxitoxin. Tetrodotoxin is usually generated by symbiotic bacteria, whereas saxitoxin is created by free-living dinoflagellates [714, 995]. Because tetrodotoxin comes from a reliable source, its use is often accompanied by striking aposematic coloration in the hosts of the bacteria [585, 953]. The dinoflagellates that generate saxitoxin are not symbiotic and often exhibit major fluctuations in annual abundance. Animals that sequester saxitoxin for defense cannot count on a steady supply of dinoflagellate prey, and thus usually do not have aposematic coloration. Sequestering the defenses of prey or symbionts imposes costs and thus trade-offs with other fitness functions [953]. For example, certain arctiid moth larvae sequester glyosides from their host plants to provide chemical defenses for both larvae and adults. Larvae feeding on plants with higher glycoside dosages do not acquire better protection, and in fact suffer higher handling costs that reduce adult fecundity and warning coloration levels [527]. Similar trade-offs are seen in other taxa [345].

Aposematic signals are widespread among animal taxa. Gorgonians (sea fans, sea whips) are cnidarians that are unpalatable to fish due to concentrated quinones and terpenoids [655]; most are brightly colored in conspicuous reds, oranges, yellows, and purples. A variety of diurnal marine flatworms (Polycladida) are conspicuously colored and where examined, are unpalatable [12, 879]. Starfish and fish with sharp or toxic spines often show starkly colored spine patterns [432, 830]. Not all aposematic signals rely on vision. Both sponges and the dorid nudibranchs that eat them exude chemicals that appear to warn off predators [655, 715]. Spiny lobsters advertise their prickly nature by stridulating when harassed [835]. Many species of puffer fish (Tetraodontidae) sequester significant amounts of tetrodotoxin from ingested prey [629]. Instead of wearing bright colors, these fish grossly inflate their stomachs with water when threatened; this behavior both serves as a warning signal and makes them harder to bite.

Similar examples can be found in terrestrial contexts (**Figure 14.7**). Insects that feed on plants often defend themselves against predators by sequestering secondary metabolites that their host plants create (in these cases, unsuccessfully) to deter herbivores. Major groups exhibiting both plant-derived chemical defenses and conspicuouos aposematic signals include Lepidoptera (butterflies and moths), beetles (Coleoptera), and true bugs (Heteroptera) [642]. Like the marine forms described above, arthropods that eat toxic arthropod prey may themselves be toxic and show aposematic coloration [232, 653]. Also like marine species, insects that are active at night may use modalities other than

(A)

(B)

(C)

(D)

(E)

(F)

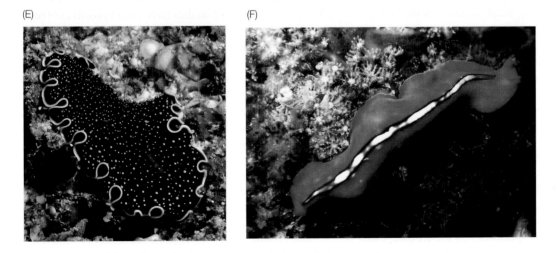

FIGURE 14.6   Aposematic coloration in tropical Pacific Ocean invertebrates Examples of nudibranchs (Mollusca): (A) *Hypselodoris apolegma*; (B) *Chromodoris leopardus*; (C) *Flabellina* sp.; and (D) *Nembrotha cristata*. Examples of flatworms (Platyhelminthes): (E) *Thysanozoon nigropapillosus* and (F) *Pseudoceros bifurcus*. Advertised defenses are obtained by storing toxins from eating sponges (A,B) or colonial ascidians (D–F), or by retaining undischarged nematocysts from hydroid or sea anemone prey (C).

(A)

(B)

(C)

(D)

FIGURE 14.7   Aposematic signals in terrestrial animals (A) The giant leopard moth (*Hypercompe scribonia*) is inactive during the day. If disturbed, it releases noxious chemicals from its thorax. Its distinctive visual markings act as a clear aposematic warning to diurnal predators. (B) Silkworm moth caterpillars (*Antheraea polyphemus*) produce audible clicking with their mandibles and regurgitate a noxious liquid when threatened. The sounds appear to serve as aposematic signals to nocturnal predators such as bats [110]. (C) The Panamanian poison dart frog (*Dendrobates pumilio*; top) and the Madagascar yellow frog (*Mantella baroni*) both display exotic aposematic warnings. These species sequester toxic alkaloids from their insect prey in their bodies [147]. (D) Two species of *Pitohui*, the hooded pitohui (*P. dichrous*, shown here) and *P. kirhocephalus*, both concentrate highly toxic alkaloids from their beetle prey in their skin and feathers. Interestingly, the two species have convergent aposematic coloration [222, 225].

vision for aposematic signals. Toxic artiid moths produce ultrasonic sounds to warn off attacking bats, and a variety of unpalatable caterpillars produce aposematic sound signals [110, 118, 425, 426, 701]. Most members of the Hymenoptera defend themselves with venomous stings or spray offensive chemicals such as alkaloids [512]; while diurnal bees and wasps host bright aposematic coloration, less visible ants vibrate the substrate as an aposematic signal [465, 699, 737, 744].

Amphibians are notorious for hosting chemicals that deter predators; most terrestrial species with such defenses are conspicuously colored [188, 189, 373, 898]. At least 28 species of frogs, toads, and salamanders are protected by tetrodotoxin or its analogs. Dendrobatid and mantellid frogs are protected by other alkaloids obtained from arthropod prey (mostly ants, beetles, mites, and millipedes) [147, 192, 868]. While it is not clear whether amphibians collect or synthesize tetrodotoxins, many species clearly secrete their own amine, peptide, protein, or steroid defensive chemicals [189, 373]. Poisonous snakes warn off predators using conspicuous markings (e.g., coral and keelback snakes), rattling of the tail (vipers), or erecting the upper body and expanding the neck (cobras) [310, 332, 429, 430, 627]. Many birds eat insects that have chemical defenses. These compounds may

accumulate in the birds' tissues and in the secretions of the uropygial gland, where they may be modified by bacterial action and later preened over the entire plumage [48, 365, 557, 734, 943]. Several studies have found positive correlations between the conspicuousness of bird plumages and the unpalatability of the birds as assessed by human researchers [169, 170, 324]. Birds in the New Guinea genera *Pitohui* and *Ifrita* concentrate the alkaloid toxins of their beetle prey and are themselves highly toxic to predators. All species in these genera are conspicuously colored [218-225, 454]. Among mammals, members of the families Mustelidae, Viverridae, and Herpestidae often have glands producing extremely malodorous secretions; many have distinctive black and white markings, although the correlation with defensive chemicals is weak [129, 617, 643, 644]. Porcupines, hedgehogs, and tenrecs advertise their spiny defenses with sounds, visual markings, or odors [127].

The evolutionary economics of aposematic signals are quite complicated [728]. Most aposematic signals are conventional in form: they have no direct dependence on the traits that they advertise [356]. Predator receivers must attack some of the aposematic prey before they can learn to associate the signal with the prey's unpalatability or other defenses. While other prey benefit from this training of the predators, it is not

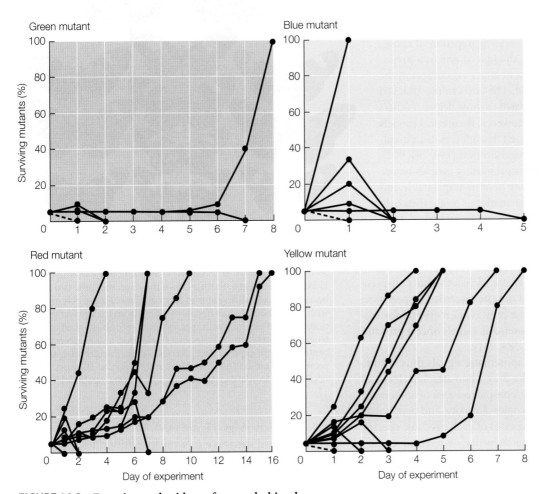

FIGURE 14.8 **Experimental evidence for neophobic advantages of aposematic mutants** During an initial training period, the daily food of wild-caught European robins (*Erithacus rubecula*) was augmented for a 2 hour period with 20 brown-colored pastry "prey." On the first experimental day, one "mutant" pastry prey of a different color (red, yellow, blue, or green) replaced one of the brown ones. On the next and successive days, 20 prey were again provided, but now in the ratio of colors seen at the end of the prior day's experiment. Over a 1–2 week period, the brightly colored prey either went "extinct" or went to 100% of the prey population (fixation). Graphs show the percentages of brightly colored prey over time for mutants of each color. The robins took significantly longer to sample the first red and yellow mutants than to try the blue or green ones. Blue and green mutants that went extinct did so much faster than did red or yellow mutants, and the latter went to fixation much more often than blue and green mutants. Since the pastries were fully palatable, these effects suggest dietary neophobia in the robins, which show stronger biases against red and yellow than against green and blue prey. Red and yellow are by far the most common colors seen in aposematic animals. (After [893].)

clear what benefits might compensate for the costs suffered by the attacked prey, particularly if prey who benefit are not close kin of the victims. If the aposematic signals are conspicuous, naïve predators will be more likely to detect and attack them, and while such prey remain rare, there will continue to be more naïve predators than experienced ones and attack rates will remain high. A mutant that is aposematic must thus reach

some threshold abundance before the benefits can begin to outweigh the costs [681]. Even after reaching threshold abundances, some aposematic prey will be attacked as each new generation of predators learns the conventions by attacking the prey. This recurrent "testing" of the aposematic signals is one force that keeps these conventions honest [357].

A great deal of theoretical effort has gone into understanding how aposematic signals might get through the disadvantageous rare stage [552, 728]. One contributing factor may be limited prey mortality. If prey are not badly injured or killed, but dropped as soon as the predator detects that they are distasteful or armed, mutant aposematic forms could steadily increase past the threshold [101, 238, 807, 809, 951]. Prey with externally located bitter tastes or rank odors are more likely to survive attacks than those with internally stored toxic compounds that affect predators later. A second factor is that many predators are neophobic and avoid novel prey, or are hesitant to switch from a familiar prey (**Figure 14.8**) [513, 555, 556, 893]. In some cases, these biases might be innate, whereas in others, they must be learned [166, 248, 721, 808]. One possible reason to avoid conspicuous novel prey is that they should not be encountered frequently unless they were toxic or defended; hence it may be prudent to avoid them [777]. Any such preexisting predator biases might allow mutant aposematic forms to accumulate until they reach the threshold. A third factor facilitating mutant proliferation past the threshold might be social aggregation: if aposematic prey cluster spatially, the loss of one to a predator might minimize subsequent attacks on others in the group by that predator [391]. The costs to the attacked individual might be compensated to some degree if others in the group are close kin. However, theoretical models and relevant experiments suggest that aposematism usually evolves before, not with, gregariousness [790, 791, 902]. A fourth possibility is that unpalatibility evolves before aposematic signals. Once prey became unpalatable and common, some distinguishing markings might then help reduce attack rates [370]. Initially cryptic and unpalatable species might have to acquire aposematic signals if they subsequently invade more conspicuous habitats [579, 831]. Finally, early stages in the evolution of aposematic signals might be favored by intermediate levels of signal conspicuousness: markings that could only be discerned by predators at close range would help the prey remain cryptic for distant predators, but act as aposematic signals once a predator approached the prey [73, 86, 384, 731].

Evolutionary models of aposematic signals thus need to combine mortality risk to the prey when attacked, levels of prey toxicity or unpalatibility, conspicuousness of aposematic signals, predator density, predator diversity, and neophobia, learning rates, and forgetting rates of each predator species. Not surprisingly, these models show that only certain combinations of the included variables allow for the successful evolution of aposematic signals [355, 384, 514, 728, 767, 829]. However, there are multiple combinations that are sufficient because values in one variable that hinder aposematic evolution may well be compensated by values of other variables that favor it. Therefore, what initially looked like a difficult type of signal to evolve can in fact evolve by a number of routes.

Predators often have little time in which to decide whether a prey is palatable or not. The most expedient, and widely confirmed, solution is a quick categorization of the prey based on a few salient stimuli. Put another way, predators generalize and classify prey into categories [299]. Where generalization rules are sufficiently broad, different aposematic prey species may benefit from each other's abundances by converging on a common set of warning signals. This is usually envisioned as a two-step process in which an initial mutation changes one element in the aposematic signal of one unpalatable species sufficiently that a common predator now assigns it to the same category to which it assigns another unpalatable species [41, 455, 728, 778, 905, 906]. The two species benefit by being treated similarly by the predator, since their combined abundances increase the speed at which naïve predators learn the shared signals. These advantages favor successive mutations in one or both aposematic species that modify other elements in their warning signals until they are increasingly identical. Unpalatable or otherwise unsuitable prey species that show such convergence in their aposematic signals are called **Müllerian mimics**. Often, more than two different species converge on common aposematic signals; these groups of species are

FIGURE 14.9 Müllerian mimicry ring in neotropical Lepidoptera Top row shows the convergence between *Heliconius melpomene* (left) and *H. erato* (right) at a common site on the upper Huallaga River, Peru. Farther down the river, the same two species become part of a different mimicry ring (lower three columns) with marked differences in color pattern from the upriver populations but convergence with species in the second site. Left column from top to bottom: *H. melpoene, H. elevatus,* and *H. demeter;* center column: *Laparus doris, Neruda aoede, Eueides tales,* and a periocopine moth (*Dysschema* sp.); right column: *H. erato, H. burneyi,* and *H. xanthocles.* All of these species are unpalatable.

called **mimicry rings** (**Figure 14.9**). They may consist of many closely related species (e.g., sympatric butterflies in the genus *Heliconius*) or a diverse array of taxa (e.g., one classic mimicry ring contains similar beetles, wasps, moths, true bugs, and flies) [528, 545, 672]. Some species of toxic New Guinea birds in the genus *Pitohui* also appear to be a convergent Müllerian ring [222, 224].

It is also possible for palatable species to mimic the aposematic signals of unpalatable species. This is called **Batesian mimicry**, and aposematic species that are mimicked are called the **models** for the emulation. In addition to exhibiting the markings and shapes of their models, Batesian mimics may also imitate the models' behaviors: droneflies colored like honeybees mimic the hovering flight of their bee models; some octopuses color themselves like poisonous sea snakes and lionfish and move in similar fashions; and salticid spiders that mimic ants emulate moving antennae with special leg movements [134, 317, 613, 631]. Burrowing owls imitate the rattling of rattlesnakes when defending their nests against predators [648, 722].

Whereas Müllerian rings benefit by adding new members, abundant Batesian mimics undermine the aposematic effectiveness of their models because naïve predators will frequently encounter palatable mimics sporting the warning signals. This generates selection on the aposematic models to distinguish themselves from the mimics, and on the predators to find ways to discriminate between the two [278, 904]. An abundance of Batesian mimics also works against itself: there may be selection within such a species to mimic multiple models (e.g., evolve polymorphism) [912]. The situation is complicated by mimics that are only slightly unpalatable, or

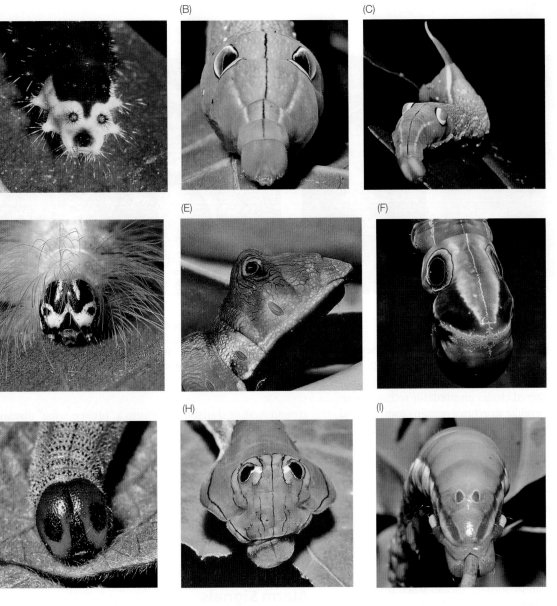

(A) (B) (C)
(D) (E) (F)
(G) (H) (I)

FIGURE 14.10   Eyespots on nine species of caterpillars from the Area Conservación Guanacaste, Costa Rica The convergence of eyespots among these and the much larger survey of taxonomically unrelated larval Lepidoptera at this site and the augmentation of surrounding structures into facelike structures support the idea that eyespots exploit an innate fear of paired eyes by avian predators which are themselves vulnerable to cryptically hidden snakes, lizards, and small mammals [438].

unpalatable only to certain predators, when compared to their models: it is not clear whether these should be called Müllerian or Batesian mimics [102]. Müllerian rings are often accompanied by numerous Batesian mimics, making the entire ring a complex mixture of the two processes. A plethora of evolutionary models have been put forward to explain various aspects of Müllerian and Batesian mimicry (using every type of modeling approach outlined in Chapter 9; see excellent summary in [728]). Key factors that are predicted to affect the number and diversity of mimicry rings include the number of relevant predator species, the generalization functions for each predator, the absolute abundances of models and mimics, and the abundances of alternative prey for the predators. Batesian mimics can be imperfect imitations of their models as long as the model is abundant, particularly well defended, or both; the special case of Batesian polymorphism seems to require particular preadaptations in the relevant genetic architecture [136–138, 625, 728].

Last-ditch prey signals to predators

Once a predator initiates an attack, a prey has fewer options. One obvious strategy is to flee if possible. Many cephalopods hide their departure direction in a cloud of black ink [968]. When flight is no longer feasible, prey may produce a **deimatic display**—a loud sound or a sudden exposure of visual markings that startles the predator and interrupts its attack [228, 480, 728]. Cephalopods, fish, and a variety of insects, especially larval and adult Lepidoptera, exhibit eyespots to nearby predators (**Figure 14.10**). Depending on the species, these may be permanently on display, or suddenly exposed when a predator is sufficiently close. Eyespots do

(A)

(B)

FIGURE 14.11 **Vertebrate decoys** (A) Male killdeer (*Charadrius vociferus*) performing a broken-wing display to decoy a predator away from its nest. (B) Skink (*Plestiodon septentrionalis*) next to the recently autotomized tip of its tail.

cause certain predators to hesitate or withdraw, and they can reduce predation rates on their bearers [483, 500, 728, 908, 909]. Whether these effects are due to simple novelty, pattern contrast of dark-centered circles on predator retinas, or an innate fear of paired eyelike structures remains controversial [71, 438, 742, 743, 849, 853, 854].

Various marine organisms, insects, amphibians, birds, and mammals produce **distress signals** when physically contacted by a predator. Acoustic distress signals are invariably loud and harsh with wide bandwidths, many harmonics, and often chaotic components that propagate well and whose source is easily localized [34, 35, 562, 918]. In some species, such signals warn nearby kin. They may also attract other nearby prey that then mob the predator, actions we will review in the section on **victim calls**. However, distress signals are often given by solitary species when no kin are present and no mobbing is likely. Such signals are clearly directed at the immediate or other more distant predators. Several studies have examined the distribution of distress calls in birds and concluded that they function largely, albeit not very effectively, to startle the predator [158, 339, 615, 962]. An alternative, and not exclusive, possibility is that distress signals alert other predators who might interfere with the attack or even attack the predator, thus giving the prey a chance to escape [315]. Some distress signals are visual; one example (often likened to a burglar alarm) is the flashing by bioluminescent dinoflagellates when they are attacked by nocturnal crustaceans. The light flashes attract larger fish predators, which then hunt the crustaceans [273, 362, 575]. Distress calls may also serve as signals of condition; a strongly delivered signal might cause a predator to evaluate a prey as being too much trouble: short-toed larks (*Calandrella rufescens*) that are healthy give longer and harsher distress calls than do diseased birds [496, 497].

Another last-minute prey strategy is the deployment of a decoy that redirects an attacking predator's attention away from the immediate prey (**Figure 14.11**). Several species of

ground-nesting shorebirds perform a "broken-wing" display to distract an approaching predator away from their nests [17, 18, 314]. A more radical decoy is an autotomized body part whose continued twitching distracts the predator while the rest of the prey escapes (see Chapter 9). Bright colors on some lizards' tails appear to serve as amplifiers to hold a predator's attention after tail autotomy [161, 162]. Last-ditch signals have received only spotty field or laboratory attention, and at this point, the assignment of functions should be considered tentative [728].

Alarm Signals

Like predator deterrent signals, alarm signals can be emitted at any of the five stages of a predator/prey interaction. However, the two types of signals differ in the identity of the primary receivers and in the relevant economics. The intended receivers of predator deterrent signals are predators, whereas the intended receivers of alarm signals are fellow prey. In both cases, the sender pays some costs to generate the signals. If a predator deterrent signal is honest and accepted by the predator, both parties can benefit. The key question is what guarantees sender honesty: as we saw, some predator notifications are indices (snorts followed by staring at the predator); others are handicaps (stotting); and still others are conventional signals periodically tested for honesty by predators (aposematic signals).

While nearby prey usually benefit from receiving alarm calls, the sender must also gain some benefit to justify its signaling costs. Each of the classical benefits that might explain what otherwise might seem an altruistic act (see Chapters 9 and 13) has been invoked for some alarm signal. When a school of fish, flock of birds, or herd of antelope flees from a predator, the predator has to select which one to attack. If all fleeing prey act and look exactly the same, each individual's risk is simply chance (here equal to the reciprocal of the group size). Being in a group "dilutes" risk to any one individual and the larger the group, the more diluted that risk [372]. However, if a member of a group acts differently from the rest, that individual may be singled out for attack by the predator. Coordination is critical to preserve the dilution effect of group flight, and alarm signals can ensure such coordination. In this case, alarm signal coordination benefits both the sender (which does not want to flee by itself) and all receivers (which do not want to be left behind). It thus can be explained as mutualism. Where a threatened prey is going to emit a given signal anyway, and nearby conspecifics eavesdrop on that signal, the economics can be interpreted as by-product mutualism. Predator detection signals benefit their senders directly, but any nearby prey will also benefit by attending to the signal. This may in fact be a common starting point for the evolution of alarm signals [773], and is a likely explanation for foot thumping in kangaroos and related macropods [60, 61, 718]. Another example is the escape response of cephalopods and sea hares (*Aplysia* spp.) to inks released by a fleeing conspecific [469, 968]. Alarm signals in some taxa are preferentially given in the presence of close kin; indirect nepotistic benefits may compensate for sender costs in these cases [390, 775]. Where groups are relatively stable, group members may alternate in the role of alarm signaller, and signaling may then be a form of reciprocity [900]. Alarm signaling is disproportionately undertaken by males in some species: this may in part be explained by consequent mating advantages and sexual selection [291, 959, 982]. Some authors have suggested that undertaking sentinel and alarm sender roles provides individuals in a group with prestige and status that pays off in other contexts [989, 990]. Finally, alarm senders may manipulate fellow prey, either by frightening them off from new food finds or by causing them to take flight at times or in directions that make them, rather than the sender, the most likely victims of a predator [139, 711]. There is considerable debate about which, if any, of the above benefits is the more widespread justification for alarm signaling [127, 151, 216, 1005]. Below, we review a wide range of animal alarm signals and indicate likely or argued compensatory benefits to senders where data are available. We then examine interesting examples of interspecific coalitions, the effects of false alarms, and the processes by which receivers might learn the more complex alarm coding schemes. It will be useful to categorize alarm signals into three classes, roughly based on the temporal sequencing of a typical predator/prey interaction. Surveillance signals are usually given before a predator makes contact with prey, and elicit receiver flight, entry into a refuge, or increased crypsis. Mobbing signals are given in similar contexts, but the receiver response is to aggregate conspicuously and harass, threaten, or attack the predator. Victim signals are alarms that are produced only after a predator has physically contacted a prey. We take up each category in turn.

Surveillance signals

BASIC ECONOMICS   Species differ in their use of **surveillance signals** along two independent axes. The first is the degree to which the members of a group act as **sentinels**, setting aside other activities to watch for predators and produce alarm signals when threats are detected (**Figure 14.12**). Differences between individuals in their assumption of sentinel roles can vary discretely (e.g., only certain subclasses take on this role) or continuously, in which the fraction of time spent on guard duty varies quantitatively with sex, age, status, recent fecundity, recent foraging success, or other factors.

In the simplest case, all members of the group are, in principle, equally likely to detect predators and emit alarm signals. The usual context occurs when all are engaged in a common activity, such as foraging, and each member must trade off the time spent in foraging with time spent in

(A)

(B)

FIGURE 14.12   Alternative strategies for predator surveillance (A) Herd of impalas (*Aepyceros melampus*) in which neither a permanent nor a rotating division of labor is present. Any member can stop foraging to look for predators, and usually when one does, many others do so also to avoid being uninformed if the herd starts to flee. (B) Group of banded mongooses (*Mungos mungo*) in which members take turns as sentinels, allowing the others to concentrate on foraging, or in this case, moving between feeding sites.

FIGURE 14.13 Simulations of optimal vigilance in populations without assigned sentinels All examples assume a fitness trade-off between foraging and predator vigilance. Each animal is assumed to have 12 relevant neighbors. The blue line in the left-hand graphs indicates the optimal fraction of time that an average animal should be vigilant (vertical axis) as a function of the number of neighbors that are also currently vigilant (horizontal axis). Stable equilibria (red circles and red lines) and unstable equilibria (blue circles) occur whenever this function is intersected by the dashed line indicating the current number of group members that are also vigilant. (A) If increased vigilance by neighbors significantly decreases a focal animal's risk, and foragers are no more likely than vigilant members to be taken during a predator attack, the optimal time the focal animal should spend in vigilance decreases and then levels off as the number of vigilant neighbors increases. The only equilibrium is stable and occurs at an intermediate value of individual vigilance. A typical aerial snapshot (right) of which animals are vigilant (black) and which are foraging (background color) shows little clumping of vigilant animals. (B) If increased vigilance by neighbors again reduces a focal animal's risk, but foragers are more at risk than vigilant animals, the optimal vigilance first decreases as the number of vigilant neighbors increases, and then increases again because it becomes increasingly risky to forage if nearly all of your neighbors are vigilant. There are two stable equilibria: one at a low level of individual vigilance when few neighbors are vigilant, and one at a maximal level when all the neighbors are vigilant. The aerial snapshot thus shows large clumps of neighbors that are all vigilant at the same time, and intervening pockets where all neighboring animals have low vigilance. (C) If increased vigilance by neighbors does not reduce a focal animal's risk, and foragers are more at risk than vigilant animals, the optimal individual vigilance is similar to the fraction of vigilant neighbors. The equilibrium occurs when both the focal animal and the neighbors adopt a low level of vigilance. Clumping of vigilant individuals is then intermediate between (A) and (B) [806].

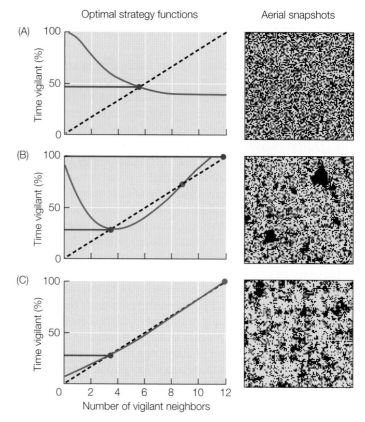

predator surveillance [422, 520]. Although early models of the optimal trade-offs between foraging and vigilance assumed independent scanning for predators by group members [678], subsequent studies have shown that the timing and duration of predator surveys are often not independent: whether a given animal stops to survey may depend on how recently it last did so or how much food it has recently eaten, and a survey by one member will often trigger concurrent surveys by other group members [53, 55, 255, 656–659].

Whereas coordinated flight ensures equal risk for all prey in the group, delayed flight by a subset allows a predator to simplify its target choice by focusing on that subset. Thus it may pay for prey to initiate a predator survey when they detect that adjacent prey are doing so; this ensures that they will be as informed as their neighbors and thus be able to flee at the same time if necessary [176, 523, 806]. The result is clumped instead of randomly timed surveillance in many species (**Figure 14.13**). Individuals also may differ in their optimal trade-off between time spent foraging and time spent in predator surveys: effective foragers may be able to watch more, heavier prey may need to compensate for slower flight should a predator appear, and personality differences may generate persistent differences in surveillance tendencies

[177, 450, 451, 696]. Even when there are no assigned sentinel roles, which individuals in a group detect predators and produce alarms will rarely be random.

Heterogeneities can also arise during flight from a predator. The coordination of a fleeing group may be achieved by signals as simple as a change in heading or velocity by a member of a fleeing group [36, 40, 116, 172, 651, 897]. When one member changes its trajectory or velocity, simple rules about near neighbor spacing may cause this change to propagate to all in the group (see Chapter 15 for details). However, it is often noted that "leaders" tend to be individuals that have more experience and less to lose by going their own way; their changes in direction are likely to be followed by other group members [159, 173, 269, 609]. This further contributes to heterogeneities in who surveys, who starts flight, and who leads once flight is initiated.

In other species and contexts, predator surveillance is unequally distributed among group members, raising the question of what compensates such sentinel duty. As with ostensibly homogeneous groups, it is not always clear whether alarm signals are not also predator deterrent signals and thus possibly justified by a direct benefit to the caller. However, other factors may contribute to caller benefits. In Siberian jays (*Perisoreus infaustus*) and some semicolonial and colonial squirrels, adult females are most likely to produce alarm signals if they have dependent young or close kin nearby [335, 419, 536, 760, 775, 776]. For most altricial birds, alarm calling by the parents causes nestlings to become silent and cryptic [251, 363]. While this suggests nepotistic benefits to alarm signaling, the evidence for nepostism beyond immediate offspring is not compelling: sample sizes in studies differentiating levels of kinship are often small, and preferential

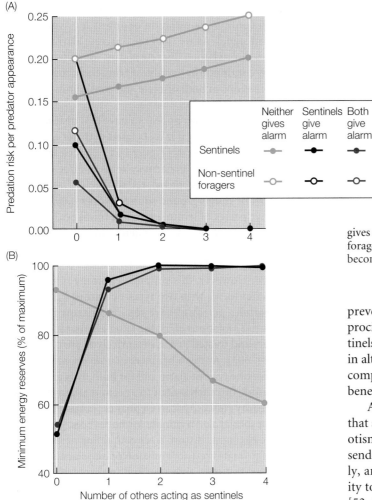

(A)

(B)

Number of others acting as sentinels

FIGURE 14.14 Dynamic game models of optimal sentinel activity (A) The model assumes a group of five individuals, some of whom are sentinels and some foragers. Sentinels (closed circles) have consistently lower risk than foragers (open circles) because they detect predators and respond early, and predators tend to favor the slower responders. If nobody gives alarm calls, increasing the number of sentinels increases the risk for everyone since there will be fewer slow members to distract the predator. If individuals give alarm calls, risk decreases as the number of sentinels increases, and the more animals that can give alarms, the greater the risk reduction. (B) An animal must balance the time it spends acting as sentinel with the time it needs to forage. This model predicts the minimum threshold energy reserves that an individual should have before it pays to act as a sentinel. When nobody gives alarms, the threshold for acting as a sentinel decreases, since foragers become increasingly at risk as more group members become sentinels [52].

prevented from giving alarm calls, suggesting possible reciprocity among males [641]. Although many species with sentinels show rotations of guard duty, tabulations of time spent in alternative roles argue against reciprocity as a widespread compensation [151, 152]. Other studies have excluded status benefits as likely compensations [975].

A dynamic game analysis of typical examples suggests that sentinel behavior can be favored without invoking nepotism, sexual selection, status modulation, or reciprocity if senders can reduce costs and risks of sentinel duty sufficiently, and there are selfish antipredator benefits of this activity to senders that exceed the reduced costs (**Figure 14.14**) [52, 54]. Minimization of sender costs and risks appears to be widespread in sentinel species [480]. Meerkats (*Suricata suricatta*) that have recently fed to satiation, and thus have nothing else to do, are the most likely individuals to take over a sentinel role when another abandons its station to forage [152]. Individual Florida scrub jays (*Aphelocoma coerulescens*) and Arabian babblers (*Turdoides squamiceps*) are more likely to become sentinels if experimentally provisioned with food, and nonsupplemented individuals are then more likely to give up being sentinels [56, 976]. Willow tits (*Parus montanus*) give alarms only when hawks are still far enough away for birds to get to cover [7]; starlings (*Sturnus vulgaris*) are most likely to give alarms when their escape path is over thick foliage [152]; and meerkat sentinels usually take up stations near to burrow holes [204]. Great gerbils (*Rhombomys opimus*) also give their alarm signals at the edge of their burrows as long as the relevant predator is too large to fit through the entry hole [690]. A variety of prey species produce alarm signals that are difficult for predators to detect or localize at a distance: ground squirrels (*Spermophilus spp.*) and rats (*Rattus norvegicus*) produce ultrasonic alarm calls that hawks cannot hear [231, 529, 958]; rats produce short-range alarm pheromones [431, 479]; and small birds produce alarm calls whose sources are difficult for approaching hawks to localize [453, 458, 481, 554, 969]. Some birds can vary the shape of

signaling when kin are present usually applies only to terrestrial predators and not to aerial predators. In addition, studies on rodents such as marmots have found no support for alarm signaling to kin other than current offspring [80]. Evidence in support of nepotistic benefits in other taxa is also mixed. Adult males are the most active sentinels in family groups of Siberian jays (*Perisoreus infaustus*), groups of capybaras (*Hydrochaeris hydrochaeris*), and troops of guenons, langurs, and mangabey monkeys [16, 39, 336, 947, 948, 979, 998]. These males have usually fathered offspring currently in the group. The loud alarms of gibbons may benefit recent offspring that have settled in territories near their parents [887]. In none of these cases is it clear that nepotistic benefits of alarm signaling extend beyond protecting current or recent offspring.

Some taxa show sentinel behavior that is very unlikely to be compensated by nepostistic benefits. The alarm signaling of roosters is closely linked to their mating strategies [959]. Male red-winged blackbirds (*Agelaius phoeniceus*) with adjacent nesting territories take turns as sentinels [120, 982]. Studies found no effects of kinship with male neighbors or extra-pair paternities in adjacent territories; however, males stopped sharing with those neighbors experimentally

their alarm call sound fields: narrow beams can be aimed at predators as a predator notification, whereas omnidirectional beams are most effective at warning other nearby prey [654, 986].

The potential benefits to serving as sentinel are the same ones that favor coordinated predator surveys in groups without explicit sentinel roles: sentinels are more likely than nonsentinels to know where a predator is located, initiate escape from an attack sooner, and know which direction to flee [52, 806]. The benefits to the sentinel of giving an alarm signal as it flees are also the same as for nonsentinel species: coordinated flight reduces risk and giving an alarm signal is the best way to ensure simultaneous group action. If it is so beneficial to be a sentinel, why don't all species do it? Two conditions appear to be necessary before nonsentinels give up their own predator surveys to concentrate on foraging. The first is that sentinels must be able to detect predators at a sufficient distance that nonsentinel foragers have sufficient time after an alarm to participate in any coordinated response. This may be one reason why sentinel behavior is most common in species that live high in trees or if terrestrial, in open habitats [52]. The second condition is that foraging efficiency is enhanced by not having to do surveillance. This appears to be the case in most sentinel species. Nonreproductive helpers in cooperatively breeding jays and other corvids release the breeders from predator surveillance and increase their foraging time [368]. Sentinels in a variety of species emit regular signals while on duty: these include specific sounds or body movements such as tail flicking [235, 401, 415, 697, 698, 949]. Sentinel sounds are positively correlated with the foraging efficiency of nonsentinels in meerkats and pied babblers (*Turdoides bicolor*); the "all clear call" given by Diana monkeys (*Cercopithecus diana*) sentinels may serve a similar function [415, 548, 907]. In pied babbler groups, the higher the post taken by the sentinel, and thus the longer its range of view, the more intensive the foraging by other group members [683].

AMOUNT OF INFORMATION   The second axis that can be used to classify surveillance signals concerns the amount of information that such signals provide. Again, signal variation can be discrete, continuous, or both. Some species have discrete alarm signals for two or more types of predators. This has been of great interest to researchers studying the evolution of human language, because such animal alarm calls can be considered **referential**: like human words, they identify different classes of objects [142, 210, 244, 270, 397, 434, 665]. However, it has been argued that there is little adaptive advantage to naming predators per se, and that the number of alarm signals actually reflects the number of escape strategies required [210, 535]. If the best response to one class of predators is to fly into the tree canopy and the best response to another is to hide in low bushes, two different alarm calls might be highly adaptive; where there is only one suitable response for all predators, a single type of alarm should suffice. Even when there is only a single escape strategy, it may be adaptive to be able to indicate urgency: the

level of urgency could be encoded in several discrete alarms, or through continuous variation of one or more properties in the alarm signal. As we shall see, discrete alarms for different types of predators, discrete alarms for different levels of urgency, continuously variable alarms encoding urgency, and combinations of both discrete and continuous variation have all been reported.

A wide variety of taxa appear to have two distinct alarm signals: usually one signal is associated with aerial and the other with terrestrial threats (**Figure 14.15**). Species that have both aerial and terrestrial alarm systems include chickens, trumpeters (*Psophia leucoptera*), red squirrels (*Tamiasciurus hudsonicus*), Belding's ground squirrels (*Spermophilus beldingi*), many species of lemurs, tamarin monkeys (*Saguinus fuscicollis and S. mystax*), many guenon monkeys (*Cercopithecus* spp.), and capuchin monkeys (*Cebus capucinus*) [207, 261, 264, 293, 294, 331, 360, 476, 640, 650, 763, 776, 998, 1002, 1003]. However, this simple dichotomy is not found in many other forms with two types of alarms. Yellow warblers (*Dendroica petechia*) have separate alarms for hawks and cowbirds (*Molothrus ater*); the latter are nest parasites of the warblers and treated just as defensively as are predacious hawks [304, 305]. Arabian babblers (*Turdoides squamiceps*) initially give one type of alarm call for both owls and terrestrial predators, but then switch to a second alarm if the owl remains nearby [608]. African elephants (*Loxodonta africanus*) have a special alarm rumble for honeybees, which sting them if the elephants disturb trees containing nests [473]. A number of primates restrict emission of one alarm to predator presence, but may give a second in response to a wide variety of environmental or social disturbances [259, 263, 846]. Some monkeys with multiple discrete alarms mix them in sequences whose syntax may provide additional information (**Figure 14.16**) [19–22, 645, 646, 740, 1005]. Although the most common multiple repertoire appears to be two discrete signals, vervet monkeys (*Cercopithecus aethiops*) produce five distinct alarm signals corresponding to as many types of predators and suitable responses [768–770, 864].

Independently of the number of predator types, most species also vary their alarm calling to provide urgency information. This may involve different discrete signals, continuous variation in a single alarm type, or mixtures of the two. The probability that a hen (chicken) will emit an aerial alarm call and her rate of calling depend on the size of the hawk relative to the size of her chicks [649]. The whistle produced when doves and pigeons take flight may function as a variable signal encoding sender urgency [402]. Superb fairy wrens (*Malurus cyaneus*) and white-browed scrubwrens (*Sericornis frontalis*) encode proximity and urgency of predator approach by adding additional elements to each alarm call and raising the maximal frequencies [249, 511]. Chickadees (*Poecile* spp.) and tufted titmice (*Baeolophus bicolor*) carry this one step further by varying the number of different syllable types in their alarm calls to encode and transmit a diversity of information about predator proximity, type, and risk [284, 285, 787, 827, 885].

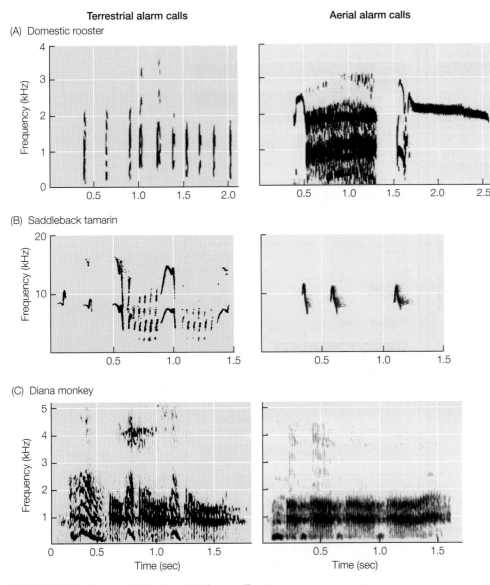

Terrestrial alarm calls   **Aerial alarm calls**

(A) Domestic rooster

(B) Saddleback tamarin

(C) Diana monkey

FIGURE 14.15   Terrestrial and aerial alarm calls Spectrograms of alarm calls of (A) domestic roosters (*Gallus domesticus*); (B) saddleback tamarins (*Saguinus fuscicollis*); and (C) Diana monkeys (*Cercopithecus diana*). (After [360, 476, 1002].)

Burrowing rodents and mongooses typically vary alarm call properties to encode urgency. In marmots (*Marmota* spp.), urgency and predator proximity are the only conditions causing signal variation and no additional information is provided about predator type [77–79, 81]. Great gerbils (*Rhombomys opimus*) produce a variably rhythmic alarm call that can encode both predator types and urgency [691, 692]. Columbian (*Spermophilus columbianus*) and Richardson (*S. richardsonii*) ground squirrels signal urgency through the degree of repetition of the same call variant or similar variants, and through the rate at which successive calls are emitted [536, 812, 934]. Perhaps the most complex rodent alarm system occurs in two species of prairie dogs (*Cynomys ludovicianus* and *C. gunnisoni*) which emit a wide variety of

alarm calls differing subtly in structure, and eliciting different receiver responses, according to equally subtle differences in predator proximity, body shape, color, and size [3, 281, 477, 668, 813, 815, 816]. Whereas banded mongooses (*Mungos mungo*) and yellow mongooses (*Cynictis penicillata*) produce several distinct alarm signals, these are largely associated with urgency and not with predator type [287, 506]. In contrast, the closely related meerkats (*Suricata suricatta*) produce a complex lexicon of discrete and graded alarm signals that encode both predator type and urgency (**Figure 14.17**; also see Figure 8.16) [549–551]. Burrowing squirrels (*Xerus inauris*) sympatric with the meerkats face the same predators, but encode only urgency in their alarm calls [288]. Primates show a range of alarm signal variation similar to that seen in rodents and mongooses. Red-fronted lemurs (*Eulemur fulvus rufus*) and squirrel monkeys (*Saimiri sciureus*) both vary acoustic parameters in their calls continuously to encode urgency, but provide little information about predator type [258, 260]. In contrast, baboons vary calls continuously along

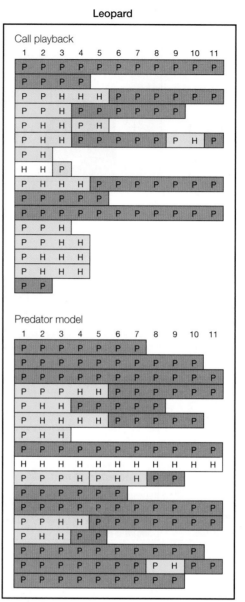

FIGURE 14.16 Alarm calls of putty-faced monkeys (*Cercopithecus nictitans*) These monkeys produce two kinds of alarm calls: hacks (H) and pyows (P). When predators are detected, alarm callers produce a long string of calls [21]. Plots show the first 11 calls given in multiple trials for different combinations of predator (crowned eagle or leopard), and modality (audio playback of the predator's call in top sections or presentation of a lifelike model of the predator in lower sections). Monkeys produce both types of calls for both predators and either kind of presentation, but more hacks are given for eagles and more pyows for leopards. Dark green indicates long strings of pyows and light green indicates mixed sequences. Additional experiments suggest that the syntax of the hack-pyow combinations provides additional information about the monkey troop's likely next movements [19, 1005].

multiple acoustic parameters, and the same graded series may be used in both social conflict and predator confrontations; subtly different versions can refer to different contexts including different types of predators [266-268, 478].

In addition to predator type and situation urgency, alarm signals often include information to identify the signaler.

Significant differences within alarm call types related to caller identity have been shown in chickens, crows, great gerbils, marmots, ground squirrels, meerkats, tamarins, and guenon monkeys [49, 82, 83, 141, 571, 675, 692, 741, 833, 878, 985]. Similar information is surely present, but unstudied, in many more of the species that produce alarm signals. Several explanations for the adaptive utility of individual identity information have been put forward. One possible benefit is that it allows receivers to discount alarms from inexperienced juveniles or known false alarmers. However, in yellow-bellied marmots (*Marmota flaviventris*), juvenile alarms elicit stronger responses than those of adults, and in Richardson ground squirrels (*Spermophilus richardsonii*) and meerkats, the age and identity of alarm callers are completely ignored by receivers [82, 741, 878]. A more likely benefit is that receivers can quickly focus on the sentinel that has given an alarm, see

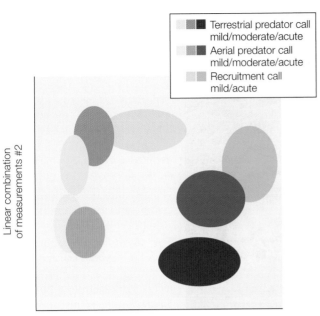

(A)

(B)

FIGURE 14.17 **Variations in meerkat (*Suricata suricatta*) alarm calls** Meerkats have separate alarm call types for large terrestrial predators such as mammals (red), aerial predators such as hawks and eagles (blue), and smaller terrestrial threats such as snakes that generate recruitment around the caller instead of flight (green). Urgency is indicated as mild (light colors), moderate (medium colors), or acute (dark colors). Data are based on 28 different acoustic measures which were then combined into two orthogonal linear combinations that best separated the 254 calls into predator-urgency classes (a method called discriminant function analysis). Ellipses show one standard deviation in each direction around the mean for each class of calls. (After [551].)

which way it is looking, and then infer the location of the predator.

Mobbing signals

Mobbing signals differ little from surveillance signals except in the response of receivers: mobbing signals attract prey which then inspect, harass, or attack the advertised predator (**Figure 14.18**). Unlike surveillance signals that are designed to obscure the location of the caller, mobbing signals make the location of the signaler easy to determine [480]. For example, "barking," with its sudden onsets and offsets and wideband frequency content, is a widely used and easily localized mobbing signal [533]. Like surveillance signals, mobbing signals may function in part as predator deterrents: they notify predators that they have been spotted and elicit predator inspections before initiating predator harassment [182, 183, 945]. Mobbing may also be induced by the distress signals of captured prey. In some species, distress signals by a member of one prey species will attract other species of prey which

FIGURE 14.18 **Examples of mobbing** (A) Starlings (*Sturnus vulgaris*) mobbing an osprey (*Pandion haliaetus*). (B) European bee-eaters (*Merops apiaster*) mobbing a ladder snake (*Elaphe scalaris*) that is trying to enter their nesting holes in a cliff face. (C) Spotted hyenas (*Crocuta crocuta*) mobbing a female lion (*Panther leo*).

(C)

FIGURE 14.19 **Meerkats mobbing a cape cobra** This photo shows meerkats engaging in predator inspection, mobbing, and instructing offspring in both predator types and predator risks.

then mob the predator as a heterospecific group [127, 150, 445, 843].

The degree to which nearby prey are recruited by mobbing or distress calls varies widely. Playbacks of mobbing calls recruited more small passerine participants if undertaken far from the nearest hawk's nest than when close to it [275]. The loud alarm calls given by blackcaps (*Sylvia atricapilla*) when cuckoo nest parasites are nearby attract a wide variety of inspecting birds, but none is induced to mob the cuckoo as a result of the blackcap calls [346]. The probability that pied flycatchers (*Ficedula hypoleuca*) will join a mobbing group upon hearing the relevant calls varies according to the current level of predation risk in that area, and the frequency with which the callers have themselves participated in mobbing groups when called [488, 489]. Chaffinches (*Fringilla coelebs*) are also more likely to recruit neighboring territory owners into mobbing groups if they have themselves joined mobs recruited by the neighbors [487]. Both examples suggest that reciprocity may play some role in mobbing economics. Meerkats mob a wide range of other species ranging from very dangerous to harmless (**Figure 14.19**); this range may arise from groups testing whether a particular species is or is not dangerous and instructing naïve members on relative risks [330].

The amount of information provided by mobbing calls shows the same levels of variation seen in surveillance signals. Tail flagging by California ground squirrels is associated with a wide variety of contexts, but becomes particularly intense and shows particular patterns when the squirrels are mobbing snakes [400, 401]. Nesting reed warblers (*Acrocephalus scirpaceus*) vary the rate and amplitude of their mobbing calls continuously according to the proximity of cuckoo nest parasites [941, 942]. Both the durations of mobbing calls and the intervals between them are modified by magpie jays

(*Calocitta formosa*) as a function of urgency; either change appears to carry the relevant information [234]. Siberian jays use the same calls for alarms and mobbing, but use different calls for hawk versus owl predators and vary these according to urgency [336, 337]. Chickadees (*Poecile* spp.) have distinctly different calls for alarms and mobbing, although in both types of signals, the call structure can be varied over a wide range to encode urgency [38, 265, 885]. Interestingly, birds that breed sympatrically with chickadees in temperate latitudes and then migrate to tropical latitudes for the winter retain their responsiveness to chickadee mobbing calls; playbacks to such migrants in Belize elicited normal mobbing responses, whereas local tropical species that have never heard chickadee calls ignored the playbacks [628].

Victim signals

Many species, even those that are normally solitary, emit signals once a predator makes contact with them. We have reviewed distress signals given in this context, in which the intended receiver is the current or other nearby predators. **Victim signals** are given in the same context, but the primary receivers are conspecifics or other nearby prey. Some signals given by attacked prey are surely directed at both predators and adjacent prey. As noted earlier, victim signals often recruit other prey to inspect or mob the attacking predator.

One likely outcome of predator contact is wounding of the prey. A large variety of aquatic species detect and respond to chemicals released into the water by damaged conspecifics. Such chemicals can be reliable cues that the animal detecting them might also be in danger, and we should not be surprised if selection favored enhanced olfactory sensitivities to particularly indicative chemicals. Clams modify the extension of their siphons and marine snails adjust investments in new shell upon exposure to chemicals released by damaged conspecifics [95, 819]. Spiny lobsters have special receptors for detecting components in the blood of injured conspecifics, and crayfish respond to compounds

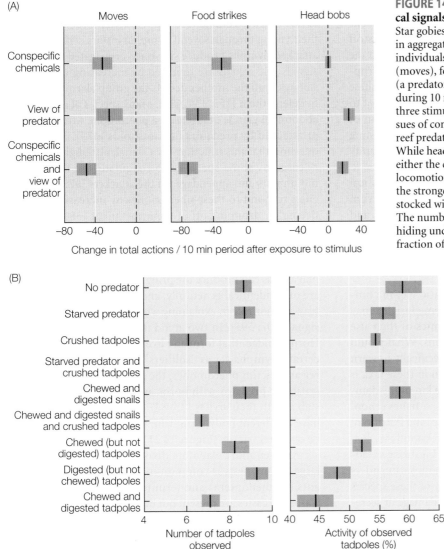

FIGURE 14.20 **Receiver responses to victim chemical signals are often augmented by other cues** (A) Star gobies (*Asterropteryx semipunctatus*) often occur in aggregations on coral reefs. Experiments using three individuals in a tank compared the rate of movement (moves), foraging (food strikes), and head bob displays (a predator deterrent and conspecific alarm signal) during 10 minute periods before and after exposure to three stimulus conditions: chemical extracts from tissues of conspecifics (a victim signal); view of a common reef predator (rock cod *Cephalopholis boenak*); or both. While head bobs are given only if the predator is visible, either the chemical signal or predator view depresses locomotion and foraging, and the combination produces the strongest effect [570]. (B) Wading pools were each stocked with 20 leopard frog tradpoles (*Rana pipiens*). The number of tadpoles that were observed (i.e., not hiding under provided leaf litter; left graph) and the fraction of visible tadpoles that were active (right) were then scored after exposure to a variety of stimuli, including the presence of a predator (a nymph of the dragonfly *Anax junius*). The strongest responses of the tadpoles were to the presence of chewed or crushed conspecific tissues (victim signal) or to digestive products indicating that the predator had recently eaten tadpoles. Starved predators did not give off digestive products that alarmed tadpoles. (After [757].)

in the urine of stressed conspecifics [750, 772]. Gammarid amphipods, blue crabs, and caddis fly larvae withdraw into refuges after detecting water containing damaged conspecifics, and both *Daphnia* and sympatric oligochaete worms avoid areas around damaged *Daphnia* [289, 460, 588, 963]. An enormous number of fish and amphibian tadpole species respond adversely to water containing chemicals from damaged conspecifics and in some cases, from taxonomically related heterospecifics [104–109, 366, 470, 570, 583, 757, 916, 965, 994]. Receiver fish can be extremely sensitive to these chemicals, called **schreckstoff** [926], although active spaces in natural contexts are limited to half a meter from the victim [504, 964]. Efforts to identify the active chemicals in fish schreckstoff have led to conflicting outcomes, although it now appears that several compounds may be involved, and these can vary with species and even different populations of the same species [105, 107, 109, 213, 583, 916]. Olfactory sensitivities specifically designed to detect tissue damage chemicals appear to have evolved in a number of fish and amphibian larvae [213, 301]. Both fish and

amphibian larvae may also be highly attentive to cues from predators and alter their responses if both predator cues and victim signals are detected (**Figure 14.20**).

There remains considerable debate about when, if ever, chemicals from damaged aquatic animals should be considered signals instead of cues [144, 213]. While some species have specialized receiver sensitivities, this could arise through either route. The critical question is whether the chemicals released when prey are damaged are specifically designed to alarm conspecifics or instead primarily fulfill other functions. In many aquatic invertebrates, it seems likely that the chemicals are simply blood or tissue compounds released inadvertently during predator attacks. Species that are unpalatable or toxic may release defensive compounds when attacked [994]. Again, it would pay conspecifics to flee when they detected these chemicals in the water. However, the vast majority of fishes that respond to conspecific schreckstoff are not unpalatable or toxic. Another possibility is that damage chemicals are distress signals designed to attract other predators that may interrupt the initial attack and allow the prey to escape

[563]. It is not clear that this type of interference occurs widely enough to explain the ubiquity of schreckstoff sensitivity in taxa such as ostariophysan fish. The schreckstoff chemicals in many fish are stored in special club cells in the skin, and can only be released externally if the cells are ruptured. While it initially appeared that the substances in these cells served antiparasitic functions, that now appears not to be the case [437]. However, experimental suppression of the immune system reduces the size of club cells, suggesting that they may have, or originally had, some other immunity function [145, 369]. Tadpoles of the frogs *Rana sylvatica, R. clamitans,* and *R. pipiens* actively secrete a special chemical when attacked; the chemical is generated as needed from two otherwise sequestered components in their skins, and although tissue damage will release the chemicals, damage is not required [277]. These chemicals do not appear to be toxic or unpalatable and no other physiological functions have yet been identified for them.

It is thus unclear which of these aquatic victim compounds might be signals instead of cues. Some insights may be gleaned by asking whether the economics of their use would support their having signal functions. What would justify the costs of producing chemicals specifically to warn conspecifics? The only likely compensation in these taxa is that the receivers of victim signals are close kin of the damaged senders. Although some ostariophysan fishes school preferentially with kin [297], this is only true for a fraction of the nearly 8000 species that respond to schreckstoff [144, 464, 823, 824]. The case for amphibian larvae is a bit stronger: tadpoles of many frog species and larvae of various salamanders aggregate preferentially with kin, and many of these species also respond to chemicals released by tissue damage or secretions of conspecifics attacked by predators [70, 328, 379, 407, 534, 687, 867, 922, 931, 936]. But there are notable exceptions: tadpoles of the common frog, *Rana pipiens,* produce their warning skin secretions despite showing no abilities to identify and preferentially aggregate with kin; and tadpoles of the wood frog, *Rana sylvatica,* produce similar secretions, but although they can identify kin, they often use this ability to *avoid* aggregating with close relatives [70, 277, 371, 757]. Tadpoles of some frog species form mixed-species aggregations and at least some of these species respond to damage-released chemicals from other taxonomically unrelated but co-aggregated species [311, 312]. These responses cannot be explained through kin effects.

The most likely explanation for these heterogeneous results is that many aquatic species have been selected to attend to cues released by damaged conspecifics or jointly aggregating species. The costs of false alarms will favor a focus on those released chemicals that are species specific as opposed to widely shared among sympatric taxa. These are receiver refinements. Where neighbors are close kin, additional selection may promote sender production of modified chemicals that are even more unique. What we see is thus a continuum of receiver preadaptations for cue detection that merges into sender refinements and true signaling only in certain cases.

The evolution of victim signals in terrestrial species is less murky. Female treehoppers (Membracidae) guard their young through several nymphal instars. Threatened or attacked nymphs produce substrate-propagated vibrational signals; the signals are repeated by increasing numbers of siblings until the mother detects the group alarm and comes to defend them [153, 154, 688]. Many species of aphids have abdominal glands that release a pheromone when harassed or attacked by predators or parasitoids [377]. There is no question that this functions as a signal, although chemicals released by some species also coagulate and immobilize mouthparts and appendages of the attackers [977]. Conspecifics respond to these pheromones by increasing crypsis, fleeing, or dropping off of their host plant [389]. In some species, the pheromone causes reproductives to reduce the number of wingless offspring they produce in favor of more winged offspring that then leave the site [492]. Because colonial aphids can produce offspring asexually, group members are often identical genetically, and nepotistic benefits surely play a major role in explaining why senders should bother to signal [507, 593]. In two aphid families (Pemphigidae and Hormaphidae) that form and live inside galls on host plants, certain nymphs act as "soldiers": when a colony member releases victim pheromones, the soldiers rush in and try to crush the predator with horns and their back legs, or they stab it with their stylets [13, 847]. Soldiers in some species have a different body form from nonsoldiers, and may even be obligately sterile (**Figure 14.21**).

An even more elaborate division of labor in defense of large colony nests is common in eusocial wasps, bees, and ants (Hymenoptera) and termites (Isoptera). In bees and wasps, no castes are present but defense is more often undertaken by nonreproducing workers than dominants and queens [97, 112]. Many ants and termites have specialized worker or soldier castes that never reproduce and regularly undertake colony defense. All such species can emit victim signals that either cause nearby conspecifics to flee (small colonies) or instead attract workers and soldiers that attack the offending entity (large colonies) [531]. Recruitment for mobbing and predator attack is most often triggered by pheromones, although termites and ants augment the pheromones with substrate vibrations, and honeybees vibrate their wings to produce a hiss [97, 246, 412]. In termites, stingless bees, and ants that defend by biting, defensive organs and the glands generating victim signal chemicals are all located in the head; in Hymenoptera where stinging is the main offensive action, the chemicals are generated by glands associated with the sting apparatus [74, 97, 112, 168, 181, 412, 446, 498, 673, 707, 708, 758, 811]. Thus, as with aphids, victim signals are often generated in the same region of the body used for defense. The entangling glues of some termites and the poison components in Hymenopteran venoms are usually nonvolatile proteins; they are thus unlikely to function as victim pheromones [74, 200]. To create victim signals, most eusocial termites and Hymenoptera must produce additional volatile chemicals that are either mixed with the defensive

(A)

(B)

(C)

(D)

FIGURE 14.21 **Soldier castes in gall-forming aphids** (A) Gall on a leaf stalk of black poplar (*Populus nigra*) housing a colony of the aphid *Pemphigus spyrothecae*. (B) Same gall as in (A) cut open to show the aphids living inside. (C) Soldier (left) and normal nymph (right) of *Pemphigus spyrothecae*. Soldiers in this species have heavier cuticles but are otherwise similar in size and potential fertility to normal nymphs. (D) Taiwanese gall-forming aphid (*Colophina monstrifica*) with sterile soldier caste (left) that is much larger than normal nymphs (right). Soldiers in most species perform housekeeping and defend against predators, but they cannot identify kin and so do not exclude unrelated individuals [1, 581, 664, 847, 972].

chemicals or released by nearby glands [75]. Victim pheromones of eusocial insects tend to have many more chemical components (30 or more in many species) and be produced in larger volumes than pheromones used for other functions. The many volatile components often have different Q/K and diffusion properties (see Chapter 6) [76, 97, 111, 191, 276, 412]. Some of this diversity is associated with coordinating different stages in colony member recruitment and attack, and in marking the target for subsequent attackers [97, 704, 811, 929]. Additional diversity may reflect the presence of components with multiple functions or nonvictim signal functions such as secondary chemical offense, antibacterial action, mate attraction, or other reasons for recruitment of colony members [76, 412, 903]. Even closely related species may differ in the composition of their victim pheromone mixtures; this presumably reflects fine-tuning of the patterns of recruitment and mobbing to specific ecological and social contexts [111, 113, 190, 276].

Other economic considerations

INTERSPECIFIC ALARMS AND EAVESDROPPING   It should not be surprising that many species respond to the alarm signals of other sympatric species. Predators can use alarm signals to locate prey: the salticid spider *Habronestes bradleyi* responds specifically to the victim signals of meat ants (*Iridomyrmex pupureus*), which it then approaches and attacks [8]. Other examples of predator eavesdropping were provided in Chapter 9.

Prey species that live in the same habitat often attend to each other's alarm signals. Although they neither make sounds nor engage in acoustic communication, marine iguanas (*Amblyrhynchus cristatus*) are very responsive to the alarm calls of sympatric Galápagos mockingbirds (*Nesomimus parvulus*) [924]. A wide variety of birds attend to distress and alarm calls of other vertebrates. Sympatric song and swamp sparrows (*Melospiza melodia* and *M. georgiana*) respond similarly to playbacks of their own and each other's distress/victim calls; the two species' calls are quite similar in temporal and frequency patterns. In contrast, neither species responds strongly to the quite different alarm calls of sympatric white-throated sparrows (*Zonotrichia albicollis*) [844]. Taxonomic relatedness is not a prerequisite for interspecific alarm responses. Zenaida doves (*Zenaida aurita*) respond to the alarm calls of sympatric carib grackles (*Quiscalus lugubris*); cardinals (*Cardinal cardinalis*) and house sparrows (*Passer domesticus*) startle at the wing whistles of fleeing mourning doves (*Zenaida macroura*) [156, 338]. Many unrelated species of bats investigate alarm and distress signals of other bat species, and Gunther's dik-dik (*Madoqua guentheri*), a small antelope, flees when it hears the alarm calls of sympatric go-away birds (*Corythaixoides leucogaster*) [508, 727]. Eastern chipmunks (*Tamias striatus*) respond strongly to alarm calls of titmice (*Baeolophus bicolor*), red squirrels (*Sciurus vulgaris*) respond to European jay (*Garrulus glandarius*) alarms, and yellowbellied marmots (*Marmota flaviventris*) and sympatric golden-mantled ground squirrels (*Spermophilus lateralis*) show similar responses to playbacks of their own and each other's alarms [694, 745, 780]. Mixed species groups are common among forest birds, forest monkeys, savannah ungulates, and cetaceans: in nearly all cases, at least some of the member species respond to alarm signals of other species in their group [50, 131, 294, 595, 834].

The response of heterospecific receivers to alarm calls is not necessarily the same as the response of conspecific

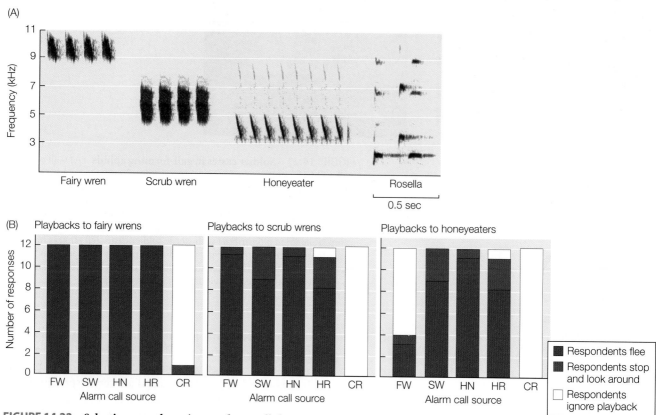

FIGURE 14.22 Selective eavesdropping on alarm calls by sympatric Australian birds (A) Spectrograms of alarm calls of superb fairy wren (*Malurus cyaneus*), white-browed scrubwren (*Sericornis frontalis*), and New Holland honeyeater (*Phylidonyris novaehollandiae*), and contact (not alarm) call of crimson rosella parrot (*Platycercus elegans*). (B) Responses of the first three species to 12 playbacks of the alarms of fairy wren (FW), scrubwren (SW), and honeyeater (HN for normal call and HR for call reduced to the lower amplitude of wren calls), and the contact call of rosella (CR). Respondents could either flee (red), stop and look around (blue), or ignore the playback (white). The small (9–10 g) fairy wrens, which are vulnerable to all of the predators of the scrubwrens and honeyeaters, fled at alarms of those species, but ignored the contact calls of the parrot. The large (19 g) honeyeaters have fewer predators and ignored many of the fairy wren alarms, but still share enough predators with scrubwrens (12–14 g) to attend to their calls. The intermediate-sized scrubwrens responded more to fairy wren calls than did honeyeaters, and about as often to honeyeater calls as the honeyeaters did. (After [542].)

receivers. Although African plovers (*Vanellus* spp.) typically encode urgency into their alarm calls, eavesdropping banded mongooses (*Mungo mungo*) ignore the urgency information and respond to plover alarms in an all-or-none manner [599]. Sympatric superb fairywrens (*Malurus cyaneus*) and white-browed scrubwrens (*Sericornis frontalis*) always respond to alarm calls of New Holland honeyeaters (*Phylidonyris novaehollandiae*); but many predators of the fairywrens are of no threat to the honeyeaters, who often ignore fairy-wren alarm calls (**Figure 14.22**). Scrubwrens and honeyeaters have intermediate predator overlap and honeyeaters respond to scrubwren alarms [541, 542].

As noted earlier, black-capped chickadees (*Poecile atri-capillus*) vary the composition of their alarm calls to encode information about the size of a predator and the level of risk; sympatric red-breasted nuthatches (*Sitta canadensis*) not only respond to chickadee alarms, but use the detailed information to adjust the nature of their responses [886]. As noted earlier, many species have distinct alarm signals for aerial versus terrestrial predators. Vervet monkeys (*Cercopithecus aethiops*) respond appropriately to the two different alarm calls of sympatric superb starlings (*Spreo superbus*) [140, 392, 864]. Ringtailed lemurs (*Lemur catta*) respond differentially to the distinct aerial and terrestrial alarm calls of sympatric Verreaux's sifaka lemurs (*Propithecus verreauxi*) [640], and Diana monkeys (*Cercopithecus diana*) show distinct responses to playbacks of leopard versus human predator calls of sympatric crested guinea fowl (*Guttera pucherani*) [1000]. Both Diana monkeys and sympatric Campbell's monkeys (*Cercopithecus campbelli*) have separate eagle and leopard alarm calls, and each species responds appropriately to the other species' alarms [1001, 1003, 1005]. These two species often form mixed species troops, with Campbell's monkeys in the canopy watching for eagles and the Diana monkeys lower down watching for leopards [967]. Interestingly, Campbell's monkeys precede an alarm with a loud "boom" if the threat is distant or of uncertain urgency: both species show reduced responses if a Campbell's monkey alarm is preceded by a boom [1004]. At least two species of hornbills (*Ceratogymna atrata* and *C. elata*) sympatric with these monkeys respond to Diana monkey alarms and differentiate between the leopard

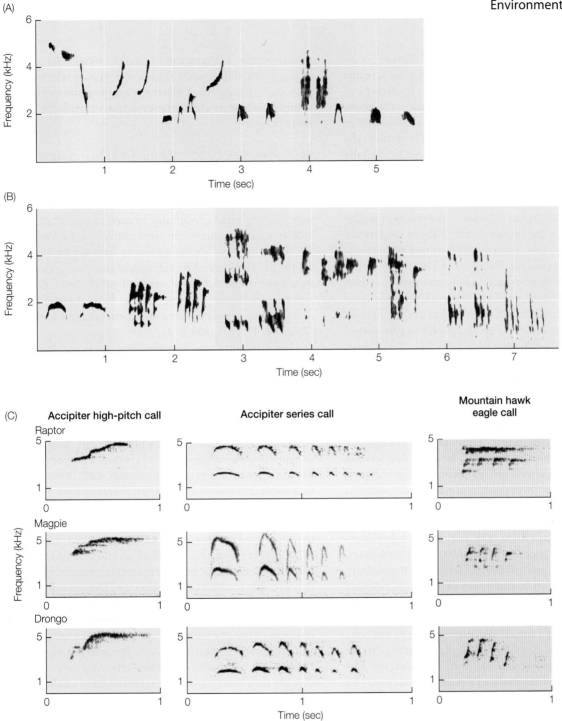

(A)

(B)

(C)

Accipiter high-pitch call

Accipiter series call

Mountain hawk eagle call

Raptor

Magpie

Drongo

Time (sec)

versus eagle calls; the smaller species (*C. atrata*) also responds to and discriminates between the eagle and leopard alarms of Campbell's monkeys [685, 686]. Greater racket-tailed drongos (*Dicrurus paradiseus*) and Sri Lankan magpies (*Urocissa ornata*) not only respond to the alarm signals of other species in their mixed insectivorous flocks, they also imitate these calls when appropriate predators are spotted (**Figure 14.23**) [320, 321]. As the drongos tend to forage in the canopy, they act as sentinels for other species that forage lower down, and these other species respond properly when the drongos give either their own or other species' alarm calls [322, 711]. The drongs and magpies also imitate the calls of avian predators

FIGURE 14.23 Vocal mimicry of alarm and predator calls by Sri Lankan birds (A) Types of own species-specific alarm calls of racket-tailed drongos (*Dicrurus paradiseus*). Alternating colors separate different alarm call types. (B) Racket-tailed drongo imitations of alarm calls of other bird species with which it often forms mixed flocks. These calls are often given to alert other mixed-flock members of danger and to incite mobbing, but may also be used deceptively to gain biased access to food finds [321, 322, 738]. (C) Calls of two avian predators sympatric with the drongos (top row): high-pitched and series calls of local accipter hawks (*Besra virgatus* and *B. trivirgatus*), and a call of the mountain hawk eagle (*Nisaetus nipalensis kelaarti*). Sympatric Sri Lankan magpies (*Uracissa ornata*; middle row) and the racket-tailed drongos (bottom row) both imitate these raptor calls as alarms. (After [702].)

and use these sounds as general alerts and alarms [702]. Phainopeplas (*Phainopepla nitens*) produce loud screams when predators are near and may then add imitations of the alarms and distress calls of sympatric species. Playbacks showed that heterospecifics attracted by these calls were more likely to mob the predator when unmodified imitation calls were present than when screams were followed by digitally scrambled versions of the imitations [146].

FALSE ALARMS   False alarms are not uncommon in animals. Some cases are natural mistakes or the results of sender inexperience. Since the signals of many alarm senders are individually distinctive, receiver discounting of untrustworthy senders may correct for this risk [141, 378]. Other false alarms are clearly deceptive. In both neotropical and Sri Lankan mixed insectivorous flocks, the usual sentinels (antshrikes or shrike-tanagers in the neotropics and drongos in Sri Lanka) occasionally give false alarm calls to gain biased access to particularly rich food finds [603, 711, 738]. Great, willow, and marsh tits (*Parus major, Poecile montanus,* and *P. palustris*), and capuchin monkeys have also been observed to emit false alarms to scare off competitors from food [364, 567, 591, 945]. False alarm rates in the tits can be as high as 50–60%, and an average wild capuchin may produce one false alarm for every two hours of observation. Young male white-throated magpie jays (*Calocitta formosa*) that serve as sentinels to breeding groups give frequent and highly diverse alarm calls about low-threat predators; evidence suggests this increases their chances of being noticed and perhaps considered for mating by breeding females [235].

High rates of false alarms (and thus low signal reliability) raise interesting questions about the relevant economics. Several authors have pointed out that the cost of unnecessary flight might be a small decrease in foraging rates, whereas failure to heed a warning when a predator is about to attack may be extremely costly [762]. However, this only considers the economics of a single false alarm, and does not consider aggregate costs. Numerous false alarms may generate a significant cumulative reduction in foraging success. The impact of false alarms depends on a number of factors. The alternative to responding to alarms is to increase individual vigilance, which also interferes with foraging. Once an alarm is given, members of a focal animal's group may begin to flee; as the residual group size decreases, the dilution effect is diminished and each remaining individual has a higher risk of being taken by the predator. At the same time, if error rates are not too high, the number of group members that elect to respond to an alarm may be a more reliable indication of risk than the alarm signal alone [472]. Finally, rates of false alarms may vary with the ease of detecting predators: this can depend on the current foraging habitat, time of day, weather, and other conditions. A number of authors have created quantitative models incorporating many of these factors to predict optimal receiver strategies [51, 521, 522, 676, 899]. The general consensus is that (1) higher rates of false alarms can be tolerated as group size increases; (2) the number of

group members fleeing can be used to weight the salience of any given alarm; and (3) receivers should rely more on personal vigilance than alarms if escape options are constrained, predators are harder to detect, or predator attacks more frequent.

ACQUIRING THE CODE   To master an alarm signal code, senders must acquire the ability to produce those signals, and receivers the ability to perceive and identify them. In addition, both parties must come to associate the same signals with the same conditions. In many cases, senders and receivers are potentially the same individual, which simplifies the tasks. Most species appear to have innate templates for the production and recognition of conspecific alarm signals; some of the individual, sex, and population differences reported for various taxa suggest that subsequent modifications of these templates based on experience are also possible [231, 262, 413, 571, 663, 814].

Learning and experience play their greatest role in helping senders associate alarm signals with appropriate conditions [417]. Young Verreaux's sifaka lemurs, meerkats, and vervet monkeys, are able to produce the different alarm signals of their species from an early age [264, 413, 771]. There is considerable evidence that they then rely on learning to refine which signals are used in which contexts. Sifakas raised in captivity exhibit the same call types as wild populations, but in response to very different stimuli [262]. Young meerkats, especially young females, learn correct associations of calls with predator conditions faster if they associate with adults [416]. Call variants associated with urgency appear to be mastered before variants associated with predator type [414]. Vervet monkeys initially produce alarms for many inappropriate objects, but like meerkats, improve quickly if they are able to observe adult models. They usually master the appropriate responses to the different vervet alarms first, and then begin to emit calls. Adult vervets that follow the response or call of a youngster with the correct call reinforce and accelerate this learning [771]. Spectral tarsier mothers (*Tarsius spectrum*) also echo their offspring's alarm calls if they are used correctly; they do not reinforce a call if it is in error [358].

Learning to respond to alarm signals is a natural subset of most animals' tendencies to associate new cues with important contexts throughout life. Innate influences can still be important constraints. For example, many fish learn to associate chemical victim signals with specific predators and contexts, whereas associations between visual cues and types of predators are largely innate [466]. Rats (*Rattus norvegicus*) are innately predisposed to associate ultrasonic (22 kHz) sounds with aversive entities, but experience is required to narrow that bias to specific stimuli [237]. Infant squirrel monkeys (*Saimiri sciureus*) respond innately and differentially to their species' aerial and terrestrial alarms; however, experience is required to associate each call type with specific types of objects [237]. As with alarm production, responses to different call types may develop sequentially: Belding's ground squirrels respond first to the species' whistle given for fast-moving predators such as hawks, and only later to the trills given for slower, usually terrestrial predators [561]. Nestling scrubwrens do not need to worry about aerial predators until they fledge; not surprisingly, they ignore their parents' aerial alarms but respond early and strongly to their terrestrial alarm calls [669]. Responses to heterospecific alarms, which often differ markedly from conspecific signals, are almost always learned. This requires both suitable exposure to heterospecific models and time to learn the signals and their associations. Wild fairy wrens attend to playbacks of the alarm calls of scrubwrens or honeyeaters if these species are currently sympatric with the fairy wrens [543]. Young vervets take considerable time and enhancement by adult reinforcement to learn the separate aerial and terrestrial predator alarm calls of sympatric starlings [392]. An interesting exception to the requirement of learning for heterospecific alarm responses is a race of common cuckoos (*Cuculus canorus*) that specializes in parasitizing the nests of reed warblers (*Acrocephalus scirpaceus*). Nestlings of this cuckoo respond innately to the alarm signals of adult reed warblers and not to the alarms of other species parasitized by other races of this cuckoo [193]. Learning does not appear to be necessary.

Food Signals

Basic economics

Signals that advertise food discoveries raise the same economic questions that alarm signals do: senders pay an immediate direct cost, whereas direct benefits of the signals largely accrue to receivers. One difference is that many alarm signals serve the dual functions of predator deterrence and fellow prey warning; food signals are usually not directed at the food and thus tend to have only a single function. This may partly explain why animal alarm signals are much more widely encountered than food signals. However, food signals do occur in a number of taxa, and this raises the economic question of why these senders should bother to signal and pay the consequent costs of sharing their food finds.

A number of models have identified various factors that singly or in combination might cause food calling to be evolutionarily stable [122, 307, 308, 619, 848, 946]. Most of the economic compensations of sending alarm signals have a counterpart for food signals. A discoverer of a new food patch that does not recruit other foragers will likely have a lower feeding rate because it cannot share vigilance with others, and it will lack the dilution of risk should a predator attack [307, 308, 520, 618]. Benefits of foraging in a group increase asymptotically with increasing group sizes, the asymptote varying with the level of predator risk [520]. These benefits must be traded off against increased competition for limited food in larger groups. There are many examples of facultative signaling depending on these trade-offs. House sparrows (*Passer domesticus*) are more likely to use their chirrup call to recruit conspecifics to food finds if they are in exposed locations and host sufficient food for multiple foragers [233]. Chickadees use serial calls with lots of "D" notes to attract conspecifics to a new food site; as the numbers of recruits increase, the number of "D" notes is reduced and recruitment slows [284, 544]. Many species vary their likelihood of giving food calls depending upon the amount and divisibility of the advertised food. Primate examples include capuchin monkeys, spider monkeys, toque macaques, and chimpanzees [135, 205, 209, 395, 633, 671]. Sender costs of sharing are also minimized if food patches are highly ephemeral, as in the case of cliff swallows (*Petrochelidon pyrrhonota*) which call to recruit colony mates to aerial aggregations of insect prey [103].

Species that require the presence of other individuals to obtain and extract food gain a mutualistic benefit by emitting recruitment signals. Some bark beetles (*Dendroctonus spp.*) can overwhelm the chemical defenses of food plants only if sufficient numbers attack the same plant; the beetles thus release a pheromone to attract conspecifics (**Figure 14.24**) [501, 502, 679, 680]. When a funnel ant (*Aphaenogaster albisetosa*) encounters a large prey item, it releases a pheromone from its poison gland that attracts colony mates from up to 2 m away [412]. The finder may produce stridulation sounds, which increase the attraction until a sufficient group is present to carry away the prey. Red Sea groupers (*Plectropomus pessuliferus*) recruit sympatric moray eels (*Gymnothorax javanicus*) to hunt prey in crevices. Groupers perform a rapid head-shaking display to eels to induce them to leave their lairs. The two then prowl the reef together: when the grouper spots a prey in a crevice, it shakes its head at the site. The eel then enters and either captures the prey or forces it to escape into open water where the grouper catches it [114]. Other species use signals to coordinate their foraging: chimpanzees emit a short bark to initiate and coordinate the hunting of red colobus monkeys; white-faced capuchins produce a regular contact call that spreads the troop out efficiently over a dense fruit patch; and spinner dolphins (*Stenella longirostris*) rely on each other's echolocation clicks to maintain encirclement of prey fish schools [62, 63, 84, 87, 178, 179, 937]. Reciprocity among specific individuals may play a role in justifying food signals in capuchins and chimpanzees [195–198, 584].

Other economics can play a role. Nepotism may help underwrite the elaborate food signals of eusocial insects [151, 960, 961]. Pied babbler (*Turdoides bicolor*) adults give food calls that differentially attract recently fledged young [682]; they provide the young with food and instruct them in new food types. Mate attraction and sexual selection explain some cases of food calls. Roosters (*Gallus gallus*) produce a soft but quickly repeated call and a characteristic body posture when they locate food ("tidbitting") [242]. The call rate is higher if hens are nearby, and also varies with the rooster's dominance status, recent sexual success, familiarity with the hens, and level of hunger [241, 667, 911]. Roosters sometimes give this call in a site where they earlier attracted hens, even if no food is currently present [913]. If they currently have no food, hens respond to these calls by looking down [243,

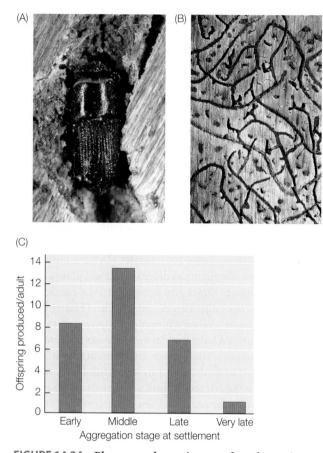

(C)

FIGURE 14.24 Pheromonal recruitment of southern pine beetles (*Dendroctonus frontalis*) on victim trees These beetles burrow into trees to feed on phloem tissues. (A) Early attackers of a tree pay the heavy cost of countering tree defenses (sticky resins) and producing a pheromone to attract conspecifics. (B) Once enough beetles have been attracted, they overwhelm the tree's defenses and create elaborate galleries in which they lay their eggs and larvae can feed and mature. (C) Reproductive success depends on time of settlement: early settlers are less successful due to the costs of overcoming tree defenses and producing pheromones; middle-stage settlers have the highest success; late settlers have lower success, as the tree dies and competition for the remaining phloem becomes acute. In later stages, beetles produce a second pheromone that discourages further recruitment. (After [679].)

244]. The full tidbitting display (visual and acoustic components) appears to be used by hens as a measure of male status and mate suitability [667, 820, 821]. Many other Galliformes (pheasants, quail, fowl, and francolins) show similar male tidbitting behaviors [860]. Male bonobos (*Pan paniscus*) use food calls to attract females as potential mates; female bonobos call in other females to help form coalitions to prevent males from dominating food finds [914].

In many species, differential relationships between senders and receivers significantly modulate food signal exchanges. Cotton-topped tamarins (*Saguinus oedipus*) are more likely to give food calls if mated but without offspring than if unmated or mated with offspring [720]. Brown-throated conures (*Aratinga pertinax*) foraging on a discovered patch

of seeds or fruit often cannot be seen from above (**Figure 14.25**). Overflying flocks emit contact calls that may or may not be answered by the foraging birds; if answered, the overflying birds descend and join the foragers. This selective recruitment is unrelated to the amount or quality of the current food patch. Why foragers should be selective in recruiting individual conspecifics is unclear. Nepotism is very unlikely in this species. Given that individual conures can recognize each other by contact calls alone, some form of reciprocity has been proposed [117]. Status often affects who emits food calls. Nonterritorial ravens (*Corvus corax*) that locate a carcass within the territory of a mated pair will often call loudly to recruit other nonterritorial birds; if sufficient numbers are recruited, the nonterritorial group may be able to take the carcass away from the usually dominant territorial pair [399, 558]. In many primates, dominant individuals are more likely than subordinates to emit food calls. This may be due to the fact that dominants will lose little of their share of any food to subordinate recruits, whereas food calling subordinates may lose a substantial share if dominants are recruited [135, 206, 348]. Many primates thus advertise food finds facultatively, depending on whether other conspecifics are nearby or are likely to have detected the same cues and whether the food can be consumed before being interrupted by a dominant animal [393, 394, 495].

FIGURE 14.25 Pair of brown-throated conures (*Aratinga pertinax*) feeding in foliage These birds are very cryptic when surrounded by vegetation. Conspecifics searching for new food sites repeatedly emit contact calls while in flight. Birds feeding in foliage may or may not call back; if they do, overflying birds will usually join them. This selective recruitment is not correlated with the amount or quality of the food site, but instead reflects the ability of conures to recognize individuals by their contact calls [117].

(A)

(B)

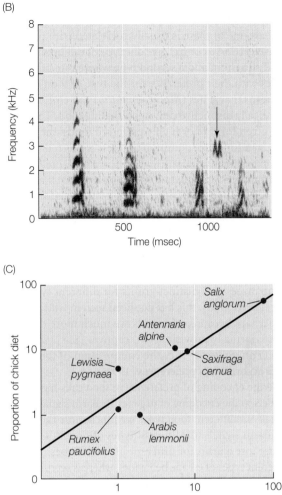

FIGURE 14.26   Dietary training of white-tailed ptarmigan (*Lagopus leucurus*) chicks by mother   (A) Hen ptarmigan foraging with chicks. When a preferred food plant is encountered, the hen performs a multimodal display beginning with a series of food calls, followed by tearing off morsels of the food and dropping them while bobbing her head. This attracts chicks, which then sample the tidbits. (B) Spectrogram of the food calling portion of the display. The first two calls were given with the hen's head up, and the last two head-down near tidbits. Arrow shows answering peep by chick. (C) Data for six major food species shows how the fraction of each species in the chicks' diet (vertical axis) increases with the fraction of tidbitting calls given by the mother to draw attention to that plant species (horizontal axis). Note that both axes are on logarithmic scales. (B,C after [148].)

Amount of information

The many parallels between the economics of alarm and food signals might suggest that the patterns of encoding information are also similar. While it is true that the encoding of quality, quantity, and type in food signals shows clear parallels with the encoding of urgency, proximity, and escape strategy in alarm signals, there is one major difference between the two types of signals: whereas alarm signals need not specify *where* a detected predator is located to be useful, this is often critical information for food signals. This imposes constraints and demands on food signals that are largely lacking in the evolution of alarms. There are four ways that senders might provide information to receivers about the location of a food find: (1) take up a position at the food and broadcast an easily localizable signal to fellow foragers; (2) return to a central location, recruit fellow foragers, and lead them back to the food site; (3) generate a chemical or visible trail between the food and a central location and then induce fellow foragers to follow this trail to the food; or (4) return to a central location and provide directions to fellow foragers on how to find the food. As we shall see, these four options differ markedly in their relevant economics. Below, we first examine the encoding of food quantity, quality, and type in food signals, and then examine the informational challenges posed by each mechanism for specifying where the food is located.

ENCODING FOOD QUALITY, QUANTITY, AND TYPE   As with alarm signals, an example of every possible mapping of discrete or continuous food variables onto discrete or continuous signal sets can be found in animals. However, the most common pattern is to produce one food signal whose emission rate or duration varies continuously with preference. Preference in this case is some weighted combination of food quality, quantity, accessibility, and type into one continuous variable. Preference in food signals thus parallels the widespread encoding of urgency in alarms. Forager stingless bees (*Melipona panamica*) return to their nest and produce stridulations that are a continuous function of food quality and proximity [160]. Rates of tidbitting in fowl varies continuously with food quantity and dispersion [939]. In contrast, rates of food calling by cotton-topped tamarins (*Saguinus oedipus*) are unaffected by the amount of food encountered, but vary according to preferences with maximal calling rates for foods of intermediate preference [720]. Female white-tailed ptarmigan (*Lagopus leucurus*) hens vary the rate of food calls to direct chicks to food plants of different suitabilities; chicks respond to these differences and the composition of their diets reflects their mother's call variations (**Figure 14.26**) [9, 148]. Similar adult recruitment of offspring to desirable foods occurs in pied babblers (*Turdoides bicolor*) and bobwhite quail (*Colinus virginianus*) [684, 955, 956].

Although some species appear to have multiple food calls, at least one of these is usually correlated with preferences. Ravens (*Corvus corax*) give one type of call upon finding a food situation, and a second type whose rate of emission varies with their preference for the type of food found [115]. Rhesus macaques (*Macaca mulatta*) can emit any of five acoustically distinctive calls upon finding food. Three of these are given for rare or highly preferred foods; the other two are given for more common and less preferred items. Playback experiments indicate that the three high-preference calls are equivalent: habituation to one by repeated playback results in habituation to the other two. The two calls for more common items are also equivalent to each other [396]. Both wild and captive chimpanzees (*Pan troglodytes*) emit rough grunts whose acoustic structure varies consistently with preferences for the current food; call rate is independent of food preference and largely reflects hunger [817]. Although captive chimpanzees give distinct grunts for different types of highly preferred foods, this has not been seen in wild populations, perhaps because wild foods are much more diverse than those normally provisioned to captives [818]. Captive bonobos (*Pan paniscus*) vary the composition of food call sequences according to preferences [149].

While the encoding of preference in food calls has obvious parallels to the encoding of urgency in alarm signals, distinct signals for different food types are much more rare than discrete alarm signals for different types of predators. This may be because alarms for different predator types usually elicit different types of escape behaviors, whereas recruitment for eating usually leads to the same behavior. One of the few examples of food type specificity occurs in trumpeters (*Psophia leucoptera*), which emit a unique call to announce the discovery of snakes [763]. Snake prey may require sufficiently different strategies from other foods in these birds' omnivorous diets to require a separate signal.

ENCODING FOOD LOCATION: BROADCAST ADVERTISEMENTS   Advertising a food find from its location is relatively easy: the signal can be conventional, and thus optimized for maximum efficacy, range, and localizability. The major cost is eavesdropping: there is nothing to prevent both intraspecific and interspecific competitors from using the signals to find the food, and predators may use them to localize signalers and any recruited conspecifics. Eavesdropping risks increase with the active space of a food signal and the fraction of time that it is being emitted, both of which vary with the taxon and context. Below we first examine some short-range broadcast signalers and then turn to longer-range examples.

Where primary receivers are already close to the sender, such as in galliform families or chickadee flocks, food calls can be of low amplitude, with minimal risk of predator eavesdropping. Bees, however, forage over very large home ranges and the likelihood that a colony mate is nearby when food is found is often low. Bees thus mark flowers they have recently sampled with short-range scents that fade away after a day or less. In many cases, these marks *repel* colony mates who then do not waste foraging time on a depleted flower; other conspecifics and even heterospecifics may benefit from eavesdropping on these marks at no additional cost to the sender [295, 309, 326, 954, 983, 984]. In other cases, bee scent marks *attract* conspecifics and eavesdropping heterospecifics that happen within the mark's limited active space [89, 254, 282, 283, 427, 519, 620, 621, 709, 747, 748]. The same mark of a honeybee (*Apis mellifera*) or a bumblebee (*Bombus* spp.) can function as a repellent or an attractant depending on the occasion [326].

There has been considerable debate about how bee scent marks might serve as both attractants and repellents, and whether they are explicit signals or instead inadvertent cues [733, 957, 966]. Some insights are provided by the relevant economics. Leaving a mark on a flower after depleting it, a sender can signal to colony mates (as well as itself) to avoid that flower until it replenishes its nectar. Maximal intake occurs after the flower is fully replenished. However, a bee that visits the flower before that point may still get significant nectar. This generates an arms race among competitors for earlier and earlier returns to a replenishing flower. For any given replenishment rate, there is an ESS waiting period below which it does not pay any forager to return to the flower [863]. If foragers could tell how long it had been since a flower was exploited, they could compare this to the threshold waiting time for that type of flower (**Figure 14.27**). Bee scent marks appear to provide the relevant information. Scent mark glands of stingless bees, bumblebees, and honeybees produce complex mixtures of alkenes and alkanes with different volatilities [254, 303, 327, 440, 749, 862]. The age of a mark can be estimated by the relative compositions of these components: less volatile components will increasingly dominate the mark over time. When fresh, the mixture repels; once the composition indicates the mark has passed the time threshold for that flower type, the mark becomes an attractant. Bumblebees appear to match quite closely the switch in appropriate behavior with the predicted ESS threshold for a variety of flower types [863].

Other taxa use food signals that must have large active spaces to be effective, and this increases the risks of eavesdropping. Bark beetles (Scotylidae) emit a communal cloud of aggregation pheromones as the infestation of a particular tree progresses. These signals have large active spaces and are emitted continuously. Not surprisingly, specialized predatory beetles (e.g., Trogositidae) home in on these pheromones to attack the bark beetles and their larvae [256, 257, 398, 705]. In contrast, food calls of birds and mammals have minimal amplitudes, durations, and emission rates. Exceptions include species in which males use food calls to attract or manipulate females. Territorial groups of birds and primates have already excluded competitive eavesdroppers from their vicinity; moderately loud food calls pose little risk from conspecific eavesdroppers although they may attract predators. One unusual example of a loud food call is the recruitment "yell" of nonterritorial ravens when they locate a new carcass: the success of this gambit depends on attracting as many nonterritorial collaborators as possible to fend off territorial pairs [399, 558].

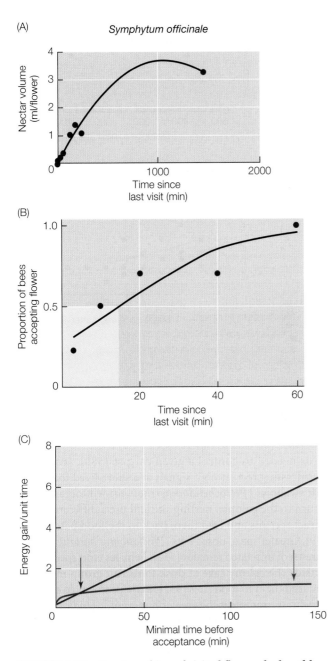

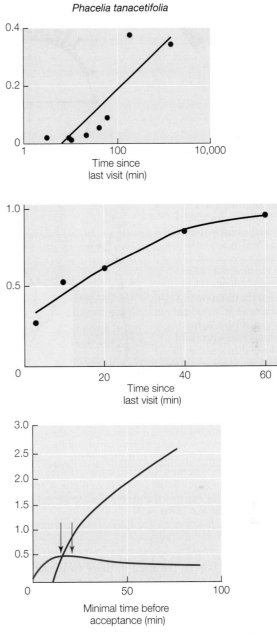

FIGURE 14.27 Scent marking of visited flowers by bumblebees (*Bombus terrestris*) (A) Rates of nectar replenishment in two species with flowers: comfrey (*Symphytum officinale*) and lacy phacelia (*Phacelia tanacetifolia*). Note that phacelia produces about a tenth as much nectar per flower as does the comfrey. (B) Wild bumblebees assess the age of scent marks on a flower left by prior foragers and either collect its nectar or reject it. Graphs here show rates of acceptance of each species by wild bumblebees as a function of time since the last forager marked the flower. Blue areas show the lag required before 50% of visitors will accept a flower (about 14–18 minutes for both species). (C) The predicted energy gain per forager, given typical search and handling times for each species as a function of the minimal time that foragers wait for a flower to replenish its nectar. Blue lines indicate rates of gain if all foragers respect a single communal waiting time; the red line shows the gain for a single "cheat" that ignores scent marks and takes nectar earlier than foragers respecting shared time. The blue arrows indicate the optimal time that bees should wait if all foragers respect the same waiting time. The red arrows indicate the ESS if cheaters are likely. Optimal and ESS waiting times are quite different for comfrey; as seen in (B), observed times are much closer to the ESS. For phacelia, the two predictions are nearly identical, and very close to the observed value. (After [863].)

ENCODING FOOD LOCATION: LEADING Leading creates fewer eavesdropper risks than broadcast signals from the food site. However, a potential leader requires a mechanism for locating likely recruits, the ability to find the food again, and sufficient compensation for the extra time and energy leading imposes on it. Potential recruits are most easily found at the communal roosts or nurseries used by many social insects, birds, and mammals. It is not clear, however,

(A)

(B)

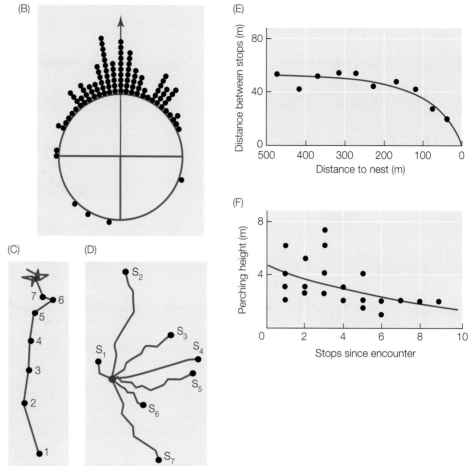

(C)

(D)

(E)

(F)

FIGURE 14.28 **Protocols by which the greater honeyguide (*Indicator indicator*) leads humans (Boran tribe in Kenya) to honeybee colonies it has found** (A) A hunt begins when a human attracts the bird with special whistle, or the bird flies close to a Boran, emitting a special call. (B) Summary of the directions of successive honeyguide flights (black dots) relative to the direction to the hive (red arrow). (C) Typical hunt showing successive flights and stops (numbered) where bird moves ahead and then waits for the human to catch up. Once at the bee colony (red dot), the bird circles the site with short flights. (D) Experiment in which the bee colony was not opened. The honeyguide then advertised the same colony in seven different hunts (S₁–S₇), always leading a Boran to the colony. (E) The bird indicates increasing proximity to the bee colony by flying shorter distances between successive stops. (F) Initially, the bird perches high during stops so the Boran can see it. As it gets closer, it drops to the height of the bee colony and emits a different call, indicating its arrival at the site. (After [433].)

that communication is even required in these contexts: if the success of returning foragers can be estimated from cues (e.g., parental birds bringing food to young or odors on the pelage of returning colonial bats), unsuccessful foragers ("scroungers") could then follow successful foragers ("producers") on their next trips [44]. The economics of following (and being followed) depend on the dispersion of the food, the size of food patches, the relative numbers of producers and scroungers in the colony, and whether successful foragers can avoid being followed [45, 46, 119, 308, 710, 933]. Where successful foragers will be followed in any case, no recruitment or leading signals are required.

Communally roosting birds and bats rarely exhibit leading behaviors. However, there are several interesting exceptions. Instead of broadcast calling at a discovered carcass, nonterritorial ravens (*Corvus corax*) sometimes return to a communal night roost and lead recruits back to the site the following day [559]. Whether immediate broadcast calling or delayed leading is the optimal strategy depends on the active space of broadcast calling relative to the typical home range size of the ravens, and the number of nonterritorial

birds needed to control a given carcass [580]. Instead of broadcasting recruitment signals at a short-lived food find, cliff swallows (*Petrochelidon pyrrhonota*) will occasionally return to the colonial nesting site and use a different call to attract recruits and lead them back to the food patch [103, 859]. Greater spear-nosed bats (*Phyllostomus hastatus*) live in harems of unrelated females [572]. Females emit special calls outside the colony cave at dusk to recruit group members and lead them to food sites [952]. Perhaps the most amazing vertebrate example is the greater honeyguide (*Indicator indicator*) which uses an elaborate set of sounds and behaviors to guide local humans to honeybee nests that it has discovered (**Figure 14.28**) [433]. The humans open the nests for the honey, and the bird eats the bee larvae and digests the leftover wax with help from its specialized gut flora. In all of these cases, mutualism or reciprocity appears to explain the costs of signaling, leading, and sharing food finds.

A combination of nepotism and mutualism likely justifies **tandem running** in a variety of ant species (Ponerinae, Myrmicinae, and Formicinae): a successful forager returns to the nest, regurgitates food it has collected, releases a pheromone from the sting, and then slowly and carefully leads a single recruit back to the food source [412]. If the follower gets behind, the leader stops until they catch up and regain antennal contact. While leading a single worker seems inefficient, this strategy is most often used for food finds by small-colony species; in the process, the follower learns the food location and can then teach additional workers how to find it [279, 280, 594].

FIGURE 14.29 Tandem running (leading) in the ponerine ant (*Pachycondyla* sp.) Note that the following ant has its antennae firmly touching the abdomen of the leader, and is not attending to any possible trail odors.

ENCODING FOOD LOCATION: TRAILS   Pheromone-marked trails between food and nests are common in ants, termites, and stingless bees. In some ants that perform tandem running, leaders use a pheromone trail laid during their return to the nest to find their way back to the food; the follower apparently ignores the trail chemical and instead uses its antennae to maintain physical contact with the leader (**Figure 14.29**). In other species, such as *Camponotus socius*, food finders lay a trail back to the nest, recruit tens of workers with a waggling display, and then lead this large group back to the food using both antennal contact and the trail pheromone. This is called **group recruitment**. In still other species, no leaders or recruitment displays are necessary: workers at the nest follow strongly marked trails to their terminus on their own. This process, called **mass recruitment**, can produce large aggregations of workers at a food find in a short period. Workers that follow the trail and find abundant food reinforce the trail pheromone on their return to the nest; those that find little food or an exhausted source mark less or not at all, and the trail dwindles or disappears. Mass recruitment thus self-adjusts the distributions of foragers among alternative trails to match changing food distributions [203, 436].

Trail pheromones are generated by glands associated with the sting in most ants except for the Formicinae, which use hind gut contents, and species in the genera *Crematogaster* and *Amblyopone*, which have trail-marking glands on the tips of the hind legs [69, 594]. Those with glands on the abdomen drag its tip or the sting on the substrate when marking. Active trail marking chemicals include terpenoids, pyrazines, coumarins, fatty acids, and moderate-chain-length alcohols, esters, or aldehydes [412, 594]. Trail pheromones of ant species can consist of single chemicals, simple mixtures of 2–3 chemicals, or complex mixtures of up to 14 different components. The composition and fade-out times depend upon a species' ecological needs. Workers of the ant *Pheidole megacephala* have two trail pheromones: a long-lasting but weakly recruiting marker of recently explored areas and a short-lived

but strongly attractive marker of new food finds. The relative use of these two trail markers allows for complex shifts in recruitment, depending on the food distribution [226]. Pharaoh ants (*Monomorium pharaonis*) mark trail branches that no longer lead to productive sites with a long-lived repellent component and those that lead to productive trails with a short-lived attractant component [436, 716, 717]. The trails branch at a standard 60° angle relative to the main trunk; this helps workers that get off and then rediscover a trail to know which direction is back to the nest [157, 435]. Each colony of harvester ants (*Pogonomyrmex barbatus*) has 30–50 "patroller" workers that mark trunk trails connecting the nest to that day's popular branch trails to steer traffic in the correct directions [333]. Harvester ants learn landmarks quickly, and once they visit food sites, often do not need trail markers for subsequent trips [66]. This is also true of garden ants (*Lasius niger*), which appear to rely about equally on landmarks and trail pheromones to relocate food patches [247]. Raiding army ants (*Eciton burchelli*) divide a single trail into an inner homeward-bound lane and two peripheral outward-bound lanes (**Figure 14.30**) [171]. Incoming and outgoing leaf-cutter ants (*Atta* spp.) routinely use a single co-mingled lane: this helps the outgoing workers to assess the number and quality of leaf pieces being brought back on that route, and the tiny "minim" caste workers, which specialize in trail pheromone reinforcement, to focus on marking only one lane [246, 250].

Termites show a diversity of food pheromones similar to that of ants. They also deposit trail pheromones, which are usually mixtures of components, by dragging the gland-bearing parts of their abdomens on the substrate [168, 459, 486, 652, 706, 788, 789, 822]. One risk of continuous ant and termite food trails is their interception by eavesdroppers: both conspecific and heterospecific eavesdropping occurs on ant trails, and ants that prey on termites may use the latter's trails to locate victims [412]. Stingless bees reduce this risk by using a broken trail between colonies and food discoveries: a successful forager uses labial glands to dab pheromone on leaves, branches, and rocks every 5–10 meters en route back to the colony [468, 525, 838]. Once back at the nest, it performs an agitated and (in some species) acoustic display to recruit foragers, and then leads recruits at least part way back to the food. Species vary in the completeness of the "dotted" trail: some completely link the food and the colony. Others terminate 10–20 meters from the food source, and the only other signals are the broadcast food marks at the food, which have active spaces of only 1–12 meters [47, 439, 620, 622, 624]. A dotted trail is more difficult for an eavesdropper to follow than a solid one, and in those species where it extends only part of the way and requires a leader to show recruits where it starts, it is even less vulnerable. Despite these precautions, the stingless bee *Trigona spinipes* is able to follow the dotted odor trails of the smaller *Melipona rufiventris* and take over the latter's food finds [623].

A few other species of animals use food trails. Colonial tent caterpillars (*Malacosoma disstria*) navigate between sites

(A)

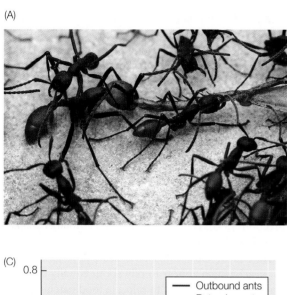

(C)

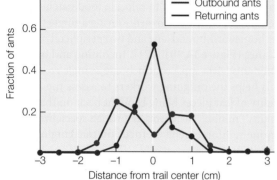

(B)

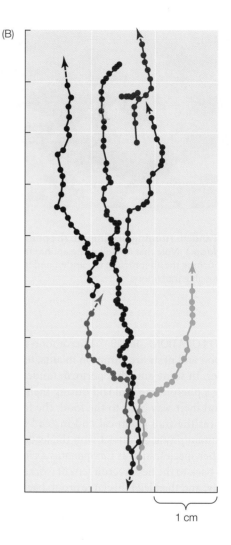

1 cm

FIGURE 14.30 **Traffic lanes of New World army ants (*Eciton burchelli*)** (A) Each night, army ants stream from the nest on raids to kill and bring back invertebrate prey. While ants in initial streams uniformly follow trails away from the nest, eventually ants with captured prey begin to return along the same trails. (B) Mapped trajectories of five outbound army ants (different shades of blue) and a concurrently returning ant (red) showing apparent separation into lanes. (C) Relative distributions of 84 outbound and 97 returning army ants relative to the center of the trail, showing clear three-lane configuration with the central return lane flanked by two outbound lanes. Trail pheromones are mostly replenished by returning ants. A two-lane stream would keep drifting to one side, because outbound ants tend to veer toward higher-pheromone concentrations in the inbound lane. The three-lane system does not drift, because the two outbound lanes both veer toward the center lane. (After [171].)

using a complex network of threads. Members of groups trace each other's recent trajectories using pheromones added to particular threads; threads not marked recently are ignored [155]. Naked mole rats (*Heterocephalus glaber*), a eusocial rodent, can track the recent tunnel trajectory of a colony member that returns to the central nest with a food sample; it is presumed that odors left by the recruiter provide the necessary information [456].

ENCODING FOOD LOCATION: DIRECTIONS Eavesdropping risks are minimized if returning foragers can provide directions to food finds. Where recruits are experienced and know many of the likely food locations, the returning forager need only announce its successful return, usually with some vibratory, acoustic, and/or pheromonal display, and share a regurgitated food sample with recruits; the latter can then check out known locations for this type of food. This, plus broadcast marking of recently visited sites, appears to work well for bumblebees and some ants [329, 589, 590, 677]. However, the small stingless bee, *Melipona mandacaia*, can forage up to 2 km away from the nest, and honeybees (*Apis mellifera*) forage up to 12 km away [227, 491, 764]: while workers in these taxa routinely mark food finds with broadcast pheromones and regurgitate samples back at the nest [350, 569], recruits have an enormous home range in which to search for the advertised food sites.

Stingless bees solve this challenge by using the dotted scent trails and initial leading of recruits discussed earlier. Honeybees have evolved a quite different solution: they perform a "waggle dance" to attending recruits in the nest that gives the approximate polar coordinates of current food sites [59, 926–928]. In the dwarf honeybee, *Apis florea*, this dance is performed on a horizontal surface at the top of the exposed combs: the distance to the site is encoded in

(A)

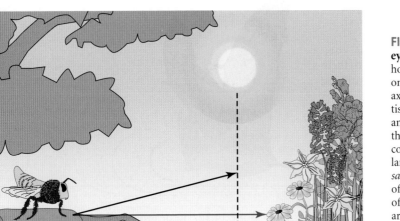

FIGURE 14.31 Waggle dances of honeybees (A) Returning foragers of the dwarf honeybee (*Apis florea*) perform waggle dances on top of the exposed single comb. The main axis of the waggles points toward the advertised food. Observer bees note the relative angle between the direction to the food and the sun, and use this to maintain a correct course even after they have passed familiar landmarks. (B) The giant honeybee (*Apis dorsata*) dances on the exposed vertical surfaces of its elaborate comb system. The main angle of the waggle dance relative to gravity (black arrow) tells recruited foragers what azimuth relative to the sun to follow to the food. The number of waggles per loop of the dance (dancers follow a figure-eight path) indicates the distance to the source. (C) Honeybees that nest and dance in dark cavities (*Apis cerana* and *A. mellifera*) augment the visual and touch components of the waggle dance with acoustical components to ensure access to directional and distance information. (After [227, 526].)

(B)   (C)

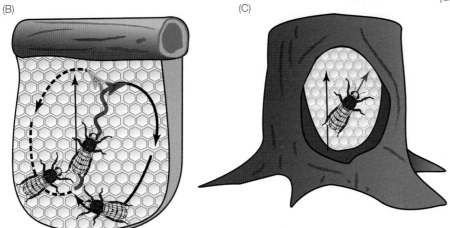

the number of waggles per dance and thus the duration of each rendition, whereas compass direction to the food in the horizontal plane—the **azimuth**—is given by the axis of the waggle axis relative to the location of the sun (**Figure 14.31A**) [524, 526]. The sun provides a reference for azimuth that can be determined from any location, even far from the nest. Because the sun moves over time, honeybees have internal clocks such that both dancers and recruits correct for shifts in the sun's position [928]. Honeybee species that dance on vertical combs (*A. dorsata, A. cerana, A. mellifera*) use gravity as a proxy for the position of the sun: a food site 10° to the right of the sun will be advertised with a waggle dance whose main axis is 10° to the right of an upward-pointing vector defined by gravity (**Figure 14.31B**). Honeybee species with nests in enclosed cavities (*A. cerana* and *A. mellifera*), and one that forages and dances at night (*A. dorsata*) add acoustic components to their dances that help recruits evaluate dance durations and angles in the dark (**Figure 14.31C**) [227]. They may also release pheromones to attract recruits [892]. The vigor and duration of a bout of dancing increases with the

profitability of the food find; foragers that return from dangerous flowers dance at lower rates or not at all [2, 765]. New colonies are formed when a subset of an existing honeybee colony leaves and bivouacs on a branch or a rock. Scouts from this swarm then leave to search for new nest sites. Returning scouts perform waggle dances on top of the perched swarm to advertise possible locations. Eventually, consensus is reached and the swarm migrates to its new nest site [766, 923].

The waggle dance minimizes the risks of eavesdropping, and the number of returning foragers advertising each site regulates the distribution of foraging effort similarly to trail reinforcement and mass recruitment in ants. The cost of this signal system is a significant cognitive burden on both senders and receivers [576]. Waggle dances are also an energetically expensive display. Multiple sensory organs and intensive brain processing must be used by recruits in the interpretation of dances [99]. Both parties use "optic flow," the rate at which images move across their visual fields, to measure the distance flown to or from a site [186, 187, 239, 881]. Both parties must learn by experience where and when the sun is located at each time of day. Honeybees can use polarized light patterns in patches of blue sky to estimate sun position when the sun itself is covered by clouds or patchy vegetation [928]. When the sky is completely cloudy, experienced honeybees can still exploit azimuth information in dances by remembering where the sun would be relative to familiar landmarks at each time of day [227]. Whereas naïve recruits clearly benefit from attending to dances [712], knowledge of local geography leads many experienced workers to ignore them and simply use floral odors on dancers, the type of food regurgitated, and their own most recent trips to direct their subsequent foraging [14, 67, 302, 351–354]. Ignoring of dance information occurs less often in tropical than in temperate habitats since food patches are smaller and less predictable in the tropics [100, 211, 212, 774].

Autocommunication

We conclude this chapter with a communication system in which sender and receiver are the same individual: autocommunication. The sender emits a signal that is altered by the immediate environment during propagation. It then retrieves the altered signal and identifies the alterations (**Figure 14.32**). As with all signals, interpretation requires a coding scheme correlating specific signal alterations with specific environmental conditions. In this signaling system, however, the sender and receiver have completely identical interests. The economics of autocommunication can thus be focused on maximizing efficacy [612]. There are two primary models: electrolocation in gymnotiform and mormyriform fish, and echolocation in bats and toothed cetaceans. Since we have already discussed electrolocation in some detail in Chapter 7, we focus here on echolocation. The basic principles are simple: a brief sound is emitted, and the time it takes for an echo to return can be used to estimate target distance. The sophisticated directionality of most animal ears facilitates estimation of target azimuth and elevation. This may be augmented by beaming of the outgoing sound using a nose leaf in bats or acoustic melon in cetaceans. If the emitted sound contains multiple frequencies, differences in the spectral compositions of emitted and reflected sounds can be used to assess target shape, size, and in water, composition.

Primitive forms of echolocation occur in a few insects, small mammals, and birds. Termites vibrate wood substrates and monitor the reverberations to assess wood quality, and some parasitic wasps induce substrate vibrations in plants to detect embedded animal prey [245, 490]. Mole rats (*Spalax ehrenbergi*) generate low frequency vibrations in the earth and use the echoes to detect obstacles ahead of tunnels they are digging [471]. Shrews with poor vision (e.g., *Sorex araneus* and *Crocidura russula*) emit repetitive "twitter" sounds to help detect nearby obstacles [786]. Swiftlets (*Aerodromus* spp.) and the oilbird (*Steatornis caripensis*) roost in caves and produce clicks whose echoes help them avoid obstacles while flying in the dark [286, 340, 343, 344, 674, 895]. However, these systems are not very elaborate: the same syringeal mechanisms are used for social calls and clicks, auditory systems are minimally modified for echolocation, and the birds only respond to echoes from large objects [343, 484, 573, 826, 876, 877, 894, 896].

In contrast, the echolocation systems of bats and toothed cetaceans (Odontocetes) are highly sophisticated. Both taxa face the challenges of rapid foraging in three dimensions,

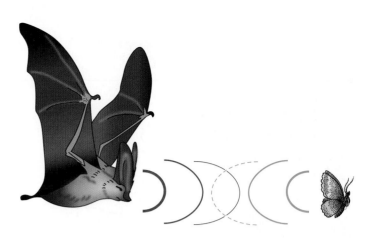

FIGURE 14.32　Basic process of echolocation The echolocator emits a short pulse of high-frequency sound that radiates away from it with usual spreading, heat, and scattering amplitude reduction (red waves). Nearby objects produce echoes (blue) that radiate back to the echolocator with second-stage spreading, heat, and scattering losses. If the echolocator can detect the resulting faint echo, it can use the time delay between emission and echo reception to estimate the distance to the object. Determination of target azimuth and altitude can be improved by emitting sounds through beaming structures such as nose leaves (see Figure 14.39).

limited visibility, and accurate interception of small moving targets. Both appear to have solved these problems in the Eocene, and then gone on to high levels of diversification and worldwide distribution [37, 229, 452, 516, 517, 547, 606, 635, 636, 804, 805, 832, 842, 882–884, 891, 910]. Whereas the two taxa independently hit on the same solutions for some problems, in other cases they required different refinements due to differences between the properties of sound in air versus water [32]. One shared refinement, limited to mammals among vertebrates, was a shift to very high frequencies in the emitted sounds. As we saw in Chapter 2, significant echoes are generated only when wavelengths of the impinging sound are the same size or smaller than the echo-producing target. Whereas low frequency sounds produce decent echoes from cave walls, echoes from small insects or fish require very small wavelengths, and thus very high frequencies: detection of a 0.5 cm moth in air demands frequencies of 69 kHz or higher. Because of different sound speeds, wavelengths for any frequency in water are 4.4 times as long as those in air, and thus require frequencies 4.4 times as high in order to produce echoes from the same-sized target. Because cetaceans are much larger than bats and feed on larger prey, they can get by with echolocation frequencies in water that are about the same as those used by bats in air: 12–200 kHz.

A second common feature is the limitation of echolocation pulse durations and emission rates to avoid overlap between outgoing pulses and returning echoes (**Figure 14.33**) [408, 756]. This is necessary because double spreading losses (first for the outgoing sound and then for the returning echo) and severe heat losses and scatter due to the high frequencies make echo amplitudes minuscule compared to those of emitted pulses. Given typical detection distances and the fact that pulse durations have to become even shorter as the echolocator approaches its target, most bats use pulses of 0.5–25 msec duration. The higher speed of sound in water requires

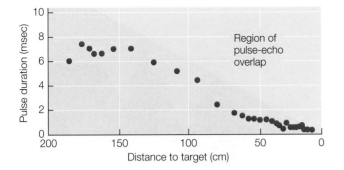

FIGURE 14.33   Avoidance of pulse-echo overlap by European pipistrelle bat (*Pipistrellus pipistrellus*) during interception of insect prey   Each data point shows the duration of an echolocation pulse emitted at that distance from the target. As the bat moves closer to the insect (left to right), it consistently reduces pulse duration so that it remains under the limit imposed by potential overlap of the outgoing pulse and the returning echo. (After [462].)

cetaceans to use much shorter pulses: typical dolphin (*Tursiops truncatus*) pulses are 0.04–0.08 msec in duration, allowing for only 4–8 cycles of the sound. This contrasts with hundreds of cycles in a typical bat pulse. Whereas bats can vary pulse fine structure to meet design challenges, cetacean pulses are much more constrained by their short durations [27]. Much of the sophistication of bat echolocation arises from their ability to vary pulse fine structure according to informational needs. Each stage in a typical predator/prey interaction, (detection, evaluation, interception, attack, and consumption), may require a different optimal pulse design, and the optimal designs at each stage can vary depending upon the habitat in which the bat is hunting. The ability to diversify their echolocation pulses may be one reason why there are currently over 760 species of echolocating bats worldwide, whereas there are only about 74 extant echolocating cetaceans [634, 784, 915].

Web Topic 14.2 *Videos of echolocating bats and cetaceans*

Here we provide web links to a variety of video sources demonstrating echolocation or showing wild bats and cetaceans hunting.

Bat echolocation

TARGET DETECTION   Bats that forage in open areas and high above the ground are not likely to get echoes from anything other than aerial prey; there are no objects nearby to generate "clutter" echoes. The optimal prey detection pulse is emitted at low rates, has a loud amplitude and long duration, and consists of a relatively low and constant frequency ([306, 405, 461, 475, 755]. Keeping emission rates low ensures that there is no pulse/echo overlap even from quite distant targets. Pulse amplitudes measured at 10 cm from a bat can be as high as

120–130 dB SPL (base reference for measuring sound amplitudes in air; see Web Topic 2.3); high amplitudes maximize the range at which echoes can be detected [409, 873]. Even at these amplitudes, insect prey are only detected at distances of 1–3 m [409, 755, 756, 935]. The long duration increases the chances that a prey item will encounter the sound beam and produce an echo. In addition, if the insect flaps its wings multiple times while in the sound beam, this will generate a string of amplitude modulations in the echo ("glints") that will be highly conspicuous [482]. The optimal search phase frequency is a compromise between the low values that minimize heat losses and extend range and the higher values required to produce echoes from small targets [252].

When bats forage closer to substrates, background clutter echoes become more intense and the problem of echo detection grades into a problem of echo discrimination (**Figure 14.34**). As we shall see in the next section, a frequency-modulated (FM) pulse design is usually better for target discrimination than a constant-frequency (CF) one. Many bats forage in intermediate zones where background echoes are detectable, but not overwhelming. Such species favor pulses with both CF and FM components. This provides a compromise that allows them to forage in both uncluttered and cluttered contexts. Examples include bats that trawl for fish just under the water's surface or insects just above it (e.g., various species in the genera *Macrophyllum*, *Myotis*, and *Noctilio*). Where the water is smooth, background echoes from the water's surface will follow trajectories away from the bat [98, 782, 785, 944]. Echoes from ripples generated by fish near the surface or insects flying just over the surface can then be used by the bats to find prey with little overlapping clutter. However, if the water's surface is irregular due to wind or currents, the levels of background echoes can become very high. Since a trawling bat might encounter successive patches of smooth and turbulent water on even a single pass along a river or over a pond, it will typically emit pulses that combine CF and FM components.

TARGET DISCRIMINATION AND CLASSIFICATION   By emitting a pulse with many frequency components and comparing the spectral compositions of echoes and emitted pulses, a bat is able to categorize, and in some cases identify, different targets [96, 201, 296, 349, 796, 798, 874, 925, 940]. This exploits the fact that objects of different sizes and shapes reflect different frequencies with different amplitudes [987, 988]. There is also considerable evidence that bats learn to recognize regularly visited routes and landscapes using some form of spectral classification [94, 443, 598, 739, 756]. The constant-frequency pulses used by many species for prey detection in uncluttered contexts cannot support spectral discriminations between prey and backgrounds, multiple landmarks, or different kinds of prey. As noted earlier, bats in moderately cluttered contexts include a frequency-modulated (FM) component either at the beginning or at the end of any CF component. Even open-country bats usually shift from CF detection signals to pure FM pulses as they

(A)

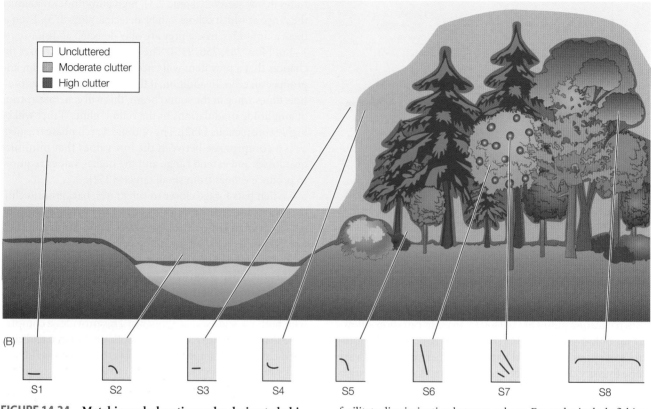

(B)

S1	S2	S3	S4	S5	S6	S7	S8

FIGURE 14.34 **Matching echolocation pulse design to habitat** (A) Typical intensities of clutter (background) echoes faced by foraging bats in different habitats. Bats foraging high above the ground and far from any background (uncluttered zones) detect echoes from aerial prey only. Along forest edges, above the forest canopy, or over smooth water surfaces (moderate clutter), background echoes are detectable, but can usually be discriminated from prey echoes. Bats that forage close to complex surfaces such as vegetation and the ground (high clutter) face a difficult challenge discriminating between background and prey echoes. (B) Spectrograms of echolocation pulses of bats foraging in each of the above habitats. Bats such as *Nyctalus noctula* (S1) that feed in uncluttered zones favor loud and constant-frequency (CF) pulses at as low a frequency as their body size permits, and durations just below that leading to pulse/echo overlap. Bats foraging in moderately cluttered zones combine a CF component to maximize detection abilities with a frequency-modulated (FM) component to facilitate discrimination between echoes. Examples include fishing bats such as *Noctilio leporinus* (S2) and under-canopy insectivores such as *Pteronotus quadridens* (S5). The tiny Kuhl's pipistrelle bat (*Pipistrellus kuhli*) often forages right at the boundary between uncluttered and moderately cluttered zones: it thus switches between CF (S3) and FM-CF (S4) pulses depending on the level of background echoes. Bats like *Myotis myotis* (S6) that glean insects right off the foliage or frugivores like *Artibeus jamaicensis* (S7) require very broadband FM pulses to allow them to discriminate between target and background echoes. Finally, horseshoe bats such as *Rhinolophus ferrumequinum* (S8) use CF pulses that are so long the echoes overlap with pulses. This allows the bats to null out echoes from very close background surfaces but detect "glints" from moving insect prey. All spectrograms with frequency scale 0–150 kHz, and time scales of 0–25 msec except for S2 (time 0–15 msec) and S8 (time 0–100 msec). (After [755].)

approach the target. As we shall see in the next section, this improves target interception accuracy, but it also provides the spectral diversity necessary to characterize targets into types. One additional strategy is found in some insectivorous bats that feed in open areas adjacent to cliffs or forest edge [457]. Echoes of the most recently emitted pulse from small nearby prey might overlap in time with delayed echoes from more distant forest or cliff surfaces. By alternating between two slightly different CF frequencies on successive pulses, the bats can tag the pulses and thus discriminate between the two sources of echoes (**Figure 14.35**) [408].

Bats that "glean" immobile foods very close to vegetation or the ground face an even more challenging discrimination

between food and inedible background echoes. Nearly all such species use pulses that are rapidly frequency-modulated from high to low frequencies and often include 2–3 harmonics to extend the range further [341, 411, 449, 592, 616, 755, 756]. Some species can vary the relative emphases of different harmonics depending upon the current discrimination task [515]. FM pulses produce a very rich set of frequency components that can be used for target classification. Some species, such as *Myotis nattereri*, which plucks immobile spiders from their webs, appear able to discriminate between prey and inedible targets largely through echolocation [781, 784]. Other species augment echolocation with other cues such as sounds or odors generated by food items: reliance

(A)

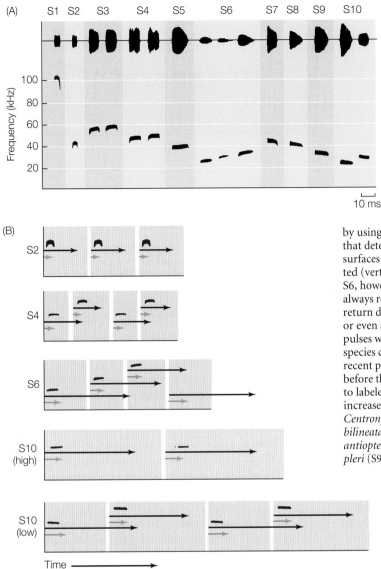

10 ms

FIGURE 14.35 Pulse labeling by neotropical emballonurid bats (A) Most of the 10 sampled species favor 5–10 msec CF pulses for hunting insect prey (presumably to detect glints from prey wing movements). Whereas species foraging in uncluttered open zones (S7–S10) or over water (S1) usually emit one pulse at a time, three of the species feeding along the forest edge (S3, S4, and S6) regularly emit pairs or triplets of pulses labeled with slightly different frequencies. The high-altitude forager S10 also shows labeled pairs on occasions. (B) Bats feeding close to large objects like trees at a forest edge or cliffs must use fairly high pulse emission rates to avoid collisions with objects. They then risk confounding echoes of their most recent pulse reflected by nearby prey with delayed echoes reflected by large but distant objects from the prior pulse. Species S2 avoids this problem by using pulse durations, amplitudes, and repetition rates such that detectable echoes from both prey (blue arrow) and large surfaces (black arrow) arrive well before the next pulse is emitted (vertical white bars indicate pulse emissions). Species S4 and S6, however, use longer and louder pulses, and while prey echoes always return before the next pulse is emitted, large surfaces may return detectable echoes after the emission of the next pulse (S4) or even after two more pulses (S6). By labeling pairs or triplets of pulses with different frequencies (red images on diagram), these species can determine whether a given echo was from the most recent pulse or from an earlier one. Species S10 receives all echoes before the next pulse is emitted when flying high, but switches to labeled pairs when it is close enough to the canopy to have to increase its pulse emission rate. Species: *Rhynchonycteris naso* (S1), *Centronycteris centralis* (S2), *Saccopteryx leptura* (S3), *Saccopteryx bilineata* (S4), *Cyttarops alecto* (S5), *Cormura brevirostris* (S6), *Balantiopteryx plicata* (S7), *Peropteryx macrotis* (S8), *Peropteryx kappleri* (S9), and *Diclidurus alba* (S10). (After [457].)

on multiple cues has been demonstrated in bats that eat ripe fruit, flower nectar, calling frogs, sleeping or migrating birds, the blood of large mammals (vampire), and small mammals including other bats [72, 240, 316, 485, 553, 746].

Bats in the families Rhinolophidae, Hipposideridae, and Mormoopidae have an alternative and highly specialized method of discriminating between prey and background echoes [361, 616, 751, 783]. Because they are often very close to backgrounds, they have given up on avoiding pulse/echo overlap and instead emit long-duration constant-frequency (CF) pulses. These pulses tend to be at much higher frequencies (80–90 kHz) than those used by open-country foragers (20–40 kHz). The overlap enables them to adjust the frequency of the emitted pulses to compensate for Doppler shifts due to their motion relative to the static background (**Figure 14.36**). All echoes from immobile backgrounds thus return at a standard reference frequency to which the bats are extremely sensitive. When insects near

such a background flutter their wings, they create echoes with Doppler shifts different from that generated by the bat's motions. These are perceived by the bats as glints with frequencies slightly different from the reference frequency. Because of the bat's sensitivity to this narrow band of frequencies, the glints are highly conspicuous and allow the bat to differentiate echoes of moving prey from those of the static background.

PREY INTERCEPTION AND ATTACK   At the same time that an open country bat begins to classify a target, it must also begin to define the target's location. Most insectivorous bats first detect prey at distances of 1–2 meters: a bat flying at a velocity of 3–8 m/sec will then have between one-quarter and two-thirds of a second to characterize the target, plot its position in space, and intercept it. Many aerial insect prey can detect bat echolocation sounds and take evasive actions such as diving to the ground or into nearby vegetation, flying erratically, emitting aposematic warnings, or even trying to jam the bat's echolocation system [42, 43, 167, 385, 424, 426, 872, 980]. Interception thus involves tracking an unpredictably moving target. Bats other than close surface gleaners typically produce a "terminal buzz" during prey interception

FIGURE 14.36 Doppler compensation by the greater horseshoe bat (*Rhinolophus ferrumequinum*) (A) Individual bats were outfitted with microphone-bearing telemetry devices and allowed to fly from one perch to another (4–5 m). A microphone recorded both the emitted pulses of the bats and the returned echoes. The plot shows how emitted frequency is decreased during flight, to compensate for Doppler shifts. (B) A plot of the measured pulse and echo frequencies of a similarly outfitted greater horseshoe bat on a 3 m flight. Like most species showing Doppler compensation, this bat undercompensates slightly relative to its resting (nonflying) frequency. (A after [406]; B after [779].)

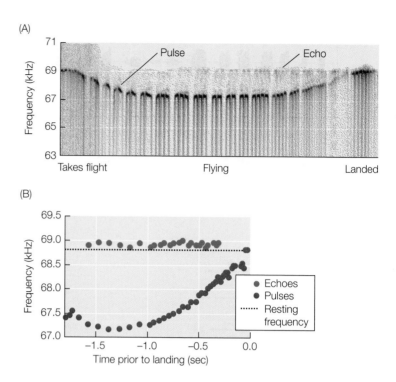

(**Figure 14.37**). As the bat approaches the target, it can increase its pulse emission rate without pulse/echo overlap if it also decreases pulse duration [341, 342, 461, 755]. This allows for rapid updating of both the remaining distance to the target and its changing angular location relative to the bat. Bats also tend to reduce pulse amplitude and auditory system sensitivity as they get closer to the target: in some cases, this maintains a constant target echo amplitude despite changing proximity, but in other cases, amplitude reduction may have other functions, such as minimizing clutter echoes from unimportant targets [90, 382, 383, 404, 406, 411, 873].

Relatively accurate location of prey is necessary to effect interception. Many bats encircle their prey with the wings, body, and tail membrane before biting it: this maneuver must be executed within 1–3 cm of insect prey to effect capture [342]. If the prey is moving, the bat needs fine updates on the prey's location in order to estimate its trajectory and time the interception actions. Finally, if the prey are near to obstacles, the bat must be able to discriminate between the echoes of the target and nearby objects. As with target evaluation, the long-duration CF pulses used by many species to detect prey are not very useful for the accurate determination of prey position. Like other mammals, bats require multiple frequencies in a sound to identify its source azimuth and altitude accurately. Estimates of target distance are similarly dependent on pulse bandwidth. If only part of a CF pulse is reflected back as an echo, the bat has no way of knowing whether it came from the beginning, middle, or end of the pulse. The uncertainty can be reduced by shortening pulse duration, but there is a better solution. An FM pulse is "labeled" in that each piece consists of a different frequency. As long as the bat knows when it emitted each frequency segment, any

short piece returning as an echo can be used to estimate target distance. If enough different pieces are returned in the echo, estimates from each can be combined into one value with low error. Once they switch to FM pulses, open-country bats can estimate target ranges to within 0.6–1.5 cm, well within the tolerances for capture maneuvers [560, 597, 793, 794]. The ability of these bats to detect *changes* in location is even more remarkable: some bats can respond to target movements as small as 0.007–0.034 cm [574, 596, 794]. Even finer discriminations have been shown (less than 0.001 cm) in big brown bats (*Eptesicus fuscus*), but such tiny differences in distance cannot be very useful for prey location; instead, this acuity appears to be required for perceptual separation of overlapping echoes of targets and clutter closer than 6 cm or concurrent reflections from different parts of the same target (e.g., wings and head of a moth) [800, 836]. Early skepticism that any mammalian auditory system could achieve these levels of accuracy and resolution [57, 574, 670, 754], which are close to the physical limits identified by sonar/radar theory [467, 474, 971], have since been allayed by experiments and a number of increasingly realistic auditory system models [10, 58, 92, 274, 564–566, 614, 616, 732, 736, 759, 795, 797–799, 802, 803, 950].

Most bats change the fine structure of their pulses as they approach a target and when they shift to different habitats [253, 342, 442, 461, 560, 592, 639, 755, 756]. One reason to do so is to correct for two sources of error in target distance estimation [91–93, 411, 449, 920, 935]. The first source of error is the fact that a rapidly flying bat emits a pulse at one location and then receives the echoes closer to an approached target. Since it relies on the delay between emission and echo to gauge distance, it will underestimate the distance between

(A)

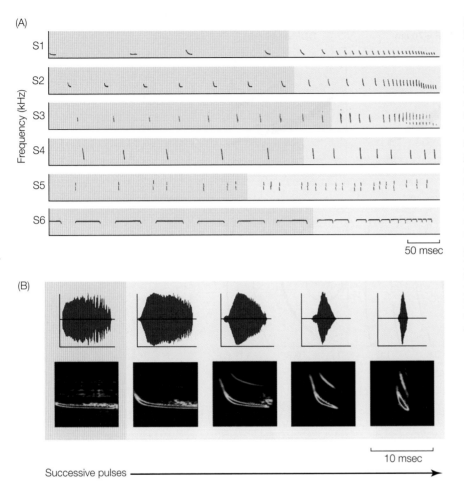

50 msec

(B)

Successive pulses ⟶

10 msec

FIGURE 14.37 Terminal buzzes by bats (A) Patterns of bat pulse emission during the transition from search (tan) to target approach and contact (blue). Insectivorous species include open country forager *Nyctalus noctula* (S1); moderate clutter forager *Pipistrellus pipistrellus* (S2); trawler over water *Myotis daubentonii* (S3); foliage gleaner *Myotis myotis* (S4); and close-range flutter specialist *Rhinolophus ferrumequinum* (S6). A frugivorous example is *Carollia perspicillata* (S5). (B) Changes in pulse fine structure as an aerial insectivore *Eptesicus bottae* searches for (tan) and approaches and captures a moth (blue). The top row shows waveforms and the bottom row corresponding spectrograms (frequency scale 0–100 kHz). Each pulse in the approach phase was emitted at a distance about half that of the prior pulse. As the bat gets closer to its prey, its pulses shift from a CF detection strategy to an FM location and discrimination strategy. (A after [755]; B after [411].)

bat and target at the time of pulse emission. A second set of errors arises from Doppler shifts due to the motion of the bat relative to the target: if the bat is approaching the target, all segments in the emitted pulse will return in the echoes with their frequencies raised by 2–5%. This complicates the strategy of tagging different parts of the pulse with different frequencies: for FM pulses that sweep from high to low frequencies, a segment in a Doppler-shifted echo with the same frequency as a focal segment in the emitted call actually corresponds to a segment in the original call emitted *after* the focal segment. As a result, not correcting for the Doppler shifts when flying toward the target causes the bat to overestimate the distance. For any given target distance, it is possible to find a pulse bandwidth, duration, modulation curvature, and harmonic weighting at which these two errors just cancel each other out. Put another way, each pulse pattern has its own **distance of focus** at which ranging errors are minimized (**Figure 14.38**). Bats producing FM pulses can use the distance estimated from one pulse to predict the most useful distance of focus in the next pulse. This results in a steady change in pulse durations, bandwidths, curvatures, and harmonic emphases as the bat moves from detection to interception. Similar adjustments may be necessary when bats forage in different habitats. Although the observed pulse structures in some bats show excellent fits to distance of

focus predictions, others fit less perfectly [91, 92, 410, 449]. This is likely due to trade-offs between maximizing distance accuracy while also classifying targets and navigating without collisions through the landscape.

Although wide-bandwidth FM pulses provide an excellent basis for the angular location of targets, many bats show additional refinements that improve angular resolutions. Open country-bats usually emit pulses from their mouths to maximize the width of the sound fields [381, 586]. They can, however, adjust their oral cavities to create moderate beams for scanning targets [300, 874, 875]. Species that forage close to vegetation favor much more narrow emission beams, and many emit their pulses through their nostrils (**Figure 14.39**). Interference between sounds emerging from these two adjacent openings can be exploited to create a narrow sound beam. Many of these bats host accessory nose leaf appendages, ridges, and furrows to create local resonances and directional reflections [313, 380, 991–993]. The size, shape, and structure of noseleaves is better correlated with habitat and diet than with the type of echolocation pulses used [15, 85, 325]. Elaboration of pinnae is also common in echolocating bats. A given species' pinnae are usually tuned to the dominant frequencies in its echolocation calls, with additional refinements that complement the particular pulse designs used for target ranging and discrimination [194, 638,

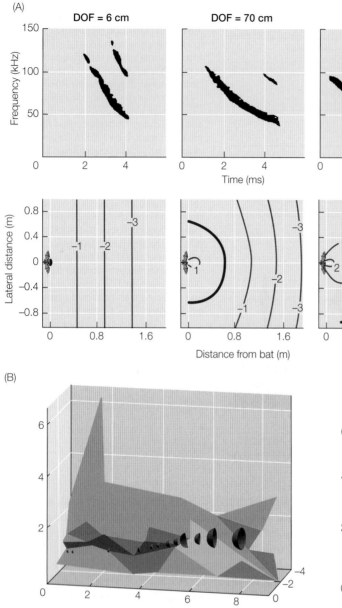

(A)

(B)

FIGURE 14.38 **Effects on target distance resolution of alternative echolocation calls of whiskered bats (*Myotis mystacinus*)** (A) Distance of focus (DOF) is the distance in front of the bat at which Doppler errors in distance estimation are just compensated by movement errors. Top row: spectrograms of three typical whiskered bat pulses. More steeply modulated pulses bring the DOF closer to the bat, and less steeply modulated pulses move it away. Lower row: Degree of error (in cm) for bat's estimation of target distance for each of three pulse shapes shown in the top row. Red line shows DOF where error is minimal. (B) Two recorded instances of whiskered bats approaching from the right and then flying very close to a hedge (green). Red areas show the location of the DOF regions in front of the bat. When far from the hedge, bats use slowly modulated pulses producing large DOF areas; as they get close to the hedge, they increase their modulation rate to reduce the DOF to the small distance needed to navigate past the surface. (After [410].)

662, 801, 981]. The tragus is often enlarged, specially shaped, and carefully located to maximize the ability to determine target elevation [143, 503, 601, 973, 974]. Whereas large neotropical gleaners can move their two ears independently, Old World equivalents tend to have a flap of skin connecting the two ears to keep their foci parallel [96, 602, 932]. Bats like rhinolophids that use CF search signals tend to combine narrow beams from their noseleaves with high directionality in their pinnae to create a single narrow perceptual focus for prey scanning [347, 752].

Cetacean echolocation

Correcting for the different acoustic impedances of air and water, the energy flux levels (joules/m^2) produced by different species of echolocating bats and odontocete cetaceans are largely overlapping [25, 29, 32, 921]; however, the loudest

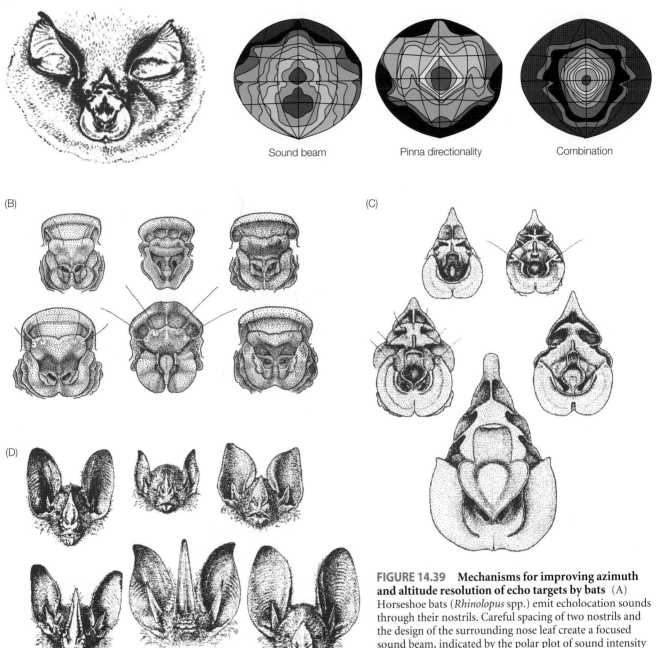

(A)

Sound beam   Pinna directionality   Combination

(B)

(C)

(D)

FIGURE 14.39   Mechanisms for improving azimuth and altitude resolution of echo targets by bats (A) Horseshoe bats (*Rhinolopus* spp.) emit echolocation sounds through their nostrils. Careful spacing of two nostrils and the design of the surrounding nose leaf create a focused sound beam, indicated by the polar plot of sound intensity in front of the bat. Red indicates the region of maximal sound intensity. Pinnae are also highly directional for sounds returning along the main body axis. The polar plot indicates zones of maximal sensitivity (red). Sound beam and pinna designs are matched so that noisy zones in one cancel out those in the other. The combination yields a highly focused zone of echo sampling. (B) Nose leaf designs for various African horseshoe bats in the genus *Hipposideros*. (C) Equivalent sample of nose leaf designs for various African *Rhinolophus* species. (D) Elongated nose leaves and matching pinna designs of six neotropical phyllostomatid bats from the neotropics. (A after [753]; B,C from [719]; D from [323].)

sounds made by any animal are the echolocation pulses of sperm whales (*Physeter macrocephalus*) [587]. Like bats, many odontocetes use low to moderate pulse emission rates when searching for prey and a "buzz" at higher rates during the interception phase [25, 202, 420, 448, 537, 539, 582, 828, 919, 938]. Also like bats, most odontocetes reduce pulse amplitudes and auditory sensitivities as they get close to a target, keeping echo amplitudes roughly constant [23, 28, 30, 441, 607, 713, 869–871].

There are also major differences between the two taxa [25, 32]. Because high-frequency sound attenuation is so much lower in water than in air, odontocetes can detect a target of a given size at a greater distance than can a bat: big brown bats

(*Eptesicus fuscus*) can just detect a 2 cm sphere at 5 m under quiet conditions, whereas a bottlenosed dolphin (*Tursiops truncatus*) can detect a similarly sized object at 73 m [25]. Many odontocetes can detect prey at 70–300 m distances [6,

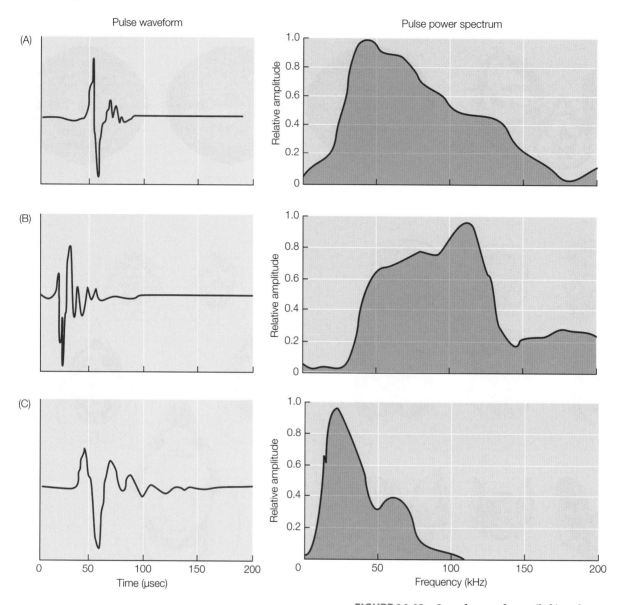

Pulse waveform   Pulse power spectrum

FIGURE 14.40 **Sample waveforms (left) and power spectra (right) of echolocation pulses of three common odontocete cetaceans** (A) Atlantic spotted dolphin (*Stenella frontalis*). (B) White-beaked dolphin (*Lagenorhynchus albirostris*). (C) Killer whale (*Orcinus orca*). (After [29].)

25, 538, 792]. Although the speed of sound in water is 4.4 times as high as that in air, odontocete pulses are only one-hundredth as long as those in bats. This is far shorter than needed to avoid pulse/echo overlap. However, the durations necessary to avoid overlap, especially when foraging near the water surface, surely preclude the serial FM fine structure used by bats to measure target distances, angles, and properties. The odontocete solution is to make pulses even shorter, so that they consist of a broad range of frequencies. As we saw in Chapter 2, extremely short duration sounds have very wide frequency compositions. Typical cetacean pulse durations of 0.03–0.10 msec are short enough to generate bandwidths of 20–40 kHz (**Figure 14.40**) [29, 538, 600]. Multiple frequencies allow for multiple measures of delays and distances even if they are emitted simultaneously (as opposed to serially as in FM bats); this greatly increases the accuracy of distance and source angle measurements [971]. Typical odontocetes, like bats, are thus able to measure target distances with accuracies of several cm [25]. All odontocete cetaceans emit their pulses through an oily melon structure on their foreheads (see Chapter 2), and retrieve echoes through fatty channels in their lower jaws (see Chapter 3). As with bat noseleaves, odonotocete melons produce echolocation beams carefully adapted to a given species' ecological tasks [25, 26, 31, 174, 175]. Most melons produce highly focused beams, at least in the forward direction [5, 24–26, 31, 632, 700, 997].Odontocetes also have a large advantage over bats when classifying targets: whereas bat targets have acoustic impedances so much higher than air that all incident energy is reflected at the target surface, odontocetes live in a world in which target impedances are sufficiently similar to water that some incident pulse energy enters the target. This can lead to successive reflections from both external and internal parts of

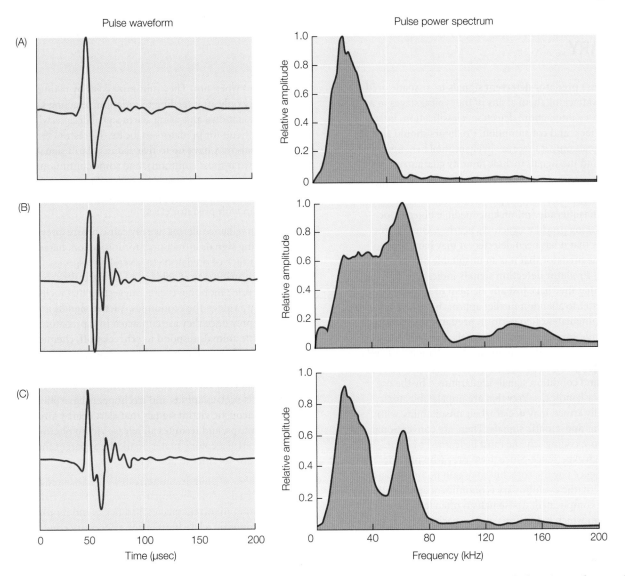

Pulse waveform

Pulse power spectrum

(A)

(B)

(C)

Time (μsec)

Frequency (kHz)

FIGURE 14.41 **Variation in echolocation pulses used by a single species of dolphin** Examples of three different echolocation pulse types produced by bottlenose dolphins (*Tursiops truncatus*), showing how subtle changes in the pulse waveforms (left) can produce major differences in the corresponding power spectra (right). (After [600].)

the target, as well as induce resonant vibrations in the target that interact with echo formation [25]. All of these effects will produce signature interactions between a given type of target and a given type of incident pulse. Bats enhance their ability to classify targets by sampling the same target with successive pulses that vary in serial fine structure: this provides multiple comparisons between the spectra of the emitted pulses and those of the echoes. Most odontocetes cannot incorporate serial frequency modulation in their pulses, but they can modulate the amplitude of successive cycles within a pulse, and this significantly changes the pulse spectrum [25, 26, 29, 30, 421]. Many species produce 4–7 different pulse types that differ in bandwidth, whether their spectrum is unimodal or bimodal, and if bimodal, which peak frequency is dominant (**Figure 14.41**). They then use all of these types in varying mixtures to obtain diverse echo spectra as they approach a target. Target classification in odontocetes is thus as good and in many cases better than that of bats [25, 33, 292, 540, 600, 841]. Two species of beaked whales that feed at great ocean depths, *Ziphius cavirostris* and *Mesoplodon densirostris*, take a different tack. They use search pulses that are 0.2–0.3

msec in duration, (3–10 times longer than other odontocetes), and frequency modulate them serially from 25–50 kHz [447, 448, 539, 996]. They can only begin emitting these longer-duration FM pulses after they have reached depths where there is no pulse/echo overlap from the surface. When they find prey, they switch to terminal buzz pulses of half the duration, twice the bandwidth (despite no FM), and amplitudes that *increase* with increasing proximity. It has been suggested that the use of serial FM pulses in the search phase is not to improve range or angle accuracy, but to enhance discrimination between different types of targets [448, 841]. Odontocetes, like bats, may interrupt pulse emission and use passive listening to detect soniferous prey, or in the case of killer whales, avoid alerting whale and dolphin prey of their approach [199, 290].

SUMMARY

1. Prey may use **predator deterrent signals** to dissuade predators from attacking them at any of the typical stages in a predator/prey interaction: detection, classification, interception, attack, and consumption. Predators should attend to such signals only if there are likely mutual benefits to doing so, and the signals include honesty guarantees.

2. Prey may try to avoid detection by using different forms of **camouflage**, including **crypsis** (blending into the background), **masquerade** (mimicking inedible objects), or **decoys** (placing preylike structures nearby). Once a prey determines that it has been detected, it may have time to signal to the predator that it is an unsuitable or less suitable target. **Predator detection signals** indicate to stalking predators that they have lost any element of surprise and may do better to abandon further approach. Fish, small birds, and ungulates often perform **predator inspection** as a group while emitting detection signals. **Prey condition signals** are given to coursing predators to demonstrate relative agility, health, or energy reserves. Honesty of predator detection and condition signals is guaranteed by the use of indices or handicaps. Prey that are unpalatable, toxic, or physically armed may declare their unsuitability with conspicuous **aposematic signals**. These are conventional signals whose coding scheme must be learned by predators through experience. Since those prey who provide that experience pay the costs but may gain no immediate direct benefit, the evolutionary economics of aposematic signals are complicated. Aposematism often leads to **mimicry rings** in which many unpalatable species (**Müllerian mimics**) and more palatable cheaters (**Batesian mimics**) converge on a common set of signals. Once under attack by a predator, some species produce **deimatic displays** that use unexpected colors, patterns, movements, or sounds to startle and distract the predator. **Distress signals** may attract other predators who interrupt the attack of the first and increase the chances that the prey can escape in the ensuing tussle.

3. **Alarm signals** warn nearby prey of the presence of predators. Possible compensatory benefits to senders to justify the costs of alarm signals include mutualistic coordination of flight, by-product mutualism of predator detection calls, nepotistic benefits to nearby kin, reciprocal surveillance by members of stable groups, subsequent mating advantages, status enhancement, or manipulation of nearby prey to minimize sender risks.

4. Most species face a trade-off between **surveillance** for predators and foraging. In some species, any member in a group may interrupt foraging to survey for predators; such species often show synchronous surveys since no member wants to be conspicuously delayed if the group flees. In other species, group members take turns as **sentinels** while others concentrate on feeding. Sentinels give alarm signals when predators are detected and may give regular "all clear" signals when not. They minimize costs by taking on sentinel duty only after feeding to satiation, selecting safe surveillance stations, and using alarm signal designs whose source is difficult for predators to locate. Terrestrial birds and mammals may have up to five distinct alarm signals (although two is most common), and some continuous variation within a signal type. Alarm signal variants are more often associated with appropriate flight strategies and urgency than with predator class.

5. Whereas surveillance alarms usually cause nearby prey to flee, **mobbing signals** attract prey for inspection, harassment, and attack of predators. In several bird species, neighbors contribute to mobbing only if the callers contributed to prior mobbing events, suggesting that reciprocity may play a role in the economics. **Victim signals** are emitted by prey once they are contacted by a predator. Many aquatic animals respond to **schreckstoff**, chemicals released from damaged conspecifics and even from heterospecifics. While these chemicals act as cues in many cases, some frogs actively secrete victim signals when attacked. Social insects such as aphids and treehoppers have pheromonal and acoustic victim signals that alert nearby kin. Bees, ants, wasps, and termites all release victim pheromones that attract colony mates, mark the apparent threat, and induce conspecifics to attack it.

6. A large number of aquatic and terrestrial species attend to the alarm signals of other sympatric species. This is particularly common in mixed-species bird flocks and monkey troops. Taxonomic relatedness is not a necessary condition for this eavesdropping, and there are many examples where reptiles, birds, and mammals attend to the alarm signals of another class. False alarms are surprisingly common in both single species and mixed-species assemblages. Models show that they can be increasingly tolerated economically as group size increases, predator attacks become less common, or group responses provide an index of veracity. Templates for the production and recognition of conspecific alarm signals are largely innate; experience is needed to refine and focus these signals on appropriate entities. Recognition and interpretation of heterospecific alarms is nearly always learned.

7. Like alarm signals, **food advertisement signals** impose immediate costs on senders and confer immediate benefits on receivers. The list of economic compensations to senders for providing food signals is nearly identical to that for alarms. Nevertheless, food signals are much less common than alarm signals, and food signalers are much more selective than alarm signalers about when and to whom they give their signals.

8. The emission rate or call structure of food signals usually reflects food preferences and only rarely food type. The location of a food find is often as important as its preferability. Many species remain at the find and broadcast a

signal to recruit fellow foragers. This type of advertisement is especially vulnerable to eavesdropping by unintended receivers. Some ants, birds, and bats minimize eavesdropping by returning to a communal roost or nest, recruiting foragers with special signals, and then leading them back to the food find. Ants, termites, and stingless bees lay pheromone trails between food finds and colony nest sites. This can lead to **mass recruitment**, in which returning workers reinforce the trail if the food find is valuable, and this in turn recruits even more workers to follow it. Trails are vulnerable to both conspecific and specialized heterospecific eavesdroppers. Honeybees do not use pheromone trails, and thus avoid any eavesdropping. Instead, they return to the nest and perform a waggle dance that provides recruits with the approximate azimuth and distance to a food find. The cost is a high cognitive burden on both senders and receivers.

9. **Autocommunication** occurs when sender and receiver are the same individual. The relevant economics can ignore honesty guarantees and focus on signaling efficacy. Examples include **electroreception** in mormyrid and gymnotid fish, and **echolocation** in bats and toothed cetaceans (some whales and all porpoises). In each of these groups, efficacy is extremely high, close to the limits imposed by the laws of physics.

10. Both bats and cetaceans echolocate with frequencies (12–200 kHz) whose wavelengths are similar to the sizes of their prey. Both use short-duration pulses to prevent an outgoing pulse from overlapping and masking a returning echo. Pulse durations of bats (0.5–25 msec) are long enough for them to modify pulse fine structure according to contextual needs: the pulses of open-country bats tend to be long in duration, with narrow bandwidths (CF) for the detection of small and distant prey; during interception, these shift to shorter-duration frequency-modulated (FM) pulses with wide bandwidths to monitor the distance, angular location, size, and shape of targets. Because sound in air does not penetrate solid targets, bats cannot determine target compositions. Bats that glean static prey close to vegetation both search for and intercept prey using FM pulses at lower amplitudes and often use additional sound, odor, or visual cues to complement echolocation information. Those hunting moving prey close to substrates exploit Doppler shifts to detect prey movements. Bats foraging near to substrates often emit their pulses through elaborate nose leaf structures to create directional sound beams and recover the echoes using elaborately directional pinnae.

11. The higher speed of sound in water requires echolocating cetaceans to use even shorter-duration pulses than those used by bats. However, most species use pulses (0.03–0.10 msec) far shorter than necessary to avoid pulse/echo overlap. These short-duration pulses generate very broad bandwidths without the detailed within-pulse regulation required by bats. The wide bandwidths are needed for the determination of target distance, angular location, size, and shape. Sound in water usually penetrates cetacean targets, and the resulting interactions during echo formation allow cetaceans to assess target composition. The reduced attenuation of sounds propagating in water allows cetaceans to detect prey at large distances without the long-duration CF pulses used by bats. All toothed cetaceans create highly directional echolocation beams using the oily melons on their foreheads, and most achieve directional reception using fatty channels in their lower jaws.

Further Reading

Ruxton et al. [728] review predator deterrent and alarm signals in animals in general, and Caro [127] provides an equivalently detailed review of such signals in birds and mammals. Both sources examine theoretical models as well as extensive examples. Ghirlanda and Enquist [299] compare the patterns of generalization used by predators to classify prey versus nonprey. Mappes et al. [552] discuss the consequences of predator generalization for the evolution of aposematism, Bond [88] relates it to color polymorphisms, and Mallet and Joron [546] show how it generates mimicry rings.

Taxon-specific reviews of predator notification and alarm signals include Wu et al. on aphids [977], Hölldobler and Wilson on ants [412], Prestwich [673] and Costa-Leonardo et al. on termites [168], Klump and Schalter on birds [480], and Zuberbühler on primates [1005]. Hollén and Radford[417] examine the ontogeny of alarm signals in birds and mammals.

Leading and trail advertisements of food finds by ants are reviewed by Hölldobler and Wilson [412], Wyatt [978], and Morgan [594]. The food trails of termites are reviewed by Costa-Leonardo et al. [168], and those of stingless bees by Nieh [622]. The food dances of honeybees are reviewed by Seeley [764] and Dyer [227]. Detrain and Deneubourg [203] compare food recruitment and group decision making in ants and bees.

The classic book by Griffin [341] describing echolocation in bats is still a valuable introduction. More recent reviews of how echolocation pulse design varies with habitat and task can be found in Schnitzler and Kalko [755] and Schnitzler et al. [756]. Jones and Holdereid [449] and Holdereid et al. [411] explain the importance of distance of focus when bats use FM pulses, and Neuweiler [616] outlines some of the auditory-system challenges faced by echolocating bats. Acoustic engineers' perspectives on bat echolocation are nicely presented in Waters [935] and Vespe et al. [920]. Teeling [884] reviews the most recent data on how and when early bats evolved echolocation. Au [25] provides an excellent synopsis of echolocation in cetaceans, and Au and Simmons [32] compare the echolocation systems of cetaceans and bats. The early evolution of cetaceans and their acquisition of relevant sound production and receiving adaptations are described in Thewissen and Williams [891], Nummela et al. [635, 636], Bajpai et al. [37], and Steeman et al. [842].

COMPANION WEBSITE

sites.sinauer.com/animalcommunication2e

Go to the companion website for Chapter Outlines, Chapter Summaries, and References for all works cited in the textbook. In addition, the following resources are available for this chapter:

Web Topic 14.1 *Models for environmental signaling*
A number of theoretical models are cited in this chapter. Here we provide a quick synopsis of the basic logic and methods used in many of these.

Web Topic 14.2 *Videos of echolocating bats and cetaceans*
Here we provide web links to a variety of video sources demonstrating echolocation or showing wild bats and cetaceans hunting.

Chapter 15

Communication Networks

Overview

This chapter looks at the complications that arise when a signal exchange involves more than a single sender/receiver pair. The many senders in a frog chorus or a grouse lek often adjust their display rates and timing according to what other males are currently doing. Signal eavesdropping in this and other contexts results in many concurrent receivers, some of whom may be of other species. The interactions between multiple senders and receivers can best be described as links in a **network**. Complex networks often show emergent properties and behaviors that would never be observed in isolated pairs of interactants. Variation in the structure of social networks can significantly modify evolutionary trajectories. Where the evolving traits affect communication, we thus need to know a lot about relevant network structure before we can fully understand how and why signals evolve as they do. This chapter reviews basic principles of network theory and then shows how they may be applied to animal communication and its evolution.

The Utility of Network Analysis

In prior chapters, we focused on the economics of a given signal exchange between one sender and one receiver (i.e., dyadic interactions). However, few pairs of senders and receivers communicate in a vacuum: other animals are usually within the active space of their signals and may both be affected by these signals and respond to them in ways that feed back on the focal pair. Eavesdropping by third parties extends the effects of communication outside the sender/receiver dyad. Males of many species eavesdrop on the courting signals of neighboring competitors; both conspecific and heterospecific eavesdroppers exploit the food trails of ants; and many species of predators detect and localize prey using the latter's conspecific signals. Territorial songbirds countersing particular motifs depending on what various neighbors are singing; this is especially noticeable during the dawn chorus. Aggressive signals by a dominant animal directed at the next most dominant individual may cause the latter to direct a behavior at the next animal in the hierarchy or even some uninvolved bystander [163]. The initial aggressive signal can thus trigger a chain of subsequent interactions, not necessarily of the same type, that propagates through the hierarchy network. Baboons and dolphins form alliances within larger groups, and responses to their signals often differ depending on whether receivers are part of an alliance or not. Schools of fish and flocks of birds maintain tight

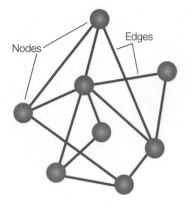

FIGURE 15.1   Sample network graph Nodes represent individual animals, and edges represent relationships between individuals. If two nodes are not connected by an edge, they do not have a significant relationship.

coordination when moving: members mimic small changes in speed or direction adopted by neighbors that then propagate to the entire group. One can diagram each of these processes by assigning each affected individual its own **node** on a graph, and linking any two nodes that pass on effects to each other with a linear **edge** (**Figure 15.1**). Such a graph is called a network, and the ways in which the **topology** of a given network modulates both the propagation of effects and the economic consequences can be studied using **network analysis**.

For some effects of a pair of communicating animals, we do not require network analyses to understand their evolution and properties. We have already noted that the average payoffs of a strategy can depend upon the relative frequencies of alternative strategies in the population. Such frequency dependence is typical of the evolutionary games that arise repeatedly in animal communication. However, the game models that we have discussed thus far tacitly assume that the population is sufficiently mixed that each strategy is encountered in proportion to its global abundance. This is most easily guaranteed if every animal in the population network is directly linked to every other animal. It is not necessary that any pair actually interact, but it is assumed that they are as likely to do so as any other pair. In real animal populations, everyone may not be connected to everyone else. Some animals may have lots of links, whereas others may have only a few. These heterogeneities violate the thorough mixing assumptions of evolutionary game theory. This raises the question of how communication and other social games will be played out when the underlying networks are heterogeneous. Does one get the same predictions?

When defining network structures, it is important to make a distinction between **relationships** and **interactions** [148, 353, 354]. Relationships generally exist prior to interactions: one animal may be dominant to another, a given male and female may be bonded mates, or two male songbirds may have adjacent territories. These relatively persistent linkages constitute relationships. Relationships define who is likely to

engage in an interaction, and who else is most likely to be affected by that interaction. The edges in a network diagram plot existing relationships between animals. While it is often the case that researchers tally interactions to identify relationships, they are tacitly assuming that these reflect some preexisting relationships. The same set of animals may be related in different ways, given different criteria: thus the relevant network for dominance relationships may be different from that for reciprocity or kinship links. Biological relationships can also change over time, often as a result of cumulative interactions: the utility of a measured network in predicting economic, informational, and evolutionary consequences is thus limited by that network's longevity. As long as a given network remains a valid representation of relationships, we can use it to predict how a given type of interaction will propagate effects and identify the likely economic consequences.

Network analysis can thus provide many new insights into signal evolution. General principles may only become apparent after one classifies networks into specific structural or dynamic categories. Such a task requires tools to characterize and compare networks. We outline some of the more widely used measures in the next section. Effects might be expected to propagate in different ways, over different distances, and with different time lags, depending on a network's structure. The second section of this chapter provides a classification of different network types and examines the role of structure in determining effect propagation. One surprising finding is that effect propagation in some networks is nonlinear, leading to the sudden emergence of properties and patterns that cannot be explained as the simple sum of dyadic interactions. The third chapter section examines evolutionary processes when the underlying social networks are heterogeneous. We find that the amount and distribution of linkage in a network can dramatically affect the predicted outcomes of evolutionary game and adaptive dynamic models. We conclude the chapter by applying this background to specific examples in which network structure seems to play an important role in social and signal evolution.

Characterizing Networks

Basic designs

Networks represent sets of relationships. Relevant relationships can be described in any of several ways. One approach lists the links between each pair of individuals in a two-dimensional **association matrix** (**Figure 15.2**). Cell values might be scored as 0 (no relationship) versus 1 (a relationship). For example, if group compositions in a species tend to be relatively stable, animals seen together in a group might be scored 1 and those in different groups scored a 0. The assumption that animals seen together have some sort of relationship is known as the **gambit of the group** [353, 354]. Depending on the relative stabilities of the groups in question, this may or may not be a valid basis for identifying relationships. While the gambit of the group tends to

(A)

Individual

	1	2	3	4	5	6	7	8
1		1	1	0	0	0	0	1
2	1		1	0	0	0	1	0
3	1	1		1	1	1	0	1
4	0	0	1		0	0	0	1
5	0	0	1	0		1	1	0
6	0	0	1	0	1		1	0
7	0	1	0	0	1	1		1
8	1	0	1	1	0	0	1	

(Individual — row axis label)

(B)

Individual

	1	2	3	4	5	6	7	8
1		.13	.65	0	0	0	0	.06
2	.13		.79	0	0	0	.37	0
3	.65	.79		.22	.87	.15	0	.08
4	0	0	.22		0	0	0	.39
5	0	0	.87	0		.23	.08	0
6	0	0	.15	0	.23		.73	0
7	0	.37	0	0	.08	.73		.31
8	.05	0	.08	.39	0	0	.31	

(Individual — row axis label)

(C)

Recipient

	1	2	3	4	5	6	7	8
1		1	1	0	0	0	0	1
2	1		1	0	0	0	0	0
3	0	0		0	1	0	0	1
4	0	0	1		0	0	0	1
5	0	0	1	0		0	0	0
6	0	0	1	0	1		1	0
7	0	1	0	0	0	1		1
8	0	0	0	0	0	0	0	

(Groomer — row axis label)

(D)

Recipient

	1	2	3	4	5	6	7	8
1		.15	.70	0	0	0	0	.23
2	.63		.65	0	0	0	0	0
3	0	0		−.32	.85	0	0	.26
4	0	0	0		0	0	0	.34
5	0	0	0	0		0	0	0
6	0	0	.17	0	0		.34	0
7	0	.33	0	0	−.19	.93		.38
8	0	0	0	0	0	0	0	

(Actor — row axis label)

FIGURE 15.2 Association matrices for a hypothetical monkey group Matrices summarize the relationships between paired individuals. If weighted, they show the strength of these relationships. Cells along the main diagonal are blue, as self-association is not usually relevant. Symmetry occurs if the cell value in column *i* and row *j* is the same as that in column *j* and row *i*. (A) Binary symmetric matrix recording whether a given individual has ever been seen foraging with another individual: (1) if it has, and (0) if it has not. (B) Weighted symmetric matrix showing the fraction of sample days that each pair of monkeys was seen foraging together. (C) Directional and binary matrix summarizing who was seen grooming whom. Note that only some relationships are symmetrical. (D) Directional and weighted matrix showing fraction of time each monkey supported the other during conflicts; negative values (red) indicate that a monkey supported the other's opponent.

network in which a stimulus signal caused a focal animal to attack some neighbors, ignore others, and help defend still others. Relationships can also be scored quantitatively such that they vary in both sign and magnitude. For example, a kinship association matrix might store the value of the coefficient of relatedness between each pair of animals, with negative values indicating relatedness less than the population average, positive values a relatedness greater than average, and the absolute magnitude a measure of how far that relationship deviated from average levels.

Any association matrix can be displayed as a **network graph** (**Figure 15.3**). Each individual is represented by one node. Discrete binary relationships either become edges between nodes if scored as 1 in the association matrix, or no edge between the nodes if scored as 0. Where the nodes are located and how the network is oriented are immaterial, as it is the relationships among the animals that are being illustrated. However, there may be dispersions of the nodes that make it easier to identify any structure in the topology of the network (e.g., spring embedding algorithms) [85, 217]. Edges in network graphs may be **undirected** or **directed**. An undirected edge implies that the effects of an interaction can propagate similarly in either direction; one or more directed edges mean that effects propagate asymmetrically between that pair of nodes. The direction is usually indicated on a network graph by an arrowhead on one end of the edge. If effects *can* move in either direction between two nodes in an otherwise directed network, this can be indicated by two parallel edges, each pointing in a different direction, or by a single edge with arrowheads on each end. A directed network can be **cyclic** if it has one or more loops that move in a consistent direction and begin and end at the same node, or **acyclic** if it lacks such loops. Ecological food webs and phylogenetic trees of animals tend to be acyclic, since edges leaving a node usually cannot get back to it; many social interaction networks of animals are at least partially cyclic. As with corresponding association matrices, edges can be **signed** (positive or negative; 0 is indicated by no edge), and they can be **weighted** according to the values of the cells in the corresponding matrix.

produce symmetrical matrices, (e.g., the relationship of animal A to animal B is the same as for B to A), discrete association matrices need not be symmetrical: animal C may always propagate signal effects to animal D, but D may never pass on effects to C. Where effects can be beneficial, neutral, or detrimental to a recipient, the relevant matrix might use relationship scores of 1, 0, and −1. An example would be a social

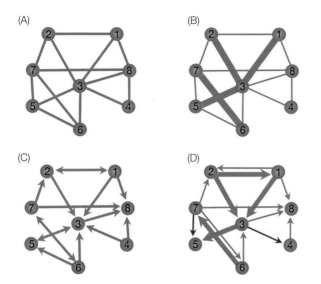

(A)

(B)

(C)

(D)

FIGURE 15.3 **Network graphs for the association matrices in Figure 15.2** Individuals were assigned the same identification numbers. (A) Undirected, unweighted, and unsigned network graph for data on joint foraging (compare to Figure 15.2A). (B) Weighted and undirected network showing the fraction of time each pair foraged together (Figure 15.2B). Weights greater than 50% are shown as thick edges; those with smaller values are shown as thin edges. (C) Directed and unweighted grooming network (Figure 15.2C). Note that many monkeys do not groom reciprocally. (D) Weighted, signed, and directed network graph for alliances during conflicts (Figure 15.2D). Thick blue edges represent support of the other monkey in 50% or more of conflicts, thin blue lines indicate support in less than 50% of conflicts, and red lines indicate support of the monkey's opponent (which never occurred more than 50% of the time). Again, note that some monkeys do not support each other symmetrically.

Assessing and diagramming real animal networks are not easy tasks, and there are many potential pitfalls [85]. Diverse indices for characterizing relationships have been invoked in studies on animals, and these vary in their resolution and general applicability [353, 354]. Insufficient sampling of the underlying relationships can lead to ambiguous or even erroneous network graphs [317]. Not all patterns of relationships fit into a standard association matrix [52]. Some relationships may involve more than simple dyads: descriptions of multiple-individual alliances as in primates and dolphins may require the drawing of **hyperedges** in graphs that have more than two dimensions [217]. Finally, if relationships change over time, some method will be required to track, quantify, and display these changes [328].

Network measures

An enormous amount of effort has been invested in defining useful measures of network structure [5, 37, 50, 51, 85, 217, 348]. The outcome is a set of measures that are broadly applicable in fields as disparate as physics, economics, cell biology, genetics, ecology, animal behavior, information technology, neurobiology, sociology, and psychology. We shall exploit this toolkit to compare observed networks to theoretical models;

compare two or more real networks for similarities in structure; assess whether characteristics of individual nodes (e.g., sex, age, dominance status, or species) are or are not randomly distributed over the network; and determine whether topology plays a role in the propagation of effects through the network or in the outcome of evolutionary processes. Many measures in this toolkit were initially developed for simple networks with undirected, unsigned, and unweighted edges. Faced with complex animal networks in which some or all of these assumptions are violated, one tactic is to reduce the initially signed and weighted association matrices to simpler formats by replacing cell values above or below some threshold value with 1s and 0s respectively (called **filtering**). In some contexts, this results in qualitatively useful analyses. For other tasks, particularly evolutionary modeling, the retention of edge signs and directionality is obligatory [233], and recent simulation studies find that retaining weights in real animal social networks gives much more reliable results than does filtering the network [117, 184]. Conveniently, most of the measures originally defined for unweighted networks now have weighted network counterparts [221].

Below, we summarize the more widely applicable measures. We divide these into measures of network **connectivity** versus those of network **centrality**. Connectivity measures reflect the degree to which the network is similar to or deviates from a **complete network**, in which every node is connected directly to every other node. Centrality measures characterize the heterogeneity among a network's nodes and edges. Many measures can be computed for each node or edge and used for comparisons with other nodes or edges; a subset of these can also be averaged across nodes or edges to give an overall measure for that network. Such averages are often used to compare networks to each other or to theoretical models. Note that few of these measures are independent of each other, and that adding, deleting, or redistributing edges can affect both connectivity and centrality measures. In addition, because cell values in the association matrix on which a network diagram is based are usually not independent, permutation methods are required to determine whether or not a given network measure deviates significantly from random expectations [85].

CONNECTIVITY MEASURES   The maximum number of edges that a network can host occurs when it is complete. In this chapter, we are largely interested in *incomplete* networks. An obvious measure of connectivity for incomplete networks is the fraction of potential edges that are actually present in a network. This measure, called **edge density**, varies between 0 and 1 (**Figure 15.4**) and is equal to the average of the cell values (including zeroes) in the relevant association matrix. If edge density is small, the network is called **sparse**; if it is at least 0.5, it is said to be **dense**. In an unweighted network, the number of edges intersecting a node is called its **degree**. If the network is directed, one can compute both "input" and "output" degrees for each node. The average degree across the network is often used as one metric of overall network

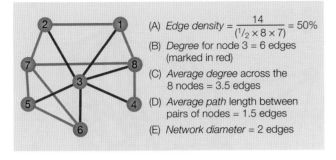

(A) *Edge density* = $\frac{14}{(1/2 \times 8 \times 7)}$ = 50%

(B) *Degree* for node 3 = 6 edges (marked in red)

(C) *Average degree* across the 8 nodes = 3.5 edges

(D) *Average path* length between pairs of nodes = 1.5 edges

(E) *Network diameter* = 2 edges

FIGURE 15.4   Some commonly used connectivity measures These measures are here computed for the network in Figure 15.3A. (A) Edge density: Each of the N nodes in a complete and directed network will connect to each of the other $N-1$ nodes. There will thus be $N(N-1)$ total edges. In an undirected and unweighted network, the link connecting a given node i to any other given node j is also the one connecting node j to node i. Thus, a complete unweighted and undirected network will have $(1/2\, N\,(N-1))$ total edges. This example has 8 nodes and thus could have up to $1/2 \times 8 \times 7 = 28$ edges. Edge density is the percentage of the edges that are present compared to the maximum possible. Here it equals 14/28 = 50%. (B) The degree of a node in an unweighted and undirected network is the number of links attached to it. (C) Average degree is the mean of the degrees of all nodes in the network. (D) Although there may be many paths connecting any two nodes in a network, we focus only on the path lengths (in edges) for the shortest routes. Average path length for a node is the average of the shortest paths between it and every other reachable node in the network. Average path length for a network is the average of the average path lengths of all of its nodes. (E) Network diameter is the longest of the shortest paths (in edges) between two nodes in the network.

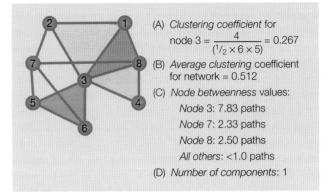

(A) *Clustering coefficient* for node 3 = $\frac{4}{(1/2 \times 6 \times 5)}$ = 0.267

(B) *Average clustering* coefficient for network = 0.512

(C) *Node betweenness* values:

Node 3: 7.83 paths

Node 7: 2.33 paths

Node 8: 2.50 paths

All others: <1.0 paths

(D) *Number of components*: 1

FIGURE 15.5   Some commonly used centrality measures These measures are here computed for the network in Figure 15.3A. (A) Clustering coefficients: Being linked to immediate neighbors that are also linked to each other is called transitivity. The clustering coefficient is the number of actual triangles in which a node is a vertex divided by the maximum possible triangles. For a node of degree k, the maximum number of triangles equals $(1/2\, k\,(k-1))$. In this sample network, node 3 is part of four triangles (colored) and since it has a degree of 6, it could have been part of 15 different triangles. Its clustering coefficient is thus 4/15 = 0.267. (B) Repeating this calculation for all nodes in this example and averaging the results gives the average clustering coefficient for the network. (C) Node betweenness is the number of shortest paths between other pairs of nodes that pass through a focal node. (D) Components: This is the number of separate regions of a network that are completely isolated from each other. If all nodes are connected by some path, there will only be a single component.

connectivity. Note, however, that this average is proportional to edge density and thus not an independent measure. If the network's edges are weighted, the **intensity** (also called **strength**) of a node is the sum of the weights of its edges. One can compute an average intensity across all of a network's nodes, and an edge density which is again equal to the average cell value in the relevant association matrix.

Another measure of connectivity is **path length**. This is the average *shortest* distance between a focal node and all other nodes in the network that can be reached by traversing edges. The **average path length** of a network is the global mean of the path lengths for all nodes in the network, and the network **diameter** is the longest of those path lengths. In unweighted networks, the unit of path length is the number of edges traversed; in weighted networks, the reciprocals of edge weights are summed along a path. Reciprocals of path lengths are also used when an unweighted network consists of separate unconnected subnetworks. Because reciprocals of path lengths measure proximity, the resultant average is alternatively called network **closeness** or **efficiency**, the latter term reflecting the likely speed with which interaction effects can propagate through the network.

CENTRALITY MEASURES   Centrality measures quantify the level of heterogeneity in a network structure. At one extreme

is a **regular** network in which all nodes have the same amount of connectivity and heterogeneity is absent. As a network is made less regular, heterogeneity can arise at different scales. This has led to a wide variety of centrality measures that focus on different scales of heterogeneity. We begin by examining centrality measures at very local scales and then move up to characterizing more global patterns.

Nodes with above-average degree or intensity are considered to be more central than those with low degree. Those with the highest degree or intensity are called **hubs**. The **reach** of a focal node is the number of other nodes that are connected to it by some small number or fewer edges. More central nodes have higher reach values than less central ones. High-degree nodes that connect to other high-degree nodes can be considered even more central. Two local centrality indices, **affinity** and **eigenvector centrality**, include the pooled degrees of a node's immediate neighbors in their computation (see formal definitions in [160, 217]). Because the average degree for a network is simply proportional to edge density, it does not provide any additional information that can be used to compare different networks. Some studies have tried to get around this limitation by comparing the maximal degrees, intensities, or eigenvector centrality values of different networks [160].

The **clustering coefficient** is a widely used measure of local heterogeneity (**Figure 15.5**). This computes the

transitivity between a focal node and its immediate neighbors. If node *A* is linked to node *B*, and node *B* is linked to node *C*, transitivity occurs if node *A* is also linked to node *C* (forming a triangle in the network graph). The clustering coefficient for a node in an undirected network equals the number of complete triangles that include that node divided by the total number that would be possible, given that node's degree. An overall measure for a network can be computed by averaging the clustering coefficients of all nodes. Both measures vary from 0 (no triangles) to 1 (all linked triplets are indeed in triangles). Most real networks fall between these extremes and are characterized as being **weakly clustered** (with low overall clustering coefficients) or **strongly clustered** (with high values) [293]. One might expect that clustering coefficients would generally increase as edge density or degree increases. The real situation, however, is more complicated. Although human and animal social networks tend to have low edge densities (0.01–0.15), they routinely have high overall clustering coefficients (0.3–0.8) [5, 37, 83, 84, 160, 181, 215, 217, 220, 255]. Many biochemical, social, and digital networks exhibit high clustering coefficients for nodes of low degree but low clustering coefficients for nodes of high degree [271, 327]. The overall level of clustering may thus depend on the relative abundances of low- versus high-degree nodes. As we shall see later, the propagation of effects in a network depends significantly on both the overall mean and the distribution of clustering coefficients among nodes [292–294].

A different type of local network heterogeneity involves the nonrandom selection of a node's topological neighbors. In a **positively assortative** network, like tends to be closely linked to like. For example, if males tend to be the nearest network neighbors of other males, and females the nearest neighbors of other females, the network is said to show positive assortativity with respect to sex (**Figure 15.6**). Alternatively, an older animal may favor contacts with younger animals and vice versa; such a network then shows **negative assortativity** (also called **disassortativity**) with respect to age. A variety of indices have been derived to characterize overall network associativity, with most acting like correlation coefficients that vary between −1 and 1 [219, 224]. One special case of assortativity occurs when the trait that defines the nonrandom patterns is node degree: if high-degree nodes (hubs) are most likely to be attached to other hubs, and low-degree nodes are linked to other low-degree nodes, the degree assortativity is positive; if hubs are mostly linked to low-degree nodes and vice versa, the degree assortativity is negative. Some human and animal social networks tend to show overall positive degree correlations, whereas the World Wide Web and biochemical networks tend to show overall negative degree correlations [217]. Note that degree assortatitivity and clustering coefficients cannot vary completely independently. In particular, high negative or low positive degree associativity restricts clustering coefficients to low values [293]. Although mean degree is not useful in comparing overall heterogeneities between networks, the relative abundances of different

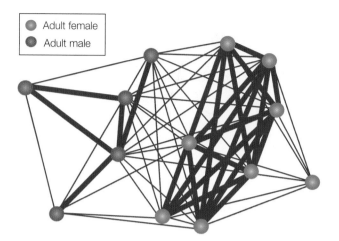

FIGURE 15.6 Network associativity in a wild spider monkey (*Ateles geoffroyi*) group Average weighted interaction network for adult males and females over one year of study. Relative interaction weights are indicated by edge thicknesses. Although some males interacted with many females, most interactions were with members of the same sex, resulting in a network with positive sex associativity. (After [268].)

degrees (or intensities) may provide a useful comparative tool. **Degree distributions** are routinely summarized in graphs, with the logarithm of increasing degree values on the horizontal axis and the logarithm of the fraction of the network's nodes having a particular degree on the vertical axis. In a regular network, all nodes have the same degree, and the degree distribution plot consists of a single thin vertical line over the ubiquitous degree value. Real distributions are more irregular, and this greater heterogeneity results in much wider degree plots with a variety of shapes. As we shall see in the next section, the shape of the degree distribution plot can be used to classify networks into types, explain effect propagation, and predict evolutionary trajectories.

There are several measures that focus on characterizing the highest levels of network heterogeneity. Network regions in which all nodes are more thoroughly linked to each other than they are to other regions are called **communities** (or **modules**) [37]. The extreme case in which within-community linkage is complete is called a **clique** (**Figure 15.7**). A wide variety of algorithms have been developed to identify communities and cliques in networks. Some use successive partitioning to divide the network into a discrete number of separate units, whereas others look for hierarchical structure in which clusters of linked nodes form subgroups, subgroups are clustered to form groups, and groups are clustered into supergroups [70, 271]. Criteria for defining large-scale patterns include the distributions of centrality measures throughout the network, the dependence of node clustering coefficients on degree, the effects of successive edge removals, and eigenanalysis of the original association matrices or derivatives of those matrices [37, 69, 70, 85, 101, 125, 175, 217, 222–224, 246, 247, 271, 292, 293]. A final measure of centrality is a node's (or edge's) **betweenness**. This equals the number of

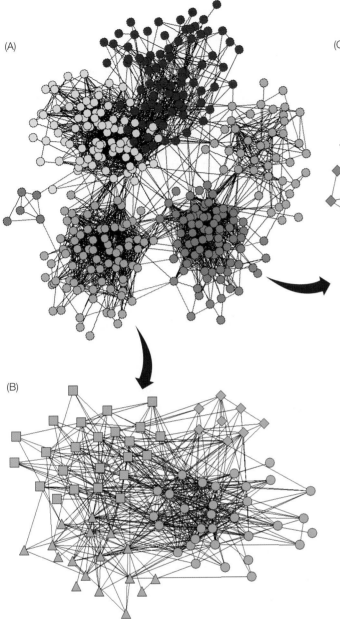

(A)

(B)

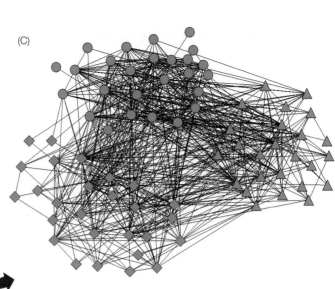

(C)

FIGURE 15.7   Community structure in a network of Galápagos sea lions (*Zalophus wollebaeki*) (A) Overall social network for 380 individuals living around Islet Caamaño, in the Galápagos Islands, based on joint sightings in the immediate locale. Separate communities are indicated by different colors. (B,C) Expanded view of the networks for two of the communities, showing constituent cliques (members of same clique have the same node symbol shape). (After [360].)

shortest paths between other nodes that pass through the focal node (or edge). A related measure, **information centrality**, is the sum of all paths through a node weighted inversely by their lengths. Because nodes or edges with relatively high betweenness or information centrality serve as the bridges between communities, their identification provides another way to define separate communities (**Figure 15.8**). Note that in some networks, there may be no bridges between separate communities. In those cases, the network is said to have multiple **components**. There can be no propagation of effects between different components in such a network.

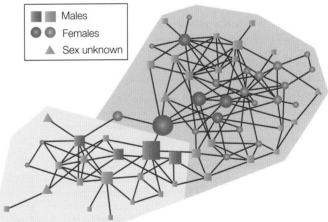

FIGURE 15.8   Betweenness and community structure in a bottlenose dolphin (*Tursiops truncatus*) social network Squares represent males, circles females, and triangles animals of unknown sex. The size of a node symbol reflects the betweenness measure for that individual. Edges are based on seeing a given pair of animals together more often than expected by chance. Note partitioning of the network into two distinct communities (pale blue and pink polygons) bridged by a few individuals (red nodes) with very high betweenness values. (After [182].)

Network Structure and Behavior

With so many variable measures, it might seem that the number of possible combinations, and thus types of networks, might be endless. In practice, many network measures are correlated with each other; for example, edge density and average degree are always positively correlated, whereas each is usually inversely correlated with average path length. The level of degree associativity can limit allowable values of cluster coefficients. Simplification due to correlated measures and careful modeling of alternative topologies have turned up several network classes that recur in fields as disparate as physics, information technology, community ecology, biochemistry, epidemiology, and human sociology. Below, we identify these classes, discuss how they differ in their behavior, particularly with regard to effect propagation, and then discuss where we might see each type in biological contexts.

Structural types of networks

Consider a regular undirected and unweighted network in which all nodes have the same degree. Assume the edge density is high enough that the network consists of a single component, but is sufficiently low that only topological neighbors are fully connected. The high local connectivity will be reflected in high clustering coefficients and will ensure rapid propagation of effects locally. However, because nodes on opposite sides of the network are connected only through a long chain of intervening edges, effect propagation between them will be slow.

Now suppose a small number of shortcut edges are added to the network so as to link some of the most distantly connected nodes. Shortcuts have a multiplicative effect on network properties: they not only reduce the path length between the two newly connected nodes, they also reduce path

lengths between the many linked neighbors of those nodes. A small number of shortcuts will eliminate many of the longest paths found in the original network and thus reduce its diameter. While eliminating longer paths will also reduce the average path length for the network, the magnitude of this reduction depends on network topology (**Figure 15.9**). Adding a few shortcuts to a ring-shaped network that lacks any central nodes will significantly reduce both the network diameter and the average path length; adding the same number of shortcuts to a full lattice will similarly reduce the diameter, but reductions in average path length will be small, because the numerous central nodes are minimally affected. Given their effects on many pairs of nodes, shortcuts also reduce the rate at which the average path length changes when the number of nodes in a network changes. They thus provide some measure of stability in network properties. Note that the number of shortcuts that must be added to provide these effects is small enough that they usually cause only minor changes in the network's edge density and average clustering coefficient.

The combination of reduced diameters and average path lengths and retained high clustering coefficients promotes rapid propagation of effects at both the local and whole network levels. Networks that exhibit these properties are called

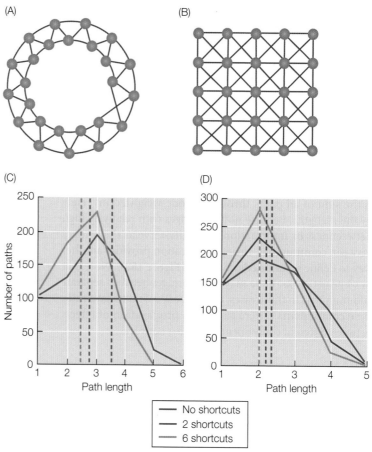

FIGURE 15.9 Small world networks (A) A 25-node ring network which is sparse since many nodes are not directly connected. All nodes in this ring have the same degree making it regular. (B) A similarly sparse 25-node rectangular lattice. Because edge and central nodes have slightly different degrees, this network is only quasi-regular. (C) Distribution of path lengths in the ring network. Adding as few as 2–6 shortcuts between the most distantly separated nodes both reduces mean path length (dashed lines) and eliminates large numbers of longer paths. (D) Similar distribution plots for lattice. Adding similar shortcuts here eliminates large numbers of longer paths, but has less of an effect on average path length, since nodes in the core of the network are already well connected. Adding six shortcuts to the ring network increased edge density from 17% to 19% while reducing the average clustering coefficient from 0.5 to 0.4; equivalent changes to the rectangular network only increased edge density from 27% to 29% and reduced the average clustering coefficient from 0.74 to 0.53, a bigger drop but still a high level of residual clustering.

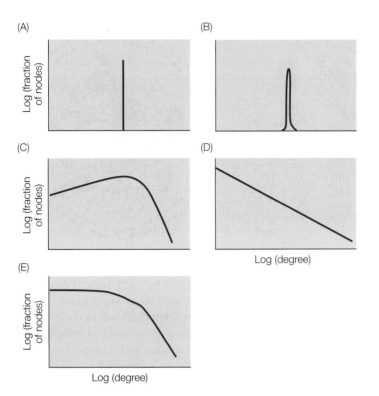

FIGURE 15.10 **Sample degree distribution graphs** (A) A fully regular network in which all nodes have the same degree. (B) A small world generated by adding a few short-cuts to a regular network. (C) A network with fully random edge assignments (Erdös-Rényi). (D) A scale-free network. (E) A real animal distribution similar to those found in stickleback, guppy, and dolphin social networks. (After [84, 181].)

the two researchers who first explored it (**Figure 15.10C**). If the probability that any pair of nodes is connected is p and the total number of nodes is N, the peak of the unimodal degree distribution plot will occur for a degree value of $p(N-1)$. Although these networks show no significant transitivity, assortativity, or community structures, they exhibit one trait that they share with more complex networks. For sparse networks at low values of p, the network is broken into many isolated components. As p is increased, the edge density in the network also increases. When the number of edges in a random network becomes greater than $N/2$, there is a sudden transition from many unconnected components into one **giant component** and a few small isolated ones. Given the low average path length at these high edge densities, effect propagation inside the giant component can be very fast. However, propagation to unconnected components is not possible until they, too, are linked to the giant component.

Suppose we now take a random network and preferentially reassign one end of a fraction of its edges to the hub nodes that already have the highest degrees. Since most of the nodes in a random network have an intermediate degree, this reassignment will remove some of their links, turning many of them into nodes of low degree while increasing the degree of existing hubs. This will raise the two ends of the degree distribution plot and depress its middle section. Because this process creates many new nodes of small degree while increasing the degree of existing hubs more often than it creates new ones, the new plot line will show a decreasing trend with increasing degree. Degree distribution plots with excess low- and high-degree nodes when compared to random networks are said to be **heavy-tailed** [95, 96]. Where the excess of high-degree nodes is large enough, the degree distribution plot is a straight line with negative slope (**Figure 15.10D**). Such a network is called **scale-free**, because the rate of decrease in numbers of nodes with increasing degree is independent of other factors [5, 23–26, 264, 265]. Those networks with the highest heterogeneity in degree values exhibit degree plots with the shallowest slopes and those with less heterogeneity have steeper slopes [226]. Most biochemical networks—as well as human networks involving sexual partners, actor groupings, the input functions of the World Wide Web, and collaborations between scientific authors—are scale-free, with slopes for their degree plots between −2 and −3 [25, 37, 217]. The few social networks of animals that have been studied and the

small worlds [315, 349, 350]. Many real networks appear to act like small worlds. For example, a now classic study provided human volunteers from cities in Nebraska and Kansas with letters addressed to a person in Boston whom they did not know. Each volunteer was asked to forward the letter to someone they thought more likely to know the recipient, who was then to forward the letter to someone they thought closer to the recipient until the letters reached the addressee [208]. The average number of transfer steps needed to deliver the letters was found to be six, leading to the phrase "six degrees of separation" for randomly selected pairs of people in human networks. Similarly low average path lengths (2–15 edges) and moderate clustering coefficient values (0.4–0.8) are found in Internet router networks, metabolic networks in biological cells, e-mail networks, and social networks in animals [5, 37, 83, 84, 217, 218].

Suppose we begin again with a regular sparse network and, holding edge density and number of nodes fixed, begin randomly reassigning the original edges. As more and more edges are relocated, the heterogeneity in the degree distribution will increase. We can compare levels of heterogeneity with a degree distribution graph. Whereas the original regular network is represented by a single vertical line on such a graph (**Figure 15.10A**), a small world network with some shortcuts added will appear as a narrow unimodal peak (**Figure 15.10B**). As more and more edges are randomly reassigned, the unimodal peak of the small world distribution will broaden. The random assignments will also disrupt local transitivity until the average clustering coefficient equals the edge density. A distribution in which all edges are assigned randomly is called an **Erdös–Rényi random network**, after

FIGURE 15.11 **Hub linkage in scale-free networks** Scale-free networks have high heterogeneity in degree distributions. They can differ in how directly their hubs are linked. (A) A simulated network in which most hubs are connected directly to other hubs, making the network a small world. (B) An interaction network of yeast nuclear proteins. The hubs in this network, unlike those in (A), are only indirectly linked to other hubs, resulting in relatively isolated modules and eliminating most small world effects. (A after [315]; B after [190].)

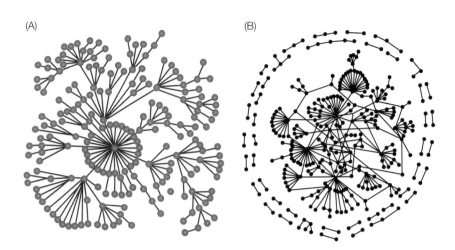

(A)　　　　　　　　　(B)

output functions of the World Wide Web show heavy-tailed but curvilinear degree distribution plots, largely because of their relative shortage of low-degree nodes when compared to scale-free systems (**Figure 15.10E**) [85, 95, 181].

Although most real networks have heavy-tailed degree distributions, they can differ significantly in how their hubs are distributed [27, 271, 280, 308, 309]. At one extreme are heavy-tailed networks in which hubs are usually linked directly to other hubs (**Figure 15.11A**). Such networks are invariably small worlds with shorter and more abundant shortcuts than in random networks [72]. As noted earlier, adding more nodes to a small world increases the network diameter and the average path length only very slowly. Human examples include Internet router networks and electrical power grids. At the other extreme are networks in which hubs are rarely linked directly but instead are connected indirectly via one or more low-degree nodes (**Figure 15.11B**). The cluster of nodes connected to such a relatively isolated hub forms a module. Adding nodes to such a network typically enlarges existing modules rather than bridging them; if anything, network growth increases the separation between hubs. Many biochemical networks show this type of modularity. Actor associations, scientific author collaborations, and some metabolic systems go one step further by having positive degree associativity among low-degree nodes, as well as negative associativity among hubs. This generates an inverse relationship between clustering coefficients and degree, and creates modules that are highly linked internally but sparsely linked to each other.

Network structure can get more complicated if linkage heterogeneity is hierarchical, with clusters of nodes at the lowest level, clusters of clusters at the next highest level, and so on up to the single cluster that is the entire population. At each level, the relevant units may show positive, negative, or no degree associativity. Different patterns of association can exist at different levels within the same network. While the potential combinations may again seem countless, variation among real networks is largely limited by the distribution of shortcut edges [121, 281, 309]. In general, the probability that a randomly added edge will connect two nodes decreases with the current topological distance between them. If this probability decreases very rapidly with increasing distance, nearby nodes will likely be connected, but there will be no shortcuts between scattered hubs at low levels or modules at higher levels. This will isolate hubs, modules, and clusters of modules from each other and create networks in which a modular pattern at the node level tends to be repeated at each successive level [308–310]. Patterns in nature that repeat at successively larger scales are called **fractals** (**Figure 15.12**). The modular networks discussed previously are usually fractal. Nonfractal networks can be generated by relaxing the rate at which the probability of a shortcut falls off with the distance between nodes. Once shortcuts can occur at the greatest possible distances, the network will become entirely small world and lose any fractal character. Most real networks have shortcut distributions that place them somewhere between the extremes of pure fractal and pure small world. Biochemical networks and the World Wide Web tend to be more fractal and modular, whereas Internet router systems are nonfractal and show little modular structure. As we shall see below, where a network falls along the fractal–small world continuum can significantly affect its properties and behavior.

Effect propagation in networks

As we mentioned at the beginning of this chapter, a signal emitted by one individual can trigger a series of effects that propagate according to the social network structure. All individuals falling within the active space of the sender's signal are potentially linked to the sender by directional edges pointing away from the sender. These individuals constitute the "nearest neighbors" of the sender in the communication network; they typically include the primary receivers and at least some eavesdroppers. Outside of this first shell of individuals around the sender is a second shell consisting of individuals that cannot themselves detect the signal but can detect responses by individuals in the first shell. These are the secondary responders around the sender. One can continue to define successive shells of potential responders until the entire network is catalogued. Note that once the signal reaches individuals in the first or outer shells, effects

(A)

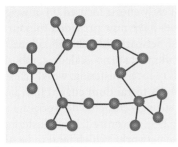

Original network

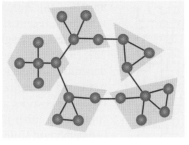

Box size = 3
Number of boxes = 5

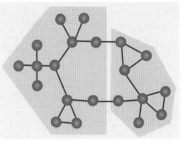

Box size = 5
Number of boxes = 2

(B)

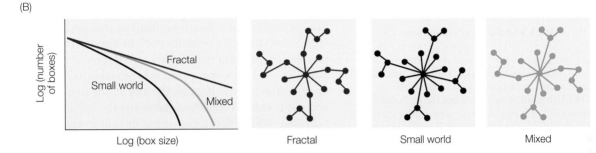

Log (number of boxes)

Fractal

Small world

Mixed

Log (box size)

Fractal   Small world   Mixed

FIGURE 15.12 **Fractal networks** The protocol for determining the fractality of a network. (A) Boxes of equal size are used to partition the network's nodes. The least number of edges connecting each pair of nodes in a box must be less than the size of the box. The goal is to find the minimum number of boxes needed to partition the entire network at a given box size. One then repeats the exercise at increasing box sizes until only one box is obtained. (B) One then plots the logarithm of the number of boxes found at each stage in the process against the logarithm of the box size (in edges). A 100% fractal network with no links between hubs will show a straight line plot on this graph. A completely small world network with close links between hubs will show a strongly concave plot. Networks that have a mix of fractal and small world regions will show curves that are intermediate [310].

could propagate in all directions (i.e., to individuals in outer shells, to other individuals in the same shell, and to those in inner shells including the sender). The relevant links can have either positive or negative signs and various weights, and different individuals may respond in different ways. While the size of the first shell will be large for broadcast signals, it will be small for low-intensity signals. However, intensity is only one factor modulating a signal's topological range. Another relevant factor is signal persistence: while a slowly vaporizing olfactory mark might have a small active space, many animals in the network may encounter it, given enough time. Depending on the lifetime of the mark and the mobility of potential receivers, a loud short sound and a slowly released olfactory mark might eventually affect the same total number of individuals in the network. What would differ is the time it took to do so. This suggests that we may need both spatial and temporal measures of effect propagation.

PERCOLATION   Let us begin with some action or perturbation at one or more nodes. This perturbation may be a

short-lived event such as a sound signal, or a persistent input such as a slowly released olfactory signal. In either case, the perturbation instigates effects that propagate through the network. Different research fields have given effect propagation different names [5, 217, 218]. For example, biochemists focus on the **transport** of metabolites through a cellular network, while sociologists and computer engineers track information **flow** through their respective networks. The theoretical network literature calls all such processes **percolation**, a term originally used in physics for the leaching of fluids through porous solids. Our communication example above suggests that we consider at least two measures: (1) **percolation fraction**, the percentage of nodes in a network that are affected by a given perturbation; and (2) **percolation time**, the time that must elapse before all the nodes in the percolation fraction are affected. How might the common types of networks defined in the prior sections differ in these two measures?

Consider first a sparse network with a very low level of connectivity and many isolated components: an event at one node may percolate only to a small fraction of those present (**Figure 15.13**). Given the short path lengths over which an effect can percolate, percolation times will likely be short. In contrast, an event at a node within the giant component of a network with high edge density is likely to percolate to a large fraction of the total nodes. As noted earlier, the giant component can appear quite suddenly in a network when edge density surpasses a threshold. Different types of networks form giant components at different threshold edge densities, and may also differ in the fraction of the network's nodes that are included in the giant component [37, 95, 96, 217]. The

(A)

(B)

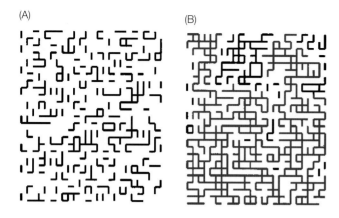

FIGURE 15.13 **Percolation and the giant component** (A) In this 25 × 25 node lattice network, the probability that any two adjacent nodes are connected is 0.315. This value is below the percolation threshold of 0.500. The network is therefore divided into many small and isolated components within which effects propagate locally. (B) This is the same lattice, but the probability that adjacent nodes are linked is now 0.525, a value slightly greater than the percolation threshold. The network now largely consists of a giant component (red) within which a perturbation at any node can, in principle, propagate to any other node. (After [5].)

percolation fraction is often equated with the relative size of a network's giant component. We discuss this further in the next section. If the network has sufficient small world shortcuts, average path lengths will be short, and thus percolation times will also be short. Fractal networks consisting of relatively isolated modules will show high percolation fractions and low percolation times within a module, but opposite conditions for percolation between modules. Overall percolation fractions and times thus depend significantly on the number and, if the network is weighted, the weights of the linkages that bridge modules [68, 120].

ROBUSTNESS   A network's **robustness** (also called **resilience**) is its ability to maintain similar percolation fractions and times after removal of some original edges or nodes [218]. For example, random deletion of a few of the edges or nodes from a network's giant component is unlikely to change the percolation properties. However, if enough edges or nodes are removed or deactivated, the giant component properties may be lost, and the percolation process switches to that of a sparse network. The threshold fraction of original nodes or edges above which a network exhibits giant component percolation and below which it acts like a sparse network is often used as a measure of its robustness [315]. A network with a high threshold is less robust to edge or node loss than is one with a low threshold. Much effort has gone into deriving robustness thresholds for different classes of networks. One approach finds the largest set of connected nodes that have a degree greater than some threshold value. This is called the **k-core**, or **backbone**, of the network. The robustness of the k-core to edge removal is then examined [29, 96, 126, 292]. Other approaches use model networks and

simulations to identify general robustness patterns [37, 87, 95, 217, 218]. Although early efforts focused on unweighted networks, recent studies have now provided measures of robustness in weighted networks [27, 28, 55, 57, 68, 153, 221, 267, 275, For networks that are not scale-free, robustness depends upon the levels of local clustering: weakly clustered networks are more robust to random edge removal than random networks, and strongly clustered networks are more robust than either [292, 293]. In contrast, local clustering has little effect on scale-free networks, which tend to be highly resilient to random edge or node removals (**Figure 15.14**) [4, 25, 129, 217, 294]. While scale-free networks are robust to random edge removal, they are quite vulnerable to selective removal of hubs. This is particularly true for those that are nonfractal; only a few hubs must be removed to change percolation measures significantly [4, 53]. Fractal networks, however, tend to be more robust to selective hub removal than nonfractal ones because the loss of one hub does not necessarily reduce connectivity between or within other hubs [120, 280, 309]. Where the strongest links in a weighted network occur in the k-core, weighting largely reinforces the above patterns. It can, however, change the levels of percolation and thus robustness in modular networks depending on whether bridges between modules are strong or weak [68].

As a general rule, robustness tends to trade off with percolation properties: fractal networks are resistant to hub removal but at the cost of reduced percolation, while nonfractal but scale-free networks show extensive and rapid percolation at the cost of reduced resistance to hub loss. As we noted earlier, most biological networks exhibit relatively high levels of local clustering and are relatively robust to random loss of nodes and edges. They differ in their relative emphasis on maximizing percolation versus maximizing robustness to hub loss. The optimal emphasis depends on the biological context [26, 37, 120, 126, 129, 281, 308, 309]. For example, high percolation fractions and low percolation times are disadvantageous when the effect spreading across the network is a disease. As a result, shortcut-rich small worlds are more susceptible to epidemics than are fractal structures or sparse networks. Cells are more likely to maintain normal functioning if their biochemical networks have a modular fractal structure than if these networks are nonmodular and nonfractal. Fractals are, in fact, the rule in cellular biochemistry networks [280]. On the other hand, where rapid and thorough percolation is beneficial, such as alarm signaling, shortcut-rich small worlds are favored over a fractal structure. The fact that different biological networks have evolved to different points along the fractal–small world continuum presumably reflects differences in the optimal emphasis on percolation versus robustness [15, 16].

EMERGENCE AND SELF-ORGANIZATION   Even though the link between any pair of nodes may propagate effects using very simple rules, the pooled interactions of large numbers of nodes, particularly effect feedbacks, can cause a network to behave in a nonlinear manner. It should not be surprising

(A)

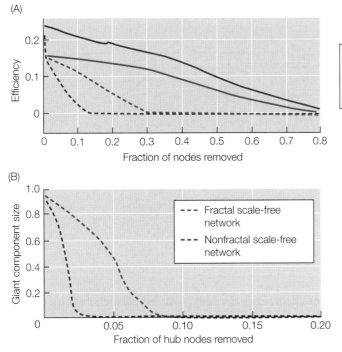

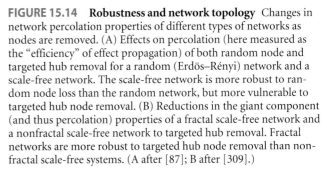

(B)

FIGURE 15.14 **Robustness and network topology** Changes in network percolation properties of different types of networks as nodes are removed. (A) Effects on percolation (here measured as the "efficiency" of effect propagation) of both random node and targeted hub removal for a random (Erdös–Rényi) network and a scale-free network. The scale-free network is more robust to random node loss than the random network, but more vulnerable to targeted hub node removal. (B) Reductions in the giant component (and thus percolation) properties of a fractal scale-free network and a nonfractal scale-free network to targeted hub removal. Fractal networks are more robust to targeted hub node removal than nonfractal scale-free systems. (A after [87]; B after [309].)

that effect propagation in large networks can trigger sudden bifurcations in global behavior, lead to different equilibria depending on starting conditions, or even generate limit cycles or chaos [96]. Network behaviors that are not simply the summed responses of all nodes but arise from nonlinear processes are said to be **emergent**. Where the nonlinear behaviors of a network create a global pattern that is not predictable given expected responses of isolated nodes, the pattern is **self-organized** [56, 92, 162]. For example, wind blowing over loose sand often causes the formation of rows of dunes. Neither the wind nor the sand grains possess blueprints for dune formation; instead, rows of dunes are a self-organized outcome of the force of the wind, gravity, and friction between individual sand grains. Self-organization is also widely observed in developmental biology. A simple network in which developing cells replicate effects from neighbors but fail to respond—or respond in an opposite fashion—when stimulated by effects from more distant cells can produce a self-organized pattern of alternating zones. Such a process appears to explain segmentation in embryos and banding patterns in animal coloration. The foraging trails of an ant colony provide a behavioral example (**Figure 15.15**). The pattern of trunk and branch trails on a given day does not follow any overall template in the ants' brains; instead, the pattern is the self-organized outcome of simple rules, such as (1) lay no trail pheromone while searching for food, (2) lay a trail en route home after food is found, and (3) switch onto existing trails and reinforce pheromone on successful return trips. When hundreds of ants in the same colony apply these rules, the result is the self-organized trunk and branch trail system seen in nature [92].

Perhaps the best studied self-organized pattern of networks is **synchrony** [14, 71, 253, 254, 314–316, 356]. Consider a network in which all nodes spontaneously perform some repetitive activity: a herd of antelopes in which each individual repeatedly lowers its head to take one or more bites and then lifts it to watch for predators while it chews; a chorus of male frogs or cicadas, each of which emits advertisement

FIGURE 15.15 **Self-organization in ant trails** Here, foraging *Lasius niger* are given access to two equally long paths between their nest and food. Although ants explore both paths, the basic rules used for creating paths between food and nest—lay a pheromone trail using any existing trails and reinforce the new trail on subsequent trips—result in one branch becoming the only used route. Different sides are chosen by chance on different trials. This self-organized pattern is an emergent property of pooled network interactions [31].

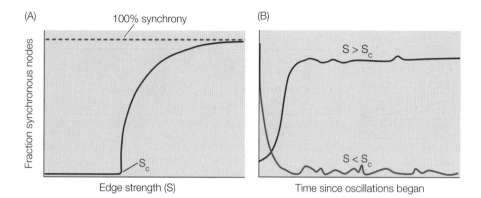

FIGURE 15.16 Synchrony thresholds (A) A model network of independently oscillating nodes will suddenly show some synchronous oscillation when edge strength (S) exceeds a threshold value (S_c). The fraction of nodes in synchrony then increases asymptotically to 100% with increasing edge strength. (B) Time course of increasing synchronization once nodes begin some common action, such as oscillation. Synchrony only increases to a stable limit if edge strength is greater than the critical value S_c (red line). If edge strength is less than the critical value (blue line), any initial synchrony decays to chance values. (After [314].)

calls at intervals; or a meadow filled with male fireflies emitting flashes to elicit responses from otherwise cryptic females. While there are no physical links between the nodes of these networks, individual animals *are* linked by the effects that the action of one individual has on the payoffs of the actions of others. We saw in Chapter 14 that random surveillance by foraging herd members puts those animals with heads down at a disadvantage when nonforaging individuals spot a predator and give an alarm. It thus pays for animals in such herds to look up when others look up and forage when others have their heads down. The fitness consequences favor their paying attention to each other, and this mutual attention creates a linked network with synchronized surveillance. Similar synchrony can be found in male calling choruses and firefly aggregations [47, 135].

A large amount of research effort has focused on how a network's topology might promote or hinder the emergence of synchrony in its behaviors [14, 96, 228]. Most theoretical models assume undirected and similarly weighted edges and begin with the performance of the same repetitive behavior by different nodes but at different rates and relative timings. They then ask what conditions are required to synchronize the shared behavior across the network. These models agree that the strength of the links between nodes is a critical factor: below a certain threshold edge strength, nodes show no synchrony at all. At strengths above this threshold, pockets of synchrony suddenly appear in the network, and as edge strength is increased further, these fuse until the entire network is synchronized (**Figure 15.16**).

Network topology can affect the initial threshold at which synchrony appears, the edge strength at which all nodes are synchronous, and whether global synchrony persists or instead decays into other patterns such as limit cycles

or chaos. Edge strength must be very high before a regular lattice network shows any synchrony at all. Adding random shortcuts to an otherwise regular lattice, e.g., converting it to a small world, reduces the threshold somewhat, making the initiation of synchrony more likely.

The increase in synchrony with increasing edge strength is more complicated in random and heavy-tailed networks [127, 128, 342]. At low edge strengths, random networks tend to form many small clusters of nodes that are internally synchronous but asynchronous with each other. Scale-free networks at low edge strengths show synchrony around the major hubs but little synchrony elsewhere. At low edge strengths, more of a scale-free network will be synchronous than in a random one. At higher edge strengths, synchronous clusters are larger in both kinds of networks. However, at a critical strength value, many of the individual clusters in the random network begin to interact, and the network contains a giant mass of synchronous nodes that is larger than any of the synchronous clusters in the scale-free network. Thus at low edge strengths, scale-free networks show more synchrony than random networks; but at higher edge strengths, random networks can show higher synchrony than scale-free ones.

Modular networks with low edge strengths show higher synchrony within modules than between them. At intermediate edge strengths, where interactions between modules first occur, they show a decrease in between-module synchrony. At high edge strengths, they finally express full synchrony within and between modules. In heavy-tailed networks, higher clustering coefficients reduce the threshold edge strength at which synchrony appears down to a specific level, after which the threshold can only be reduced further by also reducing average path length [127].

Network topology also plays a key role in the persistence of synchrony once initiated [14]. Not surprisingly, regular networks are least able to sustain synchrony, although adding shortcuts to create small worlds makes stable synchrony more likely [43, 315]. Although nonfractal scale-free networks are usually small worlds, the high-degree heterogeneity among nodes can inhibit stable synchrony [44, 91, 168, 226]. This is because the few highly connected hubs in a heterogeneous network receive so many disparate inputs from other nodes that they cannot lock into a stable and common rhythm. In undirected and unweighted scale-free networks, synchrony is thus hard to maintain. However, scale-free biological networks are usually directed *and* weighted, and this opens up solutions that can support synchrony despite degree heterogeneity.

As noted earlier, the sum of the weights of all inputs to a node is called the intensity (or **temperature**) of that node. Even if a network's degree distribution is highly heterogeneous, the intensities of all nodes can be equalized, given suitable weight combinations. A network with equalized intensities is said to be **isothermal**. Equalizing input intensities minimizes the traffic jams at hubs and significantly increases the chances that a scale-free network will sustain synchrony [212, 225, 227, 363]. It is even easier to homogenize the total node traffic and promote synchrony if some edges are inhibitory (e.g., have negative weights). Since only certain weight and sign topologies achieve isothermal conditions, it becomes easier to understand why either adding or deleting links to a scale-free network might alter its ability to sustain synchrony [136]. Above a minimal number of nodes, there may be many alternative combinations of weights and directions that promote synchrony [228]. Thus serial additions of more nodes (individuals) and edges (relations) to a network (group) might first promote synchrony, then reduce it, then promote it again, and so on in an alternating sequence (**Figure 15.17**). While the necessary conditions for sustained synchrony may seem prohibitive, many biological systems exhibit persistent synchronous behaviors. We shall return to the emergence of synchrony, particularly in the context of signaling, later in this chapter.

Modeling Evolution in Networks

As noted in Chapter 9, many evolutionary models assume complete mixing within a population: every strategy is likely to encounter every other strategy in proportion to their relative abundances. The same is assumed for genotypes. This is rarely true in nature: instead, different strategies and genotypes may be spatially separated and never interact. Put another way, well-mixed evolutionary models assume that the relevant networks are complete and that all nodes have identical connectivity, while, as we have just seen, many of the possible edges can be missing, and node degree and intensity can be highly heterogeneous. Just as network topology can have profound effects on effect propagation, we shall see below that the topology of incomplete networks can affect the processes of evolution.

Evolutionary graph theory

The modeling of evolution on incomplete networks is called **evolutionary graph theory** [233, 234]. In common with

other modeling designs, this approach usually assumes weak selection, haploid genetics, and stable population sizes. There are two relevant networks: the **interaction network** connects those individuals whose actions affect each other's fitness, and the **replacement network** connects those individuals whose offspring might provide replacements if one of them dies. To simplify the requisite calculations, most evolutionary graph models assume that these two networks are identical (for a treatment with only partial overlap, see [239]). Biological reality requires that the edges in both the interaction and replacement networks be directed and possibly weighted. Again, many evolutionary graph models simplify calculations by assuming that the two directional edges connecting any pair of individuals in the replacement network have equal weights (**dispersal symmetry**). This means that either individual's offspring, if they have them, would be able to fill a vacancy when one appears. Selection differences then are related only to who has a handy offspring and who dies.

For a population to remain stable, any fluctuation in density in one direction must be compensated by a subsequent fluctuation in the opposite direction. If an individual produces more than the average number of births (B), this will increase competition and cause a corresponding increase in deaths (D). Similarly, if an individual dies, this

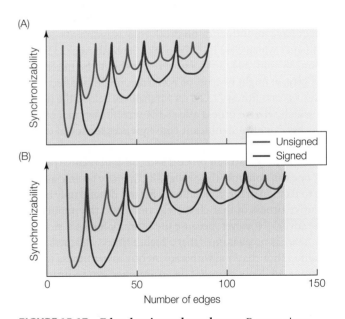

FIGURE 15.17 Edge density and synchrony For any given edge density in an undirected network, there is usually an optimal edge distribution that maximizes the chances of network synchrony. Removing successive edges from an initially complete network will alternately decrease and increase the network's synchronizability, depending on whether the current density is or is not an integer multiple of a critical value. (A) Synchronizability as a function of edge density for a network of 10 nodes (maximum of 90 edges in the complete network). Binary and unsigned networks show synchrony at more edge densities than equivalent signed networks. (B) Similar contrast for a network with 12 nodes (maximum of 132 edges). These complex patterns explain why adding (or deleting) edges in a network sometimes increases synchronizability, and at other times decreases it. (After [228].)

allows survivors to increase birth rates to fill the vacancy. A demographic history can thus be broken down into a series of birth/death (*BD*) and death/birth (*DB*) events. At the finest accounting level, each event will involve only two individuals: one that dies and one that provides a replacement offspring. A demographic model assuming single death/single replacement events is called a **Moran process** [211]. In the absence of consistent fitness differences between individuals, the probability that a given individual is chosen for either role in a Moran process is assumed to be random. If there are identifiable classes in this population—e.g., all individuals have one of several alternative alleles or adopt one of several alternative strategies—the probability that any class is chosen for either role in a Moran event depends only on the relative abundance of that class. Since an individual of the more common class will most often replace a dead individual of any class, the more common class will tend to increase until the other classes disappear. Changes in class abundances in this population over evolutionary time will then be solely due to random drift.

Now consider a population in which there *are* consistent fitness differences between classes. The probability that any given class is selected for a particular role in a Moran event will be biased by its fitness; this selection combined with drift will determine the population's evolutionary trajectory. Many evolutionary graph models assume that fitness differences have no influence on death role assignments, but do influence which class is likely to take the birth role [233, 234, 236–238, 240, 334, 335]. However, it should be noted that the reverse might be true, or both roles might be affected by relative class fitness [132]. As we shall see below, the degree to which relative fitness affects the second step in a Moran process can have significant consequences for social evolution on graphs.

Assuming a Moran process and constant relative fitnesses, one can compute the probability that any given class will go to fixation in the population. This will depend only on the relative fitness of that class (selection) and the population size (drift). However, because the Moran process is stochastic, there is no guarantee that a higher fitness class *will* go to fixation, nor that one with consistently lower fitness will be eliminated. Note also that by assuming a Moran process, a model tacitly assumes an overlap of generations, since only one individual dies per time interval. This can be contrasted with alternative approaches, such as a **Fisher-Wright process**, in which the entire adult population dies at the same time and is replaced by the next generation [105, 132, 333]. Assuming different demographic schedules can lead to quite different evolutionary trajectories and fixation predictions [132]. Although the traditional approach for evolutionary games assumed a Fisher-Wright process of generational replacement, it soon became clear that overlapping generation models were more appropriate for social evolution on graphs [154, 155, 333]. Thus most evolutionary graph models assume a Moran process.

The Moran process was originally designed to track evolution in well-mixed populations, or equivalently, in a complete network with undirected and equally weighted edges between all nodes. The networks studied with evolutionary graph theory are invariably incomplete, in that only a subset of individuals can provide the replacement for a given individual that has died. One major finding of evolutionary graph theory is that as long as the population network is isothermal, (i.e., the sum of the input weights is identical for all nodes), there is sufficient homogeneity that Moran process predictions about fixation probabilities still hold [233]. However, this will not be the case for nonisothermal networks. For example, network nodes that have only output links and no inputs are called **roots**. A root has a lower intensity than the rest of the network nodes. Evolving networks with multiple roots experience a reduced role of selection and an amplified role of drift relative to a well-mixed population. At the other extreme are scale-free networks in which hubs have much higher intensities than other nodes; these conditions amplify the effects of selection and suppress the effects of drift when compared to well-mixed networks. If stochasticity in event outcomes is minimized, so that a lower-fitness player is nearly always replaced by a higher-fitness player, network evolution can become highly nonlinear and exhibit frequent bifurcations, multiple equilibria, limit cycles, and chaos [230]. The population may even evolve into a complex spatial mosaic or a fractal distribution of classes [233]. Given these complications, most evolutionary graph models assume that the evolving network has some minimal levels of node homogeneity and stochasticity [132, 234, 334].

Social evolution on graphs

COMPENSATION ON GRAPHS   A key component of a Moran process is **compensation**: this is the birth after a death, or the death after a birth that is required to keep a population stable. While a Moran process in a well-mixed population allows the parties experiencing the triggering event and the consequent compensation to be anywhere, social interactions rarely involve random and distant individuals. In many natural populations, both social interaction and demographic consequences tend to be localized. As a result, compensation is most likely to fall on the immediate neighbors of the animal that dies or produces a bumper number of offspring (**Figure 15.18**). We shall refer to an incomplete network in which both social interactions and demographic consequences are spatially local as a **structured population.** Although *BD* and *DB* sequences lead to similar fixation predictions in well-mixed populations, they can produce quite different evolutionary trajectories in structured populations [131, 132, 233, 234, 236, 237, 332]. Why is this so?

Consider a focal animal in a structured population and a Moran process of incremental evolution. If an immediate neighbor has above-average fecundity, this will negatively impact the survival of the focal animal (a *BD* event). If survival probabilities prior to the birth of a neighbor's supernumerary offspring are the same for all individuals, the focal animal is stuck carrying its share of the compensation. But

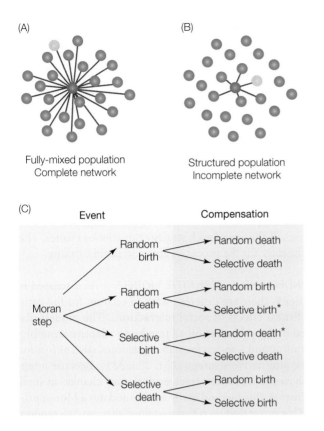

(A)

Fully-mixed population
Complete network

(B)

Structured population
Incomplete network

(C)

Event

Compensation

Moran step

Random birth → Random death
Random birth → Selective death

Random death → Random birth
Random death → Selective birth*

Selective birth → Random death*
Selective birth → Selective death

Selective death → Random birth
Selective death → Selective birth

FIGURE 15.18 Moran processes In a Moran process, the population size must return to its stable value after each step: each death must be compensated by a corresponding birth to replace the mortality; any birth must be compensated by a subsequent death. (A) In a fully mixed population, a focal individual (red node) is potentially connected to every other individual in the population. A birth or death at a focal node can be compensated by any other node, including very distant ones (e.g., yellow node). (B) In structured populations, the only network links are local: a focal animal that dies is mostly likely to be replaced by the offspring of neighbors (e.g., yellow node), and the offspring of the focal animal are most likely to replace their dead parent or their neighbors. (C) There are eight possible Moran sequences, depending on whether the initial event is a death or a birth and random or biased by selection, and whether the ensuing compensation is random or based on selective differences. Contrasts between the two sequences marked with asterisks indicate that they differ in their sensitivity to network topology. Although early interpretations focused on the type of initial event, it now appears that it is the type of compensation that modulates sensitivity to network topology (see text and Figure 15.19 for details).

if some of the fecund animal's other neighbors had survival probabilities lower than that of the focal animal prior to the births, they are more likely to die when competition increases, and the mortality risk to the focal animal will be reduced. When the compensatory costs in a *BD* event are randomly distributed among neighbors, details about other neighbors are irrelevant to the focal animal. Put another way, the relevant economic neighborhood for the focal animal is limited to it and its fecund neighbor. If the compensatory deaths are not randomly assigned to the fecund animal's neighbors, but

instead depend on their relative survival probabilities prior to the event, then the neighbors of the fecund animal must be included in the relevant economic neighborhood of the focal animal. The economic neighborhood of a focal animal for a *BD* event is thus smaller and harsher if compensation is randomly distributed, and larger and possibly less harsh if compensation is not distributed equally among the neighbors [132, 237, 240].

A parallel pattern occurs for *DB* events. When the neighbor of a focal animal dies, the focal animal will compete with the other neighbors of the dead individual to provide the replacement offspring. If the winner of the competition is randomly determined, the relevant economic neighborhood is again limited to the focal animal and its dead neighbor. If relative fecundities prior to the event bias the odds of which neighbor's offspring replaces the mortality, the relevant economic neighborhood for the focal animal must be expanded to include all of the dead individual's neighbors. While *BD* events emphasize relative survival probabilities and *DB* events emphasize relative fecundities, the two demographic scenarios do share one major prediction: the relevant economic neighborhood is small if the compensation role is randomly assigned among neighbors but larger if it depends on relative fitness components of neighbors before the event [132, 332]. For various historical reasons, many evolutionary graph models assume random assignments of compensation in *BD* events and fitness-biased assignments in *DB* events [233, 236, 237, 335]. Although many publications link evolutionary outcomes to the type of demographic event, the critical factors shaping social evolution turn out to be the degree to which compensation is selective and the consequent size of the economic neighborhood.

SOCIAL GAMES ON GRAPHS   In a typical social interaction, such as communication, an actor does something that affects the subsequent behavior or physiology of one or more recipient animals. The actor's behavior will usually affect its own survival or fecundity directly, and it will affect the survival or fecundity of the recipient. In a stable population, any effects of the action on the recipient's fitness will feed back on the actor in the form of compensation. In well-mixed populations, compensation can be spread over the whole population; in structured populations with limited mixing, compensation is localized and thus felt more intensely by neighbors of the recipient, including the actor. We can reframe such social interactions using the logic of the prior section. The actor is our focal animal and the recipient of the action is a linked neighbor whose fecundity (*B*) or survival (*D*) is altered by the focal animal's action. In this case, the focal animal is not a passive participant in the occurrence of a demographic fluctuation; instead, it actively triggers a *BD* event (if its action alters the neighbor's fecundity) or a *DB* event (if it alters the neighbor's survival). The question then is: Should the actor perform the action or not? If several actions are possible, which is optimal? As in prior chapters, these questions are best answered using evolutionary game theory. However, this time the relevant games are played by small numbers of linked neighbors in the social network.

The most widely modeled network game explores the evolution of altruism. When is the provision of a benefit to another at some personal cost (the altruist strategy) evolutionarily favored over an alternative strategy of accepting such gifts but never providing anything to another if there is a personal cost (the defector strategy)? We discussed this question in Chapters 9 and 13, and concluded that altruism is favored only if enough of the original benefit trickles back to the donor to more than compensate for the costs. How much the original benefit can be discounted during the trickle-back and still compensate for costs can usually be computed in well-mixed populations. The consequent criterion requires that the ratio of recipient benefit to altruist cost be larger than some function of the discounting fraction. However, the localization of interactions in structured populations makes such calculations much more challenging. Modeling such a population as an incomplete network has greatly helped to clarify the conditions that favor or disfavor the evolution of altruistic behaviors. As noted earlier, even these models are only analytically tractable if the network is assumed to be reasonably homogeneous. However, simulations for less homogeneous network structures suggest that many homogeneous network predictions may have considerable generality.

A widely replicated finding of this approach is that sustained altruism is more likely to evolve in a structured population than in a well-mixed one if most demographic events are *DB* and local compensation is nonrandomly distributed [233, 234, 236–238, 334]. Conversely, altruism is less likely to evolve in a structured than in a well-mixed population if most demographic events are *BD* and compensation is randomly distributed. Although less widely cited, a *BD* model in which compensation is nonrandom is also predicted to favor altruism in structured populations more than in a well-mixed system, whereas a *DB* model with random compensation reduces the likelihood of altruism below that predicted for the well-mixed case [132, 332]. In short, altruism is more likely to evolve in structured than in well-mixed populations as long as compensation is nonrandom. Why?

We provide a simple example of evolution given different combinations of birth/death and random/selection sequences in **Figure 15.19**. We can interpret the results as differences in the relevant economic neighborhood. In a well-mixed population, the fitness of an evolutionary game strategy depends on the overall relative abundances of the various strategies. In structured populations, the fitness of an individual adopting a given strategy is mostly dependent on the relative abundances of the different strategies among its neighbors. Altruists in structured populations have higher fitness if their immediate neighbors are altruists than if their neighbors are defectors. The larger economic neighborhood as a result of nonrandom compensation increases the chances that an altruist will have neighbors that are also altruists. When enough neighbors are altruists, a focal altruist will be less likely to die when its neighbor has enhanced fecundity

and more likely to provide the replacement offspring when its neighbor dies. The potential reduction in compensation costs reduces the fraction of benefit that has to trickle back to an altruist to justify its action. Note that once altruism is fixed in the population, there is no longer an advantage to nonrandom compensation, since now everyone bears compensation equally. However, at that point, selection is irrelevant except to weed out the occasional defector mutant. Note, also, that a minimal number of altruists must be present before the nonrandom assignment can exert any selective advantages to altruists. However, in the stochastic environment of a Moran process, even a single altruist mutant may, by chance, produce the offspring that replaces a defector neighbor if it dies. Thus small clusters of each strategy can arise by chance. This sets the stage for the processes above to begin working.

EXTENDED AND ALTERNATIVE MODELS   As discussed in Chapter 9, there are different ways to account for the complex fitness effects of social interactions. The models discussed in the prior section all invoke an adaptive dynamics accounting and a search for stable outcomes, such as fixation of one alternative strategy [231, 233, 331]. Because adaptive dynamics equations track evolutionary changes in small increments, this method easily accommodates a Moran process. One novel finding is that sustained altruism in a random death/selective birth world requires that the benefit-to-cost ratio must be larger than a simple function of the average degree of the network: as more and more edges are added to each node, it becomes harder and harder for altruism to evolve [236, 237]. In the limit, every node is connected to every other node, the population is well-mixed, and a strategy is favored only if there is a net direct benefit to the actor. Increasing network degree dilutes the indirect benefits to actors until they are insufficient to compensate for the costs.

The adaptive dynamic approach has been taken one step further by examining a general model of two-strategy symmetric evolutionary games on structured networks. Again, the relevant networks are assumed to be relatively homogeneous. Interestingly, a general ESS can be predicted using only the payoff matrix and a single parameter, σ, that combines the effects of network structure, event type (*BD* or *DB*), and the mutation rate (**Figure 15.20**) [330]. This parameter essentially measures the degree to which individuals adopting a given strategy are likely to interact with other players adopting the same strategy. The altruism-defector game described above can be seen as a special case [234]. The model has been useful for game theoretic analyses of the evolution of reciprocity and greenbeards as well as the dynamics of social evolution when strategies are learned and players belong to overlapping associations [10, 237, 238, 240, 329]. Some related models examine what happens if offspring can disperse and establish interaction links assortatively. These models find that a structured and incomplete network can arise in an otherwise well-mixed population as long as individuals are highly selective about whom they select as neighbors and interaction partners. Such biased dispersal facilitates

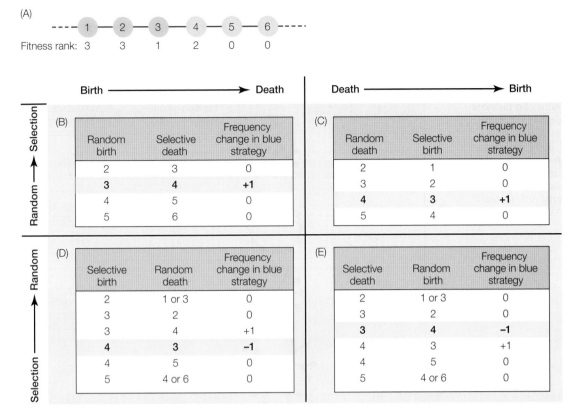

FIGURE 15.19   Social evolution on a simple network (A) Linear replacement network with a uniform degree of 2 and no transitive triangles (i.e., clustering coefficient = 0). Two strategies are present (blue and yellow). An individual's fitness depends on its and its neighbors' strategies: the fitness of blues surrounded by blues is greater than the fitness of a yellow with blue and yellow neighbors, which is greater than the fitness of a blue with blue and yellow neighbors, which is in turn greater than the fitness of yellows surrounded by yellow neighbors. This pattern fits most altruist (blue) versus defector (yellow) evolutionary games and many communication games. In a Moran process on this network, a death can only be compensated by the birth of offspring of immediate neighbors. How does the frequency of the blue strategy change over time given different combinations of random and selection-based events? (B) Random birth/selective death: All individuals are equally likely to give birth. If individual 3 gives birth, either individual 2 or 4 dies to compensate; 4 is more likely to die as its initial fitness is lower than that of 2. Births by others do not affect blue frequency. Over time, blue strategists will likely increase to fixation. (C) Random death/selective birth: The only event that changes blue frequency is death by individual 4; individual 3 has higher fitness than individual 5 and is more likely to supply the replacement offspring. Again, blue strategies will likely increase to fixation. (D) Selective birth/random death: Once an individual has a birth, its two neighbors have equal chances of dying. Only births by individuals 3 and 4 affect blue frequency, and they do so in opposite directions. Since 4 has higher fitness, it will give birth more often than 3, and when it does, blue frequency decreases. Yellow strategy will likely go to fixation. (E) Selective death/random birth: Similarly, deaths of only individuals 3 and 4 affect blue frequency, again in opposite directions. Since 3 is more likely to die than 4, decreases in blue strategy occur more often than increases. Yellow strategy is more likely to be fixed over time. In general, blue strategy increases if compensation is selective and decreases if it is random.

the evolution of altruism while also generating interaction networks with a low average degree and heavy-tailed degree distributions [284–287].

An alternative accounting for the evolution of altruism invokes inclusive fitness measures (see Chapter 9). Most structured population models assume that neighbors supply the replacements for an individual that dies. Over time, neighbors then become increasingly related genetically, and the computation of relative fitness takes this sharing of alleles into account. The relevant inclusive fitness models make the same predictions noted earlier: well-mixed populations are more likely to sustain altruism than *BD* worlds with random

compensation, and *DB* worlds with nonrandom compensation are even more likely to sustain altruism than well-mixed populations [131, 132, 174, 332, 334, 335]. The congruence in predictions using the two accounting methods is reassuring, but not surprising. Some recent work suggests the two methods may even measure the same thing [361].

In Chapter 9, we noted that proponents of inclusive fitness and group selection accounting methods often disagree about how best to track social evolution: each group claims that their method is more general than that of the other. A similar dispute exists between proponents of adaptive dynamics on networks and those advocating inclusive

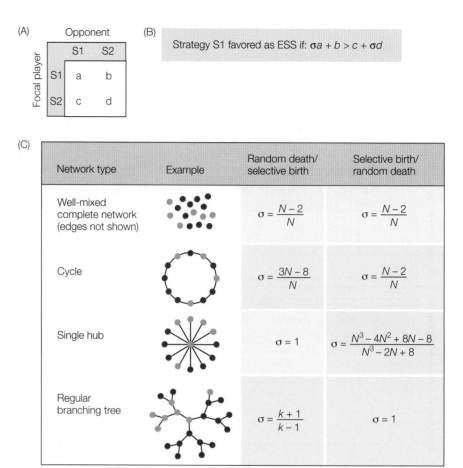

FIGURE 15.20 **ESS for symmetric game on network** (A) Payoff matrix for discrete symmetric evolutionary game with two alternative strategies, S1 and S2. (B) The general condition for strategy S1 to be an ESS depends on the network-specific parameter σ and the values in the payoff matrix. (C) The value of σ for some simple network topologies and different Moran process sequences. N is the number of nodes in network, and k is the average degree. The node colors correspond to different strategies. The results in tan cells apply for any mutation rate, whereas the results in blue cells are valid only for low mutation rates. (After [330].)

fitness to model evolution in structured populations. Each group claims that the simplifying assumptions of the other are more limiting than their own [131, 132, 234, 235, 332, 336]. It is instructive that so far, the two approaches have produced largely similar predictions, and challenges by one group that the other's method cannot deal with a particular phenomenon are invariably followed by a suitable extension of the challenged method. All models make simplifications: this is the whole point of the exercise. Our sense, as with the earlier conflict, is that having multiple accountings with different model assumptions invariably provides complementary insights. This certainly appears to be the case for studies of social evolution on networks.

Web Topic 15.2 *Additional information on evolutionary graph theory*
Here we provide links to several sites that give more detailed explanations and examples of Moran processes and the modeling of social evolution on networks.

Animal Communication Networks

We now turn to examples of animal communication in which the previous discussions of network measurement, behavior, and evolution may provide new insights. Of particular interest are social behaviors, including communication, that show emergent properties and self-organized patterns that can only be understood using network models [300]. The dissemination of new information about predators and foraging options often percolates along a population's social network. Increases in population density can trigger bifurcations in social structure when an initially sparse signaling network suddenly acquires a giant component. Locomotion synchrony in flocks and schools is often an emergent property of links between adjacent members, requiring no leaders. Below, we reexamine these and other topics raised in prior chapters, this time with a network perspective.

Linkage patterns

Many of the evolutionary models of communication discussed in prior chapters either focused on single sender/receiver dyads (optimization economics) or assumed a well-mixed population in which each strategy is encountered in proportion to its relative abundance (evolutionary game theory). These two cases are network extremes: the first assumes a sparse network of isolated pairs with each node linked to only one other node, and the second assumes a complete network. They are theoretical favorites because their respective homogeneities simplify model computations. But most real communication networks are neither dyadic nor complete; they are incomplete, with edges that are directed and unequally weighted [85, 354]. As we have seen, the heterogeneous topologies of such intermediate cases can play a major role in how far and fast effects propagate, and what evolutionary outcomes are most likely when alternative strategies are available.

FIGURE 15.21 **Percolation in a communication network** Consider signaling in a baboon group. The brown female calls to solicit a nearby subordinate male, who then approaches. The active space of her call is shown in blue. A dominant male (green) is inside her active space, hears her call, and moves to intercept the subordinate male. Another male (dark blue), outside the active space, doesn't hear the female call, but seeing the dominant male preoccupied, attacks another individual usually protected by the dominant. Other baboons close to this conflict (orange zone), including the female who called, move away. Individuals in the yellow zone see movements of orange zone individuals toward them and also move away, but not as far. Individuals outside any zone are unaffected. The female's calls thus trigger a cascade of effects that feed back on her, radiate away, and gradually fade out at sufficient topological distances. The relevant percolation fraction is much larger than the active space of the initial sounds, but does not include the entire interaction network.

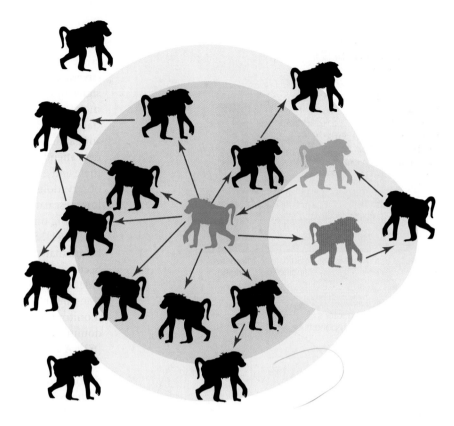

What fraction of an overall interaction network is relevant to communication? Some authors have defined a communication network as the group of animals within signaling and receiving distance of each other [201–203]. An even narrower definition identifies a focal individual's communication network as the set of conspecifics within the active space of its signals [191]. A limitation of these definitions is that they exclude network links between individuals that are not mediated by signals but which could be activated by signals. When we reviewed percolation on networks in a prior section, we noted that effect propagation does not require that effects at successively more distant nodes be of identical type: the response of a primary receiver or eavesdropper to a focal animal's signal need not be to produce the same or even another signal. Individuals outside the original signal's active space, yet in the same network, may be affected by the response of a receiver within the active space. Thus a signal exchange between two individuals can initiate a radiating chain of effects that may include no further signals at all. Because edge weights are rarely maximal, effect amplitudes usually decay at successively more distant nodes from the original stimulus. However, the percolation fraction of the overall interaction network that is affected by an initial signal stimulus is likely to be much larger than that enclosed by the active space of that signal (**Figure 15.21**). Remember also that once effects propagate several links away from the stimulus, propagation can be in both directions, resulting in feedback to the sender and primary receiver from responses by nodes well outside the signal's active space. Such feedback is often essential to the appearance of emergent properties and self-organization.

The size of the percolation network following a given signal exchange depends on two factors. The first is whether a dyad can manage a signal exchange without triggering propagation in their interaction network. Although dyads can attempt to restrict the active space to a minimum, they usually cannot control the presence of eavesdroppers. Eavesdroppers provide a bridge between isolated signaling dyads and the rest of the interaction network. Thus eavesdropping can play an important role in determining percolation fractions. The second factor is the topology of the preexisting interaction network. Even if links exist between a signaling dyad and the network, the percolation fraction and percolation times will depend critically on the density, weighting, and distribution of edges in the interaction network. We take up each of these factors below.

EAVESDROPPING   Animals eavesdrop all the time: they eavesdrop on conspecifics, and they eavesdrop on other species. The risk of eavesdropping can significantly alter the economics of communication for both senders and receivers, and one or both parties may exert considerable effort to minimize these risks [159]. Their ability to minimize eavesdropping depends on the usual distance between senders and receivers (i.e., whether they use broadcast or short-range signals); how well senders can regulate active spaces, given signal modality and propagation conditions; and the density of likely eavesdroppers in that habitat [88, 157, 357]. For many signaling contexts, eavesdropping is inevitable.

Eavesdroppers obtain information by attending to signals. Two classes of eavesdropping have been proposed, based on when an eavesdropper acts on the information it obtains

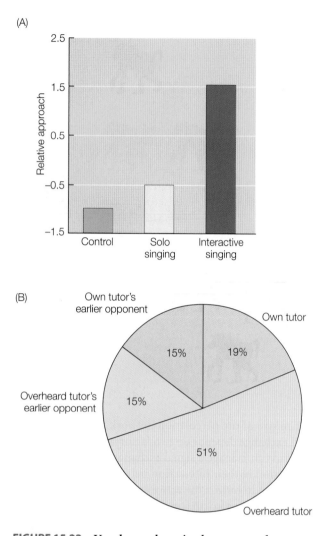

FIGURE 15.22 Vocal eavesdropping by young male song sparrows (*Melospiza melodia*) (A) Juvenile male song sparrows were tracked with radio telemetry during field playbacks of black-capped chickadee (*Poecile atricapilus*) song (control); solo songs of one adult male song sparrow; and songs recorded during competitive interactions between pairs of adult male song sparrows. The juveniles preferentially approached playbacks of interacting conspecific males over either solo conspecific or heterospecific song. However, they did not approach closely, nor did they vocalize, so it is likely that they were eavesdropping on the adults' song contests. (B) Hand-raised juveniles that had not yet begun to sing were allowed to hear two pairs of adults countersinging. Later, the juveniles were allowed to interact vocally with one of the adults, and also to overhear vocal interactions between another of the adults and another juvenile. The pie chart shows that the young birds learned songs from each of the four adults to which they were exposed, but as many from the adult they heard interacting with another juvenile as from the other three tutors combined. Again, this suggests that eavesdropping on vocal contests plays a major role in song acquisition in these birds. (After [32, 337].)

[252]. **Interceptive eavesdroppers** respond to the signal very quickly and usually interfere in some way with the sender, the receiver, or both parties. Interspecific examples include the many predators, parasites, and parasitoids that use a prey's signals to find and attack it. Intraspecific examples include a

territorial songbird's immediate eviction of a singing intruder, or a parent's disruption of a fight among offspring. **Social eavesdroppers** incorporate the information provided by the signal into the adjustment of long-term strategies. There are many demonstrated examples. Young songbirds preferentially learn song types heard while eavesdropping on adult song contests (**Figure 15.22**). Female canaries and chickadees attend to male countersinging contests and adjust future mate and extra-pair copulation probabilities based on these observations [6, 173, 206, 244, 245]. Male songbirds raise or lower escalation levels in contests with neighbors after eavesdropping on an opponent's contests with other neighbors [213, 251]. Social eavesdropping significantly affects both future mate choices and fighting strategies in guppies [102]. Scent marking, particularly the levels of overmarking, provide rodents with detailed information about recent interactions among other residents [158]. Social primates regularly monitor interactions among other group members to assess dominance status, kinship relationships, and leadership roles [67]. A variety of species decrease cooperative interactions with neighbors that are observed failing to fulfill their share of some joint task; eavesdropping is thus a key component in the evolution of indirect reciprocity [3, 108, 176, 232, 241, 261, 278, 343]. Nest parasitic birds compare the courtship and territorial signals of potential hosts before selecting a nest to parasitize [248]. Client fish often scrutinize the signals and interactions between cleaner fish and previous clients before exposing themselves for a cleaning [46].

Given the ubiquity of eavesdropping, it is not surprising that senders may adjust their signaling strategies depending upon who is likely to be in the neighborhood [88]. This is called the **audience effect**. There are two classes of audience effect (**Figure 15.23**). First-order effects reflect a sender's estimation that a primary receiver is within the active space of its signal: if that probability is sufficiently low, it may not be worth the costs to emit the signal. Simple time and energy considerations may also explain omitted or reduced signaling when primary receivers are unlikely to be present. First-order examples include roosters that are less likely to give food calls when hens are absent; yellow mongooses that give alarm calls only when conspecifics are nearby; drongos that produce terrestrial predator alarm calls only when terrestrially foraging babblers are members of their mixed flocks; and zebra finch males that infuse more carotenoids into their bills when they are housed with females [123, 172, 189, 276].

Second-order audience effects occur when primary receivers are likely to be present, but senders modify their signaling because of the risks of eavesdropping [195]. Again, there are numerous examples. Fiddler crabs, mollies, Siamese fighting fish, and salamanders alter their courtship displays to females if other males are nearby [13, 17, 97, 195, 196, 259, 260, 262]. Female chimpanzees avoid giving pant-grunt greetings to subordinate males when dominant males are within earshot [171]. In a variety of songbirds, males emit unusually low-intensity songs in the final stages of escalated fights [7, 8, 20, 149, 167, 169, 289]. While the latter case may be due to the imposition of

(A)

Drongo

Babbler

(B)

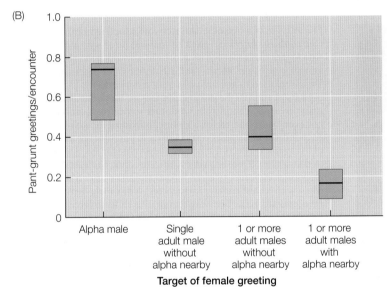

FIGURE 15.23   Audience effects (A) First-order audience effect, in which the sender gives a signal only if the appropriate receiver is present. Canopy-feeding fork-tailed drongos (*Dicrurus adsimilis*) routinely emit alarm calls to other mixed flock members when aerial predators appear, but rarely emit alarms for terrestrial predators unless terrestrially foraging pied babblers (*Turdoides bicolor*) are present in same flock (see Chapter 14 for more details). (B) Second-order audience effect, in which the sender alters signal emission rate or content depending on whether eavesdroppers are present. Female chimpanzees (*Pan troglodytes*) usually give pant-grunt greetings to the group's alpha male more often than to other adult males encountered when the alpha is not present. When the alpha male *is* present, the female is much less likely to give greetings to other adult males. (After [171, 276].)

vulnerability handicaps, (e.g., a combatant makes itself vulnerable by giving a low-intensity threat that can be heard only at close range), there are clear examples which appear to be due to minimization of eavesdropping risks.

SOCIAL NETWORKS   Individuals in nearly any signaling species are part of some preexisting interaction network. The structure of these networks determines who signals to whom and how far a given effect will propagate (see illustrative human example in [63]). Interactions in structured populations are limited to spatial neighbors. As noted in prior chapters, kinship often determines the location and strength of suitable interaction links. Positive unisexual alliances are widespread in mammals: male coalitions are found in chimpanzees, bottlenose dolphins, and bachelor zebras, and female coalitions occur in hyenas and a variety of primates such as baboons [74, 107, 183, 188, 210, 268, 296, 301, 302, 305, 312, 351]. The need to display in an aggregation and an animal's dominance status can also shape its interaction network.

While widely used for human studies [57, 156], network metrics have only recently been recruited for studies of animal social dynamics and evolution [85, 164, 325, 352, 354]. Some specific examples are instructive, as they show the types of interaction networks that are being found in animals. Unfortunately, the field is still so new, and the number of possible metrics so great, that different authors use different subsets of the possible measures. On top of this diversity, different authors use different names for the same metric. Luckily, a number of the metrics that we defined earlier are now sufficiently widely used that some comparisons are feasible. Below, we compare these metrics in species with one of three different social organizations: those that are largely solitary, those that live in fission–fusion societies, and those that interact in highly cohesive groups.

The networks of two relatively solitary species have been examined quantitatively. Sleepy lizards (*Tiliqua rugosa*) have stable individual home ranges. However, these overlap enough that a typical lizard makes contacts with 12 others

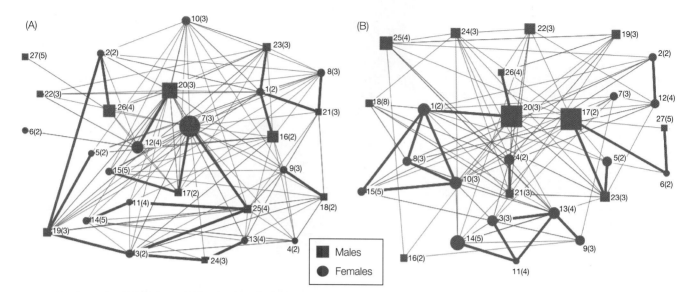

FIGURE 15.24 **Association networks for Tasmanian devils** (***Sarcophilus harrisii***) Each animal was fitted with a proximity data logger that recorded each occasion in which another outfitted animal was within 30 cm—sufficiently close for potential contact. Data were used to generate association networks for (A) the mating season and (B) the postmating season. Individual females and males are each identified by a number so relative locations and links can be compared for the same individual in two seasons. Each individual's age in years is given in parentheses, and the size of the symbol reflects the betweenness value for that individual in that season. Thin-line edges indicate contact at least once, and thick lines indicate preferred links that remain after filtering out edges with lowest contact frequencies. While male–female links predominate in the breeding season, this shifts to largely female–female associations postbreeding. (After [137].)

(the average degree) in the premating season, and 10 in the postmating season [177]. If the relevant association matrix is filtered to focus only on strongly "preferred" links, the average degree values drop to 1.1 in the mating season and 1.0 in the postmating season, and there is significant sexual assortativity: male–female pairs are much more likely to associate regularly than any other combination. For Tasmanian devils (*Sarcophilus harrisii*), which also live relatively solitary lives but with overlapping home ranges [137], the average degree based on the presence or absence of dyadic encounters is 12.7 during the mating season and 10.4 in the nonmating season; the values for preferred links with more than a threshold number of encounters drop to 1.8 and 1.6, respectively (**Figure 15.24**). Male–female links dominate the mating season network, while female–female associations are significantly more common in the nonmating season. The degree distributions for the Tasmanian devils were significantly heavy-tailed, but they were not scale-free. Some individuals had very high betweenness values (25–26 shortest paths between other dyads passing through them in both seasons); clustering coefficients were moderate (0.40–0.47), but significantly

higher than for a random network. These two studies clearly show that even relatively solitary animals interact via networks. However, neither of these networks is highly structured, with hierarchical or modular clusterings.

Animals that live in fission–fusion societies are more social than the species discussed above, but their frequent mingling and repartitioning into groups would seem to preclude significant structure in their interaction networks. Surprisingly, this has not turned out to be the case. Metrics for association networks for shoaling guppies (*Poecilia reticulata*) and sticklebacks (*Gasterosteus aculeatus*) based on samples of 110–147 fish/shoal include edge densities that are moderate (15–30%), but still large enough to host a giant component; intermediate average degrees (14–28 links/node); short average path lengths (2–2.3 links); and high clustering coefficients (0.57–1.0) [83–86]. The short path lengths and high clustering coefficients imply small world behavior in these networks. Degree distributions are heavy-tailed, due to a shortage of high-degree hubs, but are not scale-free. The networks also exhibit positive assortativity for degree, body size, and some behavioral measures (such as boldness versus shyness).

(A)

(B)

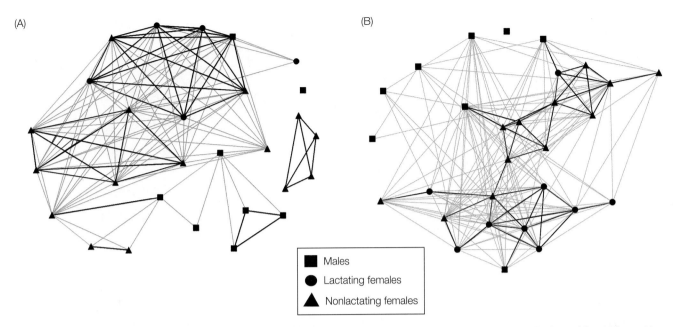

◼ Males

● Lactating females

▲ Nonlactating females

No assortativity by sex was found. Community structure was present in the networks of the guppies, and this was clearly correlated with each individual favoring one of the adjacent pool basins in the river sampled.

Bottlenose dolphins (*Tursiops* spp.) also live in fission–fusion societies [74]. A variety of network metrics have been obtained for one New Zealand population of *T. truncatus* containing 62–67 individuals [181, 182, 185]. These include edge density (8%), average degree (4.97), average path length (3.36 links), a moderately high clustering coefficient (0.30), and high betweenness for hub nodes (7.3 and higher). Although the network is definitely sparse, the low path length and moderate clustering coefficient argue for some small world character to this network. Degree distributions were heavy-tailed, but not scale-free due to a shortage of low-degree nodes. Although they often exchanged members, two distinct communities were identified (see Figure 15.8). Significantly positive assortativity was noted for sex and age; degree correlations were not significant. A slightly larger population (N = 95–120) of *T. aduncus* on Australia's eastern coast was also found to contain two interacting communities [358]. Average intensities (measured as the sum of the weights of all links to a node), ranged from 6.4 to 9.3 if links between communities are included, and from 4.5 to 7.7 if only links within a community were considered. Affinities (the average intensities of a node's neighbors) were very similar to a node's own intensity. Interestingly, the clustering coefficients in the Australian population (0.13–0.21 for links inside a community) were significantly less than for a random network and opposite in direction from the New Zealand samples. While each Australian dolphin has many connections, there is less local interconnectivity than for their counterparts in the New Zealand community.

Fission–fusion herds are very common in hoofed mammals. One study used network metrics to compare wild zebras (*Equus grevyi*) in Kenya brushland to onagers (*Equus*

FIGURE 15.25 Association networks for wild equids Wild herds of Grevy's zebras (*Equus grevyi*) and onagers (*Equus hemionus khur*) were censused repeatedly for individual composition, using natural markings. Resulting association networks for zebras (A) and onagers (B) are marked here to distinguish males, lactating females, nonlactating females, and any association (thin edges) from preferred associations after filtering (thick edges). Results show the presence of cliques and high clustering coefficients in zebras, but much looser connectivity patterns in the onagers. (After [326].)

hemionus) in the Indian desert [326]. These two species have herd sizes with similar means (5.1 and 4.1 animals respectively) and ranges (1–15 and 1–25). However, they differ subtly in network topology (**Figure 15.25**). Both networks consist mostly of one giant component (containing 23 and 28 individuals respectively) and 1–2 additional components consisting of a few individuals each. Average degrees are low (1–6) for both species. If only preferred (heavily weighted) links are considered, the zebra network breaks up into 11 components (several of which qualify as cliques), while the onager network breaks up into 6 components (with only loose communities and no cliques). Average degree is higher for onagers than for zebras, indicating more general mixing in herds. The clustering coefficients are high for both networks (0.91 and 0.71 respectively). Average path lengths are 1.9 and 1.6 links for the overall network, but 1.0 and 3.5 for preferred connections. Both networks act as small worlds, but there are longer shortcuts in the preferred network in onagers. Overall, zebras form networks with more consistent modularity than do onagers.

We finally consider species in which group composition is relatively stable and separate groups interact infrequently. The most relevant metrics here involve network structure inside an average group. Social wasps and bees live in highly cohesive colonies that are largely kin. One recent

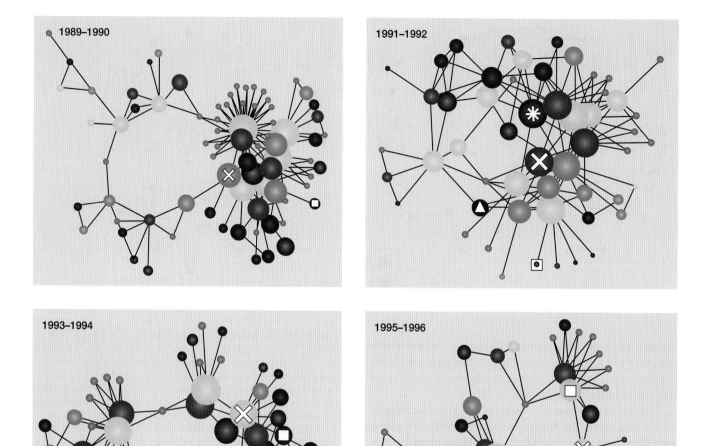

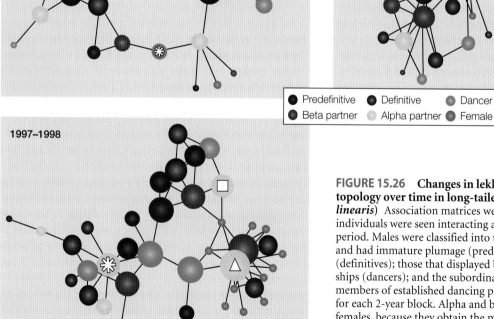

FIGURE 15.26 Changes in lekking status and network topology over time in long-tailed manakins (*Chiroxiphia linearis*) Association matrices were constructed by noting which individuals were seen interacting at a given lek site over a 10-year period. Males were classified into those that had yet to display and had immature plumage (predefinitives) or mature plumage (definitives); those that displayed but had no established partnerships (dancers); and the subordinate (beta) and dominant (alpha) members of established dancing pairs. Networks were created for each 2-year block. Alpha and beta males show many links to females, because they obtain the most copulations. The size of the nodes indicates the information centrality metric for that individual in that 2-year block. Males with a high information centrality value early in their dancing careers (such as the ones marked with an X, asterisk, triangle, or square) were most likely to move up to alpha status and high mating success later in life. (After [199].)

study examined colonies of the wasp *Rhopalida marginata* ranging from 8–40 individuals [215]. The modeled network contained weighted edges based on the relative number of positive interactions between dyads; dominance interactions were not scored. Resulting metrics included edge densities (10–30%), average path lengths (2–4 links), and clustering coefficients (0.5–0.8). Degree distributions were random for small colonies but became increasingly heavy-tailed (but never scale-free) at larger colony sizes. Average path length also increased with colony size, while edge density and clustering coefficients both decreased in large colonies. Given the short average path lengths and high clustering coefficients, these colonies surely act as small worlds. The networks are also highly robust to random removal of nodes: neither path lengths nor clustering coefficients changed significantly when 3–6 random colony members were removed experimentally.

Males of most manakin species (Pipridae) display on leks [307]. While individual display is the most common pattern, pairs of males perform joint displays in several species in the genus *Pipra* and in all species of the genus *Chiroxiphia* [100, 114, 140, 198, 277, 306]. These birds tend to be long-lived, and local populations are relatively stable in composition. This facilitates the establishment of complex social networks based in part on age and in part on display partnerships [99, 115, 197]. Paired display is facultative in *Pipra filicauda*. An examination of joint display networks in this species based on who displayed with whom provided a variety of metrics: edge density (8–9%), average degree (2.6–3.9), network diameter (8.3 links), average path length (3.9 links), betweenness (6–13 paths), and clustering coefficient (0.39) [283]. Despite the low edge density, networks at three study populations all consisted of a single component. Degree distributions were heavy-tailed but not scale-free. Older males tended to be the higher-degree hubs. A given male's reproductive success was correlated with degree and betweenness measures, as well as with his tenure at a display perch. A parallel study on *Chiroxiphia linearis*, in which joint display is an obligate condition for matings, resulted in shorter average path lengths (2.5 links) and a smaller network diameter (6 links) (**Figure 15.26**) [199, 200]. In addition, the best predictor of a male's rise in dominance status, a key factor in his reproductive success, was his information centrality metric soon after he began displaying. As noted earlier, information centrality is an extended version of the betweenness metric. Interestingly, cooperative display in these and other manakins appears to be unrelated to kinship [116, 180, 200]. The low average degree and heavy-tailed degree distributions observed in these species are thus consistent with evolutionary graph theory models that predict cooperative signaling without kin altruism when individuals are free to select interaction partners [284–287].

The highly cohesive groups of social primates have been extensively studied in both the field and captivity. One recent review tabulated network metrics for 30 species (5 families and 17 genera) [160]. Only positive interactions within a group (grooming, body contact, social play, etc.) were scored

to define network edges; dominance, aggression, and inter-group interactions were not considered. About half of the groups sampled had group sizes of less than 10, and the other half ranged from 10–35 individuals. In general, edge densities inside primate groups are very high (49–93%). As a result, average path lengths are nearly 1, no two nodes are more than 2–3 links apart, and the network always consists of a single component. Because nearly all nodes have the same high degree, only those additional metrics that use edge weight were computed. Average intensities and weighted clustering coefficients varied between groups but were largely independent of group size; weighted eigenvector centrality increased somewhat with increasing group size, and larger groups had higher levels of modularity. The distributions of intensities were heavy-tailed but not scale-free. Some positive degree associativity was found in most groups, although it was not a strong effect. More significant was associativity based on sex and age: such links are well-known components of primate sociality. The most striking finding is the extremely high social connectivity within these primate groups—far higher than that found in wasp colonies of similar group sizes. This should promote much faster and more extensive percolation in primates than in wasps, with consequent differences in the selection pressures on the signaling that mediates these social interactions.

Just because most primate groups are cohesive does not necessarily mean that their networks are static. Groups that live in highly seasonal habitats might be expected to show subtle changes in network topology and associated measures, based on changing environmental conditions. This is in fact the case: the social networks of female baboons (*Papio hamadryas ursinus*) show cyclic changes in a number of network metrics that are well correlated with seasonal changes (**Figure 15.27**).

Synchrony

To what degree are the theoretical models reviewed earlier useful in understanding how and when behavioral synchrony arises in social animals? These models all focused on synchrony as an emergent property of local interactions between many network nodes. The network was assumed to be static in structure, and all nodes were already performing some common action. The models then sought to determine when the nodes might perform this action in a coordinated way. The most striking prediction was a sharp bifurcation between synchronous and asynchronous behaviors: below a threshold link strength or edge density, behaviors would be asynchronous, and above that critical value, a certain degree of network synchrony would appear spontaneously. The fraction of nodes that would eventually synchronize, and how long it would take to achieve that fraction, vary with link strength and network topology.

Behavioral synchrony in animals is more complicated than these models would suggest. In real animals, synchrony requires two steps: first, a significant number of animals in a group must adopt the same behavior; and second, those

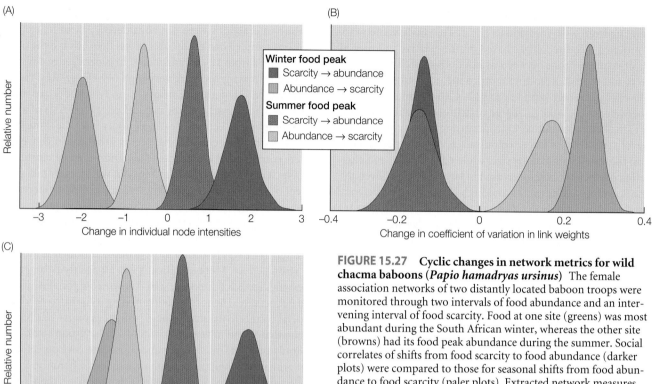

(A)

Relative number

Change in individual node intensities

Winter food peak
- Scarcity → abundance
- Abundance → scarcity

Summer food peak
- Scarcity → abundance
- Abundance → scarcity

(B)

Change in coefficient of variation in link weights

(C)

Relative number

Change in individual clustering coefficients

FIGURE 15.27 Cyclic changes in network metrics for wild chacma baboons (*Papio hamadryas ursinus*) The female association networks of two distantly located baboon troops were monitored through two intervals of food abundance and an intervening interval of food scarcity. Food at one site (greens) was most abundant during the South African winter, whereas the other site (browns) had its food peak abundance during the summer. Social correlates of shifts from food scarcity to food abundance (darker plots) were compared to those for seasonal shifts from food abundance to food scarcity (paler plots). Extracted network measures included the distributions of (A) network intensities, (B) edge weight variation within each node (as measured by a coefficient of variation), and (C) individual clustering coefficients. At both sites, the seasonal decrease in food abundance led to increased linkage intensity, reduced heterogeneity in link weights, and higher clustering coefficients; subsequent food increases produced the opposite effects. (After [146].)

adopting the behavior must coordinate their performance of it with each other [257]. Theoretical models of emergent synchrony ignore the first step. In addition, behavioral synchrony in animals can arise for reasons other than network emergence: the behavior of all group members may be entrained to some external stimulus (e.g., time of day or the appearance of a predator), or a group may have one (despotic) or more (oligarchic) leaders whose signals and cues orchestrate behavioral synchrony [56]. In most animals, behavioral synchrony arises from a mixture of external cues, leader signals, and network interactions, with the relative weightings of the three factors varying both among and within species. Whereas most theoretical models for synchrony assume undirected and identical link strengths between each pair of nodes, social networks often have different strengths for input and output links. Input link strength affects how sensitive a node is to the actions of other nodes in the vicinity, and output node strength affects how far away a focal node's actions will be felt. Selection can tune these two link strengths independently and thus affect the propensity for emergent synchrony, whether the asymptotic limit over time is partial or complete synchrony, and how rapidly this limit is reached. The degree to which the interaction network in a group is modular or more homogeneous also contributes to the likelihood of synchrony. Finally, interaction networks may not be static in real

animals: the intensities of a given node's input links, output links, or both can vary over time, and links can be added or removed from the network. Despite all these complications, the key predictions of theoretical models are largely preserved in animal systems: synchrony appears as a behavioral bifurcation when either link strength or edge density increases, and both the maximal fraction of synchronized nodes and the time to reach that level are functions of linkage strength and network topology [62].

Consider a simple example. Desert locusts (*Schistocerca gregaria*) move frequently to find food. At low densities, individual locusts are cryptically colored and actively avoid each other. Above a threshold density, nymphs experiencing frequent physical contact molt and acquire a more conspicuous body color, an attractive odor, and a propensity to coordinate their movements with those of others (**Figure 15.28**) [303, 304]. This sharp bifurcation from asynchronous to highly synchronous behavior can be seen as a shift from a sparse network of many components to a highly connected giant single component with much stronger links between nodes [9, 48]. These are precisely the theoretical conditions favoring the emergence of synchrony.

(A)

(B)

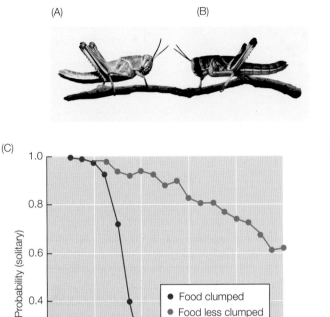

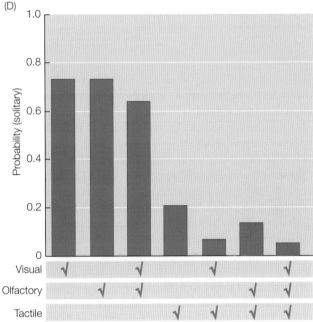

FIGURE 15.28 **Behavioral bifurcation in desert locusts (*Schistocerca gregaria*)** (A) Cryptic morph, living solitarily. (B) Gregarious morph, living in a high-density population. (C) The probability that a cryptic morph will continue to show solitary behaviors after 8 hours of exposure to different densities of conspecifics. Locusts show a sudden change (bifurcation) in behavior at a threshold density if food is provided in a clumped distribution; the change still occurs if food is less clumped, but more gradually. (D) Comparison of the effectiveness of visual, olfactory, and tactile cues from conspecifics as triggers for the shift from solitary to gregarious behaviors. Each initially cryptic animal was exposed to single or combined stimuli for 4 hours and then tested for behavioral traits. Tactile cues had by far the greatest single effect, although adding concurrent olfactory and visual stimuli augmented the effect. (After [73, 279, 303].)

Several theoretical treatments have extended the classical models to consider mixtures of leadership and emergent synchrony. Where a group is faced with isolated decisions, the role of expert leaders can be favored over the collective pooling of information across all group members; when the same decision occurs repeatedly, so that the success of previous decisions can be evaluated, emergent synchrony is favored over leadership [161]. Evolutionary game models for groups in which there are conflicts of interest predict that group choices among discrete options (e.g., to which of several foraging sites should a group move next) are best made by leaders, whereas those made among continuous choices (e.g., what time to leave) are best made by emergent consensus [78, 79]. Leadership is not necessarily based on relative access to information, but can itself be an emergent property

arising from differences in nutritional needs or previous status interactions [76, 257, 266, 269, 321, 324]. The different origins of group leadership are discussed in more detail in Chapter 13. In this chapter, we are most interested in behavioral synchrony that is an emergent property of pooled network interactions. Below we review three common contexts in which emergent synchrony appears to play the dominant role. However, even here, some effects of external cues or leadership may contribute to the relevant thresholds.

COLLECTIVE DECISION MAKING   As discussed in Chapter 13, most social groups of animals make periodic group decisions [77]. These often begin with one or more informed individuals advertising their bias toward a particular choice. Advertisement can be as simple as a short intention movement

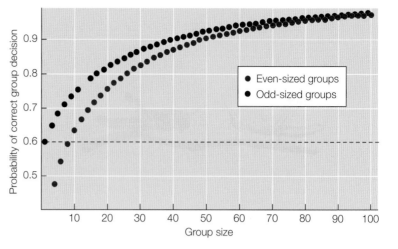

FIGURE 15.29 **The wisdom of the crowd** Suppose an isolated animal makes the correct choice between two alternatives 60% of the time (i.e., somewhat better than the 50% expected by chance). What is the probability that a group of such individuals will make a correct decision by pooling their individual decisions and honoring majority rule? The graph shows that the probability of such a group making correct decisions increases with group size, an effect called the wisdom of the crowd. Because even-sized groups can have ties, they do slightly worse than odd-sized groups. Very small groups may do worse than the individual value of 60%. (After [179].)

or as complex as a honeybee dance that transmits the polar coordinates of the advertised location in each dance and the site's quality by the number of dances performed (see Chapter 14). Synchrony increases locally as observers indicate their support of particular advertisers. A tendency to support the currently most popular options can act as a positive feedback, accelerating the reaching of consensus [178, 323]. Advertisers who fail to attract supporters may give up advertising, or may themselves switch to currently popular options. Once a sufficient number of individuals indicates support for the same option, the decision is made and the group takes the relevant action. As long as the average individual decision is more likely to be correct than not, communal decisions based on the polling of many individual "votes" can significantly improve the chances that a communal choice will be the correct one. This potential benefit is called the wisdom of the crowd (**Figure 15.29**). However, this benefit comes at a cost if achieving consensus takes longer than would an individual decision. There is thus a trade-off between exploiting the crowd's pooled wisdom and minimizing the costs of delayed decisions [118]. The optimal trade-off will differ for different taxa and can be achieved by appropriate adjustments in the topology of the signaling network and the threshold level of consensus at which polling ends.

The choice of a new nest site by a bivouacked swarm of honeybees (*Apis mellifera*) provides a useful example [290]. This process is outlined in some detail in Figure 13.31; here we review it again with an emphasis on those components that are relevant to the network properties. Scouts leave the bivouac and look for possible nest sites within a several-kilometer range. If they find one, they return and dance the location to group members. The number of dances initially performed reflects the scout's evaluation of the cavity's suitability based on its volume, height above the ground, entrance-hole size, and other features. The dancer then makes additional trips to and from the site, but regardless of the site quality, the number of dances that it performs on each return decreases with successive trips [291]. Other swarm members that observe these dances then fly out to inspect the advertised sites; the number of recruits at each site is proportional to the number of dances recently performed for that site. If recruits also find the site to be of high quality, they, too, return and dance for that site. When scouts

that have yet to search encounter no strongly advertised sites before leaving, they search for additional ones. Because a dancer for a high-quality site incites multiple recruits to inspect its site, the number of scouts visiting the most popular sites can increase multiplicatively. At the same time, the number of sites being currently advertised gradually decreases. The decision is made when scouts simultaneously visiting a prospective site perceive that their numbers have reached or exceeded a threshold value (**quorum**); when these individuals return to the bivouac, they now produce a set of special signals that activate the swarm and trigger a coordinated departure to the chosen site [270].

There are several properties of this interaction network that help achieve an optimal trade-off between picking the best site and minimizing decision time [216, 250]. A returning scout can dance to only a tiny fraction of the bivouacked swarm. It thus acts as a node in a very sparse network. However, the fact that it can incite multiple recruits to visit a site (i.e., act as a network hub), amplifies the rate at which a threshold buildup can occur and thus shortens the decision time. The inevitable decay of a scout's dancing rate regardless of site quality (i.e., a decrease in her output link intensity) acts as a negative feedback eliminating previous sites from consideration if they fail to increase in popularity. It also ensures that any one site will be inspected by many different individuals and thus increases the cumulative accuracy of pooled assessments. Since this is a one-shot decision for which retrospection will not help future decisions, oligarchic leadership is favored over a full-group consensus. However, the assignment of leadership to the quorum oligarchy is itself an emergent property of the networked interactions that preceded the decision: few if any of the bees in the oligarchy will have visited multiple sites and thus be in a position to compare them. Any consensus is an emergent function of the dynamics of pooled site advertisements. Finally, the value of the quorum number at which a decision is reached can be fine-tuned to optimize the trade-off between accuracy and timing: a low value minimizes decision times, but at the cost of reduced accuracy; a high value has the opposite effect. Honeybees use a value that appears to optimize the relevant costs and benefits [250].

Rock ants (*Temnothorax albipennis*) move their nests at intervals and appear to use a process similar to that of the

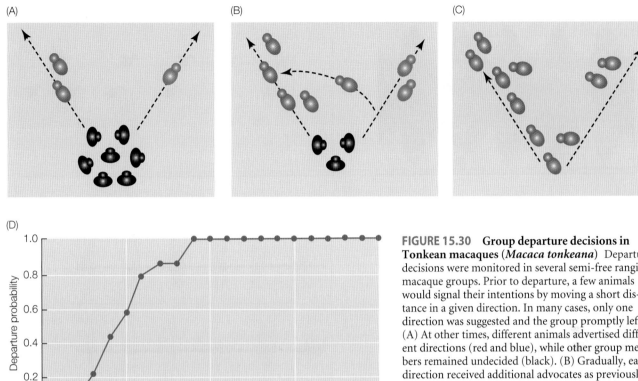

(A)

(B)

(C)

(D)

FIGURE 15.30 Group departure decisions in Tonkean macaques (*Macaca tonkeana*) Departure decisions were monitored in several semi-free ranging macaque groups. Prior to departure, a few animals would signal their intentions by moving a short distance in a given direction. In many cases, only one direction was suggested and the group promptly left. (A) At other times, different animals advertised different directions (red and blue), while other group members remained undecided (black). (B) Gradually, each direction received additional advocates as previously undecided animals joined one group of advertisers, and some individuals that had advertised one direction switched to advertising the other. (C) At a critical point, an initiator led the entire group in the direction currently advertised by the majority of individuals. (D) Pooled decisions by one group of 22 macaques show that the critical point for a final decision is reached when a threshold fraction of the group is participating in the process by advertising one direction or the other. A similar threshold was found for a second group. (After [320].)

honeybees [118]. Accuracy is maximized by looking for potential sites ahead of time, and then having scouts fan out in all directions when migration is imminent [103, 313]. Instead of dances, returning scouts use tandem running to lead recruits from the old nest to an advertised site. As with the honeybees, when the number of ants inspecting a site exceeds a quorum threshold, these scouts suddenly shift tactics: they return to the nest, and instead of tandem running, begin transporting other adult ants and pupae to the new site [263]. To expedite the transport phase, scouts use tandem running to show recently transported adults the route back to the old nest site [119]. This constitutes another positive feedback loop in the process.

Although quorum formation occurs away from the main group in social insects, it occurs within the main group for many other taxa. This is particularly evident in the decisions exhibited by most social vertebrates about where and when to move next [34, 75, 77, 166, 256, 257, 318–320, 347]. Typically, various group members perform intention movements or equivalent signals prior to nonurgent departures (**Figure 15.30**). Individuals observe the "votes" of others and may then shift their own vote to a more popular option; these shifts may or may not be biased by the animal's stronger interaction links [319]. Dominants or regular group leaders are not necessarily the initiators nor the moderators of this process; and even when they are, they appear to take the votes of other group members into account [257]. Synchrony is, again, an emergent

property of the many interactions between group members and can arise even in the absence of significant leadership.

COORDINATED ADVERTISEMENT Emergent synchrony has been observed in a variety of species in which animals display competitively. Higher levels of escalation in song contests are marked by the matching of song frequency in species with single song types, such as chickadees (*Poecile atricapillus*), or by matching of song types in species with large song repertoires, such as banded wrens (*Thryothorus pleurostictus*) [112, 341, 345]. Contests between a particular dyad can trigger propagated responses that radiate out through many successive rings of territories in the network. This is particularly common during the dawn chorus. Because links between neighbors can be both directional and weighted, effect propagation need not be uniform, but may follow multiple and divergent tracks through the territorial array [49, 109, 112, 113].

A different level of competition exists among displaying males in swarms, choruses, and leks. Here, males emit one or more signals repeatedly over long periods, and females select mates based on their relative display properties. A male bird or mammal that is displaying on a lek typically increases his

(A)

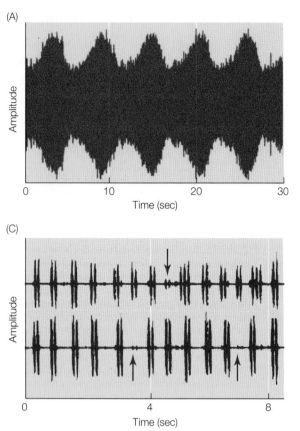

(B)

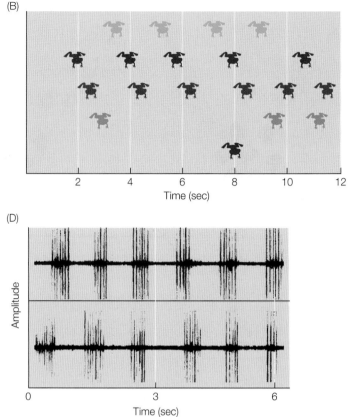

(C)

(D)

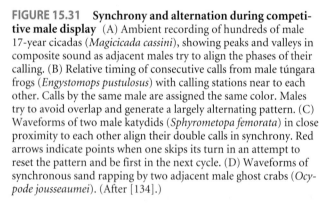

FIGURE 15.31 **Synchrony and alternation during competitive male display** (A) Ambient recording of hundreds of male 17-year cicadas (*Magicicada cassini*), showing peaks and valleys in composite sound as adjacent males try to align the phases of their calling. (B) Relative timing of consecutive calls from male túngara frogs (*Engystomops pustulosus*) with calling stations near to each other. Calls by the same male are assigned the same color. Males try to avoid overlap and generate a largely alternating pattern. (C) Waveforms of two male katydids (*Sphyrometopa femorata*) in close proximity to each other align their double calls in synchrony. Red arrows indicate points when one skips its turn in an attempt to reset the pattern and be first in the next cycle. (D) Waveforms of synchronous sand rapping by two adjacent male ghost crabs (*Ocypode jousseaumei*). (After [134].)

display rate when a female enters his display territory [150]. This causes his male neighbors to increase their display rates, and the perturbation can radiate out and affect males many territories away. Birds and mammals usually do not take the second step of synchronizing the timing of their displays (the long-tailed manakins discussed earlier are a rare exception). In contrast, certain species of fiddler and sand bubbler crabs, chorusing arthropods and anurans, and fireflies show both steps of emergent synchrony: they begin signaling or increase their signaling rate when their neighbors do, and they then coordinate the timing of their signal emissions with their neighbors [2, 17, 18, 124, 133–135]. Coordination can be in the form of *alternation*, which maximizes the time intervals between the signal emissions of different males, or *synchrony*, which maximizes signal overlap (**Figure 15.31**). Species differ in whether coordination is spatially localized (fiddler

crabs) or global (southeast Asian fireflies), and transient (some frogs) or persistent (cicadas). Some species of katydids maintain their own display rate but use short "skips" in the pattern to provide temporary coordination with neighbors; others modify both the delays and their own rate of signal emission to achieve long-term coordination [104, 135]. For a few species, coordinated display may be directly adaptive: some fireflies may exhibit synchrony to ensure that a female can recognize their species-specific flash patterns despite the jumbled flashing in multispecies communities [209]. On the other hand, no obvious adaptive benefit has been identified for many other taxa. Synchrony is then most likely an emergent property of displaying male interactions. How might this arise? One widespread correlate of coordination in invertebrate and anuran display aggregations is a female mating bias favoring a leading male over a following male when male signals show partial overlap [33, 45, 124, 130, 133–135, 274]. This is called a **precedence effect**. It may exist because females are able to localize the leading signal more accurately, or because those males that can sustain the highest display rates are more likely to be leading signalers [133, 229]. Where females do express precedence biases, alternation and synchrony are two ways that males can attempt to equalize their attractiveness to females. Both patterns are thus emergent properties of interactions that involve feedback among the various participants in the displaying network.

COORDINATED LOCOMOTION   Finally, we consider the cohesive locomotion of schools of squid and fish; swarms

(A)

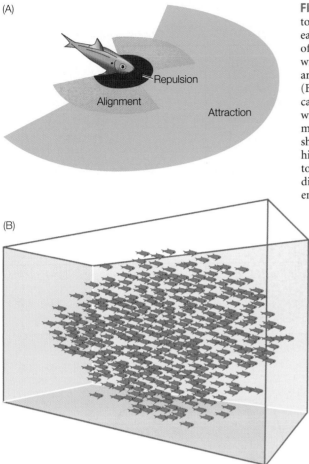

(B)

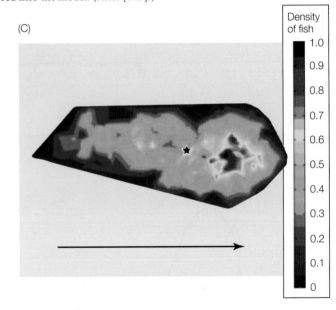

FIGURE 15.32 **Synchronous schooling in fish** (A) A model seeking to explain emergent school synchrony in fish defines three zones around each individual: individuals move apart when one enters the *repulsion zone* of another; each individual aligns its direction of movement and velocity with neighbors in a slightly more distant *alignment zone*; and individuals are attracted to fish in an outer *attraction zone*, keeping the school together. (B) By combining measured sizes of zones of real fish with known physical parameters of swimming in water, the model predicts an ovoid school with a tapering tail end as shown here swimming from left to right. (C) The model also predicts variations in the density of individuals. The arrow here shows the direction of movement of the school. The density of fish is much higher in the forward end of the school, causing the center of gravity (star) to be farther forward than the geometrical center of the school. Both predictions match patterns seen in schools of the fish whose parameters were entered into the model. (After [143].)

of foraging ants and termites; flocks of birds in flight; and herds of large mammals in migration. Some locomotory synchrony is due to group members following leaders. The leaders may simply be those most anxious to reach a goal, such as a feeding site, or for other reasons those least motivated to ensure group cohesion [76]. Lactating females routinely provide the leadership role for moving harems and multiharem herds of plains zebras (*Equus quagga*) [107]. Dominance hierarchies often form within pigeon flocks, and the dominants routinely take up critical locations during flight from which they dictate flock direction and flight velocity [214]. Experienced male bottlenose dolphins (*Tursiops truncatus*) perform displays to initiate group departures and changes in activity; experience plays less of a role in displays by females that counter the male proposals [185]. Leaders with different credentials (e.g., some based on experience or knowledge and some based on dominance) can coexist given oversight of different aspects of a group's synchronized activities [346].

Of greater interest in this chapter are the highly coordinated movements of aquatic animal schools and flying bird flocks. Because they do not require recurrent substrate contact and can move in three dimensions (instead of two), the locomotory movements of these groups can be extremely rapid. Although there are often no nearby obstacles to act as

external synchronizing cues, these groups often exhibit highly coordinated turns, accelerations, and reversals in direction. They also exhibit highly characteristic group shapes while in motion. A wide variety of theoretical and field measurement studies have demonstrated that the observed synchrony and group geometry do not require leaders or external cues, but are instead emergent properties of the networked linkages between individuals [1, 11, 19, 21, 22, 58–62, 80–82, 106, 141, 143, 147, 249].

Remarkably simple network models can explain fish school and bird flock synchrony [11, 80, 143, 249]. Because there are no leaders, the relevant networks lack hubs. Three topological zones can be identified around each nodal individual (**Figure 15.32**). The most immediate zone involves negatively weighted links: when a neighbor intrudes into this zone, both parties move apart until the nodal individual is again the only occupant. The next concentric zone involves positively weighted links: the nodal individual aligns its direction of motion with that of neighbors at this intermediate distance. Finally, the nodal individual is positively attracted to neighbors in a third outermost zone. This attraction reflects each animal's motivation to remain in the group. Outside of the third zone, links to the nodal individual are weak or missing. However, because each individual attends

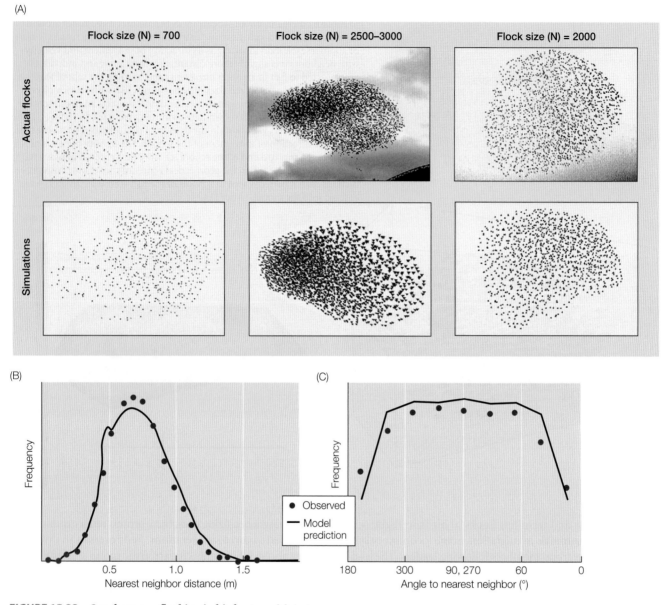

FIGURE 15.33 Synchronous flocking in birds A model similar to that of Figure 15.32 was applied to flying flocks of wild starlings (*Sturnus vulgaris*), correcting for the different physical properties of the medium (air versus water). (A) Visual comparisons between the shapes and internal structure of three real starling flocks (top) in flight and model predictions (below). The flocks in the middle column were making a turn. (B) Comparison between the observed nearest neighbor distances in starling flocks and model predictions. (C) Comparison of observed angles between the forward movement vector of individual birds and the location of their nearest neighbor to model predictions. As predicted, few birds had neighbors directly in front of or behind them; most nearest neighbors were to one side or the other. (From [147].)

to the same three zones in the same way, the entire school or flock forms a network with one giant component.

Given this network architecture, cohesive group movement is an emergent property. The shape and cohesiveness of the group depends on the relative sizes of each of the three zones: depending upon the combination of zone sizes, the group may form a loose swarm with only moderate orientation synchrony; a torus, in which the group moves in a circular path around an empty center; or a highly synchronized group that can move in any direction with the same orientation and angular momentum [80]. Fish and birds have somewhat differently coordinated groups. Fish schools tend to be oblong in shape, with a higher density of individuals in the forward end of the school and tighter spacing as group size increases. Flocks of birds can take any of a variety of shapes. The inter-bird spacing is largely independent of group size, though there tend to be higher densities of birds at the flock margins than in the interior (**Figure 15.33**) [21, 22, 58, 61, 143, 145, 147]. Although not necessary to generate locomotory synchrony, external cues such as nearby predators, obstacles, or feeding sites can affect zone sizes and thus influence the shape, spacing, and direction of the motion of synchronized groups [38, 338–340].

In fast-moving fish schools and bird flocks, the links between individuals usually do not involve signaling, but instead reflect a very high sensitivity to visual and proprioceptive cues emitted by neighbors. Squid, which advertise alarm or other behavioral states via variable color patterns, may use these patterns to coordinate school movements [138, 193, 194]. Signals play a much more important role in promoting slower synchronous movements. As discussed in Chapter 13, contact calls are widely used by members of bird flocks to coordinate movements in dense vegetation. In Chapter 14, we discussed the role of trail pheromones in channeling ant and termite movements along specific routes. These trails are self-organized patterns emerging from the pheromone contributions of many successful foragers; they expire when foragers that have found no food at the terminus return without reinforcing the marks. The complex network interactions between many individual ants or termites can lead to a network of trails exhibiting bifurcations, multiple equilibria, limit cycles, and chaos, depending on the parameters examined [30, 40, 41, 90, 93, 122].

Other self-organized patterns

There are many other aspects of animal social behavior that exhibit self-organized patterns due to networked interactions. Below we briefly review two of these that typically include signaling as part of the network linkages.

DOMINANCE   We saw in Chapter 11 that the linearity and stability of dominance hierarchies are emergent properties of the network of interactions between members within a group [12, 39, 64–66, 152, 170, 282, 344]. It would be impossible to generate a linear hierarchy of the members of a group based only on comparisons of fighting abilities, since these abilities are usually normally distributed. Instead, each individual adjusts its tendencies to escalate or retreat, and these responses to the outcomes of recent contests result in emergent linearity [242, 243]. Both the contests that lead up to a stable hierarchy and the interactions that maintain it are typically mediated by signals. While earlier models of this process focused on pooling the dyadic interactions of all group members, a network approach suggests that the dynamics of dominance are much more complicated [142, 144]. Audiences of dyadic contests can adjust their own tendencies to escalate or retreat based on what they observe [98]. Losers of a recent contest may attack other nearby individuals in front of a local audience to restore their status [163]. Dominants may attack subordinates in front of audiences, as hyenas do, to reinforce status [305]. Even with audiences nearby, the effects of a given contest may only propagate short distances within a social network. One behavior pattern that can link these many interacting neighborhoods into a small world is policing: here, a few dominant individuals roam the group and enforce minimal fighting. The small world connectivity and the hub roles of the policing individuals lead to more integrated social networks; in the absence of the policers, the social networks tend to fragment into very local communities and cliques (**Figure 15.34**). This fragmentation can hinder valuable social interactions such as play, allogrooming, infant care, and shared surveillance [110, 111].

PERSONALITY   In several prior chapters, we noted that many animal populations contain multiple personality types. Personality traits are behavioral propensities that are relatively consistent within individuals but variable between individuals [94, 272, 273, 297]. Most personality traits can be treated as continuous variables, with each animal showing a particular value within a bounded range. Examples include an animal's position along a bold/shy axis, an inquisitive/neophobic axis, an active/inactive axis, a sociable/unsociable axis, and an aggressive/meek axis [272, 273]. While the values of these and additional personality traits can vary independently, those in real populations are often correlated: thus one set of individuals might be consistently bold, exploratory, and aggressive whereas another set is shy, neophobic, and meek. Significantly correlated suites of personality traits are called **behavioral syndromes** [297–299].

Although originally ignored as "noise" in the social system, animal personalities and behavioral syndromes are now acknowledged to be widespread and can have significant effects on fitness (**Figure 15.35**). This raises a number of questions. Why are personality traits so consistent within individuals over time and across multiple contexts? Why might different personality traits be correlated, creating behavioral syndromes? And what evolutionary processes allow multiple personalities and syndromes to coexist in the same population [94, 359]? The answers to all of these questions are related to the fact that personalities and syndromes are, at least in part, properties emerging from the animals' social networks [165, 300]. Other factors may play a role, but network structure is often a key force promoting distinct personalities in animals.

To see why this is true, we first note that optimal strategies for any animal depend upon its current state [151]. This state includes its prior history and its current anatomy, physiology, and social and ecological environments. Different state variables change values over different time scales. For many animals, body size and sex are immutable after a given age. Some state variables such as growth rate and resting metabolic rate are set early in life and are difficult to change later. State variables such as energy reserves, residual reproductive value, accumulated knowledge, and physical condition can be changed given the right behavioral choices, but usually only slowly (see Chapter 9 for details). Other variables, such as membership in a particular herd or swarm, might be altered rapidly and reversibly.

Because interacting animals differ in genotype and prior history, they are likely to differ in state and thus may differ in optimal behaviors. Where an optimal behavior is largely dictated by an immutable state variable like body size, sex, or growth rate, it is not surprising that individuals are consistent in their performance of that behavior and that different individuals repeatedly perform that behavior differently

FIGURE 15.34 Hubs and network stability in monkeys Normal grooming and play networks (top row) for a group of captive pig-tailed macaques (*Macaca nemestrina*). Three adult males served as "policers" by interceding in selected conflicts and maintaining general peace. While these males contributed many direct links to the network, they also had indirect effects on network topology. This was demonstrated by comparing topological removal of these males (removing them from the relevant association matrices and then reconstructing the network) to actual experimental removal. Visual inspection of the networks shows that topological removal of males has no effect on play networks, but experimental removal significantly changed the play network topology. More subtle effects of the males' presence were shown using network metrics. Without the males, the remaining group members significantly reduced their number of grooming and play connections (lower degree), limited their range of interactions (lower reach), shifted to interactions with animals of similar degree (more positive degree associativity), and favored more links to topologically close neighbors (increased clustering coefficient). This study shows that networks are not simply a linear sums of their parts, but instead can show emergent properties based on pooled interactions. (After [110].)

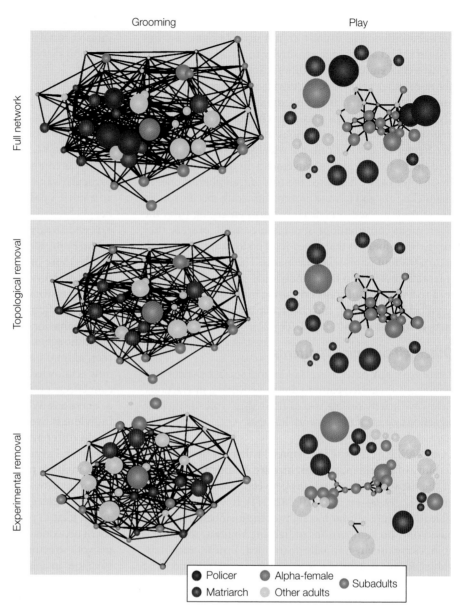

Grooming | Play

Full network

Topological removal

Experimental removal

● Policer   ● Alpha-female   ● Subadults
● Matriarch   ○ Other adults

or even adopt a totally different optimal behavior [54, 311]. Where an immutable or slowly changing state variable affects multiple optimal strategies and thus multiple personality traits, behavioral syndromes are a direct consequence [35, 36]. Immutable states can thus be one cause of behavioral consistency along various personality axes. However, where states can be changed by behavioral actions, consistent performance within individuals and consistent differences in performance between individuals are likely to persist only if there are positive feedbacks between the performance of a behavior and the state that favors that particular performance (**Figure 15.36**) [94, 186, 299, 359, 362]. Such feedbacks amplify initially small individual differences in state into large and persistent personality differences. For example, an antelope whose initial state makes it a faster-than-average runner may benefit by foraging in a group because a predator is more likely to select the slower group members as prey.

While in a group, the fast antelope can afford to spend more time feeding and less time in surveillance than other group members. This increases differences in state between group members and makes faster members even more likely to be cavalier about surveillance and slower members to be even more wary. Positive feedbacks thus amplify initial state differences into emergent personality differences.

Note in the antelope example that it is relative and not absolute state in the group that generates the positive feedback and emergence of personality differences. Although fast, a solitary antelope may gain no advantage from its initial state, as a predator has no alternative prey to consider. This reinforces a common theme in this book: the payoffs of a given strategy are often contingent upon what strategies are being performed by others in a focal animal's vicinity. The social network in which an animal is embedded can thus modulate the levels of amplification and personality

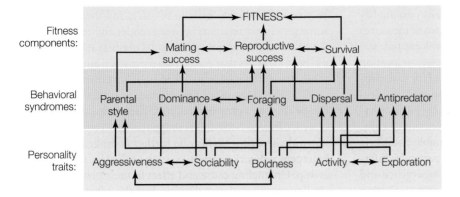

FIGURE 15.35 Links between personality traits and fitness Personality traits affect an animal's fitness through multiple indirect paths, many of which interact with each other. Here is one proposed set of the more obvious linkages. (After [272].)

emergence. When social networks are incomplete, the composition of topological neighborhoods and thus the relative state of a focal animal can vary. A given personality or behavioral syndrome may emerge in one neighborhood, but not in another. Once personality differences have emerged, these can affect the emergence of additional personality differences. Heterogeneity in incomplete social networks may be a reason for the persistence of multiple personalities and syndromes in a given population.

While an animal that finds its personality poorly adapted to a given neighborhood may be unable to change that personality, given its physiological and anatomical state, it may be able to relocate or adjust its network linkages to reduce the discordance [285, 286]. Personality traits are correlated with network measures in a variety of taxa [139, 187, 192, 207, 243]. For example, bold members of schools of three-spined sticklebacks (*Gasterosteus aculeatus*) were found to have fewer links (lower degree) than shy fish, but to spread these out more evenly across the entire school (low clustering coefficients); shy fish were highly selective in their interactions, with a bias toward links with spatial neighbors [258]. Similar results were obtained for guppies (*Poecilia reticulata*) with the additional finding that linkages were associative for personality type [86]. A combination of immutable state

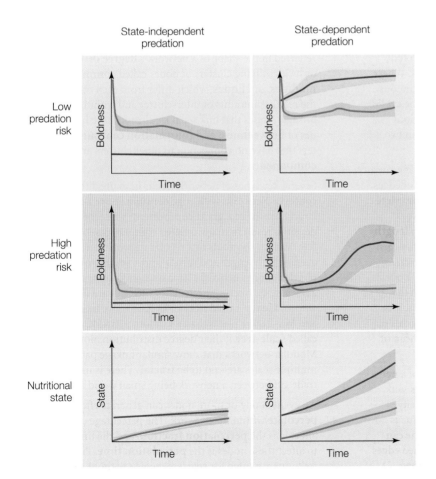

FIGURE 15.36 Simulation models of personality evolution In these computer simulations, 1000 model foragers are allowed to adopt any level of boldness over a fixed range. Bolder foragers take more predation risks but obtain food faster than less bold foragers. Blue foragers begin each simulation with a lower nutritional state than red foragers. Each combination of conditions was run 1000 times and the results averaged (solid lines in graphs with variations as colored zones). In all runs, blue foragers are initially very bold but gradually reduce boldness as their state improves. If predation depends only on boldness and average predator density and is independent of forager state (left column), the optimal boldness for blue foragers remains higher than that for red foragers. Blue foragers thus improve state faster than red foragers, if they are not killed by predators, and eventually all foragers converge on the same state and the same boldness. Personality differences are not favored in this context. If predation *is* dependent on state (right column), red foragers can afford to be bolder, improve their state faster, and this positive feedback amplifies the initial differences in state and consequently, the optimal boldnesses. A stable mix of personalities is then the expected outcome. (After [186].)

differences, dominance status, and emergent personality traits was required to explain observed network measures in cooperatively breeding cichlids (*Neolamprologus pulcher*) [288]. Where social networks are nearly complete, relocation and redistribution of linkages provide no relief for maladaptive personalities. These will eventually be removed by selection. However, even when a high level of mixing in complete networks forces evolutionary games to be played out globally, some conditions are predicted to support stable equilibria with multiple personalities [89, 205, 355]. Of particular interest is a game theoretic model predicting the emergence and stable persistence of multiple personalities in communication contexts [42]. The model assumes that senders have imperfect information about the state that they are advertising to receivers; signals are costly to senders; and there are at least some genetic correlations between sender and receiver codes. Given these assumptions, the model predicts stable mixtures of "bold" senders that exaggerate their state, and "shy" senders that are more conservative in their advertisement.

The gist of the preceding discussion is that personalities can be emergent properties of social networks, and once generated, can feed back and reshape the structure of the social network. The process may be further complicated by other emergent network properties, such as dominance and leadership. Untangling cause and effect in these processes can be challenging in real populations, but the advent of suitable network and personality measurement tools now makes that goal feasible [165].

SUMMARY

1. **Network analysis** examines the pattern of pooled **relationships** between all animals in a group or population. Relationships can be estimated by tallying **interactions** between each pair; inferring relationships through joint attendance in specified contexts is called **the gambit of the group**. All combinations of relationships can be tabulated in an **association matrix** or visualized in a two-dimensional **network graph** in which each individual is assigned one **node**, and **edges** link all pairs with recorded relationships. **Weighted** edges can be used to indicate relationship strengths, **signed** edges to indicate beneficial versus detrimental relationships, and pairs of **directed** edges to accommodate asymmetrical relationships. Where directed edges of three or more nodes form a continuous path in the same direction, the network is said to be **cyclic**. Complex relationships such as alliances and coalitions may require **hyperedges** on graphs with more than two dimensions.

2. **Connectivity measures** reflect the amount of linkage in a network. **Edge density** is the ratio of the number of edges in a network to that expected were the network **complete** (every node connected to every other node). Networks with low edge densities are said to be **sparse**. **Path length** is the average of the shortest topological distance between a focal node and all other nodes in the network; **average path length** is the mean path length across all nodes; and network **diameter** is the largest path length in the network. Topological distance is measured as the number of traversed edges in unweighted networks; reciprocals of weights are used for weighted networks, and the mean of reciprocal sums is called network **closeness**.

3. **Centrality measures** indicate the heterogeneity of a network's linkage. A network with no heterogeneity is said to be **regular**. Irregular networks show heterogeneity at different scales, and different centrality measures focus on these different levels. At the most local level, one can compare nodes according to the number of attached edges (**degree**) in unweighted networks or the sum of their attached edge weights (**intensity**) in weighted networks. **Reach** is the number of nodes within a fixed number of edges of a focal node. Central nodes have greater reach than noncentral nodes. Nodes of high degree or intensity (**hubs**) whose linked neighbors are also hubs will exhibit high **eigenvector centrality**. **Clustering coefficients** measure the level of local connectivity in adjacent trios of nodes. Network heterogeneity can also be generated if individuals selectively form links according to type: such **assortativity** may be based on sex, age, alliances, or even node degree. The highest levels of heterogeneity will be reflected in the shape of a network's **degree distribution**, or by identifying clusters of nodes called **communities**, **modules**, or **cliques**, which differ from each other in the relative amounts of inter-cluster and within-cluster linkage. Nodes that link separate clusters have high values of **betweenness** and **information centrality**. Clusters that have no connection to other clusters are called **components**.

4. Several key types of networks recur in many different contexts. Networks with edge densities above a threshold largely consist of a **giant component**, while lower edge densities result in many isolated components. Adding a few random shortcuts to a regular network can reduce average path length and turn it into a **small world**. Completely random networks have no significant transitivity, associativity, or community structure, but are usually small worlds. **Heavy-tailed networks** have more high-degree and low-degree nodes than a random network, and are called **scale-free** if their degree distribution plot is linear. Modular networks that show similar linkage patterns at multiple scales are said to be **fractal**. There is usually a trade-off between a network being small world and fractal.

5. An event at one or more nodes can generate effects that **percolate** within the network. The percentage of nodes affected is the **percolation fraction**, and the time it takes to affect these nodes is the **percolation time**. Percolation fractions are small for sparse networks without a giant

component and high for those with a giant component. Percolation times are short for small worlds but longer for fractal networks with limited between-module connectivity. **Robust networks** retain initial percolation properties even after some original edges or nodes are removed. Networks that are heavy-tailed are usually robust to random edge or node removal, but not to targeted hub removals unless they are also fractal and modular in structure. In general, network structures that favor high percolation fractions and low percolation times have low robustness, and vice versa. Most biological systems appear to be selected for an optimal weighting in this trade-off.

6. Effect propagation in many networks is nonlinear, and this results in **emergent properties** and **self-organized patterns** that are not the linear sum of each node's actions. An example is the emergent **synchrony** of repetitive actions by a network's nodes. Sufficient edge strength is a key condition for the emergence of synchrony as well as its spread throughout a network. Even for a given edge strength, some network structures promote synchrony more than others. Regular networks usually fail to synchronize unless sufficient shortcuts are present to make them into small worlds. Heavy-tailed networks start to synchronize at lower edge strengths than random networks, but have trouble both achieving full synchrony and maintaining it for long periods at higher edge strengths. Full synchrony can be achieved in such networks if edges are weighted and input intensities are equalized (**isothermal**) for all nodes.

7. **Evolutionary graph models** of social evolution typically assume weak selection, haploid genetics, stable population sizes, and similar **interaction networks** and **replacement networks**. Most assume a **structured population** in which both social and demographic events are localized. They also invoke a **Moran process** that breaks evolutionary trajectories into a series of events in which either a birth results in a compensating death (*BD*) or a death results in a compensating birth (*DB*). The evolution of cooperative behaviors such as altruism and communication in a structured population requires that the fulfillment of the compensation step be based not on chance (drift) but on the prior fitness differences between players (selection). Models based on the assumptions that death is random but births reflect selective differences routinely predict the evolution of cooperation for *DB* events but not for *BD* events. However, real systems may vary as to which parts of a Moran event are based on selection and which on random assignments. Interestingly, inclusive fitness models for the evolution of cooperation in structured populations provide similar predictions despite using different accountings.

8. **Eavesdroppers** often provide a link between a communicating dyad and the larger social network outside the active space of the signals. Propagated effects need not be signals and may feed back on the original communicators. **Interceptive eavesdroppers** respond to signals immediately, while **social eavesdroppers** exploit the signal information for future decisions and actions. Senders often anticipate eavesdropping and avoid emitting signals unless primary receivers are sure to be present (**first-order audience effects**) or modify their signaling strategies when potential eavesdroppers are present (**second-order audience effects**).

9. The social networks of real animals tend to be small worlds with short path lengths, high clustering coefficients, and heavy-tailed but not scale-free degree distributions. Most fission–fusion societies have sparse networks with low edge densities and average degrees, whereas networks within cohesive groups such as monkey troops or wasp colonies have higher edge densities and average degrees. Networks in both types of society contain multiple communities if sufficiently large: the degree to which these are modular or overlapping depends upon the species. The presence and direction of associativity varies widely among species and across network topologies: when present, the most common criteria for associative linking are sex, age, and body size. Degree assortativity only occurs sporadically and may show different patterns for different subsets of the network.

10. Behavioral synchrony in animals requires two steps: individuals must adopt the same action, and they must then perform it in a coordinated way. Either step can be achieved through entrainment to external cues, attention to leader signals, as an emergent property of interactions within the network, or as some mix of these factors. A group's **collective decisions** usually follow the theoretically predicted sequence of increasing emergent synchrony: advertisement of different options by different informed individuals; accretion of supporters around particular advertisers generating local pockets of synchrony; and iterative redistributions of support until some threshold synchrony is reached. Switching support may be biased by existing network affiliations and often includes mechanisms with positive feedbacks such as shifting to the currently most popular option. Network topology may be selected to produce an optimal trade-off between making correct collective choices and minimizing time spent on decisions.

11. **Coordinated advertisement** is an emergent pattern of behavioral synchrony seen in aggregations of animals, usually males, that display competitively. Males on avian and mammalian leks coordinate their rates of display when females are present, but they usually do not coordinate display timing. Some fiddler crabs, chorusing insects and anurans, and fireflies, however, coordinate both rate of display and the timing of signal emissions. This is apparently due to a **precedence effect**, in which females favor the first signaling male that they detect. Two emergent patterns that minimize the effects of precedence biases are alternation and synchrony in male signal emission.

12. A widespread case of emergent behavioral synchrony is the **coordinated locomotion** of schools of squid, fish, and cetaceans; flocks of flying birds; trails of ants and termites; and migrations of large mammals. The cohesion of rapidly moving bird flocks or fish schools does not require leaders, but instead relies on the existence of different types of links

between a focal animal and three successively more distant shells of neighbors. The shape of the moving group and the spacing of its members depend on the relative sizes of the three zones around each individual; the zones and thus the group shape and spacing can change suddenly when the group encounters a predator or an obstacle. Rapidly moving animals rely only on cues from neighbors to pursue zoning rules, but slowly moving groups may use coordinating signals as well.

13. **Dominance hierarchies** and the existence of **multiple personalities** are additional self-organized patterns that emerge from the cumulative interactions within social networks. The linearity of dominance hierarchies arises from positive feedback loops in which individuals adjust the weights and signs of the links in their interaction network according to recent contests. In addition to their own contests, these individuals can also make adjustments after eavesdropping on the contests of others in the network. Policing by dominants can add further feedbacks that reinforce dominance differences. Multiple personalities can emerge in a similar fashion: small differences among group members or topological neighbors in physical and physiological states may favor differences in actions; positive feedbacks from the consequences of the actions then exacerbate the state differences. One result is persistent differences in personality traits such as boldness, neophobia, activity level, sociability, and aggressivity. Where an initial state difference affects multiple personality traits, one can also observe stable **behavioral syndromes** in animals. If animals can relocate or be selective in associations, personality differences can reshape an initial network structure into a new one. Only certain combinations of personalities may be stable equilibria (ESSs) in this case.

Further Reading

Whitehead [354] reviews the collection, preparation, and filtering of the data that are tabulated into association and interaction matrices, and then visualized as network graphs. An excellent introduction to network analysis measures with sample applications in behavioral ecology can be found in Croft et al. [85]. These authors also outline some of the methodological problems associated with network studies. More detailed and technical summaries of network measures and types of networks can be found in Strogatz [315], Newman [217], Barabási and Bonabeau [25], and Boccaletti et al. [37].

Percolation and robustness are compared in many types of networks by Dorogovtsev et al. [96], while Rozenfeld and Makse [280] and Rozenfeld et al. [281] examine the trade-off between percolation and robustness in fractal versus small world networks. Strogatz [316] provides a highly readable overview of emergent synchrony in networks, and Camazine et al. [56] discuss what does and does not qualify as self-organization in biology.

Nowak [233] presents a general introduction to the application of adaptive dynamics models in structured populations, and both this book and Nowak et al. [234] review examples of stochastic (Moran) and deterministic evolutionary graph theory. Grafen and Archetti [132] and Taylor [332] expand on the interpretations of birth/death versus death/birth evolutionary processes, and examine the similarities between predictions of evolutionary graph theory and inclusive fitness accountings.

Eavesdropping as a network process is treated in detail in McGregor [204]. Sumpter provides two broad reviews of collective animal behaviors [322, 325]. Useful summaries of collective decision making in animals can be found in Conradt and Roper [75], Petit and Bon [257], and Couzin [82]. Greenfield [134] compares display synchrony in aggregations of male arthropods and anurans, and Couzin et al. [80] define the three-zone model of local interaction that is now widely used to explain locomotory coordination in fish and birds. Flack and Krakauer [111] show how the emergence of dominance interacts with network structure, and Réale et al. [273] and Sih et al. [300] discuss the evolution of multiple personalities in the context of social networks.

COMPANION WEBSITE
sites.sinauer.com/animalcommunication2e

Go to the companion website for Chapter Outlines, Chapter Summaries, and References for all works cited in the textbook. In addition, the following resources are available for this chapter:

Web Topic 15.1 *Software for network measures*
A number of free and commercial computer packages are available for calculating the network measures defined in this chapter. Here we present a sampling of the options.

Web Topic 15.2 *Additional information on evolutionary graph theory*
Here we provide links to several sites that give more detailed explanations and examples of Moran processes and the modeling of social evolution on networks.

Chapter *16*

The Broader View: Microbes, Plants, and Humans

Overview

In the previous chapters we have developed an array of principles of animal communication, ranging from the physics, chemistry, and physiology of signal production, transmission, and reception in different modalities to the conditions under which it pays senders to send signals and for receivers to incorporate them into their decision-making, and finally to the design of signals that provide mostly reliable information in a variety of social and environmental contexts. In this chapter we ask whether these principles can be applied to an odd assortment of other organisms, namely microbes, plants, and humans. For each of these taxa, we describe the degree to which individuals can be claimed to communicate with and obtain information from others, and identify the similarities and differences between their communication systems and those of the invertebrate and vertebrate animals we have been discussing.

Microbial Communication

We begin our discussion of microbial communication with a look at the tree of life in order to define microbes—single-celled organisms—and how they are related to each other. **Figure 16.1** shows the currently accepted tree, which is separated at the highest level into three domains—Bacteria, Archaea, and Eukaryota—and at the next level into the major kingdoms. There is general agreement that these domains all evolved from a common ancestor, with archaea and eukaryotes possessing some important biological similarities despite the morphological similarity of archaea and bacteria [609]. **Bacteria** and **archaea**, together called **prokaryotes**, are characteristically small single-celled organisms with one circular chromosome attached to the plasma membrane and no nuclear membrane, organelles, or cytoskeleton. Most possess a cell wall to maintain a rigid shape and flagella for movement. Both reproduce primarily by **binary cell fission** following chromosomal duplication (**mitosis**) to form genetic clones. They differ in the chemical structure of the cell wall—archaea lack the compound **peptidoglycan** and instead have a surface layer of proteins, which provides strong chemical and physical protection against extreme environmental conditions. The process of chromosomal replication varies with regard to the structure of the DNA polymerases used, and the process of transcription and translation of the DNA code into proteins via RNA polymerases and tRNA also varies, with archaea being more similar to eukaryotes than to bacteria in these respects. Archaea inhabit a wide range of habitats, including some with extreme

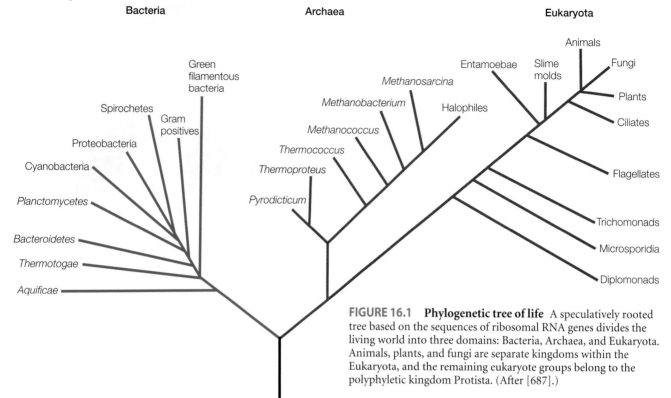

FIGURE 16.1 **Phylogenetic tree of life** A speculatively rooted tree based on the sequences of ribosomal RNA genes divides the living world into three domains: Bacteria, Archaea, and Eukaryota. Animals, plants, and fungi are separate kingdoms within the Eukaryota, and the remaining eukaryote groups belong to the polyphyletic kingdom Protista. (After [687].)

temperatures, salinity, acidity, and alkalinity. They are common in the oceans, marshes, and soils, as are bacteria. There are no known pathogenic archaea, but numerous bacteria are responsible for diseases in multicellular organisms. Archaea use diverse sources of energy, including light, inorganic compounds such as sulfur and ammonia, and various organic compounds. No archaea employ photosynthesis, a process found in some bacteria and eukaryotes.

Eukaryotes differ from prokaryotes in that their cells are larger, with multiple linear chromosomes enclosed in a nuclear envelope; organelles such as mitochondria, plastids, Golgi apparatus, and endoplasmic reticulum; and an organizing cytoplasmic skeleton of microtubules and microfilaments. In addition to fissioning by mitosis, most eukaryotes reproduce by **meiosis** with recombination to create **haploid gametes**, which later fuse to form genetically novel **zygotes**. Eukaryotes are believed to have arisen from a series of symbiotic mergers among various archaea and bacteria cells [95, 120, 125, 384, 639]. Eukaryotic taxa with a unicellular level of organization, without differentiation into tissues, are often lumped into one polyphyletic kingdom called the **Protista** [6]. The multicellular kingdoms include plants, fungi, and animals. In this section we shall focus on the single-celled taxa: bacteria, archaea, protists, and some fungi.

Mate choice

The process of natural selection favors individuals that maximize the genetic diversity of their offspring so that at least some of them are well adapted to current and changing environmental conditions. In Chapter 12, we learned that animals achieve this goal not only by reproducing sexually, but also by choosing mates carefully. Preferred mates may be well

adapted as indicated by their survival and vigorous displays and provide additive genetic benefits, or they may be genetically complementary and provide nonadditive genetic benefits in the form of optimally outbred offspring. This **mate choice** principle applies equally well to microorganisms, and they possess various mechanisms to achieve this end. The mere fact that these ancient organisms still exist despite enormous changes in the earth's environment during the past two billion years attests to their ability to adapt successfully.

Asexual reproduction by binary fission in archaea and bacteria generates two identical daughter cells. Genetic variability is generated in other ways. First, mutations occur at a low rate during every chromosomal duplication event, approximately 10^{-8} per base pair per generation [142]. In a culture of *Escherichia coli* bacteria that has divided 30 times, an estimated 1.5 percent of the cells carry mutations, many of which are deleterious. When this mutation rate is multiplied by the very rapid generation time of prokaryotes (10 minutes to 2 days), the opportunity for new adaptations to arise is high. Prokaryotes can also acquire new genetic material by directly taking up free DNA fragments from the environment and by infection with a virus. However, the most important source of potentially beneficial genetic variation comes from the process of **conjugation**, in which two prokaryotic cells make contact through a bridgelike connection and exchange genetic material. This exchange is unidirectional, from a cell that possesses a plasmid or conjugative transposon element to another cell that lacks this element (**Figure 16.2**). Most conjugative plasmids and elements carry genes for a mechanism that allows their host cell to detect the presence or absence of identical elements in the recipient cell—a simple mate choice mechanism [13, 381, 566].

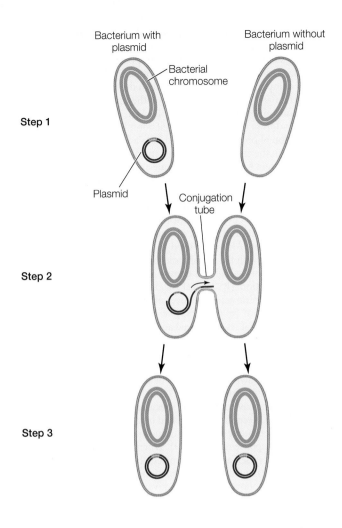

FIGURE 16.2 **Prokaryotic conjugation** Conjugation is a mechanism for horizontal gene transfer, which can occur between two cells of the same species or between cells of different species, even of different kingdoms and domains. In step (1), the donor cell with a plasmid detects another cell that does not have this plasmid. The plasmid (red) has a region of DNA for the origination of replication (green). In step (2), the donor cell produces a conjugation tube that connects it to the recipient. The mobile plasmid is nicked, and a single strand of its DNA peels away and is transferred to the recipient cell. In step (3), both cells recircularize their plasmids and synthesize second strands; each is now a viable donor. Some small plasmids are able to integrate into the bacterial chromosome. Beneficial plasmids can provide antibiotic resistance, tolerance to extreme abiotic conditions, or the ability to use a new metabolite. Other plasmids may be more like genetic parasites—mobile elements that spread themselves into new hosts.

called **isogamy**, but in most cases there are in fact two **cryptic mating types**. Only opposite types can fuse to form a zygote, with recognition being mediated by mating-type-specific pheromones and pheromone receptors. This two-mating-type system is theoretically very stable and is illustrated by various yeasts (unicellular fungi) and green algae [293, 310]. The mating types may be genetically determined by a single locus with alternate alleles that specify the mating-type pheromone, the mating-type receptor, or both. In some cases, two loci specify the mating type, one for the pheromone and the other for the receptor. This configuration can present a problem if having two mating types is favored, and generally leads to the translocation of these loci so that they are in close proximity on the same chromosome and strongly linked into a supergene. Recombination between them is often suppressed by a chromosome inversion, so that a given gamete releases one type of pheromone and is receptive only to the opposite type [205, 294, 430]. These two-mating-type systems are believed to be the precursors for the anisogamous sexual gamete systems of more advanced multicellular eukaryotic organisms [426].

Some protists and fungi have complex life cycles characterized by an **alternation of generations** between diploid and haploid multicellular forms. Each phase consists of a distinct organism—the two forms may look very similar, or be quite different. The diploid form produces spores by meiosis, and the haploid form produce gametes by mitosis (**Figure 16.3**). All gametes are identical in size and appearance, a condition

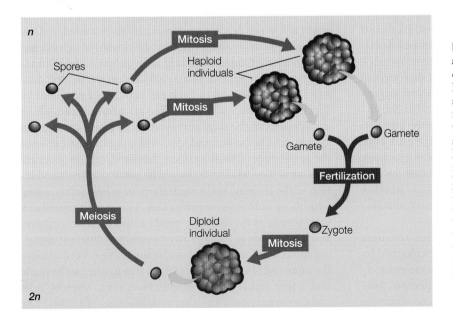

FIGURE 16.3 **Eukaryotic life cycle with alternation of generations** In eukaryotic organisms, cell division by meiosis produces haploid gametes, while division by mitosis sustains somatic growth. Cycling meiosis and fertilization events result in a series of transitions between alternating haploid and diploid states. In some species the primary multicellular organism is the diploid form; in others it is the haploid form; and in still others the living organism alternates between multicellular states, as illustrated here. This type of cycle is also termed sporic, because meiosis in the diploid (2n) individual produces spores that disperse and grow into haploid (n) individuals. The spores, haploid individuals, and gametes are very often comprised of two isogamous (same-sized) and cryptic mating types, depicted here as red and blue. Only gametes of opposite types can fuse to form a zygote.

A few isogamous species have more than two mating types. An increase in the number of mating types has two effects. First, it increases the probability that an encountered conspecific is an allowable mating partner from 0.5 (50% chance) with 2 types to $(n − 1)/n$ with n types (e.g., 90% with 10 types). More types would therefore be advantageous if the encounter rate is low or if the time frame for seeking a conjugation partner is limited. Second, an increase in the number of mating types increases outbreeding by creating a mechanism to detect and reject very close relatives or members of one's own clone. Ciliates are an interesting taxon of unicellular protists with different species having from 2 to 12 mating types. They inhabit aquatic environments and are covered by hairlike cilia that enable them to swim around. Ciliates possess two nuclei: a smaller micronucleus that divides by meiosis into haploid daughter nuclei, which are exchanged with those of another individual during conjugation; and a larger macronucleus responsible for general cell maintenance, which is regenerated following each sexual micronuclear fusion event to match the new genetic makeup of the cell. Cells undergo many generations of duplication by binary fission between these sexual conjugation events. Both factors mentioned above may play a role in determining the number of mating types in ciliates. Species with more than two types are generally unable to self-fertilize, whereas those with two types do have this ability, presumably as a backup strategy when they can't find an acceptable partner [153, 489]. Some fungus species also have more than two mating types, occasionally numbering in the hundreds or thousands. In this case, the driving force behind the proliferation of mating types is a **self-recognition/self-incompatibility system** that prevents an organism from fusing with its own gametes of different types and maximizes outbreeding [93, 144, 366, 429]. We shall learn later in this chapter that plants employ a similar mechanism to achieve optimal outbreeding.

Cooperation and conflict

Microbes engage in a surprising array of social behaviors [112, 674]. They cooperatively forage for resources and attack prey and hosts, defend themselves against competitive groups, construct protective shelters, and sometimes disperse and reproduce in a group. Some species also develop mutualistic relationships with multicellular eukaryotic organisms. In animals, such activities require signals for coordination, synchronization, and recognition. Over the past two decades, microbiologists have discovered tantalizing evidence of potential chemical signals that mediate these interactions in microbes. We shall examine some of these examples and attempt to evaluate whether they are truly signals, or whether they may be better described as either cues or manipulators.

QUORUM SENSING   Many of the cooperative activities undertaken by bacteria are effective only when performed by large or dense populations of cells. Bacteria release extracellular enzymes to digest prey, toxins to kill competitors, protein surfactants to facilitate group movement, and virulence

FIGURE 16.4   Quorum-sensing signals and genetic architecture for gram-negative and gram-positive bacteria (A) Typical system observed in gram-negative bacteria. Left, the signal molecule, or autoinducer, is an acyl homoserine lactone (AHL). The tail component (R group) varies with the species of bacteria. In the schematic illustration of a cell on the right, LuxI (blue) is the enzyme protein that synthesizes the AHL molecules. The AHL molecules diffuse freely across the cell boundary and thus move out as a function of production rate and in as a function of their concentration in the environment. LuxR (purple) is the cytoplasmic receptor protein for the AHL molecule and is unstable and rapidly degraded when not bound with AHL. When AHL accumulates in the cell and becomes bound, the LuxR-AHL complex recognizes a consensus binding sequence upstream of the operon that activates target gene (yellow) expression for the public goods product. This also activates LuxI production, so when the quorum-sensing circuit engages, more AHL is induced and the surrounding environment is flooded with the signal molecule. The positive feedback loop operates only at high cell density and induces public goods product production. (B) Typical system observed in gram-positive bacteria. Left, amino acid sequences of three precursor peptides in *Bacillus subtilis* and *Streptococcus pneumoniae* (in 1-letter abbreviations), and the final set of cyclized autoinducer peptides (AIPs) for *Staphylococcus aureus* (in 3-letter abbreviations). These peptides are genetically encoded; thus each bacteria species can produce signals with a unique sequence. Schematic cell diagram of the production process is shown on the right. The peptides are processed and cyclized into autoinducer peptides (gold heptagons with tails) and released into the environment via specialized transporters (T) in the membrane. Reception is mediated by a two-component system, including a membrane-bound histidine kinase receptor (H) and a cytoplasmic response regulator (D). When the receptor binds AIP, it phsphorylates itself and then transfers the phosphate group (P) to the response regulator, which functions as a transcriptional activator in its phosphorylated form. Activation of the target genes that code for the production of the public goods products also induces production of the AIP precursor and the histidine kinase receptor, resulting in a positive feedback loop at high cell densities, as in the AHL system. (After [458, 592].)

factors to overcome a host's immune defenses, but these cell products must be released in large quanitites, and thus require the concerted effort of a large number of bacterial cells to be effective. Since the entire population benefits from the effects of the cell products in the local environment, these effects are known as a **public goods benefit**. But synthesis of such beneficial chemicals is presumably costly and not worth doing at low cell densities, so a mechanism is needed to allow individual cells to assess whether the threshold density has been attained. **Quorum sensing** is the proposed mechanism that enables individuals to assess local conspecific cell density [148, 214, 662, 680]. Each cell produces and releases a specific signaling molecule, called an **autoinducer**, which accumulates in the local environment. A specific reception system detects the autoinducer. When the concentration of the autoinducer crosses a critical threshold, certain genes are then turned on to initiate the public goods activity (production of the enzyme, toxin, surfactant, or other factor), along with more autoinducer molecules. This positive feedback loop generates more public goods activity. Gram-negative and gram-positive bacteria have developed different signaling molecules and signal transduction pathways to achieve the same end (**Figure 16.4**) [458]. A strong case can be made that a true signal has evolved in both taxa, whereby cells can communicate with other nearby cells [149, 333]. The

(A)

(B)

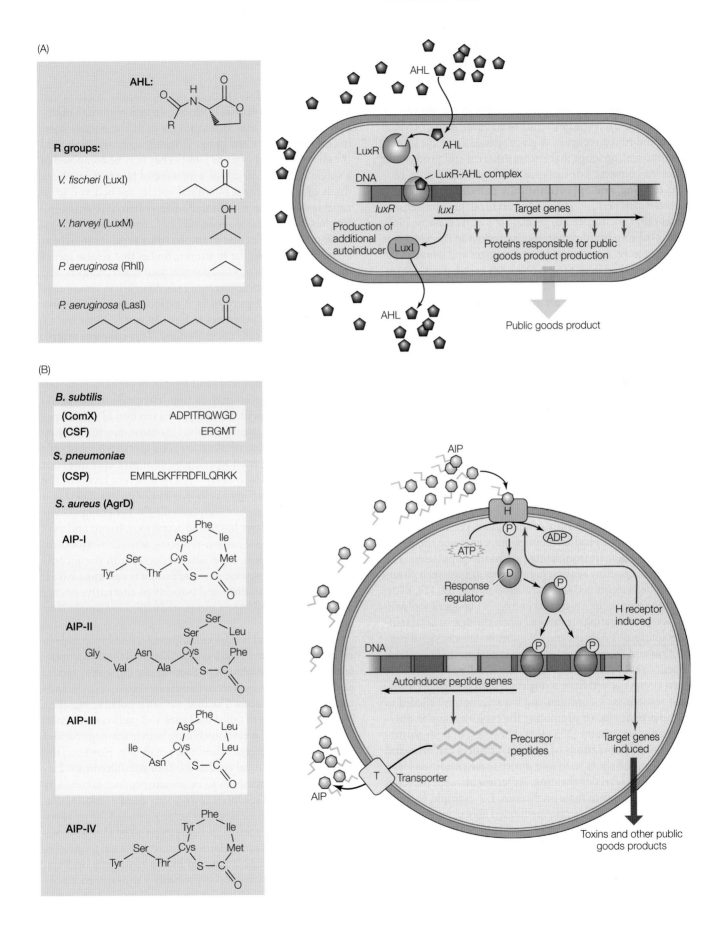

chemicals are small and soluble and therefore diffuse readily in water. Experimental addition of the chemical to a low-density population of bacterial cells induces expression of the genes that code for synthesis of the public goods product. The compounds are sufficiently complex to encode species identity and each species tends to have its own "private" signal and tuned receptor. A given species can have several quorum-sensing loops that release different end products in different contexts. For instance, the highly adaptable, bio-film-forming, and opportunistically pathogenic bacterium *Pseudomonas aeruginosa* has a hierarchical quorum-sensing system that is estimated to regulate 6% of its genome [568, 650]. One important public goods product in this species is a **siderophore**, a compound that scavenges and converts the essential mineral iron into a biologically metabolizable form. Accelerated bacterial growth due to the production of this siderophore is a major factor in the development of virulence in this species [255, 672].

Synthesis of both quorum-sensing signals and public goods products requires significant metabolic energy and is therefore costly [333]. Cooperative quorum-sensing systems are potentially invadable by cheaters that avoid these costs by exploiting the systems in various ways [148, 674]. One possible type of cheater could fail to produce the signal while still monitoring the ambient level of signal molecules produced by other cells, thus benefiting from public goods while avoiding the cost of disseminating information. Another type of cheater could produce the signal but not the product, thereby inciting others to participate in the public goods activity while avoiding the cost of synthesis itself and reducing the reliability of the signal. Yet another type of cheater might produce neither the signal nor the product, yet still benefit from being part of the population. The stability of quorum-sensing populations against such cheaters has been examined in several models [68, 69, 114, 672]. The models concur on several points: (1) cooperation with low levels of cheating are favored when the degree of relatedness among cells is high; and (2) quorum-sensing signal production and reception are also favored when there is a high benefit to cooperating. Because bacterial cell populations are often clones derived from a single cell with limited dispersal, high relatedness is common and kin selection is believed to be the primary factor promoting the evolution and maintenance of cooperation [255, 368]. However, in both natural and clinical populations of bacteria, cheaters can arise and prosper in the short term and under some conditions lead to moderate levels of cheating. Experiments with bacterial populations confirm these predictions. For example, in the *P. aeruginosa* siderophore case mentioned above, nonsignaling or nonresponding mutant "cheaters" in populations increase relative to fully cooperative and signaling individuals because their costs are lower [316], and their presence reduces the virulence benefit of the infecting population [150, 535, 540, 640]. If the cheaters succeed in completely outcompeting the cooperators, the host ultimately benefits from reduced bacterial virulence. New knowledge of these bacterial interactions could lead to novel medical intervention strategies against bacterial infection [70].

BIOFILMS **Biofilms** are dense aggregations of bacteria that form on nearly any biotic or abiotic surface. Common examples include the plaque that forms on teeth, the slippery coating on river stones, the gunge clogging up water pipes, and the goo in infected wounds [360]. They may be composed of a single bacterial species or a mixture of two to many hundreds of species. Most microbial species are believed to be capable of forming a biofilm. Biofilms often have structural features that suggest the occurrence of coordination or cooperation among cells. For example, some form multicellular aerial structures similar to fruiting bodies that release spores capable of dispersing, some contain fluid-filled channels, and most possess multiple layers of cells differing in appearance and metabolism (**Figure 16.5**). A defining feature of biofilms is the secretion of **polymers** to form an extracellular matrix within which the cells are embedded. This **cohesive matrix** promotes surface attachment, provides structural support, and can protect the cells against external threats such as antimicrobial compounds and predators [451]. Cell-to-cell signaling is known to play an important role in the initial attraction, settlement, and polymer excretion of single-species films, and frequently involves the same quorum-sensing chemical signals described earlier. The specialization of cells within a growing monospecific biofilm probably does not involve signals. More likely, cells in different layers of the film experience different resource levels (microniches) and modify their own metabolism accordingly. For example, cells in the outermost layer have good access to nutrients and oxygen and can employ carbon-reducing aerobic metabolism; those in the interior experience a decrease in these commodities and an increase in waste products and must switch to anaerobic metabolism and possibly to alternative energy sources such as sulfate or nitrate reduction, or simply stop growing [593].

Do the different species in multispecies biofilms engage in reciprocal signaling, or **cross-talk**? Kin selection, an important force for the evolution of quorum sensing within species and clones, cannot operate to maintain honesty in communication between species. The discovery of a putative signaling and reception system called the LuxS/AI-2 pathway, common to a wide range of species, including both gram-negative and gram-positive bacteria, raised this possibility. However, the autoinducer chemical (called AI-2, or 4,5-dihydroxy-2,3-pentanedione) turned out to be a waste product excreted by many bacteria, so it cannot convey very precise information [333, 686]. Nevertheless, it may serve as a cue of the presence of other bacterial species, as demonstrated by *Pseudomonas aeruginosa*, which upregulates its virulence factor promotors in the presence of AI-2 released by the host's oral bacterial flora [158]. Cue detection is probably a widespread mechanism by which multispecies biofilms form, and the ensuing interaction may be mutually beneficial, manipulative, or relatively neutral. An example of a beneficial interaction

(A)

(B)

(C)

(D)

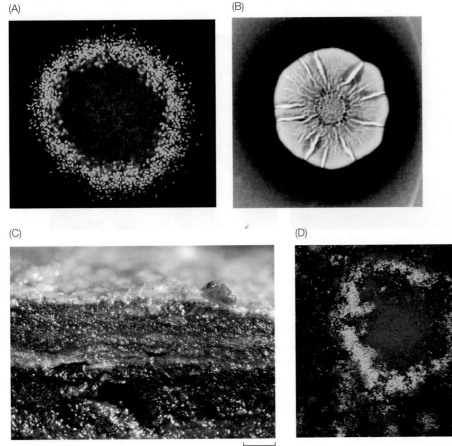

2 mm

FIGURE 16.5   Structure in biofilms (A) Monospecific biofilm of *Pseudoalteromonas tunicata* with a subpopulation of cells differentiating into diverse phenotypes that lead to localized cell death; green flourescent cells are viable, whereas red fluorescent cells have a compromised cell membrane and are dead [402, 663]. (B) A colony of *Bacillus subtilis* from a natural isolate stock grown for five days on an agar plate forms a structured biofilm. Eventually, aerial structures, or fruiting bodies, are formed that serve as sporulation sites [61]. (C) Different species in a salt marsh microbial mat separate into discrete layers according to their metabolic properties; the top yellow-green layer is formed by unicellular cyanobacteria, the darker green layer below contains mostly filamentous cyanobacteria, the purple layer is anoxygenic purple sulfur bacteria, the white layer is colorless sulfur bacteria, and the black layer is formed by sulfate-reducing bacteria [529, 530]. (D) When benzyl alcohol is the sole carbon source, *Acinetobacter* sp. (red) is invaded by wild-type *Pseudomonas putida* (green). *Acinertobacter* initially grows quickly, but after three days, as shown here, *P. putida* begins to invade and eventually overgrows the *Acinetobacter* colony. Under different conditions, the two species can achieve a stable mixture [267].

was found in two unrelated soil-inhabiting bacteria, *Acinetobacter* sp. and *Pseudomonas putida*. Simple mutations in the genome of *P. putida* can adapt it to the presence of *Acinetobacter*, allowing formation of a close-knit biofilm with mutual exploitation that is more productive than the biofilm of either species growing alone [267]. A manipulative interaction between two members of the human oral biofilm community has been described: *Streptococcus gordonii* ferments carbohydrates to form lactic acid, a preferred substrate for *Veillonella atypica*; a "signal" from *Veillonella* to *Streptococcus* causes the latter to alter its gene expression and metabolism—to its detriment and *Veillonella*'s benefit [165].

KIN DISCRIMINATION   Kin selection can operate to favor cooperative behaviors in most bacteria without the need for a kin recognition mechanism, because of bacteria's clonal growth pattern and low dispersal rate, discussed in Chapter 9. In contrast, more mobile, dispersing, and solitary-living microbes may require kin discrimination abilities if they are to cooperate. A few microbial species have evolved **kin discrimination** abilities in conjunction with truly altruistic social systems. **Myxobacteria** are soil-dwelling gram-negative bacteria with several social developmental stages and an exceptionally large genome to regulate developmental changes and cell-to-cell communication (**Figure 16.6**). During vegetative growth, the rod-shaped cells aggregate and glide together in swarms in search of rich food patches. These large foraging groups have been likened to wolf packs because they prey on other bacteria by releasing extracellular digestive enzymes (a public goods product). Individual cells are free to leave the group and establish colonies of their own. When conditions become poor and cells reach a certain starvation level, they migrate (swarm) toward focal points and form **fruiting bodies**. Only about 10–20% of the original swarm cells become spores; 10% become

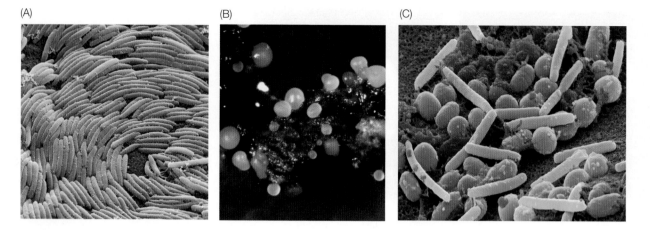

FIGURE 16.6 Myxobacteria life cycle stages (A) *Myxococcus xanthus* rod-shaped vegetative cells engaging in coordinated motility. (B) Fruiting bodies emerging from a soil particle. Each fruiting body contains up to 100,000 spores. (C) Spherical spore form (green) and vegetative rod form (yellow). The process of sporulation involves encasing the chromosomal DNA with protein to protect it from damaging UV radiation and heat, and thickening the cell wall to resist toxins, so that the cell can go dormant and survive a period of dry and harsh conditions. Only some bacteria species can sporulate [636–638].

peripheral rods that protect the dormant fruiting bodies, and the remaining 80% of cells die. In *Myxococcus xanthus*, experimental mixing of vegetative cells from different clones results in antagonism, exclusion, and exploitation. Discrimination between clones is apparently mediated by a secreted extracellular protein blend, which differs even among closely related clones [197, 243, 361, 572, 636–638, 645].

The **social amoebae**, including the well-studied *Dictyostelium discoideum*, are eukaryotic protists with a convergent life cycle (**Figure 16.7**). They live in the soil and feed on bacteria as solitary haploid individuals. When conditions deteriorate, the cells aggregate and merge into a multicellular slug. The slug moves to an open well-lit location and there about 20% of the cells die in the formation of a fruiting body—a stalk that lifts the remaining cells up to an optimal location for sporulation and dispersal (see Web Topic 16.1). Although many genetic lines may be present in a small sample of soil, the cells in a slug are usually members of the same clone. Chimeras composed of more than one clone sometimes occur in nature and can be created in the lab. Mixtures often result in unequal representation of clone members in the spore and stalk, shorter stalks, and reduced migration distances. The greater the genetic dissimilarity between clones, the more likely it is that they will segregate into clone-specific slugs. Communication among amoeba cells during these aggregation episodes occurs in stages. Initial aggregation is stimulated by the release of cyclic AMP from a focal core of cells. Once cells are close, cohesion proteins on the surface of the cell membranes, including the greenbeard gene product mentioned in Chapter 13, cause the cells to stick together. Slug formation requires the final stage of kin recognition, which

is mediated by two adjacent hyperallelic genes called *lagB1* and *lagC1*. These genes encode two transmembrane proteins with extracellular loops possessing highly variable amino acid sequences. The similarity of these sequences determines the likelihood that adjacent amoeba cells will merge [43, 94, 203, 237, 343, 348, 419, 471, 507, 594]. As we discussed in Chapter 9, such genetic kin recognition systems require a special set of conditions to maintain the genetic marker variability [534]. First, there must be a significant benefit to cooperating. Second, frequency-dependent selection favoring rare alleles (which are better indicators of close relatedness if shared between individuals) must be coupled with a high mutation rate at these loci to maintain a balanced polymorphism among alternative alleles. Third, a low rate of recombination helps to prevent frequent breakup of genetic linkages between the marker and the altruistic genes. Finally, the continual presence and threat of cheaters may foster the ongoing evolution of novel resistance and cheater-detection strategies [231, 248, 348, 594, 618]. Soil-dwelling bacteria and social amoebae appear to meet all of these conditions, and show convergent evolution in their altruistic fruiting bodies and kin recognition capabilities.

SPITE   As mentioned in Chapter 9, spiteful acts are those that decrease the fitness of both the actor and the recipient, but benefit relatives of the actor if directed toward less-related recipients [228, 382, 675]. **Spiteful behavior** thus also requires discrimination between kin and non-kin. Bacteria may be particularly amenable to spiteful interactions due to their clonal reproduction and widespread production of **bacteriocins** [673]. Bacteriocins are small to large proteins that differ from traditional antibiotics in one critical way: they are narrowly and lethally toxic only to competing strains of conspecifics. The producing strain is immune to its effects. In many Gram-negative bacteria such as *Escherichia coli*, production of the toxin is particularly costly for the producing cells because they must lyse (commit suicide) to release the protein [229, 523]. Evidence of spiteful dynamics among bacterial strains is growing. Inoculation of moth larvae with equal amounts of two strains of the pathogenic bacterium

(A)   (B)   (C)

(D)   (E)

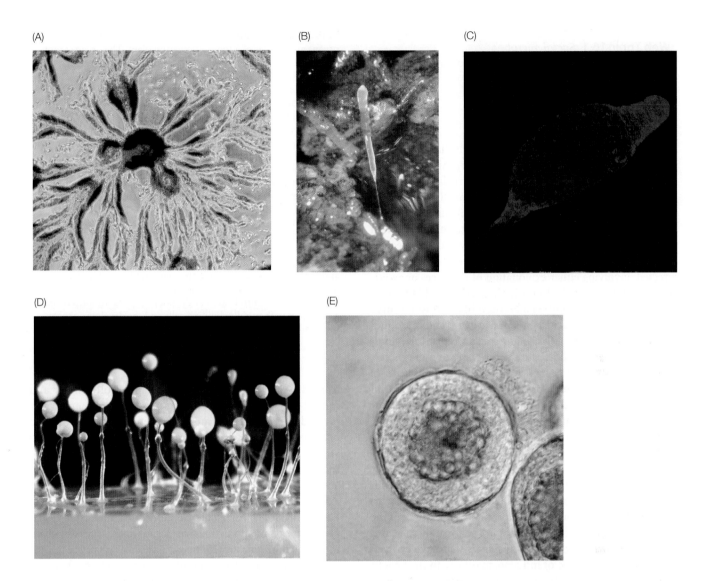

FIGURE 16.7   Life stages of the social amoeba *Dictyostelium discoideum* (A) Aggregating amoeba cells. Solitary amoeba cells that first begin to starve become the focal point for attracting others. They release a long-distance pheromone, identified as cyclic adenosine monophosphate (cAMP). The chemical is synthesized and secreted at 6-minute intervals to create oscillating concentration waves that propagate through the cell population. The leading edge of each wave provides a cAMP concentration gradient that directs cells toward the center of the aggregation [370, 614]. (B) Slug. Slugs contain 10^4–10^6 individual cells; larger slugs can travel farther [204, 490, 526]. (C) Culminant. The cells bunch together and start to form a fruiting body. A fluorescent dye marking cells that express a cytoskeletal gene indicates the parts of the mound that will become the stalk and cap structures, while unmarked cells will become spores [526]. (D) Fruiting bodies. The front of the culminant attaches to an optimal site and elongates to lift the sorus containing spores aloft as high as possible. The spores disperse by sticking to passing animals [343]. (E) A maturing (40-hour-old) macrocyst. During a sexual cycle, two haploid amoeba cells of opposite mating type fuse to form a giant cell. They release cAMP to attract other individuals, which are then cannibalized. Some of the prey cells form a cellulose wall around the entire group as shown here. The macrocyst then undergoes recombination and meiosis, divides multiple times by mitosis, and hatches hundreds of recombinants [420].

Pseudomonas aeruginosa, one a producer and the other a sensitive nonproducer, resulted in maximum growth of the producer but lower virulence for the host, presumably because the nonproducer strain was killed [308]. In the soil-dwelling pathogenic bacterium *Xenorhabdus bovienii*, samples collected at various distances from each other (from within several centimeters to several meters apart) showed increasing genetic differences with distance, and pairwise inhibition assays revealed a range of among-strain antagonistic effects ranging from resistance at short distances to strong inhibition at longer distances [277]. As mentioned earlier, similar antagonistic effects were found between competing strains of *Myxococcus xanthus*, and this species is known to possess a self/nonself recognition system [645]. Chemical warfare is the consequence, and in the case of pathogenic bacteria, the host might actually benefit as the bacteria impede each other's growth [70]. A few examples of spite have been described in animals (see Chapter 9). We should expect to observe spite wherever individuals interact with kin and non-kin in highly competitive environments [230].

Plant Communication

Plants have more in common with animals than first meets the eye. They are complex multicellular organisms with differentiated cells and tissues, organs, and circulatory systems. Although they are rooted to a given location, like sessile animals, they have remarkable powers of movement, as first studied by Charles Darwin himself [119]. This movement is slow, but with time-lapse photography becomes readily apparent to us (see Web Topic 16.2). It is quite clearly goal-directed, involving either exploitation of resources or avoidance of competitors and herbivore predators. Plants face the same trade-offs between allocation of effort to resource acquisition, defense against enemies, growth, and reproduction that animals do. Most plant species reproduce sexually—hence they are subject to the processes of mate choice, sexual selection, and sexual conflict. Plants must acquire information about their abiotic and biotic environment in order to make decisions about where to seek resources, how to avoid competitors and herbivores, and when to reproduce. They therefore possess sensory systems for the detection of environmental information, and they can transmit this information to other parts of the plant through their circulatory network [63, 619]. Finally, plants directly communicate with conspecifics, other plant species, and nonplant species with which they have mutualistic relationships [328]. Much of the information plants acquire comes in the form of cues, but there are some clear examples of signals that have evolved for the express purpose of influencing others, to the benefit of the sender. The signaling modalities employed are chemical, visual, and tactile; as far as we know, acoustic signals are not used. This section begins with an overview of plant function, sensory systems, and behavior, and then moves on to examine plant communication in four contexts: competitive interactions, mate choice, defense against herbivores, and mutualistic interactions.

Overview of plant function, sensory systems, and behavior

The primordial plant cell solved the problem of acquiring energy by engulfing photosynthesizing **cyanobacteria** and converting them into organelles called **chloroplasts**. Chloroplasts contain the pigment chlorophyll, which along with the other components of the photosynthetic apparatus, traps sunlight energy and stores it as ATP and NADPH while splitting water molecules and releasing oxygen. The energy is then used to convert carbon dioxide into sugars and other organic compounds that fuel cellular metabolism. Since sunlight and carbon dioxide are ubiquitous on the Earth's surface, the need for wide-ranging movement was not an evolutionary

FIGURE 16.8 Anatomy of a vascular plant The principle organs and tissues of a typical flowering plant. The organs (root, stem, and leaf) are composed of tissues containing groups of specialized cells with distinct structures and functions. The root system (below ground level) includes the central taproot and the lateral roots. The shoot system (above ground level) includes the stems and leaves; shoots are organized into nodes (where one or more leaves are attached) and internodes (stem section between two nodes). Axillary buds, which can give rise to shoot branches, are found in the axils, the upper angle between leaf and stem. Lateral root branches arise from the inner tissues of the roots. The apical meristems provide new cells for growth in length, and the lateral meristems allow increases in the thickness of stems and roots. (A) The mesophyll tissue of leaves is specialized for photosynthesis. Gas exchange occurs through the stomata (singular *stoma*), which open or close as a result of changes in guard cell turgor pressure. (B) In the stem, the vascular tissues, xylem and phloem, occur together in a ring of bundles just inside the cortex and form a continuous system throughout the plant body, including the veins of leaves. (C) In the root, the vascular tissues are contained within a central vascular cylinder. (D) The root tip contains a meristem with rapidly dividing cells and the developing seive tube and vessel elements of the vascular system.

imperative for plants. As plants became multicellular and moved onto land, their needs for light, minerals, and water favored the evolution of morphologies that allowed them to occupy and exploit local resources. A **modular branching structure**, with multiple shoot growth tips and leaves above the ground surface to gather sunlight and carbon dioxide, and multiple root growth tips below ground to acquire water and minerals, emerged as the most biologically efficient strategy. **Figure 16.8** illustrates the anatomy of a modern terrestrial vascular plant and defines the organs and tissues discussed in this section. The growth tips of a vascular plant are known as **meristems**—clusters of undifferentiated cells that have the ability to develop into organs with a variety of characteristics. Each meristem and its associated **root** or **shoot** and auxiliary structures (e.g., leaves, buds, and secondary meristem for lateral tissue growth) is a semiautonomous unit that can sense local resources and grow in optimal ways to maximally exploit them. For example, if a root or shoot encounters a poor resource patch, it can grow in length until it encounters a rich patch, whereupon it can branch profusely [292, 305]. This repetitive modular structure also enables plants to recover from herbivore and other damage [619, 678].

Plants must be sensitive to a variety of environmental cues in order to respond appropriately to changing conditions and they have evolved specific sensory systems for detecting these cues. **Light detection** is, of course, crucial. Plants can respond not only to the total intensity of light received, but also to its spectral composition, the direction from which it is coming, the spatial gradient of its intensity, and the amount of time that it is available. Three photoreceptor systems outside of the chloroplast/chlorophyll photosynthesis system, in conjunction with plant hormones and biochemical signaling cascades, control aspects of light-dependent movement, growth, and reproduction.

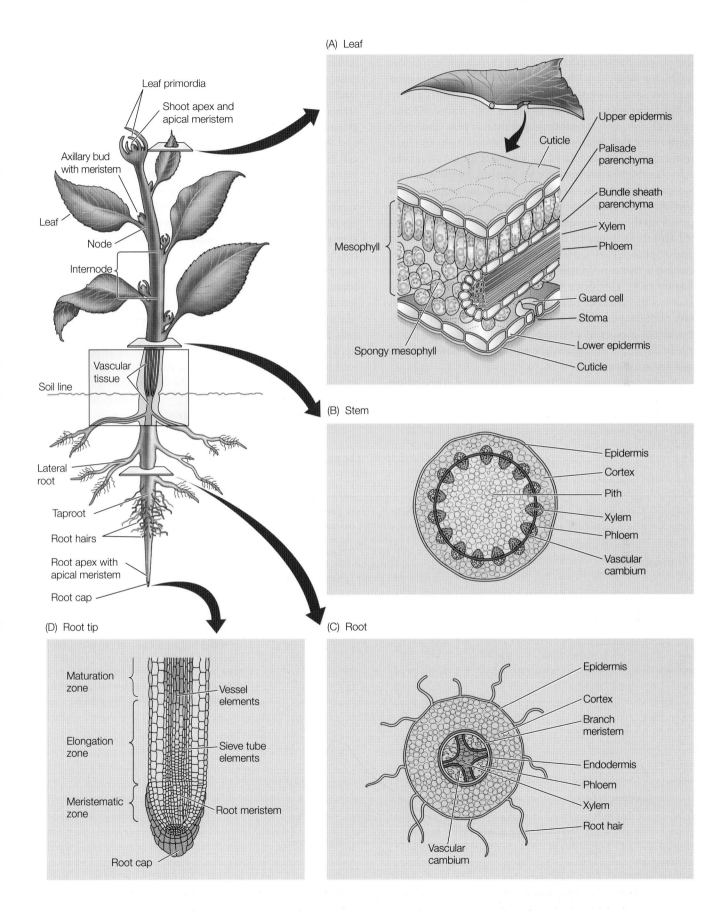

(A) Leaf

Upper epidermis
Cuticle
Palisade parenchyma
Bundle sheath parenchyma
Xylem
Phloem
Mesophyll
Guard cell
Stoma
Lower epidermis
Cuticle
Spongy mesophyll

(B) Stem

Epidermis
Cortex
Pith
Xylem
Phloem
Vascular cambium

(C) Root

Epidermis
Cortex
Branch meristem
Endodermis
Phloem
Xylem
Root hair
Vascular cambium

(D) Root tip

Maturation zone
Vessel elements
Elongation zone
Sieve tube elements
Meristematic zone
Root meristem
Root cap

Leaf primordia
Shoot apex and apical meristem
Axillary bud with meristem
Leaf
Node
Internode
Vascular tissue
Soil line
Lateral root
Taproot
Root hairs
Root apex with apical meristem
Root cap

Cryptochromes are a class of blue/UV-A light receptors composed of a protein with a flavin chromophore (color-producing molecule). They play a key role in the regulation of growth and flowering times of plants, and contribute to the entrainment of circadian clocks in both plants and animals [92, 349, 394]. **Phototropin** is another blue-absorbing flavoprotein that detects persistent directional light and light gradients, and causes plants to bend and grow toward the light. It achieves this bending by causing the growth hormone auxin to be shunted to the side of the plant or stem opposite to the illumination, so that growth is asymmetrically greater on the shaded side [103]. **Phytochromes** are red/far-red-absorbing pigments composed of a protein plus a bilin chromophore. They sense the ratio of red to far-red light by shifting between two forms with different absorption peaks (**Figure 16.9**). This shade-sensing system is cleverly designed to detect the presence of nearby plants and, in conjunction with phototropin, causes growth to occur away from shade-generating competitors [30, 584]. Phytochromes also play a role in daylength evaluation and regulation of flowering and seed germination.

Plants perceive several types of **mechanical stimulation** and respond over different time scales [59, 437]. *Gravity* is a constant force and a critical environmental cue for plants to detect—the main stem of a plant must grow straight up to keep the weight of the crown balanced, and roots must grow down into the soil. Gravity perception occurs in specialized column-shaped cells surrounding the vascular tissues of stems and in the tips of roots, which contain dense starch grains. The grains gravitate to the lowest point in the cell, and the position of the sedimentation at the "bottom" of the cell indicates the cell's deviation from vertical. Exactly how plant cells sense the movement and position of the grains and direct auxin to induce compensatory growth remains controversial [353, 447]. *Wind* significantly affects the growth strategy of plants, causing them to develop thicker, more compact stems, trunks, and roots in order to withstand the additional mechanical force stresses [49]. Perception of wind occurs via two possible mechanisms: (1) tension on the plasma membrane from stem bending opens stretch-activated calcium ion channels, or (2) specialized proteins situated between the plasma membrane and the more rigid cell wall of all plant cells change their conformation in response to the sheer that occurs during bending [437]. Either of these effects could lead to biochemical cascades that cause more lateral stem growth and less stem elongation. *Physical wounding* of a plant causes a rapid loss of hydrostatic pressure and generates an electrical perturbation called a **variation potential** to travel

(A) Phytochromobilin (P$_r$ form)

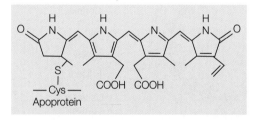

FIGURE 16.9 **The phytochrome shade-sensing system** (A) Phytochromes are photoreceptor pigments consisting of a bilin chromophore attached to a protein. (B) They exist in two interconvertible forms: red-absorbing P$_r$ (red curve), with a peak absorption at 668 nm and visibly reflecting a blue-green color, and far-red-absorbing P$_{fr}$ (black curve), with a peak absorption at 730 nm and appearing green. When P$_r$ absorbs red light it converts to P$_{fr}$, and when P$_{fr}$ absorbs far-red light it reverts back to P$_r$. P$_{fr}$ is the biologically active form, and the plant responds to the ratio of P$_r$ to P$_{fr}$. In open habitat (full sunlight, blue irradiance spectrum), the P$_r$/P$_{fr}$ ratio is around 50% because both forms are equally activated. In shaded habitat under the canopy (green irradiance spectrum), light is filtered through photosynthesizing leaves, which absorb most of the red radiation below 700 nm but pass and reflect the far-red radiation. The low ratio of red light to far-red light in shade increases the proportion of phytochrome that is in the P$_r$ form. High P$_r$/P$_{fr}$ ratios thus indicate the presence of shading plants and stimulate light-seeking growth strategies in shade-intolerant plants.

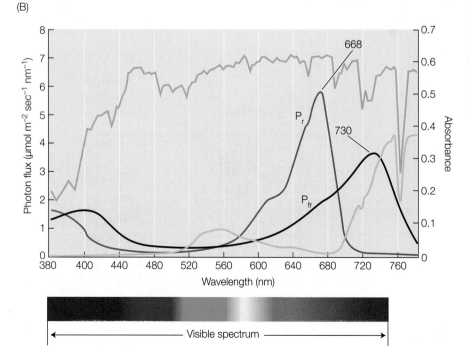

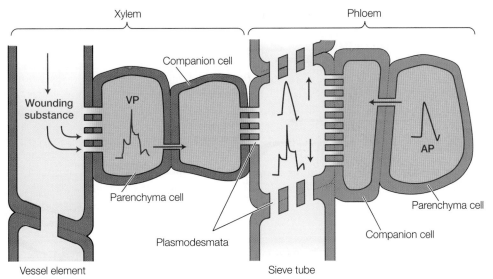

FIGURE 16.10 Electrical signals in plant tissues Figure shows a schematic longitudinal slice through the vascular bundle of a stem, with xylem tissues on the left and phloem tissues on the right. Each type of tissue contains long tube-like cells, parenchyma cells, and companion cells, which are connected laterally by small gaps called plasmodesmata that permit transfer of some cell constituents. An action potential (AP) generated in a phloem parenchyma cell can propagate laterally over short distances to a sieve tube, whereupon it can travel vertically over long distances. In contrast to animal neurons, where the ionic mechanism for action potentials depends on inward-flowing Na^+ (depolarization) and outward-flowing K^+ (repolarization), plant cells use Ca^{2+}, Cl^-, and K^+ ions. A variation potential (VP) is generated in the parenchyma cells adjacent to xylem vessel elements by a hydraulic wave or a wounding substance. Variation potentials may also travel through plasmodesmata to the phloem pathway, but they do not travel very far because their amplitude decreases rapidly with distance.

a short distance beyond the wound site through the xylem (**Figure 16.10**). This internal signal can lead to the release of the stress hormone **jasmonate**, which mobilizes the plant's chemical defenses against herbivores [398, 585]. Specialized **touch receptor systems** in some species generate true electrical **action potentials** (voltage spikes) that are rapidly propagated over greater distances through the sieve tube cells in the phloem tissue (see Figure 16.10). This nervelike network is responsible for the rapid closing of the Venus flytrap when an insect has stimulated the hairs on its specialized leaves, for the rapid folding of leaves by touch-sensitive *Mimosa*, and for the tentacle movement of carnivorous sundews (**Figure 16.11**) [59, 63, 213]. Web Topic 16.2 shows some video clips of these actions. Action potentials are also generated by pollination, light changes, heat and cold shock, and sudden irrigation, leading to adaptive whole-plant changes in metabolism [213].

Chemical detection operates rather differently in plants than in animals, in that there are no olfactory organs or collections of cells with diverse ligand receptors. In a sense, most plant cells can "smell." Plants primarily require inorganic

chemicals for their growth. **Root hair cells** actively take up soil nutrients such as sulfur, phosphorus, magnesium, calcium, nitrogen, potassium, and other trace elements, using membrane-embedded **carrier proteins** and **channel proteins** specific to each type of ion [514]. Carbon dioxide gas is taken up through the stomatal openings in leaf epidermis, where it is first dissolved in water before passing into **mesophyll** cells by diffusion (see Figure 16.8). The mesophyll tissue within the leaf adjacent to the **stoma** (plural *stomata*) is a spongy

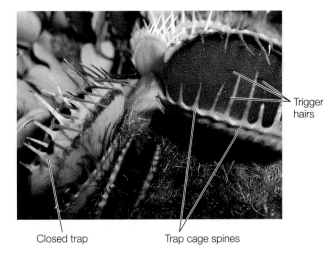

FIGURE 16.11 Prey capture by the carnivorous Venus flytrap *Dionaea muscipula* A potential insect prey approaches an open trap formed by specialized bilobed leaves. Each leaf has three trigger hairs, emerging perpendicularly from the center of the red leaf pad. The insect must bump into at least two hairs within 40 seconds for the trap to close. Stimulating the hairs activates mechanosensitive ion channels and generates a single action potential, which causes rapid loss of osmotic pressure in cells at the base of the lobes. The trap closes in less than a second. The sharp spines on the leaf margins interlace when the trap is closed, barring the insect in the cage. The movements of the agitated struggling insect tighten the trap closure, and the leaves then secrete digestive enzymes that kill and digest the prey. Carnivorous plants tend to evolve in environments low in nitrogen [59, 644].

FIGURE 16.12   A modular view of plant foraging Plants consist of independent aboveground and belowground modules that sense environmental cues and signals, take up resources, and decide where and how to grow. The diagram at the left shows a plant made up of root and shoot units (rectangles). The units are connected to other units via vascular connections. Roots take up water and minerals, and these resources flow upward from the root modules and into the growing aboveground shoot units via the xylem (red arrows). Above-ground units manufacture sugars and other organic compounds, which are then transported to other modules via the phloem (blue arrows). Each module has its own receptors to detect local environmental quality, and internal signal transduction components (hormones, proteins, and metabolites) that lead to foraging responses. The lower right diagram depicts a root module with its receptors, internal components, and resource output. This module also interacts with more distant modules (top right rectangle) in the plant, including other root modules as well as shoot modules. Distant modules detect their own local environmental signals and cues, such as water or light; this information, along with distantly produced resources, feeds back to the lower module through the vascular system to modulate its foraging responses. Thus the global cocktail of cues derived from independent modules circulates within the plant and affects the responses and decisions of distant modules. (After [129].)

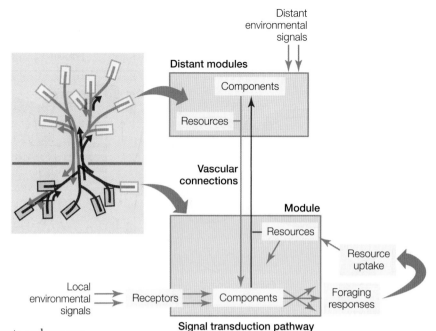

layer of thin-walled cells interspersed with intercellular air spaces that can passively assimilate hydrophilic and hydrophobic chemicals. Thus other small volatile compounds can also enter the leaf through the stomata. Finally, the epidermal surface of leaves is covered with a **waxy cuticle** that reduces water loss, but volatile lipophilic chemicals can adhere to this substrate and diffuse into the interior leaf tissue [521]. Whether and how plants evolve specific responses to these chemical cues depends on the costs and benefits of doing so. For example, seedlings of the parasitic plant called dodder (*Cuscuta pentagona*), which does not photosynthesize but rather extracts nutrients from its host, use volatile plant cues to seek out the healthiest host plants. The fact that they grow differentially toward certain plant chemicals suggests that they have evolved a tuned chemical reception mechanism [536].

In the past decade, the responses of plants to environmental events have come to be viewed as **plant behavior**, and the application of animal behavior principles, especially the individual fitness-optimizing approach, has led to new insights into the evolution of plasticity (indeterminate, flexible development) in plants [328, 619, 620]. For instance, the perception of resources and the growth and placement of plant organs where they can best harvest them is viewed as *foraging behavior*. Plant growth in heterogeneous environments can be examined from the perspective of marginal value models of optimal patch staying time, and optimal diet

or prey choice models have been applied to ask which forms of nutrients plants should take up and whether it is worth the cost of collaborating with a mutualist. A useful animal model analogous to plant foraging is actually the honeybee colony (see Chapter 14), in which modular units can independently sense cues and perform different behaviors simultaneously, but where coordination is achieved by perceiving and responding to cues from the circulating nutrients (**Figure 16.12**) [129, 305, 418]. The observation of trade-offs among different growth, reproductive, and defensive strategies in plants leads to the notion of cost-benefit analysis and *decision-making*. For example, plants that are simultaneously challenged by pathogen infection and herbivore damage must decide how to optimally allocate available resources between conflicting defense mechanisms [54]. Plants can *anticipate future conditions* by accurately perceiving reliable environmental cues, such as day length; for instance, they respond to the shortening days of autumn by dropping leaves that would be damaged by cold winter weather. Plants exhibit *conditional learning and memory*, in the sense that they alter their behavior depending on experience. For example, once an individual has been attacked by a parasite and mounted a defensive immune response, it responds faster if attacked again [108]. Rapid physical movements, as described earlier, have been viewed as goal-oriented *intentional behaviors* [59, 164, 621]. We turn now to *communication behavior*, whereby plants produce signals and receive information from other organisms.

Web Topic 16.2 *Plant movement*
We typically think of plants as passive organisms. In fact, they move quite extensively, but at their own time scales and in their own ways. Here are some examples of plants that move for a variety of reasons, including foraging, defense, and reproduction.

Competitive plant interactions

Plants compete fiercely with neighbors for access to light and soil nutrients. Unlike animals, they have little ability to choose their neighbors; but like animals, they can decide to minimize competitive encounters by avoiding them, maximize their competitive effects by aggressively confronting them, or they can simply tolerate them [460]. Because plants can adjust the positions of their organs only slowly (by moving or by differential growth), and in the case of woody plants invest heavily in irreversible secondary lateral growth, it is important for them to detect competitors and evaluate their relative competitive potential well ahead of time, so they can build the organs relevant for functioning in their future environment.

The shoot system has several strategies for dealing with light competitors. We mentioned above the phytochrome red/far-red shade-sensing system, which is well-suited to the detection of green plant competitors. Plants can actually detect a neighbor with this sensory system well before the neighbor has shaded them. The vertical stems of a plant are sensitive to the horizontal flux of red/far-red radiation reflecting off of the neighbor's leaves. This information provides the plant with an early warning system and initiates an **aggressive response syndrome** of stem elongation, reduced branching, and increased apical dominance to increase plant height [18, 28, 29]. Putting collars around the stems that block only far-red radiation to this part of the plant abolishes the elongation response. Light for photosynthesis is a directional resource coming from above. By preemptively investing in growing taller, a plant can strive to get ahead of its neighbors and eventually deny them the light they need [668]. However, elongation comes at the cost of thinner stems with risk of breakage, less horizontal growth, and reduced root development, and hence should be undertaken only if there is a benefit to the plant [104]. (Vines have circumvented these costs entirely by growing long, thin stems and using other plants for support, a highly aggressive strategy.) An alternative strategy upon detecting a close neighbor on one side is to **avoid competition** by reducing the growth rate of branches close to the neighbor and increasing the growth of branches away from the neighbor. For some plants, such as trees, such asymmetric growth risks a strongly imbalanced crown, which could, in extreme cases, lead to the tree falling over [695]. In highly competitive environments, such as tropical forests, some tree species develop buttresses to counteract crown asymmetry [696]. Finally, plants can **tolerate** the shade of competitors with a variety of adjustments. Shade causes a reduction in the rate of photosynthesis. This reduced production causes plants to become more photosynthetically efficient by increasing chlorophyll concentration and reducing respiratory loss of CO_2 through the stomata, and over the longer term to maximize exposure to sun flecks by increasing leaf size and reducing leaf clumping [18, 460, 629]. These light-competition strategies all involve responses to actively detected cues from other plants.

Root systems also engage in competitive interactions with neighbors [128]. Some plant species employ **aggressive deterrence** against the root growth of adjacent plants by exuding toxic compounds from their roots, leaves, or decomposing tissue. Known **phytotoxins** (plant toxins) include a range of mostly aromatic chemical structures (e.g., flavonoids, quinones, quinolines, and hydroxamic acids) with different modes of action. They can negatively affect metabolite production, photosynthesis, respiration, membrane transport, germination, or root and shoot growth in susceptible plants [26, 669]. Secondary plant compounds that have evolved for the primary purpose of inhibiting neighbor growth qualify as **territorial signals**. In some cases, only heterospecific plants are affected. For example, black walnut trees (*Juglans nigra*) produce a quinone that impairs water uptake by roots and limits growth in several other plant species [280, 681]. In other cases, adult plants primarily inhibit the growth of young conspecifics close to them but do not affect other species [664]. Still other toxins are detrimental to both conspecifics and heterospecifics [258, 400, 486]. The role of toxins is often demonstrated experimentally by adding activated carbon to the soil, which adsorbs most organic compounds and ameliorates the detrimental effects. This chemical deterrent strategy, called **allelopathy**, is not common and tends to be observed in plants inhabiting resource-poor environments such as deserts and in invasive non-native plants (**Figure 16.13**) [26, 87, 550].

It is more common to observe that roots *avoid* growing into areas containing other roots. Roots absorb water and nutrients from the soil and generate a depletion zone around them. Adjacent plant roots may simply detect the declining gradient of resources when they approach the roots of another individual and reduce their growth in that direction in favor of growth into less depleted areas [550, 551]. Some species appear to *tolerate* the roots of other individuals, especially if they typically have neighbors on all sides; they cope with the depleted resources using some optimal allocation of investment among roots, shoots, and leaves. Water and soil nutrients, with the exception of phosphorus, are mobile resources and likely to be distributed in temporary three-dimensional patches, so root zone size is less important to the relative competitive ability of the root system than is height for the shoot system [569]. The question arises: Can plants detect the presence of root competitors, as they can detect shoot competitors, and adopt preemptive strategies? Several laboratory studies addressed this question on a variety of species by quantifying growth and allocation responses of plants to the presence or absence of a root competitor while carefully controlling the amount of nutrients. Most of the studies found that the plants increased their root growth in the presence of a competitor, but consequently reduced their shoot growth and net reproductive fitness. These results were touted not only as evidence for **self/non-self detection**, but also as an example of a **Tragedy of the Commons game**, in which competitors selfishly use up a nonrenewable resource and all players pay a cost [186, 187, 234, 401, 463]. These studies were subsequently criticized for failing to take into

FIGURE 16.13 **Territorial roots** The creosotebush (*Larrea tridentate*) occurs in nearly monospecific stands in the desert habitat of the southwestern United States. Individual plants are more uniformly spaced than would be expected by chance. (A) A map of the excavated root systems in a small population of creosotebushes. (B) A map representing the root area of each plant in (A) as a polygon. Shaded areas represent the surface where there is extensive overlap of at least four root polygons. (C) A map of hypothetical circular root systems centered on each plant location and equal in area to its polygon area; the overlap area is much greater. The asymmetrical root distribution results from toxic compounds exuded by the roots, which inhibit growth of nearby individuals. (After [65, 400].)

account the larger pot volume provided to the competing pairs, but some results were consistent with strategic growth allocation in response to a competitor. More compelling evidence for root detection are studies in which plants were placed in competition with kin versus non-kin and exhibited different growth allocation responses (**Figure 16.14**) [159, 335, 449], although similar studies on other species found no effects of kinship [99, 431]. It is not clear how plants would recognize other plant individuals. Root activity generates ionic currents, so plants may be able to detect ion and pH gradients. Roots also release digestive agents, antimicrobial

FIGURE 16.14 **Kin cooperation in plants** Pale jewelweed (*Impatiens pallida*) is an annual herbaceous plant found in moist shady woods in eastern North America. Seed dispersal occurs by explosive dehiscence, which allows seeds to land close to the mother plant and creates dense stands of closely related individuals. Greenhouse experiments compared the growth patterns of plants potted together with either close kin or strangers to the growth patterns of solitary potted plants from the same stock as the kin and stranger treatments. Root, stem, and leaf areas differed significantly between plants sharing pots with kin and those sharing pots with strangers, as indicated by the asterisks. When strangers were potted together, both plants reduced root allocation and invested in leaf area, a carbon-maximizing strategy. When kin were potted together, the plants elongated their stems, became taller, and increased their branches, a strategy that reduces self-shading and mutual shading. Stands of kin had higher fitness than stands of strangers. (After [335, 449].)

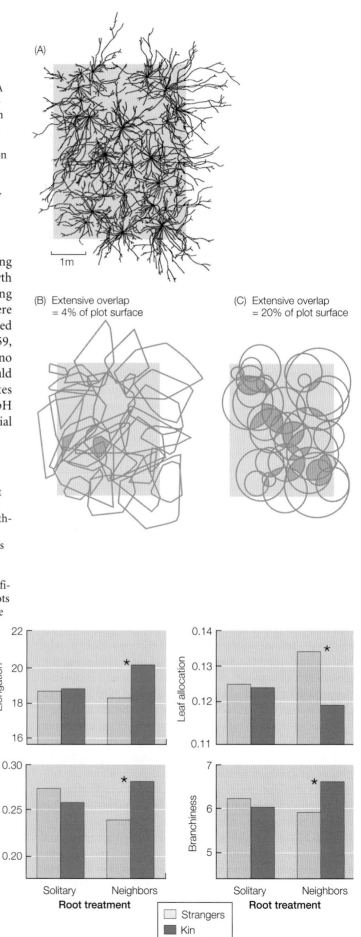

defense chemicals, mutualist attraction allomones, and waste products [25]. These compounds most certainly can be species-specific and provide a rich source of chemical information to other roots that can detect them.

Sexual selection in plants

Like the majority of animal species, most plant species reproduce sexually. Haploid sperm and eggs are produced via the process of meiosis (followed by mitosis), and the fusion of an egg with the sperm of (usually) another individual forms a diploid zygote that develops into a new adult individual. Plants thus obtain the same evolutionary advantages of genetic recombination as other eukaryotic organisms. However, in contrast to animals, most plant species are **hermaphroditic**, meaning that each individual produces both male and female gametes. We include both **monoclinous** species (male and female organs contained in the same flower) and **monoecious** species (male and female organs in separate structures within the individual) in our definition of hermaphrodite, because these two strategies face the same genetic constraints, as discussed below. Rather than referring to males and females, as we do when the sexes are separated into different individuals, we use the terms **male function** and **female function** when referring to the sperm- and egg-producing organ components of hermaphrodites. A few animal species are also hermaphrodites, and some plant species have separate sexes like most animals and are referred to as **dioecious**.

Figure 16.15 illustrates the reproductive organs and fertilization process of a typical monoclinous **flowering plant**. **Pollen grains**, produced on the **anthers**, contain the male gametes. The **pistil**, comprising the **stigma**, **style**, and **ovary**, serves the female function. For fertilization to occur, pollen grains must be released from the anther and land on the stigma of a flower of the same species. The pollen grains germinate on the stigma and produce a **pollen tube**, which grows down the style and carries the sperm to the ovules, where fertilization takes place. The zygotes mature into embryos enclosed within seeds, which are then dispersed through various means. Some plants self-fertilize—that is, their zygotes are formed from sperm and eggs from the same individual. However, many other plants tend to outcross (cross-fertilize) and some species are unable to self-fertilize (self-incompatible) due to various barriers. In these cases, pollen grains from the flower of one plant must be removed and transported to the stigma of the flower of another plant, usually by some type of animal (insect, bird, or bat) or by wind. Flowering plant species are collectively called **angiosperms**, the largest and most diverse taxon of land plants. **Gymnosperms** (conifers, cycads, gnetophytes, and ginkgos), the other main land plant taxon, have a similar sexual seed-producing system, but the conelike reproductive organs leave the ovules directly exposed to the air, rather than enclosed as they are in angiosperms. Coniferous pines are monoecious, and ginkgos and cycads are dioecious. Transfer of pollen in gymnosperms is usually via wind.

Although it was once believed that sexual selection could not occur in hermaphrodites, in fact the underlying driving force for sexual selection—**anisogamy**—is just as prevalent in plants as it is in animals. Seed maturation and fruit growth require a huge resource investment, so female-function fitness is limited more by access to resources than it is by access to pollen. Male-function fitness in cross-fertilizing, animal-pollinated species is limited by access to pollinators, through which pollen gains access to the stigmas and ovules on other plant individuals. These sexual function differences set up the conditions for strong **male–male competition** (intrasexual selection) and **female choice of mates** (intersexual selection). Features of male function that increase the competitive advantage of a given pollen donor relative to other individuals are strongly favored, including: production of large amounts of pollen; strategies for attracting and attaching pollen to pollinators; rapid pollen germination upon arrival at a stigma; and rapid pollen tube growth down the style to win the race to the ovules. The stigma has no control over whose pollen reaches it, but it typically receives far more pollen grains than it has ovules available for fertilizing. Female function thus favors strategies that allow the plant to assess and select high-quality sperm, by selectively nurturing the pollen tube growth of certain pollen donors and selectively aborting the seeds or fruit sired by less preferred pollen donors [582]. Nevertheless, hermaphroditism may curtail opportunities for sex-specific trait expression. Any increase in allocation of resources to one sexual function comes at the cost of reduced investment in the other sexual function, so selection will lead to an optimal level of sex allocation. This allocation level could vary depending on the absolute size or resource availability of the plant, such as a small or poor-condition individual investing more in the less costly male function [16, 97, 98]. In dioecious (separate-sex) species, the female can carry the genes for the male trait without expressing it, and they therefore do not bear the survival cost of possessing genes for extreme male traits. Sex chromosomes have even evolved in some dioecious species to resolve **sexual conflict** [40]. Hermaphrodites, however, cannot avoid this cost, since survival reflects the balance of an individual's male and female function traits. Thus runaway Fisherian selection for exaggerated male traits is expected to be rare in hermaphrodites [445]. Moreover, in the absence of sex chromosomes and sex-limited trait expression, sexually antagonistic traits are exposed to selection in every individual. As a consequence, cycles of sexually antagonistic conflict should be curtailed, but not eliminated [40].

What is the evidence for female mate choice of pollen from certain pollen donors? The stigma performs the first level of filtering by blocking the germination of both nonconspecific pollen and self-pollen (in self-incompatible species). The mechanisms for discriminating against these two classes of pollen are different [295]. Discrimination against clearly heterospecific pollen is believed to be a passive process called **incongruity**, in which there are simply too many biochemical differences between the pollen and pistil factors for

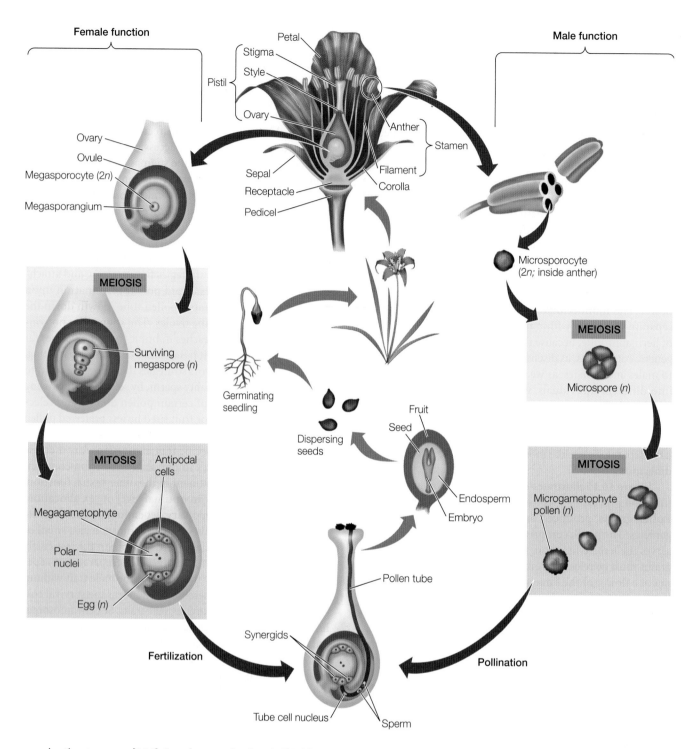

Female function

Male function

Petal
Stigma
Style
Pistil
Ovary
Sepal
Receptacle
Pedicel
Anther
Stamen
Filament
Corolla

Ovary
Ovule
Megasporocyte (2*n*)
Megasporangium

Microsporocyte
(2*n*; inside anther)

MEIOSIS

Surviving
megaspore (*n*)

MEIOSIS

Microspore (*n*)

Germinating
seedling

Dispersing
seeds

Fruit
Seed

Endosperm
Embryo

MITOSIS

Antipodal
cells

Megagametophyte

Polar
nuclei

Egg (*n*)

MITOSIS

Microgametophyte
pollen (*n*)

Pollen tube

Synergids

Fertilization

Tube cell nucleus
Sperm

Pollination

germination to occur [296]. In other words, the pistil is blind to such pollen. If a plant species has encountered a closely related species or strain that yields inviable zygotes or sterile offspring, it is likely to evolve one or more recognition molecules for pollen and stigma surface secretions and require a match of alleles between the pollen and stigma versions before germination can proceed [331, 342]. Self-pollen, on the other hand, is identified via a different self-recognition mechanism, similar to the immune system of animals, mediated by a single hypervariable locus called the **S-locus**. When a match between pollen and pistil factors is detected, the

pollen is actively rejected in the case of a self-incompatible species, a process called **incompatibility** [266, 304, 453]. Some species allow selfing under some circumstances (e.g., insufficient amount of outcrossed pollen received, varying environmental conditions, and age of the plant and flower [409]).

Once these two filters have been passed, the question then remains: Is there female choice among acceptable pollen for sires that are themselves high-quality plants, superior competitors against other sires, or otherwise more genetically compatible with the pollen-receiving plant? And if there is

◀ **FIGURE 16.15** **Sexual organs and reproduction in flowering plants (angiosperms)** Top center: The flower of a monoclinous hermaphroditic species contains functioning male and female organs surrounded by the petals, which together with nectar and olfactory signals serve to attract pollinators. Female function is shown on the left side. The female organ is the pistil, which comprises the stigma, style, and ovary. Each ovule contains a diploid cell (the megasporocyte) that undergoes meiosis to produce four haploid cells, only one of which develops into the gametophyte. The mature gametophyte includes a single haploid egg cell and a large central cell that contains two haploid nuclei (the polar nuclei). Male function is shown on the right. Multiple anthers, producing large numbers of microsporocytes, constitute the male organ. Each microsporocyte undergoes meiosis and mitosis, and each resulting pollen grain contains a generative (sperm) cell (which divides one more time to produce two sperm) and a tube cell, both of which are haploid. The pollen grains disperse via wind or animal vectors. To achieve pollination, a pollen grain lodges onto the stigma and becomes hydrated. The pollen grain extrudes a pollen tube, which penetrates the cuticle of the stigma if the species recognition criterion has been met and the self-identity alleles are different. The tube grows down into the style toward the ovule, receiving nutrients from the pistil. When the pollen tube reaches an ovule, one of the two sperm cells inside the tube fuses with the egg to produce a fertilized diploid embryo. The embryo receives nutrient support from the endosperm, which is derived from a second fertilization event between the second pollen sperm cell and the two female polar nuclei (and is therefore triploid). The endosperm and embryo are enclosed in a hard seed coat for protection during dispersal. Dispersal may be facilitated by floating devices that promote wind transport; by a dehiscing seedpod; or by the development of a fleshy fruit that is eaten by an animal and later dropped or passed through the digestive system. The seeds germinate and develop into the next generation of plants.

evidence of nonrandom mate choice, does it lead to higher offspring fitness? There is evidence for finer-scale selection of the most genetically compatible pollen. Experimental field studies of the alpine wildflower *Delphinium nelsonii* involving hand-pollination from pollen donors at varying distances from the female plant found the highest offspring survival of seeds from sires located intermediate distances away. This result suggests that local genetic adaptations are very important, and that the female function in plants benefits by selecting optimally outcrossed sires [658, 659]. Similar nonadditive genetic benefits and interaction effects between the male and female genome on offspring fitness components have been found in other plant species [48, 260, 408, 488, 571, 590]. On the other hand, evidence of *consistent pollen donor performance effects* across maternal plants implies potential additive genetic benefits to females [407, 683]. Much research has focused on the male performance trait of pollen tube growth rate. The pistil is an arena for intrasexual competition among pollen donors and for intersexual assessment of male quality by the female. Pollen tube growth rate is both heritable and affected by environmental factors and the condition of the pollen donor plant [9, 46, 138, 218, 374, 375, 560, 694]. Several studies that compared the siring ability of competing pollen donors by simultaneously applying pollen from two different plants to a single stigma found that individuals with faster-growing pollen tubes exhibited greater siring success than their rivals [477, 580, 588, 589]. However, tube growth rate reflected the quality of the individual sire in only a few cases. One example, illustrated in **Figure 16.16**, is the violet

FIGURE 16.16 **Evidence of sexual selection for good genes** (A) *Viola tricolor* is an outcrossing annual plant that is common in Europe. Pollen from pairs of individuals possessing distinctive genetic markers was hand-pollinated onto the stigmas of a series of maternal plants. The seed production rate per capsule of the pollen donors was quantified, as well as the pollen tube growth rate and fertilization success of the pollen donors. The offspring (seeds) sired by the superior and inferior pollen competitors were then allowed to germinate and grow, and the seed production rate, pollen tube growth rate, and competitive siring ability of these second generation individuals were also measured. A third generation of offspring from the superior and inferior pollen donors was similarly monitored for seed production and pollen tube growth rate. The seed production rate of a plant was significantly correlated with its pollen tube growth rate, indicating that donor quality is reflected in the vigor of its pollen tube growth. The difference in pollen tube growth rate of two competitors was significantly correlated with the difference in seed production rate. The pollen tube growth rate of a plant was also positively correlated with the pollen tube growth rate of its offspring, indicating that siring ability is somewhat heritable. (B) The siring ability of a pollen donor, shown as pollen donor rank, was correlated with its offspring's seed production. The female plant function thus obtains two benefits from the receipt of fast-growing pollen: offspring of higher seed fecundity and offspring of higher siring ability. Pollen tube growth rate could therefore function as a good-genes cue for female choice. (After [581].)

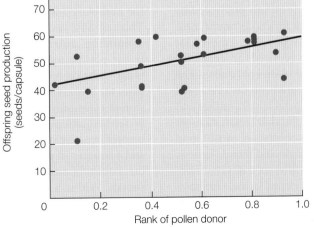

(A)   (B)

- Reduced pistil
- Stamen

- Pistil
- Reduced stamen

FIGURE 16.17 **Dioecious flowers of holly (*Ilex aquifolium*)** (Top) Flowers from a male plant showing pronounced stamens bearing pollen and reduced, sterile pistil. (Bottom) Flowers from a female plant showing prominent pistil and reduced stamens with no pollen.

(*Viola tricolor*), in which pollen tube growth rate is correlated with seed production rate (female function) of the pollen donor plant, as well as with the male- and female-function fitness of the offspring [581]. Offspring fitness (survival rate, germination rate, or growth rate) appears to be enhanced when there is a high degree of competition among pollen genotypes [121, 373, 448, 508, 685]. However, it is difficult to sort out whether these effects are caused primarily by male competitive differences, female promotion of tube growth by preferred pollen genotypes, or male–female interactions such as selective maternal provisioning of ovules fertilized first [46, 139, 414, 478, 582]. As with cryptic female choice and post-fertilization zygote survival in animals, we do not yet understand the level of information available to maternal plants at the gamete level regarding sire characteristics [47, 406].

It is somewhat easier to sort out male- and female-function effects on floral characteristics and prepollination processes in self-incompatible angiosperms pollinated by animals [140]. Here, fitness for both functions can be measured in the same currency—amount of pollen removed for male fitness, and amount of pollen received for female fitness. Various field and experimental studies have quantified male- and female-function selection gradients on a variety of floral traits, such as corolla width, petal length, anther and pistil length and position, flower height, stalk length, number of flowers, flowering date, and flower lifespan. Most found no conflict or trade-off between male and female fitness, because the two processes favored different traits that were compatible in the same flower. For example, male function typically selects for larger flower size, while female function selects for longer pistils [22, 217]. This result implies that any

trait that attracts more pollinators favors both sexual functions [464]. Some trade-offs were found for the timing variables. For instance, floral lifespan is often divided temporally into a male phase (anthers mature) first followed by a female phase (stigma matures). Shifts in the transition point always favor one sex to the detriment of the other (e.g., earlier or longer female phase increases female fitness and lowers male fitness). Some species do show opposing selection gradients for male- and female-function flower traits [88, 446]. Not surprisingly, this result is more likely in species that are partially or completely dioecious, implying that the two sexual functions cannot be maximized in the same flower or plant [21, 137]. **Figure 16.17** shows the typical morphological differences between the male and female flowers of a dioecious plant species. As in animals, sexual conflict is expected to be stronger in separate-sex species compared to hermaphrodites because aggressive sex-specific traits can arise and flourish once the same-body constraint of hermaphroditism has been bypassed [40] (**Figure 16.18**).

Defensive alarm signals

Because sessile plants cannot flee to escape from mobile predators such as chewing or sucking insects and mites and larger herbivores, they have evolved a set of defense mechanisms to prevent or limit the damage [345, 383, 585, 657]. The first line of defense is preexisting (called **constitutive**) **physical barriers** and **deterrents**, such as the tough cuticle of leaf surfaces; hard woody covers (bark) that can withstand the attack of small herbivores; sharp silica crystals and thorns that deter large vertebrate herbivores, **trichomes** (hairs on the stem and leaf surfaces that make it difficult for insects to walk on the plant and sometimes exude toxic or sticky substances to further impede progress [481]); and other specialized organs that further restrict herbivore access to the more nutritious parts of the plant. Once an attack has occurred, the plant mounts a complex chemical response to *directly* impact the herbivore and strengthen the plant's internal defenses. This response begins with the perception of tissue damage. **Wounding** causes hydrostatic changes in the cell, a change in the plasma membrane potential that induces a short-range electrical response, and the release of oligosaccharides from the broken membranes [399]. Also quickly released are stored (constitutive) **secondary plant compounds** (monoterpenes, sesquiterpenes, and aromatics) that are toxic to herbivores; reactive oxygen species such as hydrogen peroxide; and a bouquet of **green leaf volatiles**, derived from oxidative breakdown of membrane lipids, whose role we will describe shortly. Within a few hours, the phytohormone **jasmonate** is produced from linolenic acid, an essential fatty acid and abundant component of cell membranes. Jasmonate, in

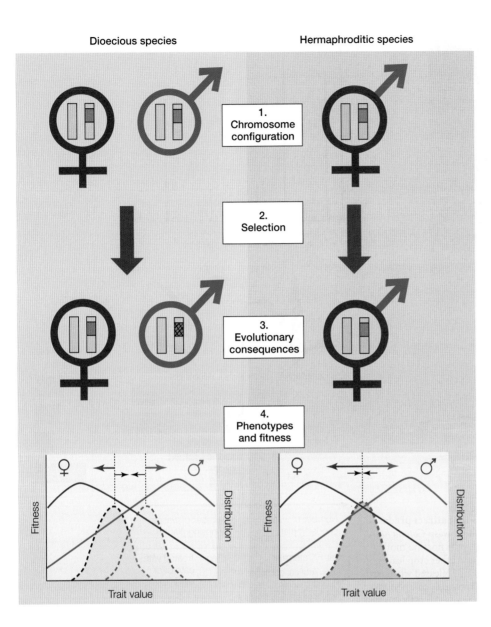

FIGURE 16.18 Sexually antagonistic conflict in dioecious and hermaphroditic species This figure illustrates the consequences of intralocus conflict, where a mutant allele at a given locus (green section) increases female fitness while decreasing male fitness. Step (1) shows the chromosomal configuration for a dioecious species with separate sexes and a hermaphroditic species with combined sexes. Selection (Step 2) leads to sex chromosomes, sex-specific regulation mechanisms, and sexual masking (hatched section) in the dioecious species. These regulatory and compensatory mechanisms cannot occur in the hermaphrodite, and the allele cannot be masked, so it is exposed to selection at each generation. The evolutionary consequence (Step 3) for the dioecious species is the development and maintenance of sexual polymorphism, because male and female fitnesses peak at different trait values (solid red and blue lines) and the distributions for male and female phenotypes diverge (dashed red and blue lines), as shown in Step 4. The hermaphrodite, on the other hand, cannot develop sexual polymorphism, despite fitness peaks at different trait values and strong selection to diverge (red and blue arrows). Thin black arrows show the constraint owing to the male-female correlation, which counterbalances the divergent selection in the dioecious species to some extent. (After [40].)

conjunction with several other compounds such as ethylene, abscisic acid, and salicylic acid, activate the wound-response genes that code for the production of a second round of **wound-induced chemical defenses**. These compounds may have local effects such as repairing the damaged tissue, reducing pathogen infection, or producing additional toxins that inhibit the predator's growth or impair its ability to digest plant tissue. In addition, jasmonate moves throughout the plant to activate systemic responses, priming the plant's immune system and building up defensive toxins in the event of herbivore attack elsewhere.

A final and *indirect* line of defense against herbivores involves attracting mobile **bodyguards**, such as insect predators, parasitoids, entomophagous nematodes, and even birds, which then attack the herbivores. Plants can attract bodyguards by providing shelter, such as the hollow thorns on the ant acacia, or by providing alternative food, such as

extrafloral nectaries. Another attraction mechanism is to emit chemical cues or signals that notify natural enemies of the herbivores about the location of an ongoing infestation. This strategy is described as plants **crying for help**, and occurs in at least 13 plant families in response to a wide range of herbivorous insects and mites [147, 473]. Several sources of volatile chemical compounds can be employed to notify the bodyguards. One is the green leaf volatiles mentioned above. These are compounds constitutively present in plant tissues and released immediately upon wounding. These chemical *cues* are highly volatile 6-carbon alcohols, aldehydes, and esters that disperse short distances by diffusion, probably not much farther than to other branches of the emitting plant. Other volatile chemicals released within a few hours of herbivore attack include indoles and terpenoids such as linalool and farnesene, slightly larger molecules that are probably transported farther by turbulent flow in fragmented plumes

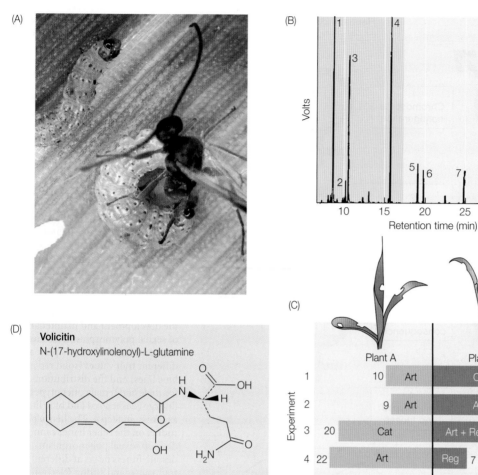

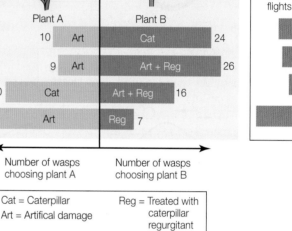

FIGURE 16.19   Plant defensive signals that attract predators of damaging herbivores (A) Parasitoid wasp *Cotesia marginiventris* ovipositing on first instar larva of the noctuid moth *Spodoptera littoralis*, which feeds on corn and other crop leaves. (B) Gas chromatogram of volatile compounds released when corn leaves are damaged by *Spodoptera* larvae. The small compounds (1–4; green region) are green leaf volatiles released locally immediately upon damage, whereas the larger compounds (5–11; yellow region) are mainly terpenoids induced by larval damage and released several hours later from both the damaged and undamaged leaves of an attacked plant (i.e., systemically). See Figure 16.21 for the chemical structures of a few examples of green leaf volatiles and terpenoids. (C) Responses during two-choice flight tunnel tests by *C. marginiventris* females to corn seedings that underwent various treatments. Artificially damaged plants release the green leaf volatiles in large amounts but only trace amounts of the induced compounds. Addition of the regurgitant to artificial damage causes the plant to release all of the induced compounds in similar amounts as with caterpillar damage. Wasps prefer plants with caterpillar damage over those with artificial damage (Experiment 1), artificial damage with added regurgitant over artificial damage alone (Experiment 2), and artificial damage over regurgitant alone (Experiment 4); they show no preference between caterpillar damage and artificial damage plus regurgitant (Experiment 3). Some wasps did not fly to either plant ("Incomplete flights"). (D) The active component of caterpillar regurgitant is volicitin, a conjugated form of linolenic acid. Application of this compound to artificially damaged plants causes the release of the full complement of induced compounds and full preference by the wasp. (After [10, 623–626].)

(see Chapter 6) [27, 626]. Synthesis of these compounds is induced only after herbivore damage and is mediated by the jasmonate/ethylene/salicylic acid signal transduction pathways. These **herbivore-induced plant volatiles** are emitted in large amounts through the leaves of the entire plant; in other words, their emission is a systemic response. The blend of herbivore-induced volatile chemicals is often specific to the particular herbivore, and thus primarily attracts those predator species that prey on the specific herbivore species. This specificity is achieved by **elicitors**, chemicals in the saliva or oviposition fluids of herbivorous insects, which combine with the damaged plant tissue to induce the volatile products. A compelling example of such a tritrophic (three-species) interaction, in which most of the components have been worked out, is shown in **Figure 16.19**. Other good examples exist [130, 131, 157, 261, 306, 320, 321, 564, 579, 643].

Several lines of evidence indicate that plants have evolved this chemical response for the express purpose of **signaling to predators**, as opposed to passively sending cues through the release of direct defense chemicals [473, 603]. First, the chemicals are emitted only at those times when the predators are active (usually daytime). Second, the chemicals are released systemically in large volumes, not just locally, to deter currently active herbivores. Third, some plants only induce the chemicals when being consumed by certain developmental stages of the herbivore. For instance, early instar caterpillar larvae cause a far stronger release of chemicals, and predation by the parasitoid is not only easier at this stage, but it also more effectively reduces damage to the plant, compared to older instar larvae [602, 693]. Thus plants evolve specific detect-and-signal mechanisms for those contexts in which they can gain a significant benefit. Field studies have shown that plants increase their fitness by calling for help when attacked [147, 291, 344]. A selective herbivore-induced signaling strategy is more effective against herbivores than less specific constitutive chemical defenses, against which insect herbivores can evolve mechanisms to detoxify and sequester the toxins or even co-opt them for their own defense [244, 631, 676]. A highly specific signal blend also has the advantage of preventing the attraction of other predators that attack the desirable bodyguards [239]. Nevertheless, some plants produce a similar chemical blend regardless of the herbivore, and some herbivore species attack several different plant species. In this latter situation, the insect predators must learn to associate the odor with their prey and host plant [124, 156, 583, 604, 641].

Herbivore-induced volatile chemicals may also be employed in communication with other plants. Some plants possess such a highly sectorial structure that systemic signals are restricted to independent stems. In this case, volatile signals arising from local herbivore damage alert other branches of the same plant and prime them for defense, a form of **autocommunication** [278, 279, 327]. Chemical defense signals, targeted either to other branches of the same plant or to insect bodyguards, can be detected by **eavesdropping** neighbors and used to prime their own defenses. Even other plant species can eavesdrop on these signals [27, 323-326, 346]. A few studies have claimed that uninfested plants might detect the presence of nearby herbivory via eavesdropping and then secondarily emit the volatile chemical signal themselves to aid nearby relatives, a putative case of kin-altruistic signaling [329, 358]. Most studies of plant volatile signals examine only tritrophic relationships in controlled laboratory contexts, but natural communities consist of complex trophic interactions among guilds of plants, herbivores, and their predators, which may aid, interfere, or exploit each other. It will be important to address these interactions in natural communities [147].

Attraction of animal mutualists

Many plants rely on mobile animals to disperse their pollen and seeds. A common strategy is to attract the animals with a food reward—nectar or pollen in the case of flowers, and fleshy nutritious fruit in the case of seeds—in order to obtain these **dispersal services**. A few plants offer other rewards such as shelter, breeding sites, and heated roost sites. Both flowers and fruits have evolved *multimodal signals* (visual and olfactory) for the express purpose of attracting their mutualists: nectarivore and pollenivore pollinators for flowers and frugivore seed dispersers for fruits. Like animal mate attraction signals, flower and fruit signals are designed to be maximally conspicuous against the background to desirable **animal vectors** (effective couriers), while minimizing their detectability to undesirable animals (herbivores). We can therefore address many of the same issues we confronted with the evolution of long-distance mate attraction signals in animals, such as whether signal form arises from the exploitation of preexisting receiver biases or by selection to provide reliable information about reward type and quality. But there are some important differences we should keep in mind. Because a plant and an animal vector do not share a gene pool, one party can readily exploit the other without suffering serious consequences, so continued interaction requires stabilizing mutualistic benefits with some sanctions against cheating [350]. A second important difference is the presence of multiple receiver species that may exert different selective pressures on plant signal design and force the plant to optimize and compromise [194, 542]. Plants could opt to produce generalized signals to attract a large number of diverse vectors, or to fine-tune the signal to target a small number of more specialized vectors. All of these points bear on a controversial topic in plant signaling biology, the existence of flower and fruit syndromes—suites of plant species with similar floral or fruit traits held in common because they attract a particular set of pollinator or frugivore species with different body designs and sensory capabilities. We shall touch on this question as we discuss the production of fruit and flower signals, strategies for dealing with mutual receivers, evidence for honest signaling, the role of sensory exploitation, and some fascinating cases of cheating.

Flowers and fruits employ the same classes of *pigments* to produce the color components of their displays. **Carotenoids** are responsible for yellow, orange, and ultraviolet hues, whereas **anthocyanins** produce red, blue, violet, and black hues (**Figure 16.20**). Black is generated by a subtractive mixture of red and blue layers on top of each other [185]. Additional pigments include ivory and white **flavonols** and red **betacyanins**. Fruits are usually unicolored, and their signal and reward derive from the same structure, so the potential for honest signaling of fruit quality is high. Fruits can add accessory structures, such as colored bracts and stems, to amplify their display by increasing background contrast [53]. Flowers, on the other hand, are more variable in color and shape, because they are comprised of ontogenetically differentiated structures that serve multiple purposes. The signaling structures (petals) are separate from the structures containing the rewards (nectar and pollen), so there is no physiological constraint to guarantee signal honesty. Pollinators therefore cannot visually assess reward quality from a distance unless some other mechanism favors reliable signaling [542]. The chemicals responsible for the *olfactory*

(A)

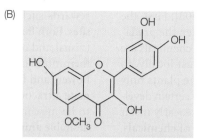

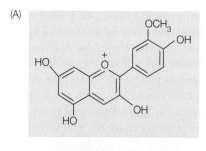

(B)

(C)

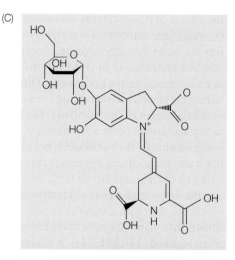

FIGURE 16.20   Plant color pigments (A) Peonidin, a common anthocyanidin responsible for red, violet, and blue colors of fruit and flowers. Different colors are produced by differences in the numbers and locations of hydroxyl and methoxyl groups on the anthocyanidin molecule, and by pH. This pigment becomes anthocyanin when a glucose moiety is attached. Photos show *Vaccinium* spp., blueberries (top) and cranberries (bottom). (B) An example of a flavonol, a class of compounds responsible for the color of ivory, white, and yellow flowers, here *Rhododendron*. (C) Betacyanin, responsible for the particular red color of beets, rhubarb, and flowers such as *Amaranthus caudatus*, shown here. Betacyanins are not related to the flavonoids, but are aromatic nitrogen-containing indole derivatives synthesized from tyrosine with a sugar portion attached. Synthesis is promoted by light. Plants also produce carotenoid pigments, which generate yellow, orange, and red colors (see Figure 4.14).

signaling component of fruits and flowers are much more diverse than the color pigments, and include aliphatics, benzenoids, phenylpropanoids, mono- and sesquiterpenes, aromatic alcohols and esters, nitrogenous compounds, and even sulfuric compounds such as dimethyl disulfide, which smells like a rotting carcass (**Figure 16.21**) [357]. Odor signals have a far greater capacity to generate **private wavelengths** for targeting specific animal vectors [511]. Although chemical signals are often detected at relatively short ranges, in cases of coevolved insect–plant pollination mutualisms such as that of figs and their wasps, the primary attraction signal is an extraordinarily long-range olfactory signal [299].

Flower pollination and fruit dispersal differ in their strategies for dealing with multiple receivers. To achieve

successful cross-pollination, an animal vector must visit at least two different plant individuals of the same species within a relatively short period of time, once to remove pollen and later to deposit it. Plants therefore benefit by attracting animal vectors that are fairly faithful to the plant species, thereby maximizing the deposition and receipt of conspecific pollen while reducing the deposition and receipt of heterospecific pollen. This pollinator behavior is called **flower constancy**, and is facilitated by flowers being easily recognizable and distinctive from other simultaneously flowering species. Selection for species distinctiveness is responsible for the complexity and diversity of floral displays. Selection pressure for faithful pollinators also leads to floral traits that are detectable and selectively accessible to a limited range of animal vectors. This principle is responsible for the so-called classical **pollinator syndromes** shown in **Figure 16.22**. For instance, flowers adapted to hummingbird pollinators tend to have red flowers with a tubular corolla, no scent, and copious nectar. Bee-attracting flowers have violet, blue or yellow flowers with ultraviolet guides for directing the bee to the nectar source. Butterfly-specific flowers have a landing platform and long narrow nectaries that can be accessed with a butterfly's long tongue. In each case, the anthers and stigmas are positioned for advantageous removal and deposition of pollen by the preferred visitor. Fruit-based seed dispersal, on the other hand, does not require such strong **consumer constancy**, because a frugivore that restricts its foraging to a single plant species would tend to deposit the seeds under

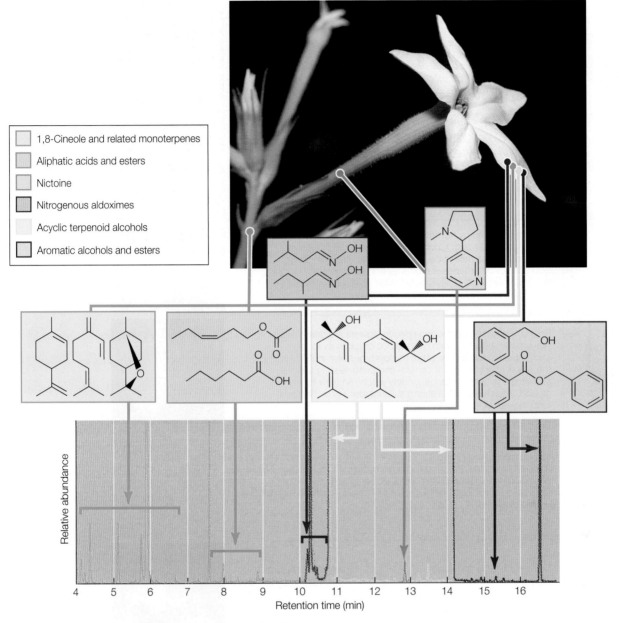

FIGURE 16.21   Diverse chemical components in floral scents This gas chromatograph analysis of the volatile compounds released by tobacco plants (*Nicotiana*) shows 12 different compounds, grouped by chemical class and site of production. During the night, the petals emit highly volatile cineole and related monoterpenes (blue), along with nitrogenous aldoximes (purple), acyclic terpenoid alcohols (yellow), and aromatic alcohols and esters (red). Trichome hairs on the leaves and on the flower's calyx (fused sepals) emit aliphatic acids and esters (green), which function as a constitutive defense against wounding. Nicotine (orange) is a component of the nectar, but is also released as part of the induced wound defense system. (After [511].)

conspecific plants, where seedling mortality is high. Thus most plants benefit from attracting a wide variety of fruit consumers. Selection pressure for diverse fruit dispersers should favor maximal fruit display conspicuousness and convergence of numerous plant species on a smaller number of **fruit syndromes**. For instance, 50–70% of fruits in tropical habitats are either red or black and are dispersed primarily by birds, while the remainder are green, yellow, or orange and are dispersed by primates and other mammals [542, 544, 677, 682]. Even after removing effects of plant phylogeny, comparative studies support the notion of strong convergence on different fruit color syndromes for these two groups of consumers [396]. Primate-dispersed fruits do not contrast with the background as strongly as bird-dispersed fruits, but they are likely to emit a stronger odor signal component.

The occurrence of flower and fruit syndromes is consistent with the notion that plant signals evolved to exploit the receiver capabilities of the most effective animal vectors. However, it is also possible that animal color vision might have evolved to detect flower and fruit signals. Several studies have attempted to resolve the cause-and-effect relationship between plant signals and animal sensory systems. Flowering plants first appeared in the early Cretaceous period about 130

Bee	Flowers are often yellow or purple with UV nectar guides; very variable in size and shape, can be open and bowl-shaped or more complex, with either radial or bilateral symmetry. Pollen, nectar, or both may be offered as reward. The sugar in nectar is predominantly sucrose. Bumblebee on purple *Geranium platypetalum* flower.	
Butterfly	Flowers tend to be large and showy, pink or lavender in color, frequently have a landing area, and are usually scented. Simple nectar guides may be present. Nectaries are usually hidden in narrow tubes reached by the butterfly's long tongue. Fritillary butterfly (*Speyeria cybele*) on purple coneflower (*Echinacea purpurea*).	
Moth	Flowers pollinated by hovering hawk moths tend to be large, showy, and white, with tubular corollas. The flowers open at night and release their strong sweet scent at this time. Plants pollinated by other types of moths tend to have smaller flowers. Hummingbird hawkmoth (*Macroglossum stellatarum*) on star jasmine (*Trachelospermum jasminoides*) flower.	
Fly	For nectar- and pollen-feeding species, flowers are often purple, violet, blue, and white, with open dishes or tubes as for bees above. Such flowers have either no scent, or scent that mimics the fly sex pheromone. For carrion- and dung-feeding species, flowers are brownish, orange, or dirty yellow, and low to the ground, often hairy, with a strong sulfuric odor. Fly on *Rafflesia keithii* flower.	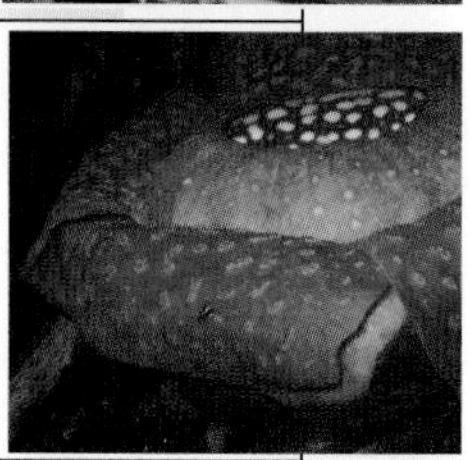
Bird	Hummingbirds hover in front of flower, so flowers are long and tubular, often red or orange. Perching nectarivorous birds require a substantial landing platform. Flowers offer odorless, copious dilute nectar dominated by fructose and glucose. Female ruby-throated hummingbird (*Archilochus colubris*) on cardinal flower (*Lobelia cardinalis*).	
Bat	Flowers are large and showy, white or light-colored, opening at night. They are sometimes bell-shaped, hanging, or located on long stalks for access by hovering bats. They have a strong odor, sometimes based on sulfuric compounds, and especially large pollen organs. Mexican long-tongued bat (*Choeronycteris mexicana*) on *Agave* flower.	
Beetle	Flowers are large, greenish or off-white, flattened or dish-shaped, and heavily scented with spicy, fruity, or decaying-material odor. The plant's ovaries are usually well protected from the biting mouthparts of beetles. Cerambycid beetle on *Trillium ovatum*.	

FIGURE 16.22 Classical pollinator syndromes Descriptions of syndromes and illustrations of pollinators and flowers that fit the traditional scheme. In a large-scale quantitative analysis of flower characteristics and their associated primary pollinator species from six habitats, only about one-third of the plant species fit these categories. Possible reasons for the lack of fit: (1) many plants are generalists visited by several types of pollinator species; (2) color vision capabilities of birds and insects are broad; (3) each animal group contains numerous specialists with different, sometimes unusual, pollinating tactics; (4) some important flower traits may not have been included in the original analyses; (5) floral adaptations might also be influenced by antagonistic floral visitors [194, 470].

million years ago and achieved worldwide dominance about 90 million years ago. Insects existed long before this point (570 mya). The vast majority of insects, and the crustaceans before them, all possess the trichromatic color perception system we see in bees (see Figure 5.26B), with photoreceptor peaks at about 350 nm (UV), 440 nm (blue), and 520 nm (green). Bee- and beetle-pollinated flower colors clearly cluster in the color space defined by this color vision system [100, 101]. Hummingbirds evolved relatively recently, so red

(A)

(B)

(C)

(D)

FIGURE 16.23   Honest signals of nectar reward Young, sexually viable and rewarding flowers are visually distinguishable from old, spent flowers via the appearance or loss of anthocyanin or carotenoid pigments. (A) The central eye of primrose (*Androsace langinosa*) changes from yellow to red. (B) In Spanish flag (*Lantana camara*) the whole flower changes from yellow to red or magenta. (C) In lupin (*Lupinus nanus*) the banner petal spot changes from white to purple or magenta. (D) Central petals of clove currant (*Ribes odoratum*) change from yellow to red.

or nectar **reward quality**? Plants definitely benefit by signaling the availability and quality of their fruit rewards. Immature fruits are an inconspicuous green and lack olfactory bouquets and sweet taste in order to avoid being consumed too early, but once they ripen, their color changes to a contrasting hue and the flesh is charged with sugars and fruity odors. In a survey of 45 Neotropical rainforest fruits, red fruits and black fruits contrasted with the background but the precise hue was not correlated with the amount of carbohydrate, protein, lipid, phenol, or tannin content. However, among fruits of other colors, hue was correlated with nutritive content: saturated orange and yellow fruit signal high protein; saturated blue fruit signal high carbohydrate and low tannin content; and white fruit are intermediate in terms of protein, carbohydrate, and tannin content [543, 545, 562]. In black elder (*Sambucus nigra*), individual variation in fruit quality (sugar content) is correlated with a visual signal—not the color of the fruit, but the color of the pedicels that support the fruit cluster. Red pedicels generate an overall greater contrast between the total fruit cluster display and the green leaf background. The anthocyanin-based color of the pedicels is light-induced, and plants growing in a well-illuminated habitat produce both redder pedicels and fruit with much higher sugar content than those growing in shade [546]. This is an example of an environmentally based quality cue or index signal. As mentioned earlier, flower petal signals are decoupled from the nectar reward and thus are not constrained to provide direct information about nectar quality. Nevertheless, the flowers of numerous plants change color once they have been pollinated (**Figure 16.23**) [462, 670, 671]. These unrewarding flowers still contribute to attracting pollinators to the remaining unpollinated flowers on the same plant. Moreover, some plants imbue their nectar with scents distinct from the floral scent, so that pollinators can remotely assess the presence of nectar in a given flower [509]. These honest nectar signals increase the foraging efficiency of the

flower colors that attract birds probably evolved much later. Hummingbird-pollinated flowers are not specifically tuned to hummingbird color sensitivity but instead appear to be designed for decreased conspicuousness to bees and other insects that have poorer visual sensitivity to long-wavelength color [11, 102]. A nice example of the tight correspondence between flower color and pollinator type was shown experimentally in two sister species of monkeyflower (*Mimulus*). A pink-flowered species is pollinated by bumblebees and a red-colored species by hummingbirds. Switching a single allele between the two plant species caused the pink species to produce darker red flowers that attract hummingbirds and the red species to produce orange flowers that attract bees [60]. The best case for the reversed cause and effect, with animals evolving new color sensitivity capabilities in order to detect plant signals, has been proposed for primates. The evolution from the ancestral mammalian dichromatic system to a trichromatic visual system significantly improved the ability of Old World monkeys and some New World monkeys to distinguish red fruit and leaves from a green background [599] (see also Web Topic 5.3).

Can pollinators and fruit consumers extract reliable information in the form of cues about the quality of the reward plants have to offer, and are there contexts in which plants benefit from providing reliable signals of their fruit

(A)

(B)

(C)

FIGURE 16.24 Sexual deception in an orchid The orchid
Chiloglottis trapeziformis provides no nectar reward but deceives
male *Neozeleboria cryptoides* wasps by mimicking a female wasp.
The deception is both visual and olfactory. Females are wingless.
They emerge from the ground, climb to the top of a stem, and
call by emitting a sex attraction pheromone. Males transport the
females to a nectar source during copulation. (A) The labellum, a
petal specialized for pollination in all orchids, visually resembles
a female wasp (arrow) and emits the female wasp pheromone,
chiloglottone (2-ethyl-5-propyl cyclohex-1,3-dione), shown in
(B). The labellum is hinged, so that when the male grasps it he flies
into the column and pollinates the flower. (C) Male wasp attempting to copulate with a plastic bead infused with the pheromone.
The labellum is larger than a normal female wasp and produces
10 times as much pheromone; males prefer the labellum to a real
female. Once males have been tricked, they learn to avoid the area,
so if they succumb to a second orchid individual some distance
away, the plant benefits from outcrossing [479, 510, 556–558].

pollinators once they arrive by directing them toward the
rewarding flowers and away from the depleted ones, benefiting both the pollinator and the plant [247, 300, 403]. There
are no known examples of flowers signaling the quality of
their nectar in terms of sugar concentration, but inbreeding
depression in some plants reduces the amount and chemical composition of the floral volatile blend and could reduce
visitation rates as well as plant defenses [141, 196].

Many plants do not provide pollinators with any type of
remote signal or cue of the presence of nectar in individual
flowers. This opens up an opportunity for plants to cheat
by not producing nectar in some proportion of its flowers,
thereby conserving a significant amount of energy. In fact,
cheater plants should evolve mechanisms to conceal any cues
of nectar availability. The risk of attempting this cheater strategy is that pollinators may learn to associate a low reward rate
with the particular plant species, and subsequently avoid it

[195]. Experiments with bumblebees found that they switched
to an alternative flower type when the proportion of nectarless
flowers exceeded 40% [586]. Several models have addressed
the problem of optimizing the proportion of nectarless and
nectarful flowers and find that the optimal strategy varies with
conditions such as plant clustering, pollinator density, number of flowers on a plant, and the costs of learning to assess
nectar presence remotely [41, 42, 586, 608]. Field data from
these studies showed that species with a higher proportion of
nectarless flowers had a larger number of flowers per inflo-
rescence and higher local densities, contexts in which pollinators have fewer options and a more difficult discrimination
task. One hundred percent nectarful flowers, along with honest signaling of nectar content, is favored when plants are far
apart, and in this context pollinators restrict their visits to the
rewarding and honest plant species.

In contrast to this relatively common partial cheating
strategy, about a third of all orchid species provide absolutely
no reward to pollinators. The orchid family is renowned for
its large diversity of pollination mechanisms and unusually
high occurrence of **nonrewarding** and **deceptive strategies**.
These strategies include food deception, food-deceptive floral
mimicry of sympatric rewarding species, brood site imitation,
shelter imitation, pseudoantagonism (stimulating defensive
behavior by territorial bees), rendezvous attraction, and
sexual deception [315]. Generalized food deception is the
most common strategy, and these orchids succeed because
they mimic the appearance of sympatric rewarding species
[236]. Sexual deception is the second most common strategy.
The deceived pollinators are typically various hymenopteran species. An example of sexual deception is illustrated in
Figure 16.24. Cheating appears to be the ancestral condition in orchids, and mutualism with real nectar reward for

pollinators has evolved independently numerous times [111, 317, 587]. Contrary to expectations, lineages with cheating and those with true rewards do not differ in extinction rates or speciation rates. Cheating may be most profitable for very low-density orchid populations, where the duped male insect does not encounter the trap very often in relation to real females and thus does not learn to avoid it. Reproductive success for both male and female function in rewardless orchids is often highly skewed, with a few individuals producing a very large number of seeds [212]. Selection on deceptive orchids may be analogous to the evolution of cheating and exaggerated sexual selection traits in male animals [587].

Human Communication

Spoken language sets our species' communication system well apart from the animal communication systems we have been discussing in this book. **Human speech**, characterized by voluntary utterances of arbitrary and learned symbolic sounds that are syntactically ordered and recombined according to grammatical rules, enables us to make strategic plans, imagine the future, negotiate exchanges of goods, cooperate to achieve common goals, and develop loyalties and close social bonds. Animal communication systems show some very rudimentary precursors of these abilities, but they do not provide much insight into the evolution of human speech. Humans also have a rich repertoire of **nonverbal communication signals**, and some of these displays exhibit more obvious parallels with animal signals. Verbal and nonverbal signals are closely integrated. In this section we shall begin with a brief overview of human evolutionary history, behavioral ecology, and social organization, before moving on to discuss nonverbal communication signals and their integration with speech, social interaction, and mating.

Human evolutionary history

Chimpanzees and bonobos are our closest extant primate relatives. Archeological discoveries during the past several decades have enabled us to piece together a good picture of the intervening evolutionary steps [417, 605, 688]. Early hominins diverged from the last common great ape ancestor between 5 and 7 million years ago. Drought conditions at that time reduced the African equatorial rainforest belt and favored the exploitation of the expanding savannah and woodland habitat by ground-dwelling, upright-walking apes. They likely retained the basic social organization of most primates with small, closed, territorial troops. Bipedal locomotion freed the hands to become more dexterous, and simple tool fabrication and cooperative hunting probably evolved along with increasing brain size. Transitional hominoids such as *Homo habilis* appeared about 2.5 million years ago. The highly successful *H. erectus*, which apparently used fire, migrated out of Africa and spread throughout the Old World around 1.8 million years ago. Neanderthals (*H. neanderthalensis*) arose about 400,000 years ago and moved into Europe. This robust species was primarily, although not

entirely, carnivorous [286]. Using sharpened stone-arrow-tipped stabbing spears, males and females cooperated in hunting large mammals [367]. They made a variety of sharp stone hand tools for butchering the carcasses. Neanderthals likely lived in small closed troops within a confined area and did not interact much with other troops [372]. Their lifespans were short, about 35–40 years, owing to their dangerous mode of hunting, a periodically harsh climate, unpredictable resources, and apparently high levels of interpersonal violence. There is evidence that they practiced cannibalism [135, 628]. They sometimes buried their dead, but did so with little or no ritual. Neanderthals used body paint, but did not fabricate personal ornaments or produce symbolic artwork [698]. Although their brains were as large as ours, and they were clever tacticians while hunting, they showed no evidence of innovation or long-range planning, basically using the same kinds of tools throughout their 350,000 year existence. They must have possessed a repertoire of voluntarily produced gestures and vocalizations, and possibly a declarative, imperative, and exclamatory mode of speech, to coordinate hunting activities and teach these skills to the young, but they probably lacked a syntactical speech system [436, 689].

Anatomically modern *Homo sapiens* appeared in Africa about 200,000 years ago. They had an omnivorous diet and both hunted game and gathered plant foods, much like the earlier *Homo* ancestors. Archeological sites of *H. sapiens* dating from 150,000 years ago to the present show a steady increase in technology and in complexity of social organization [107, 201, 415]. Tools became more sophisticated and included axes, hammers, bone tools, barbed points, arrows, and needles. Dwelling sites were structurally organized for different tasks, and there is evidence of planned land use, which eventually led to agriculture. These sites also contained fabricated multicomponent artifacts such as clothing, personal ornaments, shell beads, abstract figurines, and cave artwork, which we associate with symbolic thinking. New modes of hunting were developed, such as shellfish hunting, fishing, projectile arrow hunting, and the use of animal snares. Group size increased (the **troop-to-tribe transition**) and a hunter-gatherer lifestyle with sexual division of labor was adopted (males hunted, females gathered). This lifestyle is associated with a high rate of interaction among groups, including **seasonal gatherings** and **trading networks** with exchange of information, technologies, and goods [369]. Evidence for trading networks as early as 140,000 years ago exists in the form of shell beads and tool materials that clearly originated from sites quite distant from where they were found [415, 635]. All of these adaptations created a more stable and secure environment for raising offspring and increasing adult longevity, and thus the human population flourished. Between 90,000 and 60,000 years ago, there was a major migration period of *H. sapiens* out of Africa into the Levant (Middle East), which later split into European and Asian migrations. The latter eventually spread across the Bering Strait land bridge to the Americas [31, 424]. *Homo sapiens* overlapped with Neanderthals in the Middle East and later in Europe, probably replacing or interbreeding with them and any other remaining hominins or humans from earlier migrations [254, 515, 653]. Modern human races diverged during this brief period, as evidenced by the generally high level of genetic homogeneity among the races and the gradually increasing genetic distance of human populations as a function of distance from Africa [33, 136, 303, 319, 354, 513, 516, 607].

These technological and social organizational changes required a symbolic language that could express abstract concepts and refer to past and future events. The key to this advanced level of cognitive thinking and speaking was the evolution of **enhanced short-term working memory capacity** in the brain of early *Homo sapiens*. Working memory is comprised of three components: (1) the *executive decision-making center* in the dorsolateral prefrontal cortex that controls attention and maintains relevant stimuli in the face of interference; (2) the *phonological loop* between the cortex and auditory centers that stores short-term verbal and acoustic stimuli, rehearses this input, and transfers it to long-term memory; and (3) the *visuospatial sketchpad* that processes and stores spatial location and orientation input [23]. These centers together allow an individual to receive and process long combinatorial strings of abstract sounds along with visual stimuli, to hold and store this input, to evaluate it in relation to previously acquired information, and to retrieve and articulate it later [123, 393]. This **mental time travel** facilitated long-range strategic planning and enhanced collective planning through information sharing with others. It also promoted numerous other distinctively human activities: the design and fabrication of multicomponent artifacts and interpretive artwork; the establishment of interaction networks and social bonds with non-kin based on reciprocity and trust; the teaching of language, technology, and social skills to children; experimentation and innovation to find new solutions to problems; and the development of population-specific cultures, mores, and rituals [4, 14, 199, 518, 533, 690].

The maintenance of successful relationships with others depends not only on verbal exchanges, but also upon the ability to send and receive nonverbal signals. Human nonverbal communication spans all the major modalities: visual signals such as facial expressions; body postures and hand movements; gazing; acoustic signals such as cries, shrieks, and various speech modifiers; olfactory cues and signals; and tactile signals. Nonverbal signals and cues are pervasive and omnipresent in our social lives. By some measures, more than half (60–65%) of the meaning in our communication exchanges is nonverbal, although this varies with context, receiver age,

TABLE 16.1 *Gestures*

Gesture category[a]	Definition	Examples
EMBLEMS	Intentional signals with a clear symbolic meaning and direct verbal translation can substitute for the words they represent. Understood by all members of a social group, sometimes specific to a culture.	• Hand wave: "Hello, here I am" • Beckoning with palms up, curling in toward body: "Come here" • Two-finger "V" sign: "Victory" • Shoulder shrug: "I don't know" • Stomach rub: "I'm hungry" • Index finger drawn across throat: "Death"
ILLUSTRATORS	Movements accompanying speech used to illustrate what is being said, trace direction of speech, set the rhythm, and hold the listener's attention. May accent, complement, repeat, or contradict what is being said.	• Pointing with head, mouth, or index finger • Pictorial gestures bearing resemblance to object being described • Pictorial gestures representing abstract concept • Hand movements following rhythm of speech • Enumerating points with serial finger-pulling
REGULATORS	Behaviors designed to maintain or regulate turn-taking during conversations. Generally not intentional, but almost involuntary, highly overlearned habits. Rules are very strict; failure to follow them is considered rude.	• Turn-yielding by gazing, turning toward, reducing or increasing intonation, pausing • Turn-denying by not pausing, using connecting words like "and" • Turn-requesting by raising hand, pointing index finger up, breathing in and straightening back • Backchanneling signals to encourage speaker, such as nodding, smiling, "mm-hmm," "right"
ADAPTORS	Behaviors that satisfy physical or psychological needs, manage emotions, help individuals adapt to stresses. Not intended to communicate but do reveal sender's internal state.	• Covering/touching mouth, ears, or head • Scratching, playing with hair, working the jaw, biting lip • Cleaning or wiping dirt from another person • Manipulation of objects such as pencil • Leg movements, kicking, foot tapping
AFFECT DISPLAYS	Displays that reveal emotions, mainly with universal facial expressions and associated vocalizations and body gestures.	• Expression of positive emotions with smiling, laughing, hugging • Expression of negative emotions with frowning, clenching fists, screaming, crying

Source: After [170].

[a]Definitions and examples of the main categories of gestures involving hand, arm, face, head, and body movements. Affective displays are described in greater detail in Table 16.2, and postural signals are illustrated in Figure 16.26.

gender, and individual personality differences [356]. Nonverbal communication is the first form of communication we employ as infants, when we interact with our caregivers through nursing, grasping, rocking, holding, crying, cooing, singing, and eventually recognition of faces and facial expressions. Nonverbal cues and signals also precede verbal communication in the first minutes of interacting with a new individual; our first impressions are formed before we open our mouths to speak. When verbal and nonverbal information conflicts, adults are more likely to believe the nonverbal information. Young children actually rely more on verbal information [76]. This prevailing faith in nonverbal communication likely occurs because of the phylogenetic inheritance of such signals from our primate ancestors. We may trust the authenticity and honesty of nonverbal signals more because we intuitively know that talk is cheap and easily twisted from the truth. Nevertheless, nonverbal signals can be recruited to deceive, and decoders must be very sensitive to subtle cues to extract useful information. Humans have an estimated 1,000 different postures; 6,000–8,000 distinct gestures; and 20,000 facial expressions [51, 289, 363, 364, 480]. **Table 16.1** lists the main categories of human gestures—emblems, illustrators, regulators, adaptors, and affect displays—which we refer to in subsequent sections. Below we shall describe the nonverbal signals used to express emotions and dominance and review how people evaluate the veracity of a speaker. We conclude with a discussion of multimodal human courtship behavior and mate choice.

Emotional expression

Humans are very emotional primates. **Emotion** is notoriously difficult to define precisely [110, 177, 209, 443, 538, 552]. A dictionary defines emotion as a mental state or feeling that arises spontaneously in response to certain events. As Charles Darwin first noted in his insightful book on this topic, different emotional responses have evolved via natural selection in both animals and humans because of their adaptive benefits [118, 181]. Since then, investigators from a wide range of disciplines, including psychologists, philosophers, neurobiologists, clinical practitioners, and behavioral ecologists, have contributed theories and evidence for the mechanisms and functions of emotional responses. Emotions stimulate cognition and prioritize information so that people behave in ways that ensure survival and help them get along with others. A more scientifically rigorous definition has emerged, which defines emotions as episodic, short-term, biologically-based patterns of perception, experience, physiology, communication, and action that occur in response to specific external events [339, 386, 552]. The **component process approach** to emotions, illustrated in **Figure 16.25**, encompasses many of the earlier theories. The point here is that the perception and appraisal of certain events triggers a complex of neurophysiological, neuroendocrine, cortical, and somatic motor patterns that produce the feelings (called **affects**), expressions (cues and signals), and motivated actions characteristic of the various emotions [537, 553]. Cognitive appraisal plays a central role in both evaluating the importance of external events and regulating the expression of emotions [552].

FIGURE 16.25 The component process approach to emotional experience Emotional feelings and their associated actions are triggered by a stimulus event, usually an external event such as a natural phenomenon or the behavior of another individual that has significance for one's well-being. The event can also be an internal stimulus, such as a physiological change or memory image. This event is then cognitively evaluated and appraised for its relevance and consequences for the salient needs, desires, or goals of the appraiser. Four additional organismic subsystems are simultaneously activated, each of which can lead to observable responses. The neurophysiological component integrates the central nervous system, neuroendocrine system, autonomic nervous system, and somatic nervous system, and can cause various bodily symptoms (e.g., heart rate changes) and physiologically-driven cues (e.g., pupil dilation). The motivational component is involved in the executive function of the central nervous system and results in taking relevant actions. The motor expression component controls the somatic nervous system and results in stereotyped facial and vocal expressions for communication purposes. The subjective feeling component monitors the internal stage and leads to the sensation of emotional experience or affect. (After [386, 552].)

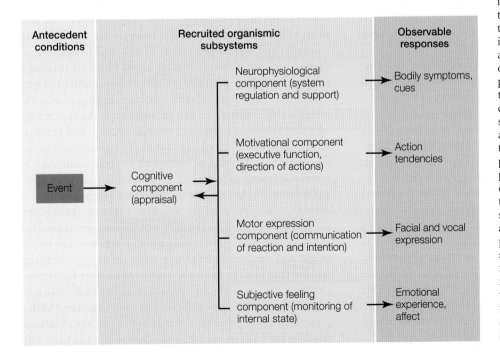

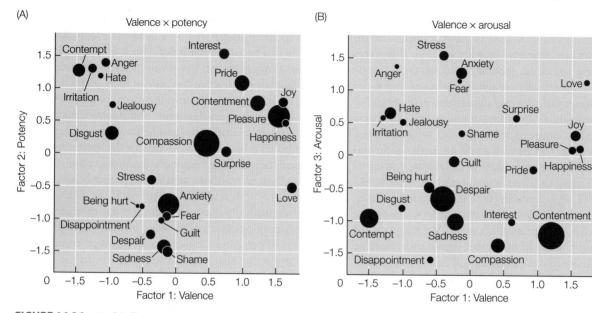

FIGURE 16.26 Multi-dimensional categorization of emotions Results of a quantitative analysis to determine the number of dimensions needed to describe the range of emotional experience. Subjects in three countries were asked to rate 24 common emotions on the basis of a panel of 144 emotional features. These features were derived from the components shown in Figure 16.25 and included types of facial, vocal, and gestural expression, types of appraisal, bodily experiences, subjective feelings, actions taken, and sense of control over the situation. The rating scores were subjected to a factor analysis, which finds the dimensions of greatest variance in the data set and represents each emotion by its coordinates along these dimension axes. The two graphs here show the positions of the 24 emotions along three of the dimensions (valence, potency, and arousal factors) two at a time. Circle diameter indicates degree of variation among samples from different countries. (Axes here have been transformed from the original analysis so that larger valence values correspond to positive emotional valence, and larger potency values correspond to greater control.) The fourth dimension, unpredictability (not shown here), primarily separated surprise from the other emotions. (After [202].)

Conveniently, most languages have specific words for different emotional affects, so we can ask people to describe what they are feeling and relate these self-reported emotions to specific antecedents, physiological responses, expressions, and actions [110, 210]. While animals have only a few basic emotions (anger, fear, subordinance, frustration, and distress; see Chapter 10 and Web Topic 10.2), humans have 6–28 basic emotions depending on the authority [146, 176, 178, 202, 312, 531]. There are numerous ways to categorize them. The simplest model arranges the emotions around a two-dimensional wheel, where the orthogonal axes are **valence** (pleasantness of the emotion, or positive versus negative emotions) and **arousal** (degree of activation produced by the emotion [692]. However, at least one more dimension is needed to fully separate out the emotions, namely **potency** (degree of control over the situation) [470], and some studies suggest a fourth dimension, **unpredictability** [202]. **Figure 16.26** shows the results of a quantitative analysis in which the number of dimensions was allowed to emerge from the data. Four dimensions were found, with the following order of importance: valence, potency, arousal, and unpredictability. A similar type of study investigated quantitative measures of the acoustic properties of speech given by subjects experiencing different emotions [379]. This study found three main dimensions, and the acoustic properties associated with the main emotions have been incorporated into the summary below (see Table 16.2).

AUDIOVISUAL SIGNALS Whether and how we communicate emotional affects depends on several factors. Only some emotions are associated with definitive signals, presumably because there is a selective advantage to revealing one's current feelings. Seven emotions—**fear, anger, happiness, sadness, surprise, disgust,** and **contempt**—are signaled with characteristic facial expressions (**Table 16.2**). These facial expressions are universally recognized by all cultures and occur in response to specific universal antecedent events [177, 531]. For example, sadness occurs in response to the loss of a significant other, fear in response to the appearance of a threatening predator or human enemy, and surprise in response to an unexpected event. These expressions are involuntary and innate, as evidenced by their occurrence in infants or young children and in adults who have been blind since birth and lacked any opportunity to learn by observation [166, 173, 176, 182, 207, 216, 311, 425]. Moreover, the facial muscles critical for expression are specialized for rapid movements by possessing a high fraction of fast twitch fibers, and they are under strong selection to be present and symmetrical in all individuals, compared to other facial muscles that vary more among and within individuals [246, 525, 591, 656]. Each emotional facial pattern is associated with a prescribed set of physiological manifestations, vocalizations, acoustic patterns of speech, and whole-body gestures, also

TABLE 16.2 *Affect Displays*

Facial expression[a]	Physiological effects[b]	Acoustic components[c]	Gestures[d]
FEAR Eyes widen and *upper lids rise*, as in surprise, but *brows are drawn together*; *lips stretch horizontally*	• Vasoconstriction, blanching • Piloerection/goosebumps • Pupil dilation • Bulging eyelid muscles • Sweating, clamminess • High heart rate • Low skin temperature	• Vocalization: screaming, whimpering, moaning • Speech modulation: loud volume; fast tempo; pitch level high if in attack mode, quiet if in escape mode; small pitch variation	• Crouching posture, turned away from object • Head sunken between shoulders • Hands twitching, opening and closing
ANGER Both lower and upper eyelids tighten and brows lowered and drawn together; intense anger also raises upper eyelids. The jaw thrusts forward, *lips press together*, and *lower lip may push up a little.*	• Vasodilation, reddening, bulging blood vessels • Salivary gland activation, foaming at the mouth • Piloerection/goosebumps • Pupil constriction • Bulging eyelid muscles • High heart rate, skin temp • Increased adrenalin, noradrenalin	• Vocalization: growling, yelling, shouting • Speech modulation: loud volume; fast tempo; pitch level high and rising if frustrated, low if annoyed or threatened; small pitch variation	• Erect posture, shoulders up, chest expanded • Head facing toward • Arms stretched up, bent against chest, or rigid against sides • Fists clenched, pointing • Rapid gait, pacing, stomping feet • Trembling, high movemen movement dynamics
HAPPINESS The corners of the mouth lift in a smile. As the eyelids tighten, the *cheeks rise* and the *outside corners of the brows pull down.*	• Lacrimal gland activation plus contraction of orbicularis occuli, causing eye twinkling • Reduced heart rate	• Vocalization: laughter • Speech modulation: moderately loud and varied; fast tempo; moderately high pitch; large pitch variation	• Erect posture, thrown backward with extreme laughter • Head upright • Jumping, clapping, dancing, high movement dynamics
SADNESS The eyelids droop as the *inner corners of the brows rise* and in extreme sadness *draw together*. The corners of the lips pull down and the lower lip may push up in a pout.	• Lacrimal gland activation, tearing and crying • High heart rate • Reduced skin temperature	• Vocalization: crying, sobbing, wailing • Speech modulation: low volume; slow tempo; high pitch if crying; low monotone pitch if speaking	• Collapsed posture • Head hanging on chest • Motionless, low movement dynamics
SURPRISE Eyebrows rise and jaw drops open; *upper eyelids rise*	• Reduced heart rate	• Vocalization: rapid intake breath, upward inflected "oh" • Speech modulation: fast tempo; high pitch level	• Erect posture • Head tilted back slightly • Arms stretched out frontally • Lateralized movements
DISGUST The nose wrinkles and the upper lip lower rises while the *lower lip protrudes*	• Salivary gland activation, drooling • Reduced heart rate	• Vocalization: downward inflected "eeew" or "ick" • Speech modulation: slow tempo; small pitch variation with rising pitch	• Collapsed posture • Shoulders backward or forward • Head downward • Arms crossed in front of chest or pushing away object
CONTEMPT The only asymmetric facial expression, one half of the *upper lip tightens upward*	• Reduced heart rate	• Vocalization: "humpf," "pah" • Speech modulation: slow tempo	• Turning whole body away from object • Snapping fingers • Low movement activity

Sources: After [62, 173, 387, 565, 654].

[a]Each of the seven basic human emotions—fear, anger, happiness, sadness, surprise, disgust, and contempt—is expressed with a specific facial expression. Italicized components are moderately or very difficult to fake.

[b]The autonomic system manifestations associated with each emotion.

[c]Specific vocalizations and typical modifications of speech associated with each emotion.

[d]Body postures and gestures associated with each emotion.

listed in Table 16.2. The linkages between emotions, physiological changes, and expressions are tight and **bidirectional**. Not only does the physiological state of an individual drive the production of unbidden facial motor patterns, but if a neutral person is asked to enact a particular facial expression, they also report feeling that emotion and show some of the associated physiological manifestations, a phenomenon called **facial feedback** [173, 387]. Emotional expressions are characterized by quick onset and brief duration. This feature distinguishes emotions from moods, which persist longer and have less specific intentional antecedents. Finally, receiver recognition of facial expressions and some of the key vocalizations, such as screams and cries, is hard-wired in the brain and exceptionally rapid (150 msec), implying the presence of specialized **emotion-detecting systems** [168, 541]. Many of these facial expressions have clear precursors in our closest primate ancestors, especially chimpanzees and bonobos [83, 132, 235, 474, 476]. Loss of facial hair and specialized facial muscle structures and innervation in these ancestors evolved to support the production of similar signals [475, 642].

Other emotions are similarly communicated with multimodal mixtures of nonverbal signals. For example, **embarrassment** is signaled by blushing while smiling and averting the gaze; **pride from victory** with raised arms; **liking/interest** with pupil dilation; **awe** with widened eyes; **frustration** with lashing out; **anxiety** and **stress** with furrowed brows and tense body postures; **disliking/hate** with intense staring and constricted pupils; **relief** with sighing and a relaxed posture; and **shame**, **remorse**, and **guilt** with slumped shoulders and bowed head [12, 337, 338, 412, 427, 617]. **Pain** is not considered an emotion because, like hunger and thirst, it is initiated by internal homeostatic conditions, but pain is clearly signaled with a distinctive facial and vocal expression. Interestingly, some very strong emotions have no particular signal but clearly lead to highly motivated behaviors. These include **jealousy**, which induces mate-defense behaviors, and **envy**, which may lead to aggression. Feelings of **regret** lead to doing something differently, **love** and **infatuation** to romantic acts, and **hope** to positive thinking [177, 531]. Societies have different cultural norms for managing and regulating affective expressions, a phenomenon called **cultural display rules**. For example, in the United States, boys are taught to hold back tears if they are injured, and women often learn that they will face negative consequences at work if they display negative emotions too aggressively. In **collectivist cultures**, such as most Eastern societies and some South American countries, people particularly value harmony, loyalty, and tradition and discourage or mask the expression of negative emotions such as anger, contempt, and disgust. This suppression probably operates by modifying the perception and appraisal component of emotional expression [313, 387, 410, 411, 554]. On the other hand, **individualist cultures**, including most Western societies, particularly value personal space, autonomy, privacy, and freedom, and encourage the right to express oneself verbally and emotionally.

FUNCTIONS   As noted above, the function of these emotions is to *adaptively mobilize* the organism to deal quickly with important interindividual interactions. The physiological effects of some emotions are designed to prepare an individual for certain types of directly beneficial actions. For example, the fear response causes the release of adrenalin, increased heart rate, shunting of blood to the peripheral muscles, and excess perspiration to support fight-or-flight behavior. Similarly, emotions with no associated signal nevertheless strongly motivate people to take future fitness-enhancing actions. For those emotions with a closely linked facial expression or other signal component, the benefit to the sender involves influencing receivers to respond in ways that benefit the sender. As mentioned earlier, human receivers are extraordinarily sensitive to facial expressions and certain vocalizations. In addition, humans, primates, and some birds possess **mirror neurons** that fire both when an individual performs a particular act and when it observes or hears the same action performed by another [527]. This system is believed to facilitate learning by imitation, including vocal and language learning as well as action pattern learning. Similarly, when a receiver observes an emotional facial expression by a sender, it tends to perform the same expression [37, 151, 606]. The mirror system, combined with the facial feedback phenomenon described earlier that induces affective feelings merely by making the expression, leads to **synchronized dyadic sharing** of the emotion (i.e., putting one's self in someone else's shoes). By vicariously experiencing the sender's emotion, the receiver may understand the sender better and offer comfort, empathy, or sympathy [7, 90, 211, 347, 474, 497]. Thus a sad, despairing, anxious, psychologically hurt, or injured sender may benefit from encoding a facial expression linked to his/her affective state by eliciting support from the receiver [242].

Sharing of positive emotions can signal mutual liking between dyads and foster friendships and bond formation. For instance, a person who is attracted to another individual involuntarily opens his/her pupils wider, and this signal of interest may make them more attractive to the receiver who in turn widens his/her pupils [288]. Emotion sharing also operates at the group level through the **emotional contagion effect** [36, 272]. Sharing of positive emotions among working groups of people via happy facial expressions enhances cooperation and improves group efficiency, productivity, and a sense of belonging [336, 630]. Likewise, sharing of negative emotions such as anger, contempt, and fear can mobilize and synchronize cooperative defensive actions against targeted enemies. Chimpanzees and bonobos exhibit some of these behaviors, including dyadic emotion sharing, emotional contagion, reconciliation, consolation, and simple affective empathy, and they can even collaborate to achieve a shared goal (joint territorial aggression), but they do not exhibit **cognitive empathy** and the ability to jointly create or negotiate shared intentions and goals. This requires a **theory of mind**, the ability to attribute mental states—beliefs, intentions, desires, and knowledge—to oneself and to others and to understand that others have beliefs, desires, and intentions that are different from one's own [132, 362, 434, 435, 474, 496, 497].

The unique human ability to create shared goals, along with spoken language, is responsible for the phenomenon of **human cultures**. Human tribes and societies develop **cultural rules**, **mores**, and **rituals** to foster group cohesion, loyalty, and mutually beneficial cooperation [57]. Some uniquely human emotions are the consequence of cultural rules [456]. Critical for achieving and maintaining cooperation is the **punishment** of transgressors, defectors, and free-riders [56, 58, 227, 285, 506]. Several unique emotions have evolved in humans as a consequence of this socially imposed cost. Numerous studies have shown that the punishment received by a person who has been caught cheating is significantly reduced if they express signals of embarrassment, shame, guilt, and remorse. For example, embarrassment can be experimentally induced by requiring an adult to suck on a pacifier. Observers respond with expressions of surprise, ridicule, contempt, or laughter, making the subject feel ashamed and belittled. If the subject blushes, they are far more likely to be consoled, forgiven, and reassured of their value and trustworthiness than if they do not blush [127, 337, 338]. Likewise, the nonverbal expression of shame and remorse by an individual found guilty of a serious moral transgression leads to some degree of sympathy and a lighter punishment [126, 337, 427]. The flip side of punishment for cheaters is **reward** for extreme altruists [57]. Altruists are honored as heroes, accorded a high status or prestigious reputation, and looked up to with expressions of awe. They are also likely to receive gifts and aid from others themselves, a cooperative process called **indirect reciprocity** [461]. Rules for **fair trading** and **reciprocity** select for honest signals of trustworthiness, such as genuine smiles and the offering of gifts and donations. Recipients of a gift then feel **beholden** to return a similar favor to the gift-giver, and tend to feel **guilty** if they fail to reciprocate [622].

Dominance, power, and influence

While the emotional interactions between humans described above can be viewed as a *horizontal* dimension of interpersonal relationships, interactions related to dominance, power, and status represent a *vertical* dimension of interpersonal relationships. In our discussion of dominance relationships in animals, we learned that individuals establish their status, or position in the hierarchy, largely on the basis of their ability to win aggressive encounters in competition for access to resources. More dominant individuals are those with greater fighting ability and motivation. Humans can certainly display aggression; fighting occurs both within and between groups, and hierarchies exist within groups. However, one's status is not necessarily determined by one's fighting ability. Rather, **social power** largely regulates vertical interactions among individuals in humans. Power is defined in terms of three specific abilities: (1) to do what one wants without interference from others, (2) to influence other people to act in ways

that benefit oneself, and (3) to resist the attempts of others to influence oneself [78, 81]. People with greater power tend to control valuable resources, such as time, money, and affection. According to the classic taxonomy of vertical relationships, there are five sources, or bases, of power [208]. **Coercive power** derives from a person's ability to punish others or take away a valuable resource, such as a boss who can demote an employee or a dominating husband who controls his wife's activities and access to funds. This type of power can sometimes result in aggression toward and abuse of subordinates. **Legitimate power** is based on the notion that a person has the vested right to influence others, who are then obligated to comply. Chiefs, policepersons, and judges are good examples. **Reward power** is the ability to dispense benefits or rewards to others beneath one in the hierarchy. For example, supervisors can give raises and promotions to employees; teachers can give good grades to pupils; and parents can give love, affection, and material goods to their children. **Expert power** arises from possession of specialized skills and knowledge. Often the most knowledgeable person in a workgroup becomes the de facto leader of the group, and an individual in possession of critical information similarly is perceived as having power. Finally, **referent power** is the influence granted to some people because they are liked, respected, and admired by others. **Credibility** can enhance the perception of all of these forms of power; a person who demonstrates truthfulness, accuracy, and follow-through of promised rewards or threats is taken more seriously, and can therefore exert more influence and obtain greater compliance than a person of low credibility [455].

Dominance in human societies refers to communication behaviors that are effective in exerting power and influence. Aggressiveness and strength play minor roles in the characterization of a dominant individual, although male–female interactions are affected by this asymmetry [67, 282, 283]. More commonly, dominant individuals exude confidence, poise, and assertiveness, and develop the **social skills** to communicate these traits more effectively. Specific nonverbal signals and cues expressed by dominant individuals include an open body posture, relaxed facial expression, the use of hand gestures, a close approach to targets, a direct gaze, and conversational control (by interrupting more often and speaking more, faster, and louder). Subordinate individuals, in contrast, show more closed body positions, limited use of space, low gazing rates, and shorter turns at speaking [44, 78, 79, 161, 264]. A very robust difference between dominant and subordinate individuals in dyadic interactions is the **visual dominance ratio**: dominants more often look directly at the subordinate while speaking but look away while listening, whereas subordinates do the reverse: look at the dominant while listening and look away while speaking [155, 184, 359]. Some communication strategies depend on the type of power base. Individuals with high referent power tend to be very charismatic, dynamic, enthusiastic and expressive, and communicate these traits by gesturing actively while speaking, smiling broadly, nodding and shaking the head while listening, and initiating

FIGURE 16.27 Common body postures and gestures (A) Open posture with relaxed stance, arms spread, palms up, indicates positive attitude. (B) Closed posture with arms and legs crossed indicates disagreement or defensiveness. (C) Slumped shoulders, head down, hands in pocket, indicates depression. (D) Hands on hips, legs apart, indicates readiness to take action. (E) Leaning forward, gazing at speaker, indicates great interest. (F) Head resting on heel of hand, slouched posture, looking down or away, indicates boredom. (G) Sitting with tense posture, symmetric arm and leg positions, ankles crossed or feet pressed together, hands together or clasped, indicates anxiety or nervousness. (H) Sitting relaxed, with one leg crossed, hands behind head, ultimate signal of dominance and control. (I) Hand on cheek, gazing at speaker, indicates interest and evaluation. (J) Hand on forehead indicates forgetfulness. (K) Hand on chin indicates negative thoughts, and fingers over mouth indicates that either the speaker or the receiver might be lying. (L) Steepled fingers indicate self-confidence or self-satisfaction [307, 459].

activities [78, 79]. At the other extreme, individuals with high coercive power may threaten with a glaring stare and unsmiling facial expression, and assume a more erect, tense, body posture [8, 332]. Furthermore, greater height and lower voice pitch are perceived as indicating greater power [264]. Height has a direct effect on behavior and confidence. A clever experiment using digital avatars found that participants assigned to taller avatars behaved more confidently in a negotiation task than those assigned to shorter avatars [691]. Taller individuals also earn higher salaries [270]. These threatening and body-size-related signals and cues are obviously inherited from our primate relatives and still play a role in interpersonal interactions. **Figure 16.27** shows some of the body postures and gestures employed to express dominance, interest, confidence, tension, and negative thoughts; further examples are shown in Web Topic 16.3.

As mentioned above, conversational control is one subtle way that dominance interactions may unfold. **Conversation management** is an extremely important facet of human life that is very precisely regulated by a series of nonverbal signals and rules [340]. Conversations are preceded by a **greeting**. Typically an individual sights another individual he or she wishes to meet, waits for that individual to return the gaze, and one or both may give an eyebrow flash greeting signal, head dip, or hand wave from a distance. This interaction initiates smiling and approach, and upon closing the distance gap to less than 10 feet, a tactile salutation gesture such as a handshake or embrace concludes the greeting [167]. Conversation then begins by each individual taking turns speaking. **Regulators** (see Table 16.1) organize the verbal traffic during conversations. **Turn-taking** is managed with signals by both individuals [24, 162, 322, 532, 539, 679]. To maintain possession of the floor, speakers use **turn-suppressing** signals such as audible inhalation of breath, continuation of a gesture, facing away, sustained intonation, and connecting words like "and." To yield the floor to the other individual, the speaker can terminate gesturing, face the receiver and make eye contact, and either raise the intonation if asking a question or lower the intonation if making a declarative statement. Slowed tempo and silence also signal **turn-yielding**. Listeners can give **turn-requesting** signals, such as raising the hand in a classroom context, gazing at the speaker, nodding the head, leaning forward, raising the index finger, or inhaling the breath while straightening the back. **Turn-forfeiting** signals function to reject the partner's offer to take the floor. **Backchanneling** gives the listener a way to tell the speaker to continue speaking, and includes nodding or shaking the head, smiling, frowning, raising eyebrows, or uttering short vocalizations such as "uh-huh," "right," and "I see." Floor exchanges are smoothest when speakers display many turn-yielding signals, and multiple displays also make listeners more likely to take the floor [162]. Conversations are terminated with verbal expressions of supportiveness, summarization, and closure, along with nonverbal signals such as leaning forward, moving the legs and trunk, standing up, slapping the thigh, breaking eye contact, and shaking

hands [334, 355]. As with facial expressions, two conversing individuals may show convergence and matching of behaviors cause by one or both parties adapting to the style of the other (body posture feedback and mirroring) [77, 351]. But a dominant individual can control the conversation by talking more, initiating topic changes, and interrupting the other person [161, 190, 309, 468]. Interruptions and overlapping in some contexts are violations of the rules and lead to a variety of verbal and nonverbal strategies by the interrupted person to resolve the disruption and resume the flow, including laughter, nodding, self-touch, and talking faster; in affiliative contexts, interruptions can indicate excitement and participation [50, 77, 191, 457, 548, 684].

> **Web Topic 16.3** *Body language*
> Postures and movements are an important component of human communication. This online unit summarizes attempts to quantify these signals and the perception of them by receivers, focusing on the use of computer animated models.

Honesty and deceit

We humans readily form impressions of other people based on verbal and nonverbal information, and we also try to manage our own image to affect how others view us. Given the strong drive to build a positive reputation and increase power, and the fairly automatic responses by receivers to emotional facial signals described earlier, what prevents senders from sending false signals to manipulate receivers? Deception is widespread—from one-quarter to one-third of all conversations involve some form of deception [81]. The arms race between manipulating senders and mind-reading receivers envisioned by Dawkins and Krebs [122] (see Chapter 10) is probably more relevant to human communication than to animal communication. Although people look to nonverbal signals, especially facial expressions, body language, and voice cues, to evaluate the veracity and sincerity of spoken words, these signals can also be recruited to shape a deceptive image. Some expressions are easier to fake than others. Facial muscles, like peripheral muscles elsewhere in the body, are innervated by two motor nerve tracts: the **pyramidal tract**, which originates in cortex brain nuclei and connects via the spinal chord to motor neurons that directly control voluntary movements including speech; and the **extrapyramidal tract**, which originates in subcortical nuclei and modulates motor neuron activity. This latter system is largely involuntary and indirectly regulates complex patterns of movement such as coordination of locomotion, postural control, and emotional expressions. The lower half of the face is under stronger voluntary and contralateral control (allowing independent movements of each side of the face). The upper half of the face is less controllable in general and innervation is more bilateral (both sides of the face fire synchronously) [525]. It turns out that each of the basic emotional expressions except surprise has at least one critical facial movement component

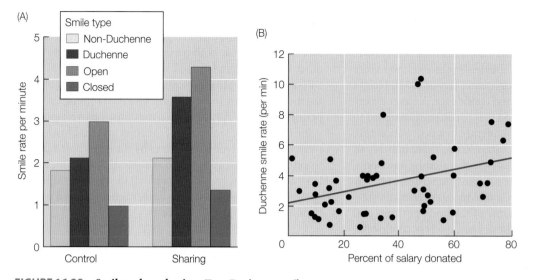

FIGURE 16.28 **Smiles when sharing** True Duchenne smiles may be honest signals of altruistic dispositions. In an experimental study, 48 pairs of friends were covertly filmed while interacting in two contexts, a control context in which they conversed to decide on pseudonyms of famous people that they would use during the experimental context, and an experimental context in which they were given an amount of money and told to decide how to divide it between them. After each conversation context, the participants filled out questionnaires to self-report their emotional feelings, and after the second experimental context, each participant also reported how much of their salary they would be willing to give to a friend in need. The number of Duchenne versus non-Duchenne, and openmouthed versus closemouthed smiles were scored from the videos taken during both conversation periods. (A) Both open and Duchenne smiles were given at higher rates during the sharing period compared to the control period, while non-Duchenne and closed smiles were given at similar and lower rates in both contexts. (B) Individuals that were willing to donate more to a needy friend also gave more Duchenne smiles during the sharing condition. Emotional feelings were not correlated with one's own smiling rate, but the friend reported feeling happier if the partner gave more open smiles during the sharing period, evidence for the contagion effect. Open and Duchenne smiles may play an important signaling role in the formation and maintenance of cooperative relationships. (After [421].)

that is very difficult for most people to produce deliberately (i.e., dishonestly) [171, 179] (see Table 16.1). Truly felt emotions thus lead to facial expressions that are more complete, more symmetrical, smoother in onset and offset, and shorter in duration than faked expressions, especially when the faked expression is used to mask a very different underlying emotion (a lie of commission). Individuals attempting this masking sometimes show **leakage** of their true emotions in fleeting **microfacial expressions** [169, 493, 494]. An individual attempting to hide a strongly felt emotion does better to adopt a neutral expression (a lie of omission).

Smiling is a well-studied example of a highly nuanced human expression. Its phylogenetic precursor is the silent bared-teeth display (see Figure 13.26), which in most primates is an appeasement display given by subordinates to dominants. In chimpanzees, bonobos, and a few macaque

species, it is given not only bidirectionally but also in a wider range of social contexts such as reconciliation, greeting, grooming, embracing, and huddling [487, 498, 499, 610, 634, 655]. In humans it has been combined with the primate relaxed openmouthed play face to form a complex graded signal that includes laughter. There are two basic forms of smiling, a fully felt expression of happiness which entails the upturned mouth corners and the orbital eye muscle contraction and raised cheeks (called a **Duchenne smile** after the French doctor who described it), and a partial smile with the easy-to-control upturned mouth corners but not the difficult-to-fake eye and cheek components (called a **social smile**) [75, 172, 174, 175, 365]. Social smiles are performed for longer durations than full smiles and are used by the sender to project a positive image to receivers. Only very discerning receivers can differentiate the two, but full smiles are more commonly given in friendly cooperative contexts and elicit stronger positive receiver scores and greater levels of trust, sharing, and perceived generosity [330, 365, 421–423, 428, 466, 467, 561, 567, 598]. **Figure 16.28** summarizes a study of the relationship between the rate of Duchenne smiling and the tendency to donate resources, and indicates that full smiles are a reliable indicator of altruistic intentions.

We turn now to the dark side of human nature, lying and deceit. People may lie to improve their reputation; retain or gain resources; avoid abuse, conflict, and punishment; or protect a partner from worry, embarrassment, hurt, or fear. For **liars** to succeed, they must not only speak words that are untruthful, but also convince the listener to believe what they are saying by projecting credible nonverbal signals and cues. Listeners must evaluate both the words and the nonverbal information to judge the speaker's honesty. Most people believe that gaze aversion, a nervous voice, and fidgeting hands and legs are reliable indicators of lying, but in fact, these are some of the least useful cues [701]. Four decades of intensive research have uncovered a far better mix of strategies for detecting deceit [648]. This work has been driven by three theories about how lying affects the sender's behavior and leads to cue leakage [81, 647, 700]. (1) The **arousal** hypothesis proposes that a liar is more emotional than a truth-teller, because he or she is feeling guilty about lying, afraid of being caught, or excited about having the opportunity to fool someone. This higher emotional level involves increased autonomic responses such as dilated pupils and rapid blinking, along with a higher pitched voice, less facial pleasantness, and more use of **adaptors** (see Table 16.1). (2) The **cognitive processing** hypothesis argues that much cognitive effort is required to concoct a plausible lie, avoid contradicting oneself, and tell a lie that is consistent with everything the listener knows or might find out, while avoiding making slips of the tongue. This high level of concentration by liars should result in longer response latencies, more pauses, more speech disturbances, and fewer gestures. (3) The **attempted control** hypothesis suggests that deceivers are aware that their nonverbal behavior might give them away, so they overcontrol their behavior. This idea predicts stiff, rigid, awkward movements, inexpressive facial patterns, and lack of spontaneity. These hypotheses are not exclusive. To complicate matters, truth-tellers confronted with accusatory questions may also experience arousal, fear, anxiety, and cognitive effort, so presence of these cues does not necessarily distinguish deceptive from honest individuals [416]. Variation due to context, seriousness of the accusation (motivation to lie), gender, dominance, and personality differences makes detecting liars accurately an even more difficult task [105]. Nevertheless, the most reliable cues projected by deceivers turn out to be, by body region: (1) eyes—increased pupil dilation, and less blinking while lying but increased blinking afterward; (2) head—reduced head movement, less facial expressiveness, increased lip presses and other lip adaptors; (3) body—fewer **illustrators** (see Table 16.1), more adaptor gestures, and greater postural rigidity; (4) voice—elevated pitch, increased response latency, shorter-duration responses, increased speech errors, more speech fillers such as "ah," "um," and "er," and increased vocal tension and uncertainty; and (5) overall demeanor—uncooperative, increased general nervousness, and reduced spontaneity [52, 81, 143, 380, 645, 699]. Thus elements of all three hypotheses play a role, and voice cues are the most reliable channel for assessing sender veracity. Quantitative statistical analysis of combined video, acoustic, and dialog data can bring the accuracy rate of liar and truth-teller detection up to around 80%, but normal receivers at best achieve a 55% correct rate, barely above the 50% chance level. Lie detection is clearly fraught with receiver biases (see Chapter 8), especially a **truth bias** of assuming that communicators are usually honest [80].

Sexual selection, mate attraction, and courtship

Humans are very **sexually dimorphic** mammals, more so than our closest living primate relatives. Men are taller than women, and develop deep voices, facial hair, broad shoulders, and muscular strength under the influence of a testosterone surge at puberty. Women are shorter with higher-pitched voices, have less body hair, and develop wider hips

accentuated by fat accumulation in the thighs and buttocks and fatty breasts larger than required for nursing infants under the influence of an estrogen surge at puberty (**Figure 16.29**). Paternal investment in offspring, long-term bonds between men and women, and a predominantly monogamous mating system mean that both sexes are expected to compete for high-quality mates.

Our terrestrial primate relatives show intense male competition for females, polygyny, sperm competition, physical coercion of females, and relatively low freedom of female choice (see Table 12.4). By contrast, humans are characterized as having mutual mate choice, as we described for some animals in Chapter 12, where the sexual dimorphism is partly a consequence of males and females preferring different kinds of traits in mates of the opposite sex. However, most research in the field of evolutionary psychology focuses on modern-day developed societies, where women are relatively free to choose their mates. This is not the case in other societies, where parents decide whom their children shall marry, or males raid neighboring villages and kidnap women for mates. Below we shall first describe nonverbal courtship signals in societies with free choice of mates, and then investigate the signals and cues males and females use to attract and choose mates; the possible adaptive significance of these traits; and the relative roles of inter- and intrasexual selection in generating sexually dimorphic traits.

COURTSHIP   Courtship behavior in humans is rich in nonverbal signals, owing to its more powerful means of expressing fundamental contingencies in social relationships such as liking, disliking, superiority, timidity, and fear. As in

FIGURE 16.29 Human sexual dimorphism A young couple, taking a jog on the beach, illustrating the key differences in male and female physical features.

FIGURE 16.30 Courtship in humans (A) The flirtatious glance, with shy smile and hands playing with hair to indicate ambivalence. (B) Typical body postures of a courting couple. Man has confident open posture and, in this example, is employing creative courtship serenading; woman is expressing coyness by leaning back against a structure with partially closed arm position and asymmetric leg position.

(A)

(B)

many animals, human courtship proceeds through a series of stages. Five stages have been described, each employing a different set of nonverbal signals [238, 442, 484]. In the first **attention stage**, the sexes use different signals to get the attention of a desired individual in such a way as to convey interest, but not excessive interest (**Figure 16.30A**). Women giggle, primp, and toss their hair back, while men engage in glancing and adopt open body postures with arms or legs extended to move closer to the woman's personal space zone [439, 517]. This stage is full of ambivalent and contradictory signals, such as looking at the desired person for a few seconds but then covering one's face with the hands, and adaptors and self-grooming behaviors such as twisting a ring, touching or biting lips, or brushing hair out of one's eyes. After securing someone's attention, the function of the next phase, called the **recognition stage**, is to determine how receptive the desired person is to interacting further. If the person looks away, yawns, or frowns, this is interpreted as a rejection. Availability and interest are signaled with eyebrow raises, gazing, direct body orientation, a forward lean, and smiling [438, 440, 485]. For both sexes, light touching indicates stronger courtship intentions [257]. Although men seem to initiate courtship by approaching women, they will not do so unless the woman has first signaled her interest through subtle solicitation behaviors such as darting glances, smiling, nodding, and tossing hair. The woman thus facilitates and controls these early stages of courtship [250, 256]. In more modern times, women also initiate courtship by approaching men directly [145, 649]. Women are generally better at both encoding and decoding nonverbal behaviors than men [263]. Although men are very sensitive to female solicitation signals, they are less likely to pick up on rejection behaviors by women [284, 441].

In the third **interaction stage**, nonverbal behaviors become more intimate and active conversation commences (**Figure 16.30B**). People place themselves in close face-to-face positions so they can converse while shutting other people out. Conversation is highly animated, with longer gazes, more nodding, and intent listening. The couple's behaviors may become more synchronized, and the ease of conversation likely determines the fate of the interaction. There is less ambivalence than before, but both individuals may give submissive signals, such as demure downward glances, shoulder shrugging, and head tilting [484]. Men switch to a softer, slower, low-pitched, and warmer speech pattern, a major component of their **seduction strategy** [17, 302]. If these first three stages have proceeded successfully, the couple may progress to the fourth **sexual arousal stage**. These behaviors include showing affection, intimate touching, and kissing. Finally, a couple may enter the **resolution stage**, characterized by copulation. One party may decide not to take this final step while the other wishes to, which indicates that the terminator was engaging in flirtatious behavior or quasi-courtship. Men often overinterpret a woman's friendly gestures as sexual interest [1, 2, 250, 371]. Thus humans fit the basic expectations of sexual selection well, namely **persuasive males** and **coy females**.

SEXUALLY-SELECTED TRAITS AND THE BASIS OF MATE CHOICE   We focus first on male sexually-selected traits and the potential roles of female preferences and male–male competition in driving their evolution, then turn to female sexually-selected traits and the potential roles of male preferences and female–female competition.

As mentioned earlier, human males differ from females in a number of physical and behavioral traits. Men are on average 8% taller than women and 15–20% heavier in overall body weight. However, fat-free body mass is 40% greater in men, and lean muscle mass is 60% greater. These differences in muscle mass translate into significant differences between the sexes in strength and speed. Men have 90% greater upper body strength and 65% greater lower body strength. Top sprint speed is 22% faster, and vertical leap 45% higher [3, 378, 413]. Aggressive and violent behavior is well known to be greater in men, and this tendency is seen early on in the play behavior of boys [19, 20, 117]. Male faces have heavier brows and more robust jaws, which are accentuated with the growth of hair in these regions during puberty. Men also develop more body hair on the chest, legs, and arms. The pitch of men's voices lowers during puberty as a consequence of an increase in the size and thickness of the vocal folds [39, 297]. Contrary to common belief, among males there is little correlation between weight or height and vocal pitch [39, 245]. Finally, compared to our primate relatives, the human penis is longer and wider. However, testis size and sperm motility relative to body size are closer to the expectation for a single-male or monogamous species, not the larger predicted size for a polygamous species with intense sperm competition, implying a minor role of sperm competition in humans [116, 268, 269, 452, 573].

Many studies have quantified women's preferences for various male traits. Some studies ask women to rank the importance of attributes such as wealth, status, paternal commitment, sexual fidelity, religious affiliation, intelligence, sense of humor, facial attractiveness, and body build. Other studies present women with alternative trait stimuli and ask them to identify the more versus less preferred variant, investigating traits such as degree of facial masculinity, height, body shapes, voice pitch, dancing performance, and the odors of t-shirts worn by men with different attributes. Good studies are careful to clarify whether the context of the preference is for a long-term partner versus a short-term affair, and

to note whether subjects are currently in the fertile or non-fertile stage of their cycle and what method of birth control they are using, because results are often dependent on these contingencies. Women seeking long-term partners overwhelmingly assert a preference for men with high resources (earning capacity) and a willingness to invest in family [84, 91, 290]. Facial attractiveness and skin condition are two other commonly preferred traits that have been suggested to be good genes indicators of health and immunocompetence [520, 597]. Women's attractiveness ratings of photos of male faces result in their selection of individuals with high MHC genetic diversity, implying the potential for both direct and good genes benefits [389, 390, 392, 528]. Women are also sensitive to olfactory cues of MHC similarity and find the t-shirt odor of more genetically dissimilar men more attractive, a potential complementary genes benefit [96, 287, 528, 665, 666].

In contrast, women in their fertile phase prefer a different set of male traits. Normally ovulating, non-pill-using women in the fertile phase (about 9–16 days after the onset of menses) prefer a number of masculine male traits, including the scent of social dominance, facial masculinity, more masculine body shapes, lower-pitched voices, masculine behavioral displays, and taller height [225]. The same women in their non-fertile phase do not express such strong preferences for masculine traits (**Figure 16.31**). Women also state preferences for such masculine traits in the context of seeking a short-term relationship, but not when seeking a long-term relationship. These observations lead to some interesting conclusions. First, human females do possess an **estrous phase** during

(A)

−50%   −30%   0   30%   50%

(B)

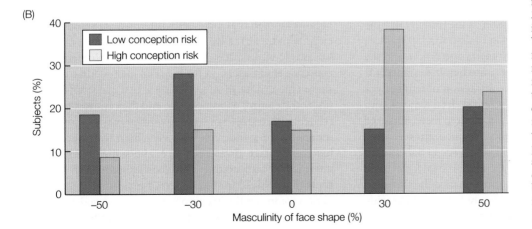

FIGURE 16.31 Change in women's preference for facial masculinity as a function of their menstrual cycle phase (A) Average male faces were generated by blending photos of about 20 men (center), and then variants were generated that were 30% and 50% feminized (toward the left), and 30% and 50% masculinized (toward the right). (B) Women were shown all five facial variants and asked to report which one was the most attractive. Women in their fertile phase (high conception risk) preferred 30% masculinized faces. The same women in their nonfertile phase (low conception risk) showed no significant preference for a particular stimulus but tended to prefer 30% feminized faces more. (After [385, 482, 483].)

the days just prior to and shortly after ovulation that functions in a similar way to that observed in other female vertebrates. Hormonal changes motivate females to attract and seek out high-quality males and increase their choosiness for particular "sexy" or dominant male characteristics. Female chimpanzees, for instance, become choosier during their fertile phase and copulate selectively with dominant males in their group, whereas they mate at high rates with many other group males when not fertile as a strategy to confuse paternity and prevent infanticide [596]. Second, unlike chimps, human females have **concealed estrus**; they have no obvious signals of cycling and receptivity, such as inflated red bottoms or estrus-related chemical cues. Women and their partners are not consciously aware of estrus. Nevertheless, a woman's behaviors and preferences clearly change during estrus, and information about her estrous status "leaks" out in a variety of cues. These cues include an increase in voice pitch, more attractive body odor, subtle gait differences, dressing in sexy clothes, increased flirting, and seeking social interactions with available men [74, 226, 252, 275, 491, 501, 612]. Men in turn pick up on these cues, and a primary or committed partner increases his mate-guarding intensity and copulation frequency when his mate is in estrus [82, 240, 241, 271, 570]. Third, paired women express greater extra-pair desires during their fertile period if they rate their partner's sexual attractiveness as low [223, 271, 492]. These male partners respond by bestowing even more love and attention on their mates during estrus. This putative **dual mating strategy** in women—pairing socially with a man who can provide good resources but seeking extra-pair matings with higher-quality sires when fertile, as commonly observed in birds—places a couple in a conflict-of-interest situation. In fact, most women do not act on these desires because of the severe social costs, and extra-pair paternity rates in human societies are fairly low, typically around 2% (but may reach 10% in some places) [15, 573]. Men, on the other hand, are more likely to have extra-pair sex, and more attractive men may sometimes expend greater effort seeking additional mates than investing in a single mate [505].

Did the masculine traits discussed above as the targets of estrous female preferences evolve primarily via intersexual selection for good genes in sires, or via male–male competition for access to resources and females? A well-reasoned case has been made for male competition as the primary driving force for most, but not all, human male traits [505]. Male traits such as size, muscularity, strength, aggression, and the manufacture and use of weapons, are better designed for contest competition than for the attraction of females per se. Traumatic injuries found in ancient skeletal remains, and higher incidents of facial fractures in modern men compared to women, attest to the occurrence of much interpersonal violence among men [5, 274, 555, 652]. Facial hair (beard and heavy eyebrows) emphasize the more robust facial bone structure of men, and may not only enhance the threat or dominance effect of facial expressions, but also could provide some protection from injury during a physical fight [259,

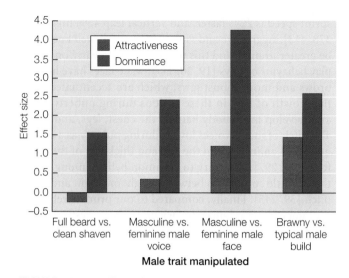

FIGURE 16.32 **Effect of masculine features on perception of dominance versus perception of attractiveness** Four studies examined the relative magnitude of the perception of dominance versus attractiveness for variants of different male characteristics: facial hair, voice pitch, facial features, and body muscularity. Receivers were asked to either rate the stimulus variants on both a dominance and an attractiveness scale, or they were asked to select the most dominant and most attractive among the stimulus set. Receivers were women in all but the voice pitch study, which used male receivers to assess physical dominance versus social dominance (attractiveness). In all studies, the more masculine trait has a stronger association with perceived dominance than with perceived attractiveness. Methodologies differed for these studies, so absolute comparisons of effect sizes among studies are not meaningful. (After [134, 206, 454, 503, 505].)

450, 454]. Low-pitched voices are perceived as more dominant and more threatening, and fundamental frequency is negatively correlated with testosterone levels, suggesting that voice pitch is a reliable indicator of male hormonal status [73, 115, 183, 504]. Deep voices and facial hair signal dominance more effectively than they increase attractiveness to women. **Figure 16.32** presents a graphic summary of studies that compared the attractiveness to females versus the perception of dominance for a few male traits. Masculine voice, masculine faces, and brawny body types show far stronger effect sizes for signaling dominance than for attracting females, because women find males with intermediate values of masculinity more attractive than those with extreme masculine variants [134, 192, 206, 259, 450, 454, 503]. Women actually express a slight aversion for bearded men relative to clean-shaven men. Since these secondary sexual male traits develop under the control of testosterone and their expression is correlated with testosterone level, women may be using these indicators to avoid men with very high testosterone and high tendencies toward violence, and settle for males with intermediate testosterone. Thus females secondarily use male traits that evolved in the context of male–male signaling to obtain a competitively competent mate, as we described in some animals [45]. Contest competition appears to be the

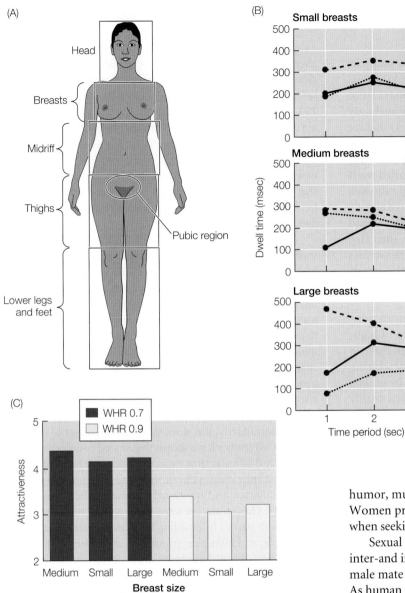

(A)

Head

Breasts

Midriff

Thighs

Pubic region

Lower legs
and feet

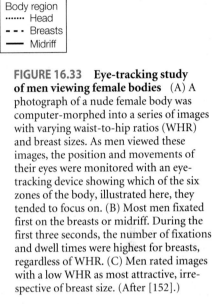

(B)

Small breasts

Medium breasts

Dwell time (msec)

Body region
······· Head
- - - Breasts
——— Midriff

Large breasts

Time period (sec)

(C)

■ WHR 0.7
□ WHR 0.9

Attractiveness

Medium Small Large Medium Small Large
Breast size

FIGURE 16.33 Eye-tracking study of men viewing female bodies (A) A photograph of a nude female body was computer-morphed into a series of images with varying waist-to-hip ratios (WHR) and breast sizes. As men viewed these images, the position and movements of their eyes were monitored with an eye-tracking device showing which of the six zones of the body, illustrated here, they tended to focus on. (B) Most men fixated first on the breasts or midriff. During the first three seconds, the number of fixations and dwell times were highest for breasts, regardless of WHR. (C) Men rated images with a low WHR as most attractive, irrespective of breast size. (After [152].)

humor, music, poetry, and other artistic courtship displays. Women prefer creativity and intelligence in men, especially when seeking long-term relationships [64, 224, 341, 500].

Sexual selection on female traits also shows a mixture of inter-and intrasexual components, but is primarily driven by male mate choice preferences for reproductively fit women. As human males evolutionarily increased their investment in the offspring of one or a few females, females were selected to advertise their reproductive value in order to attract a high-quality male. Men the world over prefer attractive, youthful, and sexually mature women [84]. The reproductive value of women peaks around 20 years of age and decreases thereafter. Women advertise their youth with **neotenous traits** (retained juvenile characteristics) such as high-pitched voices, reduced body hair, large wide-set eyes, full lips, and fine facial features, all traits preferred by men [34, 106, 113, 318, 505]. Reproductive capability is advertised with fatty breast and hip deposits. No other primate has the extreme sexually dimorphic body fat deposits found in women, and men strongly prefer women with optimal fat in these body regions [89, 160, 404]. In particular, it is the **hourglass figure**, or the **waist-to-hip ratio** (**WHR**), and the closely related **body mass index** (**BMI**) that men evaluate with great precision (**Figure 16.33**) [109, 152, 215, 405, 524, 575, 577, 578, 595, 615, 616]. Contrary to earlier suggestions that these fat deposits are deceptive structures that do not reflect a woman's true capacity to produce an ample supply of milk [397], recent evidence shows that hip and breast deposits, but not abdominal fat, are required for regular endocrine cycling and ovulation, are associated

primary driving force for the evolution of male aggressiveness and signals of fighting ability, both in terms of defending resources (usually in male coalitions) and in terms of direct control of women. In societies with strong physical control over women, women will have little opportunity for mate choice, compared to societies in which male aggression is primarily channeled into resource acquisition [505].

In contrast, a few male traits may have been directly selected via female preferences. The fleshy human male penis is one. Women report greater satisfaction with larger penises, and a man's ability to stimulate orgasm in a woman boosts sperm activation and retention and encourages further copulations [219, 388, 502]. The second proposed female-choice-driven trait is enhanced ability to produce **creative displays** designed to court and attract women, the **mating mind** hypothesis [432, 433]. As also proposed for the evolution of complex song mimicry in birds like mockingbirds [55], this idea suggests that creative displays indicate high neurophysiological efficiency and possibly high intelligence, and thereby advertise both direct parental investment abilities as well as indirect heritable fitness benefits. Creative displays include

with a significantly higher probability of conception per cycle, and are mobilized during pregnancy and lactation [314, 352, 376, 444, 661, 667, 697]. Gluteofemoral (lower body) fat in particular stores essential long-chain polyunsaturated fatty acids such as omega-3 docosahexaenoic acid (DHA) that are critical for fetal and infant brain development. The offspring of women with WHR values above 0.7 show proportionally decreasing scores on cognitive tests [377]. The small waist circumference is an important component of this ratio as a fitness indicator because it allows for the detection of already pregnant women, women with low estrogen, and obese and diseased individuals. In addition to body shape, men also express preferences for facial and body odor characteristics that result in the selection of women with greater MHC genetic dissimilarity to themselves and with greater general genetic heterozygosity [276, 389, 392, 612, 666].

As far as intrasexual selection is concerned, women compete strongly among themselves for the attention of high-quality men by enhancing facial attractiveness with makeup that accentuates eye size and skin condition, by wearing clothing that accentuates the hourglass figure, and by smiling and appearing chaste and faithful [89, 600]. Women usually do not engage in direct contest competition over access to men, but instead often derogate their competitor's chastity and attractiveness to improve their own relative value [91, 198, 563].

A contentious aspect of the perception of facial and body **beauty** in humans is the role of **symmetry**. As we discussed in Chapter 12, fluctuating asymmetry is putatively correlated with developmental instability caused by genetic and environmental stress, but there is very little evidence in animals that asymmetry of body parts is either detected by receivers or associated with heritable mate quality. Some research on humans suggests that fluctuating asymmetry might be more important in our species. Numerous studies in which receivers evaluated attractiveness based on facial cues, voice cues, olfactory cues, and dancing performance find that more attractive individuals are more symmetrical in one or more body measurements [71, 72, 133, 221, 222, 232, 249, 251, 298, 301, 522, 549, 574, 576, 611, 612]. However, much of the human evidence has been called into question. One criticism is that humans vary in degree of left-right handedness, which in turn affects limb development. Correcting for this effect, there is no evidence that asymmetry reflects developmental instability [632, 633]. Nevertheless, humans do find symmetrical bodies and faces more beautiful, although the effect size is usually small [667]. People also find well-proportioned faces and bodies more attractive, as well as computer-averaged stimuli, which conform better to idealized proportions [66, 154, 188, 189, 395]. A deeply rooted inherent sensory bias for symmetry and proportion is one

alternative explanation [273, 395, 472, 651]. Attractive people clearly obtain social benefits from this bias in terms of higher salaries, power, and influence, called the **halo effect** [163, 193, 265, 298, 520, 549, 627]. Attractiveness definitely affects sexual success, measured as number of sexual partners and age at onset of sexual activity [519]. The evidence for facial attractiveness as an indicator of health and longevity remains weak and controversial, but some evidence does now suggest that MHC heterozygosity, which is correlated with attractiveness, is associated with greater disease resistance [253, 281, 390, 391, 520, 667].

Finally, there are many sexually monomorphic human traits that are mutually selected mate preferences expressed by both sexes. We mentioned facial attractiveness as a trait preferred by both sexes, although more so by males choosing females. Couples mate assortatively with respect to this trait. Attractive females in particular are far choosier of mates than less attractive females. They strive to attract those rare males that rank highly in all four key desirable male mate characteristics, namely good genes, good resources, parenting proclivities, and emotional commitment [85, 86, 613]. We also mentioned above that both sexes are attracted to facial and olfactory cues that indicate genetic heterozygosity and MHC dissimilarity, but does this really play a role in mate choice? Two studies have shown that mated couples are more genetically dissimilar than expected by chance [96, 465]. Perhaps most importantly, both men and women value kindness, intelligence, and emotional commitment in a long-term mate [84]. These traits are important, as they increase the fitness of both parties by enabling families to maintain a stable, supportive, and loving environment for children [200, 233]. Information about these traits must obviously be obtained over some period of time, after initial attraction and during the interaction stage of courtship. Women in particular show ambivalence during courtship as they attempt to assess the veracity of a man's persuasive courtship displays.

In conclusion, human mate attraction is complicated. Most male sexually selected traits have evolved under *intrasexual* selection pressure, as men compete for access to high-quality attractive women and the resources they require. Most female sexually selected traits evolved under *intersexual* selection pressure, as men who will parentally invest in the offspring of just one or a few women focus on traits that reliably signal female fecundity. Mated pairs cooperate to raise children, often with some division of labor. But pairs live in close proximity to other pairs, often linked through male kinship, and dual mating strategies by both men and women create a conflict between the sexes. Through the lens of the principles of animal communication, we are gaining a better understanding of our own signals and behaviors.

SUMMARY

1. This chapter extends the principles of animal communication to microbes, plants, and humans. The current tree of life recognizes three primary domains: **Bacteria**, **Archaea**, and **Eukaryota**. Single-celled microbes include bacteria, archaea, the polyphyletic eukaryotes lumped within the **Protista**, and some fungi. These organisms do not have male and female sexes, but they possess equivalent **mate choice** mechanisms to achieve genetically diverse offspring. Archaea and bacteria reproduce by **binary fission** and **mitosis** (simple chromosomal duplication) into two equivalent daughter cells, but they occasionally undergo the process of **conjugation**, in which two cells connect and unidirectionally transfer some genetic material. Ciliates (Protista) and fungi produce gametes via **meiosis** that are all of the same size (**isogamy**) but comprise two or more **cryptic mating types**. Only different types can fuse to produce a **zygote**. Many species have two mating types, and some of these species show an **alternation of generations** between clonal multicellular diploid and haploid organisms. Other species have several to many mating types, a strategy that primarily serves a **self-recognition/self-incompatibility** function to reduce inbreeding.

2. Microbes cooperate in several ways, using signals or cues to mediate the interaction. Bacterial populations must attain a certain density in order to obtain the **public goods benefits** of releasing prey digestion enzymes, toxins, surfactants, or virulence factors. They have evolved **quorum sensing** mechanisms to communicate population density. Each cell releases low levels of an **autoinducer** signal molecule that accumulates in the environment, and has tuned signal receptors on the cell surface to monitor autoinducer levels. When the concentration reaches a certain threshold, production of the public good product is triggered. This altruistic behavior is vulnerable to mutant cheaters that do not signal or produce the product, but high relatedness among nearby cells tends to keep cheating at a low level. **Biofilms** are dense aggregations of bacteria that form on many kinds of surfaces. Their tight adhesion protects them from antimicrobial toxins and predators. Extracellular chemical signals are responsible for attraction, settlement, and the polymer production that maintains the strong **cohesive matrix**. Both beneficial and manipulative interactions may occur in multispecific biofilms, and are probably mediated by cues rather than signals. Soil-dwelling **myxobacteria** and **social amoebae** have independently evolved **kin discrimination** mechanisms that enable more closely related cells to converge under poor conditions and form multicellular **fruiting bodies**. Only a fraction of the cells survive and contribute genes to the next generation. Some bacteria produce **bacteriocins** that are selectively toxic to less-related conspecific individuals, an example of **spiteful behavior**.

3. Plants are sessile multicellular organisms, and their **modular branching structure** enables them to forage, move, and behave in ways that maximize resource intake and avoid competitors. Sensory systems for detecting environmental information have been co-opted for communication among conspecifics, with other plant species, and with animal mutualists. Plants have three **light detection** systems: **cryptochromes** for regulating flowering time and circadian rhythms, **phototrophin** for sensing the incoming direction of light and light gradients, and **phytochromes** for detecting the presence of nearby plants. Plants detect several types of **mechanical stimulation**, including gravity, wind, physical wounding, and touch. **Chemical detection** does not involve an olfactory organ, as it does in animals, but occurs in a variety of tissues: **root hair cells** contain membrane-embedded **carrier** and **channel proteins** that take up specific ions in the soil; the **stomata** on the underside of the leaf takes in volatile airborne chemicals; and the **waxy cuticle** of the leaf surface absorbs lipophilic compounds.

4. Plants compete with adjacent neighbors for access to light and soil nutrients using several strategies. The phytochrome red/far-red **shade-sensing system** can detect the horizontal flux of light reflected of adjacent green plants before they cause real shade. Activation of this system leads to an **aggressive response syndrome** of stem elongation, reduced branching, and increased apical dominance to increase plant height. Alternatively, plants can use this detector system to **avoid competition** by reducing growth in the direction of the neighbor and increasing branch growth away from the neighbor. Other plants **tolerate** the shade of neighbors with adjustments in leaf area. Analogous strategies are used to avoid or tolerate root competition. Some species employ **aggressive deterrence** against root growth of competitors with toxic **territorial signals**.

5. Most plants reproduce sexually, but are **hermaphroditic** in the sense that **male** and **female function** reside in the same individual. In **monoclinous flowering plants**, male and female reproductive organs are contained in the same flower. **Pollen grains** on the **anthers** are the male gametes, which, like sperm, are mobile and move via wind or animal vectors among plant individuals. The **pistil**, comprising **stigma**, **style**, and **ovary**, contains the eggs. Pollen grains land on the stigma, germinate, and grow a pollen tube down the style into the ovary. There is both **male–male competition** among pollen for access to ovaries, and the potential for **female choice of mates**, as pollen tube growth requires interaction with the style. Many plants have a **self-recognition system**, the hypervariable **S-locus**, that blocks self-pollen from fertilizing the eggs in **self-incompatible** species. **Incongruity** is the process by which nonconspecific pollen is rejected. There is limited evidence that females promote the pollination success of genetically high-quality

mates by sensing pollen tube growth performance. **Sexual conflict** is expected to be stronger in **dioecious** species (in which the sexes are on separate individuals) than in hermaphroditic species.

6. Plants employ several defensive strategies against herbivorous predators and parasites. **Constitutive physical barriers** and **deterrents** are the first line of defense, and include a tough cuticle, hard woody covers, sharp silica crystals, thorns, and **trichomes** (hairs) to directly impede herbivory. **Secondary plant compounds** that are toxic to herbivores are another line of defense. **Wounding** by an herbivore releases the plant hormone **jasmonate**, which is circulated through the plant's vascular system to build up toxic defenses elsewhere. A final strategy involves plants **crying for help** by producing signals that attract predators of the herbivore. These **herbivore-induced plant volatiles** are specific to the herbivore, and predators learn to associate these volatile chemical mixes with their prey species. In highly sectorial plant species without a continuous vascular system, a similar chemical warning system may be used to prepare defenses in other parts of the same plant. **Eavesdropping** neighbors may detect these **autocommunication** signals.

7. Many plants rely on mobile **animal vectors** to disperse their pollen and seeds. In addition to offering rewards such as shelter, breeding sites, and roost sites, plants typically display multimodal visual and olfactory signals and offer food rewards to obtain these **dispersal services**. Flowers and fruits use pigments such as **carotenoids**, **anthocyanins**, **flavonols**, and **betacyanins** to produce colorful displays. Olfactory signaling components are more diverse and have the potential to target specific pollinators. Pollination requires that visitors show **flower constancy** by visiting multiple individuals of the same species to pick up and then unload pollen, so flower appearance varies significantly among simultaneously flowering species. Flowers show some evidence of **pollinator syndromes**—convergent appearance, color, and structure among species designed to attract the same type of pollinator (e.g., bee, beetle, butterfly, or bird). Fruit dispersal requires that visitors move the seeds far from the parent plant, so fruit **consumer constancy** is not favored. Plants have thus converged on a smaller number of **fruit syndromes**. Because fruits are unicolored and their signal and reward are derived from the same structure, there is greater potential for signal variants to honestly reflect **reward quality**. Flowers, on the other hand, have separately derived signaling (petals) and reward (nectar and pollen) components, so there is no constraint on honesty. Plants may cheat by failing to fill some portion of their flowers with nectar. Orchid species are the ultimate plant cheaters, employing a variety of **nonrewarding** and **deceptive strategies** to lure pollinators.

8. **Human speech**, characterized by voluntary utterances of arbitrary and learned symbolic sounds that are syntactically ordered and recombined with grammatical rules, has no animal precedent. It enables people to make strategic plans, imagine the future, negotiate exchanges of goods, and cooperate to achieve common goals. It likely evolved gradually in our *Homo* ancestors as social organization underwent the **troop-to-tribe transition**, and **seasonal gatherings** and **trading networks** increased among-group interactions and the exchange of technical information. **Nonverbal communication signals**, on the other hand, are more obviously derived from ancestral primate signals. The human nonverbal repertoire has expanded to meet the needs of more complex social interactions and is strongly integrated with speech, by some measures encoding 65% of the meaning of communication exchanges.

9. **Emotions** are episodic, short-term, biologically based patterns of perception, experience, physiology, communication, and action that occur in response to specific external events. Universal emotions such as **fear**, **anger**, **happiness**, **sadness**, **surprise**, **disgust**, and **contempt** are signaled with characteristic **affect displays**, including facial expressions and associated vocalizations. Such expressions are both **bidirectional** (event-triggered feelings produce the expression, called **feedback**, and enacting the expression induces some physiological responses) and regulated by **mirror neurons** (observing expression in another individual activates the same emotion in receiver). These combined processes **synchronize dyadic sharing** of emotions and produce **emotional contagion** effects in group situations, which mobilizes adaptive cooperative and defensive responses that benefit all parties. Other emotions in humans such as **embarrassment**, **shame**, **remorse**, and **guilt** promote reciprocity and honesty and reduce cheating and dishonesty.

10. **Social power** in human interactions refers to the ability to do what one wants without interference from others, to exert influence on others, and to avoid being influenced. Five bases of power are traditionally recognized: **coercive power**, **legitimate power**, **reward power**, **expert power**, and **referent power**. **Dominance** refers to the communication behaviors that are effective in exerting power and influence. Physical fighting ability rarely determines dominance in human societies, but taller height and deeper voice pitch do contribute to the perception of dominance. For most power bases, dominance is communicated by gestures of confidence, relaxed and open body postures, and conversational control. **Turn-taking** in conversations is managed by a set of **regulators**, including patterns of intonation, gazing, and body language that two people employ unconsciously to avoid speaking at the same time.

11. Deception is widespread in human interactions. Although people tend to believe the nonverbal cues and signals over verbal messages when information from the two sources is contradictory, nonverbal signals can also be used deceptively. Some facial expressions are difficult to fake, and true emotions may **leak out** in the form of fleeting **microfacial expressions**. There are two forms of smiling, the truly felt **Duchenne smile** and the **social smile**, distinguished by the eye and cheek components of the display. Duchenne smiling appears to be honestly related to altruistic intentions.

Liars give deceitful or misleading verbal information to improve their reputation, gain resources, or avoid punishment, and simultaneously attempt to manage their nonverbal signals to promote the deception. The consequences of high **arousal**, intense **cognitive processing**, and **attempted control** efforts from telling a concocted story often yield leaked cues of lying to discerning receivers.

12. Nonverbal signals play an important role during mate attraction and courtship in humans. Humans are more **sexually dimorphic** than their closest primate relatives: males have greater height, more lean muscle mass and strength, more facial and body hair, and deeper voices than women. Women have fat deposits on breasts and hips, higher voices, and **neotenous** (youthful) facial features. Human females have lost specific signals of estrus (**concealed estrus**), but cues of estrus nevertheless leak out. Women show stronger preferences for slightly more masculine facial and body features when they are in the fertile phase of their cycle, but in general prefer average male features. Male facial and body characteristics most likely have been selected via intrasexual selection (male–male competition and dominance interactions) rather than via intersexual female choice. Female traits, on the other hand, have been selected primarily via intersexual male mate choice for fertile mates. The **hourglass figure**, a low **waist-to-hip ratio** in particular, is correlated with greater endocrine cycling and ovulation regularity, overall health, and ability to support fetal brain development, growth, and lactation. The initial stages of courtship in humans rely on assessments of **beauty** in women and **resources** and **confidence** in men, regulated by nonverbal signals such as darting glances, body postures, and touching. During the interaction phase of courtship, couples converse intimately. Men employ **creative displays** and **seduction strategies** to persuasively court the woman, while women are ambivalently interested and **coy** and attempt to learn about the resourcefulness, status, and paternal commitment of the man. Mate choice is thus usually **mutual**, except in societies where parents select mates for their offspring. Both sexes value intelligence, kindness, and a sense of humor in a long-term mate.

Further Reading

A recent analysis of the universal common ancestry of life forms can be found in the article by Theobald [609]. For a detailed description of conjugation in bacteria see Alvarez-Martinez and Christie [13]. Phadke [489] presents a recent overview of mating types in ciliates. For fungi, see Metin et al. [429]. West and colleagues have written several accessible reviews, models, and experimental tests of quorum sensing, cooperation, and spite in microbes [149, 150, 229, 255, 535, 672-675]. Staley et al. [592] is a good general microbiology textbook.

Raven et al. [514] and Taiz and Zeiger [601] are recommended as general textbooks on plant biology and evolution.

Reviews of plant behavior can be found in several recent articles: Karban [328], Novoplansky [460], and Trewavas [619-621]. For foraging behavior specifically see de Kroon et al. [129] and McNickle et al. [418]; for photoreceptor sensing see Ballaré [30] and Batschauer [38]; for plant neurobiology see Baluška and Mancuso [32], Barlow [35], and Brenner et al. [63]; and for defenses against herbivores see Baldwin et al. [27], Dicke [147], Halitschke et al. [262], and Kessler and Baldwin [345]. Skogsmyr and Lankinen [582] provide an excellent review of sexual selection in plants, and other related articles include Bedhomme et al. [40], Delph and Ashman [140], and Prasad and Bedhomme [495]. Informative articles on multimodal fruit and flower signals include those by Raguso [509-512] and Schaefer [542, 543, 545]. For perspectives on the controversial topic of frugivore and pollinator syndromes see Fenster [194], Waser [660], Schaefer [544], Ollerton [470] and Lomáscolo [396]. Deception and mimicry are discussed by Schaefer [547], Jersáková [315] and Schiestl [559].

For general information on human nonverbal communication, see recent editions of the textbooks by Burgoon et al. [81] and Knapp and Hall [356]. Recent reviews on facial expressions and emotions by the classical authors, Ekman [179-181] and Scherer [552, 553] provide a good entry into this topic. Several articles discuss the difficulty of detecting deceit in lying humans: Bond and De Paulo [52], Burgoon et al. [80], and Vrij [646-648]. Puts [504] and Gallup and Frederick [220] present excellent overviews of sexual selection and the basis of mate choice in humans.

COMPANION WEBSITE
sites.sinauer.com/animalcommunication2e

Go to the companion website for Chapter Outlines, Chapter Summaries, and References for all works cited in the textbook. In addition, the following resources are available for this chapter:

Web Topic 16.1 *Social microbes*
View videos of the remarkable aggregation and fruiting body formation in social amoebae *Dictyostelium* and social *Myxobacteria*.

Web Topic 16.2 *Plant movement*
We typically think of plants as passive organisms. In fact, they move quite extensively, but at their own time scales and in their own ways. Here are some examples of plants that move for a variety of reasons, including foraging, defense, and reproduction.

Web Topic 16.3 *Body language*
Postures and movements are an important component of human communication. This online unit summarizes attempts to quantify these signals and the perception of them by receivers, focusing on the use of computer animated models.

Credits

CHAPTER 4   4.8B: Courtesy of NASA/nasaimages.org. 4.13B: © PhotosByNancy/Shutterstock. 4.13C: Courtesy of Tam Stuart, tamstuart.com, tmstuart@comcast.net. 4.13D: © Darren J. Bradley/Shutterstock. 4.13E: Courtesy of Cameron Rognan and The Cornell Lab of Ornithology. 4.14A *left*: Courtesy of David Louis Burton. 4.14A *right*: © Tim Laman, 19 Woodpark Circle, Lexington MA 02421, timlaman.com. 4.14B *left*: Courtesy of Sandra Vehrencamp. 4.14B *right*: Courtesy of Sascha Schulz, saschas31@gmail.com. 4.14C *left*: Courtesy of Mathew Brust, Assistant Professor, Chadron State College, Nebraska, flickr.com/photos/24608578@N00/. 4.14C *right*: Courtesy of Patrick Steinberger, salamandro.de. 4.14D *left*: Courtesy of Dino E. Cardone, Ph.D., for cdd.lint-ernet.com. 4.14D *right*: Courtesy of Jennifer Mizen Malpass, jennmizen@gmail.com. 4.14E *left*: © AZPworldwide/Shutterstock. 4.14E *right*: Courtesy of Juliann Schamel, arcticwarbler@hotmail.com. 4.15B *left*: Courtesy of Mathew L. Brust, Assistant Professor, Chadron State College, Nebraska, flickr.com/photos/24608578@N00/. 4.15B *middle*: Courtesy of Dr. Kenneth E. Clifton, Associate Professor of Biology, Lewis and Clark College, Portland, OR, 97219, clifton@lclark.edu. 4.15B *right*, 4.15C *middle*: Courtesy of Tam Stuart, tamstuart.com, tmstuart@comcast.net. 4.15C: *left*: © Tim Laman. 4.15C *right*: © Juniors Bildarchiv/Alamy. 4.16B,C,D: From Lim, M. L. M., Land, M. F., and Li, D. Q. 2007. *Science* 315: 481. Courtesy of Matthew L. M. Lim. 4.16E,F: From Mazel, C. H., Cronin, T. W., et al. 2004. *Science* 303: 51. Courtesy of Roy L. Caldwell. 4.17B: Courtesy of Richard Prum. 4.17C: Courtesy of Tam Stuart, tamstuart.com, tmstuart@comcast.net. 4.18D *left*: Courtesy of Ainsley Seago. 4.18D *right*: From Neville, A. C. 1977. *J Insect Physiol* 23: 1267–1274. Courtesy of Ainsley Seago, CSIRO Entomology, GPO Box 1700, Canberra ACT 2601 Australia, Ainsley.seago@csiro.au, (02) 6246–4103. 4.18E *left*: Courtesy of Mathew L. Brust, Assistant Professor, Chadron State College, Nebraska, flickr.com/photos/24608578@N00/. 4.18E *right*: From Seago, A. E., Brady, P., et al. 2009. *J R Soc Interface* 6: S165–S184. Courtesy of Tom D. Schultz. 4.18F *left*: From Kinoshita, S., Yoshioka, S., and Kawagoe, K. 2002. *Proc R Soc Lond B Biol Sci* 269: 1417–1421. Courtesy of Shuichi Kinoshita. 4.19B: From Parker, A. R. 1995. *Proc R Soc Lond B Biol Sci* 262: 349–355. Courtesy of Andrew Parker. 4.19C,D: Courtesy of Mohan Srinivasarao. 4.21A *left*: Courtesy of Ainsley Seago. 4.21A *right*: From Parker, A. R., Welch, V. L., et al. 2003. *Nature* 426: 786–787. Courtesy of Andrew Parker. 4.21B *left*: Courtesy of Steeve Collard, steevecollard@yahoo.com. 4.21B *right*: From Ghiradella, H. 2010. *Adv In Insect Phys* 38: 135–180. Courtesy of H. Ghiradella. 4.22A: From Prum, R. O., and Torres, R. 2003. *J Exp Biol* 206: 2409–2429. © Kenneth W. Fink, 1209 Hueneme St., Apt. 12, San Diego, CA, 92110. 4.22B,C,D: From Prum, R. O., and Torres, R. 2003. *J Exp Biol* 206: 2409–2429. Courtesy of Richard Prum. 4.23B *left*: Courtesy of Richard Prum. 4.23B *right*: © Ch'ien C. Lee, Rainforest Pictures of Tropical Asia, mail@wildborneo.com.my, wildborneo.com.my. 4.23C *left*: Courtesy of Matthew Shawkey. 4.23C *right*: Courtesy of Cameron Rognan and The Cornell Lab of Ornithology. 4.24A *left*: From Shawkey, M. D., Morehouse, N. I., and Vukusic, P. 2009. *J R Soc Interface* 6(Suppl 2): S221–S231. Courtesy of Matthew Shawkey. 4.24A *right*: Courtesy of Juan D. Ramirez Rpo., colombiabirding.com. 4.24B *left*: Courtesy of Richard Prum. 4.24B *right*: © Glenn Bartley/photolibrary. 4.24C *left*: From Durrer, H., and Villiger, W. 1970b. *Journal fur Onithologie* 111: 133–153. 4.24C *right*: Courtesy of Dr. Kenneth E. Clifton, Associate Professor of Biology, Lewis and Clark College, Portland, OR, 97219, clifton@lclark.edu. 4.24D *left*: From Zi, H., Yu, X. D., et al. 2003. *PNAS USA* 100: 12576–12578. Courtesy of Jian Zi. 4.24D *right*: Courtesy of Jian Zi. 4.25B: © JTB Photo/photolibrary. 4.25C,D,E: Courtesy of Steven Haddock, biolum@lifesci.ucsb.edu. 4.26: © Arco Images GmbH/Alamy. 4.27A: Courtesy of Tamara Dial Gray, flickr.com/photos/love_notes_from_home/, GrayMDiv@aol.com. 4.27B,D: Courtesy of Mohan Srinivasarao, Jung Ok Park, and Matija Crne. 4.28A: Courtesy of Tom D. Schultz. 4.28B: From Seago, A. E., Brady, P., et al. 2009. *J R Soc Interface* 6: S165–S184. Courtesy of Ainsley Seago. 4.28C: Courtesy of Tom D. Schultz. 4.29A: From Dyck, J. 1987. *Biol Skrift* 30: 2–43. 4.29A: © Nik Borrow. 4.31A: Courtesy of Sandra Vehrencamp. 4.31B: Courtesy of Roger E. Carpenter. 4.31C,D: Courtesy of Richard Prum. 4.32A: Courtesy of Sandra Vehrencamp. 4.32B *middle*, *bottom*: Courtesy of Christopher V. Anderson, chamaeleonidae.com. 4.33B,C: Courtesy of Kendra Buresch and Roger Hanlon. 4.34A: © Mark Levy/Alamy. 4.34B: From Vigneron, J. P., Pasteels, J. M., et al. 2007. *Phys Rev E Stat Nonlin Soft Matter Phys* 76(3): 031907. Courtesy of Jean Pol Vigneron. 4.34C: © Gregory Guida, gguida.com. 4.34D: Courtesy of Doris Evans, flickr.com/dorisevans. 4.35A: © Tim Laman, 19 Woodpark Circle, Lexington MA 02421, timlaman.com. 4.35B: Courtesy of Dr. Kenneth E. Clifton, Associate Professor of Biology, Lewis and Clark College, Portland, OR, 97219, clifton@lclark.edu. 4.35C: © Paddy Ryan. 4.35D: © Mariko Yuki/Shutterstock. 4.36B: Courtesy of Chen Ching-Fu, freebsd.tspes.tpc.edu.tw/~afu/. 4.36C: © David Wrobel. 4.36D: Courtesy of Will Smith and Mayang Adnin, mayang.com/textures. 4.36E: Courtesy of Jeffrey Jeffords/Divegallery.com.

CHAPTER 5   5.2A: Courtesy of Andrew M. Snyder, andrewsnyder87@gmail.com. 5.2B,C: Courtesy of Dr. Kenneth E. Clifton, Associate Professor of Biology, Lewis and Clark College, Portland, OR, 97219, clifton@lclark.edu. 5.2D: © Tim Laman, 19 Woodpark Circle, Lexington MA 02421, timlaman.com. 5.7B: Courtesy of Atle Grimsby. 5.7D: Courtesy of David Fotheringham, Blue Leaf Natural Resources, dave@lauderstewart.co.uk. 5.8A: Courtesy of Dr. Kenneth E. Clifton, Associate Professor of Biology, Lewis and Clark College, Portland, OR, 97219, clifton@lclark.edu. 5.8B: © Steve Parish Publishing, steveparish.com.au. 5.8C: Courtesy of Razvan Marescu, Vancouver, Canada, copyright@marescu.net. 5.9A: Courtesy of Ryan Shaw. 5.9B: Courtesy of Diana Giffin. 5.9C: Courtesy of D. Westerman Bowen. 5.9D: Courtesy of Gary A. Garner, 2204 Bonita, Austin, TX 78703. 5.10: © Steve Parish Publishing, steveparish.com.au. 5.11A–D:

Courtesy of John G. Blake. 5.14A *left, right*: Courtesy of Dr. R. D. Young, Cardiff University. 5.14B: From Stavenga, D. G. and Arikawa, K. 2006. *Arthropod Struct Dev* 35(4): 307–318. Courtesy of Doekele Stavenga. 5.16A,B,D: Courtesy of Roger E. Carpenter. 5.16C: Courtesy of Sandra Vehrencamp. 5.22A: Courtesy of Tam Stuart, tamstuart.com, tmstuart@comcast. net. 5.23A: © David Haring/Duke Lemur Center. 5.23B: Courtesy of Doris Evans, flickr.com/photos/dorisevans. 5.23C: Courtesy of Steven Haddock/MBARI, lifesci.ucsb. edu/~biolum/. 5.27A,C: Courtesy of Roy Caldwell. 5.30: Courtesy of Jack Bradbury.

CHAPTER 6 6.2A: Courtesy of Dave Rudie, dave@catalinaop.com. 6.2B: Courtesy of Laura Davidson, Dept. of Biological Sciences, The University of Hull, Hull England, HU6 7RX. 6.2C *left*: From Dreanno, C., Kirby, R.R., and Clare, A. S. 2006. *Biol Lett* 2: 423–425. Courtesy of Anthony S. Clare and Catherine Dreanno. 6.2C *right*: Courtesy of Alan Cressler. 6.2D: Courtesy of Roger Steeb. 6.5A: Courtesy of Scott Bauer, USDA Agricultural Research Service, bugwood.org. 6.5D: Courtesy of Charles Linn and Callie Musto. 6.6B,D: Courtesy of Jack Bradbury. 6.6E: Courtesy of Felice Bond, Botanimal Images, fbond203@gmail.com. 6.6F: © Allison C. Alberts, Ph.D., Chief Conservation and Research Officer, San Diego Zoo Global, Institute for Conservation Research, 15600 San Pasqual Valley Road, Escondido, CA, 92027, aalberts@sandiegozoo.org. 6.9B: Courtesy of Fiene van Beelen. 6.10: Courtesy of Rob Beynon. 6.11: © Bernhard Jacobi, www.flickr.com/photos/29697818@N03/, h.b.jacobi@ gmx.de. 6.12A: Courtesy of Thomas Eisner, Cornell University. 6.13A: Courtesy of Joaquín Ramírez López, c/Escritora Luciana Narváez, nº 3-4º A, 29011, Málaga, España, jrlmalaga@hotmail.com. 6.13B: Courtesy of Emily Wroblewski. 6.14B: © Warren Photographic. 6.14C: Courtesy of Angie Lott. 6.15A *left*: Courtesy of Christian C. Voigt. 6.15A *right*: Courtesy of Michael Stifter, michael-stifter.de. 6.15B: Courtesy of Paolo Mazzei, herp.it. 6.16A: Courtesy of Ian Jones. 6.16B: Courtesy of Xavier Bayod Farré. 6.16C: From Müller-Schwarze et al. 1977. *J Ultrastruct Res* 59(3): 223–230. Courtesy of D. Müller-Schwarze. 6.16D: Courtesy of Manvendra Bhangui. 6.17A: Courtesy of Jack Bradbury. 6.17B: © Nigel Dennis, nigeldennis.com. 6.17C: © Gregory Guida, gguida. com. 6.20A: Courtesy of Russ Hopcroft and Thomas Kiørboe. 6.21B: Courtesy of Carl Rettenmeyer, armyantbiology. com. 6.24A: Courtesy of Don Webster, Webster and Weissburg (2001), *Limnology and Oceanography*. 6.24B: Courtesy of Chun-Ho Liu. 6.25B: Courtesy of Kenneth Catania. 6.25D: Courtesy of M. Koehl, J. Koseff, and J. Crimaldi. 6.26A: From Ozaki, M., Wada-Katsumata, K., et al. 2005. *Science* 309: 311–314. Courtesy of Mamiko Ozaki and Masayuki Iwasaki. 6.26B: From Filoramo, N. I. and Schwenk, K. 2009. *J Exp Zool A Ecol Genet Physiol* 311(1): 20–34. Courtesy of Nirvana Filoramo and Kurt Schwenk. 6.26C: Courtesy of D. Muller-Schwarze, *Chemical Ecology of Vertebrates*. 6.31B: From Shaheen, N., Patel, K., et al. 2005. *Anim Behav* 70(5): 1067–1077. Courtesy of Melissa A. Harrington, Delaware

State University. 6.33A: Courtesy of Brian DeMeester, brian.d.demeester@gmail.com, 924 Rockledge Dr., Carlisle, PA 17013.

CHAPTER 7 7.1A: © Brian Mayes, flickr.com/photos/ brianmayes/. 7.1B: Courtesy of Mace Hack. 7.2A: Courtesy of Javier Torrent, flickr.com/photos/seta666, seta666@yahoo. es. 7.2B: Courtesy of Piero Cravedi, piero.cravedi@unicatt. it, taken in the Institute of Entomology and Plant Pathology of the Universitá Cattolica del Sacro Cuore, Piacenza, Italy. 7.2C: © J.G. Hall and The Mammal Image Library of the American Society of Mammalogists. 7.3A: © Alex Wild, alexanderwild.com. 7.3B: Courtesy of Richard Wrangham. 7.3C: Courtesy of Phyllis Keating, Mississauga, Ontario, flickr.com/photos/flipkeat/. 7.4A–C: Courtesy of Terry Fitzgerald, Social caterpillars, web.cortland.edu/fitzgerald/. 7.10: From Takahashi-Iwanaga, H., Maeda, T., and Abe, K. 1997. *J Comp Neurol* 389(1): 177–184. Courtesy of Hiromi Takahashi-Iwanaga. 7.16A: Courtesy of Derek Haslam, flickr.com/photos/dirks_images/. 7.16B: Courtesy of Claudia Domenig, cdomenig@gmail.com, flickr.com/photos/ cdomenig/, cdomenig@gmail.com. 7.17 A–D: © Kenneth C. Catania, Vanderbilt University, Dept. of Biological Sciences, VU Station B, Box 35-1634, Nashville, TN, 37235–1634; parts B and D drawn by Lana Finch. 7.18: From Hanke, W., Brücker, C., and Blekmann, H. 2000. *J Exp Biol* 203: 1193–1200, available at jeb.biologists.org/cgi/reprint/203/7/1193. 7.21: From Jacobs, G. A., and Theunissen, F. E. 2000. *J Neurosci* 20(8): 2934–2943. 7.26A: Courtesy of Northeast Fisheries Science Center Aquarium. 7.29: Courtesy of Franz Uhl. 7.39: © S. Karger AG, Basel. Nelson, M. E., MacIver, M. A., and Coombs, S. 2002. *Brain Behav Evol* 59: 199–210.

CHAPTER 8 8.1: Courtesy of Fernando A. Campos, Dept. of Anthropology, University of Calgary, 2500 University Drive N.W., Calgary, Alberta T2N 1N4 Canada, camposfa@ gmail.com. 8.8: Courtesy of Ingrid Taylar, ingridtaylar.com. 8.17A: Courtesy of Grant Brummett, grantbrummett.com. 8.17B: Courtesy of Kathy Goodson, kathygoodson@gmail. com. 8.17C: Courtesy of C. van Maarth, Caliente, CA, cvanmaarth@yahoo.com, flickr.com/photos/clarasbell. 8.18A: © R. A. Behrstock/Naturewide Images, naturewideimages.com. 8.18B: Courtesy of Brad and Lynn's Field Photography, 8641 Dallas St., La Mesa, CA 91942, brad.weinert@cox.net, herpindiego.com. 8.18C: © Paul Lathbury, Latherspics, flickr.com/ photos/latherspics, paul.lathbury@gmail.com. 8.18D: Courtesy of Marie-Louise Hagen. 8.19: Courtesy of B. Caspers.

CHAPTER 9 9.1 *top*: Courtesy of Ken Clifton. 9.1 *middle, bottom*: Courtesy of Jack Bradbury. 9.2A: © Gerrit Vyn Photography; gerritvynphoto.com. 9.2B: © Merlin D. Tuttle, Bat Conservation International, batcon.org. 9.2C: Courtesy of Ronald R. Hoy. 9.3: Artwork by W. T. Cooper. 9.10 *left*: Courtesy of Maciek Wrzyszcz, flickr.com/photos/maciunio/. 9.10 *right*: Courtesy of Janet Hughes, flickr.com/photos/ janethughes/. 9.14: © Karel Broz/iStockphoto. 9.17A–C:

whitinglab.com, martin.whiting@mq.edu.au. 11.25B: Courtesy of Hans Joachim Kaiser. 11.25C: Courtesy of Gabri Mtnez, gabrimtnezmarmol@yahoo.es. 11.26A: Courtesy of Andrew M. Durso, amdurso@gmail.com. 11.29: Courtesy of Jorge Martin Silva Rivera, San Cristobal de las Casas, Chiapas, Mexico, jmsilva_r@hotmail.com. 11.30: From Corl, A., Davis, A. R., et al. 2010. *PNAS USA* 107(9): 4254–4259. Courtesy of Barry Sinervo and Ammon Corl. 11.31: Courtesy of Johan van Rensburg, jjvrensburg52@gmail.com, flickr.com/photos/johanvanrensburg/.

CHAPTER 12 12.10B: Courtesy of Thomas Eisner, Cornell University. 12.11A: From Maklakov, A. A., and Arnqvist, G. 2009. *Curr Biol* 19(Suppl): 1903–1906. 12.13A: © Sam Cotton, 2010, flickr.com/photos/samcotton/. 12.15: Courtesy of Alexander Badyaev, tenbestphotos.com, Ecology and Evolutionary Biology, University of Arizona, Tuscon, AZ 85721-0088, abadyaev@email.arizona.edu. 12.16A: Courtesy of P.-A. Ravussin. 12.16B: From Punzalan, D., Cooray, M., et al. 2008. *J Evol Biol* 21: 1297–1306. Courtesy of David Punzalan. 12.17A,B: Kemp, D. J., and Rutowski, R. L. 2007. *Evolution* 61: 168–183. Courtesy of Ronald L. Rutowski and Darrell Kemp. 12.18A: © Clyde Barrett. 12.18B: © Angela Arenal/iStockphoto. 12.18C: Courtesy of Danny Gibson, dgpix.org.uk, danny.gibson@btconnect.com. 12.18D: © Carlos Palacín, cpalacin@mncn.csic.es. 12.19A: Courtesy of Euan Reid, 1005 Queen St. West, Mississauga Ontario, Canada, ereid@rogers.com. 12.19B: © INTERFOTO/Alamy. 12.19C: Courtesy of Pravin Indrekar. 12.20A: Courtesy of H. Carl Gerhardt, gerhardth@missouri.edu. 12.21A: Courtesy of Oldcar Lee (Hong Kong). 12.21B: Courtesy of David Miguel González Martins, www.flickr.com/photos/david94/. 12.22A: Courtesy of Mark Blows. 12.23A: Courtesy of Jörg Vierke, fischverhalten.de. 12.24C *left, middle, right*: Courtesy of Stephanie Doucet. 12.24D: Courtesy of Dr. Kenneth E. Clifton, Lewis and Clark College, Portland, OR, 97219, clifton@lclark.edu. 12.25A: Courtesy of Wesley Bicha. 12.27: Courtesy of Tam Stuart, tamstuart.com, tmstuart@comcast.net. 12.31A: Courtesy of Szűts Tamás. 12.31B: © Helen Glazer, 2007, helenglazer.com, helen@helenglazer.com. 12.31C: Courtesy of Hannah and Craig Meddaugh, Hannah and Craig Meddaugh Nature Photography, meddaughphotography.com, admin@meddaughphotography.com. 12.32A: © Judd Patterson, juddpatterson.com, judd@juddpatterson.com. 12.32B: © imagebroker/Alamy. 12.32C: Courtesy of Thomas Schultz. 12.33A,D: Courtesy of Paolo Mazzei, www.herp.it. 12.33B,C: Courtesy of Stevan J. Arnold. 12.35B: Courtesy of Gail Patricelli, published in Patricelli, G. L., Coleman, S. W., and Borgia, G. 2006. *Anim Behav* 71: 49–59. 12.36A: Courtesy of Darryl Gwynne. 12.36B: Courtesy of Andreas Svensson. 12.36C: Courtesy of Natalia J. Demong, njd7@cornell.edu.

CHAPTER 13 13.3A: Courtesy of Jill M. Mateo, The University of Chicago, jmateo@uchicago.edu. 13.5A: From Tibbetts, E. A., and Dale, J. 2007. *Trends Ecol Evol* 22: 529–537. Photos by Max Holdt, courtesy of James Dale. 13.5B:

Courtesy of Claire Spottiswoode, Dept. of Zoology, University of Cambridge, Downing St., Cambridge CB2 3EJ, UK, spottiswoode@cantab.net. 13.5C: Courtesy of Kim Wolhuter, wildhoot@mweb.co.za. 13.6A: Courtesy of Karen McComb. 13.7C: Courtesy of Eric Hilton. 13.7D: Courtesy of Sarah Huber. 13.8: Courtesy of H. Carl Gerhardt, gerhardth@missouri.edu. 13.9: © Yamil Saenz, 2010, flickr.com/photos/ysaenz. 13.11A: © Sue Thatcher, bakalo@btinternet.com. 13.11B: © Pablo Cervantes, pablocervan@gmail.com. 13.12: Courtesy of Terri Eagle, randomletters@gmail.com. 13.16A: © Bill Lloyd/iStockphoto. 13.17A: Courtesy of Isabelle Charrier, Centre Neurosciences Paris Sud, UMR 8195 CNRS, Université Paris Sud, bat. 446, 91405 Orsay, France, isabelle.charrier@u-psud.fr. 13.20: Courtesy of Rebecca Kilner. 13.21: Courtesy of Per Smiseth. 13.22: Courtesy of R. B. Cocroft. 13.23A,B: Courtesy of Matthew Johnson. 13.24A: Courtesy of Katherine McAuliffe, Dept. of Human Evolutionary Biology, Harvard University, 11 Divinity Ave., Cambridge, MA 02138. 13.25: Courtesy of Elizabeth Tibbetts. 13.26A,C,D: Courtesy of Julia Fischer, Cognitive Ethology Laboratory, German Primate Center, www.cog-ethol.de. 13.26B,E,F: Courtesy of Andreas Ploss, andreasploss@googlemail.com. 13.27A: © Chris van Rooyen. 13.27B: Courtesy of Fernando A. Campos, Dept. of Anthropology, University of Calgary, 2500 University Dr. N.W., Calgary, Alberta T2N 1N4, Canada. camposfa@gmail.com. 13.27C: Courtesy of Tony Johnson, England, flickr.com/photos/tony_6. 13.28A: Courtesy of S. P. Henzi. 13.29A: Courtesy of Frans de Waal. 13.29B: Courtesy of Michael R. Reilly. 13.29C: Courtesy of Fernando Colmenares. 13.29D: Courtesy of Kay E. Holekamp, Dept. of Zoology, Michigan State University, East Lansing, MI 48824-1115. 13.30A: Courtesy of Dorothy Delina Porter, Oregon, flickr.com/photos/delina. 13.31B: Courtesy of Thomas D. Seeley. 13.32: From King, A. J., Johnson, D. D. P., and van Vugt, M. 2009. *Curr Biol* 19: R911–R916. Courtesy of Susan M. Lusseau, Lighthouse Field Station, University of Aberdeen. 13.33A–D: Courtesy of Frode Jacobsen, flickr.com/photos/frodejacobsen/. 13.33E: Courtesy of Muriel Neddermeyer, birdybird2010@yahoo.com. 13.33F: © Lois Manowitz, Tuscon, Arizona, loismanow@gmail.com. 13.35A: Courtesy of Roger Sargent. 13.35B,C: Courtesy of Ria Tan, wildsingapore.com.

CHAPTER 14 14.1A: Courtesy of Edward Curmi, Malta, edward_curmi@yahoo.com, edwardcurmi.com. 14.1B: © Ch'ien C. Lee, Rainforest Pictures of Tropical Asia, mail@wildborneo.com.my, wildborneo.com.my. 14.1C: Courtesy of David Evans, Australia, funkyfoton@yahoo.com, Australia. 14.1D: Courtesy of Nicolas Moulin Macrophotography, Spain, flickr.com/photos/nimou/. 14.2A–D: © Ch'ien C. Lee, Rainforest Pictures of Tropical Asia, mail@wildborneo.com.my, wildborneo.com.my. 14.3A: Courtesy of Joseph W. Adair, joe@joeadair.com. 14.3B: Courtesy of Troy G. Murphy, Dept. of Biology, Trinity University, San Antonio TX 78212. 14.3C: Courtesy of Stephen Zozaya, stephen.zozaya@my.jcu.edu.au. 14.3D: © Rita Summers, Charles and Rita

Summers, 23463 East Moraine Place, Aurora, CO 80016, wildimages.biz, T: 303-840-3355. 14.4A: Courtesy of Laura Thrower, desk@laurathrower.com. 14.4B: Courtesy of Ashley F. Grooms, ashleyfgrooms@gmail.com. 14.4C: © Michael P. O'Neil/Photoshot, photoshot.com, T: 44-207-421-6000, sales@photoshot.com. 14.5A: © Nigel Dennis, Nigel Dennis Wildlife Photography, nigeldennis.com. 14.5B: Courtesy of Steve Miller, srmillerphotography.co.uk. 14.5C: © Elliott Neep, neepimages.com, info@enwp.co.uk, T: +44 (0)7788 577404. 14.5D: © Taco Meeuwsen, esox@pn.nl. 14.6A–F: Courtesy of David Evans, Australia, funkyfoton@yahoo.com. 14.7A: Courtesy of Ronnie Pitman. 14.7B: Courtesy of Alan Cressler. 14.7C: Courtesy of Valerie C. Clark, frogcaller.com, frogchemistry@gmail.com. 14.7D: © W. Peckover/VIREO. 14.9: © J. Mallet. 14.10: From Janzen, D. H., Hallwachs, W., and Burns, J. M. 2010. *PNAS USA* 107: 11659–11665. Courtesy of Daniel H. Janzen, djanzen@sas.upenn.edu. 14.11A: Courtesy of Jason M. Hogle, jmhogle@gmail.com. 14.11B: © Allen B. Sheldon. 14.12A: Courtesy of Casey A. Parker, kzintoscana.blogspot.com. 14.12B: Courtesy of Michael R. Reilly. 14.18A: Courtesy of John Dunstan. 14.18B: © Mathieu Bruno/biosphoto, biosphoto.com/. 14.18C: © Harvey Martin/biosphoto, biosphoto.com/. 14.19: Courtesy of Marta Manser, Kalahari Meerkat Project, mkproj@mweb.co.za, photo © Robert Sutcliffe. 14.21A,B: Courtesy of Jojanneke Bijkerk, plantengallen.com, the Netherlands. 14.24A: Courtesy of the USDA Forest Service - Region 8 - Southern Archive, USDA Forest Service, bugwood.org. 14.24B: Courtesy of Ronald. F. Billings, Texas Forest Service, bugwood.org. 14.25: © 2010 Steve Martin, World Parrot Trust, parrots.org/. All rights reserved. 14.26A: Courtesy of Bryan King, b.j.king@telus.net. 14.29, 14.30A: © Alex Wild, alexanderwild.com.

CHAPTER 15   15.15: © Emmanuel Perrin, L'équipe Photothèque, CNRS Images - Photothèque, 1 Place Aristide Briand, 92195 MEUDON CEDEX, France, T: 33-01-45-07-57-90, phototheque.cnrs.fr/, phototheque@cnrs-bellevue.fr. 15.23A *drongo, babbler insets*: Courtesy of Andy Radford, Senior lecturer, School of Biological Sciences, University of Bristol, andy.radford@bristol.ac.uk, T: 44-117-9288246. 15.24A: Courtesy of Carl Johan Heickendorf. 15.28A,B: Courtesy of Stephen J. Simpson. 15.33A: From Hildenbrandt, H., Carere, C., and Hemelrijk, C. K. 2010. *Behav Ecol* 21: 1349–1359.

CHAPTER 16   16.5A: Courtesy of Jeremy S. Webb, Anne Mai-Prochnow, and Staffan Kjelleberg. 16.5B: From Branda, S. S., Gonzalez-Pastor, J. E., et al. 2001. *PNAS USA* 98: 11621–11626. Courtesy of Roberto Kolter. 16.5C: Courtesy of Guus Roeselers, Harvard University. 16.5D: Courtesy of Janus Haagensen and Søren Molin. 16.6A–C: From Velicer, G. J., and Vos, M. 2009. *Annu Rev Microbiol* 63: 599–623. Courtesy of Greg Velicer. 16.7A,B,D: Courtesy of Owen Gilbert, owen.gilbert@gmail.com. 16.7C: From Rivero, F., Kuspa, A., et al. 1998. *J Cell Biol* 142: 735–750. Courtesy of Francisco Rivero, Centre for Cardiovascular and Metabolic Research, The Hull York Medical School, University of Hull, Cottingham Rd., Hull HU6 7RX, UK, francisco.rivero@hyms.ac.uk. 16.7E: Courtesy of Kei Inouye. 16.11: Courtesy of Keith Povall, Twitter @soxer99. 16.14: Courtesy of John M. Hagstrom. 16.16: Courtesy of Jaap Cost Budde. 16.17 *both*: Courtesy of Dr. Ahlert Schmidt, flickr.com/photos/bienenwabe. 16.19A: Courtesy of Sergio Rasmann, Dept. of Ecology and Evolution, Lausanne, Switzerland, sergio.rasmann@unil.ch. 16.20A *top*: Courtesy of Lloyd W. White, Bay Roberts NL, Canada. 16.20A *bottom*: Courtesy of Otto Loesel. 16.20B: Courtesy of Marek Trylinski, marek.trylinski@gmail.com. 16.20C: Courtesy of Robert Silverwood, flickr.com/photos/silverwood. 16.21: From Raguso, A. 2008. *Annu Rev Ecol Evol Syst* 39: 549–569. Courtesy of Robert A. Raguso. 16.22 *bee*: Courtesy of Bill Wakefield. 16.22 *butterfly*: Courtesy of Thomas O'Brien. 16.22 *moth*: Courtesy of Ron James. 16.22 *fly*: Courtesy of Emerald Leung, eleungda@gmail.com. 16.22 *bird*: © Steve Byland, stevebyland.com. 16.22 *bat*: © Joe Coelho, Quincy University, 1800 College Ave., Quincy, IL, 62301, coelhjo@quincy.edu. 16.22 *beetle*: Courtesy of John Brew, 2734 NE 184th Pl., Lake Forest Park, WA, 98155. 16.23A: © Emilio Esteban-Infantes, emilesteban@gmail.com, flickr.com/photos/96454410@N00. 16.23B: Courtesy of J. Ramírez, jrlmalaga@hotmail.com, naturalezadeandalucia.blogspot.com. 16.23C: © Photomajik by Marilyn, iammarilynt@earthlink.net. 16.23D: Courtesy of Kristina Medve, suricatasuricatta@gmail.com. 16.24A: Courtesy of Rod Peakall, Evolution, Ecology and Genetics, Research School of Biology, The Australian National University, Canberra ACT 0200. 16.24C: Courtesy of F. P. Schiestl. 16.29: © Asia Images Group Pte Ltd/Alamy. 16.30A: © GoGo Images Corporation/Alamy. 16.30B: © Suprijono Suharjoto/iStockphoto. 16.31A: From Johnston, V. S., Hagel, R., et al. 2001. *Evol Hum Behav* 22: 251–267. Table 16.2 *happiness*: © Daniel Laflor/iStockphoto.

Index

Page numbers in *italic* type indicate the information will be found in an illustration.